Essentials of Medical Biochemistry

With Clinical Cases

ELSEVIER
science &
technology books

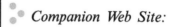

:• *Companion Web Site:*

http://booksite.elsevier.com/9780124166875

Essentials of Medical Biochemistry, Second edition
N. V. Bhagavan and Chung-Eun Ha

Resources for Professors:

• Multiple choice questions and answers.

• All figures and tables from the book available as PowerPoint slides.

ELSEVIER

ACADEMIC
PRESS

Essentials of Medical Biochemistry

With Clinical Cases

Second Edition

N. V. Bhagavan
Chung-Eun Ha

ELSEVIER

AMSTERDAM • BOSTON • HEIDELBERG • LONDON • NEW YORK • OXFORD
PARIS • SAN DIEGO • SAN FRANCISCO • SINGAPORE • SYDNEY • TOKYO

Academic Press is an imprint of Elsevier

Academic Press is an imprint of Elsevier
525 B Street, Suite 1800, San Diego, CA 92101-4495, USA
32 Jamestown Road, London NW1 7BY, UK
225 Wyman Street, Waltham, MA 02451, USA

First edition 2011
Second edition 2015

British Library Cataloguing-in-Publication Data
A catalogue record for this book is available from the British Library

Library of Congress Cataloging-in-Publication Data
A catalog record for this book is available from the Library of Congress

ISBN: 978-0-12-416687-5

For information on all Academic Press publications
visit our website at store.elsevier.com

Publisher: Janice Audet
Acquisition Editor: Jill Leonard
Editorial Project Manager: Pat Gonzalez
Production Project Manager: Melissa Read
Designer: Maria Ines Cruz

Contents

16. Lipids I: Fatty Acids and Eicosanoids

17. Lipids II: Phospholipids, Glycosphingolipids, and Cholesterol

18. Lipids III: Plasma Lipoproteins

19. Contractile Systems

20. Perturbations of Energy Metabolism: Obesity and Diabetes Mellitus

21. Structure and Properties of DNA

Preface

Much of the content and concepts for the second edition of *Essentials of Medical Biochemistry: With Clinical Cases* were derived from the first edition, but this latest edition has been extensively re-designed in an effort to better integrate the principles of biochemistry into disciplines of basic and clinical sciences. The goal is to foster a more comprehensive understanding of health and disease. We hope that we have accomplished this task by making the principles of biochemistry and molecular biology more clinically relevant and applicable.

Unique features of this textbook include: for each chapter, a list of key points which outline the main concepts or questions that will be discussed in the chapter, followed by the text and a list of required readings in some chapters. The text also includes supplemental references where appropriate, to encourage readers to continue learning beyond the chapter material. Most chapters include actual clinical cases with teaching points that pertain to the subject matter of the chapter. This textbook has also been revised to include the latest research on various aspects of biochemistry and molecular biology topics that have emerged since the first edition.

Every effort has been made to create an enjoyable and educational learning experience for the student. Biochemistry has often been perceived as a topic of rote memorization of reactions and pathways, but here we present a well-organized and a systematic mosaic of basic and clinical sciences, which altogether clearly illustrates the interconnectedness between biochemistry and molecular biology as they relate to health and disease. Biochemistry, therefore, is not a topic limited to textbooks or laboratories; life depends on it. This is illustrated by the clinical cases that we have presented immediately following the chapter material. These cases were abstracted from previously published articles that describe actual clinical conditions. This serves as a reminder of the importance of medical biochemistry, and it aids the student in developing their skills in the diagnostic acuity and subsequent treatment.

We hope students cultivate a full understanding of the pathophysiologic processes of the body by examination at the molecular level, in an effort to accomplish the overall goal of logical clinical problem-solving, with a compassionate understanding of human ailment.

We also promote problem-solving skills that are based on accurate and current information. In almost every chapter, we have listed required reading references as well as a enrichment reference list at the end of the text, so that the student may refer to the original literature. We hope that students can develop skills in the understanding of evidence-based medicine.

As clinicians, academicians, and researchers, the authors and their associates in the production of this textbook firmly believe that the new format of our text encourages students to embrace medical biochemistry and apply its principles to all areas of medicine. The versatility of this textbook also makes it appropriate for use by students pursuing various disciplines in basic and clinical sciences.

Acknowledgments

We are grateful to our collaborative authors for their contributions and sharing their knowledge and insights. We also appreciate their cooperation during the lengthy writing and production process.

The preparation of this text required collective efforts and collaboration of many individuals. We are indebted and appreciative of several individuals who contributed to the book as reviewers of selected chapters, served as consultants, provided constructive suggestions and academic support. These individuals are Stacey A.A. Honda, Jodi Matsuura-Eaves, Brock Kaya, Stanley Loo, Carlos Rios, Lourna M. Murakami, Leslie Y. Uyeno, Karen Higa, Gale Fugitani, Charity Saludares, Miki Loscalzo, Ji-Sook Ha, Scott Lozanoff, Karen Yamaga, Kumar Bhagavan, Kimberly Yamauchi, Celeste Wong, Winona K. Lee, Keawe'aimoku Kaholokula, Mari C. Kuroyama, and Michael D. Black. We are grateful to Kenton Kramer for reviewing and proof reading several chapters and providing many constructive suggestions for improving the text. One of the authors, Craig M. Jackson, particularly thanks M. Peter Esnouf for always constructive and valuable advice.

The production of this manuscript, which involved a myriad of processes, required the contributions from many dedicated individuals. James Ha and Jennifer Ha were responsible for transforming the rough manuscript into its final form and we appreciate their vital contribution. Other individuals who assisted in this process are Michael Tabacchini, Teri Laupola, Anastacia Guittap, and Tricia Yamaguchi.

The acquisition of current information in the process of development of our manuscript was aided and assisted by the staff of the John A. Burns School of Medicine Health Science Library. The persons involved in this process were Virginia Tanji, Leah Gazan, and Hilda Baroza.

From Elsevier staff we received continued encouragement and support from Elizabeth Gibson, Jill Leonard, and Pat Gonzalez. They helped us in maintaining schedules for the manuscript preparation, and assisted us in many ways, which moved the project forward.

LIST OF AUTHORS

1. Nadhipuram V. Bhagavan, Ph.D., F.A.C.B.
   Emeritus Professor
   Department of Anatomy, Biochemistry, and Physiology
   John A. Burns School of Medicine
   University of Hawaii at Manoa
   Chapters 1−4, 8−18, 20, 25−32, and 35−37

2. Chung-Eun Ha, Ph.D.
   Associate Professor
   Department of Native Hawaiian Health
   John A. Burns School of Medicine
   University of Hawaii at Manoa
   Chapters 21−24

3. William Gosnell, Ph.D.
   Assistant Professor
   Department of Tropical Medicine, Medical Microbiology, and Pharmacology
   John A. Burns School of Medicine
   University of Hawaii at Manoa
   Chapter 33

4. Craig M. Jackson, Ph.D., FNACB, FAAAS(Chem)
   Formerly Professor, Biological Chemistry
   Associate Professor, Washington University,
   St. Louis, MO
   San Diego, CA
   Chapters 5, 6, 34

5. Nicole Mahealani Lum, D.O.
   Resident Physician
   Family Medicine Residency Program
   Department of Family Medicine and Community Health
   John A. Burns School of Medicine
   University of Hawaii at Manoa
   Chapter 19

Dr. Lum also prepared the following clinical cases:
CC 4.1, 4.2; CC 7.1−3; CC 10.1−10.3;
CC 11.1−11.5; CC 17.1; CC 19.1, 19.2; CC 26.1, 26.3;
CC 27.1; CC 29.1; CC 30.1; CC 31.1, 31.2;
CC 32.1−32.3; CC 35.1−35.3; CC 37.1, 37.2.

Chapter 1

The Human Organism: Organ Systems, Cells, Organelles, and Our Microbiota

Key Points

1. The human organism is hierarchically organized; at the highest level, it is organized into organ systems classically related to functions and anatomical structures.
2. Distinctions among organs are the consequence of their specialized tissues and cells that are produced during embryonic differentiation, a process that begins after fertilization of an ovum by a spermatozoon and continues for some organs for a time post-partum.
3. Cells are the basic building blocks for organs and tissues in all living systems.
4. Biochemical processes within cells, and thus within organ systems, include metabolism, growth, reproduction, mutation, response, self-destruction, and evolution.
5. The body of an adult consists of more than 200 differentiated and thus specialized cell types.
6. Specialized structures with functional distinctions, organelles, exist with cells and constrain particular biological processes that create unique functions that confer metabolic efficiency and cellular integrity. Intracellular organelles work interdependently to degrade, synthesize, transport, and excrete intracellular products.
7. Cellular shape and ability to exchange substances with the circulatory and other transport systems are derived from the cellular membranes and intracellular membranous structures.
   a. Mitochondria are the primary sources of energy for cells, organs, and the living animal.
   b. The nucleus provides the genetic and genetic expression systems that enable growth, cell division, and cell differentiation.
8. Stem cells, a unique precursor to all cell types, tissues, and organ types, are capable of repairing damaged or defective cells, tissues, and organs.
9. The human body is colonized by microorganisms at several locations. The microbiota is primarily in the distal intestine within the gastrointestinal tract. Gastrointestinal microbes contribute to the host as both symbiotic and pathogenic agents in many physiological processes.

ORGAN SYSTEMS: INTEGRATED FUNCTION AT THE HIGHEST LEVEL

The highest level of integrated functionality in humans is the organ system; these systems have been structurally recognized for thousands of years but recognized for their functional importance for only a few hundred. The 11 organ systems of humans are as follows:

- Skeletal—bones, cartilage, and ligaments. This system provides the framework and physical form for the body.
- Integumentary—skin, hair, and nails. This system provides a barrier between the outside world and the body.
- Muscular—skeletal, cardiac, and smooth muscle. This system enables movement of the body, skeleton, and internal organs.
- Digestive—mouth, stomach, small and large intestines, colon, and anus. This system provides the pathway for food ingestion and processing to extract nutrients and thus energy, and an environment in which symbiosis occurs with microorganisms that convert foodstuffs into nutrients absent from the foodstuffs but required for survival.
- Cardiovascular or circulatory—heart, arteries, veins, and lymphatic vessels. This is the transportation system by which nutrients and oxygen are exchanged with both the lungs and cells within organs.
- Respiratory—lungs with their alveolar sacs, trachea, nasal orifices, and diaphragm. This is the system for oxygen exchange with the atmosphere and specialized muscle that enables breathing.
- Excretory—kidneys and ureters, bladder, and urethra. This is the system for waste removal from nutrient utilization and organ and cell renewal as well as for maintaining electrolyte balance.

N. V. Bhagavan and C.-E. Ha: Essentials of Medical Biochemistry, Second edition. DOI: http://dx.doi.org/10.1016/B978-0-12-416687-5.00001-4

- Nervous—brain, spinal cord, nerves, and some receptors. This system provides cognition and electrical signal/information processing and transmission, and acts as the control center for the other organ systems of the human animal.
- Endocrine—glands and secretory tissues. This system provides for chemical signaling via transport using the circulatory system, a second control system.
- Reproductive—testes, ovaries, uterus, genitalia. This is the system by which the species ensures continuation of itself.
- Immune—white blood cell types, thymus, spleen. This is the system by which protection from pathogens enables survival.

Organs are composed of cells that compartmentalize the lower-level functions that are the basis for organ function in living organisms. These are, in turn, composed of organelles that further compartmentalize the biochemical reactions and processes that are highly evolved and specialized to produce the chemical substances needed to create these structures and to extract energy to drive the chemical reactions and mechanical actions which the various organ systems provide. Normal health and disease diagnosis are related to the organ systems of humans, and thus the primary sections and chapters of this biochemical text are similarly organized.

CELLS: STRUCTURES AND FUNCTIONS

The unifying principle of biology is that all living organisms from the smallest and least complex (bacteria) to the largest (whales) and most complex (humans) are composed of cells. The precise location of cells in the multicellular organisms and the location of intracellular organelles within cells are vital in normal development and function. During injury, wound repair, or morphogenesis, the precise location and migratory patterns of cells in multicellular organisms involve several strategies, which include establishment of gradient of small molecules, regulatory networks, and genetic diversity [1]. The membrane trafficking and metabolites to correct intracellular locations are precisely regulated. Defects in the membrane trafficking lead to pathological consequences [2].

In the simplest forms of life, such as bacteria, cellular organization and biochemical functions are relatively uncomplicated and are primarily devoted to growth and reproduction. As a consequence, bacteria have evolved to survive and thrive in the widest range of environments imaginable—soil, rivers and oceans, hot springs, and frozen land, as well as in most areas of the human body. The only regions of the body that are normally sterile are the respiratory tract below the vocal chords; the sinus and

middle ear; the liver and gall bladder; the urinary tract above the urethra; bones; joints; muscles; blood; the linings around the lungs; and cerebrospinal fluid. Intestinal colon contains numerous microorganisms, and they are collectively known as **microbiota**. Humans and microorganisms have a symbiotic relationship. A population of normal healthy microbiota is essential for the maintenance of optimal host physiology by providing nutrients and energy balance. In germ-free mice, changes in the microbial flora have ameliorated obesity. Fecal flora from healthy persons has been used to re-establish normal microbiota in a recurrent *Clostridium difficile* enteric infection. Diet, the quality of the gut microbiota, and genetic makeup of the host all may play a role in producing a proatherosclerotic metabolite known as trimethylamine-N-oxide (TMAO) leading to cardiovascular disease. Recent studies have revealed that dietary sources of carnitine and phosphatidylcholine (lecithin) are converted to trimethylamine by gut microbes, which in turn enter the enterohepatic blood circulation followed by its conversion to TMAO by hepatic flavin-containing monooxygenases. Colon epithelial barrier damage can provoke infection and inflammation and may lead to a spectrum of diseases. (These aspects, along with required reading references, are discussed in Chapter 11).

All bacteria belong to the super kingdom called **prokaryotes**. Yeasts, molds, and protozoa are also single-celled organisms, but their cellular structures and functions are more complex than those of bacteria. These organisms belong to the other super kingdom called **eukaryotes**, along with all higher plants and all multicellular animals. A prokaryote cell has no true nucleus or specialized organelles in the cytoplasm. Bacteria reproduce asexually by cell division (fission). Because mitochondria (discussed later) have many properties in common with bacteria, it suggests that bacteria-like organisms were assimilated into eukaryotic cells early in their evolution.

All eukaryotic cells have a well-defined nucleus surrounded by a nuclear membrane and cytoplasm containing organelles that perform specialized functions. All eukaryotic somatic cells reproduce by the complex mechanisms of mitosis and cytokinesis. Germinal cells (sperm and ova) are formed by a slightly different mechanism called **meiosis**.

Although the size and complexity of eukaryotic organisms differ enormously (e.g., amoeba, fly, worm, crab, bird, dog, dolphin, chimpanzee, human being), the basic organization and chemistry of their individual cells are quite similar. Sequencing of the nuclear DNA of many different organisms has shown remarkable conservation of key genes and proteins among widely dissimilar organisms. In some cases a protein produced by a human gene will function just as well when the human gene is swapped for the comparable gene in yeast.

Properties of "Living" Cells

The human body actually contains billions of both prokaryotic and eukaryotic cells that perform metabolic functions, many of them synergistic. Thus, bacteria in the human body are just as essential to the health and survival of a person as are her or his own cells (discussed earlier). Although no single definition serves to distinguish "living" from "nonliving," both prokaryotic and eukaryotic cells share certain properties that distinguish them from nonliving matter.

Metabolism

The sum of all chemical reactions in cells that maintain life is metabolism. Living cells extract energy from the environment to fuel their chemical reactions; the metabolism of dead or dying cells is significantly different from that of healthy cells. Bacteria generally extract the chemicals they need from their immediate environment; human cells obtain essential chemicals and nutrients via the circulatory system. The ultimate source of energy for all plants and animals is sunlight, which is used directly by plants and indirectly by animals that consume plants.

Growth

As a result of the utilization of energy and the synthesis of new molecules, cells increase in size and weight.

Reproduction

All cells reproduce by giving rise to identical copies of themselves. As a result of growth, cells reach a size that triggers reproduction and the production of progeny cells. For example, prokaryotes divide asexually while eukaryotes are capable of sexual reproduction.

Mutation

In the process of growth and reproduction, cells occasionally undergo a mutation that is a permanent heritable change in the genetic information in the cell's DNA. Such mutations that arise in sperm or ova are called **germinal mutations**, and these may lead to hereditary disorders. Those that arise in body cells other than sperm or ova are called **somatic mutations**; these mutations may alter normal cell growth and reproduction, and may underlie the development of cancer (unregulated cell growth), aging, or other derangements of cellular functions.

Response

All living cells and multicellular organisms respond to environmental stimuli that change chemical reactions and behaviors. Stimuli that evoke cellular responses include light, nutrients, noxious chemicals, stress, and other environmental factors.

Evolution

As a result of mutation and other genetic mechanisms, the genetic information, chemical reactions, and other properties of organisms change (evolve) over time. Some of the inherited changes that occur in organisms make some individuals better able to survive and reproduce in particular environments. Evolution causes populations of organisms to evolve over time; some become extinct and others develop into new species.

First proposed in the 1800s, the cell theory of life is now well integrated into biological sciences and medicine. In general, the **cell theory** states that (1) all organisms consist of one or more cells; (2) cells are the smallest units characteristic of life; and (3) all cells arise from pre-existing cells. How the first cells(s) arose on Earth is still matter of intense scientific research, speculation, and controversy.

Structures and Organelles in Eukaryotic Cells

Cytosol

The intracellular aqueous compartment that surrounds all of the subcellular organelles is known as **cytosol**. The content of cytosol includes ionic compartment and macromolecular enzyme units. The cytosolic components participate in the osmoregulation, extracellular transduction events, and transport and delivery of metabolites to selected cellular locations and several metabolic pathways. The metabolic pathways of cytosol involve interdependence of organelles. For example, synthesis of glucose (gluconeogenesis), heme, urea, and pyrimidines requires enzymes located in both cytosol and mitochondria. Some metabolic pathways occur entirely in the cytosol, such as glycolysis and hexose monophosphate (HMP) shunt pathways.

The metabolic pathways are organized and located at specific sites of the cytosol. The organization and integration of feedback loops of proteins that participate in specific pathways and signaling networks are facilitated by scaffolding proteins [3,4]. Cytosolic metabolic pathways are regulated depending on the availability of nutrients and oxygen supply. This is illustrated in the circulating red blood cells (RBCs). During the relative deoxygenated state of hemoglobin, glycolysis operates to produce ATP and 2,3-diphosphoglycerate (2,3-DPG). ATP provides energy and 2,3-DPG facilitates oxygen delivery to tissues from oxyhemoglobin. During the oxygenated state of hemoglobin, glucose oxidation is shunted to the HMP pathway for the production of NADPH and glutathione (GSH), both of which are required to protect the RBCs from oxidative stress (discussed in Chapters 12 and 26).

All eukaryotic cells possess characteristic structures and organelles (see Figure 1.1). The shape and size of eukaryotic cells differ markedly depending on their functions, but all are much larger than even the largest prokaryote. Most eukaryotic cells contain all of the structures shown in Figure 1.1, but there are exceptions. Non-motile cells usually lack a flagellum or cilia; the nuclei of red blood cells are extruded after being synthesized in bone marrow before they enter the circulation; the nuclei also are digested in the outermost layer of skin cells. Cilia are sensory organelles and possess hair-like microtubular structures. They participate in signaling pathways, which include detections of external signals and their integrations into metabolic functions. Ciliary dysfunction due to genetic defects can result in a wide variety of developmental and degenerative disorders known as **ciliopathies** [5]. Most cells in the body are in a dynamic state of degradation and renewal. Skin cells and gastrointestinal epithelial cells, for example, are destroyed and replaced on a regular basis. If portions of the liver are damaged by disease or surgical removal, the organ will regrow to its original size. Table 1.1 lists the primary functions of the specialized structures and organelles in a generalized animal cell. The size of cellular organelles is subject to nutrient availabilities, metabolic demands, and stressful conditions [6].

Cells of every adult multicellular organism trace their ancestry to the **zygote**, the first cell of a new individual that is formed by union of sperm and egg. During development, cells undergo repeated mitosis and division and ultimately differentiate into specialized cells that have structures and functions specific to the needs of each tissue or organ in the body.

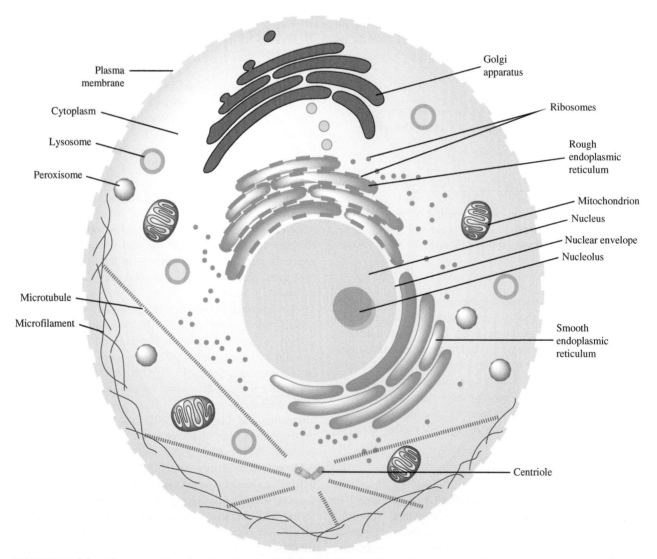

FIGURE 1.1 Schematic representation of a eukaryotic animal cell showing characteristic structures and organelles.

TABLE 1.1 Functions of Components and Organelles in Eukaryotic Cells

Component	Function
Plasma membrane	Lipid bilayer in which proteins are embedded. Regulates influx and efflux of nutrients and chemicals; also cell–cell recognition
Cytoskeleton	An array of microfilaments and microtubules in dynamic rearrangement
Flagellum and cilia	Motility of cell or movement of fluids over tissues
Endoplasmic reticulum	Internal membranes (rough and smooth) that serve as sites of synthesis for proteins, lipids, and carbohydrates; also modify and transport proteins
Nucleus	Structure with double membrane that contains chromosomes (DNA) that express genes and control chemical activities of the cell
Nucleolus	Site of ribosome synthesis and assembly
Golgi apparatus	Site of packaging of proteins for transport from the cell
Lysosomes	Structures containing enzymes that degrade unwanted cellular products
Mitochondria	Organelles that produce cellular energy (ATP); contain small DNA molecules (mitochondrial DNA)
Ribosomes	Sites of protein synthesis in cells; active ribosomes are bound to rough endoplasmic reticulum
Microfilaments	Long, thin filaments of actin; form and degrade in response to changes in intracellular ion concentrations
Microtubules	Composed of tubulin; found in flagella, cilia, and spindle fibers used to separate chromosomes in mitosis and meiosis
Intracytoplasmic lipid droplets	Sites of triacylglycerol and cholesterol esters, structurally organized by protein known as perilipins

The Plasma Membrane and Cytoskeleton

The plasma membrane and cytoskeleton determine a cell's morphology and transport of molecules. The **plasma membrane** consists of a complex lipid bilayer, phosphate and carbohydrate components, and a large number of proteins embedded in the membrane that connect the inner milieu of the cell with the external environment. The plasma membrane maintains the physical integrity of the cell and prevents the contents of the cell from leaking into the fluid environment. At the same time, it facilitates the entry of nutrients, ions, and other molecules from the outside. The pattern of proteins and carbohydrate–lipid complexes exposed on the cell surface also are specific to a particular cell type and individual. For example, skin cells of one individual are recognized as "foreign" by another individual's immune system, although cells from both individuals function as skin cells.

The functions of the plasma membrane are coordinated by specialized adhesion receptors called **integrins**. The complex modular structure of the extracellular portion of an integrin is shown in Figure 1.2. Integrins represent important cell receptors that regulate fundamental cellular processes such as attachment, movement, growth, and differentiation. Because integrins play such vital roles in the functions of the plasma membranes, they are also involved in many disease processes: they contribute to the initiation and progression of neoplasia, tumor metastasis, immune dysfunction, ischemia-reperfusion injury, viral infections, and osteoporosis.

The **cytoskeleton** of a cell is a constantly changing array of components (**microfilaments** and **microtubules**) that give a cell its structure and motility. Defects in microtubule transport lead to neurological and ciliary diseases [7]. The cytoskeleton also plays an important role in cell division and the transport of molecules across the plasma membrane. Microfilaments consist of long very thin strands of the protein **actin**, which is also a major component of muscle. Strands of microfilaments form spontaneously in high concentrations of Ca^{2+} and Mg^{2+} within the cell. Microtubules are long, thin tubes composed of the protein tubulin. They also assemble and disassemble in response to the ionic environment. Microtubules comprise the **spindle fibers** that separate chromosomes prior to cell division. **Centrioles** are composed of microtubules and function as the organizing center for the formation of spindle fibers.

Structures and Organelles Involved in Synthesis, Transport, and Degradation of Molecules

The **endoplasmic reticulum (ER)** appears as an intricate, complex, folded net in the cytoplasm. A portion of the ER has ribosomes bound to it, which give it a "rough"

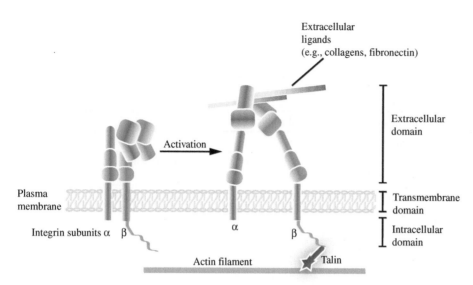

FIGURE 1.2 Diagram of integrin heterodimer modules. The extracellular domain upon activation recognizes extracellular ligands such as fibronectin to induce cell migration and extracellular matrix remodeling (signal transduction from inside to outside the cell, or inside-out signal transduction). Also, the intracellular domain upon activation interacts with intracellular components such as a cytoskeleton protein, talin, to regulate cellular processes (signal transduction from outside to inside the cell, outside-in signal transduction).

appearance when viewed by electron microscopy. The rough ER is the site for synthesis of proteins that are destined to be exported from the cell. The ER also has mechanisms for maintaining the quality of the proteins synthesized, especially those destined for transport. The ER has sensor molecules that monitor the amounts of unfolded or improperly folded proteins that accumulate. **Parkinson's disease** and **Alzheimer's disease** are both characterized by synthesis of excessive amounts of mutant proteins that are folded incorrectly. In general, accumulation of nonfunctional proteins in the ER contributes to a situation referred to as ER stress, which is associated with diabetes and cancer, as well as neurodegenerative diseases. Morphological links and metabolite and ion (e.g., Ca^{2+}) exchange occur between ER and mitochondria. ER stress can initiate programmed cell death known as **apoptosis** by mitochondria.

Other ribosomes are not attached to the ER and are "free" in the cell, although they are usually attached to the cytoskeleton. These ribosomes synthesize proteins used intracellularly. The smooth ER is the site of lipid synthesis, which does not require ribosomes.

The **Golgi apparatus** (named for its discoverer, Camillo Golgi) is a specialized organelle in which proteins are processed, modified, and prepared for export from the cell. The Golgi apparatus resembles a stack of 10 to 20 hollow, flat structures with the smallest being attached to the plasma membrane. Proteins are received from the ER and passed through layers of the Golgi apparatus where polysaccharides are synthesized and attached to proteins to make **glycoproteins**, or to lipids to make **glycolipids**.

The **lysosomes** are membrane-bounded sacs containing enzymes capable of destroying the cell if they were not confined to lysosomes. Lysosomes contain as many as 40 different hydrolytic enzymes, and a eukaryotic cell

(especially liver and kidney) may contain several hundred lysosomes. The hydrolytic enzymes found in lysosomes include proteases, nucleases, glycosidases, lipases, phosphatases, and sulfatases; all of these enzymes function at the acidic pH maintained in the lysosome.

Lysosomes assist in cell renewal by digesting old or damaged cellular components. During development, lysosomes play an important role in the formation of specialized tissues such as fingers and toes. For example, lysosomes digest the webbed tissues that join fingers and toes in the embryo. White blood cells protect the body from infectious disease by engulfing pathogenic microorganisms and isolating them in a membranous sac called a **phagosome**. A lysosome then fuses with the phagosome and digests the pathogen.

Peroxisomes are similar to lysosomes, in that they are single membranous sacs containing enzymes. However, peroxisomes contain enzymes that are used for detoxification rather than for hydrolysis. One of the most important functions of peroxisomes is the detoxification of alcohol in liver cells. Other peroxisome enzymes remove the amine group from amino acids and convert it to ammonia prior to excretion. Liver peroxisomes contain three important detoxification enzymes: **catalase**; **urate oxidase**; and **D-amino acid oxidase**. These enzymes use molecular oxygen to remove hydrogen atoms from specific substrates in oxidation reactions. The enzyme content of cellular peroxisomes varies according to the needs of the tissue. Peroxisomes also participate in degradation of long chain fatty acids. Peroxisome defects lead to disorders such as **adrenoleukodystrophy**, **Zellweger's syndrome**, and **Refsum's disease** [8].

The **ubiquitin-proteasome system** and **autophagosomes**, which are located in the cytosol, are required for intracellular proteolysis. The ubiquitin-proteasome system, which is also present in the nucleus, consists of organelles called proteasomes; an individual proteasome is a multiprotein subunit barrel-like structure with a central hollow pore. Ubiquitin, so named because it occurs ubiquitously in all eukaryotes, is a small 76 amino acid residue protein (8.5 kDa), which tags misfolded proteins as well as normal proteins destined for destruction. The polyubiquitinated protein is eventually surrendered to proteasomes for proteolysis, while ubiquitin molecules escape proteolysis and are reused. The overall process is regulated and requires several enzymes, and energy is provided by ATP hydrolysis. The autophagosome (self-digestion) system consists of sequestration of misfolded proteins, followed by their integration with lysosomes and proteolysis by lysosomal enzymes. The ubiquitin-proteasome and autophagosome systems are required for several normal cellular functions (e.g., cell cycle and division, immune response, biogenesis of organelles). Disruption of the ubiquitin-proteasome and autophagosome systems is associated with several disorders. A chemotherapeutic agent, bortezomib, inhibits ubiquitin-proteasomal proteolysis and promotes apoptosis of rapidly growing monoclonal plasma cells and thus is used in the treatment of multiple myeloma (Chapter 4).

Almost all vertebrate cells have a primary or many specialized projections on the cell surface that are called **cilia**; some cells also have a larger, more complex projection called a **flagellum** (plural flagella) that is used for cell movement. Some cilia are also involved in motility, but most have other functions and are immotile. Cilia are also specialized cellular organelles.

Cilia play essential roles in the olfactory system, in visual photoreceptors, and in mechanosensation. Cilia play vital roles in intercellular communication and signaling. Ciliary defects are associated with many human disorders, including retinal degeneration, polycystic kidney disease, and neural tube defects.

Mitochondria Provide Cells with Energy

Mitochondria are organelles in eukaryotic cells that supply energy for all cellular metabolic activities. The number of mitochondria in cells varies as do their energy needs; muscle cells, especially those in the heart, contain the largest number of mitochondria. The overall production of energy in body cells is expressed by the following equation:

$$\text{glucose} + \text{oxygen} \Rightarrow \text{carbon dioxide} + \text{water} + \text{energy}$$

Mitochondria are characterized by an outer membrane and an inner membrane that is intricately folded and organized for the transport of electrons along the **respiratory chain**. The inner membrane is the site of **oxidative phosphorylation**, which generates most of a cell's energy in the form of ATP. Mitochondria are in a dynamic state in the cells; they are mobile and constantly change from oval- to rod-shaped.

In addition, mitochondria are in a continual state of fission and fusion, so that the identity of any given mitochondrion is transient. At any point in the cell's cycle, a mitochondrion may undergo fission to give rise to two separate mitochondria. Concurrently, other mitochondria are undergoing fusion in which both the inner and outer membranes of the mitochondria break and rejoin to form a single intact mitochondrion. At least three different GTPases are involved in mitochondrial fusion. For fission, another GTPase called **dynamin-related protein 1 (DRP1)** is required. At least one serious disease involving microcephaly and metabolic abnormalities has been ascribed to mutations in DRP1 (see Chapter 13 and Clinical Case Study 13.1).

Mitochondria contain small, circular DNA molecules (mtDNA) containing 37 genes that are essential for mitochondrial functions. The ribosomes in the mitochondria also resemble prokaryotic ribosomes rather than eukaryotic ribosomes found in the cytoplasm, which further supports their derivation from bacteria that were assimilated early in the evolution of eukaryotic cells.

Mutations in mtDNA are responsible for a number of diseases (**mitochondriopathies**) that can be inherited, although not in the Mendelian pattern that is characteristic of nuclear genes (see Clinical Case Study 13.2 for an exception). Inheritance of mtDNA defects is through a maternal lineage, as mitochondria are present in large numbers in the ovum that forms the zygote. Male mitochondria are not present in the head of the sperm that fertilizes the ovum. In some patients, exercise intolerance and muscle fatigue are due to mutations in mtDNA. Mitochondria also participate in other metabolic pathways in conjunction with the cytoplasmic enzymes; these include heme biosynthesis, urea formation, fatty acid oxidation, and initiation of apoptosis by release of cytochrome C.

Lipid Droplets

Lipid droplets store triacylglycerol (triglyceride) and cholesterol esters as discrete organelles. Adipocytes and other cells that store triacylglycerols and cholesteryl esters contain lipid droplets, which are organized by proteins known as perilipins. The perilipin family of proteins and other associated proteins provide structural organization of scaffolding and metabolic functions. In the white adipose tissue cells, 90% of the cell volume consists of unilocular lipid droplets (Chapter 18).

The Nucleus Controls a Cell's Development and Chemical Activities

The nucleus is the control center of the cell; it contains the chromosomes that carry all of the individual's genetic information. The nucleus is encased in a double membrane called the **nuclear envelope**. The outer membrane is fused with the ER at multiple sites and forms **nuclear pores** that facilitate transport of molecules between the cytoplasm and the nucleus. The inner membrane encases the chromosomes and usually two or more organelles called **nucleoli** that surround a region of the DNA containing genes for making ribosomal RNA (rRNA). The many copies of rRNA synthesized in the nucleoli are bound to ribosomal proteins that are transported to the nucleus from the cytoplasm. After assembly of ribosomes in nucleoli, they are exported through the nuclear pores back into the cytoplasm.

During most of the cell cycle, the chromosomes are in a diffuse, extended state in the nucleus and are not visible. During mitosis when a cell is preparing to divide, the chromosomes condense into visible structures whose movements can be observed in the light microscope. All of the genes whose products are essential for the functions of the particular cell are regulated and expressed (transcribed) as messenger RNA (mRNA) molecules that are subsequently transported to the cytoplasm, where they are translated into proteins. However, prior to being transported to the cytoplasm, the mRNAs are extensively processed. RNA-splicing enzymes remove segments of the mRNA (introns) and rejoin segments that are to be translated (exons). Nucleotides are also added to both ends of the mRNA that facilitate its transport and prevent premature degradation.

Each of the 46 chromosomes in a human cell contains a linear, continuous strand of double helical DNA. Each DNA molecule contains multiple sites of replication, a centromere that functions in separating chromosomes during mitosis, and repeated nucleotide sequences at both ends of the DNA called **telomeres**. These special structures at the ends of DNA molecules prevent the loss of genetic information at each cycle of DNA replication. Because of the mechanics of DNA replication, some bases are lost at the ends of each molecule of DNA after each cycle of DNA replication. Thus, the telomeres protect against the loss of genetic information at each cycle of replication, while the length of the telomeric repeats becomes shorter.

The DNA in telomeres can be restored and elongated by an enzyme called **telomerase**. Most somatic cells in an adult no longer produce active telomerase. As a result, telomeres are gradually lost during repeated cell divisions until the cell reaches a stage when it is no longer viable. Recent studies demonstrate that telomeres and telomerase play essential roles in the biology of cancer, stem cells, aging, and an inherited disorder, **dyskeratosis congenita**. Cells that contain serious defects in telomere length or in telomerase activity are destroyed by **apoptosis** (programmed cell death). Apoptosis is believed to be necessary during normal differentiation of tissues in development and for replacement of aging cells in the adult. Abnormalities in apoptosis may play a role in cancer; as many as 90% of cancer cells

have reactivated telomerase, which may contribute to the unregulated growth of tumors.

Apoptosis is a normal part of embryonic and postnatal development of the nervous system. Many cells become "gratuitous" after serving their function during development. After performing their necessary functions, they are removed by apoptosis. Thus, apoptosis is essential for normal development and for the removal of aging cells in the adult. However, failure to inhibit apoptotic activity in adult neural cells that are not replaced may be one of the causes of neurological diseases. Many chronic, neurological diseases are associated with neural cell death, including **amyotrophic lateral sclerosis** (**ALS**), **Alzheimer's disease**, **Parkinson's disease**, and **Huntington's disease**.

The Body of an Adult Consists of More Than 200 Specialized Types of Cells

During the development of a human being (and other vertebrates), cells undergo **differentiation** and become specialized in both form and function as they assemble into specialized tissues and organs. Examples of major tissues in the body are epithelial tissue, connective tissue, muscle tissue, and nervous tissue. **Epithelial cells** aggregate into sheets of cells that line the inner and outer surfaces of the body. Depending on their location in the body, epithelial cells differentiate into secretory cells, ciliated cells, absorptive cells, and other cell types. The spaces between tissues and organs are filled with **connective cells** that provide a matrix for other organs and tissues. **Nervous tissue** consists of highly specialized cells in the brain and central nervous system that control all of the organism's activities and capabilities. **Muscle tissue** provides the mechanical force that allows movement.

In addition to these tissue systems, organs such as the liver, heart, kidney, ovary, testis, and pancreas possess specialized cells whose biochemical activities are specific to each organ. All cellular activities must be regulated and integrated with the functions of other tissues and organs for optimal health of the individual. **Understanding the chemical activities of specialized cells and tissues, their molecular interactions with other cells and tissues, and how aberrations produce disease is the subject of medical biochemistry.**

Stem Cells Are a Renewable Source of Specialized Cells

Embryonic stem cells are derived from the inner cell mass of 5- to 10-day-old blastocysts. These cells are pluripotent and are capable of differentiating into any specialized cell in the body. Human embryonic stem cells have been obtained by transferring a somatic cell (skin

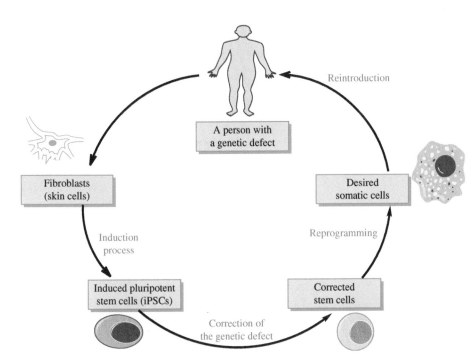

FIGURE 1.3 A model for correction of monogenic defect using patients' own fibroblasts. This method uses iPSCs derived from human α_1-antitrypsin (α_1-AT) fibroblasts after the correction of the genetic defect and converts them to hepatocytes, the physiologic source of α_1-AT. The reprogrammed hepatocytes injected into mice were found in the liver and synthesized α_1-AT [9].

cell) nucleus into denucleated human oocytes (see Enrichment references 6,7). Stem cells can also be derived from fetuses, umbilical cord blood, the placenta, or from adult tissues. Stem cells from these sources may be less than pluripotent. When cultured under favorable conditions, embryonic stem cells have the potential to develop into stable lines of differentiated cell types. The major goal of all stem cell research is to develop techniques for propagating specific cells from an individual that can be used to treat diseases, for example, replacing pancreatic islet cells from a person with type 1 diabetes.

Stem cells have three characteristics that distinguish them from other cells: (1) the capacity for self-renewal; at cell division one or both daughter cells have the same biological properties as the parent cell; (2) the capacity to develop into multiple types of cell lineages; (3) the potential to proliferate indefinitely. Hematopoietic stem cells migrate through the blood, across endothelial vasculature, and end up in specific organs and bone marrow. Migration of stem cells from blood to destined tissue is a complex process called **homing**, which involves stress signals and other factors.

The controversy over stem cell research derives from the origin of cells used for research or treatment, namely embryos created *in vitro* or aborted fetuses. However, a noncontroversial source of stem cells is umbilical cord blood, which is usually discarded after birth. Stem cells harvested from cord blood have been used in the treatment of leukemia and lymphoma. Multifaceted and multidisciplinary approaches are underway to produce pluripotent stem cells from noncontroversial sources.

One of the methods involves the following steps: the induction of pluripotent stem cells from fibroblasts (iPSCs) obtained from a person with a genetic defect, correction of the genetic defect, reprogramming of the corrected stem cells to the desired somatic cells, and finally the reintroduction of repaired somatic cells to the original person from which the fibroblasts were obtained (Figure 1.3). This concept has been accomplished in part in α_1-antitrypsin (α_1-AT) deficiency. Fibroblasts obtained from an α_1-AT deficient human are converted to iPSCs and, after correction of the genetic defect, the cells are reprogrammed to hepatocytes and injected into mice. In the injected mice with corrected human hepatocytes, they express normal α_1-AT [9].

Another avenue of research involves obtaining stem cells from human tooth pulp. Injection of these cells into transected rat spinal cord showed functional improvements [10].

In studies using rodents [11,12], it has been shown that the fibroblasts of injured myocardium can be reprogrammed into becoming normal cardiomyocytes by administration of exogenous transcription factors and microRNA (Chapter 24).

All of these studies are aimed at obtaining readily available, immunologically compatible stem cells by various methods for the therapy of many diseases.

REQUIRED READING

[1]   A.D. Lander, How cells know where they are, Science 339 (2013) 923–927.

[2]   M.A. De Matteis, A. Luini, Mechanism of disease: Mendelian disorders of membrane trafficking, N. Engl. J. Med. 365 (2011) 927–938.

[3] M.C. Good, J.G. Zalatan, W.A. Lim, Scaffold proteins: Hubs for controlling the flow of cellular information, Science 332 (2011) 680–686.

[4] F. Shi, M.A. Lemmon, KSR plays CRAF-ty, Science 332 (2011) 1043–1044.

[5] F. Hildebrandt, T. Benzing, N. Katsanis, Ciliopathies, N. Engl. J. Med. 364 (2011) 1533–1543.

[6] Y.M. Chan, W.F. Marshall, How cells know the size of their organelles, Science 337 (2012) 1186–1189.

[7] J.M. Gerdes, N. Katsanis, Microtubule transport defects in neurological and ciliary disease, CMLS Cell. Mol. Life Sci. 32 (2005) 1556–1570.

[8] H.R. Waterham, M.S. Ebberink, Genetics and molecular basis of human peroxisome biogenesis disorders, Biochim. Biophys. Acta 2012 (1822) 1430–1441.

[9] R.A. Sandhaus, Gene therapy meets stem cells, N. Engl. J. Med. 366 (2012) 567–569.

[10] E.Y. Snyder, Y.D. Teng, Stem cells and spinal cord repair, N. Engl. J. Med. 366 (2012) 1940–1942.

[11] Y. Nam, K. Song, E.N. Olson, Heart repair by cardiac reprogramming, Nat. Med. 19 (2013) 413–415.

[12] P. Bostrom, J. Frisén, New cells in old hearts, N. Engl. J. Med. 368 (2013) 1358–1360.

ENRICHMENT READING

[1] S.E. Senyo, M.L. Steinhauser, C.L. Pizzimenti, V.K. Yang, L. Cai, M. Wang, et al., Mammalian heart renewal by pre-existing cardiomyocytes, Nature 493 (2013) 433–437.

[2] S.A. Goldman, M. Nedergaard, M.S. Windrem, Glial progenitor cell-based treatment and modeling of neurological disease, Science 338 (2012) 491–495.

[3] A.M.K. Choi, S.W. Ryter, B. Levine, Autophagy in human health and disease, N. Engl. J. Med. 368 (2013) 651–662.

[4] K. Yusa, S.T. Rashid, H. Strick-Marchand, I. Varela, P. Liu, D.E. Paschon, et al., Targeted gene correction of α_1-antitrypsin deficiency in induced pluripotent stem cells, Nature 478 (2011) 391–396.

[5] K. Sakai, A. Yamamoto, K. Matsubara, S. Nakamura, M. Naruse, M. Yamagata, et al., Human dental pulp-derived stem cells promote locomotor recovery after complete transection of the rat spinal cord by multiple neuro-regenerative mechanisms, J. Clin. Invest. 122 (2012) 80–90.

[6] D. Cyranoski, Human stem cells created by cloning, Nature 497 (2013) 295–296.

[7] M. Tachibana, P. Amato, M. Sparman, N.M. Gutierrez, R. Tippner-Hedges, H. Ma, et al., Human embryonic stem cells derived by somatic cell nuclear transfer, Cell 153 (2013) 1–11.

Chapter 2

Water, Acids, Bases, and Buffers

Key Points

1. Water plays a major role in all aspects of metabolism, and life is not possible without water.
2. Micelles are submicroscopic spherical aggregates of amphipathic molecules that contain large nonpolar hydrocarbon chains (hydrophobic groups) and polar or ionic groups (hydrophilic groups).
3. The reversible dissociation of water, although very weak, is important in maintaining and regulating the body's acid—base homeostasis.
4. An optimal acid—base balance is maintained in body fluids and cells despite large fluxes of metabolites. A buffer system protects the body from fluctuations in pH.
5. Metabolism produces both inorganic and organic acids. Acids generated from metabolites other than CO_2 are nonvolatile and are excreted via the kidney. Nonvolatile acids are lactic acid, acetoacetic acid, β-hydroxybutyrate, and acids derived from sulfur-containing amino acids and phosphorus-containing compounds.
6. CO_2 produced by metabolism is transported as bicarbonate ion (HCO_3^-) in the bicarbonate-carbonic system in plasma. CO_2 is eliminated in the lungs. Perturbations in this system can lead either to retention of CO_2 (acidosis) or to excessive loss of CO_2 (alkalosis).
7. Transport of CO_2 involves red blood cell carbonic anhydrase, which catalyzes the reversible reaction of hydration of CO_2 to H_2CO_3. Hemoglobin in tissue capillaries binds H^+ and releases oxygen to the tissues.
8. The H^+ ion concentration of the extracellular fluid, including plasma, is maintained at pH 7.37—7.44 (36—43 nmol/L).

PROPERTIES OF WATER

Acid and base concentrations in living systems are carefully regulated to maintain conditions compatible with normal life. Biochemical reactions involving acids and bases occur in the body water, whereas buffer systems protect the body from significant variations in the concentrations of acids and bases. This chapter introduces basic concepts of the properties of water, acids, bases, and buffers.

Life cannot be sustained without water. Water constitutes 45%—73% of total human body weight. It is distributed in intracellular (55%) and extracellular (45%) compartments and provides a continuous solvent phase between body compartments. As the biological solvent, water plays a major role in all aspects of metabolism: absorption, transport, digestion, and excretion of inorganic and organic substances, as well as maintenance of body temperature. The unique properties of water are inherent in its structure.

Hydrogen Bonding

Water (H_2O) is a hydride of oxygen in which the highly electronegative oxygen atom attracts the bonding electrons from two hydrogen atoms. This leads to polar H—O bonds in which the hydrogen atoms have a slightly positive charge (δ^+) and the oxygen atom has a slightly negative charge (δ^-) (Figure 2.1). Water molecules have a relatively high dipole moment because of the angle (104.5°) of the H—O—H bond and the polarity of the bonds. Neighboring liquid water molecules interact with one another to form an extensive lattice-like structure, similar to the structure of ice. The intermolecular bonding between water molecules arises from the attraction between the partial negative charge on the oxygen atom and the partial positive charge on the hydrogen atom of adjacent water molecules. This type of attraction involving a hydrogen atom is known as a **hydrogen bond** (Figure 2.2).

Hydrogen bonds contain a hydrogen atom between two electronegative atoms (e.g., O and N). One is the formal hydrogen donor; the other is the hydrogen acceptor. The amount of energy required to break a hydrogen bond (bond energy) is estimated to be 2—5 kcal/mol (8.4—20.9 kJ/mol) in the gas phase. Covalent bonds have bond energies of 50—100 kcal/mol (209—418 kJ/mol). The cumulative effect of many hydrogen bonds is equivalent to the stabilizing effect of covalent bonds. In proteins, nucleic acids, and water, hydrogen bonds are essential to stabilize overall structure. In ice, each water molecule forms a hydrogen bond with four other water molecules, giving rise to a rigid tetrahedral arrangement (Figure 2.2). In the liquid state, water maintains a tetrahedrally coordinated structure over short ranges and for short time periods.

N. V. Bhagavan and C.-E. Ha: Essentials of Medical Biochemistry, Second edition. DOI: http://dx.doi.org/10.1016/B978-0-12-416687-5.00002-6

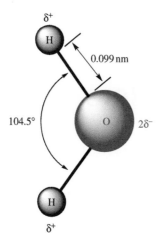

FIGURE 2.1   Structure of the water molecule.

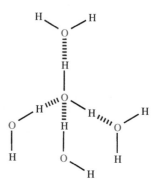

FIGURE 2.2 Tetrahedral hydrogen-bonded structure of water molecules in ice. The tetrahedral arrangement is due to the fact that each water molecule has four fractional charges: two negative charges due to the presence of a lone pair of electrons on the oxygen atom and two positive charges, one on each of the two hydrogen atoms. In the liquid phase, this tetrahedral array occurs transiently.

Physical Properties

Properties of water uniquely suited to biological systems include melting point, boiling point, heat of vaporization (quantity of heat energy required to transform 1 g of liquid to vapor at the boiling point), heat of fusion (quantity of heat energy required to convert 1 g of solid to liquid at the melting point), specific heat (the amount of heat required to raise the temperature of 1 g of substance by 1°C), and surface tension. All these values for water are much higher than those for other low molecular weight substances because of the strong intermolecular hydrogen bonding of water. These properties contribute to maintenance of temperature and to dissipation of heat in living systems. Thus, water plays a major role in thermoregulation in living systems. The optimal body temperature is a balance between heat production and heat dissipation. Impaired thermoregulation causes either **hypothermia** or **hyperthermia**, and has serious metabolic consequences; if uncorrected, impaired thermoregulation may lead to death. Maintenance of therapeutic hypothermia at 32°C−34°C for 12−24 hours has been employed to improve neurologic outcomes in patients with cardiac arrhythmia [1].

Water is transported across cell membranes in one of two ways: by simple diffusion through the phospholipid bilayer and by the action of membrane-spanning transport proteins known as **aquaporins**.

Thus, the concentration of water is in thermodynamic equilibrium across the cell membrane. In the renal collecting duct, water is reabsorbed through a specific aquaporin channel protein (aquaporin 2). This reabsorption of water is regulated by antidiuretic hormone (also known as **vasopressin**). A defect or lack of functional aquaporin 2, vasopressin, or its receptor leads to enormous loss of water in the urine, causing the disease known as **diabetes insipidus** (Chapter 37). Water plays a significant role in enzyme functions, molecular assembly of macromolecules, and allosteric regulation of proteins. For example, the effect of protein solvation in allosteric regulation is implicated in the transition of deoxyhemoglobin to oxyhemoglobin. During this process, about 60 extra water molecules bind to oxyhemoglobin.

Solutes, Micelles, and Hydrophobic Interactions

Water is an excellent solvent for both ionic compounds (e.g., NaCl) and low molecular weight nonionic polar compounds (e.g., sugars and alcohols). Ionic compounds are soluble because water can overcome the electrostatic attraction between ions through solvation of the ions. Nonionic polar compounds are soluble because water molecules can form hydrogen bonds to polar groups (e.g., −OH).

Amphipathic compounds, which contain both large nonpolar hydrocarbon chains (hydrophobic groups) and polar or ionic groups (hydrophilic groups), may associate with each other in submicroscopic aggregations called **micelles**. Micelles have hydrophilic (water-liking) groups on their exterior (bonding with solvent water) and hydrophobic (water-disliking) groups clustered in their interior. They occur in spherical, cylindrical, or ellipsoidal shapes. Micelle structures are stabilized by hydrogen bonding with water, by van der Waals attractive forces between hydrocarbon groups in the interior, and by energy of hydrophobic reactions. As with hydrogen bonds, each hydrophobic interaction is very weak, but many such interactions result in formation of large, stable structures.

Hydrophobic interaction plays a major role in maintaining the structure and function of cell membranes, the activity of proteins, the anesthetic action of nonpolar compounds such as chloroform and nitrous oxide, the absorption of digested fats, and the circulation of hydrophobic molecules in the interior of micelles in blood plasma.

Colligative Properties

The colligative properties of a solvent depend on the concentration of solute particles. These properties include freezing point depression, vapor pressure depression, osmotic pressure, and boiling point elevation. The freezing point of water is depressed by 1.86°C when 1 mol of nonvolatile solute, which neither dissociates nor associates in solution, is dissolved in 1 kg of water. The same concentration of solute elevates the boiling point by 0.543°C. Osmotic pressure is a measure of the tendency of water molecules to migrate from a dilute to a concentrated solution through a semipermeable membrane. This migration of water molecules is termed **osmosis**. A solution containing 1 mol of solute particles in 1 kg of water is a 1-osmolal solution. When 1 mol of a solute (such as NaCl) that dissociates into two ions (Na^+ and Cl^-) is dissolved in 1 kg of water, the solution is 2-osmolal.

Measurement of colligative properties is useful in estimating solute concentrations in biological fluids. For example, in blood plasma, the normal total concentration of solutes is remarkably constant (275−295 milliosmolal). Pathological conditions (e.g., dehydration, renal failure) involving abnormal plasma osmolality are discussed in Chapter 37.

Dissociation of Water and the pH Scale

Water dissociates to yield a hydrogen ion (H^+) and a hydroxyl ion (OH^-).

$$H_2O \rightleftharpoons H^+ + OH^- \qquad (2.1)$$

The H^+ bonds to the oxygen atom of an undissociated H_2O molecule to form a hydronium ion (H_3O^+).

$$H_2O + H_2O \rightleftharpoons H_3O^+ + OH^-$$

Thus, water functions as both an acid (donor of H^+ or proton) and as a base (acceptor of H^+ or proton). This description of an acid and a base follows from the Bronsted−Lowry theory. According to the Lewis theory, acids are electron pair acceptors and bases are electron pair donors. The equilibrium constant, K, for the dissociation reaction in Equation (2.1) is

$$K = \frac{[H^+][OH^-]}{[H_2O]} \qquad (2.2)$$

where the square brackets refer to the molar concentrations of the ions involved. K can be determined by measurement of the electrical conductivity of pure water, which has the value of 1.8×10^{-16} M at 25°C, indicative of a very small ion concentration, where M (molar) is the units of moles per liter. Therefore, the concentration of undissociated water is essentially unchanged by the dissociation reaction.

Since 1 liter of water weighs 1000 g and 1 mol of water weighs 18 g, the molar concentration of pure water is 55.5 M. Substitution for K and [H_2O] in Equation (2.2) yields

$$[H^+][OH^-] = (55.5 \text{ M}) \times (1.8 \times 10^{-16} \text{ M})$$

$$[H^+][OH^-] = 1.0 \times 10^{-14} \text{M}^2 = K_w$$

K_w is known as the **ion product of water**. In pure water, [H^+] and [OH^-] are equal, so that

$$[OH^-] = [H^+] = 1.0 \times 10^{-7} \text{ M}$$

pH is employed to express these ion concentrations in a convenient form, where the "p" of pH symbolizes "negative logarithm (to the base 10)" of the concentration in question (see Table 2.1). Thus,

$$pH = -\log_{10}[H^+] = \log \frac{1}{[H^+]}$$

TABLE 2.1 The pH Scale

[H^+] M	pH	[OH^-] M
Acidic		
10.0	−1	10^{-15}
1.0	0	10^{-14}
0.1	1	10^{-13}
0.01 (10^{-2})	2	10^{-12}
10^{-3}	3	10^{-11}
10^{-4}	4	10^{-10}
10^{-5}	5	10^{-9}
10^{-6}	6	10^{-8}
Neutral		
10^{-7}	7	10^{-7}
Basic		
10^{-8}	8	10^{-6}
10^{-9}	9	10^{-5}
10^{-10}	10	10^{-4}
10^{-11}	11	10^{-3}
10^{-12}	12	0.01
10^{-13}	13	0.1
10^{-14}	14	1
10^{-15}	15	10

Similarly,

$$pOH = -\log_{10}[OH^-] = \log\frac{1}{[OH^-]}$$

Therefore, for water

$$\log[H^+] + \log[OH^-] = \log 10^{-14}$$

or

$$pH + pOH = 14.$$

The pH value of 7 for pure water at 25°C is considered to be neutral, values below 7 are considered acidic, and values above 7 are considered basic. It is important to recognize that as the pH decreases, $[H^+]$ increases. A decrease in one pH unit reflects a 10-fold increase in H^+ concentration. In discussions of acid–base problems in human biochemistry, it is often preferable to express H^+ concentration as nanomoles per liter (nmol/L).

BUFFERS

Buffers resist change in pH in solutions when acids or bases are added. They are either a mixture of a weak acid (HA) and its conjugate base (A^-) or a mixture of a weak base (B) and its conjugate acid (HB^+).

Henderson–Hasselbalch Equation

The Henderson–Hasselbalch equation was developed independently by the American biological chemist L. J. Henderson and the Swedish physiologist K. A. Hasselbalch, for relating the pH to the bicarbonate buffer system of the blood (see below). In its general form, the Henderson–Hasselbalch equation is a useful expression for buffer calculations.

It can be derived from the equilibrium constant expression for a dissociation reaction of the general weak acid (HA) in Equation (2.3):

$$K = \frac{[H^+][A^-]}{[HA]} \tag{2.3}$$

where K is the equilibrium constant at a given temperature. For a defined set of experimental conditions, this equilibrium constant is designated as K' (K prime) and referred to as an apparent dissociation constant. The higher the value of K', the greater the number of H^+ ions liberated per mole of acid in solution and hence the stronger the acid. K' is thus a measure of the strength of an acid. Rearrangement of Equation (2.3) yields

$$[H^+] = \frac{K'[HA]}{[A^-]} \tag{2.4}$$

Taking logarithms of both sides of Equation (2.4) and multiplying throughout by -1 gives

$$-\log[H^+] = -\log K' - \log[HA] + \log[A^-] \tag{2.5}$$

Substituting pH for $-\log[H^+]$ and pK' for $-\log K'$ yields

$$pH = pK' + \log\frac{[A^-]}{[HA]} \tag{2.6}$$

or

$$pH = pK' + \log\frac{[conjugate\ base]}{[acid]} \tag{2.7}$$

This relationship represents the Henderson–Hasselbalch equation.

Since a buffer is intended to give only a small change in pH with added H^+ or OH^-, the best buffer for a given pH is the one that gives the smallest change. As may be seen from the Henderson–Hasselbalch equation, when the pH of the solution equals the pK' of the buffer, [conjugate base] = [acid], the buffer can respond equally to both added acid and added base. Similarly to the weak acid dissociation reaction, the Henderson–Hasselbalch relationship for weak base can be applied.

For a weak acid, if the pH is one unit below its pK' value, the solution contains approximately 91% of the unionized species (protonated, HA). Conversely, if the pH is one unit above its pK' value, the solution contains 90% of the ionized species (unprotonated, A^-). For a weak base, if the pH is one unit below its pK' value, the solution contains approximately 91% of the ionized species (protonated, BH^+). Conversely, if the pH is one unit above its pK' value, the solution contains approximately 90% of unionized species (unprotonated, B) (Table 2.2).

pH Partitioning and Ion Trapping across Cell Membranes of Various Body Compartments of Weak Acidic and Weak Basic Drugs

Many therapeutic agents are weak acids or bases (Table 2.3). Unionized species of weak electrolytes (HA or B) are usually lipid soluble and diffuse through cell membranes. In contrast, the ionized species (A^- or BH^+) do not permeate freely across cell membranes. Thus, the partitioning of an ionized drug and unionized drug across cell membranes is governed by the pK' of the drug and the prevailing pH of a particular body compartment. For example, acetylsalicylate (aspirin), an acidic drug (pK' = 3.5), is absorbed more rapidly in the un-ionized form in the acidic conditions of the stomach compared to the small intestine where the pH is alkaline. However, due to the availability of a very large absorptive surface area of the villi and microvilli of the intestine, the absorption of acidic drugs occurs

TABLE 2.2 Percent Unprotonated Species and Ratio of Unprotonated Forms Relative to the Difference between pH and pK′

%A⁻	[A⁻] /[HA]	$\text{Log} \frac{[\text{A}^-]}{[\text{HA}]} (\text{pH} - \text{pK}')$
"100"	999/1	3.00
99	99/1	2.00
98	98/2	1.69
96	96/4	1.38
94	94/6	1.20
92	92/8	1.06
91	91/9	1.00
90	90/10	0.95
80	80/20	0.60
70	70/30	0.37
60	60/40	0.18
50	50/50	0.00
40	40/60	− 0.18
30	30/70	− 0.37
20	20/80	− 0.60
10	10/90	− 0.95
8	8/92	− 1.06
6	6/94	− 1.20
4	4/96	− 1.38
2	2/98	− 1.69
1	1/99	− 2.00
"0"	1/999	− 3.00

Reproduced, with permission, from Aronson, J. N. (1981). The Henderson−Hasselbalch equation revisited. *Biochemical Education* 11 (2), 68.

TABLE 2.3 pK′ Values of Selected Weak Acidic and Basic Drugs

	pK′
Weak Acids	
Ampicillin	2.5
Aspirin	3.5
Chlorpropamide	5.0
Furosemide	3.9
Ibuprofen	4.4, 5.2
Levodopa	2.3
Methotrexate	4.8
Tolbutamide	5.3
Warfarin	5.0
Weak Bases	
Albuterol	9.3
Allopurinol	9.4, 12.3
Amiloride	8.7
Amphetamine	9.8
Atropine	9.7
Codeine	8.2
Isoproterenol	8.6
Methadone	8.4
Methyldopa	10.6
Morphine	7.9
Procainamide	9.2
Propranolol	9.4

predominantly at this site of the gastrointestinal tract. The pH partitioning also occurs at several other sites in the body, and these include renal tubule, biliary tract, lactating mammary gland, and blood−brain barrier. In the renal system, the weak acidic and basic drugs filtered at the glomeruli are excreted. The urine pH determines the amount of the drug excreted, and therefore the manipulation of urine pH has therapeutic implications. In general, weak acidic drugs are excreted more rapidly in alkaline urine and less rapidly in acidic urine. This effect of urine pH has a converse effect on weak basic drugs. Manipulation of urine pH can be accomplished by the administration of sodium bicarbonate and carbonic anhydrase inhibitors, affecting blood levels of weak acidic and weak basic drugs.

The transport of drugs is a passive process and is not energy-dependent, although the establishment of the pH gradient across the cell membrane is an energy-dependent active process. There are many other transport systems that are utilized for movement of molecules across cell membranes. They include carrier-mediated membrane transport systems that can either be an active-energy-dependent or a non-energy-dependent facilitated diffusion process, endocytosis or exocytosis. A transport system that participates primarily in the efflux of components from cells through cell membranes is mediated by P-glycoprotein or multidrug-resistant type 1 (MDR1) transporter. Neoplastic (cancer) cells can develop resistance to chemotherapeutic agents by expelling the drug via MDR1 transporters.

Buffer Systems of Blood and Exchange of O_2 and CO_2

If the H^+ concentration deviates significantly from its normal level in blood, the health and survival of the human body are in jeopardy. H^+ is the smallest ion, and it combines with many negatively charged and neutral functional groups. Changes of $[H^+]$, therefore, affect the charged regions of many molecular structures, such as enzymes, cell membranes, and nucleic acids, and dramatically alter their physiological activity. If the plasma pH reaches either 6.8 or 7.8, death may be unavoidable. Despite the fact that large amounts of acidic and basic metabolites are produced and eliminated from the body, buffer systems maintain a fairly constant pH in body fluids.

The major metabolic product from oxidation of ingested carbon compounds is CO_2. Hydration of CO_2 dissolved in water yields the weak acid H_2CO_3 (carbonic acid). Depending on the type of food ingested and oxidized, 0.7–1.0 mol of CO_2 is produced per mole of O_2 consumed. This results in the metabolic production of about 13 mol of hydrated CO_2 each day in a normal person.

For efficient transport of relatively insoluble CO_2 from the tissues where it is formed to the lungs where it must be exhaled, the carbonic anhydrase of red blood cells convert CO_2 to the very soluble anionic form HCO_3^- (bicarbonate ion). The principal buffers in blood are bicarbonate–carbonic acid in plasma, hemoglobin in red blood cells, and protein functional groups in both. The normal balance between rates of elimination and production of CO_2 yields a steady-state concentration of CO_2 in the body fluids and a relatively constant pH.

Other acids that are products of metabolism are lactic acid, acetoacetic acid, β-hydroxybutyric acid, phosphoric acid, sulfuric acid, and hydrochloric acid. The organic acids (e.g., lactate, acetoacetate, and β-hydroxybutyrate) are normally oxidized further to form CO_2 and H_2O. The hydrogen ions and anions contributed by mineral acids and any unmetabolized organic acids are eliminated via the excretory system of the kidneys. Thus, although body metabolism produces a large amount of acid, a constant pH is maintained by transport of H^+ ions and other acid anions in buffer systems, and by elimination of CO_2 through alveolar ventilation in the lungs and excretion of aqueous acids in the urine.

Metabolic activities continuously release CO_2 to the blood (Figure 2.3), and the lungs continuously eliminate CO_2 (Figure 2.4). As oxygen is consumed in peripheral tissues, CO_2 is formed and its pressure (P_{CO_2}) builds to about 50 mmHg, whereas the blood entering the tissue

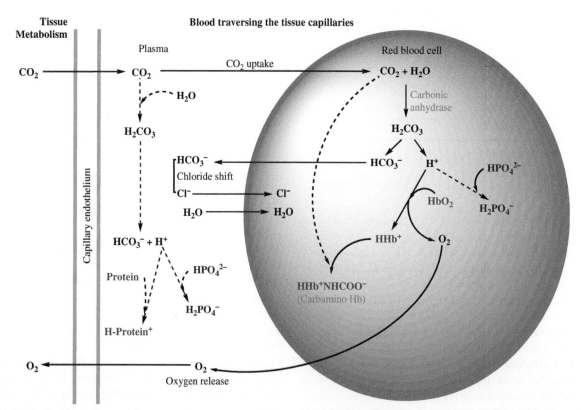

FIGURE 2.3 Schematic representation of the transport of CO_2 from the tissues to the blood. Note that most of the CO_2 is transported as HCO_3^- in the plasma, and that the principal buffer in the red blood cell is hemoglobin. Solid lines refer to major pathways, and broken lines refer to minor pathways. Hb = hemoglobin.

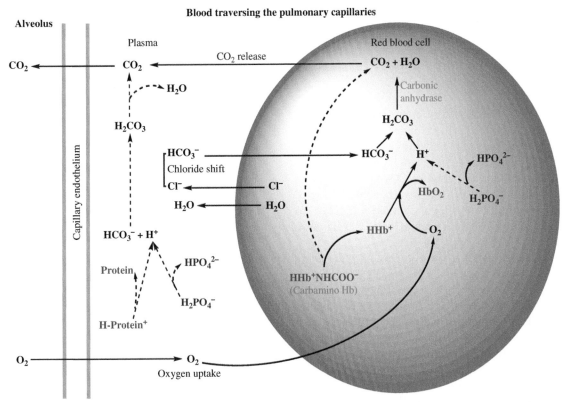

FIGURE 2.4 Schematic representation of the transfer of CO_2 from the alveolus (and its loss in the expired air in the lungs) and oxygenation of hemoglobin. Note that the sequence of events occurring in the pulmonary capillaries is the opposite of the process taking place in the tissue capillaries (Figure 2.3). Solid lines indicate major pathways and broken lines indicate minor pathways. Hb = hemoglobin.

capillaries has a P_{CO_2} of about 40 mmHg. Because of this difference in P_{CO_2} values, CO_2 diffuses through the cell membranes of the capillary endothelium and the blood P_{CO_2} rises to 45−46 mmHg. Despite this increase in P_{CO_2}, the blood pH value drops by only about 0.03 during the flow from the arterial capillary (pH 7.41) to the venous capillary (pH 7.38) as a consequence of buffering.

About 95% of the CO_2 entering the blood diffuses into the red blood cells where the enzyme carbonic anhydrase catalyzes conversion of most of the CO_2 to H_2CO_3:

$$CO_2 + H_2O \rightleftharpoons H_2CO_3$$

H_2CO_3 dissociates into H^+ and HCO_3^-. Although H_2CO_3 is a weak acid, its dissociation is essentially 100% because of removal of H^+ ions by the buffering action of hemoglobin. The presence of CO_2 and the production of H^+ cause a reduction in the affinity of hemoglobin for oxygen. Oxyhemoglobin (HbO_2) consequently dissociates into oxygen and deoxyhemoglobin (Hb). This effect of pH on the binding of O_2 to hemoglobin is known as the **Bohr effect**.

Oxygen diffuses into the tissues because the P_{O_2} in blood is greater than the P_{O_2} in tissue cells, and because protonated deoxyhemoglobin (HHb^+) is a weaker acid than HbO_2 and thereby binds H^+ more strongly than

HbO_2. When purified HbO_2 dissociates at pH 7.4 to yield oxygen and Hb, the Hb binds 0.7 mol of H^+ per mole of oxygen released. However, under physiological conditions in whole blood, the Hb combines 0.31 mol of H^+ per mole of oxygen released. This process is reversible. The remainder of the H^+ is buffered by phosphate and proteins other than hemoglobin. The main buffering group involved in the transport of H^+ is an imidazolium group of a histidine residue in hemoglobin. The imidazolium group has a pK' value of about 6.5 (Chapter 3). The difference in acid−base properties between the two forms of hemoglobin molecules is explained by the conformational change that accompanies conversion from HbO_2 to HHb^+ (Chapters 6 and 26).

As the concentration of HCO_3^- (i.e., of metabolic CO_2) in red blood cells increases, an imbalance occurs between the bicarbonate ion concentrations in the red blood cell and plasma. This osmotic imbalance causes a marked efflux of HCO_3^- to plasma and a consequent influx of Cl^- from plasma, in order to maintain the balance of electrostatic charges. The latter osmotic influx, known as the **chloride shift**, is accompanied by migration of water to red blood cells. Thus, transport of metabolic CO_2 in the blood occurs primarily in the form of plasma bicarbonate formed after CO_2 diffuses into red blood cells.

A small percentage of CO_2 entering the red blood cells combines reversibly with an un-ionized amino group ($-NH_2$) of hemoglobin:

$$\begin{array}{ccc} \text{hemoglobin} & & \text{hemoglobin} \\ | & + CO_2 \longleftrightarrow & | & + H^+ \\ NH_2 & & NH-COO^- \end{array}$$

Hemoglobin$-NH-COO^-$, commonly known as **carbaminohemoglobin**, is more correctly named hemoglobin carbamate. Formation of this compound causes a lowering of the affinity of hemoglobin for oxygen. Thus, an elevated concentration of CO_2 favors dissociation of oxyhemoglobin to oxygen and deoxyhemoglobin. Conversely, CO_2 binds more tightly to deoxyhemoglobin than to oxyhemoglobin. All of these processes occurring in the red blood cells of peripheral capillaries are functionally reversed in the lungs (Figure 2.4). Since alveolar P_{O_2} is higher than that of the incoming deoxygenated blood, oxygenation of hemoglobin and release of H^+ occur. The H^+ release takes place because HbO_2 is a stronger acid (i.e., has a lower pK') than deoxyhemoglobin. The released bicarbonate, which is transported to the red blood cells with the corresponding efflux of Cl^-, combines with the released H^+ to form H_2CO_3. Cellular carbonic anhydrase catalyzes dehydration of H_2CO_3 and release of CO_2 from the red blood cells.

Thus, **red blood cell carbonic anhydrase, which catalyzes the reversible hydration of CO_2, plays a vital role in carbon dioxide transport and elimination**. Carbonic anhydrase is a monomeric (M.W. 29,000) zinc metalloenzyme and is present in several different isoenzyme forms. The most prevalent red blood cell isoenzyme of carbonic anhydrase is type I (CAI). Zinc ion, held by coordinate covalent linkage by three imidazole groups of three histidine residues, is involved in the catalytic mechanism of carbonic anhydrase. Water bound to zinc ion reacts with CO_2 bound to the nearby catalytic site of carbonic anhydrase to produce H_2CO_3. The action of carbonic anhydrase is essential for a number of metabolic functions. Some include the formation of H^+ ion in stomach parietal cells (Chapter 11), in bone resorption by the osteoclasts (Chapter 35), and reclamation of HCO_3^- in renal tubule cells (Chapter 37). In osteoclasts and in the renal tubule cells, the isoenzyme CAII catalyzes the hydration reaction of CO_2. A deficiency of CAII caused by an autosomal recessive disorder consists of osteopetrosis (marble bone disease), renal tubular acidosis, and cerebral calcification.

The diffusion of CO_2 from venous blood into the alveoli is facilitated by a pressure gradient of CO_2 between the venous blood (45 mmHg) and the alveoli (40 mmHg), and by the high permeability of the pulmonary membrane to CO_2. Blood leaving the lungs has a P_{CO_2} of about 40 mmHg; thus, essentially complete equilibration occurs between alveolar CO_2 and blood CO_2.

Blood Buffer Calculations

Carbonic acid has the following pK values:

$$H_2CO_3 \rightleftharpoons H^+ + HCO_3^- \quad pK'_1 = 3.8 \qquad (2.8)$$

$$HCO_3^- \rightleftharpoons CO_3^{2-} + H^+ \quad pK'_2 = 10.2 \qquad (2.9)$$

It is apparent from the pK' values that neither equilibrium can serve as a buffer system at the physiological pH of 7.4. However, carbonic acid (the proton donor) is in equilibrium with dissolved CO_2, which in turn is in equilibrium with gaseous CO_2:

$$H_2O + CO_2(\text{aqueous}) \rightleftharpoons H_2CO_3 \qquad (2.10)$$

The hydration reaction (2.10), coupled with the first dissociation of carbonic acid (2.8), produces an apparent pK' of 6.1 for bicarbonate formation. Thus, the summation of Equations (2.8) and (2.10) yield

$$H_2O + CO_2 \rightleftharpoons H^+ + HCO_3^- \qquad (2.11)$$

$$pK'(\text{apparent}) = \frac{[HCO_3^-][H^+]}{[H_2CO_3]} = 6.1 \qquad (2.12)$$

The ratio of HCO_3^- to H_2CO_3 at a physiological pH of 7.4 can be calculated by means of employing the Henderson$-$Hasselbalch equation:

$$7.4 = 6.1 + \log\frac{[HCO_3^-]}{[H_2CO_3]} \qquad (2.13)$$

$$\log\frac{[HCO_3^-]}{[H_2CO_3]} = 1.3$$

Taking antilogarithms:

$$\frac{[HCO_3^-]}{[H_2CO_3]} = \frac{20}{1} = \frac{\text{proton acceptor}}{\text{proton donor}}$$

This ratio is large because the pH is greater than the pK' (see Table 2.3). At pH 7.4, the bicarbonate system is a good buffer toward acid (i.e., it can neutralize large amounts of acid), but a poor buffer for alkali. However, blood H_2CO_3 is in rapid equilibrium with a relatively large (about 1000 times as much) reservoir of cellular CO_2, and can function as an effective buffer against increases in alkalinity. The HCO_3^-/H_2CO_3 ratio in blood is coupled to the partial pressure of CO_2, i.e., to the metabolic production of CO_2 and to the loss of CO_2 during respiration. In the equilibrium expression for the bicarbonate$-$carbonic acid buffer system at pH 7.4, the carbonic acid term can be replaced by a pressure term because the carbonic acid concentration is proportional to P_{CO_2} in the blood.

$$pH = 6.1 + \log\frac{[HCO_3^-]}{aP_{CO_2}} \qquad (2.14)$$

where a, a proportionality constant (for normal plasma at 37°C, 0.0301), is defined by the equation

$$[H_2CO_3] = aP_{CO_2} \qquad (2.15)$$

The HCO_3^-/H_2CO_3 buffer system effectively maintains a constant blood pH of 7.4 if bicarbonate and H_2CO_3 concentrations are maintained at a ratio of 20:1. The concentration of HCO_3^- is regulated by its selective excretion and reclamation by the membranes of the renal tubular epithelial cell. Levels of P_{CO_2} and $[H_2CO_3]$ in the blood can be altered by changes in the rate and depth of respiration. Respiration is controlled by a skeletal muscle known as a diaphragm. It contains fibers consisting of both type I and type IIa. The former is fatigue-resistant, and the latter is fast-twitch fibers (Chapter 19). Diaphragm structure is dome shaped and separates thoracic and abdominal cavities. Phrenic nerves whose roots originate in the neck region (c3−c5) regulate diaphragm muscle and thus respiration. Dysfunctional breathing can occur due to many causes and affects blood P_{CO_2} [2]. For example, **hypoventilation** (slow, shallow breathing) leads to increased blood P_{CO_2}, whereas **hyperventilation** (rapid, deep breathing) has the opposite effect. P_{CO_2} changes mediated by the lungs are more rapid than $[HCO_3^-]$ changes affected through the kidneys (Chapter 37).

Non-Bicarbonate Buffers in Blood

Other important non-bicarbonate blood buffers are protein and phosphate. The predominant buffer system in the red blood cells is hemoglobin. Protein amino acid side chains (R-groups) that act as buffers are carboxylate groups of glutamate and aspartate, and the weakly basic groups of lysine, arginine, and histidine. To be effective, the pK' value of a buffer should be close to the pH of the system to be buffered. Except for the R-group of histidine, which is an imidazolium group, the pK' values of the other amino acids mentioned previously are not close enough to the physiological pH of blood to be effective buffers. The imidazolium group has a pK' value of 6.5, but it can vary from 5.3 to 8.3 depending on differences in electrostatic environment, either within the same protein molecule or in different proteins. Another potential buffering group in protein is the α-amino group of the amino acid residues at the amino terminus of the protein. This group has a pK' value ranging from 7.8 to 10.6, with a typical value of about 8.

In plasma, the protein buffer system has a limited role; the principal plasma buffer is the bicarbonate−carbonic acid system, and in the red blood cells hemoglobin buffers a major portion of the hydrogen ions produced by the dissociation of H_2CO_3 generated by the hydration of CO_2.

H^+ CONCENTRATION AND pH

The use of pH to designate $[H^+]$ is due to the fact that a broad range of $[H^+]$ can be compressed within the numerical scale of 0−14. However, in clinical acid−base problems, use of the pH scale has some disadvantages. Since the pH is the logarithm of the reciprocal of $[H^+]$, significant variations of $[H^+]$ in a patient may not be fully appreciated. For example, if the blood pH decreases from 7.4 to 7.1, $[H^+]$ is doubled, or if the pH increases from 7.4 to 7.7, $[H^+]$ is halved (Figure 2.5). In addition, the use of the pH scale masks the relationship between $[H^+]$ and the concentrations of other cations, e.g., Na^+ and K^+. Thus, in clinical situations, it is preferable to express $[H^+]$ directly as nanomoles per liter in order to better evaluate acid−base changes and interpret laboratory tests.

A blood pH of 7.4 corresponds to 40 nM $[H^+]$, which is the mean of the normal range (Figure 2.5). The normal range is 7.36−7.44 on the pH scale, or 36−44 nM $[H^+]$. If the pH of blood falls below pH 7.36 ($[H^+] > 44$ nM), the condition is called **acidemia**. Conversely, if the pH rises above pH 7.44 ($[H^+] < 36$ nM), the condition is called **alkalemia**. The suffix "-emia" refers to blood, and usually to an abnormal concentration in blood. Over the pH range of 7.20−7.50, for every change of 0.01 pH unit, there is a change of approximately 1 nM $[H^+]$ in the opposite direction.

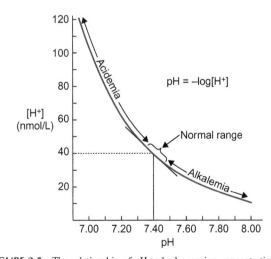

FIGURE 2.5 The relationship of pH to hydrogen ion concentration (in nanomoles per liter). The normal blood pH of 7.4 corresponds to 40 nmol/L of H^+. The solid straight line is drawn to show the linear relationship between the concentration of H^+ and pH, over the pH range of 7.2−7.5. A 0.01-unit change in pH is equivalent to about 1.0 nmol/L change in the opposite direction.

Since the Henderson−Hasselbalch equation uses pH terms, its utility in clinical situations is less than optimal. Kassirer and Bleich have derived a modified Henderson−Hasselbalch expression that relates $[H^+]$, instead of pH, to P_{CO_2} and HCO_3^-, as follows:

$$H^+ = \frac{K'a \times P_{CO_2}}{[HCO_3^-]} \qquad (2.16)$$

In Equation (2.16), K' and a are constants, and the numerical value of $K'a$ is 24 when P_{CO_2} is expressed in mmHg, $[HCO_3^-]$ in mM, and $[H^+]$ in nM. Therefore,

$$H^+ = \frac{24 \times P_{CO_2}}{[HCO_3^-]} \qquad (2.17)$$

This equation expresses the interdependence of three factors; if two of them are known, the third can be calculated. For example, at a blood $[HCO_3^-]$ of 24 mM and a P_{CO_2} of 40 mmHg, $[H^+]$ is 40 nM. Clinical applications of Equation (2.17) are discussed in Chapter 37.

REQUIRED READING

[1] D.F.M. Brown, F.A. Jaffer, J.N. Baker, M.E. Gurol, Case 28-2013: A 52-year-old man with cardiac arrest after an acute myocardial infarction, N. Engl. J. Med. 369 (2013) 1047−1054.

[2] F.D. McCool, G.E. Tzelepis, Dysfunction of the diaphragm, N. Engl. J. Med. 366 (2012) 932−942.

Chapter 3

Amino Acids

Key Points

1. The basic structure of an amino acid consists of an α-carboxylic acid group, an α-amino group (or an imino group), a hydrogen atom, and a unique side chain known as an R-group. Thus, the α-carbon atom is linked to four substituent groups.

2. If the four substituent groups linked to an α-carbon atom are all different, then the α-carbon becomes an asymmetric center (also known as chiral center) giving rise to two optically active isomers designated as D- and L-isomers.

3. All amino acids have at least one asymmetric center, with the exception of glycine, which has two hydrogen atoms linked to the α-carbon atom.

4. L-amino acids are used in mammalian metabolism, whereas D-amino acids are used in bacterial metabolism,

5. In humans, L-amino acids are exclusively used in protein synthesis. However, in long-lived extracellular proteins, some amino acid residues (e.g., aspartate) undergo conversion from L-form to D-form, very slowly with respect to time (aging). This process of conversion of an L-isomer to D-isomer is known as racemization.

6. The R-group linked to the α-carbon of an amino acid defines its unique properties as well as its role in protein structure and function.

7. The R-group of amino acids, after their incorporation into proteins, undergoes numerous modifications during normal and abnormal metabolism.

8. Amino acids are classified based on the properties of the R-group: hydrophobic (nonpolar), uncharged hydrophilic (polar), or charged hydrophilic (polar).

9. Monocarboxylic and monoamino acids exist as the dipolar ionic (zwitterionic) form at physiologic pH.

10. In addition to being building units for peptides and proteins, amino acids, either directly or in their modified form, participate in numerous metabolic functions. These include neurotransmitter function, and acting as precursors for the synthesis of non-protein nitrogen containing metabolites (e.g., heme, creatine, purines, and pyrimidines).

11. Some amino acids are not synthesized in the body; they are known as essential (indispensable) amino acids and are supplied in the diet to maintain normal growth and metabolism.

12. Either a deficiency or an excess quantity of some amino acids can cause metabolic abnormalities.

13. Some amino acids and their derivatives are used therapeutically.

14. Some plants, fungi, and bacterial-derived amino acids can be toxic to the human body.

Proteins are the most abundant class of organic compounds in the healthy, lean human body, constituting more than half of its cellular dry weight. Proteins are polymers of amino acids and have molecular weights ranging from approximately 10,000 to more than one million. Biochemical functions of proteins include catalysis, transport, contraction, protection, structure, and metabolic regulation.

Amino acids are the monomeric units, or building blocks, of proteins joined by a specific type of covalent linkage. The properties of proteins depend on the characteristic sequence of component amino acids, each of which has distinctive side chains.

Amino acid polymerization requires elimination of a water molecule as the carboxyl group of one amino acid reacts with the amino group of another amino acid to form a covalent amide bond. The repetition of this process with many amino acids yields a polymer, known as a **polypeptide**. The amide bonds linking amino acids to each other are known as **peptide bonds**. Each amino acid unit within the polypeptide is referred to as a **residue**. The sequence of amino acids in a protein is dictated by the sequence of nucleotides in a segment of the DNA in the chromosomes, and the uniqueness of each living organism is due to its endowment of specific proteins.

L-α-AMINO ACIDS: STRUCTURE

Almost all of the naturally occurring amino acids in proteins are L-α-amino acids. The principal 20 amino acids in proteins have an amino group, a carboxyl group, a hydrogen atom, and an R-group attached to the α-carbon (Figure 3.1).

N. V. Bhagavan and C.-E. Ha: Essentials of Medical Biochemistry, Second edition. DOI: http://dx.doi.org/10.1016/B978-0-12-416687-5.00003-8

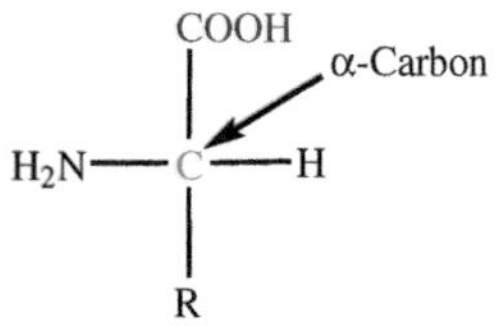

FIGURE 3.1 Basic structure of an α-amino acid.

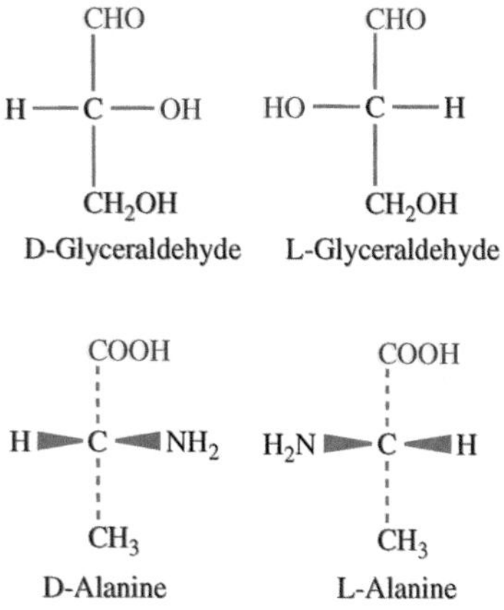

FIGURE 3.2 Stereoisomers of alanine.

Proline is an exception, because it has a cyclic structure and contains a secondary amine group (called an **imino group**) instead of a primary amine group (called an **amino group**). Amino acids are classified according to the chemical properties of the R-group. Except for glycine (R = H), the amino acids have at least one asymmetrical carbon atom (the α-carbon). The absolute configuration of the four groups attached to the α-carbon is conventionally compared to the configuration of L-glyceraldehyde (Figure 3.2). The D and L designations specify absolute configuration and not the dextro (right) or levo (left) direction of rotation of plane-polarized light by the asymmetrical carbon center.

CLASSIFICATION

A useful classification of the amino acids is based on the solubility (i.e., ionization and polarity) characteristics of the side chains (R-groups). The R-groups fall into four classes:

1. Nonpolar (hydrophobic);
2. Polar negatively charged (acidic);
3. Polar positively charged (basic); and
4. Polar neutral (un-ionized).

Within each class, R-groups differ in size, shape, and other properties. Figure 3.3 shows the structure of each amino acid according to this classification, with the R-group outlined. Ionizable structures are drawn as they would exist at pH 7.0. The three-letter and one-letter abbreviations for each amino acid are given in Table 3.1.

A "-yl" ending on an amino acid residue indicates that the carboxyl group of an amino acid is linked to another functional group (e.g., in a peptide bond).

The eight **essential amino acids** (Table 3.1) are those which humans cannot synthesize and which must be supplied in the diet. The remaining amino acids are synthesized in the body by various biochemical pathways.

Nonpolar Amino Acids

Glycine

Glycine is the smallest amino acid and has an H atom as its R-group. It is the only α-amino acid that is not optically active. The small R-group provides a minimum of steric hindrance to rotation about bonds; therefore, glycine fits into crowded regions of many peptide chains. Collagen, a rotationally restricted fibrous protein, has glycyl residues in about every third position in its polypeptide chains. Glycine is used for the biosynthesis of many nonprotein compounds, such as porphyrins and purines.

Glycine and taurine are conjugated with bile acids, products derived from cholesterol, before they are excreted into the biliary system. Conjugated bile acids are amphipathic and are important in lipid absorption. Glycine is also a neurotransmitter; it is inhibitory in the spinal cord, and excitatory in the cerebral cortex and other regions of the forebrain. Nonketotic hyperglycinemia (NKH) is an inborn error of glycine degradation in which a large amount of glycine accumulates throughout the body. NKH gives rise to severe consequences in the central nervous system (CNS) and leads to death.

Alanine

The side chain of alanine is a hydrophobic methyl group, CH_3. Other amino acids may be considered to be chemical derivatives of alanine, with substituents on the β-carbon. Alanine and glutamate provide links between amino acid and carbohydrate metabolism.

Valine, Leucine, and Isoleucine

The branched-chain aliphatic amino acids valine, leucine, and isoleucine contain bulky nonpolar R-groups and participate in hydrophobic interactions. All three are essential amino acids. A defect in their catabolism leads to **maple syrup urine disease**. Isoleucine has asymmetrical centers at both the α- and β-carbons, and four stereoisomers, only one of which occurs in protein. The bulky side chains tend to associate in the interior of water-soluble globular proteins. Thus, the hydrophobic amino acid residues stabilize the three-dimensional structure of the polymer.

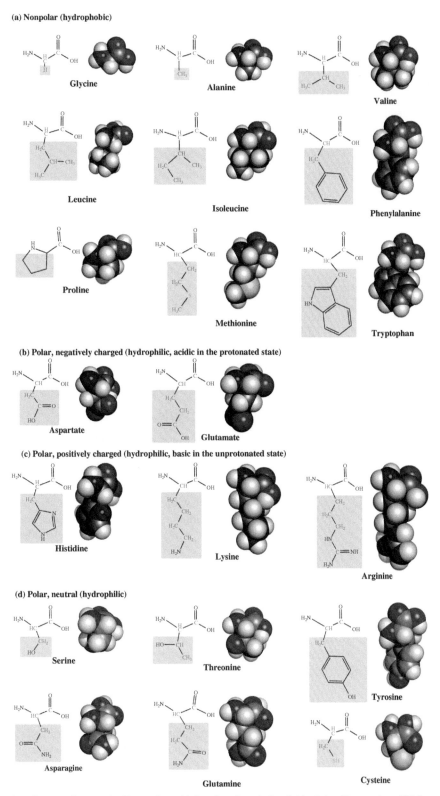

FIGURE 3.3 Classification of commonly occurring L-α-amino acids based on the polarity of side chains (R-groups) at pH 7.0.

TABLE 3.1 Amino Acid Abbreviations and Nutritional Properties*

Alanine	Ala	A	nonessential
Arginine	Arg	R	conditionally essential
Asparagine	Asn	N	nonessential
Aspartic acid	Asp	D	nonessential
Cysteine	Cys	C	nonessential
Glutamic acid	Glu	E	nonessential
Glutamine	Gln	Q	conditionally essential
Glycine	Gly	G	nonessential
Histidine	His	H	conditionally essential
Isoleucine	Ile	I	essential
Leucine	Leu	L	essential
Lysine	Lys	K	essential
Methionine	Met	M	essential
Phenylalanine	Phe	F	essential
Proline	Pro	P	nonessential
Serine	Ser	S	nonessential
Threonine	Thr	T	essential
Tryptophan	Trp	W	essential
Tyrosine	Tyr	Y	nonessential
Valine	Val	V	essential

The eight essential amino acids are not synthesized in the body and have to be supplied in the diet. The conditionally essential amino acids, although synthesized in the body, may require supplementation during certain physiological conditions such as pregnancy. The nonessential amino acids can be synthesized from various metabolites.

Phenylalanine

A planar hydrophobic phenyl ring is part of the bulky R-group of phenylalanine. It is an essential amino acid whose metabolic conversion to tyrosine is defective in **phenylketonuria**. Phenylalanine, tyrosine, and tryptophan are the only α-amino acids that contain aromatic groups, and consequently are the only residues that absorb ultraviolet (UV) light.

Tryptophan

A bicyclic nitrogenous aromatic ring system (known as an indole ring) is attached to the β-carbon of alanine to form the R-group of tryptophan. Tryptophan is a precursor of serotonin, melatonin, nicotinamide, and many naturally occurring medicinal compounds derived from plants. It is an essential amino acid. The indole group absorbs UV light at 280 nm—a property that is useful for spectrophotometric measurement of protein concentration.

Methionine

Methionine contains an R-group with a methyl group attached to sulfur. The essential amino acid serves as the donor of a methyl group in many transmethylation reactions, e.g., in the synthesis of epinephrine, creatine, and melatonin. Almost all of the sulfur-containing compounds of the body are derived from the sulfur atom of methionine.

Proline

Proline contains a secondary amine group, called an **imine**, instead of a primary amine group. For this reason, proline is called an **imino acid**. Since the three-carbon R-group of proline is fused to the α-nitrogen group, this compound has a rotationally constrained rigid-ring structure. As a result, prolyl residues in a polypeptide introduce restrictions on the folding of chains. In collagen, the principal protein of human connective tissue, certain prolyl residues are hydroxylated. The hydroxylation occurs during protein synthesis and requires ascorbic acid (vitamin C) as a cofactor. Deficiency of vitamin C causes formation of defective collagen and **scurvy**.

Acidic Amino Acids

Aspartic Acid

The β-carboxylic acid group of aspartic acid has a pK′ of 3.86 and is ionized at pH 7.0 (the anionic form is called **aspartate**). The anionic carboxylate groups tend to occur on the surface of water-soluble proteins, where they interact with water. Such surface interactions stabilize protein folding. Aspartic acid is required for the *de novo* synthesis of purine and pyrimidine nucleotides.

Glutamic Acid

The γ-carboxylic acid group of glutamic acid has a pK′ of 4.25 and is ionized at physiological pH. The anionic groups of glutamate (like those of aspartate) tend to occur on the surfaces of proteins in aqueous environments. Glutamate is the primary excitatory neurotransmitter in the brain, and these neurons are known as glutamatergic neurons. Its levels are regulated by clearance that is mediated by glutamate transfer protein in critical motor control areas in the CNS.

In some proteins, the γ-carbon of glutamic acid contains an additional carboxyl group. Residues of **γ-carboxyglutamic acid** (Gla) bear two negative charges. γ-Carboxylation of glutamic acid residues is a post-translational modification and requires **vitamin K** as a cofactor. Residues of γ-carboxyglutamate are present in a number of blood coagulation proteins (factors II, VII, IX, and X) and anticoagulant proteins C and S. Osteocalcin, a protein present in the bone, also contains γ-carboxyglutamate residues.

Basic Amino Acids

Lysine

Lysine is an essential amino acid. The long side chain of lysine has a reactive amino group attached to the ε-carbon. The ε-NH_2 (pK′ = 10.53) is protonated at physiological pH. The lysyl side chain forms ionic bonds with negatively charged groups of acidic amino acids. The ε-NH_2 groups of lysyl residues are covalently linked to biotin (a vitamin), lipoic acid, and retinal, a derivative of vitamin A and a constituent of visual pigment.

In collagen and in some glycoproteins, δ-carbons of some lysyl residues are hydroxylated (Figure 3.4), and sugar moieties are attached at these sites. In elastin and collagen, some ε-carbons of lysyl residues are oxidized to reactive aldehyde (−CHO) groups, with elimination of NH_3. These aldehyde groups then react with other ε-NH_2 groups to form covalent cross-links between polypeptides, thereby providing tensile strength and insolubility to protein fibers. Examples of cross-linked amino acid structures are desmosine, isodesmosine (Figure 3.4), dehydrolysinonorleucine, lysinonorleucine, merodesmosine, and dehydromerodesmosine. Lysyl R-groups participate in a different type of cross-linking in the formation of fibrin, a process essential for the clotting of blood. In this reaction, the ε-NH_2 group of one fibrin polypeptide forms a covalent linkage with the glutamyl residue of another fibrin polypeptide.

Histidine

The imidazole group attached to the β-carbon of histidine has a pK′ value of 6.0. The pK′ value of histidyl residues in protein varies, depending on the nature of the neighboring residues. The imidazolium−imidazole buffering pair has a major role in acid−base regulation (e.g., in hemoglobin). The imidazole group functions as a nucleophile, or a general base, in the active sites of many enzymes, and may bind metal ions. Histidine is nonessential in adults but is essential in the diet of infants and individuals with **uremia** (a kidney disorder). Decarboxylation of histidine to yield **histamine** occurs in mast cells present in loose connective tissue and around blood vessels, basophils of blood, and **enterochromaffin-like** (**ECL**) **cells** present in the acid-producing glandular portion (oxyntic cells) of the stomach.

The many specific reactions of histamine are determined by the type of receptor (H_1 or H_2) present in the target cells. The contraction of smooth muscle (e.g., gut and bronchi) is mediated by H_1 receptors and antagonized by diphenhydramine and pyrilamine. H_1-receptor antagonists are used in the treatment of allergic disorders. Secretion of HCl by the stomach and an increase in heart rate are mediated by H_2 receptors. Examples of H_2-receptor antagonists of histamine action are cimetidine and ranitidine, agents used in the treatment of gastric ulcers.

Arginine

The positively charged guanidinium group attached to the δ-carbon of arginine is stabilized by resonance structures between the two NH_2 groups and has a pK′ value of 12.48. Arginine is utilized in the synthesis of creatine, and it participates in the urea cycle.

The nitrogen of the guanidino group of arginine is converted to nitric oxide (NO) by nitric oxide synthase. NO is unstable, highly reactive, and has a lifespan of only a few seconds. However, NO affects many biological activities including vasodilation, inflammation, and neurotransmission.

Post-translational modifications of guanidinium groups of peptidyl arginine residues by deimination produce citrulline residues. These citrulline residues occur in some target proteins and connective tissue proteins. Autoantibodies directed against citrullinated proteins are utilized as a serological marker for a chronic inflammatory disease, namely, rheumatoid arthritis.

Neutral Amino Acids

Serine

The primary alcohol group of serine can form esters with phosphoric acid (Figure 3.4) and glycosides with sugars. The phosphorylation and dephosphorylation processes regulate the biochemical activity of many proteins. Active centers of some enzymes contain seryl hydroxyl groups and can be inactivated by irreversible derivatization of these groups. The −OH group of serine has a weakly acidic pK′ of 13.6.

Threonine

The essential amino acid threonine has a second asymmetrical carbon atom in the side chain and therefore can have four isomers, only one of which, L-threonine, occurs in proteins. The hydroxyl group, as in the case of serine, participates in reactions with phosphorylation and dephosphorylation and with sugar residues.

Cysteine

The weakly acidic (pK′ = 8.33) sulfhydryl group (−SH) of cysteine is essentially undissociated at physiological pH. Free −SH groups are essential for the function of many enzymes and structural proteins. Heavy metal ions, e.g., Pb^{2+} and Hg^{2+}, inactivate these proteins by combining with their −SH groups. Two cysteinyl −SH groups can be oxidized to form **cystine**. A covalent disulfide bond of cystine can join two parts of a single polypeptide chain or two different polypeptide chains through crosslinking of cysteine residues. These −S−S− bonds are essential both for the folding of polypeptide chains and for the association of polypeptides in proteins that have more than one chain, e.g., insulin and immunoglobulins.

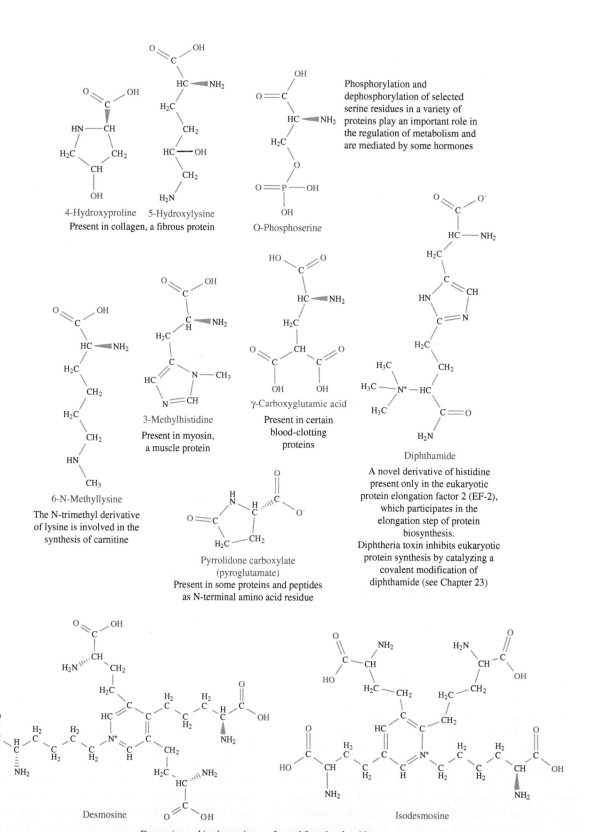

4-Hydroxyproline   5-Hydroxylysine
Present in collagen, a fibrous protein

O-Phosphoserine

Phosphorylation and
dephosphorylation of selected
serine residues in a variety of
proteins play an important role in
the regulation of metabolism and
are mediated by some hormones

6-N-Methyllysine
The N-trimethyl derivative
of lysine is involved in the
synthesis of carnitine

3-Methylhistidine
Present in myosin,
a muscle protein

γ-Carboxyglutamic acid
Present in certain
blood-clotting
proteins

Pyrrolidone carboxylate
(pyroglutamate)
Present in some proteins and peptides
as N-terminal amino acid residue

Diphthamide
A novel derivative of histidine
present only in the eukaryotic
protein elongation factor 2 (EF-2),
which participates in the
elongation step of protein
biosynthesis.
Diphtheria toxin inhibits eukaryotic
protein synthesis by catalyzing a
covalent modification of
diphthamide (see Chapter 23)

Desmosine

Isodesmosine

Desmosine and isodesmosine are formed from lysyl residues
of the polypeptide chains of elastin, a fibrous protein

FIGURE 3.4 Post-translationally modified derivatives of certain amino acids that are found in proteins.

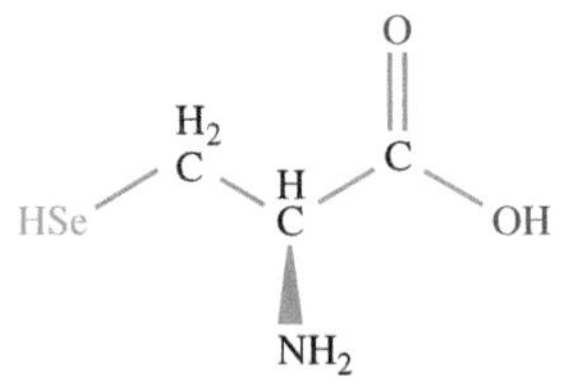

FIGURE 3.5 Structure of selenocysteine.

The loops of disulfide linkages of Cys-Cys pairs are required for the function of ligand-gated ion channel superfamily of neurotransmitter proteins.

Selenocysteine

Selenocysteine is an analogue of cysteine (Figure 3.5). In place of a sulfur atom in the cysteine, selenocysteine contains selenium (Se). Several human proteins and enzymes are selenoproteins. Selenocysteine is located in the active sites of enzymes that participate in oxidation–reduction reactions. These include glutathione peroxidase, thioredoxin reductase, and iodothyronine deiodinase. The incorporation of selenocysteine into the growing peptide chain occurs by a unique suppressor tRNA and a stop codon. The physiological importance of selenium as an essential trace element is exemplified in a disorder of congestive cardiomyopathy known as *Keshan* disease. This disease is prevalent mainly in children and young women in China who have markedly low levels of selenium in their water and food, due to its low levels in the soil.

Tyrosine

The phenolic hydroxyl group of this aromatic amino acid has a weakly acidic pK′ of about 10, and therefore is un-ionized at physiological pH. In some enzymes, the hydrogen of the phenolic hydroxyl group can participate in hydrogen bond formation with oxygen and nitrogen atoms. The phenolic hydroxyl group of tyrosine residues in protein can be sulfated (e.g., in gastrin and cholecystokinin), or phosphorylated by a reaction catalyzed by the tyrosine-specific protein kinase that is a product of some oncogenes. Phosphorylation and dephosphorylation of selected tyrosyl residues in specific protein domains participate in the pathways of signal transduction. Tyrosine kinase activity also resides in a family of cell surface receptors that includes receptors for such anabolic polypeptides as insulin, epidermal growth factor, platelet-derived growth factor, and insulin-like growth factor type 1. All of these receptors have a common motif of an external ligand-binding domain, a transmembrane segment, and a cytoplasmic tyrosine kinase domain. Tyrosine accumulates in tissues and blood in **tyrosinosis** and **tyrosinemia**, which are due to inherited defects in catabolism

of this amino acid. Tyrosine is the biosynthetic precursor of thyroxine, catecholamines, and melanin.

Asparagine

The R-group of asparagine, an amide derivative of aspartic acid, has no acidic or basic properties, but it is polar and participates in hydrogen bond formation. It is hydrolyzed to aspartic acid and ammonia by the enzyme asparaginase. In glycoproteins, the carbohydrate side chain is often linked through the amide group of asparagine.

Glutamine

An amide of glutamic acid, glutamine has properties similar to those of asparagine. The γ-amido nitrogen, derived from ammonia, can be used in the synthesis of purine and pyrimidine nucleotides, converted to urea in the liver, or released as NH_3 in the kidney tubular epithelial cells. The latter reaction, which is catalyzed by the enzyme glutaminase, functions in acid–base regulation by neutralizing H^+ ions in the urine (Chapter 37).

Glutamine is the most abundant amino acid in the body. It comprises more than 60% of the free amino acid pool in skeletal muscle. It is metabolized in both liver and gut tissues. Glutamine, along with alanine, is a significant precursor of glucose production during fasting. It is a nitrogen donor in the synthesis of purines and pyrimidines required for nucleic acid synthesis. Glutamine synthesized in the astrocytes is the precursor of glutamate neurotransmitter in the glutamatergic neurons. Astrocytes are supporting cells for neurons and also are the source for neurotransmitter γ-aminobutyrate. Deficiency of astrocyte glutamine due to lack of glutamine synthetase can cause severe neurological disturbances. Glutamine is enriched in enteral and parenteral nutrition to promote growth of tissues; it also enhances immune functions in patients recovering from surgical procedures. Thus, glutamine may be classified conditionally as an essential amino acid during severe trauma and illness.

Unusual Amino Acids

Several L-amino acids have physiological functions as free amino acids, rather than as constituents of proteins. Examples are as follows:

1. β-Alanine is part of the vitamin pantothenic acid.
2. Homocysteine, homoserine, ornithine, and citrulline are intermediates in the biosynthesis of certain other amino acids.
3. Taurine, which has an amino group in the β-carbon and a sulfonic acid group instead of COOH, is present in the CNS, and, as a component of certain bile acids, participates in digestion and absorption of lipids in the gastrointestinal tract.
4. γ-Aminobutyric acid is an inhibitory neurotransmitter.

Amino Acids and Derivatives Used as Drugs

Amino acids and their derivatives are used as drugs; they include D-penicillamine for chelation therapy in Wilson's disease; N-acetylcysteines for treating acetaminophen toxicity (see Clinical Case Study 7.3); cysteamine for the therapy of cystinosis; and gabapentin as an anticonvulsive agent. Some nonprotein α-amino acids that are associated with toxic and abnormal metabolic changes can be found online.

ELECTROLYTE AND ACID−BASE PROPERTIES

Amino acids are ampholytes; i.e., they contain both acidic and basic groups. Free amino acids can *never* occur as nonionic molecules. Instead, they exist as neutral **zwitterions** that contain both positively and negatively charged groups. Zwitterions are electrically neutral and so do not migrate in an electric field. In an acidic solution (below pH 2.0), the predominant species of an amino acid is positively charged and migrates toward the cathode. In a basic solution (above pH 9.7), the predominant species of an amino acid is negatively charged and migrates toward the anode.

The **isoelectric** point (pI) of an amino acid is the pH at which the molecule has an average net charge of zero and therefore does not migrate in an electric field. The pI is calculated by averaging the pK′ values for the two functional groups that react as the zwitterion becomes alternately a monovalent cation or a monovalent anion.

At physiological pH, monoaminomonocarboxylic amino acids, e.g., glycine and alanine, exist as zwitterions. That is, at a pH of 6.9−7.4, the α-carboxyl group (pK′ = 2.4) is dissociated to yield a negatively charged carboxylate ion (−COO⁻), while the α-amino group (pK′ = .7) is protonated to yield an ammonium group (−NH₃⁺). The pK′ value of the α-carboxyl group is considerably lower than that of a comparable aliphatic acid, e.g., acetic acid (pK′ = 4.6). This stronger acidity is due to electron withdrawal by the positively charged ammonium ion and the consequent increased tendency of a carboxyl hydrogen to dissociate as an H⁺. The α-ammonium group is correspondingly a weaker acid than an aliphatic ammonium ion, e.g., ethylamine (pK′ = 9.0), because the inductive effect of the negatively charged carboxylate anion tends to prevent dissociation of H⁺. The titration profile of glycine (Figure 3.6) is nearly identical to the profiles of all other monoaminomonocarboxylic amino acids with

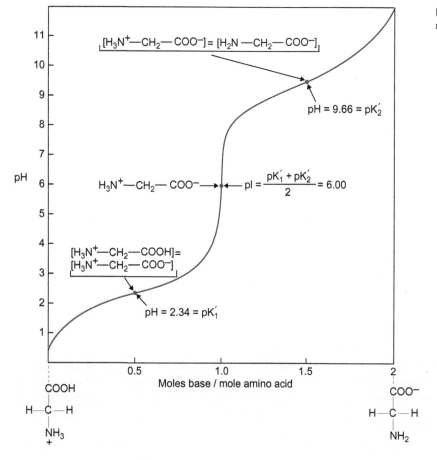

FIGURE 3.6 Titration profile of glycine, a monoaminomonocarboxylic acid.

TABLE 3.2 pK' and pI Values of Selected Free Amino Acids at 25°C*

Amino Acid	pK$_1'$ (α-COOH)	pK$_2'$	pK$_3'$	pI
Alanine	2.34	9.69 (β-NH$_3^+$)		6.00
Aspartic acid	2.09	3.86 (γ-COOH)	9.82 (α-NH$_3^+$)	$2.98\left(\dfrac{pK_1'+pK_2'}{2}\right)$
Glutamic acid	2.19	4.25 (γ-COOH)	9.67 (α-NH$_3^+$)	$3.22\left(\dfrac{pK_1'+pK_2'}{2}\right)$
Arginine	2.17	9.04 (α-NH$_3^+$)	12.48 (Guanidinium)	$10.76\left(\dfrac{pK_2'+pK_3'}{2}\right)$
Histidine	1.82	6.00 (Imidazolium)	9.17 (NH$_3^+$)	$7.59\left(\dfrac{pK_2'+pK_3'}{2}\right)$
Lysine	2.18	8.95 (α-NH$_3^+$)	10.53 (ε-NH$_3^+$)	$9.74\left(\dfrac{pK_2'+pK_3'}{2}\right)$
Cysteine	1.71	8.33 (—SH)	10.78 (α-NH$_3^+$)	$5.02\left(\dfrac{pK_1'+pK_2'}{2}\right)$
Tyrosine	2.20	9.11 (α-NH$_3^+$)	10.07 (Phenol OH)	$5.66\left(\dfrac{pK_1'+pK_2'}{2}\right)$
Serine	2.21	9.15 (α-NH$_3^+$)	13.6 (Alcohol OH)	$5.68\left(\dfrac{pK_1'+pK_2'}{2}\right)$

*The pK' values for functional groups in proteins may vary significantly from the values for free amino acids.

nonionizable R-groups (Ala, Val, Leu, Ile, Phe, Ser, Thr, Gln, Asn, Met, and Pro).

The titration of glycine has the following major features. The titration is initiated with glycine hydrochloride, $Cl^-(H_3^+NCH_2COOH)$, which is the fully protonated form of the amino acid. In this form, the molecule contains two acidic functional groups; therefore, two equivalents of base are required to completely titrate 1 mol of glycine hydrochloride. There are two pK' values: pK$_1'$ due to reaction of the carboxyl group and pK$_2'$ due to reaction of the ammonium group. Addition of 0.5 eq of base to 1 mol of glycine hydrochloride raises the pH 2.34 (pK$_1'$), whereas addition of 1.5 eq further increases the pH to 9.66 (pK$_2'$). At low pH values (e.g., 0.4), the molecules are predominantly cations with one positive charge; at pH values of 5−7, most molecules have a net charge of zero; at high pH values (e.g., 11.7), all of the molecules are essentially anions with one negative charge. The midpoint between the two pK' values [i.e., at pH = (2.34 + 9.66)/2 = 6.0] is the pI. Thus, pI is the arithmetic mean of pK$_1'$ and pK$_2'$

values and the inflection point between the two segments of the titration profile.

The buffering capacities of weak acids and weak bases are maximal at their pK' values. Thus, monoamino monocarboxylic acids exhibit their greatest buffering capacities in the two pH ranges near their two pK' values, namely, pH 2.3 and pH 9.7 (Figure 3.6). Neither these amino acids nor the α-amino or α-carboxyl groups of other amino acids (which have similar pK' values) have significant buffering capacity in the neutral (physiological) pH range. The only amino acids with R-groups that have buffering capacity in the physiological pH range are histidine (imidazole; pK' = 6.0) and cysteine (sulfhydryl; pK' = 8.3). The pK and pI values of selected amino acids are listed in Table 3.2. The pK' values for R-groups vary with the ionic environment.

The pK' values for functional groups in proteins may vary significantly from the values for free amino acids. The R-groups are ionized at physiological pH and have anionic and cationic groups, respectively.

Chapter 4

Three-Dimensional Structure of Proteins and Disorders of Protein Misfolding

Key Points

1. Proteins and polypeptides consist of about 20 different amino acids. The amino acids are linked by peptide bonds, and the amino acids in a polypeptide are known as amino acid residues.

2. The protein structure is defined based on the four hierarchical structures: primary, secondary, tertiary, and quaternary.

3. Primary structure: linear sequence of specific amino acid residues, determined by the coding DNA sequence (exon).

4. Secondary structure: consists of four structural motifs: α-helix, β-pleated sheet (also known as β-structures and β-sheet), β-turns, and nonrepetitive structure (also known as loop sections or random coil).

5. Supersecondary structure consists of packing of secondary structures such as helix-turn-helix, β-α-β units are considered as structures intermediate between secondary and tertiary structures.

6. Tertiary structure: defines the overall three-dimensional conformation; consists of folding of the motifs to bring together disparate amino acid residues in the assembly of a functional unit. Three-dimensional structure is stabilized by the nature of the R-groups, which allows for hydrophobic interactions, to achieve maximum van der Waals forces, electrostatic (ionic) interactions, and H-bonds. Formation of disulfide (−S−S−) covalent bonds between two strategically located cysteine residues stabilizes the folded conformation (e.g., insulin).

7. Quaternary structure refers to those proteins which consist of more than one polypeptide chain, known as subunit (monomer) and their assembly into a functional molecule (oligomers). Subunits are held together by noncovalent interactions that exist between complementary surfaces (e.g., hemoglobin).

8. Globular proteins, in aqueous medium, have their hydrophilic R-groups of amino acid residues on the surface and the hydrophobic R-groups in the interior. Opposite locations of R-groups occur in the hydrophobic environment (e.g., cell membrane proteins).

9. Different functional units of a protein such as binding sites for substrates, cofactors, and modifiers are known as domains. Some proteins contain many domains including tandem repeats.

10. Proteins may exist in two or more different well-defined, energetically favorable conformations to meet the needs of their biological functions (e.g., hemoglobin, immunoglobulins, many enzymes).

11. Defects in the synthesis, assembly, storage, and disposal of proteins have pathological consequences.

12. Protein degradation (proteolysis) occurs in the body by different processes: proteolysis in the gastrointestinal tract by pancreatic and gastrointestinal enzymes, intracellular lysosomal and autophage-phagosome-lysosomal proteolysis, and cytosolic and nuclear proteolysis by the energy-dependent ubiquitin−proteasome complex.

13. Defective proteins synthesized in the mRNA-ribosomal pathway and the cytosolic regulatory short-lived proteins are degraded in the ubiquitin−proteasome complex. This pathway of proteolysis is an essential component of cellular metabolism. The inhibition of this pathway by specific and potent chemical inhibitors has therapeutic applications.

14. Aberrantly folded proteins, which escape degradation, can give rise to proteolytic-resistant conformations leading to pathological consequences (e.g., prion diseases and Alzheimer's disease).

Proteins that consist of a single polypeptide chain are generally considered at three levels of organization: **primary**, **secondary**, and **tertiary structure**. For proteins that contain two or more polypeptide chains, each chain is a subunit and there is a **quaternary** level of structure. The primary structure is the unique sequence of amino acids that make up a particular polypeptide; primary structure is maintained by covalent bonds. Secondary, tertiary, and quaternary structures are maintained principally by noncovalent bonds; disulfide bridges may also be considered at the secondary and tertiary levels. Secondary structure arises from repeated hydrogen bonding within a chain, as in the α-**helix**, the β-**pleated sheet**, and β-**turns**

N. V. Bhagavan and C.-E. Ha: Essentials of Medical Biochemistry, Second edition. DOI: http://dx.doi.org/10.1016/B978-0-12-416687-5.00004-X

(discussed later). Tertiary structure describes the three-dimensional stereo-chemical relationships of all of the amino acid residues in a single protein chain. *Folding* of a polypeptide is an orderly sequential process by which the polypeptide attains the lowest possible state of energy. The folding of the polypeptide into its secondary structure is determined primarily by the primary structure. Once the secondary structures are in place, a tertiary structure is formed and stabilized by interactions among amino acids that may be far from each other in the primary sequence but that are close to each other in the three-dimensional structure.

In a discussion of protein structure, it is necessary to differentiate the terms **configuration** and **conformation**. *Configuration* refers to the absolute arrangement of atoms or substituent groups in space around a given atom. Configurational isomers cannot be interconverted without breaking one or more covalent bonds. For example, D- and L-amino acids (Chapter 3), which have different amino acid configurations around the asymmetrical carbon atom, are not interconvertible without the breaking and remaking of one or more covalent bonds. Conformation refers to a three-dimensional arrangement of groups of atoms that can be altered without breaking any covalent bonds. For example, rotation around single bonds allows molecules to undergo transitions between conformational isomers (conformers), as in the eclipsed and staggered conformers of

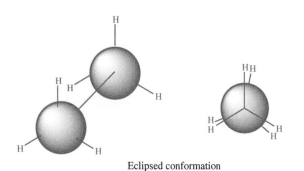

Eclipsed conformation

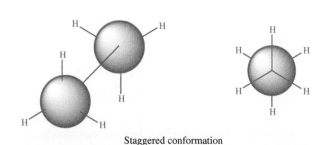

Staggered conformation

FIGURE 4.1 Conformational isomers of ethane (H_3C-CH_3). Eclipsed and staggered conformations for ethane are possible by virtue of the unrestricted rotation around the carbon–carbon single bond. There is a potential energy difference between the two forms, the staggered form being at the minimum and the eclipsed form at the maximum.

ethane (Figure 4.1). Since rotation is relatively unrestricted around the H_3C-CH_3 bond, the two conformers rapidly interconvert.

Proteins contain many single bonds capable of free rotation. Theoretically, therefore, proteins can assume an infinite number of possible conformations, but under normal biological conditions, they assume only one or a very small number of "most stable" conformations. Proteins depend on these stable conformations for their specific biological functions. A functional protein is said to be in its *native* form, usually the most stable one. The three-dimensional conformation of a polypeptide chain is ultimately determined by its amino acid sequence (primary structure). Changes in that sequence, as they arise from mutations in DNA, may yield conformationally altered (and often less stable, less active, or inactive) proteins. Since the biological function of a protein depends on a particular conformation, changes such as **denaturation** (protein unfolding) can lead to loss of biological activity.

COVALENT AND COORDINATE COVALENT BONDS IN PROTEIN STRUCTURE

Covalent bonds involve the equal sharing of an electron pair by two atoms. Examples of important covalent bonds are **peptide** (amide) and **disulfide** bonds between amino acids, and C–C, C–O, and C–N bonds within amino acids.

Coordinate covalent bonds involve the unequal sharing of an electron pair by two atoms, with both electrons (originally) coming from the same atom. The electron pair donor is the ligand, or Lewis base, whereas the acceptor is the central atom (because it frequently can accept more than one pair of electrons), or Lewis acid. These bonds are important in all interactions between transition metals and organic ligands (e.g., Fe^{2+} in hemoglobin and the cytochromes).

NONCOVALENT INTERACTIONS IN PROTEIN STRUCTURE

Both attractive and repulsive interactions occur among different regions of polypeptide chains and are responsible for most secondary and tertiary structures.

Attractive Forces

Ionic interactions arise from electrostatic attraction between two groups of opposite charge. These bonds are formed between positively charged (α-ammonium, ε-ammonium, guanidinium, and imidazolium) side chains and negatively charged (ionized forms of α-carboxyl, β-carboxyl, γ-carboxyl, phosphate, and sulfate) groups.

Hydrogen bonds involve the sharing of a hydrogen atom between two electronegative atoms that have unbonded electrons. These bonds, although weak compared to the bonds discussed previously, are important in water−water interactions, and their existence explains many of the unusual properties of water and ice (Chapter 2). In proteins, groups possessing a hydrogen atom that can be shared include −NH (peptide nitrogen, imidazole, and indole), −SH (cysteine), −OH (serine, threonine, tyrosine, and hydroxyproline), −NH$_2$ and −NH$_3^+$ (arginine, lysine, and alpha-amino), and −CONH$_2$ (carbamino, asparagine, and glutamine). Groups capable of sharing a hydrogen atom include −COO− (aspartate, glutamate, and alpha-carboxylate), −S−CH$_3$ (methionine), −S−S− (disulfide), and −C=O (in peptide and ester linkages).

Van der Waals attractive forces are due to a fixed dipole in one molecule that induces rapidly oscillating dipoles in another molecule through distortion of the electron cloud. The positive end of a fixed dipole will pull an electron cloud toward it; the negative end will push it away. The strength of these interactions is strongly dependent on distance, varying as $1/r^6$, where r is the interatomic separation. The van der Waals forces are particularly important in the nonpolar interior structure of proteins, where they provide attractive forces between nonpolar side chains.

Hydrophobic interactions cause nonpolar side chains (aromatic rings and hydrocarbon groups) to cling together in polar solvents, especially water. These interactions do not produce true "bonds," since there is no sharing of electrons between the groups involved. The groups are pushed together by their "expulsion" from the polar medium. Such forces are also important in lipid−lipid interactions in membranes.

Repulsive Forces

Electrostatic repulsion occurs between charged groups of the same charge and is the opposite of ionic (attractive) forces. This kind of repulsion follows Coulomb's law: $q_1 q_2 / r^2$, where q_1 and q_2 are the charges and r is the interatomic separation.

Van der Waals repulsive forces operate between atoms at very short distances from each other and result from the dipoles induced by the mutual repulsion of electron clouds. Since there is no involvement of a fixed dipole (in contrast to van der Waals attractive forces), the dependence on distance in this case is even greater ($1/r^{12}$). These repulsive forces operate when atoms not bonded to each other approach more closely than the sum of their atomic radii and are the underlying forces in steric hindrance between atoms.

PRIMARY STRUCTURE

Peptide Bond

Peptide bonds have a planar *trans* configuration and undergo very little rotation or twisting around the amide bond that links the α-amino nitrogen of one amino acid to the carbonyl carbon of the next amino acid (Figure 4.2). This effect is due to amido−imido tautomerization. The partial double bond character of the N−C bond in the transition state probably best represents what exists in nature. Electrons are shared by the nitrogen and oxygen atoms, and the N−C and C−O bonds are both (roughly) "one-and-one-half" bonds (intermediate between single and double). The short carbonyl carbon−nitrogen bond length, 0.132 nm (the usual carbon−nitrogen single bond length is 0.147 nm), is consistent with the partial double-bond character of the peptide linkage. The planarity and rigidity of the peptide bond are accounted for by the fact that free rotation cannot occur around double bonds.

Whereas most peptide bonds exist in the *trans* configuration to keep the side chains (R-groups) as far apart as possible, the peptide bond that involves the −NH group of the rigid pyrrolidone ring of proline can occur in both *trans* and *cis* arrangements (Figure 4.3). However, X-ray data suggest that the *trans* form occurs more frequently in proteins than does the *cis* form. It has been further postulated that some proline residues (known as "permissive"

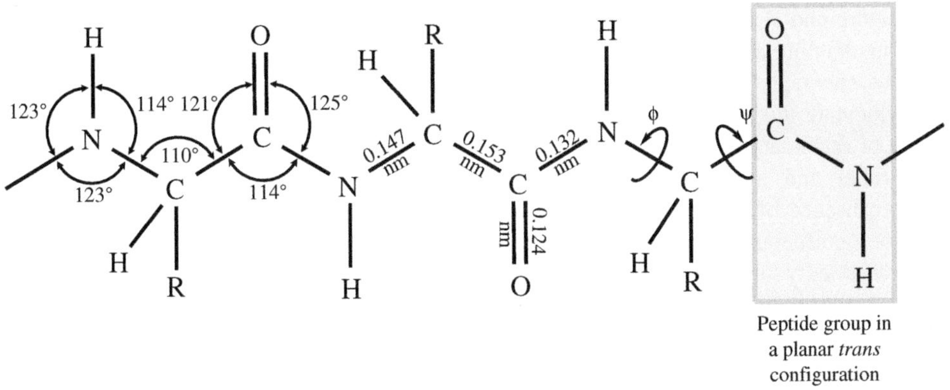

FIGURE 4.2 Geometry of a peptide (amide) linkage.

Peptide group in a planar *trans* configuration

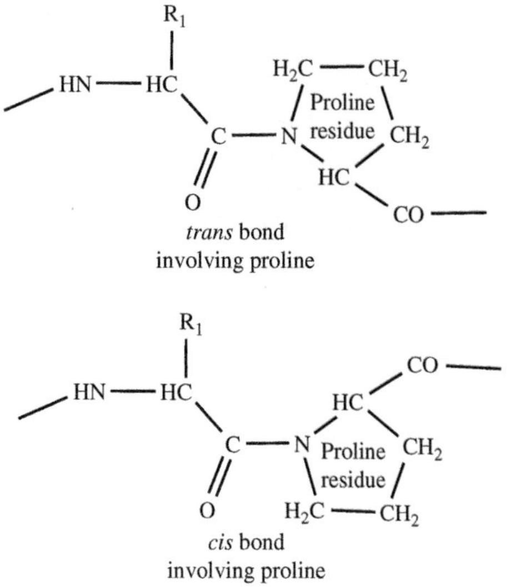

FIGURE 4.3 The *trans* and *cis* configurations of peptide bonds involving proline.

proline residues) can exist in either the *cis* or *trans* configuration. For example, of the four proline residues of ribonuclease A, at least two are thought to be in the *trans* configuration in order to form a native structure, whereas one or both of the other residues may be accommodated in either the *cis* or *trans* configuration.

The bonds on either side of the α-carbon (i.e., between the α-carbon and the nitrogen, and between the α-carbon and the carbonyl carbon) are strictly single bonds. Rotation is possible around these single bonds. The designation for the angle of rotation of the α-carbon-nitrogen bond is Φ, whereas that of the α-carbon-carbonyl carbon bond is ψ. Although theoretically an infinite number of Φ and ψ angles are possible around single bonds, only a limited number of Φ and ψ angles are actually possible in proteins. A polypeptide has specific Φ and ψ values for each residue that determines its conformation.

SECONDARY STRUCTURE

The folding of polypeptide chains into ordered structures maintained by repetitive hydrogen bonding is called **secondary structure**. The chemical nature and structures of proteins were first described by Linus Pauling and Robert Corey, who used both fundamental chemical principles and experimental observations to elucidate the secondary structures. The most common types of secondary structure are the right-handed α**-helix**, parallel and antiparallel β**-pleated sheets**, and β**-turns**. The absence of repetitive hydrogen-bonded regions (sometimes erroneously called **random coil**) may also be part of secondary structure. A protein may possess predominantly one kind of secondary structure (α-helix of hair and fibroin of silk contain mostly α-helix and β-pleated sheet, respectively), or a

protein may have more than one kind (hemoglobin has both α-helical and non-hydrogen-bonded regions). **Globular proteins** usually have mixed, and fibrous proteins have predominantly one kind of secondary structure.

α-Helix

The rod-shaped right-handed α-helix, one of the most common secondary structures found in naturally occurring proteins, consists of L-α-amino acids (Figure 4.4). In the right-handed α-helix, the helix turns counterclockwise (C-terminal to N-terminal), and, in the left-handed, it turns clockwise. A left-handed α-helix is less stable than a right-handed α-helix because its carbonyl groups and the R-groups are sterically hindered. The helical structure is stabilized by intrachain hydrogen bonds involving each —NH and —CO group of every peptide bond. These hydrogen bonds are parallel to the axis of the helix and form between the amido proton of the first residue and the carbonyl oxygen of the fourth residue, and so on, producing 3.6 amino acid residues per turn of the helix. The rise per residue is 0.15 nm, and the length of one turn is 0.54 nm (Figure 4.5).

In some proteins, α-helices contribute significantly to the secondary structure (e.g., α-keratin, myoglobin, and hemoglobin), whereas in others, their contribution may be small (e.g., chymotrypsin and cytochrome c) or absent (e.g., collagen and elastin). Whether a polypeptide segment forms an α-helix depends on the particular R-groups of the amino acid residues. Destabilization of an α-helix may occur for a variety of reasons: electrostatic repulsion between similarly charged R-groups (Asp, Glu, His, Lys, Arg); steric interactions due to bulky substitutions on the β-carbons of neighboring residues (Ile, Thr); and formation of side-chain hydrogen or ionic bonds. Glycine residues can be arranged in an α-helix; however, the preferred and more stable conformation for a glycine-rich polypeptide is the β-pleated sheet because the R-group of glycine (—H) is small and gives rise to a large degree of rotational freedom around the α-carbon of this amino acid. Prolyl and hydroxyprolyl residues usually create a bend in an α-helix because their α-nitrogen atoms are located in rigid ring structures that cannot accommodate the helical bonding angles. Moreover, they do not have an amido hydrogen and therefore can form neither the necessary hydrogen bond nor the usual planar peptide bond. However, some proteins such as rhodopsin do contain proline residues embedded in α-helical segments.

In some proteins, the α-helices twist around each other to form rope-like structures (coiled coils) to give rise to a **supersecondary structure**. Examples of such proteins are the α-keratins, which are major protein components of hair, skin, and nails. These proteins are rich in amino acid residues that favor the formation of an α-helix.

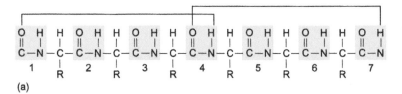

(a)

FIGURE 4.4 Hydrogen bonds in the α-helix. (a) Each peptide group forms a hydrogen bond with the fourth peptide group in each direction along the amino acid chain. (b) Coiling of an amino acid chain brings peptide groups into juxtaposition so that the hydrogen bonds shown in (a) can form. The multiple hydrogen bonds (indicated by the three dots) stabilize the helical configuration.

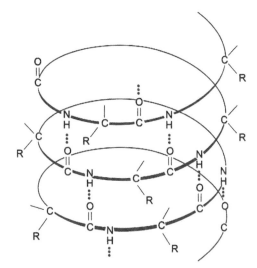

(b)

FIGURE 4.5 Average dimensions of an α-helix. The rise per residue and the length of one turn are 0.15 and 0.54 nm, corresponding to minor and major periodicity, respectively.

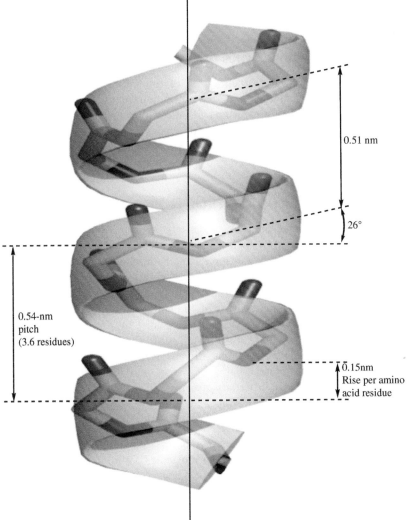

0.51 nm

26°

0.54-nm pitch (3.6 residues)

0.15nm Rise per amino acid residue

In addition, consistent with their properties of water insolubility and cohesive strength, α-keratins are rich in hydrophobic amino acid residues and disulfide cross-linkages. The α-helices are arranged parallel to their length with all the N-terminal residues present at the same end. Three α-helical polypeptides are intertwined to form a left-handed supercoil, called a **protofibril** (the α-helix itself is right-handed). Eleven protofibrils form a microfibril. The polypeptides within the supercoil are held together by disulfide linkages and are also stabilized by van der Waals interactions between the nonpolar side chains. The number of disulfide cross-linkages in α-keratins varies from one source to another. Skin is stretchable because of fewer cross-links, whereas nails are inflexible and tough because of many more cross-links.

β-Pleated Sheet

The β-structure has the amino acids in an extended conformation with a distance between adjacent residues of 0.35 nm (in the α-helix, the distance along the axis is 0.15 nm). The structure is stabilized by intermolecular hydrogen bonds between the —NH and —CO groups of *adjacent* polypeptide chains. The β-structure can occur between separate peptide chains (e.g., silk fibroin) or between segments of the same peptide chain, where it folds back on itself (e.g., lysozyme). Two types of β-pleated sheets exist: **parallel** and **antiparallel**. In the parallel sheet structure, adjacent chains are aligned in the same direction with respect to N-terminal and C-terminal residues, whereas in the antiparallel sheet structure, the alignments are in the opposite directions (Figure 4.6). Some amino acid residues such as glycine and alanine promote the formation of β-pleated sheets.

The β-pleated sheet occurs as a principal secondary structure in proteins found in persons with **amyloidosis**. The generic name **β-fibrilloses** has been suggested for this group of disorders. The proteins that accumulate are called **amyloid** and are aggregates of twisted β-pleated sheet fibrils. They are derived from endogenous proteins (e.g., immunoglobulins) on selective proteolysis and other chemical modifications. The fibrillar proteins are insoluble and relatively inert to proteolysis. Their accumulation in tissues and organs can severely disrupt normal physiological processes. The amyloid deposit, which occurs in several different tissues, is produced in certain chronic inflammatory diseases, in some cancers, and in the brain with some disorders, e.g., **prion diseases, Alzheimer's disease** (discussed later).

β-Turns

β-Turns, which are stabilized by a hydrogen bond, cause polypeptide chains to be compact molecules (e.g., globular proteins of spherical or ellipsoidal shape). The four amino acid residues of a β-turn form a hairpin structure in a polypeptide chain, thus providing an energetically economical and space-saving method of turning a corner. Two tetrapeptide conformations can accomplish a β-turn that is stabilized by a hydrogen bond (Figure 4.7).

Random Coil

Certain regions of peptides may not possess any definable repeat pattern in which each residue of the peptide chain interacts with other residues, as in an α-helix. However, a given amino acid sequence has only one conformation, or possibly a few, into which it coils itself. This conformation has minimal energy. Since energy is required to bring about change in protein conformation, the molecule may remain trapped in a conformation corresponding to minimal energy, even though it is not at absolute minimum internal energy. This concept of a molecule seeking a preferred, low-energy state is the basis for the tenet that the primary amino acid sequence of proteins determines the secondary, tertiary, and quaternary structures [1].

Other Types of Secondary Structures

Other distinct types of protein secondary structures include the type present in collagen, a fibrous connective tissue protein and the most abundant of all human proteins. Collagen peptide chains are twisted together into a three-stranded helix. The resultant "three-stranded rope" is then twisted into a **superhelix** (Chapter 10).

TERTIARY STRUCTURE

Three-dimensional tertiary structure in proteins is maintained by ionic bonds, hydrogen bonds, —S—S— bridges, van der Waals forces, and hydrophobic interactions.

The first protein whose tertiary structure was determined is **myoglobin**, an oxygen-binding protein of muscle cells that consists of 153 amino acid residues; its structure was deduced from X-ray studies by Kendrew, Perutz, and co-investigators (Figure 4.8). Their studies provided not only the first three-dimensional representation of a globular protein but also insight into the important bonding modes in tertiary structures. The major features of myoglobin are as follows:

1. Myoglobin is an extremely compact molecule with very little empty space, accommodating only a small number of water molecules within the overall molecular dimensions of $4.5 \times 3.5 \times 2.5$ nm. All of the peptide bonds are planar with the carbonyl and amide groups in *trans* configurations to each other.

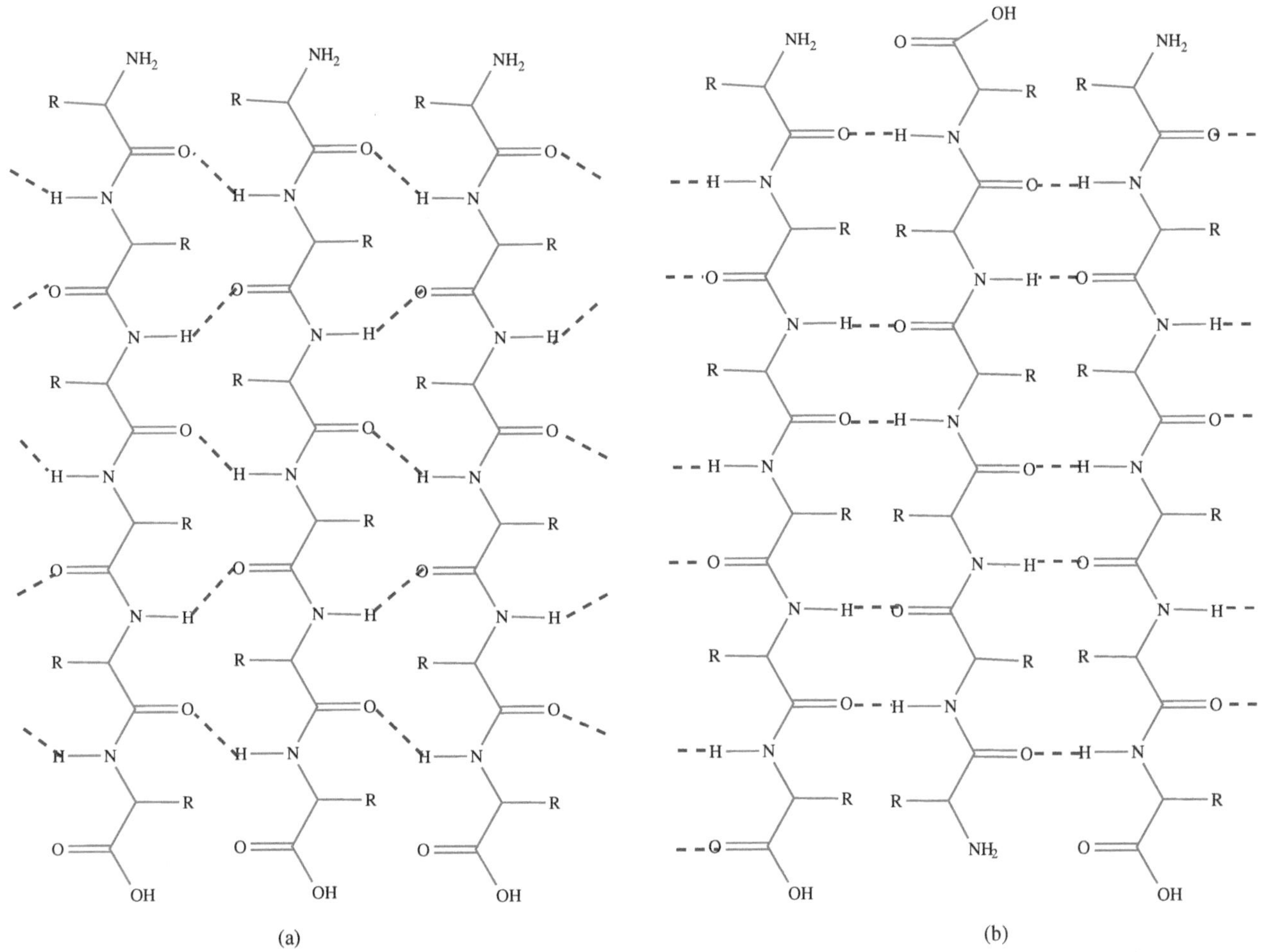

FIGURE 4.6 Hydrogen bonding pattern of parallel (a) and antiparallel (b) β-pleated sheet structures.

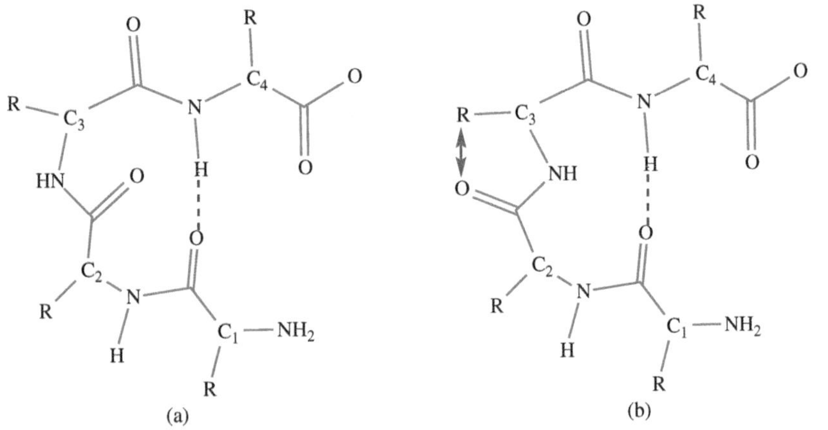

FIGURE 4.7 Two forms of β-turns. Each is a tetrapeptide and accomplishes a hairpin turn. The amino acid residues are identified by numbering the α-carbons 1–4. The CO group of residue 1 is hydrogen-bonded to the NH group of residue 4. Structure (b) is stable only if a glycine (R = H) residue is present as the third residue because of steric hindrance between the R-group and the carbonyl oxygen (double-headed arrow).

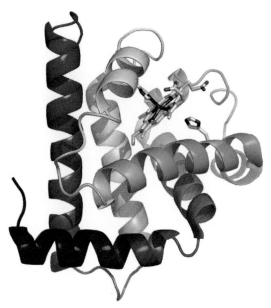

FIGURE 4.8 A high-resolution, three-dimensional structure of myoglobin.

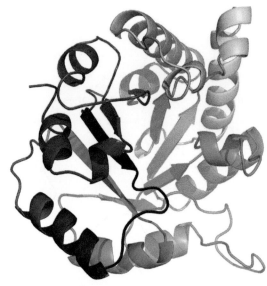

FIGURE 4.9 The αβ-barrel structure seen in triose phosphate isomerase. This structure consists of eight β-pleated sheets (represented by arrows) arranged in a cylindrical manner and surrounded by eight α-helices.

2. Eight right-handed α-helical segments involve approximately 75% of the chain. Five nonhelical regions separate the helical segments. There are two nonhelical regions: one at the N terminus and another at the C terminus.
3. Eight terminations of the α-helices occur in the molecule: four at the four prolyl residues and the rest at residues of isoleucine and serine.
4. Except for two histidyl residues, the interior of the molecule contains almost all nonpolar amino acids. The two histidyl residues interact in a specific manner with the iron of the heme group required for myoglobin activity (Chapter 26). The exterior of the molecule contains polar residues that are hydrated and nonpolar amino acid residues. The amino acid residues whose R-groups contain both polar and nonpolar portions (e.g., threonine, tyrosine, and tryptophan) are oriented so that the nonpolar group faces inward and the polar group faces outward, allowing only the polar portion to come in contact with water.
5. Myoglobin has no S—S bridges that generally help stabilize the conformation formed by amino acid interactions.

Three-dimensional analyses of proteins other than myoglobin reveal the incorporation of every type of secondary structure into tertiary globular arrays. For example, triose phosphate isomerase, a glycolytic enzyme (Chapter 12), contains a core of eight β-pleated sheets surrounded by eight α-helices arranged in a symmetrical cylindrical barrel structure known as αβ-barrels (Figure 4.9). This type of structure has also been noted among other glycolytic as well as nonglycolytic enzymes (e.g., catalase, peroxidase). A few generalizations can be made. Larger globular proteins tend to have a higher percentage of hydrophobic amino acids than do smaller proteins. Membrane proteins tend to be rich in hydrophobic amino acids that facilitate their noncovalent interaction with membrane lipids. In both large globular and membrane proteins, the ratio of aqueous solvent-exposed surface to volume is small; hence, the need for polar, ionic, and hydrogen bonding groups is minimized, whereas the need for hydrophobic interaction is maximized. Small globular proteins tend to have more S—S bonds, which contribute significantly to tertiary structure stability, than do large globular proteins. These S—S bonds may be critical for proteins that lack sufficient stabilization from hydrophobic interactions.

Hydrophobic interactions are considered to be a major determinant in the maintenance of protein conformation. Hydrophobic interactions are due to the tendency of nonpolar side chains to interact with other nonpolar side chains rather than with water. Such interactions may involve the side chains of different adjacent molecules, or they may occur between side chains on the same protein molecule.

QUATERNARY STRUCTURE

Quaternary structure exists in proteins consisting of two or more identical or different polypeptide chains (**subunits**). These proteins are called **oligomers** because they have two or more subunits. The quaternary structure describes the manner in which subunits are arranged in the native protein. Subunits are held together by **noncovalent forces**; as a result, oligomeric proteins can undergo rapid conformational changes that affect biological activity. Oligomeric proteins include the hemoglobins, allosteric enzymes responsible for the regulation of metabolism, contractile proteins such as actin and tubulin, and cell membrane proteins. Several proteins inserted into the membrane participate in the transport of metabolites and ions. Their structure consists of twists and turns in the lipid bilayers of membranes and can undergo topological rearrangements [2]. The subunits of neurotransmitters consist of N-terminal cysteine–cysteine disulfide linkages forming a characteristic loop structure (see the enrichment references).

DENATURATION

Denaturation of a native protein may be described as a change in its physical, chemical, or biological properties. Mild denaturation may disrupt tertiary or quaternary structures, whereas harsher conditions may fragment the chain. Mild denaturation is normally a reversible process.

CONVERSION OF PRECURSOR PROTEINS TO ACTIVE PROTEINS BY PROTEOLYSIS

Many proteins undergo post-translational endoproteolysis at the target sites by proteases (also called **proteinases**). (See also Chapter 4.) Activation of precursor proteins at the target sites is vital to maintain homeostasis. These precursor proteins include enzymes, protein hormones, and receptors, and they are discussed throughout the text in the appropriate chapters. The majority of the proteases (e.g., trypsin and chymotrypsin) contain a triad of amino acid residues—serine, histidine, and aspartate—in variable sequences at the active site.

A small group of serine endoprotease, known as proprotein convertases, are calcium-dependent and located in several different intracellular locations [3]. Proprotein convertases participate in several physiological processes (e.g., processing of pro-opiomelanocortin pathway, and conversion of proinsulin to insulin, proglucagon to glucagon).

Intramembrane proteins located within the lipid bilayer are also targets for proteolysis. They include presenilins, which comprise γ-secretase catalytic subunits and endoplasmic signal peptide peptidase. Presenilin γ-secretase is responsible for the production of amyloid-β-peptide, which forms plaques in the brains of Alzheimer's disease patients (discussed later). The intramembrane proteases have the GxGD motif at the catalytic site, and the catalysis also requires a second aspartyl residue [4].

PROTEASE INHIBITORS USED AS THERAPEUTIC AGENTS FOR HEPATITIS C AND HUMAN IMMUNODEFICIENCY VIRUS INFECTIONS

Protease inhibitors are used as therapeutic agents in the treatment of hepatitis C virus (HCV) and human immunodeficiency virus (HIV).

Both HCV and HIV infectious diseases are major global public health problems. HCV infection can lead to cirrhosis and hepatocellular carcinoma. The replication of HCV is mediated by decoding its single-stranded RNA into a large polyprotein. The polyprotein is cleaved initially into subunits by host proteases. These subunits are assembled to promote replication of the virus. The protease inhibitors *boceprevir* and *telaprevir* are used along with traditional therapeutic agents pegylated interferon and ribavirin [5,6]. One major limitation of currently used protease inhibitors is that they exhibit antiviral activity only against predominant HCV genotype 1; other therapeutic agents are under study [7].

Another example of the use of site-specific protease inhibitors is in the treatment of acquired immunodeficiency syndrome (AIDS). AIDS is caused by human immunodeficiency virus (HIV) infection. HIV causes AIDS by destroying $CD4^+$ T lymphocytes which have CD4 antigen on their cell membrane (see Chapter 33). Susceptibility to opportunistic infections and cancer are complications of AIDS. HIV is a retrovirus and consists of two copies of positive single-stranded RNA and several enzymes required for its replication in the host cell. The propagation of the virus in the host $CD4^+$ T lymphocytes (which can be interrupted at several points of the viral replication cycle, such as reverse transcriptase) consists of the following steps: synthesis of complementary DNA (cDNA) using viral RNA as a template by viral transcriptase, and integration of cDNA with the host DNA followed by its transcription and translation into a polyprotein. The propagation of the virus can be interrupted by inhibitors at different target sites of the replication. One of the sites of inhibition occurs through inhibition of HIV-protease, which cleaves the precursor polyprotein into functional proteins required for assembly and maturation into the virion. Several antiretroviral protease inhibitors are currently used in the management of AIDS (see the Enrichment references 7 and 8).

Several protease inhibitors protect physiological systems from indiscriminate action that can be caused by proteases. Proteases with physiological function are compartmentalized (e.g., lysosomes) or activated only at the appropriate site (e.g., trypsin). α_2-Macroglobulins, α_1-antitrypsin, antithrombin, and complement-1 inhibitor are examples of protease inhibitors. With the exception of macroglobulins, the protease inhibitors mentioned here are known as *serpins*

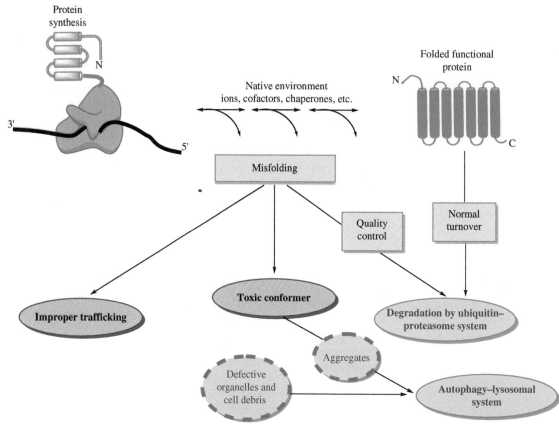

FIGURE 4.10 Pathway of protein folding. Normal folding occurs with the help of chaperones and other factors. Misfolding of polypeptides can lead to targeting to inappropriate cellular locations, which are often resistant to proteolysis and form aggregates, such as amyloid plaques.

because they inhibit proteases that have serine at their catalytic site. Loss of function mutations of serpins can have serious pathological consequences (discussed later in the appropriate sections of text).

PROTEIN FOLDING AND INTRACELLULAR DEGRADATION OF MISFOLDED AND DEFECTIVE PROTEINS

Proteins are synthesized on ribosomes as nascent polypeptides in the lumen of the endoplasmic reticulum (ER). The amino acid sequence of proteins that determines the secondary and tertiary structures is dictated by the nucleotide sequence of mRNA. In turn, mRNA sequences are determined by DNA sequences (see Chapters 21–23). As discussed earlier, the classic experiments of Pauling and Anfinsen led to the concepts that certain key amino acids at the proper positions are essential for the folding of proteins into a functional, unique three-dimensional conformation. It is amazing that of hundreds of millions of conformational possibilities, only a single conformational form is associated with a functional protein. The process of directing and targeting the folding of intermediate polypeptides to the fully folded structures is aided, in some instances, by proteins known as **molecular chaperones** (also called **chaperonins**) (Figure 4.10).

Chaperones bind reversibly to unfolded polypeptide segments and prevent their misfolding and premature aggregation. This process involves energy expenditure provided by the hydrolysis of ATP. A major class of chaperones is **heat-shock proteins** (HSP), which are synthesized in both prokaryotic and eukaryotic cells in response to heat shock or other stresses (e.g., free-radical exposure). There are many classes of heat-shock chaperones (HSP-60, HSP-70, and HSP-90) that are present in various organelles of the cell. HSP-70 chaperones contain two domains: an ATPase domain and a peptide-binding domain. They stabilize nascent polypeptides and also are able to reconform denatured forms of polypeptides. The HSP-70 family of chaperones shows a high degree of sequence homology among various species (e.g., E. coli and human HSP-70 proteins show 50% sequence homology).

Another chaperone, **calnexin**, is a 90 kDa Ca^{2+}-binding protein and is an integral membrane phosphoprotein of the ER. Calnexin monitors the export of newly synthesized glycoproteins by complexing with misfolded glycoproteins that have undergone glycosylation. If a protein cannot be folded

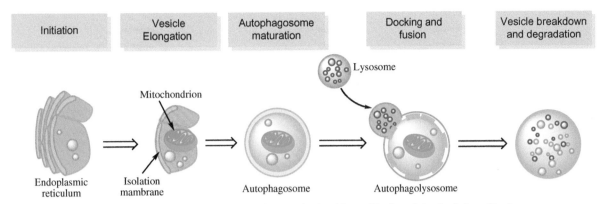

FIGURE 4.11 The pathway of the process of autophagy of a defective mitochondrion and its degradation by fusion with a lysosome.

into its proper conformation, chaperones assist in its destruction. The process of folding is also facilitated by the ionic environment, cofactors, and enzymes. For example, folding is affected by protein disulfide isomerase, which catalyzes the formation of correct disulfide linkages, and by peptidyl prolyl isomerases, which catalyze the *cis—trans* isomerization of specific amino acid—proline peptide bonds.

AUTOPHAGY

Intracellular degradation of short-lived and regulatory proteins and dysfunctional or damaged organelles takes place through two processes to maintain cellular homeostasis. The two pathways are autophagy ("self-eating") and the ubiquitin—proteasome system [8—10]. The major role of autophagy is removal of cellular debris.

Autophagy is an essential component of cellular function, and it occurs in the cytosol of all eukaryotic cells. The process of autophagy begins with targeting defective and nonfunctional organelles, misfolded proteins, and their aggregates, long-lived and short-lived regulatory proteins. The targeted components are sequestered into double-membrane organelles known as autophagosomes. The ultimate degradation of autophagosomal contents occurs when autophagosomes fuse with lysosomes, followed by cleavage by lysosomal acid hydrolases. These products, which are metabolic precursors, are utilized in the resynthesis of macromolecules (Figure 4.11).

Autophagy participates in many physiological processes (e.g., nutrient supply, immune function, inflammation), and defects in autophagy are implicated in several diseases. Autophagy is regulated by several genes and their products. Starvation and energy deprivation prompt autophagy to provide metabolites to maintain vital cellular functions. Insulin and growth factors are negative regulators of autophagy. Disruption of autophagy is implicated in several protein-aggregating diseases (e.g., polyglutamine disorders such as Huntington's disease, Alzheimer's disease, amyotrophic lateral sclerosis, familial Parkinson's disease).

A beneficial effect of promoting degradation of intracellular lipid droplets by stimulation of the autophagy—lysosomal system has been observed in bovine aortic endothelial cells. The stimulant is derived from a polyphenol known as epigallocatechin gallate (EGCG), which is an abundant constituent of green tea. The presumed mechanism consists of elevation of cytosolic Ca^{2+}, followed by activation in succession of calmodulin-dependent protein kinase kinase and AMP-activated protein kinase (see the Enrichment references for more details). Another mechanism by which defective proteins have a diminished capacity to properly fold into three dimensions in the endoplasmic reticulum—Golgi pathway consists of O-mannosylation. The mannosylated defective proteins are rerouted for proteolysis in the cytosol [11,12].

Proteolysis (Protein Degradation) by ATP-Dependent Ubiquitin—Proteasome Complex

If a protein cannot be folded into its proper conformation, chaperones assist in its destruction (proteolysis) in the ubiquitin—proteasome complex. The chaperone-dependent ubiquitin—proteasome system is involved in the regulation of a wide array of cellular processes: quality control of newly synthesized protein, and cell-cycle and division; DNA repair; growth and differentiation; regulation of membrane receptors and ion channels. Failure of the ubiquitin—proteasome system can lead to severe pathological consequences (e.g., many neurodegenerative diseases). Modulation of activity of the proteasome system is an emerging therapeutic target for cancer and some inherited disorders.

Components and the Properties of the Ubiquitin—Proteasome System

1. The process is highly conserved evolutionarily. (See Figure 4.12.)
2. The process is regulated, specific, and selective, and it consumes energy.

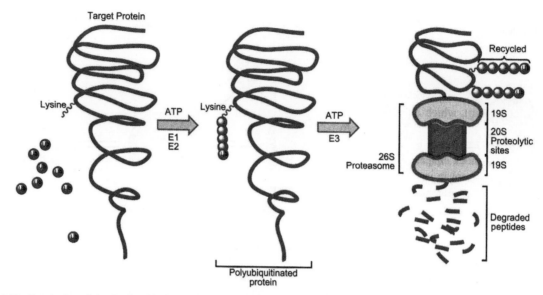

FIGURE 4.12   Protein degradation by the ubiquitin–proteasome pathway. The first step in the proteolysis consists of multiple ATP-dependent steps of conjugation of several ubiquitin (Ub) molecules to the target protein. Polyubiquitination to target protein involves three enzymes: E1 (ubiquitin-activating enzyme); E2 (ubiquitin-conjugating enzyme); and E3 (ubiquitin–protein ligase). During the second step, the protein is stripped of its ubiquitin (which is reused), and in the 19S regulatory component, which consists of ring ATPases, the protein is unfolded and degraded by the proteolytic activity within the core of the proteasome. Bortezomib is an inhibitor of the proteolytic catalytic activity of the proteasome (see Figure 4.13).

3. Selection of proteins that are targeted for proteolysis is complex and may be based on N-terminal amino acid residues, whether they are basic amino acids or large hydrophobic amino acids, and specific internal sequences.
4. Many housekeeping cellular functions are mediated by this system.
5. Degradation of the target protein involves concerted action of three enzymes (E1, E2, and E3), with the beginning of conjugation of several ubiquitin molecules with the targeted protein, and degradation of the tagged protein in the 26S barrel-shaped multicomponent protease that recognizes ubiquitin-tagged substrate.
6. Peptides released from the proteasome are hydrolyzed to amino acids by cytosolic proteases.
7. The attached ubiquitin molecules do not enter the core of the 26S particle, and, after conversion to monomers, they are reused.
8. Proteasomal proteolysis can also begin with monoubiquitination (e.g., transmembrane proteins, histones) or without any ubiquitination (e.g., ornithine decarboxylase).

Bortezomib, an inhibitor of catalytic activity of proteosome, is used in the treatment of multiple myeloma. Bortezomib contains a boronate moiety linked to dipeptide (Figure 4.13). Multiple myeloma is a malignancy of plasma cells (Chapters 15 and 33). It affects the bone marrow and skeleton. Many cytokines, most notably IL-6 and growth factors, promote the proliferation and survival of plasma cells. The antitumor effect of bortezomib may involve the modulation of several regulatory pathways of apoptosis and survival. One of the important pathways involves nuclear factor-κB (NF-κB) and its inhibitory partner protein (IκB). The NF-κB/IκB system is a transcriptional regulatory system for cell death (apoptosis). In the cytosol the NF-κB/IκB complex is in the inactive state. When the cell receives extra- or intracellular signals, kinase cascades are activated, leading to phosphorylation of IκB. This phosphorylation of IκB causes its dissociation from NF-κB and subsequent degradation in the polyubiquitination–proteasome complex. The unbound NF-κB is translocated to the nucleus, where it induces transcription of genes whose protein products maintain cell survival. Inhibition of proteolysis of IκB by bortezomib allows NF-κB to remain in cytosol complexed with IκB, preventing its action on nuclear DNA and promoting tumor cell death (Figure 4.13).

The selectivity and specificity of bortezomib's inhibitory action on the proteasome of multiple myeloma is not completely understood. It should be emphasized that normal proteasomal function is essential for cellular metabolic functions. For example, in some neurodegenerative diseases (e.g., Alzheimer's disease, prion diseases), the pathogenesis may involve the inhibition of proteolysis by the proteasome; furthermore, the toxic microaggregates of protein may cause the inhibition. Thus, bortezomib can potentially accelerate neurodegenerative diseases; however, it does not cross the blood–brain barrier, sparing the central nervous system.

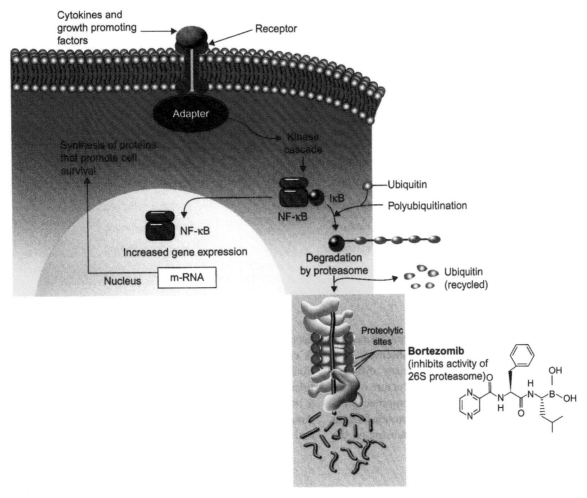

FIGURE 4.13 Inhibition of proteasome proteolytic activity by bortezomib. A major therapeutic effect of bortezomib is the treatment of multiple myeloma mediated by its inhibition of proteolysis of the inhibitory nuclear factor κB (IκB). This leads to IκB remaining bound to the transcription factor nuclear factor-κB (NF-κB) in the cytosol, preventing its translocation to the nucleus. In tumor cells, IκB dissociates from the NF-κB/IκB complex initiated by cell signaling pathways which involve phosphorylation, and IκB is degraded in the proteasome. The released NF-κB in the nucleus induces transcription of genes whose products promote cell survival and proliferation.

PROTEIN MISFOLDING AND FIBRILLOGENIC DISEASES

Misfolded proteins that cannot be removed by ubiquitination or autophagy accumulate both intracellularly and in soft tissues. The misfolded proteins are aligned into characteristic β-sheet structures of a thermodynamically stable pathogenic fibril structure. These structures are protease resistant and are known as amyloid (a misnomer, because they are not starch). Amyloidosis is a diverse group of diseases with fatal consequences caused by several amyloidogenic proteins. Their diagnosis and treatment are based on the type of amyloidogenic protein. Two disorders that cause progressive neurologic damage, along with cognitive function decline, are discussed next.

Transmissible Spongiform Encephalopathies

Transmissible spongiform encephalopathies (TSEs), also known as prion diseases, are caused by neuronal proteins that undergo abnormal limited proteolysis followed by an accumulation of misfolded proteins into amyloid fibrils. They cause progressive neurodegeneration. The most common human prion disease is Creutzfeldt–Jakob disease (CJD). CJD is caused by a conformational transformation of a normal protein of the central nervous system with yet unknown function (PrP^c) into a pathologic isoform (PrP^{Sc}). A protease-resistant polypeptide (PrP^{TSE}) that has the propensity to form amyloid structure is generated from limited proteolysis of PrP^{Sc}. The accumulation of PrP^{TSE} and its products leads to progressive neurological abnormality, including cognitive dysfunction. The

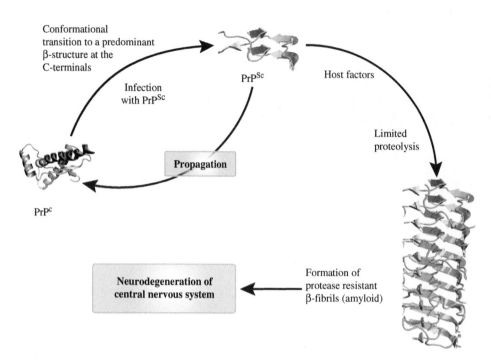

FIGURE 4.14 Diagrammatic representation of transformation of a normal endogenous cellular prion protein (PrP^c) into the transmissible and pathological midfolded isoform (PrP^{Sc}).

transmission of CJD and other prion diseases is propagated by the abnormal prion protein itself, and thus it has the properties of an infectious agent (see Figure 4.14 and see Clinical Case Study 4.1).

Protein folding disorders of an unusual nature may account for a group of TSEs involving proteins called **prions (PrP)**. These disorders, known as prion diseases, are all characterized by amyloid deposition in the brain of animals and humans. The clinical features include neurological symptoms with loss of motor control, dementia, paralysis, and wasting. Incubation periods for prion diseases are months in animals and years in humans. No treatments are available for any of these diseases. TSEs occur in several species of animals and humans, and animal models have been essential in deciphering the molecular basis of these diseases. Some examples of prion diseases occurring in animals and humans are:

- Cats: transmissible feline encephalopathy
- Cows: bovine spongiform encephalopathy (BSE)
- Mink: transmissible mink encephalopathy
- Mule, deer and elk: chronic wasting disease
- Sheep: scrapie
- Humans: Creutzfeld−Jakob disease (CJD), kuru, fatal-familial insomnia syndrome, and Gerstmann−Straussler−Schenker disease

TSEs can exhibit inherited, infectious, and sporadic presentations. Additionally, the inherited disease can also be infectious. CJD occurs both as an inherited autosomal-dominant disorder and in a transmissible form. In the "protein only" hypothesis, the abnormal prion protein, either introduced from external sources or produced by the mutated prion protein gene, affects normal protein folding and shifts the prion protein folding toward the formation of abnormal prion protein. The conversion of the normal prion protein, whose function is unknown, to an aberrant form involves a conformational change rather than a covalent modification. The abnormal prion protein functions as a seed that induces the change of normal cellular prion protein toward the abnormal amyloidogenic rich, β-structure proteins, which can be propagated and transmitted to other cells. The aggregated form of prion protein-forming amyloid is resistant to proteolysis.

The conversion of naturally occurring protease-sensitive prion protein to a protease-resistant form occurs *in vitro* by mixing the two proteins. However, these protease-resistant prion proteins are not infectious. Thus, in the "protein-only" hypothesis of prion infection, the acquisition of an aberrant conformation is not sufficient for the propagation of infectivity.

In the last 25 years, a serious public health problem has arisen by showing that prion disease in cattle can cross species barriers and infect humans. This occurred when cattle were fed meal made from sheep infected with scrapie. The cattle developed BSE (commonly called "mad cow disease"). Subsequently, when people consumed prion-contaminated beef, a small number, primarily in Great Britain, developed a variant of CJD (vCJD) approximately 5 years afterward. The variant form of CJD is a unique form of prion disease occurring in a much younger population than would be expected from inherited or sporadic CJD. Both BSE and vCJD share many similar pathologic characteristics, suggesting an etiologic link between human vCJD and cattle BSE.

The tumor suppressor protein $p53$ provides yet another example of protein misfolding that can lead to pathological effects, in this case cancers (p stands for protein and 53 for its approximate molecular weight of 53,000). The gene for $p53$ is located on the short arm of chromosome 17 (17p) and codes for a 393-amino-acid phosphoprotein. In many cancers, the $p53$ gene is mutated, and the lack of normal $p53$ protein has been linked to the development of as many as 40% of human cancers.

Normal $p53$ functions as a tumor suppressor and is a transcription factor that normally participates in the regulation of several genes required to control **cell growth**, **DNA repair**, and **apoptosis** (programmed cell death). Normal $p53$ is a tetramer, and it binds to DNA in a sequence-specific manner. One of the $p53$-regulated genes produces a protein known as $p21$, which interferes with the cell cycle by binding to cyclin kinases. Other genes regulated by $p53$ are MDM2 and BAX. The former gene codes for a protein that inhibits the action of $p53$ by functioning as a part of a regulatory feedback mechanism. The protein made by the BAX gene is thought to play a role in $p53$-induced apoptosis.

Most mutations of $p53$ genes are somatic missense mutations involving amino acid substitutions in the DNA binding domain. The mutant forms of $p53$ are misfolded proteins with abnormal conformations and the inability to bind to DNA, or they are less stable. Individuals with the rare disorder **Li−Fraumeni syndrome** (an autosomal dominant trait) have one mutated $p53$ gene and one normal $p53$ gene. These individuals have increased susceptibility to many cancers, such as leukemia, breast carcinomas, soft-tissue sarcomas, brain tumors, and osteosarcomas.

Clinical trials are underway to investigate whether the introduction of a normal $p53$ gene into tumor cells by means of gene therapy has beneficial effects in the treatment of cancer. Early results with $p53$ gene therapy indicate that it may shrink the tumor by triggering apoptosis.

Several disorders of protein folding with many different ethologies are known; they have the characteristic pathological hallmark of protein aggregation and deposits in and around the cells. A list of protein folding disorders is given in Table 4.1; they are discussed in subsequent chapters.

Transthyretin amyloidosis (also called **familial amyloid polyneuropathy**) is an autosomal-dominant syndrome characterized by peripheral neuropathy. This disease results from one of five mutations identified thus far in the gene for transthyretin. Transthyretin is also called **prealbumin** (although it has no structural relationship to albumin) because it migrates ahead of albumin in standard electrophoresis at pH 8.6. Transthyretin is synthesized in the liver and is a normal plasma protein with a concentration of 20−40 mg/dL. It transports thyroxine and retinol binding protein (Chapter 15). The concentration of transthyretin is significantly decreased in malnutrition, and plasma levels are diagnostic of disorders of malnutrition (Chapter 15). The gene for transthyretin resides on chromosome 18, and it is expressed in a constitutive manner. The primary structure of transthyretin consists of 127 amino acid residues and 8 β-sheet structures arranged in an antiparallel conformation on parallel planes (Figure 4.15).

Alzheimer's Disease

A dementia syndrome characterized by an insidious progressive decline in memory, cognition, behavioral stability, and independent function was described by **Alois Alzheimer** and is known as **Alzheimer's disease (AD)** [13−16; see also Clinical Case Study 4.2]. Age is an important risk factor for AD; it affects 10% of persons over 65 years of age and about 40% of those over age 85. The characteristic neuropathological changes include formation of extracellular neuritic plaques and intraneuronal tangles with associated neuronal loss in hippocampus and neocortex (Figure 4.16).

The major constituent of the extracellular plaques is **amyloid β-protein (Aβ)**, which aggregates into 8 nm filaments. Aβ is a peptide of 42 amino acid residues and is proteolytically derived from a transmembrane glycoprotein known as **β-amyloid precursor protein (βAPP)**. The enzymes that cleave βAPP to Aβ are known as secretases. In the amyloidogenic pathway, the APP is initially cleaved by β-secretase followed by γ-secretase (presenilin), an intramembrane enzyme producing $A\beta_{42}$ (Figure 4.17). βAPP is widely expressed, particularly in the brain, and its gene has been localized to chromosome 21q. Two major observations have aided in understanding the role of Aβ peptides in the pathology of Alzheimer's disease. The first is that patients with **Down's syndrome** have trisomy 21 (i.e., three chromosome 21s instead of two), exhibit Aβ deposits, and develop classical features of Alzheimer's disease at age 40 years or earlier. Second, several missense mutations in βAPP have been identified in cases of autosomal dominant Alzheimer's disease. These dominant mutations in βAPP adversely affect the action of secretases either by increasing the absolute rate of Aβ excretion (N-terminal mutations) or by increasing the ratio $A\beta_{42}$ to $A\beta_{40}$ (C-terminal mutations).

Inherited disorders of Alzheimer's disease represent less than 1% of all cases. The Aβ peptides aggregate to form β-structures, leading to fibrils. The $A\beta_{42}$ peptides

TABLE 4.1 Examples of Protein Misfolding Diseases

Disease	Mutant Protein/Protein Involved	Molecular Phenotype
Inability to fold		
Cystic fibrosis	CFTR	Misfolding/altered HSP-70 and calnexin interactions
Marfan syndrome	Fibrillin	Misfolding
Amyotrophic lateral sclerosis	Superoxide dismutase	Misfolding
Scurvy	Collagen	Misfolding
Maple syrup urine disease	α-Ketoacid dehydrogenase complex	Misassembly/misfolding
Cancer	*p*53	Misfolding/altered
		HSP-70 interaction
Osteogenesis imperfecta	Type I procollagen proα	Misassembly/altered
		BiP expression
Toxic folds		
Scrapie/Creutzfeldt—Jakob/Familial insomnia	Prion protein	Aggregation
Alzheimer's disease	β-Amyloid	Aggregation
Familial amyloidosis	Transthyretin/lysozyme	Aggregation
Cataracts	Crystallins	Aggregation
Parkinson's disease	α-Synuclein	Aggregation (Lewy bodies)
Mislocalization owing to misfolding		
Familial hypercholesterolemia	LDL receptor	Improper trafficking
α_1-Antitrypsin deficiency	α_1-Antitrypsin	Improper trafficking
Tay—Sachs disease	β-Hexosaminidase	Improper trafficking
Retinitis pigmentosa	Rhodopsin	Improper trafficking
Leprechaunism	Insulin receptor	Improper trafficking

Reproduced with permission from Thomas, P. J., Qu, B-H., & Pedersen, P. L. (1995). *Trends in Biochemical Sciences* 20, 456.

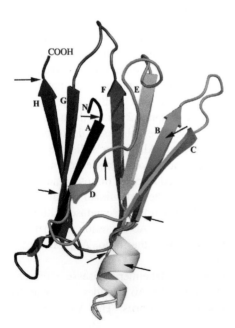

FIGURE 4.15 The structure of transthyretin. The molecule contains eight antiparallel β-strands (A–H) arranged in two parallel planes. The circulating form of transthyretin is a tetramer. Some mutations in the transthyretin gene lead to amyloidosis; eight of the amino acid alterations causing this disease are indicated by arrows. In plasma, transthyretin is a tetramer composed of identical monomers. It appears that mutations cause the monomeric unfolded intermediate of transthyretin to aggregate into an insoluble β-amyloid fibril formation.

are neurotoxic and produce toxic effects through many interrelated mechanisms. These may involve oxidative injury, changes in intracellular Ca^{2+} homeostasis, cytoskeletal reorganization, and actions by cytokines.

The intraneuronal tangles found in AD are bundles of long paired helical filaments that consist of the microtubule-associated protein *tau*. The normal function of tau protein is to stabilize microtubules in neurons by enhancing polymerization of *tubulin*. Normally, tau protein is soluble; however, when it is excessively phosphorylated, it turns into an insoluble filamentous polymer. The dysregulation of phosphorylation/dephosphorylation events has been attributed to an enhanced activity of certain kinases

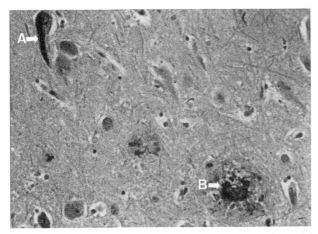

FIGURE 4.16 Section of cerebral cortex from a patient with Alzheimer's disease containing neurofibrillary tangle (A) and neuritic plaque (B). The section was processed with Bielschowsky's stain. *Courtesy of John A. Hardman.*

and a diminished activity of certain phosphatases. Whereas plaques are pathognomonic for AD, tangles are found in etiologically different neurological diseases. Disorders of abnormal hyperphosphorylation and aberrant aggregation of tau protein into fibrillar polymers are known as **taupathies**. Examples of taupathies in addition to AD are *progressive supranuclear palsy*, *Pick's disease, corticobasal degeneration*, and *frontotemporal dementias*.

Two other genes in addition to βAPP have been implicated in the early onset of autosomal-dominant forms of Alzheimer's disease. The other two causative genes are located on chromosomes 14 and 1 and code for transmembrane proteins **presenilin 1** (consisting of 467 amino acid residues) and **presenilin 2** (consisting of 448 amino acid residues). These proteins are synthesized in neurons, but their functions are not known. However, mutations in the presenilin genes lead to excessive production of $A\beta_{42}$ peptides.

Sporadic forms of Alzheimer's disease, responsible for 90% of all cases, are complex diseases and may represent the combined action of both environmental factors and genetic traits that manifest over long time spans. Various forms of a polymorphic gene for **apolipoprotein E (apo E)**, which is on chromosome 19, have been found to occur in higher frequency in persons with Alzheimer's disease. There are three alleles of the apo E gene with six combinations: ε2/ε2, ε3/ε3, ε4/ε4, ε2/ε3, ε2/ε4, and ε3/ε4. Apo E is a lipid carrier protein that is primarily synthesized in the liver; however, it is also synthesized in astrocytes and neurons. The function of apo E proteins in lipoprotein metabolism and their relation to premature atherosclerosis are discussed in Chapter 18.

Of the several genotypes for apo E, the acquisition of two apo E ε4 alleles may increase the risk for

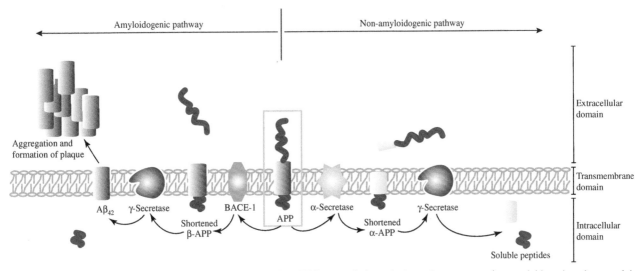

FIGURE 4.17 Proteolytic processing of amyloid precursor protein (APP) by proteolysis catalyzing various secretases into amyloidogenic and nonamyloidogenic pathways in the central nervous system. In the nonamyloidogenic pathway of proteolysis, APP is cleaved by α-secretase followed by γ-secretase releasing nonaggregable soluble peptides; whereas in the amyloidogenic pathway, APP is cleaved at a different site by β-secretase (also known as beta-site amyloid precursor protein-cleaving enzyme, BACE-1) followed by γ-secretase releasing amyloidogenic peptide ($A\beta_{42}$). It should be noted that the proteolytic cleavage occurs in the intramembrane, ultimately releasing either soluble peptides or amyloidogenic peptide.

Alzhiemer's disease up to eightfold. Each copy of the apo E gene increases the risk and shifts the onset to lower ages. The biochemical mechanism by which apo E ε4 protein participates in formation of tangles and plaques is unclear. Several mechanisms have been suggested, namely, interaction with tau protein and generation, and clearance of Aβ peptides.

Pharmacological therapy for Alzheimer's disease consists of correcting the **cholinergic deficit** by administering **acetylcholinesterase inhibitors** (e.g., tacrine, donepizil). **Estrogen therapy** in women with Alzheimer's disease has been associated with improved cognitive performance. Estrogen's beneficial effect may be due to cholinergic and neurotrophic actions. Other therapeutic strategies are directed at inhibiting or decreasing the formation of neurotoxic peptides. In addition, drugs that selectively digest the aggregated peptides may prove useful.

When a patient is evaluated for Alzheimer's disease, it is essential that other treatable causes of dementia be excluded by determining critical biochemical and clinical parameters. Some of the treatable, relatively common abnormalities that produce dementia include drug abuse, electrolyte imbalance, thyroid abnormalities, and vitamin B_{12} deficiency; less common abnormalities are tumor, stroke, and **Wernicke's encephalopathy**.

Sports-related traumatic injuries, either acute or chronic, can potentially increase the risk of neurodegenerative disorders, known as *dementia pugilistica*, resembling Alzheimer's disease [17,18].

CLINICAL CASE STUDY 4.1 89-Year-Old Man with Rapidly Progressive Decline in Cognitive Function

This case was abstracted from case reference 1.

Case Presentation

An 89-year-old man was brought to the emergency department owing to abrupt changes in cognition and personality that began 6 months previously and continued to worsen to the point of functional impairment. The patient's symptoms started with new short-term memory loss, irritability, confusion, and bizarre behavior such as inappropriately undressing or intentionally pouring milk onto the table. Within only a few months, the patient steadily deteriorated, until he could no longer live independently or perform his activities of daily living. He did not have any individual incidences of impaired cognition suggestive of stroke, but rather had an uninterrupted course of decline. Evaluation in the ED revealed no substance abuse, electrolyte abnormalities, or vitamin B_{12} deficiency. Physical examination was most significant for widespread neurologic dysfunction, with upper motor neuron signs and impairment most prominent in the frontal network. These findings were highly suggestive of prion disease, the most common cause of rapidly progressive dementia, but routine lab work was performed to identify any potentially reversible causes of cognitive decline. The serum chemistry, complete blood count, vitamin B_{12} level, and thyroid function were normal. HIV serology, urinary toxicology test, serum IgG and IgA levels, and serum viscosity were also normal. Imaging of the brain showed no evidence of hemorrhage, atrophy, or masses. MRI with T_2-weighted and fluid-attenuated inversion recovery (FLAIR) revealed widespread hyperintensity in multiple focal areas of the cortex, classic for Creutzfeldt–Jakob disease (CJD). Electroencephalogram findings were consistent with encephalopathy. To rule out any other conditions that may imitate CJD, such as autoimmune encephalitides, an array of antibodies were tested and found undetectable. This patient was thus concluded to have CJD by clinical diagnosis. Definitive diagnosis of CJD requires pathological findings of abnormal prion protein PrP^{Sc} in the brain tissue, although confirmation by brain biopsy seldom changes management because there is no effective treatment for CJD. This patient was treated empirically with IV methylprednisolone for a possible glucocorticoid-sensitive encephalopathy. However, within 4 weeks, the patient's condition continued to worsen with complications of prominent myoclonus, dysphagia, and aspiration. With the subsequent withdrawal of further testing and treatment, this patient expired shortly thereafter, just 6 months after his onset of symptoms. Upon autopsy, CJD was confirmed by the presence of prion-containing brain tissue, likely of sporadic etiology.

Teaching Points

1. Creutzfeldt–Jakob disease (CJD) belongs to a family of prion diseases, which are rare, rapidly progressive, and fatal neurodegenerative disorders that affect humans and other animals. In prion diseases, the normal prion protein (PrP^C), which naturally occurs in the central nervous system and has unknown function, undergoes a spontaneous conformational change induced by a prion protein infection into a pathologic, infectious form (PrP^{Sc}), which then has the capability to induce the same transformation in other normal prion proteins. The prion, or proteinaceous infectious particle, differs from conventional infectious agents in that it is merely a misfolded protein, and its infectivity lies in the misfolding of other proteins. The accumulation and deposition of this abnormal isoform in the brain leads to ultimate widespread neurologic dysfunction, followed by neuronal degeneration.

2. CJD is the most common cause of rapidly progressive dementia, but it can also present with a wide variety of other neurologic dysfunctions. Myoclonus is the most frequent physical sign, appearing in 90% of affected patients. Death is invariable and usually occurs within one year of onset of symptoms. Reversible causes of

(Continued)

CLINICAL CASE STUDY 4.1 (Continued)

rapidly progressive dementia must be ruled out, such as substance abuse, electrolyte abnormalities, B_{12} deficiency, and hypothyroidism.

3. CJD is not the same as "mad cow disease" or bovine spongiform encephalopathy (BSE), which is a prion disease in cattle. Classic CJD is caused by sporadic occurrence, familial transmission, or iatrogenic contamination with infected human tissue. The consumption of infected cow meat may cause a variant Creutzfeldt–Jakob disease (vCJD), as studies have shown a strong causal association in which the agent responsible for BSE is the same as that in vCJD. The significance of classic CJD and variant CJD is that they differ clinically and pathologically. Patients with classic CJD have dementia and early neurologic signs as in the case presented here, whereas patients with variant CJD have delayed neurologic signs. The median duration of illness in classic CJD is 5 months, compared to 14 months in vCJD. The median age at death in classic CJD is 68 years old, compared with 28 years old in vCJD. There is no effective treatment for either type of CJD.

Case Reference
[1] M.L. Rinne, S.M. McGinnis, M.A. Samuels, J.T. Katz, J. Loscalzo, A startling decline, N. Engl. J. Med. 366 (2012) 836–842.

Additional Case References
[1] R.T. Johnson, R.G. Gonzalez, M.P. Frosch, An 80-year-old man with fatigue, unsteady gait, and confusion, N. Engl. J. Med. 353 (2005) 1042–1050.

[2] N. Venna, R.G. Gonzalez, S.I. Camelo-Piragua, A 69-year-old woman with lethargy, confusion, and abnormalities on brain imaging, N. Engl. J. Med. 362 (2010) 1431–1437.

References for Supplemental Reading
[1] J.J. Hauw, P.Y. Naccache, D. Seilhean, S. Camilleri, K. Mokhtari, C. Duyckaerts, Neuropathology of non conventional infectious agents or prions, Pathol. Biol. 43 (1995) 43–52.

[2] E.D. Belay, Transmissible spongiform encephalopathies in humans, Annu. Rev. Microbiol. 53 (1999) 283–314.

[3] S.B. Prusiner, Prions and neurodegenerative diseases, N. Engl. J. Med. 317 (1987) 1571–1581.

[4] E. Belay, L. Schonberger, Variant Creutzfeldt–Jakob disease and bovine spongiform encephalopathy, Clin. Lab. Med. 22 (2002) 849–862.

CLINICAL CASE STUDY 4.2 82-Year-Old Male with Slowly Progressive Memory Impairment

An 82-year-old male, a retired high school teacher, was referred by his primary care physician to a neurologist regarding his progressive memory impairment over the past 6 years. After performing physical examinations, tests for neurologic and mental status, and magnetic resonance imaging (MRI) studies, the neurologist gave the diagnosis of dementia and probable Alzheimer's disease. In arriving at this diagnosis, the neurologist eliminated other causes of dementia by appropriate testing. These included depression, electrolyte abnormalities, and drug abuse including alcoholism, thyroid disorders (e.g., hypothyroidism), and vitamin B_{12} deficiency. The subject's symptoms showed only a mild improvement with cholinesterase inhibitors. The subject was admitted to a supervised care home. During the next 3 years, his mental status progressively declined; he declined to eat and passed away due to nutritional failure at age 82. The family requested an autopsy limited to the brain, which showed extracellular neuritic plaques and interneuronal tangles associated with neuronal loss in the hippocampus and neocortex, confirming the diagnosis of Alzheimer's disease.

Teaching Points

1. The neurotoxic plaques in Alzheimer's disease consist of nondigestible, aggregation-prone β-structure amyloid peptides of 42 amino acids ($A\beta_{42}$). $A\beta_{42}$ is normally derived from the cleavage of transmembrane glycoprotein β-amyloid precursor protein (βAPP). βAPP undergoes both nonamyloidogenic and amyloidogenic proteolytic pathways. The

former yields $A\beta_{40}$ and the latter $A\beta_{42}$, which is less prevalent but more prone to aggregation and cellular damage. The amyloidogenic pathway that forms $A\beta_{42}$ is initiated by a proteolytic enzyme known as β-secretase beta site amyloid precursor protein-cleaving enzyme 1 (BACE-1), followed by the action of γ-secretase. The steady-state levels of $A\beta_{40}$ and $A\beta_{42}$ are normally maintained by a proteolytic enzyme known as neprilysin. Any error in this complex cascade of events causes an imbalance between the synthesis and degradation of Aβ, causing an accumulation of the β-amyloid peptides into neurotoxic plaques. At the level of the synapse, Aβ impairs synaptic plasticity and impairs synaptic transmission. It is synaptic dysfunction that best correlates with cognitive decline in AD.

2. The cytotoxic neurofibrillary tangles of AD are aggregates of hyperphosphorylated tau proteins. The normal function of tau protein is to promote intracellular vesicle transport by promoting the assembly of microtubules. The extent of tau protein phosphorylation is controlled by regulatory enzymes. When tau protein becomes hyperphosphorylated, it loses affinity for microtubules, becomes insoluble, and forms filamentous inclusions.

3. Currently, definitive diagnosis of AD is only made at autopsy, when characteristic histopathologic hallmarks of brain reveal neurofibrillary tangles and amyloid plaques. However, measurement of $A\beta_{42}$, total tau proteins (T-tau), and phosphorylated tau protein 181 (P-tau 181) in cerebrospinal fluid (CSF) may provide a clinical diagnosis of

(Continued)

CLINICAL CASE STUDY 4.2 (Continued)

AD in subjects with progressive cognate deficit. Compared to healthy control subjects, the $A\beta_{42}$ level is decreased (because it is consumed in the formation of plaques) and both T-tau and P-tau 181 are increased (because they are released due to neurodegeneration), in potential subjects who may develop AD.

4. The understanding of $A\beta_{42}$ peptide in the pathology of AD is aided by studies with patients with Down's syndrome and several missense mutations in the gene for βAPP.

5. As discussed previously, $A\beta$ and tau are particularly useful for the clinical and pathological diagnosis of AD, but these are not the only abnormalities that occur in the disease process. The pathogenesis of AD is multifold, and there is no single linear sequence of events. However, from synaptic failure to mitochondrial dysfunction to oxidative and inflammatory damage, studies have shown that the underlying cause for this widespread dysfunction is an abnormal accumulation of misfolded proteins.

6. Protein folding disorders can occur with many other proteins due to excessive production and/or mutations. These

are prions (see reference 3); immunoglobin light chains (AL, elevated in some forms of multiple myeloma; see reference 4); serum precursor protein (SAA, elevated in inflammation); and mutations in transthyretin, lysozyme, fibrinogen, and some other proteins. The amyloid formation due to these proteins may occur throughout the body.

Selected References for Further Study and Enrichment

[1] C.H. Kawas, Early Alzheimer's disease, N. Engl. J. Med. 349 (2003) 1056–1063.

[2] H.W. Querfurth, F.M. LaFerla, Alzheimer's disease, N. Engl. J. Med. 362 (2010) 329–344.

[3] R.T. Johnson, R.G. Gonzalez, M.P. Frosch, Case 27-2005: an 80-year-old man with fatigue, unsteady gait, and confusion, N. Engl. J. Med. 353 (2005) 1042–1050.

[4] L.M. Dember, J.O. Shepard, F. Nesta, J.R. Stone, Case 15-2005: an 80-year-old man with shortness of breath, edema, and proteinuria, N. Engl. J. Med. 352 (2005) 2111–2119.

[5] A.Z. Herskovits, J.H. Growdon, Sharpen that needle, Arch. Neurol. 67 (2010) 918–920.

[6] G.D. Meyer, F. Shapiro, H. Vanderstichele, et al., Diagnosis-independent Alzheimer disease biomarker signature in cognitively normal elderly people, Arch. Neurol. 67 (2010) 949–956.

REQUIRED READING

[1] K.A. Dill, J.L. MacCallum, The protein-folding problem, 50 years on, Science 338 (2012) 1042–1046.

[2] J.U. Bowie, Membrane protein twists and turns, Science 339 (2013) 398–399.

[3] A.W. Artenstein, S.M. Opal, Proprotein convertases in health and disease, N. Engl. J. Med. 365(26) (2011) 2507–2518.

[4] T. Tomita, T. Iwatsubo, Structural biology of presenilins and signal peptide peptidases, J. Biol. Chem. 288(21) (2013) 14673–14680.

[5] J.P.H. Drenth, HCV treatment—No more room for interferonologists? N. Engl. J. Med. 368(20) (2013) 1931–1932.

[6] T.J. Liang, M.G. Ghany, Current and future therapies for hepatitis C virus infection, N. Engl. J. Med. 368(20) (2013) 1907–1917.

[7] N. Alkhouri, N.N. Zein, Protease inhibitors: silver bullets for chronic hepatitis C infection? Cleve. Clin. J. Med. 79(3) (2012) 213–222.

[8] A.M.K. Choi, S.W. Ryter, B. Levine, Autophagy in human health and disease, N. Engl. J. Med. 368(7) (2013) 651–662.

[9] I. Koren, A. Kimchi, Promoting tumorigenesis by suppressing autophagy, Science 338 (2012) 889–890.

[10] R.A. Nixon, The role of autophage in neurodegenerative disease, Nat. Med. 19(8) (2013) 983–997.

[11] B. Kleizen, I. Braakman, A sweet send-off, Science 340 (2013) 930–931.

[12] C. Xu, S. Wang, G. Thibault, D.T.W. Ng, Futile protein folding cycles in the ER are terminated by the unfolded protein O-mannosylation pathway, Science 340 (2013) 978–981.

[13] H.W. Querfurth, F.M. LaFerla, Alzheimer's disease, N. Engl. J. Med. 362(4) (2010) 329–344.

[14] S. Gandy, Lifelong management of amyloid-beta metabolism to prevent Alzheimer's disease, N. Engl. J. Med. 367(9) (2012) 864–866.

[15] R.J. Bateman, C. Xiong, T.L.S. Benzinger, A.M. Fagan, A. Goate, N.C. Fox, et al., Clinical and biomarker changes in dominantly inherited Alzheimer's disease, N. Engl. J. Med. 367(9) (2012) 795–804.

[16] A. Aguzzi, T. O'Connor, Protein aggregation diseases: pathogenicity and therapeutic perspectives, Nat. Rev. Drug. Discov. 9 (2010) 237–248.

[17] B.D. Jordan, The clinical spectrum of sport-related traumatic brain injury, Nat. Rev. Neurol. 9 (2013) 222–230.

[18] K. Blennow, J. Hardy, H. Zetterberg, The neuropathology and neurobiology of traumatic brain injury, Neuron 76 (2012) 886–899.

ENRICHMENT READING

[1] F.A. Stephenson, Introduction to thematic minireview series on celebrating the discovery of the cysteine loop ligand-gated ion channel superfamily, J. Biol. Chem. 287(48) (2012) 40205–40206.

[2] J. Changeux, The nicotinic acetylcholine receptor: the founding father of the pentameric ligand-gated ion channel superfamily, J. Biol. Chem. 287(48) (2012) 40207–40215.

[3] S. Dutertre, C. Becker, H. Betz, Inhibitory glycine receptors: an update, J. Biol. Chem. 287(48) (2012) 40216–40223.

[4] E. Sigel, M.E. Steinmann, Structure, function, and modulation of $GABA_A$ receptors, J. Biol. Chem. 287(48) (2012) 40224–40231.

[5] A.J. Wolstenholme, Glutamate-gated chloride channels, J. Biol. Chem. 287(48) (2012) 40232–40238.

[6] S.C.R. Lummis, 5-HT(3) receptors, J. Biol. Chem. 287(48) (2012) 40239–40245.

[7] A. Brik, C.-H. Wong, HIV-1 protease: mechanism and drug discovery, Org. Biomol. Chem. 1 (2003) 5–14.

[8] I. Pérez-Valero, J.R. Arribas, Protease inhibitor monotherapy, Curr. Opin. Infect. Dis. 24 (2011) 7–11.

[9]   T. Johansen, T. Lamark, Selective autophagy mediated by autophagic adapter proteins, Autophagy 7(3) (2011) 279−296.

[10]   L.N. Micel, J.J. Tentler, P.G. Smith, S.G. Eckhardt, Role of ubiquitin ligases and the proteasome in oncogenesis: novel targets for anticancer therapies, J. Clin. Oncol. 31(9) (2013) 1231−1238.

[11]   J. Calise, S.R. Powell, The ubiquitin proteasome system and myocardial ischemia, Am. J. Physiol. Heart Circ. Physiol. 304 (2013) 337−349.

[12]   S. Chitra, G. Nalini, G. Rajasekhar, The ubiquitin proteasome system and efficacy of proteasome inhibitors in diseases, Int. J. Rheum. Dis. 15 (2012) 249−260.

[13]   D.H. Margolin, M. Kousi, Y.-M. Chan, E.T. Lim, J.D. Schmahmann, M. Hadjivassiliou, et al., Ataxia, dementia, and hypogonadotropism caused by disordered ubiquitination, N. Engl. J. Med. 368(21) (2013) 1992−2003.

[14]   S.M. Paul, Therapeutic antibodies for brain disorders, Sci. Transl. Med. 3(84) (2011) 1−5.

[15]   D.J. Selkoe, Preventing Alzheimer's disease, Science 337 (2012) 1488−1492.

[16]   F.M. LaFerla, Preclinical success against Alzheimer's disease with an old drug, N. Engl. J. Med. 367(6) (2012) 570−572.

[17]   J. Lowenthal, S. Chandros Hull, S.D. Pearson, The ethics of early evidence—Preparing for a possible breakthrough in Alzheimer's disease, N. Engl. J. Med. 367(6) (2012) 488−490.

[18]   T. Jonsson, H. Stefansson, S. Steinberg, I. Jonsdottir, P.V. Jonsson, J. Snaedal, et al., Variant of TREM2 associated with the risk of Alzheimer's disease, N. Engl. J. Med. 368(2) (2013) 107−116.

[19]   M.S. Willis, C. Patterson, Proteotoxicity and cardiac dysfunction—Alzheimer's disease of the heart? N. Engl. J. Med. 368(5) (2013) 455−464.

[20]   J. Li, T. Kanekiyo, M. Shinohara, Y. Zhang, M. J. Ladu, H. Xu, et al., Differential regulation of amyloid-β endocytic trafficking and lysosomal degradation by apolipoprotein E isoforms, J. Biol. Chem. 287(53) (2012) 44593−44601.

[21]   W.S.T. Griffin, Neuroinflammatory cytokine signalling and Alzheimer's disease, N. Engl. J. Med. 368(8) (2013) 770−771.

[22]   D.H. Smith, V.E. Johnson, W. Stewart, Chronic neuropathologies of single and repetitive TBI: substrates of dementia? Nat. Rev. Neurol. 9 (2013) 211−221.

[23]   T.K.H. Scheel, C.M. Rice, Understanding the hepatitis C virus life cycle paves the way for highly effective therapies, Nat. Med. 19 (7) (2013) 837−849.

[24]   H. Kim, V. Montana, H. Jang, V. Parpura, J. Kim, Epigallocatechin gallate (EGCG) stimulates autophagy in vascular endothelial cells, J. Biol. Chem. 288(31) (2013) 22693−22705.

Chapter 5

Energetics of Biological Systems

Key Points

1. Humans and other organisms require energy to live.
   a. Diet provides both sources of energy and building blocks for the synthesis of biological molecules.
   b. Energy balance is a prerequisite to health, and thus understanding the relationships between energy sources, energy requirements, and mechanisms by which energy is transformed for use in humans is necessary.
   c. Energetics in biological systems is governed by the same laws of physics and chemistry as in nonbiological systems.
      i. Thermodynamics—describes the principles by which chemical reactions provide energy or consume energy.
      ii. Kinetics—describes the principles by which chemical reactions are made to occur fast enough to meet the demands of the living organism.
2. Thermodynamics is concerned *only* with initial and final states of chemical reactions.
   a. A chemical reaction that will occur spontaneously, i.e., without the *net* input of energy is described as thermodynamically favorable.
      i. Favorable reactions can be too slow to occur in the absence of catalysts, e.g., enzymes.
   b. Chemical reactions will be favorable or not depending on the free energy difference between the initial (reactant) and final (product) states.
      i. Energy differences between initial and final states are determined by
         1. Energy necessary to break or form chemical bonds.
         2. Concentrations of the substances that are reacting.
   c. Chemical reactions that are not favorable can be made to occur by
      i. Coupling with other reactions to provide the energy necessary for reactions to become favorable, e.g., using phosphorylation by ATP.
      ii. Coupling with reactions to change concentrations so that transformation from reactant to product occurs; e.g., a product being removed from another reaction is a thermodynamic description of a metabolic pathway.
3. Interactions between biological molecules guide and modulate biological processes.
   a. Interactions between biological molecules are governed by the laws of thermodynamics and apply to reactions that involve noncovalent bond formation and covalent bond breaking and making.
   b. Binding of ligands to receptors serves to initiate, modulate, and control biological processes with very high discrimination as exhibited by hormone regulation and immune system recognitions.
4. Kinetics is concerned with the speed at which thermodynamically favorable reactions proceed from the initial to the final state.
   a. Catalysts are required for most reactions to occur at appreciable rates at body temperature and in aqueous solutions such as conditions found in cells and tissues.
   b. Enzymes (and ribozymes) are the catalysts in living systems.
   c. Enzymes determine the pathways by which thermodynamically favorable reactions occur at acceptable rates in living systems.

Living organisms such as humans depend on chemical reactions to obtain the energy required to perform the functions that distinguish living from nonliving systems. Reactions involved in performing mechanical work by muscle and chemical work in synthesis of proteins, DNA, RNA, and complex polysaccharides consume energy. Humans, and other animals, use substances from the diet to provide the energy demanded for life and require a constant input of energy-providing substances. The diet also provides chemical substances that humans cannot make and thus are obtained from plants and animals consumed as food.

The energetics of living systems, i.e., the relationships between energy available and energy used in biochemical reactions, are governed by the same laws of chemical **thermodynamics** and **chemical kinetics** as govern nonliving systems. What distinguishes living from nonliving systems are the methods and molecules that have evolved to obtain energy from chemical substances (reactants) and the use of these to create both simpler and more complex substances (products). A rudimentary knowledge of energetics of biological systems, here human biology, is thus necessary to understand energy balance and how energy is obtained and converted into other forms. Rudimentary principles of chemical thermodynamics are thus the subject of the first section of this chapter.

N. V. Bhagavan and C.-E. Ha: Essentials of Medical Biochemistry, Second edition. DOI: http://dx.doi.org/10.1016/B978-0-12-416687-5.00005-1

THERMODYNAMICS

The first law of thermodynamics states that energy can neither be created nor destroyed (i.e., the energy of the universe is constant) but can be converted from one form into another. The second law states that in all processes involving energy changes under a given set of conditions of temperature and pressure, some energy is dissipated unproductively. This is the amount of energy unavailable for performing useful work. The lost energy is called the **entropy** of the system, and it attains a maximum value at equilibrium. **Entropy (S)** may be considered a measure of disorder or randomness. However, increases in disorder can promote a chemical reaction and thus contribute to determining whether a reaction will or will not occur. An example of entropy driving an important process in biological systems is the formation of the lipid bilayer of cell membranes; however, this example is somewhat subtle because the increase in randomness is in the water around the lipid molecules when they are "tucked" into the membrane bilayer.

A powerful property of thermodynamic quantities is that they are independent of the chemical pathway taken by the reactants in their conversion to products. Thermodynamics predicts, on the basis of the known energies of reactants and products, whether a reaction can be expected to occur without input of energy from an external source. Using thermodynamic quantities, we can obtain information about reactions that cannot be studied directly in living systems and apply this information to the living system.

Thermodynamic Relationships for Chemical Reactions in Biological Systems

In biological systems, **thermodynamics** is employed to establish the possibility that chemical reactions will occur under physiological conditions, i.e., constant temperature and pressure and acidity. Included in the biochemical reactions are chemical reactions that involve covalent bond breaking and bond making and noncovalent interactions to form complexes that are involved in biochemical processes. Common examples of noncovalent reactions are the binding of metabolites, regulatory substances, and drugs to cellular and intracellular receptors. Chemical bond making and chemical bond breaking reactions in living organisms provide the energy for doing work and transforming the chemical substances of the diet into those required for life. When biochemical reactions must consume energy to make them occur—i.e., when a reaction is not favorable at reactant concentrations and conditions within the cell and/or the energy involved in the bond breaking or making process are not favorable—this is a thermodynamic "problem." Living systems have, however, evolved processes that can solve this problem; i.e., they combine or couple reactions to make the net reaction favorable. Biochemical thermodynamics is

concerned with uses of energy from other sources and reactions so that the biologically necessary reactions are made to be favorable. Metabolic pathways are a primary example of linked biochemical reactions in living systems. In applying the laws of thermodynamics to human metabolism, the primary concern is how the concentrations and energy content of reactants and products affect the **net** direction of the reactions in a metabolic pathway.

Reactions that are thermodynamically favorable may still *not* occur fast enough to be compatible with the demands of the living organism. These reactions are made to be faster by catalysts; in biological systems, enzymes are the primary catalysts. The energetics of reaction catalysis, a phenomenon totally independent of thermodynamics, is considered in the second section of this chapter.

Energetics of favorable chemical reactions can be illustrated by a diagram that shows both the thermodynamic requirements for a reaction to be favorable and how energetic barriers exist that also make kinetic considerations relevant (see Figure 5.1).

During the conversion of a reactant **B** to product **C**, there is a net energy change in the reaction; this energy is called the **Gibbs free energy (G)**. The Gibbs free energy change is what is relevant to whether the reaction will occur without net input of energy; i.e., the reaction is thermodynamically favorable. The change is designated **ΔG**, and is the difference between the energy in the final state; i.e., the product **C** and the reactant **B** are present in their equilibrium concentrations. This is illustrated by plotting the change in free energy of the participating reactant molecules as a function of the reaction progress as the reaction goes to equilibrium. The phrase "progress of reaction" on the abscissa of the graph includes all changes in bond lengths and angles (i.e., in the shapes and atomic interactions) of the reactant molecules as they are converted to product molecules. The conceptual "progress of reaction" provides the link between thermodynamic and kinetic

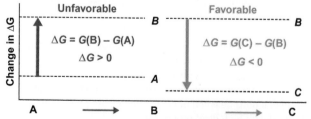

FIGURE 5.1 Energy changes in simple chemical reactions; a coupled reaction illustrating how a thermodynamically unfavorable reaction becomes favorable by coupling to a favorable reaction with a greater energy change.

characteristics of reactions, both of which are key elements in understanding biochemical processes. An isolated reaction continues until it reaches its final state, i.e., reaches equilibrium, and a difference in energy no longer exists. At equilibrium, the *change* in the Gibbs free energy, $\Delta G = 0$. The equilibrium state is dynamic, however, not static. When both reactant and product molecules are present in a reaction mixture, molecules of reactant will be converted to product and "product" molecules back to "reactant" molecules. However, at any given moment, the concentrations of the reactants and products are constant, the net energy difference is zero between reactant and product, and thus the reaction is at equilibrium.

An energy profile for a hypothetical chemical reaction, $B \rightleftharpoons C$ shown in Figure 5.1, is for a "spontaneous," i.e., thermodynamically **favorable**, reaction. The free-energy difference between reactants and products is indicated by ΔG. A negative value for ΔG for a chemical reaction (as in Figure 5.1) indicates that energy is released in the transformation of reactants to products; such a reaction is termed **exergonic**. For reactions with $\Delta G < 0$, we can predict that they are *thermodynamically favorable*, and thus no net energy input is needed for the reaction as written, the "forward reaction," to occur. An example of great relevance to living systems is the conversion of glucose + oxygen (glucose is the principal energy source for the brain) to carbon dioxide and water; in this case the reaction involves two reactants and two products:

$$C_6H_{12}O_6 + 6O_2 \rightarrow 6CO_2 + 6H_2O \qquad (5.1)$$

The free energy change for the conversion of glucose to carbon dioxide and water is very large; ~ 685 kcal/mole at 37°C under blood and tissue oxygen concentration.

An energy diagram for a thermodynamically *unfavorable* chemical reaction, $A \rightarrow B$ is shown in Figure 5.1. In this reaction the free energy difference between reactants and products, ΔG, has a positive value. If a chemical reaction has a positive ΔG value, the reaction is *thermodynamically unfavorable* and will not occur except with the assistance of an external source of energy or by coupling of the unfavorable reaction with favorable reactions. Such a reaction with $\Delta G > 0$ is termed **endergonic**. Biological systems use coupling of reactions, e.g., in metabolic pathways, to make otherwise thermodynamically unfavorable reactions favorable. Dramatically illustrative of an unfavorable reaction is the **reverse reaction to oxidation of glucose** to carbon dioxide and water, i.e., conversion of carbon dioxide and water to glucose. We know that humans and other animals don't carry out this reaction. In fact, this is the principal reaction of photosynthesis in plants, and the ultimate source of external energy is provided by sunlight.

Whether a reaction will be thermodynamically favorable or unfavorable depends not only on energy differences, but also on the concentrations of the reactants and products. The equilibrium concentrations of reactants and products in a chemical reaction are given by the mass action expression, $K_{eq} = [\text{products}]_{eq}/[\text{reactants}]_{eq}$ where the subscript "eq" means concentrations at equilibrium. For a reversible chemical reaction, the dependence of ΔG on concentration of reactants and products at constant temperature is given by the expression

$$\Delta G = \Delta G^0 + RT \ln \frac{[\text{products}]}{[\text{reactants}]}; \text{ in our examples}$$

$$\frac{[B]}{[C]} \text{ or } \frac{[B]}{[A]} \qquad (5.2)$$

where ΔG^0 is the standard free-energy change in calories per mole of reactant consumed; R is the gas constant (1.987×10^{-3} kcal·mol^{-1}·deg^{-1});[1] T is the absolute temperature in degrees Kelvin ($= °C$ plus 273); $\ln$ is the natural logarithm (logarithm to the base e); and [products] and [reactants] are the corresponding molar concentrations for B (reactant) and C (product) or A (reactant) and B (product). A standard state value (G^0 and ΔG^0) is necessary because we need a reference state or situation to which to relate our ΔG values and to calculate the equilibrium constants, which are of more direct interest because equilibrium constants enable us to calculate the concentrations of reactants and products. Moreover, by measuring the concentrations of reactants and products at equilibrium, we can calculate ΔG^0.

Since at equilibrium $\Delta G = 0$, this equation can be written

$$\Delta G^0 = -RT \ln K_{eq} \text{ where } K_{eq} = \frac{[\text{products}]_{eq}}{[\text{reactants}]_{eq}} \qquad (5.3)$$

Note that the standard free-energy change is proportional to the natural logarithm of the equilibrium constant, K_{eq}. We use ΔG and ΔG^0 in discussing energetics because free energies can be added and subtracted, whereas equilibrium constants cannot. In biological systems another convention is also used, $\Delta G^{0\prime}$ where the $\prime$ (prime) refers to values of ΔG^0 at pH 7.0.

A physical interpretation of ΔG^0 can be obtained by noting from Equation (5.2) that if [products]/[reactants] = 1, (*the ln of 1 = 0*), then $\Delta G = \Delta G^0$. ΔG^0 thus represents the amount of useful work that can be obtained from conversion of 1 mol of each reactant in its standard state to 1 mol of each product in its standard state.[2]

---

1. The preferred units are joules, rather than calories or kilocalories. However, because dietary quantities are given in calories (kilocalories, kcal), we retain their usage.

2. This equation and standard states for thermodynamic parameters which explain differences between ΔG and ΔG^0 more extensively can be found in any textbook on physical chemistry or physical biochemistry; see supplementary references.

Physiological Concentrations and How ΔG Can Be <0 When $\Delta G^0 > 0$

As noted previously, if the value ΔG^0 for a given chemical reaction is **positive**, thermodynamics of that reaction alone suggests that the reaction will not occur under standard conditions without external energy input. However, the value of ΔG (*not* ΔG^0) may be negative, indicating that the reaction can occur under the **prevailing concentrations** of reactants and products. One reaction from the metabolic pathway, glycolysis, is catalyzed by the enzyme aldolase (Chapter 13). This reaction furnishes a good illustration of how ΔG depends on actual reactant and product concentrations, and shows that physiological, e.g., intracellular or intravascular, reactant and product concentrations dictate whether reactions are or are not thermodynamically favorable.

Fructose 1,6-Bisphosphate $\rightleftharpoons$ Glyceraldehyde 3-Phosphate + Dihydroxyacetone Phosphate

The equilibrium constant K_{eq} for this reaction at 25°C is 6.7×10^{-5} M. We calculate ΔG^0 from Equation (5.2) to be +5.67 kcal/mol. The positive value for ΔG^0 indicates that the reaction is not favored in the forward direction under standard conditions. If we calculate the ΔG value using the concentration of 50 µmol/L (or 50×10^{-6} M) for both reactant and product, which is close to their physiological concentrations, then:

$$\Delta G = +5.67 \text{ kcal/mol} + RT \ln([50 \times 10^{-6}]$$
$$\times [50 \times 10^{-6}]]/[50 \times 10^{-6}])$$
$$= +5.67 \text{ kcal/mol} + (-5.85 \text{ kcal/mol})$$
$$= -0.18 \text{ kcal/mol, i.e., } \Delta G \text{ is} < 0$$

The negative value of ΔG indicates that, under physiological conditions, the forward reaction (**B** to **C**) is favorable despite the fact that ΔG^0 is positive.

In living systems, chemical reactions rarely reach true equilibrium (are invariant with time) but may attain near-equilibrium. In a metabolic pathway, a nonequilibrium reaction can become the rate-determining (slowest) step of that pathway, and its **kinetic regulation** provides directionality to the pathway. This is a second mechanism in addition to thermodynamic favorability by which living systems control reaction pathways and the direction that they take. In the reaction sequence

$$A \rightleftharpoons B \rightleftharpoons C \rightleftharpoons D \cdots \qquad (5.4)$$

the reaction $A \rightleftharpoons B$, which in our example is not thermodynamically favorable, can be made thermodynamically favorable when it is coupled with the reaction **B**. **B** can be prevented from reaching near-equilibrium by removing it—by converting it to **C** as fast as it can be made

from **A**. However, the conversion of **B** to **C** may attain near-equilibrium. The near-equilibrium reactions are reversible and can allow for reversal of steps in a biochemical pathway. All reactions in the body are interrelated, and the system as a whole is in a **steady-state condition rather than equilibrium**. In a steady state, concentrations of particular substances are kept constant by control of the rates of linked reactions in a pathway. A change in concentration of any component (product of one reaction used as a reactant in another reaction) shifts the concentration of all other components linked to it by means of a sequence of chemical reactions, resulting in the attainment of a new steady state. Metabolic pathways are regulated by the flow of metabolites through them, governed by thermodynamics and, as will be seen later, by adequate rates of reaction that are provided by enzyme catalysis of each reaction.

Reactions that may be favorable thermodynamically may not be occurring at all at an evident rate. For example, glucose, the primary energy source for the brain, is stable in a bottle or in an aqueous solution that is uncontaminated by microorganisms. The reason is that, for the chemical reaction shown in Figure 5.1 to occur at an appreciable rate, the energy of reactants has to be raised to that of the **transition state** before the products are formed; i.e., energy must be input to enable the chemical changes of the reaction to proceed at a reasonable rate. For example, energy must be provided for the breaking of chemical bonds in the reactants, but this activation energy will be recovered by the energy released in forming the new chemical bonds that exist in the product—thus the primary importance of net energy change (ΔG^0 and ΔG). The lowering of the activation energy that is the essence of reaction catalysis is illustrated in Figure 5.2. The significance of the

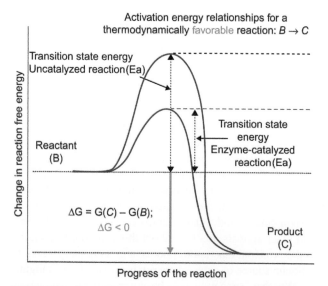

FIGURE 5.2 Energy diagram showing the free energy of activation in a chemical reaction and the activation energy reduction in an enzyme-catalyzed reaction.

transition state energy barrier is discussed further in this chapter under "Chemical Kinetics" and in Chapter 6, "Enzymes and Enzyme Regulation." Here, we are concerned only with the energy differences between initial and final states and how biological systems make unfavorable reactions become favorable by coupling.

Thermodynamic Relationships for Binding of Molecules to Each Other and to Cellular Receptors

The binding of drugs, metabolites, and other molecules regulates the behavior of the enzyme catalysts of metabolic reactions. Biochemical reactions are turned from off to on and vice versa, and their efficiencies changed by such binding. The binding of ions to macromolecules, drugs to their receptors, and antigens to antibodies is governed by the same laws of thermodynamics as govern the chemical reactions already described. In binding equilibria, however, no covalent bonds are made or broken, but the consequences of such molecular interactions can be as varied as stimulation of membrane transport, release of contents of subcellular organelles, and initiation of response of immune system cells to foreign substances. Substances that bind to receptors and cause changes in the receptor function or activity are called **agonists**; those that interfere with agonist binding are **antagonists**. These binding interactions are the subject of pharmacology and related disciplines.

Binding interactions are commonly characterized by the **specificity** of binding and the **affinity** of the substance that binds (**ligand**). The site to which the substance binds is an **acceptor** or **receptor**. More than one ligand may bind to some macromolecular acceptors or cellular receptors, or to multiple receptors on cells. Thus, a third property is required to characterize the binding process: the number of sites (**n**) on the acceptor. For example, the maximum number of ligands that can be bound to cellular receptors, L_nR (assuming the **stoichiometry** is one ligand per receptor molecule), indicates the number of receptor sites on the cell. The most widely used algebraic equation to describe ligand−acceptor binding is a rearranged form of the law of mass action, $K_d = [product]/[reactant]$. In binding equilibria, the "product" is the ligand−acceptor complex, and the reactant is the ligand. Commonly, the ligand concentration greatly exceeds the concentration of the acceptor (including the number of sites on the acceptor) so that algebraic simplifications can be made to the rearranged mass action equation. The binding process can thus be described as follows: (1) **n** = the number of sites on the acceptor or receptor OR the number of receptors on a cell; (2) **L**, the concentration of the ligand; (3) K_a, the association constant for the binding ($K_a = 1/K_d$, a nomenclature difference related to history alone); and (4) the fraction of acceptor or receptor sites

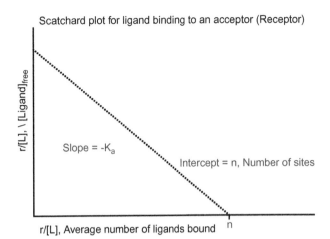

Scatchard plot for ligand binding to an acceptor (Receptor)

r/[L], \ [Ligand]$_{free}$

Slope = -K_a

Intercept = n, Number of sites

r/[L], Average number of ligands bound   n

FIGURE 5.3 Scatchard plot for the binding of a ligand to an acceptor. This is the most common graphical method for determining binding affinity and number of binding sites.

occupied by the ligand, written **r** or **r** with a bar above it. The binding ratio, **r**, is the average concentration of ligands bound to the total concentration of receptors and is calculated as $[L]_{bound}/[receptors]_{total}$.[3] If binding is to cells, the denominator may be a cell count, and thus the value of **n** may be interpretable as the number of receptors per cell. The binding equation thus can be written as follows:

$$r = \frac{nK_a[L]}{(K_a + [L])} \qquad (5.5)$$

The thermodynamic equation (5.3, earlier) is written $\Delta G^0 = -RT \ln K_a$, and the strength of the interaction between ligand and receptor determined by the value of ΔG^0.

Measurements of the binding of ligands to acceptors and receptors are analyzed to obtain values for the number of sites, **n**, and the affinity, K_a. The relative affinity of different ligands for a particular receptor is the parameter that most closely indicates the relative specificity of the ligand for the receptor. The most commonly used graphical method is the Scatchard plot; a graph of **r/[L]** versus **r** is shown in Figure 5.3.

The affinity of receptors for the ligands varies greatly; e.g., hormone (ligand) concentrations can be as low as 10^{-12} M. Other substances bind much more weakly, e.g., Ca^{2+} ions bind with low affinity, inferred to be suitably tight because their physiological concentrations are about 10^{-3} M in blood plasma. The ΔG^0 values are thus related to the affinity of a receptor for its ligand; free energies of binding can be large as evidenced by the ~ -16 kcal calculated for a K_a of 10^{12}, a value representative of a very tightly bound hormone.

---

3. Concentration appears in both numerator and denominator, so **r** is the number of ligands per acceptor or receptor.

Thermodynamic Relationships for Oxidation–Reduction (Redox) Reactions

The production of energy for living systems and their biochemical reactions depends on the availability of reactions that are thermodynamically favorable. Energy to make the necessary reactions thermodynamically favorable comes from several types of reactions although almost all proceed via the cellular energy "currency," adenosine triphosphate (ATP) and the nicotinamide–adenine nucleotide phosphates (NAD$^+$/NADH and NAD$^+$P/NADPH).[4] In the following paragraphs the principles required to understand oxidation–reduction reactions and oxidative phosphorylation are outlined. A number of metabolic reactions result in and/or depend on the transfer of electrons (oxidation and reduction), and thus, their thermodynamic properties are related to their electrochemical properties. **Oxidation** is the loss of electrons by an atom, ion, or hydride (H^-) ion or by a molecule. **Reduction** is the gain of electrons by an atom, ion, or hydride (H^-) ion by a molecule. In a chemical reaction, the transfer of one hydride ion results in the transfer of *two* electrons; $H^- \rightleftharpoons H^+ + 2e^-$. These reactions are a subset of the chemical reactions that, because they involve electron transfer, are described by a particular thermodynamic relationship (the Nernst equation) that relates electrochemical reactions to the ΔG changes.

The amount of work required to add or remove electrons is called the **electromotive potential or force** (**emf**) and is designated E. It is measured in volts (joules per coulomb, where a coulomb is a unit of electric charge or a quantity of electrons). The standard **emf**, E^0 (or as commonly used in biochemical systems for pH7, $E^{0\prime}$), is the **emf** measured when the temperature is 25°C and the materials being oxidized or reduced are present at concentrations of 1.0 M. We employ $E^{0\prime}$ because it is commonly used in biochemistry. The **emf** measured for a *redox* reaction under nonstandard concentrations is designated E_h. It is mathematically related to the **emf** measured for the same reaction under standard conditions, $E^{0\prime}$, through the Nernst equation:

$$E_h = E^{0\prime} + (RT/nF)\ln [\text{electron acceptor/electron donor}]$$

where E_h = observed potential, E^0 = standard redox potential, R = gas constant (8.31 J deg$^{-1}\cdot$mol), T = temperature in Kelvin, n = number of electrons being transferred, and F is the *Faraday constant* (96,487 J/V). It is evident from the Nernst equation that $E_h = E^{0\prime}$ when [electron acceptor] = [electron donor], analogously to the previous relationship for ΔG^0. In this manner, E^0 values for many redox reactions of

biochemical importance have been determined (Chapter 14). Electrochemical reactions are commonly written as **half-reactions**, i.e., reactions in which electrons or hydride ions are written explicitly for individual donors or acceptors.

The key substances involved are the oxidized and reduced forms of nicotinamide adenine dinucleotide, NAD$^+$ and NADH. Their structures are given in Chapter 6; their metabolic functions are discussed throughout the entirety of biochemistry as the result of their widespread involvement. The half-reaction for NAD$^+$ reduction to NADH can be written as follows:

$$NAD^+ + H^- \rightarrow NADH \quad E^{0\prime} = +0.32 \text{ V}$$

The commonly coupled reaction can be written as follows:

$$H_2O \rightarrow \tfrac{1}{2}O_2 + 2H^+ + 2e^- \quad E^{0\prime} = -0.816 \text{ V}$$

Since free electrons combine rapidly with whatever is at hand, half-reactions never occur by themselves. Something must accept electrons as fast as they are released. The substance releasing electrons (or H^-) is the **reductant**, or **reducing agent** (it is oxidized), and the substance accepting electrons is the **oxidant**, or **oxidizing agent** (it is reduced). Two half-reactions, when combined, give a redox reaction. When balanced, such reactions never show free (uncombined) electrons. For example,

$$H^+ + NADH + \tfrac{1}{2}O_2 \rightarrow NAD^+ + H_2O$$

$$\Delta E^{0\prime} = -0.32 - (+0.816) = -1.136 \text{ V}$$

$\Delta E^{0\prime}$ for this reaction is calculated according to

$$\Delta E^{0\prime} = \Delta E^{0\prime}(\text{reductant}) - \Delta E^{0\prime}(\text{oxidant})$$

Under standard conditions (and pH = 7.0), the reaction will occur as written if $\Delta E^{0\prime} < 0$.

When ΔE^0 in volts is converted to calories per mole, the result is $\Delta G^{0\prime}$ for the redox reaction being considered. This conversion is accomplished by means of the following equation:

$$\Delta G^{0\prime} = (nF/R)T \cdot \Delta E^{0\prime} = (23.061)n \cdot \Delta E^{0\prime}$$

As before, F is Faraday's constant = 96,487 J/V; n is the number of electrons transferred per mole of material oxidized or reduced; E^0 is E^0 (reductant) $-$ E^0 (oxidant) in volts; and ΔG^0 is the standard free energy change in calories per mole of material oxidized or reduced (since there are 4.184 J/cal).

In the case of the reaction just considered,

$$H^+ + NADH + \tfrac{1}{2}O_2 \rightarrow NAD^+ + H_2O; \Delta E^{0\prime} = -1.136 \text{ V}$$

---

4. The nicotinamide–adenine dinucleotides are ubiquitous cofactors in enzyme-catalyzed oxidation–reduction reactions (see Chapter 6).

n equals 4 gram-equivalents of electrons per mole of O_2 or 2 gram-equivalents of electrons per mole of H_2O, NAD^+, or NADH. Then

$$\Delta G^0 = 23.061n \times (-1.136) = -26.197n = -26.197.2$$

$$\Delta G^0 = -52.394 \text{ calories/mol of } H_2O \text{ or } NAD^+$$

formed or NADH converted, and

$$\Delta G^0 = -26.197 \times 4 = -104.788 \text{ calories/mol of}$$

O_2 transformed

As in all reactions, the actual value of ΔG^0 depends on the reactant or product for which it is calculated.

The significance of free-energy changes and redox reactions lies in the fact that life on Earth depends on the redox reaction in which CO_2 is reduced by H_2O to yield **$C_6H_{12}O_6$**, glucose, in photosynthesis using sunlight as the energy source, the reverse of the reaction described previously for the oxidation of glucose to CO_2 and H_2O. The $\Delta E^{0\prime}$ for the forward reaction is about $+1.24$ V, or as noted earlier, $\Delta G^{0\prime}$ is about $+685.8$ kcal for glucose formation.

Energy from the Sun is absorbed by the chloroplasts of green plants, causing water to be oxidized and carbon dioxide to be reduced, producing oxygen and carbohydrate, e.g., glucose. In oxidation reactions, carbohydrates (or lipids) and oxygen are consumed, and energy, water, and carbon dioxide are released. The energy is captured primarily as adenosine triphosphate (ATP), which is used as an immediate source of energy in most cellular endergonic processes.

Standard Free Energy of Hydrolysis of ATP

In the living organism, ATP functions as the most important energy intermediate, linking energy-releasing (exergonic) with energy-requiring (endergonic) processes. Exergonic processes (e.g., oxidation of glucose, glycogen, and lipids) are coupled to the formation of ATP, and endergonic processes are coupled to the consumption of ATP (e.g., in biosynthesis, muscle contraction, active transport across membranes). ATP is called a "high-energy compound" because of its large negative free energy of hydrolysis:

$$ATP^{4-} + H_2O \rightarrow ADP^{3-} + HPO_4^{2-} (\text{or } P_i) + H^+$$

$$\Delta G^{0\prime} = -7.3 \text{ kcal/mol}$$

The reason for the large release of free energy associated with hydrolysis of ATP is that the products of the reaction, ADP and P_i, are much more stable than ATP. Several factors contribute to their increased stability: relief of electrostatic repulsion, resonance stabilization, and ionization.

At pH 7.0, ATP has four closely spaced negative charges with a strong repulsion between these charges (Figure 5.4).

Upon hydrolysis, the strain in the molecule due to electrostatic repulsion is relieved by formation of less negatively charged products (ADP^{3-} and P_i^{2-}). Furthermore, since these two products are also negatively charged, their tendency to approach each other to re-form ATP is minimized. The number of resonance forms of ADP and P_i exceeds those present in ATP so that the products are stabilized by larger resonance energy. Ionization of $H_2PO_4^- \rightleftharpoons HPO_4^{2-} + H^+$ has a large negative $\Delta G^{0\prime}$, contributing the $\Delta G^{0\prime}$ for ATP hydrolysis.

FIGURE 5.4 Structural formula of adenosine triphosphate (ATP) at pH 7.0.

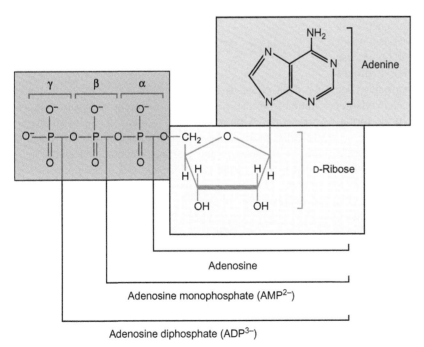

The free energy of hydrolysis of ATP is also affected by Mg^{2+} and the physiological substrate is $Mg^{2+} \cdot ATP^{4-}$ (or $Mg \cdot ATP^{2-}$). The presence of magnesium ions favors ATP hydrolysis because both ADP and P_i have higher affinity (about six times) for Mg^{2+} than ATP. *In vivo*, the ΔG^0 value for ATP hydrolysis is probably much higher than that for $\Delta G^{0\prime}$. As noted previously, the reason is the prevailing intracellular concentration of the reactants and products; thus, in erythrocytes, ΔG^0 is -12.4 kcal/mol.

In some energy-consuming reactions, ATP is hydrolyzed to AMP and pyrophosphate, with a value of free energy of hydrolysis of similar magnitude to that of ATP to ADP and P_i:

$$ATP^{4-} + H_2O \rightarrow AMP^{2-} + HP_2O_7^{3-} + H^+,$$
$$\Delta G^0 = -7.7 \text{ kcal/mol} \qquad (5.6)$$

In vivo, pyrophosphate is usually hydrolyzed to inorganic phosphate, which also has a large negative $\Delta G^{0\prime}$:

$$HP_2O_7^{3-} + H_2O \rightarrow 2HPO_4^{2-} + H^+,$$
$$\Delta G^0 = -7.17 \text{ kcal/mol} \qquad (5.7)$$

Pyrophosphate hydrolysis, although not coupled to any particular endergonic reaction, ensures completion of the forward reaction or process (e.g., activation of fatty acids and amino acids, synthesis of nucleotides and polynucleotides).

The hydrolysis of AMP to adenosine and P_i does not yield a high negative $\Delta G^{0\prime}$ because the phosphate is linked in a normal ester bond, as opposed to the other two linkages (β and γ), which are phosphoanhydride bonds (Figure 5.4). The $\Delta G^{0\prime}$ values of hydrolysis of other nucleoside triphosphates (e.g., UTP, CTP and GTP) are similar to that of ATP, and they are utilized in the biosynthesis of carbohydrates, lipids, and proteins (discussed elsewhere). In addition to undergoing hydrolysis, ATP may act as a donor of phosphoryl groups (e.g., in the formation of glucose 6-phosphate; see Chapter 12), pyrophosphoryl groups (e.g., in the formation of phosphoribosylpyrophosphate; see Chapter 25), and adenylyl groups (e.g., in the adenylylation of glutamine synthetase; or adenosyl groups in the formation of S-adenosylmethionine; see Chapter 15).

Since ATP synthesis from ADP and P_i requires energy, it requires compounds or processes that yield larger negative $\Delta G^{0\prime}$ values than the reverse reaction, ATP hydrolysis, requires. Some of the compounds that provide this energy are organic. For example, the conversion of phosphoenolpyruvate to pyruvate provides -14.8 kcal/mol (-61.9 kJ/mol), or conversion of 1,3-bisphosphoglycerate to 3-phosphoglycerate -11.8 kcal/mol (-49.3 kJ/mol), whereas formation of ATP from ADP and P_i requires just $+7.3$ kcal/mol ($+30.5$ kJ/mol). Other high-energy compounds are thioesters, aminoacyl esters, sulfonium derivatives, and sugar nucleoside diphosphates, which can also provide the energy required for ATP synthesis.

CHEMICAL KINETICS

The study of the velocity of chemical reactions is called **chemical kinetics**. Whereas thermodynamics deals with the relative energy states of reactants and products, kinetics deals with how fast a thermodynamically favorable reaction occurs and the chemical pathway it follows. The detailed chemical pathway is the reaction mechanism. Referring back to Figure 5.2, and the discussion of thermodynamics, there is an energy barrier that must be overcome for the reaction to occur: the activation energy. First, we consider how the kinetics of reactions are described and how the activation energy is related to the velocity of the reaction.

A simple chemical reaction, the conversion of a reactant **B** to a product **C**, is sufficient to illustrate the terminology of chemical kinetics:

$$B \xrightarrow{k} C \qquad (5.8)$$

The chemical changes, i.e., the conversion $B \rightarrow C$, can be represented diagrammatically by the increase in concentration of product, [C], with respect to time (where **k** is a simple proportionality constant, the **rate constant**) (Figure 5.5). As the reaction proceeds, reactant concentration decreases and product concentration increases until the reaction reaches equilibrium.

The reaction rate or velocity is related to the concentrations of the reactant **B** and the rate constant. The average velocity, v, for the period of time between t_1 to t_2 is the slope of the reaction progress curve at a time between t_1 and t_2. The rate or velocity is moles liter^{-1} time^{-1} because concentration is moles liter^{-1}:

$$v = \frac{-\Delta[B]}{\Delta t} \text{ or } v = \frac{-([B]_2 - [B]_1)}{(t_2 - t_1)} \qquad (5.9)$$

This equation represents how the velocity is determined from the measurement of product formation during a reaction. A more desirable velocity is an instantaneous velocity, particularly at the very beginning of the reaction (at $t = 0$) because the starting reactant concentration is what is used in the equation, and thus it simplifies calculations. The instantaneous velocity, v, at any time ($t = t_x$) is expressed as follows:

$$v = \frac{-d[B_{(t=x)}]}{dt} \text{(evaluated at } t = x) \qquad (5.10)$$

The equation $-d[B(t)]/dt$ is the first derivative of $[B(t)]$ with respect to t; it is the slope of the curve at the time, $t = x$. One can also consider the change in concentration of the reactant, **B**. Since the reactant concentration, as stated in the chemical equation [B], decreases with the

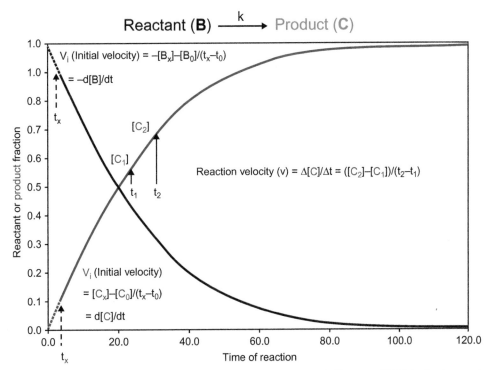

FIGURE 5.5 The time course of a chemical reaction, B→C. The instantaneous velocity, d[C]/dt or −d[B]/dt is the slope at the point labeled t_x. For illustrative purposes, t_x is shown at the start of the reaction for both C formation and B disappearance; the dotted line interval is intended to represent a very short time such that the slope between t_0 and t_x is essentially the same as the initial velocity.

same stoichiometry and at the same rate at which the product concentration, [C], increases, i.e.:

$$v = \frac{d[C(t)]}{dt} = -\frac{d[B(t)]}{dt} \qquad (5.11)$$

The rate constant (Equation 5.8) expresses the proportionality between the velocity of formation of **C** and the molar concentration of **B** and is characteristic of the reaction, its pH, and temperature. The units of **k** depend on the order of the reaction, i.e., on the number of reactants that are participating in the reaction in which they are converted to product. In this example, **k** relates the velocity of the reaction (moles liter^{-1} time^{-1}) to the concentrations of **B** (moles liter^{-1}) and t; therefore, the units of **k** are time^{-1}.

A somewhat more complicated example is

$$A + B \xrightarrow{k} C + D \qquad (5.12)$$

$$v = \frac{d[C]}{dt} = \frac{d[D]}{dt} = \frac{d[products]}{dt} = k[A][B] \qquad (5.13)$$

This reaction is first order with respect to **A** and first order with respect to **B**, and second order overall. If there are two reactants such as in the chemical equation **A + B→C + D** (Equation 5.12), then the reaction is designated second order and the units of **k** for a second order reaction are liters moles^{-1} time^{-1}. The units of **k**

are whatever is needed for $d[\mathbf{B}]/dt$ to have the units of moles liters^{-1} time^{-1}.

In some situations, the reaction velocity can be independent of the reactant concentration, a reaction designated zero order. For *zero order* reactions, the units of **k** are moles liter^{-1} time^{-1}. Time is, by international agreement, in seconds.

If the concentrations of the reactants are 1 mole liter^{-1}, then the rate constant is equal to the value of the reaction velocity at that concentration. This provides a convenient way to visualize the rate constant as a reaction rate, something that can be intuitively more meaningful.

Theoretically, all reactions are reversible, and thus, the correct way to write **B→C** is **B⇌C**, where $\mathbf{k_1}$ is the rate constant for the conversion of **A** to **B**, and $\mathbf{k_{-1}}$ is the rate constant for the reverse reaction, i.e., conversion of **B** to **A**. The reaction is effectively kinetically irreversible if $\mathbf{k_{-1} << k_1}$. The rate of a reversible reaction is written as follows:

$$v = \frac{d[B]}{dt} = k_1[B] - k_{-1}[C]. \qquad (5.14)$$

Kinetic schemes can become complicated with many linked reactions and intermediates, but are of crucial importance in biological systems. Metabolic pathways provide examples of complex kinetic schemes.

Energetic Considerations of Chemical Kinetics

Reactions involve the collisions between molecules that result in changes in the nature of the chemical substances; reactants disappear and products appear. The number of molecules colliding and reacting gives rise to the concept of reaction order.[5] In a first-order reaction, no collisions are required (since only one reactant molecule is involved). In a second-order reaction, two molecules must collide with each other for the products to form. The frequency with which molecules collide and the energy of the collisions that is required for reaction are the subjects of the paragraphs that follow.

Zero-order reactions can be explained in two ways. In a process that is truly zero order with respect to all reactants (and catalysts, in the case of enzymes), either the activation energy is zero, or every molecule has sufficient energy to overcome the activation barrier. Alternatively, a reaction can be zero order with respect to one or more (but not all) of the reactants. This kind of reaction is important in enzyme kinetics and assays of enzymes of clinical importance. If one of the reactants (or a catalyst, such as an enzyme) is limited, then increasing the availability (concentrations) of the other reactants can result in no increase in the velocity of the reaction beyond that dictated by the limiting reagent. Hence, the rate is independent of the concentrations of the nonlimiting materials and zero order with respect to that reactant.

The rate at which a reaction will proceed (characterized by the rate constant k) is directly related to the amount of energy that must be supplied before reactants and products can be interconverted. In the absence of a catalyst, this activation energy (**Ea**) comes from the energy of the reactants. As illustrated in Figure 5.2, the larger the activation energy, the slower the reaction rate and the smaller the rate constant.

The reaction rate constant is determined by two components. **Ea**, which is *analogous* to thermodynamic free energy and consists of an internal contribution and comes from the inherent energy of the reactants, e.g., vibrational and electronic, is useful when one molecule rearranges itself or eliminates some atoms to form the product. This other form of energy is kinetic (translational and rotational motion), which is necessary if two molecules collide to react. The second component is an entropic contribution, which is related to collision frequency and orientation of the colliding molecules. Not only must the reactants have enough energy to make and break the requisite chemical bonds, but the molecules must also be properly oriented with respect to each for the reaction to occur and products to form.

The simplest equation that expresses the relation between rate constants and activation energies is the Arrhenius equation:

$$k = A \cdot e^{-Ea/RT} \qquad (5.15)$$

Ea is the Arrhenius activation energy in calories per mole; R is the gas constant (1.987 calories mole^{-1} degree K^{-1}); **T** is the absolute temperature (K). **A** is the Arrhenius pre-exponential factor, a parameter that is predominantly entropic and includes collision frequency and orientation. k is the rate constant, and **e** is 2.71828, the base of the natural logarithms. From the Arrhenius equation, it is apparent that the larger the value of **Ea**, the smaller the value of k. This equation is particularly useful in describing the temperature dependence of chemical reactions.

Reactions with high activation energies, even if they are highly favorable thermodynamically, may not be observed unless a catalyst is present. For example, the conversion of glucose in the presence of oxygen to carbon dioxide and water is favored (Equation 5.1), but glucose and glucose solutions are kinetically stable. Even though energy may be provided by heating in ordinary chemical reactions, obtaining the maximum amount of work from the conversion is not possible. In biological systems, however, which operate at constant temperature, with enzyme catalysts, **Ea** is reduced so that the reactions can occur at an acceptable velocity in cells and tissues, and the maximum ΔG is obtained in the process. How enzymes accomplish the feat of lowering activation energies and directing metabolic pathways is the subject of the next chapter.

ENRICHMENT READING

[1] R.H. Abeles, P.A. Frey, W.P. Jencks, Biochemistry, Jones and Bartlett Publishers, Boston, 1992.
[2] C. Tanford, Physical chemistry of macromolecules, John Wiley & Sons, Inc., New York, 1961.
[3] G. Scatchard, The attractions of proteins for small molecules and ions, Ann. N. Y. Acad. Sci. 51 (1949) 660−672.
[4] N.S. Kresge, D. Robert, R.L. Hill, Historical perspectives on bioenergetics, J. Biol. Chem. 57 (2010).
[5] D. Wei, X. Huang, J. Liu, M. Tang, C.-G. Zhan, Reaction pathway and free energy profile for papain-catalyzed hydrolysis of N-acetyl-phe-gly 4-nitroanilide, Biochemistry 52(30) (2013) 5145−5154.
[6] A.R. Fersht, Structure and Mechanism in Protein Science, W.H. Freeman and Company, New York, 1998.
[7] S.J. Harris, C.M. Jackson, D.J. Winzor, Effects of heterogeneity and cooperativity on the forms of binding curves for multivalent ligands, J. Protein Chem. 14(6) (1995) 399−407.
[8] S.J. Harris, C.M. Jackson, D.J. Winzor, The rectangular hyperbolic binding equation for multivalent ligands, Arch. Biochem. Biophys. 316(1) (1995) 20−23.
[9] A.D. Vogt, E. Di Cera, Conformational selection is a dominant mechanism of ligand binding, Biochemistry 52(34) (2013) 5723−5729.

---

5. This description is somewhat oversimplified, but a comprehensive and more rigorous treatment is beyond the scope of understanding intended to be provided here.

Chapter 6

Enzymes and Enzyme Regulation

Key Points

1. Enzymes guide the chemical reactions in biological systems to extract energy from foodstuffs and create the substances unique to living organisms.

2. Enzymes are proteins with unique structures that are folded to create three-dimensional sites and thus enable specific recognition of the substrates that they transform into new chemical substances or from which they facilitate optimal energy extraction.

3. The specificity of an enzyme for its substrate is determined by the amino acid residues of the enzyme that create a binding site for the substrate and, as a consequence, lower the activation energy for the reaction that the enzyme catalyzes.

4. The rate of an enzyme-catalyzed reaction often is regulated by molecules that bind to the enzyme at sites distinguishable from the active or catalytic site and alter the enzyme's efficiency as a catalyst. "Allosteric" is the term that describes this type of regulation.

5. The catalysis of a chemical reaction by an enzyme can be described by a kinetic or rate equation, the Michaelis—Menten equation. This equation describes the relationship between the velocity of the reaction and (1) enzyme concentration, (2) substrate concentration, and (3) two constants, K_m and k_c.

6. Many enzymes require cofactors to catalyze their particular chemical reactions; cofactors participate in a reaction but are not consumed.

7. Some enzymes can be irreversibly inactivated to regulate their activity, often by inhibitors that act as suicide substrates. *In vitro* many substances that cause irreversible inactivation are not physiological but are foreign to living systems.

8. Metabolic pathways by which foodstuffs are transformed into energy and building blocks for the organism are determined by the enzymes that catalyze the individual reactions of the pathway and by the enzymes' kinetic properties.

9. Multiple forms of many enzymes aid in identifying the tissue source of the enzymes and thus in diagnosis of disease.

Enzymes are the catalysts that direct the pathways of cellular metabolism. They are most commonly proteins, but some RNA molecules also can act as catalysts. This and the following chapter discuss protein enzymes. Many enzymes bind nonprotein components that are essential for catalytic activity. The term **prosthetic group** is generally reserved for those nonprotein components that are bound covalently or very tightly, whereas the term **coenzyme** applies to less tightly bound nonprotein components. Coenzymes are complex organic compounds, often derived from vitamins. The term **cofactor** is used for metal ions and simple organic compounds that participate in enzyme catalysis. In enzymes with prosthetic groups or cofactors, the protein portion is the **apoenzyme**, and the fully functional enzyme with its attached nonprotein component is the **holoenzyme** (i.e., apoenzyme + coenzyme = holoenzyme).

Catalytic RNA molecules (ribozymes) recognize specific nucleotide sequences in the target RNA and hydrolyze phosphodiester bonds. Ribosomal RNA also functions as a peptidyl transferase in protein biosynthesis. Some RNA molecules also undergo a self-splicing process. One example is the transformation of pre-mRNA to mature, functional mRNA by the splicing out of intervening sequences (**introns**) and ligation of the coding sequences (**exons**). The splicing—ligation process requires **small nuclear ribonucleoprotein particles** (snRNPs) and other proteins. Catalytic RNA also can act on other RNA molecules. An example is ribonuclease P (RNase P), which is involved in the conversion of precursor tRNA to functional tRNA by generating 5-phosphate and 3-hydroxyl termini; RNase P contains both a catalytic RNA moiety and an associated protein.

NOMENCLATURE

Enzymes are commonly named for the **substrate** or chemical group on which they act, and the name takes the suffix "-ase." Thus, the enzyme that hydrolyzes urea is named urease. Exceptions to this terminology are also common, e.g., trypsin, pepsin, and papain, which are trivial names. The Enzyme Commission of the International Union of Biochemistry has developed nomenclature for enzymes. This system provides a rational and practical basis for identification of all enzymes currently known as well as for new enzymes. The systematic name describes

N. V. Bhagavan and C.-E. Ha: Essentials of Medical Biochemistry, Second edition. DOI: http://dx.doi.org/10.1016/B978-0-12-416687-5.00006-3

the substrate, the nature of the reaction catalyzed, and other characteristics. (See http://www.ebi.ac.uk/intenz/ for an up-to-date table of enzyme names and catalyzed reactions.)

CATALYSIS

Specificity of Enzyme Catalysis

Enzymes are highly specific and usually catalyze only one type of reaction. Some enzymes show specificity for a single substrate. For example, pyruvate kinase catalyzes the transfer of a phosphate group only from phosphoenolpyruvate to adenosine diphosphate during glycolysis. Other enzymes catalyze the same type of reaction but act on several, structurally related substrates. The following enzymes show less specificity:

- Hexokinase transfers the phosphate group from adenosine triphosphate to several hexoses (D-glucose, 2-deoxy-D-glucose, D-fructose, and D-mannose) at almost equivalent rates.
- Phosphatases hydrolyze phosphate groups from a large variety of organic phosphate esters.
- Esterases hydrolyze esters to alcohols and carboxylic acids, with considerable variation in chain length in both the alcohol and acyl portions of the ester.
- Proteinases (proteases) hydrolyze particular peptide bonds irrespective of the protein substrate.

Enzymes show stereoisomeric specificity. For example, human α-amylase catalyzes the hydrolysis of glucose from the linear portion of starch but not from cellulose. Starch and cellulose are both polymers of glucose, but in the starch the sugar residues are connected by $\alpha(1 \rightarrow 4)$, whereas in cellulose they are connected by $\beta(1 \rightarrow 4)$ glycosidic linkages.

ACTIVE SITE AND ENZYME–SUBSTRATE COMPLEX

An enzyme-catalyzed reaction is initiated when the enzyme binds its substrate to form an enzyme–substrate complex. In general, enzyme molecules are considerably larger than the substrate molecules. Exceptions are proteinases, nucleases, and amylases that act on macromolecular substrates. Irrespective of the size of the substrate, binding of the groups in the substrate on which the enzyme will act occurs at a specific and specialized region known as the **active site**. The active site is a cleft or pocket in the surface of the enzyme that constitutes only a small portion of the enzyme molecule. Catalytic function occurs at this site because various chemical groups important in substrate binding are brought together in a spatial arrangement that confers specificity

and catalytic ability on the enzyme. The unique catalytic properties of an enzyme are based on its three-dimensional structure and on an active site whose chemical groups may be brought into close proximity from different regions of the polypeptide chain. The specificity of an enzyme for a substrate was originally visualized as reflecting a lock-and-key relationship. This description implies that the enzyme has an active site that fits the exact dimensions of the substrate, an oversimplified but useful analogy (Figure 6.1a).

Subsequent evidence has shown that after attachment of substrate, the enzyme may undergo conformational changes that provide a more complementary fit between the active site residues and the substrate. This process has been described as **induced fit**. Evidence from contemporary physical methods has refined the induced fit explanation to a more realistic "selection of an active site conformation" visualization of enzyme–substrate binding. Actual structures of two enzyme active sites with bound substrates are shown in Figure 6.1b and c.

Factors Governing the Rate of Enzyme-Catalyzed Reactions

The overall reaction involving conversion of substrate (S) to product (P) with formation of the enzyme–substrate complex, ES, can be written

$$\text{E} + \text{S} \underset{k_{-1}}{\overset{k_1}{\rightleftarrows}} \text{ES} \underset{k_{-2}}{\overset{k_2}{\rightleftarrows}} \text{E} + \text{P} \qquad (6.1)$$

where k_1, k_{-1}, k_2, and k_{-2} are rate constants for the designated steps; the positive subscripts indicate forward reactions, and the negative subscripts indicate reverse reactions. Because an enzyme increases the rate of a particular reaction by decreasing the free energy of activation, enzyme-catalyzed reactions are invariably much faster than uncatalyzed reactions. Also, because an enzyme catalyzes the reaction without itself being consumed or permanently altered, a small number of enzyme molecules can convert an extremely large number of substrate molecules to products. Enzymes do not alter the equilibrium distribution of substrates and products in a chemical reaction because they catalyze both the forward and backward reactions. Thus, enzymes affect *only the rate* at which equilibrium is established between reactants and products in any single reaction.

Effect of Temperature

The rates of almost all chemical reactions increase with a rise in temperature. This reflects the increase in the average kinetic energy of the molecules and results in a higher frequency of reaction collisions and greater ability to get

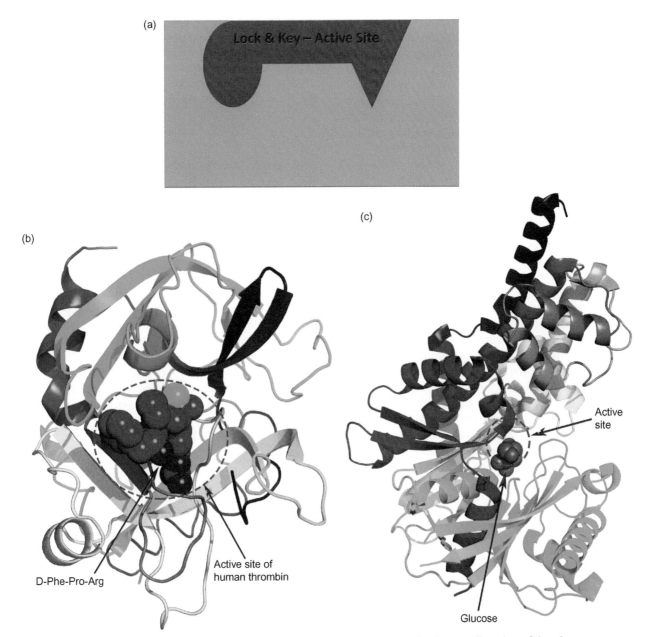

(a)

Lock & Key – Active Site

(c)

(b)

D-Phe-Pro-Arg

Active site of
human thrombin

Active
site

Glucose

FIGURE 6.1 The "lock and key" description implies that the enzyme has an active site that fits the exact dimensions of the substrate, now recognized to be an oversimplification, but a useful analogy is shown in panel (a). More realistically and supported by evidence from a contemporary state-of-the-art method, the substrate and enzyme mutually interact to form the active site. Panels (b) and (c) show an actual structure of an enzyme active site with a bound substrate analog.

over the activation energy barrier. Enzyme-catalyzed reactions, however, because of the arrangement of particular amino acid residues on the enzyme's surface and within its active site can increase reaction rates more than temperature alone because of the lower activation barrier (Chapter 5).

Enzymes can undergo denaturation (unfolding with loss of activity) at high temperatures; thus, there is an optimum temperature for enzymatic reactions. The optimal temperature for many enzymes is close to the normal temperature of the organism in which catalysis occurs. In humans, most enzymes have an optimal temperature near or slightly higher than 37°C.

Effect of pH

The activity of enzymes depends on pH; this is a consequence of the ionizable amino acid residues of the enzyme, particularly residues in the active site. The pH−enzyme activity profile commonly delineates a bell-shaped curve exhibiting a maximum pH at which activity is optimal. This pH is usually the same pH as the fluid in which the enzyme functions. Thus, most enzymes have their highest activity between pH 6 and pH 8 (the pH of human blood is about 7.4). However, pepsin, which must function at the low pH of gastric juice, has maximal activity at about pH 2; this curve is not bell shaped because pH values lower than 1 are essentially unattainable in aqueous solutions.

The pH dependence of enzyme activity is the result of several effects: ionizable groups (1) in the active site of the enzyme (or elsewhere because of changes in the enzyme conformation), (2) in the substrate, or (3) in the enzyme—substrate complex. Changes in pH affect catalysis depending on whether the protons on the reactive groups are dissociated or undissociated. Ionization of these groups depends on their pK values, the chemical properties of surrounding groups, and the pH and salt concentration of the reaction medium. Because changes in pH affect the binding of the substrate at the active site of the enzyme and also the rate of breakdown of the enzyme—substrate complex, it is possible to infer the identity of an ionizable group that participates at the active site from the pH—activity profile for a given enzyme. pH dependence is very informative in understanding enzyme mechanisms. The enzymes in living systems function at nearly constant pH because they are in an environment in which the molecules present, e.g. proteins and metabolites, are pH buffers.

Effects of Enzyme and Substrate Concentrations

Enzyme—substrate interactions obey the mass-action law, the same as other chemical reactions. Consequently, the reactions depend on enzyme and substrate concentrations. Under usual conditions (exceptions are discussed later), the reaction rate is directly proportional to the concentration of the enzyme—a consequence of substrate molecules being present in large excess relative to enzyme concentration.

For a given enzyme concentration, the reaction velocity increases initially with increasing substrate concentration. Eventually, a limiting velocity is reached, and further addition of substrate has essentially no effect on reaction velocity (v) (Figure 6.2). The shape of a plot of reaction velocity (v) versus substrate concentration, [S], is a rectangular hyperbola and is characteristic of all nonallosteric enzymes. These dependences on enzyme and substrate concentrations led to a simple algebraic equation that describes most enzyme-catalyzed reactions.

MICHAELIS—MENTEN MODEL FOR ENZYME-CATALYZED REACTIONS

Many enzyme-catalyzed reactions can be described by a simplified chemical equation in which the second step is considered to be irreversible, Equation (6.2):

$$E + S \underset{k_{-1}}{\overset{k_1}{\rightleftharpoons}} ES \overset{k_2}{\longrightarrow} E + P \qquad (6.2)$$

A simple kinetic mechanism for enzyme catalysis based on Equation (6.2) was proposed by Michaelis and Menten in 1913 and later modified by Briggs and Haldane to include a slightly more general set of relationships among the rate constants of the mechanism. The resulting algebraic equation is the classical Michaelis—Menten equation, Equation (6.3). This equation relates the initial rate of an enzyme-catalyzed reaction to the substrate concentration and to two constants, K_m, the Michaelis constant, and k_c, the catalytic constant. The two constants of the Michaelis—Menten equation can be viewed as the kinetic consequences of the structure of the active site and its ability to catalyze the designated chemical reaction. It should be noted that [S] >> [E] is assumed in the derivation of the Michaelis—Menten equation and that the last step is slower than the first two steps in the reaction.

$$v = \frac{V_{max}[S]}{K_m + [S]} = \frac{k_c[E]_t[S]}{K_m + [S]} \qquad (6.3)$$

The Michaelis—Menten equation describes the following relationships.

1. If the substrate concentration is very high, i.e. >> K_m, a limiting maximum velocity (V_{max}) is attained. The reaction velocity, $v = k_c[E] = V_{max}$. High substrate concentrations are thus optimal for measuring enzyme activity. The constant, k_c, is the proportionality constant that relates the enzyme's maximum ability to convert substrate into product, i.e., the enzyme turnover number for the substrate.

2. If the value for the velocity is one-half the value of V_{max}, algebraic rearrangement shows that the value of K_m is equal to the substrate concentration at this velocity, $v = V_{max}/2$. This relationship can be used to obtain a value for K_m and provides a rough measure of the affinity of the enzyme and the particular substrate upon which it is acting.

3. If the substrate concentration is << than K_m, the equation can be simplified to $v = k_c [E_t] \cdot [S]/K_m$; the reaction depends linearly on both [E_t] and [S] as noted previously. The ratio, k_c/K_m, is a measure of the overall catalytic efficiency of the enzyme for the substrate and is used when comparing different enzymes that act on the same substrate.

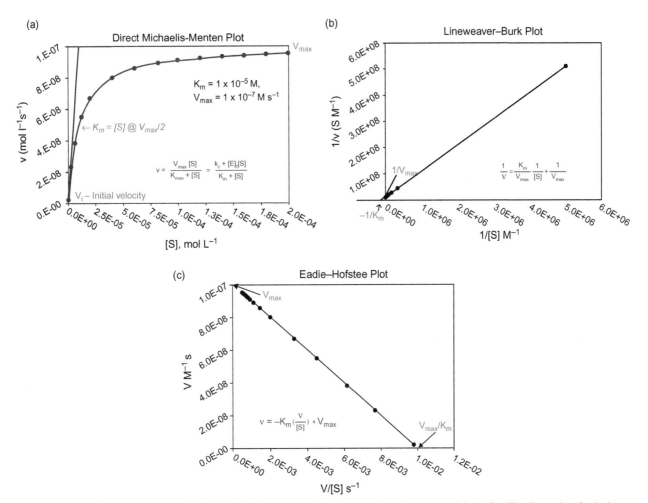

FIGURE 6.2 Graphical representations of the Michaelis−Menten equation. Axes are labeled in exponential notation. The direct plot of velocity versus substrate concentration is the most intuitive representation of the reaction behavior. Linearizations of the Michaelis−Menten equation provide convenient methods for obtaining the values for K_m and k_c, less so today than in the past because computerized curve fitting provides better estimates for the kinetic parameters. (a) Plot of the initial velocity of an enzyme-catalyzed reaction versus the substrate concentration. (b) Double reciprocal plot or Lineweaver−Burk plot. Serious limitations exist in obtaining accurate estimates for K_m and k_c using the Lineweaver−Burk plot. The disadvantage of this plot is that it depends on less well determined points obtained at low values of [S], whereas the more well determined points are obtained at high values of [S]. These points are clustered and appear less important in drawing the linear plot or in fitting the line. (c) The Eadie−Hofstee plot. The Eadie−Hofstee plot represents an improvement over the Lineweaver−Burk plot in that the experimental points are usually more equally spaced, and the drawn line is less biased by the least accurate data points.

Several particularly important points should be noted regarding enzyme-catalyzed reactions:

1. K_m is a characteristic constant for a particular enzyme and substrate and is independent of enzyme and substrate concentrations. It is a composite kinetic constant, $K_m = (k_{-1} + k_2)/k_1$; however, when k_2 is very small relative to k_{-1}, the value of K_m very closely approximates the equilibrium constant for the binding of the substrate to the enzyme, and K_m is designated K_s and $K_s = k_{-1}/k_1$. V_{max} depends on enzyme concentration, and at saturating substrate concentration, it becomes independent of substrate concentration.

2. K_m and k_c are influenced by pH, temperature, and other factors such as ionic strength (electrolyte concentration).

3. If an enzyme binds more than one substrate, the K_m values for the various substrates can be used as a relative measure of the affinity of the enzyme for each substrate (the smaller the value of K_m, the higher the affinity of the enzyme for that substrate).

4. In many reactions, the change in free energy, $-\Delta G$ is very large, and thus the reaction is effectively irreversible. When this situation exists, the reverse reaction for which the rate constant is k_{-2} (Equation 6.1) is commonly omitted from the general chemical equation.

5. In a metabolic pathway, k_c/K_m values for enzymes that catalyze the sequential reactions may indicate the rate-limiting step for the pathway (the lowest k_c/K_m corresponds roughly to the slowest step).

6. The reaction product of one step in a metabolic pathway is transformed in the second step to another product, thus providing a direction and progressive movement from one substrate to the next until the pathway is complete.

7. Turnover numbers for most enzymes usually range from 1 to 10^4 per second. A few enzymes have turnover numbers above 10^5. The ability of a cell to produce a given amount of product by an enzymatic reaction during its lifespan is proportional to the turnover number and the number of molecules of that enzyme in the cell.

8. The activity of an enzyme is expressed practically as specific activity, defined as micromoles (μmol) of substrate converted to product per minute per milligram (mg) of enzyme protein. Historically, a unit of activity is usually defined as that quantity of enzyme which catalyzes the conversion of $1\,\mu$mol of substrate to product per minute under a defined set of optimal conditions. This unit, referred to as the **International Unit** (U), is expressed in terms of U/mL of biological specimen (e.g., serum), or U/L. The International Union of Biochemistry recommends use of a unit known as **katal** (kat); one katal is the amount of enzyme that converts one mole of substrate to product per second. As noted earlier in point 3, units of activity and katals are dependent on the pH, temperature, electrolyte concentration, and other properties of the solution in which the reaction occurs.

9. In clinical disorders, the activity of a variety of enzymes is measured in biological fluids. Elevated activities of enzymes originating from the liver and myocardium are indicative of damage to these organs and elevated concentrations of substrates, e.g., creatinine of impaired renal function.

Linear Plots for Michaelis—Menten Kinetic Behavior

Before computers became widespread, particularly personal computers, the Michaelis—Menten equation was algebraically rearranged to linear forms to facilitate estimation of values for K_m and k_c. Straight-line plots, much easier to evaluate than curves, also provide some useful simplifications for estimating values of K_m and k_c, when data are not available to fully describe the entire rectangular hyperbola. Two such reformulations are historically important: the **Lineweaver—Burk plot**, which is a double-reciprocal plot (Equation 6.4a) and the

Eadie—Hofstee plot (Equation 6.4b). These plots are shown in Figure 6.2b and c. The algebraic expressions by which K_m and k_c are obtained from their respective graphs are noted within the figures.

$$\frac{1}{v} = \frac{K_m}{V_{max}}\frac{1}{[S]} + \frac{1}{V_{max}} \tag{6.4a}$$

$$v = -K_m\left(\frac{v}{[S]}\right) + V_{max} \tag{6.4b}$$

Kinetics of Enzymes Catalyzing Two-Substrate Reactions

Most enzymatic reactions involving two substrates (called **bisubstrate** or **ternary complex reactions**) show more complex kinetics than do one-substrate reactions. Examples are reactions catalyzed by dehydrogenases and aminotransferases. Although hydrolytic reactions are bisubstrate reactions in which water is one of the substrates, the change in water concentration is negligible and has no effect on the rate of reaction; thus, hydrolytic reactions are indistinguishable from single-substrate reactions.

The general two-substrate reaction can be most simply written as follows:

$$\text{Substrate A} + \text{Substrate B} \rightleftharpoons \text{Product C} + \text{Product D} \tag{6.5}$$

The enzyme—substrate interactions can proceed by either **single-displacement** or **double-displacement reactions** (also known as "ping-pong" reactions). A substrate reaction proceeding by way of a single-displacement reaction can be shown as follows:

$$\text{E} + \text{A} \rightleftharpoons \text{EA} \quad \text{or} \quad \text{E} + \text{B} \rightleftharpoons \text{EB} \tag{6.6a}$$
a random order of first substrate binding

$$\text{EA} + \text{B} \rightleftharpoons \text{EAB} \quad \text{or} \quad \text{EB} + \text{A} \rightleftharpoons \text{EAB}$$
formation of the ternary enzyme substrate complex

$$\tag{6.6b}$$

$$\text{EAB} \rightleftharpoons \text{C} + \text{D} \tag{6.6c}$$
transformation of the two substrates into two products

Note that the ternary complex EAB can be formed in two different ways. If the formation of EAB can occur with either substrate binding first, the reaction is known as a **random** single-displacement reaction. Many reactions catalyzed by **phosphotransferases** are of this type. If a particular substrate must bind first with the enzyme before the second substrate can bind, the reaction is known as an **ordered** single-displacement reaction. Many reactions catalyzed by **dehydrogenases** are of this type. The values for

K_m and V_{max} for each substrate can be obtained from experiments in which the concentration of one substance is held constant at saturating levels while the concentration of the second substrate is varied. Kinetic analyses can distinguish between these types of reactions.

In a double-displacement reaction, initially only one substrate is bound; this is followed by the release of one product. An intermediate covalently modified enzyme (E*) remains. E* then reacts with the second substrate to form the second product. These reactions are also called **substituted enzyme reactions**. Reactions catalyzed by **aminotransferases** are of this type. In this mechanism, no ternary complex EAB is formed. The double displacement reaction sequence is as follows:

$$E + A \sim X \rightleftharpoons EA \sim X; \quad E \sim X \rightleftharpoons E * X + A \quad (6.7a)$$

$$E * X + B \rightleftharpoons E * XB; \quad E * XB \rightleftharpoons E + XB \quad (6.7b)$$

INHIBITION

Enzyme inhibition is one of the ways in which enzyme activity is regulated naturally and experimentally. Most therapeutic drugs function by inhibiting a specific enzyme. Inhibitor studies have contributed much of the available information about enzyme mechanisms. In the body, some of the processes controlled by enzyme inhibition are blood coagulation (hemostasis), blood clot dissolution (fibrinolysis), complement activation, connective tissue turnover, and inflammatory reactions.

Enzyme inhibitors are classified as **reversible** or **irreversible**. Reversible inhibitors bind noncovalently; irreversible inhibitors commonly form covalent bonds with the enzyme or react with residues involved in catalysis and modify them chemically. Many irreversible inhibitors are similar to substrates, but do not dissociate from the enzyme active site; they act as suicide substrates.

Reversible Inhibition

In reversible inhibition, which is further subdivided into **competitive**, **noncompetitive**, **uncompetitive**, and **mixed** types, the activity of the enzyme is fully restored when the inhibitor is removed from the system in which the enzyme functions. In reversible inhibition, equilibrium exists between the inhibitor, I, and the enzyme, E (Equation 6.8):

$$E + I \rightleftharpoons EI \quad (6.8)$$

The equilibrium constant for the **dissociation** of the enzyme–inhibitor complex, known as the inhibitor constant K_i, is given by Equation (6.9):

$$K_i = \frac{[E][I]}{[EI]} \quad (6.9)$$

Thus, K_i is a measure of the affinity of the inhibitor for an enzyme, similarly to the way K_s is a measure of the affinity of a substrate for the enzyme.

In **competitive inhibition**, the inhibitor is commonly a structural analogue of a substrate and competes with the substrate for binding at the active site. Thus, two reactions are possible:

$$E + S \rightleftharpoons ES \rightarrow E + P \quad (6.10)$$

$$E + I \rightleftharpoons EI \quad (6.11)$$

The modified Michaelis–Menten equation that relates the velocity of the reaction in the presence of competitive inhibitor to the concentrations of substrate and inhibitor is as follows:

$$v = \frac{V_{max}[S]}{K_m \left(1 + \dfrac{[I]}{K_i}\right) + [S]} \quad (6.12)$$

In this relationship, K_m is multiplied by a term that includes the inhibitor concentration, [I], and the inhibitor constant, $(1 + [I]/K_i)$. The observed "K_m," also called $K_{m(app)}$, is $> K_m$ because S and I compete for binding at the active site, and thus a higher concentration of S is required to achieve half-maximal velocity.

The Lineweaver–Burk plot is diagnostically useful for distinguishing types of reversible inhibition. For competitive inhibition, the lines for [S] and [I] at any [I] intersect at the same point on the ordinate but have different slopes and different intercepts on the abscissa (because of differences in $K_{m(app)}$). V_{max} is unaltered. Table 6.1 shows the algebraic expressions for the apparent K_m for competitive and other types of inhibition.

In **noncompetitive inhibition**, the inhibitor does not usually bear any structural resemblance to the substrate, and it binds to the enzyme at a site distinct from the substrate-binding site. No competition exists between the

TABLE 6.1 Effects of Competitive Inhibition on K_m and V_{max} (Michaelis–Menten Kinetic Parameters)

Inhibitor Type	V_{max}	K_m
Competitive	None ($V_{max_{app}} = V_{max}$)	$K_{m_{app}} = K_m \left(1 + \dfrac{1}{K_i}\right)$
Noncompetitive	$V_{max_{app}} = \dfrac{V_{max}}{\left(1 + \dfrac{1}{K_i}\right)}$	None ($K_{m_{app}} = K_m$)
Uncompetitive	$V_{max_{app}} = \dfrac{V_{max}}{\left(1 + \dfrac{1}{K_i}\right)}$	$K_{m_{app}} = \dfrac{K_m}{\left(1 + \dfrac{1}{K_i}\right)}$

Note: $K_i = \dfrac{[E][I]}{[EI]}$ or $\dfrac{[ES][I]}{[ESI]}$. In uncompetitive inhibition, the inhibitor binds equally well to E and ES.

inhibitor and the substrate, and the inhibition cannot be overcome by increase of substrate concentration. An inhibitor may bind either to a free enzyme or to an enzyme–substrate complex; in both cases, the complex is catalytically inactive. Noncompetitive reactions are shown in Equations (6.13) and (6.14). Noncompetitive inhibition is relatively rare except for certain heavy metal inhibition of enzymes in which a reactive is involved.

$$E + I \leftrightarrow EI_{inactive} \tag{6.13}$$

$$E + I \leftrightarrow ESI_{inactive} \tag{6.14}$$

The value of V_{max} is reduced by the inhibitor because the concentration of active enzyme is reduced. K_m is unaffected because the affinity of S for E is not altered.

Enzymes with sulfhydryl groups (−SH) that participate in maintenance of the three-dimensional conformation of the molecule are noncompetitively inhibited by heavy metal ions (e.g., Ag^+, Pb^{2+}, and Hg^{2+}). Heavy metal ions react with S-containing, O-containing, and N-containing ligands, present as −OH, −COO$^-$, −OPO$_3$H$^-$, −C = O, −NH$_2$, and −NH groups. An example is shown in Equation (6.15):

$$E - SH + Hg^{2+} \rightleftharpoons E - S - Hg^+ + H^+ \tag{6.15}$$

Lead poisoning causes anemia (low levels of hemoglobin), owing to inhibition of heme synthesis at two sites at least. Porphobilinogen synthase and ferrochelatase, both of which contain sulfhydryl groups (Chapter 27), are thus inhibited.

Enzymes that are dependent on divalent metal ion (e.g., Mg^{2+} and Ca^{2+}) for activity are inhibited by chelating agents (e.g., ethylenediamine tetraacetate) that remove the metal ion from the enzyme, although these strictly involve the removal of a metal cofactor, not the binding of an inhibitor.

Enolase catalyzes a step in the metabolism of glucose, a reaction that has an absolute requirement for a divalent metal ion (e.g., Mg^{2+} or Mn^{2+}) complexed to the enzyme before the substrate is bound. (See the discussion of glycolysis in Chapter 13.)

$$2\text{-Phospho-D-glycerate} \rightarrow \text{phosphoenolpyruvate} + H_2O \tag{6.16}$$

This reaction is inhibited by fluoride ion (F^-) in a process involving the formation of a complex with phosphate, giving rise to a phosphofluoridate ion that binds Mg^{2+}. Thus, addition of fluoride ions inhibits the breakdown of glucose in the glycolytic pathway. For this reason, F^- is used as a preservative in clinical specimens (e.g., blood) in which glucose determinations are to be made.

In **uncompetitive inhibition**, inhibitor, I, combines only with ES to form an enzyme–substrate inhibitor complex:

$$E + I \rightleftharpoons ESI_{inactive} \tag{6.17}$$

The double-reciprocal plot with an uncompetitive inhibitor yields parallel lines (i.e., the slope remains constant), but the intercepts on both the x and y axes are altered by the presence of the inhibitor (Figure 6.3). The apparent V_{max} and the apparent K_m are both divided by a factor of $(1 + [I] / K_i)$ (Table 6.1). Uncompetitive inhibition is rarely observed in single-substrate reactions. A noteworthy example in clinical enzymology is the inhibition of intestinal alkaline phosphatase by L-phenylalanine. Uncompetitive inhibition is more common in two-substrate reactions with a double-displacement reaction mechanism.

In **mixed inhibition**, the inhibitor binds to either E or ES. The inhibitory effect on the activity, however, differs for K_m and V_{max} and thus two K_i values will be observed: K_i for binding to E and K_i' for binding to ES. See Table 6.1.

Competitive inhibition occurs in several different circumstances, which depend on the mechanism of the enzyme-catalyzed reaction; four are illustrated next.

Reversible Inhibition: Examples in Single-Substrate Reactions

Structural analogues compete with the substrate for binding at the active site of the enzyme. Several specific examples are described in the following sections.

Succinate Dehydrogenase

A classic example of competitive inhibition occurs with the succinate dehydrogenase reaction:

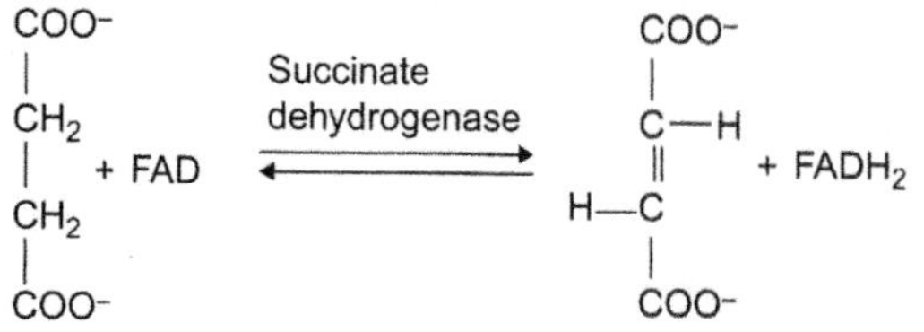

This enzyme is competitively inhibited by malonate, oxalate, or oxaloacetate, all structural analogues of succinate.

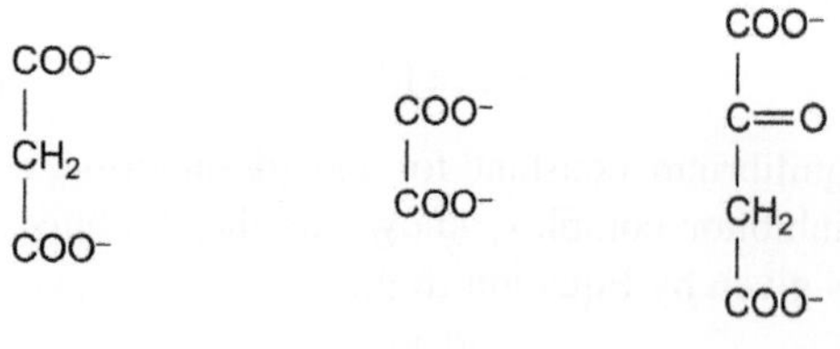

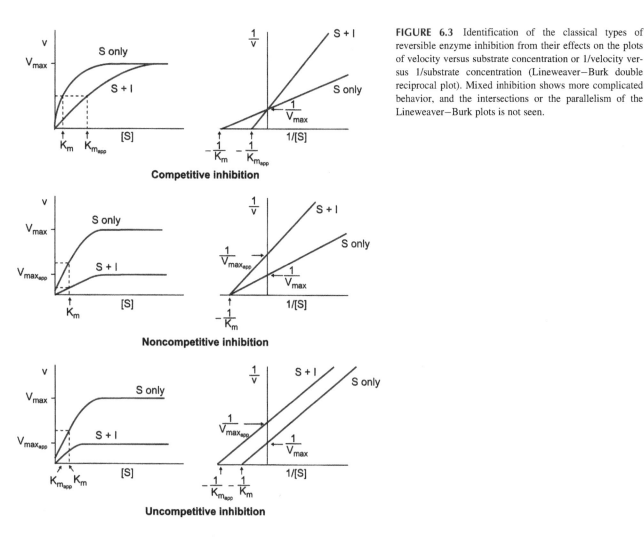

FIGURE 6.3 Identification of the classical types of reversible enzyme inhibition from their effects on the plots of velocity versus substrate concentration or 1/velocity versus 1/substrate concentration (Lineweaver−Burk double reciprocal plot). Mixed inhibition shows more complicated behavior, and the intersections or the parallelism of the Lineweaver−Burk plots is not seen.

Folate Synthesis

Competitive inhibition of a biosynthetic step in folate synthesis accounts for the antimicrobial action of sulfonamides, which are structural analogues of p-aminobenzoic acid (PABA):

p-Aminobenzoic acid   Sulfanilamide

PABA is used by bacteria in the synthesis of folic acid (pteroylglutamic acid), which functions as a coenzyme in one-carbon transfer reactions that are important in amino acid metabolism, in the synthesis of RNA and DNA, and thus in cell growth and division. Sulfonamides inhibit the bacterial enzyme responsible for incorporation of PABA into 7,8-dihydropteroic acid and lead to the inhibition of growth (bacteriostasis) of a wide range of gram-positive and gram-negative microorganisms. Microorganisms susceptible to sulfonamides are those that synthesize their own folic acid or that cannot absorb folic acid derived from the host. Sulfonamides, however, have no effects on host cells (or other mammalian cells) that require preformed folic acid.

Dihydrofolate Reductase

Folate-dependent reactions in the body are inhibited by folate analogues (or antagonists, e.g., methotrexate). Before it can function as a coenzyme in one-carbon transfer reactions, folate (F) must be reduced by dihydrofolate reductase to tetrahydrofolate (FH_4). Dihydrofolate reductase is competitively inhibited by methotrexate. Since FH_4 is needed for the synthesis of DNA precursors, a deficiency causes most harm to those cells that synthesize DNA rapidly. Certain types of cancers (e.g., the leukemias) exhibit an extremely high rate of cell division and are particularly susceptible to folate antagonists and thus the use of methotrexate to treat some types of cancer.

The reactions of one-carbon metabolism are discussed in Chapter 25.

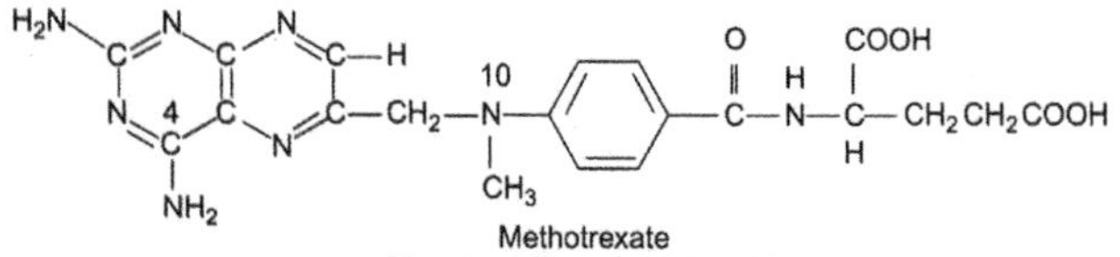

Methotrexate
(4-amino-N^{10}-methyl folic acid)

Xanthine Oxidase

Uric acid, the end product of purine catabolism in humans, is formed by the sequential oxidation of hypoxanthine, and xanthine is catalyzed by xanthine oxidase:

$$\text{Hypoxanthine} \xrightarrow{\text{xanthine oxidase}} \text{xanthine} \tag{6.19}$$
$$\xrightarrow{\text{xanthine oxidase}} \text{uric acid}$$

Allopurinol, a structural analogue of hypoxanthine, is a competitive inhibitor as well as a substrate for xanthine oxidase.

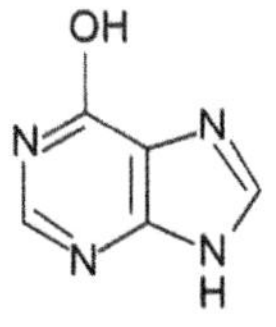

Hypoxanthine 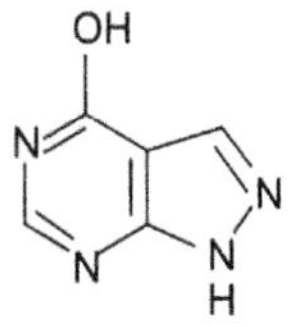   Allopurinol

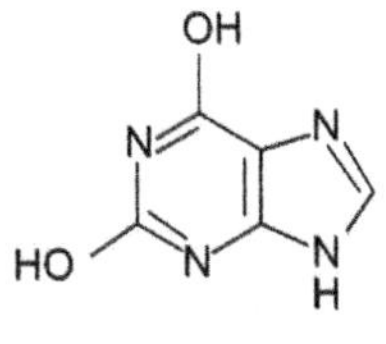

Xanthine 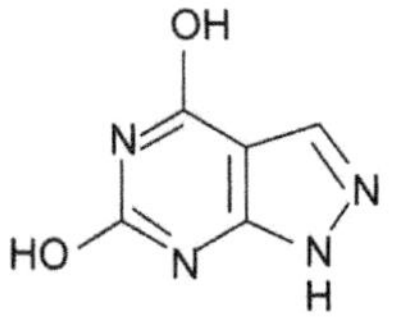   Alloxanthine

Although allopurinol is transformed into alloxanthine, it remains tightly bound to the active site of the enzyme by chelation with Mo^{4+}. Xanthine oxidase uses molybdenum in a catalytic cycle that requires the reversible oxidation and reduction of Mo^{4+} to Mo^{6+}. In the presence of alloxanthine, the reoxidation of Mo^{4+} to Mo^{6+} is very slow, and thus, the rate of the overall catalytic process is slowed. In this type of inhibition, an inhibitor bearing a particular structural similarity to the substrate binds to the active site of the enzyme and, through the catalytic action of the enzyme, is converted to a reactive compound that can form a covalent (or coordinate covalent) bond with a functional group at the active site. This type of inhibition, described as **mechanism-based enzyme inactivation**, depends on both the structural similarity of the inhibitor to the substrate and the mechanism of action of the enzyme. The substrate analogue is a **suicide substrate** because the enzyme is inactivated in one of the steps of the catalytic cycle. The suicide substrate, by virtue of its high selectivity, provides possibilities for many *in vivo* applications (e.g., development of rational drug design). Suicide substrates with potential clinical applications for several enzymes (e.g., penicillinase, prostaglandin cyclooxygenase) have been synthesized. Examples of suicide substrates harmful to the body are given in Chapter 27.

Allopurinol, which affects both the penultimate and ultimate steps in the production of uric acid, is used to lower plasma uric acid levels in conditions associated with excessive urate production (e.g., gout, hematologic disorders, and antineoplastic therapy). Sodium urate has a low solubility in biological fluids and tends to crystallize in derangements of purine metabolism that result in hyperuricemia. The crystalline deposits of sodium urate are responsible for recurrent attacks of acute arthritis or of renal colic (pain in kidney(s) due to either stone formation or acute inflammation; see also discussion of purine catabolism in Chapter 25).

Reversible Inhibition: Example in a Two-Substrate Reaction

In two-substrate enzyme-catalyzed reactions with a double-displacement reaction mechanism, high concentrations of the second substrate may compete with the first substrate for binding. For example, in the reaction catalyzed by aspartate aminotransferase

$$\text{L-Aspartate} + \alpha\text{-ketoglutarate} \rightleftharpoons \text{L-glutamate} \atop + \text{oxaloacetate} \tag{6.20}$$

The enzyme is inhibited by excess concentrations of α-ketoglutarate; inhibition is competitive with respect to L-aspartate.

Reversible Inhibition by Reaction Products

Competitive inhibition can occur in freely reversible reactions owing to accumulation of products. Even in reactions that are not readily reversible, a product can function as an inhibitor when an irreversible step precedes the dissociation of the products from the enzyme. In the alkaline phosphatase reaction, in which hydrolysis of a wide variety of organic monophosphate esters into the corresponding alcohols (or phenols) and inorganic phosphates occurs, the inorganic phosphate acts as a competitive inhibitor. Both the inhibitor and the substrate have similar enzyme-binding affinities (i.e., K_m and K_i are of the same order of magnitude).

Reversible Inhibition by Metal Ion Substitution

In reactions that require metal ions as cofactors, similar metal ions can compete for the same binding site on the enzyme. For example, Ca^{2+} inhibits some enzymes that require Mg^{2+} for catalytic function. Pyruvate kinase catalyzes the reaction for which K^+ is an obligatory activator, whereas Na^+ and Li^+ are potent competitive inhibitors:

$$\text{Phosphoenolpyruvate} + \text{ADP} \rightarrow \text{ATP} + \text{pyruvate} \quad (6.21)$$

Use of Competitive Substrates for Treatment of Intoxication

Competitive inhibition is the basis for the treatment of some types of intoxication (e.g., methyl alcohol, ethylene glycol) (see Clinical Case Study 6.1).

IRREVERSIBLE INHIBITION

Irreversible inhibition occurs when the inhibitor reacts with residues of the active site of the enzyme by covalent modification, or when the inhibitor binds so tightly that, for practical purposes, there is no dissociation of enzyme and inhibitor. The latter situation occurs in the case of some proteinase inhibitors (see following section). The irreversible inhibitor reaction is written as follows:

$$E + I \rightleftharpoons E - I \rightarrow EI \quad (6.22)$$

Following are some examples of irreversible inhibitors of enzymes:

1. Enzymes that contain free sulfhydryl groups at the active site (e.g., glyceraldehyde-3-phosphate dehydrogenase; see Chapter 11) react with an alkylating reagent, iodoacetic acid, resulting in inactivation of the enzyme.

$$\text{Enzyme-SH} + \frac{\text{ICH}_2\text{COOH}}{\text{iodoacetic acid}}$$

$$\rightarrow \frac{\text{enzyme-}S\text{-CH}_2\text{COOH}}{\text{Inactive covalent derivative of enzyme}} + \text{HI}$$

$$(6.23)$$

The imidazole ring of histidine also undergoes alkylation on reaction with iodoacetate. In ribonuclease, two residues (His 12 and His 119) are alkylated with loss of activity when the enzyme is treated with iodoacetate at pH 5.5.

2. Enzymes with seryl hydroxyl groups at the active sites can be inactivated by organophosphorus compounds. Thus, diisopropylphosphofluoridate (DPF) inactivates serine hydrolases by phosphorylation at the active site:

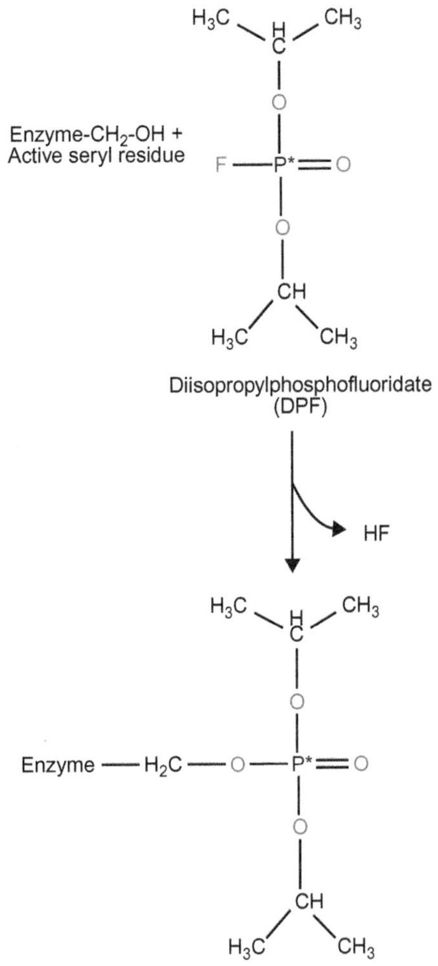

A specific example is inactivation of acetylcholinesterase, which catalyzes hydrolysis of acetylcholine to acetate and choline. Acetylcholine is a neurotransmitter, a chemical mediator of a nerve impulse at a junction—known as a **synapse**—between two neurons or between a neuron and a muscle fiber. On arrival of a nerve impulse at the ending of the neuron, acetylcholine (which is stored in the vesicles of the presynaptic nerve terminal) is released. The released acetylcholine acts on the postsynaptic membrane to increase the permeability of Na^+ entry across the membrane. Depolarization results in the inside of the membrane becoming more positive than the outside; normally, the inside of the membrane is more negative than the outside. This process may propagate an action potential along a nerve fiber, or it may lead to contraction of a muscle (Chapter 19). Acetylcholine is quickly destroyed by acetylcholinesterase present in the basal lamina of the neuromuscular junction (Figure 6.4). If, however, acetylcholine is not destroyed, as in the case of inactivation of acetylcholinesterase by DPF, its continued presence causes extended transmission of impulses.

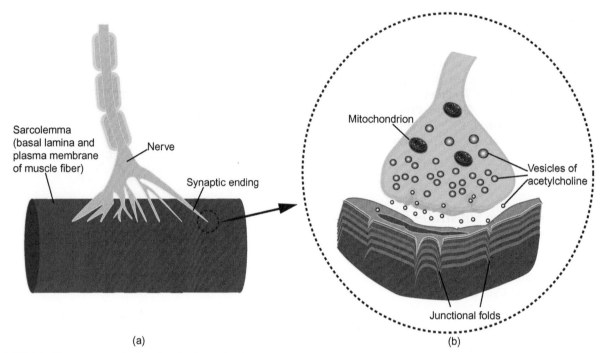

Sarcolemma
(basal lamina and
plasma membrane
of muscle fiber)

Nerve

Synaptic ending

Mitochondrion

Vesicles of
acetylcholine

Junctional folds

(a)

(b)

FIGURE 6.4   The neuromuscular junction. (a) Connection between a nerve and a muscle fiber. (b) Invagination of a nerve terminal into the muscle fiber. The acetylcholine receptor-rich domains are located on the crest of the fold.

In muscle fibers, continuous depolarization leads to paralysis. The cause of death in DPF intoxication is respiratory failure due to paralysis of the respiratory muscles (including the diaphragm and abdominal muscles). Several organophosphorus compounds are used as agricultural insecticides, improper exposure to which can result in toxic manifestations and death.

Knowledge of the mechanisms of action of acetylcholinesterase and of the reaction of organophosphorus compounds with esterases led to the development of drugs useful in the treatment of this kind of intoxication. The active site of acetylcholinesterase consists of two subsites: an esteratic site and a site consisting of a negative charge (Figure 6.5).

The esteratic site contains seryl hydroxyl group whose nucleophilicity toward the carbonyl carbon of the substrate is enhanced by an appropriately located imidazole group of histidine that functions as a general base catalyst. A side-chain carboxyl group that is suitably located apparently functions to hold the imidazole and imidazolium ion in place. The substrate is positioned on the enzyme so that its positively charged nitrogen atom is attracted to the negatively charged active site of the enzyme by both coulombic and hydrophobic forces. The acyl carbon of the substrate is subjected to nucleophilic attack by the oxygen atom of the serine hydroxyl group. This catalytic mechanism is similar to that of other serine hydrolases (esterases and the proteinases, i.e.,

chymotrypsin, trypsin, elastase, and thrombin). Although they have different functions, these enzymes have a common catalytic mechanism, which supports the view that they evolved from a common ancestor.

The inactive phosphorylated acetylcholinesterase undergoes hydrolysis to yield free enzyme, but the reaction is extremely slow. However, a nucleophilic reagent (e.g., hydroxylamine, hydroxamic acids, and oximes) can reactivate the enzyme much more rapidly than spontaneous hydrolysis. Wilson and coworkers accomplished the reactivation by use of the active-site-specific nucleophilic pralidoxime. This compound, oriented by its quaternary nitrogen atom, brings about a nucleophilic attack on the phosphorus atom, leading to the formation of an oxime-phosphonate-enzyme complex that dissociates into oxime-phosphonate and free enzyme (Figure 6.6). Pralidoxime has found use in the treatment of organophosphorus poisoning and is most active in relieving the inhibition of skeletal muscle acetylcholinesterase. The phosphorylated enzyme can also lose an isopropoxy residue; such an enzyme-inhibitor complex has been named an "aged" enzyme. The aged enzyme is resistant to regeneration by pralidoxime because the phosphorus atom is no longer an effective center for nucleophilic attack. The severity and duration of toxicity of organophosphorus compounds depend on their lipid solubility, the stability of the bond linking the phosphorus atom to the oxygen of the serine hydroxyl group, and the ease with which the aged enzyme

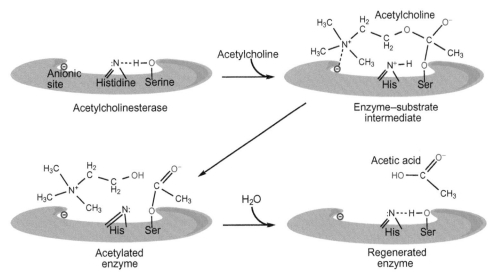

FIGURE 6.5 Hydrolysis of acetylcholine by acetylcholinesterase.

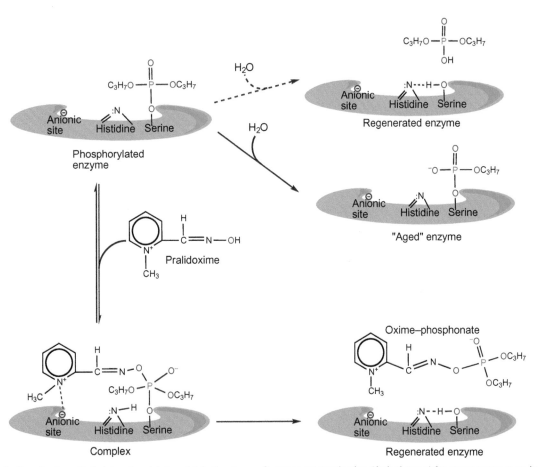

FIGURE 6.6 Reactivation of alkylphosphorylated acetylcholinesterase. Spontaneous reactivation (dashed arrow) by water occurs at an insignificant rate; however, loss of one isopropoxy group occurs at a much more rapid rate, yielding the "aged" enzyme, which is resistant to reactivation. The regeneration of the enzyme by pralidoxime is shown at bottom.

complex is formed. Some organophosphorus compounds are so toxic that they can be fatal within a few seconds of exposure. Other manifestations (e.g., excessive mucus secretion and bronchoconstriction) are reversed by administration of atropine, which does not relieve enzyme inactivation but, by binding to some of the acetylcholine receptor sites, renders ineffective the accumulated acetylcholine.

Organophosphorus compounds have been used in the identification of functional groups essential for catalysis. Another approach to identification of functional amino acid residues is **affinity labeling**. The labeling is produced by a synthetic substrate-like reagent designed to form a covalent linkage with some amino acid residue at or near the active site of the enzyme. After the labeling, the enzyme is subjected to classic techniques of degradative protein chemistry (Chapter 4) to determine the amino acid sequence of the structurally altered portion of the enzyme, which is inferred to be the active site.

Inactivation and Reactivation of Cytochrome Oxidase

Cyanides are among the most rapidly acting toxic substances. Cyanide (CN^-) inhibits cellular respiration to cause tissue hypoxia by binding to the trivalent iron of cytochrome oxidase, a terminal component of the mitochondrial electron transport chain. This chain consists of electron-transferring proteins and other carriers arranged sequentially in the inner mitochondrial membrane. The reducing equivalents, obtained from a variety of substrates, are passed through the electron transport system to molecular oxygen with the formation of water and energy (Chapter 13). Thus, cyanide severely impairs the normal energy-generating functions of mitochondria by inhibition of mitochondrial respiration, leading to cell death, particularly affecting the central nervous system. Death in acute cyanide poisoning is due to respiratory failure.

Cytochrome oxidase is a multienzyme complex that contains oxidation-reduction centers of iron-porphyrin prosthetic groups as well as centers of copper ion. Cyanide has a higher affinity for the oxidized form of cytochrome oxidase than for the reduced form. CN^- probably forms a loose complex with Fe^{2+} of porphyrin. When Fe^{2+} is oxidized to Fe^{3+}, the latter forms a stable complex with CN^-. This complex cannot by reduced, thus preventing electron flow and uptake of O_2.

Cyanide and cyanide precursors occur widely in nature. Foods that contain moderate to high levels of cyanogenic glycosides include cassava (a dietary staple in several regions of Africa), kernels of some fruits (peach, cherry, plum, and apricot), lima beans, sorghum, linseed, sweet potato, maize, millet, and bamboo shoots. **Amygdalin** (Figure 6.7) is one of the principal

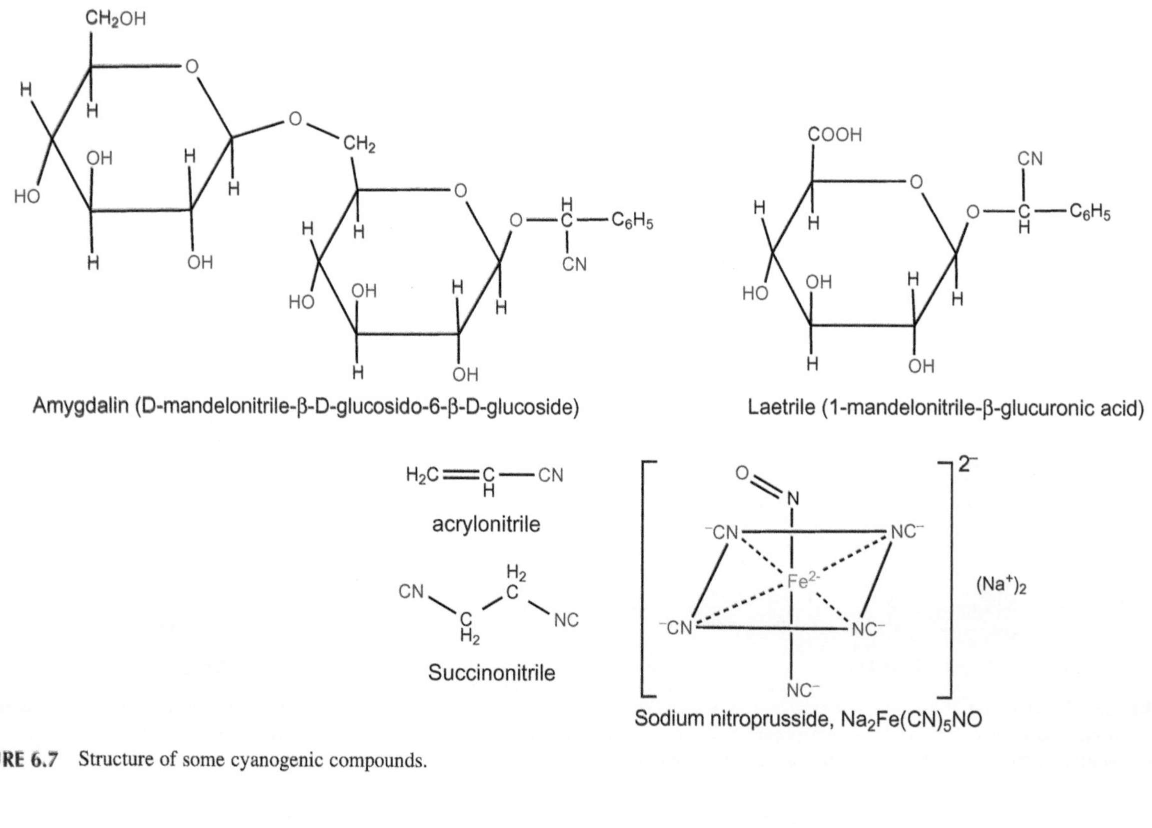

Amygdalin (D-mandelonitrile-β-D-glucosido-6-β-D-glucoside)

Laetrile (1-mandelonitrile-β-glucuronic acid)

acrylonitrile

Succinonitrile

Sodium nitroprusside, $Na_2Fe(CN)_5NO$

FIGURE 6.7 Structure of some cyanogenic compounds.

cyanogenic glycosides of dietary origin. Hydrogen cyanide is released when amygdalin undergoes enzymatic hydrolysis in the gastrointestinal tract:

$$\underset{\text{Amygdalin}}{C_{20}H_{27}NO_{11}} + 2H_2 \rightarrow 2 \underset{\text{Glucose}}{C_6H_{12}O_6}$$
$$+ \underset{\text{Benzaldehyde}}{C_6H_5CHO} + \underset{\text{Hydrogen cyanide}}{HCN} \tag{6.24}$$

The toxicity of amygdalin is directly related to the release of hydrogen cyanide. The enzymes that catalyze the hydrolysis of amygdalin are supplied by the microbial flora of the intestine, which probably explains why amygdalin is many times more toxic when taken by mouth than when given intravenously. Amygdalin and the related compound **laetrile** have been at the center of a controversy regarding their efficacy as anticancer agents. Their anticancer activity is claimed to depend on selective hydrolysis at the tumor site by β-glucuronidase or β-glucosidase, with local release of HCN to cause cell death. Presumably, normal cells are not affected by HCN because they contain CN^--inactivating enzyme (rhodanese; see following text). No objective evidence demonstrates any therapeutic value for these compounds as anticancer agents. Their ingestion has led to severe toxicity in some reported cases.

Nondietary sources of cyanide include sodium nitroprusside (a hypotensive agent), succinonitrile (an antidepressant agent), acrylonitrile (used in the plastics industry and as a fumigant to kill dry-wood termites), and tobacco smoke. Chronic exposure to cyanogenic compounds leads to toxic manifestations such as demyelination, lesions of the optic nerves, ataxia (failure of muscle coordination), and depressed thyroid functions. This last effect arises from accumulation of thiocyanate, the detoxified product of cyanide in the body (see following text). Thiocyanate inhibits the active uptake of iodide by the thyroid gland and, therefore, the formation of thyroid hormones (Chapter 31).

Acute cyanide poisoning requires prompt and rapid treatment. The biochemical basis for an effective mode of treatment consists of creating a relatively nontoxic porphyrin-ferric complex that can compete effectively with cytochrome oxidase for binding the cyanide ion. This is accomplished by the administration of nitrites ($NaNO_2$ solution intravenously, and amylnitrite by inhalation), which convert a portion of the normal oxygen-carrying hemoglobin with divalent iron to oxidized hemoglobin (i.e., methemoglobin with trivalent iron, which does not transport oxygen). Methemoglobin binds cyanide to form cyanomethemoglobin, whose formation is favored because of an excess of methemoglobin relative to cytochrome oxidase. This process may lead to the restoration of normal cytochrome oxidase activity (Figure 6.8).

Cyanomethemoglobin is no more toxic than methemoglobin and can be removed by normal processes that degrade erythrocytes and by the reaction catalyzed by rhodanese. Rhodanese (thiosulfatecyanide sulfurtransferase) catalyzes the reaction involving cyanide and thiosulfate to form thiocyanate:

$$\underset{\text{Thiosulfate}}{SSO_3^{3-}} + enzyme \leftrightarrow \underset{\text{Sulfite}}{SO_3^{2-}} + \underset{\text{Sulfur-substituted enzyme}}{enzyme\text{-}S}$$
$$Enzyme - S + \underset{\text{Cyanide}}{CN^-} \rightarrow \underset{\text{Thiocyanate}}{SCN^-} + enzyme$$

$$(6.25)$$

Rodanese is present in the mitochondria, particularly of liver and kidney cells. A double-displacement mechanism has been proposed for its biochemical action. The steps are as follows: the free enzyme reacts with a sulfane sulfur-containing compound (a sulfane sulfur is one that is divalent and covalently bonded only to other sulfurs), cleaving the S—S bond of the donor substrate (e.g., SSO_3^{2-}) to form the sulfur-substituted enzyme. The latter reacts with the cyanide (a thiophilic acceptor) to form thiocyanate in an essentially irreversible reaction.

Nitrite administration has been augmented by thiosulfate administration (intravenously) in the treatment of cyanide poisoning. Cobalt-containing compounds (e.g., cobalt chloride and cobalt ethylenediaminetetraacetate) have also been used to form complexes with cyanide, in order to decrease the amount of cyanide available for binding with cytochrome oxidase.

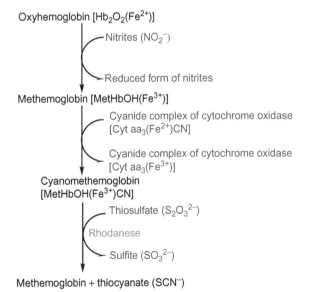

FIGURE 6.8 Reactivation of cytochrome oxidase and inactivation of cyanide.

Irreversible Proteinase Inhibitors: Inhibitors of Great Clinical Significance

In the body, enzymes are compartmentalized within cells and subcellular organelles, and thus function under highly restricted conditions. Some enzymes (e.g., digestive proteinases) are not protein substrate-specific, but rather are specific for particular amino acid residues in a protein substrate. When released in active form in an inappropriate tissue, proteinases (also called **proteases**) act indiscriminately on proteins and cause considerable damage to the tissue. Inhibitors inactivate these enzymes at sites where their action is not desired. Proteinase inhibitors, which are themselves proteins, are widely distributed in intracellular and extracellular fluids. Protein inhibitors of enzymes other than proteinases are relatively rare, although such inhibitors exist for α-amylases, deoxyribonuclease I, phospholipase A, and protein kinases.

Many irreversible proteinase inhibitors are present in blood plasma and participate in the control of blood coagulation and dissolution of blood clots (Chapter 34), inactivation of the complement cascade proteinases (Chapter 33), formation and destruction of some peptide hormones, and inactivation of proteinases released from phagocytic cells. The proteinase inhibitors rapidly combine with their target enzymes to form stable complexes that are practically nondissociable. Inhibition occurs through binding of a reactive site residue (a substrate-like region) of the inhibitor to the active site of the proteinase.

Neutrophils and macrophages function protectively against foreign organisms in an inflammatory process. In inflammation a number of proteinases escape to surrounding tissues. One such enzyme is elastase, which normally catalyzes the hydrolysis of elastin, a protein of connective tissue (Chapter 10). The activity of elastase is inhibited by the **α₁-proteinase inhibitor** (also known as **α₁-antitrypsin**), which inhibits a broad spectrum of proteinases containing serine in their active sites (e.g., elastase, trypsin, and chymotrypsin). Genetic deficiency of α₁-proteinase inhibitor strongly predisposes to pulmonary emphysema and liver disease. Emphysema (which means swelling or inflation) results from the breakdown of alveolar walls, which leads to a gradual decrease in the effectiveness of CO_2 elimination and oxygenation of hemoglobin. Oxidants in tobacco smoke convert a methionine residue in α₁-proteinase inhibitor and inactivate it, predisposing smokers to emphysema (see Clinical Case Study 6.1).

A high molecular weight proteinase inhibitor with broad specificity, known as **α₂-macroglobulin**, is present in the plasma of all mammals and exhibits an interesting inhibition pattern. α₂-Macroglobulin combines with proteinases in a cage-like complex; it inhibits only the proteolytic activity of the enzyme toward large protein substances without significantly affecting the catalysis of low molecular weight substrates.

Proteinases and their inhibitors also play a major role in the metastasis of cancer. Metastasis of tumors requires remodeling of extracellular matrix (ECM). Remodeling is a balance between proteolysis and respective proteinase inhibitors. This process is aided by proteolytic enzymes synthesized by the tumor cells. Examples of proteolytic enzymes include serine proteinases, cathepsins, and matrix metalloproteinases. Some of the **cathepsins** are cysteine proteinases, and their inhibitors belong to the cystatin superfamily. The **cystatin family** consists of three subfamilies: types 1, 2, and 3; type 3 is a group of proteins called **kininogens**. Progressive loss of expression of the proteinase inhibitors may contribute to metastasis. The loss of control over proteinase expression and their respective inhibitors is influenced by a variety of biological response modifiers such as growth factors, cytokines, tumor promoters, and suppressor genes.

Viral proteinases offer unique targets for antiviral drugs. An example is in the treatment of human immunodeficiency virus (HIV) infections by HIV protease inhibitors.

Mechanisms of Enzyme Action

The mechanism of a reaction catalyzed by an enzyme gives a detailed description of the chemical interactions occurring among the substrates, enzymes, and cofactors. For example, although the overall reaction catalyzed by acetylcholinesterase consists of hydrolysis of acetylcholine to choline and acetic acid, the detailed mechanism is a two-step displacement reaction in which an alcohol (choline) is produced first, followed by an acid (acetic acid). The same is true for the rhodanese-catalyzed reaction. Many types of experiments are carried out to arrive at the description of an enzyme mechanism, including synthesis of recombinant proteins with modifications of specific amino acid residues.

Coenzymes, Prosthetic Groups, and Cofactors

Many enzymes require the presence of low molecular weight nonprotein molecules. These small molecules may be bound to the enzyme by a covalent, tight, or noncovalent linkage. Prosthetic groups are usually bound by covalent or tight linkages and coenzymes by noncovalent linkages.

Cofactors commonly are metal ions or organic compounds (some are vitamins, e.g., ascorbic acid). Prosthetic groups and coenzymes are complex organic compounds, many of which are derived from vitamins. These compounds are recycled and are needed only in small amounts to convert a large amount of reactants to products. Coenzymes function as substrates in two-substrate reactions, being bound only momentarily to the enzyme during catalysis. They are chemically altered during the reaction and are reconverted to their original forms by the same or another enzyme (e.g., NAD1- or NADP1-dependent dehydrogenase reactions). Prosthetic groups form part of the active center and undergo cyclic chemical changes during the reaction (e.g., pyridoxal phosphate in aminotransferases). Principal coenzymes, prosthetic groups, and cofactors and their metabolic roles are discussed elsewhere, e.g., with the reactions they catalyze and the vitamins from which they are derived.

Regulation

The metabolism of key substances, which can proceed via multiple pathways, is regulated and integrated. A close interrelationship exists among products formed by different metabolic pathways from a common metabolite. For example, glucose can be consumed either by oxidation to CO_2 or by conversion to glycogen, lipids, nonessential amino acids, or other sugar molecules. The glucose supply of the body can be derived either from the diet or from the breakdown of glycogen, a polymer of glucose (primarily from the liver and the kidney), or it can be synthesized from some amino acids or lactate (predominantly in the liver). These processes of glucose utilization and synthesis are under tight regulation. In fact, the plasma glucose level is maintained at the level at which tissues that require glucose as a primary substrate (e.g., brain, erythrocytes, kidney medulla, and the lens and cornea of the eye) are not deprived of this essential fuel. Each of these metabolic pathways is mediated by enzymes that are unique for a given pathway and that are under control by the mechanisms discussed previously and in their synthesis. Because of compartmentalization, metabolic pathways do not usually compete with each other for utilization of a substrate, and operate only to serve a particular physiological need or function.

Types of Regulation

A metabolic pathway involves many enzymes functioning in a sequential manner or in some unique, branched arrangement to carry out a particular metabolic process. Control of a pathway is accomplished through modulation of the activity of only one or a few key enzymes. These **regulatory enzymes** usually catalyze either the first or an early reaction in a metabolic sequence. A regulatory enzyme catalyzes a **rate-limiting** (or rate-determining) chemical reaction that controls the overall flux of metabolites within the pathway. It may also catalyze a chemical reaction unique to that pathway, which is known as a

committed step. In the metabolic pathway for the formation of E from A

$$A \xrightarrow{E_1} B \xrightarrow{E_2} C \xrightarrow{E_3} D \xrightarrow{E_4} E \qquad (6.26)$$

the conversion of A to B, catalyzed by the enzyme E_1, is the rate-limiting step and also a committed step. The rate-limiting step need not be the same as the committed step. In the branched metabolic pathway:

$$(6.27)$$

If the conversion of A to B is the rate-limiting step, the committed step in the pathway for the formation of N is the conversion of B to L ($\mathbf{B \rightarrow L}$), catalyzed by the enzyme E_5. Those enzymes which catalyze the rate-limiting step or the committed step of a pathway are under regulation. When the end product exceeds the steady-state level concentration, it inhibits the regulatory enzyme in an attempt to normalize the overall process. This type of control, known as **feedback inhibition** (see following text), ensures a high degree of efficiency in the utilization of materials and of energy in living systems.

Regulation may be achieved in other ways. The absolute amount of a regulatory enzyme may be altered through mechanisms that control gene expression (Chapter 24). This regulation at the genetic level occurs during various phases of reproduction, growth, and development, with different metabolic pathways being turned on or off in accordance with the special requirements of each phase. In eukaryotic cells, regulation at the genetic level is a relatively long-term process. Several short-term regulatory mechanisms control metabolic activity rapidly (see following text). Substrates and some hormones play significant roles in regulating the concentration of key enzymes at this level. Many drugs or other chemicals can increase levels of enzymes that affect their own metabolism. Examples are phenobarbital and polycyclic hydrocarbons that cause an increase in the levels of microsomal enzyme systems involved in their metabolism.

Regulation of metabolic processes can be accomplished by other methods. One is the use of a **multienzyme complex** (e.g., the pyruvate dehydrogenase complex or fatty acid synthase complex). In multienzyme complexes, enzymes are organized so that the product of one becomes the substrate for an adjacent enzyme. A single polypeptide chain may contain multiple catalytic centers that carry out a sequence of transformations (e.g., the mammalian fatty acid synthase; see Chapter 16). Such multifunctional polypeptides increase catalytic efficiency by abolishing the accumulation of free intermediates and by maintaining a stoichiometry of 1:1 between catalytic centers.

Another type of regulation is accomplished by a series of **proenzymes** in which a biological signal causes activation of the first proenzyme, which then activates the second proenzyme, which, in turn, activates the third, and so on. Such an enzyme cascade process can provide great amplification in terms of the amount of final product formed. Examples are blood coagulation, the dissolution of blood clots, complement activation, and glycogen breakdown.

Regulation is further accomplished by compartmentalization of the enzyme systems involved in either anabolic or catabolic pathways into different cell organelles. For example, fatty acid synthesis occurs in the soluble fraction of the cytoplasm, whereas fatty acid oxidation takes place in mitochondria. Heme synthesis begins and is completed in mitochondria, but some of the intermediate reactions take place in the cytosol. Heme catabolism is initiated in the smooth endoplasmic reticulum. Transport of key metabolites across an organelle membrane system is also a form of regulation.

Many enzymes occur in several molecular forms called **isoenzymes** (or isozymes); those that are genetically determined may be called **primary isoenzymes**. The different isoenzymes catalyze the same chemical reaction but differ in their primary structure and kinetic properties. The tissue distribution of isoenzymes imparts distinctive properties and distinctive patterns of metabolism to particular organs. The presence of isoenzymes may reveal differences not only between organs, but also between cells that make up an organ or between organelles within a cell. During different stages of differentiation and development from embryonic life to adulthood, the isoenzyme distribution in an organ undergoes characteristic changes. When an adult organ reverts to the embryonic or fetal state (e.g., in cancer), the isoenzyme distributions change to those characteristic of that developmental state. The existence of isoenzymes in human tissues has important implications in the study of human disease and in disease diagnosis.

Zymogen (e.g., trypsinogen and chymotrypsinogen) synthesis, secretion, transport, and activation and the rate of inactivation of the active enzyme by inhibitors may all be considered means of enzyme regulation; these are irreversible chemical reactions.

Enzyme activity can be regulated by covalent modification or by noncovalent (allosteric) modification. A few enzymes can undergo both forms of modification (e.g., glycogen phosphorylase and glutamine synthetase). Some covalent chemical modifications are phosphorylation and dephosphorylation, acetylation and deacetylation, adenylylation and deadenylylation, uridylylation and deuridylylation, and methylation and demethylation. In mammalian systems, phosphorylation and dephosphorylation, reactions catalyzed by kinases and phosphatases, are commonly used as a means of metabolic control (see Chapter 28).

Allosteric Enzyme Regulation

Those enzymes in metabolic pathways whose activities can be regulated by noncovalent interactions with certain compounds at sites other than the catalytic site are known as **allosteric enzymes**. They are usually rate-determining enzymes and play a critical role in the control and integration of metabolic processes. The term *allosteric* is of Greek origin, the root word *allos* meaning *other*. Thus, an **allosteric site** is a unique region of an enzyme other than the substrate-binding site that affects catalysis. At the allosteric site, the enzyme is regulated by noncovalent interaction with specific ligands known as **effectors**, **modulators**, or **modifiers**.

The properties of allosteric enzymes differ significantly from those of unregulated enzymes. Ligands (in some instances even the substrate) can bind at remote sites by a cooperative binding process. Cooperativity describes the process by which binding of a ligand to a regulatory site affects binding of the same or of another ligand to the enzyme. Allosteric enzymes have a more complex structure than nonallosteric enzymes, and the reaction dependence on substrate concentration differs from Michaelis−Menten kinetics. An allosteric site is specific for its ligand, just as the active site is specific for its substrate. Binding of an allosteric modulator causes a change in the conformation of the enzyme (see following text) that leads to a change in the binding affinity of the enzyme for the substrate. The effect of a modulator may be positive (activatory) or negative (inhibitory). The former leads to increased affinity of the enzyme for its substrate, whereas the reverse is true for the latter. Activating sites and inhibiting sites are separate and specific for their respective modulators. Thus, if an end product of a metabolic pathway accumulates in excess of its steady-state level, it can slow down or turn off the metabolic pathway by binding to the inhibitory site of the regulatory enzyme of the pathway. As the concentration of the end product (inhibitor) decreases below the steady-state level, the number of enzymes having bound inhibitors decreases, and they revert to their active form. In this instance, the substrate and the negative modulator bear no structural resemblance. An allosteric enzyme also may be positively modulated by the substrate itself or by a metabolite of another pathway that depends on production of the end product in question for its eventual utilization (e.g., pathways of synthesis of purine and pyrimidine nucleotides in the formation of nucleic acids; see Chapter 25).

Most allosteric enzymes are **oligomers** (i.e., they consist of two or more polypeptide chains or subunits). The individual subunits are known as **protomers**. Two types of interactions occur in allosteric enzymes: **homotropic** and **heterotropic**. In a **homotropic** interaction, the same ligand positively influences the cooperativity between different modulator sites on the enzyme. An example is a regulatory enzyme modulated by its own substrate. Thus, this class of enzyme has at least two substrate-binding sites which respond to situations that lead to substrate excess by increasing its rate of removal. **Heterotropic** interaction refers to the effect of one ligand on the binding of a *different* ligand. For example, a regulatory enzyme modulated by a ligand other than its substrate constitutes a **heterotropic** system, in which the cooperativity can be positive or negative. Some allosteric enzymes exhibit mixed homotropic and heterotropic interactions.

Kinetics of Allosteric Proteins

The kinetic properties of allosteric enzymes vary significantly from those of nonallosteric enzymes, exhibiting cooperative interactions between the substrate, the activator, and the inhibitor sites. These properties are responsible for deviations from the classic Michaelis–Menten kinetics that applies to nonallosteric enzymes. Nonallosteric enzymes yield a rectangular hyperbolic curve when the initial velocity (v) is plotted against the substrate concentration [S]. For allosteric enzymes, a plot of v versus [S] yields curves of different shapes, including sigmoid-shaped curves in some cases. (A sigmoidal curve can result from other mechanisms.) The v versus [S] plot for a homotropic enzyme is shown in Figure 6.9.

The following features should be noted:

1. The substrate functions as a positive modulator; i.e., there is positive cooperativity between the substrate-binding sites so that binding of the substrate at one binding site greatly enhances binding of the substrate at the other sites. As the substrate concentration increases, there is a large increase in the velocity of the reaction.
2. Owing to the above effect, the shape of the curve is sigmoidal.
3. The value of the substrate concentration corresponding to half-maximal velocity is designated as $K_{0.5}$ and not K_m, since the allosteric kinetics do not follow the hyperbolic Michaelis–Menten relationship.
4. Maximum velocity (V_{max}) is attainable at a rather high substrate concentration, implying saturation of the catalytic site of the enzyme.

The v versus [S] plot for **heterotropic** enzymes is more complex. The kinetic profiles can be divided into

two classes, depending on whether the allosteric effector alters $K_{0.5}$ and maintains a constant V_{max} or alters V_{max} and maintains a nearly constant $K_{0.5}$. The v versus [S] profile of an allosteric enzyme that follows the former set of properties is shown in Figure 6.10.

In the absence of any modulators, the [S] profile is sigmoidal (**curve a**). In the presence of a positive modulator (**curve b**), the value for $K_{0.5}$ is decreased; i.e., a lower substrate concentration is required to attain half-maximal velocity. Curve b is more hyperbolic than sigmoidal. Curve c, obtained with a negative modulator, is more

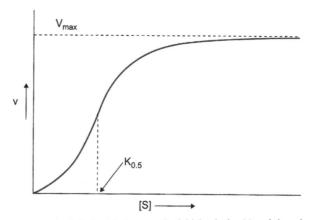

FIGURE 6.9 Relationship between the initial velocity (v) and the substrate concentration [S] for an allosteric enzyme that shows a homotropic effect. The substrate functions as a positive modulator. The profile is sigmoidal, and during the steep part of the profile, small changes in [S] can cause large changes in v. $K_{0.5}$ represents the substrate concentration corresponding to half-maximal velocity.

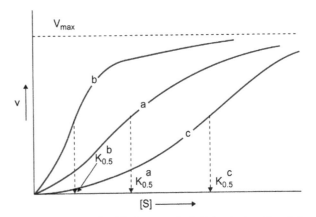

FIGURE 6.10 Relationship between the initial velocity (v) and the substrate concentration [S] for an allosteric enzyme that shows a heterotropic effect with constant V_{max} but with varying $K_{0.5}$. Curve a is obtained in the absence of any modulators, curve b in the presence of a positive modulator, and curve c in the presence of a negative modulator. Regulation is achieved by modulation of $K_{0.5}$ without change in V_{max}.

sigmoidal than curve a, and the $K_{0.5}$ value is increased, reflecting a decreased affinity for the substrate; i.e., a higher substrate concentration is required to attain half-maximal velocity. Regulation of the enzyme is achieved through positive and negative modulators. Thus, at a given substrate concentration (e.g., steady-state level), the activity of the enzyme can be turned on or off with appropriate modulators. Figure 6.11 shows v versus [S] plots for allosteric enzymes modulated by changes in V_{max} but retaining an essentially constant $K_{0.5}$. This type of modulation is less common than the two previous cases considered. The positive modulator increases V_{max} (curve b), whereas the negative modulator decreases V_{max} (curve c). Hemoglobin, the prototypical example of allosteric regulation, is considered in Chapter 26. The metabolic significance of allosteric enzymes is discussed at the appropriate places in the text.

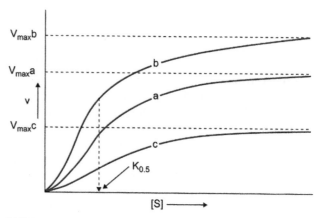

FIGURE 6.11 Relationship between the initial velocity (v) and the substrate concentration [S] for an allosteric enzyme that shows a heterotropic effect with constant $K_{0.5}$ but with varying V_{max}. Curve a is obtained in the absence of any modulator, curve b in the presence of a positive modulator, and curve c in the presence of a negative modulator. Regulation is achieved by modulation of V_{max} without change in $K_{0.5}$.

CLINICAL CASE STUDY 6.1 Consequences of α_1-Antitrypsin Deficiency and Management

Case Description

A 38-year-old male, with a history of smoking cigarettes (a pack a day), who developed a chronic progressive condition of shortness of breath (dyspnea) and incapability of sustaining any physical activity, was referred to a pulmonologist by his primary care physician. Testing for lung cancer and examination of lung biopsy yielded the diagnosis of panacinar emphysema without any evidence for cancer. A routine serum electrophoretic study showed a decreased α_1-globulin fraction, which led to the quantitative measurement of α_1-antitrypsin (α_1-AT). The serum α_1-AT level was severely deficient, and the level was 22 mg/dL (reference range: 78–220 mg/dL, to convert to μmol/L, divide by 5.2). The isoelectric focusing studies revealed that the subject phenotype was ZZ (PiZZ). The subject's liver function tests did not show any abnormalities.

The subject was counseled to stop smoking immediately, limit exposure to respiratory irritants, get vaccinations against pneumococcal and influenza infections, and initiate prompt and early antibiotic therapy in the event of any suspected bacterial infections. The subject was placed on a periodic augmentation therapy of α_1-AT infusion.

Teaching Points

1. α_1-Antitrypsin (α_1-AT) deficiency occurrence is estimated to be 1 in 3000–5000 people of Caucasian ancestry, often unrecognized, which can lead to chronic obstructive pulmonary disease (COPD) and liver disease. Thus, in these groups of patients, measurement of serum α_1-AT levels is warranted and, when these are low, assessment of phenotype and genotype.

2. Why does α_1-AT deficiency result in COPD and/or liver disease? α_1-Antitrypsin consists of a single polypeptide chain of 394 amino acid residues with three

oligosaccharide side chains, all of which are attached to asparagine residues. It contains three β sheets and eight α helices. It has an overall molecular weight of 51,000. It is a polar molecule and readily migrates into tissue fluids. It is synthesized in hepatocytes and secreted into the plasma, where it has a half-life of 6 days. Its normal serum concentration is 78–220 mg/dL (depending on the method of determination). Although the presence of α_1-AT has been noted in other cell types, the primary source in plasma is the hepatocyte. α_1-AT does not possess a propeptide but does contain a 24-residue signal peptide, which is eliminated during its passage through the endoplasmic reticulum. α_1-AT is a defense protein, and its synthesis and release into circulation increase in response to trauma and inflammatory stimulus. Thus, it is known as an acute-phase reactant. α_1-AT functions by forming a tight complex with the active site of the target enzyme that inhibits its proteolytic activity. The enzyme-inhibitor complexes are rapidly cleared by the reticuloendothelial cells. Thus, α_1-AT is a suicide protein. Many genetic variants of α_1-AT have been identified by isoelectric focusing. The α_1-AT gene is on chromosome 14, is 10.2 kb long, and contains four introns. The phenotype is based on a codominant pattern with both alleles expressed equally. Variants are classified by a system known as **Pi (Protease inhibitor)**. The most common form is PiMM (carried by 95% of the American population). Most polymorphisms observed in α_1-AT are due to single-amino acid substitutions.

α_1-AT is a broad-spectrum serine proteinase inhibitor. Its principal action is to inhibit leukocyte (neutrophil) elastase. This function appears to be most important in maintaining the integrity of the elastic fibers for the elastic

(Continued)

CLINICAL CASE STUDY 6.1 (Continued)

recoil of normal lung tissue. The elastic fibers contain an amorphous core of elastin surrounded by an envelope of microfibrils (Chapter 10). The target substrate for neutrophil elastases is elastin. The turnover rate of mature elastin is extremely low, and if it is destroyed, replacement is severely limited. Thus, the risk of development of a lung disease (**emphysema**), characterized by dilatation of air spaces distal to the terminal bronchiole and destruction of bronchiole walls, is high in α_1-AT deficiency.

A human phenotype in which a severe deficiency of α_1-AT (10%−15% of normal serum concentration) has been observed is designated PiZ. The risk of development of emphysema in PiZZ (homozygotic ZZ individuals) is 20 times that in PiMM (a normal genotype). In addition, cigarette smoking by PiZ individuals accelerates development of destructive lung disease for several reasons.

a. The levels of α_1-AT are very low, and smoking causes an increase in elastolytic load owing to a change in the phagocyte population.

b. A reduction of the proteinase inhibitory activity is due to oxidation of the reactive methionine residue of the active center by cigarette smoke, as well as by oxygen radicals produced by leukocytes and macrophages. Emphysema is a common disease, and its most common cause is cigarette smoking. Only 1%−2% of cases are due to genetic deficiency of α_1-AT.

The substitution of a lysine for a glutamic acid at position 342 leads to the deficiency of α_1-AT in PiZ individuals (Figure 6.12). This substitution of a basic for an acidic amino acid is consistent with a base change of cytosine to thymine in DNA. The amino acid change prevents normal processing of oligosaccharide side chains of the protein and therefore its secretion. As a result, Z-α_1-AT accumulates in hepatocytes. An association of hepatic disease with severe α_1-AT deficiency has been observed. The sequestered α_1-AT causes damage to hepatocytes or renders them more susceptible to injury, toxins, or viruses. Liver transplantation has been attempted with limited success in patients with hepatocellular failure. The therapy for destructive lung disease is administration of an anabolic steroid (danazol), which increases the serum levels of α_1-AT by about 40%, or direct replacement of α_1-AT by transfusion.

Another structural variant of α_1-AT that has been studied in detail is S-α_1-AT (PiSS). This protein differs from M-α_1-AT at position 264, where a glutamic acid residue is substituted by valine (Figure 6.12). The S-α_1-AT undergoes normal addition of oligosaccharides, and the protein is secreted into the blood. Affected individuals have sufficiently high levels of S-α_1-AT to protect them against neutrophil elastase and so are not at higher risk for developing emphysema.

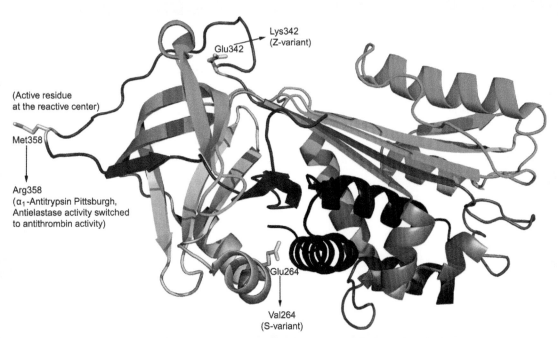

FIGURE 6.12   Structure of α_1-antitrypsin. Arrows indicate three amino acid substitution sites found in α_1-antitrypsin variants.

(Continued)

CLINICAL CASE STUDY 6.1 (Continued)

3. The reactive site of α_1-AT is methionine residue located at the 358 position. The sequence around the reactive center of several proteinase inhibitors, including α_1-AT, antithrombin III, and ovalbumin (a protein without known inhibitory function), shares extensive homologies and may all have evolved from a common ancestral protease inhibitor more than 500 million years ago. A fatal bleeding disorder due to substitution of a methionine for an arginine residue at position 358 of α_1-AT has been reported. This particular alteration brought a change in the specificity of the inhibitor. This altered protein (known as α_1-antitrypsin Pittsburgh) exhibits antithrombin activity while losing antielastase activity. Antithrombin, which has an important regulatory role in hemostasis (Chapter 34), has a reactive arginine at its active center. Since the α_1-AT and antithrombin III are structurally similar, the substitution of methionine of α_1-AT for arginine at the active center alters its specificity. This bleeding disorder is further complicated by the fact that the concentration of normal antithrombin III does not increase in response to inflammatory stress, as does the level of α_1-AT. Therefore, with each hemolytic episode, the abnormal α_1-AT increases, creating a vicious cycle resulting in uncontrolled bleeding (see case studies listed in reference 2).

4. It should be noted that in P_i^* Null/Null genotypes, α_1-AT is not synthesized, and therefore the risk of liver disease is nil, but the risk of COPD at an early age is very high.

5. The subject has requested his physician for a stem cell transplantation to achieve a permanent cure (see discussion on stem cells). How should he be advised? Your thoughts?

See Chapters 10, 23, and 34 for further enrichment.

References for Case Studies and Enrichment
[1] E.K. Silverman, R.A. Sandhaus, Alpha1-antitrypsin deficiency, N. Engl. J. Med. 360 (2009) 2749−2757.
[2] B. Hua, L. Fan, Y. Liang, Y. Zhao, E.G.D. Tuddenham, α_1-Antitrypsin Pittsburgh in a family with bleeding tendency, Haematologica 94 (2009) 881−884.

ENRICHMENT READING

[1] A.R. Fersht, Structure and Mechanism in Protein Science, W.H. Freeman and Company, New York, 1998.

[2] A. Cornish-Bowden, Principles of Enzyme Kinetics, first ed., Butterworths, London, 2003.

[3] L. Michaelis, M.L. Menten, K.A. Johnson, R.S. Goody, The original Michaelis constant: translation of the 1913 Michaelis−Menten paper, Biochemistry 50(39) (2011) 8264−8269.

[4] J.P. Richard, Enzymatic rate enhancements: a review and perspective, Biochemistry 52(12) (2013) 2009−2011.

[5] T.L. Amyes, J.P. Richard, Specificity in transition state binding: the Pauling model revisited, Biochemistry 52(12) (2013) 2021−2035.

[6] S.J. Benkovic, G.G. Hammes, S. Hammes-Schiffer, Free-energy landscape of enzyme catalysis, Biochemistry 47(11) (2008) 3317−3321.

[7] G.G. Hammes, S.J. Benkovic, S. Hammes-Schiffer, Flexibility, diversity, and cooperativity: pillars of enzyme catalysis, Biochemistry 50(48) (2011) 10422−10430.

[8] J. Monod, J. Wyman, J.P. Changeux, On the nature of allosteric transitions: a plausible model, J. Mol. Biol. 12 (1965) 88−118.

[9] D.E. Koshland Jr., G. Nemethy, D. Filmer, Comparison of experimental binding data and theoretical models in proteins containing subunits, Biochemistry 5(1) (1966) 365−385.

[10] A. Cornish-Bowden, Enzyme kinetics from a metabolic perspective, Biochem. Soc. Trans. 27(2) (1999) 281−284.

[11] A.C. Storer, A. Cornish-Bowden, The kinetics of coupled enzyme reactions. Applications to the assay of glucokinase, with glucose 6-phosphate dehydrogenase as coupling enzyme, Biochem. J. 141 (1) (1974) 205−209.

[12] J.P. Klinman, Importance of protein dynamics during enzymatic C-H bond cleavage catalysis, Biochemistry 52(12) (2013) 2068−2077.

Chapter 7

Clinical Enzymology and Biomarkers of Tissue Injury

Key Points

1. Circulating levels of enzymes, nonenzyme proteins, and peptides are used as markers of tissue injury. Their levels depend on the rate of synthesis regulated by transcriptional and translational control, removal, and degradation, and volume of distribution with respect to intracellular and extracellular space.

2. The enzymes in circulation either can be plasma-specific, meaning that they are present in plasma to perform a normal function (e.g., thrombin and plasmin), or non-plasma-specific, meaning that their concentrations are low or undetectable under normal conditions but undergo elevations due to tissue injury and damage.

3. Since many enzymes have isoenzyme forms, which have relative tissue specificity, the measurement of these isoenzyme levels facilitates the diagnosis of a specific tissue injury.

4. Factors that affect the plasma or serum enzyme activities include rate of cell turnover (e.g., infection and cancer), increased synthesis due to normal growth and repair, stimulation of synthesis or degradation by drugs (e.g., ethanol and antiepileptic agents), and the clearance rate of enzymes in circulation. Plasma levels of small molecular weight enzymes (e.g., amylase) cleared by glomerular filtration can be affected by renal failure.

5. The time course of appearance and disappearance of enzymes depend on molecular size; intracellular location (cytosol mitochondria, lysosomes, or other organelles); the gradient of enzyme levels between plasma and the cellular content; the degree, magnitude, and nature of cell injury; and half-life (e.g., AST is shorter than ALT, CK is shorter than LD).

6. The concentration of an enzyme in a biological fluid is determined by measuring either its mass by immunological techniques or its enzyme activity.

7. Measurement of serum markers in the diagnosis and prognosis of disorders, tissue dysfunction, and/or damage is a vital component of clinical management. Examples include myocardium (heart), liver, and pancreas.

8. Myocardial markers can be either myocardial infarction (MI) or heart failure due to increased volume load. MI markers are cardiac troponins, and the heart failure marker is B-natriuretic peptide (BNP).

9. BNP produced by the stretched cardiac myocytes and as an endocrine response by the heart initiates natriuresis and inhibition of the renin–aldosterone–angiotensin system to compensate for excess blood volume load.

10. Enzymes are used in the quantification of other enzymes, as well as many other metabolites and drugs, by coupled reactions to chemical tags, which are easily measurable.

11. Enzymes are used as therapeutic agents in the treatment of diseases. Both the use of recombinant DNA technology to produce human enzymes and polyethylene-coupled enzymes with properties of diminished immunogenicity and prolonged half-life have increased the therapeutic benefits of enzymes.

DIAGNOSIS AND PROGNOSIS OF DISEASE

Enzyme assays employed in the diagnosis of diseases are among the most frequently used clinical laboratory procedures. The most commonly used body fluid for this purpose is serum, the fluid that appears after the blood has clotted. The liquid portion of unclotted blood is called **plasma**. Serum is used for many enzyme assays because the preparation of plasma requires the addition of anticoagulants (e.g., chelating agents) that interfere with some assays, and elimination of fibrinogen, which can form aggregates that interfere with activity measurements. Enzymes in circulating plasma are either **plasma-specific** or **non-plasma-specific**. Plasma-specific enzymes normally present in plasma perform their primary functions in blood. Examples are those enzymes and their precursors involved in hemostasis (blood clotting) (e.g., thrombin), fibrinolysis (e.g., plasmin), and complement activation, as well as cholinesterase and ceruloplasmin. These enzymes are synthesized mainly in the liver and released into the circulation at a rate that maintains optimal steady-state concentrations.

Non-plasma-specific enzymes are intracellular enzymes normally present in plasma at minimal levels or at concentrations well below those in tissue cells. Their presence in plasma is normally due to turnover of tissue cells, but they

N. V. Bhagavan and C.-E. Ha: Essentials of Medical Biochemistry, Second edition. DOI: http://dx.doi.org/10.1016/B978-0-12-416687-5.00007-5

are released into the body fluids in excessive concentrations as a result of cellular damage or impairment of membrane function. Tissue injury and impairment of membrane function can be caused by diminished oxygen supply (e.g., myocardial infarction), infection (e.g., hepatitis), and toxic chemicals. Proliferation of cells, with consequent increased turnover, can also raise levels in plasma of enzymes characteristic of those cells (e.g., elevation of serum acid phosphatase in prostatic carcinoma). Intracellular enzymes are essentially confined to their cells of origin. A few are enzymes that are secreted by some selected tissue (salivary gland, gastric mucosa, or pancreas) into the gastrointestinal tract, where they participate in digestion of food constituents (Chapter 11). Plasma levels of secretory enzymes increase when their cells of origin undergo damage or membrane impairment, or when the usual pathways of enzyme secretion are obstructed. For example, large amounts of pancreatic amylase and triacylglycerol lipase (commonly known as lipase) enter the blood circulation in patients suffering from pancreatitis. These enzymes can digest the pancreas itself and surrounding adipose tissue in a process known as enzymatic necrosis (death of tissue cells).

The diagnosis of organ disease is aided by the measurement of a number of enzymes characteristic of that tissue or organ. Most tissues have characteristic enzyme patterns that may be reflected in the changes in serum concentrations of these enzymes in disease. The diseased tissue can be further identified by determination of the isoenzyme pattern of one of these enzymes (e.g., lactate dehydrogenase, creatine kinase) in the serum, since many tissues have characteristic isoenzyme distribution patterns for a given enzyme. For example, creatine kinase (CK) is a dimer composed of two subunits, M (for muscle) and B (for brain), that occur in three isoenzyme forms, BB(CK1), MB(CK2), and MM(CK3). CK catalyzes the reversible phosphorylation of creatine with adenosine triphosphate (ATP) as the phosphate donor:

$$\text{Creatine} + \text{ATP} \xleftrightarrow{\text{CK}} \text{Phosphocreatine} + \text{ADP}$$

This reaction provides ATP for muscle contraction (Chapter 19). Skeletal muscle contains predominantly CK3, whereas heart muscle (myocardium) contains CK3 and CK2. Serum normally contains a small amount of CK3 derived predominantly from skeletal muscle. Detection of CK2 in serum (in an appropriate clinical setting) is strongly suggestive of myocardial damage. Since an abnormal isoenzyme level may occur with apparently normal total activity of the enzyme, determination of the isoenzyme concentrations is essential in diagnostic enzymology (Table 7.1). In the diagnosis of myocardial injury, CKMB has been replaced with more specific and reliable cardiac-specific troponins I and T (discussed later).

Factors Affecting Presence and Removal of Intracellular Enzymes from Plasma

Many factors are taken into consideration in the clinical interpretation of serum enzyme levels. Membrane permeability changes and cell destruction affect the release of intracellular enzymes. These changes can result from a decrease in intracellular ATP concentration due to any of the following conditions: deficiency of one or more of the enzymes needed in ATP synthesis (e.g., pyruvate kinase in red blood cells); glucose deprivation; localized hypoxia; and high extracellular K^+ (ATP depletion results in decreased activity of the Na^+, K^+-ATPase in the cell plasma membrane required to maintain the proper K^+/Na^+ ratio between the intracellular and extracellular environments). Localized hypoxia can result from poor blood flow, the result of obstruction of blood vessels responsible for the territorial distribution of the blood (a condition known as **ischemia**). Ischemia may result from narrowing of the lumen of the blood vessels (e.g., deposition of lipids in the vessel wall, which leads to atherosclerosis; see Chapter 18) or from formation of blood clots within the vessels. Ischemic necrosis leads to infarction (death of surrounding tissue). When cells of the myocardium die as a result of severe ischemia, the lesion is known as a myocardial infarct.

The amounts of enzymes released depend on the degree of cellular damage, the intracellular concentrations of the enzymes, and the mass of affected tissue. The subcellular source of the enzymes released reflects the severity and the nature of the damage. Mild inflammatory conditions are likely to release cytoplasmic enzymes, whereas necrotic conditions also yield mitochondrial enzymes. Thus, in severe liver damage (e.g., hepatitis), the serum aspartate aminotransferase (AST) level is extremely high (much greater than that of alanine aminotransferase) because the mitochondrial isoenzyme of AST is released in addition to the corresponding cytoplasmic isoenzyme.

The amount of enzyme released into the plasma from an injured tissue is usually much greater than can be accounted for on the basis of tissue enzyme concentration and the magnitude of injury. Loss of enzymes from cells may stimulate further synthesis of enzymes. Many drugs cause an increase in drug-metabolizing enzymes as a result of an increase in *de novo* synthesis (i.e., enzyme induction). These drug-metabolizing enzymes, located in the smooth endoplasmic reticulum (microsomal fraction; see Chapter 1) of liver and other tissues, catalyze the following chemical reactions: hydroxylation; demethylation; de-ethylation; acetylation; epoxidation; deamination; glucuronidation; and dehalogenation. Thus, serum levels of some of these enzymes may be elevated following exposure to enzyme-inducing agents, such as ethanol,

TABLE 7.1 Tissues of Origin of Some Diagnostically Important Serum Enzymes

Enzyme	Principal Tissue Source
Acid phosphatase	Prostate
Alanine aminotransferase (glutamate pyruvate transaminase)	Liver
Alcohol dehydrogenase	Liver
Alkaline phosphatase	Bone, intestinal mucosa, hepatobiliary system, placenta, kidney
Amylase*	Pancreas, salivary glands
Arginase	Liver
Aspartate aminotransferase (glutamate oxaloacetate transaminase)	Heart and skeletal muscle, liver kidney, brain
Ceruloplasmin	Liver
Cholinesterase	Liver
Chymotrypsin(-ogen)*	Pancreas
Creatine kinase	Skeletal and heart muscle, brain
Fructose-bisphosphate aldolase	Skeletal and heart muscle
γ-Glutamyl transferase	Kidney, hepatobiliary system, prostate, pancreas
Glutamate dehydrogenase	Liver
Isocitrate dehydrogenase	Liver
Lactate dehydrogenase	Skeletal and heart muscle, liver, kidney, erythrocytes, pancreas, lungs
Leucine aminopeptidase	Hepatobiliary system, intestine, pancreas, kidney
Ornithine carbamoyl-transferase	Liver
Pepsin(-ogen)*	Gastric mucosa
Prostatic specific antigen (a serine proteinase)	Prostate
Sorbitol dehydrogenase	Liver
Triacylglycerol lipase* (lipase)	Pancreas
Trypsin(-ogen)*	Pancreas

Secretory enzymes.

barbiturates, phenytoin, and polycyclic hydrocarbons. While plasma levels of enzymes can become elevated because of tissue injury, the levels may drop (in spite of continued progress of the injury) to normal (or below normal) levels when the blood circulation is altered and limited, or when the functional part of the tissue is replaced by repair or nonfunctional tissue (e.g., connective tissue, as in extensive fibrosis of the liver in the disease known as cirrhosis).

Inactivation or removal of plasma enzymes may be accomplished by several processes: separation from its natural substrate or coenzyme; presence of enzyme inhibitors (e.g., falsely decreased activity of amylase in acute pancreatitis with hyperlipidemia); removal by the reticuloendothelial system; digestion by circulating proteinases; uptake by tissues and subsequent degradation by tissue proteinases; and clearance by the kidneys of enzymes of low molecular mass (amylase and lysozyme).

A schematic representation of the causes of the appearance and disappearance of non-plasma-specific enzymes is shown in Figure 7.1. Since enzymes differ in their rates of disappearance from plasma, it is important to know when the blood specimen was obtained relative to the time of injury. It is also important to know how soon after the occurrence of injury various enzyme levels begin to rise. The biological half-lives for enzymes and their various isoenzymes are different.

The use of appropriate normal ranges (also known as **reference intervals**) is important in evaluating abnormal levels of plasma enzymes. However, an abnormal

Presence

Removal

Normal turnover of tissues →

Leakage through cell membranes →

Tissue necrosis →

Increased enzyme synthesis →

Non-plasma-specific enzymes

→ Intracellular inactivation (dilution, lack of substrates and coenzymes, inhibitors, and proteinases)

→ Uptake by tissues, with subsequent inactivation

→ Removal by the reticuloendothelial system

→ Excretion in urine (of low molecular weight enzymes)

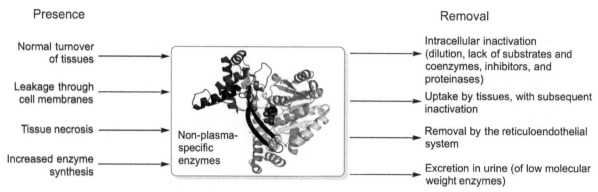

FIGURE 7.1 Factors affecting the presence and removal of non-plasma-specific enzymes.

isoenzyme pattern may occur despite normal total activity (see previous description). The standard unit for enzyme activity was discussed in Chapter 6. The normal range is affected by a variety of factors: age; sex; race; degree of obesity; pregnancy; alcohol or other drug consumption; and malnutrition. Drugs can alter enzyme levels *in vivo* and interfere with their measurement *in vitro*.

Enzyme activities may also be measured in urine, cerebrospinal fluid, bone marrow cells or fluid, amniotic cells or fluid, red blood cells, leukocytes, and tissue cells. Cytochemical localization is possible in leukocytes and biopsy specimens (e.g., from liver and muscle). Under ideal conditions, both the concentration of mass of the enzyme and its activity would be measured. Radioimmunoassay (RIA) and its alternative modes such as fluorescence immunoassay (FIA), fluorescence polarization immunoassay (FPIA), and chemiluminescence immunoassay, can be used to measure enzyme concentration, as well as other clinically important parameters.

SERUM AND PLASMA MARKERS IN THE DIAGNOSIS OF TISSUE DAMAGE

Myocardium

Laboratory tests for cardiovascular diseases are multifactorial. These tests of biomarkers include the assessment of inflammation (e.g., C-reactive protein), oxidative stress (e.g., oxidized low-density lipoproteins, myeloperoxidase), neurohormones (e.g., norepinephrine, renin, aldosterone), myocyte stress and function (B-natriuretic peptide), myocardial ischemia (e.g., ischemia-modified albumin), and myocyte injury (e.g., cardiac troponins I and T). In this section, tests for acute cardiac tissue damage are discussed and tests for other clinical conditions are discussed in Chapters 18 and 37. See also Clinical Case Study 7.1 with four vignettes.

Plasma B-Natriuretic Peptide in the Diagnosis of Acute Heart Failure also Known as Decompensated Heart Failure or Heart Failure

Three natriuretic peptides (NPs) are known. They are ANP, BNP, and CNP [1−5]. These peptides possess a common 17-amino acid ring structure formed by a disulfide linkage. Both ANP and BNP are synthesized by cardiac myocytes, and CNP is produced by the endothelial cells. Circulating ANP is primarily from atria, and BNP is from ventricles. Both ANP and BNP are released from cardiac myocytes when there is enhanced stretching due to disorders of decreased cardiac output, which results in tension of the myocardial walls (heart failure). The increased volume of blood is caused by the failure of the heart in pumping blood from the ventricles. Both sympathetic and the renin−angiotensin−aldosterone systems (Chapter 30) are activated as an immediate response to heart failure. However, activation of these systems is counterproductive, since they promote increase in salt/water retention adding to the volume load, systemic vasoconstriction, and compensatory enhanced myocardial contractility. The NPs counteract and modulate the compensatory mechanisms by promoting natriuresis; they do so by preventing renal tubular Na^+ reabsorption, increasing glomerular filtration rate, and inhibiting renin−angiotensin−aldosterone and sympathetic nervous systems. These effects of NPs cause a reduction in hemodynamic overload. Thus, the heart functions as **an endocrine organ when it is volume-stressed due to overload, with consequent production of NPs to ameliorate the deleterious effects**. Serum BNP level is used in the diagnosis and monitoring of heart failure. It is a 32-amino acid physiologically active fragment released by cardiac myocytes. An inactive larger fragment known as N-terminal (NT)-BNP has longer plasma half-life (20 minutes vs. 60−120 minutes) and is also released from stretched cardiac myocytes. NT-BNP is also used in

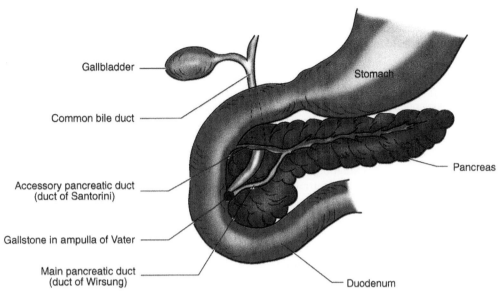

Gallbladder

Stomach

Common bile duct

Pancreas

Accessory pancreatic duct
(duct of Santorini)

Gallstone in ampulla of Vater

Main pancreatic duct
(duct of Wirsung)

Duodenum

FIGURE 7.2 Pancreatic duct obstruction by a gallstone at the ampulla of Vater. Obstruction can lead to pancreatitis, from induction of bile reflux, which eventually damages the acinar cells of the pancreas.

the clinical assessment of heart failure. The test has a negative predictive value of 99%. Since subjects with heart failure and those with pulmonary diseases (e.g., emphysema) both have symptoms of labored breathing (dyspnea), serum BNP levels are used in the differential diagnosis between the two.

The signal transduction pathways of BNP at the target sites are mediated by binding of BNP and activating membrane-bound guanylate cyclase. The activated guanylate cyclase catalyzes the conversion of GTP to cyclic GMP (cGMP) in the cytosol. The increased levels of cGMP promote biochemical pathways and ion channels that counteract cardiac volume overload. Because of the favorable actions of BNP, a synthetic recombinant BNP (nesiritide [Natrecor®]) has been used in the therapy of heart failure. It is administered intravenously.

Markers of Myocardial Infarction

Coronary artery occlusion causes heart tissue damage due to ischemia and can lead to myocardial infarction (MI) [6–11]. The immediate and common cause of artery obstruction is the formation of a thrombus or a ruptured plaque. **Antithrombolytic therapy (Chapter 34), with streptokinase or recombinant tissue plasminogen activator, protects the myocardium from permanent damage by restoring blood flow**. An early diagnosis of acute MI (AMI) is crucial for proper management. A patient's history, presence of chest pain, and electrocardiograms are problematic in the diagnosis of AMI. Therefore, measurements of circulatory proteins released from the necrotic myocardial tissue are useful in the diagnosis of AMI. The definition of MI has been elaborated in a

consensus document published in the *Journal of the American College of Cardiology* [6]. See also Clinical Case Study 7.1 with four vignettes.

Characteristics of an ideal myocardial injury marker are cardiac specificity, rapid appearance in the serum, substantial elevation for a clinically useful period of time, and ease and rapidity of the analytical assay. At present, no serum marker fulfills all of these criteria. Some cardiac markers that appear in plasma are myoglobin, LDH, CK, and cardiac troponins that have replaced all the other markers. Troponins TnI and TnT appear in the plasma within 3 hours after onset of chest pain. Troponins consist of three different proteins I, C, and T, and are expressed in both cardiac and skeletal muscle. The tripartite troponin complex regulates the calcium-dependent interaction of myosin with actin (Chapter 19). Troponins are encoded by different genes. Cardiac troponin I and T isoforms have unique structural differences from their skeletal muscle counterparts. The introduction of high-sensitive assays cardiac troponins with serial timed measurements has made significant advances in the early treatment of MI. Many nonischemic clinical conditions (e.g., heart failure, pulmonary embolism, chronic kidney disease) can give rise to elevations of cardiac troponins [7].

Pancreas

The pancreas is both an exocrine and an endocrine gland [12–15]. The exocrine function is the digestion of food substances (Chapter 11); the endocrine function involves glucose homeostasis (Chapter 20). Acute pancreatitis, which is characterized by epigastric pain, is an inflammatory process and potentially fatal. Obstruction of the pancreatic duct (Figure 7.2), which delivers pancreatic juice to the small intestine, by gallstones (Chapter 11), and alcohol abuse, are the most common causes of acute pancreatitis, representing 80% of all causes. The pathophysiology is due to inappropriate release of pancreatic enzymes and their premature activation. The principal pancreatic enzyme is trypsinogen, which, after activation to trypsin, converts many other enzymes to their active forms. Some of these are kallikrein, phospholipase A2, elastase, enzymes of blood coagulation and fibrinolysis, and complement. The effects of these abnormal processes are autodigestion of the pancreas, vasodilation, increased capillary permeability, and disseminated intravascular coagulation. These can result in circulatory collapse, renal insufficiency, and respiratory failure.

Laboratory diagnosis of acute pancreatitis involves the measurement of the pancreatic digestive enzymes, amylase and lipase (Chapter 11). Elevated serum amylase level is a sensitive diagnostic indicator in the assessment of acute pancreatitis, but it has low specificity because there are many nonpancreatic causes of hyperamylasemia. Furthermore, amylase (M.W. 55,000) is rapidly cleared by the kidneys and returns to normal levels by the third or fourth day after the onset of abdominal pain. Amylase activity in the serum appears within 2−12 hours after the onset of pain. Serum lipase is also used to assess pancreatic disorders and has a higher specificity than serum amylase. It appears in the plasma within 4−8 hours, peaks at about 24 hours, and remains elevated for 8−14 days. Measurements of amylase and lipase provide 90%−95% accuracy in the diagnosis of acute pancreatitis with abdominal pain (see Clinical Case Study 7.2).

Since exocrine cells of the pancreas contain many enzymes, attempts have been made to use markers other than amylase and lipase to diagnose acute pancreatitis. One such enzyme is trypsinogen. It is a 25 kDa protein that exists in two isoenzyme forms: trypsinogen-1 (cationic) and trypsinogen-2 (anionic). Both forms are readily filtered through the kidney glomeruli. However, the tubular reabsorption of trypsinogen-2 is less than that for trypsinogen-1; a dipstick method has been developed to detect trypsinogen-2 in the urine of patients suspected of having acute pancreatitis. The test strip contains monoclonal antibodies specific for trypsinogen-2. In the neonatal screening programs for cystic fibrosis, trypsinogen is measured in dried blood spots. In cystic fibrosis, pancreatic ducts are blocked, due to a membrane chloride ion channel defect, leading to the appearance of the trypsinogen in the blood (Chapter 11).

Liver

The liver is the largest glandular organ; its parenchymal cells are called **hepatocytes** [16,17; see Chapter 11 for a discussion on structured organization of liver and Clinical Case Study 7.3]. The liver has numerous functions, including metabolism, detoxification, formation and excretion of bile, storage, and synthesis. Liver diseases include alcohol abuse, medication, chronic hepatitis B and C infections, steatosis and steatohepatitis, autoimmune hepatitis, hemochromatosis, Wilson's disease, α_1-antitrypsin deficiency, malignancy, and poisons and infectious agents. These disorders require specific laboratory testing procedures and are discussed in the appropriate places in the text. The serum enzymes used in assessment of liver function are divided into two categories: (1) markers used in hepatocellular necrosis and (2) markers that reflect cholestasis. Serum enzymes used as markers of cholestasis include alkaline phosphatase, 5′-nucleotidase, and γ-glutamyl transferase. Cholestasis can be due either to intrahepatic or extrahepatic origin. The latter was discussed under acute pancreatitis. Intrahepatic cholestasis can result from loss of function mutations in adenosine triphosphates-binding cassette family of protein (ABCB11) [17]. Alanine aminotransferase and aspartate aminotransferases are markers for hepatocellular necrosis. Other tests used in the assessment of liver disorders include measurement of bilirubin, albumin, and α-fetoprotein.

ENZYMES AS ANALYTICAL REAGENTS

The use of enzymes as analytical reagents in the clinical laboratory has found widespread application in the measurement of substrates, drugs, and activity of other enzymes. The advantages include specificity of the substrate that is being measured and direct measurement of the substrate in a complex mixture that avoids preliminary separation and purification steps such as serum.

ENZYMES AS THERAPEUTIC AGENTS

Enzymes have found a few applications as therapeutic agents. Some examples are transfusion of fresh blood in bleeding disorders, oral administration of digestive enzymes in digestive diseases (e.g., cystic fibrosis), administration of fibrinolytic enzymes (e.g., streptokinase) to recanalize blood vessels occluded by blood clots (thrombi) in thromboembolic disorders (e.g., pulmonary embolism, acute myocardial infarction), treatment of selected disorders of inborn errors of metabolism (e.g., Gaucher's disease), and cancer therapy (e.g., L-asparaginase in acute lymphocytic leukemia). For enzymes to be therapeutically useful, they should be derived from human sources to prevent immunological problems. Although enzymes derived from human blood are readily obtainable, enzymes derived from tissues, which would be particularly useful in the treatment of inborn errors of metabolism, are difficult to obtain in adequate quantities. Transport of specific enzymes to target tissues is also a problem, but some recent advances and commercial applications (e.g., propagation of human tissue culture cell lines, isolation and cloning of specific genes) have the potential of overcoming these difficulties. Such techniques have been used in the production of peptide hormones such as somatostatin and insulin, interferon, and tissue plasminogen activator. In treatment with enzymes or proteins, covalent attachment of an inert polymer polyethylene glycol (PEG) provides many therapeutic benefits. They include slowing clearance, diminished immunogenicity, prevention of degradation, and binding to antibody. PEG-enzyme therapy is used in the treatment of the immunodeficiency disease caused by adenosine deaminase deficiency (Chapter 25), and PEG-interferon alfa complex (peginterferon alfa-2a) is used in treatment of chronic hepatitis C infection. Ultimately, however, when a gene is cloned, techniques will be developed to incorporate it into the genome of persons lacking the gene or having a mutated gene.

CLINICAL CASE STUDY 7.1 Biomarkers of Myocardial Injury

Vignette 1: Elevated Serum C-Reactive Protein in Ischemic Acute Myocardial Infarction

This clinical case was abstracted from: P.M. Ridker, Myocardial infarction in a 72-year-old woman with low LDL-C and increased hsCRP: implications for statin therapy, Clin. Chem. 55 (2009) 369–375.

Synopsis

A 72-year-old female was admitted to the cardiac unit of the local hospital with a diagnosis of acute myocardial infarction (AMI). Her coronary angiography studies showed significant stenosis, and she required angioplasty and placement of drug-eluting stents.

The patient's pertinent medical history included the following: systolic hypertension (treated with angiotensin converting enzyme inhibitor); life-long smoking habit; she was thin (BMI = 23 kg/m^2); and she had a family history of premature heart disease. Her serum lipid parameters measured 4 months prior to this AMI event were all considered optimal. They were total cholesterol = 216 mg/dL; LDL-C = 104 mg/dL; and HDL-C = 82 mg/dL. However, she had elevated serum levels of high-sensitivity C-reactive protein (hsCRP), measured at 7.7 mg/L and then 7.4 mg/L 2 weeks later. Serum hsCRP values greater than 3 mg/dL are considered higher relative risk categories for predicting abnormal vascular events of AMI and stroke (as compared with <1 mg/dL for low risk, and 1–3 mg/dL for moderate risk).

Subsequent to her treatment at the hospital, the patient was placed on pharmacotherapy consisting of a statin (an HMG-CoA reductase inhibitor used to lower cholesterol levels), aspirin (an antiplatelet agent used to prevent clotting), and clopidogrel (another antiplatelet agent used to prevent clotting). Thirty days after the initiation of the therapy, her serum LDL-C had decreased to 78 mg/dL, and her hsCRP had decreased to 4.8 mg/dL. She was entered into a clinical trial and a cardiac rehabilitation program.

Teaching Points

1. While the preferred biochemical markers for detecting myocardial injury are serum cardiac troponins T and I due to their high sensitivity and specificity, hsCRP is a biomarker of inflammation that has been shown to predict cardiovascular events such as AMI and stroke. Inflammation plays a key role in the pathogenesis of atherosclerosis, which can ultimately cause abnormal cardiovascular events. hsCRP is an acute-phase reactant released from the liver into the serum in response to inflammation and tissue injury. hsCRP is also thought to be produced by the atherosclerotic lesion itself, which actually further worsens cell dysfunction and atherogenesis. Thus, elevated measurements of serum hsCRP are indicative of a vascular event, as hsCRP is not only an indicator but also a contributor to atherosclerosis. Patients can decrease their risk of AMI and stroke with reduction of hsCRP through diet, exercise, smoking cessation, and statin therapy.

2. Abnormal lipid levels can increase a patient's risk for ischemic cardiac injury, but this case demonstrates that AMI can occur even in the absence of dyslipidemia. Although this patient did not have many traditional risk factors for coronary heart disease, she did have a life-long history of smoking and a family history of AMI. Additionally, her serum hsCRP values were elevated, which placed her in the high-risk category for the occurrence of abnormal vascular events.

Integration with Chapter 18, "Lipids III: Plasma Lipoproteins"

Vignette 2: Elevated Plasma Cardiac Troponin T in Nonischemic Acute Myocardial Injury

This case was abstracted from: J.A. de Lemos, Increasingly sensitive assays for cardiac troponins, JAMA 309, 21 (2013) 2262–2269.

Synopsis

A 62-year-old man presented to the emergency department with shortness of breath and atypical chest pain, which started after a 3-hour plane flight. His past medical history included gastroesophageal reflux disease (GERD), diabetes, hypertension, stage 2 chronic kidney disease, paroxysmal atrial flutter, and chronic heart failure. Coronary angiography done 3 years previously was reported to be normal. However, based on EKG findings and elevated cTnT of 20 ng/L, a diagnosis of non-ST-segment elevation myocardial infarction was made.

He was given aspirin (an antiplatelet agent used to prevent clotting of the blood), clopidogrel (another antiplatelet agent used for additional prevention of clotting), enoxaparin (a low molecular weight heparin used for anticoagulation), simvastatin (an HMG CoA reductase inhibitor used to lower

(Continued)

CLINICAL CASE STUDY 7.1 (Continued)

cholesterol levels), lisinopril (an angiotensin-converting enzyme inhibitor used for lowering high blood pressure and for cardioprotection), and carvedilol (a nonselective β-adrenergic inhibitor used in the treatment of heart failure). This patient was unable to undergo cardiac catheterization due to elevated INR (international normalized ratio) and increased risk for bleeding, but serial measurements of cTnT unexpectedly remained unchanged at 20 ng/L over the next 2 days, warranting risk stratification. A less invasive approach with a vasodilator perfusion scan revealed left ventricular ejection fraction of 32% and evidence of nonischemic cardiomyopathy, a potential cause for his elevated troponin levels. This patient had multiple potential nonischemic etiologies for his elevated troponin levels (acute and chronic heart failure, structural heart disease, chronic kidney disease); therefore, proper management and treatment would have required each of these possible underlying causes to be addressed.

Teaching Points

1. Cardiac troponins are highly sensitive and specific for myocardial injury, but they are not specific for any particular mechanism of injury. Serum cardiac troponins can be elevated in ischemic and nonischemic cardiac injuries. As illustrated in this case, there are various causes of elevated serum cardiac troponins, which are outlined in reference 2.
2. Patients with nonischemic myocardial injury, such as heart failure, renal failure, sepsis, and pulmonary embolism, may have a worse prognosis in the setting of elevated cardiac troponin levels. In patients with suspected ischemic injury, even slightly elevated troponin levels indicate increased risk for death and recurrent ischemic events. Thus, regardless of mechanism, elevated troponins have serious clinical implications.
3. The emergence of highly sensitive assays for cTnT and cTnI, which can detect troponin levels well below the threshold for MI but at the expense of decreased specificity, poses challenges in interpreting slightly elevated troponins and deciding how to proceed with management. As discussed in reference 2, there are multiple causes for elevated troponins besides MI; thus, detection of troponins with a poorly specific test used in the emergency setting could lead to further unnecessary testing, misdiagnoses, inappropriate treatment, and delays in treating the underlying cause of elevated troponins. The utility of highly sensitive assays in the ambulatory setting may be helpful for identifying and monitoring outpatients with stable chronic heart failure or coronary artery disease, as there is a dose-dependent relationship between cardiac troponins and risk for acute heart failure and death.

Vignette 3: Elevated Plasma Cardiac Troponin I in the Absence of Myocardial Injury
This case was abstracted from: L.A. Legendre-Bazydlo, D.M. Haverstick, J.L. Kennedy, J.M. Dent, D.E. Bruns, Persistent increase of cardiac troponin I in plasma without evidence of cardiac injury, Clin. Chem. 56 (2010) 702−707.

Synopsis
A 69-year-old male with a history of several risk factors for acute coronary syndrome (ACS, e.g., dyslipidemia, diabetes type 2, hypertension, and ischemic strokes) was brought to the emergency department twice over an interval of 3 months, with complaints of difficulty balancing. During both admissions, his serial cTnI values were elevated, despite intervention with cardiac catheterization and stenting. A comprehensive investigation later revealed the cTnI elevation to be a false positive result for ACS, and was instead due to a formation of a cTNI-IgG complex. Further cardiac evaluation was discontinued as there was no clinical evidence for AMI.

Teaching Points

1. Although measurement of cardiac troponins plays a central role in the diagnosis of AMI, elevations of cTnI can also occur in the absence of AMI, as illustrated in this case. This patient had several risk factors for ischemic myocardial injury, but his persistently elevated troponins did not match the clinical presentation required for the diagnosis of AMI, raising suspicion for other etiologies of elevated troponins.
2. The cTNI-IgG complex is not very uncommon, and in some individuals it clears slowly from the circulation, resulting in increased concentrations in the serum, without any injury to the myocardium. Therefore, while elevated troponins are highly sensitive for cardiac injury, in a patient with troponin elevations not consistent with the clinical presentation for AMI like chest pain, other etiologies should be considered.

Vignette 4: Ischemic Acute Myocardial Infarction in a Patient without Traditional Risk Factors for Ischemia
This case was abstracted from: S. Leng, B.K. Nallamothu, S. Saint, L.J. Appleman, G.M. Bump, Clinical problem-solving. Simple and complex, N. Engl. J. Med. 368 (2013) 65−71.

Synopsis
A 44-year-old man with no atherosclerotic risk factors presented to the emergency department with chest pain. EKG revealed ST-segment elevation myocardial infarction. Initial measurement of troponin level was normal but peaked later to 127 ng/mL. He was given aspirin (an antiplatelet agent used to prevent clotting), nitroglycerin (a vasodilator used as needed for relief of angina chest pain), heparin (used for anticoagulation), atorvastatin (an HMG CoA reductase inhibitor), and clopidogrel (another antiplatelet used for additional prevention of clotting). He underwent percutaneous coronary intervention and placement of two drug-eluting stents. Further investigation of the cause of AMI in this relatively young and healthy patient, with no ACS risk factors or family history, revealed thrombocytosis (increased platelets), a gene mutation of JAK2 V617F, and bone marrow biopsy findings consistent with a myeloproliferative neoplasm. After 6 months of treatment with hydroxyurea (an antineoplastic

(Continued)

CLINICAL CASE STUDY 7.1 (Continued)

agent that decreases the production of deoxyribonucleotides [Chapter 25]), the patient no longer had chest pain, his cell counts normalized, and his cardiac function returned to normal.

Teaching Points

1. Even without the traditional risk factors for ischemia, this patient suffered an AMI due to his high platelet count and resulting hypercoagulable state. Hypercoagulability can be caused by primary myelofibrosis (as in this case), polycythemia vera, essential thrombocythemia, and chronic myelogenous leukemia.

2. The JAK2 V617F is a gain of function Janus tyrosine kinase 2 [Chapter 28] mutation that results in increased proliferation and survival of proliferative cells stimulated by erythropoietin and thrombopoietin. This mutation is present not only in primary myelofibrosis, but also in polycythemia vera and essential thrombocytopenia.

3. Cardiac troponins should be interpreted with caution to time elapsed since onset of symptoms. The expected rise and fall of cardiac troponins in AMI is best appreciated by serial troponin measurements. In this case, troponin was initially normal because it was too early for detection, but later peaked to 127 ng/mL, indicating the presence of myocardial injury. An isolated measurement of troponin is only 70%–80% sensitive, so serial measurements are needed to rule out AMI.

Integration with Chapter 34, "Biochemistry of Hemostasis"

Case References

[1] P.M. Ridker, Myocardial infarction in a 72-year-old woman with low LDL-C and increased hsCRP: implications for statin therapy, Clin. Chem. 55 (2009) 369–375.

[2] J.A. de Lemos, Increasingly sensitive assays for cardiac troponins, JAMA 309(21) (2013) 2262–2269.

[3] L.A. Legendre-Bazydlo, D.M. Haverstick, J.L. Kennedy, J.M. Dent, D.E. Bruns, Persistent increase of cardiac troponin I in plasma without evidence of cardiac injury, Clin. Chem. 56 (2010) 702–707.

[4] S. Leng, B.K. Nallamothu, S. Saint, L.J. Appleman, G.M. Bump, Clinical problem-solving. Simple and complex, N. Engl. J. Med. 368 (2013) 65–71.

Additional Case References

[5] S. Devaraj, U. Singh, I. Jialal, The evolving role of C-reactive protein in atherothrombosis, Clin. Chem. 55(2) (2009) 229–238.

[6] A.D. Hingorani, T. Shah, J.P. Casas, S.E. Humphries, P.J. Talmud, C-reactive protein and coronary heart disease: predictive test or therapeutic target, Clin. Chem. 55(2) (2009) 239–255.

Additional Clinical Case Studies

[7] S.V. Arnold, M.W. Rich, Hyperlipidemia in older adults, Clin. Geriatr. December (2009) 18–24.

[8] N.J. Leeper, L.S. Wener, G. Dhaliwal, S. Saint, R.M. Wachter, One surprise after another, N. Engl. J. Med. 352 (2005) 1474–1479.

Supplemental References for Enrichment

[9] T. Reichlin, W. Hochholzer, S. Bassetti, S. Stever, C. Stelzig, S. Hartwiger, et al., Early diagnosis of myocardial infarction with sensitive cardiac troponin assays, N. Engl. J. Med. 361 (2009) 858–867.

[10] C. Laine, D. Goldmann, L.A. Kopin, T.A. Pearson, In the clinic: dyslipidemia, Ann. Intern. Med. September (2007)ITC9-1-16

[11] D.A. Morrow, Clinical application of sensitive troponin assays, N. Engl. J. Med. 361 (2009) 913–915.

[12] P.A. Kavsak, A. Worster, J.J. You, M. Oremus, A. Elsharif, S.A. Hill, et al., Identification of myocardial injury in the emergency setting, Clin. Biochem. 43 (2010) 539–544.

[13] E. Braunwald, Biomarkers in heart failure, N. Engl. J. Med. 358 (2008) 2148–2159.

[14] J.H. O'Keefe, M.D. Carter, C.J. Lavie, Primary and secondary prevention of cardiovascular diseases, a practical evidence-based approach, Mayo Clin. Proc. 84 (2009) 741–757.

[15] A. Kumar, C.P. Cannon, Acute coronary syndromes, diagnosis and management. Part II, Mayo Clin. Proc. 84 (2009) 1021–1036.

CLINICAL CASE STUDY 7.2 Biomarkers of Pancreatic Injury

Synopsis

A 65-year-old obese male (BMI = 32) developed severe epigastric pain with episodes of vomiting 3 hours after dinner. The subject was admitted to the emergency department of the local hospital. After several clinical studies including laboratory biochemical parameters and imaging studies, a diagnosis of acute pancreatitis due to gallstones was made. The serum enzyme levels were abnormal: amylase 1050 U/L (reference interval 25–125 U/L); lipase 880 U/L (reference interval 18–180 U/L); alanine aminotransferase 250 U/L (reference interval 7–40 U/L); and lactate dehydrogenase 420 U/L (reference interval 110–210 U/L). The patient was admitted to the CCU, and the following therapeutic measures were performed. Pain was managed by use of narcotic analgesics, the fluid and electrolyte balance and nutritional support were maintained by intravenous measures, and a sphincterotomy by endoscopic

retrograde cholangio-pancreatography (ERCP) was performed, followed by cholecystectomy 4 days later. The patient was placed on oral ursodiol treatment (ursodeoxycholic acid, a naturally occurring bile acid that increases the pool of bile acids and prevents the formation of cholesterol-containing gallstones) for prevention of gallstone formation, and he was referred to a clinical nutritionist for weight reduction.

Teaching Points

1. Acute pancreatitis (AP), if not diagnosed and managed promptly, has severe medical consequences with a high mortality rate even after treatment. The diagnosis consists of two of the three parameters: abdominal pain, serum amylase or lipase level three times the upper limit of the reference interval, and characteristic findings of AP on a computed tomography imaging scan (see reference 2). Either

(Continued)

CLINICAL CASE STUDY 7.2 (Continued)

serum amylase or lipase measurement is sufficient for the diagnosis of AP. Lipase elevation remains longer than amylase. Both tests have low specificity and high sensitivity. Thus, a normal serum amylase level generally excludes the diagnosis of AP, while an elevated serum amylase level can result from many clinical conditions (see reference 3). The extent of elevation of serum lipase and amylase in AP does not correlate with the severity or outcome of the disease, although some studies suggest that urinary trypsinogen-2 is a novel prognostic marker (see reference 9).

2. Amylase is mainly synthesized in the pancreas and secreted into the digestive system to break down starch and glycogen, whereas lipase normally breaks down fats. Refer to Chapters 11 and 17 to learn more about the role of the pancreas and the metabolism of lipids. In AP, amylase and lipase are released from the pancreas. The pathogenesis of AP is initiated by inappropriate and premature conversion of trypsinogen to trypsin. Trypsin promotes self-destruction of pancreatic tissue, followed by seepage of pancreatic enzymes (including amylase and lipase) into circulation, a systemic inflammatory response, and various deleterious clinical manifestations requiring prompt therapeutic interventions.

3. Gallstones (cholesterol-containing insoluble material) and alcohol abuse are the most common causes of AP.

4. Chronic pancreatitis with recurrent abdominal pain can result from structural and nonstructural abnormalities with normal serum parameters (see reference 5).

Integration with Chapters 11, "Gastrointestinal Digestion and Absorption," 15, "Protein and Amino Acid Metabolism," and 17, "Lipids II: Phospholipids, Glycosphingolipids, and Cholesterol".

References

[1] D.C. Whitcomb, Acute pancreatitis, N. Engl. J. Med. 354 (2006) 2142–2150.
[2] P.A. Banks, M.I. Freeman, Practice Parameters Committee of the American College of Gastroenterology, Practice guidelines in acute pancreatitis, Am. J. Gastroenterol. 101 (2006) 2379–2400.
[3] D. Yadav, N. Agarwal, C.S. Pitchumoni, A critical evaluation of laboratory tests in acute pancreatitis, Am. J. Gastroenterol. 97 (2006) 1309–1318.
[4] R.M.S. Mitchell, M.F. Byrne, J. Baillie, Pancreatitis, Lancet 361 (2003) 1447–1455.
[5] M.P. Callery, S.D. Freedman, A 21-year-old man with chronic pancreatitis, JAMA 299(13) (2008) 1588–1594.
[6] J.M. Beauregard, J.A. Lyon, C. Slovis, Using the literature to evaluate diagnostic tests: amylase or lipase for diagnosing acute pancreatitis, J. Med. Libr. Assoc. 95(2) (2007) 121–126.
[7] B. Yegneswaran, C.S. Pitchumoni, When should serum amylase and lipase levels be repeated in a patient with acute pancreatitis, Clev. Clin. J. Med. 77(4) (2010) 230–231.

Supplemental References for Enrichment

[8] T. Baron, Managing severe acute pancreatitis, Clev. Clin. J. Med. 80 (6) (2013) 354–359.
[9] M. Lempinen, M.L. Kylanpaa-Back, U.H. Stenman, P. Puolakkainen, R. Haapiainen, P. Finne, et al., Predicting the severity of acute pancreatitis by rapid measurement of trypsinogen-2 in urine, Clin. Chem. 47(12) (2001) 2103–2107.
[10] J. Oiva, O. Itkonen, R. Koistinen, K. Hotakainen, W.M. Zhang, E. Kemppainen, et al., Specific immunoassay reveals increased serum trypsinogen 3 in acute pancreatitis, Clin. Chem. 57(11) (2011) 1506–1513.

CLINICAL CASE STUDY 7.3 Biomarkers of Hepatic Injury

Synopsis

A 52-year-old male was brought to the emergency department with a history of chronic alcoholism and complaints of severe headache, gastrointestinal distress, and jaundice. His serum biochemical parameters revealed highly elevated transaminases, ALT and AST (aminotransferases), and bilirubin levels. On questioning, the patient revealed he had taken several acetaminophen (Tylenol, also known as paracetamol) tablets of 300 mg each for relief of his headache over the past 24 hours. His serum acetaminophen concentration was 200 mcg/mL. The patient was admitted to the hospital, and after intensive therapy with N-acetlcysteine, the patient's symptoms and laboratory findings normalized. He was discharged and referred to psychiatric care for the treatment of alcoholism.

Teaching Points

1. Alanine aminotransferase (ALT) and aspartate aminotransferase (AST) are biomarkers for the detection of hepatocellular injury. ALT is more specific than AST for liver damage. The degree of elevation of these aminotransferases can indicate the extent of disease and, in some cases, the

cause. For example, an AST:ALT ratio of 2:1 suggests alcoholic liver disease, whereas AST and ALT values as high as over 10 times normal can suggest acute viral hepatitis or drug toxicity, as can be caused by excessive acetaminophen intake. However, it is important to interpret liver enzyme levels along with the patient's clinical presentation and history, as normal values do not necessarily mean absence of disease. For example, patients with cirrhosis (an insidious disorder in which normal parenchymal liver cells are replaced with fibrous tissue) can have normal aminotransferases due to the lack of functioning tissue. However, these patients can have abnormalities in other liver function tests such as bilirubin, albumin, prothrombin time, and ammonia.

2. Acetaminophen, which is commonly used for its analgesic and antipyretic properties, can produce toxic effects with acute hepatic damage. Acetaminophen toxicity is the most common cause of acute liver failure in the United States. Acetaminophen-induced liver failure can occur particularly in patients with chronic alcoholism, malnutrition, and use of medications that induce cytochrome P-450 enzymes. In patients with cirrhosis who

(Continued)

CLINICAL CASE STUDY 7.3 (Continued)

require pain management, the current recommendation regarding acetaminophen is cautious use of 2 to 3 g/d, to avoid the risk of drug-induced hepatic injury in patients with preexisting liver disease.

3. Under normal circumstances, acetaminophen is mostly metabolized to nontoxic end products through glucuronidation and sulfation. A small fraction of the drug is converted to the extremely toxic electrophile N-acetyl-p-benzoquinone imine (NAPQI) by cytochrome P-4502E1. NAPQI is inactivated promptly by hepatocyte glutathione (gamma-glutamylcysteinylglycine). However, in situations with an activated cytochrome P-450 system, malnutrition, and pre-existing liver diseases, glutathione levels are low and NAPQI causes acute hepatic necrosis. N-acetylcysteine, the antidote, rapidly enters the hepatocytes and provides the substrate cysteine for glutathione synthesis, and rescues hepatic injury (refer to Chapters 7 and 15). See references 1 and 2 for more information.

4. N-acetylcysteine is also used as an inhalation therapy to convert thick mucus, which occurs in the airways of certain lung diseases (e.g., cystic fibrosis, emphysema) to less viscous fluid (Chapter 11).

References
[1] K.J. Heard, Acetylcysteine for acetaminophen poisoning, N. Engl. J. Med. 359 (2008) 285–292.

[2] A. Schilling, R. Corey, M. Leonard, B. Eghtesad, Acetaminophen: old drug, new warnings, Clev. Clin. J. Med 77 (2010) 19–27.

[3] N. Chandok, K.D.S. Watt, Pain management in the cirrhotic patient: the clinical challenge, Mayo Clin. Proc. 85(5) (2010) 451–458.

[4] G. Aragon, Z.M. Younossi, When and how to evaluate mildly elevated liver enzymes in apparently healthy patients, Clev. Clin. J. Med. 77(3) (2010) 195–204.

[5] D. Prati, E. Taioli, A. Zanella, E. Della Torre, S. Butelli, E. Del Vecchio, et al., Updated definitions of healthy ranges for serum alanine aminotransferase levels, Ann. Intern. Med. 137(1) (2002) 1–10.

[6] D.E. Johnston, Special considerations in interpreting liver function tests, Am. Fam. Physician 59(8) (1999) 2223–2230.

ENRICHMENT READING

[1] M.F. McGrath, M.L. Kuroski de Bold, A.J. de Bold, The endocrine function of the heart, Trends Endocrinol. Metabol. 16(10) (2005) 469–477.

[2] J.P. Goetze, B-type natriuretic peptide: from posttranslational processing to clinical measurement, Clin. Chem. 58(1) (2012) 83–91.

[3] L.R. Potter, Guanylyl cyclase structure, function and regulation, Cell. Signal. 23 (2011) 1921–1928.

[4] L. Badimon, J.C. Romero, J. Cubedo, M. Borrell-Pages, Circulating biomarkers, Thromb. Res. 130 (2012) S12–S15.

[5] H.D. Shah, R.K. Goyal, Nesiritide: a recombinant human BNP as a therapy for decompensated heart failure, Indian J. Pharmacol. 37(3) (2005) 196–197.

[6] K. Thygesen, J.S. Alpert, A.S. Jaffe, M.L. Simoons, B.R. Chaitman, H.D. White, et al., Third universal definition of myocardial infarction, J. Am. Coll. Cardiol. 60(16) (2012) 1581–1598.

[7] L.K. Newby, R.L. Jesse, J.D. Babb, R.H. Christenson, T.M. De Fer, G.A. Diamond, et al., ACCF 2012 expert consensus document on practical clinical considerations in the interpretation of troponin elevations: a report of the American College of Cardiology Foundation task force on Clinical Expert Consensus Documents, J. Am. Coll. Cardiol. 60(23) (2012) 2427–2463.

[8] T. Keller, T. Zeller, D. Peetz, S. Tzikas, A. Roth, E. Czyz, et al., Sensitive troponin I assay in early diagnosis of acute myocardial infarction, N. Engl. J. Med. 361 (2009) 868–877.

[9] T. Reichlin, W. Hochholzer, S. Bassetti, S. Steuer, C. Stelzig, S. Hartwiger, et al., Early diagnosis of myocardial infarction with sensitive cardiac troponin assays, N. Engl. J. Med. 361 (2009) 858–867.

[10] K. Thygesen, J. Mair, E. Giannitsis, C. Mueller, B. Lindahl, S. Blankenberg, et al., How to use high-sensitivity cardiac troponins in acute cardiac care, Eur. Heart J. 33(18) (2012) 2252–2257.

[11] K. Kanjwal, N. Imran, B. Grubb, Y. Janjwal, Troponin elevation in patients with various tachycardias and normal epicardial coronaries, Indian Pacing Electrophysiol. J. 8(3) (2008) 172–174.

[12] J. Smotkin, S. Tenner, Laboratory diagnostic tests in acute pancreatitis, J. Clin. Gastroenterol. 34(4) (2002) 459–462.

[13] O. Itkonen, Human trypsinogens in the pancreas and in cancer, Scand. J. Clin. Lab. Invest. 70(2) (2010) 136–143.

[14] G. Lippi, M. Valentino, G. Cervellin, Laboratory diagnosis of acute pancreatitis: in search of the Holy Grail, Crit. Rev. Clin. Lab. Sci. 49(1) (2012) 18–31.

[15] T. Baron, Managing severe acute pancreatitis, Cleve. Clin. J. Med. 80(6) (2013) 354–359.

[16] G. Aragon, Z.M. Younossi, When and how to evaluate mildly elevated liver enzymes in apparently healthy patients, Cleve. Clin. J. Med. 77(3) (2010) 195–204.

[17] S.S. Strautnieks, J.A. Byene, L. Pawlikowska, D. Cebecauerova, A. Rayner, L. Dutton, et al., Severe bile salt export pump deficiency: 82 different BCB11 mutations in 109 families, Gastroenterology 134 (2008) 1203–1214.

Chapter 8

Simple Carbohydrates

Key Points

1. Monosaccharides are polyhydroxy compounds that also contain carbonyl functional groups, namely an aldehyde or ketone. They are classified based on the number of carbon atoms they contain, e.g., trioses, tetroses, pentoses, hexoses, heptoses, and nanoses.

2. The structure of carbohydrates, with the exception of dihydroxyacetone, contains at least one asymmetrical or chiral carbon atom giving rise to stereoisomers. The number of stereoisomers or enantiomers for a given carbohydrate is determined by 2^n, where n is the number of chiral carbons.

3. D- and L- designation of a monosaccharide is based on the chiral carbon located farthest from the carbonyl (i.e., most oxidized carbon atom) functional group. In the Fischer projection, the D-monosaccharide has a hydroxyl group attached to the chiral carbon, on the right-hand side.

4. Aldoses and pentoses are the most abundant monosaccharides, and in humans, they are D-stereoisomers.

5. Stereoisomers that are not mirror images of each other are known as diastereoisomers; examples are D-ribose and D-arabinose. Diastereoisomers that differ by only a single chiral carbon atom are known as epimers; examples are D-glucose and D-galactose.

6. In aqueous solutions, stable forms of monosaccharides of five or more carbon atoms exist as cyclic structures, which are formed due to reversible chemical reactions between a hydroxyl group and carbonyl group. In the process of cyclization, the carbonyl carbon becomes a chiral atom and is known as an anomeric carbon atom. The resulting two diastereoisomers are known as anomers, designated as α- or β-structures.

7. In aqueous solutions, the α- and β-forms undergo interconversion and reach a stable ratio of anomers; this process is known as mutarotation.

8. The conformational structural formulae of ring structures of monosaccharides are accurate representations in aqueous solutions.

9. The major monosaccharides are glucose, galactose, and fructose.

10. The physiologically important monosaccharide derivatives include sugar alcohols, sugar acids, amino sugars, sugar phosphates, and deoxysugars.

11. Glycosides are formed when the hydroxyl group linked to an anomeric carbon atom condenses with the hydroxyl group of a second molecule with the elimination of water. This linkage is known as a glycosidic bond. Many glycosides serve as therapeutic agents.

12. Disaccharides contain monosaccharides linked in a glycosidic linkage. Nutritionally and physiologically important disaccharides include sucrose, maltose, and lactose. A therapeutically useful nonabsorbable disaccharide is lactulose, which is used in the management of hepatic encephalopathy due to ammonia toxicity to the central nervous system.

13. The chemosensory perception of the sweet taste of sucrose and other nonsucrose molecules is mediated by G-protein-coupled receptors located on the sensory cells of the tongue, ionotropic channels, and generation of electrical impulses. The nonsucrose synthetic sweeteners are varied in structure and have therapeutic applications in the management of diabetes and obesity.

14. Polysaccharides contain many monosaccharides linked in glycosidic linkages. If the monosaccharides are all the same, they are known as homopolysaccharides; and if they are different, they are known as heteropolysaccharides.

15. The three most important homopolysaccharides are glycogen, starch, and cellulose, and they contain glucose units. In both glycogen and starch, which is derived from plants, the glycosidic linkages are $\alpha(1 \rightarrow 4)$ and $\alpha(1 \rightarrow 6)$. In cellulose, a nondigestible carbohydrate in humans, the linkages between glucose units are $\beta(1 \rightarrow 4)$.

16. Cellulose and other indigestible polysaccharides and a nonpolysaccharide component, lignin, constitute dietary fiber and provide fecal mass. Dietary fiber is an essential component in the maintenance of optimal health and nutrition.

Carbohydrates are major functional constituents of living systems. They are the primary source of energy in animal cells; carbohydrates are synthesized in green plants from carbon dioxide, water, and solar energy. They provide the skeletal framework for tissues and organs of the human body, and serve as lubricants and support elements of

N. V. Bhagavan and C.-E. Ha: Essentials of Medical Biochemistry, Second edition. DOI: http://dx.doi.org/10.1016/B978-0-12-416687-5.00008-7

connective tissue. Major energy requirements of the human body are met by dietary carbohydrates. They confer biological specificity and provide recognition elements on cell membranes. In addition, they are components of nucleic acids and are found covalently linked with lipids and proteins.

CLASSIFICATION

Carbohydrates consist of polyhydroxyketone or polyhydroxyaldehyde compounds and their condensation products. The term *carbohydrate* literally means hydrate of carbon, a compound with an empirical formula $(CH_2O)_n$. This formula applies to many carbohydrates, such as glucose, which is $C_6H_{12}O_6$, or $(CH_2O)_6$. However, a large number of compounds are classified as carbohydrates even though they do not have this empirical formula; these compounds are derivatives of simple sugars (e.g., deoxyribose, $C_5H_{10}O_4$). Carbohydrates may be classified as monosaccharides, oligosaccharides, or polysaccharides; the term *saccharide* is derived from the Greek word for sugar. Monosaccharides are single polyhydroxyaldehyde (e.g., glucose) or polyhydroxyketone units (e.g., fructose), whereas oligosaccharides consist of 2–10 monosaccharide units joined together by glycosidic linkages. Sucrose and lactose are disaccharides, since they are each made up of two monosaccharide units. The names of common monosaccharides and disaccharides take the suffix "-ose" (e.g., glucose, sucrose). Polysaccharides, also known as **glycans**, are polymers that may contain many hundreds of monosaccharide units. They are further divided into homopolysaccharides and heteropolysaccharides. The former contain only a single type of polysaccharide unit (e.g., starch and cellulose, both of which are polymers of glucose), whereas the latter contain two or more different monosaccharide units.

Monosaccharides

Monosaccharides are identified by their carbonyl functional group (aldehyde or ketone) and by the number of carbon atoms they contain. The simplest monosaccharides are the two trioses: glyceraldehyde (an aldotriose) and dihydroxyacetone (a ketotriose). Four-, five-, six-, and seven-carbon-containing monosaccharides are called **tetroses**, **pentoses**, **hexoses**, and **heptoses**, respectively. All monosaccharides, with the exception of dihydroxyacetone, contain at least one asymmetrical or chiral carbon atom, and therefore two or more stereoisomers are possible for each monosaccharide depending on the number of asymmetrical (chiral) centers it contains. Glyceraldehyde, with one asymmetrical center, has two possible stereoisomers, designated D and L forms (Figure 8.1). A method for representing the D and L forms is the Fischer

projection formula, named for the Nobel Prize-winning German chemist, Emil Fischer. The D and L representations are used for all monosaccharides. The designation of D or L, given to a monosaccharide with two or more asymmetrical centers, is based on the configuration of the asymmetrical carbon atom located farthest from the carbonyl functional group. Thus, if the configuration at that carbon is the same as that of D-glyceraldehyde (with the hydroxyl group on the right-hand side), it belongs to the D series. A similar relationship exists between L-glyceraldehyde (with the hydroxyl group on the left-hand side) and the L series of monosaccharides. The optical rotation of a monosaccharide with multiple asymmetrical centers is the net result of contributions from the rotations of each optically active center. Thus, the prefix D or L provides no information with regard to optical rotation; it indicates only the configuration around the asymmetrical carbon atom located farthest from the carbonyl carbon.

The numbering system for monosaccharides depends on the location of the carbonyl carbon (or carbon atom in the most oxidized state), which is assigned the lowest possible number. For glucose (an aldohexose), carbon C1 bears the carbonyl group and the farthest asymmetrical carbon atom is C5 (the penultimate carbon), the configuration which determines the D and L series. For fructose (a ketohexose), C2 bears the carbonyl group, and C5 is the highest numbered asymmetrical carbon atom. Glucose and fructose have identical configurations around C3 to C6.

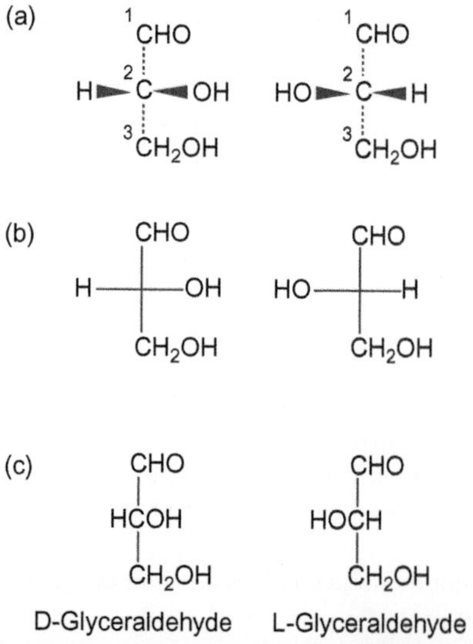

FIGURE 8.1 Stereoisomers of glyceraldehyde. (a) Perspective formulas showing tetrahedral arrangement of the chiral carbon 2 with four different substituents. (b) Projection formulas in which the horizontal substituents project forward and the vertical substituents project backward from the plane of the page. (c) A common method of representation.

The structure of glucose, written in straight-chain form (Figure 8.2), shows four asymmetrical centers. In general, the total number of possible isomers with a compound of n asymmetrical centers is 2^n. Thus, for aldohexoses having four asymmetrical centers, 16 isomers are possible, 8 of which are mirror images of the other 8 (enantiomers). These two groups constitute members of the D and L series of aldohexoses. Most of the physiologically important isomers belong to the D series, although a few L-isomers are also found. In later discussions, the designation of D and L is omitted, and it is assumed that a monosaccharide belongs to the D series unless it is specifically designated an L-isomer. Of the D series of aldohexoses, three are physiologically important: D-glucose, D-galactose, and D-mannose. Structurally, D-glucose and

D-galactose differ only in the configuration around C4; D-glucose and D-mannose differ only in the configuration around C2. Pairs of sugars (e.g., glucose and galactose, glucose and mannose) that differ only in the configuration around a single carbon are known as **epimers**. D-galactose and D-mannose are not epimers, since they differ in configurations around both C2 and C4. D-fructose, one of eight 2-ketohexoses, is the physiologically important ketohexose. Monosaccharides with five or more carbons occur predominantly in cyclic (ring) forms owing to a reaction between the carbonyl group (aldehyde or ketone) and an alcohol group:

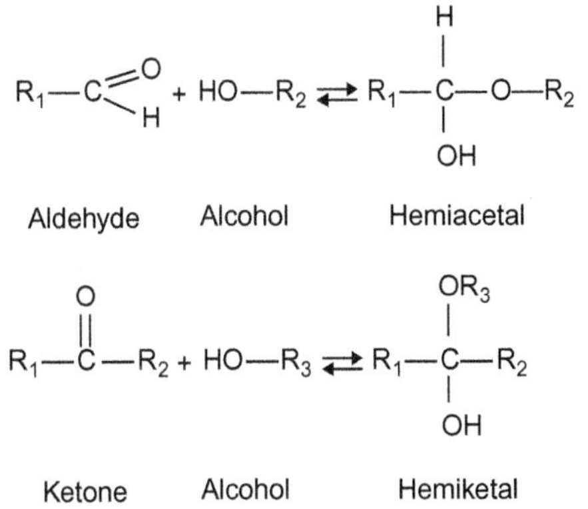

$$ (8.1) $$

Formation of the cyclic forms of monosaccharides is favored because these structures have lower energies than the straight-chain forms. Cyclic forms of D-glucose are formed by the hemiacetal linkage between the C1 aldehyde group and the C4 or C5 alcohol group. If the ring structure is formed between C1 and C4, the resulting five-membered ring structure is named D-glucofuranose because it resembles the compound furan:

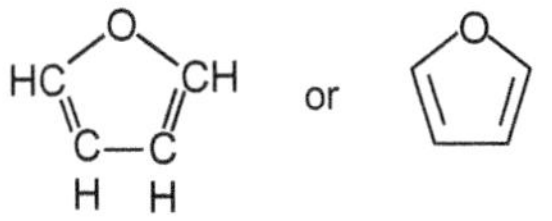

Furan structure

A linear representation of D-glucose

A modified representation of the D-glucose molecule, showing the formation of the hemiacetal linkage

If the ring structure is formed between C1 and C5, the resulting six-membered ring is named D-glucopyranose because it resembles the compound pyran:

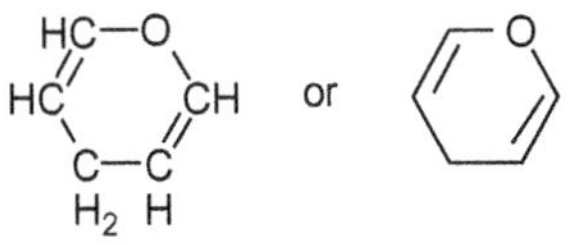

Pyran structure

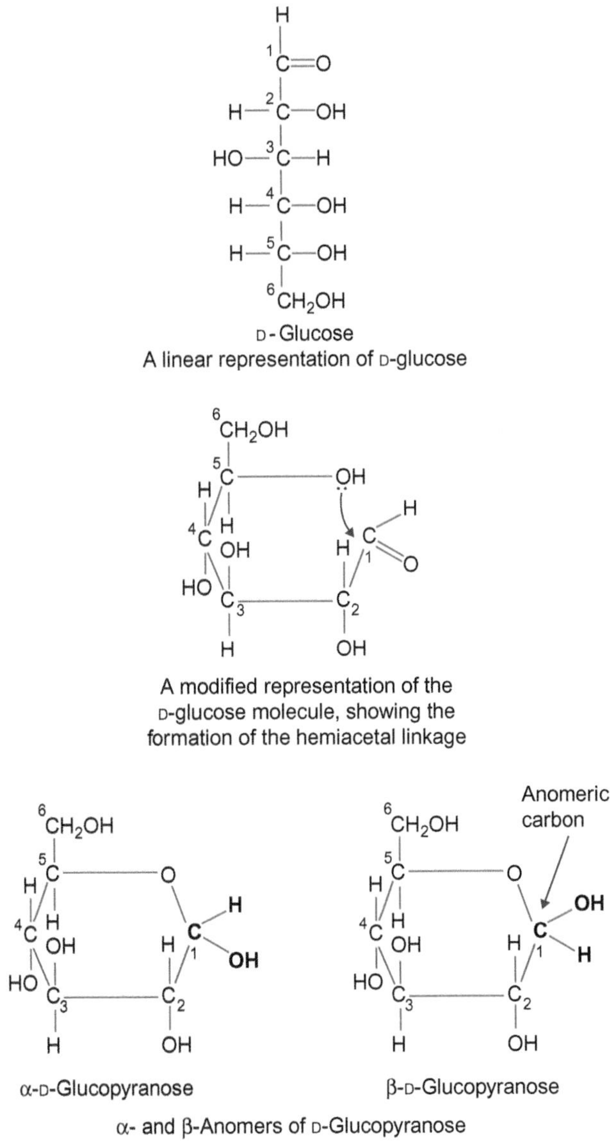

α-D-Glucopyranose   β-D-Glucopyranose

α- and β-Anomers of D-Glucopyranose

FIGURE 8.2 Formation of α- and β-anomers from D-glucose.

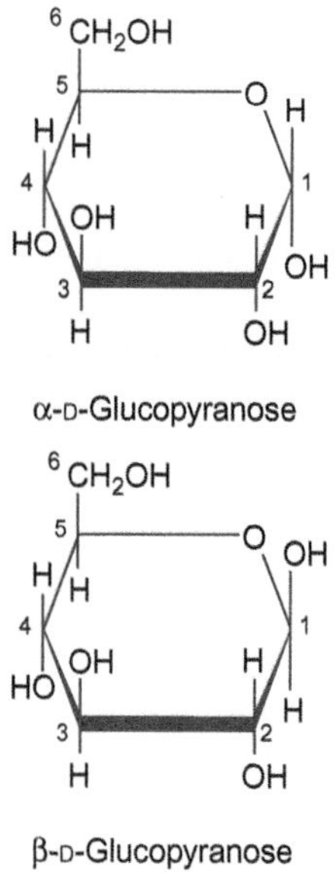

FIGURE 8.3 Haworth projection formulas of anomers of D-glucopyranose. The thick line of the structure projects out toward the observer and the upper edge (thin line) projects behind the plane of the paper.

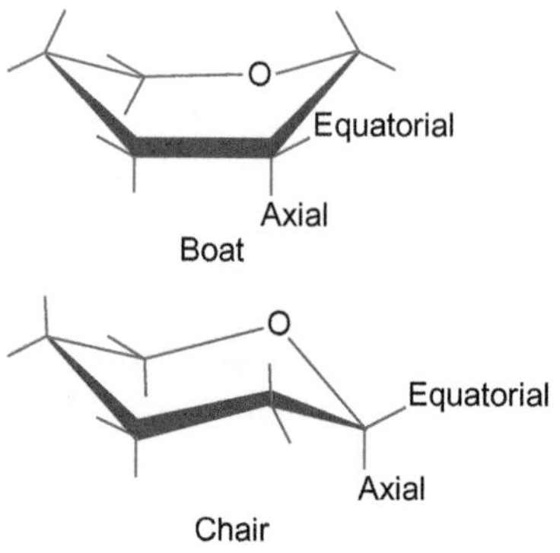

FIGURE 8.4 Conformational formulas for the boat and chair forms of pyranose.

Aldohexoses exist in solutions mainly as six-membered pyranose ring forms, since these forms are thermodynamically more stable than furanose ring forms.

Cyclization of a monosaccharide results in the formation of an additional asymmetrical center, known as the **anomeric carbon**, when the carbon of the carbonyl group reacts with the C5 hydroxyl group. The two possible stereoisomers resulting from the cyclization are called α- and β-**anomers** (Figure 8.3). Aldohexoses in their cyclic forms have five asymmetrical centers and, therefore, 32 stereoisomers. In other words, each of the 16 isomers that belong to the D or L series has two anomeric forms. The systematic names for these two anomers are α-D-glucopyranose and β-D-glucopyranose. Three-dimensional representations of ring structures are frequently shown as Haworth projection formulas (Figure 8.3), in which the lower edge of the ring is presented as a thick line, to indicate that this part of the structure projects out toward the observer, and the upper edge is shown as a thin line that projects behind the plane of the paper. Carbon atoms of the ring are not explicitly shown but occur at junctions of lines representing bonds. Sometimes the hydrogen atoms are also omitted and are presumed to exist wherever a bond line ends without a specified group.

The pyranose ring is not planar, being similar to that of cyclohexane. The conformational formulas are shown in Figure 8.4. Most pyranoses occur in the chair conformation.

Interconversion of α- and β-forms can be followed in a polarimeter by measuring the optical rotation of D-glucopyranose (D-glucose) in aqueous solutions.

Freshly prepared solutions of the α- and β-forms show specific rotations of +112.2° and +18.7°, respectively. However, over a period of a few hours at room temperature, the specific rotation of both forms in aqueous solution changes and attains a stable value of +52.7°. This change in optical rotation, known as **mutarotation**, is characteristic of sugars that form cyclic structures. It represents the interconversion of the α- and β-forms, yielding an equilibrium mixture consisting of about two-thirds β-form, one-third α-form, and a very small amount of the noncyclic form (Figure 8.5). Thus, the change in structure can occur in solution and attain equilibrium, which favors the formation of more stable (lowest energy) forms.

D-fructose, a ketohexose, can potentially form either a five-membered (furanose) or a six-membered (pyranose) ring involving formation of an internal hemiketal linkage between C2 (the anomeric carbon atom) and the C5 or C6 hydroxyl group, respectively. The hemiketal linkage introduces a new asymmetrical center at the C2 position. Thus, two anomeric forms of each of the fructofuranose and fructopyranose ring structures are possible (Figure 8.6). In aqueous solution at equilibrium, fructose is present predominantly in the β-fructopyranose form.

However, when fructose is linked with itself or with other sugars, or when it is phosphorylated, it assumes the furanose form. Fructose 1,6-bisphosphate is present in the β-fructofuranose form, with a 4:1 ratio of β- to α-anomeric forms.

Fructose is a major constituent (38%) of honey; the other constituents are glucose (31%), water (17%), maltose (a glucose disaccharide, 7%), sucrose (a glucose–fructose disaccharide, 1%), and polysaccharide (1%). The variability of these sugars in honey from different sources is quite large.

Sugars that contain a carbonyl functional group are reducing agents. Therefore, an oxidizing agent (e.g., ferricyanide, cupric ion, or hydrogen peroxide) becomes reduced with the simultaneous oxidation of the carbonyl group of the sugar. This property is exploited in the

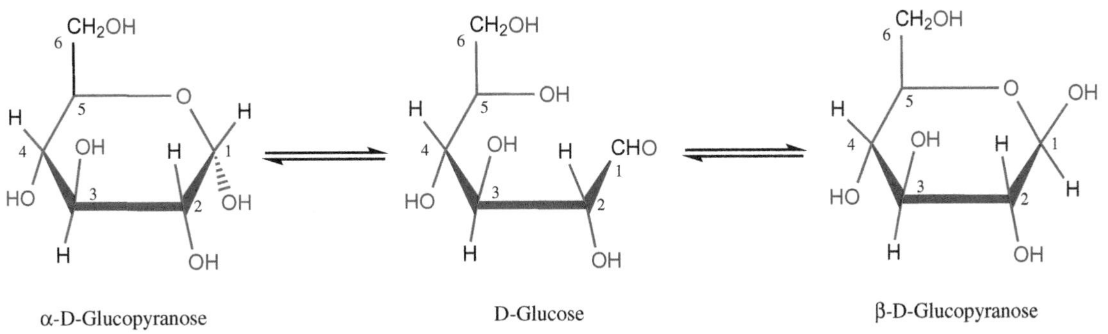

(a)

α-D-Glucose ⇌ Equilibrium mixture containing α-, β-, and open-chain forms ⇌ β-D-Glucose

Specific rotation $[\alpha]^{20}_D$:   +112.2°   +52.7°   +18.7°

$$\begin{bmatrix} \beta\text{-form:} & 63\% \\ \alpha\text{-form:} & 36\% \\ \text{open form:} & <1\% \end{bmatrix}$$

(b)

FIGURE 8.5   (a) Interconversion of the anomeric forms of D-glucopyranose via the intermediate open-chain form. (b) Mutarotation of D-glucose.

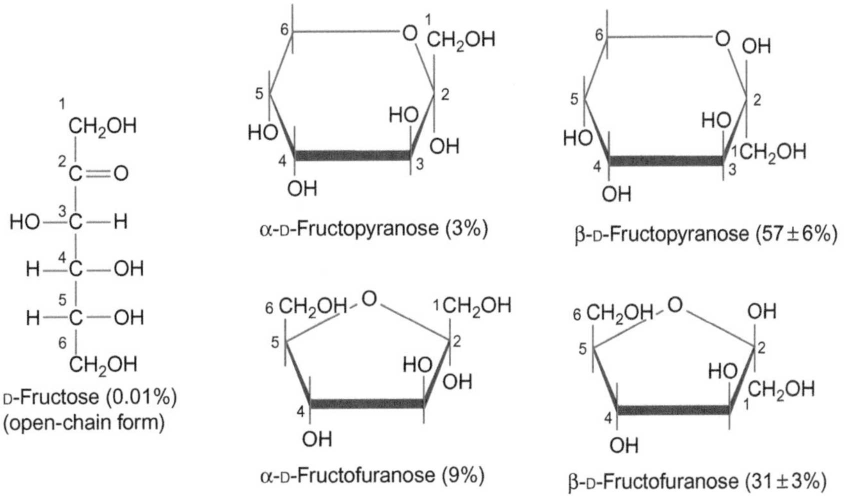

FIGURE 8.6   Structures of open-chain and Haworth projections of D-fructose and the approximate percentage of each at equilibrium in aqueous solution.

estimation of reducing sugars. The individual monosaccharides are better quantitated by specific procedures, such as an enzymatic procedure (e.g., for glucose, the glucose oxidase method).

A reaction frequently used in the determination of carbohydrate structure, and for its identification in tissue preparations, is the periodate reaction. Periodate (sodium periodate, $NaIO_4$) oxidatively cleaves carbon−carbon bonds bearing adjacent oxidizable functional groups, e.g., vicinal glycols, α-hydroxyaldehydes, and ketones. The vicinal glycols on periodate oxidation yield a dialdehyde; this reaction is quantitative.

Periodate cleavage of glycogen (a polyglucose) yields a polyaldehyde that can be coupled to a visible dye reaction. This is referred to as periodic acid-Schiff (PAS) staining. A tissue slice is treated with periodate, followed by staining with Schiff's reagent (basic fuchsin bleached with sulfurous acid). The polyaldehyde, if present, will combine with the NH_2 groups of the bleached dye to produce a magenta or purple complex. The specificity of the reaction can be confirmed by first treating another tissue slice with α-amylase (which breaks down the glycogen to smaller fragments that are removed by washing) and then staining it with PAS. Absence of color confirms that the

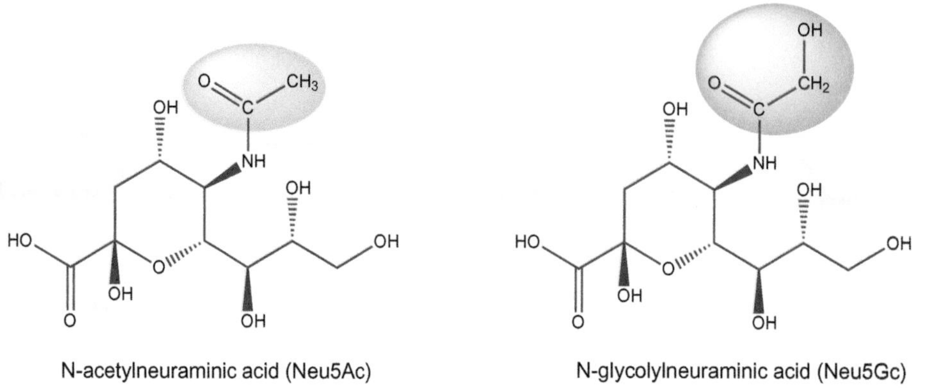

N-acetylneuraminic acid (Neu5Ac)   N-glycolylneuraminic acid (Neu5Gc)

FIGURE 8.7   Structures of the two most frequently occurring sialic acids found on the outer leaflet of mammalian plasma cell membranes.

material was glycogen, whereas the appearance of color suggests the presence of a nonglycogen carbohydrate, such as a glycoprotein.

Some physiologically important monosaccharide derivatives include sugar alcohols, sugar acids, amino sugars, sugar phosphates, deoxy sugars, and sugar glycosides. Their metabolic roles are discussed in the appropriate sections of the text.

Sialic Acids

Sialic acids are derivatives of a nine-carbon-containing monosaccharide, **ketonanose**, known as **neuraminic acid**. They are synthesized by the condensation of D-mannosamine with pyruvic acid. They typically occur as the terminating units of oligosaccharide side chains of some glycoproteins and glycolipids of mammalian cell membranes, as well as on secreted glycoproteins (Chapter 9). The two most common sialic acids found in mammals are **N-acetylneuraminic acid** and **N-glycolyl (hydroxyl acetic acid)neuraminic acid** (Figure 8.7). In humans, due to deletion of the gene for the hydroxylase enzyme, N-glycolylneuraminic acid (**Neu5Gc**) is absent. However, Neu5Gc is found in foods derived from poultry, fish, red meat, and milk products, and has been shown to be metabolically incorporated into cell surface membranes of humans. Recent studies have revealed that the incorporation of this nonhuman glycan provides high-affinity receptors for a cytotoxic protein secreted by Shiga toxigenic *Escherichia coli*. The gastrointestinal disease (see Chapter 11 for mechanism) and hemolytic uremic syndrome caused by the infection of this toxigenic *E. coli* are initiated by Neu5Gc-containing receptors present on human gut epithelia and kidney vasculature.

Disaccharides

Structures of some disaccharides are illustrated in Figure 8.8. Maltose is composed of two glucose residues

joined by an α-glycosidic linkage between C1 of one residue and C4 of the other residue [designated α(1→4)]. In maltose, the second sugar residue has an unsubstituted anomeric carbon atom and therefore can function as a reducing agent, as well as exhibit mutarotation. In trehalose, two glucose residues are joined by an α-linkage through both anomeric carbon atoms; therefore, the disaccharide is not a reducing sugar, nor does it exhibit mutarotation. Lactose, synthesized only by secretory cells of the mammary gland during lactation, is a disaccharide consisting of galactose and glucose. The glycosidic bond is a β-linkage between C1 of galactose and C4 of glucose (Figure 8.8). Lactose is a reducing sugar and exhibits mutarotation by virtue of the anomeric C1 of the glucose residue. Lactulose is a synthetic disaccharide consisting of galactose and fructose linked through a β-linkage between C1 of galactose and C4 of fructose. It is used in the treatment of some forms of chronic liver disease (such as hepatic encephalopathy) in which the ammonia content in the blood is elevated (hyperammonemia). Normally, ammonia produced in the gastrointestinal tract, principally in the colon by microbial action, is transported to the liver via the portal circulation and inactivated by conversion to urea (Chapter 15). Oral administration of lactulose relieves hyperammonemia by microfloral conversion in the colon to a variety of organic acids (e.g., lactic acid) that acidify the colonic contents. Lactulose is neither broken down nor absorbed in the small intestine. Reduction of the colonic luminal pH favors conversion of ammonia (NH_3) to the ammonium ion (NH_4^+), which is not easily absorbed, and thus its absorption is decreased. Reduction of luminal pH may additionally promote a microflora that causes a decrease in the production of ammonia, as well as an increase in its utilization. The osmotic activity of the disaccharide and its metabolites causes an osmotic diarrhea, which is useful in eliminating toxic waste products.

Another synthetic nonabsorbable disaccharide used in the treatment of hepatic encephalopathy is lactitol

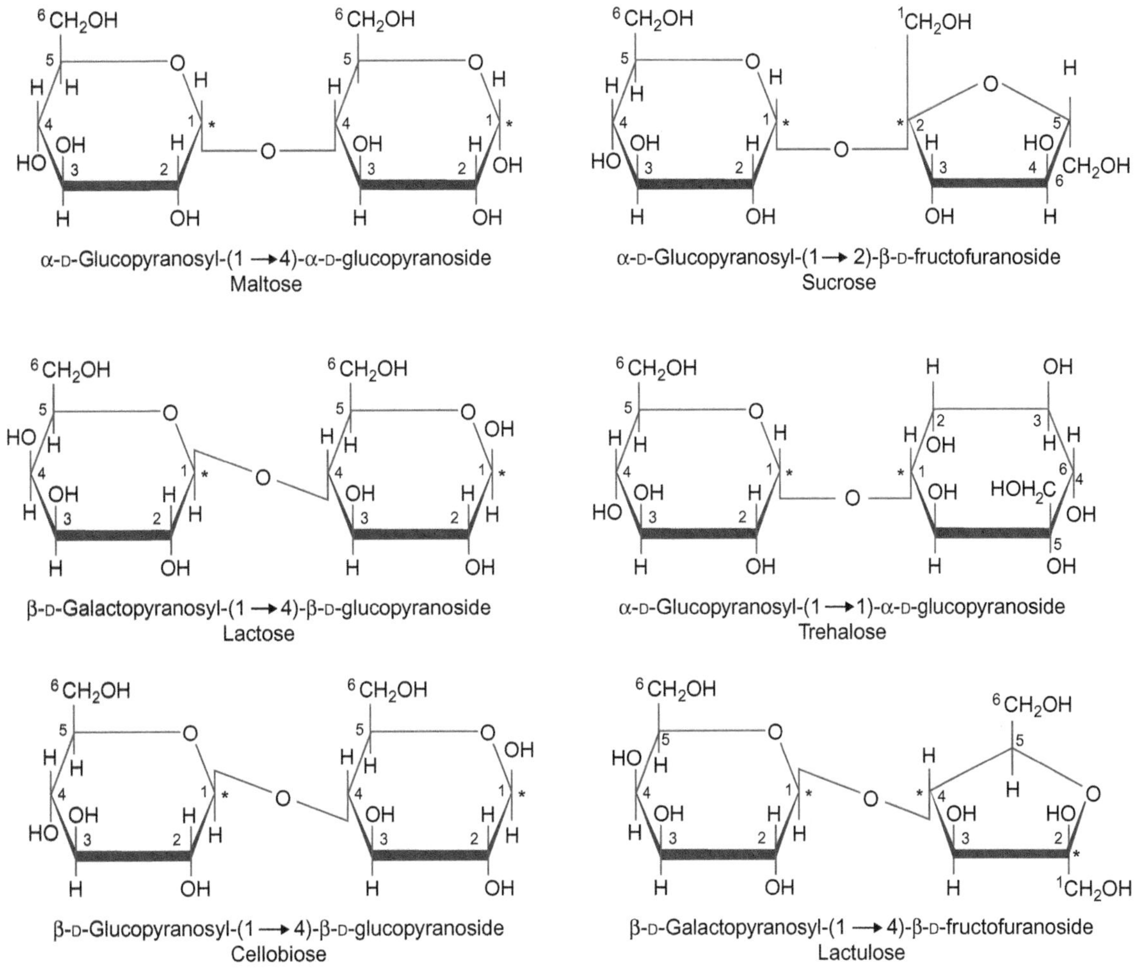

FIGURE 8.8 Structures of some disaccharides. *Anomeric carbon atoms. All structures are Haworth projections.

(β-galactosidosorbitol). Compared to lactulose, lactitol has the advantage of higher palatability and fewer side effects (e.g., flatulence). Ammonia production in the colonic lumen by urease-producing bacteria can be reduced by administering antibiotics such as neomycin or metronidazole. The therapeutic effect of the combined use of a nonabsorbable disaccharide and an antibiotic may result from the metabolism of the disaccharide by antibiotic-resistant bacteria.

Sucrose, a widely occurring disaccharide found in many plants (cane sugar and beet sugar), consists of glucose and fructose moieties linked together through C1 of glucose and C2 of fructose. Sucrose is not a reducing sugar and does not mutarotate. Because of its sweet taste, sucrose is consumed in large amounts. The perception of sweetness is mediated by taste buds submerged in the tongue and oral mucous membranes. The taste bud, a pear-like organ, consists of sensory cells (taste cells) interwoven with a branching network of nerve fibers. The taste bud contains two additional cell types: basal and supporting cells. Sensory cells have a short lifespan of

about 10 days, and new cells are derived from basal cells that continually undergo mitosis. Sensory cells contain microvilli (thin hair-like projections on the surface of the cells). The microvilli protrude through the pores of the taste buds to provide a receptor surface for the perception of taste. Highly soluble and diffusible substances, such as salt (NaCl) and sugar, enter the taste pores and produce taste sensations. The chemical stimuli received by the sensory cells are transduced into electrical impulses. These impulses, in turn, are passed on to the nerve fibers through neurotransmitters. Less soluble and diffusible compounds, such as starch and protein, produce correspondingly less taste sensation.

The five primary taste sensations, perceived by the taste-receptor cells (TRC), are sweet, salty, bitter, sour, and umami. In humans, the umami taste sensation is characterized by delicious flavor and evoked by two amino acids, monosodium glutamate (MSG) and aspartate. The chemosensory perceptions of taste received by TRC are transduced into electrical impulses, which are ultimately recognized in the brain. At the molecular level the taste perceptions are mediated by ionotropic channels and G-protein-coupled receptors (GPCRs; Chapter 28). Three families of GPCRs, namely T1R1, T1R2, and T1R3, either as homodimeric or heterodimeric receptor complexes, are receptors for sweet and umami taste ligands. The long amino-terminal extracellular domains or GPCRs, which look like a cauliflower, provide sites for ligand(s) recognition and binding. For example, the heterodimeric sweet taste receptor complex of T1R2 and T1R3 provides receptors for natural, artificial, and protein sweet tasting compounds. The umami sensing and umami taste are potentiated by purine nucleotide IMP and GMP.

In contrast to sweet and umami, the bitter taste receptors consist of many divergent GPCRs (T2Rs), encoded by a large family of genes. It is thought that the evolution of bitter taste recognition evolved to prevent ingestion of toxic compounds, regardless of their structure.

The second messenger downstream signaling pathways for both T1Rs and T2Rs consist of G-protein-mediated activation of phospholipase C. This process leads to the generation of two intracellular messengers: inositol-1,4,5-trisphosphate (IP_3) and diacylglycerol (DAG) (Chapter 28).

Both salty and sour taste perceptions are mediated by specialized ion-membrane Na^+ and H^+ channels located on the apical cell surface of TRCs. Each taste bud possesses different degrees of sensitivity for all four qualities, but it usually has greatest sensitivity to one or two. The integration of taste perception occurs in the cerebral cortex, which receives nerve signals arising from the taste buds that pass through the medulla and the thalamus. An important function of taste perception is to provide reflex stimuli that regulate the output of saliva. A pleasant taste perception increases saliva production, whereas an unpleasant taste reduces output. Taste also affects the overall digestive process by affecting gastric contractions, pancreatic flow, and intestinal motility.

Choice of food and dietary habits are influenced by taste and smell, which are interrelated. The sense of smell resides in receptors of specialized bipolar neuronal cells (olfactory cells) located on each side of the upper region of the nasal cavity. Like taste sensory cells, the receptors of the olfactory cells, with cilia protruding into the mucus covering the epithelium, also undergo continuous renewal but with a longer turnover rate of about 30 days. The receptors of olfactory cells are stimulated by volatile airborne compounds. Since perceptions of taste and smell are triggered by chemicals; they are called **chemosensory perceptions**. At the molecular level, they are mediated by ionotropic channels and G-protein-coupled receptors (Chapter 28). Chemosensory perceptions are affected by a number of factors. Normal aging leads to perceptual, as well as anatomical, losses in chemosensory processes. Increased thresholds for both taste and smell accompany aging. For example, aged persons need two to three times as much sugar or salt as young persons to produce the same degree of taste perception. The reduction in chemosensory acuity may contribute to weight loss and malnutrition in elderly people. Other causes of chemosensory disorders include aberrations of nutrition and hormones, infectious diseases, treatment with drugs, radiation, or surgery. Sugars exhibit different degrees of sweetness (Table 8.1). Sucrose is sweeter than the other common disaccharides, maltose and lactose. D-fructose is sweeter than either D-glucose or sucrose. D-fructose is manufactured commercially starting with hydrolysis of corn starch to yield D-glucose, which is subsequently converted to D-fructose by the plant enzyme glucose isomerase.

Synthetic noncarbohydrate compounds can also produce a sweet taste. Saccharin, a synthetic compound, tastes 400 times as sweet as sucrose and has the following structure:

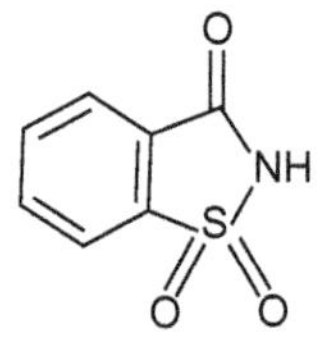

Saccharin

TABLE 8.1 The Relative Sweetness of Sugars, Sugar Alcohols, and Noncarbohydrate Sweeteners

Types of Compounds	Percent Sweetness Relative to Sucrose
Disaccharides	
Sucrose	100
Lactose	20
Maltose	30
Monosaccharides	
Glucose	50–70
Fructose	130–170
Galactose	30
Sugar alcohols	
Sorbitol	35–60
Mannitol	45–60
Xylitol	200–250
Noncarbohydrate sweeteners	
Saccharin	40,000
Aspartame	16,000

Another synthetic sweetener is aspartame (L-aspartyl-L-phenylalanine methyl ester), a dipeptide. Aspartame is 160 times as sweet as sucrose and, unlike saccharin, is said to have no aftertaste. Artificial sweeteners, because of their high degree of sweetness on a weight-for-weight basis compared to sucrose, contribute very little energy in human nutrition. They are useful in the management of obesity and diabetes mellitus. However, use of aspartame during pregnancy, particularly by individuals heterozygous or homozygous for phenylketonuria (Chapter 15), may be hazardous to the fetus.

The perception of sweet taste can be elicited by a wide range of chemical compounds. Two naturally occurring sweet proteins, thaumatin and monellin, are derived from the fruits of two African plants called katemfe and serendipity berries, respectively. These two proteins are intensely sweet and produce a perception of sweetness at a concentration as low as 10^{-8} mol/L. Despite the similarity in sweetness, thaumatin and monellin bear no significant structural similarities with respect to amino acid sequence or crystalline structure. However, they do exhibit immunological cross-reactivity, suggesting a common chemical and structural feature. A sweet-tasting, naturally occurring steviol glycoside, known as **rebaudioside A**, found in the leaves of Stevia plants which are indigenous to Paraguay and Brazil, has been recently approved for human consumption in the United States. Sweet substances may act as short-term antidepressants, presumably by raising serotonin (a metabolite of tryptophan) levels in the central nervous system. This property of sweet-tasting carbohydrates may unwittingly contribute to the development of obesity in susceptible individuals.

Polysaccharides

Polysaccharides, also known as **glycans**, contain many monosaccharide units joined together by glycosidic linkages. They may be homopolysaccharides (e.g., glycogen, starch, and cellulose), which contain only one type of monomeric residue, or heteropolysaccharides, which consist of two or more different types of monosaccharide units glycosidically joined in different ways. The heteropolysaccharides have complex structures, and they may also be found covalently linked with proteins and lipids (e.g., proteoglycans and glycosphingolipids).

Starch and glycogen are energy storage forms of carbohydrate and are thus known as **storage carbohydrates**. When the supply of carbohydrate exceeds the needs of the cell, the excess is converted to storage forms. When the situation is reversed, the storage forms are converted to usable forms of carbohydrate. Therefore, a storage carbohydrate should be capable of rapid synthesis as well as breakdown in response to the energy requirements of the cell. As monosaccharide accumulates in the cell, its rapid conversion to insoluble, high molecular weight polysaccharide prevents an osmotic imbalance and also maintains a favorable concentration gradient between the intra- and extracellular compartments, which facilitates sugar transport. Starch, the storage polysaccharide of most plants and particularly of tubers (e.g., potatoes) and seeds (corn and rice), consists of a mixture of amylose and amylopectin. Amylose is an unbranched polymer of glucose in which the glucosyl residues are linked in $\alpha(1 \rightarrow 4)$ glycosidic linkages (Figure 8.9).

Amylopectin contains glucosyl units joined together in both $\alpha(1 \rightarrow 4)$ and $\alpha(1 \rightarrow 6)$ linkages, the latter linkages being responsible for branch points (Figure 8.9). Unlike amylose, amylopectin is unable to assume a stable helical conformation because of the branching. Amylopectin complexes with iodine to a much lesser extent than amylose; therefore, the amylopectin–iodine complex has a red-violet color that is much less intense than the blue of the amylose–iodine complex. Starch from different sources contains different amounts of amylose and amylopectin. In most plants, amylopectin is the more abundant form (about 75%–80%). Virtually no amylose is found in starch obtained from some waxy varieties of maize (corn) and rice. Starch is easily digested by humans (Chapter 11).

Glycogen is the animal equivalent of starch in plants and functions as the main storage polysaccharide in humans. It is a branched polysaccharide of D-glucose and, like amylopectin, contains both $\alpha(1 \rightarrow 4)$ and $\alpha(1 \rightarrow 6)$ linkages, the latter forming branch points (Figure 8.9). Each molecule of glycogen contains one reducing glucose residue (i.e., unsubstituted C1 hydroxyl group), which is the terminal unit on one of the chains. At each of the other termini, the glucose residue has a free hydroxyl group at C4, while the C1 hydroxyl group participates in the glycosidic linkage. Synthesis and breakdown take place at these termini.

A glycogen molecule contains about 10^5 glucose units. Glycogen has no discrete molecular weight, since its size varies considerably depending on the tissue of origin and its physiological state. In the human body, most glycogen is found in liver and muscle. The functional roles of glycogen in these two tissues are quite different: in muscle, glycogen serves as an energy reserve mostly for contraction, whereas liver glycogen supplies glucose to other tissues via the blood circulatory system.

Cellulose, the most abundant carbohydrate on Earth, is an unbranched polymer with glucosyl residues joined together in $\beta(1 \rightarrow 4)$ linkages (Figure 8.10). Cellulose from different sources varies in molecular weight, and the number of glucose units lies in the range of 2500–14,000. Cellulose is the structural polysaccharide in plants. Cellulose microfibrils are tightly packed aggregates of cellulose molecules, which are chemically inert

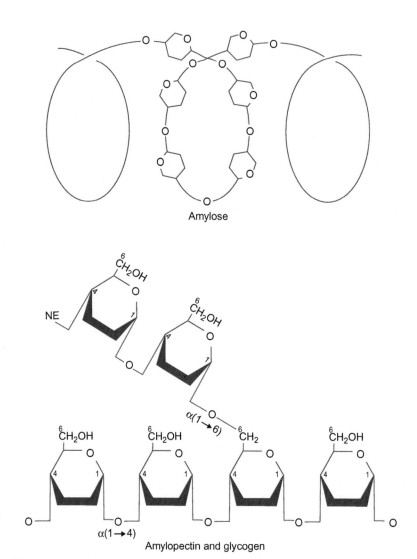

FIGURE 8.9 Structures of the starch polysaccharides amylose and amylopectin and glycogen. NE 5 nonreducing end (C4).

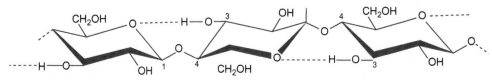

FIGURE 8.10 Conformational structure of cellulose. $\beta(1 \rightarrow 4)$ linkage of glucose residues in cellulose and showing hydrogen bonds (-----) between the ring oxygens and the C3 hydroxyl groups.

and insoluble and possess significant mechanical strength. In humans, cellulose is not digested in the small intestine but is digested in the large intestine to varying degrees by the microflora to yield short-chain fatty acids, hydrogen, carbon dioxide, and methane. Undigested cellulose forms a part of the indigestible component of the diet, known as **dietary fiber**. Ruminants and termites are able to digest cellulose, which is a primary energy source for them, because they harbor microorganisms in their intestinal tract that produce cellulase, which catalyzes hydrolysis of the $\beta(1 \rightarrow 4)$ glycosidic linkages.

Dietary fiber consists of cellulose and other polysaccharides (hemicelluloses, pectins, gums, and alginates) and a nonpolysaccharide component, lignin. These different types of dietary fiber are derived from the cell wall and the sap components of plant cells.

ENRICHMENT READING

[1] S.C. Kinnamon, Taste receptor signalling—From tongues to lungs, Acta Physiol. 204 (2012) 158–168.

[2] C.A. Nguyen, Y. Akiba, J.D. Kaunitz, Recent advances in gut nutrient chemosensing, Curr. Med. Chem. 19 (2012) 28–34.

[3] M. Aliani, C.C. Udenigwe, A.T. Girgih, T.L. Pownall, J.L. Bugera, M.N.A. Eskin, Zinc deficiency and taste perception in the elderly, Crit. Rev. Food Sci. Nutr. 53 (2013) 245–250.

[4] A. Varki, Uniquely human evolution of sialic acid genetics and biology, Proc. Natl Acad. Sci. U.S.A. 107 (2010) 8939–8946.

[5] M. Bardor, D.H. Nguyen, S. Diaz, A. Varki, Mechanism of uptake and incorporation of the non-human sialic acid N-glycolylneuraminic acid into human cells, J. Biol. Chem. 280 (2005) 4228–4237.

Chapter 9

Heteropolysaccharides: Glycoconjugates, Glycoproteins, and Glycolipids

Key Points

1. Glycoconjugates consist of covalently linked oligosaccharides with proteins and lipids.

2. Carbohydrate-linked protein conjugates can either be proteoglycans or glycoproteins. These two classes of glycoconjugates have widely different functions. Proteoglycans have a very high carbohydrate content compared to glycoproteins, and the oligosaccharides are covalently linked to small polypeptides (Chapter 10).

3. Glycoproteins are structurally diverse and have many functions. They occur both in soluble forms in extracellular fluids and also bound to membranes.

4. Nonenzymatic glycation involves condensation of amino groups of proteins with the keto or aldehyde groups of carbohydrates and yield glycoproteins that have pathologic implications in disorders such as diabetes mellitus.

5. In glycoproteins, the carbohydrates have three types of linkages: O-glycosidic linkage, N-glycosidic linkage, and glycosylphosphatidylinositol (GPI)-anchored linkage.

6. Human cell membrane constituents include lipids (phospholipids, glycosphingolipids, and cholesterol), glycoproteins, lipid-linked proteins, and proteins.

7. Membranes are phospholipid bilayers, and their composition and arrangements are unique for their specific function, such as selective permeability channels for metabolites and receptors for ligand recognition.

8. Glycolipids are predominantly located on the outer leaflet of plasma membranes. Glycosphingolipids contain a lipid moiety known as ceramide that is linked to mono, di, or tri oligosaccharides through O-glycosidic linkage. Several bacterial toxins (e.g., cholera, tetanus, and botulism) and some bacteria (e.g. *Escherichia coli*, *Streptococcus pneumoniae*, and *Neisseria gonorrhoeae*) bind to glycolipid receptors.

9. The bilayer of red blood cell (RBC) membranes is internally supported by a skeleton of interacting proteins, which are required for maintenance of morphology and function. Defects in these skeletal proteins lead to abnormal shape, mobility, and survivability of RBCs.

10. Cell surface glycoproteins are involved in many pathologic processes, and abnormalities in their synthesis and proper localization can cause diseases.

11. Glycoproteins as cell surface adhesion molecules (CAM) are central to the many physiological functions. They exist in various states of activation and transmit extracellular signals to the interior of the cell via G-proteins, kinases, and cytoskeletal components.

12. The process of activation of platelets (Chapter 34), leukocyte emigration from the blood vessels in the host-defense against infection, and extracellular matrix formation require specific glycoproteins.

13. Blood group substances present on the membrane of RBCs, other cells, and body fluids are oligosaccharides linked to proteins or lipids. Some blood group substances are made up of only proteins. All of the blood group substances are inherited according to the laws of Mendel, and their identification is essential in transfusion medicine.

14. Molecular mimicry involving a ganglioside between a pathogen and human antigen can lead to postinfectious autoimmune disease.

GLYCOPROTEINS

Glycoproteins, a wide range of compounds of diverse structure and function, are components of cell membranes, intercellular matrices, and extracellular fluids, such as plasma. They occur in soluble and membrane-bound forms (Table 9.1).

The proportion of carbohydrate varies considerably in glycoproteins derived from different tissues or from various sources. Collagen, for example, contains an amount of carbohydrate that varies with the source: skin tissue collagen, about 0.5%; cartilage collagen, about 4%; and basement membrane collagen, more than 10%. Glycoproteins containing high amounts of carbohydrate include glycophorin, a membrane constituent of human erythrocytes, about 60%, and soluble blood group substances, as much as 85%.

In glycoproteins, the protein and the carbohydrate residues are bound in covalent linkage. There are five common

N. V. Bhagavan and C.-E. Ha: Essentials of Medical Biochemistry, Second edition. DOI: http://dx.doi.org/10.1016/B978-0-12-416687-5.00009-9

TABLE 9.1 Some Functions of Glycoproteins in the Body

Function	Examples
Structure	Collagen
Lubrication and protection	Epithelial mucins, synovial fluid glycoproteins
Transport	Ceruloplasmin (copper carrier), transferrin (iron carrier)
Endocrine regulation	Thyrotropin, chorionic gonadotropin, erythropoietin
Catalysis	Proteases, nucleases, glycosidases, hydrolases
Defense against infection	Immunoglobulins, complement proteins, interferon, selectins and integrins
Membrane receptors	Receptors for hormones (e.g., insulin), acetylcholine, cholera toxin, electromagnetic radiation (e.g., rhodopsin)
Antigens	Blood group substances
Cell–cell recognition and adhesion	Fibronectin, laminin, chondronectin
Miscellaneous	Glycophorin (an intrinsic red blood cell membrane constituent), intrinsic factor (essential for absorption of dietary vitamin B_{12}), clotting factors (e.g., fibrinogen)

types of carbohydrate, four of which are common to human glycoproteins (Figure 9.1). All the linkages are either N- or O-glycosidic bonds. The amino acid residue that participates in the N-glycosidic linkage is asparagine, and the amino acid residues that participate in O-glycosidic linkage are serine, threonine, hydroxylysine, and hydroxyproline. The glycoproteins exhibit microheterogeneity, i.e., glycoproteins with an identical polypeptide sequence may vary in the structure of their oligosaccharide chains. Microheterogeneity arises from incomplete synthesis or partial degradation and poses problems in the purification and characterization of glycoproteins. For example, human serum α_1-acid glycoprotein has five linkage sites for carbohydrates and occurs in at least 19 different forms as a result of differences in oligosaccharide structures.

The oligosaccharide side chains of glycoproteins consist of only a limited number (about 11) of different monosaccharides. These monosaccharides are hexoses and their derivatives (N-acetylhexosamine, uronic acid, and deoxyhexose), pentoses, and sialic acids derived from neuraminic acid, a nine-carbon sugar. The most common of the many different types of sialic acids is N-acetylneuraminic acid. (All of these sugars are in the D-configuration, unless indicated otherwise.) The sugar residues of the oligosaccharide chains are not present in serial repeat units (unlike the proteoglycans; Chapter 10), but they do exhibit common structural features. For example, the oligosaccharide chains bound in N-glycosidic linkages may be envisaged as consisting of two domains. The inner domain, common to all glycoproteins, is attached to the protein via an N-α-glycosidic linkage between an asparaginyl residue and N-acetylglucosamine. The inner domain consists of a branched polysaccharide,

which is trimannosyl-di-N-acetylglucosamine (Figure 9.2). The peripheral mannose residue of the inner domain is linked to an oligosaccharide chain, known as the **outer domain**. The outer domain is made up of either oligosaccharides consisting of mannose residues (oligomannosidic types) or N-acetyllactosamine units.

Glycoproteins with O-glycosidic linkages do not show the common features of glycoproteins with N-glycosidic linkages. The number of sugar residues may vary from one (e.g., collagen) to many (e.g., blood group substances). Many glycoproteins, however, do show the presence of a common disaccharide constituent, namely, galactosyl-β-(1→3)-N-acetylgalactosamine, which is linked to either serine or threonine. In collagens, the O-glycosidic linkages occur via hydroxyproline or hydroxylysine residues. A given glycoprotein may contain oligosaccharide chains of both N- and O-glycosidic types. If more than one N-glycosidic carbohydrate linkage site is present in a glycoprotein, they are usually separated by several amino acid residues. In contrast, O-glycosidic carbohydrate linkages may be found in adjacent hydroxyamino acid residues, or they may occur in close proximity, e.g., in glycophorin, human chorionic gonadotropin, and antifreeze glycoprotein. The latter is found in the blood of Arctic and Antarctic fish species and other species on the eastern coast of North America. It contains a very high amount of carbohydrate, since every threonine residue of the glycoprotein is linked with a galactosyl-β-(1→3)-N-acetylgalactosamine unit. The protein consists of the repeating tripeptide sequence of alanyl–alanyl–threonine. The presence of antifreeze glycoprotein, along with high concentrations of NaCl in the blood, prevents water from freezing in blood vessels and permits survival at the low

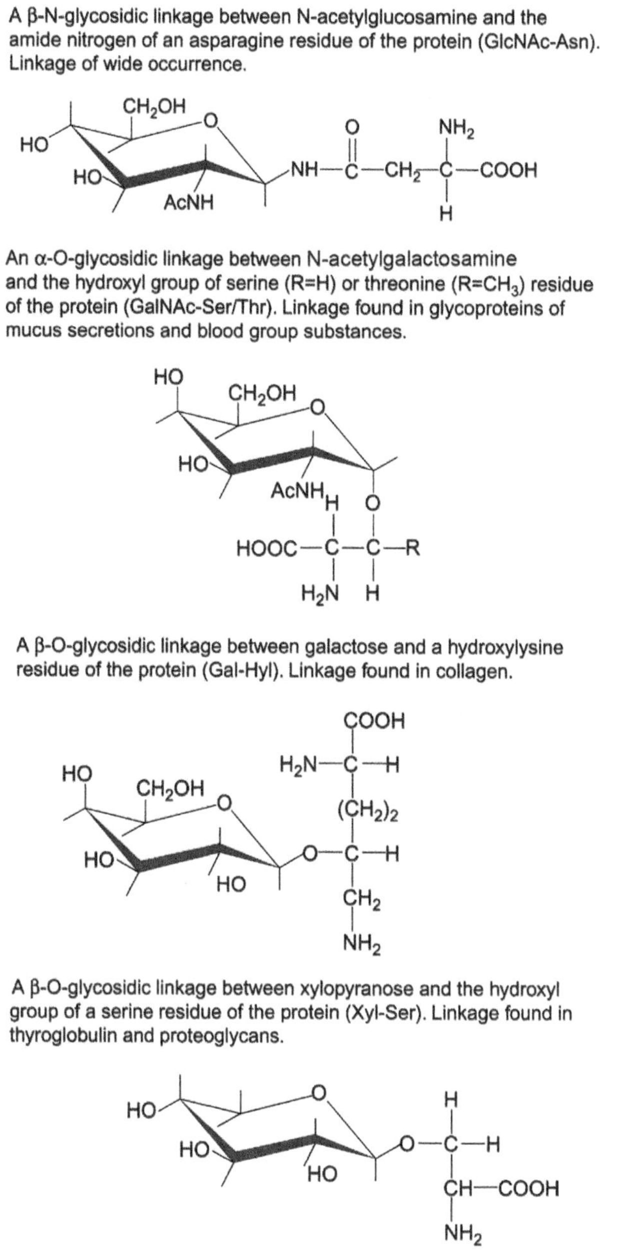

A β-N-glycosidic linkage between N-acetylglucosamine and the amide nitrogen of an asparagine residue of the protein (GlcNAc-Asn). Linkage of wide occurrence.

An α-O-glycosidic linkage between N-acetylgalactosamine and the hydroxyl group of serine (R=H) or threonine (R=CH₃) residue of the protein (GalNAc-Ser/Thr). Linkage found in glycoproteins of mucus secretions and blood group substances.

A β-O-glycosidic linkage between galactose and a hydroxylysine residue of the protein (Gal-Hyl). Linkage found in collagen.

A β-O-glycosidic linkage between xylopyranose and the hydroxyl group of a serine residue of the protein (Xyl-Ser). Linkage found in thyroglobulin and proteoglycans.

FIGURE 9.1 Carbohydrate−peptide linkages in glycoproteins.

Man $\alpha(1\rightarrow6)$
Man $\beta(1\rightarrow4)$ GlcNAc $\beta(1\rightarrow4)$ GlcNAc $\beta\rightarrow$Asn
Man $\alpha(1\rightarrow3)$

FIGURE 9.2 Structure of the common branched inner domain of oligosaccharides linked in N-glycosidic linkages with asparagine.

temperatures of polar seawater. This freezing-point depression by antifreeze glycoproteins has been attributed to their highly hydrated and expanded structure, which interferes with the formation of ice crystals.

In all glycoproteins, the polypeptide component is synthesized first on the membrane-bound ribosomes of the rough endoplasmic reticulum; carbohydrate side chains are added during passage through the endoplasmic reticulum and Golgi apparatus. The carbohydrate additions involve specific glycosyltransferases and their substrates (uridine diphosphate sugars) and, in some glycoproteins, an oligosaccharide carrier known as **dolichol** (a lipid).

Glycoproteins can also be formed by addition of carbohydrate residues without any of the complex enzymatic pathways of carbohydrate addition. This process, which is known as **nonenzymatic glycation**, proceeds by the condensation of a monosaccharide, usually glucose, with certain reactive amino groups on the protein. The initial, labile Schiff base adduct slowly rearranges to the stable ketoamine or fructosamine form. For example, a small fraction of hemoglobin A, the main hemoglobin of adult humans, is present in the red blood cells (RBCs) as glycated hemoglobin (HbA₁C). The glycation of hemoglobin is a continuous process occurring throughout the 120-day lifespan of the RBC. In HbA₁C, glucose is incorporated via an N-glycosidic linkage into the N-terminal amino group of valine of each β-chain. Enhanced levels of HbA₁C occur in individuals with diabetes mellitus, and measurement of glycated hemoglobin has been useful in monitoring the effects of therapy. Human serum albumin, which has a half-life of 19 days, is also subjected to nonenzymatic glycation producing a stable condensation product known as **fructosamine**. Fructosamine is a generic term applied to the stable condensation product of glucose with serum proteins, of which albumin is quantitatively the largest fraction. Measurement of fructosamine concentration provides a means by which short-term (1−3 weeks) plasma glucose levels can be estimated, whereas measurement of HbA₁C concentration reflects integrated plasma glucose levels over a longer period (2−3 months). Human lens proteins, α-, β-, and γ-crystallins, which have much longer lifespans than other proteins in the body, also undergo age-dependent, nonenzymatic glycation at the ε-amino groups of their lysine residues. In diabetics, this process occurs twice as often as in normal individuals of comparable age. Lens cells, like RBCs, do not require insulin for the inward transport of glucose. However, the extent of nonenzymatic glycation of crystallins is much lower than that of hemoglobin.

Crystallins constitute 90% of the soluble proteins of the lens cells (also called **fiber cells**). The human lens—a transparent, biconvex, elliptical, semisolid, avascular structure—is responsible for focusing the visual image onto the retina. The lens grows throughout life at a slowly decreasing rate, building layer upon layer of fiber cells around a central core and never shedding the cells. Crystallin turnover is very slow or nonexistent. In addition to nonenzymatic glycation, crystallins undergo other age-dependent, post-translational modifications *in vivo*: formation of disulfide bonds and other covalent cross-links; accumulation of high molecular weight aggregates; deamidation of asparagine and glutamine residues; partial proteolysis at characteristic sites; racemization of aspartic acid residues; and photo-oxidation of tryptophan. Some of these processes contribute to the increasing amount of insoluble crystallins during aging. Nonenzymatic incorporation of carbohydrates into proteins *in vivo* can be extensive and may contribute to the pathophysiology of diabetes mellitus and galactosemia.

CELL MEMBRANE CONSTITUENTS

Various aspects of the cell membrane are discussed throughout this text, and a brief introduction is presented here. The living system's ability to segregate from and protect itself against—and interact with and against—changes in the external environment is accomplished by membranes. In the body, membranes function at the level of tissues, cells, and intracellular domains. They function as protective barriers and as transducers of extracellular messages carried by chemical agents because they have recognition sites that interact with metabolites, ions, hormones, antibodies, or other cells in a specific manner. This characteristic selectivity of membranes to interact with specific molecules imparts unique properties to a given cell type. Within the cell, the membranes of organelles are highly differentiated and have properties consistent with metabolic function. Examples include electron transport and energy conservation systems in the mitochondrial membrane, protein biosynthesis in the rough endoplasmic reticulum, modification and packaging of proteins for export in the membranes of the Golgi complex, drug detoxification in the smooth endoplasmic reticulum, and light reception and transduction in the disk membranes of retinal cells.

The membrane constituents are lipids (phospholipids, glycosphingolipids, and cholesterol; Figure 9.3), carbohydrates, and proteins. The ratio of protein:lipid:carbohydrate on a weight basis varies considerably from membrane to membrane. For example, the human erythrocyte membrane has a ratio of about 49:43:8, whereas myelin has a ratio of 18:79:3. All membrane lipids are amphipathic (i.e., polar lipids). The polar heads of the phospholipids may be neutral, anionic, or dipolar. The surface of the membrane bears a net negative charge. The distribution of lipid constituents in the bilayer is asymmetrical. For example, in the erythrocyte membrane, phosphatidylethanolamine and phosphatidylserine are located primarily in the internal monolayer, whereas phosphatidylcholine and sphingomyelin are located in the external monolayer.

Lipids are organized into bilayers that account for most of the membrane barrier properties. Membrane proteins may be peripheral (extrinsic) or integral (intrinsic). Peripheral proteins are located on either side of the bilayer and are easily removed by ionic solutions, whereas integral proteins are embedded in the bilayer to varying degrees. Some integral proteins penetrate the bilayer and are exposed to both external and internal environments. By spanning both external and internal environments of the cell, these proteins may provide a means of communication across the bilayer that may be useful in the transport of metabolites, ions, and water, or in the transmission of signals in response to external stimuli provided by hormones, antibodies, or other cells. Because of hydrophobic interactions, isolation of integral proteins requires harsh methods, such as physical disruption of the bilayer and chemical extraction procedures using synthetic detergents or bile salts to disrupt the lipid–protein interactions.

Carbohydrate residues are covalently linked (exclusively on the external side of the bilayer) to proteins or lipids to form glycoproteins or glycolipids, respectively, both of which are asymmetrically distributed in the lipid bilayer (Figure 9.4). Fluidity of the membrane structure is determined by the degree of unsaturation of the hydrocarbon chains of the phospholipids and by the amount of cholesterol in the membrane. Hydrocarbon chains with *cis*-double bonds produce kinks and allow a greater degree of freedom of movement for the neighboring alkyl side chains; hence, these unsaturated chains give rise to more fluidity than do saturated alkyl chains, which associate in ordered arrays. Cholesterol, an inflexible polycyclic molecule, is packed between fatty alkyl chains, the ring bearing the polar hydroxyl group interacting with the polar groups of phospho- and glycolipids. The presence of cholesterol disrupts the orderly stacking of alkyl side chains, restricts their mobility, and causes increased membrane viscosity. Thus, the lipid composition of the membrane at physiological temperatures can have significant effects on fluidity and permeability. Some correlation appears to exist between the concentrations of sphingomyelin and cholesterol in different membranes. Plasma membranes (e.g., RBCs and myelin sheaths) are rich in both; the inner membrane of mitochondria contains neither.

Membrane proteins show considerable mobility in the plane of the bilayer (lateral motion). There is no evidence that proteins migrate from one side of the bilayer to the other. The frequency of reorientation of lipid components (flip-flop migration) is extremely slow or nonexistent.

CELL-SURFACE GLYCOPROTEINS

Glycoproteins play major roles in antigen–antibody reactions, hormone function, enzyme catalysis, and cell–cell interactions. Membrane glycoproteins have domains of

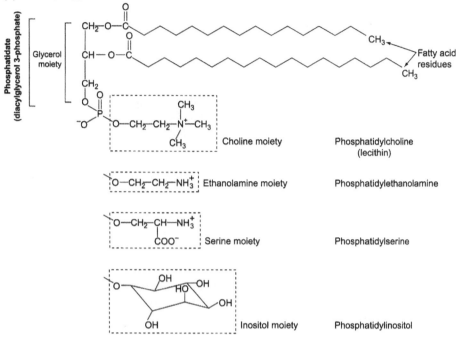

(a) **Phosphoglycerides**

Phosphatidate
(diacylglycerol 3-phosphate)

Glycerol
moiety

Fatty acid
residues

Choline moiety   Phosphatidylcholine
(lecithin)

Ethanolamine moiety   Phosphatidylethanolamine

Serine moiety   Phosphatidylserine

Inositol moiety   Phosphatidylinositol

(b)   **Phosphosphingolipids**

These lipids contain a long unsaturated hydrocarbon chain amino alcohol known as **sphingosine**:

$C_{18}H_{37}NO_2$

A derivative of sphingosine, in which a fatty acid is linked by an amide linkage, is **ceramide**:

Fatty acid
residue

Amide
linkage

Primary
alcohol group

The product obtained when the alcohol hydroxyl group of sphingosine is esterified with phosphorylcholine is **sphingomyelin**. The conformations of phosphatidylcholine and sphingomyelin are similar.

FIGURE 9.3   Structures of some membrane lipids.

(c) Glycosphingolipids (cerebrosides)

A derivative of ceramide that contains a monosaccharide unit (either glucose or galactose) linked in a β-glycosidic linkage is a cerebroside. These neutral lipids occur most abundantly in the brain and myelin sheath of nerves. The specific galactocerebroside, phrenosine, contains a 2-hydroxy-24-carbon fatty acid residue. Following is a structure of a glucocerebroside.

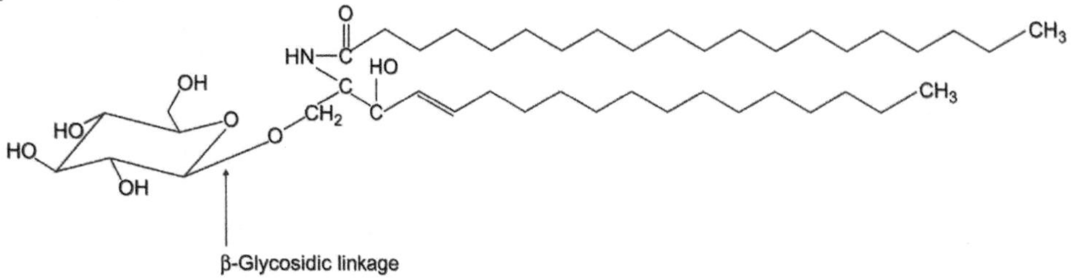

(d) Cholesterol

Cholesterol (3-hydroxy-5, 6-cholestene) belongs to a family of compounds derived from a fused, reduced, nonlinear four-ring system of cyclopenta [α]-phenanthrene. Bile acids, steroid hormones, and vitamin D metabolites are derived from cholesterol.

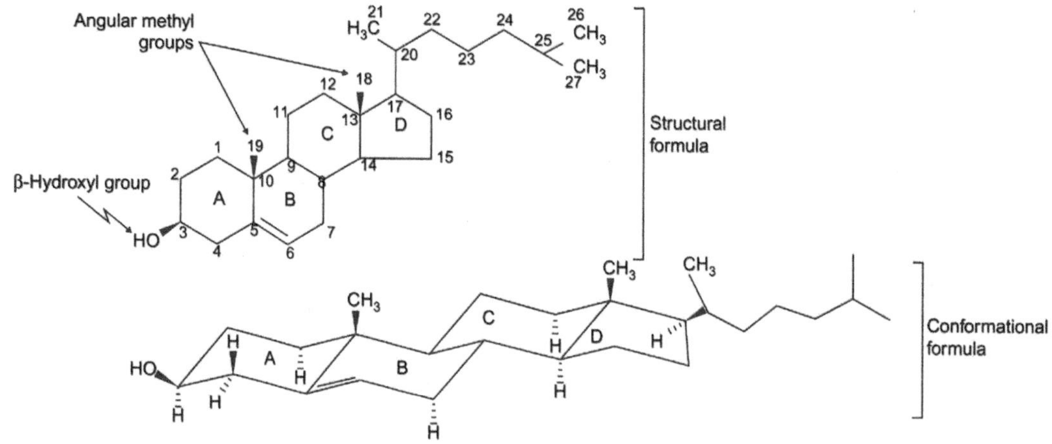

FIGURE 9.3 (Continued)

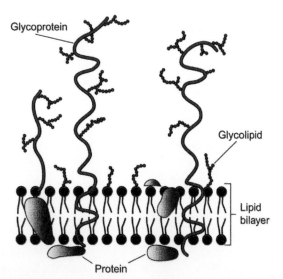

FIGURE 9.4 Membrane asymmetry with respect to location of glycoproteins and glycolipids. These carbohydrate-containing molecules are exclusively present on the external surface of the plasma membrane.

hydrophilic and hydrophobic sequences, and are amphipathic molecules. The carbohydrate moieties of glycoproteins are distributed asymmetrically in cell membranes, cluster near one end of the protein molecule (Figure 9.4), and constitute a hydrophilic domain of amino acid residues as well as carbohydrates. The hydrophobic domain of the molecule interacts with the lipid bilayer.

The role of glycoproteins in cell–cell interaction is coordination and regulation of adhesion, growth, differentiation of cells, and cell size. Disruption of these processes may lead to loss of control of cell division and growth, a property characteristic of cancer cells. Normal cells, when grown in tissue culture, show **contact inhibition**. When cells are allowed to grow in a tissue culture medium under optimal conditions on a surface, such as that of a Petri dish, they grow and divide until the surface is covered with a monolayer and further growth is inhibited. However, when these cells are transformed by treatment with certain viruses or carcinogenic chemicals, they lose contact inhibition and continue to grow beyond the

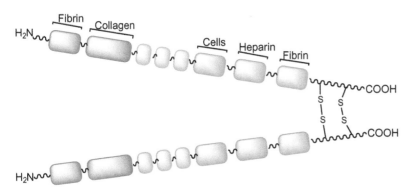

FIGURE 9.5 Schematic representation of a fibronectin molecule. It is a dimer of similar subunits (M.W. of each, ~250,000) joined by a pair of disulfide linkages. The various functional domains by which fibronectin can interact with other protein and membrane components are indicated.

monolayer to form multilayered masses of cells. Cancer cells (malignant neoplastic cells), like transformed cells, show continued growth and invasiveness (spreading) in tissue culture. In both transformed and cancer cells, cell—cell interaction has been altered. Reduced adhesion of cancer cells to a matrix has been related to a reduced synthesis of fibronectin and collagen.

Fibronectin, a glycoprotein abundant on the cell surface of normal cells, promotes that attachment and subsequent spreading of many cell types. Known also as a cell surface protein, fibronectin is a large, external, transformation-sensitive protein that binds to a number of substances (e.g., collagen and fibrin). The name derives from Latin fibra ("fiber") and Latin nectere ("to tie"). Two types of fibronectin are recognized: cell surface fibronectin, which occurs as dimers and multimers; and plasma fibronectin, which occurs primarily as dimers (Figure 9.5). Fibronectin is a multifunctional molecule containing regions that recognize glycoconjugates (e.g., glycolipids) of the cell surface, proteoglycans, and collagen fibers. Fibronectin contains about 5% carbohydrate by weight. Plasma fibronectin plays several roles in wound repair: in the formation of a fibrin clot as cross-linked fibronectin, in some reactions of platelets, in the enhancement of the opsonic activity of macrophages (important for removal of foreign material and necrotic tissue), and in attracting fibroblasts (which participate in the production of repair components such as collagen and proteoglycans in the extracellular matrix). A unique form of fibronectin, known as **fetal fibronectin**, is found in the extracellular matrix surrounding the extravillous trophoblast at the uteroplacental junction. The presence of fetal fibronectin in cervicovaginal secretions may be used as a marker in assessing the risk for preterm delivery. It is measured by enzyme immunoassay procedures. In women during 24−35 weeks of gestation with symptoms of preterm labor consisting of contractions and advanced cervical dilation, elevated fetal fibronectin levels ($\geq$50 mg/L) in the cervicovaginal fluid are indicative of an increased risk of preterm delivery. The fetal fibronectin levels have a high negative predictive value in the assessment of preterm delivery. Antenatal treatment (e.g., with corticosteroids) in women who are at high risk for delivery of premature infants significantly decreases morbidity and mortality. Inhibition of uterine myometrial contractions, which is known as **tocolysis**, by appropriate therapeutic agents is used in preventing prematurity.

In some cell types, fibronectin is not involved in cell adhesion. For example, the extracellular matrix adjacent to epithelial cells and chondrocytes does not contain fibronectin but rather two other glycoproteins, laminin and chondronectin. Laminin mediates in adhesion of epithelial cells, whereas chondronectin mediates the attachment of chondrocytes to collagen. Mutations in a laminin molecule, namely laminin 5, manifest in abnormalities of skin fragility, causing one type of disorder known as **epidermolysis bullosa**. A family of cell surface adhesion receptor proteins known as **integrins** binds with fibronectin. Integrins are a family of proteins containing $\alpha\beta$ heterodimers that possess receptors not only for fibronectin but also for collagens, laminin, fibrinogen, vitronectin, and integral membrane proteins of the immunoglobulin superfamily (Chapter 33). A sequence of three amino acids (RGD) in fibronectin is a recognition site for binding with integrin. Other amino acid sequences that are present in proteins for integrin recognition include KQAGDV, DGEA, EILDV, and GPRP. A cascade of physiologic events is required for the circulating leukocytes in the blood vessel to emigrate to inflammatory foci to fight infection (Figure 9.6). The emigration of leukocytes through the endothelial wall of a blood vessel requires coordinated activities of a group of glycoproteins, selectins, integrins, and intercellular adhesion molecules. The selectin family of glycoproteins consists of three members, P-, E-, and L-selectins, named for the initial cell type where they were discovered, platelets, endothelial cells, and lymphocytes, respectively. The quiescent rolling leukocytes mediated by L-selectins in the blood vessels are stimulated by inflammatory cytokines for their exodus process. Endothelial cell P- and E-selectins provide receptors for binding with carbohydrate ligands of leukocytes. Defects in either the selectins or integrins can lead to clinical syndromes of increased susceptibility to bacterial infection.

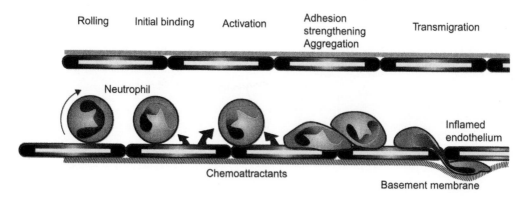

FIGURE 9.6 The multistep process of emigration of neutrophils (leukocytes) through the blood vessel endothelial barrier to sites of inflammation. The rolling neutrophils passing through inflammatory endothelium are snared, trapped, and extravasated to the surrounding tissues. The process of emigration of leukocytes requires inflammatory signals that lead to upregulation of several glycoproteins, including selectins, integrins, and adhesion molecules. *[Reproduced with permission from C. R. Scriver, A. L. Beaudet, W. S. Sly, et al. (Eds.), The Metabolic Molecular Basis of Inherited Disease, 8th ed. New York: McGraw-Hill, 2001, 4834.]*

Blood Group Antigens

Antigenic determinants (Chapter 33) including both oligosaccharides (e.g., A, B, and O) and proteins (e.g., Rh) are known as **blood group substances**. Oligosaccharides are present on the surface of RBCs, and the proteins span the RBC membrane. The blood group substances are inherited according to Mendel's law. Identification of blood group substances is essential for blood transfusions, and is valuable in forensic medicine and anthropological studies. The antigens are recognized by specific antigen−antibody interactions that produce agglutination. For example, when red blood cells containing a specific antigenic determinant are mixed with plasma containing specific antibodies to that antigen, cells will agglutinate through formation of a network of antigen−antibody linkages. More than 100 different blood group antigens have been categorized on the basis of their structural relationships into 15 independent blood group systems. The blood group substances are found on RBC membranes and in a variety of tissue cells. Soluble blood group substances are found as glycoproteins in saliva, gastric juice, milk, seminal fluid, urine, fluids produced in ovarian cysts, and amniotic fluid.

SERUM GLYCOPROTEINS

Almost all serum proteins, with the notable exception of albumin, are glycoproteins. The sugar residues found most commonly in the outer domain of the oligosaccharides of these glycoproteins are galactose, N-acetylhexosamine, and sialic acid. L-fucose is a minor constituent. The liver plays a major role in the synthesis and catabolism of these proteins. Glycoproteins lose their terminal sialic acid residues through the action of neuraminidase (sialidase) during circulation in the blood, which exposes the galactose residues. The resulting galactose-terminated glycoproteins, known as **asialoglycoproteins**, are taken up after binding to receptors on hepatocytes. The bound asialoglycoprotein is internalized by a process known as **receptor-mediated endocytosis** (Chapter 10) and subjected to lysosomal degradation. In liver disease, the plasma levels of asialoglycoproteins are elevated.

The internalized glycoproteins are catabolized to their monomeric units by the lysosomal enzymes. The oligosaccharides are degraded sequentially by specified hydrolases, starting from the nonreducing termini. Hereditary deficiency of some of these enzymes has been reported: α-D-mannosidase in mannosidosis; α-L-fucosidase in fucosidosis; glycoprotein-specific α-neuraminidase in sialidosis; and aspartylglycosaminidase in aspartylglycosaminuria. In these disorders, undigested or partially digested oligosaccharides derived from glycoproteins accumulate within the lysosomes. Keratan sulfate (a glycosaminoglycan; Chapter 10) also accumulates, presumably because the above enzymes are required for its degradation. These disorders are inherited as autosomal recessive traits and can be diagnosed prenatally. Clinically, they resemble mild forms of mucopolysaccharidosis (Chapter 10). There is no definitive treatment.

MOLECULAR MIMICRY OF OLIGOSACCHARIDES AND HOST SUSCEPTIBILITY

In the previous section we discussed the presence of oligosaccharide and protein blood group substances that cause susceptibility to certain pathogens. In this section we discuss an infectious agent that contains a common antigenic epitope with the host and that elicits antibodies that cause a disease. Guillain−Barré syndrome (GBS) is one such disease. GBS is an acute inflammatory neuropathy with progressive weakness in both arms and legs, leading to paralysis. It is a self-limited autoimmune disease. Most GBS cases have resulted from an antecedent, acute infection of bacterial or viral origin; in children GBS has also been identified following vaccination.

One major cause of bacterial gastroenteritis is *Campylobacter jejuni*, and this infection also may be followed by GBS. Some serotypes of *C. jejuni* possess liposaccharides that contain terminal tetrasaccharides identical to ganglioside G_{M1}. Antibodies made in response to the bacterial infection also attack G_{M1}, which is widely present in the nervous system. The antibodies directed against the G_{M1} epitope cause an immune-mediated destruction of nerve fibers. Thus, the sharing of homologous epitopes between bacterial liposaccharides and gangliosides is an example of molecular mimicry that causes disease. Host factors may also influence a person's susceptibility to GBS. Plasma exchange and intravenous immunoglobulin administration are used for immunomodulation in the therapy of GBS.

ENRICHMENT READING

[1] C. Has, G. Sparta, D. Kiritsi, L. Weibel, A. Moeller, V. Vega-Warner, et al., Integrin α_3 mutations with kidney, lung, and skin disease, N. Engl. J. Med. 366 (2012) 1508−1514.

[2] N. Yuki, H. Hartung, Guillain-Barre Syndrome, N. Engl. J. Med. 366 (2012) 2294−2304.

[3] E.W. Gelfand, Intravenous immune globulin in autoimmune and inflammatory diseases, N. Engl. J. Med. 367 (2012) 2015−2025.

Chapter 10

Connective Tissue: Fibrous and Nonfibrous Proteins and Proteoglycans

Key Points

1. Connective tissue that provides support and framework for the body consists of fibrous proteins and nonfibrous ground substance in varying proportions depending on their functions.

2. Collagen, which is the most abundant protein, constitutes about one-third of all body protein. More than 19 different types of collagens, encoded by 30 widely dispersed genes with characteristic distribution among tissues, are known.

3. Collagens have a common basic structure consisting of three polypeptide chains wound into a triple helix.

4. In collagen each of the three-polypeptide chains are coiled in a left-handed helix conformation with about three amino acids per turn. The three polypeptide chains are intertwined and coiled into a right-handed triple helix.

5. Collagen is stabilized through lysine-derived covalent intermolecular cross-linkages.

6. The synthesis and assembly of polypeptide chains require intracellular and extracellular processes. Acquired and inherited disorders cause several clinical disorders (Chapter 25).

7. Collagenases are a group of enzymes that belong to the family of Zn^{2+}-dependent proteolytic enzymes known as matrix metalloproteinases.

8. Fibers of the extracellular matrix confer elastic properties on tissues such as large blood vessels, lungs and skin, which contain elastin and fibrillin. Elastin is vastly different from collagen. Elastin contains mostly nonpolar amino acids, and it is insoluble.

9. Elastases belong to the family of serine proteases similar to trypsin and are found in exocrine pancreatic tissue and leukocytes. Their activity is inhibited by α_1-antitrypsin and α_2-macroglobulin. Inappropriate action of elastase can lead to disorders such as emphysema.

10. A major component of 10–12 nm extracellular microfibrils is glycoprotein fibrillin 1. It has a large molecular weight and consists of multidomains, with several cysteine-rich motifs. Genetic defects of the fibrillin 1 (FBN1) gene present on chromosome 15 cause a multisystem connective tissue disorder known as Marfan's syndrome. Mutations in a homologous gene on chromosome 5 that encode fibrillin 2 glycoprotein also cause connective tissue disorders.

11. Proteoglycans are present in extracellular matrix and on cell surfaces, and they are multifunctional. They contain covalently linked glycosaminoglycan side chains linked to core proteins by N- and O-glycosidic linkages. Their carbohydrate content is high (as much as 95%) compared to glycoproteins.

12. In connective tissue, proteoglycans interact with other fibrous structures consisting of collagen, elastin, fibronectin, and fibrillin.

13. The turnover of proteoglycans is accomplished by lysosomal enzymes. Deficiencies of these enzymes degradative of glycosaminoglycans lead to a family of heritable diseases known as mucopolysaccharidoses.

14. Peptidoglycans, also called murein, consist of short peptide units covalently linked to long carbohydrate chains and are components of bacterial cell walls. The polymeric network of peptidoglycans imparts semi-rigidity and strength to the bacterial cell walls. A bacterium is broadly classified as gram positive or gram negative based on whether or not they retain crystal violet stain. The cell walls of gram-positive and gram-negative bacteria are vastly different.

15. Lysozyme found in body fluids is microbicidal for certain gram-positive bacteria and the yeast *Candida albicans*. Their microbicidal action is due to hydrolysis of the $\beta 1 \rightarrow 4$ linkage between N-acetylmuramic acid and N-acetylglucosamine in peptidoglycan. The antibacterial activity of penicillin and other β-lactams is due to their inhibition of a transpeptidase that is required for cross-linkage of peptidoglycans. However, bacteria can develop resistance to β-lactam antibiotics by several different mechanisms.

16. Lectins are proteins with recognition domains for carbohydrates. They are involved in many biological processes such as extravasations of leukocytes from the endothelium (Chapter 9), infection by microorganisms (*Helicobacter pylori* lectins binding oligosaccharide units of blood group O of gastric mucosa), toxic manifestations of massive diarrhea of cholera toxin (Chapter 11), and the lectin pathway of complement activation (Chapter 33).

N. V. Bhagavan and C.-E. Ha: Essentials of Medical Biochemistry, Second edition. DOI: http://dx.doi.org/10.1016/B978-0-12-416687-5.00010-5

PROTEIN FIBERS

Connective tissues are composed of insoluble protein fibers (the glycoprotein collagen and the nonglycoprotein elastin) embedded in a matrix of proteoglycans (ground substance). The connective tissues bind tissues together and provide support for the organs and other structures of the body. Their properties depend on the proportion of different components present. A tissue of very high tensile strength, the Achilles tendon, is composed of about 32% collagen and 2.6% elastin, whereas an elastic tissue, the ligamentum nuchae, is composed of about 32% elastin and 7% collagen. The proteins and proteoglycans are synthesized by connective tissue cells: fibroblasts (generalized connective tissue); chondroblasts (cartilage); and osteoblasts (bone). Connective tissue also contains blood and lymphatic vessels and various transient cells including macrophages and mast cells. Adipose tissue is a specialized form of connective tissue consisting of a collection of adipocytes (stores of triacylglycerol) that cluster between the protein fibers.

Collagen

Collagens, which are extracellular proteins of connective tissue, make up about one-third of all body protein. They are a family of related glycoproteins in which hydroxylysyl residues provide the sites for attachment of glucose, galactose, or an $\alpha(1 \rightarrow 2)$ glucosylgalactose residue via an α-O-glycosidic linkage (Figure 10.1). In brief, the synthesis of collagen can be considered to occur in two stages: intracellular and extracellular. The intracellular stage consists of the production of procollagen from precursor polypeptide chains that undergo the sequence of hydroxylation, glycosylation, formation of a triple helix, and secretion. The hydroxylation requires vitamin C (see Clinical Case Study 10.1, Vignette 3). The extracellular stage consists of the conversion of procollagen to tropocollagen by limited proteolysis from the amino and carboxyl termini, self-assembly of tropocollagen molecules into fibrils, and finally cross-linking of the fibrils to form collagen fibers (Chapter 23).

Collagen exists predominantly as fibrous protein; however, in the basement membrane of many tissues, including kidney glomeruli and the lens capsule, it is present in a nonfibrous form. The unique property of each type of connective tissue (e.g., the flexibility of skin, rigidity of bone, elasticity of large arteries, and strength of tendons) depends on the composition and organization of collagen and other matrix components.

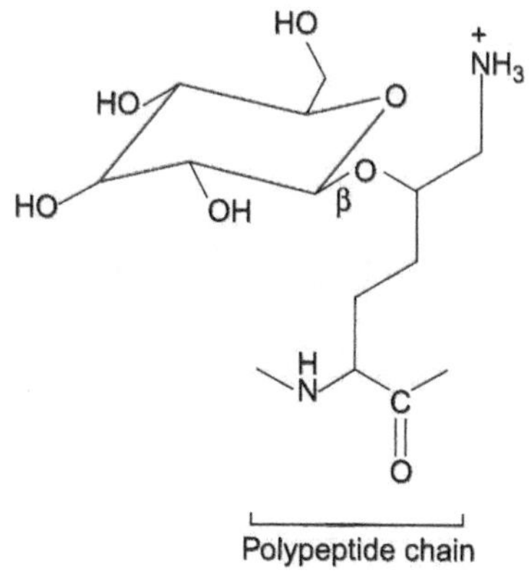

Collagen Types

More than 19 different types of collagen have been reported (see Table 10.1). They constitute the most

β-D-Glucosylhydroxylysyl residue

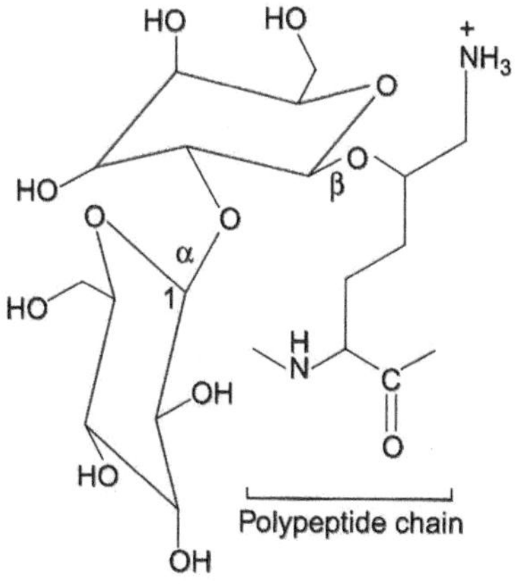

D-Glucosyl (α1→2)-O-β-D-galactosylhydroxylysyl residue

FIGURE 10.1 Glycosides of the collagen polypeptide chain.

abundant family of proteins in the human body. The collagens are encoded by 30 genes dispersed in at least 12 different chromosomes.

Type I collagen consists of two identical chains of $\alpha 1$ (I) and one chain of $\alpha 2$; it is the major connective tissue protein of skin, bone, tendon, dentin, and some other tissues. Type II collagen consists of three identical chains of $\alpha 1$(II); it is found in cartilage, cornea, vitreous humor, and neural retinal tissue. Type III collagen consists of three identical chains of $\alpha 1$(III); it is present, along with type I collagen, in skin, arteries, and uterine tissue. Type IV collagens are composed of $\alpha 1$(IV) to $\alpha 5$(IV) and are found in the basement membranes of various tissues. Unlike collagen types I, II, III, V, and XI, which are fibrillar, type IV collagen is not fibrillar in structure.

The synthesis of collagens in cultured cells has aided the understanding of collagen biochemistry. The tissue specificity of various types of collagen is also reflected in cultured cells obtained from appropriate tissues. For

TABLE 10.1 Types of Collagen, and Their Distributions in Tissue and Properties

Type*	Composition	Distribution in Tissues	Examples of Known Disorders
I	$[\alpha1(I)]_2 [\alpha2(I)]$ $[\alpha1(I)]_3$	Bone, Tendon, Skin, Dentin, Fascia, Arteries	Osteogenesis imperfecta and Ehlers–Danlos syndrome. Both syndromes are clinically heterogeneous due to genetic defects that affect biosynthesis, assembly, post-translational modification, secretion, fibrillogenesis, or other extracellular matrix components
II	$[\alpha1(II)]_3$	Cartilage, Vitreous humor	Chondrodysplasia: spondyloepiphysial dysplasia, achondrogenesis, and Stickler syndrome
III	$[\alpha1(III)]_3$	Skin, Blood vessels, Uterus	EDS type IV: mutation affects synthesis, structure, or secretion
IV	$[\alpha1(IV)]_2 [\alpha2(IV)]$; other forms are also known	Basement membrane	Alport Syndrome: characterized by nephritis and sensorineural deafness
V	$[\alpha1(V)]_3$ $[(\alpha1(V)]_2 [\alpha2(V)]$ $[\alpha1(V)][\alpha2(V)][\alpha3(V)]$	Skin, Placenta, Blood vessels, Chorion uterus	EDS types I & II
VII	$[\alpha1(VII)]_3$	Anchoring fibrils	Epidermolysis bullosa
VIII	Not yet known	Cornea, Blood vessels, Network-forming collagen	
IX	$[\alpha1(IX)][\alpha2(IX)][\alpha2(IX)]$	Cartilage, Fibril associated collagen	
X	$[\alpha1(X)]_3$	Cartilage	
XI	$[\alpha1(XI)][\alpha2(XI)][\alpha1(II)]$	Cartilage	
XII	$[\alpha1(XII)]_3$	Soft tissues	

Collagen Types I, II, III, V, and XI are fibrillar.

example, human fibroblasts and smooth muscle cells synthesize both types I and III collagen; epithelial and endothelial cells synthesize type IV collagen; and chondroblasts synthesize type II collagen.

Structure and Function

Each collagen molecule contains three polypeptide chains coiled around each other in a triple helix. (See Clinical Case Studies 10.1 and 10.2). These chains, called **α-chains**, are designated by Roman numerals according to the chronological order of their discovery. The three polypeptide chains may be identical or may consist of two identical chains and one dissimilar chain.

Each α-polypeptide chain consists of about 1000 amino acid residues, of which every third following a glycine is also a glycine. Thus, the molecular formula of an α-chain may be written as $(Gly-X-Y)_{333}$, where X and Y represent amino acid residues other than glycine. In mammalian collagens, about 100 of the X positions are

occupied by proline residues, and 100 of the Y positions are occupied by 4-hydroxyproline residues. At a few X positions, 3-hydroxyproline residues are present; however, they only occur adjacent to 4-hydroxyproline. Most hydroxyproline residues are present as the *trans* isomer. Although all collagen polypeptides have the general structure $(Gly-X-Y)_n$, differences between the various collagen types are associated with the particular sequences of amino acid residues in the X and Y positions. Hydroxyproline residues are not common in proteins; other than the collagens, hydroxyproline residues are found in elastin, acetylcholinesterase, and the C1q subcomponent of the complement system (Chapter 33).

Another unique amino acid residue found in collagen is hydroxylysine, which occurs at the Y position. The number of hydroxylysine residues per polypeptide chain lies in the range of 5 to 50. The hydroxylysine residues provide the sites for α-O-glycosidic linkage with galactose or glucose or $\alpha(1 \rightarrow 2)$ glucosylgalactose. The collagens differ in their ratio of monosaccharide to disaccharide residues, as well as in their total carbohydrate content. For example, the carbohydrate contents of collagen from skin, cartilage, and basement membrane are about 0.5%, 4%, and more than 10%, respectively. Because collagen biosynthesis involves intracellular and extracellular post-translational protein modifications, e.g., hydroxylation, glycosylation, fibril formation, and formation of cross-links, a given genetically determined collagen may show a great deal of heterogeneity. This is particularly true for type I collagen. Fibronectin and type I collagens are interacting key components of extracellular matrix. They participate in several essential cellular functions such as adhesion, migration, growth, and differentiation [1].

Collagen also contains alanine residues in relatively high quantities. The only amino acid not found in collagen is tryptophan (an essential amino acid). Collagen consists essentially of four amino acids in abundant quantities and negligible amounts of almost all other amino acids. For this reason, collagen is an inferior protein nutritionally. It is very insoluble in water and, for the most part, indigestible. However, collagen can be converted to soluble and digestible products by hydrolysis of some covalent bonds (by the use of heat or by certain plant proteinases) to yield gelatin. In the human body, collagen is resistant to most proteinases. Polymerized fibrillar collagen is a stable component of the extracellular matrix, and its turnover rate is insignificant except in areas where tissue remodeling and repair occur. However, nonfibrillar collagens (e.g., type IV) are susceptible to proteolytic attack. Specific proteases (collagenases) cleave collagen at specific sites.

The amino acid sequence of collagen produces its unique secondary and tertiary structures. A tropocollagen molecule contains three polypeptide chains, each coiled into a left-handed helix having about three amino acid residues per turn (Figure 10.2). This type of helix is unique to collagen and differs significantly from the α-helix in its periodicity and dimensions (Chapter 4). The three helical polypeptides twist tightly around each other, except for the two short nonhelical regions at the C and N termini that form a right-handed triple-stranded superhelix. Tropocollagen has a molecular weight of about 300,000, a length of 300 nm, and a diameter of 1.5 nm. The glycine residue, which occurs at every third position, has the smallest R-group and is thus able to fit into the restricted space where the three polypeptide chains are closest to each other. Since the proline and hydroxyproline residues hinder free rotation around the N−C bond (Chapter 4), the polypeptide chain has a rigid and kinked conformation. These stereochemical properties are responsible for the superhelix. Hydrogen bond formation between the NH group of a glycyl residue in one chain and the CO group of a prolyl or other amino acid residue

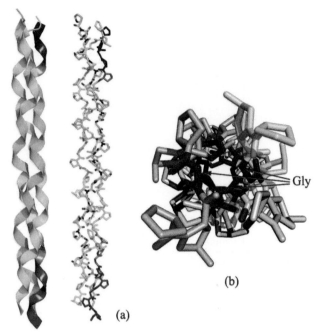

FIGURE 10.2   Structure of tropocollagen. (a) The coiling of three left-handed helices of collagen polypeptides. (b) A bird's-eye view of collagen triple helix.

in the X position of an adjacent chain stabilizes the triple helix. The hydrogen bond between the glycyl residue of one chain and the pro-lysyl residue of another chain is direct. If the prolyl residue is replaced by any other amino acid, the interchain hydrogen bonds occur through a water-bridged structure. These bonds are further stabilized by hydrogen bonding with the hydroxyl group of the trans-4-hydroxyprolyl residue, which occurs at the Y position. Additionally, the chains are held together by covalent linkages involving the lysine residues of native tropocollagen (Figure 10.3).

Polypeptide chains consisting only of glycine, proline, and hydroxyproline residues (in that order) form an extremely stable triple helix. Furthermore, the stability (thermal) of the triple helix decreases in the following order for repeating sequences of the chain: Gly−Pro−Hyp > Gly−Pro−Y > Gly−X−Pro > Gly−X−Y. In a given polypeptide chain of native tropocollagen, about one-third of the molecule contains the Gly−Pro−Hyp sequence and two-thirds involve Gly−X−Y, which decreases the stability of the triple helix. Amino acid residues other than proline and hydroxyproline that occupy the X and Y positions decrease the helix stability but are essential for the next level of organization of collagen—the formation of microfibrils.

Hydroxylysine glycosides occur at the Y position and may play a role in determining fibril diameter. The side chains of these amino acids project outward from the center of the triple helix, permitting hydrophobic and ionic

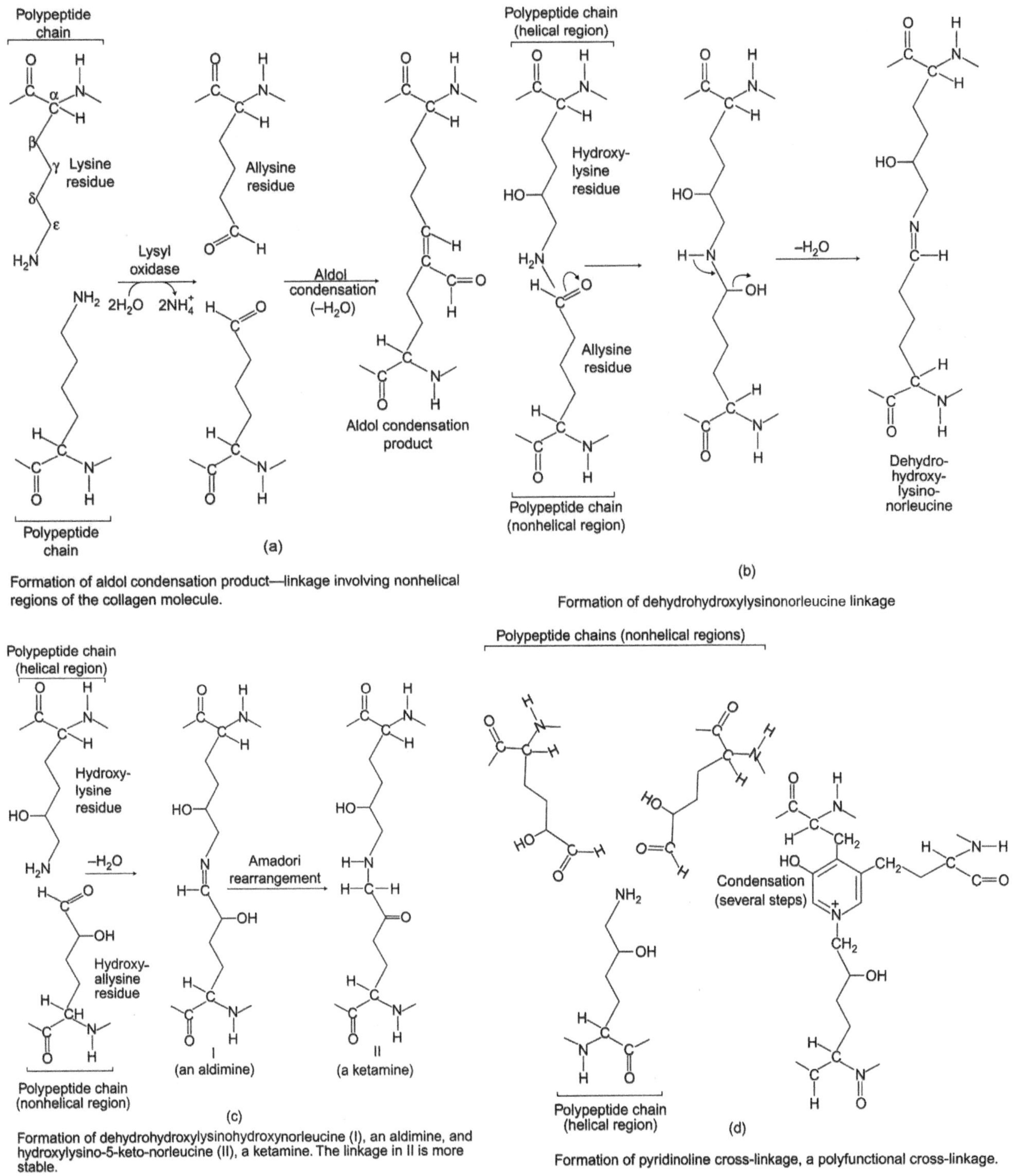

(a)

Formation of aldol condensation product—linkage involving nonhelical regions of the collagen molecule.

(b)

Formation of dehydrohydroxylysinonorleucine linkage

(c)

Formation of dehydrohydroxylysinohydroxynorleucine (I), an aldimine, and hydroxylysino-5-keto-norleucine (II), a ketamine. The linkage in II is more stable.

(d)

Formation of pyridinoline cross-linkage, a polyfunctional cross-linkage.

FIGURE 10.3 Formation of cross-links in collagen. All cross-links are derived from lysyl and hydroxylysyl amino acid residues. The initial step is the oxidative deamination of the α-NH$_2$ group of two amino acid residues located at certain strategic positions (e.g., short nonhelical segments at both ends of the collagen molecule), with the formation of corresponding aldehyde-containing residues, allysine, and hydroxyallysine. This reaction occurs extracellularly and is catalyzed by lysyl oxidase, a copper-dependent enzyme. The aldehyde groups react spontaneously with other aldehyde groups located in the adjacent chain of the same molecule or adjacent molecule. The structures do not show carbohydrate moieties for the sake of clarity.

interactions between tropocollagen molecules. These interactions determine the manner in which individual tropocollagen molecules aggregate to form microfibrils initially, then larger fibrils, and eventually fibers.

The microfibril, about 4 nm wide, consists of four to eight tropocollagen molecules that aggregate in a highly ordered and specific manner, owing to interactions of amino acid residues at the X and Y positions. In this ordered arrangement, each molecule is displaced longitudinally by about one-quarter of its length from its nearest neighbors. The longitudinally displaced tropocollagen molecules are not linked, and there is a gap of about 40 nm between the end of one triple helix and the beginning of the next. These holes may provide sites for deposition of hydroxyapatite $[Ca_{10}(PO_4)_6(OH)_2]$ crystals in the formation of bone (Chapter 35). Electron microscopic studies of negatively stained collagen fibrils reveal alternating light and dark regions. The light region, where the stain does not deposit, represents the overlapping of tropocollagen molecules; the dark region corresponds to a hole where the stain is deposited.

The tensile strength of collagen fibrils is determined by covalent cross-links involving lysyl and hydroxylysyl side chains. The packing arrangement of tropocollagen molecules provides tensile strength and also prevents sliding of molecules over one another. This organization does not allow for stretching, which does occur in elastin. The nature and extent of cross-linking depend on the physiological function and age of the tissue. With age, the density of cross-linkages increases, rendering connective tissue rigid and brittle. The arrangement of fiber bundles varies with the tissue; it is random in bone and skin, sheet-like in blood vessels, crossed in the cornea, and parallel in tendons.

Turnover of Collagen and Tissue Repair

The catabolism of collagen in the connective tissue matrix is carried out by collagenases [2]. Collagenases belong to a family of zinc-dependent enzymes known as **matrix metalloproteinases**. Part of the degradation may also involve neutral proteinases. A metalloenzyme specific for collagen catalyzes the cleavage of the triple helix at a single peptide bond, located at a distance from its N terminus corresponding to about three-quarters the length of the tropocollagen molecule. The cleavage sites in type I collagen are the peptide bonds of Gly−Ile of the α1-chain and of Gly−Leu of the α2-chain. The role and regulation of collagenase activity *in vivo* are unclear. The activity of the enzyme may be regulated via the formation of an enzyme−inhibitor complex or by the activation of a proenzyme, or both.

A substantial fraction (20%−40%) of newly synthesized polypeptide chains of collagen undergoes intracellular degradation. This degradation, which appears to occur in lysosomes, may be important in regulating the amount of collagen synthesized and in removing any defective or abnormal polypeptide chains that may be synthesized.

Elastin

Structure and Function

Elastin is a fibrous, insoluble protein that is not a glycoprotein but is present with collagen in the connective tissues. Connective tissues rich in elastic fibers exhibit a characteristic yellow color. Elastic fibers are highly branched structures responsible for physiological elasticity. They are capable of stretching in two dimensions and are found most notably in tissues subjected to continual high-pressure differentials, tension, or physical deformation. Elastin imparts to these tissues the properties of stretchability and subsequent recoil that depend only on the application of some physical force. Tissues rich in elastic fibers include the aorta and other vascular connective tissues, various ligaments (e.g., ligamentum nuchae), and the lungs. Microscopically, elastic fibers are thinner than collagen fibers and lack longitudinal striations.

Elastin fibers can be separated into amorphous and fibrillar components. The amorphous component consists of elastin, which is characterized by having 95% nonpolar amino acid residues and two unique lysine-derived amino acid residues, desmosine and isodesmosine.

Mature elastin is a linear polypeptide, tropoelastin, which has a molecular weight of about 72,000 and contains about 850 amino acid residues. Although glycine accounts for one-third of the residues, the repeat sequence Gly−X−Y characteristic of collagen is not present in elastin. Instead, glycine residues are present in the repeat units Gly−Gly−Val−Pro, Pro−Gly−Val−Gly−Val, and Pro−Gly−Val−Gly−Val−Ala. Elastin is relatively rich in the nonpolar amino acids: alanine, valine, and proline. In contrast to collagen, only a few hydroxyproline residues are present in elastin. Elastin contains no hydroxylysine or sugar residues.

A feature of mature elastin is the presence of covalent cross-links that connect elastin polypeptide chains into a fiber network. The major cross-linkages involve desmosine and isodesmosine, both of which are derived from lysine residues. Several regions rich in lysine residues can provide cross-links. Two such regions that contain peptide sequences that are repeated several times in tropoelastin have the primary structure −Lys−Ala−Ala−Ala−Lys− and −Lys−Ala−Ala−Lys−. The clustering of lysine residues with alanine residues provides the appropriate geometry for the formation of cross-links. Cross-linking elastin occurs extracellularly. The polypeptides are synthesized intracellularly by connective tissue cells (such as smooth muscle cells), according to the same principles for the formation of other export proteins (e.g., collagen; see

Chapter 23). After transport into the extracellular space, the polypeptides undergo cross-linking. A key step is the oxidative deamination of certain lysyl residues, catalyzed by lysyl oxidase (a Cu^{2+} protein). These lysyl residues are converted to very reactive aldehydes (α-aminoadipic acid δ-semialdehyde) known as **allysine residues**. Through an unknown mechanism, three allysine residues and one unmodified lysine residue react to form a pyridinium ring, alkylated in four positions, which is the basis for the cross-links (Figure 10.4). The cross-links may occur within the same polypeptide or involve two to four different polypeptide chains. Current models of elastin structure suggest that only two polypeptide chains are required to form the desmosine cross-link, one chain donating a lysine and an allysine residue, the other contributing two allysine residues. Elastin polypeptides are also cross-linked by condensation of a lysine residue with an allysine residue, followed by reduction of the aldimine to yield a lysinonorleucine residue (Figure 10.5). These linkages are present to a much lesser extent than in collagen.

Formation of cross-linkages in elastin can be prevented by inhibition of lysyl oxidase. Animals maintained on a copper-deficient diet and those administered lathyrogens such as β-aminopropionitrile or α-aminoacetonitrile develop connective tissue abnormalities. Lathyrogens are so called because of their presence in certain peas of the genus *Lathyrus*. They are noncompetitive inhibitors of lysyl oxidase, both *in vitro* and *in vivo*. Consumption of certain types of sweet pea (e.g., *Lathyrus odoratus*) can lead to lathyrism. In osteolathyrism, the abnormalities involve bone and other connective tissues, the responsible compounds being the above-mentioned nitriles. The toxic agent responsible for neurolathyrism, β-N-oxalyl-α,β-diaminopropionic acid, has been isolated from *Lathyrus sativus*, but its mode of action is not known. Thus, copper deficiency or administration of lathyrogens in animals prevents the formation of the highly insoluble cross-linked elastin, but results in the accumulation of soluble elastin. Soluble elastin obtained in this manner is particularly useful in structural studies. Both collagen and elastin are required for the functioning of the

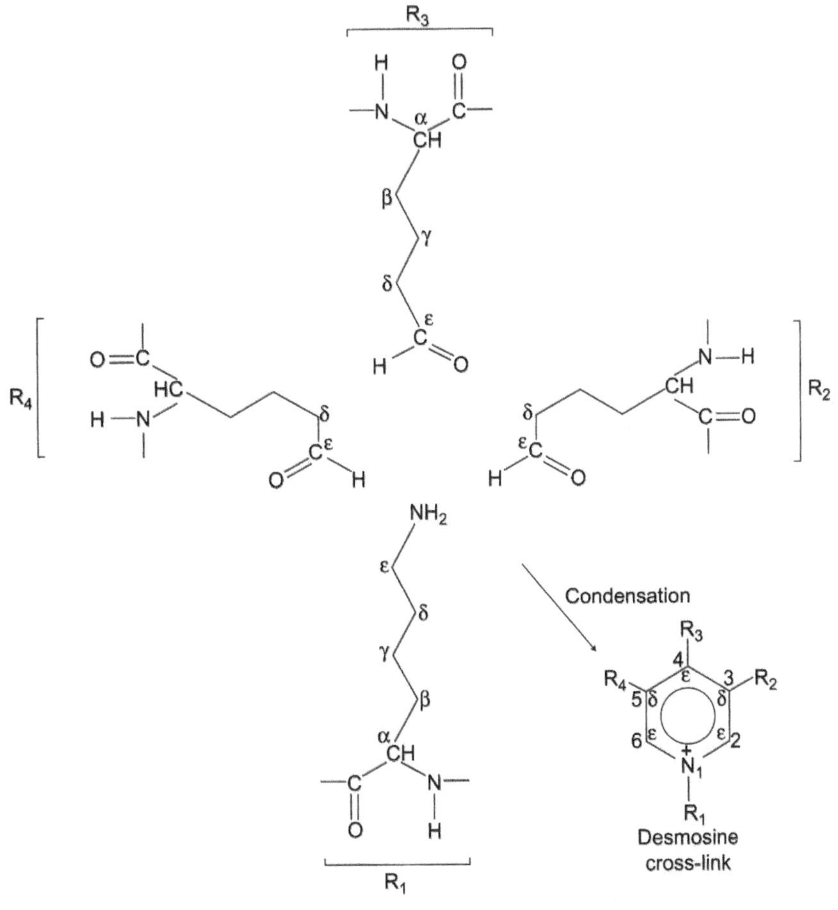

FIGURE 10.4 Formation of desmosine and isodesmosine covalent cross-links in elastin. Three allysine residues (R2, R3, and R4) and one lysyl residue (R1) condense to give a desmosine cross-link. The allysine residues (ε-aldehydes) are derived from the oxidative deamination of lysyl residues. The isodesmosine cross-link is formed similarly, except that it contains a substitution at position 2 rather than at position 4, along with substitutions at 1, 3, and 5 on the pyridinium ring.

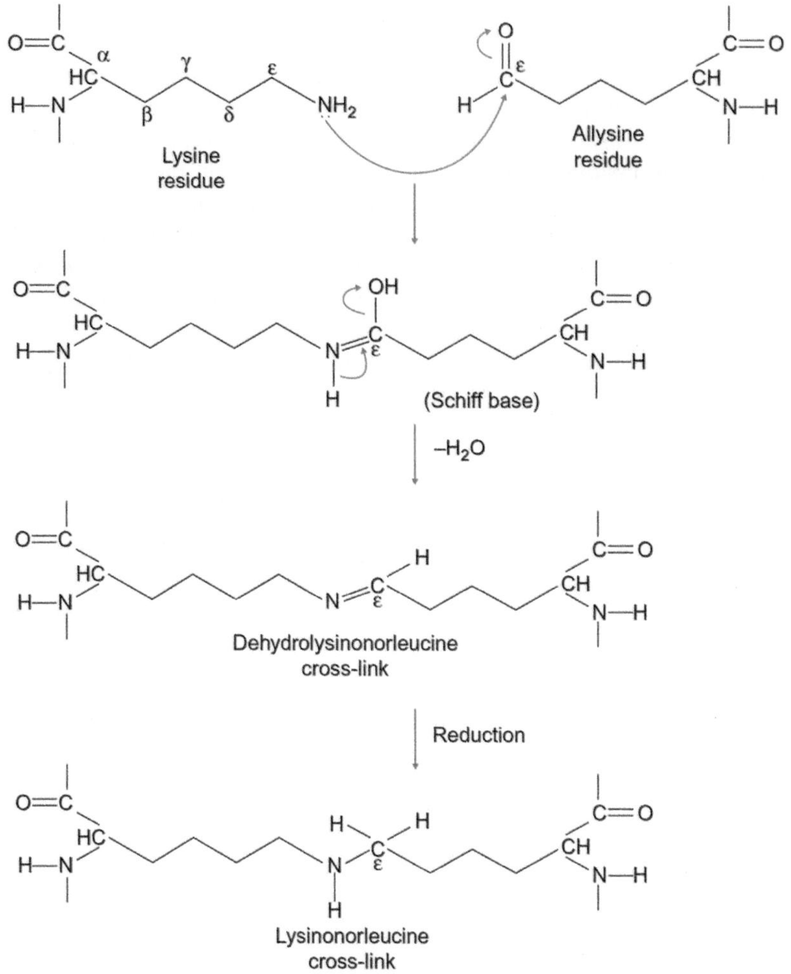

FIGURE 10.5 Formation of the lysinonorleucine cross-link between an allysine and a lysine residue in elastin.

major arteries, such as aorta, to provide elasticity and strength. Both aging and arterial diseases produce deleterious changes leading to stiffening aneurysms and rupture of the blood vessels. Ruptures of aortic aneurysms caused by inherited disorders, hypertension, and other comorbid abnormalities can have fatal outcomes [3,4].

In the degeneration connective disease of articular cartilage, administration of a heterocyclic compound known as **kartogenin** has rejuvenated the affected tissue in animal studies. The mechanism of kartogenin activity involves stimulating of endogenous progenitor chondrocytes to become matrix synthesizing (e.g., collagen type II) chondrocytes [5].

Fibrillin

A major fibrillar component is a glycoprotein called **fibrillin**. The structure of fibrillin consists of several cysteine-rich motifs and exhibits a multidomain organization similar to epidermal growth factor. The structure is stabilized by disulfide linkages. Fibrillin monomers undergo aggregation extracellularly into supramolecular structures. The fibrillar structures have a diameter of 10−12 nm and surround the amorphous component elastin. During fetal development, the fibrillar component appears first in the extracellular matrix. With increasing fetal age and with maturation of the fibers, the amorphous component is deposited within the framework of the microfibrillar bundles, producing mature, elastic microfibrils.

Fibrillin is encoded by a gene located in the long arm of chromosome 15. Mutations in the fibrillin gene lead to an autosomal dominant trait known as **Marfan's syndrome**. The incidence of this disorder is 1:10,000, and 15%−30% of cases are caused by new mutations in the fibrillin gene. Consistent with the function of fibrillin in the elastic connective tissues, the clinical manifestations present as disorders of cardiovascular, musculoskeletal, and ophthalmic systems. For example, dissecting aneurysm of the aorta, preceded by a dilatation, is a potentially fatal cardiac manifestation.

The clinical management of Marfan's syndrome requires a multidisciplinary approach. β-Adrenergic receptor and angiotensin II receptor blockades delay or prevent cardiovascular disorders. The inhibition of angiotensin II action downregulates the signaling pathway mediated by transforming growth factor β (TGFβ). The downregulation of TGFβ leads to inhibition of metalloproteinases with a reduction in extracellular matrix breakdown.

Proteoglycans

Proteoglycans are high molecular weight, complex molecules with diverse structures and functions. They are polyanionic substances containing a core protein to which at least one glycosaminoglycan (also known as **mucopolysaccharide**) chain is covalently attached. Proteoglycans are major components of connective tissue and participate with other structural protein constituents, namely, collagen and elastin, in the organization of the extracellular matrix.

Types, Structures, and Functions of Glycosaminoglycans

Six classes of glycosaminoglycans have been described. All are heteropolysaccharides and contain repeating disaccharide units. The compositions and structures of the repeating disaccharide units are shown in Table 10.2 and Figure 10.6. In five glycosaminoglycans, the disaccharide

units consist of amino sugars (usually D-galactosamine) alternating with uronic acids. In one type, keratan sulfate, the uronic acid is replaced by galactose. The amino sugars are usually present as N-acetyl derivatives. In heparin and heparan sulfate, however, most amino sugars are found as N-sulfates in sulfamide linkage, with a small number of glucosamine residues as N-acetyl derivatives. With the exception of hyaluronate, all glycosaminoglycans contain sulfate groups in ester linkages with the hydroxyl groups of the amino sugar residues.

The presence of carboxylate and sulfate groups provides the glycosaminoglycans with a high negative charge density and makes them extremely hydrophilic. Almost all water in the intercellular matrix is bound to glycosaminoglycans, which assume a random conformation in solution, occupying as much solvent space as is available by entrapping the surrounding solvent molecules. In solution, glycosaminoglycans also undergo aggregation and impart a high viscosity. In order to maintain electrical neutrality, negatively charged (anionic) groups of glycosaminoglycans, fixed to the matrix of the connective tissue, are neutralized by positively charged (cationic) groups (e.g., Na^+). Osmotic pressure within the matrix is thereby elevated, contributing turgor to the tissue. Thus, proteoglycans, because they contain glycosaminoglycans, are polyanionic and usually occupy large hydrodynamic volumes relative to glycoproteins or

TABLE 10.2 Composition, Properties, and Distribution of Glycosaminoglycans

Glycosaminoglycan (Mol Wt Range)	Amino Sugar	Uronic Acid	Type of Sulfate Linkage*	Tissue Distribution
Hyaluronate (4 to 80×10^6)	D-Glucosamine	D-Glucuronate	None	Connective tissues, cartilage, synovial fluid, vitreous humor, umbilical cord
Chondroitin sulfate (5000 to 50,000)	D-Galactosamine	D-Glucuronate	4-O- and/or 6-O-sulfate on galactosamine	Cartilage, bone, skin, cornea, blood vessel walls
Dermatan sulfate (15,000 to 40,000)	D-Galactosamine	L-Iduronate, D-Glucuronate	4-O-sulfate on galactosamine; 2-O-sulfate on iduronate	Skin, heart valve, tendon, blood vessel walls
Keratan sulfate (4000 to 19,000)	D-Glucosamine	None (but contains D-Galactose)	6-O-sulfate on both carbohydrate residues	Cartilage, cornea, intervertebral disks
Heparan sulfate (10^4 to 10^5)	D-Glucosamine	D-Glucuronate (major), L-Iduronate (minor)	6-O-sulfate and N-sulfate (or N-acetyl) on glucosamine; 2-O-sulfate on iduronate	Lung, blood vessel walls, many cell surfaces
Heparin (10^3 to 10^6)	D-Glucosamine	L-Iduronate (major) D-Glucuronate (minor)	2-O-sulfate on iduronate; 6-O-sulfate and N-sulfate (or N-acetyl) on glucosamine	Lung, liver, skin, intestinal mucosa (mast cells)

Extent of sulfation is variable.

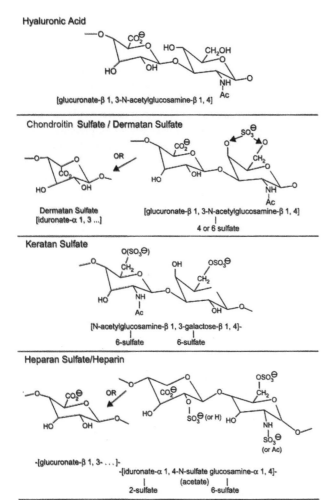

Hyaluronic Acid

[glucuronate-β 1, 3-N-acetylglucosamine-β 1, 4]

Chondroitin Sulfate / Dermatan Sulfate

Dermatan Sulfate
[iduronate-α 1, 3 ...]

[glucuronate-β 1, 3-N-acetylglucosamine-β 1, 4]
|
4 or 6 sulfate

Keratan Sulfate

[N-acetylglucosamine-β 1, 3-galactose-β 1, 4]-
| |
6-sulfate   6-sulfate

Heparan Sulfate/Heparin

-[glucuronate-β 1, 3- . . .]-
-[iduronate-α 1, 4-N-sulfate glucosamine-α 1, 4]-
| (acetate) |
2-sulfate   6-sulfate

FIGURE 10.6 Basic repeating disaccharide units of six classes of gly-cosaminoglycans. Chondroitin sulfate and dermatan sulfate differ in type of principal uronic acid residue; the former contains glucuronate and the latter, iduronate. Heparin and heparan sulfate differ in that heparin contains more iduronate 2-sulfate and N-sulfated glucosamine residues than does heparan sulfate. In the sulfate-containing glycosaminoglycans, the extent of sulfation is variable. All of these molecules have a high negative charge density and are extremely hydrophilic. *[Reproduced with permission from V.C. Hascall, J.H. Kimura, Proteoglycans: isolation and characterization. Meth. Enzymol. 82(A), 1982.]*

globular proteins of equivalent molecular mass. Usually, they have high buoyant densities, and this property has been utilized in their isolation and characterization. The physical properties of connective tissue also depend on collagen or elastin fibers. Collagen fibers resist stretching and provide shape and tensile strength; the hydrated network of proteoglycans resists compression and provides both swelling and pressure to maintain volume. Thus, the properties of these different macromolecules are complementary, and their relative compositions vary according to tissue function. For example, sulfated

glycosaminoglycans tend to provide a solid consistency, whereas hyaluronate provides a softer, more fluid consistency. The vitreous space behind the lens of the eye is filled with a viscous, gelatinous solution of hyaluronate free of fibrous proteins and transparent to light. Hyaluronate is not sulfated, and there is no evidence that it is linked to a protein molecule, as are the other glyco-saminoglycans. Hyaluronate is also present in synovial fluid in joint cavities that unite long bones, bursae, and tendon sheaths. The viscous and elastic properties of hya-luronate contribute to the functioning of synovial fluid as a lubricant and shock absorber.

Hyaluronate is depolymerized by hyaluronidase, which hydrolyzes the $\beta(1 \rightarrow 4)$ glycosidic linkages between N-acetylglucosamine and glucuronate. Some pathogenic bacteria secrete hyaluronidase, which breaks down the protective barrier of hyaluronate and renders the tissue more susceptible to infection. Hyaluronidase, found in spermatozoa, may facilitate fertilization by hydrolyzing the outer mucopolysaccharide layer of the ovum, thereby enabling the sperm to penetrate it. Hyaluronidase (prepared from mammalian testes) is used therapeutically to enhance dispersion of drugs administered in various parts of the body.

Heparin is a heterogeneous glycosaminoglycan found in tissues that contain mast cells (e.g., lungs and perivas-cular connective tissue). Its primary physiological function is unknown, but it is a powerful inhibitor of blood clotting and is used therapeutically for this purpose (Chapter 34).

Tendons have a high content of collagen and sulfated glycosaminoglycans (chondroitin and dermatan sulfates). Tendons are fibrous cords that fuse with skeletal muscle at each end and penetrate bones at the two sides of a joint. Thus, tendons are aligned along their long axis, providing flexible strength in the direction of the muscle pull.

Remarkable progress has been achieved in understand-ing the organization of the macromolecular components of cartilage. Cartilage is one type of dense connective tis-sue; bone is the other (Chapter 35). Cartilage is less resis-tant to pressure than bone, and its weight-bearing capacity is exceeded by that of bone. It has a smooth, resilient sur-face. During the growth and development of an embryo, cartilage provides the temporary framework for the for-mation and development of bone. In postnatal life, it per-mits the long bones of the extremities to grow until their longitudinal growth ceases. Its smooth surface provides for freely movable joints (e.g., knee and elbow). The properties of cartilage depend on its content of collagen (or collagen and elastin) and proteoglycans. Cells respon-sible for the synthesis and maintenance of cartilage are known as **chondrocytes**.

The proteoglycan monomer of cartilage consists of glycosaminoglycan chains of chondroitin sulfate and keratan sulfate covalently linked to a core protein about 300 nm long with a molecular weight of 250,000. Each molecule of core protein contains about 80 chondroitin sulfate and 100 keratan sulfate chains. The linkage between the oligosaccharides and the protein can be O-glycosidic between xylose and serine residues, O-glycosidic between N-acetylgalactosamine and serine or threonine residues, or N-glycosidic between N-acetylglucosamine and the amide nitrogen of asparagine. The glycosaminoglycan chains with their negatively charged groups extend out of the core protein to occupy a large volume and to entrap surrounding solvent. One end of the core protein has relatively few or no attached glycosaminoglycan chains and resembles a glycoprotein; at this end, the core protein possesses an active site that binds to a very long filament of hyaluronate by a noncovalent interaction at intervals of 30 nm. The interaction between the core protein and the hyaluronate is aided by a protein (M.W. 50,000) known as **link protein**. The macromolecular complex is negatively charged and highly hydrated; it interacts electrostatically with basic charges on collagen fibrils, which define tissue space and provide tensile strength. The interspersed proteoglycan aggregates provide a hydrated, viscous gel for resistance to compressive loads. Proteoglycans originating in different tissues and exhibiting differences in molecular structure are presumed to be products of different genes. Synthesis and secretion of proteoglycans involve several different phases of the internal membrane systems of the cell. Synthesis of the protein component occurs on ribosomes attached to the rough endoplasmic reticulum, with translocation of the newly synthesized polypeptide chains to the cisternal side of the membrane. The polypeptide chains travel through the cisternae of the endoplasmic reticulum to the Golgi apparatus, where the oligosaccharide side chains are synthesized by the addition of one sugar residue at a time. Further modification of the sugar residues (the conversion of selected residues of D-glucuronate to L-iduronate, and sulfation) takes place after completion of the oligosaccharide. The fully synthesized proteoglycans are then transferred from the Golgi apparatus, in the form of vesicles, to the cell membrane for the purpose of secretion into the extracellular spaces.

The proteoglycan content in animal tissues appears to change with aging. Tissues rich in keratan sulfate show a continual increase in the amount of this proteoglycan throughout life. The amounts of chondroitin sulfate in cartilage and in the nucleus pulposus (part of the intervertebral disk) and of hyaluronic acid in skin decrease with age. Administration of growth hormone to an older animal appears to reverse the pattern of proteoglycan synthesis, making it similar to that of a younger animal. Some of the effects of growth hormone are mediated indirectly by a group of peptides known as **somatomedins**, which are secreted by the liver (and possibly by the kidney) under the influence of growth hormone (Chapter 29). Somatomedin A, formerly known as **sulfation factor**, exhibits cartilage-stimulating activity. Somatomedin affects the incorporation of sulfate into proteoglycans. Testosterone, a steroid hormone (Chapter 32), increases the rate of hyaluronate synthesis. Insulin deficiency causes diminished synthesis of proteoglycans in rats that are rendered diabetic experimentally; synthesis is restored to normal following insulin administration. Some chronic complications of diabetes mellitus, namely, accelerated vascular degeneration, poor wound healing, and increased susceptibility to infection, may be partly attributable to altered proteoglycan metabolism.

Chondroitin sulfate and glucosamine have been used as dietary supplements in the management of knee osteoarthritis. However, randomized, double blind, placebo-controlled studies are required to support the clinical use of these supplements.

Turnover of Proteoglycans and Role of Lysosomes

Proteoglycans undergo continuous turnover at rates dependent upon the nature of the proteoglycan and tissue location. Their half-life is between one and several days. Degradation of proteoglycans is initiated by proteolytic enzymes that release glycosaminoglycans; the latter are subsequently degraded by lysosomal enzymes. Some of the products of hydrolysis (e.g., dermatan sulfate and heparan sulfate) are excreted in the urine.

Lysosomes are subcellular organelles in which a wide range of catabolic enzymes are stored in a closed, protective membrane system. They are major sites of intracellular digestion of complex macromolecules derived from both intracellular (autophagic) and extracellular (heterophagic) sources. Lysosomal enzymes show optimal activity at acidic pH. The pattern of enzymes in lysosomes may depend on the tissue of origin, as well as on the physiological or developmental state of the cells.

Lysosomal enzymes are synthesized on the ribosomes of the rough endoplasmic reticulum, passed through the Golgi apparatus, and packaged into vesicles. The hydrolyases are glycoproteins, some of which contain mannose 6-phosphate markers necessary for the normal uptake of glycoproteins into lysosomes. Thus, carbohydrates may also serve as determinants of recognition in the intracellular localization of the glycoproteins following their synthesis. In two biochemically related disorders of lysosomal function, inclusion (I)-cell disease (mucolipidosis II) and pseudo-Hurler's polydystrophy (mucolipidosis III), the lesion is in the post-translational modification step for acid hydrolases destined to be packaged into lysosomes. In normal cells, these acid hydrolyases are glycoproteins carrying mannose 6-phosphate markers that direct them to lysosomes through a receptor-mediated process. In other words, the presence of phosphomannose residues on the newly synthesized acid hydrolases and of phosphomannose receptors on selected membranes leads to the segregation of these enzymes in the Golgi apparatus, with subsequent translocation into lysosomes. In mucolipidosis II, the acid hydrolases are not phosphorylated; in mucolipidosis III, the enzymes either are not phosphorylated or have significantly diminished phosphate content. The absence of mannose 6-phosphate leads to defective localization of acid hydrolases. Instead of being packaged in lysosomes, the acid hydrolases are exported out of the cell; thus, the enzyme activity in plasma reaches high levels. These enzymes could cause indiscriminate damage. At least eight acid hydrolases (glycosidases, sulfatases, and cathepsins) appear to be affected in this manner. However, not all lysosomal enzymes and tissue cells are affected. For example, lysosomal acid phosphatase and α-glucosidase, and hepatocytes and neurons, appear to be spared from this defect. In mucolipidosis II and III, large inclusions of undigested glycosaminoglycans and glycolipids occupy almost all of the cytoplasmic space in cultured skin fibroblasts. In addition to the phosphorylation defect, the acid hydrolases are much larger than their normal counterparts, presumably owing to lack of the limited proteolysis of the hydrolases that occurs in normal lysosomes. Both disorders are inherited as autosomal recessive traits, affect primarily connective tissue, and are characterized by psychomotor retardation, skeletal deformities, and early death.

The segregation of lysosomal enzymes into lysosomes requires carbohydrate recognition markers (phosphomannose in some) and also the formation of coated vesicles into which the enzymes are sequestered. Coated vesicles shuttle macromolecules between organelles and may be responsible for selectivity in intercompartmental transport (e.g., from endoplasmic reticulum to Golgi, from Golgi to lysosomes). The major component of coated vesicles is clathrin, a nonglycosylated protein (M.W. 180,000). It is located on the outer surface (cytoplasmic side) of the coated vesicles. Clathrin and its tightly bound light chains (M.W. 33,000 and 36,000) form flexible lattices which function as structural scaffolds that surround the vesicles.

In addition to their role in intracellular transport, coated vesicles are involved in receptor-mediated endocytosis. This process accomplishes internalization of macromolecules (ligands) by binding them to receptors on the cell membranes located in specialized regions of clathrin-containing coated pits, which invaginate into the cell to form coated vesicles. Inside the cell, the vesicles lose their clathrin coat (which may be reutilized) and fuse with one another to form endosomes, whose contents are acidified by proton pumps driven by free energy of hydrolysis derived from ATP. In the endosome, the ligand and receptor undergo dissociation. The endosome fuses with the primary lysosome to form a secondary lysosome. Receptors may also be recycled or degraded. Thus, there are several pathways of receptor-mediated endocytosis. In some cells, the receptors migrate continuously to coated pits and undergo internalization whether or not ligands are bound to them (e.g., receptors for low-density lipoproteins, transferrin, and asialoglycoproteins). In other cells, the receptors are diffusely distributed and do not migrate to coated pits unless they are bound with ligands (e.g., epidermal growth factor).

Mucopolysaccharidosis

Mucopolysaccharidosis (MPS) encompasses disorders in which undegraded or partly degraded glycosaminoglycans accumulate in the lysosomes of many tissues owing to a deficiency of specific lysosomal enzymes (see Clinical Case Study 10.3). Table 10.3 lists the missing enzymes and gives relevant clinical and laboratory findings. These disorders, with the exception of Hunter's syndrome, which is an X-linked trait, are inherited in the autosomal recessive manner. All of the deficient enzymes are acid hydrolases except for acetyltransferase in Sanfilippo's syndrome type C. In mucopolysaccharidosis, the catabolism of heparan sulfate, dermatan sulfate, and keratan sulfate is affected. Their degradation proceeds from the nonreducing end of the carbohydrate chain by the sequential actions of lysosomal exoglycosidases, exosulfatases, and an acetyltransferase (Figures 10.7 through 10.9).

These disorders are rare; collectively, they may occur in 1 in 20,000 live births. Since proteoglycans are widely distributed in human tissues, the syndromes can affect a wide variety of tissues; thus, the clinical features vary considerably. All types are characterized by reduced life expectancy, with the exception of Scheie's syndrome. All types are characterized by skeletal abnormalities, which are particularly severe in types IV and VI. Types IH, IS, IV, VI, and VII usually involve clouding of the cornea. Types IH, II, III, and VIII are characterized by severe mental retardation, while types IS, IV, and VI exhibit normal intelligence. Patients with different enzyme deficiencies may exhibit phenotypic similarities (pleiotropism), and they may also exhibit variation in clinical severity

TABLE 10.3 Classification of the Mucopolysaccharidoses (MPS)*

Number	Eponym	Clinical Manifestations	Enzyme Deficiency	Glycosaminoglycan Affected
MPS I H	Hurler	Corneal clouding, dysostosis multiplex, organomegaly, heart disease, mental retardation, death in childhood	α-L-Iduronidase	Dermatan sulfate, heparan sulfate
MPS I S	Scheie	Corneal clouding, stiff joints, normal intelligence and lifespan	α-L-Iduronidase	Dermatan sulfate, heparan sulfate
MPS I H/S	Hurler/Scheie	Phenotype intermediate between I H and I S	α-L-Iduronidase	Dermatan sulfate, heparan sulfate
MPS II (severe)	Hunter (severe)	Dysostosis multiplex, organomegaly, no corneal clouding, mental retardation, death before 15 years	Iduronate sulfatase	Dermatan sulfate, heparan sulfate
MPS II (mild)	Hunter (mild)	Normal intelligence, short stature, survival to 20s to 60s	Iduronate sulfatase	Dermatan sulfate, heparan sulfate
MPS III A	Sanfilippo A	Profound mental deterioration, hyperactivity, relatively mild somatic manifestations	Heparan N-sulfatase	Heparan sulfate
MPS III B	Sanfilippo B	Phenotype similar to III A	α-N-acetylglucosaminidase	Heparan sulfate
MPS III C	Sanfilippo C	Phenotype similar to III A	Acetyl CoA:α-glucosaminide N-acetyltransferase	Heparan sulfate
MPS III D	Sanfilippo D	Phenotype similar to III A	N-acetylglucosamine-6-sulfatase	Heparan sulfate
MPS IV A	Morquio A	Distinctive skeletal abnormalities, corneal clouding, odontoid hypoplasia; milder forms known to exist	Galactose 6-sulfatase	Keratan sulfate, chondroitin 6-sulfate
MPS IV B	Morquio B	Spectrum of severity as in IV A	1-Galactosidase	Keratan sulfate
MPS V	No longer used			
MPS VI	Maroteaux–Lamy	Dysostosis multiplex, corneal clouding, normal intelligence; survival to teens in severe form; milder forms known to exist	N-acetylgalactosamine 4-sulfatase (arylsulfatase B)	Dermatan sulfate
MPS VII	Sly	Dysostosis multiplex, hepatosplenomegaly; wide spectrum of severity, including hydrops fetalis and neonatal form	β-Glucuronidase	Dermatan sulfate, heparan sulfate, chondroitin 4-, 6-sulfates
MPS VIII	No longer used			

Reproduced with permission from E. F. Neufeld and J. Muenzer, In Metabolic basis of inherited disease, 7th ed., C. R. Scriver, A. L. Beaudet, W. S. Sly, D. Valle (Eds.), New York: McGraw-Hill 1995, 2466.

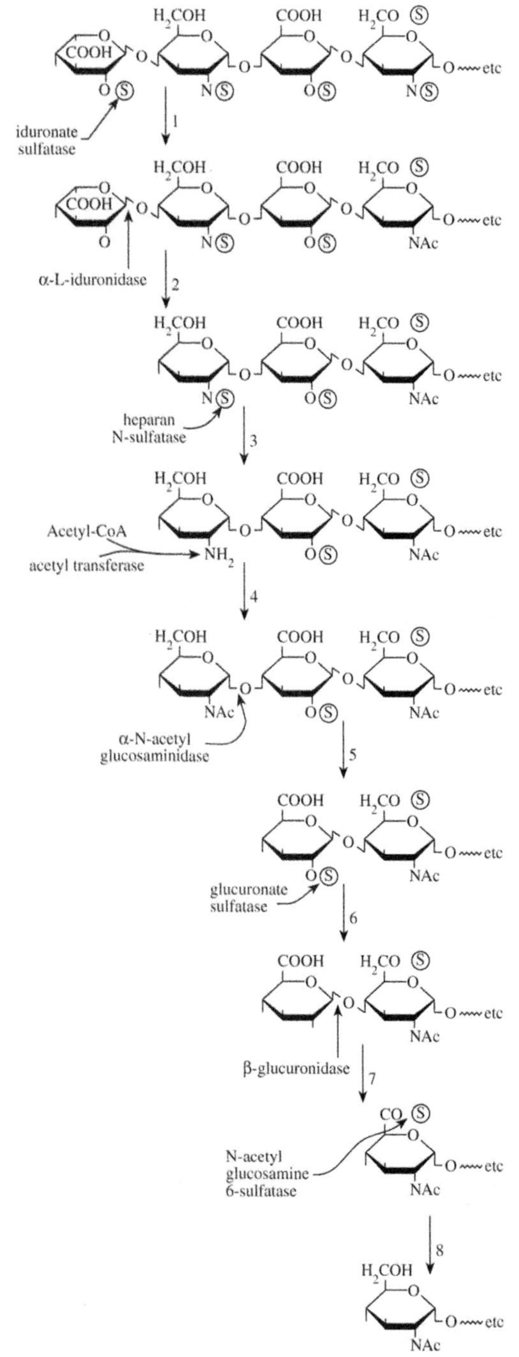

FIGURE 10.7 Stepwise degradation of heparan sulfate. The deficiency diseases corresponding to the numbered reactions are: 1 = mucopolysaccharidosis (MPS) II, Hunter's syndrome; 2 = MPS I, Hurler's, Scheie's, and Hurler–Scheie's syndromes; 3 = MPS III A, Sanfilippo's syndrome type A; 4 = MPS III C, Sanfilippo's syndrome type C; 5 = MPS III B, Sanfilippo's syndrome type B; 6 = no deficiency disease yet known; 7 = MPS VII, Sly's syndrome; 8 = MPS III D, Sanfilippo's syndrome type D. The schematic drawing depicts all structures known to occur within heparan sulfate and does not imply that they occur stoichiometrically. Very few of the glucuronic acid residues are sulfated. [*Reproduced with permission from E.F. Neufeld, J. Muenzer. In Metabolic basis of inherited disease, 7th ed., C.R. Scriver, A.L. Beaudet, W.S. Sly, D. Valle (Eds.). New York: McGraw-Hill, 1995, 2468.*]

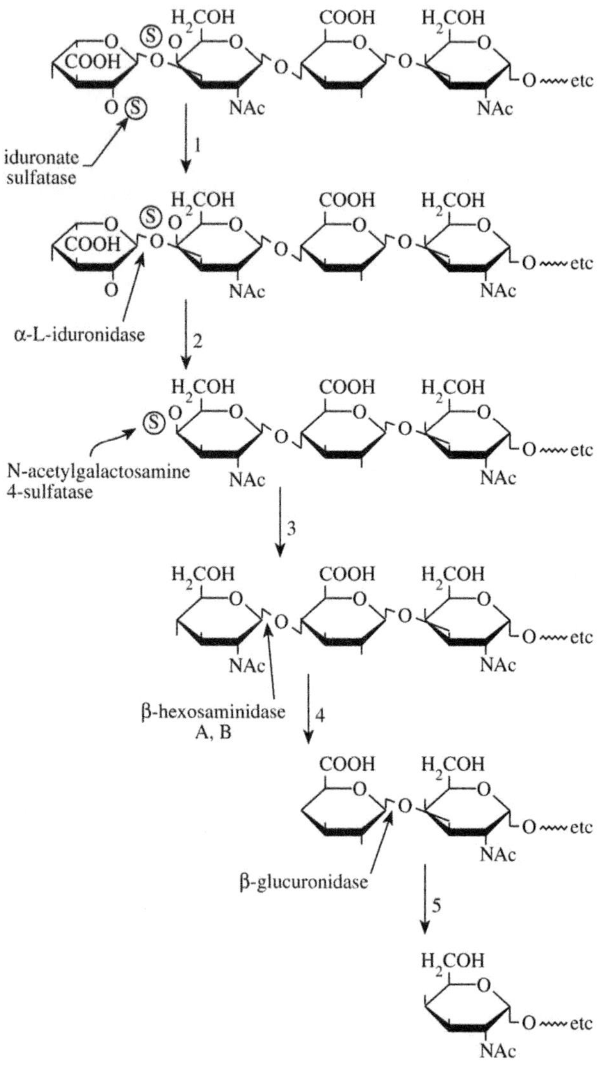

FIGURE 10.8 Stepwise degradation of dermatan sulfate. The deficiency diseases corresponding to the numbered reactions are: 1 = MPS II, Hunter's syndrome; 2 = MPS I, Hurler's, Scheie's, and Hurler–Scheie's syndrome; 3 = MPS VI, Maroteaux–Lamy syndrome; 4 = Sandhoff's disease; and 5 = MPS VII, Sly's syndrome. This schematic drawing depicts all structures known to occur within dermatan sulfate and does not imply that they occur in equal proportion. For instance, only a few of the L-iduronic acid residues are sulfated, and L-iduronic acid occurs much more frequently than glucuronic acid. [*Reproduced with permission from E.F. Neufeld, J. Muenzer. In Metabolic basis of inherited disease, 7th ed., C.R. Scriver, A.L. Beaudet, W.S. Sly, D. Valle (Eds.). New York: McGraw-Hill, 1995, 2467.*]

with the same enzyme deficiency (allelic variants). The enzyme deficiency can be established by assays on peripheral lymphocytes or cultured fibroblasts. Prenatal diagnosis is possible but requires successful culture of amniotic fluid cells and assay of specific enzymes. No specific therapy is available. Management focuses on providing prognostic information and counseling.

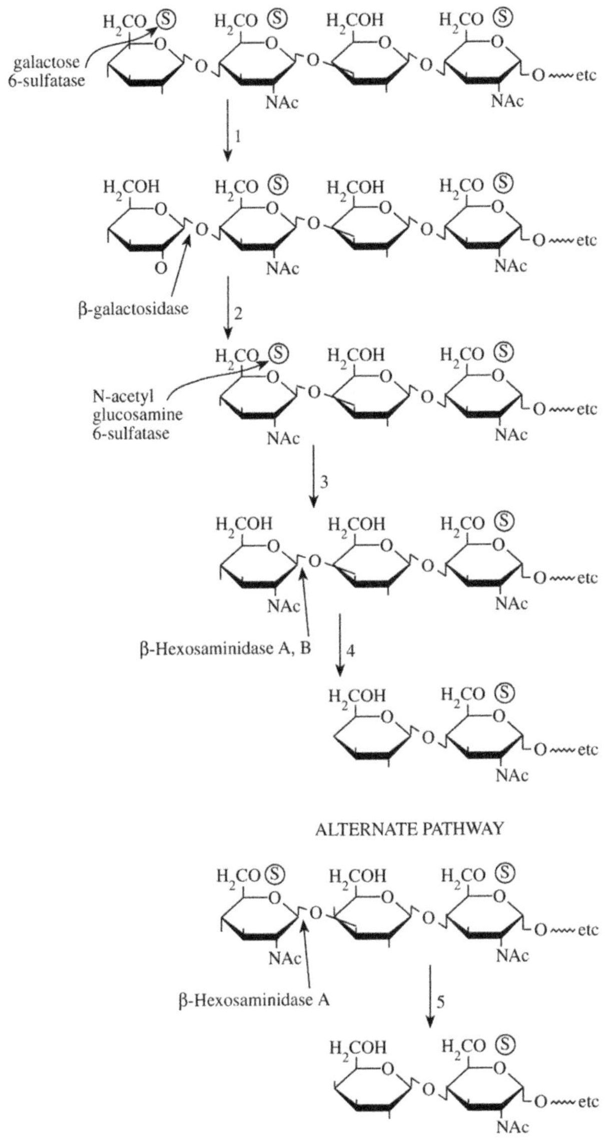

FIGURE 10.9 Stepwise degradation of keratan sulfate. The deficiency diseases corresponding to the numbered reactions are: 1 = MPS IV A, Morquio syndrome type A; 2 = MPS IV B, Morquio syndrome type B; 3 = MPS III D, Sanfilippo's syndrome type D; 4 = Sandhoff's disease; and 5 = Tay–Sachs and Sandhoff's disease. The alternate pathway releases intact N-acetylglucosamine-6-sulfate, a departure from the usual stepwise cleavage of sulfate and sugar residues. *[Reproduced with permission from E.F. Neufeld, J. Muenzer. In Metabolic basis of inherited disease, 7th ed., C.R. Scriver, A.L. Beaudet, W.S. Sly, D. Valle (Eds.). New York: McGraw-Hill, 1995, 2468.]*

Peptidoglycans

Peptidoglycans are components of bacterial cell walls and consist of cross-linked heteropolysaccharide chains; they may also exhibit variation in short peptide chains. These cell walls bear the antigenic determinants; when exposed to them, humans develop specific antibodies to defend against

bacteria. Bacterial virulence is also related to substances associated with the cell wall. Bacterial cell wall synthesis is the target for the action of the penicillins and cephalosporins.

Bacterial cell walls are rigid and complex, enable the cells to withstand severe osmotic shock, and survive in a hypotonic environment. The contents of a bacterium can exert an osmotic pressure as high as 20 atm. At cell division, the cell walls rupture and reseal rapidly.

Lectins

The lectins are a group of proteins, originally discovered in plant seeds (now known to occur more widely), which bind carbohydrates and agglutinate animal cells. They have two or more stereospecific sites that bind noncovalently with the terminal (and often penultimate) residue at the nonreducing end of an oligosaccharide chain. A number of plant lectins have been purified and their binding properties investigated. Wheat germ agglutinin binds to N-acetylglucosamine and its glycosides; concanavalin A from jack beans binds to mannose, glucose, and glycosides of mannose and glucose; peanut agglutinin binds to galactose and galactosides; and red kidney bean lectin binds to N-acetylglucosamine. Since lectins have a high affinity for specific sugar residues, they have been used to identify specific carbohydrate groups and used in the purification of carbohydrate-containing compounds. Lectins may be involved in carbohydrate transport, specific cellular recognition, embryonic development, cohesion, or binding of carbohydrates. As noted in Chapter 9, hepatocytes bind to serum glycoproteins with exposed galactose residues. This binding is thought to be a lectin-mediated clearance of partially degraded glycoproteins. Hepatocytes also contain a Ca^{2+}-dependent fucose-binding lectin. A lectin that binds to N-acetylglucosamine- and mannose-terminated glycoproteins in reticuloendothelial cells has been identified. Many lectins are glycoproteins.

Some lectins cause agglutination of red blood cells and can be used in typing of blood groups. Soybean lectin, a galactose-binding protein, binds selectively to T lymphocytes and causes their agglutination. Thus, it has been used in selective removal of mature T lymphocytes from bone marrow preparations. In treatment of disorders such as immunodeficiencies, blood cancers, and some hemoglobinopathies, bone marrow grafts are made after ensuring that donor and recipient are histocompatible. If they are not, the donor cells destroy the recipient's cells and graft-versus-host disease (GVHD) ensues. However, GVHD does not occur if the T lymphocytes are removed from the donor marrow, even without histocompatibility. A highly selective method of removal of T lymphocytes uses monoclonal antibodies (Chapter 33).

CLINICAL CASE STUDY 10.1 Disorders of Collagen Biosynthesis

Vignette 1: Ehlers–Danlos Syndrome

This clinical case was abstracted from: D. Buettner, S.H. Fortier, Ehlers–Danlos syndrome in trauma: A case review, J. Emerg. Nurs. 35 (2) (2009) 169–170.

Synopsis

An 18-year-old man was brought to the emergency room with a head injury. He was unconscious and pulseless. Emergency room physicians were told that the patient had a history of diagnoses with Ehlers–Danlos syndrome (EDS), a connective tissue disorder. CPR and other resuscitation efforts such as thoracotomy, open cardiac massage, and internal cardiac defibrillation were performed but failed to revive the patient. The autopsy report confirmed that the patient had EDS type IV, known as the vascular or arterial-ecchymotic type, which is the most serious of the six EDS types. The autopsy result indicated that the patient had bilateral hemothoraces (collections of blood in the pleural cavity) and a large left retroperitoneal hematoma due to thoracic aortic dissection.

Teaching Points

1. Ehlers–Danlos syndromes (EDS) are a group of genetic disorders caused by defects in collagen biosynthesis. The majority of EDS are caused by a defect in the primary structure of fibrillar collagen.
2. There are six major types of EDS: classical, hypermobility, vascular, kyphoscoliosis, arthrochalasis, and dermatosparaxis. All types involve the skin and joints, to some extent.
3. EDS type IV, the vascular variant, is caused by a mutation of the gene COL3A1, which encodes type III procollagen. Type III collagen is important in the structure of skin and blood vessels; thus, the most critical complication in EDS is arterial tears.

Supplemental References

[1] C.W. Chen, S.W. Jao, Ehlers–Danlos syndrome, N. Eng. J. Med. 357 (2007) e12.
[2] A.E. Sareli, W.J. Janssen, D. Sterman, S. Saint, R.E. Pyeritz, What's the connection, N. Eng. J. Med. 358 (2008) 626–632.

Vignette 2: Osteogenesis Imperfecta

This case was abstracted from: A.M. Barnes, W. Chang, R. Morello, W.A. Cabral, M. Weis, D.R. Eyre, et al., Deficiency of cartilage-associated protein in recessive lethal osteogenesis imperfect, N. Eng. J. Med. 355 (2006) 2757–2764.

Synopsis

The first child of consanguineous parents was born at 35 weeks' gestation; prenatal ultrasonography showed severe micromelia of the arms and legs. The newborn baby had a wide-open anterior grannell, eye proptosis, and white sclera. He also had pectus excavatum and rhizomelic limb shortening, and a radiographic survey showed over 20 fractures of the long bones and ribs. He died at 10 months of age. mRNA samples were purified and quantified from the patient's fibroblasts and tested for cartilage-associated protein (CRTAP), which forms a complex with prolyl 3-hydroxylase 1. Prolyl 3-hydroxylase 1 is required for hydroxylation of the proline residue in the 986 position of type 1 collagen. CRTAP mRNA was not detected from the patient's fibroblasts, and genomic DNA sequencing of exons of the CRTAP gene revealed a homozygous mutation in the splice donor site of exon 1. The parents of the patient were heterozygous for this mutation. This type of mutation induces premature termination of protein synthesis since stop codons existed in the retained intron 1 sequence.

Teaching Points

1. Classic osteogenesis imperfect is an autosomal dominant genetic disease caused by mutations in type 1 collagen genes. An autosomal recessive form of this disorder can be caused by mutations in the CRTAP gene, which leads to the dysfunction of prolyl 3-hydroxylase required for the hydroxylation of prolyl 3-hydroxylation of type 1 collagen.

Supplemental Reference

[1] F. Rauch, F.H. Glorieux, Osteogenesis imperfect, Lancet 363 (2004) 1377–1385.

Vignette 3: Scurvy

This case was abstracted from: J.M. Olmedo, J.A. Yiannias, E.B. Windgassen, M.K. Gomet, Scurvy: a disease almost forgotten, Intern. J. Dermatol. 45 (2006) 909–913.

Synopsis

A 77-year-old woman presented with fatigue, bruising, gingival bleeding, and anemia. She had been eating only bread, olive oil, and red meat for two years since she believed that her food allergies were due to all types of fruits and vegetables. She also reported that recently she had experienced occasional nosebleeds, bleeding of the gums while brushing her teeth, and soreness of the tongue. On the day of her clinic visit, she noticed bright red blood in her stool. Laboratory tests reported normocytic anemia with a hemoglobin of 8.8 g/dL, hematocrit of 26.4%, and mean corpuscular volume of 86.1 fL. Iron studies, prothrombin time, and international normalized ratio were within normal ranges. Values for vitamin B_{12} and folate were within normal ranges, whereas vitamin C was not detected. A diagnosis of vitamin C deficiency scurvy was made, and the patient was given 100 mg of oral vitamin C three times a day for 2 weeks.

Teaching Points

1. Vitamin C, ascorbic acid, is required for the hydroxylation of prolyl and lysyl residues of collagen biosynthesis, which are catalyzed by prolyl and lysyl hydroxylases. Defects in collagen biosynthesis disrupt the integrity of blood vessels, leading to their fragility and subsequent bleeding. Vitamin C is also required for the biosynthesis of carnitine and norepinephrine, metabolism of tyrosine, and amidation of peptide hormones, and aids in iron absorption by converting ferric state iron (Fe^{3+}) to ferrous state iron (Fe^{2+}) in the gastrointestinal tract.

Integration with Chapter 36

Supplemental Reference

[1] M. Weinstein, P. Babyn, S. Zlotkin, An orange a day keeps the doctor away: scurvy in the year 2000, Pediatrics 108 (2001) e55. 1–5.

CLINICAL CASE STUDY 10.2 Goodpasture's Syndrome, a Disorder of Collagen Metabolism

This clinical case was abstracted from: H. Bazari, A.R. Guimaraes, Y.B. Kushner, Case 20-2012: A 77-year-old man with leg edema hematuria, and acute renal failure, N. Eng. J. Med. 366 (2012) 2503–2015.

Synopsis

A 77-year-old man who presented to the emergency department with dyspnea and bilateral leg swelling was admitted to the hospital for further evaluation and found to have hematuria and acute renal failure. The patient was a smoker who also had had hemoptysis for several months. This patient's constellation of symptoms of hemoptysis, leg edema, and progressive renal failure met the criteria for rapidly progressive glomerulonephritis (defined as more than 50% loss of renal function in less than 3 months, with findings of glomerular injury). Serum analysis for various antibodies was performed. Results were negative, including antibodies to glomerular basement membrane (GBM), but renal biopsy showed IgG deposition along the GBM. Biopsy also revealed mild proliferative glomerulonephritis with crescents and segmental nodular glomerulosclerosis. Although serum anti-GBM antibodies were undetectable, the presence of IgG deposition on the GBM suggested that this patient had high-affinity antibodies that bound tightly to the kidney and circulated freely in very low levels. The patient was thus concluded to have anti-GBM disease, or Goodpasture's syndrome. The patient underwent therapy with glucocorticoids, cyclophosphamide, and plasmapheresis.

Teaching Points

1. Goodpasture's syndrome involves type IV collagen of the kidney's glomerular basement membrane (GBM). The development of autoantibodies to the noncollagenous 1 (NC1) domain of type IV collagen in the GBM results in hematuria, oliguria, and acute renal failure. Smokers can also have lung involvement, presenting with hemoptysis. Detection of the antibodies in the serum, along with deposition of IgG on the GBM on renal biopsy, confirms the diagnosis.

CLINICAL CASE STUDY 10.3 Mucopolysaccharidosis

This clinical case was abstracted from: G.S. Shah, T. Mahal, S. Sharma, Atypical clinical presentation of mucopolysaccharidosis type II (Hunter syndrome): a case report, J. Med. Case Rep. 4 (2010) 154.

Synopsis

A 10-year-old boy from East Asia presented to the pediatric clinic with a chief complaint of abdominal distension and intermittent dyspnea for 5 years. Review of systems revealed intermittent joint pain and below-average performance in school for 3 years. Physical exam showed mild mental retardation, short stature, an enlarged and pointed skull, a depressed nasal bridge, abdominal distension, and hepatosplenomegaly. A diagnosis of mucopolysaccharidosis type II (Hunter syndrome) was made based on clinical presentation and characteristic radiologic findings of vertebral breaking, paddle ribs, and J-shaped sella turcica (a variant configuration in which the sphenoid bone of the skull is flattened rather than saddle-shaped). This patient did not undergo measurement of urinary glycosaminoglycans, nor enzyme assay of iduronate sulfatase, as indicated to confirm the diagnosis, due to lack of testing capabilities at the medical facility. History, physical examination, and skeletal survey were sufficient for confirmation of the diagnosis.

Teaching Points

1. Mucopolysaccharidosis (MPS) is a group of disorders in which a deficiency of certain lysosomal enzymes normally responsible for the breakdown of glucosaminoglycans results in an accumulation and deposition of undegraded or partially degraded glucosaminoglycans in the lysosomes of many tissues. In MPS type II (Hunter syndrome), there is a deficiency of the lysosomal enzyme iduronate-2-sulfatase.

2. MPS type II has a mild and severe form, based on CNS involvement and length of survival. In both types, patients appear normal at birth. In the severe type, clinical features start to appear at age 2 to 4 years old, and patients have severe mental retardation and loss of function. In patients with the mild type, symptoms appear in their twenties, and they have normal intelligence with only mild mental retardation.

3. Patients found to have characteristic physical findings and radiologic evidence for MPS can undergo screening by measurement of the glucosaminoglycans heparin and dermatan sulfate in the urine. Excess urinary glucosaminoglycans can occur in MPS type I, II, or VII. The diagnosis of MPS type II is confirmed by enzyme assay showing low or absent iduronate-2-sulfatase lysosomal enzyme activity.

4. Emerging therapy for MPS type II is the replacement of the deficient enzyme using idursulfase (Elaprase), which is a recombinant human iduronate-2-sulfatase enzyme. However, the definitive treatment for MPS is transplantation of compatible bone marrow or stem cells from umbilical cord blood.

Supplemental References

[1] J. Muenzer, A. Fisher, Advances in the treatment of mucopolysaccharidosis type I, N. Engl. J. Med. 350 (2004) 1932–1934.

[2] R. Martin, M. Beck, C. Eng, R. Giugliani, P. Harmatz, V. Mufioz, et al., Recognition and diagnosis of mucopolysaccharidosis II (Hunter syndrome), Pediatrics 121(2) (2008) 377–386.

REQUIRED READING

[1] M.C. Erat, B. Sladek, I.D. Campbell, I. Vakonakis, Structural analysis of collagen type 1 interactions with human fibronectin reveals a cooperative binding mode, J. Biol. Chem. 288 (2013) 17441−17450.

[2] G.B. Fields, Interstitial collagen catabolism, J. Biol. Chem. 288 (2013) 8785−8793.

[3] A. Tsamis, J.T. Krawiec, Elastin and collagen fibre microstructgure of the human aorta in ageing and disease: a review, J. R. Soc. Interface 10 (2013) 20121004.

[4] F. Schoenhoff, J. Schmidli, M. Czerny, T.P. Carrel, Management of aortic aneurysms in patients with connective tissue disease, J. Cardiovasc. Surg. 54 (2013) 125−134.

[5] J.C. Marini, A. Forlino, Replenishing cartilage from endogenous stem cells, N. Engl. J. Med. 366 (2012) 2522−2524.

ENRICHMENT READING

[1] B.L. Loeys, G. Mortier, H.C. Dietz, Bone lessons from Marfan syndrome and related disorders: Fibrillin, TGF-B and BMP at the balance of too long and too short, Ped. Endocrinol. Rev 16 (2013) 417−423.

[2] A.D. Irvine, W.H. I. McLean, D.Y.M. Leung, Filaggrin mutations associated with skin and allergic diseases, N. Engl. J. Med. 365 (2011) 1315−1327.

[3] C. Has, G. Sparta, D. Kiritsi, L. Weibel, A. Moeller, V. Vega-Warner, et al., Integrin α_3 mutations with kidney, lung, and skin disease, N. Engl. J. Med. 366 (2012) 1508−1514.

[4] H.H. Freeze, Understanding human glycosylation disorders: biochemistry leads the charge, J. Biol. Chem. 288 (2013) 6936−6945.

[5] S.H. Hussain, B. Limthongkul, T.R. Humphreys, The biomechanical properties of the skin, Dermatol. Surg. 39 (2013) 193−203.

Chapter 11

Gastrointestinal Digestion and Absorption

Key Points

1. The gastrointestinal (GI) system consists of accessory organs and structures, namely salivary glands, teeth, and tongue in the mouth, and liver, gall bladder and pancreas in the abdomen. The GI tract is a continuous tube and consists of four concentric layers.

2. The organs of the GI system secrete many hormones and digestive enzymes.

3. Saliva provides fluids and enzymes and aids in the swallowing process through the esophagus.

4. The stomach stores food temporarily and prepares the food for digestion and absorption in the small intestines. Parietal cells secrete HCl and intrinsic factor required for eventual vitamin B_{12} absorption in the ileum of the small intestine, and pepsinogen is secreted by chief cells. The product of gastric digestion is known as chyme.

5. Most digestion of food occurs in the small intestine. It is a major site for absorption of water and electrolytes. Its finger-like projections, known as villi, contain numerous microvilli that provide a vast surface area for digestion and absorption, and they also retard the flow of chyme.

6. Digestion in the small intestine requires biliary and pancreatic secretions that are emptied into the duodenum, usually at the common site. The liver and pancreas facilitate digestion and absorption by providing juices containing bile and enzymes required for these processes.

7. Bile is formed in the hepatocytes (functional cells in the liver), then stored and concentrated in the gall bladder, and its release is mediated by cholecystokinin. Bile contains bile salts that emulsify fat in the preparation for its absorption. In the hepatocytes, bile salts are synthesized from cholesterol.

8. Pancreatic secretions contain enzymes necessary for the breakdown and processing of lipids, carbohydrates, and proteins, the products of which are absorbed by the GI epithelial cells. Pancreatic secretions also neutralize acidic chyme. Their secretions are stimulated by the cholecystokinin and secretin.

9. The large intestine (colon) harbors a large collection of microorganisms known as microbiota. They play a major role in health and disease.

10. GI function is regulated by the nervous system, which includes intrinsic and CNS neurons and GI hormones.

11. The GI tract produces many peptide and protein hormones that are integral in digestion and function in autocrine, paracrine, and endocrine fashion. For the most part, their actions are mediated by binding to hormone-specific G-protein-coupled receptors (Chapter 30).

12. Eating provokes secretion of GI hormones. Some promote plasma glucose-lowering effects through insulin secretion from the β-cells of the pancreas. These are known as incretins. Incretin mimetics are used in the treatment of type 2 diabetes.

13. GI hormones, in addition to their digestive function, regulate satiety and hunger at the CNS level and thus play a major role in energy homeostasis (Chapter 20).

14. In terms of energy homeostasis, ghrelin produced by the stomach is the only hormone that stimulates food intake (appetite), whereas the other hormones suppress food intake by promoting satiety.

15. Carbohydrate digestion involves participation of the pancreatic enzyme amylase and several intestinal brush border cell oligosaccharidases. The final product, glucose, enters the GI cells via the Na^+-glucose cotransporter (SGLT). Glucose enters the portal capillary blood circulation.

16. Lactose, the principal carbohydrate in milk, is hydrolyzed by brush border enzyme lactase to glucose and galactose. Galactose enters the cell by the same glucose transport process.

17. Deficiency of lactase or one of the brush border oligosaccharides may cause GI distress due to conversion of lactose by colonic bacteria to a large number of osmotically active metabolites.

18. Protein digestion begins in the stomach with proteolysis by pepsin, which is derived from the inactive precursor pepsinogen. The polypeptides are further hydrolyzed in the small intestine by trypsin, chymotrypsin, and carboxypeptidase, all derived from the pancreas. These enzymes arrive in the small intestine as inactive precursors (zymogens). Their optimal activities are at neutral pH. Trypsinogen is activated via proteolytic cleavage by a brush border cell enzyme, enteropeptidase, to trypsin; trypsin then converts the other zymogens to their active forms. The peptides are further hydrolyzed by brush border cell aminopeptidases and dipeptidases. The products of protein digestion yield amino acids, dipeptides, and tripeptides that undergo active carrier-mediated, Na^+-dependent transport into the brush border cell, where they are all converted to amino acids by intracellular peptidases. The amino acids enter portal blood circulation.

N. V. Bhagavan and C.-E. Ha: Essentials of Medical Biochemistry, Second edition. DOI: http://dx.doi.org/10.1016/B978-0-12-416687-5.00011-7

19. A defect in the transport of neutral amino acids and basic amino acids causes metabolic abnormalities (e.g., Hartnup disease, cystinuria).

20. Triglycerides (triacylglycerols) are absorbed into GI cells with the aid of bile acids and the pancreatic enzymes colipase and lipase. In the mucosal cells, triglycerides are incorporated into a lipoprotein known as a chylomicron that enters lacteals and eventually appears in the blood. Medium-chain fatty acyl triglycerides are absorbed in different pathways independent of bile acid and lipase–colipase participation.

21. Malabsorption of lipids leads to steatorrhea and can be caused by impairment of lipolysis, micelle formations, absorption, chylomicron formation, or its transport.

22. General malabsorption can be caused by the following disorders.
    a. Celiac disease—an immune-mediated disorder provoked by undigested peptides in cereals (wheat, rye, and barley) in susceptible individuals.
    b. Cystic fibrosis—an autosomal recessive disorder, caused by a defective chloride transport, is a multi-organ system disease.
    c. Another disorder of chloride transport involving a small intestinal infection with *Vibrio cholerae*, which secretes an exotoxin that activates the chloride channel, with loss of large volumes of watery stool containing electrolytes.

23. Food intake results in energy expenditure and heat production mediated by the sympathetic nervous system and thyroid hormones.

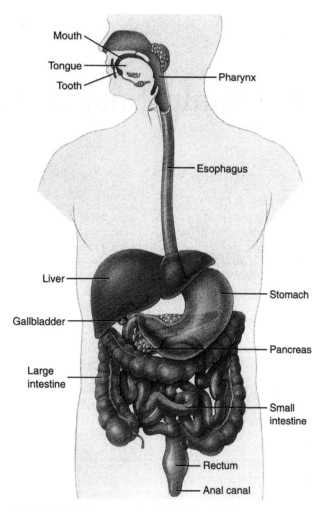

FIGURE 11.1   The organs of the human gastrointestinal system.

ANATOMY AND PHYSIOLOGY OF THE GASTROINTESTINAL TRACT

The gastrointestinal (GI) tract consists of mouth and esophagus, stomach, small intestine (duodenum, jejunum, and ileum), and large intestine (colon and rectum) (Figure 11.1).

Mouth and Esophagus

In the mouth, food is mixed with saliva, chewed to break up large particles, and propelled into the esophagus by swallowing. Saliva is secreted by three pairs of glands (parotid, submaxillary, and sublingual) and by numerous small buccal glands, and is under autonomic nervous system control. Parasympathetic nervous stimulation causes profuse secretion of saliva. Thus, atropine and other anticholinergic agents diminish salivary secretion and make the mouth dry. The presence of food in the mouth and the act of chewing stimulate secretion of saliva by reflex nervous stimulation. Salivary secretion can also be a conditioned response, so that the sight, smell, and even thought of food can elicit salivary secretion; however, in humans, this type of conditioned response is weak.

Stomach

The stomach stores food temporarily, retards its entry into the small intestine, and secretes pepsin to begin the digestion of protein. Hydrogen ions in the stomach activate pepsinogen to form pepsin and aid in maintaining the sterility of the upper GI tract. The stomach also secretes intrinsic factor (a glycoprotein), which is required for vitamin B_{12} absorption in the ileum (Chapter 36), helps prepare some of the essential minerals (e.g., iron) for absorption by the small intestine (Chapter 27), and provides mucus for protective and lubricative functions.

Small Intestine

Most digestion and absorption of food constituents occur in the small intestine. The small intestine is also the major site for absorption of water and electrolytes. It is a convoluted tube about 6 meters long, beginning at the pylorus and ending at the ileocecal valve. It is divided into the duodenum (a C-shaped, short, fixed segment), the mobile jejunum, and ileum. Its wall consists of the same four concentric layers as in other parts of the GI tract: **mucosa**, **submucosa**, **muscularis externa**, and **serosa**. Intricate arrays of finger-like projections known as **villi** that arise from the luminal side of the mucosal membrane provide a large surface area in the small intestine. Each villus consists of a single layer of tall, columnar epithelial absorptive cells (**enterocytes**) that contain **microvilli** which further increase the absorptive surface and give it a brushlike appearance (brush border). The surface membrane of the microvillus contains digestive enzymes and transport systems. The core of the microvillus consists of a bundle of filamentous structures that contain actin and associated proteins (e.g., myosin, tropomyosin, villin, fimbrin; see Chapter 19), which may have a structural role. Enterocytes are abundantly supplied with mitochondria, endoplasmic reticulum, and other organelles. The outermost surface of the microvillus is the carbohydrate-rich **glycocalyx**, which is anchored to the underlying membrane. The glycocalyx is rich in neutral and amino sugars and may protect cells against digestive enzymes. Villi also contain **goblet cells**, which secrete mucus. The absorptive cells are held together by junctional complexes that include "tight junctions." However, the tight junctions contain pores that permit the transport of water and small solutes into the intercellular spaces according to the osmotic pressure gradient.

Endocrine cells are distributed throughout the small intestine and other sections of the GI tract. About 20 different cell types have been identified. Some, easily identifiable by their staining characteristics, are argentaffin or enterochromaffin cells containing 5-hydroxytryptophan, the precursor of serotonin (Chapter 15). Neoplastic transformation of these cells leads to excessive production and secretion of serotonin, which causes diarrhea, flushing, and bronchoconstriction (**carcinoid syndrome**). Other types of cells lack 5-hydroxytryptophan but take up amine precursors (e.g., amino acids) to synthesize and store biologically active peptides (e.g., hormones).

The mucosal cells of the small intestine and stomach are among the most rapidly replaced cells of the human body, being renewed every 3–5 days. Millions of cells are sloughed off (or exfoliated) from the tips of villi every minute; about 17 billion such cells are shed per day. These mucosal cells amount to about 20–30 g of protein, most of which is reclaimed as amino acids after hydrolysis in the lumen. These cells are replaced through division of undifferentiated stem cells in the crypts of the villi (crypts of Lieberkühn) and subsequent migration to the top of the villus, followed by maturation. Paneth cells also reside in the intestinal crypt (Figure 11.2). They respond to the availability of the food by providing growth signals for stem cell production. During nutritional (caloric) deprivation, Paneth cells promote the enlargement of stem cell pool with a concurrent decrease in numbers of enterocytes. The opposite occurs during the presence of abundance of nutrition. It is thought that this process is a mechanism to adopt for situations of nutritional scarcity and abundance. The drug sirolimus mimics the condition of nutritional restriction [1]. Mitotic poisons (e.g., antineoplastic agents), injury, and infection can adversely affect mucosal cell proliferation, producing a flat mucosa devoid of villi and leading to severe malnutrition.

Digestion in the small intestine requires the biliary and pancreatic secretions that are emptied into the duodenum (usually at a common site).

Formation, Secretion, and Composition of Bile

Bile is formed and secreted continuously by polygonally shaped liver parenchymal cells called **hepatocytes**. An aqueous buffer component (e.g., HCO_3^-) is added to the bile by the hepatic bile duct cells that carry the secretion toward the common bile duct. The membrane of the hepatocytes in contact with the blood has microvilli that facilitate the exchange of substances between plasma and the cells. Hepatocytes are rich in mitochondria and endoplasmic reticulum. Hepatic bile flows into the gallbladder, where it is concentrated and stored; it is emptied into the duodenum when the partially digested contents of the stomach enter the duodenum. This movement is accomplished by contraction of the gallbladder mediated by cholecystokinin, a GI hormone.

The **liver** is the largest gland and second largest organ in the body and weighs about 1.5 kg in an adult. It functions in synthesis, storage, secretion, excretion, and specific modification of endogenously and exogenously derived substances. The liver takes up material absorbed from the small intestine, since it is mainly supplied by venous blood arriving from the GI tract. It is uniquely susceptible to changes in alimentation and to toxic substances. The liver participates in numerous metabolic functions.

The liver is organized into **lobules**, functional aggregates of hepatocytes arranged in radiating cords or plates surrounding a central vein (Figure 11.3). In adults, the branching plates are usually one cell thick, and the vascular channels between them are known as **sinusoids**. The blood in the sinusoids passes slowly between the rows of cells, facilitating the exchange of substances between the cells and the plasma. The blood supply of the sinusoids is furnished by the portal vein (75%) and by the right and left hepatic arteries (25%). Blood flows from the

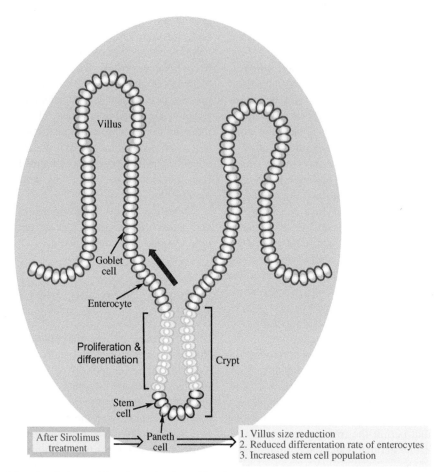

FIGURE 11.2 Role of Paneth cells in providing signals for the proliferation of stem-cell pool during nutritional deprivation. Sirolimus mimics the effects of nutritional deprivation, by the same mechanism.

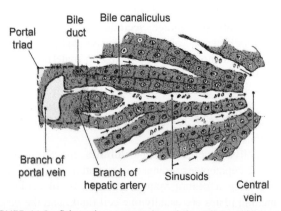

FIGURE 11.3 Schematic representation of the radial architecture of the plates formed by the liver cells. Blood from the portal vein and the hepatic artery flows into sinusoids and eventually enters the central vein. The bile canaliculi are located between the liver cells. Bile flows in the opposite direction and empties into the bile duct in portal triads. *[Reproduced with permission from A. W. Ham, Histology, 8th ed., Philadelphia, PA: J. B. Lippincott, 1979.]*

periphery of the lobules into the central vein. Sinusoids anastomose to form central veins that join to form larger veins which feed through the hepatic vein into the inferior vena cava. Sinusoids are larger than capillaries and are lined by reticuloendothelial cells, including the phagocytic **Küpffer cells**. This layer contains numerous small fenestrations that provide access of sinusoidal plasma to the surface microvilli of hepatocytes via the intervening space, called the **space of Disse**. Fine collagen fibers (reticulin) within this space provide the support for the liver cell plates (Figure 11.4).

Bile produced by hepatocytes is secreted into the bile canaliculi between adjacent hepatic cells. The wall of the canaliculus is formed by the plasma membrane of the hepatocytes, which are held together by tight junctions. Canaliculi arise near central veins and extend to the periphery of the lobules. The direction of bile flow in the canaliculi is centrifugal, whereas that of the blood flow is centripetal. Canaliculi coalesce to form ducts, which are

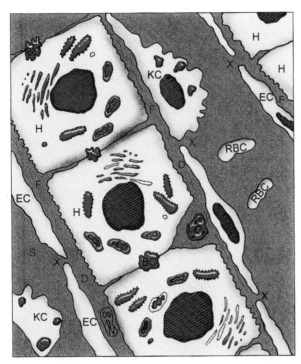

FIGURE 11.4 Schematic representation of structures within the hepatic lobule. H = hepatocyte; BC = bile canaliculus; KC = Küpffer cell; EC = endothelial cell; N = nerve fiber; F = reticulin fiber; S = sinusoid; D = space of Disse; X = gap between sinusoid lining cells; RBC = red blood cell.

lined by epithelium, and the ducts coalesce to form the right and left hepatic ducts. Outside the liver, these ducts form the common hepatic duct.

The volume of hepatic bile varies in a normal human adult from 250 to 1100 mL, depending on the rate at which bile acids recirculate in the enterohepatic cycle (see below). Some GI hormones increase the volume and bicarbonate content of bile through stimulation of the duct epithelium. Hepatic bile has a pH of 7.0−7.8 and is isotonic with plasma. During the interdigestive period, it is diverted into the gallbladder, since the **sphincter of Oddi** is closed. Bile in the gall bladder is concentrated 12- to 20-fold by absorption of electrolytes and water. Nonetheless, gallbladder bile is isotonic with plasma because cations (mainly Na^+) associate with the osmotically inactive bile acid micellar aggregates. Gallbladder capacity is 50−60 mL. Emptying requires a coordinated contraction of the gallbladder and relaxation of the sphincter of Oddi, both mediated by cholecystokinin.

Bile contains bile acids, bile pigments, cholesterol, phosphatidylcholine (lecithin), and electrolytes. Bile pigments are glucuronide conjugates of bilirubin and biliverdin derived from the degradation of heme, a prosthetic group of many proteins. Most of the bilirubin formed in the body comes from hemoglobin catabolized in reticuloendothelial cells (liver is one of the sites where these cells

are found). The bilirubin is bound to plasma albumin and transported to hepatocytes, where it is converted to glucuronide conjugates and secreted into the bile, giving it a golden yellow color. If bilirubin remains in the gallbladder, it is oxidized to biliverdin, which is greenish brown. Bile pigments are toxic metabolites (Chapter 27).

The bile acids are 24-carbon steroid derivatives. The two primary bile acids, cholic acid and chenodeoxycholic acid, are synthesized in the hepatocytes from cholesterol by hydroxylation, reduction, and side chain oxidation. They are conjugated by amide linkage to glycine or taurine before they are secreted into the bile (see cholesterol metabolism, Chapter 17). The mechanism of secretion of bile acids across the canalicular membrane is poorly understood.

Bile acids (also known as **bile salts**) are present as anions at the pH of the bile, and above a certain concentration (critical micellar concentration), they form polyanionic molecular aggregates, or micelles. Bile salts exported from hepatocytes are mediated by an ATP binding cassette protein (ABCB11). A deficiency of ABCB11 leads to cholestatic liver disease (see Chapter 7). The critical micellar concentration for each bile acid and the size of the aggregates are affected by the concentration of Na^+ and other electrolytes and of cholesterol and lecithin. Thus, bile consists of mixed micelles of conjugated bile acids, cholesterol, and lecithin. While the excretion of osmotically active bile acids is a primary determinant of water and solute transport across the canalicular membrane, in the canaliculi they contribute relatively little to osmotic activity because their anions aggregate to form micelles.

Bile is also a vehicle for excretion of other endogenous (e.g., bilirubin and steroids) and exogenous (e.g., drugs and dyes) compounds. The dye sulfobromophthalein is removed from the bloodstream by the hepatocytes, excreted in the bile, and eliminated through the GI tract. Its rate of removal from the circulation depends on the functional ability of the hepatocytes and of the hepatic blood flow, and so it is used to assess hepatic function.

Cholesterol is virtually insoluble in water but is solubilized through the formation of micelles, which also contain phosphatidylcholine and bile acid anions in specific proportions. Cholesterol can be precipitated to form stones (known as **gallstones**) when the critical concentration of micellar components is altered, i.e., when the concentration ratio of cholesterol to bile acids is increased. Gallstones may also form because of precipitation of bilirubin, e.g., with increased production of bile pigments due to hemolysis, associated with infection and bile stagnation. Presumably, under these conditions the bilirubin diglucuronide is converted to free bilirubin by β-glucuronidase. Bilirubin is insoluble and forms calcium bilirubinate stones (called **pigment stones**).

Gallstones occur in 10%−20% of the adult population in Western countries. Cholesterol stones account for 80% (see Chapter 17).

More than 90% of the bile acids are reabsorbed, mostly in the ileum. About 5%−10% undergoes deconjugation and dehydroxylation catalyzed by bacterial enzymes in the colon to form deoxycholic acid and lithocholic acid. These acids are known as **secondary bile acids**.

Deoxycholic acid is reabsorbed, but lithocholic acid, which is relatively insoluble, is poorly absorbed. The reabsorbed bile acids enter the portal blood circulation, are reconjugated in the liver, and are secreted into bile. This cycling is known as the **enterohepatic circulation** and occurs about six to eight times daily. The rate of cycling increases after meals. The lithocholic acid is conjugated with either taurine or glycine, sulfated at the ring hydroxyl group, and again secreted into the bile. Sulfated lithocholate is not reabsorbed. Free lithocholic acid is highly toxic to tissues and causes necrosis.

The bile acid pool normally consists of about 2−4 g of conjugated and unconjugated primary and secondary bile acids. Daily loss of bile acids in feces, mostly as lithocholate, is about 0.2−0.4 g. Hepatic synthesis of bile acids equals this amount, so that the size of the bile acid pool is maintained at a constant level.

Exocrine Pancreatic Secretion

The pancreatic gland is 10−15 cm long and resides in the retroperitoneal space on the posterior wall of the abdomen, with its broad and flat end ("head") fitting snugly into the loop formed by the duodenum. The endocrine function of the pancreas consists of synthesis, storage, and secretion of the hormones insulin, glucagon, and somatostatin into the venous circulation that drains the pancreas. These hormones are essential in the regulation of metabolism of carbohydrates and other substrates (Chapter 20). The endocrine cells are located in the **islets of Langerhans**, distributed throughout the pancreas but separated from the surrounding exocrine−acinar tissue by collagen fibers. The exocrine function of the pancreas consists of synthesis, storage, and secretion of digestive enzymes and bicarbonate-rich fluids.

The acinar cells of the pancreas are arranged in ellipsoidal structures known as **acini**. Each acinus contains about a dozen cells arranged around a lumen, which forms the beginning of the pancreatic duct system, and centroacinar cells, which secrete a bicarbonate solution. The acinar cells are joined to one another by tight junctions, desmosomes, and gap junctions. The tight junctions probably prevent leakage of pancreatic secretions into the extracellular spaces. Acinar cells are pyramidal in shape with the nucleus located away from the lumen and secretory granules located toward the lumen. Centroacinar and acinar cells have microvilli on their free borders.

Enzymes are synthesized in the acinar cells by ribosomes situated on the rough endoplasmic reticulum and sequestered in membrane-bound granules (Chapter 23) stored in the apex of the cell. Upon stimulation by the vagus nerve or by cholecystokinin, the contents of the granules are released into the lumen.

The centroacinar and duct cells secrete bicarbonate ions under the control of gastrointestinal hormones. The bicarbonate ions are formed by the action of carbonic anhydrase:

$$CO_2 + H_2O \xrightleftharpoons{\text{carbonic anhydrase}} H_2CO_3 \rightleftharpoons H^+ + HCO_3^-$$

The exact mechanism of HCO_3^- secretion into ductules is not known.

Composition of Pancreatic Juice

Pancreatic juice is alkaline because of its high content of HCO_3^-. It neutralizes the acidic chyme from the stomach, aided by the bile and small intestinal juices. In the jejunum, chyme is maintained at a neutral pH as required for the optimal functioning of the pancreatic digestive enzymes. A normal pancreas secretes about 1.5−2 L of juice per day.

Pancreatic juice contains the proenzymes trypsinogen, chymotrypsinogen, procarboxypeptidases, and proelastase. All are activated by trypsin in the intestinal lumen. Enteropeptidase located in the brush border of the jejunal mucosa converts trypsinogen to trypsin. A trypsin inhibitor in pancreatic juice protects against indiscriminate autodigestion from intraductal activation of trypsinogen. Other enzymes of pancreatic juice and their substrates are listed below.

Enzyme	Substrate
Prophospholipase, activated to phospholipase	Phospholipids
Cholesteryl ester hydrolase	Cholesteryl esters
Lipase and procolipase, converted to colipase	Triacylglycerols
Amylase	Starch and glycogen
Ribonuclease	RNA
Deoxyribonuclease	DNA

Large Intestine

The large intestine extends from the ileocecal valve to the anus. It is wider than the small intestine except for the descending colon, which, when empty, may have the same diameter as the small intestine. Major functions of the colon are absorption of water, Na^+ and other electrolytes, and symbiotic relationship with coexisting microbial flora as well as temporary storage of excreta, followed by their elimination. The colon harbors large numbers of bacteria, known as **microbiota**, which can cause disease if they invade tissues and enter the bloodstream. Gut microbiota metabolize carbohydrates to lactate, short-chain fatty acids (acetate, propionate, and butyrate), and gases (CO_2, CH_4, and H_2). Ammonia, a toxic waste product, is produced from urea and other nitrogenous compounds. Other toxic substances are also produced in the colon. Ammonia and amines (aromatic or aliphatic) are absorbed and transported to the liver via the portal blood, where the former is converted to urea (Chapter 15) and the latter is detoxified. The liver thus protects the rest of the body from toxic substances produced in the colon. Microbiota can also be a source of certain vitamins (e.g., vitamin K, Chapter 34).

Gut Microbiota

Several recent studies have improved our understanding of the role of gut microbiota in promoting hosts' optimal nutrition and immune function. Microbiota have also been implicated using animal studies in metabolic disorders such as obesity, metabolic syndrome, diabetes, and cardiovascular diseases. These studies also provide therapeutic options in treating diseases and promoting human health [2−6]. Following are some examples of the role of diet and gut microbiota in human health:

1. Diminished energy utilization in subjects with malnutrition due to Kwashiorkor disease [7].
2. Beneficial influence of dietary components (e.g., components of cruciferous vegetables) on hosts' immune function [8].
3. Regulation of cells involved in immune effector and regulatory pathways [9].
4. The role of diet rich in choline in phosphatidyl choline (e.g., red meat) in promoting atherosclerosis in some genetically susceptible subjects [10].
5. The role of differences in microbiota that may contribute to obesity or diabetes [11−12].
6. The re-establishment of normal healthy fecal microbiota in the treatment of antibiotic-resistant intestinal *Clostridium difficile* infection [13].
7. Relationship of diminished gut microbiota integrity, inflammation, and colorectal cancer [14].

GASTROINTESTINAL HORMONES

Gastrointestinal (GI) hormones (Table 11.1) consist of a wide variety of polypeptide hormones that regulate intestinal functions (e.g., motility and secretion) in aiding absorption and digestion of nutrients. They also regulate appetite and satiety by providing signals to the CNS of the nutrient status, and modulate insulin secretion. The GI hormones that promote insulin secretion from β-cells of pancreas are known as **incretins**. The metabolism of incretins and their therapeutic applications in the glycemic management of type 2 diabetes are discussed in Chapter 20. They also have receptors on the vagus nerves. GI tract functions are regulated by the autonomic nervous system. GI hormones' biological actions are mediated, for the most part, through binding to hormone-specific G-protein-coupled receptors located on plasma cell membranes of target cells. The intracellular signal transduction involves adenylate cyclases, protein kinases, Ras family of GTP-binding proteins, and membrane ion channels.

The endocrine cells of the GI tract are interspersed with mucosal cells and do not form discrete glands. The GI tract has been described as the largest endocrine organ in the body. In addition, the GI tract is noted for its special significance in the development of endocrinology. The first hormone discovered (initially being characterized as a chemical messenger) was the GI hormone **secretin**, by Bayliss and Starling in 1902.

The GI hormones are a heterogeneous group of peptides that are released into the bloodstream in response to specific stimuli, bind to specific receptors on target cells to produce biological responses, and are subject to feedback regulation. Some may affect neighboring cells by being transported through intercellular spaces (paracrine secretion), and others serve as neurotransmitters in peptidergic neurons (Chapter 28). Peptidergic neurons are present in the GI tract and in the peripheral and central nervous systems. These peptides are synthesized in the GI tract and in the brain although in the GI tract some may originate from vagal nerve endings. This dual localization of "brain−GI tract" peptides is an example of biological conservation.

The Physiological Roles of GI Hormones

Gastrin

Gastrin is synthesized and stored in G cells in the antral mucosa of the stomach (see Table 11.1); it is also found, to a lesser extent, in the mucosa of the duodenum and jejunum. During fetal life, gastrin is also found in the pancreatic islet cells. Gastrin-secreting tumors, known as **gastrinomas** (e.g., **Zollinger−Ellison syndrome**), may develop in the pancreas. The main form of gastrin obtained from G cells is a heptadecapeptide. At the N-terminal end of the heptadecapeptide, there is a pyroglutamyl residue, and at the C-terminal end, there is a phenylalaninamide residue, both of which protect the peptide from amino and carboxy-peptidase activity, respectively. Two forms of the

TABLE 11.1 Origins and Functions of Major Gastrointestinal Hormones

Name	Sources	Functions
1. Gastrin	G-cells of gastric antrum and duodenum	Stimulates acid (HCl) secretion from parietal cells and histamine from enterochromaffin-like cells; promotes proliferation of gastric mucosa
2. Ghrelin	A-cells of gastric fundus; other segments of GI tract; hypothalamus	Promotes food intake; promotes gastric motility; stimulates growth hormone secretion
3. Cholecystokinin (CCK)	I-cells of duodenum and jejunum; central nervous system	Stimulates gall bladder and pancreatic enzyme secretions; decreases appetite
4. Glucose-dependent insulinotropic polypeptide (GIP)	K-cells of duodenum and jejunum	Promotes insulin secretion from β-cells of pancreas (incretin effect); stimulates fatty acid synthesis in adipose tissue
5. Secretin	S-cells of the upper portions of the small intestine	Stimulates the secretion of bicarbonate-rich fluids from the duct cells and biliary tract; augments the action of CCK and decreases gastric acid secretion
6. Polypeptide YY	L-cells of distal small and large intestine	Inhibits food intake
7. Somatostatin	D-cells of the GI tract and pancreas; hypothalamus	Inhibits secretion of GI tract hormones, growth hormone, and thyrotropin
Hormones Obtained from Proglucagon Cleavage by Cell-Specific Prohormone Convertases		
1. Glucagon-like peptide-1 (GLP-1)	L-cells of distal small and large intestine	Promotes insulin secretion from β-cells of pancreas (incretin effect) and promotes β-cell growth; suppresses pancreatic glucagon secretion by α-cells; delays gastric emptying
2. Glucagon-like peptide-2 (GLP-2)	L-cells of distal small and large intestine	Promotes intestinal mucosal growth and repair; inhibits gastric secretion
3. Oxyntomodulin	L-cells of distal small and large intestine	Inhibits food intake; inhibits gastric motility and acid production

heptadecapeptide exist: one in which the tyrosyl residue at position 12 is free (gastrin 1) and the other in which it is sulfated (gastrin 2). Though larger and smaller forms of gastrin have been identified, the biological activity of gastrin resides in the C-terminal tetrapeptide. A synthetic pentapeptide is available for clinical purposes.

Gastrin is released in response to chemical, mechanical, or neural stimuli on the G cell. Hypoglycemia (plasma glucose concentration ≤ 40 mg/dL), acting through the hypothalamus, causes release of gastrin mediated through the vagus nerve. A rise in calcium ion concentration in plasma causes the release of gastrin. The main physiological actions of gastrin are stimulation of acid secretion by parietal cells, secretion of pepsin by chief cells, increase in gastric mucosal blood flow, stimulation of gastric motility, and promotion of the growth of oxyntic mucosa and exocrine pancreatic tissue. Release of gastrin is suppressed by acidification of the antral mucosa. In disorders in which H$^+$ is not excreted owing to destruction or absence of functioning parietal cells (e.g., **pernicious**

anemia, **atrophic gastritis**), gastrin plasma levels are highly elevated. Plasma levels of gastrin are also elevated in the Zollinger–Ellison syndrome (see Clinical Case Study 11.1), in which hypersecretion of acid, peptic ulcer disease, and hyperplasia of the gastric mucosa occur. Gastrin release is suppressed by all members of the secretin family.

The mechanism of acid secretion by parietal cells is complex and not completely understood. These cells have many receptors and are subjected to a variety of stimuli that can act independently or modulate one another's action. If intracellular pH is assumed to be 7 and luminal pH 1, parietal cells secrete H$^+$ at a concentration a million-fold higher than that inside the cell. Parietal cells contain the largest number of mitochondria found in eukaryotic cells. They are polar; at the apical membrane, HCl is secreted into the gastric lumen, and at the basolateral membrane, HCO_3^- is secreted. The basolateral membrane contains many different receptors, ion channels, and transport pathways. The resting cell is packed with membrane-bound vesicles called **tubulovesicles** in which the membranes contain H^+,K^+-ATPase (a proton pump), the enzyme responsible for acid secretion. When the cell is stimulated, these tubulovesicles interact by means of cytoskeletal elements to form secretory canaliculi. A schematic representation of receptor systems and ion pathways is shown in Figure 11.5. The mediators that stimulate acid secretion are acetylcholine, gastrin, and histamine. Bombesin, a gastrin-releasing peptide, is also released from enteric neurons by vagal stimulation. The action of acetylcholine can be inhibited by anticholinergic agents (e.g., atropine). The action of histamine (the decarboxylated product of histidine) is mediated by H_2 receptors. H_1 receptors mediate the action of histamine on smooth muscle, causing bronchoconstriction, as in acute anaphylaxis and allergic asthma. The actions of histamine can be selectively inhibited by specific H_1- or H_2-receptor antagonists. The introduction of H_2-receptor antagonists (e.g., cimetidine and ranitidine; see Figure 11.6) helps to heal duodenal and gastric ulcers. Secretion of gastric acid can also be suppressed by prostaglandin E derivatives and substituted benzimidazoles that inhibit H^+,K^+-ATPase (e.g., omeprazole, pantoprazole). Omeprazole consists of a sulfinyl group that bridges two ring systems consisting of benzimidazole and pyridine. The active drug is formed by protonation in the stomach and rearranges to form sulfenic acid and sulfenamide. H^+,K^+-ATPase contains critical sulfhydryl groups located on the luminal side.

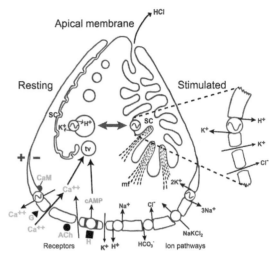

FIGURE 11.5 Schematic representation of the resting (left side) and stimulated (right side) state of the parietal cell. Basolateral membrane contains three major receptor classes: gastrin (G), acetylcholine (Ach), and histamine (H). Their actions are mediated by cAMP responses, Ca^{2+} changes, or both. In addition, there are a number of ion transport pathways. In the stimulated state, the apical membrane acquires H^+,K^+-ATPase contained in the tubulovesicles (tv) as well as the property of K^+ and Cl^- conductance, both of which are essential in the secretion of HCl. A change in cytoskeletal arrangement is also associated with stimulation. CaM = calmodulin; SC = secretory canaliculus; mf = microfilaments. *[Reproduced with permission from D.H. Malinowska, G. Sachs, Cellular mechanisms of acid secretion, Clin. Gastroenterol. 13(322) (1984) 309−326.]*

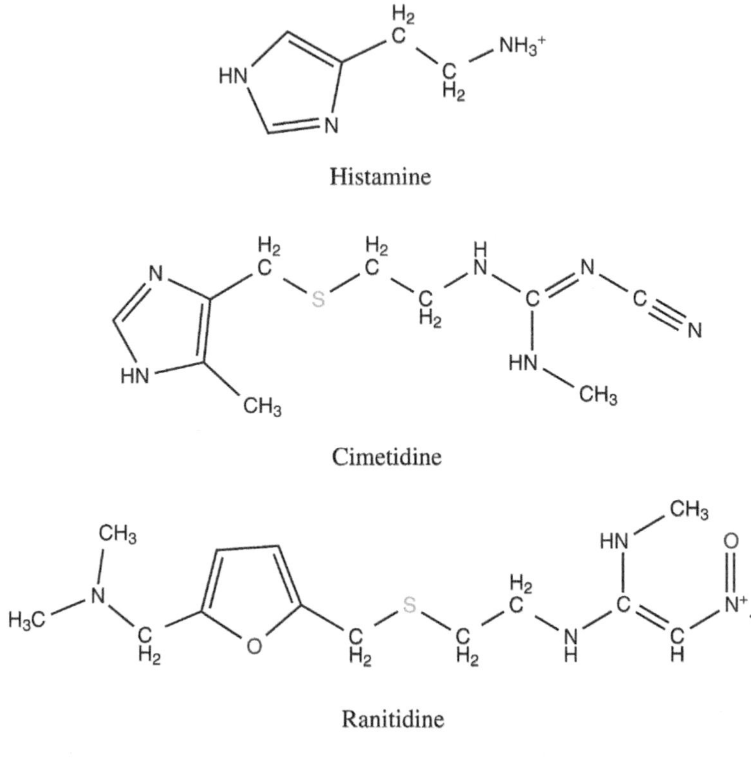

Histamine

Cimetidine

Ranitidine

FIGURE 11.6 Comparison of the structures of histamine with cimetidine and ranitidine. The latter two are H_2-receptor antagonists and act on the gastric parietal cells to inhibit gastric acid production. Ranitidine, which has a furan rather than an imidazole structure, is a more potent competitive inhibitor than cimetidine.

Sulfenamide is covalently attached to H^+,K^+-ATPase, causing an irreversible inhibition (Figure 11.7). Complete inhibition occurs when two molecules of sulfonamide bind to the enzyme. The degree of proton pump inhibition is dose-dependent, and the drug is particularly useful in patients with peptic ulcer disease now well controlled by H_2-receptor antagonists.

Although most of the action of gastrin is mediated via the H_2 receptor, residual stimulation due to gastrin alone suggests the presence of specific receptors for gastrin.

The precise intracellular events that lead to acid secretion are not clear, but the second messengers in this process appear to be cAMP and Ca^{2+}. The H_2 receptor is coupled to the adenylate cyclase system (Chapter 28), and

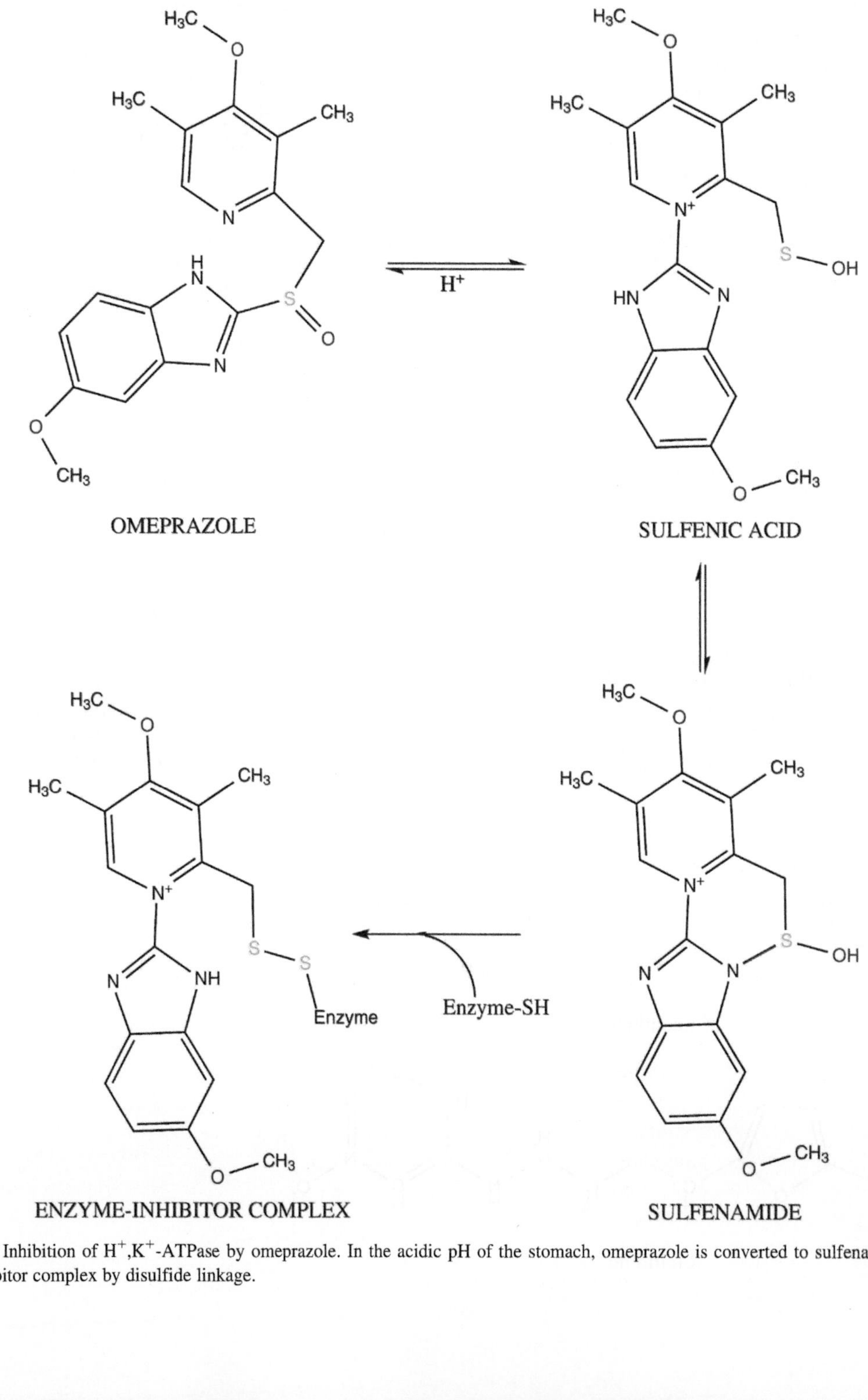

FIGURE 11.7   Inhibition of H^+,K^+-ATPase by omeprazole. In the acidic pH of the stomach, omeprazole is converted to sulfenamide, which forms an enzyme-inhibitor complex by disulfide linkage.

its activation results in the intracellular elevation of cAMP concentration. Stimulation of cholinergic receptor systems is coupled to increased Ca^{2+} permeability.

The H^+ ions in the stimulated parietal cell are produced by the action of carbonic anhydrase:

$$CO_2 + H_2O \xrightarrow{\text{carbonic anhydrase}} H_2CO_3 \rightleftarrows H^+ + HCO_3^-$$

and the H^+ ions formed are secreted with the aid of H^+,K^+-ATPase, which secretes H^+ in exchange for K^+ through the free-energy hydrolysis of ATP. K^+ is transported into the lumen via a specific ion channel. In the lumen, the accompanying anion for H^+ is Cl^- secreted from the parietal cell. The chloride ions are derived from the plasma and transported into the parietal cell via the electroneutral Cl^-/HCO_3^- exchange. Thus, for every H^+ secreted into the gastric lumen, there is a transfer of an HCO_3^- into the plasma in exchange for Cl^-, which is ultimately secreted into the gastric lumen. In instances of excessive loss of gastric fluids (e.g., persistent vomiting or nasogastric suction), metabolic alkalosis results (Chapter 37).

Peptic Ulcer Disease

Peptic ulcers are caused by an imbalance of acid secretion by the parietal cells and lack of mucosal protective barriers. There are two major causes of peptic ulcer disease. One is drug induced and the other is caused by a bacterial infection. The drugs that are related to **gastropathy** (and also possibly renal insufficiency) belong to a group of compounds known as **nonsteroidal anti-inflammatory drugs** (**NSAIDs**). NSAIDs encompass many different classes of chemical compounds including aspirin (acetyl-salicylic acid). NSAIDs exert anti-inflammatory action and are used in the treatment of many rheumatic conditions. The mechanism of action of NSAIDs involves inhibition of the cyclooxygenase group of enzymes, which are responsible for the synthesis of prostaglandins (Chapter 16). The inhibition of those enzymes by aspirin is irreversible and is caused by covalent modification involving acetylation of a key serine residue of cyclooxygenase.

At least two isoforms of cyclooxygenase (COX) are known, COX1 and COX2. COX1, a constitutive enzyme, is responsible for the synthesis of prostaglandins that are essential for normal function. Gastric prostaglandins maintain mucosal integrity by modulating parietal cell acid production, stimulate mucus and bicarbonate production in the mucous gel layer, and regulate mucosal blood flow. COX2 is induced during inflammation by cytokines and inflammatory mediators. NSAIDs inhibit both COX1 and COX2. While they produce a desirable therapeutic effect as anti-inflammatory agents, NSAIDs undesirably inhibit the glandular prostaglandin production necessary

for normal function. Thus, drugs that selectively inhibit COX2 enzymes are better anti-inflammatory agents; one such NSAID is celecoxib (Chapter 16). A stable analogue of prostaglandin E1, misoprostol has cytoprotective effects in the treatment of peptic ulcer disease. Aspirin has other pharmacological effects such as inhibition of platelet aggregation (Chapter 34).

Peptic ulcer disease is associated with *Helicobacter pylori* infection in 90% of patients with gastric and duodenal ulceration. Elimination of *H. pylori* infection with antibiotics heals the peptic ulcer and the associated symptoms. Combination therapy with antibiotics, antisecretory agents, namely H_2-receptor antagonists or proton pump inhibitors, and bismuth salts has significantly improved the clinical outcome of peptic ulcer disease, and other factors are necessary for *H. pylori* colonization and disease to occur. Flagellated motile bacteria resist peristalsis and adhere to gastric epithelium in a highly specific manner.

Several adhesins and their ligands have been identified on the host cells. Lewis blood group antigen has been identified as an epithelial cell receptor for *H. pylori* binding. Enhanced acid production due to chronic *H. pylori* infection has been attributed to the production of inflammatory mediators (e.g., cytokines, Chapter 33) and pH alterations. Urease produced by the bacteria converts urea to ammonia and carbon dioxide. Ammonia increases the pH and is essential for the survival of bacteria at acidic pH. A change in pH toward alkalinity may increase the levels of gastrin, thus causing increased acid production and a vicious cycle. Urease activity is conserved among all *H. pylori* species, as is the primary structure of the enzyme.

H. pylori organisms that produce vacuolating cytotoxin (coded by the gene *vacA*) and a high-molecular-weight protein known as **CagA** (coded by the gene *cagA*) are implicated in ulcerogenesis and gastritis. *H. pylori* strains are generically diverse, due to their ability to mutate and the ease with which they exchange genes. *H. pylori* infection may also be a risk factor in adenocarcinoma of the antrum and the body of the stomach and **non-Hodgkin's lymphoma** of the stomach.

Infection by *H. pylori* is detected by serological markers produced by host immune responses (e.g., antibodies to antigens of *H. pylori*) and a breath test. The latter, known as the **urea breath test**, consists of oral administrations of radioactively labeled urea. This is metabolized to labeled CO_2 and ammonia by the urease of *H. pylori* present in the gastric mucosa. The presence of labeled CO_2 measured in the exhaled air confirms infection.

H_2-receptor antagonists and proton pump inhibitors are used to reduce gastric acidity and are useful in the treatment of gastroesophageal reflux disease (GERD). In GERD, the most common malady of the esophagus, the contents of the stomach reflux into the esophagus. Therapy also involves increasing of lower esophageal sphincter tone. Surgical therapy involves strengthening esophageal sphincters by wrapping part of the stomach around the lower esophagus and placing pressure on the sphincter to assist its closure. This procedure, known as **fundoplication**, can be performed by abdominal surgery or, more commonly, by laparoscopy. GERD complications involve structures contiguous to the esophagus, namely laryngeal, pharyngeal, pulmonary, sinusal, and dental disorders. Lifestyle changes that can alleviate GERD include cessation of smoking and consumption of alcohol; avoidance of spicy food, coffee, and bedtime meals; correction of obesity; and elevation of the head by at least 15 cm (about 6 inches) during sleep. Therapeutic agents that contain theophylline, anticholinergic agents, and progesterone should be avoided because they delay gastric emptying and decrease lower esophageal sphincter tone. One of the changes of chronic GERD that may occur in some patients is that the healing epithelium of the esophagus is replaced not with the normal squamous cells but with specialized columnar cells. Columnar epithelial cells normally are found in the small intestine. Hence, this process is known as **intestinal metaplasia** or **Barrett's esophagus** and is regarded as a premalignant condition.

Cholecystokinin

Cholecystokinin (CCK) is found throughout the small intestine but is located predominantly in the mucosal I cells of the duodenum and jejunum. In the ileum and colon, it is localized in nerve endings, and it is widely distributed throughout the peripheral and central nervous systems. CCK consists of 33 amino acid residues (CCK-33) and shows macro- and microheterogeneity. Several other forms are known: CCK-58, CCK-39, and CCK-8. Naturally occurring CCK has a sulfated tyrosyl residue at position 27, and removal of sulfate changes the biological activity to that of gastrin. The C-terminal tetrapeptide is identical with that of gastrin. A synthetic C-terminal octapeptide is about three times more potent than CCK-33. Cerulein, a decapeptide present in the skin and GI tract of certain amphibians, has a C-terminal octapeptide sequence identical to that of CCK. These two peptides have similar biological properties, and cerulean is clinically useful for the stimulation of gallbladder contraction. CCK is secreted as a result of stimuli caused by the products of digestion of proteins and lipids. The secretion of CCK is terminated when these digestion products are absorbed or migrate into the lower portions of the GI tract. The principal physiological actions of CCK are to stimulate gallbladder contraction, to relax the sphincter of Oddi, and to stimulate secretion of pancreatic juice rich in

digestive enzymes. Other functions are stimulation of bicarbonate-rich fluid secretion, insulin secretion, and intestinal motility. CCK can induce satiety in laboratory animals and humans, the gastric vagal fibers being necessary for this effect.

Secretin

Secretin is synthesized by the S cells of the duodenum and jejunum and is also present in the brain. Its amino acid sequence is similar to that of glucagon, VIP, and GIP. All 27 amino acid residues are required for biological activity. Chyme (pH < 4.5) in the duodenum stimulates release of secretin. Secretin stimulates the secretion of pancreatic juice rich in bicarbonate, which neutralizes the acid chyme and inhibits further secretion of the hormone. This action appears to be mediated by membrane-bound adenylate cyclase and increased concentrations of intracellular cAMP.

Incretins

Gastric Inhibitory Peptide

Gastric inhibitory peptide (GIP) is found in the K cells of the duodenum and jejunum: it consists of 43 amino acid residues and occurs in multiple molecular forms. Its secretion is stimulated by the presence of glucose and lipids in the duodenum. It has two main functions:

1. To stimulate insulin secretion that prepares the appropriate tissues for the transport and metabolism of nutrients obtained from the GI tract (Chapter 20); and
2. To inhibit gastric secretion and motility.

The physiological functions of the GI hormones discussed previously are integrated and coordinated to facilitate the digestion and absorption of food (Figure 11.8).

Glucagon-like Peptide-1

Glucagon-like peptide-1 (GLP-1) is formed in the L-cells of distal small and large intestines. It promotes insulin secretion, suppresses pancreatic glucagon secretion by the α-cells, delays gastrin emptying, and promotes satiety. GLP mimetics are used in the treatment of type 2 diabetes (see Chapter 20).

DIGESTION AND ABSORPTION OF MAJOR FOOD SUBSTANCES

Carbohydrates

In most diets, carbohydrates are a major source of the body's energy requirements. The predominant digestible carbohydrates are starches (amylose and amylopectin). Depending on the diet, the digestible carbohydrates may

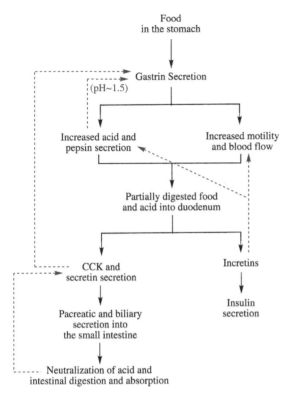

Food
in the stomach

Gastrin Secretion
(pH~1.5)

Increased acid and
pepsin secretion

Increased motility
and blood flow

Partially digested food
and acid into duodenum

CCK and
secretin secretion

Incretins

Pacreatic and biliary
secretion into
the small intestine

Insulin
secretion

Neutralization of acid and
intestinal digestion and absorption

FIGURE 11.8 Integrated function of gastrointestinal hormones in regulation of digestion and absorption of food. Dashed arrows indicate inhibition.

include glycogen, and the disaccharides sucrose and lactose. The disaccharide present in mushrooms, trehalose, is also digestible. The specific patterns of carbohydrate intake are influenced by cultural and economic factors. Many plant carbohydrates are not digestible, and they constitute the dietary fiber, which includes cellulose, hemicelluloses, pectins, gums, and alginate (Chapter 8). The digestion of starch (and glycogen) begins in the mouth during mastication and the mixing of food with salivary α-amylase, which hydrolyzes starch to some extent. The digestive action of salivary α-amylase on starch is terminated in the acidic environment of the stomach. Starch digestion is resumed in the duodenum by another α-amylase secreted by the pancreas. Salivary and pancreatic α-amylases share some properties and exist in several isoenzyme forms that are separable by electrophoresis. In pancreatic disease, the α-amylase level in the serum increases, and its measurement and isoenzyme pattern are useful in diagnosis of acute pancreatitis (Chapter 7). Carbohydrate digestion and absorption take place in a well-defined sequence:

1. Intraluminal hydrolysis of starch and glycogen by α-amylase to oligosaccharides of variable length and structure;

2. Brush-border surface hydrolysis of oligosaccharides and disaccharides (e.g., sucrose, lactose, and trehalose) to their monomers by specific oligosaccharidases that are integral to the cell membrane of the enterocyte; and

3. Transport of monosaccharides (e.g., glucose, galactose, and fructose) into enterocytes.

Digestion of Starch

α-Amylase hydrolyzes starch into α-limit dextrins (branched oligosaccharides of five to nine glucose residues), maltotriose, and maltose (Figure 11.9). The α-amylase is an endoglycosidase, and its action does not yield free glucose. It cannot catalyze the hydrolysis of $\alpha(1 \rightarrow 6)$ linkages (the branch points). α-Amylase has optimal activity at pH 7.1, an absolute requirement for the presence of Cl^-, and is stabilized by Ca^{2+}.

Brush-Border Surface Hydrolysis

Products of α-amylase digestion of starch and ingested disaccharides are hydrolyzed by oligosaccharidases on enterocyte cell membranes to yield monosaccharides that are transferred across the brush-border bilayer. The oligosaccharidases are large glycoproteins (M.W. >200,000) that are integral constituents of the cell membrane. Their active sites project toward the luminal side. The rate-determining step in absorption is the monosaccharide transport system, with the exception of mucosal lactase, which has the lowest activity of oligosaccharidases. Thus, hydrolysis of lactate rather than absorption is rate limiting. Oligosaccharide digestion is virtually complete by mid jejunum.

Transport of Monosaccharides into the Enterocyte

Glucose and galactose compete for a common transport system. This system is an active transport system; i.e., the monosaccharides are absorbed against a concentration gradient. It is saturable and obeys Michaelis–Menten kinetics, and it is carrier-mediated and Na^+-dependent. Translocation of glucose (or galactose) is shown in Figure 11.10. On the luminal side, one molecule of glucose and two sodium ions bind to the membrane carrier (presumably, Na^+ binding to the carrier molecule increases the affinity for glucose because of a conformational change). The carrier-bound Na^+ and glucose are internalized along the electrochemical gradient that results from a low intracellular Na^+ concentration. Inside the cell, the sodium ions are released from the carrier, and the diminished affinity of the carrier for glucose releases the glucose. The sodium ions that enter in this manner are transported into the lateral intercellular spaces against a concentration gradient by the free energy of ATP

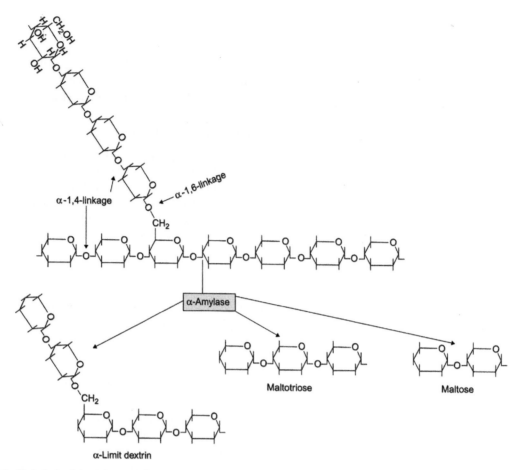

α-1,4-linkage

α-1,6-linkage

α-Amylase

Maltotriose

Maltose

α-Limit dextrin

FIGURE 11.9 Hydrolysis of starch by α-amylase.

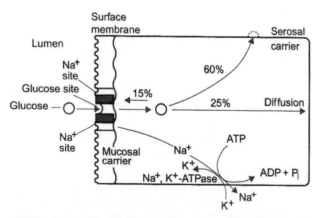

FIGURE 11.10 Schematic representation of glucose (or galactose) transport by the enterocyte. Glucose binds to the receptor, facilitated by the simultaneous binding of two Na^+ ions at separate sites. The glucose and Na^+ are released in the cytosol as the receptor affinity for them decreases. The Na^+ are actively extruded at the basolateral surface into the intercellular space by Na^+,K^+-ATPase, which provides the energy for the overall transport. Glucose is transported out of the cell into the intercellular space and hence to portal capillaries, both by a serosal carrier and by diffusion. *[Reproduced with permission from G. M. Gray, Carbohydrate absorption and malabsorption in gastrointestinal physiology. New York: Raven Press, 1981.]*

hydrolysis catalyzed by a Na^+,K^+-ATPase. Thus, glucose and Na^+ are transported by a common carrier, and energy is provided by the transport of Na^+ down the concentration and electrical gradient. The low cytoplasmic Na^+ concentration is maintained by the active transport of Na^+ out of the cell, coupled to K^+ transport into the cell by the Na^+,K^+-ATPase. Since the sugar transport does not use ATP directly, it can be considered as secondary active transport. Although this mode of glucose transport is the most significant, passive diffusion along a concentration gradient may also operate if the luminal concentration of glucose exceeds the intracellular concentration.

The intracellular glucose is transferred to the portal capillary blood by passive diffusion and by a carrier-mediated system. Intracellular glucose can be converted to lactate (Chapter 12), which is transported via the portal blood system to the liver, where it is reconverted to glucose (gluconeogenesis; Chapter 14). The quantitative significance of this mode of glucose transport is probably minimal.

Fructose transport is distinct from glucose–galactose transport and requires a specific saturable membrane carrier (facilitated diffusion).

Disorders of Carbohydrate Digestion and Absorption

Carbohydrate malabsorption can occur in a number of diseases that cause mucosal damage or dysfunction (e.g., gastroenteritis, protein deficiency, gluten-sensitive enteropathy). Disorders due to deficiencies of specific oligosaccharidases are discussed next.

Lactose Intolerance

Lactose intolerance (also known as **milk intolerance**) is the most common disorder of carbohydrate absorption, leading to diarrhea [15]. Lactase deficiency occurs in the majority of human adults throughout the world and appears to be genetically determined. The prevalence is high in persons of African and Asian ancestry (≥65%) and low in persons of Northern European ancestry. Lactase deficiency in which mucosal lactase levels are low or absent at birth is rare and is transmitted as an autosomal recessive trait. Prompt diagnosis and a lactose-free diet assure the infant's normal growth. Since hydrolysis of lactose by lactase is rate-limiting, any mucosal damage will cause lactose intolerance.

In full-term human infants, lactase activity attains peak values at birth and remains high throughout infancy. As milk intake decreases, lactase levels drop and lactose intolerance may develop. The extent of the decrease of lactase activity distinguishes lactose-tolerant from -intolerant populations.

Ingestion of milk (or lactose) by individuals who have lactose intolerance leads to a variety of symptoms (bloating, cramps, flatulence, and loose stools). Severity of the symptoms depends on the amount of lactose consumed and the enzyme activity. Symptoms disappear with elimination of lactose from the diet. Lactose-depleted milk or fermented milk products with negligible amounts of lactose are good substitutes for milk. The intestinal problems are due primarily to osmotic effects of lactose and its metabolites in the colon. The lactose not absorbed in the small intestine increases the osmolarity and causes water to be retained in the lumen. In the colon, it is metabolized by bacterial enzymes to a number of short-chain acids, further increasing osmolarity and aggravating fluid reabsorption. The bacterial fermentation also yields gaseous products (H_2, CO_2, and CH_4), hence the bloating, flatulence, and sometimes frothy diarrhea.

Intolerance of Other Carbohydrates

Intolerance to sucrose and α-limit dextrins may be due to deficiency of sucrase-α-dextrinase or to a defect in glucose–galactose transport. These disorders are rare autosomal recessive traits; clinical problems can be corrected by removing the offending sugars from the diet. Lactulose, a synthetic disaccharide consisting of galactose and fructose with a β(1→4) linkage, is hydrolyzed not in the small intestine but in the colon, and is converted to products similar to those derived from lactose fermentation. It has been used in the treatment of patients with liver disease. Normally, ammonia (NH_3) produced in the GI tract is converted in the liver to urea (Chapter 15); hence, in patients with severe liver disease, blood ammonia levels increase. Absorption of ammonia can be decreased by administration of lactulose, which acidifies the colonic constituents so that NH_3 is trapped as NH_4^+ ions (Chapter 8).

Proteins

Protein is an essential nutrient for human growth, development, and homeostasis. The nutritive value of dietary proteins depends on its amino acid composition and digestibility. Dietary proteins supply **essential amino acids**, which are not synthesized in the body. Nonessential amino acids can be synthesized from appropriate precursor substances (Chapter 15). In human adults, essential amino acids are valine, leucine, isoleucine, lysine, methionine, phenylalanine, tryptophan, and threonine. Histidine (and possibly arginine) appears to also be required for support of normal growth in children. In the absence from the diet of an essential amino acid, cellular protein synthesis does not occur. The diet must contain these amino acids in the proper proportions. Thus, quality and quantity of dietary protein consumption and adequate intake of energy (carbohydrates and lipids) are essential. Protein constitutes about 10%–15% of the average total energy intake.

Animal proteins, with the exception of collagen (which lacks tryptophan), provide all of the essential amino acids. Vegetable proteins differ in their content of essential amino acids, but a mixture of these proteins will satisfy the essential amino acid requirement. For example, the lysine lacking in grains can be provided by legumes. This combination also corrects for the methionine (which is supplied in corn) deficiency of legumes. Such combinations (e.g., lentils and rice, chick-peas and sesame seeds, spaghetti and beans, corn and beans) are widely used in different cultures to provide an optimal protein requirement. The recommended allowance for mixed proteins in an adult in the United States is 0.85 g per kilogram of body weight per day. The allowances are increased during childhood, pregnancy, and lactation.

Besides dietary protein, a large amount of endogenous protein undergoes digestion and absorption. Endogenous protein comes from three sources:

1. Enzymes, glycoproteins, and mucins secreted from the salivary glands, stomach, intestine, biliary tract, and pancreas, which together constitute about 20–30 g/day;
2. Rapid turnover of the gastrointestinal epithelium, which contributes about 30 g/day; and
3. Plasma proteins that normally diffuse into the intestinal tract at a rate of 1–2 g/day.

In several disorders of the GI tract (e.g., protein-losing gastroenteropathy), loss of plasma proteins is considerable and leads to hypoproteinemia.

Digestion

Protein digestion begins in the stomach, where protein is denatured by the low pH and is exposed to the action of pepsin. The low pH also provides the optimal H^+ concentration for pepsin activity. The zymogen precursor pepsinogen (M.W. 40,000) is secreted by the chief cells and is converted to pepsin (M.W. 32,700) in the acid medium by removal of a peptide consisting of 44 amino acid residues. This endopeptidase hydrolyzes peptide bonds that involve the carboxyl group of aromatic amino acid residues, leucine, methionine, and acidic residues. The products consist of a mixture of oligopeptides.

Chyme contains potent secretagogues for various endocrine cells in the intestinal mucosa. CCK and secretin cause release of an alkaline pancreatic juice containing trypsinogen, chymotrypsinogen, proelastase, and procarboxypeptidases A and B. Activation begins with the conversion of trypsinogen to trypsin by enteropeptidase (previously called **enterokinase**) present in the brush-border membranes of the duodenum.

Enteropeptidase cleaves between Lys-6 and Ile-7 to release a hexapeptide from the N-terminus of trypsinogen. Trypsin autocatalytically activates trypsinogen and activates the other zymogens. The importance of the initial activation of trypsinogen to trypsin by enteropeptidase is manifested by children with congenital enteropeptidase deficiency who exhibit hypoproteinemia, anemia, failure to thrive, vomiting, and diarrhea.

Trypsin, elastase, and chymotrypsin are **endopeptidases**. Carboxypeptidases are **exopeptidases**. The combined action of these enzymes produces oligopeptides having two to six amino acid residues and free amino acids. Hydrolysis of oligopeptides by the brush-border aminopeptidases releases amino acids. Leucine aminopeptidase, a Zn^{2+}-containing enzyme, is an integral transmembrane glycoprotein with a carbohydrate-rich hydrophilic portion and active site protruding into the luminal side. It cannot degrade proline-containing dipeptides, which are largely hydrolyzed intracellularly. Dipeptidases and tripeptidases are associated with the brush-border membranes, but their functions are not clearly understood. The major products of intraluminal digestion of protein are mixtures of amino acids (30%–40%) and small peptides (60%–70%).

Absorption of Amino Acids and Oligopeptides

Dipeptides and tripeptides that escape brush-border membrane peptidases are actively transported against a concentration gradient by Na^+-dependent mechanisms. Inside the cell they are hydrolyzed.

Free amino acids are transported into enterocytes by four active, carrier-mediated, Na^+-dependent transport systems remarkably similar to the system for glucose. These systems transport, respectively, neutral amino acids; basic amino acids (Lys, Arg, His) and cystine; aspartic and glutamic acids; and glycine and imino acids.

Some amino acids (e.g., glycine) have affinities for more than one transport system; within each group, amino acids compete with each other for transport. These systems are specific for L-amino acids. D-amino acids are transported by a passive diffusion process. Amino acid transport in the renal cells may use similar systems (Chapter 37). Entry of amino acids into cell compartments elsewhere in the body may require different transport systems (e.g., the γ-glutamyl cycle; Chapter 15). In the enterocyte, amino acids may be metabolized or transported to the liver. Glutamate, glutamine, aspartate, and asparagine are metabolized in the enterocyte (Chapter 15). Small amounts of protein (e.g., dietary, bacterial, and viral) may be absorbed intact by nonselective pinocytosis. This absorption may be more common in neonatal life. Absorption of food proteins (or their partially digested antigenic peptides) can cause allergic manifestations, whereas bacterial and viral antigens stimulate immunity by production and secretion of secretory IgA (Chapter 33). Since thyrotropin-releasing hormone (pGlu-His-Pro-NH$_2$) is resistant to hydrolysis, it is effective if taken orally (Chapter 31).

Disorders of Protein Digestion and Absorption

The principal causes of protein maldigestion and malabsorption are diseases of the exocrine pancreas and small intestine. Primary isolated deficiency of pepsinogen or pepsin, affecting protein assimilation, has not been described. Deficiencies of trypsinogen and enteropeptidase are rare.

Defects in neutral amino acid transport (**Hartnup disease**), in basic amino acids and cystine (**cystinuria**), dicarboxylic aminoaciduria, and aminoglycinuria have been reported. The clinical severity of these disorders is usually minimal and relates to the loss of amino acids or relative insolubility of certain amino acids in the urine. In cystinuria, for example, cystine can precipitate in acidic urine to form stones. In Hartnup disease, severe nutritional deficiencies are uncommon, since the essential amino acids are absorbed as dipeptides or oligopeptides. However, tryptophan deficiency may occur. It is a precursor of nicotinamide (a vitamin; see Chapter 36) and used in the biosynthesis of NAD^+ and $NADP^+$ (Chapter 15). Skin and neuropsychiatric manifestations characteristic of nicotinamide deficiency that occur in Hartnup disease respond to oral nicotinamide supplementation.

Lipids

Dietary fat provides energy in a highly concentrated form and accounts for 40%−45% of the total daily energy intake (100 g/day in the average Western diet). Lipids contain more than twice the energy per unit mass than carbohydrates and proteins (Chapter 5). The efficiency of fat absorption is very high; under normal conditions, almost all ingested fat is absorbed, with less than 5% appearing in the feces. The predominant dietary lipid is triacylglycerol (a **triglyceride**), which contains three long-chain (l6-carbon or longer) fatty acids (Chapter 16). The dietary lipids include essential fatty acids (Chapter 16) and the lipid-soluble vitamins A, D, E, and K (Chapters 35 and 36).

Digestion and absorption of lipids involves the coordinated function of several organs but can be divided into three phases: luminal, intracellular, and secretory.

Intraluminal Phase

Digestion of lipid in the mouth and stomach is minimal. However, lipases secreted by lingual glands at the base of the tongue are active at acid pH and initiate the hydrolysis of triacylglycerol without a requirement for bile acids. The fatty acids released stimulate the release of CCK and a flow of bile and pancreatic juice. The free fatty acids also stabilize the surface of triacylglycerol particles and promote binding of pancreatic colipase. This phase aids in the optimal action of pancreatic lipase and is particularly important in disorders of pancreatic function or secretion (e.g., in prematurity, cystic fibrosis, congenital deficiency of pancreatic lipase).

The major functions of the stomach in fat digestion are to store a fatty meal, to contribute to emulsification by the shearing actions of the pylorus, and to gradually transfer the partially digested emulsified fat to the duodenum by controlling the rate of delivery. The hydrolysis of triacylglycerol in the duodenum and jejunum requires bile and pancreatic juice. Bile acids are powerful detergents that, together with monoacylglycerol and phosphatidylcholine, promote the emulsification of lipids. Emulsification is also aided by the churning action of the GI tract, which greatly increases the area of the lipid−water interface that promotes the action of pancreatic lipase. The products of digestion are relatively insoluble in water but are solubilized in micelles. Micelles also contain lipid-soluble vitamins, cholesterol, and phosphatidylcholine. Pancreatic lipase functions at the lipid−water interface, its activity being facilitated by colipase (M.W. ~ 10,000), also secreted by the pancreas as procolipase activated by tryptic hydrolysis of an Arg−Gly bond in the N-terminal region. Colipase anchors the lipase to the triacylglycerol emulsion in the presence of bile salts by forming a 1:1 complex with lipase, and protects lipase against denaturation. Colipase deficiency (with normal lipase) is accompanied by significant lipid malabsorption, as is pancreatic lipase deficiency. Pancreatic juice contains esterases that act on short-chain triacylglycerols and do not require bile salts, as well as cholesteryl esterase. Phosphatidylcholine in the diet (4−8 g/day) and in bile secretions (17−22 g/day) is hydrolyzed to lysophosphatidylcholine and fatty acid by phospholipase A2, a pancreatic enzyme with an absolute requirement for Ca_2^+ ions and for bile acids. The secreted form, prophospholipase A2, is activated by tryptic hydrolysis of an Arg−Ala bond in the N-terminal region. Phospholipase A2 also hydrolyzes fatty acids esterified at the 2-position of phosphatidylethanolamine, phosphatidylglycerol, phosphatidylserine, phosphatidylinositol, and cardiolipin, but has no effect on sphingolipids.

Lipid absorption in the duodenum and jejunum appears to be a passive diffusion process. Lipid-laden micelles migrate to the microvilli, and the fatty acids, monoacylglycerols, and lysophosphoglycerols are transferred across the membrane according to their solubility within the lipid bilayer of the cell surfaces. Bile acids are not absorbed into the enterocyte but migrate to the ileum, where they are actively absorbed and transferred to the liver via the portal venous system. The bile acid pool is recycled several times daily (enterohepatic circulation) to meet the demands of lipid digestion, and disorders that interfere with this process lead to malabsorption of lipids. A cytoplasmic fatty acid-binding protein with high affinity for long-chain fatty acids transports fatty acids to the smooth endoplasmic reticulum for resynthesis of triacylglycerol.

Digestion and absorption of triacylglycerols with medium-chain fatty acids (≤ 12 carbons) proceed by a different pathway. Medium-chain triacylglycerols are partly water-soluble, are rapidly hydrolyzed by lingual and pancreatic lipases, and do not require the participation of bile acids. Some are absorbed intact and hydrolyzed inside the absorptive cell. Medium-chain fatty acids enter the portal vein. Thus, medium-chain triacylglycerols can be digested and absorbed in the presence of minimal amounts of pancreatic lipase and in the absence of bile salts. For this reason, they are used to supplement energy intake in patients with malabsorption syndromes. Coconut oil is rich in trioctanoylglycerol (8-carbon) and tridecanoylglycerol (10-carbon).

Intracellular (Mucosal) Phase

Fatty acids (long-chain) are activated, and monoacylglycerols are converted to triacylglycerols at the smooth endoplasmic reticulum. The steps involved are as follows:

1. Conversion of fatty acids to acyl-CoA derivatives by acyl-CoA synthetase.

$$RCOO^- + HSCoA + ATP^{4-} \xrightarrow{Mg^{2+}, K^+}$$
$$RCOSCoA + AMP^{2-} + PP_i^{3-}$$

2. Esterification of monoacylglycerol to diacylglycerol and triacylglycerol catalyzed by monoacylglycerol transacylase and diacylglycerol transacylase, respectively.

In a minor alternative pathway, triacylglycerol is synthesized from glycerol-3-phosphate and acyl-CoA by esterification at the 1,2-positions of glycerol, removal of the phosphate group, removal of the phosphate group, and esterification at C3 (Chapter 17).

The triacylglycerols are incorporated into a heterogeneous population of spherical lipoprotein particles known as **chylomicrons** (diameter 75–600 nm) that contain about 89% triacylglycerol, 8% phospholipid, 2% cholesterol, and 1% protein. Phospholipids of chylomicrons arise by *de novo* synthesis (Chapter 17) or from reacylation of absorbed lysolecithin. Cholesterol is supplied by *de novo* synthesis (Chapter 17) or is absorbed. The protein apolipoprotein B-48 (apo B-48) forms a characteristic protein complement of chylomicrons and is synthesized in enterocytes. Synthesis of apo B-48 is an obligatory step in chylomicron formation. Absence of apo B-48 synthesis, as in the rare hereditary disease **abetalipoproteinemia**, leads to fat malabsorption. Enterocytes are involved in the synthesis of other lipoproteins (Chapter 18).

Secretion

Vesicles that contain chylomicrons synthesized within the endoplasmic reticulum and the saccules of the Golgi apparatus migrate toward the laterobasal membrane, fuse with it, and extrude the chylomicrons into the interstitial fluid, where they enter the **lymphatic vessels through fenestrations**. Medium-chain triacylglycerols are absorbed and transported by portal blood capillaries without formation of micelles or chylomicrons. Chylomicrons enter the bloodstream at the left subclavian vein via the thoracic duct. In the bloodstream, they are progressively hydrolyzed by endothelial lipoprotein lipase activated by apolipoprotein C-II. Fatty acids so released are taken up by the tissues (e.g., muscle and adipose) as blood passes through them. The chylomicron remnants consist primarily of lipid-soluble vitamins and cholesterol (and its esters) and are metabolized in the liver (Chapter 18).

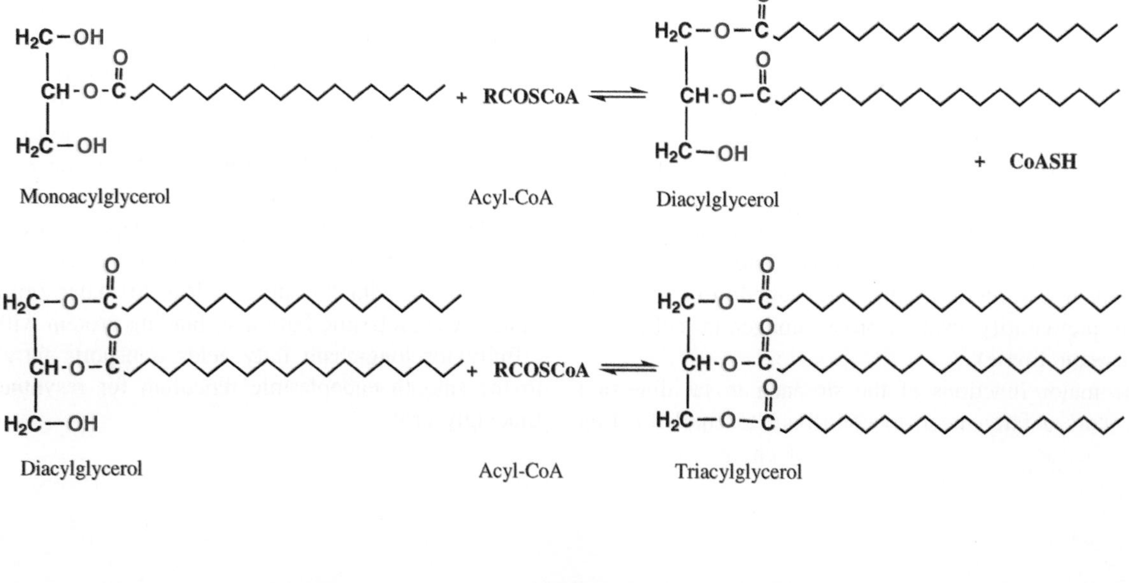

TABLE 11.2 Abnormalities of Lipid Digestion Due to Impaired Lipolysis

Type of Defect	Biochemical Disturbance	Examples of Disease States
Rapid gastric emptying	Reduction in the efficiency of lipid interaction with bile and pancreatic secretions	Gastrectomy, as in treatment of ulcers or in neoplasms of stomach
Acidic duodenal pH	Inactivation of pancreatic lipase and decreased ionization of bile acids	Zollinger–Ellison syndrome
Decreased CCK release	Deficiency of bile and pancreatic secretions	Disorders associated with mucosal destruction; regional enteritis; gluten enteropathy
Congenital lipase or colipase deficiency	Defective lipolysis	
Pancreatic insufficiency	Defective lipolysis	Chronic pancreatitis, Pancreatic duct obstruction (e.g., cystic fibrosis)
Absent or decreased bile salts	Decreased lipolysis due to impaired micelle formation	(see Table 11.3)

TABLE 11.3 Abnormalities of Lipid Digestion Due to Impaired Micelle Formation Leading to Decreased Lipolysis

Type of Defect	Examples of Disease States
Decreased hepatic synthesis of bile salts	Severe parenchymal liver disease
Decreased delivery of bile salts to the intestinal lumen	Biliary obstruction due to stone, tumor, or primary biliary cirrhosis
Decreased effective concentration of conjugated bile acids	Zollinger–Ellison syndrome (causes hyperacidity); bacterial overgrowth and stasis; administration of drugs neomycin and cholestyramine
Increased intestinal loss of bile salts	Ileal disease or resection

Chylomicrons normally begin to appear in the plasma within 1 hour after ingestion of fat and are completely removed within 5–8 hours.

Disorders of Lipid Digestion and Absorption

Normally, more than 95% of ingested lipid is absorbed. When a large fraction is excreted in the feces, it is called **steatorrhea**. Measurement of fecal lipid with adequate lipid intake is a sensitive indicator of lipid malabsorption. Malabsorption can result from impairment in: lipolysis (Table 11.2), micelle formation (Table 11.3), absorption, chylomicron formation, or transport of chylomicrons via the lymph to blood.

General Malabsorptive Problems

A malabsorptive disorder caused by proteins found in wheat, rye, and barley produces chronic sensitivity that damages the small intestine in susceptible individuals. This disorder is known as **gluten-sensitive enteropathy** or **celiac disease** (see Clinical Case Study 11.2, vignettes 1 and 2) [16]. The toxicity of the cereals is associated with a group of proteins known as gliadins, hordeins, and secalins. In susceptible individuals, peptides of 33 amino acid residues derived from gliadin survive proteolytic degradation. These 33 amino acid peptides undergo deamidation by tissue transglutaminase, leading to conversion of glutamine residues to glutamic acid. The deamidated peptides undergo proteolytic cleavage in antigen-presenting cells, producing epitopes (Chapter 33). These epitopes eventually activate T cells, leading to an undesirable immune response (Figure 11.11). The damage to the small bowel consists of conversion of normal columnar mucosal cells to cuboidal cells, villous flattening, crypt hyperplasia, and infiltration of lymphocytes and plasma cells into the lamina propria. Celiac disease has characteristics of an autoimmune disorder associated with a genetic predisposition. High-risk populations for celiac disease include patients with Down's syndrome, those with insulin-dependent diabetes, those with other autoimmune disorders such as lupus and rheumatoid arthritis, and relatives of patients with celiac disease.

The diagnosis of celiac disease includes characteristic intestinal biopsy findings (discussed previously) and serological testing for antigliadin and antiendomysial antibodies. The treatment of patients with celiac disease consists of lifelong complete abstinence from gluten-containing foods; acceptable foods are rice, potato, and maize.

Cystic fibrosis (CF) is a disease of multiple exocrinopathy and generalized malabsorption due to lack of

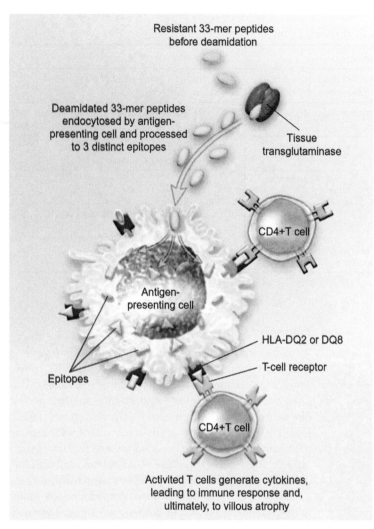

Resistant 33-mer peptides
before deamidation

Deamidated 33-mer peptides
endocytosed by antigen-
presenting cell and processed
to 3 distinct epitopes

Tissue
transglutaminase

CD4+T cell

Antigen-
presenting cell

Epitopes

HLA-DQ2 or DQ8

T-cell receptor

CD4+T cell

Activited T cells generate cytokines,
leading to immune response and,
ultimately, to villous atrophy

FIGURE 11.11 Initiation of a pathologic immunologic response to undigestible 33 amino acid peptides derived from gliadin in the lamina propria of the GI tract. *[Reprinted with permission from R. McManus, D. Kelleher, Celiac disease: the villain unmasked, N. Engl. J. Med., 348 (2003) 2573.]*

delivery of pancreatic digestive enzymes to the small intestine (see Clinical Case Study 11.3). CF is an autosomal recessive disease. Among Caucasians, the incidence is 1 in 2000–3000 births. The prevalence of heterozygous individuals is about 1 in 20. Abnormalities in affected persons are found in airways, lungs, pancreas, liver, intestine, vas deferens, and sweat glands. The primary abnormality in CF is in electrolyte transport, specifically in Cl^- secretion from the apical membrane of the epithelial cells of sweat glands, airways, pancreas, and intestine. Defective Cl^- secretion causes hyperactivity of Na^+ absorption, and these two processes cause the secreted mucus to become viscous and sticky. The most life-threatening clinical feature is related to pulmonary disease. The sticky, viscous mucus clogs the airway and compromises the normal beating of the cilia that cover the apical surface of the airway epithelium. These conditions promote lung infections by bacteria such as *Pseudomonas aeruginosa* and *Staphylococcus aureus*.

The clinical abnormalities related to the gastrointestinal tract are not life threatening and can be treated. In newborns with CF, intestinal obstruction (meconium ileus) can occur in 10%–20% of cases because of failure of digestion of intraluminal contents due to lack of pancreatic enzymes *in utero*. Exocrine pancreatic enzyme deficiency is present from birth, affecting both lipid and protein digestion. In general, carbohydrate digestion is not severely impaired.

CF is associated with a plethora of clinical conditions such as diseases of the hepatobiliary tract and genitourinary tract (e.g., male infertility). Elevation of chloride concentration in the sweat is the most consistent functional abnormality in CF and the determination of the sweat chloride concentration is used as the standard diagnostic test. The cause of elevated Cl^- concentration in the sweat is due to failure of reabsorption in the reabsorptive portion of the sweat gland. Normally, sweat is a hypotonic fluid as it emerges at the surface of the skin. This occurs because the secretory and absorptive activities of the sweat gland are located in two different regions.

The CF gene encodes a protein designated as the cystic fibrosis transmembrane conductance regulator (CFTR) and is on chromosome 7q31.2. CFTR is a glycosylated protein containing 1480 amino acid residues. It is a 170- to 180-kDa protein, and the variations in molecular weight are due to differences in glycosylation. CFTR belongs to a family of channel proteins known as **ATP-binding cassette (ABC)** transporters that are essential to virtually all cells (discussed in several sections of the text). Abnormalities in channel proteins have both inherited and noninherited causes, and associated disorders have been called **channelopathies**. Other ABC transport defects are multidrug resistance (MDR) transporter (known as P-glycoprotein) and sulfonylurea receptors (SUR1 and SUR2). P-glycoprotein gene is upregulated in its expression in response to certain chemotherapeutic drugs (e.g., vinca alkaloids). This causes the cells to become multidrug resistant because the drugs are exported out of the target cells in an ATP-dependent process. The normal physiological role of P-glycoprotein may reside in its participation in the transport of phosphatidylcholine and other phospholipids (Chapter 17). It is also thought that the P-glycoprotein functions in the detoxification process of pumping toxins that are xenobiotic out of cells. The sulfonylurea-receptor transporter is an ATP-sensitive K^+ channel found in the β-cells of the pancreatic islets of Langerhans and in cardiac and skeletal muscle. Regulation of insulin secretion from β-cells is governed by sulfonylurea receptor proteins (Chapter 20). TAP1 and TAP2, subunits of the major histocompatibility complex, are ABC transporters involved in peptide transport into the endoplasmic reticulum of antigen-presenting cells. Recurrent respiratory infections and bronchiectasis occur in patients with TAP2 defects. ABC transporters are also involved in many other biochemical processes. These include peroxisome biogenesis (**X-linked adrenoleukodystrophy, Zellweger's syndrome**, Chapter 16), cholesterol efflux (**Tangier disease**, Chapter 18), and all-*trans* retinaldehyde efflux from the inside of the disk to the cytoplasm in retinal rod cells (**early-onset macular degeneration**).

The CFTR protein consists of five domains: two membrane-spanning domains (MSDs), two nucleotide-binding domains (NBDs), and a regulatory domain (Figure 11.12). Both the amino terminus and carboxy terminus are located in the cytoplasm, and each of the two membrane-spanning domains contains six transmembrane segments. The two nucleotide-binding domains interact with ATP. The unique regulatory domain contains several consensus phosphorylation sites. The phosphorylation of the R-protein occurs both by cAMP-dependent protein kinase and by protein kinase C. CFTR protein is located in the apical cell membrane of most epithelial cells; however, it is also found in the basolateral cell membranes and presumably participates in Cl^- reabsorption. The basolateral cell membrane sites are located in sweat duct

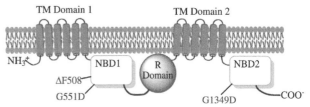

FIGURE 11.12 Diagrammatic representation of cystic fibrosis transmembrane regulatory protein (CFTR). CFTR consists of two membrane-spanning or transmembrane (TM) domains, two nucleotide-binding domains (NBDs), and one regulatory R-domain. The opening of the chloride channel requires ATP binding to at least one of the NBDs and phosphorylation of R-domain by cAMP-dependent protein kinase. There are more than 700 mutations and polymorphisms of CFTR. NBDs are hot spots for mutation; and three severe CF-causing mutations ΔF508, G551D, and G1349D, are shown. The ΔF508 mutation is the most common mutation, occurring in about 70% of CF patients.

epithelial cells and the proximal convoluted tubule and the thick ascending limbs of Henle's loop in the kidney.

The understanding of CFTR function in chloride transport has been aided by insertion of CFTR into artificial lipid membranes and by recording currents flowing through single membrane channels using patch-clamp technology. Activation of the channel for Cl^- secretion requires phosphorylation of sites located in the R-domain and binding and/or hydrolysis of ATP at the nucleotide-binding domain. Phosphorylation by protein kinases and dephosphorylation by phosphatases at the R-subunit controls Cl^- secretion. Thus, if CFTR is defective, Cl^- secretion does not occur. If CFTR is kept continually active by increasing intracellular cAMP levels, the Cl^- transport is enormously increased with accompanying fluid loss (Figure 11.13).

More than 700 mutations located throughout the gene have been identified in the CFTR gene in CF patients. The mutations include missense, nonsense, and frameshift mutations. In terms of functional defects, CFTR mutations have been grouped into four categories:

1. Protein production
2. Protein processing
3. Regulation of the channel
4. Conduction.

The most common mutation causing CF is a 3-base pair deletion resulting in loss of a phenylalanine (F) residue at position 508. This deletion (ΔF508), is found in about 70% of CF patients worldwide and is associated with severe disease. The phenylalanine-deficient CFTR protein gets trapped in the endoplasmic reticulum and is degraded, leading to net loss of chloride channel function. The G551D mutation of CF leads to a functional CFTR protein; however, it does not undergo misfolding and is translocated to the cell membrane. The mutation occurs in about 4% of CF patients. A recent therapeutic agent

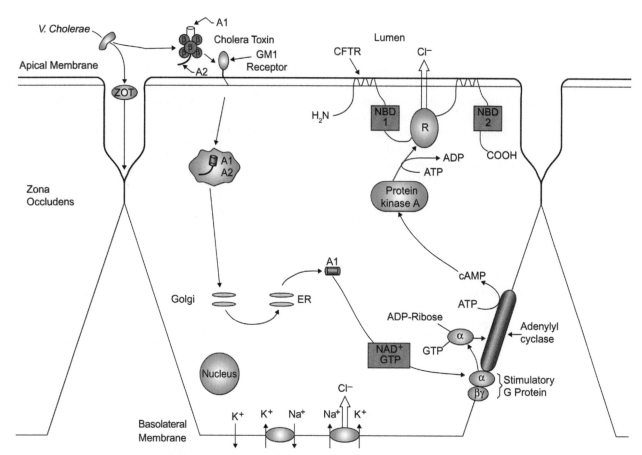

FIGURE 11.13 Mechanism of action of cholera toxin. The steps consist of cholera toxin (CT) binding to GM1 receptors anchored by its ceramide moiety; internalization of CT in endosomes; the release of the A_1-subunit of CT from the trans-Golgi network and endoplasmic reticulum (ER); ADP ribosylation of the α-subunit or stimulatory G-protein by A_1; activation of adenylyl cyclase and production of cAMP; activation of protein kinase A; phosphorylation of the regulatory (R) subunit of CFTR; and finally opening of the chloride channel. *Vibrio cholerae* also produces a zona occludens toxin (ZOT), which increases the ionic permeability of the zona occludens.

known as **ivacaftor** has been shown to promote the chloride channel activity with G551D CF patients [17].

The diagnosis of CF includes the clinical features of chronic obstructive lung disease, persistent pulmonary infection, meconium ileus, and pancreatic insufficiency with failure-to-thrive syndrome. Family history is also very helpful. The diagnosis can be confirmed by a positive sweat chloride test in which a concentration of chloride ions is greater than 60 meq/L. The genotype is confirmed by DNA analysis, and carrier detection in CF families is useful in genetic counseling. Newborn screening for CF consists of quantitation of immunoreactive trypsinogen (IRT) in dried blood spot obtained from infants a few days after birth. The IRT levels are elevated in CF, presumably due to blockage of the ducts that transport normally produced IRT and its subsequent spillage into blood circulation. However, as CF individuals age with deterioration of exocrine pancreatic function, blood

IRT levels decrease, and, therefore, IRT levels are only used during infancy. Treatment of CF patients involves a comprehensive approach. Use of antibiotics and pancreatic enzyme replacement therapy has been helpful in treating pulmonary infection and the maldigestion of food substances, respectively. Recombinant DNase (pulmozyme) and N-acetylcysteine have been used in clearing the airways in patients. A potentially useful agent in clearing airways is **recombinant gelsolin**, which degrades actin present in cell lysates and decreases viscosity. Aerosolized drugs such as amiloride, an inhibitor of the Na^+ channel, and nonhydrolyzable forms of UTP (UTPγS), an activator of non-CFTR chloride channels, are being tested clinically.

Chronic alcoholism is frequently associated with generalized malabsorption of major foods and vitamins because of liver and pancreatic involvement and mucosal dysfunction.

ABSORPTION OF WATER AND ELECTROLYTES

Fluid turnover of about 9 L occurs daily in the GI tract. Ingested water contributes about 2 L, and the remainder arises from secretions of the GI tract mucosa and associated glands. Nearly all of this water is reabsorbed, and about 200 mL (2%) or less is excreted (Table 11.4). If the amount of water excreted in feces exceeds 500 mL, diarrhea results. Similarly, only 2% of Na^+ and 10% of K^+ in gastrointestinal fluids appear in the feces. Most absorption of fluid and electrolytes occurs in the small intestine.

The gastric mucosa is relatively impermeable to water, but the small intestine is highly permeable, and water transport occurs in both directions depending on osmotic gradients. The osmolality of the duodenal contents can be low or high, depending on the food ingested. However, in the jejunum, the luminal fluid becomes isotonic and remains so throughout the rest of the small intestine. Water absorption is a passive process and is coupled to the transport of organic solutes and electrolytes. The simultaneous presence of glucose and Na^+ facilitates the absorption of all three. In the treatment of diarrhea, oral administration of solutions containing glucose and NaCl replaces fluids and electrolyte.

Na^+ is also absorbed separately from organic solutes through two coupled transport systems, one of which absorbs Na^+ in exchange for H^+, while the other absorbs Cl^- in exchange for HCO_3^-. In the lumen, H^+ and HCO_3^- combine

to give rise to CO_2 and H_2O, which enter the mucosal cell or pass through to the plasma. The Na^+ absorbed is pumped out by Na^+,K^+-ATPase in the basolateral membrane. Cl^- follows Na^+ passively, with transfer of NaCl to plasma.

In the colon Na^+, Cl^-, and water are efficiently absorbed. Na^+ absorption is regulated by aldosterone (Chapter 30). K^+ is secreted into the lumen as a component of mucus but is reabsorbed by passive diffusion. The amount of K^+ in the feces is usually far below the daily intake; however, in chronic diarrhea, the loss of ileal and colonic fluids can cause negative K^+ balance.

Disorders of Fluid and Electrolyte Absorption

If the intestinal contents become hyperosmolar, water enters the lumen to produce iso-osmolarity, and fluid and electrolyte losses occur. This condition is seen in lactose intolerance, ingestion of nonabsorbable laxatives such as magnesium salts, and ingestion of sugars such as lactulose. Bile acids inhibit the absorption of Na^+ in the colon. Under normal circumstances, bile acids do not reach the colon because they are actively reabsorbed in the ileum. When the ileum is diseased or resected, sufficient bile salts enter the colon to inhibit absorption of Na^+ and water and cause diarrhea. Hydroxylated fatty acids (e.g., ricinoleic acid, the active ingredient of castor oil) also inhibit salt and water absorption. In disorders with significant mucosal abnormalities (e.g., gluten-sensitive enteropathy), fluid and electrolyte absorption is impaired.

Loss of fluids and electrolytes in **cholera** results from stimulation of a secretory process (see Clinical Case Study 11.4, vignettes 1 and 2). The toxin secreted by *Vibrio cholerae* causes a diarrhea of up to 20 L/day, resulting in dehydration and electrolyte imbalance, which may lead to death. Cholera bacteria acquire a pathogenic protein gene by means of incorporation of a toxic gene contained in an invading bacterial virus. Bacteria in contaminated food attach chiefly to the ileal mucosa and secrete enterotoxin consisting of one fraction that binds to specific sites on the cell membrane and another responsible for the characteristic biochemical activity. The binding moiety consists of five identical polypeptide subunits (B) (M.W. 11,300) that surround the active moiety (A). The A-subunit consists of two unequally sized polypeptides, A_1 (M.W. 23,500) and A_2 (M.W. 5500), linked by a disulfide bridge. The A_2 polypeptide appears to connect the A1 polypeptide to the B-subunit. The B-subunit binds rapidly to monosialogangliosides in the membrane (G_{M1}; Chapters 9 and 17). The A_1 polypeptide then migrates through the membrane and catalyzes transfer of the ADP-ribose group from NAD^+ to the stimulatory guanine nucleotide-binding protein (G_s) that regulates adenylate cyclase activity. Adenylate cyclase, which catalyzes the conversion of ATP to cAMP (Chapter 28), is stimulated

TABLE 11.4 Daily Turnover of Water in the GI Tract

Source of Fluids	Quantity (L/d)	Composition
Input		
Diet	2.0	Variable
Saliva	1.5	Hypotonic, alkaline
Gastric juice	2.5	Isotonic, acid
Bile	0.5	Isotonic, alkaline
Pancreatic juice	1.5	Isotonic, alkaline
Intestinal juice	1.0	Isotonic, neutral
Total	9.0	
Reabsorption		
Jejunum	5.5	
Ileum	2.0	
Colon	1.3	
Total	8.8	
Lost in the stool		
(9.0 − 8.8) =	0.2	

or inhibited by the active forms of the guanine nucleotide-binding protein, G_s and G_i, respectively. The Gs protein is activated by the binding of GTP and inactivated when GTP is converted to GDP by a GTPase intrinsic to G_s protein. G proteins contain α-, β-, and γ-subunits. The ADP ribosylation of the α-subunit decreases GTP hydrolysis and thus leads to sustained activation of adenylate cyclase activity, increased intracellular levels of cAMP, and secretion of isotonic fluid throughout the entire small intestine. Cholera toxin does not cause fluid secretion in the stomach, has minimal effects on the colon, and does not affect Na^+-dependent absorption of glucose and amino acids. Its effects can be readily reversed by oral or intravenous administration of replacement fluids. The best cure for cholera is rehydration by providing uncontaminated safe drinking water and proper methods of sewage disposal [18,19].

The mechanism by which cholera toxin causes secretory diarrhea is through continuous stimulation of the CFTR-regulated Cl^- channel (Figure 11.13). In cystic fibrosis, CFTR defects cause the abolition of intestinal chloride secretion without affecting the absorptive capacity. In homozygous CF patients, the disease is eventually lethal (discussed earlier). In cholera infections, however, CFTR is over-activated with fluid and electrolyte losses that lead to intravascular volume depletion, severe metabolic acidosis, and profound hypokalemia. These metabolic changes can result in cardiac and renal failure with fatal consequences. It has been postulated that persons who are heterozygous for a CF mutation may have selective advantage during a cholera epidemic. This speculation has been tested in a mouse model; indeed, homozygous CF mice treated with cholera toxin did not show intestinal secretion of fluids despite an increase in intracellular cAMP levels. In heterozygous CF mice, the intestinal secretion is intermediate compared to that of controls. Based on these studies, CFTR chloride channel inhibitors have the potential therapeutic implications in the treatment of cholera and other enterotoxin-mediated secretory diarrheal diseases [19]. In addition, these inhibitors have potential in the treatment of fluid-filled cyst expansion, which occurs in the genetically inherited polycystic kidney disease [20]. A familial chronic diarrhea syndrome can result from an activation of a membrane-bound guanylate cyclase due to a genetic mutation. Ligands (e.g., heat-stable enterotoxin) bound to the guanylate cyclase induce the production of excessive amounts of cGMP. It is thought that cGMP may promote CFTR, resulting in chronic diarrhea [21].

A human pathogen known as *Escherichia coli* O157:H7 may cause nonbloody diarrhea, **hemorrhagic colitis**, **hemolytic uremic syndrome**, and death. The interval between exposure and illness averages only 3 days. The designation of O157:H7 derives from the fact that the bacterium expresses the 157th somatic (O) antigen and the 7th flagella (H) antigen. The pathogen is transmitted by contaminated food (e.g., ground beef, fruit and vegetables, and water) from one person to another and occasionally through occupational exposure. The pathogenicity of *E. coli* O157:H7 is due to its ability to produce a molecule composed of an enzymatic subunit (A_1) and a multimer of five receptor binding (B) subunits. The genome for the synthesis of the toxin resides on a bacteriophage inserted into *E. coli* O157 DNA. The A_1-subunit is linked to a carboxy terminal A2 fragment by a single disulfide bond.

Shiga toxin produced by *Shigella dysenteriae* has similar structural features. The toxin binds to a glycolipid (Gb3), undergoes endocytosis, and the enzymatic A_1 fragment, which is a specific N-glycosidase, removes adenine from one particular adenosine residue in the 28S RNA of the 60S ribosomal subunit. Removal of the adenine inactivates the 60S ribosome, blocking protein synthesis. **Ricin**, **abrin**, and a number of related plant proteins inhibit eukaryotic protein synthesis in a similar manner (Chapter 23).

Several *E. coli* strains also elaborate heat-labile enterotoxins that cause diarrheal disease ("traveler's" diarrhea) by similar mechanisms. In *V. cholerae*, the same enterotoxin is produced by all pathogenic strains and is chromosomally determined, whereas in *E. coli*, different enterotoxins are produced, and the toxin genes are carried on plasmids. Inflammatory GI diseases known as **Crohn's disease** and **ulcerative colitis** are widespread disorders (see Clinical Case Study 11.5, vignettes 1 and 2). The therapeutic strategies for these disorders consist of ameliorating inflammation by inhibition of proinflammatory cytokine tumor necrosis factor (TNF) and by leukocyte trafficking to the GI tract [22–25]. Recurrent abdominal discomfort known as **irritable bowel syndrome** (**IBS**) involves perturbation in motor and sensory functions of the GI tract [26].

THERMIC EFFECT OF FOOD

Energy balance depends on energy intake, energy expenditure, and existing energy stores (Chapter 5). Energy is expended in digestion, absorption, transport, metabolism, and storage of food. The level varies with type of food ingested and its metabolic rate. Energy costs for the processing of lipids, carbohydrates, and proteins are 4%, 5%, and 30% of their energy content, respectively. Lipogenesis from carbohydrate also has a high metabolic cost. Part of this energy appears as heat energy and is variously referred to as the **thermic effect of food, diet-induced thermogenesis**, or **specific dynamic action** of food. The magnitude of this thermic effect depends on the food, nutritional state, and antecedent diet. The thermic effect of food accounts for about 10% of the daily energy expenditure and exhibits interindividual variation. Activation of the sympathetic nervous system and secretion of thyroid hormone contribute significantly to diet-induced thermogenesis. Some forms of obesity may result from decreased thermic effects of food.

CLINICAL CASE STUDY 11.1 Disorders of Gastrointestinal Hormone Secretion

This clinical case was abstracted from: L.H. Simmons, A.R. Guimaraes, L.R. Zukerberg, Case 6-2013: A 54-year-old man with recurrent diarrhea, N. Engl. J. Med. 368 (2013) 757−765.

Synopsis

A 54-year-old man was admitted to the hospital owing to a one-week history of severe and worsening diarrhea, vomiting, and weight loss. This patient's past medical history was significant for recurrent abdominal cramping and nonbloody diarrhea for the past 2.5 years, for which he received multiple courses of antibiotics for presumed traveler's diarrhea, with only temporary relief. He also took omeprazole for gastroesophageal reflux. On arrival to the hospital, his lab results showed an elevated white count with hypochloremia. Stool cultures were negative. CT of the abdomen and pelvis revealed thickening of the small bowel and stomach, ulceration of the duodenum, and a pancreatic mass. EGD revealed esophagitis, excess fluid collection in the stomach, multiple ulcers of the duodenum, and an ulcer crater suspicious for perforation. Pathologic examination of the pancreatic mass revealed tumor cells positive for gastrin. With a serum gastrin level elevated 13 times the normal value, the patient was diagnosed with Zollinger−Ellison syndrome due to a pancreatic gastrinoma. In addition to surgical resection of the pancreatic mass, the patient underwent chemoradiation and then resection of a residual lymph node with no return of disease. The patient was offered sunitinib and everolimus for possible recurrence.

Teaching Points

1. This patient was diagnosed with Zollinger−Ellison syndrome, but he first required evaluation for other causes of chronic diarrhea, such as chronic inflammatory diarrhea (inflammatory bowel disease, infections, ischemic colitis, malignant tumors), chronic fatty diarrhea (malabsorption syndromes like celiac disease, and maldigestion), and chronic watery diarrhea (osmotic diarrhea as in lactose intolerance, and secretory diarrhea as in diabetes and cancer).

2. Zollinger−Ellison syndrome is a constellation of diarrhea and esophageal reflux, caused by excess gastric acid secretion and severe peptic ulcer disease. Gastric acid is secreted in response to gastrin, which is hypersecreted if a gastrinoma is present. Diarrhea in Zollinger−Ellison syndrome is caused by steatorrhea as a result of gastric acid's inactivation of pancreatic digestive enzymes. A secretory diarrhea also results from gastric acid's inhibition of sodium and water reabsorption in the small intestine.

3. An elevated fasting serum gastrin level is diagnostic of a gastrinoma; a normal level rules out Zollinger−Ellison syndrome. Proton-pump inhibitors should be discontinued for one week prior to measurement of gastrin, which may be challenging for a patient with advanced peptic ulcer disease. Alternatively, the diagnosis can also be made with a secretin stimulation test: if there is an increase of gastrin by more than 120 pg/mL in response to secretin administration, then the patient has a gastrinoma.

Reference

[1] E.C. Ellison, J.A. Johnson, The Zollinger−Ellison syndrome: a comprehensive review of historical, scientific, and clinical considerations, Curr. Probl. Surg. 46(1) (2009) 13.

CLINICAL CASE STUDY 11.2 Celiac Disease

Vignette 1

This clinical case was abstracted from: A. Fasano, C. Catassi, Celiac disease, N. Engl. J. Med. 2 (367) (2012) 2419−2426.

Synopsis

A 22-year-old woman with a history of fatigue and oral ulcers fractured her wrist while playing volleyball. Despite being relatively healthy, the patient was found to have osteopenia of the wrist on X-ray. Lab results revealed vitamin D deficiency and anemia, raising concern for a possible malabsorption syndrome like celiac disease, despite the absence of the typical symptoms of diarrhea, weight loss, or abdominal pain. Measurement of IgA anti-tissue transglutaminase antibodies was used to screen for celiac disease, with confirmation by endoscopy and biopsy of the small intestine. The mainstay of treatment for celiac disease is a lifetime gluten-free diet.

Vignette 2

This clinical case was abstracted from: S.L. Bender, N.A. Sherry, R. Masia, Case 16-2013: A 12-year-old girl with irritability, hypersomnia, and somatic symptoms, N. Engl. J. Med. 368 (2013) 2015−2024.

Synopsis

A 12-year-old girl with a history of well-controlled celiac disease, anxiety, and depression presented with chief complaints of worsening irritability, hypersomnia, and multiple somatic symptoms. Over the past 8 months, she developed nonspecific symptoms of fatigue, stomach-aches, paresthesias, dizziness, anorexia, and irritability, for which she was given bupropion and escitalopram for emotional stress related to depression and anxiety. However, upon admission, she was found to have progressive lethargy, pallor, weight loss, and prolonged QT interval. Diagnostic evaluation revealed a failed cosyntropin test, elevated plasma corticotropin, elevated plasma renin, and elevated anti-21-hydroxylase antibody, consistent with primary adrenal insufficiency, or Addison's disease. She was treated with hydrocortisone and fludrocortisone for 3 months, with resolution of her somatic symptoms. Her symptoms of depressed mood and anxiety improved after restarting bupropion.

(Continued)

CLINICAL CASE STUDY 11.2 (Continued)

Teaching Points

1. Celiac disease is a multisystem disorder that was once considered a gastrointestinal disorder for its characteristic symptoms of chronic diarrhea, weight loss, and abdominal distention. However, celiac disease can occur in the absence of gastrointestinal symptoms, as in the patient in Vignette 1. Nearly every body system can be affected by celiac disease, as manifested by systemic symptoms of iron deficiency, aphthous stomatitis (canker sores), and chronic fatigue.

2. Patients who should be screened for celiac disease have a first-degree family history of biopsy-confirmed celiac disease, a condition with a known association with celiac disease (Down or Turner syndrome), or an autoimmune disease (type 1 diabetes mellitus or autoimmune thyroiditis). There seems to be an association between celiac disease and certain autoimmune disorders, like autoimmune adrenal insufficiency as seen in the patient in Vignette 2.

3. The cornerstone treatment for celiac disease is a strict gluten-free diet for life, because gluten triggers the chain of events leading to celiac enteropathy.

Reference

[1] S.E. Crowe, In the clinic: celiac disease, Ann. Intern. Med. 154(9) (2011) ITC5-1–ITC5-15.

CLINICAL CASE STUDY 11.3 Cystic Fibrosis

This clinical case was abstracted from: U. Shah, A.S. Shenoy-Bhangle, Case 32-2011: A 19-year-old man with recurrent pancreatitis, N. Engl. J. Med. 365(16) (2011) 1528–1536.

Synopsis

A 19-year-old boy was admitted after having abdominal pain, nausea, anorexia, and lightheadedness for 3 days. This patient had a 6-month history of recurrent acute pancreatitis secondary to alcoholic binge drinking, which required three hospitalizations and resolved with intravenous (IV) fluids and supportive therapy. Upon this admission, serum lipase was elevated, but there were no abnormalities of the pancreas, pancreatic duct, or common bile duct found on abdominal MRI and magnetic resonance cholangiopancreatography. A sweat test was indeterminate, but genetic testing revealed two mutations of the CFTR gene (cystic fibrosis transmembrane conductance regulator) and a mutation of the SPINK1 gene (serine protease inhibitor Kazal type 1, also known as pancreatic secretory trypsin inhibitor). The patient was diagnosed with cystic fibrosis and recurrent acute pancreatitis due to CFTR and SPINK1 gene mutations, triggered by alcohol use. The patient did not have any findings of pulmonary involvement from cystic fibrosis, but he continued to have recurrent pancreatitis every 6 months due to his refusal to stop drinking alcohol.

Teaching Points

1. Acute pancreatitis is caused by inflammation of the pancreas with inappropriate release of pancreatic enzymes (like trypsin, which digests proteins, or lipase, which digests lipids) resulting in pancreatic autodigestion and further destruction of pancreatic tissue (Chapter 7). There are several causes of acute pancreatitis, with alcohol consumption accounting for 35%–40% of all cases. However, alcohol-induced pancreatitis is unusual in children and adolescents like the patient described here, and occurs only after many years of alcohol abuse. This history, in the absence of structural abnormality, obstruction, infection, or other metabolic disease, suggests an underlying genetic mutation. Screening for a genetic cause of recurrent pancreatitis involves detection of mutations in PRSS1, SPINK1, CFTR, and CTRC genes.

2. Mutations of the PRSS1 gene (serine protease 1) lead to elevated levels of intracellular trypsin. Mutations of the SPINK1 gene lead to pancreatic acinar cells' loss of inhibition on trypsin secretion. Mutations of the CFTR gene (cystic fibrosis transmembrane conductance regulator) lead to defective chloride secretion and excess sodium absorption, forming thick and viscous exocrine pancreatic secretions that obstruct pancreatic ducts; the same process can affect the lungs and genitourinary tract. Mutations of the CTRC gene (chymotrypsinogen C) cause a loss of destruction of activated trypsin.

3. In addition to an increased risk for developing exocrine pancreatic insufficiency, diabetes, and pancreatic cancer, this patient had a risk for pancreatitis 500 times greater than that in the general population because of his strong genetic predisposition. The treatment of uncomplicated acute pancreatitis is supportive therapy with IV fluids and pain control, but management of recurrent pancreatitis is based on minimizing all contributing factors, including the cessation of alcohol consumption.

References

[1] J. Rosendahl, O. Landt, J. Bernadova, P. Kovacs, N. Teich, H. Bodeker, et al., CFTR, SPINK1, CTRC, and PRSS1 variants in chronic pancreatitis: is the role of mutated CFTR overestimated? Gut 62(4) (2013) 582.

[2] S. Tenner, J. Baillie, J. DeWitt, S.S. Vege, American College of Gastroenterology guideline: management of acute pancreatitis, Am. J. Gastroenterol. 108 (2013) 1400–1415.

CLINICAL CASE STUDY 11.4 Cholera

Vignette 1

This clinical case was abstracted from: J.B. Harris, L.C. Ivers, M.J. Ferraro, Case 19-2011: A 4-year-old Haitian boy with vomiting and diarrhea, N. Engl. J. Med. 364 (2011) 2452−2461.

Synopsis

A 4-year-old Haitian boy was admitted to the hospital during a cholera epidemic for a chief complaint of vomiting and watery diarrhea for 10 hours. The patient was found to be thirsty, irritable, tachycardic, and tachypneic, with dry mucous membranes, weak pulses, and clammy extremities. There were no laboratory facilities available at this hospital in Haiti. The patient was treated empirically with azithromycin, and after 2.5 days of IV fluid resuscitation and oral rehydration, the patient's condition improved. He was discharged with directions for the parents to provide adequate oral rehydration, and to prevent future infection through water sterilization and hand sanitation with soap.

Vignette 2

This clinical case was abstracted from: E.T. Ryan, L.C. Madoff, M.J. Ferraro, Case 20-2011: A 30-year-old man with diarrhea after a trip to the Dominican Republic, N. Engl. J. Med. 364 (2011) 2536−2541.

Synopsis

A 30-year-old man who had just returned from the Dominican Republic presented to a U.S. emergency room with sudden onset of watery diarrhea for 2 days. The patient did not have fever, nausea, vomiting, or blood in the stool, but he had had more than 12 bouts of diarrhea since onset. The patient was informed that 13 people who had attended the same social event as he had in the Dominican Republic also had similar symptoms. The patient's stool was found to be bright green and watery, and stool culture was positive for toxigenic *V. cholerae* serotype O1. The patient was treated with one dose of azithromycin for prevention of transmission, but he did not receive any further antibiotics, given only mild dehydration and resolving loose stools. As expected for travelers' diarrhea, the patient's symptoms resolved within 3 days.

Teaching Points

1. Cholera is a diarrheal illness caused by gastrointestinal infection by the gram-negative bacterium *Vibrio cholerae*. Toxic-producing strains cause the majority of cholera infection, in which the cholera toxin binds to and continuously stimulates the CFTR-regulated chloride channel, leading to a secretory diarrhea.

2. Current recommendations state that treatment of cholera with antibiotics should be reserved only for patients with severe dehydration, in the interests of conserving resources. Otherwise, cholera is self-limiting. Management instead focuses on decreasing the risk of person-to-person transmission through hygiene education.

Reference

[1] E.J. Nelson, J.B. Harris, J.G. Morris, S.B. Calderwood, A. Camilli, Cholera transmission: the host, pathogen and bacteriophage dynamic, Microbiology 7 (2009) 693−702.

CLINICAL CASE STUDY 11.5 Inflammatory Bowel Disease

Vignette 1

This clinical case was abstracted from: O.H. Nielsen, M.A. Ainsworth, Tumor necrosis factor inhibitors for inflammatory bowel disease, N. Engl. J. Med. 369(8) (2013) 754.

Synopsis

A 35-year-old man presented with a one-week history of right lower quadrant abdominal pain and frequent bowel movements of eight to nine stools per day. The patient had an 8-year history of Crohn's disease with recurrent exacerbations requiring glucocorticoids, then azathioprine for the past year. Upon admission, the patient was found to be anemic with elevated C-reactive protein. Magnetic resonance enterography revealed distal ileal and colonic inflammation, and ileocolonoscopy showed patchy erythema and ulcerations of the hepatic flexure and terminal ileum. Biopsy of the colon revealed chronic and acute granulomatous inflammation. Based on these findings of active inflammation despite receiving standard treatment for Crohn's disease, the patient was started on a tumor necrosis factor inhibitor for remission maintenance and relapse prevention.

Vignette 2

This clinical case was abstracted from: L. Chouchana, D. Roche, R. Jian, P. Beaune, M.A. Loriot, Poor response to thiopurine in inflammatory bowel disease: how to overcome therapeutic resistance, Clin. Chem. 59(7) (2013) 1023−1027.

Synopsis

A 24-year-old woman presented to her gastroenterologist owing to recurrence of diarrheal episodes of about eight times per day. The patient had a 5-year history of ulcerative colitis that was fairly controlled with steroids. Because of her recurrence of symptoms, the patient underwent therapy with azathioprine, followed by 6-mercaptopurine, but her symptoms still persisted after 7 months. Measurement of intraerythrocyte thiopurine metabolites revealed abnormally high thiopurine S-methyltransferase activity, which inactivates thiopurines, as a cause for her resistance to two thiopurine agents (Chapter 25). The patient was then treated with a tumor necrosis factor-α antagonist.

(Continued)

CLINICAL CASE STUDY 11.5 (Continued)

Teaching Points

1. Crohn's disease and ulcerative colitis comprise the inflammatory bowel diseases (IBD), as seen in the two vignettes here. The pathophysiology of IBD is inappropriate, recurrent inflammation of the intestinal mucosa in response to normal microbial flora, due to a genetically determined disturbance in the mucosal barrier (ulcerative colitis), the sensing of microbes (Crohn's disease), or the regulation of the mucosal immune response (both). While both types of IBD have overlapping symptoms of recurrent diarrhea and abdominal pain, they are differentiated by the location of inflammation in the gastrointestinal tract, as well as the depth of involvement of the mucosal wall. Endoscopic findings confirm the diagnosis.

2. The cornerstone of treatment for maintenance of IBD is a thiopurine agent, while systemic glucocorticoids are used for active inflammation. Patients who experience recurrent inflammation despite treatment with thiopurine are good candidates for tumor necrosis factor inhibitors.

References

[1] C. Abraham, J.H. Cho, Inflammatory bowel disease, N. Engl. J. Med. 361 (2009) 2066–2078.

[2] F. Cominelli, Inhibition of leukocyte trafficking in inflammatory bowel disease, N. Engl. J. Med. 369(8) (2013) 775–776.

[3] B.G. Feagan, P. Rutgeerts, B.E. Sands, S. Hanauer, J.F. Colombel, W.J. Sandborn, et al., Vedolizumab as induction and maintenance therapy for ulcerative colitis, N. Engl. J. Med. 369(8) (2013) 699–710.

[4] W.J. Sandborn, B.G. Feagan, P. Rutgeerts, S. Hanauer, J.-F. Colombel, B.E. Sands, et al., Vedolizumab as induction and maintenance therapy for Crohn's disease, N. Engl. J. Med. 369(8) (2013) 711–721.

[5] T. Fiskerstrand, N. Arshad, B.I. Haukanes, R.R. Tronstad, K.D. Pham, S. Johansson, et al., Familial diarrhea syndrome caused by an activating GUCY2C mutation, N. Engl. J. Med. 366(17) (2012) 1586–1595.

REQUIRED READING

[1] H. Clevers, The Paneth cell, caloric restriction, and intestinal integrity, N. Engl. J. Med. 367 (2012) 1560–1561.

[2A] K. Mueller, C. Ash, E. Pennisi, O. Smith, The gut microbiota, Science 336 (2012) 1245.

[2B] M. Hvistendahl, My microbiome and me, Science 336 (2012) 1248–1250.

[3] J.I. Gordon, Honor thy gut symbionts redux, Science 336 (2012) 1251–1253.

[4] H.J. Haiser, P.J. Turnbaugh, Science 336 (2012) 1253–1255.

[5] E.K. Costello, K. Stagaman, L. Dethlefsen, B.J.M. Bohannan, D.A. Relaman, The application of ecological theory toward an understanding of the human microbiome, Science 336 (2012) 1255–1261.

[6A] J.K. Nicholson, E Holmes, J. Kinross, R. Burcelin, G. Gibson, W. Jia, et al., Host-gut microbiota metabolic interactions, Science 336 (2012) 1262–1267.

[6B] L.V. Hooper, D.R. Littman, A.J. Macpherson, Interactions between the microbiota and the immune system, Science 336 (2012) 1268–1273.

[7] W.S. Garrett, Kwashiorkor and the gut microbiota, N. Engl. J. Med. 368 (2013) 1746–1747.

[8] H. Tilg, Diet and intestinal immunity, N. Engl. J. Med. 366 (2012) 181–183.

[9] J. Bollrath, F. Powrie, Feed your Tregs more fiber, Science 341 (2013) 463–467.

[10] R.A. Koeth, Z. Wang, B.S. Levisosn, J.A. Buffa, E. Org, B.T. Sheehy, et al., Intestinal microbiota metabolism of L-carnitine, a nutrient in red meat, promotes atherosclerosis, Nat. Med. 19 (2013) 576–585.

[11] A.W. Walker, J. Parkhill, Fighting obesity with bacteria, Science 341 (2013) 1069–1070.

[12] S. Devaraj, P. Hemarajata, J. Versalovic, The human gut microbiome and body metabolism: implications for obesity and diabetes, Clin. Chem. 59 (2013) 617–628.

[13] M.D. Agito, A. Atreja, M.K. Rizk, Fecal microbiota transplantation for recurrent *C. difficile* infection: ready for prime time? Clev. Clin. J. Med. 80 (2013) 101–108.

[14] A.M. Gallimore, A. Godkin, Epithelial barriers, microbiota, and colorectal cancer, N. Engl. J. Med. 368 (2013) 282–284.

[15] F.J. Suchy, P.M. Brannon, T.O. Carpenter, J.R. Fernandez, V. Gilsanz, J.B. Gould, et al., National Institutes of Health Consensus Development Conference: lactose intolerance and health, Ann. Intern. Med. 792 (2010) 792–796.

[16] S.E. Crowe, In the clinic. Celiac disease, Ann. Intern. Med. 154 (2011) ITC5-1–ITC5-15

[17] P.D.W. Eckford, C. Li, M. Ramjeesingh, C.E. Bear, Cystic fibrosis transmemebrane conductance regulator (CFTR) potentiator VX-770 (Ivacaftor) opens the defective channel gate of mutant CFTR in a phosphorylation-dependent but ATP-independent manner, 287 (2012) 36639–36649.

[18] R.J. Waldman, E.D. Mintz, H.E. Papowitz, The cure for cholera—improving access to safe water and sanitation, 368 (2013) 592–594.

[19] E.J. Nelson, J.B. Harris, J.G. Morris, S.B. Calderwood, A. Camilli, Cholera transmission: the host, pathogen and bacteriophage dynamic, Nat. Rev. Microbiol. 693 (2009) 693–702.

[20] A.S. Verkman, D. Synder, L. Tradtrantip, J.R. Thiagarajah, M.O. Anderson, CFTR inhibitors, Curr. Opin. Pharmacol. 13 (2013) 888–894.

[21] T. Fiskerstrand, N. Arshad, B. Haukanes, R.R. Tronstad, K.D. Pham, S. Johansson, et al., Familial diarrhea syndrome caused by an activating GUCY2C mutation, N. Engl. J. Med. 366 (2012) 1586–1595.

[22] F. Cominelli, Inhibition of leukocyte trafficking in inflammatory bowel disease, N. Engl. J. Med. 369 (2013) 775–776.

[23] O.H. Neilsen, M.A. Ainsworth, Tumor necrosis factor inhibitors for inflammatory bowel disease, N. Engl. J. Med. 369 (2013) 754–762.

[24] B.G. Feagan, P. Rutgeerts, B.E. Sands, S. Hanauer, J. Colombel, W. J. Sandborn, et al., Vedolizumab as induction and maintenance therapy for ulcerative colitis, N. Engl. J. Med. 369 (2013) 699–710.

[25] W.J. Sandborn, B.G. Feagan, P. Rutgeerts, S. Hanauer, J. Colombel, B.E. Sands, et al., Vedolizumab as induction and maintenance therapy for Crohn's disease, N. Engl. J. Med. 369 (2013) 711–721.

[26] M. Camilleri, Peripheral mechanisms in irritable bowel syndrome, N. Engl. J. Med. 367 (2012) 1626–1635.

Chapter 12

Carbohydrate Metabolism I: Glycolysis and the Tricarboxylic Acid Cycle

Key Points

1. Glycolysis occurs virtually in all human cells.

2. Glucose transport to the cells is mediated by a family of transmembrane facilitative glucose transporters (GLUTs). In the GI tract and renal tubule cells, glucose is actively transported against the concentration gradient along with Na^+ mediated by sodium-dependent transporter (SGLT). The major glucose transporter (GLUT4) of muscle and adipose tissue cells is regulated by insulin.

3. Glycolysis pathway occurs in cytoplasm and consists of 10 steps. The first five steps result in one molecule of glucose being converted to two glyceraldehyde-3-phosphate molecules at the expense of two molecules of ATPs. The second five steps result in the formation of four ATP molecules with the net production of two ATP molecules per one molecule of glucose.

4. Under anaerobic and limited oxygen supply, the end product of glycolysis is lactate. In aerobic cells, the end product is pyruvate, which is oxidized in the mitochondria.

5. Fructose, galactose, and glycerol enter the glycolytic pathway via specific intermediates.

6. Three allosteric enzymes—hexokinase, phosphofructokinase, and pyruvate kinase mediated by positive and negative effectors—regulate glycolysis. Insulin promotes glycolysis, and glucagon has opposite effects. AMP-activated protein kinase (AMPK) promotes glycolysis.

7. The glycolysis pathway requires the participation of the inorganic phosphate and vitamin niacin, which is used in the synthesis of NAD^+.

8. The accumulation of L-lactate in the body either due to its overproduction (anoxic conditions) or lack of utilization (liver diseases) causes metabolic acidosis. D-lactic acid, a bacterial metabolite produced in some intestinal disorders, can also cause acidosis.

9. Positron-emitting isotope ^{18}F tagged to 2-deoxyglucose [^{18}F-2-deoxyglucose] has been used in the metabolic assessment of normal and abnormal tissue cells such as cancer by positron emission tomography.

10. In cells that contain mitochondria, pyruvate enters mitochondria mediated by mitochondrial pyruvate transporter and is converted to acetyl-CoA by oxidative decarboxylation. This reaction requires five coenzymes, four of which are derived from vitamins: thiamine, pantothenic acid, nicotinamide, and riboflavin.

11. The TCA cycle consists of a series of biochemical reactions in which two carbon atoms of acetyl-CoA are oxidized to CO_2, and the energy stored in the two carbon atoms is transferred to NAD^+ and FAD, yielding NADH and $FADH_2$.

12. One molecule of acetyl-CoA oxidation in the TCA cycle yields two CO_2, three NADH, one $FADH_2$, and one GTP molecule.

13. The TCA cycle is a multifunctional pathway and participates in both catabolism and anabolism.

14. The depletion of TCA cycle intermediates due to their use in anabolic pathways is replenished by several reactions known as anaplerotic reactions.

15. The TCA cycle is regulated based on cellular energy demands coupled to substrate levels and allosteric effectors at three irreversible enzymes: citrate synthase, isocitrate dehydrogenase, and α-ketoglutarate dehydrogenase.

16. Dysregulation of the TCA cycle due to loss of function mutations of succinate dehydrogenase and fumarate hydratase can initiate tumorigenesis. Mutations of nonparticipant isocitrate dehydrogenase isoforms 1 and 2 in the TCA cycle acquire new gain of function in which an oncometabolite is generated leading to cancer.

Carbohydrates are metabolized by several metabolic pathways, each with different functions. Although these pathways usually begin with glucose, other sugars may enter a pathway via appropriate intermediates. Glucose can be: stored as glycogen (glycogenesis), which, in turn, can be broken down to glucose (glycogenolysis); synthesized from noncarbohydrate sources (gluconeogenesis); converted to nonessential amino acids; used in the formation of other carbohydrates or their derivatives (e.g., pentoses, hexoses, uronic acids) and other noncarbohydrate metabolites; converted to fatty acids (the reverse process does not occur in humans) and stored as triacylglycerols; used in the biosynthesis of glycoconjugates (e.g., glycoproteins, glycolipids, proteoglycans); or catabolized to provide energy for cellular function (glycolysis, tricarboxylic

N. V. Bhagavan and C.-E. Ha: *Essentials of Medical Biochemistry, Second edition.* DOI: http://dx.doi.org/10.1016/B978-0-12-416687-5.00012-9

165

TABLE 12.1 Properties of Selected Members of Human Glucose Transporters

Transporters	Major Tissue Distribution	Properties
GLUT1	Brain, microvessels, red blood cells, placenta, kidney, and many other cells	Low K_m (about 1 mM), ubiquitous basal transporter
GLUT2	Liver, pancreatic β-cell, small intestine	High K_m (15–20 mM)
GLUT3	Brain, placenta, fetal muscle	Low K_m, provides glucose for tissue cells metabolically dependent on glucose
GLUT4	Skeletal and heart muscle, fat tissue (adipocytes)	K_m of 5 mM, insulin responsive transporter
GLUT5	Small intestine, testes	Exhibits high affinity for fructose
SGLT1	Small intestine and renal tubules	Low K_m (0.1–1.0 mM)
SGLT2	Renal tubules	Low K_m (1.6 mM)

acid cycle [TCA], and electron transport and oxidative phosphorylation).

GLYCOLYSIS

Glycolysis is common to most organisms and in humans occurs in virtually all tissues. Ten reactions culminate in the formation of two pyruvate molecules from each glucose molecule. All 10 reactions occur in the cytoplasm and are anaerobic. In cells that lack mitochondria (e.g., erythrocytes) and in cells that contain mitochondria but under limiting conditions of oxygen (e.g., skeletal muscle during heavy exercise), the end product is lactate. Under aerobic conditions in cells that contain mitochondria (i.e., most cells of the body), pyruvate enters the mitochondria through a mitochondrial pyruvate carrier, where it is oxidized to acetyl-coenzyme A (acetyl-CoA), which is then oxidized through the tricarboxylic acid (TCA) cycle. Thus, the pathway is the same in the presence or absence of oxygen, except for the end product formed.

Source and Entry of Glucose into Cells

Glucose can be derived from exogenous sources by assimilation and transport of dietary glucose (Chapter 11) or from endogenous sources by glycogenolysis or gluconeogenesis (Chapter 14). The blood circulation transports glucose between tissues (e.g., from intestine to liver, from liver to muscle). Control and integration of this transport are discussed in Chapter 20. Glucose transport across cell membranes can occur by carrier-mediated active transport or by a concentration gradient-dependent facilitated transport that requires a specific carrier. The latter type can be either insulin-independent or insulin-dependent. The active transport system is Na^+-dependent and occurs in intestinal epithelial cells (Chapter 11) and epithelial cells

of the renal tubule (Chapter 37). They are designated as SGLT1 and SGLT2, respectively.

The properties of glucose transporter (GLUT) proteins consist of tissue specificity and differences in functional properties reflected in their glucose metabolism. Five transporter proteins, GLUTs 1–5, have been identified (Table 12.1). The transporters combine with glucose and facilitate its transport across the intervening membrane for entry into the cells. The GLUT has features characteristic of Michaelis–Menten kinetics, namely, bidirectionality and competitive inhibition. GLUT proteins belong to a family of homologous proteins coded by multiple genes. They vary from 492 to 524 amino acids and share 39%–65% identity among primary sequences. All GLUT proteins are single-polypeptide chains and contain 12 transmembrane α-helical domains with both amino and carboxy termini extending into the cytoplasm.

SGLT1 protein consists of 664 amino acids with 14 transmembrane helices. SGLT1 and SGLT2 proteins are not homologous with GLUT proteins. Insulin-stimulated glucose uptake in muscle and adipose tissue cells is mediated by GLUT4. Insulin's role in recruiting GLUT4 proteins from intracellular vesicles to the plasma membrane consists of trafficking through multiple intracellular membrane compartments (Chapter 20).

Defects in GLUT4 can result in insulin resistance. GLUT5 is located both at the luminal and basolateral sides of the intestinal epithelial cells. At the luminal side, it functions in tandem with Na^+ glucose symporter, and at the basolateral site, it is involved in the transport of glucose from the absorptive epithelial cells into portal blood circulation. GLUT2 is located in the liver and pancreatic β-cell membranes. It has a high K_m for glucose, and therefore, the entry of glucose is proportional to blood glucose levels. In the liver, glucose can be stored as glycogen or converted to lipids when the plasma glucose levels are high (**hyperglycemia**), and during low levels of plasma glucose (**hypoglycemia**), the liver becomes a provider of glucose to extrahepatic tissues by glycogenolysis, gluconeogenesis, or both (Chapter 14). In pancreatic β-cells, plasma membrane GLUT2 participates in insulin secretion (Chapter 20). GLUT1 and 3 are present in many cell membranes and are basal transporters of glucose at a constant rate into tissues that are metabolically dependent on glucose (e.g., brain and red blood cells). These transporters have a lower K_m for glucose than GLUT2 and, therefore, transport glucose preferentially.

The importance of GLUT1 in brain metabolism is illustrated in a report of two infants with a syndrome of poorly controlled seizures and delayed development, caused by a genetic defect in the GLUT1 protein. Glucose is an essential fuel for the brain and is transported by GLUT1 across the plasma membranes of the brain endothelial cells of the blood–brain barrier system. In these two patients, despite normal blood glucose levels, low levels of glucose in cerebrospinal fluid (CSF) (**hypoglycorrhachia**), as well as low levels of CSF lactate, were observed. Since GLUT1 is present in both red blood cells and brain endothelial cells, the more accessible GLUT1 in red blood cells is used in clinical studies of brain glucose transport disorders. Two patients with a primary defect of glucose transport into the brain were treated with a ketogenic diet. The metabolism of ketogenic substrates does not depend on the glucose transporter and thus can provide a large fraction of the brain's energy requirement.

Reactions of Glycolysis

The overall pathway is shown in Figure 12.1, and some properties of these reactions and the enzymes involved are listed in Table 12.2. Glycolytic enzymes can be classified into six groups according to the type of reaction catalyzed: kinase, mutase, dehydrogenase, cleaving enzyme, isomerase, and enolase.

Phosphorylation of Glucose

Glucose is phosphorylated by hexokinase (in extrahepatic tissues) or glucokinase (in the liver):

$$\alpha\text{-D-Glucose} + \text{ATP}^{4-} \xrightarrow{\text{Mg}^{2+}}$$
$$\alpha\text{-D-glucose-6-phosphate}^{2-} + \text{ADP}^{3-} + \text{H}^+$$

The Mg^{2+} is complexed with ATP^{4-} and is present as MgATP^{2-}. This reaction, essentially an irreversible (nonequilibrium) reaction, is accompanied by a substantial loss of free energy as heat. This phosphorylation initiates glycolysis and leads to intracellular trapping of intermediates (because the plasma membrane is not permeable to phosphate esters).

Four isoenzymes of hexokinases (types I–IV) constitute a family of enzymes that probably arose from a common ancestral gene-by-gene duplication and fusion events. Hexokinases I–III (M.W. ∼ 100,000) have a K_m of about 0.1 mM and are allosterically inhibited by glucose-6-phosphate. Due to their low K_m, hexokinases I–III are saturated with the substrate glucose at concentrations found in plasma. Thus, the overall control of glucose-6-phosphate formation resides in the rate of glucose transport across the plasma membrane. Hexokinase I is expressed in many tissues and is considered a "housekeeping" enzyme. Hexokinase II is found primarily in insulin-sensitive tissues, namely, heart, skeletal muscle, and adipose tissue. In these tissues, the glucose transporter GLUT4 and hexokinase II function in concert in glucose utilization. Abnormalities in insulin, GLUT4, or hexokinase II production may be associated with disorders of insulin resistance, obesity, and diabetes mellitus (Chapter 20).

Hexokinase IV, known as **glucokinase**, differs functionally from hexokinases I–III. Glucokinase, a monomeric protein (M.W. 50,000), has a higher K_m for glucose (5 mM) compared to hexokinases I–III, is found only in hepatocytes (liver parenchymal cells) and pancreatic (islet) β-cells, and is not inhibited by glucose-6-phosphate. The functions of glucokinase in hepatocytes and β-cells are different, and this difference is consistent with their metabolic function. During postprandial and hyperglycemic periods, hepatocyte glucose uptake is increased due to increased levels of glucokinase. Thus, the hepatocyte glucokinase initiates the metabolism of glucose and maintains a gradient of inward flow with glucose being converted to other metabolic products. During periods of hypoglycemia (e.g., starvation), the hepatocyte glucokinase levels are reduced and the liver becomes a provider of glucose.

In β-cells, glucokinase functions as a glucose sensor and modulates insulin secretion. As the plasma levels of glucose increase, with the resultant increase in cytosolic glucose concentration in β-cells, glucokinase determines the rate of glucose phosphorylation, glycolytic flux, and hence, the ATP/ADP ratio. As ATP levels increase, glucokinase binds to the ATP-binding K^+ cassette transporter located in the plasma membrane and blocks the K^+ efflux from the cell. This ATP-dependent K^+ channel is known as the **sulfonylurea receptor**, because sulfonylurea drugs used in the treatment of diabetes mellitus type 2 bind and block the channel. Inhibition of K^+ efflux and the resulting cell depolarization opens the voltage-sensitive Ca^{2+} channels. The increased cytosolic Ca^{2+} triggers the secretion of insulin (Chapter 20).

Tissue-specific glucokinases expressed in liver and pancreatic islet β-cells are differentially regulated. The glucokinase gene consists of two different transcription control regions. Other examples of genes that contain different transcription control regions (promoters) yet produce mRNAs that code for identical proteins are α-amylase and

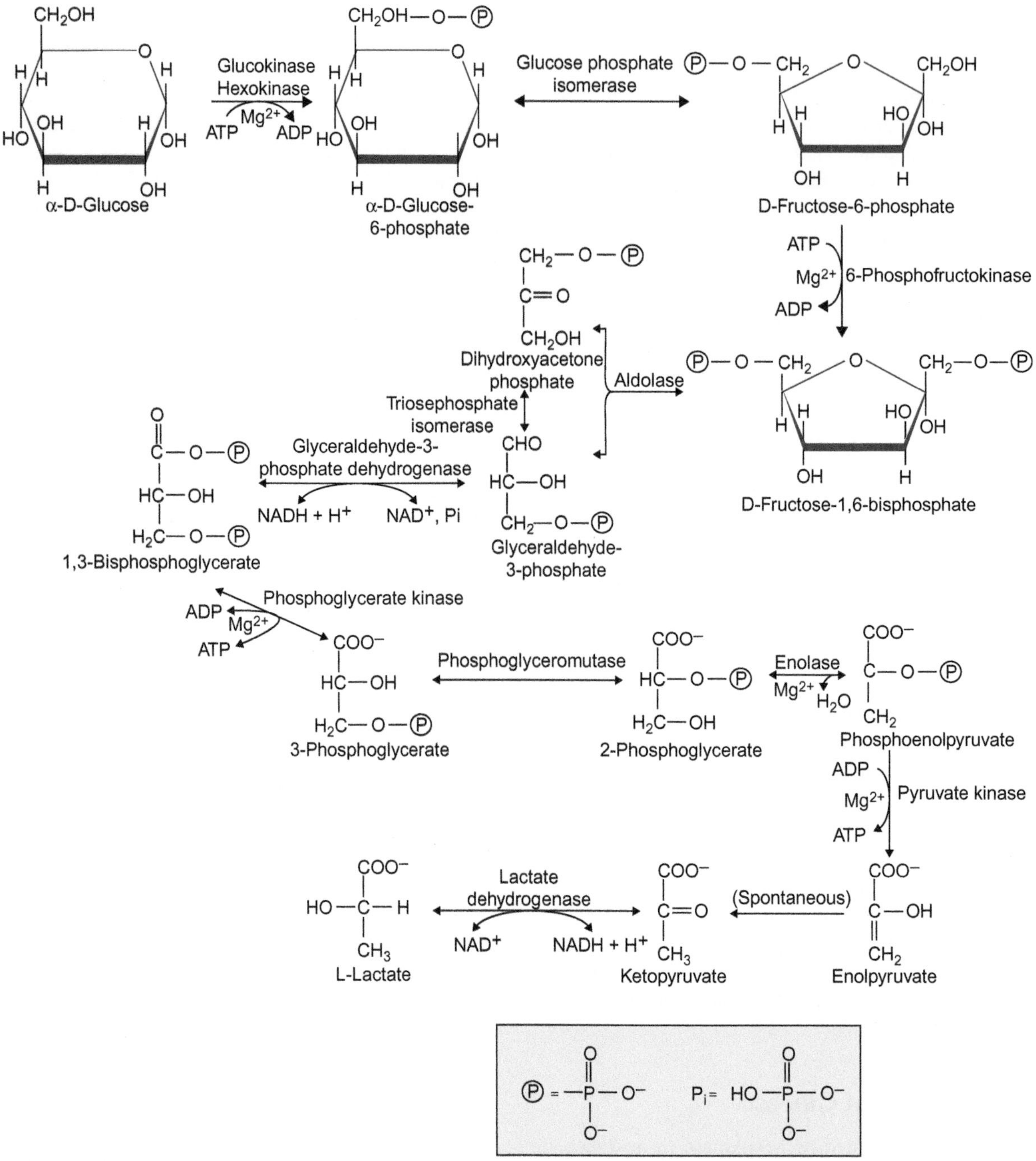

FIGURE 12.1 Pathway of glycolysis.

α_1-antitrypsin genes. In the hepatocytes, insulin promotes glucokinase expression, whereas in β-cells insulin has no effect, but glucose promotes enzyme expression. In the hepatocytes, cAMP (and thus glucagon) and insulin have opposing effects on the expression of glucokinase (Chapter 20). Tissue-specific glucose transporters and glucokinases play critical roles in glucose homeostasis (Figure 12.2). Mutations that alter these proteins may lead to inherited forms of diabetes mellitus (Chapter 20).

Isomerization of Glucose-6-Phosphate to Fructose-6-Phosphate

This freely reversible reaction requires Mg^{2+} and is specific for glucose-6-phosphate and fructose-6-phosphate. It is catalyzed by glucose-phosphate isomerase:

$$\alpha\text{-D-glucose-6-phosphate} \leftrightarrow$$
$$\alpha\text{-D-fructose-6-phosphate}$$

TABLE 12.2 Properties of Enzymes of Glycolysis

Enzyme	Coenzymes and Cofactors	Allosteric Modulators		Equilibrium Constant at pH 7.0 (K'_{eq})	$\Delta G^{\circ\prime}$ kcal/mol (kJ/mole)
		Positive	Negative		
Hexokinase	Mg^{2+}	ATP, P_i	Glucose-6-phosphate	650	−4.0 (−16.7) (nonequilibrium)*
Glucokinase	Mg^{2+}	−	−	−	(nonequilibrium)
Glucose-phosphate isomerase	Mg^{2+}	−	−	0.5	+4.0 (+1.7) (near-equilibrium)†
6-phosphofructokinase	Mg^{2+}	Fructose-2,6-bisphosphate, ADP, AMP, P_i, K^+, NH_4^+	ATP, citrate	220	−3.4 (−14.2) (nonequilibrium)*
Fructose-bisphosphate aldolase		−	−	0.001	+5.7 (+ 23.8) (near-equilibrium)‡
Triose-phosphate isomerase	Mg^{2+}		−	0.075	+1.8 (+7.5) (near-equilibrium)†
Glyceraldehyde-3-phosphate dehydrogenase	NAD	−	−	0.08	+1.5 (+ 6.3) (near-equilibrium)†
Phosphoglycerate kinase	Mg^{2+}	−	−	1500	−4.5 (−18.8) (near-equilibrium)†
Phosphoglyceromutase	Mg^{2+}, 2,3-bisphosphoglycerate	−	−	0.02	+1.1 (+4.6) (near-equilibrium)†
Enolase	Mg^{2+}	−	−	0.5	+0.4 (+1.7) (near-equilibrium)†
Pyruvate kinase§	Mg^{2+}, K^+	Fructose-1,6-bisphosphate	ATP, alanine, acetyl-CoA	200,000	−7.5 (−31.4) (nonequilibrium)*
Lactate dehydrogenase	NAD	−	−	16,000	−6.0 (−25.1) (near-equilibrium)†

Physiologically irreversible reactions.
†*Physiologically reversible reactions.*
‡*This reaction, despite a high positive $\Delta G^{\circ\prime}$ value, is reversible under in vivo conditions.*
§*Pyruvate kinase is also regulated by cAMP-dependent phosphorylation. The dephosphorylated form is more active, and the phosphorylated form is less active.*

Phosphorylation of Fructose-6-Phosphate to Fructose-I,6-Bisphosphate

This second phosphorylation reaction is catalyzed by 6-phosphofructokinase (PFK-1):

$$\text{D-Fructose-6-phosphate}^{2-} + \text{ATP}^{4-} \xrightarrow{\text{Mg}^{2+}}$$
$$\text{D-Fructose-1,6-bisphosphate}^{4-} + \text{ADP}^{3-} + \text{H}^+$$

The reaction is essentially irreversible under physiological conditions and is a major regulatory step of glycolysis. PFK-1 is an inducible, highly regulated, allosteric enzyme. In its active form, muscle PFK-1 is a homotetramer (M.W. 320,000) that requires K^+ or NH_4^+, the latter of which lowers K_m for both substrates. When adenosine triphosphate (ATP) levels are low during very active muscle contraction, PFK activity is modulated positively despite low concentrations of fructose-6-phosphate. Allosteric activators of muscle PFK-1 include adenosine monophosphate (AMP), adenosine diphosphate (ADP), fructose-6-phosphate, and inorganic phosphate (P_i); inactivators are citrate, fatty acids, and ATP.

The most potent regulator of liver PFK-1 is fructose-2,6-bisphosphate, which relieves the inhibition of PFK-1 by ATP and lowers the K_m for fructose-6-phosphate.

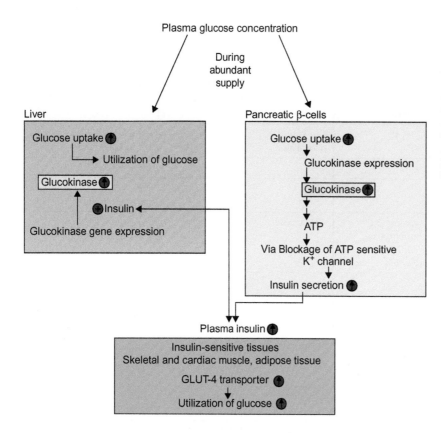

FIGURE 12.2 The role of tissue-specific glucokinases of liver and pancreatic β-cells. During the periods of abundant supply of glucose (e.g., postprandial conditions), the plasma levels of glucose are increased. This induces glucokinase expression in pancreatic islet β-cells with eventual insulin secretion. In the hepatocytes, insulin induces the glucokinase expression with accompanying increased metabolism. Insulin also enhances glucose utilization by recruiting glucose transporter GLUT4 in insulin-sensitive tissues. ⊕ = increased; ⊕ = positive effect.

Fructose-2,6-bisphosphate is a potent inhibitor of fructose-1,6-bisphosphatase (which is important in gluconeogenesis) and thus ensures the continuation of glycolysis. Metabolism of fructose-2,6-bisphosphate, its role as activator of PFK-1 and inhibitor of fructose-1,6-bisphosphatase, and its allosteric and hormonal regulation are discussed in Chapter 14.

Cleavage of Fructose-1,6-Bisphosphate into Two Triose Phosphates

In this reversible reaction, aldolase (or fructose-1,6-bisphosphate aldolase) cleaves fructose-1,6-bisphosphate into two isomers:

$$\text{D-Fructose-1,6-bisphosphate}^{4-} \leftrightarrow$$
$$\text{D-Glyceraldehyde-3-phosphate}^{2-} +$$
$$\text{dihydroxyacetone phosphate}^{2-}$$

Isomerization of Dihydroxyacetone Phosphate to Glyceraldehyde-3-Phosphate

In this reversible reaction, triose-phosphate isomerase converts dihydroxyacetone phosphate to D-glyceraldehyde-3-phosphate, which is the substrate for the next reaction. Thus, the net effect of the aldolase reaction is to yield two molecules of D-glyceraldehyde-3-phosphate from one molecule of fructose-1,6-bisphosphate. This reaction concludes the first phase of glycolysis, during which one molecule of glucose yields two triose phosphates and two ATP molecules are consumed. The second phase of glycolysis consists of energy conservation and begins with the oxidation of glyceraldehyde-3-phosphate.

Dehydrogenation of Glyceraldehyde-3-Phosphate

In a reversible reaction, glyceraldehyde-3-phosphate is converted by glyceraldehyde-phosphate dehydrogenase to an energy-rich intermediate, 1,3-bisglycerophosphate (or 3-phosphoglyceroyl-phosphate):

$$\text{D-Glyceraldehyde-3-phosphate}^{2-} + \text{NAD}^+ + \text{P}_i^{2-} \leftrightarrow$$
$$\text{1,3-Bisphosphoglycerate}^{4-} + \text{NADH} + \text{H}^+$$

The rabbit skeletal enzyme is a homotetramer (M.W. 146,000). Each subunit contains a binding region for glyceraldehyde-3-phosphate and another for nicotinamide adenine dinucleotide (NAD⁺). NAD⁺, a cosubstrate of this reaction, participates in many hydrogen transfer reactions. NAD⁺ (Figure 12.3) contains the vitamin nicotinamide. Nicotinamide can also be obtained from the amino

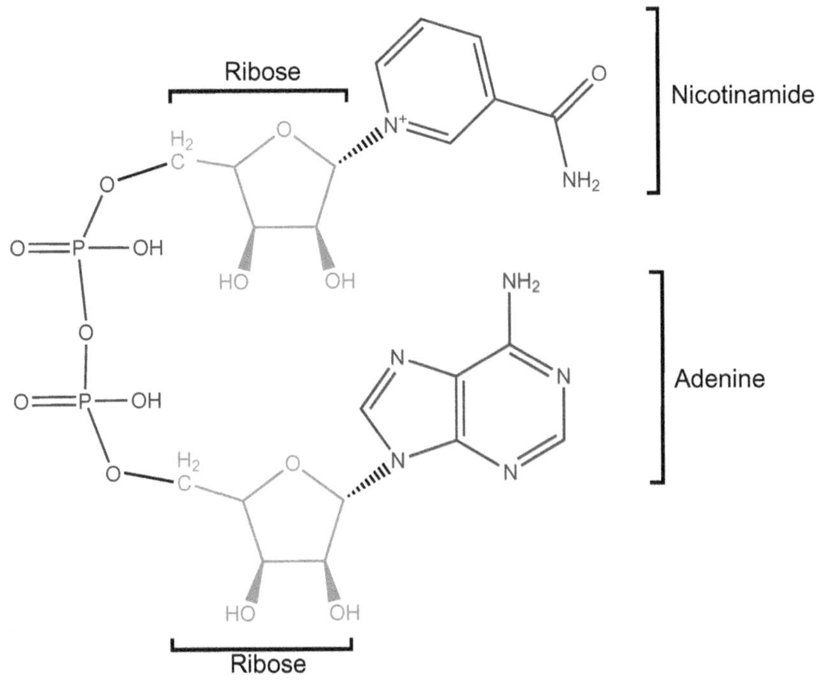

FIGURE 12.3 Nicotinamide adenine dinucleotide (NAD^+).

acid tryptophan. An $-SH$ group present at the active site plays a prominent role in the catalysis, and iodoacetate (ICH_2COO^-) and heavy metals (e.g., Hg^{2+}, Pb^{2+}), which react with sulfhydryl groups covalently, inactivate the enzyme (E):

$$E-SH + ICH_2COO^- \rightarrow E-S-CH_2COO^- + HI$$

Phosphorylation of ADP from 1,3-Bisphosphoglycerate

In a reversible reaction, the phosphoryl group of the acid anhydride of glycerol phosphate is transferred to ADP by phosphoglycerate kinase:

$$1,3\text{-Bisphosphoglycerate}^{4-} + ADP^{3-} \xrightarrow{Mg^{2+}}$$
$$3\text{-Phosphoglycerate}^{3-} + ATP^{4-}$$

The net result of this reaction is the formation of two molecules of ATP per molecule of glucose. This formation of ATP is known as **substrate-level phosphorylation** to distinguish it from the oxidative phosphorylation coupled to electron transport that occurs in mitochondria (Chapter 13). In erythrocytes, 1,3-bisphosphoglycerate is converted to 2,3-bisphosphoglycerate (also called 2,3-diphosphoglycerate or 2,3-DPG) by 1,3-bisphosphoglycerate mutase, which possesses 2,3-bisphosphoglycerate phosphatase activity and is responsible for the formation of 3-phosphoglycerate (which re-enters glycolysis) and

inorganic phosphate. In this bypass pathway, the free energy associated with the anhydride bond of 1,3-bisphosphoglycerate is dissipated as heat. Erythrocytes contain a high level of 2,3-DPG, which functions as an allosteric modulator of hemoglobin and decreases affinity for oxygen (Chapter 26). 2,3-DPG is present in trace amounts in most cells, functions as a cofactor in the next reaction, and is synthesized by 3-phosphoglycerate kinase in the reaction:

$$3\text{-Phosphoglycerate}^{2-} + ATP^{4-} \rightarrow$$
$$2,3\text{-Bisphosphoglycerate}^{4-} + ADP^{3-} + H^+$$

In the next set of reactions of glycolysis, the remaining phosphate group of glycerate is elevated to a high-energy state, at which it can phosphorylate ADP in another substrate-level phosphorylation reaction. The formation of the high-energy intermediate occurs by way of the following two reactions.

Isomerization of 3-Phosphoglycerate to 2-Phosphoglycerate

In this reversible reaction, the phosphate group is transferred to the 2-position by phosphoglycerate mutase:

$$3\text{-Phosphoglycerate}^{2-} \xrightarrow{Mg^{2+}} 2\text{-Phosphoglycerate}^{2-}$$

The reaction is similar to the interconversion catalyzed by phosphoglucomutase (Chapter 14). The reaction is primed by phosphorylation of a serine residue of the enzyme by 2,3-bisphosphoglycerate and occurs in the following steps:

$$
\begin{array}{l}
COO^- \\
| \\
CHOH \\
| \\
CH_2OPO_3^{2-}
\end{array}
\;+\; E\!-\!OPO_3^{2-}
\quad\rightarrow\quad
\begin{array}{l}
COO^- \\
| \\
CHOPO_3^{2-} \\
| \\
CH_2OPO_3^{2-}
\end{array}
\;+\; E\!-\!OH
$$

3-Phosphoglycerate   2,3-Bisphos-phoglycerate

$$
E\!-\!OPO_3^{2-} \;+\;
\begin{array}{l}
COO^- \\
| \\
HCOPO_3^{2-} \\
| \\
CH_2OH
\end{array}
$$

2-Phospho-glycerate

Dehydration of 2-Phosphoglycerate to Phosphoenolpyruvate

In this reversible reaction, dehydration of the substrate causes a redistribution of energy to form the high-energy compound phosphoenolpyruvate:

$$
\text{2-Phosphoglycerate} \xrightarrow{\ Mg^{2+}\ } \text{phosphoenolpyruvate} + H_2O
$$

Enolase (2-phospho-D-glycerate hydrolyase) is a homodimer (M.W. 88,000) that is inhibited by fluoride, with formation of the magnesium fluorophosphate complex at the active site. This property of fluoride is used to inhibit glycolysis in blood specimens obtained for measurement of glucose concentration. In the absence of fluoride (or any other antiglycolytic agent), the blood glucose concentration decreases at about 10 mg/dL (0.56 mM/L) per hour at 25°C. The rate of decrease is more rapid in blood from newborn infants owing to the increased metabolic activity of the erythrocytes, and in leukemia patients because of the larger numbers of leukocytes.

Neuron-specific and non-neuron-specific enolase isoenzymes have been used as markers to distinguish neurons from non-neuronal cells (e.g., glial cells that are physically and metabolically supportive cells of neurons) by immunocytochemical techniques. Neuron-specific enolase is extremely stable and resistant to a number of *in vitro* treatments (e.g., high temperature, urea, chloride) that inactivate other enolases. The functional significance of these isoenzymes is not known.

Phosphorylation of ADP from Phosphoenolpyruvate

In this physiologically irreversible (nonequilibrium) reaction, the high-energy group of phosphoenolpyruvate is transferred to ADP by pyruvate kinase, thereby generating ATP (i.e., two molecules of ATP per molecule of glucose). This reaction is the second substrate-level phosphorylation reaction of glycolysis:

$$
\text{Phosphoenolpyruvate}^{3-} + \text{ADP}^{3-} + \text{H}^+ \rightarrow
$$
$$
\text{pyruvate}^- + \text{ATP}^{4-}
$$

The pyruvate kinase reaction has a large equilibrium constant because the initial product of pyruvate, the enol form, rearranges nonenzymatically to the favored keto form:

$$
\begin{array}{l}
COO^- \\
| \\
COH \\
\| \\
CH_2
\end{array}
\quad\longrightarrow\quad
\begin{array}{l}
COO^- \\
| \\
C\!=\!O \\
| \\
CH_3
\end{array}
$$

Enol-pyruvate   Keto-pyruvate

Several isoenzyme forms of pyruvate kinase are known (M.W. 190,000–250,000, depending on the source). Each is a homotetramer exhibiting catalytic properties consistent with the function of the tissue in which it occurs. Enzyme activity is dependent on K^+ (which increases the affinity for phosphoenolpyruvate) and Mg^{2+}.

Pyruvate kinase is an allosteric enzyme regulated by several modifiers. The liver isoenzyme (L-type) shows sigmoidal kinetics with phosphoenolpyruvate. Fructose-1,6-bisphosphate is a positive modulator and decreases K_m for phosphoenolpyruvate; ATP and alanine are negative modulators and increase K_m for phosphoenolpyruvate. The former is an example of positive feed-forward regulation, and the latter are examples of negative feedback regulation. Alanine, a gluconeogenic precursor, is obtained by proteolysis or from pyruvate by amino transfer (Chapter 15). The modulation of pyruvate kinase is consistent with the function of the liver; when glucose abounds, its oxidation is promoted, and when glucose is deficient, its formation is favored by gluconeogenesis (Chapter 14).

The L-type is also regulated by diet and hormones. Fasting or starvation decreases activity, whereas a carbohydrate-rich diet increases it. Insulin increases activity, whereas glucagon decreases it. These hormones also have reciprocal effects on gluconeogenesis, which insulin inhibits and glucagon promotes. Glucagon action is dependent on the cAMP-mediated cascade process of phosphorylation and dephosphorylation of the enzyme by the protein kinase and phosphatase, respectively (Chapter 28). The cAMP cascade begins with stimulation of membrane-bound adenylate cycles by glucagon to form

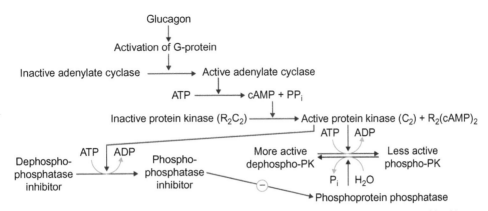

FIGURE 12.4 Action of glucagon on the liver pyruvate kinase (PK) via the cAMP-cascade system (see also Chapter 28). Glucagon, by combining at specific receptor sites on the hepatocyte plasma membrane, activates adenylate cyclase, which converts ATP to cAMP. The latter activates (cAMP-dependent) protein kinase, which phosphorylates PK (phospho-PK), converting it to a less active form and making it more susceptible to allosteric inhibition by alanine and ATP. Phospho-PK is converted to the more active form by removal of phospho groups by a phosphoprotein phosphatase, whose activity is also regulated by cAMP-dependent protein kinase via the phosphorylation of phosphatase inhibitor. The overall effect of glucagon is to diminish glycolysis, whereas the effect of insulin is to promote glycolysis. PP_i = pyrophosphate; Θ = inhibition.

cAMP and is followed by activation of cAMP-dependent protein kinase, which phosphorylates pyruvate kinase. The phospho enzyme is less active than the dephospho form, has a higher K_m for phosphoenolpyruvate, has a higher susceptibility to inhibition by the negative modulators alanine and ATP, and has virtually no activity in the absence of fructose-1,6-bisphosphate. The phospho enzyme is converted to the dephospho form by a phospho-protein phosphatase, the activity of which is also regulated by cAMP-dependent protein kinase (Figure 12.4).

The muscle and brain isoenzyme (M-type) shows hyperbolic kinetics with phosphoenolpyruvate and is inhibited by ATP, with an increase in K_m for phosphoenolpyruvate and development of sigmoidal kinetics. Fructose-1,6-bisphosphate and alanine have no effect on this isoenzyme.

The overall reaction for glycolysis in cells that possess mitochondria and are under aerobic conditions is as follows:

$$Glucose + 2NAD^+ + 2ADP^{3-} + 2HPO_4^{2-} \rightarrow$$
$$2pyruvate^- + 2NADH + 2H^+ + 2ATP^{4-} + 2H_2O$$

Pyruvate is transported into mitochondria via the mitochondrial pyruvate carrier, where it is oxidized in the TCA cycle (see next section). Although NADH is not transported into mitochondria, its reducing equivalent is transported into mitochondria (Chapter 14), where it is oxidized in the electron transport system coupled to oxidative phosphorylation (Chapter 13).

Cells that lack mitochondria (e.g., erythrocytes) or contain mitochondria but under hypoxic conditions (e.g., vigorously and repeatedly contracting muscle) reduce pyruvate to lactate by lactate dehydrogenase (LDH), which uses NADH as a reductant. This reduction regenerates NAD^+, which is required for continued oxidation of glyceraldehyde-3-phosphate.

Reduction of Pyruvate to Lactate

This reversible reaction is the final step of glycolysis and is catalyzed by (LDH):

$$Pyruvate^- + NADH + H^+ \leftrightarrow L\text{-lactate}^- + NAD^+$$

The overall reaction of glycolysis becomes:

$$Glucose + 2ADP^{3-} + 2P_i^{2-} \rightarrow$$
$$2lactate^- + 2ATP^{4-} + 2H_2O$$

LDH (M.W. 134,000) occurs as five tetrameric isoenzymes composed of two different types of subunits. Subunits M (for muscle) and H (for heart) are encoded by loci in chromosomes 11 and 12, respectively. Two subunits used in the formation of a tetramer yield five combinations: H_4(LDH-1), H_3M(LDH-2), H_2M_2(LDH-3), HM_3(LDH-4), and M_4(LDH-5). The tissue distribution of LDH isoenzymes is variable. For example, LDH-1 and LDH-2 are the principal isoenzymes in heart, kidney, brain, and erythrocytes; LDH-3 and LDH-4 predominate in endocrine glands (e.g., thyroid, adrenal, pancreas), lymph nodes, thymus, spleen, leukocytes, platelets, and nongravid uterine muscle; and LDH-4 and LDH-5 predominate in liver and skeletal muscle. In tissue injury or insult, the appropriate tissue isoenzymes appear in plasma; thus, determination of LDH isoenzyme composition has diagnostic significance.

Alternative Substrates of Glycolysis

The glycolytic pathway is also utilized by fructose, galactose, mannose, glycogen, and glycerol. The metabolism of the monosaccharides and glycogen is discussed in Chapter 14. Glycogenolysis catalyzed by phosphorylase in a nonequilibrium reaction yields glucose-1-phosphate, which is then converted to glucose-6-phosphate

(Chapter 14), bypassing the initial phosphorylation reaction of glycolysis. Therefore, the conversion of a glucosyl unit of glycogen to two lactate molecules yields three ATP, which is 50% more than the yield from a glucose molecule. Glycerol released by hydrolysis of triacylglycerol enters the glycolytic pathway by way of dihydroxyacetone phosphate, as follows:

$$\text{Glycerol} + \text{ATP}^{4-} \xrightarrow{\text{glycerol kinase, Mg}^{2+}}$$
$$\text{Glycerol 3-phosphate}^{2-} + \text{ADP}^{3-} + \text{H}^+$$

$$\text{Glycerol 3-phosphate}^{2-} + \text{NAD}^+ \xleftarrow{\text{glycerol-3-phosphate dehydrogenase}}$$
$$\text{dihydroxyacetone phosphate}^{2-} + \text{NADH} + \text{H}^+$$

Role of Anaerobic Glycolysis in Various Tissues and Cells

Red Blood Cells

In tissues that lack mitochondria or function under limiting conditions of oxygen (e.g., lens and erythrocytes), glycolysis is the predominant pathway providing ATP. The lifespan of **red blood cells** (**RBCs**, also known as **erythrocytes**) is about 120 days; they lack mitochondria, nucleus, and other organelles required for protein synthesis. Thus, their ATP source is glycolytic pathway. RBCs possess, in addition to glycolysis, a competing metabolic pathway that also utilizes glucose-6-phosphate: the hexose monophosphate (HMP) shunt pathway (also known as **pentose phosphate pathway**, discussed in Chapter 14). The glycolytic pathway and the HMP pathway serve different functions in the RBCs. Glycolysis provides ATP and 2,3 DPG, and the HMP pathway provides antioxidant NADPH, which reduces glutathione (GSH). During low pO_2, glycolysis predominates, whereas during high pO_2, the HMP pathway predominates to provide antioxidants that are required for the protection of the RBCs exposed to high pO_2. These competing pathways are regulated by an underlying cytoskeleton protein of the RBCs, known as **Band 3** [1]. Glycolytic enzymes bound to Band 3 are released and become active due to the presence of deoxyhemoglobin, and opposite conditions prevail during the presence of oxyhemoglobin, promoting the HMP pathway (Figure 12.5).

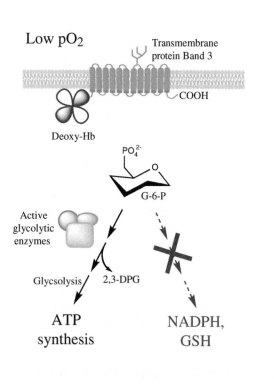
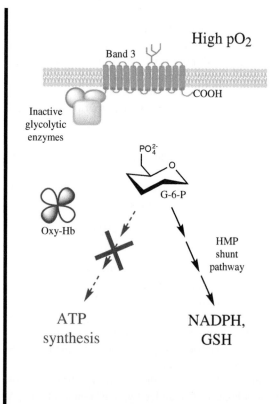

FIGURE 12.5 In circulating RBCs, the glycolytic pathway and HMP shunt pathway prevail during low pO_2 and high pO_2 conditions, respectively, to meet functional and metabolic needs. During low pO_2, deoxygenated hemoglobin binds to the amino terminal of transmembrane cytoskeletal protein Band 3 and this releases the glycolytic enzymes required for glycolysis, whereas during high pO_2 conditions, the HMP-shunt pathway is operational due to the inactive glycolytic enzymes. *[Adapted with permission from W.H. Dzik, The air we breathe: three vital respiratory gases and the red blood cell: oxygen, nitric oxide, and carbon dioxide, Transfusion 51 (2011) 676–685.]*

On the basis of speed of contraction and metabolic properties, skeletal muscle fibers may be classified into three types: type I (slow-twitch, oxidative); type IIA (fast-twitch, oxidative-glycolytic); and type IIB (fast-twitch, glycolytic). Short-term, sudden energy output is derived from glycolytic fibers. The glycolytic fibers (particularly type IIB) have few mitochondria, with relative enrichment of myofibrils and poor capillary blood supply. Thus, these fibers can perform a large amount of work in a short period of time, in contrast to slow-twitch fibers, which are rich in mitochondria, have good capillary blood supply, and power sustained efforts (Chapter 19).

A high rate of glycolysis also occurs in lymphocytes, kidney medulla, skin, and fetal and neonatal tissues. The anaerobic capacity of lymphocytes is increased for growth and cell division. Kidney medulla (Chapter 37), which contains the loops of Henle and collecting tubules, receives much less blood than the cortex and is rich in glycogen. In contrast, kidney cortex, which contains the glomeruli, proximal tubules, and parts of the distal tubules, receives a large amount of blood through the renal arterioles. The renal cortex derives its energy from oxidation of glucose, fatty acids, ketone bodies, or glutamine to CO_2 and H_2O.

Some rapidly growing tumor cells exhibit a high rate of glycolysis. Inadequate oxygen supply (hypoxia) causes tumor cells to grow more rapidly than the formation of blood vessels that supply oxygen. Thus, in hypoxic tumor cells, glycolysis is promoted by increased expression of glucose transporters and most glycolytic enzymes by hypoxia-inducible transcription factor (HIF-1).

Summary of Regulation of Glycolysis

1. Glycolysis is regulated by three enzymes, all of which catalyze irreversible reactions: hexokinases, phosphofructokinase-1, and pyruvate kinase. The allosteric effectors are listed in Table 12.1 and discussed with the enzyme reactions.
2. Insulin promotes glycolysis, whereas glucagon has the opposite effect.
3. The cellular energy supply is reflected in the ATP: AMP ratio: when the ATP level is low, the AMP level is high and AMP activates a protein kinase (AMPK). AMPK stimulates glycolysis by enhancing fructose-2,6-bisphosphate synthase, which is a positive allosteric modulator of 6-phosphofructokinase and glucose uptake by promoting biogenesis of GLUT4. AMPK also participates in other pathways of energy homeostasis (see Chapters 19 and 20).

Glycolytic Enzyme Deficiencies in Erythrocytes

Deficiency of hexokinase, glucose-phosphate isomerase, 6-phosphofructokinase, aldolase, triose-phosphate isomerase, or phosphoglycerate kinase is associated with hemolytic anemia, whereas lactate dehydrogenase deficiency is not. The most common deficiency is that of pyruvate kinase (PK), which is inherited as an autosomal recessive trait. PK deficiency occurs worldwide; however, it is most commonly found in kindreds of Northern European ancestry. Erythrocytes of PK individuals have elevated levels of 2,3-DPG and decreased ATP levels. Several mutations of the PK gene have been identified. Individuals with PK deficiency suffer from lifelong chronic hemolysis to a varying degree. Splenectomy ameliorates the hemolytic process in severe cases.

Some enzymopathies of erythrocytes may be associated with multisystem disease (e.g., aldolase deficiency with mental and growth retardation). Individuals with 6-phosphofructokinase deficiency exhibit hemolysis and myopathy and have increased deposition of muscle glycogen (a glycogen storage disease; see Chapter 14). The myopathy is usually characterized by muscle weakness and exercise intolerance (see also Chapter 19).

Use of Positron-Emitting Tracer Glucose Analogue 2-[^{18}F]-Fluoro-2-Deoxyglucose (FDG) to Identify Metabolically Active Tissue Cells by Sodium-Independent Glucose Transporters (GLUT1–4)

Inside the cell, FDG is converted to FDG-6-phosphate by hexokinase and is not further metabolized. The intracellular trapped FDH-6-phosphate emits positrons. Positrons are subatomic particles of mass equivalent to electrons but have positive charges and annihilate electrons with the formation of high-energy photons. The photons are detected by scanning devices. Since FDG-PET monitors only glucose uptake, the nature of the tissue is assessed by combining PET scans with computed tomography (CT). CT provides spatial resolution of tissues with their density and composition. Clinically, FDG-PET scans are used in localizing metabolically active tumors. However, it is not tumor specific, and metabolically active inflammatory cells, hyperplastic bone marrow, and thymic cells result in enhanced uptake of FDG. Metabolically active brown adipose tissue also gives rise to positive PET scans. In the tumor diagnostic PET scan protocols, the metabolic activity of brown adipose tissue is quenched by use of benzodiazepines and beta blockers.

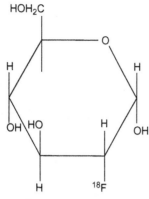

2-[^{18}F] Fluoro-2-deoxyglucose

PYRUVATE METABOLISM

Pyruvate occupies a central hub in several metabolic pathways. It can be reduced to lactate, converted to oxaloacetate in a reaction important in gluconeogenesis (Chapter 14) and in an anaplerotic reaction of the TCA cycle (see following text), transaminated to alanine (Chapter 15), or converted to acetyl-CoA and CO_2. Acetyl-CoA is utilized in fatty acid synthesis, cholesterol (and steroid) synthesis, acetylcholine synthesis, and the TCA cycle (Figure 12.6).

Conversion of pyruvate to ethanol by certain yeast strains occurs in two steps. It is first decarboxylated to acetaldehyde by pyruvate decarboxylase, which utilizes thiamine pyrophosphate (TPP) as coenzyme:

$$CH_3COCOOH \xrightarrow[\text{TPP}]{\text{Mg}^{2+}} CH_3CHO + CO_2$$

In the second step, acetaldehyde is reduced to ethanol by alcohol dehydrogenase, an NAD-dependent enzyme:

$$CH_3CHO + NADH + H^+ \leftrightarrow CH_3CH_2OH + NAD^+$$

The NADH is derived from glyceraldehyde-3-phosphate dehydrogenation. Thus, yeast fermentation yields ethanol rather than lactate as an end product of glycolysis. Small amounts of ethanol are produced by the microbial flora of the gastrointestinal tract. Other types of fermentation using similar reactions occur in microorganisms and yield a variety of products (e.g., acetate, acetone, butanol, butyrate, isopropanol, hydrogen gas).

Lactic Acidemia and Lactic Acidosis

As discussed earlier, some tissues produce lactate as an end product of metabolism [2]. The lactate produced is L-lactate and is commonly referred to simply as **lactate**. As we will see in the following discussion, D-lactate is also produced under certain pathological conditions, which presents a unique clinical problem.

Under normal conditions, lactate is metabolized in the liver, and the blood lactate level is between 1 and 2 mM.

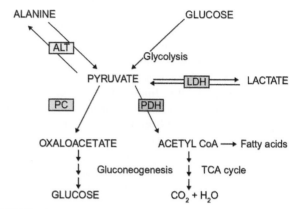

FIGURE 12.6   Major pathways of pyruvate metabolism. Pyruvate is metabolized through four major enzyme pathways: lactate dehydrogenase (LDH), pyruvate dehydrogenase complex (PDH), pyruvate carboxylase (PC), and alanine aminotransferase (ALT). Multiple arrows indicate multiple steps.

Lactate accumulation in body fluids can be due to increased production and/or decreased utilization. A blood lactate-to-pyruvate ratio below 25 suggests defects in a gluconeogenic enzyme (Chapter 14) or pyruvate dehydrogenase (discussed later). A common cause of **lactic acidosis** is tissue hypoxia caused by shock, cardiopulmonary arrest, and hypoperfusion. Inadequate blood flow leads to deprivation of oxygen and other nutrients to the tissue cells, as well as to the removal of waste products. Oxygen deprivation leads to decreased ATP production and accumulation of NADH, which promotes conversion of pyruvate to lactate. Lactic acidosis is also associated with thiamine deficiency, malignancy, hepatic toxins, and some therapeutic agents (e.g., metformin used in the treatment of type 2 diabetes mellitus).

A major cause of acidosis that occurs during inadequate cellular oxygen delivery is continual hydrolysis of the available supply of ATP that releases protons:

$$ATP^{4-} + H_2O \rightarrow ADP^{3-} + HPO_4^{2-} + H^+$$

Laboratory assessment includes measurements of blood lactate, pyruvate, β-hydroxybutyrate, and acetoacetate (Chapter 37). The primary treatment of lactic acidosis involves correcting the underlying cause, such as reversal of circulatory failure.

D-Lactic Acidosis

The unusual form of lactic acidosis known as **D-lactic acidosis** is due to increased production and accumulation of D-lactate in circulation. The normal isomer synthesized in the human body is L-lactate, but the D-lactate isomer can occur in patients with jejunoileal bypass, small bowel resection, or other types of short bowel syndrome. In these patients, ingested starch and glucose bypass the normal metabolism in the small intestine and lead to increased delivery of nutrients to the colon, where gram-positive, anaerobic bacteria (e.g., *Lactobacillus*) ferment glucose to D-lactate. The D-lactate is absorbed via the portal circulation.

A limited quantity of D-lactate is converted to pyruvate by the mitochondrial flavoprotein enzyme D-2-hydroxy acid dehydrogenase. Thus, the development of D-lactate acidosis requires excessive production of D-lactate and impairment in its metabolism. The clinical manifestations of D-lactic acidosis are characterized by episodes of encephalopathy after ingestion of foods containing carbohydrates.

The diagnosis of D-lactic acidosis is suspected in patients with disorders of the small intestine causing malabsorption, and when the serum anion gap (Chapter 37) is elevated in the presence of normal serum levels of L-lactate and other organic acids. Measurement of serum D-lactate requires special enzymatic procedures utilizing D-lactate dehydrogenase and NADH. As D-lactate is converted to pyruvate, NADH is oxidized to NAD^+, which is detected spectrophotometrically.

The treatment of D-lactic acidosis consists of oral administration of antibiotics, limitations of oral carbohydrate intake, and recolonization of the colon by bacterial flora that does not produce D-lactate.

Oxidation of Pyruvate to Acetyl-CoA

Pyruvate is transported into mitochondria by a specific carrier known as **mitochondrial pyruvate carrier** [3–5]. A limited passive transport of pyruvate across mitochondrial membrane can also occur. Defects in the mitochondrial pyruvate transport system can lead to severe clinical manifestations such as dysmorphism and lactic acidemia [6]. Inside mitochondria, pyruvate undergoes oxidative decarboxylation by three enzymes that function sequentially and are present as a complex known as the **pyruvate dehydrogenase complex**. The overall reaction is physiologically irreversible, has a high negative $\Delta G^{0\prime}$

(-8.0 kcal/mol, or -33.5 kJ/mol), and commits pyruvate to the formation of acetyl-CoA:

$$Pyruvate + NAD^+ + CoASH \rightarrow$$
$$acetyl\text{-}CoA + NADH + H^+ + CO_2$$

In the preceding reaction, CoASH stands for coenzyme A (Figure 12.7), and it contains the vitamin pantothenic acid (Chapter 36). Coenzyme A functions as a carrier of acetyl (and acyl groups) by formation of thioesters, and the acetyl-sulfur bond has a $\Delta G^{0\prime}$ comparable to that of the phosphoanhydride bond of ATP.

Dehydrogenation and decarboxylation of pyruvate occur in five reactions requiring three enzymes and six coenzymes, prosthetic groups, and cofactors (Mg^{2+}, thiamine pyrophosphate [TPP], lipoic acid, CoASH, flavin adeninedinucleotide [FAD], and NAD^+) (Figure 12.8). A kinase and a phosphatase are tightly bound to the pyruvate dehydrogenase subunit and participate in the regulation of the activity.

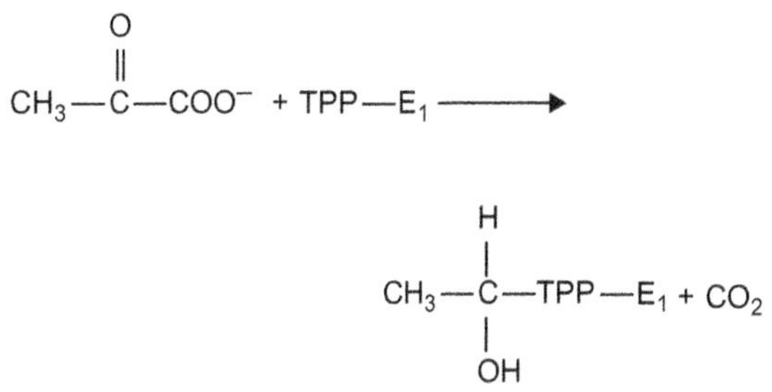

$$\underset{\overset{\|}{O}}{CH_3-C-COO^-} + TPP-E_1 \longrightarrow$$

$$CH_3-\underset{\overset{|}{OH}}{\overset{\overset{H}{|}}{C}}-TPP-E_1 + CO_2$$

The structure of FAD that contains the vitamin riboflavin is given in Figure 12.9. In the last step, catalyzed by dihydrolipoyl dehydrogenase, the hydrogens (or reducing equivalents) are transferred to NAD^+, with the formation of NADH and the oxidized flavoprotein:

$$E_3\text{-}FADH_2 + NAD^+ \rightarrow E_3\text{-}FAD + NADH + H^+$$

Thus, the flow of reducing equivalents in the pyruvate dehydrogenase complex is from pyruvate to lipoyl to

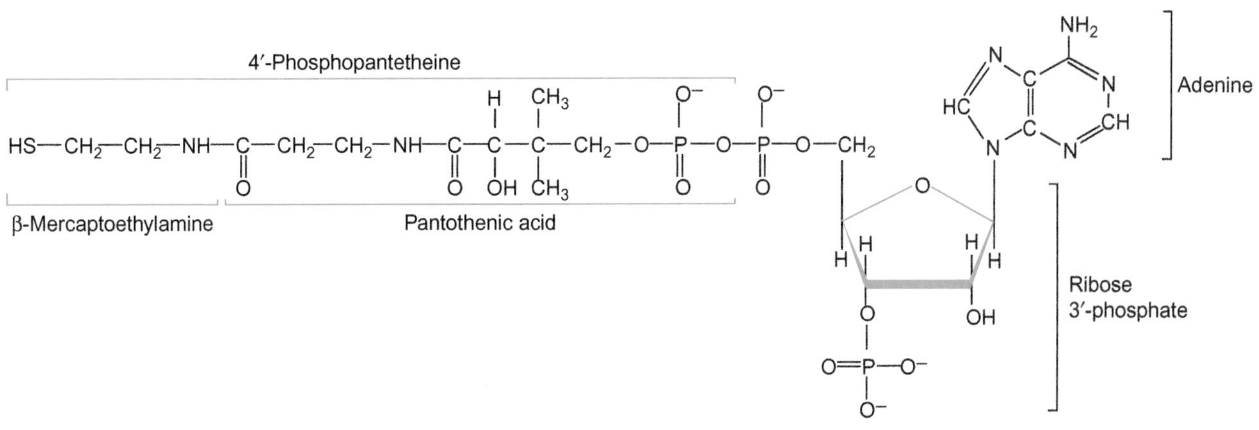

FIGURE 12.7 Coenzyme A. The terminal sulfhydryl group is the reactive group of the molecule.

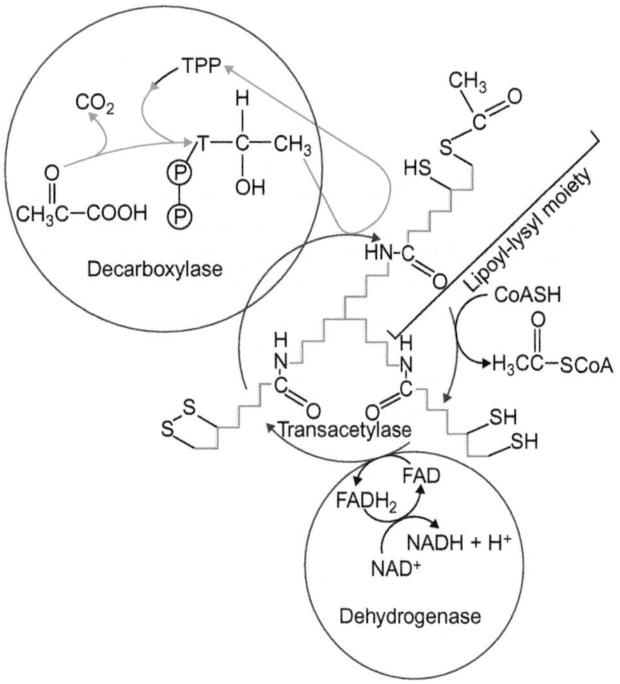

FIGURE 12.8 Schematic representation of the relationship between the three enzymes E1, E_2, and E3, of the pyruvate dehydrogenase complex. The lipoyl-lysyl moiety of the transacetylase delivers the acetyl group to CoA and the reducing equivalents to FAD. TPP = Thiamine pyrophosphate.

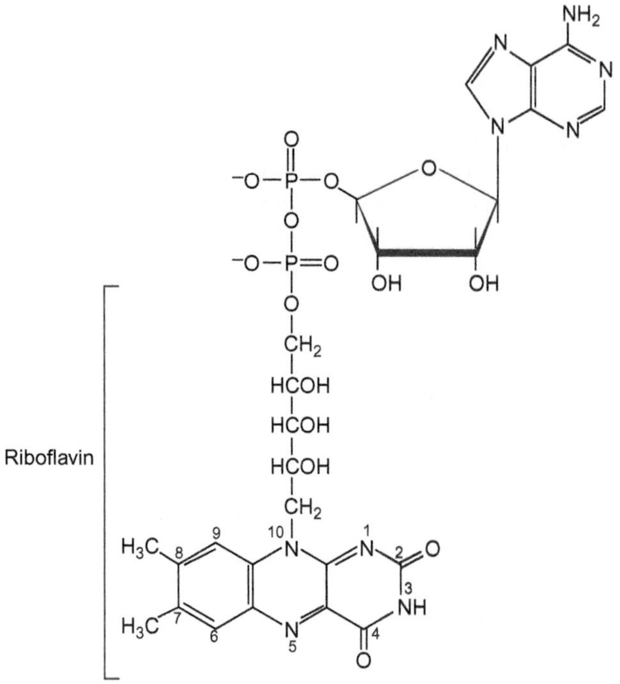

FIGURE 12.9 Flavin adenine dinucleotide (FAD).

FAD to NAD^+. Conversion of pyruvate to acetyl-CoA requires four vitamins: thiamine, pantothenic acid, riboflavin, and niacin. In contrast, in glycolysis, niacin is the only vitamin used. NADH generated in this reaction is oxidized to NAD^+ in the electron transport chain, with the production of ATP (Chapter 13). Dihydrolipoyl transacetylase plays a central role in transferring hydrogen atoms and acetyl groups from one enzyme to the next in the pyruvate dehydrogenase complex. This is made possible by the lipoyl-lysyl swinging arm of the transacetylase, which is about 1.4 nm long (Figure 12.8).

In a similar way, other α-keto acids, e.g., α-ketoglutarate (in the TCA cycle; see below) and branched-chain α-keto acids derived by transamination from the branched-chain amino acids valine, leucine, and isoleucine (Chapter 15), undergo decarboxylation and dehydrogenation catalyzed by enzyme complexes. These enzyme complexes differ in specificity of E1 and E_2, but all contain the same E3 (the dihydrolipoyl dehydrogenase).

An autoimmune mitochondrial antibody-mediated destruction of the intrahepatic bile ducts, known as **primary biliary cirrhosis**, is directed against the E_2 subunit of pyruvate dehydrogenase and α-ketoglutarate dehydrogenase. Some bacterial enzymes exhibit structural homology with human E_2 enzyme, and this molecular mimicry is suggested as the trigger for the autoimmune destruction in primary biliary cirrhosis. The antimitochondrial antibodies in serum and elevated liver enzymes are used in the diagnosis of primary biliary cirrhosis. It is treated with **ursodeoxycholic acid** to increase bile flow and cytoprotection.

Regulation of Pyruvate Dehydrogenase Activity

The pyruvate dehydrogenase complex catalyzes an irreversible reaction that is the entry point of pyruvate into the TCA cycle (see following text) and is under complex regulation by allosteric and covalent modification of the pyruvate dehydrogenase component of the complex. The end products of the overall reaction (NADH and acetyl-CoA) are potent allosteric inhibitors of the pyruvate dehydrogenase component of the complex. They also function as effectors in a non-cAMP-dependent reversible phosphorylation and dephosphorylation cycle of the dehydrogenase. Phosphorylation occurs by an ATP-specific pyruvate dehydrogenase kinase at three serine residues of the α-subunit of the enzyme and leads to inactivation. The kinase is activated by elevated [acetyl-CoA]/[CoA], [NADH]/[NAD^+], and [ATP]/[ADP] and is inhibited by increases in [pyruvate], [Ca^{2+}], and [K^+]. The phospho enzyme is converted to the active dephospho enzyme by a pyruvate dehydrogenase phosphatase, an Mg^{2+}/Ca^{2+}-stimulated enzyme that is also stimulated by insulin in adipocytes. The kinase and phosphatase are associated with pyruvate dehydrogenase. The regulation of pyruvate dehydrogenase is shown in Figure 12.10. In general, when

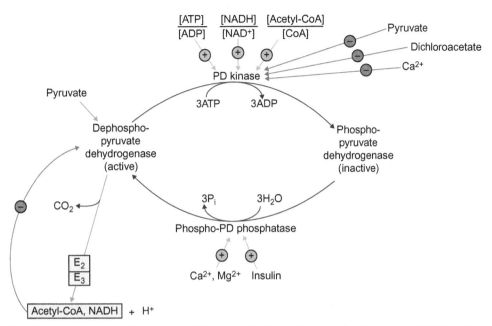

FIGURE 12.10 Regulation of pyruvate dehydrogenase (PD) by inactivation and reactivation by a non-cAMP-dependent phosphorylation–dephosphorylation cycle. Although PD kinase phosphorylates three specific seryl residues in the α-subunit of PD, phosphorylation at any of these sites inactivates PD. The kinase and the phosphatase are under the influence of several regulators, and the dephospho-active PD is also regulated by end products. $\oplus$ = Activation; $\ominus$ = inhibition; E_2 = dihydrolipoyl transacetylase; E_3 = dihydrolipoyl dehydrogenase.

the levels of ATP, NADH, or acetyl-CoA are high, the oxidation of pyruvate to acetyl-CoA is markedly decreased. For example, fatty acid oxidation (Chapter 17) provides all three metabolites and thus decreases the need for pyruvate oxidation.

An analogue of pyruvate, dichloroacetate ($CHCl_2COO^-$), inhibits pyruvate dehydrogenase kinase and maintains the pyruvate dehydrogenase complex in the active state. In experimental animals and in humans with lactic acidemia due to a variety of causes, dichloroacetate administration lowers the concentration of lactate through promotion of pyruvate oxidation. Compounds like dichloroacetate have potential applications in disorders associated with lactic acidemia (e.g., diabetes mellitus).

Abnormalities of Pyruvate Dehydrogenase Complex

Thiamine deficiency causes decreased pyruvate oxidation, leading to accumulation of pyruvate and lactate, particularly in the blood and brain, and is accompanied by impairment of the cardiovascular, nervous, and gastrointestinal systems. Inherited deficiency of pyruvate dehydrogenase complex is accompanied by lactic acidemia and abnormalities of the nervous system (e.g., ataxia and psychomotor retardation). Pyruvate carboxylase deficiency causes similar abnormalities (Chapter 14). Both inherited disorders of pyruvate utilization are autosomal recessive.

Nervous system abnormalities may be attributed in part to diminished synthesis of neurotransmitters, rather than to inadequate synthesis of ATP. In pyruvate dehydrogenase complex deficiency, diminished levels of acetyl-CoA cause decreased production of acetylcholine; in pyruvate carboxylase deficiency, decreased production of oxaloacetate may lead to deficiency of amino acid neurotransmitters (e.g., γ-aminobutyrate, glutamate, aspartate). The interrelationships of these amino acids are discussed in Chapter 16. Although no particular therapeutic method is well established in the treatment of inherited disorders of pyruvate metabolism, ketogenic diets have been beneficial in pyruvate dehydrogenase complex deficiency, since they provide the product of the deficient reaction (acetyl-CoA). Administration of large doses of thiamine may be of benefit because mutations of pyruvate dehydrogenase complex may give rise to decreased affinity for thiamine pyrophosphate. In one patient, administration of dichloroacetate was beneficial even in the absence of a ketogenic diet.

In pyruvate carboxylase deficiency, administration of diets supplemented with aspartate and glutamate demonstrated sustained improvement in neurological symptoms. These two amino acids cross the blood−brain barrier after amidation to asparagine and glutamine in non-neural tissues. The toxicity of organic arsenicals and arsenite (ASO_3^{3-}) is due to their ability to bind functional sulfhydryl groups of enzymes. One important target is the dithiol form of the lipoyl group of pyruvate dehydrogenase and α-ketoglutarate dehydrogenase complexes. In earlier times arsenicals were used in the treatment of syphilis; however, due to their toxicity, they have been replaced with better drugs such as penicillin.

TRICARBOXYLIC ACID CYCLE

The tricarboxylic acid (TCA) cycle (also called the **citric acid cycle** or **Krebs cycle**) consists of eight enzymes that oxidize acetyl-CoA with the formation of carbon dioxide, reducing equivalents in the form of NADH and FADH$_2$, and guanosine triphosphate (GTP):

$$CH_3COSCoA + 3NAD^+ + FAD + GDP^{3-} + HPO_4^{2-}$$
$$+ 2H_2O \rightarrow CoASH + 2CO_2 + 3NADH + FADH_2$$
$$+ GTP^{4-} + 2H^+$$

The reducing equivalents are transferred to the electron transport chain, ultimately to reduce oxygen and generate ATP through oxidative phosphorylation (Chapter 13). Thus, oxidation of acetyl-CoA yields a large supply of ATP.

Since the intermediates in the cycle are not formed or destroyed in its net operation, they may be considered to play catalytic roles. However, several intermediates are consumed because they are biosynthetic precursors of other metabolites (amphibolic role) and hence may become depleted. They are replenished (anaplerotic process) by other reactions to optimal concentrations.

The TCA cycle is the major final common pathway of oxidation of carbohydrates, lipids, and proteins, since their oxidation yields acetyl-CoA. Acetyl-CoA also serves as the precursor in the synthesis of fatty acids, cholesterol, and ketone bodies. All enzymes of the cycle are located in the mitochondrial matrix except for succinate dehydrogenase, which is embedded in the inner mitochondrial membrane. Thus, the reducing equivalents generated in the cycle have easy access to the electron transport chain. TCA cycle enzymes, with the exception of α-ketoglutarate dehydrogenase complex and succinate dehydrogenase, are also present outside the mitochondria. The overall TCA cycle is shown in Figure 12.11.

Reactions of TCA Cycle

Condensation of Acetyl-CoA with Oxaloacetate to Form Citrate

The first reaction of the TCA cycle is catalyzed by citrate synthase and involves a carbanion formed at the methyl group of acetyl-CoA that undergoes aldol condensation with the carbonyl carbon atom of the oxaloacetate:

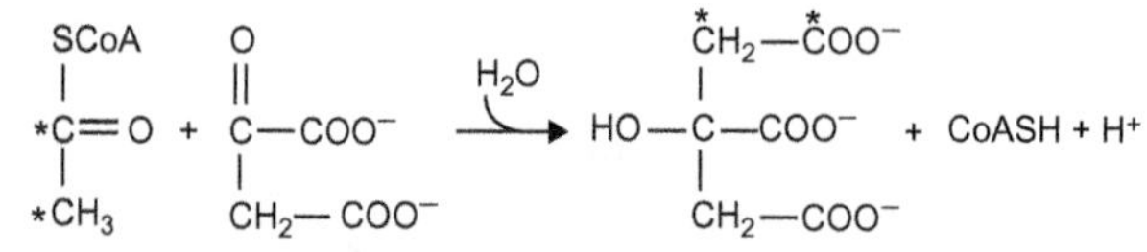

Acetyl-CoA   Oxaloacetate   Citrate

The condensation reaction is thought to yield a transient enzyme-bound intermediate, citroyl-CoA, which undergoes hydrolysis to citrate and CoASH with loss of free energy. This reaction is practically irreversible and has a $\Delta G^{0\prime}$ of -7.7 kcal/mol (-32.2 kJ/mol). Formation of citrate is the committed step of the cycle and is regulated by allosteric effectors. Depending on the cell type, succinyl-CoA (a later intermediate of the cycle), NADH, ATP, or long-chain fatty acyl-CoA functions as the negative allosteric modulator of citrate synthase. Citrate formation is also regulated by availability of substrates, and citrate is an allosteric inhibitor.

Citrate provides the precursors (acetyl-CoA, NADPH) for fatty acid synthesis and is a positive allosteric modulator of acetyl-CoA carboxylase, which is involved in the initiation of long-chain fatty acid synthesis (Chapter 16). It regulates glycolysis by negative modulation of phosphofructokinase activity (see glycolysis section). All of the preceding reactions occur in the cytoplasm, and citrate exits from mitochondria via the tricarboxylate carrier.

Isomerization of Citrate to Isocitrate

In the isomerization reaction of citrate to isocitrate, the tertiary alcoholic group of citrate is converted to a secondary alcoholic group that is more readily oxidized. The isomerization catalyzed by aconitate dehydratase (or aconitase) occurs by removal and addition of water with the formation of an intermediate, cis-aconitate:

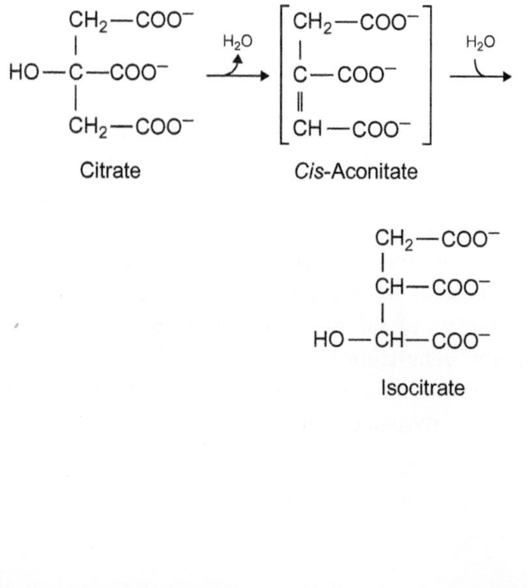

Citrate   Cis-Aconitate

Isocitrate

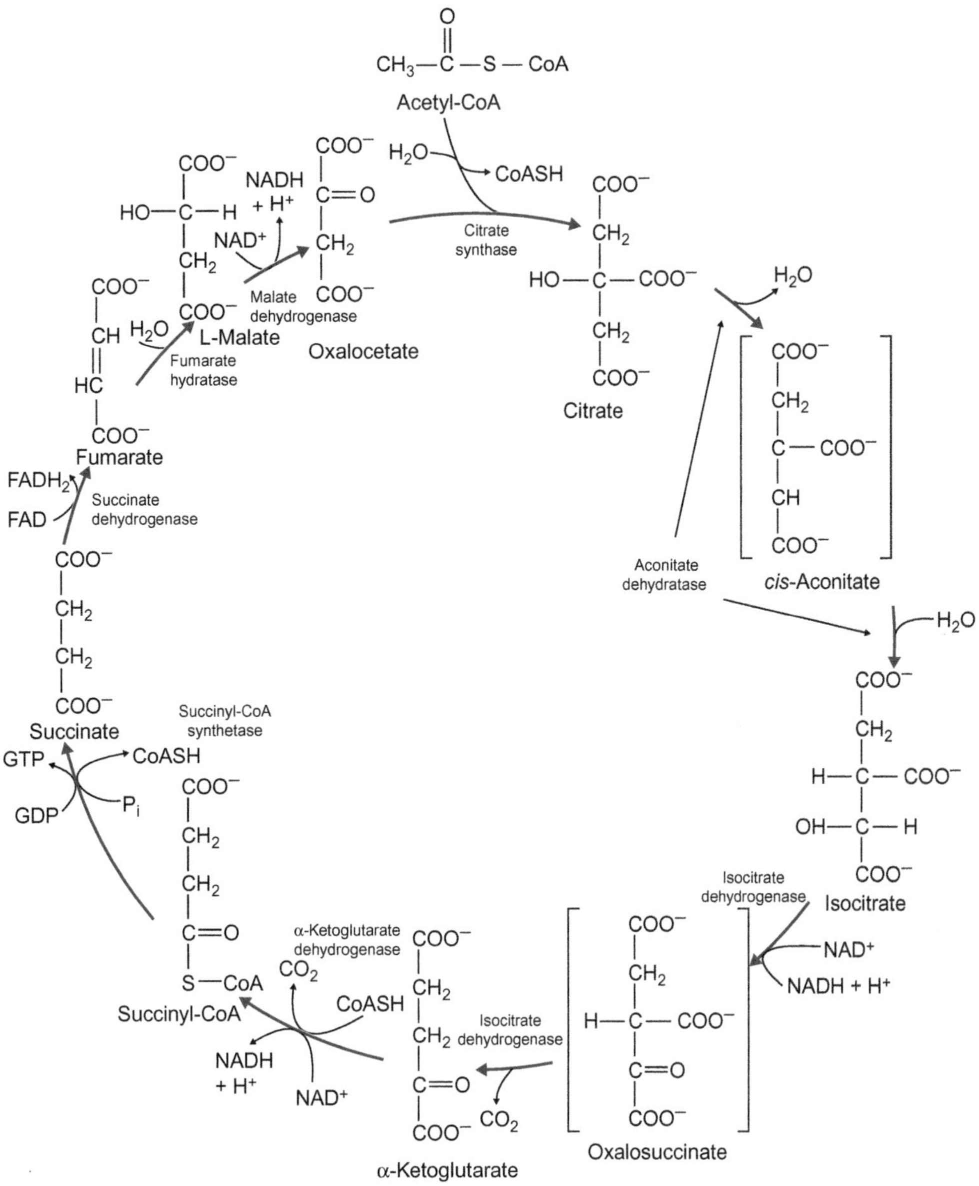

FIGURE 12.11   Reactions of the tricarboxylic acid (TCA) cycle.

The *cis*-aconitate may not be an obligatory intermediate, since the enzyme presumably can isomerize citrate to isocitrate directly. Aconitate dehydratase has an iron-sulfur center, the exact function of which is not known. The reaction is considered to be near-equilibrium, even though under physiological conditions the ratio of citrate to isocitrate is about 9:1 owing to rapid conversion of isocitrate to subsequent products of the cycle.

Although citrate is a symmetrical molecule, its two −CH$_2$COOH groups are not identical with regard to the orientation of the −OH and −COOH groups. Thus, citrate reacts in an asymmetrical manner with the enzyme, so that only the −CH$_2$COOH group derived from oxaloacetate is modified. A three-point attachment of the enzyme to the substrate accounts for the asymmetrical nature of the reaction.

A powerful competitive inhibitor (highly toxic) of aconitate dehydratase is fluorocitrate, an analogue of citrate:

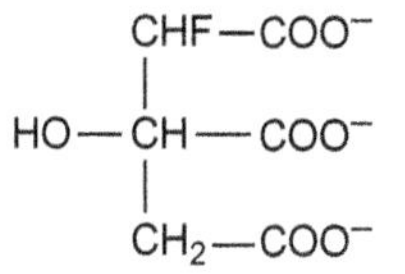

$$CHF-COO^-$$
$$HO-CH-COO^-$$
$$CH_2-COO^-$$

In vivo, fluorocitrate can be formed by condensation of fluoroacetyl-CoA with oxaloacetate by the action of citrate synthase. Fluoroacetyl-CoA itself is synthesized from fluoroacetate by acetyl-CoA synthase.

Oxidative Decarboxylation of Isocitrate to α-Ketoglutarate

Isocitrate dehydrogenase (IDH) catalyzes the first of two decarboxylations and dehydrogenations in the cycle. Three different isocitrate dehydrogenases are present: one is specific for NAD^+ (IDH3) and found in mitochondrial matrix; the other two specific IDH1 and IDH3 for $NADP^+$ are found in mitochondria and cytoplasm. The NAD^+-specific enzyme IDH3, a tetramer, is the primary enzyme with regard to the TCA cycle and plays a central role in aerobic energy production. All three require Mg^{2+} or Mn^{2+}. The reaction yields α-ketoglutarate (2-oxoglutarate), NAD(P)H, and CO_2 and involves enzyme-bound oxalosuccinate as an intermediate:

CH₂—COO⁻ CH—COO⁻ HO—CH—COO⁻ (with NAD⁺ → NAD + H⁺)

Isocitrate

CH₂—COO⁻ CH—COO⁻ O=C—COO⁻ (with CO₂) → H₂C—COO⁻ CH₂ O=C—COO⁻

Oxalosuccinate (enzyme-bound)   α-Ketoglutarate (2-oxoglutarate)

The reaction is nonequilibrium in type and has a $\Delta G^{0\prime}$ of -5.0 kcal/mol (-20.9 kJ/mol). Although in mitochondria both NAD^+- and $NADP^+$-linked enzymes are involved in the cycle, the NAD^+-linked enzyme, which is also under allosteric regulation, is the more predominant. Positive effectors are ADP and Ca^{2+}, and negative effectors are ATP and NADH. Thus, under conditions of abundance of energy, the enzyme is inhibited, and under

conditions of low energy, the enzyme is stimulated. IDH1 and IDH2 are dimers and not under allosteric regulation. They catalyze the reversible NADPH-dependent reductive carboxylation of α-ketoglutarate to isocitrate. The mutant forms of IDH1 and IDH2 found in cancer cells catalyze a new gain-of-function reaction in which α-ketoglutarate is converted to R(-)-2-hydroxyglutarate (2HG) utilizing NADPH as reductant. The formation of 2HG, an oncometabolite with depletion of NADPH due to a futile metabolic cycle, perturbs a number of processes involved in tumorigenesis and proliferation (Figure 12.12). The perturbations include impairment of redox status, epigenetic landscape (e.g., histone demethylation), and other cellular processes promoting malignant transformation [7−9].

Oxidative Decarboxylation of α-Ketoglutarate to Succinyl-CoA

The α-ketoglutarate dehydrogenase complex is analogous to the pyruvate dehydrogenase complex and consists of three enzymes: α-ketoglutarate dehydrogenase, dihydrolipoyl transsuccinylase, and dihydrolipoyl dehydrogenase (the latter dehydrogenase is identical in both complexes). The cofactor, prosthetic groups, and coenzyme requirements are identical to those of pyruvate oxidation. The overall reaction is

CH₂—COO⁻ CH₂ O=C—COO⁻ + CoASH + NAD⁺ → (with CO₂)

α-Ketoglutarate

CH₂—COO⁻ CH₂CO—SCoA + NADH + H⁺

The reaction is of the nonequilibrium type, with a $\Delta G^{0\prime}$ of -8 kcal/mol (-33.5 kJ/mol). Unlike the

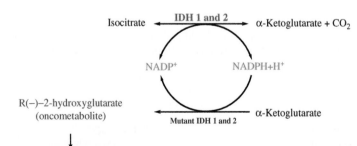

Isocitrate ⇌ (IDH 1 and 2) α-Ketoglutarate + CO_2
($NADP^+$ → $NADPH + H^+$)
R(−)-2-hydroxyglutarate (oncometabolite) ← (Mutant IDH 1 and 2) α-Ketoglutarate

↓

Impairs epigenetic landscape (e.g., impairs histone demethylation) and redox status, and promote oncogenesis

FIGURE 12.12 Reactions of normal isocitrate dehydrogenase (IDH) isoenzymes 1 and 2 and their mutants. The mutant forms generate an oncometabolite 2-hydroxyglutarate (see text for details).

pyruvate dehydrogenase complex, the α-ketoglutarate dehydrogenase complex does not possess a complex regulatory mechanism involving a kinase and a phosphatase. However, activity is inhibited by high ratios of [ATP]/[ADP], [succinyl-CoA]/[CoASH], and [NADH]/[NAD$^+$], and stimulated by Ca^{2+}. α-Ketoglutarate is reversibly transaminated to glutamate, and it is utilized in the hydroxylation of prolyl or lysyl residues of collagen (Chapter 23).

Conversion of Succinyl-CoA to Succinate Coupled to Formation of GTP

In this complex reaction catalyzed by succinyl-CoA synthase (succinate thiokinase), the energy-rich thioester linkage of succinyl-CoA is hydrolyzed with release of free energy that is conserved in the substrate phosphorylation of GDP with phosphate to form GTP:

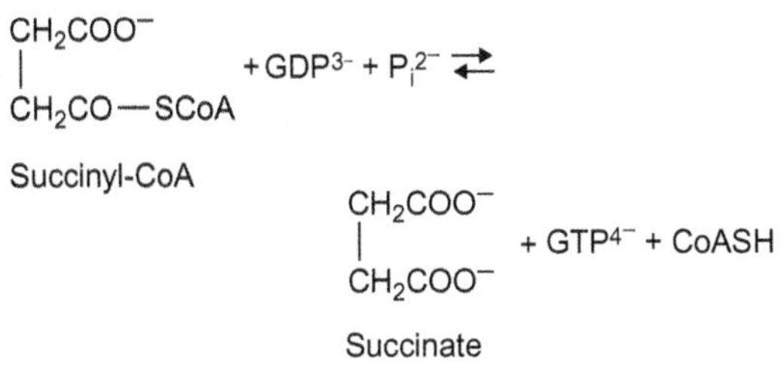

$$CH_2COO^- \\ | \qquad\qquad +GDP^{3-} + P_i^{2-} \rightleftarrows \\ CH_2CO-SCoA$$

Succinyl-CoA

$$CH_2COO^- \\ | \qquad\qquad + GTP^{4-} + CoASH \\ CH_2COO^-$$

Succinate

This reaction is readily reversible (a near-equilibrium reaction) and has a $\Delta G^{0\prime}$ of -0.7 kcal/mol (-2.9 kJ/mol). It proceeds with the formation of intermediates of the enzyme with succinyl phosphate and with phosphate (P_i), the latter being linked to a histidyl residue of the enzyme (E):

E + succinyl-CoA + $P_i^{2-} \leftrightarrow E$-succinyl phosphate + CoASH

E-succinyl phosphate $\leftrightarrow E$-phosphate + succinate

E-phosphate + GDP $\leftrightarrow$ E + GTP

GTP is converted to ATP by nucleoside-diphosphate kinase:

GTP + ADP $\leftrightarrow$ GDP + ATP

Succinyl-CoA can also be synthesized from propionyl-CoA by way of methylmalonyl-CoA, which is formed in the oxidation of branched-chain amino acids (e.g., valine, isoleucine) and in the terminal stage of oxidation of odd-chain-length fatty acids (Chapter 16). Succinyl-CoA is utilized in the activation of acetoacetate (Chapter 16) and

the formation of Δ-aminolevulinate, a precursor of porphyrin (Chapter 27).

Dehydrogenation of Succinate to Fumarate

Succinate is dehydrogenated to the trans-unsaturated compound fumarate by succinate dehydrogenase, an FAD enzyme:

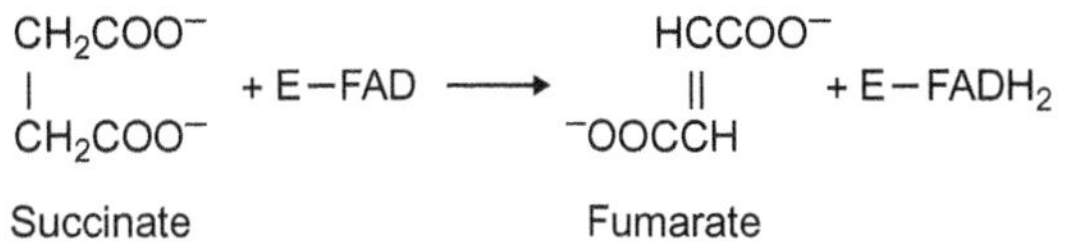

$$CH_2COO^- \qquad\qquad\qquad HCCOO^- \\ | \qquad + E-FAD \longrightarrow \quad || \qquad + E-FADH_2 \\ CH_2COO^- \qquad\qquad ^-OOCCH$$

Succinate   Fumarate

This reaction is the only dehydrogenation reaction of the TCA cycle in which NAD$^+$ does not mediate the transport of reducing equivalents. The FAD is covalently linked to a histidyl residue of the enzyme, which is an integral protein of the inner mitochondrial membrane and contains iron-sulfur centers that undergo redox changes (Fe^{3+} + e$^- \leftrightarrow$ Fe^{2+}). Reducing equivalents from FADH$_2$ enter the electron transport chain at the coenzyme Q level, bypassing one of the sites of oxidative phosphorylation (Chapter 13). The enzyme is stereospecific for the *trans* hydrogen atoms of the methylene carbons of the substrate, so that only the *trans* isomer is produced (the *cis* isomer is maleate). Succinate dehydrogenase is competitively inhibited by malonate, the next lower homologue of succinate (Chapter 6).

Mutations of succinate dehydrogenase with loss of function lead to accumulation of precursors of the TCA cycle, which may lead to oncogenesis [9].

Hydration of Fumarate to Malate

In this reaction, water is added across the double bond of fumarate by fumarate hydratase (fumarase) to yield the hydroxy compound L-malate:

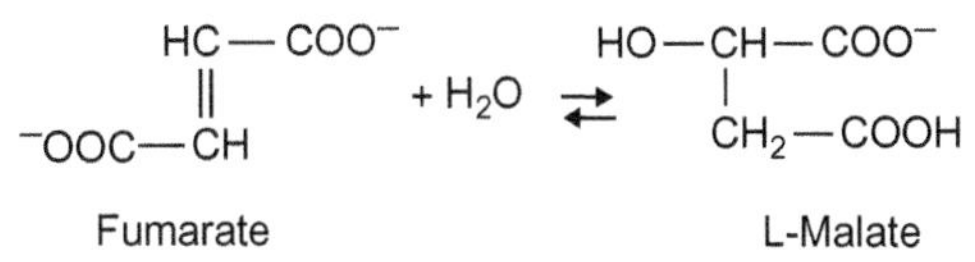

$$HC-COO^- \qquad\qquad HO-CH-COO^- \\ || \qquad\quad + H_2O \rightleftarrows \qquad | \\ ^-OOC-CH \qquad\qquad CH_2-COOH$$

Fumarate   L-Malate

The enzyme has absolute specificity for the double bond of the *trans*-unsaturated acid and for the formation of L-hydroxy acid. The reaction is freely reversible (near equilibrium). Homozygous loss of function mutations of fumarate dehydrogenase with consequent TCA cycle intermediates is associated with some cancers [9].

Dehydrogenation of Malate to Oxaloacetate

In the last reaction of the cycle, L-malate is oxidized to oxaloacetate by malate dehydrogenase, an NAD^+-linked enzyme:

$$HO-CHCOO^-$$
$$| \quad\quad + NAD^+ \rightleftarrows$$
$$CH_2-COO^-$$

L-Malate

$$\begin{array}{c} O \\ \| \\ C-COO^- \quad + NADH + H^+ \\ | \\ CH_2-COO^- \end{array}$$

Oxaloacetate

Although the equilibrium of this reaction favors malate formation, *in vivo* the reaction proceeds toward the formation of oxaloacetate, since the latter is rapidly removed by the citrate synthase reaction to initiate the next round of the cycle.

Stereochemical Aspects of the TCA Cycle

Aconitate dehydratase, succinate dehydrogenase, and fumarase yield stereospecific products. Labeling experiments with ^{14}C in methyl and carboxyl carbons of acetyl-CoA or in all of the carbons of oxaloacetate yield the following results in terms of the intermediates or product formed.

When only labeled acetyl-CoA is used, the label does not appear in CO_2 during the first revolution of the cycle, and upon completion of the first revolution, two of the four carbon atoms of oxaloacetate are labeled. Thus, during the first revolution, neither of the carbon atoms of acetyl-CoA is oxidized to CO_2; the carbon atoms of the CO_2 produced are derived from oxaloacetate. This finding is due to discrimination of the two $-CH_2COO^-$ groups of citrate by aconitate dehydratase, so that citrate is treated as a chiral molecule at the asymmetrical active center of the enzyme. However, during the succeeding revolutions of the cycle, the label is randomized because succinate, a symmetrical compound, is treated as such by the enzyme without discriminating between its two carboxyl groups. If only labeled oxaloacetate is used, half of the label is retained in the oxaloacetate at the end of the first revolution of the cycle.

Amphibolic Aspects of the TCA Cycle

At each turn of the TCA cycle, oxaloacetate is regenerated and can combine with another acetyl-CoA molecule. The TCA cycle is **amphibolic**; i.e., it serves as a catabolic and an anabolic pathway. Reactions that utilize intermediates of the cycle as precursors for the biosynthesis of other molecules are as follows. Some of these reactions occur outside the mitochondria.

1. Citrate + ATP + CoA → acetyl-CoA + oxaloacetate + ADP + P_i. This reaction takes place in the cytoplasm and is a source of acetyl-CoA for fatty acid biosynthesis.
2. α-Ketoglutarate + alanine ↔ glutamate + pyruvate.
3. α-Ketoglutarate → succinate + CO_2. This reaction is involved in the hydroxylation of prolyl and lysyl residues of protocollagen, a step in the synthesis of collagen.
4. Succinyl-CoA + glycine → Δ-aminolevulinic acid (ALA). ALA is then utilized for the synthesis of heme.
5. Succinyl-CoA + acetoacetate → acetoacetyl-CoA + succinate. This reaction is important in the activation of acetoacetate (a ketone body) and hence for its utilization in extrahepatic tissues.
6. Oxaloacetate + alanine ↔ aspartate + pyruvate.
7. Oxaloacetate + GTP $\xrightarrow{Mg^{2+}}$ phosphoenolpyruvate + GDP + CO_2

Utilization of intermediates in these reactions leads to their depletion and hence to a slowing down of the cycle unless they are replenished by the replacement or anaplerotic reactions listed here:

1. Pyruvate + CO_2 + ATP $\xrightarrow{biotin, Mg^{2+}}$ oxaloacetate + ADP + P_i

   This reaction, catalyzed by pyruvate carboxylase, is the most important anaplerotic reaction in animal tissues and occurs in mitochondria. Pyruvate carboxylase is an allosteric enzyme that requires acetyl-CoA for activity (see gluconeogenesis; Chapter 14).
2. Pyruvate + CO_2 + NADPH + H^+ ↔ L-malate + $NADP^+$.
3. Oxaloacetate and α-ketoglutarate may also be obtained from aspartate and glutamate, respectively, by aminotransferase (transaminase) reactions.
4. Oxaloacetate is obtained in the reverse of reaction 7, above.

Regulation of the TCA Cycle

The TCA cycle substrates oxaloacetate and acetyl-CoA and the product NADH are the critical regulators. The availability of acetyl-CoA is regulated by the pyruvate dehydrogenase complex. The TCA cycle enzymes citrate synthase, isocitrate dehydrogenase, and the α-ketoglutarate dehydrogenase complex are under regulation by many metabolites (discussed previously) to maintain optimal cellular energy needs. Overall fuel homeostasis is discussed in Chapter 20.

Dysregulation of some of the enzymes in the TCA cycle can initiate and promote cancer formation.

Energetics of the TCA Cycle

Oxidation of one molecule of acetyl-CoA in the cycle yields three NADH, one $FADH_2$, and one GTP. Transport of reducing equivalents in the electron transport chain from one NADH molecule yields three ATP and from one $FADH_2$, two ATP (see Chapter 13). Thus, oxidation of one molecule of acetyl-CoA yields 12 ATP. Since one glucose molecule yields two acetyl-CoA, the yield of ATP is 24. In addition, in the oxidation of glucose to acetyl-CoA, two other steps yield four NADH (i.e., glyceraldehyde-3-phosphate dehydrogenase and pyruvate dehydrogenase reactions), since one glucose molecule yields two triose phosphate molecules, which accounts for 12 more ATP. Oxidation of one molecule of glucose to two of pyruvate yields two ATP (see under Glycolysis) by substrate-level phosphorylation. Thus, complete oxidation of one molecule of glucose can yield 38 ATP. In cells dependent only on anaerobic glycolysis, glucose consumption has to be considerably greater to derive an amount of energy equal to that obtainable from aerobic oxidation (2 ATP versus 38 ATP per glucose). However, in cells capable of both aerobic and anaerobic metabolism, glycolysis, the TCA cycle, and oxidation in the electron transport chain are integrated and regulated so that only just enough substrate is oxidized to satisfy the energy needs of a particular cell. In the presence of oxygen, a reduction in glucose utilization and lactate production takes place, a phenomenon known as the **Pasteur effect**. The depression of rate of flux through glycolysis can be explained in part by the accumulation of allosteric inhibitors (e.g., ATP) of phosphofructokinase and pyruvate kinase, together with the supply of cofactors and coenzymes of certain key enzymes (e.g., glyceraldehyde-3-phosphate dehydrogenase).

In addition, to meet the energy demands during hypoxic stress, tissues undergo changes in gene expression, which result in enhanced angiogenesis, red blood cell production, and glycolytic enzymes. The enhanced synthesis of glycolytic enzymes is mediated by the DNA-binding protein known as **hypoxia-inducible factor** (Chapter 26).

REQUIRED READING

[1] W.H. Dzik, The air we breathe: three vital respiratory gases and the red blood cell: oxygen, nitric oxide, and carbon dioxide, Transfusion 51 (2011) 676−685.

[2] L.W. Andersen, J. Mackenhauer, J.C. Roberts, K.M. Berg, M.N. Cocchi, M.W. Donnino, Etiology and therapeutic approach to elevated lactate levels, Mayo Clin. Proc. 88 (2013) 1127−1140.

[3] A.S. Divakaruni, A.N. Murphy, A mitochondrial mystery, solved, Science 337 (2012) 41−43.

[4] S. Herzig, E. Raemy, S. Montessuit, J. Veuthey, N. Zamboni, B. Westermann, et al., Identification and functional expression of the mitochondrial pyruvate carrier, Science 337 (2012) 93−96.

[5] D.K. Bricker, E.B. Taylor, J.C. Schell, T. Orsak, A. Boutron, Y. Chen, et al., A mitochondrial pyruvate carrier required for pyruvate uptake in yeast, *Drosophila*, and humans, Science 337 (2012) 96−100.

[6] M. Brivet, A. Garcia-Cazorla, S. Lyonnet, Y. Dumez, M.C. Nassogne, A. Slama, et al., Impaired mitochondrial pyruvate importation in a patient and a fetus at risk, Mol. Genet. Metab. 78 (2003) 186−192.

[7] J. Kim, R.J. DeBerardinis, Silencing a metabolic oncogene, Science 340 (2013) 558−561.

[8] Z.J. Reitman, H. Yan, Isocitrate dehydrogenase 1 and 2 mutations in cancer: alterations at a crossroads of cellular metabolism, J. Natl Cancer Inst. 102 (2010) 932−941.

[9] D.C. Wallace, Mitochondria and cancer, Nat. Rev. Cancer 12 (2012) 685−698.

Chapter 13

Electron Transport Chain, Oxidative Phosphorylation, and Other Oxygen-Consuming Systems

Key Points

1. Energy requirements of aerobic cells are met by oxidation of carbohydrates, lipids, and proteins, which yields reduced coenzymes NADH and $FADH_2$ (Chapters 12, 15, and 16).

2. The oxidation of NADH and $FADH_2$ proceeds in a highly organized electron transport chain (ETC) mitochondrial system, which consists of complexes of proteins and coenzymes. The ultimate electron acceptor is oxygen, which is converted to water. The energy is conserved with the transport of electrons in the ETC as an electrochemical proton gradient, because protons (H^+) are pumped out of the matrix to the intermembranal space. The electrochemical proton gradient drives the conversion of ADP to ATP, when protons are channeled back into the matrix via ATP synthase.

3. Electron transport chain and oxidative phosphorylation (OXPHOS) systems are organized into five complexes: I, II, III, IV, and V. Each complex consists of unique species of molecules that include flavoproteins with tightly bound FAD or FMN, coenzyme Q, cytochromes, iron−sulfur (Fe-S) proteins, and copper-bound proteins. The five complexes are coupled and function in a dynamic manner. The direction of electron flow in the ETC is from a higher to a lower energy level.

4. Complex I is a flavoprotein and an NADH dehydrogenase. It transfers electrons from NADH to hydrophobic mobile electron carrier CoQ via FMN and iron−sulfur clusters, leading to the formation of $CoQH_2$. The passage of electrons by complex I is coupled to the transport of protons from the matrix to the intermembrane space of mitochondria. Complex I is inhibited by rotenone and Amytal.

5. Complex II, also known as succinate dehydrogenase, is a member of the TCA cycle (Chapter 12). It oxidizes succinate to fumarate, which is coupled to the reduction of FAD to $FADH_2$. The electrons from $FADH_2$ are transferred to CoQ via Fe-S centers with the formation of $CoQH_2$.

6. CoQ accepts electrons not only from both complex I and II but also from other flavoprotein-linked dehydrogenases, namely fatty acyl-CoA dehydrogenase (Chapter 16) and glycerophosphate dehydrogenase.

7. Complex III catalyzes the electron transport from $CoQH_2$ to cytochrome c and also is the second site of proton transport from matrix to intermembrane space. Complex III consists of three different cytochromes and an Fe-S protein. Cytochromes contain the same iron−protoporphyrin IX heme prosthetic groups found in hemoglobin and myoglobin (Chapters 26 and 27). Unlike hemoglobin and myoglobin, the iron atoms of the heme groups of cytochromes undergo oxidation and reduction cycles facilitating electron transport. The terminal electron acceptor of electrons from $CoQH_2$ is cytochrome c. It is a mobile electron carrier like CoQ, but is water soluble, loosely bound to the inner mitochondrial membrane, and under certain signals leaks into cytosol and initiates programmed cell death known as apoptosis.

8. Complex IV is cytochrome c oxidase, and it is the final destination for the electrons. It catalyzes the reduction of O_2 by using electrons obtained from reduced cytochrome c, with the formation of the end product, water; and it is the third proton pumping station. It consists of four redox centers: cytochrome a, cytochrome a_3, and two Cu ions. Inhibitors of cytochrome c oxidase are carbon monoxide, azide, and cyanide.

9. The energy conserved by the electron transport chain as an electrochemical proton gradient between the inner-membranal space and matrix is utilized in the formation of ATP from ADP. This process is known as Mitchell's Chemiosmotic hypothesis and is catalyzed by ATP synthase, also known as complex V. It consists of spheres known as F_1 that are attached to an integral membrane protein, F_0. The translocation of protons occurs via the transmembrane channel F_0. ATP synthesis occurs on F_1 subunits, which function as a rotating motor coupled to the formation of an anhydride bond between ADP and P_i. Transport of two electrons from NADH yields three ATP (P/O = 3) and transport of two electrons from $FADH_2$ yields two ATP (P/O = 2).

10. Uncoupling agents such as thermogenin of brown adipose tissue dissipate energy conserved in the proton gradient as heat (Chapter 20).

N. V. Bhagavan and C.-E. Ha: Essentials of Medical Biochemistry, Second edition. DOI: http://dx.doi.org/10.1016/B978-0-12-416687-5.00013-0

11. Mitochondrial energy production is regulated by the availability of substrates, primarily ADP. ATP−ADP exchange is mediated by a translocase located in the inner membrane.

12. Cytosolic NADH produced in the glycolytic pathway is transported by two different shuttle pathways and is oxidized in the electron transport system.

13. Mitochondria contain their own genome; they are derived from oocyte cytoplasm during fertilization, and therefore, their disorders are maternally inherited. Mitochondrial DNA is a covalently linked circular molecule and encodes for only 13 proteins required for OXPHOS. Therefore, mitochondrial replication and function require proteins encoded in nuclear DNA; thus, their overall function is under dual genetic control. The insertion of OXPHOS polypeptides synthesized in mitochondrial ribosomes (mitoribosomes) is guided by a mitoribosome protein. Mitochondrial DNA also codes for 2 mitoribosomal RNA and 22 transfer RNAs. Defects due to either mitochondrial DNA or nuclear DNA, which supports mitochondrial function, can give rise to disorders of clinical significance.

14. Under certain signals, cytochrome c that is released to the cytosol via the mitochondrial permeability transition pores formed in the mitochondrial membranes initiates apoptosis. Apoptosis is a critical component of normal development and function. However, inappropriate stimulation by external or internal stimuli can cause acute and chronic diseases.

15. Oxygen is consumed as substrate in a number of reactions, other than oxidative phosphorylation. Single-electron transfers to oxygen create reactive oxygen species (ROS), which include superoxide radical, hydrogen peroxide, the hydroxyl radical, and singlet oxygen. ROS production is required in killing bacteria in the phagosomes of neutrophils. Defects in ROS production lead to chronic granulomatous disease. However, excessive ROS production in mitochondria can cause tissue injury. Oxygen is required for many essential hydroxylation reactions, such as the cytochrome P-450 monooxygenase system.

The energy requirements of aerobic cells are met by the energy released in the oxidation of carbohydrates, fatty acids, and amino acids by molecular oxygen. In these oxidation processes, the reducing equivalents from substrates are transferred to NAD^+, flavin mononucleotide (FMN), or flavin adenine dinucleotide (FAD). The hydrogens, electrons, or hydride (H^-) ions are removed catalytically from NADH and reduced flavin nucleotides, and transferred through a series of coupled reduction−oxidation (redox) reactions to oxygen, which serves as the ultimate electron acceptor. The reduced oxygen is then converted to water. The entire process is known as **cell respiration**. The numerous coupled redox reactions are tightly linked and take place in the inner mitochondrial membrane. This reaction sequence is called the **respiratory electron transport chain**. Electrons in substrate or cofactor begin with a high potential energy and end at oxygen with a lower potential energy. During this electron flow, a portion of the free energy liberated is conserved by an energy-transducing system (by which electrical energy is changed to chemical energy). Since the energy is conserved in the terminal phosphoanhydride bond of ATP through phosphorylation of ADP to ATP, the overall coupled process is known as **oxidative phosphorylation**.

The following reactions yield NADH or $FADH_2$ during the breakdown of glucose by glycolysis and in the TCA cycle (Chapter 12). The NADH-producing reactions are

$$\text{Glyceraldehyde 3-phosphate} \rightarrow \text{1,3-bisphosphoglycerate}$$

$$\text{Pyruvate} \rightarrow \text{acetyl-CoA}$$

$$\text{Isocitrate} \rightarrow \alpha\text{-ketoglutarate}$$

$$\alpha\text{-Ketoglutarate} \rightarrow \text{succinyl-CoA}$$

$$\text{Malate} \rightarrow \text{oxaloacetate}$$

The $FADH_2$-producing reaction is

$$\text{Succinate} \rightarrow \text{fumarate}$$

Except for the conversion of glyceraldehyde 3-phosphate to 1,3-bisphosphoglycerate occurring in the cytoplasm, all of the preceding reactions take place in the mitochondria. Since mitochondria are not permeable to NADH, two shuttle pathways transport the reducing equivalents of cytoplasmic NADH into the mitochondria.

The TCA cycle functions for both the oxidation and the production of reducing equivalents from a variety of metabolites such as carbohydrates, amino acids, and fatty acids. In the mitochondria, fatty acids undergo successive removal of two carbon units in the form of acetyl-CoA, which is converted in the TCA cycle to CO_2 and reducing equivalents found in NADH and $FADH_2$. The latter two substances are then fed into the electron transport chain. The successive removal of each acetyl-CoA molecule in the β-oxidation of fatty acids requires the removal of four hydrogen atoms and is catalyzed by two dehydrogenases (Chapter 16). One dehydrogenase yields $FADH_2$; the other yields NADH.

The term *one reducing equivalent* means that 1 mol of electrons is present in the form of one equivalent of a reduced electron carrier. One mol of NAD^+, when reduced to 1 mol of NADH, utilizes 2 moles of electrons from 1 mol of substrate (Chapter 12):

$$NAD^+ + SH_2 \text{ (reduced substrate)} \rightleftharpoons NADH + H^+ + S$$
(oxidized substrate)

In the forward reaction, the 2 moles of electrons are transferred in the form of 1 mol of hydride ion $(H:)^-$.

MITOCHONDRIAL STRUCTURE AND PROPERTIES

Mitochondria are present in the cytoplasm of aerobic eukaryotic cells. They are frequently found in close proximity to the fuel sources and to the structures that require ATP for maintenance and functional activity (e.g., the contractile mechanisms, energy-dependent transport systems, and secretory processes). The number of mitochondria in a single cell varies from one type of cell to another; a rat liver cell contains about 1000, while one giant amoeba has about 10,000. In a given cell, the number of mitochondria may also depend on the cell's stage of development or functional activity.

The size and shape of mitochondria vary considerably from one cell type to another. Even within the same cell, mitochondria can undergo changes in volume and shape depending on the metabolic state of the cell. In general, they are $0.5-1.0\,\mu m$ wide and $2-3\,\mu m$ long and are known to aggregate end to end, forming long filamentous structures. They undergo fusion and fission processes (Chapter 1). Defective mitochondria can be rescued by functionally active mitochondria by fusion. This can occur during energy demand and stressful cellular conditions [1]. Dynamic metabolic networks determine mitochondrial fusion and fission processes [2]. All of the proteins required for fusion and fission are encoded by nuclear genes. Mutations in these genes give rise to serious metabolic derangements and are associated with clinical sequelae (see Clinical Case Study 13.1).

Mitochondria consist of two membranes, one encircling the other, creating two spatial regions: the intermembrane space and the central space, called the **matrix**. The outer membrane is $6-7\,nm$ thick, smooth, unfolded (Figure 13.1), and freely permeable to molecules with molecular weights below 10,000. It contains a heterogeneous group of enzymes that catalyze certain reactions of lipid metabolism as well as hydroxylation reactions. The intermembrane space ($5-10\,nm$) contains the enzymes that catalyze interconversion of adenine nucleotides.

The inner membrane ($6-8\,nm$ thick) has many folds directed toward the matrix. These invaginations, known as **cristae**, increase the surface area of the inner membrane. The lipid component, almost all of which is phospholipid, constitutes $30\%-35\%$ by weight of the inner membrane. Phospholipids are asymmetrically distributed in the lipid bilayer, with phosphatidylethanolamine predominating on the matrix side and phosphatidylcholine on the cytoplasmic side. Seventy-five percent of the cardiolipin is present on the matrix side of the membrane. The fatty acid composition of the phospholipids depends on the species, tissue, and diet. In all cases, sufficient unsaturated fatty acids are contained in the

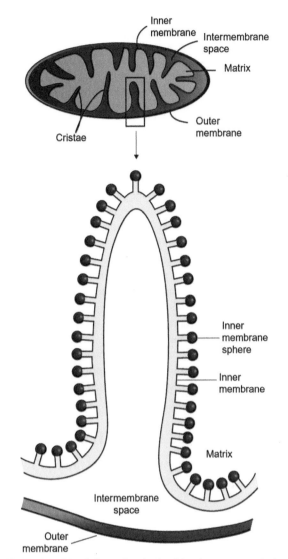

FIGURE 13.1 Morphology of a mitochondrion (transverse section).

phospholipids to provide a highly fluid membrane at physiological temperatures.

The inner membrane is studded with spheres, each $8-10\,nm$ in diameter, which are attached via stalks $4-5\,nm$ in length. These inner membrane spheres are present on the matrix side (M-side) but absent from the cytoplasmic side (C-side). The components of the inner membrane include respiratory chain proteins, a variety of transport molecules, and a part of the ATP-synthesizing apparatus (the base piece of ATP synthase). The ATP—ADP translocase and the ATP synthase together make up at least two-thirds of all protein within the inner mitochondrial membrane. Small, uncharged molecules (e.g., water, oxygen, carbon dioxide, ammonia, and ethanol) can diffuse through the inner membrane, but all other molecules that pass through require specific transport systems. The mitochondrial outer and inner membranes are vastly different in their constituents and function:

1. The outer membrane contains two to three times more phospholipids per unit of protein;
2. Cardiolipin is localized in the inner membrane; and
3. Cholesterol is found predominantly in the outer membrane.

The permeability of the outer membrane to charged or uncharged substances up to a molecular weight of 10,000 is due to pore-like structures that consist of a protein (M.W. 30,000) embedded in the phospholipid bilayer.

The matrix is viscous and contains all TCA cycle enzymes except succinate dehydrogenase, which is a component of electron transport complex II and is located within the inner membrane. Enzymes of the matrix or the inner mitochondrial membrane mediate reactions of fatty acid oxidation; ketone body formation and oxidation; and biosynthesis of urea, heme, pyrimidines, DNA, RNA, and protein. Mitochondria are involved in programmed cell death apoptosis (discussed later).

The structure, mechanism of replication of mitochondrial DNA, and processes of transcription and translation are unique in several respects (discussed later).

Components of the Electron Transport Chain

The electron transport system consists of four discrete enzyme complexes that catalyze the following four reactions:

1. $\text{NADH} + \text{H}^+ + \text{coenzyme Q (Q)} \xrightarrow{\text{NADH-CoQ reductase}}$
   $\text{NAD}^+ + \text{QH}_2$

2. $\text{Succinate} + \text{Q} \xrightarrow{\text{Succinate-CoQ reductase}} \text{fumarate} + \text{QH}_2$

3. $\text{QH}_2 + 2 \text{ ferricytochrome c} \xrightarrow{\text{CoQ cytochrome c reductase}}$
   $\text{Q} + 2 \text{ ferrocytochrome c} + 2\text{H}^+$

4. $2 \text{ Ferrocytochrome c} + 2\text{H}^+ + \frac{1}{2}\text{O}_2 \xrightarrow{\text{cytochrome c oxidase}}$
   $2 \text{ ferricytochrome c} + \text{H}_2\text{O}$

Each complex can be considered as a functional unit composed of a fixed number of electron carriers. The individual components of the four complexes are firmly bonded together and are not dissociated by mild fractionation procedures, whereas the bonds holding unlike complexes are relatively weak and can be dissociated. The functional organization of the four complexes in the inner mitochondrial membrane is shown in Figure 13.2. In addition to these complexes, the F_0/F_1-ATPase, which is required for ATP synthesis, is considered as complex V. The relative ratios of complexes I, II, III, IV, and V have been estimated to be 1:2:3:6:6. Complexes I, II, III, and IV can be combined in the presence of cytochrome c

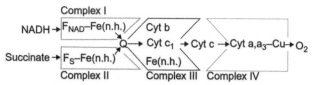

FIGURE 13.2 Diagram of the functional complexes of the electron transport system within the respiratory chain. F_{NAD} = NADH dehydrogenase flavoprotein; F_s = succinate dehydrogenase flavoprotein; Fe(n.h.) = nonheme iron.

(which separates during fractionation) to form a single unit with all of the enzymatic properties of the intact electron transport system except coupled phosphorylation.

The individual electron carriers of the four complexes of the respiratory chain, shown in Figure 13.3, are arranged in accordance with their redox potentials, with the transfer of electrons from NADH to oxygen associated with a potential drop of 1.12 V, and that of succinate to oxygen of 0.8 V. In the electron transport system, the electrons can be transferred as hydride ions $(\text{H:})^-$ or as electrons (e.g., in the cytochromes).

Electron Transport Complexes

Complex I

Complex I catalyzes an NADH-CoQ reductase reaction, and it contains the NADH dehydrogenase flavoprotein. It has two types of electron-carrying structures: FMN and several iron−sulfur centers. FMN is a tightly bound prosthetic group of the dehydrogenase enzyme, and it is reduced to $FMNH_2$ by the two reducing equivalents derived from NADH:

$$\text{NADH} + \text{H}^+ + \text{E} - \text{FMN} \leftrightarrow \text{NAD}^+ + \text{E} - \text{FMNH}_2$$

The electrons from $FMNH_2$ are transferred to the next electron carrier, coenzyme Q, via the iron−sulfur centers of the NADH-CoQ reductase. The iron−sulfur centers consist of iron atoms paired with an equal number of acid-labile sulfur atoms. The respiratory chain iron−sulfur clusters are of the Fe_2S_2 or Fe_4S_4 type. The iron atom, present as nonheme iron, undergoes oxidation−reduction cycles $(Fe^{2+} \leftrightarrow Fe^{3+} + e^-)$. In the Fe_4S_4 complexes, the centers are organized such that iron and sulfur atoms occupy alternate corners of a cube. The four iron atoms are covalently linked via the cysteinyl sulfhydryl groups of the protein (Figure 13.4). The iron−sulfur cluster assembly is mediated by a conserved mitochondrial protein known as **lataxin**. An inherited neurodegenerative disorder **Friedreich ataxia** is caused by lataxin deficiency. Complex I is one of three sites that is involved in the transport of protons from the matrix side to the intermembranal space. Complex I is inhibited

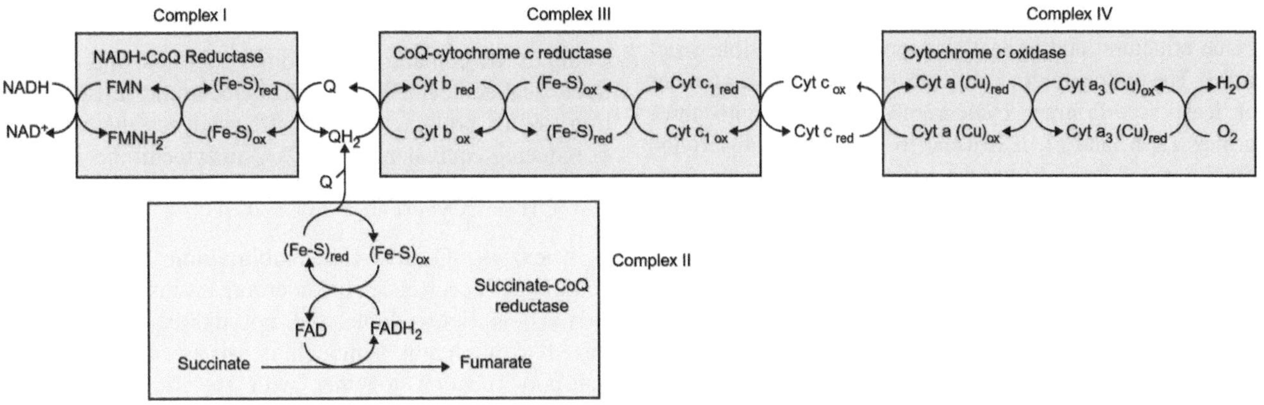

FIGURE 13.3   Mitochondrial electron transport system, ox = oxidized; red = reduced.

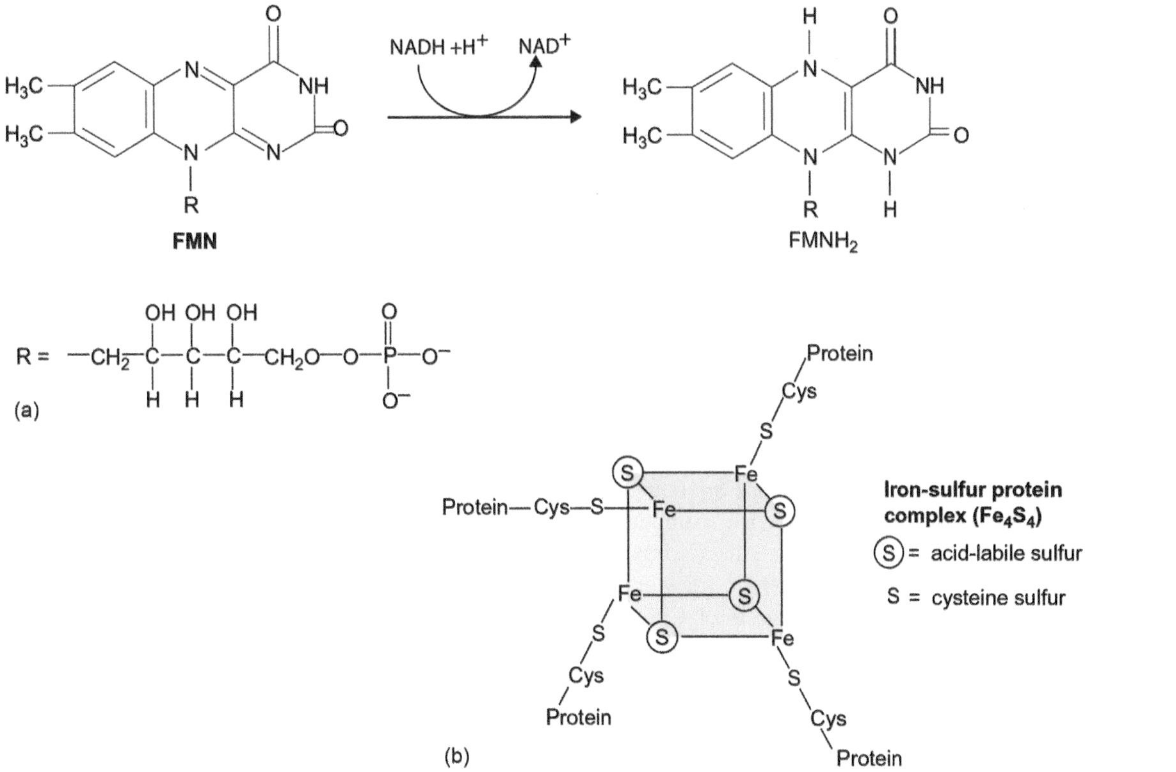

FIGURE 13.4   (a) Transport of reducing equivalents from NADH to FMN and (b) structure of the iron−sulfur protein complex that mediates electron transport from $FMNH_2$ to CoQ. Both FMN and the iron−sulfur centers are components of NADH-CoQ reductase.

by rotenone (a natural toxic plant product), amobarbital (a barbiturate), and piericidin A (an antibiotic), all of which act at specific points and are useful in the study of electron transport. Mutations of nuclear DNA-encoded subunit proteins of complex I cause a number of defects associated with neurodegenerative disorders. The aberrant biochemical effects of complex I include a reduction in functional complex I, accumulation of NADH, elevated levels of reactive oxygen species, and other deleterious effects [3,4].

Complex II

Complex II contains succinate dehydrogenase and its iron−sulfur centers and links succinate oxidation to ubiquinone reduction of electron transport chain. The complex II enzymes have four subunits consisting of a flavoprotein subunit, Fe-S protein subunit, and two integral membrane proteins (Chapter 12). The complex also contains a specific cytochrome, cytochrome b558. **Cytochromes** are heme proteins that undergo oxidation−reduction reactions and are differentiated on the basis of their apoprotein structure, heme structure, and optical absorption in the visible spectrum. The mitochondrial electron transport chain contains at least six different cytochromes classified into three groups (a, b, and c). It is usual to indicate the absorption maximum of the α-band of a particular cytochrome (e.g., cytochrome b558). Succinate dehydrogenase, an FAD-containing enzyme, is part of the TCA cycle and catalyzes the *trans* elimination of two hydrogens from succinate to form fumarate (Chapter 12).

The hydrogens are accepted by FAD, which is covalently bound to the apoprotein via a histidine residue. In many flavoproteins, the flavin nucleotide is bound to the apoprotein not covalently but rather via ionic linkages with the phosphate group. The reducing equivalents of $FADH_2$ are passed on to coenzyme Q (CoQ or Q) via the iron−sulfur centers. Thus, the overall reaction catalyzed by complex II is as follows:

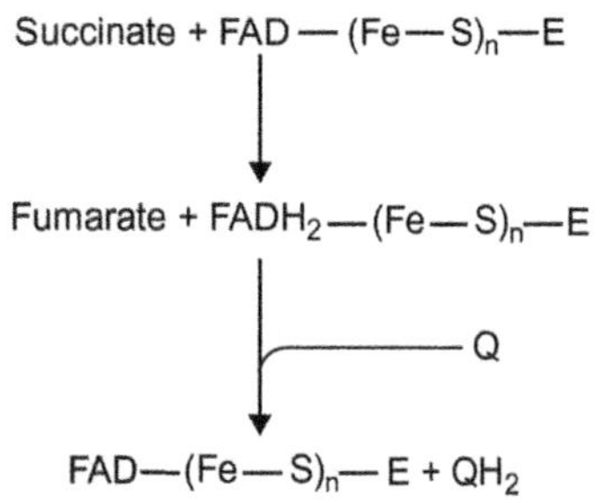

$$Succinate + FAD - (Fe-S)_n - E$$

$$\downarrow$$

$$Fumarate + FADH_2 - (Fe-S)_n - E \quad (13.1)$$

$$\downarrow \qquad Q$$

$$FAD - (Fe-S)_n - E + QH_2$$

During the terminal stages of electron transfer in complex II, cytochrome b558 is involved and provides binding sites for CoQ. Oxaloacetate and malonate are competitive inhibitors of succinate dehydrogenase and compete with the substrate for binding at the active site. Carboxin and thenoyltrifluoroacetone inhibit electron transfer from $FADH_2$ to CoQ. Defects in the synthesis of CoQ in the multistep mevalonate pathway (Chapter 17) due to mutations lead to functionally impaired CoQ and cause multiple-system atrophy [5]. Mutations of the complex II subunit are associated with neurodegeneration and/or tumor formation [6; also see Chapter 12].

Complex III

Complex III contains cytochromes b_{562} and b_{566} (collectively called cytochrome b), cytochrome c_1, and an iron−sulfur protein. Complex III catalyzes the transport of reducing equivalents from CoQ to cytochrome c:

$$QH_2 + 2Cyt\ c(Fe^{3+}) \rightarrow Q + 2Cyt\ c(Fe^{2+}) + 2H^+$$

Coenzyme Q, also called **ubiquinone** because of its ubiquitous occurrence in microorganisms, plants, and animals, is lipid-soluble and not tightly or covalently linked to a protein, although it carries out its electron transport function together with specific CoQ-binding peptides. It plays a central role in the electron transport chain because it collects reducing equivalents from NADH- and $FADH_2$-linked dehydrogenases and passes them on to the terminal cytochrome system. CoQ is a substituted 1,4-benzoquinone containing a polyisoprenoid side chain at C6 (Figure 13.5). In bacteria, CoQ usually contains 6 isoprenoid units (Q6), whereas in most mammalian mitochondria it has 10 (Q_{10}). The reduction of Q to QH_2 (a hydroquinone) requires two electrons and two protons, and probably occurs via a one-electron intermediate as shown here:

$$Q \underset{1e^-,1H^+}{\overset{1e^-,1H^+}{\rightleftharpoons}} QH^\cdot \underset{1e^-,1H^+}{\overset{1e^-,1H^+}{\rightleftharpoons}} QH_2 \quad (13.2)$$

where the dot associated with QH represents an unpaired electron (a free radical). Similarly to complex I, complex III also translocates protons from the matrix side to intermembranal space. Antimycin A (a *Streptomyces* antibiotic) inhibits the transfer of electrons from QH_2 to cytochrome c.

Cytochrome c transfers electrons from complex III to complex IV, and cytochromes a and a_3 transfer electrons to oxygen in complex IV. The structure of the heme prosthetic group (iron−protoporphyrin IX) in cytochromes b, c, and c_1 is the same as that present in hemoglobin and myoglobin but differs from the heme group (heme A) of cytochromes a and a_3 (Figure 13.6). The heme groups in

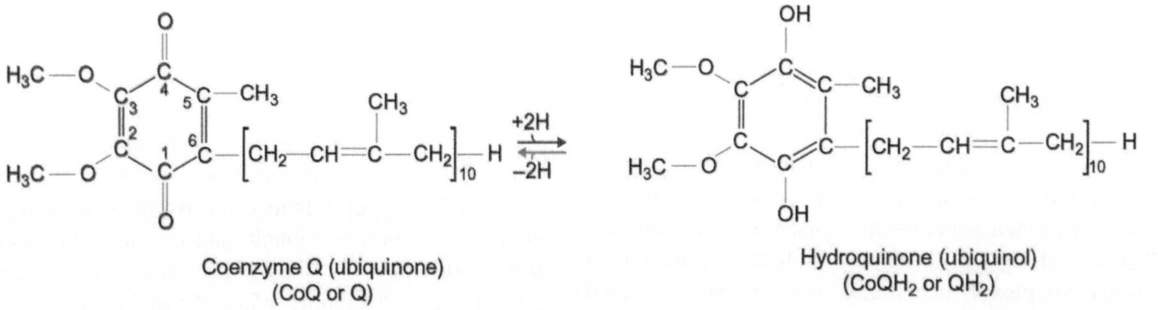

Coenzyme Q (ubiquinone)
(CoQ or Q)

Hydroquinone (ubiquinol)
(CoQH₂ or QH₂)

FIGURE 13.5 Structure and redox reaction of coenzyme Q. In most mammalian tissues, CoQ has 10 isoprenoid units. CoQ collects reducing equivalents from NADH dehydrogenase and from other flavin-linked dehydrogenases.

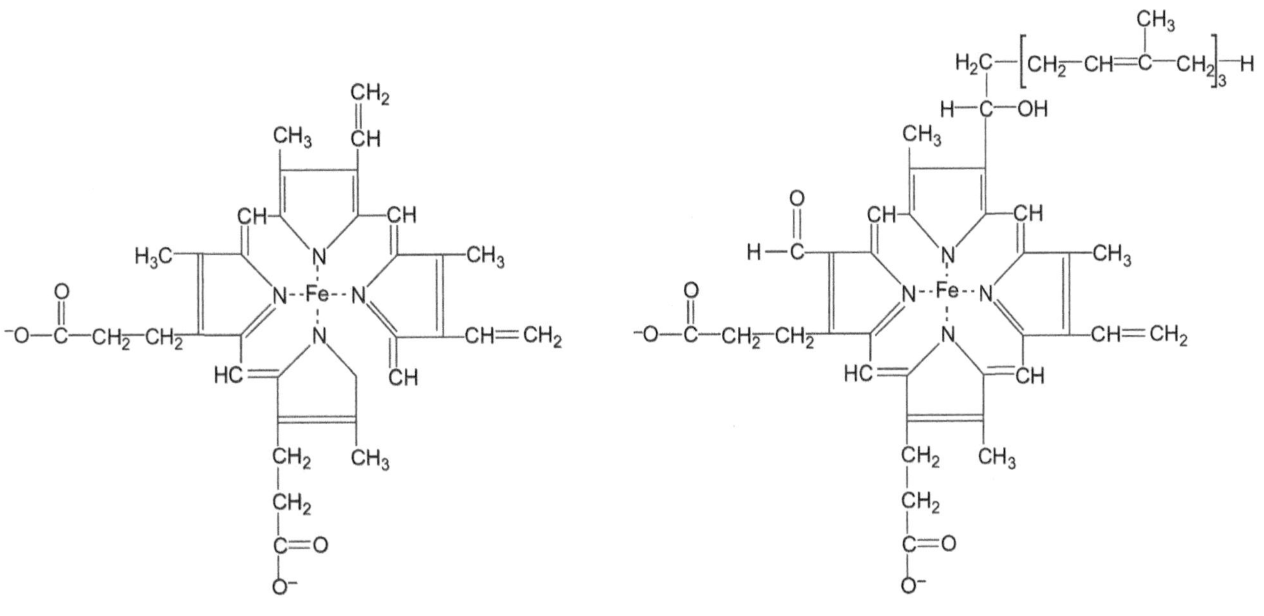

Heme
(iron-protoporphyrin IX)

Heme A

FIGURE 13.6 Structure of heme (present in cytochromes b, c, and c_1) and of heme A (present in cytochromes a and a_3).

cytochromes c and c_1 are covalently linked to the apoprotein by thioether bonds between sulfhydryl groups of two cysteine residues and the vinyl groups of the heme. The heme of cytochromes b and a_3 is bound by strong hydrophobic interactions between the heme and the apoprotein.

Cytochromes b, c_1, a, and a_3 are integral membrane proteins, whereas cytochrome c is a peripheral protein located on the C-side of the membrane and is easily isolated from mitochondria.

Complex IV

Complex IV, also called **cytochrome c oxidase**, is the terminal component of the respiratory chain. It consists of four redox centers: cytochrome a, cytochrome a_3, and two Cu ions. Like the heme iron atoms, the copper ions function as one-electron carriers:

$$Cu^{2+} + e^- \leftrightarrow Cu^+$$

Cytochrome c transfers electrons from cytochrome c_1, the terminal component of complex III, to the four redox centers of the cytochrome oxidase complex. The transfer of four electrons from each of the four redox centers of the cytochrome oxidase complex to an oxygen molecule occurs in a concerted manner to yield two molecules of water:

$$4e^- + O_2 + 4H^+ \rightarrow 2H_2O$$

Accompanying the preceding reactions, cytochrome c oxidase also drives the transport of protons from the matrix side to intermembranal space. Cyanide, carbon monoxide, and azide inhibit cytochrome oxidase. Treatment for cyanide toxicity with administration of nitrite and thiosulfate is discussed in Chapter 6.

Formation of Reactive Oxygen Species

More than 90% of metabolic oxygen is consumed in the cytochrome oxidase reaction. Oxygen contains an unconventional distribution of its two valence electrons. These two electrons occupy different orbitals and are not spin paired; thus, oxygen is a diradical. The reduction of an oxygen molecule with fewer than four electrons results in the formation of an active oxygen species. One electron transfer yields superoxide radical (O_2^-) and the two-electron transfer yields hydrogen peroxide (H_2O_2). Hydroxyl free radical ($HO^\bullet$) formation can take place from hydrogen peroxide in the presence of ferrous iron or cuprous chelates. Both O_2^- and $HO^\bullet$ free radicals are cytotoxic oxidants. In the mitochondrial electron transport system, leakage of electrons at any one of the redox centers due to aging or pathological conditions results in the formation of superoxide. Antioxidant enzymes, namely superoxide dismutases (SOD), catalase, and glutathione peroxidase, participate in the elimination of toxic oxygen metabolites. Three different SODs are present in human cells; they are located in mitochondria, cytosol, and extracellular fluid. The importance of mitochondrial SOD (labeled as SOD2), which is a manganese-containing enzyme, is exemplified in the homozygous SOD2 knockout mice. These SOD2 knockout mice have low birth weights, and they die shortly after their birth from dilated cardiomyopathy.

Redox reactions are a required part of normal metabolism. In neutrophils, for example, the killing of the invading microorganisms requires the formation of reactive oxygen metabolites (discussed later).

Oxidants are also involved in gene expression (e.g., the variety of protein kinases) and in the regulation of redox homeostasis. Perturbation of redox homeostasis causes oxidative stress and may contribute to chronic inflammatory diseases and malignancy.

Organization of the Electron Transport Chain

The arrangement of components of the electron transport chain was deduced experimentally. Since electrons pass only from electronegative systems to electropositive systems, the carriers participate according to their standard redox potential.

Complexes I–IV of the respiratory chain are organized asymmetrically in the inner membrane (Figure 13.7) as follows:

1. The flavin prosthetic groups of NADH dehydrogenase and succinate dehydrogenase face the M-side of the membrane.

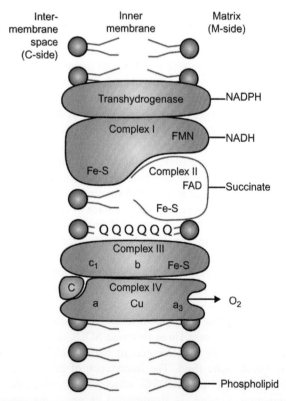

FIGURE 13.7 Orientation of the components of the electron transport complexes within the inner mitochondrial membrane. Fe-S = Iron-sulfur center; b, c, c_1, a, and a_3 = cytochromes; Cu = copper ion.

2. CoQ and cytochrome b of complex III are probably inaccessible from either side of the membrane.
3. Cytochrome c interacts with cytochrome c_1 and cytochrome a, all located on the C-side.
4. Complex IV spans the membrane, with cytochrome a oriented toward the C-side; copper ions and cytochrome a_3 are oriented toward the M-side.
5. Nicotinamide nucleotide transhydrogenase, which catalyzes the reaction $NADPH + NAD^+ \leftrightarrow NADH + NADP^+$, spans the membrane, but its catalytic site faces the M-side.

The anisotropic organization of electron carriers across the membrane accounts for the vectorial transport of protons from the inside to the outside of the membrane, which occurs with the passage of electrons. The coupling of this proton gradient to a proton-translocating ATP synthase (also known as **ATP synthetase**) accounts for the chemi-osmotic coupling in oxidative phosphorylation.

OXIDATIVE PHOSPHORYLATION

ATP is synthesized from ADP and phosphate during electron transport in the respiratory chain. This type of phosphorylation is distinguished from substrate-level phosphorylation, which occurs as an integral part of specific reactions in glycolysis and the TCA cycle. The free energy available for the synthesis of ATP during electron transfer from NADH to oxygen can be calculated from the difference in the value of the standard potential of the electron donor system and that of the electron acceptor system. The standard potential of the $NADH/NAD^+$ redox component is -0.32 V and that of $H_2O/\frac{1}{2}O_2$ is $+0.82$ V; therefore, the standard potential difference between them is as follows:

$$\Delta E^0 = (\text{electron acceptor} - \text{electron donor}) =$$
$$0.82 - (-0.32) = +1.14 \text{ V}$$

The standard free energy is calculated from the expression

$$\Delta G^{0\prime} = -n\text{F} \, \Delta E^{0\prime}$$

where n is the number of electrons transferred, and F is the **Faraday constant**, which is equal to 23,062 cal/eV. Thus, the standard free energy for a two-electron transfer from NADH to molecular oxygen is

$$\Delta G^0 = -n\text{F} \, \Delta E^{0\prime} = -2 \times 23,062 \times 1.14$$
$$= -52.6 \text{ kcal/mol}(-220 \text{ kJ/mol})$$

Although 52.6 kcal of free energy is available from the reaction, only 21.9 kcal is conserved in the formation of phosphoanhydride bonds at ATP. Formation of each phosphoanhydride bond requires 7.3 kcal (30.5 kJ), and

21.9 kcal accounts for three ATPs synthesized. The remainder of the energy is assumed to be dissipated as heat. In the mitochondria of brown adipose tissue, very little ATP is synthesized, and most of the energy liberated in the electron transport system is converted to heat.

The conservation of energy in the electron transport chain occurs at three sites coupled to proton transport from the matrix to the intermembrane space, where there is a large decrease in free energy: between NADH and FMN; between cytochromes b and c_1; and between cytochromes $(a + a_3)$ and molecular oxygen. Because electrons from succinate oxidation to fumarate bypass complex I and enter at the CoQ level, only two moles of ATP are synthesized per mole of succinate. The number of ATP molecules synthesized depends on where the reducing equivalents enter the respiratory chain, and is indicated by the ratio (or quotient) of phosphate esterified (or ATP produced) to oxygen consumed (P/O or ATP/O) per two-electron transport. The NAD^+-linked substrates (e.g., malate, pyruvate, α-ketoglutarate, isocitrate, β-hydroxybutyrate, and β-hydroxyacyl-CoA) have P/O values of 3, and some flavin-linked substrates (e.g., succinate, glycerol phosphate, and fatty acyl-CoA) have P/O ratios of 2. The complete oxidation of 1 mol of glucose yields either 36 or 38 mol of ATP, depending on which shuttle pathway is used in the transport of cytoplasmic NADH to mitochondria. A list of the energy-yielding reactions in glucose oxidation is shown in Table 13.1.

TABLE 13.1 Energy-Yielding Reactions in the Complete Oxidation of Glucose

Reaction	Net Moles of ATP Generated per Mole of Glucose
Glycolysis (phosphoglycerate kinase, pyruvate kinase; two ATPs are expended)	2
NADH shuttle glycerol–phosphate shuttle (or malate–aspartate shuttle)	4 (6)
Pyruvate dehydrogenase (NADH)	6
Succinyl CoA synthetase (GTP is equivalent to ATP)	2
Succinate dehydrogenase (succinate → fumarate + FADH₂)	4
Other TCA cycle reactions (isocitrate → α-ketoglutarate, α-ketoglutarate → succinyl CoA, malate → oxaloacetate; total of 3 NADH generated)	18
Total	36 (38)

Mechanisms of Oxidative Phosphorylation

According to the **chemiosmotic hypothesis**, developed by Peter Mitchell, an electrochemical gradient (pH gradient) generated across the inner mitochondrial membrane by the passage of reducing equivalents along the respiratory chain provides the driving force for the synthesis of ATP. There are three prerequisites for achieving oxidative phosphorylation according to this hypothesis:

1. An anisotropic (direction-oriented) proton-translocating respiratory chain capable of vectorial transport of protons across the membrane;

2. A coupling membrane impermeable to ions except via specific transport systems; and

3. An anisotropic ATP synthase whose catalytic activity is driven by an electrochemical potential.

The transport of reducing equivalents in the respiratory chain generates a proton gradient across the membrane by virtue of the specific vectorial arrangement of the redox components within the inner mitochondrial membrane. The proton gradient is generated by ejection of protons from the matrix into the intermembrane space during proton-absorbing reactions, which occur on the M-side of the inner membrane, and proton-yielding reactions, which occur on the C-side, to form redox loops.

According to the chemiosmotic hypothesis, ejection of two or more protons occurs at each of three sites in complexes I, III, and IV. Thus, in the transfer of two reducing equivalents from NADH to oxygen, about $6-10$ protons (2×3) are ejected. In respiring mitochondria, the intramembrane space is more acidic than the matrix space by about 1.4 pH units, and the transmembrane potential is about $0.180-0.220$ V. Thus, the basic premise of the chemiosmotic hypothesis is

$$\text{Transport of reducing equivalents} \rightarrow \Delta\tilde{\mu}_{H^+} \rightarrow \text{ATP synthesis}$$

The proton-motive force drives ATP synthesis via the re-entry of protons. The ATP synthase (F_0/F_1 complex), which consists of three parts, is driven by this vectorial transfer of protons from the intramembrane space into the matrix. ATP synthesis occurs in the inner membrane spheres (Figure 13.8).

Uncoupling Agents of Oxidative Phosphorylation

Uncoupling agents dissociate ATP synthesis and other energy-dependent membrane functions from the transport of reducing equivalents in the respiratory chain by abolishing the proton gradient. Normally, these processes are tightly coupled. These uncoupling agents cause a several-fold stimulation of respiration with unimpeded

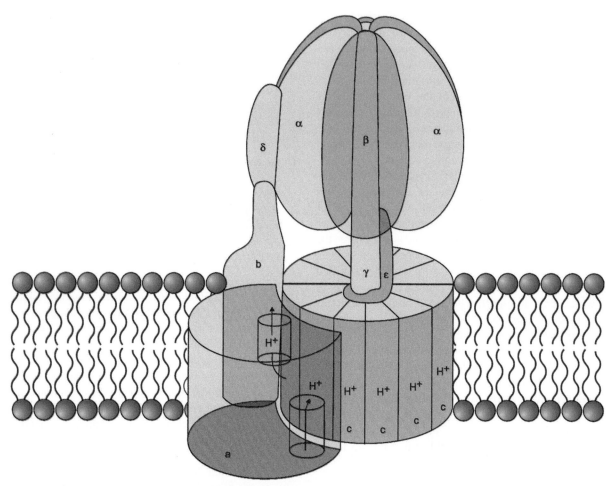

FIGURE 13.8 Structure of ATP synthase. The F_1 complex situated above the membrane on the matrix side consists of three $\alpha\beta$ dimers and single γ, δ, and ε subunits. The membrane segment F_0, also known as a stalk, consists of H^+ channels. Protons are conducted via the c-subunit channels. The rotation of c-subunits relative to a subunit of the stalk drives the rotation of the γ-subunit. *[Reproduced with permission from Y. Zhou, T. M. Duncan, R. L. Crass, Subunit rotation in Escherichia coli F_0F_1-ATP synthase during oxidative phosphorylation. Proc. Natl Acad. Sci. USA 94, (1997) 10583].*

utilization of substrate and dissipation of energy as heat. A classic example of an uncoupling agent is 2,4-dinitrophenol. Endogenous uncoupling proteins known as **thermogenin** found in the mitochondrial inner membrane dissipate proton gradient by creating passive proton channels (discussed later).

Mitochondrial Energy States

In normal mitochondria in the presence of nonlimiting amounts of respiratory chain components, the rate of oxygen consumption is affected by changes in the NADH/NAD$^+$ ratio, the phosphorylation ratio (or phosphate potential, [ATP]/[ADP][P$_i$]), and pO_2. In resting (unstimulated) mitochondria, in the presence of nonlimiting concentrations of substrate (reductant), oxygen, and phosphate, respiratory activity is controlled by the availability of ADP. In this state, the rate of endogenous energy dissipation is low and is regulated by ADP. Dinitrophenol abolishes respiratory control because it uncouples the transport of reducing equivalents in the respiratory chain from the phosphorylation of ADP.

In the oxidation of carbohydrates, energy is conserved at all three stages: glycolysis, TCA cycle, and oxidative phosphorylation. Each of these stages is regulated at specific sites, and all three are coordinated in such a manner that ATP is synthesized only to the extent that it is needed in the cell. For example, when the [ATP]/[ADP] ratio is high, glycolysis, TCA cycle, and oxidative phosphorylation are inhibited. The key regulatory enzyme, 6-phosphofructokinase, is activated by ADP and inhibited by ATP (see also Chapters 12, 14, and 20).

Energy-Linked Functions of Mitochondria Other Than ATP Synthesis

The energy derived from the respiratory chain, although primarily coupled to the formation of ATP, is also utilized for purposes such as heat production, ion transport, and the transhydrogenase reaction.

Brown adipose tissue is responsible for nonshivering thermogenesis, which is important in the arousal of hibernating animals and for maintaining body temperature in hairless neonates of mammals, including humans. Diet-induced thermogenesis also occurs in brown adipose tissue (Chapter 11). Brown adipose tissue, in contrast to white adipose tissue, consists of small heat-generating adipocytes that are rich in mitochondria and lipid dispersed in the cytoplasm as distinct droplets. Mitochondria of brown adipose tissue are naturally uncoupled, so oxidation of substrates generates heat rather than a proton-motive force. Stimulation of thermogenesis in brown adipose tissue is initiated by stimulation of the sympathetic nervous system, which releases norepinephrine at nerve endings. Norepinephrine combines with the β-adrenergic receptors of the brown adipose tissue cells and initiates cAMP-dependent activation of triacyl-glycerol lipase (Chapter 20), leading to the elevated intracellular concentrations of fatty acids that are oxidized in the mitochondria. Promoting the metabolic activities of brown adipocytes has been a therapeutic target for treating obesity and related metabolic diseases [7]. Fatty acids are also thought to activate uncoupling protein 1 (UCP1). Two other uncoupling proteins, UCP2 and UCP3, possess metabolic roles, UCP2 in the regulation of insulin secretion from the pancreatic β-cells (Chapter 20) and UCP3 in the thermogenic effect of tetraiodothyronine (T_3, Chapter 31).

A transhydrogenase catalyzes the transfer of reducing equivalents from NADH to $NADP^+$ to form NADPH by utilizing energy captured by the energy conservation mechanisms of the respiratory chain without the participation of the phosphorylation system. Similarly, ion translocation occurs at the expense of energy derived from substrate oxidation in the respiratory chain. For example, Ca^{2+} is transported from the cytoplasm to the mitochondrial matrix, through the inner mitochondrial membrane, at the expense of proton-motive force. The Ca^{2+} efflux from mitochondria is regulated so that levels of cytoplasmic Ca^{2+} that are optimal for function are achieved. Increased cytoplasmic Ca^{2+} levels initiate or promote muscle contraction (Chapter 19), glycogen breakdown (Chapter 14), and oxidation of pyruvate (Chapter 12). Decreased levels of Ca^{2+} have the opposite effect.

Translocation systems of the inner mitochondrial membrane are responsible for electroneutral movement of dicarboxylates, tricarboxylates, α-ketoglutarate, glutamate, pyruvate, and inorganic phosphate. Specific electrogenic translocator systems exchange ATP for ADP, and glutamate for aspartate, across the membrane. The metabolic function of the translocators is to provide appropriate substrates (e.g., pyruvate and fatty acids) for mitochondrial oxidation that is coupled to ATP synthesis from ADP and P_i.

The ATP/ADP translocator, also called **adenine nucleotide translocase**, plays a vital role in the metabolism of aerobic cells because mitochondrial ATP is primarily consumed outside the mitochondria to support biosynthetic reactions. The translocator is the most abundant protein in the mitochondrion; two molecules per unit of respiratory chain are present. The translocator consists of two identical hydrophobic peptides. The export of ATP^{4-} is necessarily linked to the uptake of ADP^{3-}. This transport system is electrogenic owing to the transport of nucleotides of unequal charge, the driving force being the membrane potential. Although made up of two identical subunits, the translocator is asymmetrical in its orientation: on the C-side it binds with ADP; and on the M-side it binds with ATP. This asymmetry of nucleotide transport has been demonstrated by use of the inhibitors **atractyloside** and **bongkrekic acid**. Atractyloside is a glycoside found in the rhizomes of a Mediterranean thistle; bongkrekic acid is a branched-chain unsaturated fatty acid synthesized by a fungus found in decaying coconut meat. The former inhibitor binds to the C-side of the translocator at the ADP site, whereas the latter binds to the M-side of the translocator at the ATP site.

Transport of Cytoplasmic NADH to Mitochondria

NADH generated in the cytoplasm during glycolysis must be transported into the mitochondria if it is to be oxidized in the respiratory chain. However, the inner mitochondrial membrane is not only impermeable to NADH and NAD^+; it also contains no transport systems for these substances. Impermeability to NADH is overcome by indirect transfer of the reducing equivalents through shuttling substrates that undergo oxidation−reduction reactions. This process consists of the following steps: a reaction in which NADH reduces a substrate in the cytoplasm, transport of the reduced substrate into mitochondria, and oxidation of the reduced substrate in the respiratory chain. Two pathways that transport reducing equivalents from NADH into mitochondria have been characterized and are known as the **glycerol−phosphate shuttle** and the **malate−aspartate shuttle**.

Role of Mitochondria in the Initiation of Programmed Cell Death Known as Apoptosis

Cell death occurs by either **necrosis** or **apoptosis**. Necrosis is a pathologic process (e.g., trauma, ischemia, infection), initiated by external signals, whereas apoptosis is part of both normal development and disease. Regulation of cell death and cell proliferation is an essential part of homeostasis. There are many proapoptotic signals, and they can be either extracellular or intracellular. The Bcl-2 family of proteins (Bcl is for B-cell lymphoma/leukemia), free radicals, and elevated levels of intracellular Ca^{2+} are examples of proapoptotic signals. The Bcl-2 family of proteins consists of both pro- and anti-apoptotic signals, and they maintain homeostatic balance between cell death and survival. Dysregulation in this process can cause carcinogenesis and neurodegenerative disease.

The central pathway of apoptosis begins with the attack of proapoptotic signals on the mitochondria. These insults result in the opening of the mitochondrial permeability-transition pores, with the release of cytochrome c and other mediators of cell death into the cytosol. Cytosolic cytochrome c initiates the apoptotic pathway by interaction with the multidomain seven-subunit protein known as **apoptotic protease-activating factor 1 (APaf-1)**. APaf-1—cytochrome c complex and proteolytic enzyme procaspase 9 form the apoptosome, which, in the presence of ATP, activates procaspase 9 to caspase 9. This initiates a cascade of activation of execution caspases, all of which leads to cytoskeleton catabolism, genomic degradation, nuclear condensation, cytoplasmic blebbing, cell contraction, and ultimately to cell death (Figure 13.9). The dead cells and their fragments contain marker molecules that are recognized by the surrounding phagocytes, which facilitate their removal. Caspases that have cysteine at the active site and cause proteolysis of the target protein with aspartic acid specificity have a pivotal role in apoptosis. The execution phase of caspases can also be activated by many other stimuli in addition to cytochrome c. The therapeutic agent **cyclosporine**, an immunosuppressive drug used in organ transplantation, inhibits the mitochondrial transition pore causing inhibition of cytochrome c release to cytosol. This effect of cyclosporine has been used in preventing the detrimental consequences of myocardial reperfusion injury.

THE MITOCHONDRIAL GENOME

The human mitochondrion contains its own genome: a circular double-stranded DNA molecule of 16,569 base pairs.

A typical human cell contains many mitochondria with each mitochondrion having more than one mtDNA.

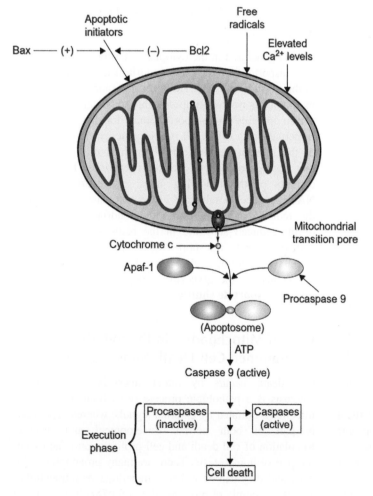

FIGURE 13.9 Schematic representation of cytochrome c initiated pathway of apoptosis. (+) = stimulation; (−) = inhibition (see text for details).

Cytoplasmic location and multiple mtDNAs give mitochondria special characteristics:

- The mtDNA is transmitted through the oocyte cytoplasm and is maternally inherited.
- Mitochondrial DNA encodes for only a limited number of proteins, 13 of which are required for OXPHOS. Thus, nuclear DNA-mediated cytoplasmic synthesis of proteins is required for mitochondrial function and regulation. Mitochondrial ribosomes (mitoribosomes) differ from both bacterial and cytoplasmic ribosomes. The mitochondrial ribosome is composed of 28S and 39S subunit particles with a composite sedimentation coefficient of 55S, whereas the bacterial ribosome and cytoplasmic ribosome have sedimentation coefficients of 70S and 80S, respectively (see Chapter 23).
- The insertion of mtDNA-encoded proteins synthesized in the mitoribosomes into the functional location requires participation of specific proteins [8].
- The occurrence of an mtDNA mutation creates a mixed intracellular population of mutant and nonmutant (original) molecules (heteroplasmy). As the heteroplasmic cells divide during mitosis or meiosis, the mutant and original mtDNAs are randomly distributed to the daughter cells. This results in drifting of the mitochondrial genotype. The replicative segregation ultimately results in cells with either mutant or normal mtDNAs (homoplasmy).
- Severe mtDNA defects reduce cellular energy outputs until they decline below the minimum energy level (energetic threshold) for normal tissue function. Such energetic thresholds differ among tissues, with the brain, heart, muscle, kidney, and endocrine organs being most reliant on mitochondrial energy.
- The mtDNA has a very high mutation rate, some 10–17 times higher than that of nDNA, which affects both germline and somatic tissue mtDNAs. Germline mutations result in maternally transmitted diseases or predispose individuals to late-onset degenerative diseases. Somatic mutations accumulate in somatic tissues and exacerbate inherited oxidative phosphorylation defects. The high mtDNA mutation rate may be due to the high concentration of oxygen radicals at the mitochondrial inner membrane, the lack of efficient mtDNA repair mechanisms, and/or the absence of histones.
- Cellular oxygen utilization decreases as a function of age, and the ATP-generating capacity of the cell is a function of age. The attenuation of ATP synthesis is associated with an age-related increase in somatic mtDNA damage in postmitotic tissue. The "normal" decline in ATP-generating capability may facilitate disease occurrence when it is associated with an inherited oxidative phosphorylation mutation.

NUCLEAR CONTROL OF RESPIRATORY CHAIN EXPRESSION

Nuclear genes contribute the majority of respiratory subunits and all of the proteins required for mtDNA transcription, translation, and replication. It is estimated that more than 80% of the genes encoding the subunits of the respiratory chain are located in the cell nucleus.

MITOCHONDRIAL DISEASES

The bioenergetic defects resulting from mtDNA mutations may be a common cause of degenerative diseases (Table 13.2). Defects in nuclear–cytoplasmic interaction are generally the result of autosomal dominant mutations; complex disease states result from depletion of mtDNAs from tissues. More than 200 mitochondrial DNA mutations are associated with a broad spectrum of chronic degenerative diseases with a variety of clinical presentations [9–11]. MtDNA mutations are detected in about 1 in 200 live births. However, symptoms due to these mutations are found in about 1 out of 10,000 adults. Identical mutations are associated with very different phenotypes, and the same phenotype can be associated with different mutations. Therapeutic measures to ameliorate

TABLE 13.2 Correlations Found between Mitochondrial DNA Mutations and Human Diseases*

Clinical Features	
A. Nucleotide Substitutions	
1. Mildly deleterious base substitutions	Familial deafness, Alzheimer's disease, Parkinson's disease
2. Moderately deleterious nucleotide substitutions	Leber's hereditary optic neuropathy (LHON), myoclonic epilepsy and ragged-red fiber disease (MERRF)
3. Severe nucleotide substitutions	Leigh's syndrome, dystonia
B. (mt)DNA Rearrangements	
1. Milder rearrangements (duplications)	Maternally inherited adult-onset diabetes and deafness
2. Severe rearrangements (deletions)	Adult-onset, chronic progressive external ophthalmoplegia (CPEO), Kearns-Sayre syndrome (KSS), lethal childhood disorders, Pearson's marrow/pancreas syndrome

*Modified from D. C. Wallace, Mitochondrial DNA mutations in diseases of energy metabolism, J. Bioenergetics Biomembranes, 26 (1994), 241–250.

mtDNA-mediated diseases are primarily palliative and supportive of control of symptoms. However, a novel and a potential method of prevention of transmission of mtDNA-mediated diseases consists of the following steps [12]: (1) The nucleus from an oocyte of the mtDNA mutation-containing mother is retrieved at any early developmental stage. (2) The retrieved nucleus is then inserted into an anucleated egg obtained from a healthy female donor. This procedure results in an oocyte containing normal donor mitochondria with the affected mother's nucleus. (3) *In vitro* fertilization and implantation of procedure can result in a normal embryo without any mtDNA mutations.

OTHER REDUCING-EQUIVALENT TRANSPORT AND OXYGEN-CONSUMING SYSTEMS

In most cells, more than 90% of the oxygen utilized is consumed in the respiratory chain that is coupled to the production of ATP. However, electron transport and oxygen utilization occur in a variety of other reactions, including those catalyzed by oxidases or oxygenases. Xanthine oxidase, an enzyme involved in purine catabolism (Chapter 25), catalyzes the oxidation of hypoxanthine to xanthine, and of xanthine to uric acid. In these reactions, reducing equivalents are transferred via FAD, Fe^{3+}, and Mo^{6+}, while the oxygen is converted to superoxide anion (O_2^-):

$$\text{Hypoxanthine} \xrightarrow[\text{H_2O,O_2}]{\text{H^+,O_2^-}} \text{Xanthine} \xrightarrow[\text{H_2O,O_2}]{\text{H^+,O_2^-}} \text{Uric acid} \quad (13.3)$$

D-amino acid oxidase is a flavoprotein located in peroxisomes. D-amino acid oxidases catalyze the oxidation of a D-amino acid to the corresponding keto acid:

$$\underset{\text{D-Amino acid}}{\text{H}-\overset{\text{COO}^-}{\underset{\text{R}}{\text{C}}}-\text{NH}_3^+ + \text{H}_2\text{O}} \xrightarrow[\text{H_2O_2}]{\text{O_2}} \underset{\text{Keto acid}}{\text{R}-\overset{\text{O}}{\text{C}}-\text{COO}^- + \text{NH}_4^+} \quad (13.4)$$

The O_2^- and H_2O_2 produced in the preceding two reactions are potentially toxic. Superoxide anion is also produced by oxygen-reducing enzymes of phagocytes (neutrophils, eosinophils, and mononuclear phagocytes) that defend the host against invading organisms by producing reactive oxidants from oxygen. The reducing equivalents are provided by NADPH derived from the hexose monophosphate shunt (Chapter 14).

Oxygen Consumption Linked to Production of Microbicidal Oxygen Species in Phagocytes

Neutrophils, monocytes, and macrophages are equipped with NADPH oxidase systems, which are essential in eradicating bacterial infections. Neutrophils are the most active participants in **phagocytosis**. Unstimulated neutrophils circulate in the bloodstream with a lifespan of a few days. When bacterial infection occurs, neutrophils actively migrate to the infected site, where they kill bacteria by the process of phagocytosis. During phagocytosis, a large amount of oxygen is consumed by the neutrophils in a reaction termed the **respiratory burst**. The burst was initially thought to represent an energy requirement for phagocytosis; however, it was subsequently found to be insensitive to inhibitors of the mitochondrial respiratory chain (cyanide, antimycin) and associated with an increased turnover of the hexose monophosphate shunt (Chapter 14).

The oxygen consumed during phagocytosis is utilized by a unique enzyme system termed the **respiratory burst oxidase** or NADPH oxidase. The oxidase generates superoxide anion (O_2^-), a one-electron reduced species, driven by intracellular NADPH:

$$\text{NADPH} + 2O_2 \rightarrow \text{NADP}^+ + 2O_2^- + \text{H}^+$$

Superoxide anion is also the source of a number of other microbicidal oxidants. It functions as an electron donor and is converted to hydrogen peroxide by superoxide dismutase:

$$2O_2^- + 2H^+ \rightarrow H_2O_2 + O_2$$

A number of highly reactive oxidizing agents are produced from H_2O_2 and O_2^-, such as the conjugate acid of superoxide, the hydroperoxy radical ($HO_2^\bullet$), hydroxyl radical ($OH^\bullet$), and hypochlorite (OCl^-). These reactive agents are powerful microbicides:

$$O_2^- + H_2O_2 \rightarrow OH^\bullet + OH^- + O_2$$

and

$$Cl^- + H_2O_2 \rightarrow OCl^- + H_2O$$

The former reaction is promoted by metal ions (e.g., Fe^{2+}), and the latter is catalyzed by myeloperoxidase present in neutrophil granules. By acquiring a single electron, oxygen can give rise to a variety of toxic products. For example, tissue destruction is enhanced when X-ray treatment is used in conjunction with hyperbaric oxygen.

The NADPH oxidase consists of six components. In the unstimulated cell, two of the components are membrane-bound (p22^{phox} and gp91^{phox}), and the small GTPase (Rac) and three components are present in cytosol (p47^{phox}, p67^{phox}, and p40^{phox}). The designations gp, phox, and p represent glycoprotein, phagocytic oxidase, and protein, respectively. The membrane-bound components occur as a heterodimeric flavo-hemocytochrome b.

The NADPH oxidase complex is dormant in resting phagocytes and becomes assembled and activated for superoxide formation upon bacterial invasion (Figure 13.10). The respiratory burst is stimulated *in vitro* as well as *in vivo* by a variety of reagents, among which are phorbol esters (PMA, phorbol 12-myristate 13-acetate), heat-aggregated IgG, unsaturated fatty acids, and analogues of bacterial peptides (FMLP, formylmethionyl-leucyl phenylalanine).

The importance of NADPH oxidase is highlighted by **chronic granulomatous disease** (**CGD**), a rare inherited disease in which affected individuals are incapable of mounting a sustained **respiratory burst**. As a consequence, people with CGD are incapable of generating the reactive oxygen compounds necessary for the intracellular killing of phagocytosed microorganisms. The CGD patient suffers from severe and recurrent bacterial and fungal infections. Studies using CGD patients' neutrophils as well as a cell-free system have helped identify the cellular components responsible for O_2^- generation.

Four different mutations give rise to CGD, and these cause defects in different components of the NADPH oxidase: the two membrane-bound components of flavo cytochrome b and two cytosolic factors p47^{phox} and p67^{phox}. The loci for the four proteins have been mapped in chromosomes X, 16, 7, and 1, respectively. CGD patients are therefore classified as being either of an X-linked type or autosomal recessive. More than 70% of CGD cases are X-linked disorders due to mutations in gp91^{phox} gene. All reported cases of CGD may be explained by abnormalities in the genes coding for one or more of these four

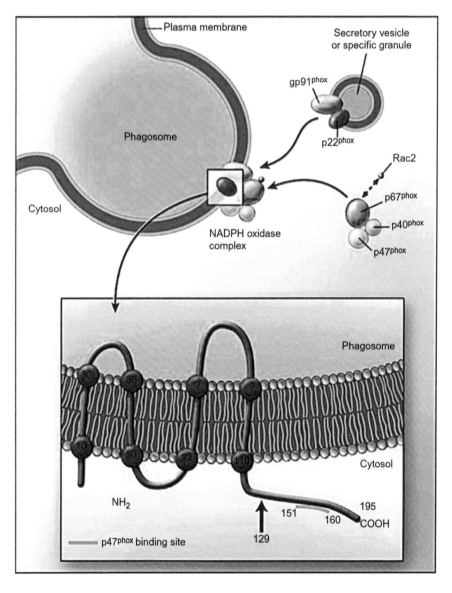

FIGURE 13.10 The assembly of the NADPH oxidase system. Neutrophils (phagocytes) upon ingestion of pathogens form phagosomes, and the dormant NADPH oxidase is converted to an active state by assembling membrane and cytosolic bound components (see text for details). *[Reprinted with permission from J. B. Harris, I. C. Michelow, S. J. Westra, R. L. Kradin. Case 21-2008: an 11-month-old boy with fever and pulmonary infiltrates. New Engl. J. Med. 359, (2008) 178].*

components of the oxidase complex. Treatment for CGD patients consists of prophylaxis with antimicrobial interferon-γ chemotherapy (Chapter 33) and granulocyte transfusion (see Clinical Case Study 13.2).

Protective Mechanisms for Oxygen Toxicity

Unsaturated fatty acid residues of membrane lipids form α-hydroperoxyalkenes by reacting with oxygen:

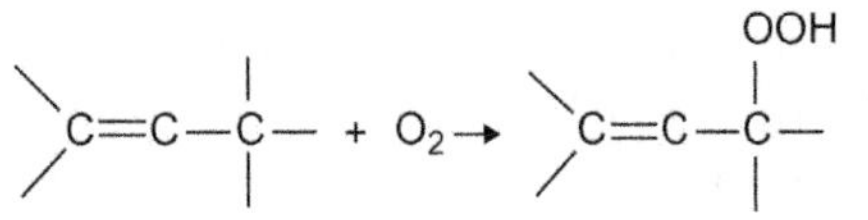

$$\begin{array}{c}\\ \diagup\!\!\!\diagup C{=}C{-}\underset{\displaystyle |}{\overset{\displaystyle |}{C}}{-} + O_2 \rightarrow \diagup\!\!\!\diagup C{=}C{-}\overset{\displaystyle OOH}{\underset{\displaystyle |}{C}}{-} \end{array} \qquad (13.5)$$

These hydroperoxides undergo cleavage to give rise to short-chain aldehydes. Peroxidation can also cause damage to DNA and proteins; in the latter, sulfhydryl groups give rise to disulfide linkages. Vitamin E functions as an antioxidant and can prevent oxygen toxicity (Chapter 36). Glutathione peroxidase, a selenium-containing enzyme, inactivates peroxides:

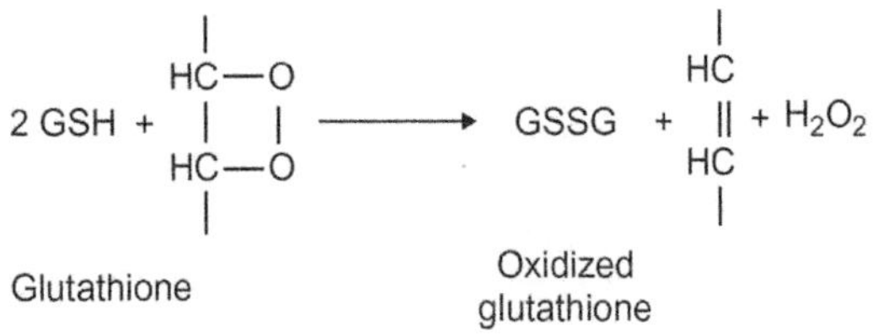

$$2\,GSH + \begin{array}{c} | \\ HC{-}O \\ |\;| \\ HC{-}O \\ | \end{array} \longrightarrow GSSG + \begin{array}{c} | \\ HC \\ \| \\ HC \\ | \end{array} + H_2O_2 \qquad (13.6)$$

Glutathione   Oxidized glutathione

Glutathione, which protects sulfhydryl groups of proteins, is regenerated by glutathione reductase, which uses reducing equivalents from NADPH:

$$GSSG + NADPH + H^+ \rightarrow 2GSH + NADP^+$$

The hydrogen peroxide formed is decomposed by catalase:

$$2H_2O_2 \rightarrow 2H_2O + O_2$$

Aerobic organisms are protected from oxygen toxicity by three enzymes: glutathione peroxidase, catalase, and superoxide dismutase. Superoxide dismutases are metalloenzymes (the cytoplasmic enzyme contains Cu^{2+} and Zn^{2+}, whereas the mitochondrial enzyme contains Mn^{2+}) widely distributed in aerobic cells. The role of superoxide

dismutases in preventing oxygen toxicity is still controversial. For example, some aerobic cells (e.g., adipocytes and some bacteria) lack superoxide dismutase, whereas some strict anaerobes possess this enzyme.

Human superoxide dismutase has been prepared in large quantities by recombinant DNA methods using *Escherichia coli*. It has potential clinical use in preventing oxygen toxicity, for example, during the re-establishment of blood flow through dissolution of a blood clot by thrombolytic agents after a myocardial infarction, during reperfusion after kidney transplant, and in the lungs of premature infants. Allopurinol, a xanthine oxidase inhibitor (Chapter 25), is potentially useful in preventing the tissue injury brought about by ischemia followed by reperfusion. During ischemia, hypoxanthine production increases by enhanced adenine nucleotide catabolism (ATP $\rightarrow$ ADP $\rightarrow$ AMP $\rightarrow$ adenosine $\rightarrow$ inosine $\rightarrow$ hypoxanthine, providing a larger supply of substrate for xanthine oxidase and thus increased formation of superoxide anions. Complete reduction of O_2 by single-electron transfer reactions (the univalent pathway) yields superoxide anion, hydrogen peroxide, and hydroxyl radical as intermediates:

$$O_2 \xrightarrow{-e^-} {}^\cdot O_2{}_- \xrightarrow{-e^-,2H^+} H_2O_2 \xrightarrow{-e^-,H^+} {}^\cdot OH \xrightarrow{-e^-,H^+} H_2O \qquad (13.7)$$

In contrast, most O_2 reduction in aerobic cells occurs via cytochrome c oxidase, which prevents the release of toxic oxygen intermediates:

$$O_2 + 4H^+ + 4e^- \rightarrow 2H_2O$$

Mitochondrial superoxide dismutase maintains intramitochondrial superoxide anion at very low steady-state concentrations.

Paraquat, a herbicide that is highly toxic to humans, causes respiratory distress that can lead to death. Damage to membranes of the epithelial cells lining the bronchioles and alveoli that occurs with paraquat poisoning has been ascribed to excessive production of superoxide anion. Paraquat readily accepts electrons from reduced substrates of high negative potential, while the reduced paraquat reacts with molecular oxygen to form superoxide anion and an oxidized paraquat molecule (Figure 13.11).

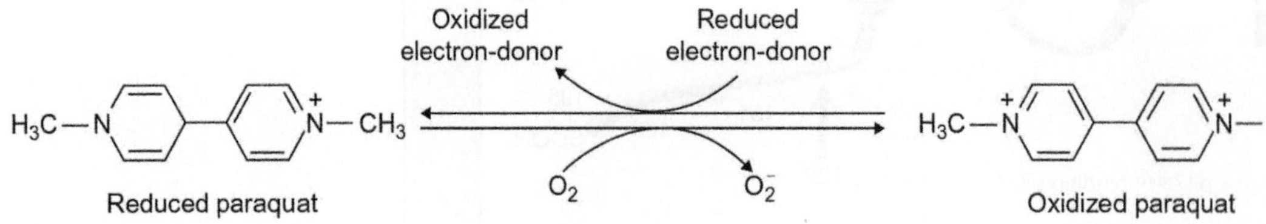

FIGURE 13.11 Oxidation and reduction of paraquat.

Monooxygenases and Dioxygenases

Oxygenases catalyze reactions in which an oxygen atom or molecule is incorporated into organic substrates. They may therefore be **monooxygenases** or **dioxygenases**. An example of a dioxygenase is tryptophan-2,3-dioxygenase (a heme enzyme), which participates in the catabolism of tryptophan (Chapter 15).

Monooxygenases, also called **hydroxylases** or **mixed-function oxidases**, catalyze the incorporation of one atom of an oxygen molecule into the substrate, while the other atom is reduced to water. A second reductant is required (e.g., NADH, NADPH, or tetrahydrobiopterin); the overall reaction is

$$RH + AH_2 + O_2 \rightarrow ROH + A + H_2O$$

where AH_2 is the second reductant.

There are many classes of hydroxylases, depending on the nature of the second reductant. One group, which acts on a variety of substrates including foreign compounds (xenobiotics), contains **cytochrome P-450**. The latter derives its name from the absorption maximum at 450 nm of its carbon monoxide adduct, the P standing for pigment. The cytochrome P-450 monooxygenase system is widely distributed in nature. In humans, it is present in most organs, but the highest concentrations are found in the liver. It is found in membranes of the endoplasmic reticulum (microsomal fraction of the cell), mitochondria, and nucleus. The role of cytochrome P-450 enzymes in the metabolism of steroids, eicosanoids, and other substrates is discussed in the appropriate sections of the text (also see enrichment reading references included in this chapter).

CLINICAL CASE STUDY 13.1 A Disorder of Fission of Organelles, Mitochondria, and Peroxisomes

This clinical case was abstracted from: H.R. Waterhan, J. Koster, C.W. van Roermund, P.A. Mooyer, R.J. Wanders, J.V. Leonard, et al., A lethal defect of mitochondrial and peroxisomal fission, N. Engl. J. Med. 356 (2007) 1736–1741.

Synopsis
A full-term female newborn infant showed feeding difficulties, several neurological abnormalities, lactic academia, and elevated serum long-chain fatty acids. Histochemical and electron microscopy studies of fibroblasts revealed fewer peroxisomes with abnormal morphology and elongated mitochondria. The infant died at the age of 37 days. The morphological abnormalities were attributed to defects in the fission of intracellular organelles, mitochondria, and peroxisomes.

The molecular studies revealed mutation in the dynamin-like protein 1 gene, which is one of the proteins involved in the mitochondrial fission.

Teaching Points
1. An intracellular organization of mitochondria and peroxisomes requires both fusion and fission processes.
2. Defects in these processes lead to serious clinical consequences consistent with their various metabolic roles (see Chapters 1, 13, and 16).

Selected Reference for Further Study and Enrichment
[1] D.C. Chan, Mitochondrial dynamics in disease, N. Engl. J. Med. 356 (2007) 1707–1709.

CLINICAL CASE STUDY 13.2 Defects in Intracellular Killing by Phagocytes

This clinical case was abstracted from: R. Fehon, S. Mehr, E. La Hei, D. Isaacs, M. Wong, Two-year-old boy with cervical and liver abscesses, J. Paediatr. Child Health 44 (2008) 670–672.

Case Description
A 2-year-old white male was referred to the Clinical Immunology service with the chief complaint of recurrent skin and soft tissue infections. The patient's past medical history was significant for recurrent perianal abscesses beginning at 4 months of age. Due to his poor response to oral antibiotics, at 9 months of age, he underwent a surgical procedure to incise, drain, and repair multiple perianal fistulas associated with the abscess progression. At 1 year of age, he had a right cervical lymphadenitis due to *Staphylococcus aureus* that was again treated with incision and drainage and a course of IV cefazolin. A cellulitis of his left elbow, which grew *Serratia marcescens* on culture, was the patient's last significant infection in the last 3 months. The WBC count, hemoglobin level, and erythrocyte sedimentation rate were normal. Screening tests for serum antibody levels, anergy skin testing, and serum complement component levels were normal. A sample of the patient's blood was sent to a neutrophil research laboratory with results that demonstrated no reduction of nitroblue tetrazolium (NBT). Testing of the patient's mother demonstrated two populations of neutrophils: one population that reduced NBT (a normal reaction) and a population that did not reduce

(Continued)

CLINICAL CASE STUDY 13.2 (Continued)

NBT. A diagnosis of chronic granulomatous disease was made, and the patient was placed on subcutaneous injections of gamma interferon (IFN-γ) and daily doses of oral trimethoprim-sulfamethoxazole.

Teaching Points

This case illustrates the importance not only of phagocytosis, but also of the proper functioning of intercellular killing mechanisms for the control of normal flora organisms. Chronic granulomatous disease (CGD) is the name given to a collection of inherited genetic disorders that disrupt the NADPH oxidase complex and are characterized by the inability of the phagocyte to produce microbial reactive superoxide anion and associated metabolites such as hydrogen peroxide. While ingestion of bacteria, degranulation, and phagolysosome formation is normal in CGD patients, regardless of the exact genetic form of CGD, the end result is that superoxide cannot be produced, and in turn, bacterial killing is

ineffective. The presence of recurrent, indolent, pyogenic infections due to *Staphylococcus aureus* and gram-negative bacteria, such as *Pseudomonas*, *Klebsiella*, and *Serratia*, beginning in the first year of life, is pathognomonic for CGD. This disease may be inherited as either X-linked-recessive (as in this case) or as autosomal-recessive. A standard screening test for CGD is the nitroblue tetrazolium (NBT) test, in which neutrophils are incubated with NBT and examined microscopically for color changes. Normal neutrophils will readily reduce NBT (which appears to color the cells blue) due to their ability to produce O_2^-; CGD patients' neutrophils will show little or no color changes in their cells. Aggressive prophylactic antibiotic therapy and IFN-γ therapy have been very effective in limiting the frequency of infections in these patients. Bone marrow transplants can cure the disease, but are limited by availability of matching donors.

Also refer to Chapter 14.

REQUIRED READING

[1] R.J. Youle, A.M. van der Bliek, Mitochondrial fission, fusion, and stress, Science 337 (2012) 1062–1065.

[2] S.L. Archer, Mitochondrial dynamics—mitochondrial fission and fusion in human diseases, N. Engl. J. Med. 369 (2013) 2236–2251.

[3] L. Blanchet, L.M.C. Buydens, J.A.M. Smeitink, P.H.G.M. Willems, W.J.H. Koopman, Isolated mitochondrial Complex I deficiency: explorative data analysis of patient cell parameters, Curr. Pharm. Des. (2011) 4023–4033.

[4] W.J.H. Koopman, P.H.G.M. Willems, J.A.M. Smeitink, Monogenic mitochondrial disorders, N. Engl. J. Med. 366 (2012) 1132–1141.

[5] The Multiple-System Atrophy Research Collaboration, Mutations in *COQ2* in familial and sporadic multiple-system atrophy, N. Engl. J. Med. 369 (2013) 233–244.

[6] T.M. Iverson, E. Maklashina, G. Cecchini, Structural basis for malfunction in Complex II, J. Biol. Chem. 287 (2012) 35430–35438.

[7] M. Harms, P. Seale, Brown and beige fat: development, function and therapeutic potential, Nat. Med. 19 (2013) 1252–1263.

[8] S.J. Kim, M. Kwon, M.J. Ryu, H.K. Chung, S. Tadi, Y.K. Kim, et al., CRIF1 is essential for the synthesis and insertion of oxidative phosphorylation polypeptides in the mammalian mitochondrial membrane, Cell Metab. 274 (2012) 274–283.

[9] A.H.V. Schapira, Mitochondrial diseases, Lancet 379 (2012) 1825–1834.

[10] S.R. Piecznik, J. Neustadt, Mitochondrial dysfunction and molecular pathways of disease, Exp. Mol. Pathol. 83 (2007) 84–92.

[11] C.M. Verity, A.M. Winstone, L. Stellitano, D. Krishnakumar, R. Will, R. Mcfarland, The clinical presentation of mitochondrial diseases in children with progressive intellectual and neurological deterioration: a national, prospective, population-based study, Dev. Med. Child Neurol. 52 (2010) 434–440.

[12] A.J. Tanaka, M.V. Sauer, D. Egli, D.H. Kort, Harnessing the stem cell potential: the path to prevent mitochondrial disease, Nat. Med. 19 (2013) 1578–1579.

ENRICHMENT READING

[1] F. Peter Guengerich, New trends in cytochrome P450 research at the half-century mark, J. Biol. Chem. 288 (2013) 17063–17064.

[2] F. Peter Guengerich, A.W. Munro, Unusual cytochrome P450 enzymes and reactions, J. Biol. Chem. 288 (2013) 17065–17073.

[3] C.M. Krest, E.L. Onderko, T.H. Yosca, J.C. Calixo, R.F. Karp, J. Livada, et al., Reactive intermediates in cytochrome P450 catalysis, J. Biol. Chem. 288 (2013) 17074–17081.

[4] E.F. Johnson, C. David Stout, Structural diversity of eukaryotic membrane cytochrome P450s, J. Biol. Chem. 288 (2013) 17082–17090.

[5] I.A. Pikuleva, M.R. Waterman, Cytochrome P450: roles in diseases, J. Biol. Chem. 288 (2013) 17091–17098.

Chapter 14

Carbohydrate Metabolism II: Gluconeogenesis, Glycogen Synthesis and Breakdown, and Alternative Pathways

Key Points

1. Gluconeogenesis, synthesis of new glucose from noncarbohydrate precursors, provides glucose when dietary intake is insufficient or absent (fasting). It also is essential in the regulation of acid–base balance, amino acid metabolism, and synthesis of carbohydrate-derived structural components.

2. Gluconeogenesis occurs in the liver and kidneys. The precursors of gluconeogenesis are lactate, glycerol, and amino acids, with propionate making a minor contribution.

3. The gluconeogenesis pathway consumes ATP, which is derived primarily from the oxidation of fatty acids. The pathway uses several enzymes of glycolysis with the exception of enzymes of the irreversible steps, namely pyruvate kinase, 6-phosphofructokinase, and hexokinase. The irreversible reactions of glycolysis are bypassed by four alternate, unique reactions of gluconeogenesis.

4. The four unique reactions of gluconeogenesis are (a) pyruvate carboxylase, a biotin (a vitamin)-dependent enzyme, located in the mitochondrial matrix; (b) phosphoenolpyruvate carboxykinase located in both mitochondrial matrix and cytosol; (c) fructose-1,6-bisphosphatase located in the cytosol; and (d) glucose-6-phosphatase located in the endoplasmic reticulum (ER).

5. Glucose-6-phosphatase, which catalyzes the conversion of glucose-6-phosphate to glucose in the final step of gluconeogenesis, is also the final step in the conversion of glycogen to glucose. The enzyme is present only in the ER of liver and kidney.

6. Gluconeogenesis is regulated by the overall energy demands of the body, allosteric effectors, and hormones. Glycolysis and gluconeogenesis are reciprocally regulated by allosteric effectors so that both pathways do not occur simultaneously. Acetyl-CoA inhibits pyruvate kinase and activates pyruvate carboxylase; fructose-2,6-bisphosphate activates 6-phosphofructokinase and inhibits fructose-1,6-bisphosphatase; and AMP and citrate inhibit 6-phosphofructokinase and activate fructose-1,6-bisphosphatase. Hormones regulate gluconeogenesis by means of phosphorylation and dephosphorylation of target proteins and by gene expression. Insulin and glucagon have opposing effects on glycolysis and gluconeogenesis: insulin promotes glycolysis and inhibits gluconeogenesis, and the opposite is true for glucagon (Chapter 20). Cortisol, a steroid hormone released from the adrenal cortex as a physiological response to stress, promotes gluconeogenesis by stimulating the enzyme synthesis required for the pathway.

7. Abnormalities in gluconeogenesis cause hypoglycemia with severe metabolic consequences. These abnormalities can result from a genetic deficiency of enzymes of gluconeogenesis or of fatty acid oxidation (Chapter 16) pathways, ethanol abuse, or the plant-derived toxin hypoglycin.

8. Glucose is stored as glycogen when glucose levels are high in virtually every cell of the body. Glycogen is especially abundant in liver and skeletal muscle. Glycogen is degraded when the glucose supply is low. It is converted to glucose-6-phosphate (e.g., in muscle) or glucose, depending on the presence of the tissue-specific glucose-6-phosphatase. The liver and kidneys are able to convert glycogen to glucose. Quantitatively, the liver plays a major role in the maintenance of optimal blood glucose levels.

9. Glycogenesis and glycogenolysis are reciprocally regulated by several allosteric modulators and hormones (insulin, glucagon, epinephrine, and cortisol). The two key enzymes that are regulated are glycogen synthase in glycogenesis and glycogen phosphorylase in glycogenolysis. Glucagon and epinephrine activate a cell-membrane-mediated signal transduction process resulting in production of cAMP, which, in turn, activates protein kinase. Phosphorylation activates glycogen phosphorylase and inactivates glycogen synthase. Muscle is insensitive to glucagon's action. Insulin promotes glycogen synthesis by activating tyrosine kinase-mediated amplification systems. The effects of these hormones are tissue specific.

10. Deficiencies of enzymes in glycogenolysis can result in glycogen storage disorders, hypoglycemia, or exercise intolerance, depending on the tissue that is lacking the enzyme.

N. V. Bhagavan and C.-E. Ha: Essentials of Medical Biochemistry, Second edition. DOI: http://dx.doi.org/10.1016/B978-0-12-416687-5.00014-2

11. The terminal CH_2OH group of the activated form of glucose (UDP-glucose) is oxidized to form UDP-glucuronic acid (UDP-GA). UDP-GA is essential for the detoxification reactions in the liver for both endogenous metabolites (e.g., bilirubin, Chapter 27) and exogenous compounds (e.g., drugs).

12. The monosaccharides fructose and galactose are converted to glucose by different pathways. Fructose enters the pathway in the liver via fructokinase and in the muscle and kidney via hexokinase. Galactose is converted to glucose in a circuitous route involving galactokinase and galactose-1-phosphate uridyltransferase. Deficiencies of the enzymes involved in fructose and galactose metabolism can result in serious clinical manifestations.

13. Amino sugars, constituents of glycoproteins and glycolipids, are synthesized in pathways that originate from glucose-6-phosphate.

14. Oxidation of glucose in the pentose phosphate pathway (also called hexose monophosphate shunt) produces NADPH and ribose-5-phosphate. The pathway consists of two oxidative reactions followed by several nonoxidative reactions. The intermediates of the pathway can be shuttled back and forth with some of the intermediates of glycolysis. NADPH is required for several metabolic reactions such as fatty acid biosynthesis, steroid biosynthesis, hydroxylations, and antioxidant and oxidant generation reactions. The pathway is active in tissue cells such as adipose tissue, liver, mammary glands, adrenal cortex, red blood cells, and leukocytes. The pentose phosphate pathway occurs in cytosol and begins with the first of two oxidation steps initiated by glucose-6-phosphate dehydrogenase (G6PD). In red blood cells, G6PD deficiency can cause hemolysis, and in leukocytes, deficiency disrupts the phagocytosis of bacteria.

In this chapter, carbohydrate metabolism is discussed in terms of nondietary sources of glucose and nonglycolytic pathways. Gluconeogenesis and glycogen synthesis and breakdown make up the first category, while the pentose phosphate pathway (also called **hexose monophosphate shunt**) and the glucuronic acid pathway make up the second. Metabolic pathways of fructose, galactose, and some other sugars belong in both categories.

GLUCONEOGENESIS

Metabolic Role

Gluconeogenesis (literally, "formation of new sugar") is the metabolic process by which glucose is formed from noncarbohydrate sources, such as lactate, amino acids, and glycerol. Gluconeogenesis provides glucose when dietary intake is insufficient to supply the requirements of the brain and nervous system, erythrocytes, renal medulla,

testes, and embryonic tissues, all of which use glucose as a major source of fuel. Gluconeogenesis has three additional functions:

1. *Control of acid—base balance.* Production of lactate in excess of its clearance causes metabolic acidosis, and resynthesis of glucose from lactate is a major route of lactate disposal. Since glycolysis is almost totally anaerobic in erythrocytes, renal medulla, and some other tissues, even under normal conditions, lactate is continually released. Other tissues, particularly muscle during vigorous exercise, can produce large amounts of lactate, which must be removed; otherwise, lactic acidosis will result (Chapter 19).

    The continuous conversion of lactate to glucose in the liver, and of glucose to lactate by anaerobic glycolysis, particularly in muscle, forms a cyclical flow of carbon called the **Cori cycle**. Deamination of amino acids prior to gluconeogenesis in the kidney also provides a supply of NH_3 to neutralize acids excreted in the urine (Chapter 37).

2. *Maintenance of amino acid balance.* Metabolic pathways for the degradation of most amino acids and for the synthesis of nonessential amino acids involve some steps of the gluconeogenic pathway. Imbalances of most amino acids, whether due to diet or to an altered metabolic state, are usually corrected in the liver by degradation of the excess amino acids or by synthesis of the deficient amino acids through gluconeogenic intermediates.

3. *Provision of biosynthetic precursors.* In the absence of adequate dietary carbohydrate intake, gluconeogenesis supplies precursors for the synthesis of glycoproteins, glycolipids, and structural carbohydrates.

Gluconeogenesis from pyruvate is essentially the reverse of glycolysis, with the exception of three nonequilibrium reactions (Figure 14.1). These reactions are

$$\text{Glucose} + \text{ATP}^{4-} \xrightarrow{\text{hexokinase or glucokinase}} \text{glucose-6-phosphate}^{2-} + \text{ADP}^{3-} + \text{H}^+ \quad (14.1)$$

$$\text{Fructose-6-phosphate}^{2-} + \text{ATP}^{4-} \xrightarrow{\text{6-phosphofructokinase}} \text{fructose-1,6-bisphosphate}^{4-} + \text{ADP}^{3-} + \text{H}^+ \quad (14.2)$$

$$\text{Phosphoenolpyruvate}^{3-} + \text{ADP}^{3-} + \text{H}^+ \xrightarrow[\text{pyruvate kinase}]{} \text{pyruvate}^- + \text{ATP}^{4-} \quad (14.3)$$

In gluconeogenesis, these reactions are bypassed by alternate steps also involving changes in free energy and also physiologically irreversible reactions.

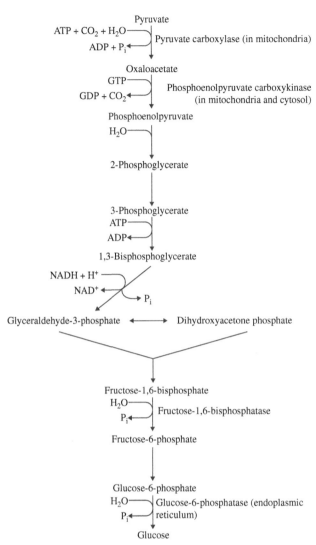

FIGURE 14.1 Pathway of gluconeogenesis from pyruvate to glucose. Only enzymes required for gluconeogenesis are indicated; others are from glycolysis. The overall reaction for the synthesis of one molecule of glucose from two molecules of pyruvate is $2 \text{ Pyruvate} + 4\text{ATP}^{4-} + 2\text{GTP}^{4-} + 2\text{NADH} + 2\text{H}^+ + 6\text{H}_2\text{O} \rightarrow \text{Glucose} + 2\text{NAD}^+ + 4\text{ADP}^{3-} + 2\text{GDP}^{3-} + 6\text{P}_i^{2-} + 4\text{H}^+$.

Conversion of pyruvate to phosphoenolpyruvate involves two enzymes and the transport of substrates and reactants into and out of the mitochondrion. In glycolysis, conversion of phosphoenolpyruvate to pyruvate results in the formation of one high-energy phosphate bond. In gluconeogenesis, two high-energy phosphate bonds are consumed ($\text{ATP} \rightarrow \text{ADP} + \text{P}_i$; $\text{GTP} \rightarrow \text{GDP} + \text{P}_i$) in reversing the reaction. Gluconeogenesis begins when pyruvate, generated in the cytosol, is transported into the mitochondrion—through the action of a specific carrier—and converted to oxaloacetate:

$$\text{Pyruvate}^- + \text{HCO}_3^- + \text{ATP}^{4-} \xrightarrow[\text{pyruvate carboxylase}]{\text{acetyl-CoA, Mg}^{2+}}$$
$$\text{oxaloacetate}^{2-} + \text{ADP}^{3-} + \text{P}_i^{2-} + \text{H}^+$$

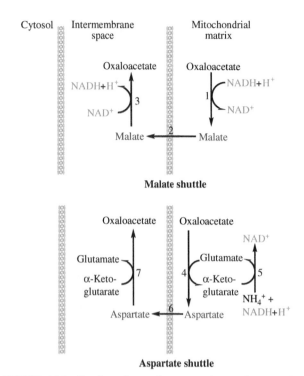

Malate shuttle

Aspartate shuttle

FIGURE 14.2 Shuttle pathways for transporting oxaloacetate from mitochondria into the cytosol. The shuttles are named for the molecule that actually moves across the mitochondrial membrane. 1 and 3 = malate dehydrogenase; 2 = malate translocase; 4 and 7 = aspartate aminotransferase; 5 = glutamate dehydrogenase; 6 = aspartate translocase.

Like many CO_2-fixing enzymes, pyruvate carboxylase contains **biotin** bound through the ε-NH_2 of a lysyl residue (Chapter 16).

The second reaction is the conversion of oxaloacetate to phosphoenolpyruvate:

$$\text{Oxaloacetate}^{2-} + \text{GTP}^{4-}(\text{or ITP}^{4-})$$

$$\xrightarrow{\text{phosphoenolpyruvate carboxykinase (PEPCK)}}$$

$$\text{phosphoenolpyruvate}^{3-} + \text{CO}_2 + \text{GDP}^{3-}(\text{or IDP}^{3-})$$

In this reaction, inosine triphosphate (ITP) can substitute for guanosine triphosphate (GTP), and the CO_2 lost is the one gained in the carboxylase reaction. The net result of these reactions is

$$\text{Pyruvate} + \text{ATP} + \text{GTP (or ITP)} \rightarrow$$
$$\text{phosphoenolpyruvate} + \text{ADP} + \text{P}_i + \text{GDP (or IDP)}$$

Pyruvate carboxylase is a mitochondrial enzyme in animal cells. In humans (and guinea pigs), PEPCK occurs in both mitochondria and cytosol.

In humans, oxaloacetate must be transported out of the mitochondrion to supply the cytosolic PEPCK. Because there is no mitochondrial carrier for oxaloacetate and its diffusion across the mitochondrial membrane is slow, it is transported as malate or aspartate (Figure 14.2). The malate shuttle carries oxaloacetate and reducing equivalents, whereas the aspartate shuttle, which does not require a preliminary reduction step, depends on the availability of glutamate and α-ketoglutarate in excess of tricarboxylic acid (TCA) cycle requirements. The proportion of oxaloacetate carried by each shuttle probably depends on the redox state of the cytosol. If most of the pyruvate is derived from lactate, the NADH/NAD$^+$ ratio in the cytosol is elevated. In this situation, there is no need to transport reducing equivalents out of the mitochondria, and the aspartate shuttle predominates. However, if alanine is the principal source of pyruvate, no cytosolic reduction occurs, and the glyceraldehyde-phosphate dehydrogenase reaction (which requires NADH) requires transport of reducing equivalents via the malate shuttle. In species in which oxaloacetate is converted in mitochondria to phosphoenolpyruvate (which is readily transported to the cytosol, perhaps via its own carrier system), no transport of oxaloacetate or reducing equivalents is required.

Phosphoenolpyruvate is converted to fructose-1,6-bisphosphate by reversal of glycolysis in the cytosol via reactions that are at near-equilibrium and whose direction is dictated by substrate concentration. Conversion of fructose-1,6-bisphosphate to fructose-6-phosphate is a nonequilibrium step, catalyzed by fructose-1,6-bisphosphatase:

$$\text{Fructose-1,6-bisphosphate}^{4-} + H_2O \xrightarrow{Mg^{2+}}$$
$$\text{fructose-6-phosphate}^{2-} + P_i^{2-}$$

Fructose-6-phosphate is then converted to glucose-6-phosphate by reversal of another near-equilibrium reaction of glycolysis. In the last reaction in gluconeogenesis, glucose-6-phosphate is converted to glucose by glucose-6-phosphatase:

$$\text{Glucose-6-phosphate}^{2-} + H_2O \rightarrow \text{glucose} + P_i^{2-}$$

Glucose-6-phosphatase is part of a multicomponent integral membrane protein system that consists of six different proteins located in the endoplasmic reticulum of liver, kidney, and intestine, but not of muscle or adipose tissue. Deficiency of any one of these five proteins leads to glycogen storage disease (discussed later).

Thus, gluconeogenesis requires the participation of enzymes of the cytosol, mitochondrion, and smooth endoplasmic reticulum, as well as of several transport systems, and it may involve more than one tissue. The complete gluconeogenic pathway, culminating in the release of glucose into the circulation, is present only in liver and kidney. Most tissues contain only some of the necessary enzymes. These "partial pathways" are probably used in glycerogenesis and in replenishing tricarboxylic acid (TCA) intermediates. Muscle can also convert lactate to glycogen, but this probably takes place only in one type of muscle fiber and only when glycogen stores are severely depleted and lactate concentrations are high, such as after heavy exercise.

Under normal conditions, the liver provides 80% or more of the glucose produced in the body. During prolonged starvation, however, this proportion decreases, while that synthesized in the kidney increases to nearly half of the total, possibly in response to a need for NH_3 to neutralize the metabolic acids eliminated in the urine in increased amounts (Chapter 20).

Gluconeogenesis is a costly metabolic process. Conversion of two molecules of pyruvate to one of glucose consumes six high-energy phosphate bonds ($4ATP + 2GTP \rightarrow 4ADP + 2GDP + 6P_i$) and results in the oxidation of two NADH molecules (Figure 14.1). In contrast, glycolytic metabolism of one molecule of glucose to two of pyruvate produces two high-energy phosphate bonds ($2ADP + 2P_i \rightarrow 2ATP$) and reduces two molecules of NAD$^+$. For gluconeogenesis to operate, the precursor supply and the energy state of the tissue must be greatly increased. Using some gluconeogenic precursors to provide energy (via glycolysis and the TCA cycle) and to convert the remainder of the precursors to glucose would be inefficient, even under aerobic conditions. Usually, the catabolic signals (catecholamines, cortisol, and increase in glucagon/insulin ratio) that increase the supply of gluconeogenic precursors also favor lipolysis, which provides fatty acids to supply the necessary ATP.

When amino acid carbons are the principal gluconeogenic precursors, the metabolic and physiological debts are particularly large compared to those incurred when lactate or glycerol is used. Amino acids are derived through breakdown of muscle protein, which is accompanied by a loss of electrolytes and tissue water.

Gluconeogenic Precursors

Gluconeogenic precursors include lactate, alanine and several other amino acids, glycerol, and propionate (Chapter 16).

Lactate, the end product of anaerobic glucose metabolism, is produced by most tissues of the body, particularly skin, muscle, erythrocytes, brain, and intestinal mucosa. In a normal adult, under basal conditions, these tissues produce 1300 mM of lactate per day, and the normal serum lactate concentration is less than 1.2 mM/L. During vigorous exercise, the production of lactate can increase several-fold. Lactate is normally removed from the circulation by the liver and kidneys. Because of its great capacity to use lactate, liver plays an important role in the pathogenesis of lactic acidosis, which may be thought of as an imbalance between the relative rates of production and utilization of lactate (Chapter 37).

Alanine, derived from muscle protein and also synthesized in the small intestine, is quantitatively the most important amino acid substrate for gluconeogenesis. It is converted to pyruvate by alanine aminotransferase (Chapter 15):

$$\text{Alanine} + \alpha\text{-ketoglutarate}^{2-} \xleftrightarrow{B_6PO_4} \text{pyruvate}^-$$
$$+ \text{glutamate}^-$$

where B_6-PO_4 = pyridoxal phosphate.

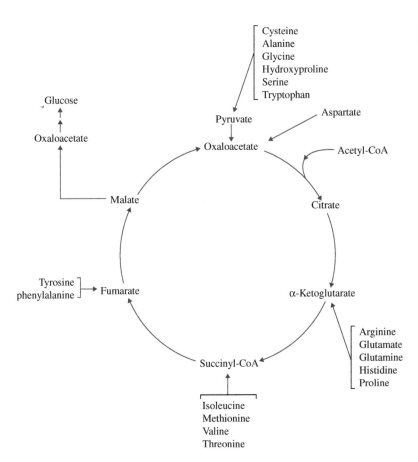

FIGURE 14.3 Points of entry of amino acids into the gluconeogenesis pathway.

An amino acid is classified as ketogenic, glucogenic, or glucogenic/ketogenic depending on whether feeding it to a starved animal increases plasma concentrations of ketone bodies (Chapter 16) or of glucose. Leucine and lysine are purely ketogenic because they are catabolized to acetyl-CoA, which cannot be used to synthesize glucose (Chapter 15). Isoleucine, phenylalanine, tyrosine, and tryptophan are both glucogenic and ketogenic, and the remaining amino acids (including alanine) are glucogenic. Entry points of the amino acids into the gluconeogenesis pathway are shown in Figure 14.3.

Glycerol is more reduced than the other gluconeogenic precursors, and it results primarily from triacylglycerol hydrolysis in adipose tissue. In liver and kidney, glycerol is converted to glycerol 3-phosphate by glycerol kinase:

$$
\begin{array}{ccc}
\text{CH}_2\text{OH} & & \text{CH}_2\text{OH} \\
| & & | \\
\text{CHOH} + \text{ATP}^{4-} \longrightarrow & & \text{CHOH} + \text{ADP}^{3-} + \text{H}^+ \\
| & & | \\
\text{CH}_2\text{OH} & & \text{CH}_2\text{OPO}_3^{2-} \\
\end{array}
$$

Glycerol   Glycerol 3-phosphate

Glycerol 3-phosphate is oxidized to dihydroxyacetone phosphate by glycerol-3-phosphate dehydrogenase:

$$
\begin{array}{ccc}
\text{CH}_2\text{OH} & & \text{CH}_2\text{OH} \\
| & & | \\
\text{CHOH} + \text{NAD}^+ \longrightarrow & & \text{C} = \text{O} + \text{NADH} + \text{H}^+ \\
| & & | \\
\text{CH}_2\text{OPO}_3^{2-} & & \text{CH}_2\text{OPO}_3^{2-} \\
\end{array}
$$

Glycerol 3-phosphate   Dihydroxyacetone phosphate

Dihydroxyacetone phosphate is the entry point of glycerol into gluconeogenesis. Glycerol cannot be metabolized in adipose tissue, which lacks glycerol kinase, and the glycerol 3-phosphate required for triacylglycerol synthesis in this tissue is derived from glucose (Chapter 20). During fasting in a resting adult, about 210 mM of glycerol per day is released, most of which is converted to glucose in the liver. During stress or exercise, glycerol release is markedly increased (Chapter 20).

Propionate is not a quantitatively significant gluconeogenic precursor in humans, but it is a major source of glucose in ruminants. It is derived from the catabolism of isoleucine, valine, methionine, and threonine; from β-oxidation of odd-chain fatty acids; and from the degradation of the side chain of cholesterol. Propionate enters gluconeogenesis via the TCA cycle after conversion to succinyl-CoA (Chapter 16).

Regulation of Gluconeogenesis

Gluconeogenesis is regulated by the production and release of precursors and by the activation and inactivation of key enzymes. Glucagon and glucocorticoids stimulate gluconeogenesis, whereas insulin suppresses it. See Chapter 20 for a discussion of the overall metabolic control and physiological implications of the several seemingly competing pathways.

Carboxylation of Pyruvate to Oxaloacetate

The pyruvate carboxylase reaction is activated by Mg^{2+} and, through mass action, by an increase in either the [ATP]/[ADP] or the [pyruvate]/[oxaloacetate] ratio. It is virtually inactive in the absence of acetyl-CoA, an allosteric activator. The enzyme is allosterically inhibited by glutamate, since oxaloacetate formed in excess would flood the TCA cycle and result in a buildup of α-ketoglutarate and glutamate.

Conversion of Oxaloacetate to Phosphoenolpyruvate

Short-term regulation of this reaction is accomplished by changes in the relative proportions of substrates and products. Increased concentrations of oxaloacetate and GTP (or ITP) increase the rate, and accumulation of phosphoenolpyruvate and GDP (or IDP) decreases it. The cytosolic PEPCK is also under long-term regulation by hormones. Its synthesis is increased by corticosteroids. Starvation and diabetes mellitus increase the synthesis of cytosolic PEPCK, whereas refeeding and insulin have the opposite effect.

Conversion of Fructose-1,6-Bisphosphate to Fructose-6-Phosphate

This reaction is catalyzed by fructose-1,6-bisphosphatase (FBPase-l). It is inhibited by AMP, inorganic phosphate, and fructose-2,6-bisphosphate, all of which are allosteric

activators of 6-phosphofructo-1-kinase (PFK-1), the competing glycolytic enzyme (Figure 14.4). This inhibition prevents the simultaneous activation of the two enzymes and helps prevent a futile ATP cycle.

Conversion of Glucose-6-Phosphate to Glucose

Glucose-6-phosphatase is regulated only by the concentration of glucose-6-phosphate, which is increased by glucocorticoids, thyroxine, and glucagon.

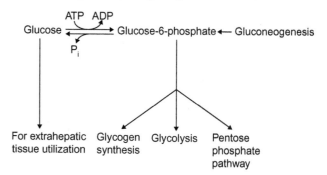

Abnormalities of Gluconeogenesis

Glucose is the predominant fuel for cells that depend largely on anaerobic metabolism, cells that lack mitochondria, and tissues such as brain that normally cannot use other metabolic fuels. An adult brain represents only 2% of total body weight, but it oxidizes about 100 g of glucose per day, accounting for 25% of the basal metabolism. Brain cannot utilize fatty acids because they are bound to serum albumin and so do not cross the blood–brain barrier. Ketone bodies are alternative fuels, but their concentration in blood is negligible, except in prolonged fasting or diabetic ketoacidosis (Chapters 16 and 20). Brain and liver glycogen stores are small relative to the needs of the brain, and gluconeogenesis is essential for survival. Decreased gluconeogenesis leads to hypoglycemia and may cause irreversible damage to the brain. Blood glucose levels below 40 mg/dL (2.2 mM/L) in adults or below 30 mg/dL (1.7 mM/L) in neonates represent severe hypoglycemia. The low levels may result from increased glucose utilization, decreased glucose production, or both. Increased glucose utilization can occur

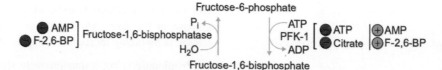

FIGURE 14.4 Regulation of liver 6-phosphofructokinase and fructose-1,6-bisphosphatase. These multimodulated enzymes catalyze nonequilibrium reactions, the former in glycolysis and the latter in gluconeogenesis. Note the dual action of fructose-2,6-bisphosphate (F-2,6-BP), which activates phosphofructokinase (PFK-1) and inactivates fructose-1,6-bisphosphatase. The activity of F-2,6-BP is under hormonal and substrate regulation. ⊕ = positive effectors; ⊖ = negative effectors.

in hyperinsulinemia secondary to an insulinoma. In a diabetic pregnancy, the mother's hyperglycemia leads to fetal hyperglycemia and causes fetal and neonatal hyperinsulinemia. Insufficiency of glucocorticoids, glucagon, or growth hormone and severe liver disease can produce hypoglycemia by depressing gluconeogenesis (Chapters 20 and 30). Ethanol consumption can cause hypoglycemia, since a major fraction of ethanol is oxidized in the liver by **cytosolic alcohol dehydrogenase**, and the NADH and acetaldehyde generated inhibit gluconeogenesis:

$$CH_3CH_2OH + NAD^+ \rightarrow CH_3CHO + NADH + H^+$$

Excessive NADH decreases the cytosolic $[NAD^+]/[NADH]$ ratio, increasing lactate dehydrogenase activity and thereby increasing conversion of pyruvate to lactate. The decrease in pyruvate concentration inhibits pyruvate carboxylase. Acetaldehyde inhibits oxidative phosphorylation, increasing the [ADP]/[ATP] ratio, promoting glycolysis, and inhibiting gluconeogenesis. Hypoglycemia and lactic acidosis are common findings in chronic alcoholics.

Pyruvate carboxylase deficiency can cause intermittent hypoglycemia, ketosis, severe psychomotor retardation, and lactic acidosis. The deficiency of either cytosolic or mitochondrial isoenzyme forms of phosphoenolpyruvate carboxykinase (PEPCK) is characterized by a failure of gluconeogenesis. Both these disorders are rare, and the mitochondrial PEPCK deficiency is inherited as an autosomal recessive trait. The major clinical manifestations include hypoglycemia, lactic acidosis, hypotonia, hepatomegaly, and failure to thrive. The treatment is supportive and is based on symptoms. Fructose-1,6-bisphosphatase deficiency, inherited as an autosomal recessive trait, severely impairs gluconeogenesis, causing hypoglycemia, ketosis, and lactic acidosis. Glucose-6-phosphatase deficiency, also an autosomal recessive trait, causes a similar condition but with excessive deposition of glycogen in liver and kidney (discussed later).

Hypoglycin A (2-methylenecyclopropylalanine), the principal toxin of the unripe akee fruit indigenous to western Africa that also grows in Central America and the Caribbean, produces severe hypoglycemia when ingested. Hypoglycin A causes hypoglycemia by inhibiting gluconeogenesis. In the liver, hypoglycin A forms nonmetabolizable esters with CoA (Figure 14.5) and carnitine, depleting the CoA and carnitine pools, and thereby inhibiting fatty acid oxidation (Chapter 16). Since the principal source of ATP for gluconeogenesis is mitochondrial oxidation of long-chain fatty acids, gluconeogenesis is stopped. Intravenous administration of glucose relieves the hypoglycemia. Inherited fatty acid oxidation defects that lead to a deficit in ATP production can cause hypoglycemia (Chapter 16).

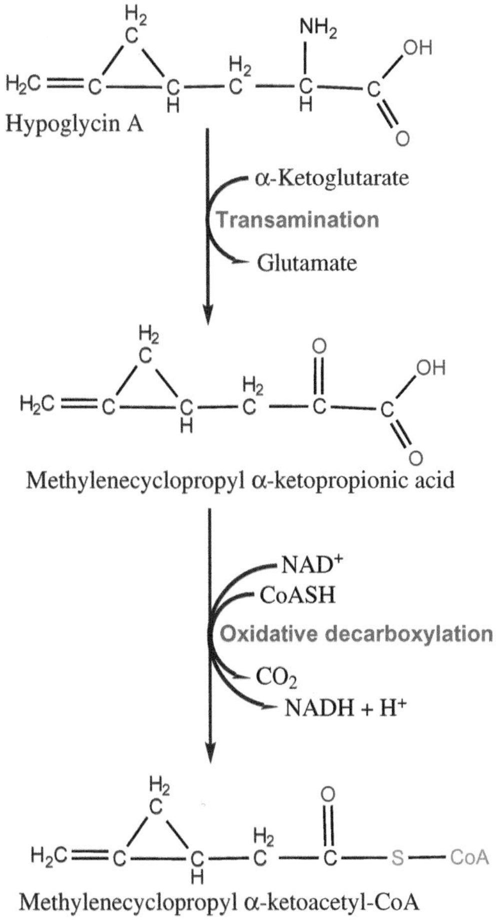

FIGURE 14.5 Hypoglycin A metabolism in the liver. Hypoglycin A is converted to a toxic metabolite, methylenecyclopropyl α-ketopropionyl-CoA, which inhibits several acyl-CoA dehydrogenases and thereby inhibits mitochondrial fatty acid oxidation. Inhibition of fatty acid oxidation causes decreased rates of gluconeogenesis due to short supply of ATP, which induces hypoglycemia.

Overactivity of gluconeogenesis due to increased secretion of catecholamines, cortisol, or growth hormone, or an increase in the glucagon/insulin ratio (Chapter 20) leads to hyperglycemia and causes many metabolic problems.

Metformin, a commonly used drug to treat type 2 diabetes, inhibits hepatic gluconeogenesis, leading to reduction in plasma glucose levels. The mechanism of action of metformin in suppressing hepatic gluconeogenesis is not completely understood. Several mechanisms of action may be contributory to metformin's hypoglycemic effect. They include (1) decreased ATP production required for gluconeogenesis by inhibiting complex I of OXPHOS, (2) decreased glucagon-stimulated cAMP production, and (3) activation of the AMP-activated protein kinase (AMPK) system. AMPK orchestrates cellular energy homeostasis (Chapter 20).

GLYCOGEN METABOLISM

Animals have developed a method of storing glucose that reduces the need to catabolize protein as a source of gluconeogenic precursors between meals. The principal storage form of glucose in mammals is glycogen. Storage of glucose as a polymer reduces the intracellular osmotic load, thereby decreasing the amount of water of hydration and increasing the energy density of the stored glucose.

In muscle, glycogen is stored in the cytosol and endoplasmic reticulum as granules, called β-**particles**, each of which is an individual glycogen molecule. In the liver, the β-particles aggregate, forming larger, rosette-shaped α-particles that can be seen with the electron microscope. The average molecular weight of glycogen is several million (10,000−50,000 glucose residues per molecule). The storage granules also contain the enzymes needed for glycogen synthesis and degradation, and for regulation of these two pathways.

Glycogen is present in virtually every cell in the body, but it is especially abundant in liver and skeletal muscle. The amount stored in tissues varies greatly in response to metabolic and physiological demands, but in a resting individual after a meal, liver usually contains roughly 4%−7% of its wet weight as glycogen and muscle about 1%. Since the body contains 10 times more muscle than hepatic tissue, the total amount of glycogen stored in muscle is greater than that in liver.

Muscle needs a rapidly available supply of glucose to provide fuel for anaerobic glycolysis when, during bursts of muscle contraction, blood may provide inadequate supplies of oxygen and fuel. The liver, under most conditions, oxidizes fatty acids for this purpose; however, liver is responsible for maintaining blood glucose levels during short fasts, and it integrates the supply of available fuels with the metabolic requirements of other tissues in different physiological states. Liver glycogen content changes primarily in response to the availability of glucose and gluconeogenic precursors. Muscle glycogen stores vary less in response to dietary signals, but they depend on the rate of muscular contraction and of oxidative metabolism (tissue respiration).

Glycogenesis (glycogen synthesis and storage) and glycogenolysis (glycogen breakdown) are separate metabolic pathways that have only one enzyme in common, namely, phosphoglucomutase. Glycogen synthesis and breakdown are often reciprocally regulated, so that stimulation of one inhibits the other. The control mechanisms are more closely interrelated than those for glycolysis and gluconeogenesis, perhaps because the glycogen pathways ensure the availability of only one substrate, whereas intermediates of several other metabolic pathways are produced and metabolized in glycolysis and gluconeogenesis.

Glycogen Synthesis

Glycogenesis begins with the phosphorylation of glucose by glucokinase in liver and by hexokinase in muscle and other tissues (Chapter 12):

$$\text{Glucose} + \text{ATP}^{4-} \xrightarrow{\text{Mg}^{2+}} \text{glucose-6-phosphate}^{2-} + \text{ADP}^{3-} + \text{H}^+$$

The second step in glycogenesis is conversion of glucose-6-phosphate to glucose-1-phosphate by phosphoglucomutase in a reaction similar to that catalyzed by phosphoglyceromutase.

In the third step, glucose-1-phosphate is converted to uridine diphosphate (UDP)-glucose, the immediate precursor of glycogen synthesis, by reaction with uridine triphosphate (UTP). This reaction is catalyzed by glucose-1-phosphate uridylyltransferase (or UDP-glucose pyrophosphorylase) (Figure 14.6).

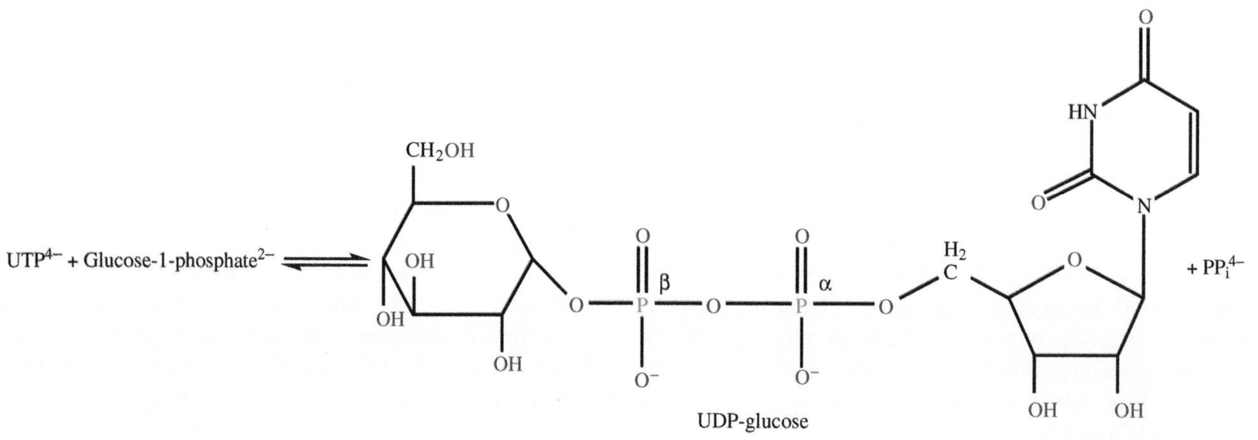

FIGURE 14.6 Formation of UDP-glucose for glycogenesis.

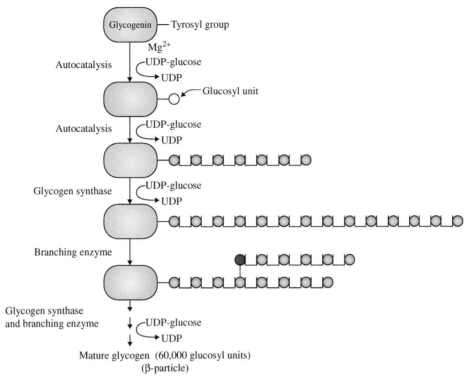

FIGURE 14.7 Schematic representation of glycogen biosynthesis. ◑—●, Glucosyl units in (1→6) linkage.

The reaction is freely reversible, but it is rendered practically irreversible by hydrolysis of the product pyrophosphate (PP_i) through the action of pyrophosphatase:

$$PP_i + H_2O \rightarrow 2P_i$$

Thus, two high-energy phosphate bonds are hydrolyzed in the formation of UDP-glucose. Nucleoside diphosphate sugars are commonly used as intermediates in carbohydrate condensation reactions. The high-energy bond between the sugar and the nucleoside diphosphate provides the energy needed to link the new sugar to the nonreducing component of another sugar or polysaccharide.

The next step in the biogenesis of glycogen is addition of a glucosyl residue at the C-1 position of UDP-glucose to a tyrosyl group located at position 194 of the enzyme glycogenin, which is a Mg^{2+}-dependent autocatalytic reaction. Glycogenin (M.W. 37,000) extends the glucan chain, again by autocatalysis, by six to seven glucosyl units in α(1→4) glycosidic linkages using UDP-glucose as substrate (Figure 14.7). The glucan primer of glycogenin is elongated by glycogen synthase using UDP-glucose. Initially, the primer glycogenin and glycogen synthase are firmly bound in a 1:1 complex. As the glucan chain grows, glycogen synthase dissociates from glycogenin. Deficiency of glycogenin leads to lack of glycogen synthases in muscle tissues and is associated with clinical manifestations including cardiac malfunction (see Clinical Case Study 14.1).

The branching of the glycogen is accomplished by transferring a minimum of six α(1→4) glucan units from the elongated external chain into the same or a neighboring chain by α(1→6) linkage. This reaction of α(1→4) to α(1→6) transglucosylation creates new nonreducing ends and is catalyzed by a branching enzyme. A mature glycogen particle is spherical, containing one molecule of glycogenin and up to 60,000 glucosyl units (β-particles). In the liver, 20−40 β-particles are aggregated into rosettes, known as **α-particles**.

As discussed previously, the glycogen synthesis catalyzed by glycogen synthase consists of a glycosyl moiety from UDP-glucose being transferred to a nonreducing end of a growing glycan chain with the release of UDP. In a very rare instance, presumably as a catalytic error, a β-phosphate residue of the substrate UDP-glucose (see Figure 14.6) is added to a glucose residue of a growing glycan chain by the glycogen synthase. In this reaction, the product is UMP. However, a phosphatase enzyme known as **laforin** subsequently removes the inappropriately incorporated phosphate group. The deficiency of laforin leads to serious clinical consequences [1−3].

Glycogen Breakdown

Glycogenolysis is catalyzed by two enzymes unique to the pathway: glycogen phosphorylase and debranching enzyme. The former normally regulates the rate of

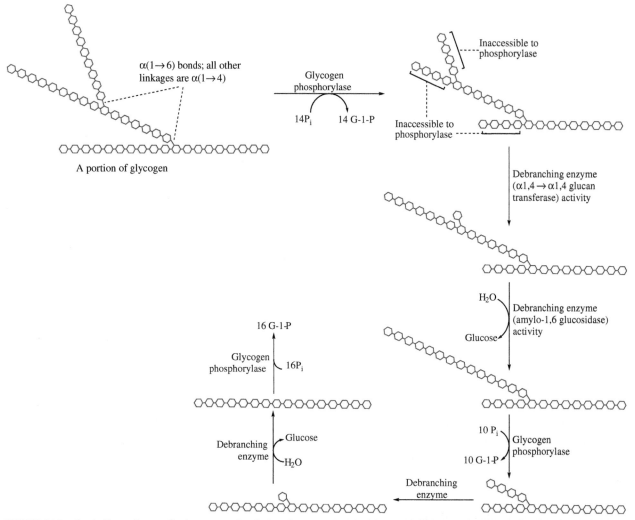

FIGURE 14.8 Catabolism of a small glycogen molecule by glycogen phosphorylase and debranching enzyme. P_i = inorganic phosphate; G-1-P = glucose-1-phosphate.

glucose release from glycogen. The progressive degradation of glycogen is illustrated in Figure 14.8. Glycogen phosphorylase catalyzes the release of glucose-1-phosphate from the terminal residue of a nonreducing end of a glycogen branch by means of phosphorolysis. A molecule of inorganic phosphate attacks the C_1 side of an $\alpha(1\rightarrow4)$ glycosidic bond, leaving a hydroxyl group on C_4 that remains in the glycogen polymer:

$$\text{Glycogen (glucosyl residues } n) + P_i \xrightarrow{\textit{Phosphorylase}}$$
$$\text{glycogen } (n-1 \text{ glucosyl residues}) + \text{glucose-1-phosphate}$$

The energy stored in the $\alpha(1\rightarrow4)$ glycosidic bond during the condensation reaction in glycogen synthesis is sufficient to permit the formation of a glucose–phosphate bond without using ATP.

Glucose-1-phosphate is next converted by phosphoglucomutase to glucose-6-phosphate. The latter may then enter the glycolytic pathway, but if glucose-6-phosphatase is present, free glucose can be formed.

Regulation of Glycogen Metabolism

Metabolism of glycogen in muscle and liver is regulated primarily through reciprocal control of glycogen synthase and glycogen phosphorylase. The activities of these enzymes vary according to the metabolic needs of the tissue (as in muscle) or of other tissues that use glucose as a fuel (as in liver). Proximal control is exerted on synthase and phosphorylase by phosphorylation/dephosphorylation and by allosteric effectors such as glucose, glucose-6-phosphate, and several nucleotides (ATP, ADP, AMP, and UDP). Glucagon and epinephrine activate a cell membrane–mediated signal transduction cascade resulting in production of cAMP, which, in turn, activates protein kinase followed by phosphorylation of target proteins. Phosphorylation activates glycogen phosphorylase, whereas it inactivates glycogen synthase.

Glycogen Storage Diseases

The group of glycogen storage diseases is characterized by the accumulation of normal or abnormal glycogen due to a deficiency of one of the enzymes of glycogen metabolism. Although all are rare (overall incidence of $\sim 1{:}25{,}000$ births), they have contributed greatly to the understanding of glycogen metabolism. They are summarized in Table 14.1 and Figure 14.9.

Infants with type I disease (subtypes Ia, Ib, Ic, and Id) develop hypoglycemia even after feeding because of their inability to convert glucose-6-phosphate to glucose. Lactate is produced at a high rate in extrahepatic tissues and is transported to the liver for gluconeogenesis.

Many characteristics of type I disease are attributed to the attendant hypoglycemia, and patients have been treated with frequent daytime feedings and continuous nocturnal intragastric feeding with a high-glucose formula. This regimen produces substantial improvement in growth, reduction in hepatomegaly, and normalization of other biochemical parameters. Feeding uncooked cornstarch every 6 hours resulted in normoglycemia, resumption of normal growth, and reduction in substrate cycling and liver size. The success of this simple nutritional therapy is thought to depend on slow hydrolysis of uncooked starch in the small intestine by pancreatic amylase, with continuous release and absorption of glucose. (Cornstarch that is cooked or altered in other ways is ineffective, presumably because of too rapid hydrolysis.) The therapy is ineffective in patients with low pancreatic amylase activity due, for example, to prematurity.

Defects in the glucose-6-phosphatase system are associated with severe chronic neutropenia due to altered glucose metabolism in neutrophils.

Myophosphorylase deficiency (also known as **McArdle's disease**) is a common form of glycogen storage disease. It exhibits pain, cramps, and fatigue during exercise. Sucrose ingestion before exercise in these subjects, the sucrose being rapidly hydrolyzed to glucose and fructose in the gastrointestinal tract, alleviates symptoms of exercise intolerance by providing glucose for muscle contraction.

ALTERNATIVE PATHWAYS OF GLUCOSE METABOLISM AND HEXOSE INTERCONVERSIONS

Glucuronic Acid Pathway

The **glucuronic acid pathway** is a quantitatively minor route of glucose metabolism. Like the pentose phosphate pathway (discussed later), it provides biosynthetic precursors and interconverts some less common sugars to ones that can be metabolized. The first steps are identical to those of glycogen synthesis, i.e., formation of glucose-6-phosphate, its isomerization to glucose-1-phosphate, and activation of glucose-1-phosphate to form UDP-glucose. UDP-glucose is then oxidized to UDP-glucuronic acid by NAD^+ and UDP-glucose dehydrogenase (Figure 14.10).

UDP-glucuronic acid is utilized in biosynthetic reactions that involve condensation of glucuronic acid with a variety of molecules to form an ether (glycoside), an ester, or an amide, depending on the nature of the acceptor molecule. As in condensation reactions that use nucleotide sugars as substrates, the high-energy bond between UDP and glucuronic acid provides the energy to form the new bond in the product. UDP is hydrolyzed to UMP (uridine monophosphate) and inorganic phosphate, further ensuring the irreversibility of the coupling reaction.

Because glucuronic acid is highly polar, its conjugation with less polar compounds such as steroids, bilirubin, and some drugs can reduce their activity and make them more water-soluble, thus facilitating renal excretion. Glucuronic acid is a component of the structural polysaccharides called **glycosaminoglycans** (hyaluronic acid and other connective tissue polysaccharides; see Chapter 10). Glucuronic acid is usually not a component of glycoproteins or glycolipids.

Fructose and Sorbitol Metabolism

Fructose is a ketohexose found in honey and a wide variety of fruits and vegetables. Combined with glucose in an $\alpha(1 \rightarrow 2)\beta$ linkage, it forms sucrose. It makes up one-sixth to one-third of the total carbohydrate intake of most individuals in industrialized nations.

Sorbitol, a sugar alcohol, is a minor dietary constituent. It can be synthesized in the body from glucose by NADPH-dependent aldose reductase (Figure 14.11). It is clinically important because of its relationship to cataract formation in diabetic patients. Fructose and sorbitol are catabolized by a common pathway (Figure 14.11). Fructose transport and metabolism are insulin-independent; only a few tissues (e.g., liver, kidney, intestinal mucosa, and adipose tissue, but not brain) can metabolize it. Fructose metabolism, which is much less tightly regulated, is more rapid than glucose metabolism. The renal threshold for fructose is very low, and fructose is more readily excreted in urine than glucose. Despite these differences, the metabolic fates of glucose and fructose are closely related because most fructose is ultimately converted to glucose. Fructose metabolism (Figure 14.11) starts with phosphorylation and formation of fructose-1-phosphate, and this reaction is the rate-limiting step. The K_m of hexokinase for fructose is several orders of magnitude higher than that for glucose, and at the

TABLE 14.1 Glycogen Storage Diseases

Type	Eponym	Deficiency	Glycogen Structure	Comments
Ia	von Gierke's disease (hepatorenal glycogenosis)	Glucose-6-phosphatase in liver, intestine, and kidney	Normal	Hypoglycemia; lack of glycogenolysis by epinephrine or glucagon; ketosis, hyperlipidemia, hyperuricemia; hepatomegaly; autosomal recessive. Galactose and fructose not converted to glucose.
Ib	...	Glucose-6-phosphate transporter in hepatocyte microsomal membrane	Normal	Clinically identical to type Ia.
Ic	...	Phosphate transporter in hepatocyte microsomal membrane	Normal	Clinically identical to type Ia.
Id	...	Glucose transporter GLUT7	Normal	Clinically identical to type Ia.
II	Pompe's disease (generalized glycogenosis)	Lysosomal α-1,4-glucosidase (acid maltase)	Normal	How this deficiency leads to glycogen storage is not well understood. In some cases, the heart is the main organ involved; in others, the nervous system is severely affected; autosomal recessive.
III	Forbes's disease, Cori's disease (limit dextrinosis)	Amylo-1,6-glucosidase (debranching enzyme)	Abnormal; very long inner and outer unbranched chain	Hypoglycemia; diminished hyperglycemic response to epinephrine or glucagon; normal hyperglycemic response to fructose or galactose; autosomal recessive. Six subtypes have been defined based on relative effects on liver and muscle, and on properties of the enzyme.
IV	Andersen's disease (branching deficiency, amylopectinosis)	$(1,4 \rightarrow 1,6)$-transglucosylase (branching enzyme)	Abnormal; outer chains missing or very short; increased number of branched points	Rare, or difficult to recognize; cirrhosis and storage of abnormal glycogen; diminished hyperglycemic response to epinephrine; abnormal liver function; autosomal recessive.
V	McArdle's disease	Muscle glycogen phosphorylase (myophosphorylase)	Normal	High muscle glycogen content (2.5%–4.1% versus 0.2%–0.9% normal); fall in blood lactate and pyruvate after exercise (normal is sharp rise) with no postexercise drop in pH; normal hyperglycemic response to epinephrine (thus normal hepatic enzyme); myoglobinuria after strenuous exercise; autosomal recessive.
VI	Hers' disease	Liver glycogen phosphorylase (hepatophosphorylase)	Normal	Not as serious as glucose-6-phosphatase deficiency; liver cannot make glucose from glycogen but can make it from pyruvate; mild hypoglycemia and ketosis; hepatomegaly due to glycogen accumulation; probably more than one disease; must be distinguished from defects in the glycogen phosphorylase-activating system.
VII	Tarui's disease	Muscle phosphofructokinase	Normal	Shows properties similar to type V; autosomal recessive; it is not completely clear why this defect results in increased glycogen storage.
VIII	...	Reduced activation of phosphorylase in hepatocytes and leukocytes	Normal	Hepatomegaly; increased hepatic glycogen stores; probably X-linked, but there may be more than one type, with some autosomally inherited; must be distinguished from glycogen phosphorylase deficiency.

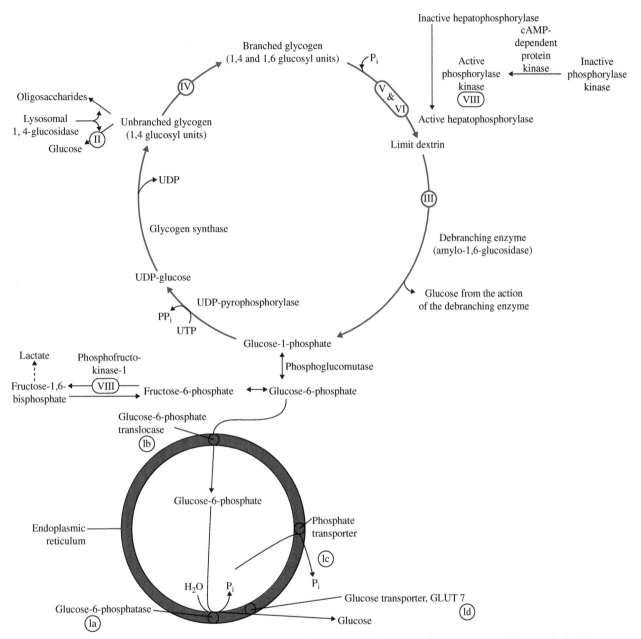

FIGURE 14.9 Locations of the glycogen storage disease enzyme defects in the overall scheme of glycogenesis and glycogenolysis. The numbers correspond to the disease types listed in Table 14.1.

concentrations of glucose found in most tissues, fructose phosphorylation by hexokinase is competitively inhibited. In tissues that contain fructokinase, such as liver, the rate of fructose phosphorylation depends primarily on fructose concentration, and an increase in fructose concentration depletes intracellular ATP. **Essential fructosuria** is caused by the absence of hepatic fructokinase activity. The condition is asymptomatic but, as with pentosuria, may be misdiagnosed as diabetes mellitus.

The next step in fructose metabolism is catalyzed by aldolase B (fructose-1-phosphate aldolase), which cleaves fructose-1-phosphate to the trioses dihydroxyacetone 3-phosphate and glyceraldehyde. Glyceraldehyde is phosphorylated by triosekinase, and glyceraldehyde 3-phosphate and dihydroxyacetone 3-phosphate can either enter the glycolytic pathway or be combined to form fructose-1,6-bisphosphate by the action of fructose-1,6-bisphosphate aldolase. Thus, fructose metabolism bypasses phosphofructokinase, the major regulatory site of glycolysis.

Most dietary fructose is converted to glucose by way of gluconeogenesis, through condensation of the triose phosphates to fructose-1,6-bisphosphate. However, administration of large doses of fructose (i.e., by intravenous feeding) may lead to hypoglycemia and lactic

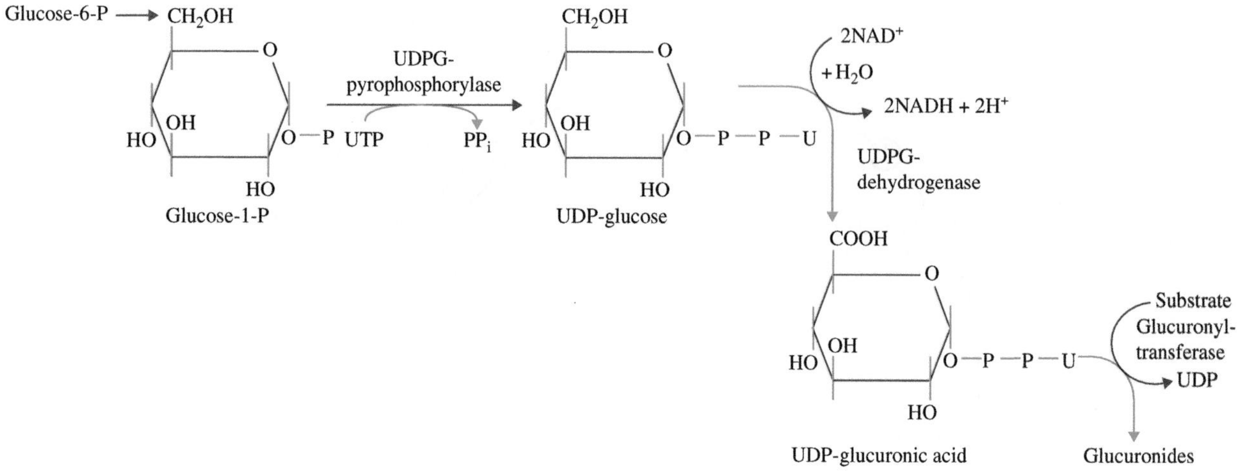

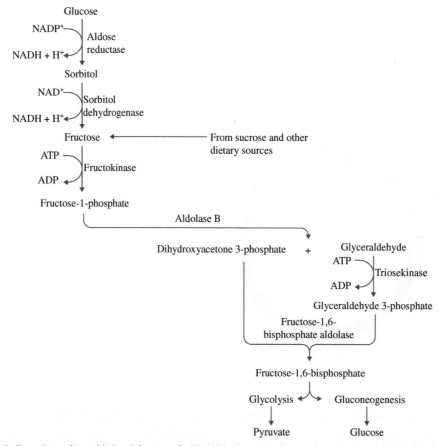

FIGURE 14.10 Synthesis of UDP-glucuronic acid. P = phosphate.

FIGURE 14.11 Metabolic pathway for sorbitol and fructose. Sorbitol dehydrogenase is sometimes known as iditol dehydrogenase. Aldolase B is also called fructose-1-phosphate aldolase, in contrast to fructose-1,6-bisphosphate aldolase.

acidosis because of saturation of aldolase B, causing accumulation of fructose-1-phosphate, and depletion of intracellular ATP and inorganic phosphate. This situation may be likened to **hereditary fructose intolerance**, which is caused by inadequate amounts of aldolase B activity. Although normally asymptomatic, individuals with this condition exhibit hypoglycemia, metabolic acidosis, vomiting, convulsions, coma, and signs of liver failure following ingestion of fructose or sucrose. The fructose-induced hypoglycemia arises from accumulation of fructose-1-phosphate, reduction in the [ATP]/[ADP] ratio, and depletion of inorganic phosphate.

Fructose-1-phosphate depresses gluconeogenesis and promotes glycolysis, while inorganic phosphate depletion inhibits ATP synthesis (Chapter 13). Glycogenolysis is inhibited by fructose-1-phosphate at the level of phosphorylase. If sucrose, fructose, and sorbitol are eliminated from the diet, complete recovery occurs. High levels of fructose also increase purine turnover owing to enhanced ATP utilization, which leads to increased production of purine degradation products: inosine, hypoxanthine, xanthine, and uric acid (Chapter 25).

Galactose Metabolism

Most galactose ingested by humans is in the form of lactose, the principal sugar in human and bovine milk. Milk sugar other than lactose is found in the sea lion and marsupials, whose first pouch milk contains a trisaccharide of galactose. Lactose is hydrolyzed to galactose and glucose by lactase, located on the microvillar membrane of the small intestine (Chapter 11). Following absorption, galactose is transported to the liver, where it is converted to glucose (Figure 14.12). The enzymes required are found in many tissues, but the liver is quantitatively the most important site for this epimerization.

Galactose is a poor substrate for hexokinase; it is phosphorylated by galactokinase. Galactose-1-phosphate is converted to UDP-galactose by galactose-1-phosphate uridylyltransferase. This enzyme may be regulated by substrate availability, since the normal hepatic concentration of galactose-1-phosphate is close to the K_m for this enzyme. The transferase is inhibited by UDP, UTP, glucose-1-phosphate, and high concentrations of UDP-glucose. Deficiency of galactokinase or transferase can cause galactosemia.

Galactose is isomerized to glucose by UDP-galactose-4-epimerase in what may be the rate-limiting step in galactose metabolism. The reaction, which is freely reversible, converts UDP-glucose to UDP-galactose for the synthesis of lactose, glycoproteins, and glycolipids. The epimerase requires NAD^+ and is inhibited by NADH. It may also be regulated by the concentrations of UDP-glucose and other uridine nucleotides. Because of this epimerase, preformed galactose is not normally required in the diet. However, infants with deficiency of epimerase in hepatic and extrahepatic tissues require small quantities of dietary galactose for normal development and growth. An isolated deficiency of epimerase in erythrocytes, which is clinically benign, has been described.

Genetically determined deficiencies of galactokinase and galactose-1-phosphate uridylyltransferase cause clinically significant galactosemia. Galactokinase deficiency is a rare, autosomal recessive trait in which high concentrations of galactose are found in the blood, particularly after a meal that includes lactose-rich foods, such as milk and nonfermented milk products. Patients frequently develop cataracts before one year of age because of the accumulation of galactitol in the lens. Galactose diffuses freely into the lens, where it is reduced by aldose reductase, the enzyme that converts glucose to sorbitol. Galactitol may cause lens opacity by the same mechanisms as for sorbitol. Galactose in the urine is detected by nonspecific tests for reducing substances, as are fructose and glucose.

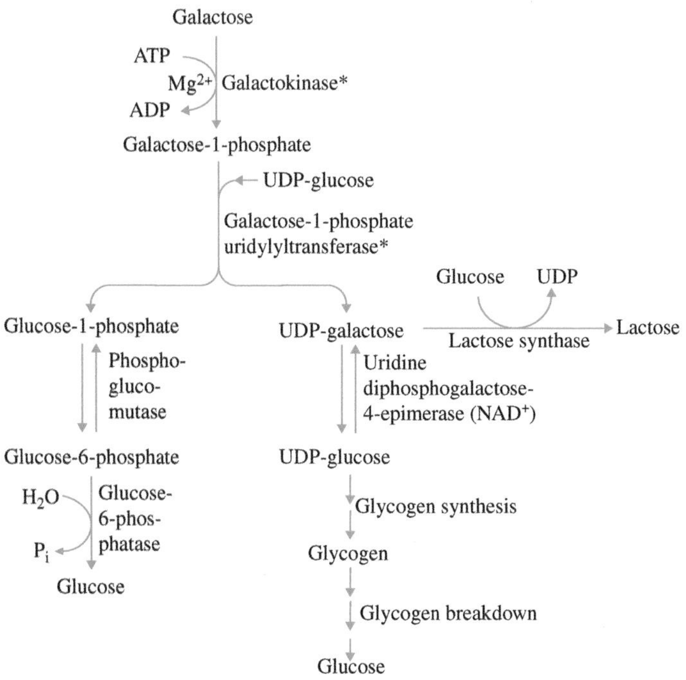

FIGURE 14.12 Metabolic pathway of galactose. *Inherited defects that lead to galactosemia.

Deficiency of the transferase causes a more severe form of galactosemia. Symptoms include cataracts, vomiting, diarrhea, jaundice, hepatosplenomegaly, failure to thrive, and mental retardation. If galactosuria is severe, nephrotoxicity with albuminuria and aminoaciduria may occur. The more severe clinical course may be due to accumulation of galactose-1-phosphate in cells. Measurement of transferase activity in red blood cells is the definitive test for this disorder. In about 70% of transferase alleles in the white population, the DNA in cells of transferase-deficient patients possesses an A-to-G transition that leads to the Q186R mutation.

Both forms of galactosemia are treated by rigorous exclusion of galactose from the diet. In pregnancies where the family history suggests that the infant may be affected, the best outcomes have been reported when the mother was maintained on a galactose-free diet during gestation.

Condensation of glucose with UDP-galactose, catalyzed by lactose synthase (Figure 14.12), forms lactose, the only disaccharide made in large quantities by mammals. This enzyme is a complex of galactosyltransferase, a membrane-bound enzyme that participates in glycoprotein synthesis (Chapter 10), and α-lactalbumin, a soluble protein secreted by the lactating mammary gland. Binding of α-lactalbumin to galactosyltransferase changes the K_m of the transferase for glucose from $1-2$ mol/L to 10^{-3} mol/L. Synthesis of α-lactalbumin by the mammary gland is initiated late in pregnancy or at parturition by the sudden decrease in progesterone levels that occurs at that time. Prolactin promotes the rate of synthesis of galactosyltransferase and α-lactalbumin.

Metabolism of Amino Sugars

In amino sugars, one hydroxyl group, usually at C_2, is replaced by an amino group. Amino sugars are important constituents of many complex polysaccharides, including glycoproteins and glycolipids, and of glycosaminoglycans (Chapter 10). *De novo* synthesis of amino sugars starts from glucose-6-phosphate, but salvage pathways can also operate.

Pentose Phosphate Pathway

The series of cytoplasmic reactions known as the **pentose phosphate pathway** are also called the **hexose monophosphate** (HMP) **shunt** (or cycle) or the **phosphogluconate pathway**. The qualitative interconversions that take place are summarized in Figure 14.13, in which stoichiometry is ignored.

The pentose phosphate pathway can be thought of as two separate pathways:

1. The oxidative conversion of glucose-6-phosphate to ribulose 5-phosphate and CO_2; and
2. Resynthesis of glucose-6-phosphate from ribulose 5-phosphate. Because the pathway begins and ends with

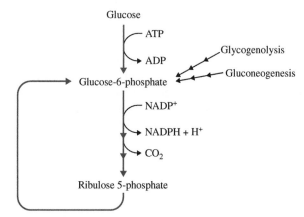

FIGURE 14.13 Summary of the pentose phosphate pathway. This diagram is intended to show the two major parts of the pathway: oxidation and decarboxylation of glucose-6-phosphate to ribulose-5-phosphate; and resynthesis of the former from the latter. The stoichiometry of the pathway is ignored.

glucose-6-phosphate, it is a cycle, or shunt. For every six molecules of glucose-6-phosphate that enter the pathway, five molecules of glucose-6-phosphate + $6CO_2$ are produced. Thus, there is a net loss of one carbon from each glucose-6-phosphate that enters the cycle.

The pathway performs a variety of functions. Production of ribose 5-phosphate for nucleotide synthesis by *de novo* and salvage pathways (Chapter 25), generation of NADPH for biosynthetic reactions in the cytosol, and interconversion of pentoses and hexoses are more valuable processes. Thus, the pathway provides a means for generating glucose from ribose and other pentoses that can be converted to ribose 5-phosphate.

The pentose phosphate pathway is active in a wide variety of cell types, particularly those that have a high rate of nucleotide synthesis or that utilize NADPH in large amounts. Nucleotide synthesis is greatest in rapidly dividing tissues, such as bone marrow, skin, and gastric mucosa. NADPH is needed in the biosynthesis of fatty acids (liver, adipose tissue, lactating mammary gland), cholesterol (liver, adrenal cortex, skin, intestine, gonads), steroid hormones (adrenal cortex, gonads), and catecholamines (nervous system, adrenal medulla), and in other reactions that involve tetrahydrobiopterin. NADPH is also needed for maintenance of a reducing atmosphere in cells exposed to high concentrations of oxygen radicals including erythrocytes, lens and cornea of the eye, and phagocytic cells, which generate peroxide and superoxide anions during the process of killing bacteria.

Oxidative Phase

The first reaction in the oxidative phase is the oxidation of glucose-6-phosphate to 6-phosphoglucono-δ-lactone and reduction of $NADP^+$ to NADPH, catalyzed by glucose-6-phosphate dehydrogenase (G6PD):

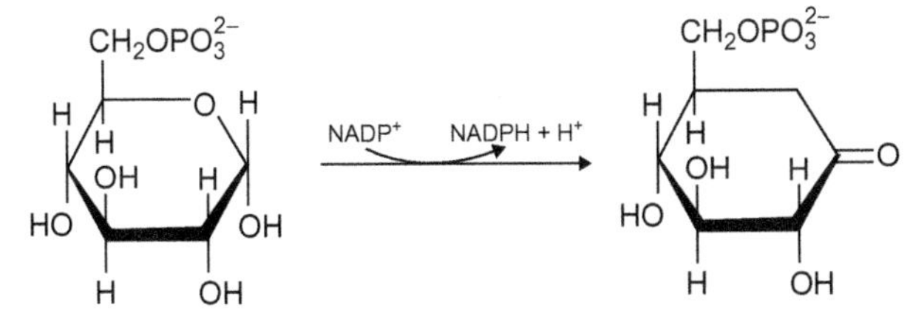

Glucose-6-phosphate → 6-phosphoglucono-δ-lactone

In the second step, 6-phosphoglucono-δ-lactone is hydrolyzed to 6-phosphogluconate by 6-phosphogluconolactonase:

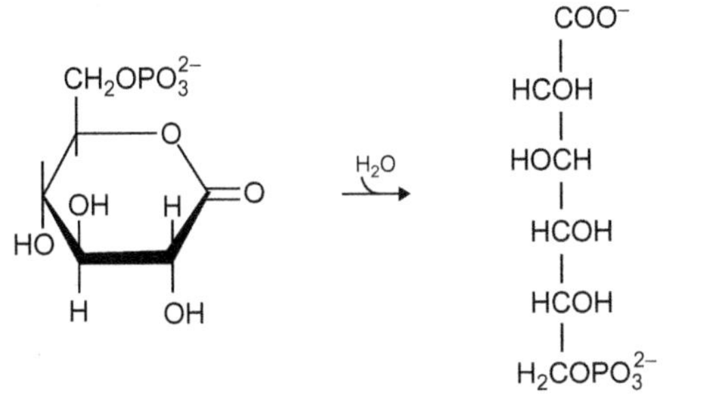

6-phosphoglucono-δ-lactone → 6-phosphogluconate

In the third step, 6-phosphogluconate is oxidatively decarboxylated to ribulose-5-phosphate in the presence of $NADP^+$, catalyzed by 6-phosphogluconate dehydrogenase:

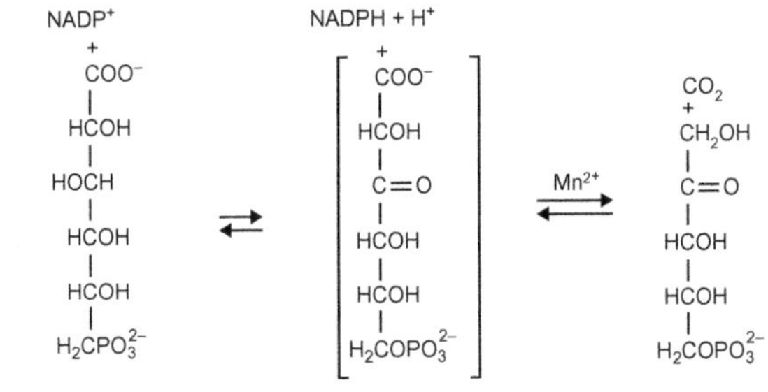

6-phosphogluconate → 3-Keto-6-phosphogluconate → D-ribulose-5-phosphate

Nonoxidative Phase

In the nonoxidative phase, ribulose-6-phosphate is converted to glucose-6-phosphate. Stoichiometrically, this process requires the rearrangement of six molecules of ketopentose phosphate to five molecules of aldohexose phosphate. One of the steps is catalyzed by transketolase, which requires thiamine pyrophosphate and Mg^{2+} as cofactors.

The versatility of this phase of the pathway allows for interconversion of a number of sugars and glycolytic intermediates, in part because of the ready reversibility of the reactions and regulation of the enzymes by substrate availability. Thus, the pathway modulates the concentrations of a number of sugars rather than acts as a unidirectional anabolic or catabolic route for carbohydrates.

Pentose Phosphate Pathway in Red Blood Cells

In the erythrocyte, glycolysis, the pentose phosphate pathway, and the metabolism of 2,3-bisphosphoglycerate (see Chapters 12 and 26) are the predominant pathways of carbohydrate metabolism. Glycolysis supplies ATP for membrane ion pumps and NADH for reoxidation of methemoglobin. The pentose phosphate pathway supplies NADPH to reduce glutathione for protection against oxidant injury. Glutathione (γ-glutamylcysteinylglycine; GSH) is synthesized by γ-glutamylcysteine synthase and GSH synthase (Figure 14.14). In the steady state, 99.8% of the glutathione is in the reduced form (GSH), and only 0.2% is in the oxidized form (GSSG), because of the NADPH-dependent reduction of oxidized glutathione, catalyzed by glutathione reductase:

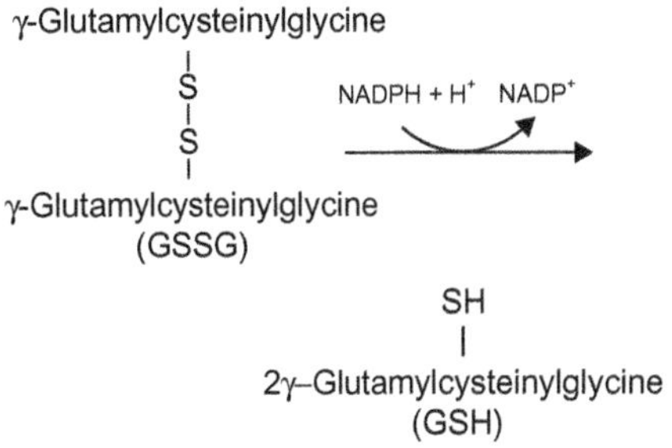

Glutathione reductase also catalyzes reduction of mixed disulfides of glutathione and proteins (Pr):

$$Pr - S - SG + NADPH + H^+ \rightarrow Pr - SH + GSH + NADP^+$$

Thus, the active sulfhydryl groups of hexokinase, glyceraldehyde−phosphate dehydrogenase, glutathione reductase, and hemoglobin (−SH group at the β-93 position) are maintained in the reduced form. Glutathione reductase is an FAD enzyme composed of two identical subunits encoded by a single locus on the short arm of human chromosome 8.

GSH is used in the inactivation of potentially damaging organic peroxides (e.g., a peroxidized unsaturated fatty acid) and of hydrogen peroxide. Hydrogen peroxide is formed through the action of superoxide dismutase (Chapter 13):

$$2O_2^- + 2H^+ \rightarrow H_2O_2 + O_2$$

Peroxide inactivation is catalyzed by glutathione peroxidase, a selenium-containing enzyme, in the following reactions:

$$2GSH + ROOH \rightarrow GSSG + H_2O + ROH$$

and

$$2GSH + H_2O_2 \rightarrow GSSG + 2H_2O$$

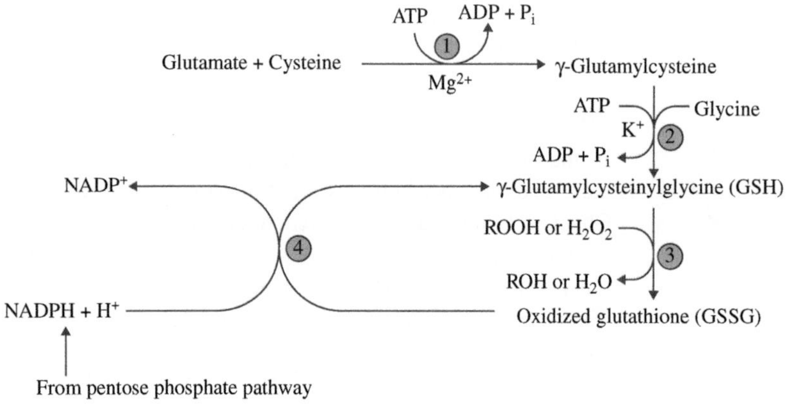

FIGURE   14.14 Metabolism   of   glutathione.
(1) γ-Glutamylcysteine synthase; (2) GSH synthase;
(3) glutathione peroxidase (Se-containing enzyme);
(4) glutathione reductase (an FAD enzyme).

The concentrations of substrates for glutathione reductase and peroxidase determine the rate at which the pentose phosphate pathway operates in erythrocytes.

Genetically determined deficiencies of γ-glutamylcysteine synthase and glutathione synthase can cause hemolytic anemia. Abnormalities of glutathione metabolism can also result from nutritional deficiencies of riboflavin or selenium. Glutathione reductase is an FAD enzyme that requires riboflavin for activity, and riboflavin deficiency can cause hemolytic anemia. An inherited lack of glutathione reductase apoenzyme has been described. Selenium deficiency diminishes activity of glutathione peroxidase and can lead to peroxidative damage. This can be partially ameliorated by vitamin E, an antioxidant (Chapters 35 and 36).

Glucose-6-Phosphate Dehydrogenase Deficiency

G6PD deficiency is the most common inherited enzyme deficiency known to cause human disease, occurring in about 100 million people (see Clinical Case Study 14.2). Most clinical manifestations are related to hemolysis, which results from impaired ability to produce cytosolic NADPH. Over 150 variants of the G6PD structural gene are known, many of which show either abnormal kinetics or instability of the enzyme. The erythrocytes are most severely affected because of their long half-lives and inability to carry out protein synthesis. Since persons with G6PD deficiency can usually make an adequate supply of NADPH under normal conditions, the defect may not become apparent until the patient takes a drug, such as primaquine, that greatly increases the demand for NADPH. The severity of the reaction depends, in part, on the particular inherited mutation.

In most persons with G6PD deficiency, hemolysis is observed as an acute phenomenon only after severe oxidative stress, leading to loss of perhaps 30%−50% of the circulating red cells. The urine may turn dark, even black, from the high concentration of hemoglobin, and a high urine flow must be maintained to prevent damage to the renal tubules by the high protein load. Because G6PD activity is highest in reticulocytes and decreases as the cell ages, the older erythrocytes (more than 70 days old) are destroyed. For this reason, measurement of red cell G6PD activity following a hemolytic crisis can lead to a spuriously high value, even within the normal range. If the patient survives the initial crisis, with or without transfusions, recovery usually occurs as the reticulocyte count increases.

The characteristics of the disorder were elucidated during investigations into the hemolytic crises observed in some patients following administration of 8-aminoquinoline derivatives, such as primaquine and pamaquine, used for prophylaxis and treatment of malaria. The relatively high frequency of such crises in some geographic areas is due to the extensive overlap in the distributions of endemic malaria and G6PD deficiency. The geographic distribution of G6PD variants, like that of sickle cell trait (Chapter 26), suggests that heterozygosity for G6PD deficiency may confer some protection against **falciparum malaria**. A number of other drugs that cause oxidative stress also cause hemolytic crises.

Oxidation of glutathione by these drugs beyond the capacity of the cell to generate NADPH for GSSG reduction causes the acute crisis. Since these drugs do not cause hemolysis *in vitro*, additional steps must take place *in vivo*. Following administration of a drug known to promote hemolysis, Heinz bodies are seen in erythrocytes in the peripheral blood. These also occur in some **thalassemias** and consist of oxidized and denatured forms of hemoglobin known as **hemichromes** (Chapter 27). Heinz bodies impair movement of the cells through the splenic pulp and probably are excised there, presumably together with the adjacent piece of plasma membrane, leaving a red cell that is more susceptible to destruction by the reticuloendothelial system. Infection and diabetic ketoacidosis can also cause hemolysis in persons with G6PD deficiency, possibly through depletion of NADPH.

Favism is characterized by acute hemolysis following ingestion of fava beans (*Vicia fava*) by people with G6PD deficiency. The fava bean is a vegetable staple of the Mediterranean region, an area in which G6PD deficiency is endemic. Infants are especially susceptible to favism. The disorder is frequently fatal unless a large amount of blood is transfused rapidly.

Studies of the genetics of human G6PD variants have contributed to the understanding of G6PD deficiency and of more general aspects of human genetics. G6PD deficiency is inherited as an X-linked trait, as are hemophilia (Chapter 34) and color blindness (Chapter 36). If the X-chromosome carrying an abnormal G6PD allele is designated X*, then the three possible genotypes containing X are

1. X*Y-hemizygous male, with full phenotypic expression of the abnormal allele;
2. XX*-heterozygous female, with a clinically normal phenotype in spite of the abnormal allele expressed in about half her cells; and
3. X*X*-homozygous female, with full phenotypic expression of the abnormal allele. Sons of affected males are usually normal (because they receive their X-chromosome from their mothers), and daughters of affected males are usually heterozygotes (because they receive one X-chromosome from their father). The rarest genotype is that of the homozygous female, since it requires that both parents have at least one abnormal X-chromosome.

Females heterozygous for a G6PD variant are phenotypic mosaics. They have two erythrocyte populations: one containing normal G6PD, the other containing the variant. In fact, in heterozygotes, every tissue has some cells expressing the normal, and some the abnormal, G6PD gene. Random X-chromosome inactivation early in embryonic development causes only one of the two X-chromosomes to be active (**Lyon's hypothesis**).

Severity of the disorder in homozygotes and hemizygotes depends on a number of factors. G6PD variants are classified into five groups (class I being the most severe) depending on the presence or absence of chronic anemia and on the amount of enzyme activity present in the erythrocytes. The most common normal activity (class IV) allele is G6PD B. Another common electrophoretic variant with normal activity is G6PD A. Among north American blacks, the gene for the absence of G6PD A (G6PD A$^-$; class III) has an incidence of about 11%. Hemizygous males have only 5%−15% of normal erythrocyte G6PD activity and exhibit a mild hemolytic anemia following an oxidative insult, such as primaquine

administration. Hemolysis may cease even with continued administration of the drug because the reticulocytes, which increase in proportion following hemolysis, have adequate G6PD activity. In persons of Mediterranean and Middle Eastern ancestry, the most common abnormal allele is G6PD M (G6PD Mediterranean; class II) associated with severe hemolysis following administration of an appropriate drug. The average incidence of this allele is approximately 5%−10%, but a subpopulation of Kurdish Jews is reported to have an incidence of 50%. Enzyme activity in erythrocytes from these patients is often less than 1% of normal, and transfusion is usually required following a hemolytic crisis. The difference in the clinical severity of the diseases associated with G6PD A$^-$ is a normal K_m for glucose-6-phosphate (50−70 µmol/L) and for NADP$^+$ (2.9−4 µmol/L), but it exhibits an abnormal pH activity curve. The deficiency is due to an accelerated rate of inactivation of G6PD A$^-$ protein. Bone marrow cells and reticulocytes have normal amounts of G6PD activity, while activity in older red cells is very low. However, the residual enzyme seems to be more resistant to inhibition by NADPH. In contrast, G6PD M molecules have an intrinsically lower catalytic activity, and both reticulocytes and older red cells have decreased amounts of G6PD activity. Consequently, when one of the drugs is given, more cells are susceptible, and hemolysis is greater than with G6PD A$^-$. The low K_m of G6PD M for G6P and NADP$^+$ may account for the near-normal survival of erythrocytes in the absence of oxidative stress.

A number of other enzymopathic disorders (e.g., pyruvate kinase, Chapter 12; and pyrimidine-5′-nucleotidase, Chapter 25), abnormal hemoglobins (Chapter 26), and abnormalities of the erythrocyte cytoskeleton (Chapter 10) may cause hemolytic anemia. Because many enzymes in the red cell are identical to those in other tissues, defects in these enzymes may have pleiotropic effects. Thus, in addition to hemolytic anemia, triose phosphate isomerase deficiency causes severe neuromuscular disease, and phosphofructokinase deficiency causes a muscle glycogen storage disease. Mutations that result in decreased enzyme stability are usually most strongly expressed in erythrocytes because of their inability to synthesize proteins.

Phagocytosis and the Pentose Phosphate Pathway

The pentose phosphate pathway is crucial to the survival of erythrocytes because of its ability to provide NADPH for reduction of toxic, spontaneously produced oxidants. In phagocytic cells, the pentose phosphate pathway generates oxidizing agents using molecular oxygen and NADPH oxidase that participate in the killing of bacteria and abnormal cells engulfed by phagocytes (see Chapter 13 for a discussion on this topic).

CLINICAL CASE STUDY 14.1 Deficiency of Glycogenin-1 Leading to Glycogen Depletion in Skeletal Muscle Fibers and Cardiac Myocytes

This clinical case was abstracted from: A.-R. Moslemi, C. Lindberg, J. Nilsson, H. Tajsharghi, B. Andersson, A. Oldfors, Glycogenin-1 deficiency and inactivated priming of glycogen synthesis, N. Engl. J. Med. 362 (2010) 1203–1210.

Synopsis

A 27-year-old man, after exercise, experienced dizziness and cardiac symptoms and was brought to emergency medical services at the hospital by ambulance. In the ambulance, the subject required cardiac defibrillation to establish normal sinus rhythm. After several laboratory tests, echocardiography, and coronary angiography, the subject's cardiac problems were corrected by cardioverter-defibrillator implantation, and pharmacotherapy with a β_1-adrenergic-receptor blocker and an angiotensin-converting enzyme inhibitor.

In order to definitively establish the biochemical basis of the subject's abnormalities, biopsies of deltoid muscle and endomyocardium were obtained. The histochemical and electron microscopic studies revealed depleted glycogen levels in both tissues, and mitochondrial proliferation and hypertrophy of cardiomyocytes. Relevant molecular biochemical studies revealed a missense mutation (Thr83Met) in the glycogenin-1 gene, and Western blotting showed the accumulation of unglycosylated defective glycogenin-1 protein.

Teaching Points

1. Glycogen, a branched polymer of glucose with $\alpha(1\to4)$ and $\alpha(1\to6)$ linkages, serves as an energy store for many tissues. Its synthesis requires three enzymes: autocatalytic glucosylation of glycogenin, which provides a priming oligosaccharide chain; glycogen synthase, which extends the oligosaccharide chain; and branching enzyme, which is responsible for the synthesis of highly branched polymers.

2. Glycogenin-1, the predominant isoform of skeletal muscle, is also present in cardiac muscle, along with the glycogenin-2 isoform. The deficiency of glycogenin-1 leads to lack of glycogen energy stores and is accompanied by metabolic abnormalities as discussed in this case. It should be emphasized that, despite the fact that cardiac muscle is capable of the oxidation of fatty acids as an energy source, glucose is an important energy source to meet its metabolic demands.

3. Glycogen synthase deficiency also leads to metabolic abnormalities similar to glycogenin-1 deficiency (see reference 1).

4. Deficiency of glycogenin-1 or glycogen synthase affects plasma glucose clearance. This observation has been attributed to a compensatory mitochondrial proliferation, which augments skeletal muscle glucose oxidation.

Integration of this topic with Chapters 8, 14, and 19.

Reference
[1] G. Kollberg, M. Tulinius, T. Gilljam, I. Ostman-Smith, G. Forsander, P. Jotorp, et al., Cardiomyopathy and exercise intolerance in muscle glycogen storage disease, N. Engl. J. Med. 357 (2007) 1507–1514.

CLINICAL CASE STUDY 14.2 Glucose-6-Phosphate Dehydrogenase (G6PD) Deficiency

This clinical case was abstracted from: A. Puig, A.S. Dighe, Case 20-2013: A 29-year-old man with anemia and jaundice, N. Engl. J. Med. 368 (2013) 2502–2509.

Synopsis

A 29-year-old man from North Africa was admitted to hospital with key complaints and clinical features of fatigue, jaundice, and dark urine. His pertinent laboratory findings included normocytic and normochromic anemia, elevated serum unconjugated bilirubin, and decreased haptoglobin levels. Other laboratory investigations such as liver function tests and hemoglobin electrophoresis to detect abnormal variants were all negative. Based on history, laboratory findings led to investigation of the G6PD level in the subject's red blood cells. The G6PD level was significantly decreased to 1.9 U per gram hemoglobin (reference interval 8.8 to 13.4). The subject's diagnosis of intravascular hemolysis was attributed to G6PD deficiency.

Resolution of the case: The subject's clinical findings resolved without any intervention after a few days.

Teaching Points

1. Ten to fifteen percent of the subject's black African decent ancestry have G6PD deficiency designated as A⁻. The mutation causes the production of an unstable G6PD whose levels also decrease as the lifespan of the red blood cells (about 128 days) declines. Thus, in an acute episode with a preponderance of young red blood cells entering the circulation, the G6PD level may be normal. In that case, the G6PD level should be determined after about 4–6 weeks.

2. Any oxidative stress or illness renders red blood cells susceptible to a hemolysis episode in G6PD deficiency.

3. The key laboratory indications for the *in vivo* hemolysis in this subject were normocytic normochromic anemia, elevated serum level of unconjugated bilirubin, decreased haptoglobin level due to its consumption with hemoglobin, and elevated lactate dehydrogenase level (see also Chapter 27).

4. The subject's black urine was caused by filtered hemoglobin and its oxidized products.

REQUIRED READING

[1] C.A. Worby, J.E. Dixon, Glycogen synthase: an old enzyme with a new trick, Cell Metab. 13 (2011) 233–234.

[2] V.S. Tagliabracci, C. Heiss, C. Karthik, C.J. Contreras, J. Glushka, M. Ishihara, et al., Phosphate incorporation during glycogen synthesis and lafora disease, Cell Metab. 13 (2011) 274–282.

[3] V.S. Tagliabracci, J. Turnbull, W. Wang, J. Girard, X. Zhao, A.V. Skurat, et al., Laforin is a glycogen phosphatase, deficiency of which leads to elevated phosphorylation of glycogen in vivo, Proc. Natl Acad. Sci. USA 104 (2007) 19262–19266.

Chapter 15

Protein and Amino Acid Metabolism

Key Points

1. Proteins constantly undergo breakdown and synthesis, which is known as protein turnover that changes during different physiologic and pathologic states of life. Serum protein analysis provides valuable information on nutritional status and in the diagnosis of various abnormalities.

2. Proper maintenance of the turnover rates of proteins requires that all their constituent 20 amino acids be available at optimal concentrations. Some of the amino acids are not synthesized in the body; they are called essential amino acids. Eight essential amino acids (isoleucine, leucine, lysine, methionine, phenylalanine, threonine, tryptophan, and valine) and two amino acids (arginine and histidine) are conditionally essential because of their limited supply during certain physiologic and pathologic states.

3. The body's requirements for amino acids are met by dietary sources, which provide both essential and nonessential amino acids. Nonessential amino acids can also be obtained from adding amino groups to endogenous α-ketoacid analogs, which are derived from carbohydrate metabolites. Ingestion of appropriate amounts of animal proteins (e.g., meat and milk) satisfies the requirement for all of the essential amino acids. Although some vegetable proteins may lack one or more essential amino acids, a proper combination of sources to complement each of the deficient amino acids provides all of the essential amino acids.

4. Protein and energy malnutrition results in severe clinical manifestations. Kwashiorkor disorder is due to inadequate intake of protein, and Marasmus disorder is due to deficiency of both protein and energy intake. Measurement of serum transthyretin (also called prealbumin) serves as a marker for protein malnutrition.

5. Several membrane-bound transport carrier systems, both specific and group specific, are required for amino acid transport across cell membranes. They include the Na^+-dependent carrier-mediated active transport systems in intestinal mucosal cells and in the kidney. Na^+-independent systems also participate in amino acid transport. Translocation of amino acids is also accomplished by the γ-glutamyl cycle, which involves six enzymes (one membrane bound, the rest cytosolic), glutathione, and the expenditure of ATP.

6. The major reactions of amino acids are:
   a. Oxidative and nonoxidative reactions convert amino acids to their respective α-keto acids, which can be oxidized in the TCA cycle (Chapter 12).
   b. The transamination reaction involves transfer of an amino group of an amino acid to an α-ketoacid with participation of a glutamate/α-ketoglutarate pair. The reaction is catalyzed by pyridoxal phosphate-dependent transaminases (also known as aminotransferases). Aminotransferases are strategically located in the cytosol and in mitochondria, and they play an important role in integrating various metabolic pathways.
   c. Three enzymes are involved in the incorporation of ammonium (ammonia) into organic metabolites: (1) carbamoyl phosphate synthase I, a mitochondrial enzyme, initiates the metabolic pathway for arginine biosynthesis and urea formation in the liver; (2) glutamate dehydrogenase, found in both cytosol and mitochondria, catalyzes the reversible addition of NH_4^+ (amination) to α-ketoglutarate and provides an integration between an amino acid and carbohydrate metabolism; and (3) glutamine synthetase catalyzes an ATP-dependent reaction, the amidation of the γ-carboxyl group of glutamate to yield glutamine.

7. Both glutamate dehydrogenase and glutamine synthetase are regulated by allosteric effectors. Both enzymes play important metabolic roles in the astrocyte–neuronal functional unit of the nervous system. The amide nitrogen of glutamine is required in the biosynthesis of purines, pyrimidines, and amino acids. In the kidneys, glutamine, upon the action of glutaminase, yields NH_3, which is required to maintain acid–base balance (Chapter 37).

8. The toxic waste product NH_3, which is formed predominantly in the large intestine due to the action of bacterial enzymes on nitrogen metabolites, is converted to the nontoxic water-soluble product urea in the liver. Urea is formed in the urea cycle, and it consists of six biochemical reactions located in both mitochondria and cytosol. Urea production requires expenditure of ATP, and it is eliminated from the body via the kidneys. Urea cycle enzymes are also utilized in the *de novo* synthesis of arginine.

N. V. Bhagavan and C.-E. Ha: Essentials of Medical Biochemistry, Second edition. DOI: http://dx.doi.org/10.1016/B978-0-12-416687-5.00015-4

9. NH_3 is particularly toxic to the central nervous system. Inherited metabolic disorders of the urea cycle, which result in severe hyperammonemia, are treated with sodium phenylacetate and sodium benzoate, which eliminate ammonia using alternate pathways. Hyperammonemia due to liver disease is managed by promoting loss of NH_3 as NH_4^+ ions in the GI tract by oral administration of lactulose (Chapter 8).

10. Arginine is the precursor of nitric oxide (NO). NO has many biological functions, including neurotransmission, blood vessel relaxation, immune response, and phagocytosis. NO synthesis is catalyzed by a family of constitutive and transcriptionally regulated (inducible) nitric oxide synthases (NOS). Both neuronal and endothelial NOS are calmodulin-dependent constitutive enzymes and are stimulated by intracellular Ca^{2+}. NOS found in phagocytic cells is calcium independent and is an inducible enzyme. NOS requires five enzyme cofactors. Nitric oxide activates guanylate cyclase, which produces cGMP from GTP, which mediates smooth muscle relaxation. Inhibition of cGMP phosphodiesterase by therapeutic agents potentiates the action of NO. Overproduction of NO can cause severe metabolic abnormalities (e.g., hypotension, stroke). Therapeutic actions of nitrovasodilators (e.g., nitrites, organic nitrates, sodium nitroprusside) are mediated by production of NO in the body. Inhaled nitric oxide therapy has a potential application in the treatment of pulmonary hypertension. In some proteins, arginine residues are post-translationally converted to citrulline residues by peptidylarginine deiminase. Autoantibodies directed against citrullinated proteins are used as serum markers for rheumatoid arthritis.

11. Glycine participates in one-carbon metabolism and is a neurotransmitter. It is utilized in the synthesis of glutathione and creatine.

12. Creatine synthesis requires glycine, arginine, and the methyl donor S-adenosyl methionine. Creatine is converted to phosphocreatine by ATP and creatine kinase. Phosphocreatine serves as an energy reservoir in tissues. Phosphocreatine undergoes a slow nonenzymatic cyclization to form the end product creatinine. Creatinine is eliminated by the kidneys, and its measurements in urine and serum are used in the assessment of renal function as glomerular filtration rate (GFR). GFR can be estimated from equations developed using serum creatinine levels. These equations have been improved using combined values for both serum creatinine and cystatin C. Cystatin C is a small molecular protein completely filtered in the kidneys and not reabsorbed.

13. Autosomal recessively inherited defects in the oxidative decarboxylation of the branched-chain amino acids leucine, isoleucine, and valine cause metabolic crisis and neurologic deficits. Early diagnosis and dietary restriction of these amino acids are used in the management of those disorders.

14. The sulfur-containing amino acids methionine, cysteine, and cystine are interrelated. Methionine is converted to S-adenosylmethionine with the adenosyl moiety provided by ATP. S-adenosylmethionine, which is a sulfonium compound, serves as a methyl donor to several metabolites and in the methylation of some arginine and lysine residues of histones and cytosine residues of DNA, both of which are involved in epigenetic effects (Chapter 24). Some of these reactions require vitamin B_6, folate, and vitamin B_{12}.

15. Phenylalanine and tyrosine are involved in the synthesis of neurotransmitters (dopamine, epinephrine, and norepinephrine), hormones (thyroxine and triiodothyronine), and the skin pigment melanin. Defects in the conversion of phenylalanine to tyrosine catalyzed by phenylalanine hydroxylase can lead to mental retardation and related disorders. Early identification of a disorder and dietary restriction of phenylalanine are required in its management.

16. Tryptophan is the precursor for the synthesis of the neurotransmitter serotonin, and the hormone melatonin.

No special storage forms of either the nitrogen or the amino acid components of proteins exist, in contrast to the case of lipids and carbohydrates. Dietary protein in excess of the body's requirement is catabolized to provide energy and ammonia, a toxic metabolite that is converted to urea in the liver and excreted by the kidneys. All body proteins serve a specific function (e.g., structural, catalytic, transport, regulatory) and are potential sources of carbon for energy production.

Proteins constantly undergo breakdown and synthesis. During growth, even though there is net deposition of protein, the rates of synthesis and breakdown are increased. Total protein turnover in a well-fed adult human is estimated at about 300 grams per day, of which approximately 100 g are myofibrillar protein, 30 g are digestive enzymes, 20 g are small intestinal cell protein, and 15 g are hemoglobin. The remainder is accounted for by turnover of cellular proteins of various other cells (e.g., hepatocytes, leukocytes, platelets) and oxidation of amino acids, and a small amount is lost as free amino acids in urine. Protein turnover rates vary from tissue to tissue, and the relative tissue contribution to total protein turnover is altered by aging, disease, and changes in dietary protein intake. Several proteins (e.g., many hepatic enzymes) have short turnover times (less than 1 hour), whereas others have much longer turnover times (e.g., collagen >1000 days). Turnover of myofibrillar protein can be estimated by measurement of 3-methylhistidine in the urine. Histidyl residues of actin and myosin are released during catabolism of these proteins and are excreted in the urine.

Protein turnover is not completely efficient in the reutilization of amino acids. Some are lost by oxidative catabolism, while others are used in synthesis of nonprotein metabolites. For this reason, a dietary source of proteins is needed to maintain adequate synthesis of protein.

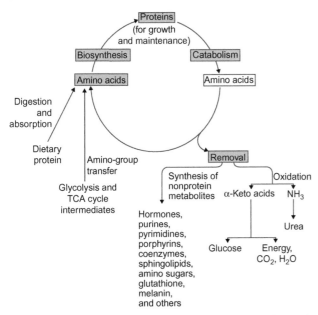

FIGURE 15.1 Overall metabolism of proteins and amino acids. Body protein is maintained by the balance between the rates of protein synthesis and breakdown. These processes are influenced by hormones and energy supply.

During periods of growth, pregnancy, lactation, or recovery from illness, supplemental dietary protein is required. These processes are affected by energy supply and hormonal factors. An overview of amino acid metabolism is presented in Figure 15.1.

Serum and urine protein analysis are widely used in the diagnosis of several disorders. A discussion of this aspect is included with the clinical case studies.

ESSENTIAL AND NONESSENTIAL AMINO ACIDS

Plants and some bacteria synthesize all 20 amino acids (see also Chapter 3). Humans (and other animals) can synthesize about half of them (the nonessential amino acids) but require the other half to be supplied by the diet (the essential amino acids). Diet must also provide a digestible source of nitrogen for synthesis of the nonessential amino acids. The eight **essential amino acids** are isoleucine, leucine, lysine, methionine, phenylalanine, threonine, tryptophan, and valine. In infants, histidine (and possibly arginine) is required for optimal development and growth, and is thus essential. In adults, histidine is nonessential, except in uremia. Under certain conditions, some nonessential amino acids may become essential. For example, when liver function is compromised by disease or premature birth, cysteine and tyrosine become

HO N OH

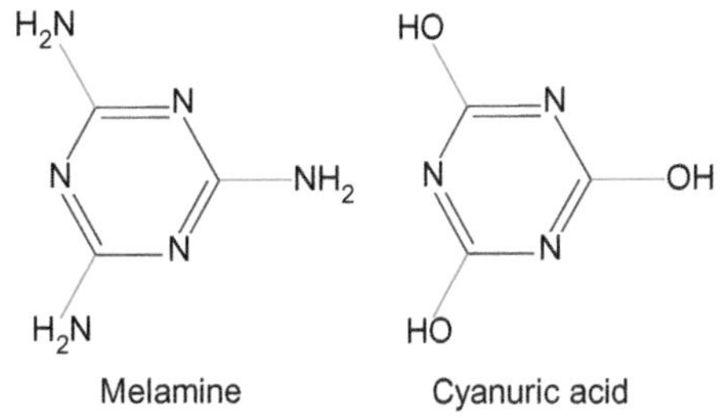

FIGURE 15.2 Structures of melamine and its metabolite cyanuric acid.

essential because they cannot be formed from their usual precursors (methionine and phenylalanine).

Glutamine—a nitrogen donor in the synthesis of purines and pyrimidines required for nucleic acid synthesis (Chapter 25)—aids in growth, repair of tissues, and promotion of immune function. Enrichment of glutamine in enteral and parenteral nutrition augments the recovery of seriously ill patients. Arginine may be considered as a semi-essential amino acid. It participates in a number of metabolic pathways, namely, formation of urea and ornithine, creatine and creatinine, spermine, agmatine and citrulline, and nitric oxide (NO). The endothelial cells lining the blood vessels produce NO from arginine, which has a major role in vasodilator function (discussed later). Dietary arginine supplementation may improve coronary blood flow, reduce episodes of angina, and help in patients with walking pain due to claudication.

For protein synthesis to occur, all 20 amino acids must be present in sufficient quantities. Absence of any one essential amino acid leads to cessation of protein synthesis, catabolism of unused amino acids, increased loss of nitrogen in urine, and reduced growth (negative nitrogen balance).

Quality and Quantity of Dietary Protein Requirement

Dietary protein provides organic nitrogen and the essential amino acids. The quantitative estimation of protein requirement must take into account the quality of protein, as determined by its essential amino acid composition. Dietary protein should provide all of the essential amino acids in the appropriate quantities.

Melamine, a high nitrogen-containing compound (Figure 15.2) used in the production of plastic and other synthetic products, has been used as an adulterant to milk and milk products. Since nitrogen is used as surrogate marker for protein quantity in foods, melamine is intentionally added to these food substances to increase their nitrogen content. Melamine and its metabolite cyanuric acid (Figure 15.2) form complexes with matrix proteins, uric acid, and phosphate to form stones in the kidneys, leading to catastrophic clinical problems including acute renal failure, especially in infants.

Protein Energy Malnutrition

Two disorders of protein energy nutrition that are widespread among children in economically depressed areas are Kwashiorkor (in Ghana, "the disease the first child gets when the second is on the way") and marasmus (from the Greek "to waste away").

Kwashiorkor occurs after weaning and is due to inadequate intake of good-quality protein and a diet consisting primarily of high-starch foods (e.g., yams, potatoes, bananas, maize, cassava) and deficient in other essential nutrients. Victims have decreased mass and function of heart and kidneys; decreased blood volume, hematocrit, and serum albumin concentration; atrophy of pancreas and intestines; decreased immunological resistance; slow wound healing; and abnormal temperature regulation. Characteristic clinical signs include edema, ascites, growth failure, apathy, skin rash, desquamation, pigment changes, and ulceration, loss of hair, liver enlargement, anorexia, and diarrhea.

Marasmus results from deficiency of protein and energy intake, as in starvation, and results in generalized wasting (atrophy of muscles and subcutaneous tissues, emaciation, loss of adipose tissue). Edema occurs in Kwashiorkor but not in marasmus; however, the distinction between these disorders is not always clear. The treatment of marasmus requires supplementation of protein and energy intake.

Protein energy malnutrition occurs with high frequency (30%−50%) in hospitalized patients as well as in populations in chronic care facilities as either an acute or a chronic problem. These individuals suffer from inadequate nutrition due to a disease or depression, and are susceptible to infections due to impaired immune function. Surgical patients with protein energy malnutrition exhibit delayed wound healing with increased length of stay in the hospital. Thus, protein energy malnutrition can cause morbidity and mortality; it also has economic consequences. Acute stressful physiological conditions such as trauma, burn, or sepsis can also precipitate protein energy malnutrition due to hypermetabolism caused by the neuroendocrine system. Prompt diagnosis and appropriate nutritional intervention are required in the management of patients with protein energy malnutrition.

Measurements of the levels of serum proteins, such as albumin, transthyretin (also known as **prealbumin**), transferrin, and retinol-binding protein are used as biochemical parameters in the assessment of protein energy malnutrition. An ideal protein marker should have rapid turnover and be present in sufficiently high concentrations in serum to be measured accurately. Transthyretin has these properties; it is a sensitive indicator of protein deficiency and is effective in assessing improvement with refeeding. The plasma half-life of transthyretin is about 1−2 days, whereas albumin has a half-life of 15−19 days.

Transport of Amino Acids into Cells

The intracellular metabolism of amino acids requires their transport across the cell membrane. Transport of L-amino acids occurs against a concentration gradient and is an active process usually coupled to Na^+-dependent carrier systems, such as for transport of glucose across the intestinal mucosa (Chapter 11). At least five transport systems for amino acids (with overlapping specificities) have been identified in kidney and intestine. They transport neutral amino acids, acidic amino acids, basic amino acids, ornithine and cystine, and glycine and proline, respectively. Within a given carrier system, amino acids may compete for transport (e.g., phenylalanine with tryptophan). Na^+-independent transport carriers for neutral and lipophilic amino acids have also been described. D-amino acids are transported by simple diffusion favored by a concentration gradient.

Inherited defects in amino acid transport affect epithelial cells of the gastrointestinal tract and renal tubules. Some affect transport of neutral amino acids (**Hartnup disease**); others that of basic amino acids and ornithine and cystine (**cystinuria**), or of glycine and proline (Chapter 11). **Cystinosis** is an intracellular transport defect characterized by high intralysosomal content of free cystine in the reticuloendothelial system, bone marrow, kidney, and eye. After degradation of endocytosed protein to amino acids within lysosomes, the amino acids are normally transported to the cytosol. The defect in cystinosis may reside in the ATP-dependent efflux system for cystine transport, and particularly in the carrier protein.

A different mechanism for translocation of amino acids in some cells is employed in the γ-**glutamyl cycle** (Figure 15.3). Its operation requires six enzymes (one membrane-bound, the rest cytosolic), glutathione (GSH; γ-glutamylcysteinylglycine present in all tissue cells), and ATP (three ATP molecules are required for each amino acid translocation). In this cycle, there is no net consumption of GSH, but an amino acid is transported at the expense of the energy of peptide bonds of GSH, which has to be resynthesized. Translocation is initiated by membrane-bound γ-glutamyltransferase (γ-GT, γ-glutamyl transpeptidase), which catalyzes formation of a γ-glutamyl amino acid and cysteinylglycine. The latter is hydrolyzed by dipeptidase to cysteine and glycine, which are utilized in resynthesis of GSH. The

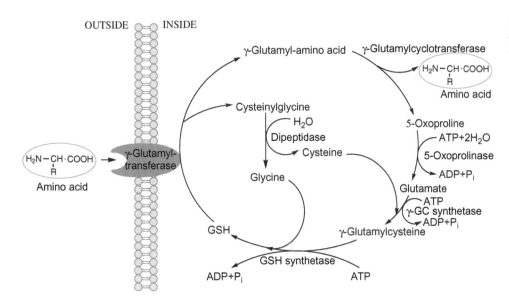

OUTSIDE   INSIDE

former is cleaved to the amino acid and 5-oxoproline (pyroglutamic acid) by γ-glutamylcyclotransferase. The cycle is completed by conversion of 5-oxoproline to glutamate and by resynthesis of GSH by two ATP-dependent enzymes: γ-glutamylcysteine synthetase and glutathione synthetase, respectively. GSH synthesis appears to be regulated by nonallosteric inhibition of γ-glutamylcysteine synthetase.

GSH has several well-established functions: it provides reducing equivalents to maintain −SH groups in other molecules (e.g., hemoglobin, membrane proteins; Chapter 14); it participates in inactivation of hydrogen peroxide, other peroxides, and free radicals (Chapter 14); it participates in other metabolic pathways (e.g., leukotrienes; Chapter 16); and it functions in inactivation of a variety of foreign compounds by conjugation through its sulfur atom. The conjugation reaction is catalyzed by specific glutathione S-transferases and the product is eventually converted to mercapturic acids and excreted.

Inherited deficiencies of GSH synthetase, γ-glutamylcysteine synthetase, γ-glutamyltransferase, and 5-oxoprolinase have been reported. Red blood cells, the central nervous system, and muscle may be affected. In GSH synthetase deficiency, γ-glutamylcysteine accumulates from lack of inhibition of γ-glutamylcysteine synthetase by glutathione synthetase by glutathione, and it is converted to 5-oxoproline and cysteine by γ-glutamylcyclotransferase. The 5-oxoproline is excreted so that GSH synthetase deficiency causes 5-oxoprolinuria (or pyroglutamic aciduria).

Measurement of serum γ-GT activity has clinical significance. The enzyme is present in all tissues, but the highest level is in the kidney; however, the serum enzyme originates primarily from the hepatobiliary system. Elevated levels of **serum γ-GT** are found in the following disorders: intra- and posthepatic biliary obstruction (elevated serum γ-GT levels indicate cholestasis, as do those of leucine aminopeptidase, 5′-nucleotidase, and alkaline phosphatase); primary or disseminated neoplasms; some pancreatic cancers, especially when associated with hepatobiliary obstruction; alcohol-induced liver disease (serum γ-GT may be exquisitely sensitive to alcohol-induced liver injury); and some prostatic carcinomas (serum from normal males has 50% higher activity than that of females). Increased activity is also found in patients receiving phenobarbital or phenytoin, possibly due to induction of γ-GT in liver cells by these drugs.

General Reactions of Amino Acids

Some general reactions that involve degradation or interconversion of amino acids provide for the synthesis of nonessential amino acids from α-keto acid precursors derived from carbohydrate intermediates.

Deamination

Removal of the α-amino group is the first step in catabolism of amino acids. It may be accomplished oxidatively or nonoxidatively.

Oxidative deamination is stereospecific and is catalyzed by L- or D-amino acid oxidase. The initial step is removal of two hydrogen atoms by the flavin coenzyme, with formation of an unstable α-amino acid intermediate. This intermediate undergoes decomposition by addition of water and forms the ammonium ion and the corresponding α-keto acid: L-amino acid oxidase occurs in the liver and kidney only.

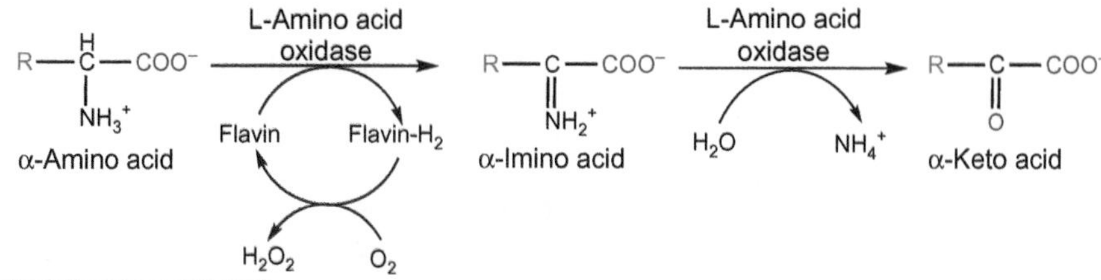

It is a flavoprotein that contains flavin mononucleotide (FMN) as a prosthetic group and does not attack glycine, dicarboxylic, or β-hydroxy amino acids. Its activity is very low.

High levels of D-amino acid oxidase are found in the liver and kidney. The enzyme contains flavin adenine dinucleotide (FAD) and deaminates many D-amino acids and glycine. The reaction for glycine is analogous to that for D-amino acids.

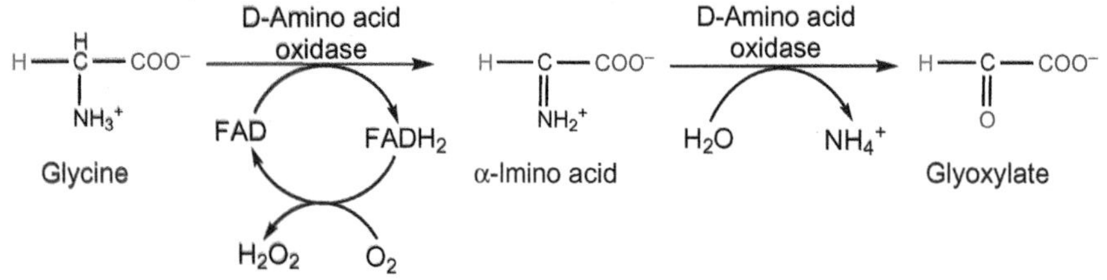

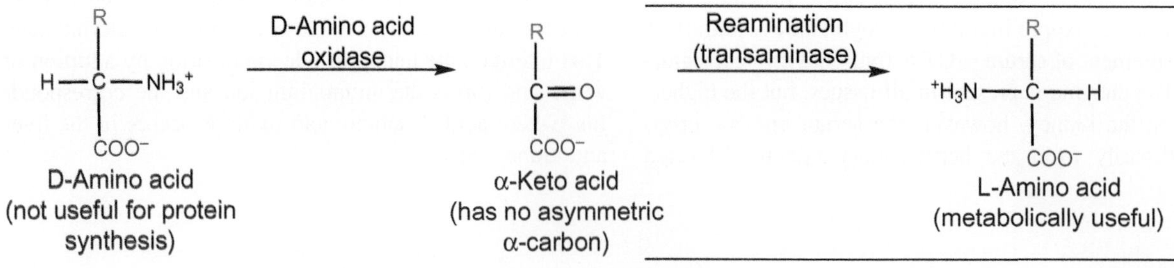

Nonoxidative deamination is accomplished by several specific enzymes.

Amino acid dehydratases deaminate hydroxyamino acids:

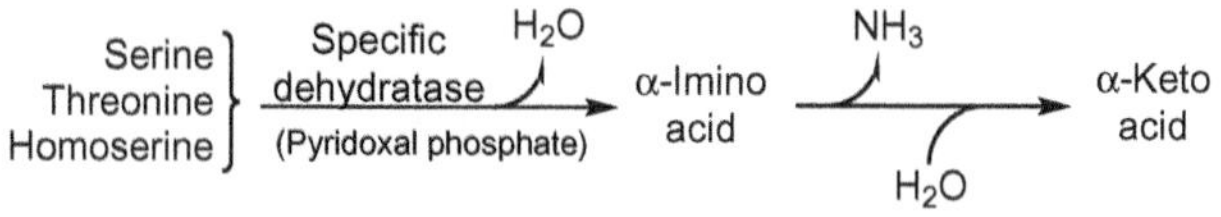

D-amino acid oxidase occurs in peroxisomes containing other enzymes that produce H_2O_2 (e.g., L-α-hydroxy acid oxidase, citrate dehydrogenase, and L-amino acid oxidase) and catalase and peroxidase, which destroy H_2O_2. In leukocytes, killing of bacteria involves hydrolases of lysosomes and production of H_2O_2 by NADPH oxidase (Chapter 13). Conversion of D-amino acids to the corresponding α-keto acids removes the asymmetry at the α-carbon atom. The keto acids may be aminated to L-amino acids. By this conversion from D- to L-amino acids, the body utilizes D-amino acids derived from the diet:

For serine dehydratase, the reaction is

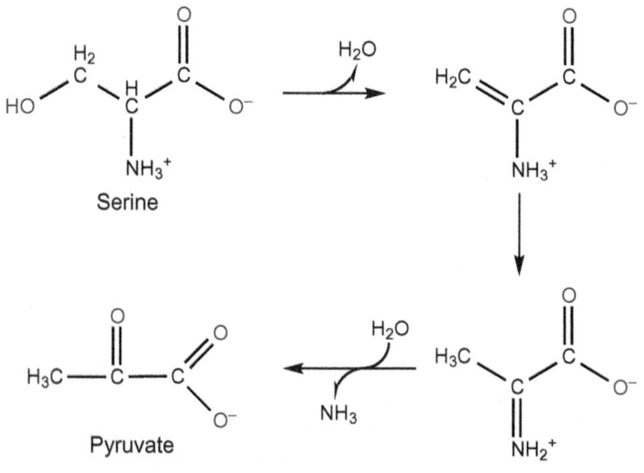

Dehydrogenation of L-Glutamate

Glutamate dehydrogenase plays a major role in amino acid metabolism. It is a zinc protein; requires NAD1 or NADP1 as a coenzyme; and is present in high concentrations in the mitochondria of liver, heart, muscle, and kidney. It catalyzes the (reversible) oxidative deamination of L-glutamate to α-ketoglutarate and NH_3. The initial step probably involves formation of α-iminoglutarate by dehydrogenation. This step is followed by hydrolysis of the imino acid to a keto acid and NH_3.

Glutamate dehydrogenase is an allosteric protein modulated positively by ADP, GDP, and some amino acids, and negatively by ATP, GTP, and NADH. Its activity is affected by thyroxine and some steroid hormones *in vitro*. Glutamate dehydrogenase is the only amino acid dehydrogenase present in most cells. It participates with appropriate transaminases (aminotransferases) in the deamination of other amino acids shown here:

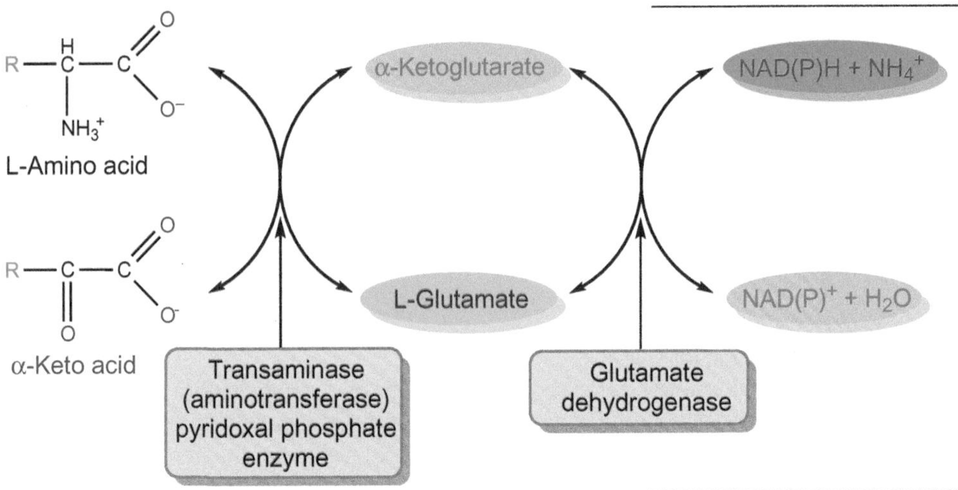

synthesized by the amidation of glutamate catalyzed by glutamine synthetase using ATP as the energy source. Glutamine synthetase is a cytoplasmic enzyme; it is both developmentally regulated and allosterically regulated by metabolites and hormones (e.g., insulin, thyroid hormone, and corticosteroids). Glutamine synthetase participates in the detoxification of ammonia, interorgan nitrogen transport, and acid−base regulation.

In the central nervous system, glutamine and glutamate cycle between the astrocytes and neurons, and serve a vital role in excitatory neurotransmission. Astrocytes are non-neuronal cells that have supporting and metabolic functions. Astrocytes are located in proximity to blood vessels, and their cellular processes surround the neurons. Astrocytes synthesize glutamine and supply it to the glutamatergic neurons. The neurons convert glutamine into glutamate using glutaminase. Excess glutamate in the synaptic cleft is taken up by the astrocytes for recycling (Figure 15.4).

These reactions are at near-equilibrium, so their overall effect depends on concentrations of the substrates and products. Aminotransferases occur in cytosol and mitochondria, but their activity is much higher in cytosol. Since glutamate dehydrogenase is restricted to mitochondria, transport of glutamate (generated by various transaminases) into mitochondria by a specific carrier becomes of central importance in amino acid metabolism. The NH_3 produced in deamination reactions must be detoxified by conversion to glutamine and asparagine or to urea, which is excreted in urine. The NADH generated is ultimately oxidized by the electron transport chain.

Formation of Glutamine and the Glutamate−Glutamine Cycle

Glutamine is the most abundant amino acid, and it participates in many essential metabolic reactions. It is

Deficiency of glutamine synthetase in the astrocytes results in deprivation of glutamate and leads to neurological disease. Deficiency in uptake of excess glutamate from the synaptic cleft is thought to result in continued excito-toxicity, resulting in neuronal death as it may occur in amyotrophic lateral sclerosis.

Transamination−Aminotransfer Reactions

Transamination reactions combine reversible amination and deamination, and they mediate redistribution of amino groups among amino acids. Transaminases (aminotransferases) are widely distributed in human tissues and are particularly active in heart muscle, liver, skeletal muscle, and kidney. The general reaction of transamination is

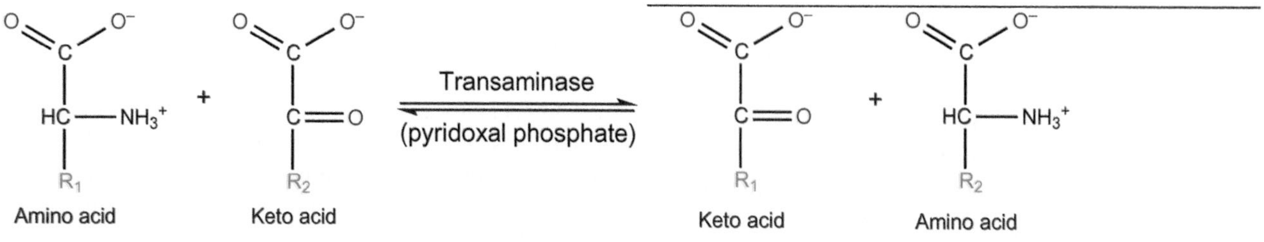

The following reaction scheme shows the general transaminase reaction:

$$\text{Amino acid} + \text{Keto acid} \xrightleftharpoons[\text{(pyridoxal phosphate)}]{\text{Transaminase}} \text{Keto acid} + \text{Amino acid}$$

The α-ketoglutarate/L-glutamate couple serves as an amino group acceptor/donor pair in transaminase reactions. The specificity of a particular transaminase is for the amino group other than the glutamate. Two transaminases whose activities in serum are used as indices of liver damage catalyze the following reactions:

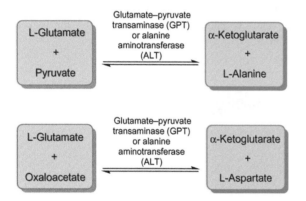

All of the amino acids except lysine, threonine, proline, and hydroxyproline participate in transamination reactions. Transaminases exist for histidine, serine, phenylalanine, and methionine, but the major pathways of their metabolism do not involve transamination. Transamination of an amino group not at the α-position can also occur. Thus, transfer of a δ-amino group of ornithine to α-ketoglutarate converts ornithine to glutamate-γ-semialdehyde.

All transaminase reactions have the same mechanism and use pyridoxal phosphate (a derivative of vitamin B$_6$; Chapter 36). Pyridoxal phosphate is linked to the enzyme by formation of a Schiff base between its aldehyde group and the ε-amino group of a specific lysyl residue at the active site and held noncovalently through its positively charged nitrogen atom and the negatively charged phosphate group (Figure 15.5). During catalysis, the amino acid substrate displaces the lysyl ε-amino group of the enzyme in the Schiff base. An electron pair is removed from the α-carbon of the substrate and transferred to the positively charged pyridine ring but is subsequently returned to the second substrate, the α-keto acid. Thus, pyridoxal phosphate functions as a carrier of amino groups and as an electron sink by facilitating dissociation of the α-hydrogen of the amino acid (Figure 15.6). In the overall reaction, the amino acid transfers its amino group to pyridoxal phosphate and then to the keto acid through formation of pyridoxamine phosphate as intermediate.

Pyridoxal phosphate is also the prosthetic group of amino acid decarboxylases, dehydratases, desulfhydrases, racemases, and aldolases, in which it participates through its ability to render labile various bonds of an amino acid molecule (Figure 15.7). Several drugs (Figure 15.8) inhibit pyridoxal phosphate-dependent enzymes. Isonicotinic acid hydrazide (used in the treatment of tuberculosis) and hydralazine (a hypertensive agent) react with the aldehyde group of pyridoxal (free or bound) to form pyridoxal hydrazones, which are eliminated in the urine. Isonicotinic acid hydrazide is normally inactivated in the liver by acetylation; some individuals are "slow acetylators" (an inherited trait) and may be susceptible to pyridoxal deficiency from accumulation of the drug. Cycloserine (an amino acid analogue and broad-spectrum antibiotic) also combines with pyridoxal phosphate.

Role of Organs in Amino Acid Metabolism

In the postabsorptive state, maintenance of steady-state concentrations of plasma amino acids depends on release of amino acids from tissue protein. After a meal, dietary amino acids enter the plasma, replenish the tissues, and are metabolized during fasting.

Liver plays a major role, since it can oxidize all amino acids except leucine, isoleucine, and valine. It also produces the nonessential amino acids from the appropriate carbon precursors. Ammonia formed in the gastrointestinal tract or from various deaminations in the liver is converted to urea and excreted in urine (discussed later).

Skeletal muscle tissue constitutes a large portion of the body weight and accounts for a significant portion of nonhepatic amino acid metabolism. It takes up the amino acids required to meet its needs for protein synthesis, and metabolizes alanine, aspartate, glutamate, and the branched-chain amino acids. Amino acids are released from muscle during the postabsorptive state (i.e., in fasting or starvation). Alanine and glutamine constitute more than 50% of the α-amino acid nitrogen released. During starvation, the total amino acid pool increases from

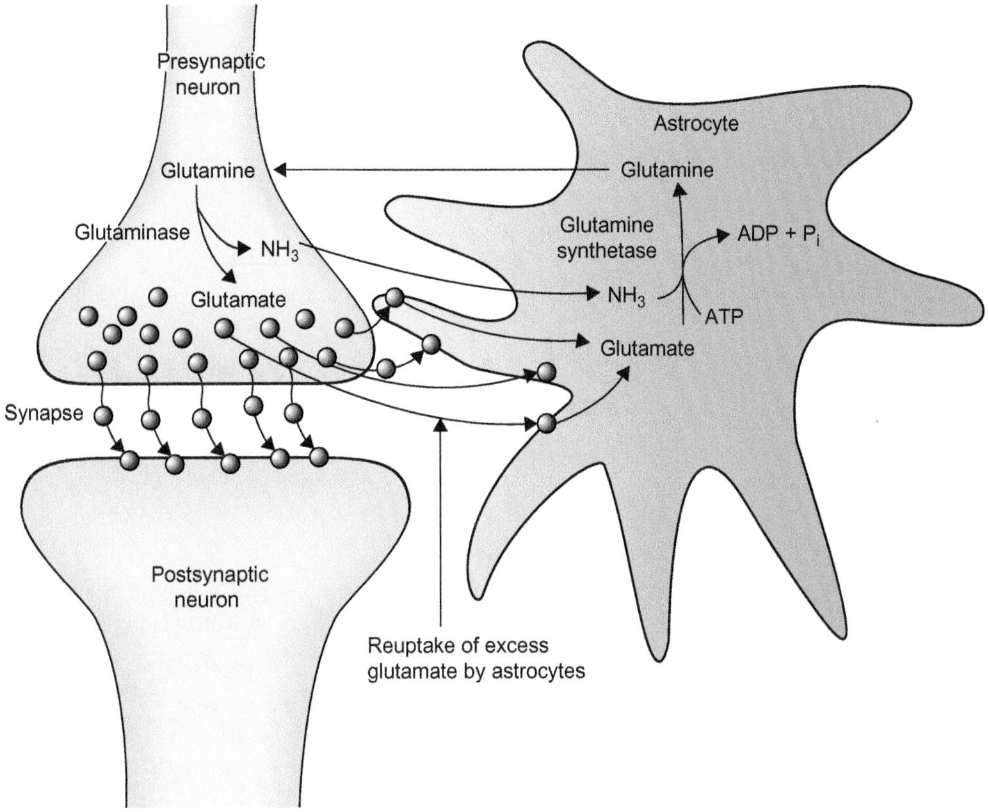

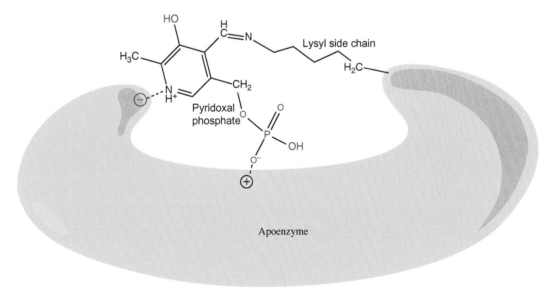

FIGURE 15.4 Glutamate—glutamine cycle between astrocytes and excitatory glutamatergic neurons. Excitatory neurotransmitter glutamate required for glutamatergic neurons is provided by the astrocytes in the form of glutamine. Glutaminase within the neurons hydrolyzes glutamine to glutamate. Astrocyte glutamine synthetase converts the excess neurotransmitter glutamate taken up from the synaptic cleft and glutamate obtained from the capillaries to glutamine. Either deficiencies of astrocyte glutamine synthetase or reuptake of glutamate by the Na^+-dependent glutamate transporters from the synaptic cleft by the astrocytes can lead to neurological and neurodegenerative diseases.

FIGURE 15.5 Binding of pyridoxal phosphate to its apoenzyme. The carbonyl carbon reacts with the ε-amino group of the lysyl residue near the active site to yield a Schiff base. Ionic interactions involve its positively charged pyridinium ion and negatively charged phosphate group.

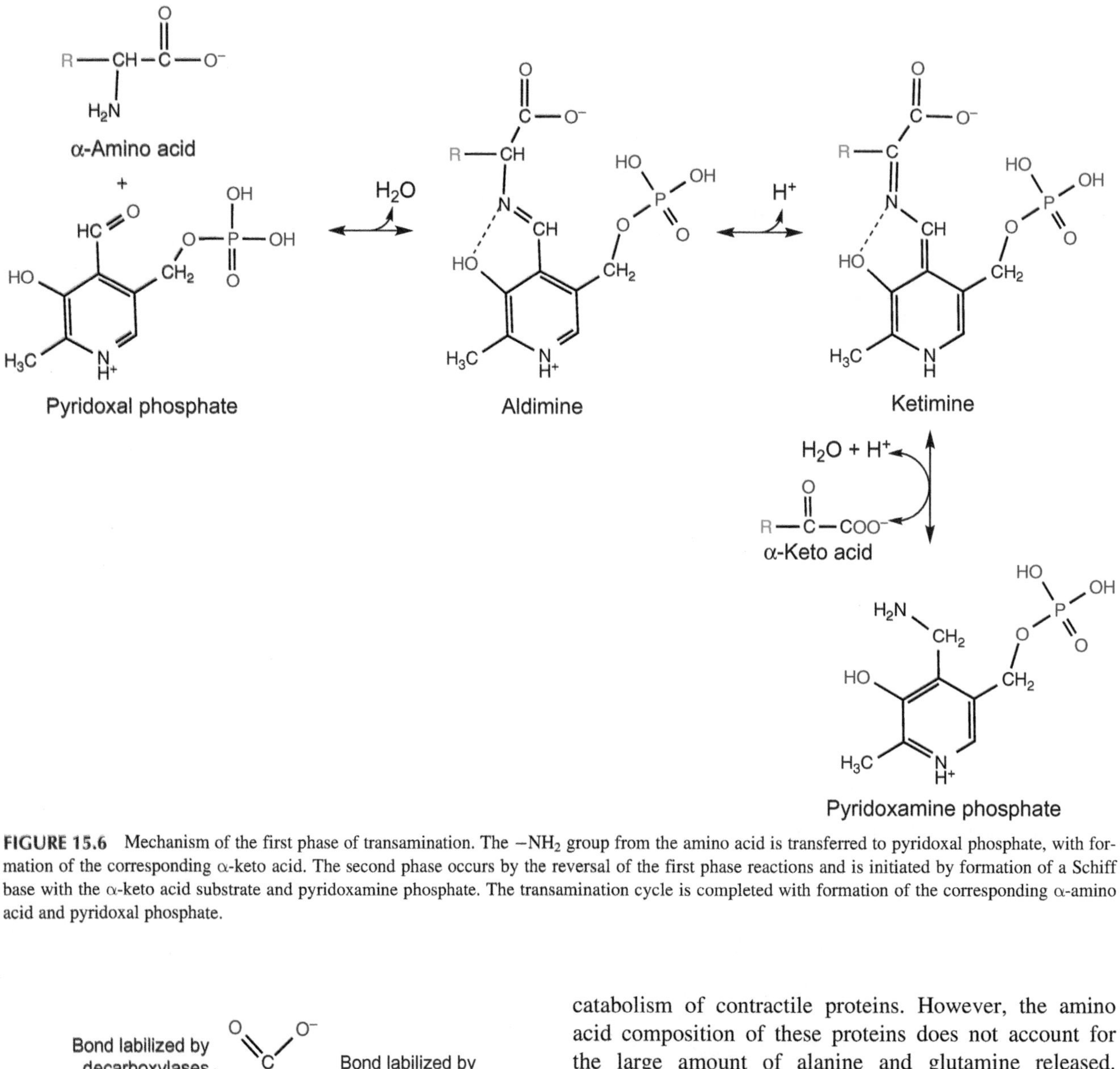

FIGURE 15.6 Mechanism of the first phase of transamination. The $-NH_2$ group from the amino acid is transferred to pyridoxal phosphate, with formation of the corresponding α-keto acid. The second phase occurs by the reversal of the first phase reactions and is initiated by formation of a Schiff base with the α-keto acid substrate and pyridoxamine phosphate. The transamination cycle is completed with formation of the corresponding α-amino acid and pyridoxal phosphate.

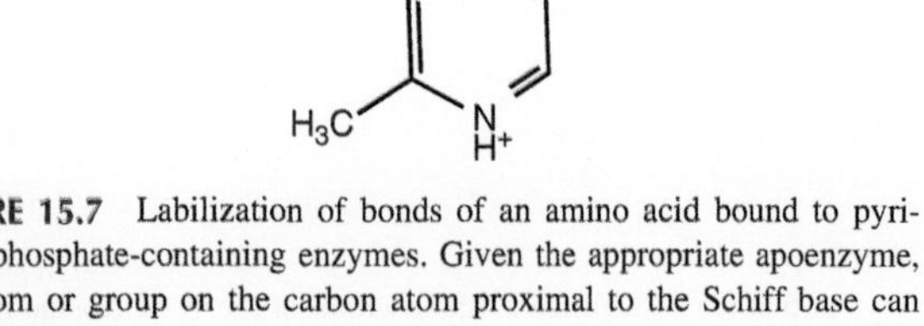

FIGURE 15.7 Labilization of bonds of an amino acid bound to pyridoxal phosphate-containing enzymes. Given the appropriate apoenzyme, any atom or group on the carbon atom proximal to the Schiff base can be cleaved.

catabolism of contractile proteins. However, the amino acid composition of these proteins does not account for the large amount of alanine and glutamine released. Amino acids that give rise to pyruvate can be transaminated to alanine. For example, aspartate can be converted to alanine as follows:

$$\text{Aspartate} \xrightarrow{\text{transaminase}} \text{Oxaloacetate} \xrightarrow{\text{PEPCK}}$$

$$\text{Phosphoenolpyruvate} \xrightarrow{\text{pyruvate kinase}}$$

$$\text{Pyruvate} \xrightarrow{\text{transaminase}} \text{Alanine}$$

Similarly, amino acids that produce tricarboxylic acid (TCA) cycle intermediates (Chapter 14) produce alanine by conversion to oxaloacetate. During starvation or intake of a carbohydrate-poor diet, conversion of pyruvate to alanine is preferred because pyruvate dehydrogenase is

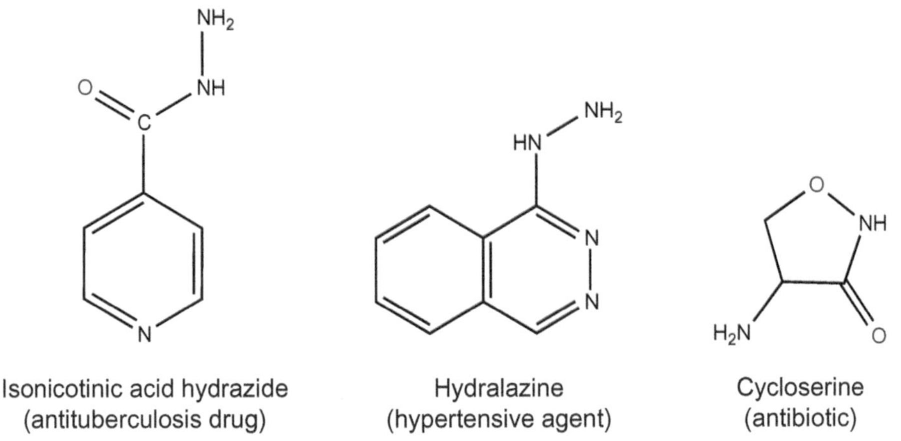

FIGURE 15.8 Structures of compounds that inhibit pyridoxal phosphate-containing enzymes.

Isonicotinic acid hydrazide
(antituberculosis drug)

Hydralazine
(hypertensive agent)

Cycloserine
(antibiotic)

inactivated by oxidation of fatty acids and ketone bodies (Chapters 12 and 16).

Glutamine is synthesized from glutamate and ammonia by glutamine synthase.

Glutamate is derived by transamination of α-ketoglutarate produced in the TCA cycle from citrate via oxaloacetate and acetyl-CoA (Chapter 12). All of the amino acids can produce acetyl-CoA. All except leucine and lysine (which are oxidized solely to acetyl-CoA) can be used in net synthesis of α-ketoglutarate to enhance glutamate synthesis. Ammonia is generated in glutamate dehydrogenase and AMP deaminase reactions.

The mucosa of the **small intestine** metabolizes dietary glutamine, glutamate, asparagine, and aspartate by oxidation to CO_2 and H_2O, or by conversion to lactate, alanine, citrulline, and NH_3. These intermediates and the unmetabolized dietary amino acids are transferred to the portal blood and then to the liver for further metabolism.

In the fasting state, the intestinal mucosa depends on other tissues for metabolites to provide energy and precursors for protein and nucleotide synthesis to maintain the rapid cell division characteristic of that tissue. Glutamine, released from liver and muscle, is utilized for purine nucleotide synthesis (Chapter 25), is oxidized to provide energy, and can be converted to aspartate for pyrimidine

nucleotide synthesis (Chapter 25). Thus, glutamine is important in cells undergoing rapid division. Intestine can also oxidize glucose, fatty acids, and ketone bodies to provide energy.

Kidney releases serine and small (but significant) quantities of alanine into the blood, and takes up glutamine, proline, and glycine. Amino acids filtered in the glomeruli are reabsorbed by renal tubule cells. Glutamine plays an important role in acid−base regulation by providing ammonia, which incorporates H^+ to form the NH_4^+ ion and which is eliminated in urine (Chapter 37). It can provide two ammonia molecules, by glutaminase and glutamate dehydrogenase, respectively, in renal tubular mitochondria. Its carbon skeleton can be oxidized or converted to glucose, since renal tissue is capable of gluconeogenesis (Chapter 14).

Brain takes up significant quantities of valine and may be a major (if not primary) site of utilization of branched-chain amino acids. The role of the glutamate−glutamine cycle between neuron−astrocyte in the excitation glutamatergic pathway was discussed previously. Aspartate and glycine are **neurotransmitters**. Glutamate is a precursor of γ-aminobutyrate; tyrosine of dopamine, norepinephrine, and epinephrine; and tryptophan of serotonin, all of which are neurotransmitters. Inactivation of

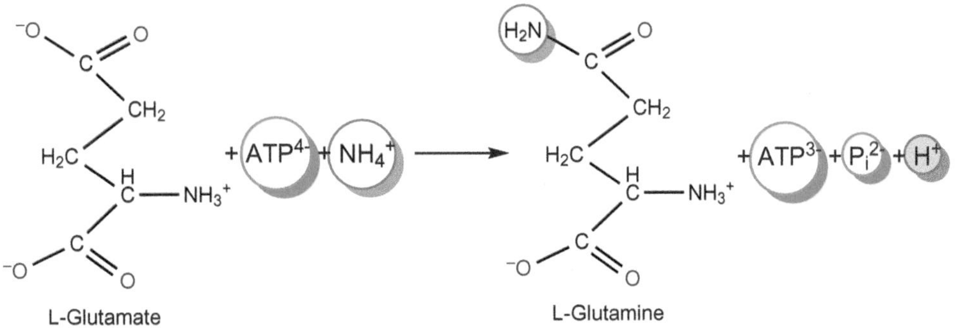

L-Glutamate   L-Glutamine

neurotransmitters involves deamination with production of ammonia, which is removed by formation of glutamine. N-acetylaspartate occurs in high levels in the brain but its function is not completely understood. It may provide acetate for myelin lipid synthesis and participate in the synthesis of neuronal dipeptide N-acetylaspartylglutamate. It is synthesized from acetyl-CoA and aspartic acid catalyzed by acetyl-CoA aspartate N-acetyl transferase. Aspartoacylase catalyzes the hydrolysis of N-acetylaspartate to acetate and aspartic acid. The deficiency of aspartoacylase, which is inherited as an autosomal recessive trait, is associated with degenerative brain changes. Patients with this disorder, also known as **Canavan dystrophy**, are usually of Eastern European Jewish heritage.

METABOLISM OF AMMONIA

Ammonia (at physiological pH, 98.5% exists as NH_4^+), the highly toxic product of protein catabolism, is rapidly inactivated by a variety of reactions. Some products of these reactions are utilized for other purposes (thus salvaging a portion of the amino nitrogen), while others are excreted. In humans, ammonia is excreted mostly as urea, which is highly water-soluble, is distributed throughout extracellular and intracellular body water, is nontoxic and metabolically inert, has a high nitrogen content (47%), and is excreted via the kidneys.

Ammonia is produced by deamination of glutamine, glutamate, other amino acids, and adenylate. A considerable quantity is derived from intestinal bacterial enzymes (e.g., urease) acting on urea and other nitrogenous compounds. The urea comes from body fluids that diffuse into the intestine, and the other nitrogenous products are derived from intestinal metabolism (e.g., glutamine) and ingested protein. The ammonia diffuses across the intestinal mucosa to the portal blood and is converted to urea in the liver.

Ammonia is particularly toxic to brain but not to other tissues, even though levels in those tissues may increase under normal physiological conditions (e.g., in muscle during heavy exercise, and in kidney during metabolic acidosis). Several hypotheses have been suggested to explain the mechanism of neurotoxicity.

In brain mitochondria, excess ammonia may drive the reductive amination of α-ketoglutarate by glutamate dehydrogenase. This step may deplete a key intermediate of the TCA cycle and lead to its impairment, with severe inhibition of respiration and considerable stimulation of glycolysis. Since the $[NAD^+]/[NADH]$ ratio will be high in mitochondria, there will be a decrease in the rate of production of ATP. This hypothesis does not explain why the same result does not occur in tissues that are not affected by ammonia. A more plausible hypothesis is depletion of glutamate, which is an excitatory neurotransmitter. Glutamine, synthesized and stored in the astrocytes and glial cells, is the precursor of glutamate. It is transported into the neurons and hydrolyzed by glutaminase. Ammonia inhibits glutaminase and depletes the glutamate concentration. A third hypothesis invokes neuronal membrane dysfunction, since elevated levels of ammonia produce increased permeability to K^+ and Cl^- ions, while glycolysis increases H^+ ion concentration (stimulates 6-phosphofructokinase; Chapter 12). Encephalopathy of hyperammonemia is characterized by brain edema and astrocyte swelling. Edema and swelling have been attributed to intracellular accumulation of glutamine, which causes osmotic shifts of water into the cell.

Behavioral disorders such as **anorexia**, sleep disturbances, and **pain insensitivity** associated with hyperammonemia have been attributed to increased tryptophan transport across the blood−brain barrier and the accumulation of its metabolites. Two of the tryptophan-derived metabolites are serotonin and quinolinic acid (discussed later). The latter is an excitotoxin at the N-methyl-D-aspartate (NMDA) glutamate receptors. Thus, the mechanism of the ammonium-induced neurological abnormalities is multifactorial. Normally, only small amounts of NH_3 (i.e., NH_4^+) are present in plasma, since NH_3 is rapidly removed by reactions in tissues of glutamate dehydrogenase, glutamine synthase, and urea formation.

Urea Synthesis

Ammonia contained in the blood flowing through the hepatic lobule is removed by the hepatocytes and converted into urea. Periportal hepatocytes are the predominant sites of urea formation. Any ammonia that is not converted to urea may be incorporated into glutamine catalyzed by glutamine synthase located in pericentral hepatocytes. Formation of urea requires the combined action of two enzymes to produce carbamoyl phosphate and of four enzymes that function in a cyclic manner in the urea cycle (Figure 15.9). Although some of these enzymes occur in extrahepatic tissues and urea formation has been shown to occur in several cell lines in tissue culture, the most important physiological site of urea formation is the liver. In hepatocytes, the first three enzymes are mitochondrial and the others are cytosolic. A citrulline−ornithine antiport is located in the inner mitochondrial membrane.

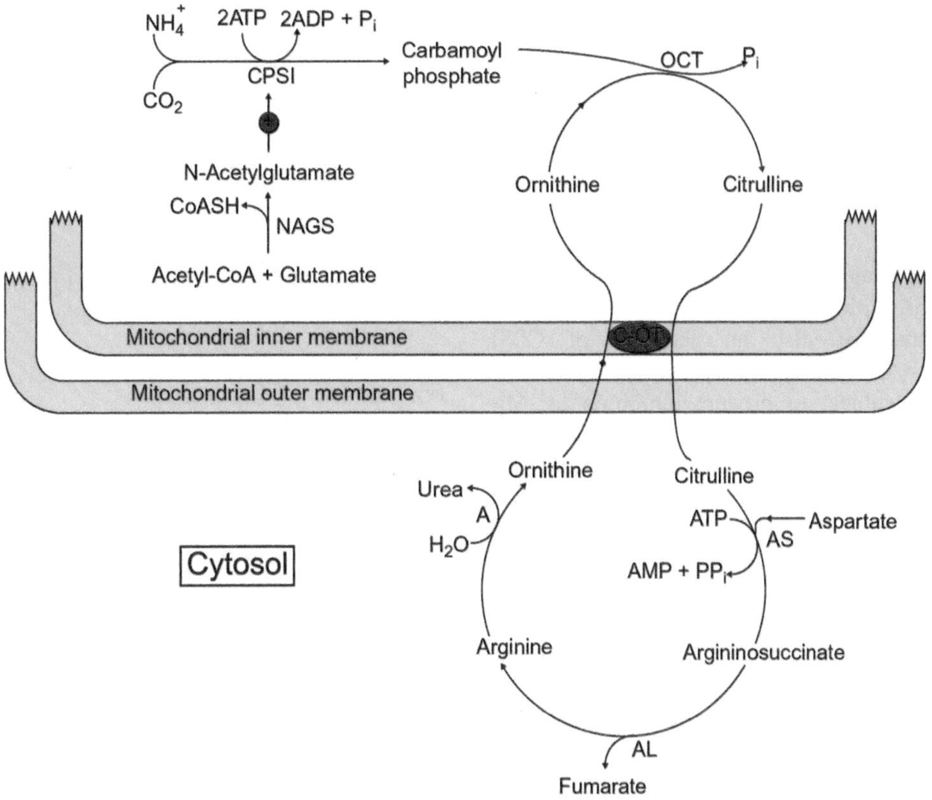

FIGURE 15.9 Formation of urea in hepatocytes. NAGS = N-acetylglutamate synthase; CPSI = carbamoylphosphate synthase I; OCT = ornithine carbamoyltransferase; C-OT = citrulline−ornithine translocase; AS = arginosuccinate synthase; AL = arginosuccinate lyase; A = arginase. —⊕→ indicates the absolute requirement of N-acetylglutamate for CPSI activity.

Formation of Carbamoyl Phosphate

Carbamoyl phosphate synthesis requires amino acid acetyltransferase (N-acetylglutamate synthase, mitochondrial) and carbamoyl phosphate synthase I (CPSI). N-Acetylglutamate (NAG) is an obligatory positive effector of CPSI. NAG synthase is under positive allosteric modulation by arginine and product inhibition by NAG. Depletion of CoASH decreases NAG synthesis and ureagenesis. This situation can occur in **organic acidemias** (e.g., propionic acidemia; Chapter 16), in which organic acids produced in excess compete for CoASH for formation of CoA derivatives that are also competitive inhibitors of N-acetylglutamate synthase and inhibitors of CPSI. **Hyperammonemia** often accompanies organic acidemias.

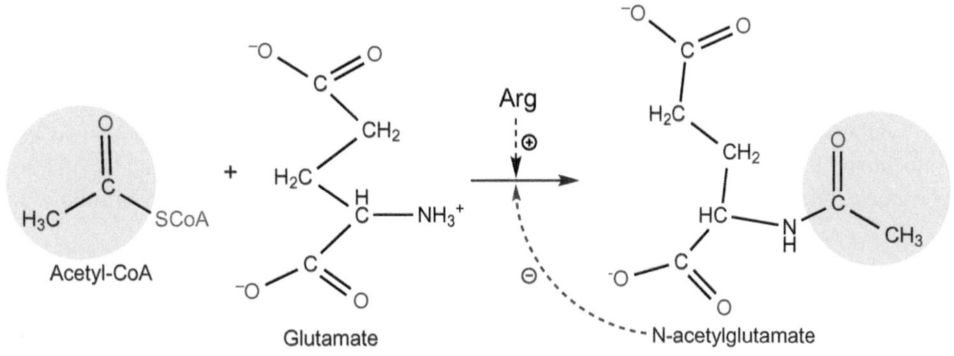

CPSI catalyzes the reaction:

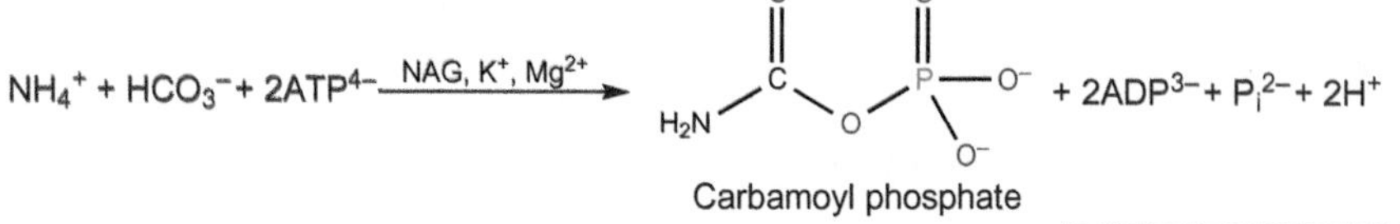

$$NH_4^+ + HCO_3^- + 2ATP^{4-} \xrightarrow{\text{NAG, K}^+\text{, Mg}^{2+}} \text{Carbamoyl phosphate} + 2ADP^{3-} + P_i^{2-} + 2H^+$$

Carbamoyl phosphate

NAG binding changes the conformation and subunit structure of CPSI, with preponderance of the monomers. Carbamolyglutamate is also an activator of CPSI. Glutamate and α-ketoglutarate compete with NAG for binding. CPSI is subject to product inhibition by Mg-ADP. It possesses two binding sites for ATP. One ATP is utilized in activation of bicarbonate by forming an enzyme-bound carboxyphosphate that reacts with phosphate spills into the cytosol and promotes synthesis of orotic acid and uridine 5′-phosphate.

Formation of Citrulline

Ornithine carbamoyltransferase (ornithine transcarbamoylase) catalyzes the condensation between carbamoyl phosphate and ornithine to yield citrulline in mitochondria:

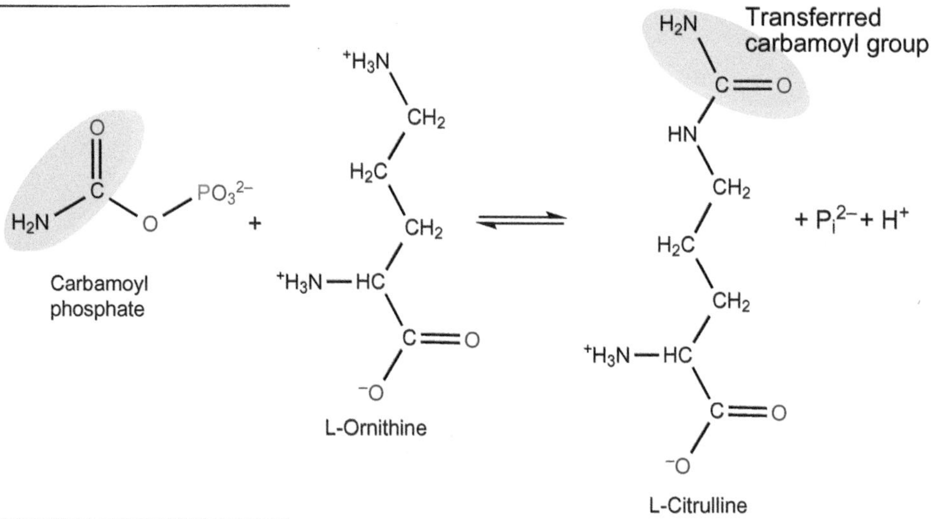

Carbamoyl phosphate

L-Ornithine

Transferrred carbamoyl group

L-Citrulline

ammonium ion to form an enzyme-bound carbamate, with elimination of inorganic phosphate. Carbamoyl phosphate is generated when the second ATP reacts with the enzyme-bound carbamate, with release of ADP and free enzyme.

In humans, there are two immunologically distinct carbamoyl phosphate synthases: one mitochondrial (CPSI) and the other cytosolic (CPSII). CPSI is involved in ureagenesis, uses NH_3 exclusively as the nitrogen donor, and requires binding of NAG for activity. CPSII uses glutamine as substrate, is not dependent on NAG for activity, and is required for synthesis of pyrimidines (Chapter 25). Normally, the mitochondrial membrane is not permeable to carbamoyl phosphate, but when the concentration increases, carbamoyl

Although the equilibrium constant strongly favors citrulline formation, the reaction is reversible. Citrulline is transported out of the mitochondria by the citrulline—ornithine antiporter.

Formation of Argininosuccinate

The condensation of citrulline and aspartate to argininosuccinate is catalyzed by argininosuccinate synthase in the cytosol and occurs in two steps. In the initial step, the ureido group is activated by ATP to form the enzyme-bound intermediate adenylylcitrulline. In the second step, nucleophilic attack of the amino group of aspartate displaces AMP and yields argininosuccinate. The overall reaction is as follows:

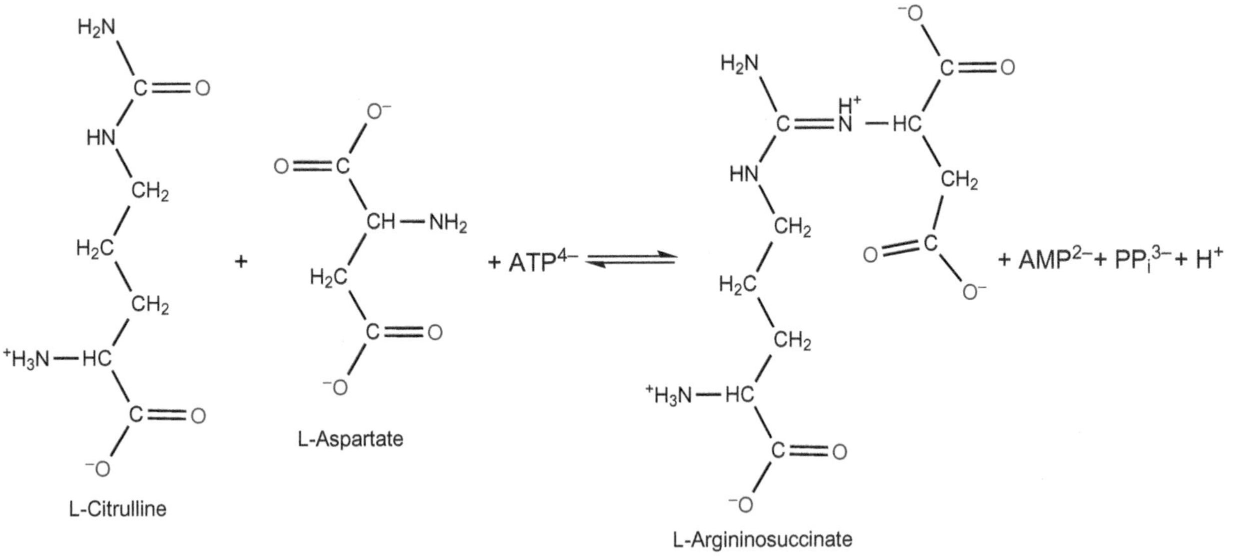

L-Citrulline + L-Aspartate + ATP^{4-} ⇌ L-Argininosuccinate + AMP^{2-} + PP$_i^{3-}$ + H$^+$

The reaction is driven forward by hydrolysis of pyrophosphate to inorganic phosphate. Argininosuccinate formation is considered as the rate-limiting step for urea synthesis. This reaction incorporates the second nitrogen atom of the urea molecule donated by aspartate.

Formation of Arginine and Fumarate

Argininosuccinate lyase in cytosol catalyzes cleavage of argininosuccinate to arginine and fumarate:

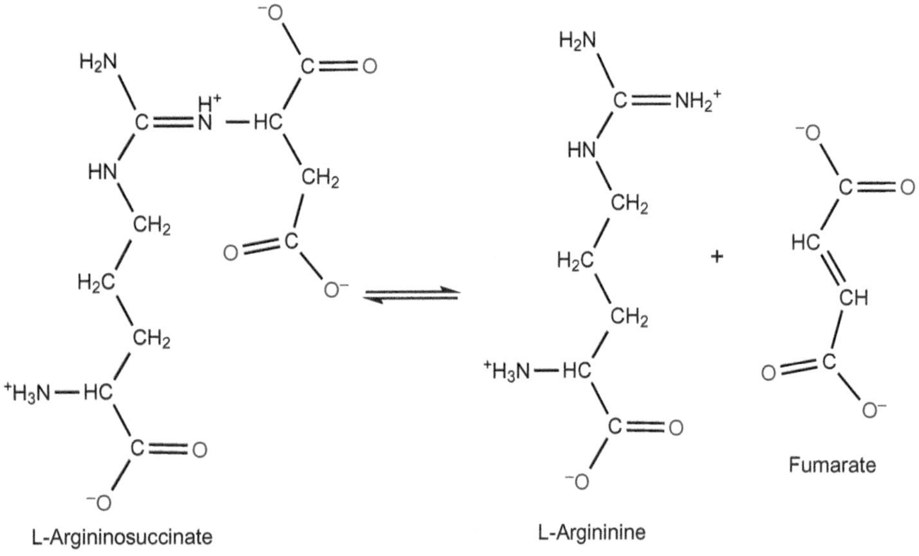

L-Argininosuccinate ⇌ L-Argininine + Fumarate

This is the pathway for synthesis of arginine, a nonessential amino acid; however, in the event of physiological deficiency, as in premature infants, or a defect in any of the enzymes discussed previously, an exogenous supply of arginine is required.

Formation of Urea and Ornithine

The formation of urea and ornithine, an irreversible reaction, is catalyzed by arginase in the cytosol:

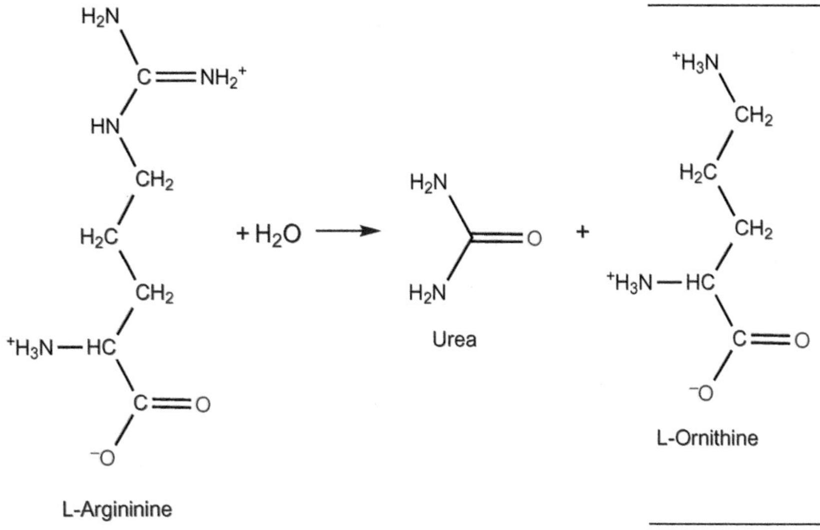

L-Argininine

   + H$_2$O ⟶

Urea

L-Ornithine

The urea formed this way is distributed throughout the body water and excreted. The renal clearance of urea is less than the glomerular filtration rate because of passive tubular back-diffusion. Diffusion of urea in the intestine leads to formation of ammonia, which enters the portal blood and is converted to urea in the liver. Re-entry of ornithine into mitochondria initiates the next revolution of the urea cycle. Ornithine can be converted to glutamate-γ-semialdehyde (which is in equilibrium with its cyclic form Δ′-pyrroline-5-carboxylate) by ornithine aminotransferase and decarboxylated to putrescine by ornithine decarboxylase. Ornithine is also produced in the arginine-glycine transaminase reaction.

The availability of substrates (ammonia and amino acids) in the liver determines the amount of urea synthesized. Urea excretion increases with increased protein intake and decreases with decreased protein intake.

Energetics of Ureagenesis

The overall reaction of ureagenesis is

$$NH_3 + HCO_3^- + aspartate + 3ATP \rightarrow urea + fumarate + 2ADP + 4P_i + AMP$$

Hydrolysis of four high-energy phosphate groups is required for the formation of one molecule of urea. If fumarate is converted to aspartate (by way of malate and oxaloacetate), one NADH molecule is generated that can give rise to three ATP molecules through the electron transport chain, so that the energy expenditure becomes one ATP molecule per molecule of urea.

Hyperammonemias

Hyperammonemias are caused by inborn errors of ureagenesis and organic acidemias, liver immaturity (transient hyperammonemia of the newborn), and liver failure (hepatic encephalopathy). Neonatal hyperammonemias are characterized by vomiting, lethargy, lack of appetite, seizures, and coma. The underlying defects can be identified by appropriate laboratory measurements (e.g., assessment of metabolic acidosis if present and characterization of organic acids, urea cycle intermediates, and glycine).

Inborn errors of the six enzymes of ureagenesis and NAG synthase have been described (Figure 15.9). The inheritance pattern of the last is not known, but five of the urea cycle defects are autosomal recessive, and ornithine carbamoyltransferase (OCT) deficiency is X-linked (see Clinical Case Study 15.2 for OCT deficiency). Antenatal diagnosis for fetuses at risk for urea cycle enzyme disorders can be made by appropriate enzyme assays and DNA analysis in cultured amniocytes.

Acute neonatal hyperammonemia, irrespective of cause, is a medical emergency and requires immediate and rapid lowering of ammonia levels to prevent serious effects on the brain. Useful measures include hemodialysis, exchange transfusion, peritoneal dialysis, and administration of arginine hydrochloride. The general goals of management are to

1. decrease nitrogen intake so as to minimize the requirement for nitrogen disposal;
2. supplement arginine intake; and
3. promote nitrogen excretion in forms other than urea.

The first goal can be accomplished by restriction of dietary protein and administration of α-keto analogues of essential amino acids. Arginine supplementation as a precursor of ornithine is essential to the urea cycle. Alternate-pathway therapy for hyperammonemia is achieved by administration of sodium benzoate or sodium (or calcium) phenylacetate. Administration of benzoate leads to elimination of hippurate (benzoylglycine):

Hippurate is rapidly secreted since its clearance is five times greater than its glomerular filtration rate. The glycine nitrogen is derived from ammonia in a complex reaction that uses CO_2, NADH, N^5,N^{10}-methylenetetrahydrofolate (a source of a single carbon unit; Chapter 25) and pyridoxal phosphate, catalyzed by mitochondrial glycine synthase (glycine cleavage enzyme).

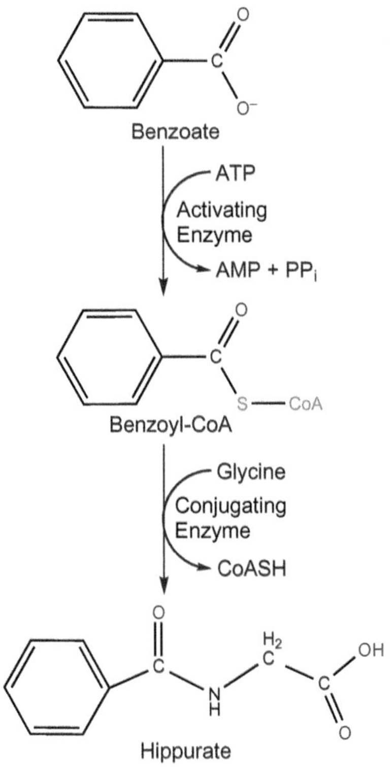

Phenylacetate or phenylbutyrate administration increases excretion of phenylacetylglutamine as follows:

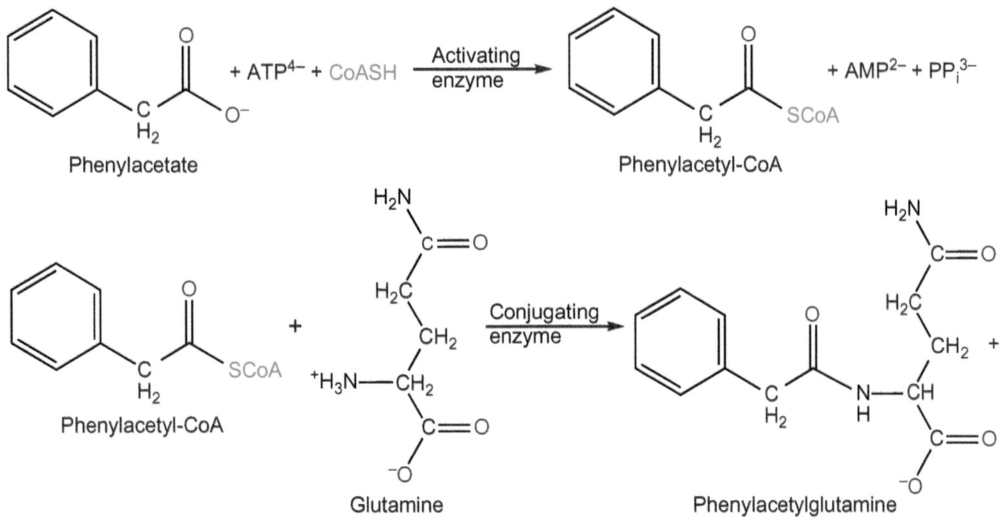

The excretion of phenylacetylglutamine produces loss of two nitrogen atoms.

NAG synthase deficiency cannot be treated by administration of NAG, since NAG undergoes cytosolic inactivation by deacylation and is not readily permeable across the inner mitochondrial membrane. An analogue of NAG, N-carbamoylglutamate, activates CPSI, does not share the undesirable properties of NAG, and has been effective in management of this deficiency.

The most common cause of hyperammonemia in adults is disease of the liver (e.g., due to ethanol abuse, infection, or cancer). The ability to detoxify ammonia is decreased in proportion to the severity of the damage. In advanced disease (e.g., cirrhosis), hyperammonemia is augmented by shunting of portal blood that carries ammonia from the intestinal tract and other splanchnic organs to the systemic blood circulation (bypassing the liver) and leads to portal-systemic encephalopathy. In addition to dietary protein restriction, colonic growth of bacteria must be suppressed by antibiotics (e.g., neomycin) and administration of lactulose (Chapter 8), a nonassimilatable disaccharide. Enteric bacteria catabolize lactulose to organic acids that convert NH_3 to NH_4^+, thereby decreasing absorption of NH_3 into the portal circulation. Catabolism of lactulose also leads to formation of osmotically active particles that draw water into the colon; produce loose, acid stools; and permit loss of ammonia as ammonium ions.

METABOLISM OF SOME INDIVIDUAL AMINO ACIDS

Mammalian tissues synthesize the nonessential amino acids from carbon skeletons derived from lipid and carbohydrate sources or from transformations that involve essential amino acids. The nitrogen is obtained from NH_4^+ or from that of other amino acids. Nonessential amino acids (and their precursors) are glutamic acid (α-ketoglutaric acid), aspartic acid (oxaloacetic acid), serine (3-phosphoglyceric acid), glycine (serine), tyrosine (phenylalanine), proline (glutamic acid), alanine (pyruvic acid), cysteine (methionine and serine), arginine (glutamate-γ-semialdehyde), glutamine (glutamic acid), and asparagine (aspartic acid).

Amino acids may be classified as ketogenic, glucogenic, or glucogenic and ketogenic, depending on whether feeding of a single amino acid to starved animals or animals with experimentally induced diabetes increases plasma or urine levels of glucose or ketone bodies (Chapter 16). Leucine and lysine are ketogenic; isoleucine, phenylalanine, tyrosine, and tryptophan are glucogenic and ketogenic; and the remaining amino acids are glucogenic. Points of entry of amino acids into the gluconeogenic pathway are discussed in Chapter 14.

Arginine

Arginine participates in a number of metabolic pathways depending on the cell type. It is synthesized as an intermediate in the urea cycle pathway and is also obtained from dietary proteins. A number of key metabolites such as nitric oxide, phosphocreatine, spermine, and ornithine are derived from arginine. During normal growth and development, under certain pathological conditions (e.g., endothelial dysfunction) and if the endogenous production of arginine is insufficient, a dietary supplement of arginine may be required. Thus, arginine is considered a semi-essential amino acid.

Metabolism and Synthesis of Nitric Oxide

Nitric oxide (NO) is a reactive diatomic gaseous molecule with an unpaired electron (a free radical). It is lipophilic and can diffuse rapidly across biological membranes. NO mediates a variety of physiological functions, such as endothelial-derived relaxation of vascular smooth muscle, inhibition of platelet aggregation, neurotransmission, and cytotoxicity. The pathophysiology of NO is a double-edged sword. Insufficient production of NO has been implicated in the development of hypertension, impotence, susceptibility to infection, and atherogenesis. Excessive NO production is linked to septic shock, inflammatory diseases, transplant rejection, stroke, and carcinogenesis.

NO is synthesized from one of the terminal nitrogen atoms of the guanidino group of arginine, with the concomitant production of citrulline. Molecular oxygen and NADPH are cosubstrates, and the reaction is catalyzed by nitric oxide synthase (NOS). NOS consists of several isoforms and is a complex enzyme containing bound FMN, FAD, tetrahydrobiopterin, heme complex, and nonheme iron. A calmodulin binding site is also present. NO formation from arginine is a two-step successive monooxygenation process requiring five-electron oxidations. The first step is the formation of N^G-hydroxylarginine (N^G denotes guanidinium nitrogen atom):

$$\text{Arginine} + O_2 + \text{NADPH} + H^+ \rightarrow \text{HO} - N^G - \text{Arg} + \text{NADP}^+ + H_2O$$

This step is a mixed-function oxidation reaction similar to the one catalyzed by cytochrome P-450 reductase, and there is considerable homology between NOS and cytochrome P-450 reductase. In the second step, further oxidation of NO-hydroxyl arginine yields NO and citrulline:

$$\text{HO-}N^G\text{-Arg} + O_2 + \frac{1}{2}(\text{NADPH} + H^+) \rightarrow \text{citrulline} + \text{NO} + H_2O + \frac{1}{2}\text{NADP}^+$$

The overall reaction is

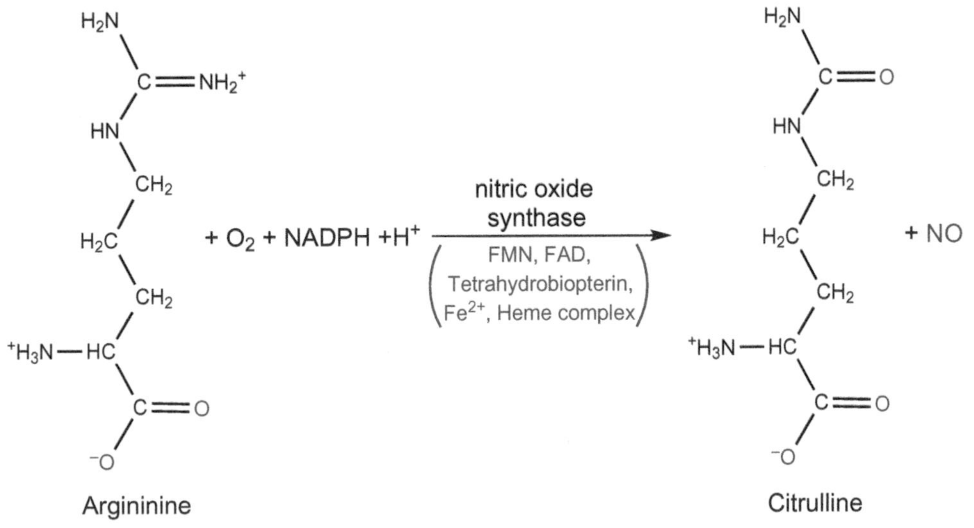

Argininine   Citrulline

The NOS activity is inhibited by N^G-substituted analogues of arginine, such as N^G-nitroarginine and N^G-monomethyl-L-arginine.

Isoforms (Also Known as Isozymes) of Nitric Oxide Synthase

There are three major isoforms of **nitric oxide synthase** (**NOS**) ranging in molecular size from 130 to 160 kDa. Amino acid similarity between any two isoforms is about 50%−60%. Isoforms of NOS exhibit differences in tissue distribution, transcriptional regulation, and activation by intracellular Ca^{2+}. Two of the three isoforms of NOS are constitutive enzymes (cNOS), and the third isoform is an inducible enzyme (iNOS). The cNOS isoforms are found in the vascular endothelium (eNOS), neuronal cells (nNOS), and many other cells, and are regulated by Ca^{2+} and calmodulin. In the vascular endothelium, agonists such as acetylcholine and bradykinin activate eNOS by enhancing intracellular Ca^{2+} concentrations via the production of inositol 1,4,5-trisphosphate, which activates the phosphoinositide second-messenger system. The NO produced in the vascular endothelium maintains basal vascular tone by vasodilation, which is mediated by vascular smooth muscle cells. Organic nitrates used in the management of ischemic heart disease act by denitration with the subsequent formation of NO. Sodium nitroprusside, an antihypertensive drug, is an NO donor. Thus, organic nitrates and sodium nitroprusside are prodrugs; the exact mechanism by which these prodrugs yield NO is not yet understood. Inhaled NO can produce pulmonary vasodilation. This property of NO has been used in the management of hypoxic respiratory failure associated with primary pulmonary hypertension in neonates. NO produced by cNOS in neuronal tissue functions as a neurotransmitter.

The inducible class of NOS (iNOS) is found in macrophages and neutrophils and is Ca^{2+}-independent. Bacterial endotoxins, cytokines (e.g., interleukin-l, interferon-γ), or bacterial lipopolysaccharides can induce and cause expression of NOS in many cell types. Glucocorticoids inhibit the induction of iNOS. In stimulated macrophages and neurophils, NO and superoxide radical ($^{•}O_2^{-}$) react to generate peroxynitrite, a powerful oxidant, and hydroxyl radicals. These reactive intermediates are involved in the killing of phagocytosed bacteria (Chapter 14). Excessive production of NO due to endotoxinemia produces hypotension and vascular hyporeactivity to vasoconstrictor agents, and leads to septic shock. NOS inhibitors have potential therapeutic application in the treatment of hypotensive crises.

Signal Transduction of NO

NO is lipophilic and diffuses readily across cell membranes. It interacts with molecules in the target cells producing various biological effects. One mechanism of action of NO is stimulation of guanylate cyclase, which catalyzes the formation of cyclic guanosine monophosphate (cGMP) from GTP, resulting in increased intracellular cGMP levels (see Figure 15.10). NO activates guanylate cyclase by binding to heme iron. The elevation of cGMP levels may activate cGMP-dependent protein kinases.

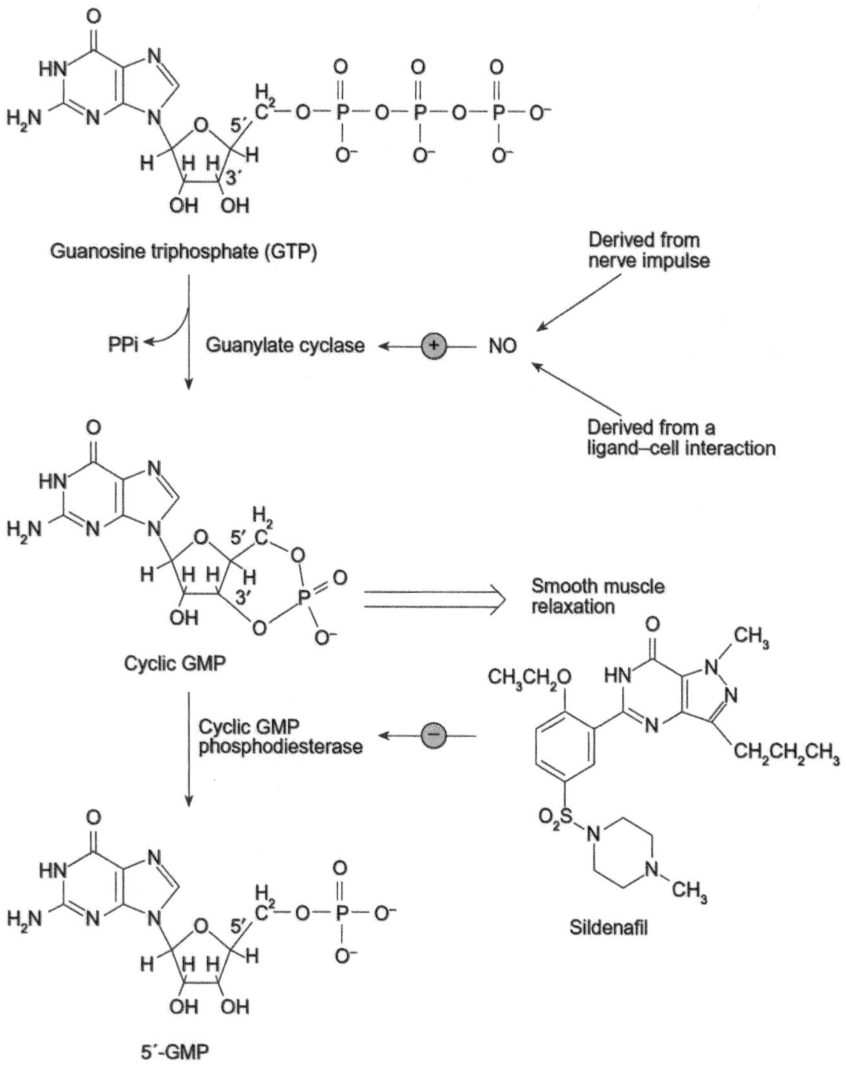

FIGURE 15.10 NO-mediated synthesis of cGMP from GTP in the corpus cavernosum that leads to smooth muscle relaxation. Sildenafil potentiates the effects of NO by inhibiting cGMP phosphodiesterase.

These kinases phosphorylate specific proteins that may be involved in removal or sequestration of Ca^{2+} or other ions, resulting in physiological stimuli. The physiological actions of cGMP are terminated by its conversion to 5'-GMP by cGMP-phosphodiesterase. Inhibitors of cGMP phosphodiesterase promote the actions of NO.

Sildenafil is a selective inhibitor of a specific cGMP phosphodiesterase (type 5) present in the corpus cavernosum. This compound (structure shown in Figure 15.10) is used orally in the therapy of some types of **erectile dysfunction**. NO is the principal transmitter involved in the relaxation of penile smooth muscle. During central or reflex sexual arousal, NO production is enhanced leading to increased production of cGMP. Smooth muscle relaxation permits the corpus cavernosum to fill with blood. Since the therapeutic effect of sildenafil potentiates the action of cGMP, the drug is ineffective in the absence of

sexual arousal. The relaxation of cavernosal smooth muscle caused by cGMP involves inhibition of Ca^{2+} uptake. Prostaglandin E_1 (alprostadil) inhibits the uptake of Ca^{2+} smooth muscle by a separate mechanism and causes erections in the absence of sexual arousal. Blood flow through the corpus cavernosum may also be increased by α-adrenergic blocking agents (e.g., phentolamine mesylate). Co-administration of NO donor drugs with NO potentiation drugs (e.g., sildenafil) may cause severe hypotension in the cardiovascular system.

Signal transduction of NO by cGMP-independent mechanisms include ADP-ribosylation of glyceraldehyde-3-phosphate dehydrogenase (GAPDH), an enzyme of the glycolytic pathway (Chapter 12), and interactions with many heme-containing and nonheme iron−sulfur containing proteins. NO activates ADP-ribosyltransferase, which catalyzes the transfer of ADP-ribose from NAD^+ to

GADPH. This results in the inactivation of GADPH, causing inhibition of glycolysis and decreased ATP production.

The anti-aggregability of platelets and the neurotoxicity of NO have been attributed to inhibition of glycolysis by NO.

Glycine

Glycine participates in a number of synthetic pathways and is oxidized to provide energy (Figure 15.11). The interconversion of glycine and serine by serine hydroxymethyltransferase is as follows:

$$NH_3^+\text{-}CH_2\text{-}COO^- + N^5, N^{10}\text{-methylene-FH}_4 + H_2O$$
glycine

$$\xrightleftharpoons[\text{pyridoxal phosphate}]{} HOH_2C\text{-}CHNH_3^+\text{-}COO^- + FH_4$$
serine

The one-carbon carrier N^5, N_{10}-methylenetetrahydrofolate is derived from reactions of the one-carbon pool (Chapter 25). [The term **one-carbon pool** refers to all single-carbon-containing metabolites (e.g., $-CH_3$, $-CHO$, $NH=C-$, etc.) that can be utilized in biosynthetic reactions such as formation of purine and pyrimidine.] These reactions include oxidation of glycine by glycine cleavage enzyme complex (glycine synthase):

$$NH_3^+\text{-}CH_2\text{-}COO^- + FH_4 + NAD^+$$
glycine

$$\xrightleftharpoons[\text{pyridoxal phosphate}]{} NH_4^+ + CO_2 + NADH$$
$$+ N^5, N^{10}\text{-methylene-FH}_4$$

This reaction favors glycine degradation, but the formation of glycine may also occur. The enzyme complex is mitochondrial and contains a pyridoxal phosphate-dependent glycine decarboxylase, a lipoic acid-containing protein that is a carrier of an aminomethyl moiety, a tetrahydrofolate-requiring enzyme, and lipoamide

dehydrogenase. The reactions of glycine cleavage resemble those of oxidative decarboxylation of pyruvate (Chapter 12).

Disorders of Glycine Catabolism

Nonketotic hyperglycinemia is an inborn error due to a defect in the glycine cleavage enzyme complex in which glycine accumulates in body fluids, especially in cerebrospinal fluid. It is characterized by mental retardation and seizures. Glycine is an inhibitory neurotransmitter in the central nervous system, including the spinal cord. Strychnine, which produces convulsions by competitive inhibition of glycine binding to its receptors, gives modest results in treatment but is not very effective. Sodium benzoate administration reduces plasma glycine levels but does not appreciably alter the course of the disease. Exchange transfusion may be useful. **Ketotic hyperglycinemia** also occurs in propionic acidemia, but the mechanism has not been established.

Type 1 primary hyperoxaluria is a genetic disorder caused by a defective enzyme, alanine glyoxylate aminotransferase (AGT). AGT is a liver-specific peroxisomal enzyme that converts a metabolite of glycine and serine metabolism, glyoxylate, to glycine. Deficiency of this enzyme leads to conversion of glyoxylate to oxalate, which is a highly insoluble product excreted via the kidneys (Figure 15.12). Overproduction of oxalate forms precipitates of calcium oxalate, causing renal damage. Hyperoxaluria, both primary and secondary, can occur due to other causes [1].

Creatine and Related Compounds

Phosphocreatine serves as a high-energy phosphate donor for ATP formation (e.g., in muscle contraction; see Chapter 19). Synthesis of **creatine** (methyl

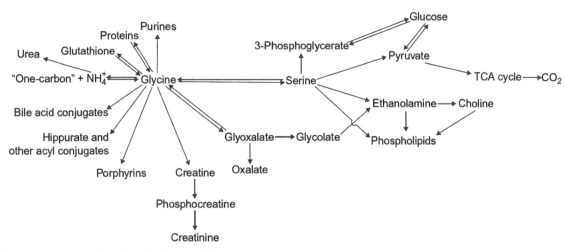

FIGURE 15.11 Overview of glycine and serine metabolism.

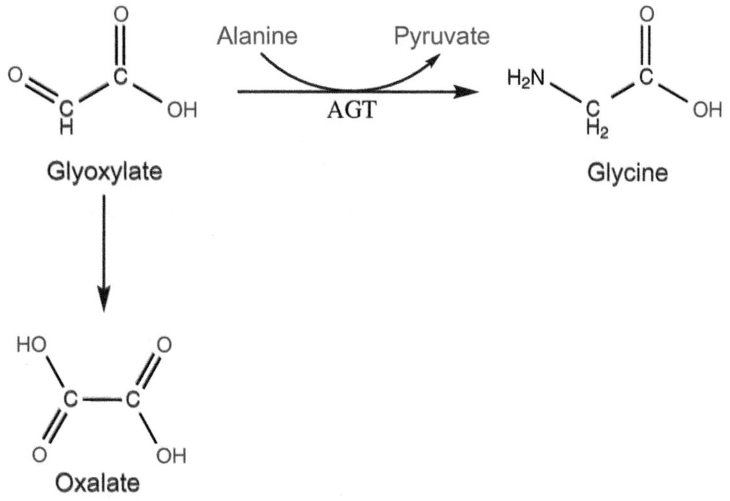

FIGURE 15.12 Conversion of glyoxylate to glycine and oxalate. The deficiency of alanine glyoxylate aminotransferase (AGT) leads to overproduction of the insoluble product oxalate.

guanidinoacetate) requires transamidination, i.e., transfer of a guanidine group from arginine to glycine, to form guanidinoacetate (glycocyamine) by mitochondrial arginine—glycine amidinotransferase:

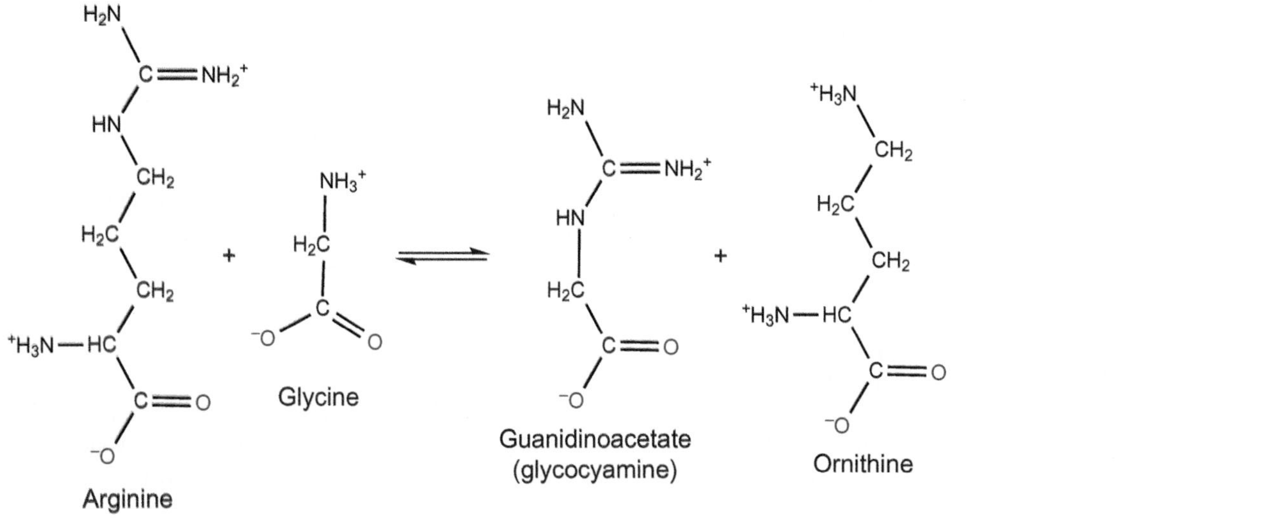

In the next step, guanidinoacetate is methylated by S-adenosylmethionine by cytosolic S-adenosylmethionine guanidinoacetate-N-methyltransferase to form creatine:

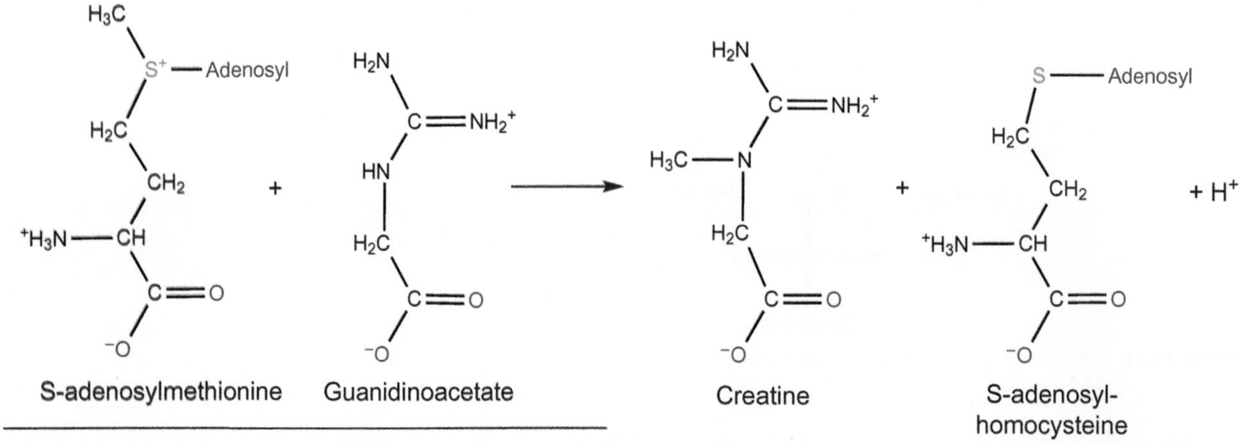

These reactions occur in liver, kidney, and pancreas, from which creatine is transported to organs such as muscle and brain. Creatine synthesis is subject to negative modulation of amidinotransferase by creatine. Phosphocreatine production is catalyzed by creatine kinase:

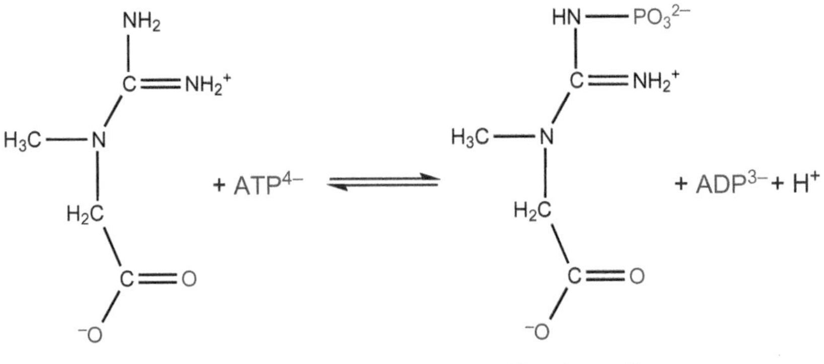

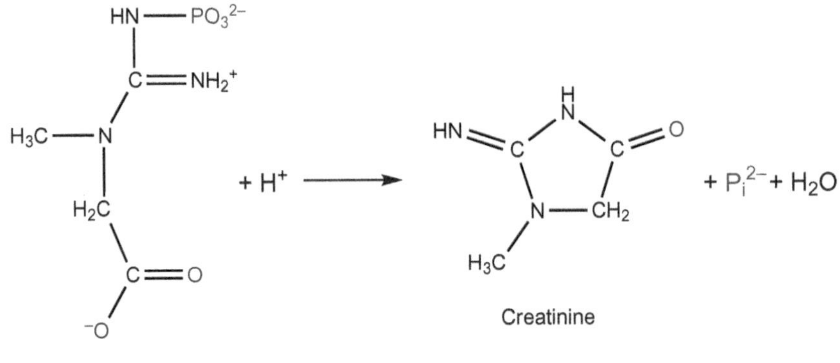

Creatinine has no useful function and is eliminated by renal glomerular filtration and to a small extent by renal tubular secretion. Creatinine clearance approximately parallels the **glomerular filtration rate (GFR)** and is used as a kidney function test. Creatinine clearance is expressed as milliliters of plasma cleared per minute per

Creatine ⇌ Phosphocreatine

This kinase is a dimer of M and B (M = muscle, B = brain) subunits produced by different structural genes. Three isozymes are possible: BB (CK-1), MB (CK-2), and MM (CK-3). Another isozyme differs immunologically and electrophoretically and is located in the intermembrane space of mitochondria. Tissues rich in CK-1 are brain, prostate, gut, lung, bladder, uterus, placenta, and thyroid; those rich in CK-3 are skeletal and cardiac muscle. Cardiac muscle contains significant amounts of CK-2 (25%−46% of total CK activity, as opposed to less than 5% in skeletal muscle), so that in myocardial infarction the rise in serum total CK activity is accompanied by a parallel rise in that of CK-2 (Chapter 7).

Phosphocreatine undergoes a slow and nonenzymatic cyclization to creatinine:

standard surface area of 1.73 square meters. It is calculated as follows:

$$\text{Creatinine clearance} = \frac{U_{cr} \times V}{P_{cr}} \times \frac{1.73}{A}$$

where

U_{cr} = Concentration of creatinine in urine;
P_{cr} = Concentration of creatinine in plasma;
V = Volume of urine flow in mL/minute;
A = Surface area of the subject that is derived from a nomogram based on height and weight.

Creatinine concentrations are measured from a precisely timed urine specimen (e.g., 4-hour, 24-hour), and a plasma specimen drawn during the urine collection period. Excretion of creatinine depends on skeletal muscle

Phosphocreatine → Creatinine

mass and varies with age and sex. However, day-to-day variation in a healthy individual is not significant. **Creatinuria**, the excessive excretion of creatine in urine, may occur during growth, fever, starvation, diabetes mellitus, extensive tissue destruction, muscular dystrophy, and hyperthyroidism.

In the clinical assessment of patients with chronic kidney disease (CKD), GFR can be estimated by using serum creatinine levels and equations that have been derived by using the inverse relationship between measured GFR and serum creatinine levels. Estimated GFR rates are used for CKD patients 18 years of age and older, under steady-state conditions.

Serum levels of cystatin C, a polypeptide of 120 amino acid residues, have also been used in the assessment of GFR and kidney function. Cystatin C is a cysteine protease inhibitor, synthesized in all nucleated cells, filtered freely in the renal glomeruli, and metabolized by the proximal tubules. Unlike creatinine, cystatin appears to be less dependent on age, sex, and muscle mass. Measurements of both serum cystatin C and creatinine in calculating estimated GFR serve as better predictors of chronic renal diseases. [2,3]

Histidine

Histidine is not essential for adults except in persons with uremia. It is essential for growth in children. **Histamine** is decarboxylated histidine.

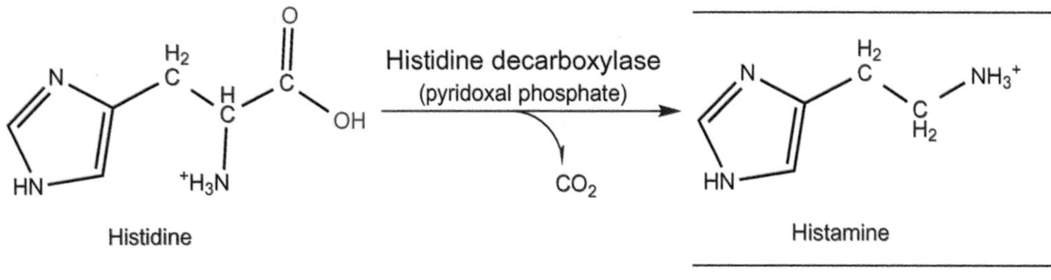

acetylcholine (released by parasympathetic nerve stimulation) and gastrin, and stimulates secretion of hydrochloric acid (Chapter 11). Histamine causes contraction of smooth muscle in various organs (gut, bronchi) by binding to H_1 receptors. The conventional antihistaminic drugs (e.g., diphenhydramine and pyrilamine) are H_1-receptor antagonists and are useful in the management of various allergic manifestations. However, in acute anaphylaxis, bronchiolar constriction is rapidly relieved by epinephrine (a physiological antagonist of histamine). Its effect on secretion of hydrochloric acid is mediated by H_2 receptors. H_2-receptor antagonists are cimetidine and ranitidine (Chapter 11), which are useful in treatment of gastric ulcers. Histamine is rapidly inactivated by methylation.

Branched-Chain Amino Acids

Leucine, isoleucine, and valine are essential amino acids that can be derived from their respective α-keto acids. A single enzyme may catalyze transamination of all three. The α-keto acids, by oxidative decarboxylation, yield the acyl-CoA thioesters, which, by α,β-dehydrogenation, yield the corresponding α,β-unsaturated acyl-CoA thioesters. The catabolism of these thioesters then diverges. Catabolism of leucine yields acetoacetate and acetyl-CoA via β-hydroxy-β-methylglutaryl-coenzyme A (HMG-CoA), which is also an intermediate in the biosynthesis of cholesterol and other isoprenoids (Chapter 17).

Histamine occurs in blood basophils, tissue mast cells, and certain cells of the gastric mucosa and other parts of the body (e.g., anterior and posterior lobes of the pituitary, some areas of the brain). Histamine is a neurotransmitter in certain nerves ("histaminergic") in the brain. In mast cells found in loose connective tissue and capsules, especially around blood vessels, and in basophils, histamine is stored in granules bound by ionic interactions to a heparin—protein complex and is released (by degranulation, vacuolization, and depletion) in immediate hypersensitivity reactions, trauma, and nonspecific injuries (infection, burns). Degranulation is affected by oxygen, temperature, and metabolic inhibitors. Release of histamine from gastric mucosal cells is mediated by

Catabolism of isoleucine yields propionyl-CoA (a glucogenic precursor) and acetyl-CoA. Catabolism of valine yields succinyl-CoA (Figure 15.13). Thus, leucine is ketogenic, and isoleucine and valine are ketogenic and glucogenic.

Oxidative decarboxylation of the α-keto acids is catalyzed by a branched-chain keto acid dehydrogenase (BCKADH) complex analogous to that of the pyruvate dehydrogenase and α-ketoglutarate dehydrogenase complexes. BCKADH is widely distributed in mammalian tissue mitochondria (especially in liver and kidney). It requires Mg^{2+}, thiamine pyrophosphate, CoASH, lipoamide, FAD, and NAD_1, and contains activities of α-keto acid decarboxylase, dihydro-lipoyl transacylase, and

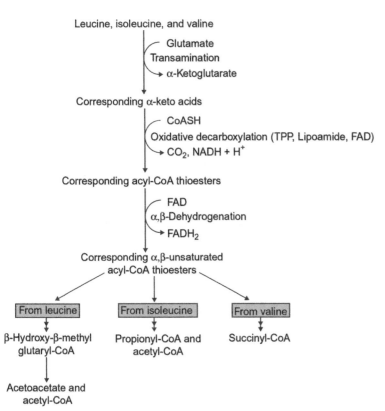

Leucine, isoleucine, and valine

Glutamate
Transamination
α-Ketoglutarate

Corresponding α-keto acids

CoASH
Oxidative decarboxylation (TPP, Lipoamide, FAD)
CO_2, NADH + H$^+$

Corresponding acyl-CoA thioesters

FAD
α,β-Dehydrogenation
$FADH_2$

Corresponding α,β-unsaturated
acyl-CoA thioesters

From leucine | From isoleucine | From valine

β-Hydroxy-β-methyl
glutaryl-CoA

Propionyl-CoA and
acetyl-CoA

Succinyl-CoA

Acetoacetate and
acetyl-CoA

FIGURE 15.13 Overview of the catabolism of branched-chain amino acids. TPP = thiamine pyrophosphate.

dihydrolipoyl dehydrogenase. Like the pyruvate dehydrogenase complex, BCKADH is regulated by product inhibition and by phosphorylation (which inactivates) and dephosphorylation (which activates).

Branched-chain ketoaciduria (maple syrup urine disease), an autosomal recessive disorder characterized by ketoacidosis, starts early in infancy and is due to a defect in the oxidative decarboxylation step of branched-chain amino acid metabolism. The name derives from the characteristic odor (reminiscent of maple syrup) of the urine of these patients. Five different variants (classic, intermittent, intermediate, thiamine-responsive, and dihydrolipoyl dehydrogenase deficient) are known, of which the first, which is due to deficiency of branched-chain α-keto acid decarboxylase, is the most severe. The incidence of maple syrup urine disease in the general population is 1 in 185,000 live births. In Mennonite populations, the incidence is extremely high (1 in 760). Neonatal screening programs consist of measuring leucine levels in dried blood spots.

Many aminoacidurias and their metabolites give rise to abnormal odors; maple syrup urine disease is one example. Some of the others are **phenylketonuria** (musty odor), **tyrosinemia type I** (boiled cabbage), **glutaric aciduria** (sweaty feet), **3-methylcrotonyl glycinuria** (cat's urine), and **trimethylaminuria** (fish). In patients with trimethylaminuria, the compound responsible for the fish odor is trimethylamine, which is a byproduct of protein catabolism by the large intestinal bacterial flora. Normally, trimethylamine is inactivated by hepatic flavin monooxygenases. Several different mutations in the gene for flavin monooxygenases have been identified in trimethylaminuric patients. An inhibitor of flavin monooxygenases is indole-3-carbinol, found in dark green vegetables (e.g., broccoli). The amelioration of symptoms of bad odor in trimethylaminuria may be achieved by limiting intake of dark green vegetables and protein, and by administering low doses of antibiotics to reduce intestinal bacterial flora.

Sulfur-Containing Amino Acids

Methionine and cysteine are the principal sources of organic sulfur in humans. Methionine is essential (unless adequate homocysteine and a source of methyl groups are available), but cysteine is not, since it can be synthesized from methionine.

Methionine

Methionine is utilized primarily in protein synthesis, providing sulfur for cysteine synthesis, and is the body's

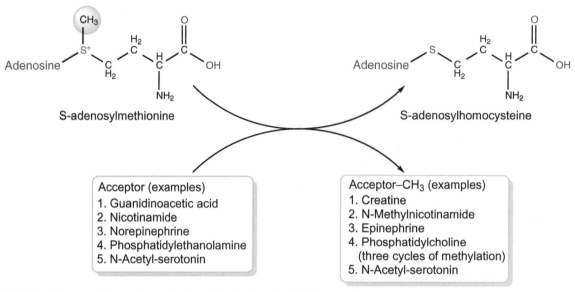

FIGURE 15.14 Selected methyl transfer reactions involving S-adenosylmethionine.

principal methyl donor. In methylation reactions, S-adenosylmethionine (SAM) is the methyl group donor. SAM is a sulfonium compound whose adenosyl moiety is derived from ATP as follows:

The methyl group is transferred to appropriate acceptors by specific methyltransferases with production of S-adenosylhomocysteine (Figure 15.14), which is hydrolyzed to homocysteine and adenosine by adenosylhomocysteinase:

Homocysteine can be recycled back to methionine either by transfer of a methyl group from betaine catalyzed by betaine–homocysteine methyltransferase, or from N^5-methyltetrahydrofolate (N^5-methyl-FH$_4$) catalyzed by N^5-methyl-FH$_4$-methyltransferase, which requires methyl cobalamin:

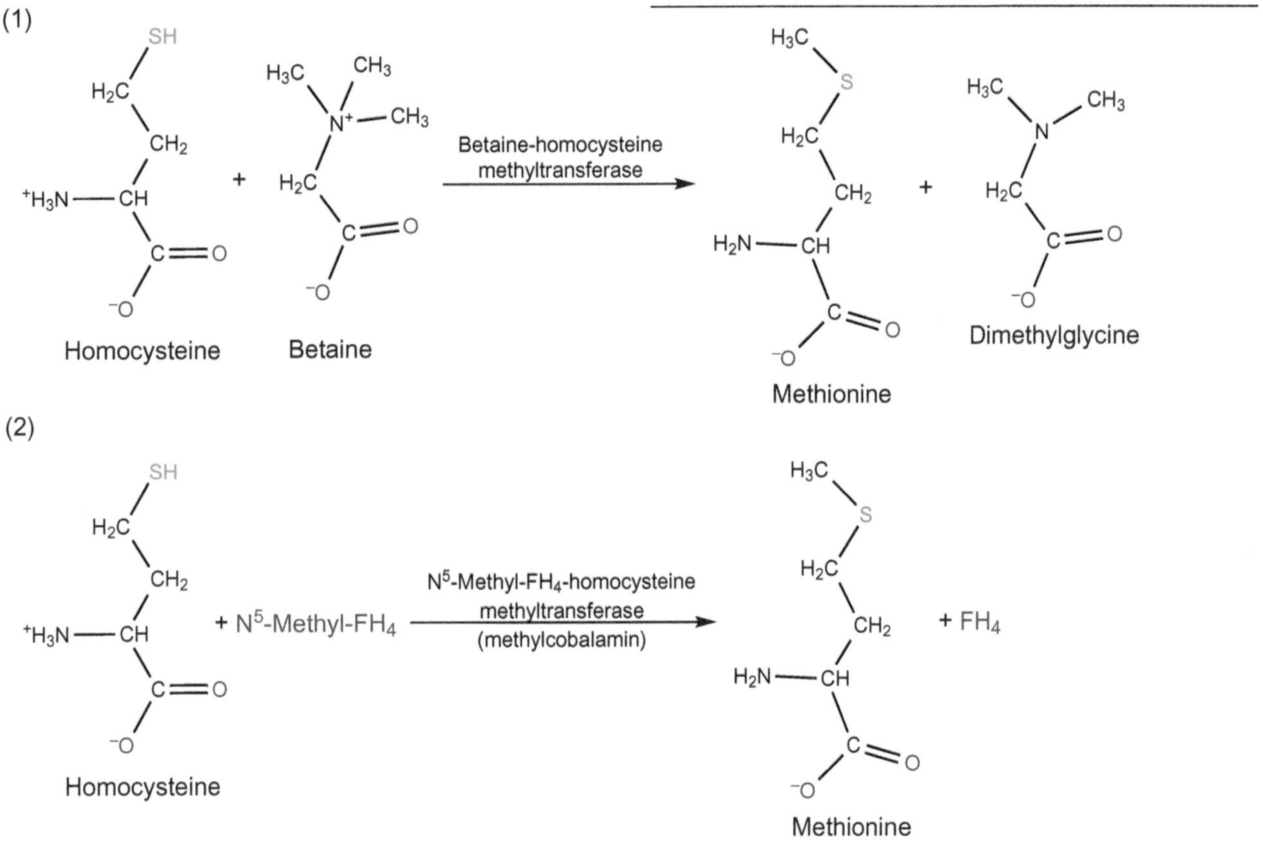

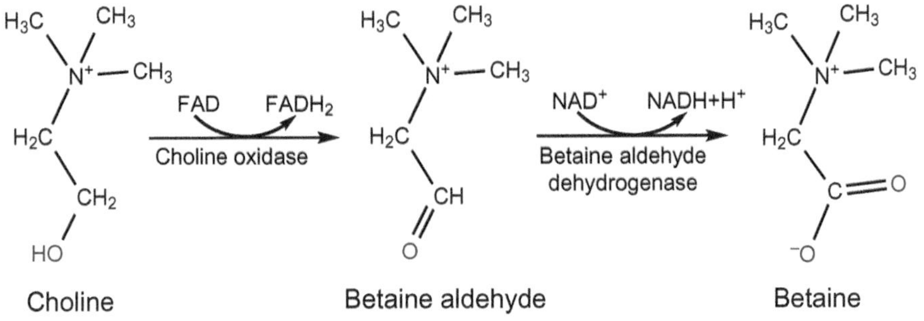

Cysteine

In the biosynthesis of cysteine, the sulfur comes from methionine by trans-sulfuration, and the carbon skeleton and the amino group are provided by serine. Cysteine regulates its own formation by functioning as an allosteric

(1)

Homocysteine   Betaine   Betaine-homocysteine methyltransferase   Methionine   Dimethylglycine

(2)

Homocysteine   + N^5-Methyl-FH$_4$ N^5-Methyl-FH$_4$-homocysteine methyltransferase (methylcobalamin)   Methionine   + FH$_4$

Betaine (an acid) is obtained from oxidation of choline (an alcohol) in two steps:

Choline   FAD   FADH$_2$ Choline oxidase   Betaine aldehyde   NAD$^+$ NADH+H$^+$ Betaine aldehyde dehydrogenase   Betaine inhibitor of cystathionine γ-lyase. α-Ketobutyrate is metabolized to succinyl-CoA by way of propionyl-CoA and methylmalonyl-CoA (Figure 15.15).

Cysteine is required for the biosynthesis of gluta-thione and of CoA-SH. A synthetic derivative, N-acet-ylcysteine, is used to replenish hepatic levels of glutathione and prevent hepatotoxicity due to overdos-age with acetaminophen. When high concentrations of acetaminophen are present in the liver, the drug under-goes N-hydroxylation to form N-acetyl-benzoquinonei-mine, which is highly reactive with sulfhydryl groups of proteins and glutathione and causes hepatic necro-sis. N-Acetylcysteine is used as a mucolytic agent (e. g., in cystic fibrosis) because it cleaves disulfide linkages of mucoproteins. Cysteine and cystine are interconverted by NAD-dependent cystine reductase and nonenzymatically by an appropriate redox agent (e.g., GSH).

The major end products of cysteine catabolism in humans are inorganic sulfate, taurine, and pyruvate. Taurine is a β-amino acid that has a sulfonic acid instead of a carboxylic acid group. Taurine is conjugated with bile acids in the liver (Chapter 17) and is readily excreted by the kidney. It is a major free amino acid of the central nervous system (where it may be an excitatory neuro-transmitter) and the most abundant in the retina; it also occurs in other tissues (e.g., muscle, lung).

Sulfate can be converted to the sulfate donor com-pound 3′-phosphoadenosine-5′-phosphosulfate (PAPS) in a two-step reaction (Figure 15.16). PAPS participates in

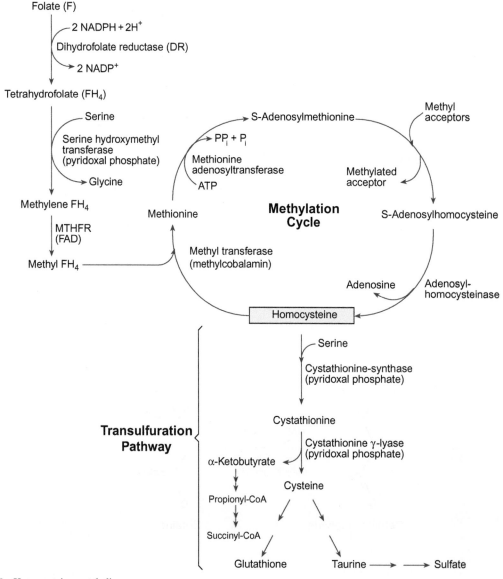

FIGURE 15.15 Homocysteine metabolism.

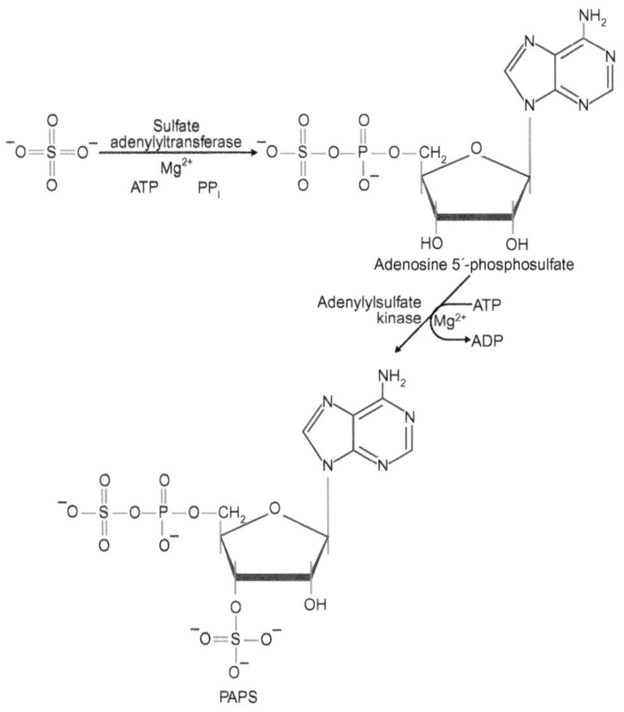

FIGURE 15.16 Formation of 3′-phosphoadenosine-5′-phosphosulfate (PAPS).

the sulfate esterification of alcoholic and phenolic functional groups (e.g., in synthesis of sulfolipids and glycosaminoglycans).

Abnormalities Involving Sulfur-Containing Amino Acids

Deficiencies of methionine adenosyltransferase, cystathionine β-synthase, and cystathionine γ-lyase have been described. The first leads to **hypermethioninemia** but no other clinical abnormality. The second leads to **hypermethioninemia**, **hyperhomocysteinemia**, and **homocystinuria**. The disorder is transmitted as an autosomal recessive trait. Its clinical manifestations may include skeletal abnormalities, mental retardation, ectopia lentis (lens dislocation), malar flush, and susceptibility to arterial and venous thromboembolism. Some patients show reduction in plasma methionine and homocysteine concentrations and in urinary homocysteine excretion after large doses of pyridoxine. Homocystinuria can also result from a deficiency of **cobalamin** (vitamin B_{12}) or folate metabolism. The third, an autosomal recessive trait, leads to cystathioninuria and no other characteristic clinical abnormality.

Hereditary sulfite oxidase deficiency can occur alone or with xanthine oxidase deficiency. Both enzymes contain molybdenum (Chapter 25). Patients with sulfite oxidase deficiency exhibit mental retardation, major motor seizures, cerebral atrophy, and lens dislocation. Dietary

deficiency of molybdenum (Chapter 35) can cause deficient activity of sulfite and xanthine oxidases.

Cystinuria is a disorder of renal and gastrointestinal tract amino acid transport that also affects lysine, ornithine, and arginine. The four amino acids share a common transport mechanism (discussed previously). Clinically, it presents as urinary stone disease because of the insolubility of cystine. In cystinosis, cystine crystals are deposited in tissues because of a transport defect in ATP-dependent cystine efflux from lysosomes (discussed previously).

Homocysteine

Homocysteine is an amino acid not found in proteins. Its metabolism involves two pathways; one is the methylation of homocysteine to methionine using N^5-methyltetrahydrofolate (N^5-methyl-FH_4) catalyzed by a vitamin B_{12}-dependent enzyme. The second is the trans-sulfuration pathway where homocysteine condenses with serine to form cystathionine; this is catalyzed by cystathionine β-synthase (CBS), which is a pyridoxal-5′-phosphate enzyme. End products of the trans-sulfuration pathway are cysteine, taurine, and sulfate (Figure 15.15). The methyl donor N^5-methyl-FH_4 is synthesized from N^5,N^{10}-methylene-FH_4, and the reaction is catalyzed by N^5,N^{10}-methylenetetrahydrofolate reductase (MTHFR). MTHFR is a FAD-dependent enzyme. Thus, the metabolism of homocysteine involves four water-soluble vitamins: folate, vitamin B_{12}, pyridoxine, and riboflavin. Any deficiencies or impairment in the conversion of the four vitamins to their active coenzyme forms will affect homocysteine levels. Several cases of **hyperhomocysteinemia** occur due to deficiencies of enzymes in the homocysteine remethylation or transsulfuration pathways. Individuals with a homozygous defect in cystathionine β-synthase have severe hyperhomocysteinemia (plasma concentrations $>50\,\mu M/L$) and their clinical manifestations are premature atherosclerosis, thromboembolic complications, skeletal abnormalities, ectopia lentis, and mental retardation.

In plasma, homocysteine is present in both free ($<1\%$) and oxidized forms ($>99\%$). The oxidized forms include protein (primarily albumin) −bound homocysteine mixed disulfide (80%−90%), homocysteine−cysteine mixed disulfide (5%−10%), and homocystine (5%−10%). Several studies have shown the relationship between homocysteine and altered endothelial cell function leading to thrombosis. Thus, hyperhomocysteinemia appears to be an independent risk factor for occlusive vascular disease. Five to ten percent of the general population have mild hyperhomocysteinemia.

It has been shown that a thermolabile form of MTHFR is a major cause of mildly elevated plasma homocysteine levels, which have been associated with coronary heart disease. The thermolabile MTHFR gene has a mutation of C to T at nucleotide position 677, which causes an alanine-to-valine amino acid substitution in the protein. The mechanism by which homocysteine mediates vascular pathology remains to be understood. The targets for homocysteine damage are connective tissue, endothelial cells, smooth muscle cells, coagulation factors, nitric oxide metabolism, plasma lipids, and their oxidized forms (Chapter 18). Vitamin supplementation with B_{12}, folate, and B_6 has reduced total plasma homocysteine levels. Vitamin supplementation may decrease the morbidity and mortality from atherosclerotic vascular disease due to hyperhomocysteinemia. However, further studies are required to assess the utility of vitamin supplementation (see Clinical Case Study 15.3).

Phenylalanine and Tyrosine

Phenylalanine is an essential amino acid. Tyrosine is synthesized by hydroxylation of phenylalanine and therefore is not essential. However, if the hydroxylase system is deficient or absent, the tyrosine requirement must be met from the diet. These amino acids are involved in synthesis of a variety of important compounds, including thyroxine, melanin, norepinephrine, and epinephrine (Figure 15.17). The conversion of phenylalanine to tyrosine and its degradation to acetoacetate and fumarate are shown in Figure 15.18.

The phenylalanine hydroxylase reaction is complex, occurring principally in liver but also in kidney. The hydroxylating system is present in hepatocyte cytosol and contains phenylalanine hydroxylase, dihydropteridine reductase, and tetrahydrobiopterin as coenzyme. The hydroxylation is physiologically irreversible and consists of a coupled oxidation of phenylalanine to tyrosine and of

tetrahydrobiopterin to a quinonoid dihydro derivative with molecular oxygen as the electron acceptor:

$$\text{Phenylalanine} + O_2 + \text{tetrahydrobiopterin}$$
$$\xrightarrow{\text{phenylalanine hydroxylase}} \text{tyrosine} + H_2O$$
$$+ \text{quinonoid dihydrobiopterin}$$

The tetrahydrobiopterin is regenerated by the reduction of the quinonoid compound dihydrobiopterin in the presence of NAD(P)H by dihydropteridine reductase:

$$\text{Quinonoid dihydrobiopterin} + \text{NAD(P)H} + H^+$$
$$\xrightarrow{\text{dihydropteridine reductase}} \text{NAD(P)}^+ + \text{tetrahydrobiopterin}$$

NADH exhibits a lower K_m and higher V_{max} for the reductase than does NADPH. Thus, the pterin coenzyme functions stoichiometrically (in the hydroxylase reaction) and catalytically (in the reductase reaction). Deficiency of dihydropteridine reductase causes a substantial decrease in the rate of phenylalanine hydroxylation. Dihydropteridine reductase and tetrahydrobiopterin are involved in hydroxylation of tyrosine and of tryptophan to yield neurotransmitters and hormones (dopamine, norepinephrine, epinephrine, and serotonin). Unlike phenylalanine hydroxylase, dihydropteridine reductase is distributed widely in tissues (e.g., brain, adrenal medulla). Tetrahydrobiopterin is synthesized starting from GTP and requires at least three enzymes. The first committed step is GTP-cyclohydrolase, which converts GTP to dihydroneopterin triphosphate. 6-Pyruvoyltetrahydrobiopterin synthase transforms dihydroneopterin triphosphate into 6-pyruvoyltetrahydrobiopterin. The latter is reduced to tetrahydrobiopterin by NADPH-dependent sepiapterin reductase. Deficiency of GTP-cyclohydrolase and sepiapterin reductase and 6-pyruvoyltetrahydrobiopterin synthase leads to **hyperphenylalanemia**, which exhibits severe clinical

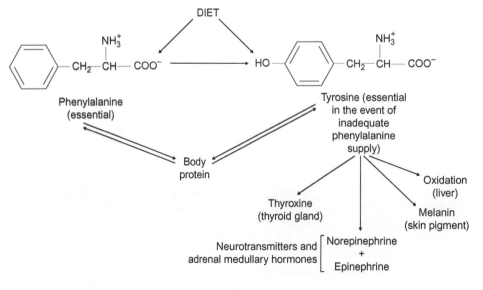

FIGURE 15.17 Overview of the metabolism of phenylalanine and tyrosine.

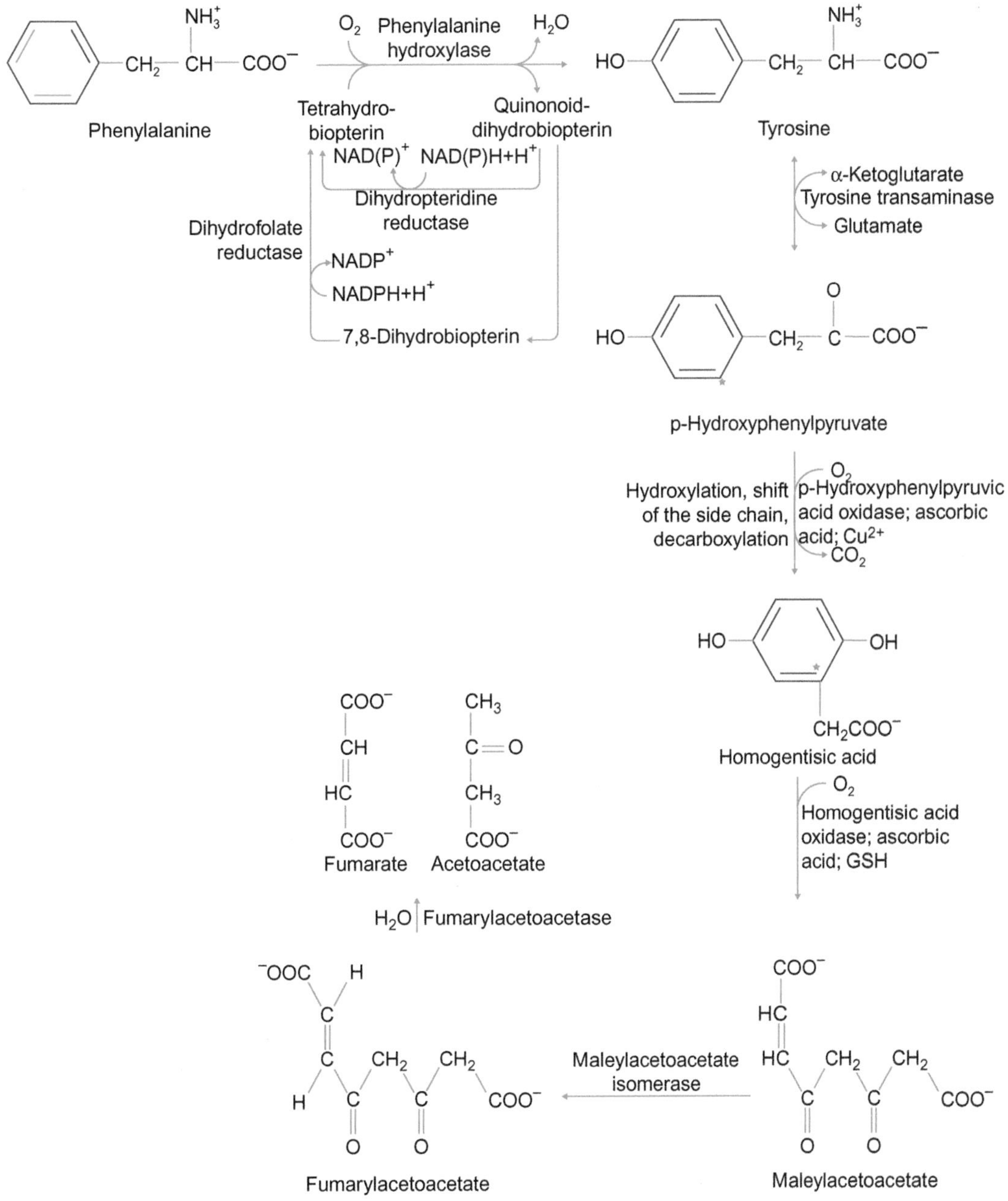

FIGURE 15.18 Conversion of phenylalanine to tyrosine and the oxidative pathway of tyrosine.

manifestations due to neurotransmitter deficiencies (see Clinical Case Study 15.4 with two vignettes).

Human liver phenylalanine hydroxylase is a multimeric homopolymer whose catalytic activity is enhanced by phenylalanine and has a feed-forward metabolic effect. Phosphorylation of phenylalanine hydroxylase by cAMP-dependent kinase leads to increased enzyme activity; dephosphorylation has an opposite effect. Thus, glucagon and insulin have opposing effects on the catalytic activity of phenylalanine hydroxylase.

Phenylketonuria (PKU)

Deficiency of phenylalanine hydroxylase, tetrahydrobiopterin, or dihydropteridine reductase results in phenylketonuria (PKU), an autosomal recessive trait. Because phenylalanine accumulates in tissues and plasma (hyperphenylalaninemia), it is metabolized by alternative pathways, and abnormal amounts of phenylpyruvate appear in urine. Phenylalanine hydroxylase deficiency may be complete (classic PKU, type I) or partial (types II and III). Many mutations of the phenylalanine hydroxylase gene have been identified (missense, nonsense, insertions, deletions, and duplications) leading to PKU or non-PKU hyperphenylalaninemia.

The incidence of classic PKU is about 1 in 10,000–20,000 live births and exhibits considerable geographic variation (the incidence in Ireland is 1 in 4000, whereas the condition is rare among blacks and Asians). About 2% of hyperphenylalaninemic infants have a deficiency of biopterin or biopterin reductase. The most important clinical presentation is mental impairment. Diagnosis can be made early in the neonatal period by measurement of phenylalanine concentration in blood collected from a heel prick onto filter paper. Treatment of phenylalanine hydroxylase deficiency consists of a diet low in phenylalanine, but which maintains normal nutrition. This diet is effective in preventing mental retardation, and its continuation throughout the first decade, or for life, may be necessary.

Treatment of biopterin and biopterin reductase deficiency consists not only of regulating the blood levels of phenylalanine, but also of supplying the missing form of coenzyme and the precursors of neurotransmitters, namely, dihydroxyphenylalanine and 5-hydroxytryptophan, along with a compound that inhibits peripheral aromatic decarboxylation. This compound is necessary because the amine products do not cross the blood–brain barrier.

Successfully treated females who have reached reproductive age may expose their offspring (who are obligate heterozygotes) to abnormal embryonic and fetal development. These effects include spontaneous abortion, microcephaly, congenital heart disease, and intrauterine growth retardation, and they correlate with the plasma level of phenylalanine in the pregnant mother. Thus, reinstitution of a low-phenylalanine diet during pre- and postconception periods may be necessary. The diet should also restrict intake of phenylalanine-containing substances, such as the synthetic sweetener aspartame, which is L-aspartyl-L-phenylalanyl methyl ester [4]. Because defective myelination occurs in the brain in PKU, there is an increased incidence of epileptic seizures, and abnormal electroencephalograms are common. The biochemical basis for the severe mental impairment is not understood.

High phenylalanine levels may disturb the transport of amino acids, in particular tryptophan required for the formation of serotonin (see next section), into cells of the central nervous system. Variations in the clinical manifestations of PKU may reflect differences in this disturbance.

Defects in pigmentation of skin and hair (light skin and hair) may be caused by interference in melanin formation by phenylalanine and its metabolites, and also by lack of tyrosine.

Melanin

Melanin is an insoluble, high molecular weight polymer of 5,6-dihydroxyindole, which is synthesized from tyrosine. It is produced by pigment cells (melanocytes) in cytoplasmic organelles (melanosomes). In the epidermis, melanocytes are associated with keratinocytes, which contain melanosomes supplied by melanocytes via dendritic processes. Color variation in human skin reflects the amount of melanin synthesized in melanosomes. Melanin synthesis is apparently under hormonal and neural regulation.

Tryptophan

Tryptophan is an essential amino acid involved in synthesis of several important compounds. Nicotinic acid (amide), a vitamin required in the synthesis of NAD^+ and $NADP^+$, can be synthesized from tryptophan. About 60 mg of tryptophan can give rise to 1 mg of nicotinamide.

5-Hydroxytryptamine (serotonin) is found in enterochromaffin cells, brain, and platelets. In the former two, it is produced from tryptophan, whereas in platelets, serotonin is taken up from plasma. Synthesis involves hydroxylation of tryptophan by tryptophan 5-hydroxylase and decarboxylation by aromatic L-amino acid decarboxylase (Figure 15.19). Hydroxylation is the rate-limiting reaction, is analogous to that of phenylalanine, and requires molecular oxygen and tetrahydrobiopterin. Serotonin is a powerful vasoconstrictor and stimulator of smooth muscle contraction. In the brain, it is a neurotransmitter, and in the pineal gland, it serves as a precursor of melatonin. Synthesis of melatonin requires N-acetylation of serotonin, followed by methylation (Figure 15.20). N-acetyltransferase is activated by increased concentrations of cytosolic cyclic AMP and Ca^{2+} that is mediated by the activation of adrenergic receptors of the pineal gland. In humans, melatonin synthesis and its release follow a circadian rhythm, which is stimulated by darkness and inhibited by light. The blood levels of melatonin increase by passive diffusion from the central nervous system after the onset of darkness, reaching a peak value during the middle of the night and declining during the second half of the night. Melatonin secretion is also regulated endogenously by signals from the suprachiasmatic nucleus. Since melatonin's peak concentration in plasma coincides with sleep, exogenous administration of the hormone can affect the circadian rhythm. Melatonin supplementation has been used to ameliorate subjective and

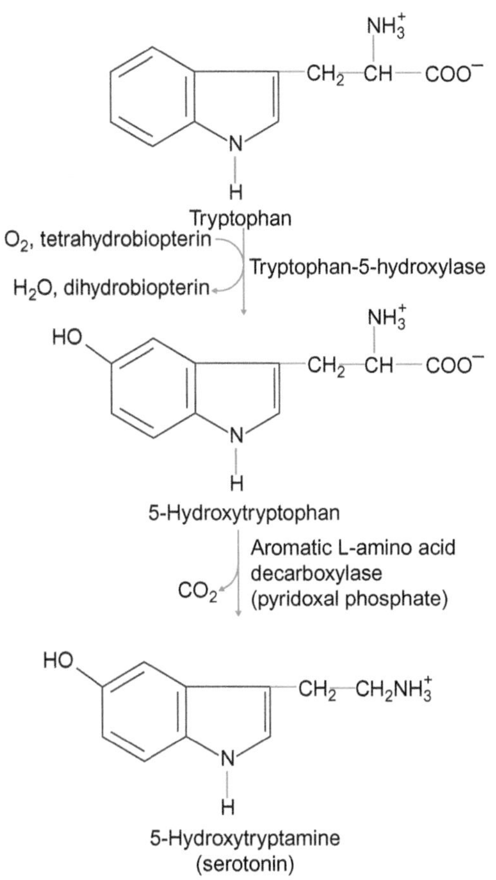

FIGURE 15.19 Biosynthesis of serotonin from tryptophan.

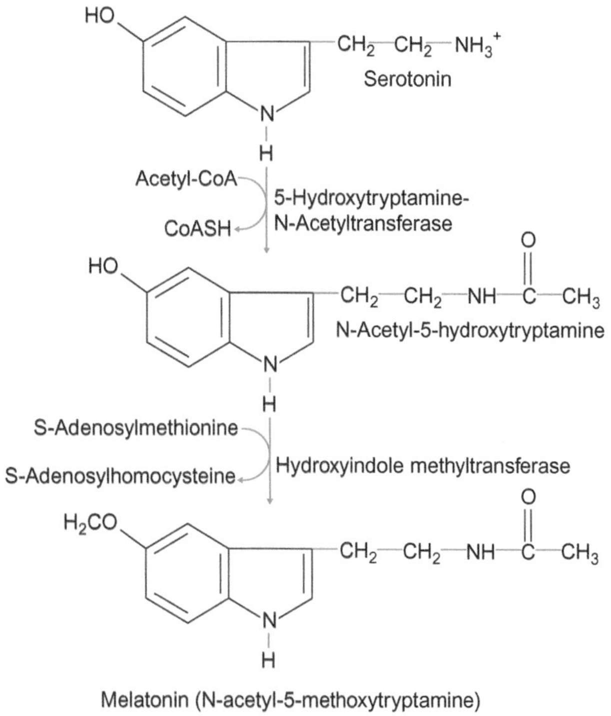

FIGURE 15.20 Biosynthesis of melatonin from serotonin.

objective symptoms of jet lag caused by travel across time zones. At high levels, melatonin promotes sleep. Short-term and long-term biological effects of melatonin supplementation have yet to be determined.

Serotonin is degraded to 5-hydroxyindoleacetic acid (5-HIAA) by monoamine oxidase and aldehyde dehydrogenase acting in sequence. 5-HIAA is excreted in the urine. Its excretion is markedly increased in subjects with carcinoid tumor (found most frequently in the gastrointestinal tract and lungs). **Carcinoid tumors** are frequently indolent and asymptomatic; however, a significant number of these tumors can manifest as metabolic problems. Although increased production of serotonin is a characteristic feature of the carcinoid tumor, cells also synthesize other substances. They include kinins, prostaglandins,

substance P, gastrin, somatostatin, corticotropin, and neuron-specific enolase. Carcinoid tumor cells are known as **enterochromaffin** cells because they stain with potassium dichromate; they are also known as argentaffin cells because they take up and reduce silver salts. The symptoms of the carcinoid tumor are due to synergistic biochemical interactions between serotonin and the previously mentioned active metabolites. Characteristics of the carcinoid syndrome are flushing, diarrhea, wheezing, heart valve dysfunction, and pellagra. Lifestyle conditions that precipitate symptoms include intake of alcohol or spicy foods, and strenuous exercise. The treatment options for the carcinoid syndrome are multidisciplinary, including lifestyle changes, inhibitors of serotonin release and serotonin receptor antagonists, somatostatin analogues, hepatic artery embolization, chemotherapy, and surgery.

Hartnup disease is a disorder of renal tubular and intestinal absorption of tryptophan and other neutral amino acids.

CLINICAL CASE STUDY 15.1 Use of Serum Electrophoretic and Related Studies for the Diagnosis of Multiple Myeloma and Small B-cell Neoplasms (Waldenström Macroglobulinemia)

Neoplasms of terminally differentiated plasma cells are known as multiple myeloma (MM). MM accounts for 1% of all malignancies and more than 10% of hematological malignancies. Serum and urine electrophoretic studies followed by appropriate additional studies are used in the diagnosis and prognosis of these disorders. The following clinical vignettes, teaching

points, and pertinent supplemental additions and references provide valuable insights into these disorders.

Vignette 1: IgG(κ) MM
A 72-year-old man with complaints of back pain, shortness of breath, dyspnea on physical exertion, and weight loss saw his

(Continued)

CLINICAL CASE STUDY 15.1 (Continued)

physician. The laboratory investigations revealed anemia with rouleaux formation. The routine serum electrophoresis showed a prominent monoclonal spike in the γ-globulin region that was IgG(κ) monoclonal protein, confirmed by immunofixation. His serum IgA and IgM levels were decreased. Urine electrophoretic studies were unremarkable; serum calcium level was increased. Radiological investigations showed multiple skeletal lytic bone lesions. Bone marrow studies showed marked plasmacytosis. His diagnosis was MM. After additional studies performed by an oncologist, the subject was considered as a candidate for autologous stem cell transplantation. His treatment began with several cycles of melphalan, prednisone, and thalidomide administration. Subsequently, the serum IgG(κ) levels were used as a surrogate marker of the tumor burden and effectiveness of the therapy. Bisphosphonate therapy to inhibit the skeletal osteolytic lesions was employed (see Chapter 35).

Teaching Points

1. The routine serum electrophoretic study, which revealed the presence of a paraprotein, was crucial in initiating the diagnostic workup, which ultimately was MM.
2. The diagnosis of IgD MM using serum protein electrophoresis and immunofixation presents unique challenges; see additional case study reference 3.

Vignette 2: Nonsecretory MM

A 56-year-old female with a history of osteoporotic compression fractures was referred to an internist. Her serum calcium level was elevated; her serum creatinine was within the reference range, and she was anemic. Her routine serum electrophoretic pattern showed only a decreased γ-globulin fraction without any monoclonal bands. Her urine electrophoretic studies did not reveal any monoclonal immunoglobulin or light chains. Her bone marrow examination revealed marked plasmacytosis, and the immunohistochemical study showed the presence of predominant cytoplasmic κ-light chain containing proteins. The concentration of serum free κ-light chain was 940 mg/L (reference interval 3.3–19.4), and λ-light chain was 10 mg/L (reference interval 5.7–26.3). The κ/λ ratio was 94 (reference interval 0.26–1.65). Based on all the studies, the subject's diagnosis of nonsecretory multiple myeloma and treatment was similar to the one discussed in Vignette 1. The effectiveness of the treatment was followed by measuring serum free κ-light chains.

Teaching Points

Nonsecretory MM, which accounts for 1%–5% of all MM cases, may not show any detectable monoclonal proteins in the serum or urine. However, some nonsecretory MM patients do secrete free light chains, and their quantification in serum aids in diagnosis and treatment. Elevated monoclonal light chains in the serum can be transported to heart and/or kidney. In these tissues, they can undergo aggregation-forming amyloid deposits affecting the organ functions (see additional case study reference 4). The clinical relevance in the use of

quantitation of serum free light chains is discussed in references 14 and 15 of the supplemental reading list.

Vignette 3: Waldenström's Macroglobulinemia

A 42-year-old woman with a history of nose bleeds, chest pain with exertion, headaches, and confusion was examined by her internist. The subject's studies revealed anemia and thrombocytopenia; a peripheral blood smear showed rouleaux formation and plasmacytoid lymphocytes. Serum electrophoretic pattern and immunofixation studies revealed a monoclonal Ig band, with highly elevated monoclonal IgM(λ) and increased serum viscosity. Bone marrow evaluation revealed hypercellular marrow with a predominant population of lymphocytes and plasmacytoid lymphocytes. After additional studies, including cytogenetic studies, the subject's diagnosis was Waldenström's macroglobulinemia (WM). The subject's immediate clinical consequences were attributed to hyperviscosity syndrome contributed by IgM. Thus, plasmapheresis was initiated to reduce IgM levels. The chemotherapy consisted of chlorambucil (a DNA alkylating agent), fludarabine (a nucleoside analogue), and rituximab (a chimeric monoclonal antibody against CD20 protein located on B-lymphocyte cell membrane).

Teaching Points

1. WM is a B-lymphocyte disorder in which monoclonal lymphoplasmacytic cells secrete IgM. Elevated levels of IgM, a pentameric immunoglobulin, can give rise to hyperviscosity syndrome requiring urgent plasmapheresis therapy.
2. Chemotherapy requires cytotoxic drugs, immunomodulator agents, and monoclonal antibodies directed against B-lymphocytes.

Integration with Chapters 15, 22, and 33.

Additional Case Study References

[1] J.L. Harousseau, P. Moreau, Autologous hematopoietic stem-cell transplantation for multiple myeloma, N. Engl. J. Med. 360 (2009) 2645–2654.
[2] B. Faiman, A.A. Licata, New tools for detecting occult monoclonal gammopathy, a cause of secondary osteoporosis, Clev. Clin. J. Med 77 (2010) 273–278.
[3] L. Harle, C. Chan, N.V. Bhagavan, C.N. Rios, C.E. Sugiyama, M. Loscalzo, et al., A healthy young man presenting with multiple rib fractures, Clin. Chem. 56 (2010) 1390–1393.
[4] A.L. Miller, R.H. Falk, B.D. Levy, J. Loscalzo, A heavy heart, N. Engl. J. Med. 363 (2010) 1464–1469.

References for Supplemental Reading

[5] A. Dispenzieri, R. Kyle, G. Merlini, J.S. Miguel, H. Ludwig, R. Hajek, et al., Spotlight review: international myeloma working group guidelines for serum-free light chain analysis in multiple myeloma and related disorders, Leukemia 23 (2009) 214–224.
[6] M. Drayson, L.X. Tang, R. Drew, G.P. Mead, H. Carr-Smith, A.R. Bradwell, Serum-free light chain measurements for identifying and monitoring patients with nonsecretory multiple myeloma, Blood 97 (2001) 2900–2902.
[7] S.P. Treon, How I treat Waldenstrom macroglobulinemia, Blood 114 (2009) 2375–2385.

(Continued)

CLINICAL CASE STUDY 15.1 (Continued)

[8] J.P. Bida, R.A. Kyle, T.M. Therneau, L.J. Melton III, M.F. Plevak, D. R. Larson, et al., Disease associations with monoclonal gammopathy of undetermined significance, a population-based study of 17,398 patients, Mayo Clin. Proc. 84 (2009) 685–693.

[9] S.K. Kumar, J.R. Michael, F.K. Buadi, D. Dingli, A. Dispenzieri, R. Fonseca, et al., Management of newly diagnosed symptomatic multiple myeloma, updated Mayo stratification of myeloma and risk-adapted therapy (mSMART) consensus guidelines, Mayo Clin. Proc. 84 (2009) 1095–1110.

[10] R.A. Kyle, S.V. Rajkumar, Treatment of multiple myeloma: a comprehensive review, Clin. Lymphoma Myeloma 9 (2009) 278–288.

[11] J.A. Burger, P. Ghia, A. Rosenwald, F. Caligaris-Coppio, The microenvironment in mature B-cell malignancies, a target for new treatment strategies, Blood 114 (2009) 3367–3375.

[12] I. Turesson, R. Velez, S.Y. Kristinsson, O. Landgren, Patterns of multiple myeloma during the past 5 decades: Stable incidence rates for all age groups in the population but rapidly changing age distribution in the clinic, Mayo Clin. Proc. 85 (2010) 225–230.

[13] J.L. Harousseau, M. Attal, H. Avet-Loiseau, The role of complete response in multiple myeloma, Blood 114 (2009) 3139–3146.

[14] E. Jenner, Serum free light chains in clinical laboratory diagnostics, Clin. Chem. 427 (2014) 15–20.

[15] M.P. McTaggart, J. Lindsay, E.M. Kearney, Replacing urine protein electrophoresis with serum free light chain analysis as a first-line test for detecting plasma cell disorders offers increased diagnostic accuracy and potential health benefit to patients, Am. J. Clin. Pathol. 140 (2013) 890–897.

[16] C. Ha, N.V. Bhagavan, Novel insights into the pleiotropic effects of human serum albumin in health and disease, Biochim. Biophys. Acta 2013 (1830) 5486–5493.

Vignette 4: Congestive Heart Failure: A Case of Protein Misfolding

This case study vignette was abstracted from: C.E. Ha, N.V. Bhagavan, M. Loscalzo, S.K. Chan, H.V. Nguyen, C.N. Rios, S.A.A. Honda, Congestive heart failure: a case of protein misfolding, Hawaii J. Med. Public Health, 73 (2014) 172–174.

Synopsis

A 65-year-old man came to his primary care physician for his complaints of shortness of breath and dyspnea on exertion (DOE). His past medical history included coronary artery disease, obstructive pulmonary disease, and surgically treated prostate cancer.

The subject's current DOE was investigated by EKG, MRI, and laboratory studies. Based on these studies, a diagnosis of restrictive cardiomyopathy due to an infiltrative process was considered. This diagnosis led to further laboratory studies that involved serum and urine protein electrophoresis and respective immunofixation electrophoresis (IFE) studies. The serum and urine IFE studies did not show the presence of any monoclonal immunoglobulin bands. However, the serum levels of IgG, IgA, and IgM levels were decreased, and also the serum-free κ-light chains were highly elevated. These studies were followed by abdominal fat pad and bone marrow biopsy studies, which led to the diagnosis of κ-light chain amyloidosis. The subject's clinical condition deteriorated, and he expired 6 months after the diagnosis of amyloidosis, despite treatment.

Teaching Points

1. This case illustrates the importance of measuring free κ and λ light chains in the serum in the workup of potential myeloma cases. In addition, hypogammaglobinemia is also an indication for further studies to establish its cause.

2. It is important to point out that many light-chain myeloma cases do not lead to amyloidosis, but a few light chain myeloma with appropriate mutations have the propensity to form β-sheet secondary structures that aggregate to form amyloid fibrils at the target sites such as heart and kidney.

3. With normal kidney function, unaggregated light chains are cleaved in the kidneys. However, aggregated light chains are not filtered in the glomeruli and do not appear in the urine, as was the situation in Vignette 4.

Discussion: Serum Protein Electrophoresis and Its Diagnostic Significance

Serum contains more than 100 different proteins, each under separate genetic control. They are transport proteins for hormones, vitamins, lipids, metals, pigments, and drugs; enzymes; enzyme inhibitors (proteinase inhibitors); hormones; antibodies; clotting factors; complement components; tumor markers; and kinin precursors. Quantitation (by radial immunodiffusion, electroimmunoassay, nephelometric methods, enzyme-linked immunological methods, and radioimmunoassay) of the various constituents of serum is of value in diagnosing and following the course of certain diseases. Several of these proteins are discussed elsewhere in the text.

A simple and useful technique involves separation of serum proteins by an electric field at pH 8.6, using agarose gel electrophoresis. Separation of these proteins is possible because each carries a different charge and hence migrates at a differing rate when subjected to an electric potential. Serum is generally used instead of plasma because the fibrinogen found in plasma appears as a narrow band in the β region, which may be mistaken for signs of monoclonal paraproteinemia (see below). The support medium, cellulose acetate, possesses several advantages over paper: the time required to separate the major proteins is short; albumin trailing is absent; and rapid quantitative determination of protein fractions by photoelectric scanning (after a suitable staining procedure) is possible. The factors that affect electrophoresis are the ionic strength of the buffer, voltage, temperature, application width, and staining.

Figure 15.21 shows normal and some abnormal patterns of serum protein electrophoresis. The electrophoretic patterns obtained are not indicative of any one disease or class of disease. Furthermore, a characteristic pattern may be obscured or not found in a disease entity where normally such a pattern would be expected. Serum electrophoretic patterns provide only a general impression of the disorder and require confirmation by other procedures. An alteration (depression or elevation) in a given fraction should be quantitated by more sensitive and specific methods.

The five major fractions seen in cellulose acetate serum protein electrophoresis are albumin, α_1-globulin, α_2-globulin,

(Continued)

CLINICAL CASE STUDY 15.1 (Continued)

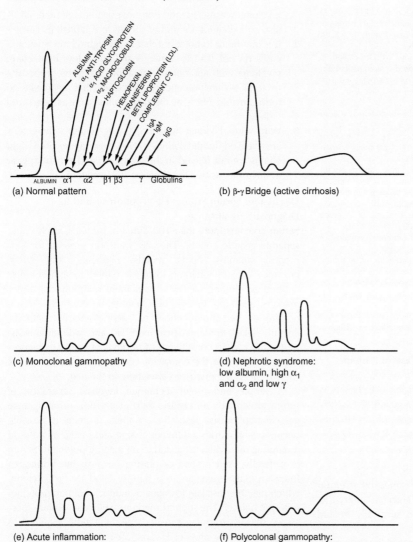

FIGURE 15.21 Serum protein electrophoresis patterns.

(a) Normal pattern

(b) β-γ Bridge (active cirrhosis)

(c) Monoclonal gammopathy

(d) Nephrotic syndrome: low albumin, high α₁ and α₂ and low γ

(e) Acute inflammation: high α₁ and α₂

(f) Polycolonal gammopathy: high γ, polyclonal elevation of immunoglobulins

β-globulin, and γ-globulin. Adult reference ranges for these five fractions, expressed as grams per 100 mL, are 3.2−5.6 for albumin, 0.1−0.4 for α_1-globulin, 0.4−0.9 for α_2-globulin, 0.5−1.1 for β-globulin, and 0.5−1.6 for γ-globulin. With the exception of γ-globulin, adult values are attained at 3 months of age for all fractions. Cord blood γ-globulin is largely of maternal origin and undergoes catabolism, reaching its lowest value at about 3 months. Adult levels for γ-globulin are not reached until about 2−7 years. After age 40, the level of albumin gradually declines and the β-globulin fraction increases.

Albumin

Albumin is synthesized in the liver. Since albumin accounts for about 75% of the oncotic pressure in blood, it is responsible for the stabilization of blood volume and regulation of vascular fluid exchange. Therefore, hypoalbuminemia can give rise to edema. **Hyperalbuminemia** is uncommon, but many types of abnormalities lead to **hypoalbuminemia**. Some hypoalbuminemic conditions include nephrotic syndrome, protein-losing enteropathies, cystic fibrosis (hypoalbuminemia with edema may be an early abnormality in infants with this disease), glomerulonephritis, cirrhosis, carcinomatosis, bacterial infections, viral hepatitis, congestive heart failure, rheumatoid arthritis, uncontrolled diabetes, intravenous feeding (the hypoalbuminemia is due to a deficiency of amino acids in portal blood), and dietary deficiency of proteins containing essential amino acids.

Hypoalbuminemia also occurs in acute stress reactions. The sharp drop in albumin is essentially due to adrenocortical stimulation, which gives rise to enhanced catabolism of albumin and sodium retention. The latter is responsible for hemodilution and expansion of extracellular fluid. The α_1- and

(Continued)

CLINICAL CASE STUDY 15.1 (Continued)

α_2-globulin fractions (see later) are also increased in an acute stress reaction. The immunoglobulins (γ-fraction) do not undergo significant alterations during the acute phase of a stress reaction but are elevated during the chronic phase. Some examples of situations in which acute stress patterns may be seen are acute infections in early stages, tissue necrosis (myocardium, renal, tumor), severe burns, rheumatoid disease of acute onset, surgery, and collagen disorders. Albumin has pleiotropic physiological effects, and the hypoalbuminemic conditions can affect several physiological conditions (see reference 16 in the supplemental reading list).

Analbuminemia is a rare autosomal recessive disorder. Affected individuals do not exhibit serious clinical symptoms, not even edema. The lack of clinical edema is presumably due to osmotic compensation by mildly elevated globulins. Osteoporosis in analbuminemia has been corrected using administration of human serum albumin. Affected females exhibit minimal pretibial edema, mild anemia, normal liver function tests, absence of proteinuria, lowered blood pressure, elevated serum cholesterol levels, and lipodystrophy. Despite elevated plasma cholesterol levels, severe atherosclerosis was not present.

Bisalbuminemia is due to albumin polymorphism. On serum electrophoresis, a double albumin band is seen. These double peaks are due either to differences in electrical charge (11 electrophoretically distinct forms have been reported) or to albumin dimers. Both forms are expressed as autosomal dominant traits and apparently present no significant clinical abnormalities. However, acquired bisalbuminemia may be associated with diabetes mellitus, the nephrotic syndrome, hyperamylasemia, and penicillin therapy. In these instances, the bisalbuminemia disappears after correction of the underlying disorder.

α_1-Globulins

The proteins that migrate in this region are α_1-antitrypsin (which accounts for 70%–90% of the α_1-peak, α_1-acid glycoprotein, or orosomucoid (which accounts for 10%–20% of the α_1-peak), α_1-lipoprotein, prothrombin, transcortin, and thyroxine-binding globulin. In acute phase reactions, elevation of the α_1-peak is due to increases in α_1-antitrypsin and α_1-acid glycoprotein, the two principal constituents of the peak. α_1-Peak elevations are also seen in chronic inflammatory and degenerative diseases, and are highly elevated in some cancers. When the α_1-peak is depressed, α_1-antitrypsin deficiency (Chapter 6) must be ruled out by direct measurements using sensitive quantitative methods. The presence of α_1-fetoprotein (AFP) in nonfetal serum is of clinical significance because of its close association with primary carcinoma of the liver; however, during pregnancy, AFP synthesized by the fetus gains access in small but measurable amounts to the maternal circulation. Measurement of maternal serum AFP (MSAFP) levels during the second trimester is widely used to detect fetal abnormalities. In neural tube defects, MSAFP levels are elevated, and in Down's syndrome, they are decreased (Chapter 36). Routine serum electrophoresis seldom gives such quantitative information, and specific immunochemical methods should be employed to measure α_1-fetoprotein in the serum.

α_2-Globulin

The α_2-fraction includes haptoglobin, α_2-macroglobulin, α_2-microglobulin, ceruloplasmin, erythropoietin, and cholinesterase. Haptoglobin is nonspecifically increased in an acute stress reaction in the presence of inflammation, tissue necrosis, or tissue destruction. A function of haptoglobin is to combine with hemoglobin to remove it from the circulation. Thus, during an episode of intravascular hemolysis, haptoglobin is depleted and may require a week or more to return to normal serum levels. Haptoglobin levels in these instances should be quantitated by immunochemical procedures. α_2-Macroglobulin functions as a protease inhibitor (Chapter 6). In the nephrotic syndrome, there is a characteristic elevation of both α_2 and β peaks with hypoalbuminemia (Figure 15.21d). The elevation in α_2 and β fractions is due to higher very-low-density lipoprotein (VLDL) and low-density lipoprotein (LDL) levels. Thus, characteristics of nephrotic syndrome include proteinuria (>3.5 g per 24 hours), hypoalbuminemia (<3 g/dL), hyperlipidemia with elevated triglyceride and cholesterol levels, lipiduria, and generalized edema. The exact mechanism causing higher VLDL and LDL levels is not understood. In part, these elevations may be due to increased synthesis of VLDLs in the liver and then conversion to LDLs in the peripheral circulation (Chapter 18), as well as their decreased catabolism. It is thought that the decreased plasma oncotic pressure due to hypoalbuminemia may stimulate increased hepatic synthesis of VLDLs.

Increased glomerular permeability for protein due to derangement in capillary walls is an early event in the nephrotic syndrome that has several causes, as listed here with examples:

1. Primary glomerular disease: membranous glomerulonephritis, lipid nephrosis;
2. Systemic disease: diabetes mellitus, amyloidosis, systemic lupus erythematosus;
3. Nephrotoxins and drugs: gold, mercury, penicillamine;
4. Infections: poststreptococcal glomerulonephritis;
5. Allergens, venoms, and vaccines;
6. Miscellaneous: toxemia of pregnancy, malignant hypertension.

The most common cause of primary nephrotic syndrome in adults is membranous nephropathy. It is characterized by the deposition of immune complex in the subepithelial portion of glomerular capillary walls.

β-Globulins

Quantitatively significant β-globulins are composed of transferrin, β-lipoproteins, complement C3, and hemopexin. Transferrin, synthesized in the liver, has a half-life of about 8.8 days and is an iron transport protein in serum that accounts for about 60% of the β peak. It is increased in iron deficiency anemia and pregnancy (Chapter 27). Transferrin levels are decreased in metabolic (e.g., liver, kidney) and

(Continued)

CLINICAL CASE STUDY 15.1 (Continued)

neoplastic diseases. Transferrin and transthyretin are both decreased in protein–calorie malnutrition. A pattern often observed in advanced cirrhosis is lack of separation of the β peak with the γ region, described as β-γ bridging (Figure 15.21b). The pattern is presumably due to elevated levels of IgA. Hypoalbuminemia is also found. Complement levels (C3 and C4) are decreased (owing to their consumption) in diseases associated with the formation of immune complexes, e.g., glomerulonephritis (acute and membranoproliferative) and systemic lupus erythematosus.

γ-Globulins

The five major classes of immunoglobulins in descending order of quantity are IgG, IgA, IgM, IgD, and IgE (see also Chapter 33). C-reactive protein migrates with the γ-globulins and is increased in trauma and acute inflammatory processes (discussed later). During an intense acute inflammatory process, its concentration may be highly elevated, giving rise to a sharp protein band that may be mistaken for monoclonal gammopathy. Variations in electrophoretic patterns due to γ-globulins can be categorized into the following three groups.

1. **Agammaglobulinemia** and **hypogammaglobulinemia**: May be primary or secondary. Secondary forms may be found in chronic lymphocytic leukemia, lymphosarcoma, multiple myeloma, nephrotic syndrome, long-term steroid treatment, and occasionally, overwhelming infection.

2. **Polyclonal gammopathy**: A diffuse polyclonal increase in γ-globulin, primarily in the IgG fraction (Figure 15.21f). Recall that in chronic stress γ-globulins are elevated. The major causes of polyclonal gammopathy are as follows:

   a. **Chronic liver disease**: An increase in polyclonal IgG and IgA with a decrease in albumin levels is a characteristic finding, regardless of the cause. The β-γ bridging observed in cirrhosis is presumably due to elevated levels of IgA that migrate to a position between the β and the γ region, increasing the trough between these bands.

   b. **Sarcoidosis**: The electrophoretic pattern shows a stepwise descent of globulin fractions starting from the γ-globulins.

   c. **Autoimmune disease**: This group includes rheumatoid arthritis (increased IgA levels), systemic lupus erythematosus (increased IgG and IgM levels), and others.

   d. **Chronic infectious disease**: This group includes bronchiectasis, chronic pyelonephritis, malaria, chronic osteomyelitis, kala-azar, leprosy, and others.

3. **Monoclonal gammopathy**: The presence of a narrow protein band in the β-γ region may indicate the presence of a homogeneous class of protein (Figures 15.21c and 15.22). Such a pattern should be further investigated using other laboratory and clinical findings to rule out multiple myeloma.

Means of investigation include quantitation of immunoglobins, immunofixation electrophoresis to identify the class and type of monoclonal immunoglobulin (Figure 15.23), bone marrow aspiration and biopsy to determine plasmacytosis,

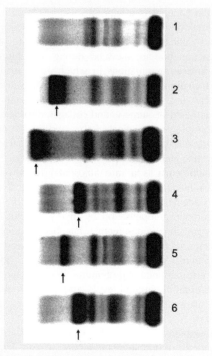

FIGURE 15.22 Serum electrophoretic protein patterns consisting of monoclonal immunoglobulins identified by arrows. Normal serum pattern is identified as 1.

and radiological examination to identify osteolytic lesions. Quantitation of serum-free light chains (κ and λ) is useful in the diagnosis of some cases of nonsecretory multiple myeloma. The normal ratio of κ/λ is 1.65. Either high or low levels suggest abnormality in immunoglobulin synthesis; however, this does not confirm the monoclonality of these proteins.

Multiple myeloma is a disorder of neoplastic proliferation of a single clone of plasma cells in the bone marrow that leads to overproduction of any one of the five immunoglobulins (IgG, IgA, IgM, IgD, or IgE) or light chains (κ or λ). Less commonly, biclonal or triclonal gammopathy also occurs. If the light chains are overproduced, they pass through the glomeruli and appear in the urine; this is known as **Bence Jones proteinuria**. In a given individual, Bence Jones proteinuria refers to either κ or λ light chains, but not both. Persons over the age of 40 may be affected, and initial findings may include spontaneous fractures, bone pain, anemia, or infections. Hypercalcemia may be found after bone destruction. Multiple myeloma is a progressive disorder and is usually fatal due to renal damage and/or recurrent infections. Cytotoxic measures (e.g., melphalan, a nitrogen mustard), immunomodulatory agents (thalidomide and lenalidomide), and ubiquitination inhibitors (bortezomib) are used in the treatment of multiple myeloma, and serum and/or urine monoclonal immunoglobulin or light chain levels serve in monitoring therapy.

(Continued)

CLINICAL CASE STUDY 15.1 (Continued)

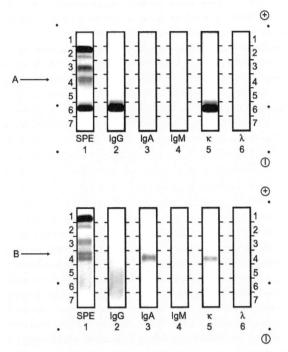

FIGURE 15.23 Serum immunofixation electrophoresis of two patients with monoclonal gammopathies. Pattern A represents IgG(κ) monoclonal gammopathy, and pattern B represents IgA(κ) monoclonal gammopathy, as indicated by arrows. The procedure consists of serum protein electrophoresis (SPE) separation, reaction of each track, with the exception of SPE, with specific respective antiserum, followed by protein staining to make visible the respective bands.

Although the monoclonal spike is usually observed in the β-γ region, it is occasionally seen in the α_2 region. In pregnancy, a monoclonal spike in the β region may be due to elevated transferrin levels. A serum electrophoretic pattern of hypogammaglobulinemia requires further studies of serum and urine (e.g., immunoglobulin measurement and immunofixation electrophoresis) to rule out multiple myeloma. The presence of monoclonal immunoglobulin in the serum is a characteristic feature of a majority of multiple myeloma cases; however, 1%–5% of patients do not show any detectable monoclonal protein in serum or urine. These disorders are known as **nonsecretory myeloma**. They fall into two classes: (1) producers of a paraprotein that remains in the cytosol of the malignant plasma cells due to secretory defects or (2) nonproducers of a paraprotein. Nonsecretory myelomas are characterized by immunoperoxidase staining methods using antibodies directed against light chain-specific immunoglobulins.

Monoclonal immunoglobulins in serum without significant clinical illness have also been found, and the incidence increases with age. Up to 3% of patients older than 70 demonstrate the presence of monoclonal immunoglobulin (benign

monoclonal proteinemia, also known as **monoclonal gammopathy of undetermined significance**, or **MGUS**). These individuals require periodic serum and urine studies. In some rare instances, monoclonal gammopathies of undetermined significance are associated with neuropathies, and the monoclonal immunoglobulin possesses an antinerve activity. A lymphoproliferative disorder characterized by monoclonal lymphocytes that produce monoclonal IgM is known as **Waldenström's macroglobulinemia**. In this disorder, the elevated serum viscosity is palliated by plasmapheresis and cytotoxic treatment (e.g., chlorambucil and nucleoside analogues, respectively) due to the presence of large quantities of IgM and the tumor burden.

It should be noted that pseudomonoclonal bands appearing in the β-γ region may be misidentified as authentic monoclonal gammopathy. Pseudomonoclonal bands occur in those serum specimens owing to hemolysis or to the presence of fibrinogen due to inappropriate blood coagulation techniques. Serum immunofixation studies should clarify these problems.

Secondary paraproteinemias may be seen in association with hematopoietic cancers (e.g., lymphomas and leukemias), other neoplasms (e.g., colon carcinoma), long-standing chronic urinary or biliary tract infection, rheumatoid factor related to IgM monoclonal protein, and amyloidosis.

Changes in Serum Proteins During Acute Phase of Disorders of Tissue Injury

The alterations in serum proteins (acute phase response) that occur within a few hours to a few days after tissue injury due to infection, trauma, burns, surgery, or infarction and inflammatory conditions are divided into two categories. The first category includes those proteins that are increased by at least 25% (positive acute phase proteins), and the second category includes proteins that are decreased by at least 25% (negative acute phase proteins). During acute stress, macrophages and monocytes are activated and produce several intercellular signaling polypeptides, known as **cytokines** (Chapter 33). Some examples of the released cytokines are interleukin-6, interleukin-1β, tumor necrosis factor-α, interferon-γ, and transforming growth factor-β. One of the functions of these cytokines is to alter the synthesis of acute phase proteins in hepatocytes by regulating expression of acute phase protein genes by both transcriptional and post-transcriptional mechanisms. Glucocorticoids stimulate the action of cytokines by promoting the production of some acute phase proteins. The postulated function of the acute phase response is to protect the body from injurious processes. One action is activation of the complement system to fight infection or to antagonize the activity of proteolytic enzymes. Some actions initiate or sustain inflammation, whereas others reflect anti-inflammatory and antioxidant properties.

Examples of positive and negative acute phase proteins are given in Table 15.1. C-reactive protein and serum amyloid A are elevated in serum as much as 1000-fold from their basal values. Serum amyloid A is an apolipoprotein; it is synthesized in hepatocytes in response to inflammatory stimuli and

(Continued)

CLINICAL CASE STUDY 15.1 (Continued)

TABLE 15.1 Examples of Positive and Negative Acute Phase Proteins

Positive

1. Several members of complement system (e.g., C3, C4)
2. Proteins of coagulation and fibrinolytic systems (e.g., fibrinogen, plasminogen)
3. Antiproteases (e.g., α_1-antitrypsin)
4. Transport proteins (e.g., haptoglobin, ceruloplasmin, hemopexin)
5. Inflammatory response modulators and others (e.g., phospholipase A_2, C-reactive protein, amyloid A, fibronectin, ferritin)

Negative

Albumin

Transferrin

Transthyretin

Thyroxine-binding globulin

associated with HDL. The function of amyloid A is not clear, and it is not commonly measured as an acute phase reactant. However, C-reactive protein, a serum protein (so named because it reacts with pneumococcal C-polysaccharide), is measured. It binds with the phosphocholine of pathogens and with the phospholipid constituents of aged blood cells and phagocytic cells, and it activates the complement system. All of the functions of C-reactive protein modulate inflammatory conditions of the body. These include induction of inflammatory cytokines, prevention of adhesion of neutrophils to endothelial cells by inhibiting the surface expression of L-selectin, and inhibition of superoxide production by neutrophils. Measurement of serum C-reactive protein is useful in differentiating an acute inflammatory condition from a noninflammatory one, as well as in the assessment of the severity of inflammation and its prognosis. Another widely used test in the assessment of acute phase disorders is the erythrocyte sedimentation rate (ESR). ESR determines the rate at which erythrocytes fall through the plasma to the bottom (sediment) of the test tube, and it depends largely on the plasma concentration of fibrinogen, an acute phase reactant (Table 15.1). Thus, ESR is increased during acute phase disorder. Compared to ESR measurement, serum C-reactive protein measurement has several advantages. ESR changes occur relatively slowly and increase with age, whereas C-reactive protein concentrations change rapidly and levels do not change with age. Since inflammation may play a role in cardiovascular disorders, measurement of a serum inflammation marker, such as C-reactive protein, is useful as a predictor of subsequent coronary events (Chapter 18). In one prospective study, healthy subjects with elevated baseline serum high sensitivity C-reactive protein levels had increased risk of myocardial infarction and ischemic stroke. The use of aspirin, an anti-inflammatory agent (Chapter 16), was also associated with a reduction in the risk of myocardial infarction. Another inflammatory marker, lipoprotein-associated phospholipase A_2, is an independent predictor of risk for abnormal cardiovascular events.

Acute phase phenomena also include a variety of metabolic and neuroendocrine changes. For example, fever is a neuroendocrine change that occurs as an acute phase response.

CLINICAL CASE STUDY 15.2 A Case of Ornithine Carbamoyltransferase Deficiency

This case was abstracted from: P.M. Jones, Altered mental status in a teenager, Clin. Chem. 59 (2013) 1442–1446.

Synopsis

A 13-year-old male was presented to the emergency department with altered mental status and a history of nausea and vomiting. The subject's clinical laboratory parameters such as serum electrolytes, hepatic and renal injury markers, and complete blood count were unremarkable. The subject was treated with intravenous fluids and antiemetic medication and was discharged. However, the subject's medical problems continued and, upon his readmission to the hospital and with measurement of blood ammonia level of 308 µmol/L (reference interval <50 µmol/L), led eventually to the diagnosis of ornithine carbamoyltransferase (OCT) deficiency. Genetic analysis confirmed the diagnosis.

Resolution of the case: The subject's dietary nitrogen intake was monitored and reduced to maintain a positive nitrogen balance. The blood level of ammonia decreased by administration of sodium phenylbutyrate and citrulline.

Teaching Points

1. OCT deficiency is an X-linked disorder and is the most common of the urea cycle defects.
2. NH_3 is neurotoxic, and it affects many systems including neurotransmitters, membrane potential, and mitochondrial function.
3. Glutamine levels are increased due to excessive levels of NH_3, and this has deleterious effects on astrocytes.
4. Treatment for OCT deficiency requires reducing the blood ammonia level by restricting dietary nitrogen intake, removing glutamine, and facilitating the urea cycle by providing its intermediates (e.g., L-citrulline, L-arginine).
5. OCT deficiency subjects develop a natural tendency toward avoiding proteins derived from animal origin.
6. In the diagnosis of OCT, family history is also valuable. In this case, one of the subject's siblings had died in the neonatal period with an unknown diagnosis.
7. Female carriers of the mutated gene for the OCT deficiency may exhibit mild symptoms.

CLINICAL CASE STUDY 15.3 A Case of Homocysteinemia

Synopsis

Two siblings, a 16-year-old boy and younger 14-year-old girl, developed walking difficulties, cognitive dysfunction, and seizure disorders. Initial laboratory investigations did not provide any diagnostic clues. Electroencephalography and brain-computed tomography of the elder sibling revealed cerebral abnormalities consistent with disorders of seizure and cognitive dysfunction. Further studies were initiated for evaluation of potential inherited juvenile-onset progressive neurocognitive disorders. These studies revealed a highly elevated level of plasma homocysteine of >150 μmol/L (reference interval 2.1–15.7 μmol/L). After several causes of hyperhomocysteinemia, such as cystathionine β-synthase and vitamin B_{12} deficiencies, were excluded, methylenetetrahydrofolate reductase (MTHFR) deficiency was considered. Accordingly, subjects' cultured skin fibroblasts showed that the MTHFR activity was severely decreased compared with control specimens.

This case was abstracted from: D. Haarburger, R. Renison, S. Meldau, R. Eastman, G. van der Watt, Teenaged siblings with progressive neurocognitive disease, Clin. Chem. 59 (2013) 1160–1165.

Resolution of the case: The subjects were treated with high-dose folate, which resulted in the resolution of the seizure disorders. However, central nervous system disorders showed only modest improvement.

Teaching Points

1. MTHFR, a cytoplasmic enzyme, catalyzes the conversion of methylenetetrahydrofolate to methyltetrahydrofolate (methyl THF). Methyl THF provides the methyl group for the conversion of homocysteine to methionine, which is part of methylation–remethylation cycle of methionine and homocysteine. Methionine after conversion to S-adenosylmethionine (SAM) is the donor of methyl group for many methylation reactions involved in the synthesis of neurotransmitters, myelin, and epigenetic regulation of gene expression (see text for more details).
2. MTHFR deficiency is an autosomal recessive disorder and is the most common inborn defect of the folate metabolism.
3. An early diagnosis followed by implementation of proper treatment is essential to prevent vascular and neurological complications.

CLINICAL CASE STUDY 15.4 L-DOPA and 5-Hydroxytryptophan Responsive Dystonia

Vignette 1: L-Dihydroxy-Phenylalanine (L-DOPA, also Known as Levodopa) Responsive Dystonia

This case was abstracted from: N. Venna, K.B. Sims, P.E. Grant, Case 26-2006: a 19-year-old woman with difficulty walking, N. Engl. J. Med. 355 (2006) 831–839.

Synopsis

A 19-year-old woman was brought to a neurology clinic with a history of neurological movement disorder consisting of painful involuntary muscle contractions. After a comprehensive clinical study and evaluation, the diagnosis of the patient's disorder was dystonia. On consideration of the differential causes of dystonia, which include many primary and secondary causes (see reference 1), the patient's dystonia was found to be a dopamine-responsive disorder. Accordingly, the patient's symptoms improved dramatically after oral administration of carbidopa and levodopa within a day, and she was maintained on levodopa therapy. After the patient's consent was obtained, genetic testing revealed a loss-of-function deletion mutation in the gene (GCH1), which codes for guanosine triphosphate cyclohydrolase 1 (GTPCH1), which is the first committed step in the formation of tetrahydrobiopterin (THB).

Teaching Points

1. Tetrahydrobiopterin is synthesized starting from GTP and initiated by GTPCH1. THB is the essential cofactor required for the synthesis of several neurotransmitters, namely dopamine, norepinephrine, epinephrine, serotonin, and nitric oxide (see also Chapter 30 for metabolic pathways).
2. Levodopa, which serves as a precursor for the synthesis of dopamine and other related neurotransmitters, is used because it readily crosses the blood–brain barrier.
3. Parkinson's disease therapy also consists of levodopa, with variable responses.
4. It is important to emphasize that this case illustrates the enormous value of understanding basic science disciplines in resolving seemingly complicated clinical problems with beneficial results.
5. Deficiency of sepiapterin reductase (SPR), which participates in the synthesis of THB, results in dystonia. Some of these conditions require supplementation of 5-hydroxytryptophan, a serotonin precursor.

Reference

[1] D. Tarsy, D.K. Simon, Dystonia, N. Engl. J. Med. 355 (2006) 818–829.

Vignette 2: Whole-Genome Sequencing for Optimized Patient Management

This case was abstracted from: M.N. Bainbridge, W. Wiszniewski, D.R. Murdock, J. Friedman, C. Gonzaga-Jauregui, I. Newsham, et. al., Whole-genome sequencing for optimized patient management, Sci. Transl. Med. 3 (2011) 87ps3.

(Continued)

CLINICAL CASE STUDY 15.4 (Continued)

Synopsis

A 14-year-old fraternal twin pair diagnosed with DOPA-responsive dystonia required additional genetic evaluation because the disorder did not respond optimally in the reduction of clinical symptoms. The whole genome sequencing identified compound heterozygous loss of function mutations for the gene encoding for sepiapterin reductase. Treatment with both L-DOPA therapy and 5-hydroxytryptophan resulted in improvement in clinical manifestations.

Teaching Points

1. Deficiency of the tetrahydrobiopterin synthetic pathway should be considered in evaluating clinical symptoms of dystonia.

2. A whole genome sequencing approach is necessary in establishing a clinical diagnosis for the initiation of optimal treatment.

Reference

[1] S.F. Kingsmore, C.J. Saunders, Deep sequencing of patient genomes for disease diagnosis: when will it become routine? Sci. Transl. Med 3 (2011) 87ps23.

REQUIRED READING

[1] P. Cochet, G. Rumsby, Primary hyperoxaluria, N. Engl. J. Med. 369 (2013) 649−658.

[2] J.R. Ingelfinger, P.A. Marsden, Estimated GFR and risk of death—is cystatin C useful? N. Engl. J. Med. 369 (2013) 974−975.

[3] M.G. Shlipak, K. Matsushita, J. Ärnolöv, L.A. Inker, R. Katz, K.R. Polkinghorne, et al., Cystatin C versus creatinine in determining risk based on kidney function, N. Engl. J. Med. 369 (2013) 932−943.

[4] N.V. Bhagavan, Letter: hazards in indiscriminate use of sweeteners containing phenylalanine, N. Engl. J. Med. 292 (1975) 52−53.

Chapter 16

Lipids I: Fatty Acids and Eicosanoids

Key Points

1. Fatty acids obtained from the diet (Chapter 11) or mobilized from hormone-sensitive lipase-mediated triglyceride hydrolysis by adipocytes are bound to albumin during circulation in the blood and then oxidized in tissues to provide energy.

2. Fatty acid oxidation occurs in mitochondria. It consists of the uptake of fatty acids into the cell, activation by conversion to acyl-CoA derivatives, transport to mitochondria involving the carnitine shuttle, and β-oxidation.

3. β-Oxidation consists of a repeated sequence of four reactions and ultimately yields acetyl-CoA, which is oxidized in the TCA cycle-electron transport system and coupled to ATP formation from ADP (Chapters 12 and 13).

4. Fatty acid oxidation is the primary source of energy for many tissues with the exception of the central nervous system and circulating red blood cells. It is regulated by metabolites and hormones. Fatty acid oxidation defects can cause metabolic abnormalities.

5. Propionyl-CoA is obtained from the oxidation of odd-chain fatty acids and some amino acids and is converted to succinyl-CoA involving biotin and vitamin B_{12} dependent enzymes.

6. Very-long-chain fatty acids are shortened initially by oxidation in peroxisomes, and defects in this process lead to neurological abnormalities.

7. Fatty acids (e.g., phytanic acid) with methyl groups blocking the β-carbon are oxidized by α-oxidation.

8. In liver mitochondria, the acetyl-CoA obtained from fatty acid oxidation is converted, in part, to ketone bodies (acetoacetate, β-hydroxybutyrate, and acetone). Both acetoacetate and β-hydroxybutyrate are oxidized in extrahepatic tissues providing energy. Excessive production of ketone bodies, as occurs in uncontrolled diabetes mellitus, leads to severe metabolic acidosis (Chapters 20 and 37).

9. Fatty acid synthesis occurs in the cytosol and requires NADPH and acetyl-CoA. Mitochondria are the major source of acetyl-CoA, which is transported to the cytosol via the citrate–malate–pyruvate shuttle. Acetyl-CoA is converted to malonyl-CoA by a biotin-dependent acetyl-CoA carboxylase. This step is the committed step of fatty acid biosynthesis. Malonyl-CoA and NADPH are used by the multienzyme fatty acid synthase to yield palmitate.

10. Fatty acid oxidation in mitochondria and fatty acid synthesis in cytosol are reciprocally regulated by metabolites and hormones.

11. Eicosanoids and leukotrienes are hormone-like metabolites which participate in many physiological functions such as cytoprotection, contraction, blood clotting, and inflammation, and are synthesized from arachidonic acid or other polyunsaturated fatty acids. These fatty acids are derived from membrane phospholipids by the action of phospholipases. Inhibitors of the synthesis of eicosanoids and leukotrienes are used as therapeutic agents in disorders of inflammation and blood coagulation.

Lipids (or fats) are a heterogeneous group of organic compounds defined by their solubility in nonpolar solvents such as chloroform, ether, and benzene, and by their poor solubility in water. Lipids may be polar or nonpolar (amphipathic). Polar lipids have limited solubility in water because they are amphipathic; i.e., they possess hydrophilic and hydrophobic regions in the same molecule. The major polar lipids include **fatty acids**, **cholesterol**, **glycerophosphatides**, and **glycosphingolipids**. Very-short-chain fatty acids and ketone bodies are readily soluble in water. Nonpolar lipids serve principally as storage and transport forms of lipid and include **triacylglycerols** (also called **triglycerides**) and **cholesteryl esters**.

Lipids have numerous functions including the following: thermal insulation; energy storage (as triacylglycerol); metabolic fuels; membrane components (phospholipids and cholesterol; Chapter 9); hormones (steroids and vitamin D metabolites; Chapters 30 and 36, respectively); precursors of **prostanoids** and **leukotrienes** (discussed later); vitamins A, C, D, E, and K (Chapters 34–36); emulsifying agents in the digestion and absorption of lipids (bile acids; Chapters 11 and 18); surfactants in the alveolar membrane (phosphatidylcholine; Chapter 17); and participation in the signal transduction pathways (e.g., inositol-1,4,5-trisphosphate and diacylglycerol; Chapter 28). The metabolism of fatty acids (saturated and unsaturated) is discussed in this chapter. The metabolism of phospholipids, glycosphingolipids, and cholesterol is considered in Chapter 17.

N. V. Bhagavan and C.-E. Ha: Essentials of Medical Biochemistry, Second edition. DOI: http://dx.doi.org/10.1016/B978-0-12-416687-5.00016-6

Fatty acids that contain no carbon–carbon double bonds are saturated, and those with carbon–carbon double bonds are unsaturated. Fatty acids that contain an even number of carbon atoms and are acyclic, unbranched, nonhydroxylated, and monocarboxylic make up the largest group. The most abundant saturated fatty acids in animals are palmitic and stearic acids. The melting point of fatty acids increases with increasing chain length, the even-numbered saturated fatty acids having higher melting points than the odd-numbered. Among the even-numbered fatty acids, the presence of *cis* double bonds lowers the melting point significantly. Free fatty acids at physiological pH are mostly ionized ($pK_a \sim 4.85$) and exist only in small quantities; in plasma, they are typically bound to albumin.

Digestion and absorption of lipids are discussed in Chapter 11. The Western diet contains about 40% fat, mostly as triacylglycerol (100–150 g/day). Triacylglycerols are packaged as chylomicrons in the intestinal epithelial cells, are delivered to the blood circulation via the lymphatic system, and hydrolyzed to glycerol and fatty acids by endothelial lipoprotein lipase. Fatty acids are taken up by the cells of the tissue, where the hydrolysis occurs, whereas glycerol is metabolized in the liver and kidney. Another means of triacylglycerol transport is very-low-density lipoprotein (VLDL), which is synthesized in the liver. VLDL triacylglycerol is also hydrolyzed by endothelial lipoprotein lipase. The metabolism of plasma lipoproteins is discussed in Chapter 18. Fatty acids are released by hydrolysis of triacylglycerol in adipocytes by hormone-sensitive lipase, particularly during starvation, stress, and prolonged exercise.

OXIDATION OF FATTY ACIDS

The overall fatty acid oxidation process in mitochondria consists of the cellular uptake of fatty acids, their activation to acyl-CoA and then to thioesters, and finally translocation into mitochondria, which involves a carnitine transesterification shuttle, and β-oxidation. Fatty acids released from chylomicrons and VLDLs are transferred across cell membranes into the cytosol facilitated by tissue-specific fatty acid-binding proteins (FABPs). Fatty acids are also obtained from the hydrolysis of triacylglycerol stored in adipose tissue, which are bound to albumin and transported in the blood. Fatty acids serve as substrates for energy production in the liver, and in skeletal and cardiac muscle during periods of fasting. Although the brain does not utilize fatty acids for generating energy directly, brain cells can utilize ketone bodies synthesized from acetyl-CoA and acetoacetyl-CoA. The latter two are obtained from β-oxidation of fatty acids in the liver. All of the enzymes involved in mitochondrial fatty acid β-oxidation are encoded by nuclear genes. After their synthesis in the cytosolic endoplasmic reticulum, the enzymes are transported into mitochondria. The transport of the enzymes into mitochondria requires, in many instances, the presence of N-terminal extensions to guide the protein across the mitochondrial membrane, receptor-mediated ATP-dependent uptake, and proteolytic processing to form fully assembled mature enzymes. During β-oxidation of an acyl-CoA, the chain length of the substrate is shortened by two carbon atoms (acetyl-CoA) each cycle. **Thus, β-oxidation requires a group of enzymes with chain length specificity.**

Activation of Fatty Acids

At least three acyl-CoA synthases exist, each specific for a particular size of fatty acid: acetyl-CoA synthase acts on acetate and other low molecular weight carboxylic acids; medium-chain acyl-CoA synthase on fatty acids with 4–11 carbon atoms; and acyl-CoA synthase on fatty acids with 6–20 carbon atoms. The activity of acetyl-CoA synthase in muscle is restricted to the mitochondrial matrix. Medium-chain acyl-CoA synthase occurs only in liver mitochondria, where medium-chain fatty acids obtained from digestion of dietary triacylglycerols and transported by the portal blood are metabolized. Acyl-CoA synthase, the major activating enzyme, occurs on the outer mitochondrial membrane surface and in the endoplasmic reticulum. The overall reaction of activation is as follows:

$$RCOO^- + ATP^{4-} + CoASH \rightleftharpoons R \overset{\overset{\displaystyle O}{\|}}{C} - SCoA + AMP^{2-} + PP_i^{3-}$$

Acyl-CoA
(a thioester)

The reaction favors the formation of fatty acyl-CoA, since the pyrophosphate formed is hydrolyzed by pyrophosphatase: $PP_i + H_2O \rightarrow 2P_i$. Thus, activation of a fatty acid molecule requires expenditure of two high-energy phosphate bonds. The reaction occurs in two steps (E = enzyme):

$$RCOO^- + ATP + E$$

$\downarrow$ Step 1

$$E - AMP - \overset{\overset{\displaystyle O}{\|}}{C} - R + PP_i$$

CoASH $\downarrow$ Step 2

$$E + R - \overset{\overset{\displaystyle O}{\|}}{C} - SCoA + AMP$$

Transport of Acyl-CoA to the Mitochondrial Matrix

Transport of acyl-CoA to the mitochondrial matrix is accomplished by carnitine (L-β-hydroxy-γ-trimethylammonium butyrate), which is required in catalytic amounts for the oxidation of fatty acids (Figure 16.1). Carnitine also participates in the transport of acetyl-CoA for cytosolic fatty acid synthesis. Two carnitine acyltransferases are involved in acyl-CoA transport: carnitine palmitoyltransferase I (CPTI), located on the outer surface of the inner mitochondrial membrane; and carnitine palmitoyltransferase II (CPTII), located on the inner surface.

The overall translocation reaction is as follows:

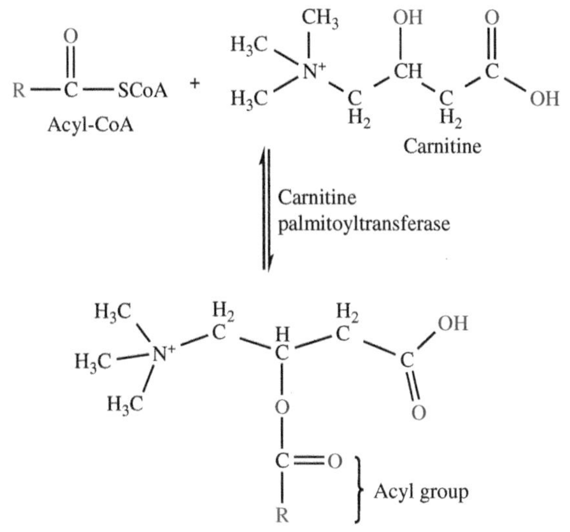

The standard free-energy change of this reaction is about zero, and therefore, the ester bond of acylcarnitine may be considered as a high-energy linkage. Malonyl-CoA, a precursor in the synthesis of fatty acids, is an allosteric inhibitor of CPTI in liver and thus prevents a futile cycle of simultaneous fatty acid oxidation and synthesis.

Carnitine is synthesized from two essential amino acids: lysine and methionine. S-adenosylmethionine donates three methyl groups to a lysyl residue of a protein with the formation of a protein-bound trimethyl-lysyl moiety. Proteolysis yields trimethyl-lysine, which is converted to carnitine. In humans, liver and kidney are major sites of carnitine production; from there, it is transported to skeletal and cardiac muscle, where it cannot be synthesized. Carnitine can also be obtained from food, and it is particularly rich in red meat.

Four inherited defects of carnitine metabolism lead to impaired utilization of long-chain fatty acids for energy production. These include defects of plasma membrane carnitine transport, carnitine palmitoyltransferase I (CPTI), carnitine palmitoyltransferase II (CPTII), and

carnitine-acylcarnitine translocase. Clinical manifestations of disorders of carnitine metabolism and fatty acid oxidation disorders (discussed later) span a wide spectrum and can be affected by the severity and site of the defect (e.g., muscle, liver, and kidney). The disorders may be characterized by hypoketotic hypoglycemia, hyperammonemia, liver disease, skeletal muscle weakness, and cardiomyopathy. In some instances, dietary intervention brings about marked improvement in the clinical manifestations. For example, patients with a carnitine transport defect respond well to carnitine therapy (see Clinical Case Study 16.1).

β-Oxidation

The main pathway for fatty acid oxidation, β-oxidation (Figure 16.2), involves oxidation of acyl-CoA at the β-carbon, and removal of two carbon fragments as acetyl-CoA; this takes place entirely in the mitochondrial matrix.

Oxidation of a saturated acyl-CoA with an even number of carbon atoms to acetyl-CoA requires repeated sequential action of four enzymes:

1. Acyl-CoA dehydrogenase dehydrogenates acyl-CoA at the α- and β-carbon atoms to yield the α,β-unsaturated acyl-CoA (or Δ^2-unsaturated acyl-CoA). Each one of four distinct dehydrogenases is specific for a given range of fatty acid chain length. All four are flavoproteins and contain a tightly bound molecule of flavin adenine dinucleotide (FAD). The electrons from the acyl-CoA dehydrogenase are transferred to the main respiratory chain via $FADH_2$ through mitochondrial electron transfer flavoprotein (ETF) and ETF-ubiquinone oxidoreductase (ETF-QO) (Figure 16.3). Both ETF and ETF-ubiquinone oxidoreductase are nuclear-encoded proteins. They also mediate transfer of electrons from dimethyl glycine dehydrogenase and sarcosine dehydrogenase.

   Inherited defects in ETF and ETF-QO cause accumulation of organic acids (acidemia) and their excretion in the urine (acidurias); examples of these disorders are glutaric acidemia type I and type II, which are inherited as autosomal recessive traits. Glutaric acid is an intermediate in the metabolism of lysine, hydroxy lysine, and tryptophan. **Glutaric acidemia type I** is caused by deficiency of glutaryl-CoA dehydrogenase, which catalyzes the conversion of glutaryl-CoA to crotonyl-CoA. **Glutaric acidemia type II** is caused by defects in the ETF/ETF-QO proteins. The clinical manifestations of these disorders are similar to medium-chain acyl-CoA dehydrogenase deficiency (discussed later). The Δ^2 double bond formed by the acyl-CoA dehydrogenase has a *trans* configuration. The double bonds in naturally occurring fatty acids generally have the *cis* configuration. The

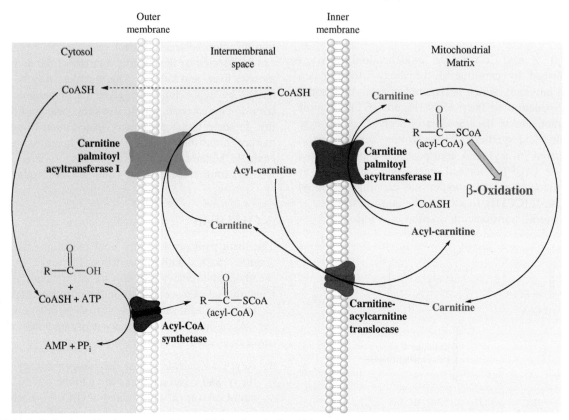

FIGURE 16.1   Role of carnitine in transport of fatty acids into the mitochondrial matrix. Entry of acylcarnitine is linked to the exit of carnitine, both mediated by a translocase.

oxidation of unsaturated *cis*-fatty acids requires two auxiliary enzymes: enoyl-CoA isomerase and 2,4-dienoyl-CoA reductase.

Acyl-CoA dehydrogenase (especially butyryl-CoA dehydrogenase) is irreversibly inactivated by methylene cyclopropylacetyl-CoA through the formation of a covalent adduct with the FAD of the enzyme. The inhibitor is derived by transamination and oxidative decarboxylation of the toxic amino acid hypoglycin (Chapter 14). Ingestion of hypoglycin causes severe hypoglycemia due to the inhibition of β-oxidation and corresponding decrease in ATP synthesis. Gluconeogenesis, which is important in maintaining fasting glucose levels, is dependent on adequate supplies of ATP. The action of hypoglycin thus serves to emphasize the importance of β-oxidation in gluconeogenesis under normal circumstances.

Among the fatty acid oxidation disorders, **medium-chain acyl-CoA dehydrogenase deficiency (MCAD)** is the most common, and its frequency is similar to that of phenylketonuria. The disorder can be identified by detecting mutant alleles and also some key abnormal metabolites in the blood specimens. An A→G substitution mutation is found at position 985 of MCAD-cDNA in about 90% of cases. This mutation leads to replacement of lysine with glutamate at position 329 (K329E) of the polypeptide.

MCAD results in elevated circulatory levels of C_6–C_{10} acylcarnitines. This observation has been adopted in newborn screening protocols for MCAD by tandem mass spectrometry in order to initiate prompt therapeutic intervention (see Clinical Case Study 16.2).

The MCAD deficiency primarily affects hepatic fatty acid oxidation, and the most common clinical presentation is episodic hypoketotic hypoglycemia initiated by fasting. The main metabolic derangement in MCAD is an inadequate supply of acetyl-CoA (Figure 16.3). The deficiency of acetyl-CoA leads to a decreased flux through the tricarboxylic acid (TCA) cycle, causing diminished ATP production, decreased ketone body formation (the ketone bodies are

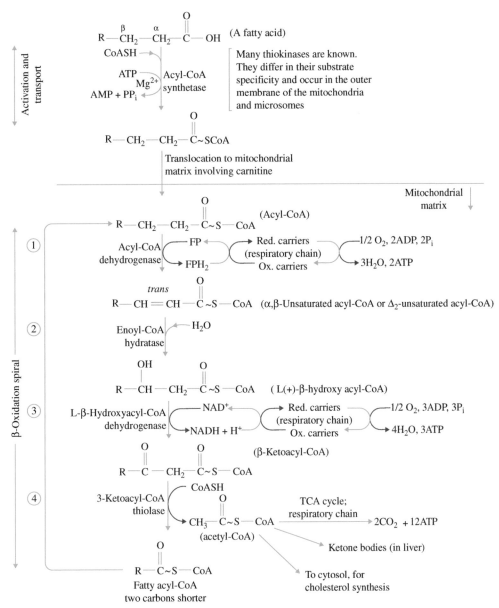

FIGURE 16.2 Fatty acid activation, transport, and β-oxidation. The shortened fatty acyl-CoA from one cycle is further oxidized in successive passes until it is entirely converted to acetyl-CoA. Odd-chain fatty acids produce one molecule of propionyl-CoA. Ox. = oxidized; Red. = reduced; respiratory chain = oxidative phosphorylation and electron transport; $\sim$ = high energy bond; FP = flavoprotein.

metabolites used by the extrahepatic tissues), decreased citrate synthesis, and decreased oxaloacetate synthesis from pyruvate (catalyzed by pyruvate carboxylase, for which acetyl-CoA is the primary activator). The decreased flux through the TCA cycle during deficiency of acetyl-CoA causes a diminished synthesis of citrate from oxaloacetate and acetyl-CoA, as well as inhibition of α-ketoglutarate dehydrogenase due to an elevated ratio of fatty acyl-CoA to CoA. The formation of oxaloacetate is crucial for

gluconeogenesis (Chapter 14). Accumulation of octanoate, which occurs in MCAD, may be responsible for encephalopathy and cerebral edema. In Reye's syndrome, octanoate is also elevated and may be responsible for its phenotypical similarity with MCAD. MCAD is managed with the avoidance of fasting, stress, and treatment with intravenous glucose during acute episodes (see Clinical Case Study 16.2).

2. Enoyl-CoA hydratase catalyses the hydration of Δ^2 unsaturated acyl-CoA. This enzyme has broad

specificity and can act on α-, β- (or Δ^2) unsaturated CoA in either the *trans* or *cis* configuration. The following products are formed:

2-*trans*-Enoyl-CoA → L(+)-β-hydroxyacyl-CoA [or L(+)-3-hydroxyacyl-CoA]

2-*cis*-Enoyl-CoA → L(−)-β-hydroxyacyl-CoA [or L(−)-3-hydroxyacyl-CoA]

The latter reaction occurs in the oxidation of natural unsaturated fatty acids, and an epimerase converts the product to the L-isomer, which is the substrate of the next enzyme.

3. β-Hydroxyacyl-CoA dehydrogenase oxidizes β-hydroxyacyl-CoA by an NAD^+-linked reaction that is absolutely specific for the L-stereoisomer. The electrons from the NADH generated are passed on to NADH dehydrogenase of the mitochondrial respiratory chain.

4. 3-Ketoacyl-CoA thiolase (β-ketothiolase) catalyzes a thiolytic cleavage, has broad specificity, and yields acetyl-CoA and an acyl-CoA shortened by two carbon atoms. The reaction is highly exergonic ($\Delta G^{0'} = -6.7$ kcal/mol) and favors thiolysis. The enzyme has a reactive −SH group on a cysteinyl residue, which participates as follows (E = enzyme):

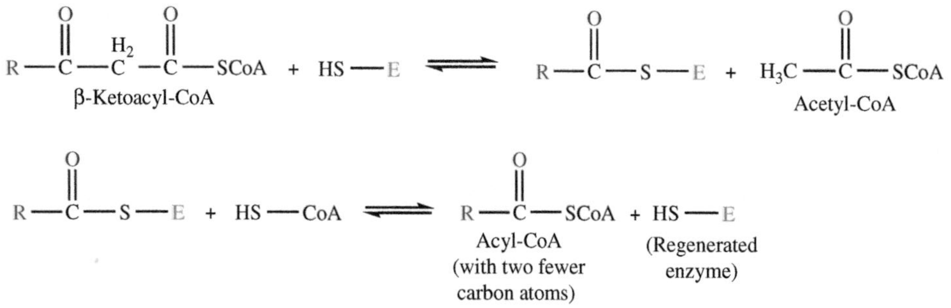

deficiency of long-chain β-hydroxyacyl-CoA dehydrogenase may themselves develop acute liver disease, hemolysis, and a low platelet count. This clinical disorder is associated with a high risk of maternal and neonatal morbidity and is known as **HELLP syndrome** (hemolysis, elevated liver enzyme levels, and low platelet count). A mutation of Glu 474 to Gln (E474Q) in the α-subunits of the trifunctional protein has been identified in three unrelated children whose mothers had an acute fatty liver episode, or HELLP syndrome, during pregnancy [1].

The fetal−maternal interaction that leads to toxic effects in women during pregnancy may be due to the transport of long-chain β-hydroxyacyl metabolites produced by the fetus and placenta to the maternal liver.

Other inborn errors of fatty acid oxidation include defects in short-chain β-hydroxyacyl-CoA dehydrogenase and medium-chain β-ketoacyl-CoA thiolase. The spectrum of clinical findings in these and other fatty acid oxidation defects are variable. Typical symptoms include fasting intolerance, cardiomyopathy, and sudden death. Children with long-chain fatty acid oxidation disorders are treated with frequent feeding of a low-fat diet consisting of medium-chain triglycerides. This dietary regimen can prevent hypoketotic hypoglycemic liver dysfunction.

Three enzyme activities—namely, of long-chain enoyl-CoA hydratase, β-hydroxyacyl-CoA dehydrogenase, and long-chain β-ketoacyl-CoA thiolase (reactions 2−4, Figure 16.2) —are associated with a trifunctional protein complex consisting of four α- and four β-subunits bound to the inner mitochondrial membrane. Each of the four α-subunits possesses hydratase and dehydrogenase enzyme activities at the N-terminal and C-terminal domains, respectively. The active site for the thiolase activity resides in the four β-subunits of the protein complex. Deficiencies of the dehydrogenase activity, and of all of the three enzyme activities for oxidation of long-chain fatty acids, have been described. These deficiencies can cause nonketotic hypoglycemia during fasting, hepatic encephalopathy, and cardiac and skeletal myopathy. In some instances, women carrying fetuses with a

Energetics of β-Oxidation

Two high-energy bonds are consumed in the activation of a fatty acid molecule. Every mole of fatty acyl-CoA that cycles through reactions 1−4 produces 1 mol of $FADH_2$, 1 mol of NADH, and 1 mol of acetyl-CoA. On the last pass of an even-chain-length fatty acid, 2 mol of acetyl-CoA are formed; and the final pass of an odd-chain-length molecule releases 1 mol of propionyl-CoA. The amount of ATP formed from complete oxidation of a hexanoic acid is calculated as shown in Table 16.1.

Fatty acid oxidation produces more moles of ATP per mole of CO_2 formed than carbohydrate oxidation. In this case, oxidation of 1 mol of hexose produces at most (assuming malate shuttle operation exclusively) 38 mol of ATP.

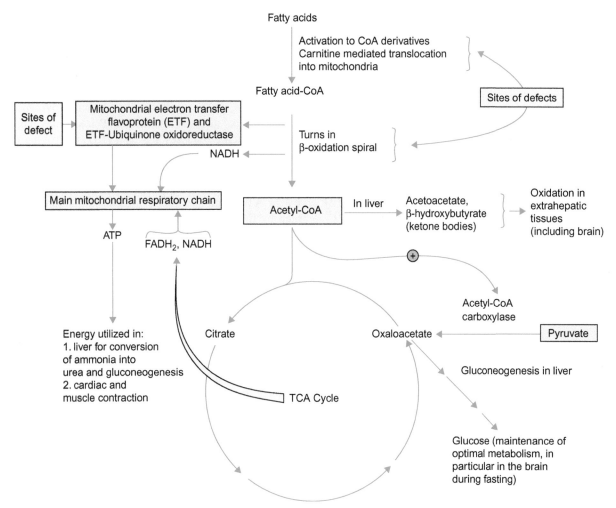

FIGURE 16.3 Spectrum of consequences of defects in fatty acid oxidation. The primary effect is inadequate production of acetyl-CoA, which leads to decreased flux through the TCA cycle and lack of ketone body synthesis in the liver. Both of these events cause energy deficits and changes in metabolic regulatory processes. Alterations in hepatic metabolism lead to hypoglycemia and hyperammonemia. Abnormalities also occur in skeletal and cardiac muscle and in the central nervous system.

TABLE 16.1 Net Amount of ATP Synthesis from Oxidation of Hexanoic Acid

Reaction	Direct Consequences of the Reaction	Moles of ATP Gained or Lost per Mole of Hexanoic Acid
Activation reaction	Hexanoic acid → Hexanoyl-CoA	−2
First dehydrogenation	Dehydrogenation of acyl-CoA; (2FAD → FADH$_2$)	+4
Hydration	Hydration of α,β-unsaturated fatty acyl-CoA	0
Second dehydrogenation	Dehydrogenation of β-hydroxy-acyl-CoA 2(NAD$^+$ → NADH + H$^+$)	+6
Oxidation of acetyl-CoA	Formation of 3 mol of acetyl-CoA, and their oxidation in the TCA cycle and electron transport system	+(3 × 12) = +36
		Total ATP = +44

Palmitoyl-CoA yields 8 acetyl-CoA molecules and 14 pairs of hydrogen atoms, via seven cycles through the β-oxidation system. Acetyl-CoA can be oxidized in the TCA cycle, used for the synthesis of cholesterol, or used for the formation of ketone bodies in liver. β-Oxidation of an acyl-CoA with an uneven number of carbon atoms yields propionyl-CoA during the acetyl-CoA acyltransferase reaction of the last cycle. Propionyl-CoA oxidation is discussed in the next section.

Complete oxidation of one molecule of palmitic acid yields 129 ATP molecules:

$$C_{15}H_{31}COOH + 8CoASH + ATP + 7FAD + 7NAD^+ + 7H_2O$$
$$\rightarrow 8CH_3COSCoA + AMP + PP_i + 7FADH_2 + 7NADH + 7H^+$$

Each molecule of acetyl-CoA yields 12 ATP ($12 \times 8 = 96$); each $FADH_2$ yields 2 ATP ($7 \times 2 = 14$); each NADH yields 3 ATP ($7 \times 3 = 21$); and two high-energy bonds are consumed (-2; ATP→AMP + PP$_i$). Thus, net ATP production is 129. It should be noted that oxidation of glucose gives a favorable yield of ATP per mole oxygen consumed of 6.33 (38 ATP/6 O_2) compared to palmitate oxidation of 5.61 (129 ATP/23 O_2). This observation has been used therapeutically in treating heart failure (ischemic cardiomyopathy) due to a lack of oxygen supply. A therapeutic agent, **trimetazidine**, is a selective inhibitor of long-chain 3-ketoacyl-CoA thiolase activity of β-oxidation and has been used as an adjunctive treatment of **angina pectoris** in Europe and Asia. Improving the efficiency of ATP production per mole of oxygen consumed by modifying the substrate utilization (i.e., glucose oxidation in favor of fatty acid oxidation) and by using therapeutic agents in heart failure of different etiologies is under investigation [2,3].

Regulation of Fatty Acid Oxidation

Regulation of fatty acid oxidation involves diet, cofactors, and competing substrates and hormones of fatty acid mobilization. Adipose tissue is one of the major sites of regulation of triacylglycerol synthesis and lipolysis. The other site is CPTI. The latter is inhibited by malonyl-CoA, which is involved in fatty acid biosynthesis (discussed later). Thus, fatty acid oxidation and synthesis do not occur simultaneously. Insulin inhibits fatty acid oxidation by blocking lipolysis in adipose tissue, and it stimulates lipogenesis and synthesis of malonyl-CoA. Glucagon stimulates fatty acid oxidation by inhibiting synthesis of acetyl-CoA carboxylase, which leads to decreased synthesis of malonyl-CoA. This causes enhanced activity of CPTI and promotion of fatty acid oxidation. In the fed state, the glucagon/insulin ratio is low, and fatty acid synthesis is promoted in the liver. In the fasting state, the glucagon/insulin ratio is high, and mobilization of free fatty acids from adipose tissue and mitochondrial fatty acid oxidation is augmented.

Peroxisomal Fatty Acid Oxidation

Peroxisomes are single membrane organelles occurring in many eukaryotic cells. They are present in many mammalian cells and are particularly abundant in liver and kidney. A normal hepatocyte contains about a thousand peroxisomes, whose proliferation is inducible by hypolipidemic drugs such as clofibrate. Peroxisomes contain H_2O_2-producing oxidases and also H_2O_2 destroying catalase. Oxidation of very-long-chain, saturated, unbranched fatty acids ($C_{24}-C_{26}$) takes place in peroxisomes after the acyl-CoA derivatives are transported across the membrane by an ATP-binding cassette protein. Oxidation is mediated by flavoprotein dehydrogenases that yield H_2O_2 and acetyl-CoA and terminates with octanoyl-CoA. Octanoyl- and acetyl-CoA are transferred to mitochondria for further oxidation. Peroxisomal oxidation does not yield ATP. All the energy produced appears as heat.

Some genetic disorders (e.g., **Zellweger's syndrome**) exhibit defective formation of peroxisomes (in Zellweger's syndrome, no morphologically detectable peroxisomes are present) or deficiency of one or more constituent enzymes. These disorders are characterized by a marked accumulation of very-long-chain, saturated, unbranched fatty acids (tetracosanoic and hexacosanoic acids) in liver and central nervous system tissues, severe neurological symptoms, and early death.

The biochemical defect in the inherited disorders known as **adrenoleukodystrophy** (**ALD**) is an abnormality in the peroxisomal ATP-binding cassette (ABC) transport membrane protein, ABCD1, for very-long-chain fatty acids (VLCFAs). There are two types of ALD: X-linked and autosomal recessive (neonatal ALD). It is estimated that the incidence of X-ALD is about 1:17,000. X-ALD is a progressive disorder in which VLCFAs accumulate in the brain, adrenals, and testes, causing neurodegeneration, adrenocortical insufficiency, and hypogonadism. Increased plasma levels of VLCFAs (e.g., C_{26}:0) are used in the diagnosis and management of affected male ALD subjects. Treatment of X-ALD consists of adrenal hormone replacement therapy (Chapter 30), hematopoietic stem cell transplantation, and dietary therapy. The goal of dietary therapy is primarily used to decrease endogenous synthesis of VLCFA. The latter is accomplished by feeding a 4:1 mixture of oils (glyceryl-trioleate and glyceryl-trierucate) known as **Lorenzo's oil**. Oleic and erucic acids are C_{18} and C_{22} *cis*-long-chain fatty acids, respectively. Lorenzo's oil normalizes plasma VLCFA levels, and the therapy is recommended for asymptomatic children.

Peroxisomes contain dihydroxyacetone phosphate acyltransferase and alkyldihydroxyacetone phosphate synthase, which are involved in the synthesis of the plasmalogens (Chapter 17). Peroxisomes may also participate in the biosynthesis of bile acids. The conversion of trihydroxyl-cholestanoic acid to cholic acid (Chapter 17) is localized in peroxisomes.

OTHER PATHWAYS OF FATTY ACID OXIDATION

Propionyl-CoA Oxidation

β-Oxidation of fatty acids with an odd number of carbon atoms yields propionyl-CoA. Since the concentration of such fatty acids in the diet is small, little propionyl-CoA is produced. Catabolized isoleucine, valine, methionine, and threonine (Chapter 15) are important sources of propionyl-CoA. The oxidation of the side chain of cholesterol also yields propionyl-CoA. Thus, propionyl-CoA is derived from the catabolism of lipids and proteins.

Propionyl-CoA is converted to succinyl-CoA, which is oxidized or converted to glucose by way of oxaloacetate and pyruvate (gluconeogenesis; Chapter 14). Succinyl-CoA may also form Δ-aminolevulinate, a precursor of porphyrin biosynthesis (Chapter 17). Formation of succinyl-CoA from propionyl-CoA requires three mitochondrial enzymes and two vitamins: biotin and cobalamin (Figure 16.4).

Inborn errors of metabolism may be due to propionyl-CoA carboxylase deficiency, defects in biotin transport or

metabolism, **methylmalonyl-CoA mutase deficiency**, or defects in adenosylcobalamin synthesis. The former two defects result in **propionic acidemia**, the latter two in methylmalonic acidemia. All cause metabolic acidosis and developmental retardation. Organic acidemias often exhibit hyperammonemia, mimicking ureagenesis disorders, because they inhibit the formation of N-acetylglutamate, an obligatory cofactor for carbamoyl phosphate synthase (Chapter 15). Some of these disorders can be partly corrected by administration of pharmacological doses of vitamin B_{12} (Chapter 36). Dietary protein restriction is therapeutically useful (since propionate is primarily derived from amino acids). Propionic and methylmalonyl acidemia (and aciduria) results from vitamin B_{12} deficiency (e.g., pernicious anemia; Chapter 36).

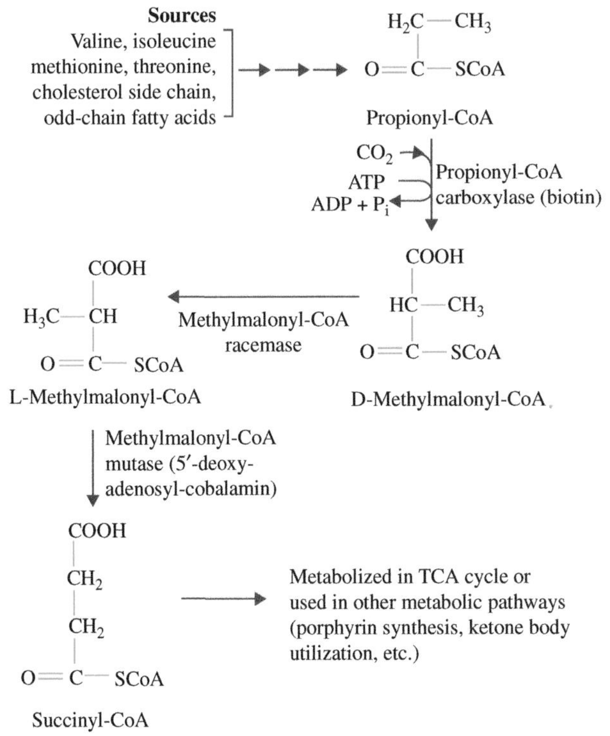

α-Oxidation

α-Oxidation is important in the catabolism of branched-chain fatty acids. In **Refsum's disease**, an autosomal recessive disorder, the defect is the deficiency of phytanoyl-CoA hydroxylase (Figure 16.5). Phytanic acid is a 20-carbon, branched-chain fatty acid derived from the polyprenyl plant alcohol phytol, which is present as an ester in chlorophyll. Thus, its origin in the body is from dietary sources. The oxidation of phytanic acid is shown in Figure 16.5. The clinical characteristics of Refsum's disease include peripheral neuropathy and ataxia, retinitis pigmentosa, and abnormalities of skin and bones. Significant improvement has been observed when patients are kept on low-phytanic acid diets for prolonged periods (e.g., diets that exclude dairy and ruminant fat).

METABOLISM OF KETONE BODIES

Ketone bodies consist of acetoacetate, D-β-hydroxybutyrate (D-3-hydroxybutyrate), and acetone. They are synthesized in liver mitochondria. The overall steps involved in the formation of ketone bodies include the mobilization of fatty acids by lipolysis from adipose tissue triacylglycerol by hormone-sensitive triacylglycerol lipase, plasma fatty acid transport, fatty acid activation, fatty acid transport into mitochondria (with acylcarnitine as an intermediate), and β-oxidation. The regulatory reactions are those of lipolysis and of acyl-CoA transport across the mitochondrial membrane (CPTI). Synthesis of ketone bodies from acetyl-CoA consists of three steps: formation of acetoacetyl-CoA; formation of acetoacetate; and reduction of acetoacetate to β-hydroxybutyrate. Nonenzymatic decarboxylation of acetoacetate yields acetone, which is eliminated via the lungs.

The pathways of formation of ketone bodies are shown in Figure 16.6. The major pathway of

FIGURE 16.4 Metabolism of propionyl-CoA.

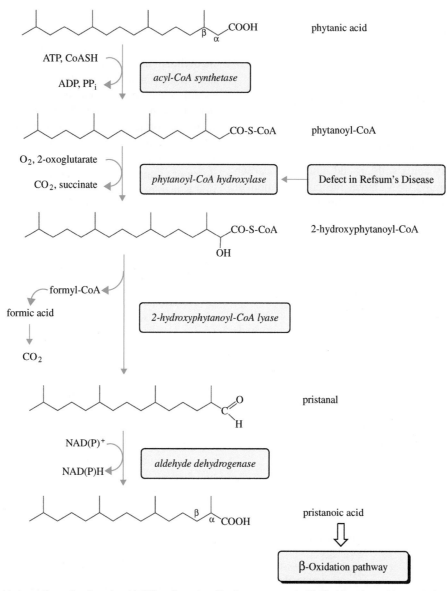

FIGURE 16.5 α-Oxidation pathway for phytanic acid. This pathway is utilized to overcome the blocked β-carbon with a methyl group so that β-oxidation can proceed. *[Reproduced, with slight modifications, with permission from R. J. A. Wanders et al. Refsum disease, in C. R. Scriver, A. L. Beaudet, W. S. Sly, D. Valle, B. Childs, K. W. Kinzler, et al. (Eds.), The metabolic and molecular bases of inherited disease, 8th ed., Chapter 132, p. 3315, New York: McGraw-Hill, 2001]*

production of acetoacetate is from β-hydroxy-β-methylglutaryl-CoA (HMG-CoA). Hydrolysis of acetoacetyl-CoA to acetoacetate by acetoacetyl-CoA hydrolase is of minor importance because the enzyme has a high K_m for acetoacetyl-CoA. HMG-CoA is also produced in the cytosol, where it is essential for the synthesis of several isoprenoid compounds and cholesterol (Chapter 17). The reduction of acetoacetyl-CoA to β-hydroxybutyrate depends on the mitochondrial $[NAD^+]/[NADH]$ ratio.

Ketone bodies are oxidized primarily in extrahepatic tissues (e.g., skeletal muscle, heart, kidney, intestines,

brain) within mitochondria. β-Hydroxybutyrate is oxidized to acetoacetate by NAD^+-dependent β-hydroxybutyrate dehydrogenase by reversal of the reaction that occurred during ketogenesis:

Activation of acetoacetate requires the transfer of coenzyme A from succinyl-CoA, derived from the TCA

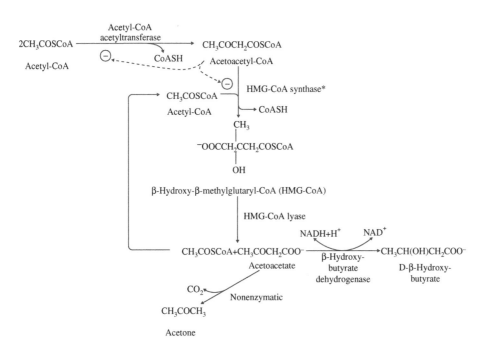

FIGURE 16.6 Ketogenesis in the liver. All reactions occur in mitochondria; the rate-controlling reactions (not shown) are release of fatty acids from adipose tissue and uptake of acyl-CoA into mitochondria, in particular, the CPTI reaction (see Figure 16.1). Acetoacetyl-CoA may regulate ketogenesis by inhibiting the transferase and the synthase. *This enzyme is similar to citrate synthase (Chapter 12), which catalyzes an analogous reaction.

cycle, by succinyl-CoA-acetoacetate-CoA transferase (thiophorase):

$$
\begin{array}{cccc}
\mathrm{CH_2COSCoA} & \mathrm{CH_2COO^-} & \mathrm{CH_2COO^-} & \mathrm{CH_2COSCoA} \\
| & + \quad | & \longrightarrow \quad | & + \quad | \\
\mathrm{CH_2COO^-} & \mathrm{COCH_3} & \mathrm{CH_2COO^-} & \mathrm{COCH_3} \\
\text{Succinyl-CoA} & \text{Acetoacetate} & \text{Succinate} & \text{Acetoacetyl-CoA}
\end{array}
$$

The activation occurs at the expense of conversion of succinyl-CoA to succinate in the TCA cycle and formation of GTP (Chapter 12). Acetoacetyl-CoA is cleaved to two molecules of acetyl-CoA by acetyl-CoA acetyltransferase, the same enzyme involved in the synthesis of acetoacetyl-CoA. Acetyl-CoA is oxidized in the TCA cycle. Thus, formation of ketone bodies in the liver and their oxidation in extrahepatic tissues are dictated by the [substrates]/[products] ratio.

Physiological and Pathological Aspects of Metabolism of Ketone Bodies

Acetoacetate and β-hydroxybutyrate are products of the normal metabolism of fatty acid oxidation and serve as metabolic fuels in extrahepatic tissues. Their level in blood depends on the rates of production and utilization. Oxidation increases as their plasma levels increase. Some extrahepatic tissues (e.g., muscle) oxidize them in preference to glucose and fatty acid. Normally, the serum concentration of ketone bodies is less than 0.3 mM.

The rate of formation of ketone bodies depends on the concentration of fatty acids derived from hydrolysis of adipose tissue triacylglycerol by hormone-sensitive lipase. Insulin depresses lipolysis and promotes triacylglycerol synthesis and storage, while glucagon has the opposite effects. Thus, insulin is antiketogenic and glucagon is ketogenic (Chapter 20). Uncontrolled insulin-dependent diabetes may result in fatal ketoacidosis (Chapters 29 and 37). Although ketonemia and ketonuria are generally assumed to be due to increased production of ketone bodies in the liver, studies with depancreatized rats have shown that ketosis may also arise from their diminished oxidation.

Ketosis can occur in starvation, in ethanol abuse, and following exercise, the latter because of a switch in blood flow. During sustained exercise, blood flow to the liver, intestines, and kidneys is substantially decreased, with a corresponding increase in blood flow to working muscles, so that more fatty acids mobilized from adipose tissue are delivered to the muscle. Thus, the formation of ketone bodies is severely curtailed. But during the post-exercise period, with the resumption of normal blood flow to the liver, ketone bodies are generated as a result of increased mobilization of fatty acids. Reduced ketone body utilization in the extrahepatic tissues can occur due to deficiency of either succinyl-CoA-acetoacetate-CoA transferase or acetyl-CoA acetyltransferase. Patients are susceptible to attacks of ketoacidosis and the presence of persistent ketone bodies in the urine. Acetone is the primary metabolite produced during isopropyl alcohol toxicity and occurs with the absence of other ketone bodies.

Metabolism of Ethanol

Ethanol is consumed widely. Microbial fermentation in the large intestine of humans can produce about 3 g of ethanol per day. Ethanol is rapidly absorbed throughout the gastrointestinal tract or, when inhaled, through the lungs. It is metabolized in the liver by a process having zero-order kinetics; i.e., the rate of oxidation is constant with time. The amount metabolized per unit time depends on liver size (or body weight); the average rate in an adult is about 30 mL in 3 hours. The energy content of alcohol is about 7 kcal/g.

Ethanol oxidation begins with conversion to acetaldehyde by alcohol dehydrogenase (M.W. $\sim 85,000$), a zinc-containing, NAD^+-dependent enzyme that is a relatively nonspecific cytoplasmic enzyme with a K_m of about 1 mM/L:

$$CH_3CH_2OH + NAD^+ \rightarrow CH_3CHO + NADH + H^+$$

Acetaldehyde is rapidly converted to acetate by NAD^+-dependent aldehyde dehydrogenase:

$$CH_3CHO + NAD^+ + H_2O \rightarrow CH_3COOH + NADH + H^+$$

Ethanol is also oxidized by the mixed-function oxidase of smooth endoplasmic reticulum, which requires NADPH, oxygen, and a cytochrome P-450 electron transport system (Chapter 13):

$$CH_3CH_2OH + NADPH + H^+ + 2O_2 \rightarrow CH_3CHO + 2H_2O_2 + NADP^+$$

Many drugs are metabolized by this enzyme, hence the competition between ethanol and other drugs (e.g., barbiturates). Peroxisomal catalase catalyzes the reaction:

$$CH_3CH_2OH + H_2O_2 \rightarrow CH_3CHO + {}+2H_2O$$

The K_m for this catalase and for the mixed function oxidase is about 10 mM/L. The extent to which these two enzymes metabolize ethanol is not known.

Acetaldehyde is converted to acetate in the liver by NAD^+-linked aldehyde dehydrogenases, one in the cytosol $(K_m = 1 \text{ mM/L})$ and another in mitochondria $(K_m = 0.01 \text{ mM/L})$:

$$CH_3CHO + NAD^+ + H_2O \rightarrow CH_3COO^- + NADH + 2H^+$$

Disulfiram (tetraethylthiuram disulfide).

causes irreversible inactivation of these aldehyde dehydrogenases by reacting with sulfhydryl groups, with a buildup of acetaldehyde that produces the acetaldehyde syndrome (vasodilation; intense throbbing, pulsating headache; respiratory difficulties; copious vomiting; sweating; thirst; hypotension; vertigo; blurred vision; and confusion). Disulfiram by itself is relatively nontoxic. It is used in the treatment of chronic alcoholism but does not cure it. Disulfiram provides a willing patient with a deterrent to consumption of alcohol. A shorter-acting reversible inhibitor of aldehyde dehydrogenase is calcium carbimide, which causes accumulation of acetaldehyde and unpleasant symptoms. Thus, calcium carbimide can also be used as a deterrent to alcohol consumption.

Symptoms similar to the disulfiram—ethanol reaction occur in high proportion in certain ethnic groups (e.g., Asians and Native Americans) who are extremely sensitive to ethanol consumption. The ethanol sensitivity in these populations is accompanied by a higher acetaldehyde steady-state concentration in the blood, which may be due to a rapid rate of formation of acetaldehyde by alcohol dehydrogenase or to a decreased rate of its removal by aldehyde dehydrogenase. Both of these dehydrogenases are present in several isozyme forms and exhibit extensive polymorphism among racial groups. An alcohol dehydrogenase variant found in the ethanol-sensitive populations has a relatively higher rate of activity at physiological pH and may account for more rapid oxidation of ethanol to acetaldehyde. However, a more important cause of acetaldehyde accumulation appears to be deficiency of an isozyme of aldehyde dehydrogenase, which has a low K_m for acetaldehyde. Thus, the cause of ethanol sensitivity may be impaired rate of removal of acetaldehyde rather than its excessive formation. Individuals who are predisposed to ethanol sensitivity should avoid ethanol intake in any form.

Acetate produced from ethanol is converted to acetyl-CoA by acetyl-CoA synthase in hepatic and extrahepatic tissues.

$$CH_3COO^- + ATP^{4-} + CoASH \rightarrow CH_3COSCoA + AMP^{2-} + PP_i^{3-}$$

Acetyl-CoA is oxidized in the TCA cycle and is used in liver and adipose tissue for biosynthesis of fatty acids and triacylglycerol.

Alcoholism affects about 10% of the drinking population, and alcohol (ethanol) abuse has been implicated in at least 20% of admissions to general hospitals. This chronic disease exhibits high mortality due to a wide variety of factors. Ethanol produces effects in virtually every organ system. The biochemical effects of ethanol are due to increased production of NADH that decreases the [NAD^+]/[NADH] ratio in the cytoplasm of liver cells at least 10-fold from the normal value of about 1000. Increased production of lactate and inhibition of gluconeogenesis (Chapter 14) result. The hyperuricemia associated with ethanol consumption has been attributed to accelerated turnover of adenine nucleotides and their catabolism to uric acid (Chapter 25). Alcohol increases hepatic fatty acid and triacylglycerol synthesis and mobilization of fat from adipose tissue, which can lead to fatty liver, hepatitis, and cirrhosis. These effects are complicated by a deficiency of B vitamins and protein.

Alcohol increases the plasma level of VLDL and of HDL cholesterol (Chapter 18).

Many actions of ethanol may be attributed to a membrane-disordering effect. Changes in membrane fluidity can affect membrane-bound enzymes (e.g., Na^+, K^+-ATPase, adenylate cyclase) and phospholipid architecture. Alcohol also affects several neurotransmitter systems in the brain. These include dopamine (mediates pleasurable effects), γ-aminobutyric acid (GABA), glutamate, serotonin, adenosine, norepinephrine, and opioid peptides. Potential drug therapy for alcohol dependence consists of the use of antagonists and agonists of alcohol-affected neurotransmitter systems. For example, naltrexone, a μ-opioid antagonist, inhibits alcohol-induced dopamine release, thus minimizing the pleasurable effect of alcohol and reducing the desire to consume alcohol. Another drug, acamprosate, reduces the craving for alcohol presumably by an agonist activity at GABA receptors and an inhibitory activity at N-methyl-D-aspartate receptors. A selective antagonist of serotonin receptor 5-HT$_3$, ondansetron, reduces alcohol consumption in patients with early onset alcoholism. The 5-HT$_3$ receptors are densely distributed in mesocorticolimbic neuronal terminals and regulate dopamine release. Attenuation of dopamine release reduces alcohol consumption.

In chronic alcoholics, heavy drinking and decreased food intake lead to ketoacidosis. Accelerated lipolysis arising from reduced insulin and increased glucagon secretion caused by hypoglycemia leads to ketosis with a high [β-hydroxybutyrate]/[acetoacetate] ratio. Treatment requires normalization of fluid and electrolyte balance (Chapter 37) and of glucose level. Administration of glucose provokes insulin release and depresses glucagon release, thus suppressing the stimuli for ketogenesis. The distinction between diabetic ketoacidosis and alcoholic ketoacidosis may be difficult to determine, and in some patients plasma glucose levels may not discriminate between the two entities (although in diabetic ketoacidosis, plasma glucose levels are usually high; whereas in alcoholic ketoacidosis, these levels may be low, normal, or marginally elevated). Fluid and electrolyte replacement and glucose administration in ketoacidosis are essential regardless of etiology.

Ethanol is a teratogen partly because it inhibits embryonic cellular proliferation. Maternal alcoholism causes fetal alcohol syndrome, which is characterized by abnormal function of the central nervous system, microcephaly, cleft palate, and micrognathia.

SYNTHESIS OF LONG-CHAIN SATURATED FATTY ACIDS

The reactions of *de novo* fatty acid biosynthesis are shown in Figure 16.7. They are carried out by two multienzyme systems functioning in sequence. The first is acetyl-CoA carboxylase, which converts acetyl-CoA to malonyl-CoA. The second is fatty acid synthase (FAS), which sequentially joins two-carbon units of malonyl-CoA, eventually producing palmitic acid. Both complexes consist of multifunctional subunits. The various catalytic functions can be readily separated in plant cells and prokaryotes, but in yeasts, birds, and mammals, attempts to subdivide catalytic functions lead to loss of activity. Important features of this system are as follows:

1. *De novo* synthesis takes place in the cytosol (whereas fatty acid oxidation occurs in mitochondria).
2. All carbon atoms are derived from acetyl-CoA (obtained from carbohydrates or amino acids), and palmitate (C_{16}) is the predominant fatty acid produced. Fatty acids longer than 16 carbons, those that are unsaturated, and hydroxy fatty acids are obtained by separate processes of chain elongation, desaturation, or α-hydroxylation, respectively.
3. The committed (rate-controlling) step is the biotin-dependent carboxylation of acetyl-CoA by acetyl-CoA carboxylase. Important allosteric effectors are citrate (positive) and long-chain acyl-CoA derivatives (negative).
4. Although the initial step requires CO_2 fixation, CO_2 is not incorporated into fatty acids. The labeled carbon in $^{14}CO_2$ (as $H^{14}CO_3-$) is not incorporated into the fatty acids synthesized.
5. Synthesis is initiated by a molecule of acetyl-CoA that functions as a primer. Its two carbons become C_{15} and C_{16} of palmitate. The acetyl group is extended by successive addition of the two carbons of malonate originally derived from acetyl-CoA, the unesterified carboxylic acid group being removed as CO_2. In mammalian liver and mammary glands, butyryl-CoA is a more active primer than acetyl-CoA. Odd-chain-length fatty acids found in some organisms are synthesized by priming the reaction with propionyl-CoA instead of acetyl-CoA.
6. Release of the finished fatty acid occurs when the chain length reaches C_{16} by action of thioester hydrolase, which is specific for long-chain acyl-CoA derivatives. A thioester hydrolase of mammary gland is specific for acyl residues of C_8, C_{10}, or C_{12}.

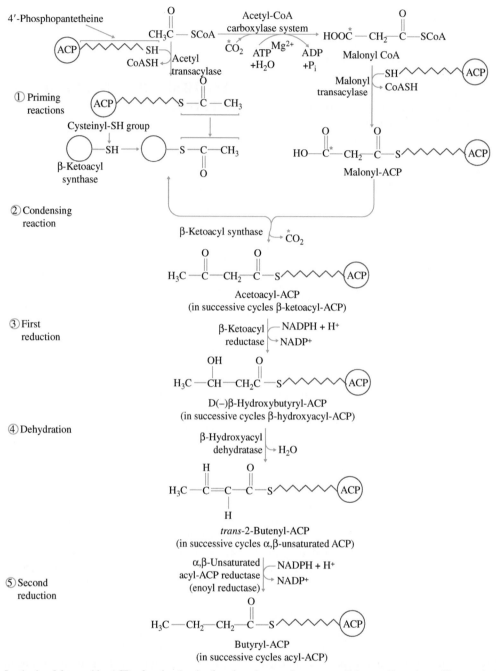

FIGURE 16.7 Synthesis of fatty acids. ACP = functional unit of acyl-carrier-protein segment of fatty acid synthase. The cysteinyl−SH group of β-ketoacyl synthase accepts the acetyl group or the acyl group, and its catalytic site, which is adjacent to the −SH group, catalyzes the condensation reaction.

The overall reaction for palmitate synthesis from acetyl-CoA is

$$8 \text{Acetyl-CoA} + 14 \text{NADPH} + 14 \text{H}^+ + 7 \text{ATP} + \text{H}_2\text{O} \rightarrow$$
$$\text{palmitate} + 8 \text{CoASH} + 14 \text{NADP}^+ + 7 \text{ADP} + 7 \text{P}_i$$

The reducing equivalents of NADPH are derived largely from the pentose phosphate pathway.

Acetyl-CoA carboxylase is a biotin-dependent enzyme. It has been purified from microorganisms, yeast, plants, and animals. In animal cells, it exists as an inactive protomer (M.W. ∼400,000) and as an active polymer (M.W. 4−8 million). The protomer contains the activity of biotin transcarboxylase, biotin carboxyl carrier protein (BCCP), transcarboxylase, and a regulatory allosteric site.

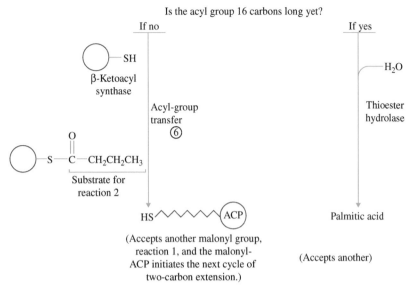

FIGURE 16.7 (Continued)

Each protomer contains a biotinyl group bound in amide linkage to the ε-amino group of a lysyl residue.

Citrate shifts the equilibrium from inactive protomer to active polymer. The polymeric form appears as long filaments in electron micrographs.

The mechanism of the carboxylation reaction consists of two half-reactions:

$$ATP + HCO_3^- + BCCP \xrightleftharpoons{\text{biotin carboxylase, Mg}^{2+}}$$
$$ADP + P_i + BCCP\text{-}COO^-$$

and

$$BCCP\text{-}COO^- + \text{acetyl-CoA} \leftrightarrow BCCP + \text{malonyl-CoA}$$

The overall reaction is

$$ATP + HCO_3^- + \text{acetyl-CoA} \leftrightarrow \text{malonyl-CoA} + ADP + P_i$$

Other biotin-dependent enzymes include propionyl-CoA carboxylase and pyruvate carboxylase (Chapter 14). The latter, like acetyl-CoA carboxylase, is subject to allosteric regulation. Pyruvate carboxylase, a mitochondrial enzyme, is activated by acetyl-CoA and converts pyruvate to oxaloacetate, which, in turn, is converted to glucose via the gluconeogenic pathway or combines with acetyl-CoA to form citrate. Some of the citrate is transported to the cytosol, where it activates the first step of fatty acid synthesis and provides acetyl-CoA as substrate (see below). Other carboxylation reactions use bicarbonate but are dependent on vitamin K, the acceptor being glutamyl residues of the glycoprotein clotting factors II, VI, IX, and X, and the anti-clotting factors Protein C and Protein S (Chapter 34).

Acetyl-CoA carboxylase (ACC) is under short- and long-term control. Allosteric modulation functions as a short-term regulator. Positive modulators are citrate and isocitrate; negative modulators are long-chain acyl-CoA derivatives. The binding of citrate increases the activity by polymerization of the protomers, whereas negative modulators favor dissociation of active polymers to inactive monomers. ACC is also regulated by covalent modification by phosphorylation, which inhibits activity, and by dephosphorylation by phosphatase, which restores activity. Phosphorylation is mediated by AMP-activated protein kinase (AMPK). Glucagon and β-adrenergic agonists also inhibit ACC activity, by inhibiting the phosphatase by an active inhibitor protein. The activation of the inhibitor protein by phosphorylation is mediated by cAMP-protein kinase A phosphorylation. This mechanism of ACC regulation is analogous to HMG-CoA reductase regulation in isoprenoid biosynthesis (Chapter 17). Insulin suppresses cAMP levels and promotes activity of acetyl-CoA carboxylase. Thus, common mediators (e.g., insulin, glucagon, and catecholamines) regulate fatty acid, carbohydrate, and isoprenoid metabolism.

Long-term regulation of acetyl-CoA carboxylase involves nutritional, hormonal (e.g., insulin, thyroxine), and other factors. The activity is high in animals on high-carbohydrate diets, fat-free diets, or undergoing choline or vitamin B_{12} deprivation. However, fasting, high intake of fat or of polyunsaturated fatty acids, and prolonged biotin deficiency leads to decreased activity. In diabetes, the enzyme activity is low, but insulin administration raises it to normal levels.

Fatty acid synthesis is also carried out by a multi-enzyme complex and uses acetyl-CoA, malonyl-CoA, and NADPH to produce palmitic acid. The overall process is presented in Figure 16.7.

The assembly of rat liver FAS involves three stages: synthesis of the multifunctional polypeptide chains, formation of the dimer, and attachment of a 4'-phosphopantetheine group by an enzyme-catalyzed reaction. This assembly process is influenced by changes in developmental, hormonal, and nutritional states. The FAS complex provides considerable catalytic efficiency, since free intermediates do not accumulate and the individual activities are present in equal amounts.

The central role of the acyl carrier domain is to carry acyl groups from one catalytic site to the next. The 4'-phosphopantetheine (20 nm long) derived from coenzyme A is bound as a phosphodiester through the hydroxyl group of a specific seryl residue. The acyl intermediates are in thioester linkage with the −SH of the prosthetic group, which serves as a swinging arm to carry acyl groups from one catalytic site to the next. The structure of 4'-phosphopantetheine attached to the serine residue of ACP is shown here:

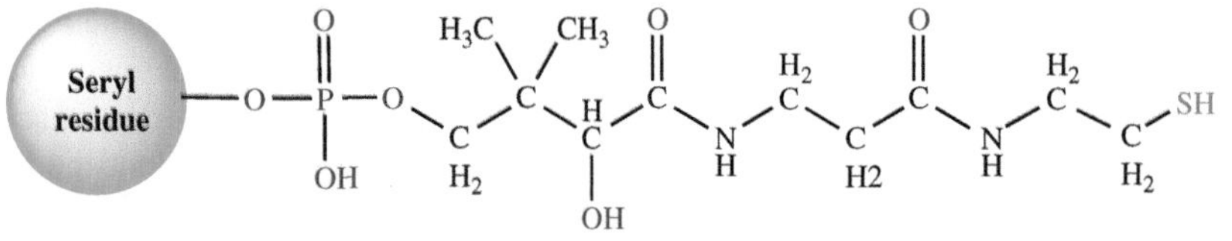

to the cytosol, its carbon atoms are transferred by two transport mechanisms:

1. Transport dependent on carnitine: Carnitine participates in the transport of long-chain acyl-CoA into the mitochondria and plays a similar role in the transport of acetyl-CoA out of mitochondria. However, carnitine acetyl transferases have a minor role in acetyl-CoA transport.
2. Cytosolic generation of acetyl-CoA (citrate shuttle): This pathway is shown in Figure 16.8. Citrate synthesized from oxaloacetate and acetyl-CoA is transported from mitochondria to the cytosol via the tricarboxylate anion carrier system and cleaved to yield acetyl-CoA and oxaloacetate.

$$\text{Citrate}^{3-} + \text{ATP}^{4-} + \text{CoA} \xrightarrow[\text{(or citrate cleavage enzyme)}]{\text{ATP citrate-lyase}}$$
$$\text{acetyl-CoA} + \text{oxaloacetate}^{2-} + \text{ADP}^{3-} + \text{P}_i^{2-}$$

Thus, citrate not only modulates the rate of fatty acid synthesis but also provides carbon atoms for the synthesis. The oxaloacetate formed from pyruvate may eventually be converted (via malate) to glucose by the gluconeogenic pathway. The glucose oxidized via the pentose phosphate

Sources of NADPH for Fatty Acid Synthesis

The reducing agent for fatty acid synthesis is NADPH, which is supplied primarily by the pentose phosphate pathway (Chapter 14). These enzymes, like the fatty acid synthase complex, are located in the cytosol. Active lipogenesis occurs in liver, adipose tissue, and lactating mammary glands, which contain a correspondingly high activity of the pentose phosphate pathway. Thus, lipogenesis is closely linked to carbohydrate oxidation. The rate of lipogenesis is high in humans on carbohydrate-rich diets. Restricted energy intake, a high-fat diet, and insulin deficiency decrease fatty acid synthesis.

Source and Transport of Acetyl-CoA

Acetyl-CoA is synthesized in mitochondria by a number of reactions: oxidative decarboxylation of pyruvate; catabolism of some amino acids (e.g., phenylalanine, tyrosine, leucine, lysine, and tryptophan); and β-oxidation of fatty acids (see earlier). Since acetyl-CoA cannot be transported directly across the inner mitochondrial membrane

pathway augments fatty acid synthesis by providing NADPH. Pyruvate generated from oxaloacetate can enter mitochondria and be converted to oxaloacetate, which is required for the formation of citrate.

Regulation of Fatty Acid Synthase

Like acetyl-CoA carboxylase, FAS is under short- and long-term control. The former is due to negative or positive allosteric modulation or to changes in the concentrations of substrate, cofactor, and product. The latter consists of changes in enzyme concentration as a result of rates of protein synthesis versus protein degradation. Variation in levels of hormones (e.g., insulin, glucagon, epinephrine, thyroid hormone, and prolactin), and in the nutritional state, affect fatty acid synthesis through short- and long-term mechanisms. In the diabetic state, hepatic fatty acid synthesis is severely impaired but is corrected by administration of insulin. The impairment may be due to defects in glucose metabolism that leads to a reduced level of an inducer, an increased level of a repressor of transcription of the FAS

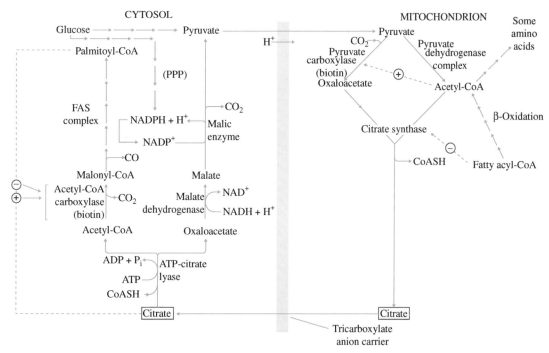

FIGURE 16.8 Cytoplasmic generation of acetyl-CoA via citrate transport and related reactions. PPP = pentose phosphate pathway; FAS = fatty acid synthase; —⊖→ = negative allosteric modifier; —⊕→ = positive allosteric modifier.

gene, or both. Glucagon and epinephrine raise intracellular levels of cAMP, and their inhibitory effect on fatty acid synthesis dephosphorylation of acetyl-CoA carboxylase maintains its inactive status. They also stimulate the action of hormone-sensitive triacylglycerol lipase and raise intracellular levels of long-chain acyl-CoA. As a result, acetyl-CoA carboxylase and citrate synthase are inhibited. The stimulatory effect of prolactin is confined to the mammary gland and may involve synthesis of the enzyme.

Fatty Acid Elongation

Cytoplasmic fatty acid synthase yields palmitate. Human triacylglycerol contains fatty acids with 18, 20, 22, and 24 carbon atoms, which are synthesized by elongation of palmitate in the endoplasmic reticulum or mitochondria. Elongation in the endoplasmic reticulum occurs mainly in liver and involves C_{10-16}-saturated and C_{18}-unsaturated fatty acids by successive addition of two-carbon groups derived from malonyl-CoA. The reductant is NADPH, and the intermediates are CoA thioesters.

The mitochondrial system uses acetyl-CoA, not malonyl-CoA, by a slightly modified reversal of β-oxidation. The substrates are saturated and unsaturated C_{12}, C_{14}, and C_{16} fatty acids, and the products are C_{18}, C_{20}, C_{22}, and C_{24} fatty acids. The first reduction step utilizes NADH, and the enzyme is β-hydroxyacyl dehydrogenase (a β-oxidation

enzyme). The second reduction step utilizes NADPH, and the enzyme is enoyl reductase.

Mitochondrial fatty acid elongation occurs primarily when the $[NADH]/[NAD^+]$ ratio is high (e.g., anaerobic conditions, excessive ethanol oxidation).

METABOLISM OF UNSATURATED FATTY ACIDS

Structure and Nomenclature of Unsaturated Fatty Acids

Unsaturated fatty acids contain one or more double bonds. A common method for designating fatty acids gives the carbon chain length, number of double bonds, and double-bond positions (in parentheses). Thus, palmitoleic acid is designated C_{16}:1(9), and linoleic acid is C_{18}:2(9,12). The location of the double bond is sometimes indicated by Δ; for example, Δ9 signifies that the double bond is between carbon 9 and carbon 10. In both methods, the carboxyl carbon is carbon 1. The double-bond position can also be related to the ω-end of the fatty acid molecule (i.e., the methyl carbon farthest from the carboxyl end). Thus, oleic acid is an ω-9 acid, and linoleic acid has double bonds at ω-6 and ω-9 carbons. The structures and names of some naturally occurring unsaturated fatty acids are given in Table 16.2.

The presence of a double bond in the hydrocarbon chain gives rise to geometrical isomerism, which is due to

TABLE 16.2 Naturally Occurring Unsaturated Fatty Acids

Common Name[‡]	Systematic Name[*]	Molecular Formula	Structural Formula	ω-Series[†]	Melting Point (°C)
Palmitoleic[‡]	9-Hexadecenoic	$C_{16}H_{30}O_2$	$CH_3(CH_2)_5CH = CH(CH_2)_7COOH$	ω-7	0.5
Oleic[‡]	9-Octadecenoic	$C_{18}H_{34}O_2$	$CH_3(CH_2)_7CH = CH(CH_2)_7COOH$	ω-9	13
Vaccenic	trans-11-Octadecenoic	$C_{18}H_{34}O_2$	$CH_3(CH_2)_5CH = CH(CH_2)_9COOH$	ω-7	43
Linoleic[‡]	9,12-Octadecadienoic	$C_{18}H_{32}O_2$	$CH_3(CH_2)_4CH = CHCH_2CH = CH(CH_2)_7COOH$	ω-6	−5
α-Linolenic	9,12,15-Octadecatrienoic	$C_{18}H_{30}O_2$	$CH_3CH_2CH = CHCH_2CH = CHCH_2CH = CH(CH_2)_7COOH$	ω-3	−11
γ-Linolenic	6,9,12-Octadecatrienoic	$C_{18}H_{30}O_2$	$CH_3(CH_2)_4CH = CHCH_2CH = CHCH_2CH = CH(CH_2)_4COOH$	ω-6	−11
Arachidonic[‡]	5,8,11,14-Eicosatetraenoic	$C_{20}H_{32}O_2$	$CH_3(CH_2)_4-(CH = CH-CH_2)_4-(CH_2)_2-COOH$	ω-6	−50
Erucic	13-Docosenoic	$C_{22}H_{42}O_2$	$CH_3(CH_2)_7CH = CH(CH_2)_{11}COOH$	ω-9	39
Nervonic	15-Tetracosenoic	$C_{24}H_{46}O_2$	$CH_3(CH_2)_7CH = CH(CH_2)_{13}COOH$	ω-9	39

*All double bonds are in the cis geometric configuration, except where indicated.
†This series is based on the number of carbon atoms present between the terminal methyl group and the nearest double bond; ω-3 and ω-6 are essential fatty acids.
‡Most abundant unsaturated fatty acids in animal lipids.

restricted rotation around carbon−carbon double bonds and is exemplified by fumaric and maleic acids.

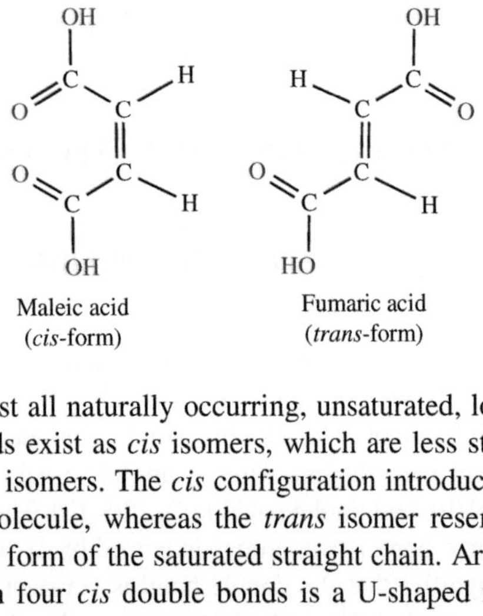

Maleic acid
(cis-form)

Fumaric acid
(trans-form)

Almost all naturally occurring, unsaturated, long-chain fatty acids exist as cis isomers, which are less stable than the trans isomers. The cis configuration introduces a bend in the molecule, whereas the trans isomer resembles the extended form of the saturated straight chain. Arachidonic acid with four cis double bonds is a U-shaped molecule. Some cis isomers are biologically active as essential fatty acids. The trans isomers cannot substitute for them but are metabolized like the saturated fatty acids.

Functions of Unsaturated Fatty Acids

The cis unsaturated fatty acids provide fluidity to triacylglycerol reserves and phospholipid membranes, and many serve as precursors of the eicosanoids (prostaglandins, prostacyclins, thromboxanes, and leukotrienes). The importance of membrane fluidity and its relationship to the membrane constituent phospholipids are discussed in Chapter 9. Eicosanoids (20 carbons) have numerous functions (see below).

Palmitoleic and oleic acids, the two most abundant monounsaturated fatty acids of animal lipids, can be synthesized from the respective saturated fatty acid coenzyme-A esters. Desaturase is a monooxygenase system present in the endoplasmic reticulum of liver and adipose tissue. The overall reaction for palmitoleic acid synthesis is as follows:

$$Palmitoyl\text{-}CoA + NAD(P)H + H^+ + O_2 \rightarrow$$
$$Palmitoleyl\text{-}CoA + NAD(P)^+ + 2H_2O$$

TRANS-FATTY ACIDS

Trans-fatty acid metabolism is similar to that of saturated fatty acids. During the partial dehydrogenation of vegetable oils (e.g., in the manufacture of margarine), the cis-fatty acids are isomerized to the trans-fatty acid forms. The "hydrogenated" margarines contain 15%−40% trans-fatty acids. The consumption of trans-fatty acids increases the risk of coronary heart disease by elevating atherogenic low-density lipoprotein (LDL) cholesterol and lowering antiatherogenic (cardioprotective) high-density lipoprotein (HDL) cholesterol (see Chapter 18).

ESSENTIAL FATTY ACIDS

The essential fatty acids (EFAs) are polyunsaturated fatty acids not synthesized in the body, but required for normal metabolism. EFAs are linoleic acid, linolenic acids (α and γ), and arachidonic acid. All contain at least one double bond located beyond C-9 or within the terminal seven carbon atoms (Table 16.2).

Introduction of double bonds does not occur in humans

Double bonds can be introduced in this region

A double bond within the terminal seven carbon atoms can be present at ω-3 or ω-6. γ-Linolenic acid is an ω-6 EFA, and α-linolenic acid is an ω-3 EFA. Other ω-3 EFAs are eicosapentaenoic acid (EPA) and docosahexaenoic acid (DCHA), both abundant in edible fish tissues. Vegetable oils are rich in ω-6 EFAs. Plants contain α-linolenic acid, which can be converted in the body to EPA and DCHA, but it is found within chloroplast membranes and not in seed oils; hence, it may not be available in significant quantities in the diet. The ω-3 and ω-6 EFAs have different metabolic effects (see below). Particularly rich sources of EPA are fish (e.g., salmon, mackerel, blue fish, herring, menhaden) that live in deep, cold waters. These fish have fat in their muscles and their skin. In contrast, cod, which has a similar habitat, stores fat in liver rather than muscle. Thus, cod liver oil is a good source of EPA, but it also contains high amounts of vitamins A and D, which can be toxic if consumed in large quantities (Chapters 35 and 36, respectively). Shellfish also contain EPA. Plankton is the ultimate source of EPA.

Linoleic acid can be converted to γ-linolenic acid and arachidonic acid in mammalian liver by the microsomal desaturation and chain elongation process. Thus, the requirement for arachidonic acid may be satisfied when the diet contains adequate amounts of linoleic acid. Similarly, α-linolenic acid is converted by desaturation and chain elongation to EPA.

Deficiency of Essential Fatty Acids

The clinical manifestations of EFA deficiency include dry, scaly skin, usually erythematous eruptions (generalized or localized and affecting the trunk, legs, and intertriginous areas), diffuse hair loss (seen frequently in infants), poor wound healing, failure of growth, and increased metabolic rate.

METABOLISM OF EICOSANOIDS

The eicosanoids include the **prostaglandins (PGs)**, **thromboxanes (TXs)**, **prostacyclins (PGIs)**, and **leukotrienes (LTs)**; they are derived from essential fatty acids and act similarly to hormones (Chapter 28). However, they are synthesized in almost all tissues (unlike hormones, which are synthesized in selected tissues) and are not stored to any significant extent. Their physiological effects on tissues occur near their sites of synthesis, rather than at a distance. They function as paracrine messengers and are sometimes referred to as autacoids. Their metabolic roles are many (e.g., regulation of vasomotor functions, smooth muscle contraction, inflammation process, hemostasis).

The four groups of eicosanoids are derived, respectively, from a 20-carbon fatty acid containing either three, four, or five double bonds; 8, 11, 14-eicosatrienoic acid (dihomo-γ-linolenic acid); 5, 8, 11, 14-eicosatetraenoic acid (arachidonic acid); and 5, 8, 11, 14, 17-eicosapentaenoic acid (Figure 16.9). In humans, the most abundant precursor is arachidonic acid. Secretion of eicosanoids in response to stimuli requires mobilization of precursor fatty acids bound by ester, ether, or amide linkages. This utilization is generally considered the rate-limiting step. Membrane phosphoglycerides contain essential fatty acids in the 2-position. In response to stimuli and after the activation of the appropriate phospholipase, the essential fatty acid is released.

The known phospholipases and their hydrolytic specificities are shown in Figure 16.10. The primary source of arachidonic acid is through action of phospholipase A_2. It may also be derived through the action of phospholipase C, which liberates diacylglycerol, which is then acted on by diacylglycerol lipase. Stimuli that increase the biosynthesis of eicosanoids cause increased mobilization of intracellular calcium, which binds to calmodulin and is then thought to activate membrane-bound phospholipases A_2 and C (phagocytosis: Chapter 14; and mechanism of hormone action: Chapter 28).

Glucocorticoids (e.g., cortisol) inhibit phospholipase A_2 activity by induction of the synthesis of a phospholipase inhibitor protein, which partly explains their anti-inflammatory effects.

The major metabolites of arachidonic acid (Figure 16.11) arise from 12-lipoxygenase, 5-lipoxygenase, and the fatty acid cyclooxygenase pathway. The 5-lipoxygenase pathway yields leukotrienes; and the cyclooxygenase pathway yields cyclic endoperoxides, which are converted to PGs, TXs, and PGIs.

Prostaglandins (PG) were discovered in human semen more than 50 years ago. Their name is derived from the

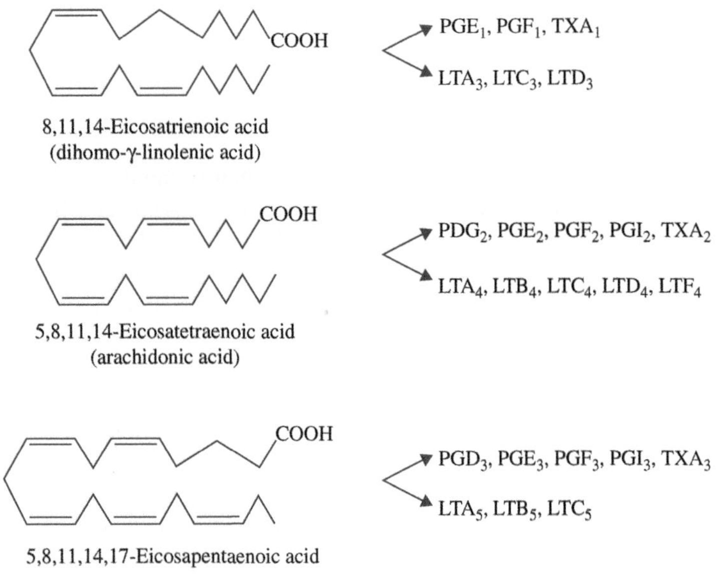

8,11,14-Eicosatrienoic acid
(dihomo-γ-linolenic acid)

→ PGE₁, PGF₁, TXA₁
→ LTA₃, LTC₃, LTD₃

5,8,11,14-Eicosatetraenoic acid
(arachidonic acid)

→ PDG₂, PGE₂, PGF₂, PGI₂, TXA₂
→ LTA₄, LTB₄, LTC₄, LTD₄, LTF₄

5,8,11,14,17-Eicosapentaenoic acid

→ PGD₃, PGE₃, PGF₃, PGI₃, TXA₃
→ LTA₅, LTB₅, LTC₅

FIGURE 16.9 Precursor and product relationships of eicosanoids: prostaglandins (PGs), prostacyclins (PGIs), thromboxanes (TXs), and leukotrienes (LTs). Arrows arising from each fatty acid indicate two different synthetic pathways: one for prostanoids (PG, PGI, TX) and the other for leukotrienes. The numerical subscript of an eicosanoid indicates the total number of double bonds in the molecule and thus the series to which it belongs. The prostanoids contain two fewer double bonds than the precursor fatty acid.

prostate gland, but they are produced in many tissues. In fact, the high concentrations found in semen arise in the seminal vesicles rather than the prostate. The chemical parent compound is a 20-carbon unnatural fatty acid known as **prostanoic acid** that contains a five-membered (cyclopentane) ring. Derivatives that contain this structure (PGs, TXs, and PGIs) are known collectively as **prostanoids**.

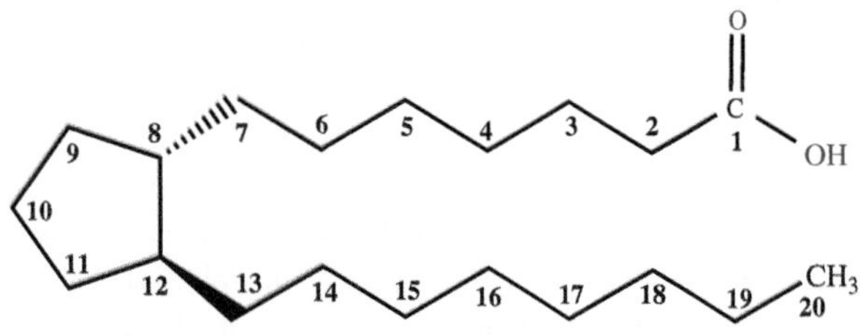

Prostanoic acid

Differences among various PGs are attributable to differences in substituents and in their positions on the five-membered ring (Figure 16.12). PGs are identified by a letter (e.g., PGE, PGF), characteristic for ring substituents, and by a numerical subscript (e.g., PGE₁, PGF₂), which indicates the number of double bonds (Figure 16.9). The location and type of double bond are as follows: PG₁, trans-C₁₃; PG₂, trans-C₁₃ cis-C₅; PG₃, trans-C₁₃ cis-C₅C₁₇. All PGs have a hydroxyl group at C₁₅ except PGG, which has a hydroperoxy group (−OOH). The hydroxyl group at C₁₅ is in the S-configuration in the naturally occurring prostaglandins. The α and β notations (e.g., PGF₂α) designate the configuration of the substituent at C₉ on the cyclopentane moiety, as used in steroid chemistry (α for below and β for above the plane of the projection of the cyclopentane ring). The natural compounds are α-derivatives.

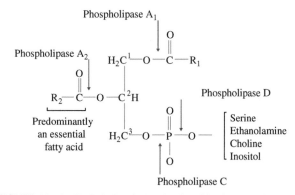

FIGURE 16.10 Hydrolysis of phosphoglycerides by phospholipases, whose cleavage sites are shown by vertical arrows.

PGs are synthesized in a stepwise manner by microsomal enzymes. The metabolic pathways discussed here use arachidonic acid as an example. Similar pathways are applicable to other polyenoic fatty acids. PG synthesis is started by microsomal prostaglandin endoperoxide synthase, which is a cyclooxygenase and a peroxidase. Cyclooxygenase activity (the rate-limiting reaction) results in 15-hydroperoxy-9,11-endoperoxide (PGG₂), which is converted to a 9,11-endoperoxide (PGH₂) by the peroxidase activity. There are two isoforms of cyclooxygenase (COX), which have been designated as COX1 and COX2. Cyclooxygenase is also known as **prostaglandin−endoperoxide synthase (PTGS)**. Both forms are membrane-associated enzymes. COX1 is constitutively expressed in many tissues, where arachidonic acid metabolites play a role in protective "housekeeping" homeostatic functions. Some of the COX1-mediated normal physiological functions include gastric cytoprotection and limiting acid secretion (Chapter 11), maintenance of renal

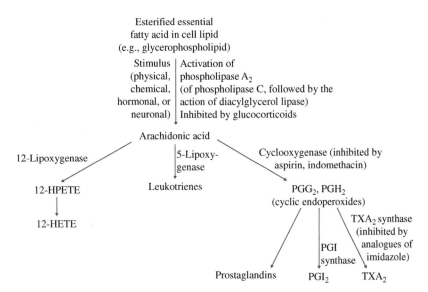

FIGURE 16.11 Pathways of arachidonate metabolism in eicosanoids synthesis and their inhibitors. HPETE = hydroperoxyeicosatetraenoic acid; HETE = hydroxyeicosatetraenoic acid; PG = prostaglandin; PGI = prostacyclin; TX = thromboxane. Conversions of arachidonic acid by various enzymes can be inhibited by analogues of the natural fatty acid, e.g., the acetylenic analogue 5,8,11,14-eicosatetraynoic acid.

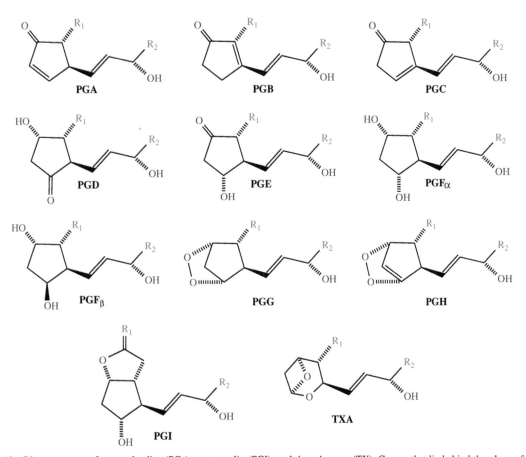

FIGURE 16.12 Ring structures of prostaglandins (PGs), prostacyclin (PGI), and thromboxane (TX). Groups that lie behind the plane of the ring are shown by ‖‖‖‖‖ and those that lie above the plane by ◀.

blood flow, vascular homeostasis, and hemostasis (e.g., antiplatelet effects, Chapter 34). COX2 activity, on the other hand, is normally undetectable in most tissues, and it is principally an inducible enzyme. In cells such as monocytes, macrophages, synoviocytes, endothelial cells, and chondrocytes, COX2 is expressed at high levels after induction by inflammatory mediators (e.g., interleukin-1 and tumor necrosis factor) and growth factors. COX2 enzymatic activity initiates the synthesis of arachidonic acid metabolites that mediate pain, inflammation, cellular differentiation, and mitogenesis. For example, PGE_2 is chemostatic for neutrophils, and PGI_2 causes changes in vascular permeability, facilitating extravasation of leukocytes. Although COX2 is generally an inducible enzyme, it is constitutively expressed in developing kidney and brain and, therefore, may be involved in their normal development and proper maturation.

The two unique isoforms, COX1 and COX2, are structurally similar, but they are encoded by separate genes differing in their tissue distribution and expression. The COX1 gene contains a promoter region without TATA sequence and is constitutively expressed. In contrast, the COX2 gene contains DNA segments that allow for rapid upregulation in response to appropriate stimuli. The anti-inflammatory action of glucocorticoids has no effect on the regulation of the COX1 gene.

The amino acid sequence homology of COX1 and COX2 is about 60%. However, in the region of the active site, the amino acid homology is about 90%, and both isoforms contain a long, narrow, largely hydrophobic channel with a hairpin bend at the end to accommodate the substrate arachidonic acid. A unique single-amino-acid difference in the wall of the hydrophobic channel (position 523) of COX1 and COX2 has been used to develop specific COX2 inhibitors. At position 523, COX1 has an isoleucine residue, whereas COX2 has a valine residue that is smaller by a single CH_2 group. The presence of the less bulky valine residue in the COX2 hydrophobic channel provides access for COX2 selective inhibitors. In COX1, the bulkier isoleucine residue prevents the entry of COX2 selective inhibitors.

In the treatment and management of pain and inflammation produced by arachidonic acid metabolites, COX inhibitors are widely used. These agents are known as **nonsteroidal anti-inflammatory drugs (NSAIDs)**. Acetylsalicylate (aspirin) is the classic anti-inflammatory and analgesic drug. Aspirin is an irreversible inhibitor of both COX1 and COX2, and it inhibits by acetylation of the hydroxyl group of the serine residue located at the active site of the enzymes. There are nonaspirin NSAIDs, the majority of which are organic acids (e.g., indomethacin, ibuprofen) that are reversible inhibitors of both COX1 and COX2. These inhibitors form a hydrogen bond with an arginine residue at position 120 of both COX1 and COX2 in the channel and block the entry of arachidonic acid. Because of their nonselectivity, aspirin and nonaspirin NSAIDs cause undesirable side effects due to inhibition of the "housekeeping" COX1 enzyme. The side effects include gastrointestinal disorders, renal dysfunction, and bleeding tendency. Thus, a COX2 selective or preferential inhibitor that spares COX1 activity is valuable in the treatment of pain and inflammation. Based on the biochemical differences between COX1 and COX2, drugs designed with COX2 inhibitor activity are associated with a markedly lower incidence of gastrointestinal injury. These drugs often possess sulfonyl, sulfone, or sulfonamide functional groups that bind with the COX2 side pocket in the hydrophobic channel. Examples of COX2 inhibitors are celecoxib, which is a 1,5-diarylpyrazole sulfonamide, and rofecoxib, which is a methylsulfonylphenyl derivative (Figure 16.13). Since nitric oxide (NO) protects gastric mucosa (Chapter 15), an NO moiety linked to conventional NSAIDs may negate the gastric toxic effects due to prostaglandin deficiency. Such drugs of NO-NSAIDs are currently being tested.

Other potential uses for COX inhibitors (in particular for COX2 inhibitors) may include the treatment of Alzheimer's disease and colon cancer. In Alzheimer's

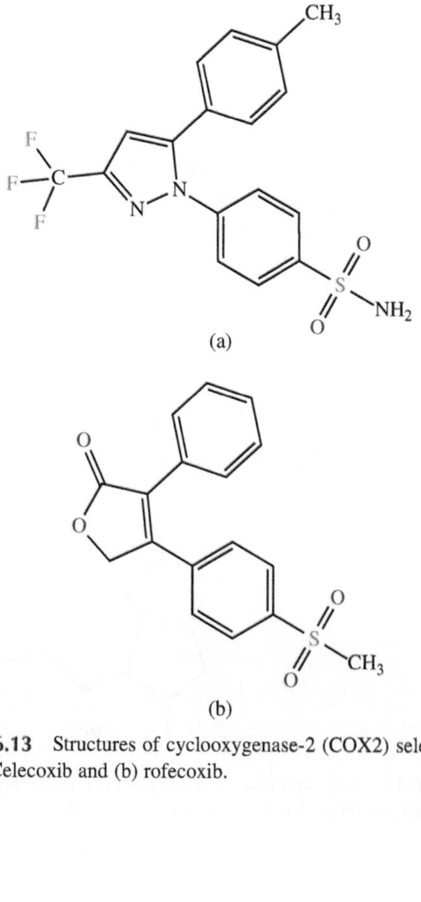

FIGURE 16.13 Structures of cyclooxygenase-2 (COX2) selective inhibitors. (a) Celecoxib and (b) rofecoxib.

disease, it is thought that an inflammatory component may lead to deposition of β-amyloid protein in neuritic plaques in the hippocampus and cortex (Chapter 4). The potential use of COX2 inhibitors such as aspirin in colon cancer arises from studies with subjects in which COX2 is overexpressed. The overexpression of the enzyme is related to the promotion and survival of intestinal adenomas and colon tumors. [4,5] The cyclooxygenase reaction is also inhibited by arachidonic acid analogues such as

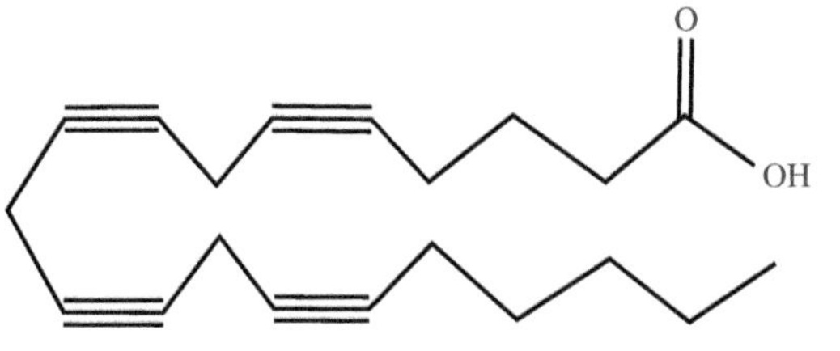

5,8,11,14-Eicosatetraynoic acid

PGH_2 is converted to PGD_2, PGE_2, $PGF_{2\alpha}$, prostacyclin (PGI_2), and thromboxane A_2 (TXA_2) by specific enzymes (Figure 16.14). PGA_2 is obtained from PGE_2 by dehydration. Since PGC_2 and PGB_2 are isomers of PGA_2, they can be synthesized by isomerases. The formation of

these compounds is shown in Figure 16.15. In some tissues, PGE_2 and PGF_2 undergo interconversion:

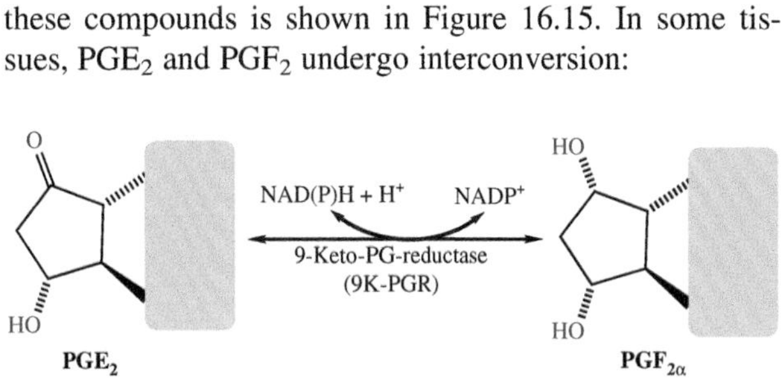

The $NAD(P)^+$ inhibits the conversion of PGE_2 to $PGF_{2\alpha}$, while reducing agents favor the formation of $PGF_{2\alpha}$. However, PGE_2 formation is favored by glutathione. In the initial catabolic reaction of both compounds by 15-hydroxy-PG-dehydrogenase (15-PGDH), the reduced NAD(P) formed in that reaction inhibits the first step. Thus, the ratio of reduced to oxidized NAD(P) may control the interconversion of PGE_2 and $PGF_{2\alpha}$ and also the first step in their catabolism. This finding is important because in many tissues PGE and PGF have opposing effects. PG biosynthesis can also be regulated by activation of latent forms of cyclooxygenase, promoted by catecholamines and serotonin. The PGs synthesized differ from tissue to

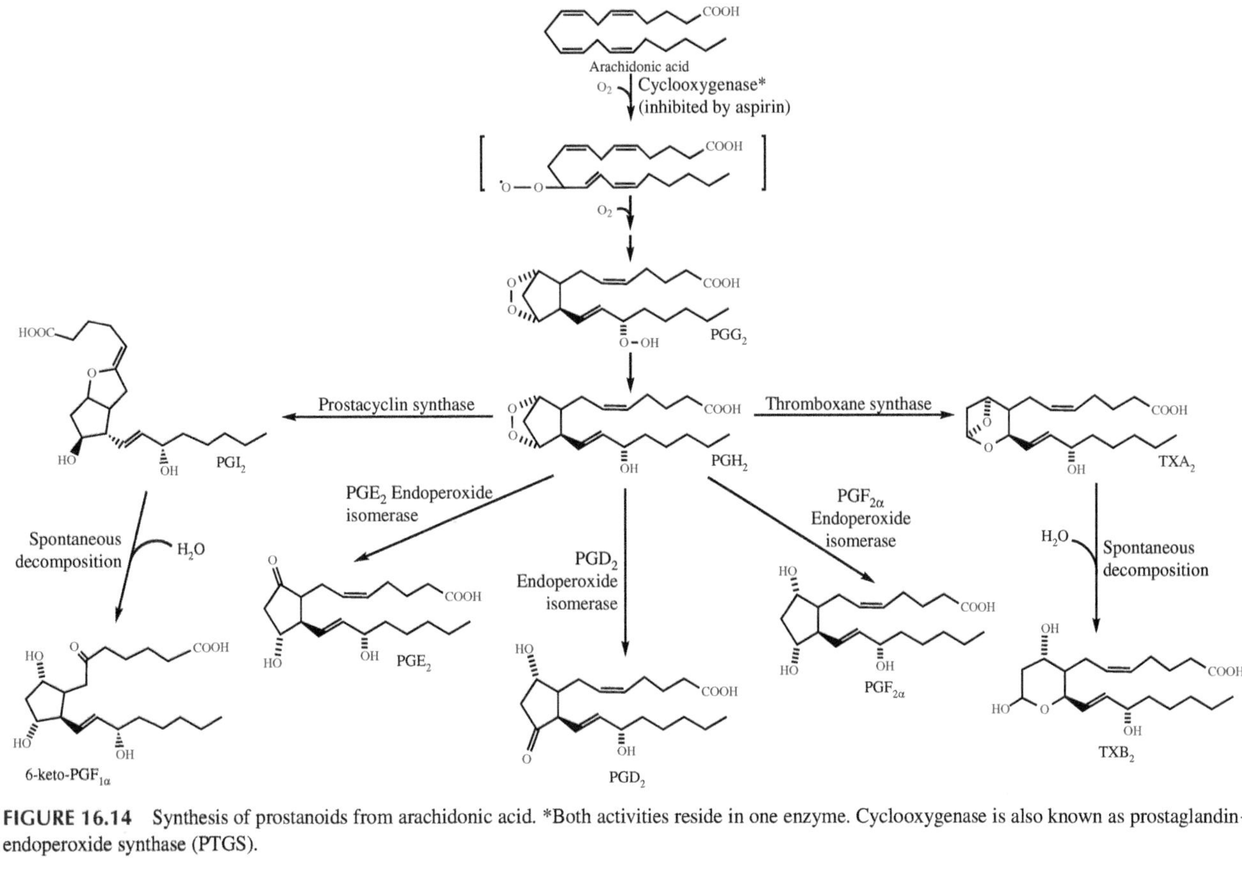

FIGURE 16.14 Synthesis of prostanoids from arachidonic acid. *Both activities reside in one enzyme. Cyclooxygenase is also known as prostaglandin—endoperoxide synthase (PTGS).

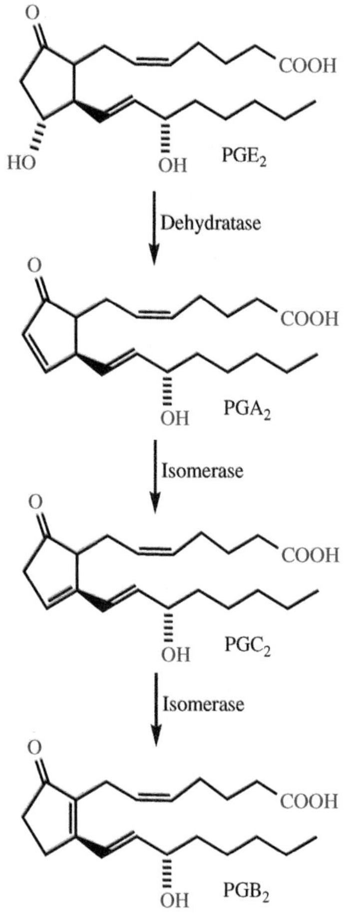

FIGURE 16.15 Conversion of PGE_2 to PGA_2, PGC_2, and PGB_2.

tissue; within the same tissue, different cells may yield products with antagonistic actions. For example, the lung parenchymal cells may produce TXA_2, while the lung vascular endothelial cells may produce PGI_2.

Catabolism of prostanoids occurs throughout the body, but the lungs can remove most of the plasma PGs during a single circulatory cycle. Despite this rapid removal, the PGs have adequate access to target organs. Catabolism starts with the reactions of 15-PGDH (oxidation of allylic-OH groups at C_{15}) and of PG reductase (reduction of the Δ^{13} double bond). 15-PGDH is found in the cytoplasm (lungs), requires NAD^+, and is specific for the $C_{15}(S)$ alcohol group. These reactions are followed by β-oxidation, ω-oxidation of the alkyl side chains, and elimination of the products. The catabolism of PGE_2 and $PGF_{2\alpha}$ is shown in Figure 16.16. The **thromboxanes (TX)**, first isolated from human and equine thrombocytes (platelets), contain an oxane ring. TXA_2 is synthesized from PGH_2 by microsomal thromboxane synthase. Thromboxane synthase is inhibited by imidazole derivatives. TXA_2 has a

very short half-life ($t_{\{1/2\}} = 30$ seconds at 37°C and pH 7.5) and undergoes rapid, nonenzymatic hydrolysis to the inactive TXB_2 (Figure 16.14).

Prostacyclin (PGI_2) is an active and unstable metabolite ($t_{\{1/2\}} = 3$ minutes at 37°C and pH 7.5) formed from PGH_2 by prostacyclin synthase. PGI_2 has a double-ring structure and is converted by nonenzymatic hydrolysis to 6-keto-$PGF_{1\alpha}$ (Figure 16.14).

Biological Properties of Prostanoids

Many effects of prostanoids are mediated through adenylate cyclase or mobilization of Ca^{2+} from intracellular stores. PGs increase cAMP in adenohypophysis, corpus luteum, fetal bone, lung platelets, and thyroid but decrease it in adipose tissue. Thromboxanes block the production of cAMP by PGs and mobilize intracellular Ca^{2+}. Thus, many endocrine glands (e.g., adrenal cortex, ovary, pancreatic islets, parathyroids) secrete hormones in response to PGs. Some of these effects are stimulation of steroid hormone production in the adrenal cortex, insulin release, thyroid hormone production, and progesterone secretion from the corpus leteum. *In vitro* PGEs, notably PGE_1, inhibit adipocyte lipolysis—the basal rate as well as that stimulated by catecholamine and other lipolytic hormones. Low doses of PGE_1 in humans tend to stimulate lipolysis through stimulation of release of catecholamines. PGEs stimulate the activity of osteoclasts with the mobilization of Ca^{2+} from bone, an effect independent of that of parathyroid hormone (Chapter 35).

Problems in delineating the primary actions of PGs arise from their frequently opposing effects and from the difficulty of distinguishing between physiological and pharmacological actions. In general, PGE_2 and $PGF_{2\alpha}$ have opposing effects on smooth muscle tone, release of mediators of immediate hypersensitivity, and cyclic nucleotide levels. Thus, the ratio between E and F compounds (due to changes in the $[NAD^+]/[NADH]$ ratio) may be a crucial factor in control of a given physiological response. The relative proportions of TXs and leukotrienes, as opposed to PGI, also appear to exert an important influence on physiological response.

In most animal species, PGI_2, PGEs, and PGAs are vasodilators, while TXA_2 is a vasoconstrictor. $PGF_{2\alpha}$ and 15-methyl $PGF_{2\alpha}$ are used for induction of mid-trimester abortions because they stimulate uterine muscle. Several PGs suppress gastric HCl production, which has therapeutic potential in the treatment of gastric ulceration and may explain the effect of aspirin to increase HCl secretion by inhibition of PG synthesis. PGs may regulate urine formation by modulating renal blood flow.

The effect of prostanoids on platelets has received considerable interest. TXA_2 synthesized in platelets

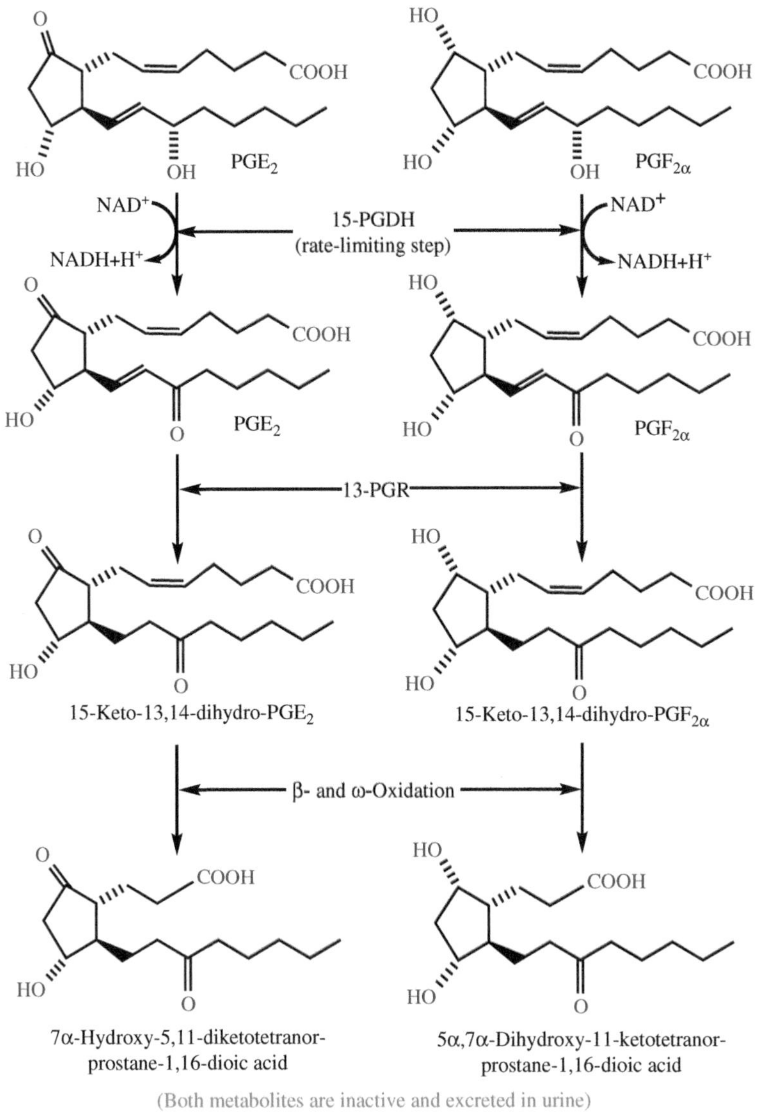

FIGURE 16.16 Catabolism of PGE$_2$ and PGF$_{2\alpha}$. 15-PGSH = 15-Hydroxy-PG-dehydrogenase; 13-PGR = 13-PG-reductase.

induces platelet aggregation, whereas PGI$_2$ generated in the vessel wall inhibits platelet aggregation. PGI$_3$ (Figure 16.9), a product of eicosapentaenoic acid (an ω-3 fatty acid), inhibits TXA$_2$ synthesis by inhibiting release of arachidonate from phospholipids and by competing for thromboxane synthase. TXA$_3$ is a much weaker aggregator of platelets than is TXA$_2$, while PGI$_3$ is a stronger antiaggregator than is PGI$_2$. The net effect is an antiplatelet effect, which may be beneficial in patients with thrombotic complications (e.g., myocardial infarction). The low incidence of coronary thrombosis in Greenland Eskimos, whose diet is almost completely derived from marine sources rich in ω-3 fatty acids, has been attributed to

antiplatelet effects. This diet is also associated with lower levels of serum cholesterol and triacylglycerol than those found in typical Western diets.

Leukotrienes

Leukotrienes (LTs) are most commonly found in leukocytes, mast cells, platelets, and vascular tissues of the lung and heart. They are formed chiefly from arachidonic acid, but they may be derived from eicosatrienoic and eicosapentaenoic acids. The name **leukotrienes** derives from their discovery in leukocytes and from the conjugated triene structure they contain. In the most active LTs, the conjugated triene is in a *trans, trans, cis* arrangement. They are distinguished by letters A—E and by a subscript that indicates the number of double bonds present.

LTs are produced in the 5-lipoxygenase pathway (Figure 16.17). Their synthesis begins with arachidonic acid obtained from cleavage of the membrane phospholipid pool due to the action of phospholipase A_2. Arachidonic acid is converted in a catalytic sequence by 5-lipoxygenase complex and its activating protein to 5-hydroperoxyeicosatetraenoic acid (5-HPETE) and then to

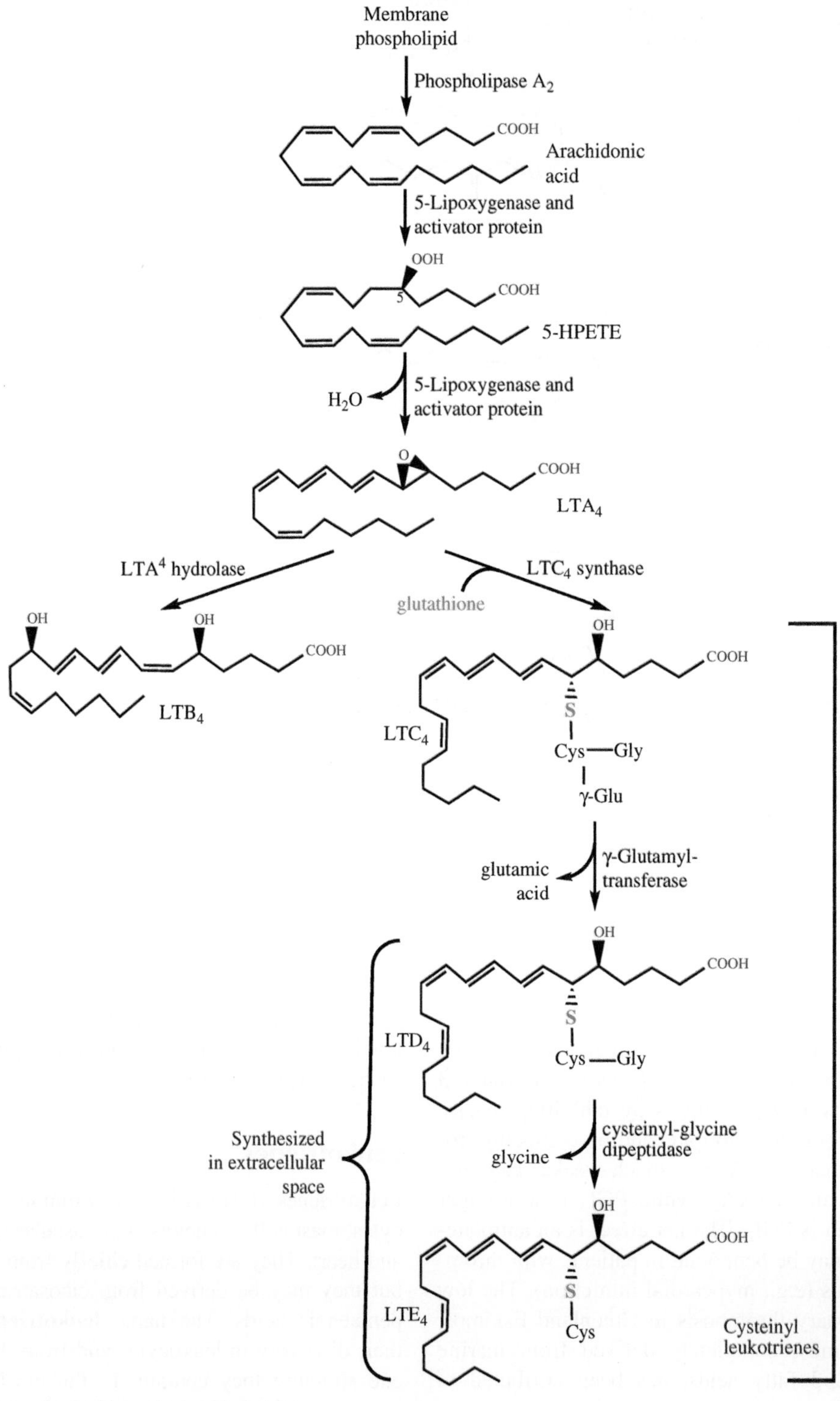

FIGURE 16.17 Pathway for the biosynthesis of leukotrienes.

leukotriene A$_4$ (5,6-oxido-7,9-*trans*-11,14-*cis*-eicosatet-raenoic acid). Leukotriene A$_4$ (LTA$_4$) is transformed by LTA$_4$ hydrolase into 5,12-dihydroxyeicosatetraenoic acid (leukotriene B$_4$, LTB$_4$) or into a glutathione adduct with the formation of a thioether linkage at C$_6$, (leukotriene C$_4$, LTC$_4$) by leukotriene C$_4$ synthase (also known as **glutathione S-transferase**). Leukotriene D$_4$ (LTD$_4$) and LTE$_4$ are synthesized in the extracellular space from LTC$_4$. A specific transmembrane transporter exports LTC$_4$ to the extracellular space. In the extracellular space, removal of the glutamyl residue from LTC$_4$ by γ-glutamyltransferase yields LTD$_4$, and the removal of the glycyl residue from LTD$_4$ by a variety of dipeptidases results in the formation of LTE$_4$ (Figure 16.17).

The three cysteinyl-linked leukotrienes—namely LTC$_4$, LTD$_4$, and LTE$_4$—are known collectively as cysteinyl leukotrienes. All three cysteinyl leukotrienes are potent mediators of inflammation and cause microvascular permeability, chemotaxis (particularly eosinophils), mucus hypersecretion, and neuronal stimulation. The potential role of LTC$_4$ as a neuromessenger or modulator has been implicated in an infant with LTC$_4$ synthase deficiency. The clinical features include muscular hypotonia, psychomotor retardation, failure to thrive, microcephaly,

and a fatal outcome. In lung tissue mast cells, eosinophils and alveolar macrophages possess the enzyme activities to synthesize cysteinyl leukotrienes and cause, in addition to the aforementioned biological actions, bronchial smooth muscle constriction and proliferation. Thus, cysteinyl leukotrienes are important mediators of immune-mediated inflammatory reactions of anaphylaxis and are constituents of substances originally called **slow-reacting substances of anaphylaxis (SRS-A)**. They are several times more potent than histamine in constricting airways and promoting tissue edema formation. The proinflammatory effect of LTE$_4$ is less than that of LTC$_4$ and LTD$_4$; it is excreted in the urine and is used as a marker of leukotriene production.

Antileukotriene agents, which can be used in treatment of allergen and exercise-induced asthma and allergic rhinitis, inhibit 5-lipoxygenase or the binding of the activator protein with 5-lipoxygenase or antagonists of leukotriene receptor at the target cell (e.g., airway epithelial cell). The traditional drugs used for treatment of asthma include inhaled corticosteroids, β$_2$-agonists, and theophyllines. Leukotriene receptor antagonists are orally active and are a new class of antiasthmatic therapeutic agents (Figure 16.18).

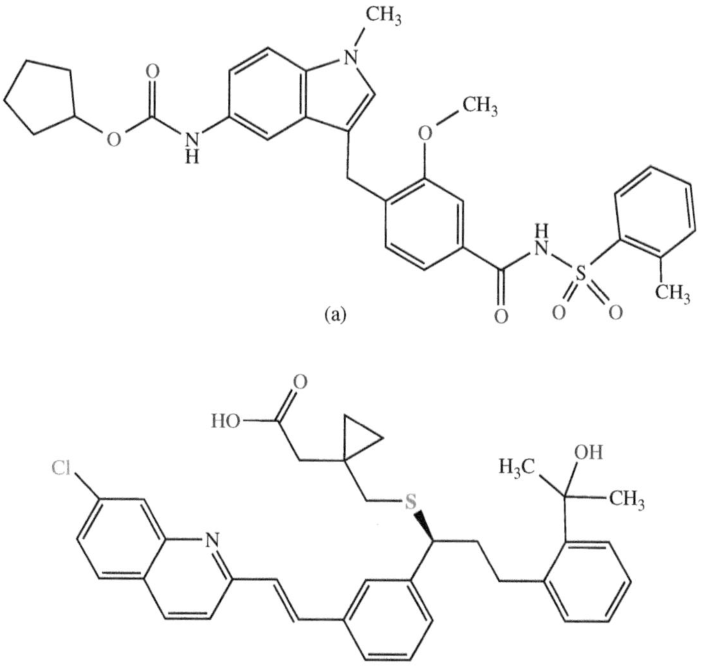

FIGURE 16.18 Structures of leukotriene receptor antagonists. (a) Zafirlukast and (b) montelukast.

CLINICAL CASE STUDY 16.1 Long-Chain Fatty Acid Oxidation Defect of Carnitine Cycling Due to Defects of Carnitine Acylcarnitine Translocase

This case was abstracted from: M.K. Fearing, E.J. Israel, I. Sahai, O. Rapalino, M. Lisovsky, Case 12-2011: a 9-month-old boy with acute liver failure, N. Engl. J. Med. 364 (2011) 1545–1556.

Synopsis

A 9-month-old infant boy was admitted to a hospital with complaints of fever, liver failure, and diarrhea. The infant had been delivered by normal vaginal delivery after a full-term gestational period. A limited newborn metabolic screening profile was unremarkable. At the current hospital admission, the clinical and laboratory studies of the infant were consistent with a working diagnosis of acute liver failure. Several causes of acute liver failure were considered, such as acetaminophen toxicity, infectious diseases, ischemic diseases, and more common metabolic disorders such as defects in long-chain fatty acid oxidation and cholesteryl-ester storage defect. Abdominal imaging and microscopic studies of liver biopsy revealed a microvascular and macrovascular steatosis. A metabolic biochemical analysis of a retrieved filter-paper blood specimen collected during newborn period revealed high levels of acylcarnitine intermediates. This result, along with clinical, imaging, and microscopic findings, led to the conclusion: metabolic defect of the infant of inborn error of CPTI and carnitine acylcarnitine translocase (CACT). The final diagnosis was CACT deficiency, which was established by gene sequencing.

Resolution of the case: The infant was placed initially on a long-chain fatty acid restricted diet with an introduction of normal diet with inclusions of carnitine and ursodeoxycholic acid. This led to progression of normal growth and development.

Teaching Points

1. The initial newborn metabolic screening panel did not detect fatty acid oxidation defects because the infant's mother had declined an optional panel due to expended costs.
2. Mitochondrial fatty acid oxidation defects are identified by elevated levels of acylcarnitines in the newborn blood metabolic profiles.
3. The clinical symptoms may vary depending on whether the defects are in oxidation of long-chain fatty acids. These disorders can cause hypoketotic hypoglycemia or acute fatty liver (see Clinical Case Study 16.2).
4. Although the defects may be asymptomatic initially, under stressful conditions, pathological conditions develop.

CLINICAL CASE STUDY 16.2 A Neonatal Death Due to Medium-Chain Acyl-CoA Dehydrogenase (MCAD) Deficiency

This clinical case was abstracted from: A.A. Manoukian, C.E. Ha, L.H. Seaver, N.V. Bhagavan, A neonatal death due to medium-chain acyl CoA dehydrogenase deficiency, Am. J. Forensic Med. Pathol. 30 (2009) 284–286.

Synopsis

A male infant born at 39 weeks of gestation via cesarean section and his mother were sent home 35 hours after delivery. At home, at 61 hours of age, the infant was found unresponsive and cyanotic. During this period, the infant had received only two feedings, and he slept most of the time. Paramedics were called, and the infant was transported by ambulance to the local hospital. Despite efforts at resuscitation, the infant was pronounced dead. A forensic examination did not reveal any foul play or accidental asphyxia. The only abnormality found on autopsy was extensive microvesicular metamorphosis of the liver. The infant's heel-stick blood had been collected on a filter paper before discharge from the hospital for the detection of inborn errors of metabolism. The specimen had been sent to a regional laboratory. The results of this metabolic profile were made available subsequent to autopsy examination and indicated MCAD deficiency. The profile showed abnormal elevation of C_6, C_8, and C_{10} fatty acyl carnitine esters. The autopsy blood acyl carnitine profile confirmed the MCAD diagnosis. The parents were carriers for two different MCAD gene mutations.

Teaching Points

1. MCAD deficiency is the most common inherited autosomal recessive metabolic defect of fatty acid oxidation. Its prevalence is between 1:8000 and 1:15,000.
2. In MCAD deficiency, the β-oxidation of fatty acids ceases at the C_6–C_{10} level. This leads to a deficiency in the production of acetyl-CoA, which is oxidized in the TCA cycle with the production of reducing equivalents as NADH and $FADH_2$. In the electron transport system, oxidation of NADH and $FADH_2$ is coupled with the formation of ATP from ADP and P_i. In the liver, ATP is required for gluconeogenesis, urea formation from ammonia, and many other vital metabolic functions. Thus, MCAD deficiency during fasting when the prevailing plasma glucose concentrations are low can result in episodic hypoketotic hypoglycemia and hyperammonemia. Recall that in the liver mitochondria, acetyl-CoA is converted to ketone bodies (acetoacetate and β-hydroxybutyrate), which serve as fuel for extrahepatic tissues (CNS may be an exception in the early stages of fasting and starvation).
3. MCAD deficiency exhibits phenotypic heterogeneity. Avoidance of fasting is essential.

ENRICHMENT READING

[1] K. Haram, E. Svendsen, U. Abildgaard, The HELLP syndrome: clinical issues and management, A Review, BMC Pregnancy and Childbirth 9 (2009) 8–23.

[2] G. Fragasso, A. Palloshi, P. Puccetti, C. Silipigni, A. Rossodivita, M. Pala, et al., A randomized clinical trial of trimetazidine, a partial free fatty acid oxidation inhibitor, in patients with heart failure, J. Am. Coll. Cardiol. 48 (2006) 992–998.

[3] L. Zhang, Y. Lu, H. Jiang, A. Sun, Y. Zou, J. Ge, Additional use of trimetazidine in patients with chronic heart failure, J. Am. Coll. Cardiol. 59 (2012) 913–922.

[4] B. Pasche, Aspirin—from prevention to targeted therapy, N. Engl. J. Med. 367 (2012) 1650–1652.

[5] X. Liao, P. Lochhead, R. Nishihara, T. Morikawa, A. Kuchiba, M. Yamauchi, et al., Aspirin use, tumor PIK3CA mutation, and colorectal-cancer survival, N. Engl. J. Med. 367 (2012) 1596–1606.

Lipids II: Phospholipids, Glycosphingolipids, and Cholesterol

Key Points

1. Phospholipids can be either glycerolipids or sphingolipids. They contain phosphate and other polar groups.

2. Phosphatidylcholine (lecithin) synthesis requires CTP for activation of intermediate metabolites.

3. Phosphatidylcholines are degraded preferentially by specific phospholipases at specific bonds. The products of hydrolysis have unique metabolic roles.

4. Sphingomyelin contains N-acylsphingosine and phosphorylcholine; the phosphorylcholine moiety is acquired from lecithin by an exchange reaction.

5. Pulmonary surfactant is a complex of phospholipids (lecithin), neutral lipids (cholesterol), and proteins. It is synthesized in alveolar type II cells and spreads at the alveolar air–liquid interface. It decreases surface tension, thus preventing alveolar collapse during the expiration cycle. Defects in surfactant synthesis, assembly, and secretion into alveolar spaces can cause respiratory distress syndrome in neonates.

6. Prenatal laboratory tests are utilized to assess surfactant deficiency and to initiate appropriate preventative measures.

7. Sphingolipids are in a continuous state of turnover. Defects in their catabolism, due to lysosomal enzyme deficiencies, lead to storage disorders with serious clinical manifestations (see Clinical Case Studies 17.1 to 17.3).

8. Cholesterol biosynthesis requires acetyl-CoA and NADPH. The enzymes are located in the cytosol and endoplasmic reticulum membrane.

9. The rate-limiting enzyme of cholesterol biosynthesis is HMG-CoA reductase. Its activity is modulated by phosphorylation (deactivation), dephosphorylation (activation), and transcriptional regulation. It is inhibited by statins that are therapeutically used in reducing cholesterol levels to ameliorate cardiovascular disorders (Chapter 18).

10. The cholesterol biosynthetic pathway is a multifunctional pathway and is utilized for the synthesis of other metabolites that contain isoprenoid units.

11. Cholesterol is the precursor for bile acid (bile salt) synthesis in the liver. Bile acids are conjugated, secreted into bile canaliculi along with cholesterol and phospholipids, and stored in the gall bladder. The formation of bile and its excretion into bile ducts by the hepatocytes is a unique exocrine function of the liver. Bile pigments (e.g., bilirubin diglucuronide), which are products of heme catabolism, are also secreted in the bile (Chapter 27). Bile salts emulsify dietary lipids in the GI tract and facilitate their absorption (Chapter 11). Primary bile acids are converted to secondary bile acids by bacterial enzymes in the GI tract. Some undergo enterohepatic circulation, and some are eliminated.

12. The presence of excess cholesterol, or deficiency of bile salts in the bile, can cause gallstones that obstruct the passage of bile into the GI tract.

13. Dissolution of gallstones may require both medical (ursodiol) and surgical interventions.

Phospholipids and glycosphingolipids are amphipathic lipid constituents of membranes (Chapter 9). They play an essential role in the synthesis of plasma lipoproteins (Chapter 18) and eicosanoids (Chapter 16). They function in the transduction of messages from cell surface receptors to the second messengers that control cellular processes (Chapter 28) and as surfactants. When the cell membrane integrity is compromised, the phospholipids appear in the plasma. This observation of changes in plasma lipids is being investigated to identify neurodegeneration in preclinical Alzheimer's disease. Cholesterol is mainly of animal origin and is an essential constituent of biomembranes (Chapter 9). In plasma, cholesterol is associated with lipoproteins (Chapter 18). Cholesterol is a precursor for bile acids formed in the liver; for steroid hormones secreted by the adrenals, gonads, and placenta; and for 7-dehydrocholesterol of vitamin D formed in the skin. In tissues, cholesterol exists primarily in the unesterified form (e.g., in the brain and in erythrocytes). However, appreciable quantities are esterified with fatty acids in the liver, skin, adrenal cortex, and blood.

N. V. Bhagavan and C.-E. Ha: Essentials of Medical Biochemistry, Second edition. DOI: http://dx.doi.org/10.1016/B978-0-12-416687-5.00017-8

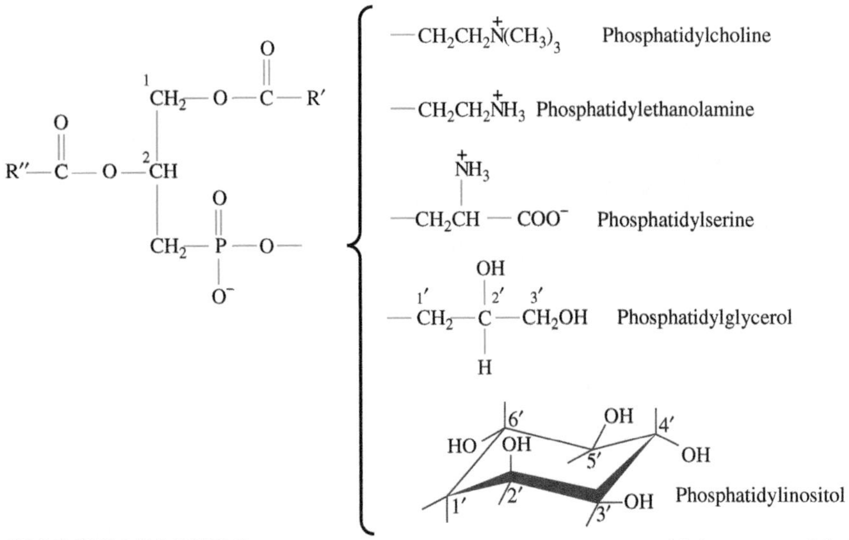

—CH₂CH₂N⁺(CH₃)₃   Phosphatidylcholine

—CH₂CH₂N⁺H₃ Phosphatidylethanolamine

—CH₂CH—COO⁻   Phosphatidylserine
(N⁺H₃)

—CH₂—C—CH₂OH   Phosphatidylglycerol
(OH, H; 1′ 2′ 3′)

Phosphatidylinositol

FIGURE 17.1 Structure of some glycerophospholipids.

PHOSPHOLIPIDS

Phospholipids can be **glycerolipids** or **sphingolipids**. Examples of glycerolipids are phosphatidylcholine, phosphatidylethanolamine, phosphatidylserine, phosphatidylinositol, and phosphatidylglycerol (Figure 17.1). To distinguish between the two primary alcohol–carbon atoms of asymmetrically substituted glycerol derivatives, the glycerol carbon atoms are numbered 1–3 from top to bottom and the C_2 hydroxyl group is written to the left. This system is known as the **stereochemical numbering** (*sn*) convention. Thus, the structural formula for *sn*-1,2-diacylglycerol is

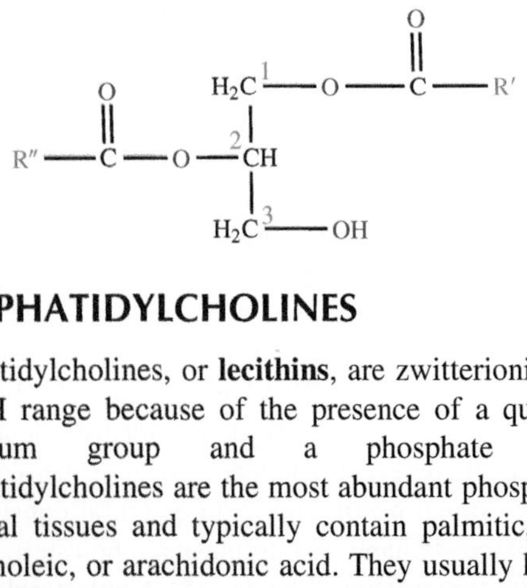

PHOSPHATIDYLCHOLINES

Phosphatidylcholines, or **lecithins**, are zwitterionic over a wide pH range because of the presence of a quaternary ammonium group and a phosphate moiety. Phosphatidylcholines are the most abundant phospholipids in animal tissues and typically contain palmitic, stearic, oleic, linoleic, or arachidonic acid. They usually have saturated fatty acids in the *sn*-1 position and unsaturated fatty acids at *sn*-2.

The *de novo* pathways for phospholipid synthesis use cytidine triphosphate (CTP) for activation of intermediate species (analogous to the role of UTP in glycogen biosynthesis; Chapter 14). The principal pathway of phosphatidylcholine biosynthesis uses cytidine diphosphate (CDP-choline) (Figure 17.2). Many reactions of phospholipid synthesis occur in the endoplasmic reticulum (ER). Choline is first phosphorylated by ATP to phosphocholine,

which reacts with CTP to form CDP-choline, from which phosphocholine is transferred to *sn*-1,2-diacylglycerol. The rate-limiting step in this pathway appears to be that catalyzed by CTP:phosphocholine cytidylyltransferase, which is activated by fatty acids.

Phosphatidylcholine can also be synthesized by the methylation pathway that converts phosphatidylethanolamine to phosphatidylcholine, principally in the liver. The methyl donor is S-adenosylmethionine (Chapter 15). Phosphatidylethanolamine-N-methyltransferase transfers three methyl groups in sequence to produce phosphatidylcholine. The fatty acid components of phosphatidylcholine can then be altered by deacylation–reacylation reactions.

Phosphatidylcholine is degraded by phospholipases that cleave preferentially at specific bonds (Chapter 16). The choline released is phosphorylated by choline kinase and reutilized in phosphatidylcholine synthesis. However, in liver mitochondria, choline is also oxidized to betaine (N-trimethylglycine):

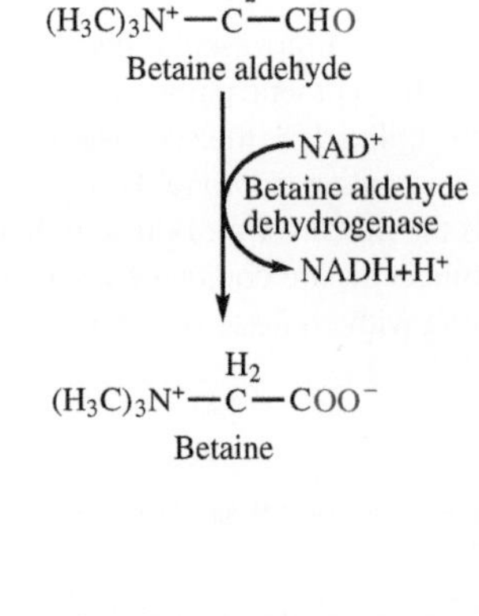

$(H_3C)_3N^+$—$\overset{H_2}{C}$—CH_2OH
Choline

↓ NAD⁺ / Choline dehydrogenase / NADH+H⁺

$(H_3C)_3N^+$—$\overset{H_2}{C}$—CHO
Betaine aldehyde

↓ NAD⁺ / Betaine aldehyde dehydrogenase / NADH+H⁺

$(H_3C)_3N^+$—$\overset{H_2}{C}$—COO⁻
Betaine

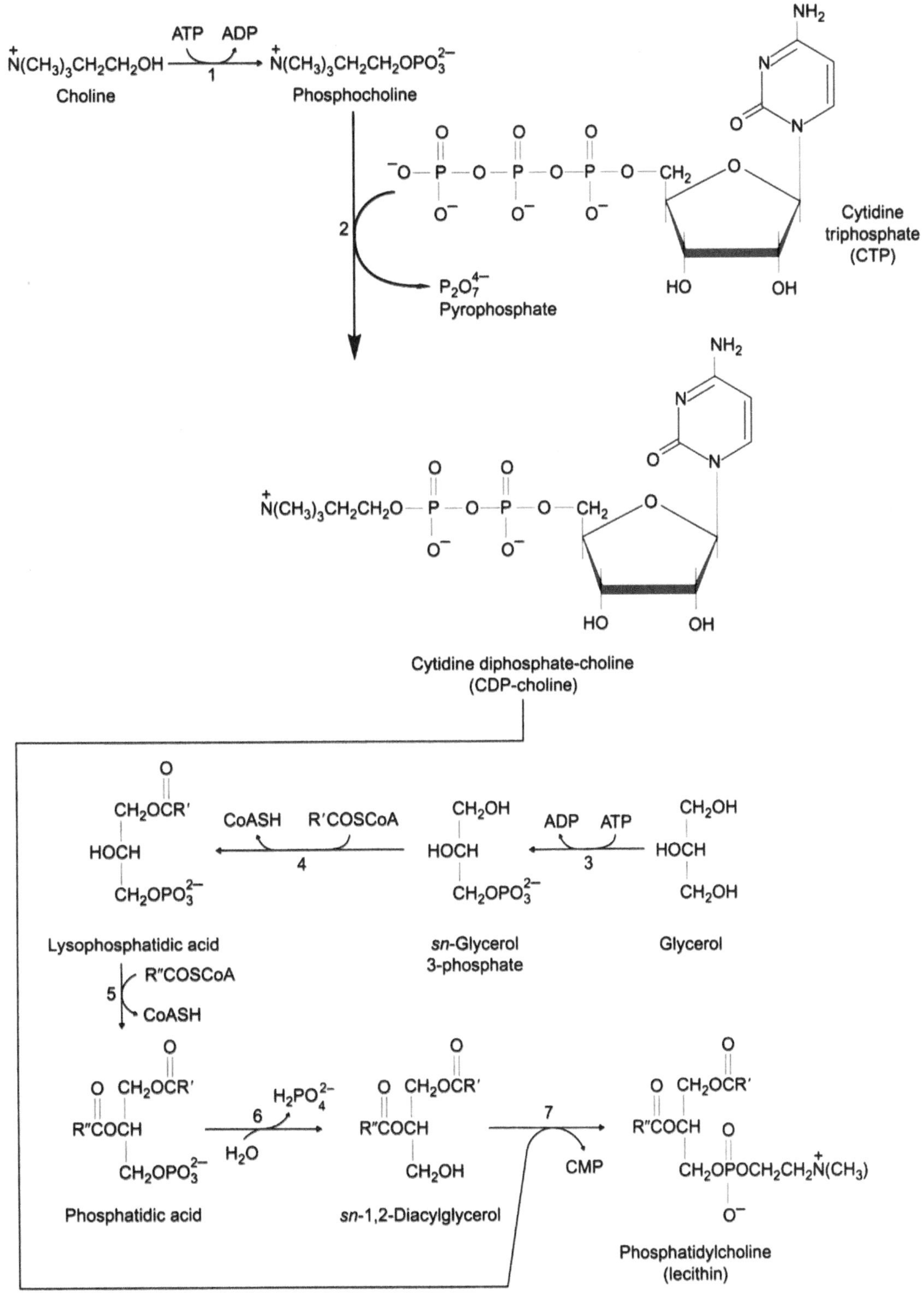

FIGURE 17.2 Synthesis of phosphatidylcholine. The rate-limiting reaction is that catalyzed by cytidylyltransferase (reaction 2), which appears to be active only when attached to the endoplasmic reticulum, although it is also found free in the cytosol. Cytidylyltransferase is inactivated by a cAMP-dependent protein kinase and activated by a phosphatase. Translocation to the endoplasmic reticulum can be stimulated by substrates such as fatty acyl coenzyme A (CoA). Choline deficiency can result in deposition of triacylglycerol in the liver and reduced phospholipid synthesis. Enzymes: (1) choline kinase; (2) CTP:phosphocholine cytidylyltransferase; (3) glycerol kinase; (4) acyl-CoA:glycerol-3-phosphate acyltransferase; (5) acyl-CoA:acyl glycerol-3-phosphate acyltransferase; (6) phosphatidic acid phosphatase; (7) CDP-choline:diacylglycerol phosphocholine transferase.

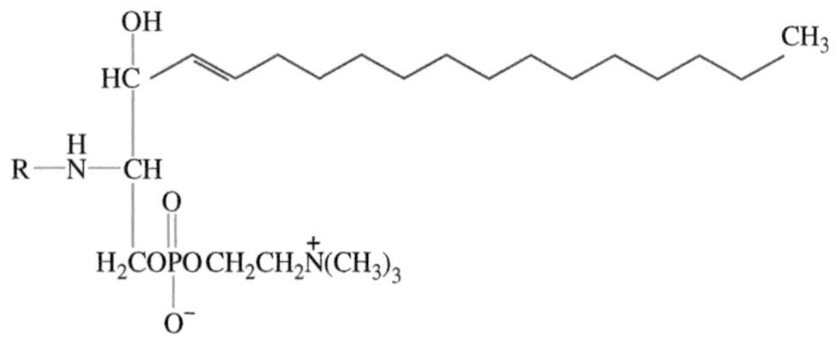

$$OH$$

$$HC$$

$$R-N-CH$$

$$H_2COPOCH_2CH_2\overset{+}{N}(CH_3)_3$$

FIGURE 17.3   Structure of sphingomyelin.

Betaine functions as a methyl donor (e.g., in methionine biosynthesis from homocysteine; Chapter 15), and it can also be converted to glycine.

PHOSPHOSPHINGOLIPIDS

The **sphingomyelins** are structurally similar to phosphatidylcholine but contain N-acylsphingosine (ceramide) instead of *sn*-1,2-diacylglycerol (Figure 17.3). They occur in high concentration in myelin and in the brain, and are a nearly ubiquitous constituent of membranes.

Sphingolipid biosynthesis is catalyzed by membrane-bound enzymes of the endoplasmic reticulum. Sphingosine, an acylaminoalcohol, is synthesized from palmitoyl-CoA and serine in a reaction that requires pyridoxal phosphate, NADPH, and Mn^{2+}.

Sphingomyelin is probably synthesized by an exchange reaction in which the phosphorylcholine moiety of phosphatidylcholine is transferred to ceramide:

$$Ceramide + phosphatidylcholine \rightarrow sphingomyelin$$
$$+ diacylglycerol$$

Synthesis of glycosphingolipids and sulfoglycosphingolipids involves the addition of sugar and sulfate residues to ceramide from UDP-sugar derivatives, or the activated sulfate donor 3′-phosphoadenosine-5′-phosphosulfate (Chapter 15) and appropriate transferases. Catabolism of sphingolipids is by specific lysosomal hydrolases.

CATABOLISM AND STORAGE DISORDERS OF SPHINGOLIPIDS

There are four groups of glycosphingolipids: cerebrosides, sulfatides, globosides, and gangliosides. Cerebrosides contain a single sugar residue linked to ceramide, which is an N-acylsphingosine. Sulfatides contain a sulfate group attached to a sugar residue. Globosides contain two or more sugar residues and an N-acetylgalactosamine group linked to ceramide. Gangliosides (G) have oligosaccharide chains that contain sialic acid residues. They are classified based on the number of sialic acid (N-acetylneuraminic acid, NANA) residues they contain and the sequence of their sugar residues. G_M, G_D, G_T, and G_Q

contain gangliosides with one, two, three, and four sialic acid residues, respectively. The number associated with M, D, T, and Q signifies the sequence of sugar residues:

1. Represents Gal−NAcGal−Gal−Glc-ceramide.
2. Represents NAcGal−Gal−Glc−ceramide.
3. Represents Gal−Glc−ceramide. Thus, the structure of ganglioside G_{M1} is:

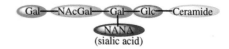

Sphingolipids are in a continuous state of turnover. They are catabolized by lysosomal enzymes through stepwise removal of sugar residues beginning at the nonreducing end of the molecule. Each sugar residue removed involves a specific exoglycosidase. The degradation of sphingolipids requires the activity of specific hydrolases, and is also dependent on nonenzymatic glycoproteins, known as **sphingolipid activator proteins (SAPs)**. The degradation of sulfolipids also requires the activity of sulfatases to remove their sulfate groups. SAP-stimulated degradation of sphingolipids is thought to involve the binding of the activator protein with the sphingolipids so that the water-soluble hydrolases can access the specific sites of hydrolysis. Genes for SAPs are located in chromosomes 5 and 10. The SAP gene that resides in chromosome 5 codes for the activator of hexosaminidase A, which hydrolyzes G_{M2}. A gene on chromosome 10 codes for a precursor that is synthesized in the endoplasmic reticulum, exported to the cell surface, and then imported into a lysosomal compartment. In the lysosomes, the precursor protein is processed to yield four mature activator proteins: sap-A, sap-B, sap-C, and sap-D. The activator functions of these proteins are as follows: sap-A stimulates glycosylceramidase and galactosylceramidase in the presence of detergents; sap-B is a nonspecific activator that stimulates hydrolysis of about 20 glycolipids as well as hydrolysis of sulfatide by arylsulfatase A; sap-C is essential for the action of glucosylceramidase; and the function of sap-D is unknown.

The importance of SAPs is exemplified in disorders where these activator proteins are not made as a result of mutations. A defect in the synthesis of the enzyme or its activator protein can both result in the same phenotype. For example, a deficiency of hexoseaminidase A deficiency or its activator protein (ganglioside G_{M2} activator) results in **Tay−Sachs disease**; a deficiency of arylsulfatase or its activator protein (sap-B) results in **juvenile metachromatic leukodystrophy**; and a deficiency of glucosylceramidase or its activator protein (sap-C) results in **Gaucher's disease**. All of these disorders are accompanied by pronounced accumulation of the respective precursor lipids in the reticuloendothelial system (see Clinical Case Studies 17.1 through 17.3). Sphingomyelin is hydrolyzed to ceramide and phosphorylcholine by sphingomyelinase:

$$\text{Sphingomyelin} + H_2O \rightarrow \text{phosphorylcholine} + \text{ceramide}$$

Sphingomyelinase deficiency leads to the development of **Niemann–Pick disease** via the accumulation of sphingomyelin (and occasionally unesterified cholesterol) in reticuloendothelial cells of the spleen, bone marrow, liver, central nervous system, and retina. This results in central nervous system damage and a cherry-red spot in the retina.

Ceramide is hydrolyzed to sphingosine and a fatty acid by ceramidase:

$$\text{Ceramide} + H_2O \rightarrow \text{sphingosine} + \text{fatty acid}$$

A nonlysosomal ceramidase in some tissues functions optimally at neutral or alkaline pH and participates in the synthesis and breakdown of ceramide. Deficiency of lysosomal (acid) ceramidase in **Farber's disease** (lipogranulomatosis) causes accumulation of ceramide. The disease is inherited as an autosomal recessive trait and is characterized by granulomatous lesions in the skin, joints, and larynx, and moderate nervous system dysfunction. It may also affect the heart, lungs, and lymph nodes and is usually fatal during the first few years of life.

Sphingosine is catabolized to *trans*-2-hexadecanal and phosphoethanolamine by way of sphingosine phosphate and is cleavage by a lyase. Catabolism of glycosphingolipids involves removal of successive glycosyl residues from their nonreducing end until ceramide is released.

Abnormalities usually involve specific exoglycosidases and their activator proteins (discussed earlier), except in metachromatic leukodystrophy, in which there is a deficiency of a sulfatase.

Catabolic pathways for the glycosphingolipids are given in Figure 17.4 and associated disorders are summarized in Table 17.1. Some comments are warranted:

1. Accumulation of a specific lipid in these disorders is frequently accompanied by deposition of one or more polysaccharides structurally related to the lipid.
2. Treatment is generally palliative or nonexistent. Enzyme replacement therapy has proven useful in some of these disorders. Because the exogenous enzymes are unable to cross the blood–brain barrier, their efficacy in the glycosphingolipidoses that involve neurological abnormalities is doubtful. Attempts to modify the enzymes to overcome this difficulty offer some hope.
3. Considerable progress has been made in the identification of carriers and in prenatal diagnosis of homozygotes. Thus, laboratory assays of enzyme activity in leukocytes or cultured skin cells using chromogenic or fluorogenic synthetic substrates have enabled the dramatical reduction of the incidence of Tay–Sachs disease.

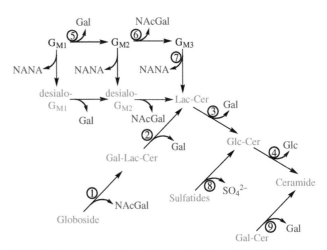

FIGURE 17.4 Degradative pathways for glycosphingolipids. $G_{M1,2,3}$ are gangliosides. Their structures and those of globosides and sulfatides should be inferred from their respective catabolic routes. The circled numbers correspond to the metabolic lesions listed in Table 17.1. Gal = Galactose; Glc = glucose; NAcGal = N-acetyl-galactose-2-amine; NANA = N-acetyl neuraminic acid (a sialic acid); Lac = lactose [galactosyl (β-1→4) glucose]; Cer = ceramide (N-acylsphingosine); and desialo = without a sialic acid (NANA) residue.

Gaucher's disease is the most common lysosomal storage disorder and also the most common inherited disease among Ashkenazi Jews, with a carrier frequency of about 1 in 14. Four mutations in the gene encoding for β-glucocerebrosidase account for at least 90% of the symptomatic patients. Gaucher's disease has three forms in which genetic effects appear to be due to errors in the same or related genetic loci. Type I, **chronic non-neuronopathic** (adult), is the most common variety. It comprises a heterogeneous group of patients characterized by the presence of hematological abnormalities (anemia, thrombocytopenia) and erosion of the cortices of long bones. Type II, **acute neuronopathic**, usually appears before 6 months of age and is fatal by 2 years. Mental damage is a primary characteristic, and the disease progresses rapidly from its onset. Type III, **subacute neuronopathic** (juvenile), comprises a heterogeneous group in which death occurs between infancy and about 30 years of age. The cerebral abnormalities usually appear 2 years postnatally at the earliest.

All three types share common features: hepatosplenomegaly, Gaucher's cells in the bone marrow (accumulation of glucocerebroside in reticuloendothelial cells in liver, spleen, and bone marrow), and autosomal recessive inheritance. Some studies have shown a correlation between the residual β-glucocerebrosidase activity and clinical severity, but the molecular basis for the genetic heterogeneity is not known. Gaucher's cells obtained from bone marrow aspirates exhibit a characteristic appearance of the cytoplasm owing to rod-shaped striated inclusion bodies composed primarily of glucocerebroside. Patients with

TABLE 17.1 Characteristics of Glycosphingolipid Storage Disorders

Disorders	Major Lipids Accumulated	Other Compounds Affected	Enzyme or Activator Protein Lacking*	Remarks
G_{M2} gangliosidosis, type II (Tay–Sachs variant; Sandhoff's disease)	Globoside and G_{M2} ganglioside		Hexosaminidases A and B (① and ⑥) G_{M2}-activator	Same clinical picture as Tay–Sachs disease but progresses more rapidly; no racial predilection; hepatosplenomegaly, cardiomyopathy.
Fabry's disease (glycosphingolipid lipidosis)	Gal-($\alpha1 \rightarrow 4$)-Lac-Cer	Gal-($4 \leftarrow 1\alpha$)-Lac-Cer accumulates.	α-Galactosidase (②)	X-linked recessive; hemizygous males have a characteristic skin lesion usually lacking in heterozygous females; pain in the extremities; death usually in the fourth decade results from renal failure or cerebral or cardiovascular disease.
Ceramide lactoside lipidosis	Gal-($\beta1 \rightarrow 4$)-Glc-Cer		Ceramide lactoside β-galactosidase (③)	Liver and spleen enlargement; slowly progressive brain damage, neurological impairment.
Gaucher's disease (glucosyl ceramide lipidosis; three types; see text)	Glc-Cer	G_{M3} ganglioside accumulates most frequently; other compounds occasionally.	β-Glucocerebrosidase (glucosylceramidase; ④) Activator protein sap-C	Hepatosplenomegaly; frequently fatal; no known treatment; occurrence of Gaucher's cells (reticuloendothelial cells that contain accumulations of erythrocyte-derived glucocerebroside).
G_{M1} gangliosidosis (two types; see text)	G_{M1}- and desialo-G_{M1}-gangliosides	Keratan sulfate-related polysaccharide accumulates.	G_{M1}-β-galactosidase (⑤)	Mental and motor deterioration; accumulation of mucopolysaccharides is as significant as accumulation of gangliosides; invariably fatal; autosomal recessive; blindness, cherry red macula (30%); hepatosplenomegaly; vacuolated lymphocytes; startle response to sound, dysostosis multiplex.
G_{M2} gangliosidosis, type I (Tay–Sachs disease; see text)	G_{M2}- and desialo-G_{M2}-gangliosides	Other desialo-hexosyl ceramides accumulate; other compounds occasionally.	Hexosaminidase A (⑥)	Red spot in retina; mental retardation; severe psychomotor retardation; seizures; blindness; startle response to sound; invariably fatal; autosomal recessive; panracial but especially prevalent among Northern European Jews.
Metachromatic leukodystrophy (MLD; sulfatide lipidoses; at least three types; see text)	3-sulfate-galactosylcerebroside	Cerebrosides other than sulfatides are decreased; ceramide dihexoside sulfate accumulates.	Sulfatidases (⑦); arylsulfatases) Activator protein sap-B	Demyelination; progressive paralysis and dementia; death usually occurs within the first decade; autosomal recessive inheritance.
Krabbe's disease (globoid cell leukodystrophy; galactosyl ceramide lipidosis)	Galactocerebroside	Sulfatides are also greatly decreased, probably as a secondary feature.	Galactocerebroside-β-galactosidase (⑧)	Mental retardation; demyelination; psychomotor retardation; failure to thrive; progressive spasticity; globoid cells in brain white matter; invariably fatal; autosomal recessive inheritance.

*The circled numbers refer to reactions in Figure 17.4. The abbreviations are the same as in that figure.

Gaucher's disease have elevated levels of acid phosphatase activity in serum and spleen, increased iron stores, increased angiotensin-converting enzyme activity, and a relative deficiency of clotting factor IX (Chapter 34).

Enzyme replacement therapy with purified macrophage-targeted human β-glucocerebrosidase in type I Gaucher's disease causes breakdown of stored glucocerebrosides. This results in a reduction in the size of the liver and spleen, improvement in hematological abnormalities (anemia and thrombocytopenia), increased bone mineralization, and decreased bone pain. Two sources of human β-glucocerebrosidase are available; one of them is derived from human placenta (aglucerase), and the other is synthesized by recombinant DNA technology (imiglucerase). The oligosaccharide chains of both enzymes are modified to expose terminal mannose residues. Macrophages, through their mannose receptors, internalize the modified enzyme (see Clinical Case Study 17.1).

Studies of cases of G_{M1}-gangliosidosis have revealed two distinct types. In **generalized gangliosidosis**, G_{M1} and desialo-G_{M1}-gangliosides accumulate in the brain and viscera. The three β-galactosidase activities isolated from normal human liver cells are absent. The disease begins at or near birth, progresses rapidly, and ends fatally, usually by 2 years of age. In **juvenile G_{M1}-gangliosidosis**, psychomotor abnormalities usually begin at about 1 year, and death ensues at 3−10 years. Two liver β-galactosidase activities are absent, possibly accounting for the lack of lipid accumulation in this organ. This enzymatic finding supports the genetic separation into two forms.

G_{M2}-**gangliosidosis** is of two types: Tay−Sachs disease, due to β-hexosaminidase A (Hex-A) deficiency, and Sandhoff's disease, due to deficiency of β-hexosaminidase A and B (Hex-A, Hex-B). The relationship between these diseases is based on the subunit composition of the two affected enzymes. Hex-A, a heteropolymer, consists of two α-chains (coded for on chromosome 15), a β-chain (coded for on chromosome 5), and an activator protein. Hex-B is a tetramer of β-chains. Mutations at the α-locus give rise to Tay−Sachs disease (see Clinical Case Study 17.2). A variant form can arise from mutation at the activator protein locus; however, it shows normal *in vitro* Hex-A activity with chromogenic substrates. Mutations at the β-locus yield **Sandhoff's disease** and affect Hex-A and Hex-B, both of which contain the β-subunit.

Treatment of sphingolipidoses is primarily symptomatic and supportive. For example, in patients with anemia due to Gaucher's disease, thrombocytopenia associated with hypersplenism is relieved by splenectomy. Infusion of appropriate purified human placental tissue enzymes in patients with Gaucher's disease and Fabry's disease reduces the accumulated glycolipids in the circulation and liver. Recent advances in the use of bacterial gene expression systems to express human genes have provided enough recombinant enzyme required for treatment. Use of the recipient's erythrocytes to deliver the enzymes is under investigation to minimize immunological complications. Exposing erythrocytes to hypotonic conditions in the presence of the enzyme causes formation of pores in the membrane that allow rapid exchange of the enzyme with the cellular contents. Restoration to isotonicity reseals the membrane and entraps some of the enzyme. Other enzyme carriers are liposomes, concentric lipid bilayers prepared from cholesterol, lecithin, and phosphatidic acid. The ideal treatment for these disorders would be addition, or replacement, of genetic material coding for the missing gene product. Replacement therapy with a polyethylene glycol-modified form of the missing enzyme, which has an extended half-life and reduced immunogenicity, may provide a promising approach to treatment.

PHOSPHOLIPIDS AND GLYCOSPHINGOLIPIDS IN CLINICAL MEDICINE

Pulmonary Surfactant Metabolism and Respiratory Distress Syndrome

Inhalation and exhalation by neonatal lungs require the lamellar bodies to reduce the surface tension of water. Lamellar bodies are made up of **pulmonary surfactant** at the lining of the air−membrane surface. Pulmonary surfactant is a complex of lipids and proteins with unique surface active properties that is synthesized exclusively in alveolar type II cells. The composition of surfactant is 90% lipids and 5%−10% surfactant-specific proteins. The lipid component is made up of dipalmitoylphosphatidylcholine (also called **lecithin**, 70%−80%) and another major phospholipid, phosphatidylglycerol (PG, 10%). The remaining lipid fraction is composed of phosphatidylinositol (PI), phosphatidylethanolamine (PE), and phosphatidylserine (PS). Immature surfactant contains higher amounts of PI compared to PG. Thus, a low ratio of PG to PI indicates lung immaturity. Cholesterol, a neutral lipid, is also a constituent of the lipid component of surfactant.

After synthesis in the various compartments of the endoplasmic reticulum of alveolar type II cells, surfactant components are assembled into lamellar bodies. The assembly and exocytosis of lamellar bodies are mediated by ATP-binding cassette (ABC) transporters (ABCA3). Mutations in the ABCA3 gene lead to surfactant deficiency with lethal consequences.

The lamellar bodies are transformed into an extracellular form of surfactant that has a quadratic lattice structure called **tubular myelin**. The three-dimensional tubulin−myelin structures spread into a monolayer at the air—liquid interface. This spreading decreases the surface tension, prevents alveolar collapse (atelectasis) at the end of the expiration cycle, and confers mechanical stability to the alveoli. The surfactant system is in a continuous state of flux. Surfactant is recycled by uptake by alveolar type II cells and is also cleared by macrophages (Figure 17.5).

Inadequate clearance of surfactants, due to defects in pulmonary macrophages, leads to accumulation of surfactants interfering with gas exchange. This disorder is known as **pulmonary alveolar proteinosis**.

The phospholipids are mainly synthesized starting from glycerol 3-phosphate, which is derived from glucose. The CDP—choline pathway is utilized in the synthesis of phosphatidylcholine or lecithin (Figure 17.2). The protein component of surfactant is lung-specific and consists of four proteins designated SP-A, SP-B, SP-C, and SP-D. These surfactant proteins perform important functions that lead to a reduction in alveolar surface tension during respiration. These include structural transformation of lamellar body to tubular myelin (SP-A and SP-B in the presence of Ca^{2+}), enhancement of surface-tension lowering properties, promotion of adsorption of surfactant phospholipids at the air—liquid interface (SP-B and SP-C), re-uptake by endocytosis of surfactant by type II cells, and the activation of alveolar macrophages to facilitate surfactant clearance. Both SP-A and SP-D possess antimicrobial properties. SP-A is chemotactic for macrophages and promotes bacterial phagocytosis.

SP-A is the most abundant, water-soluble glycoprotein. The primary structure of SP-A is highly conserved among several mammalian species. It has two domains: the N terminus is collagen-like with Gly—X—Y repeats (where Y is frequently a prolyl residue); and the C terminus has lectin-like properties. SP-B and SP-C are highly hydrophobic proteins. SP-D is a glycoprotein and has a

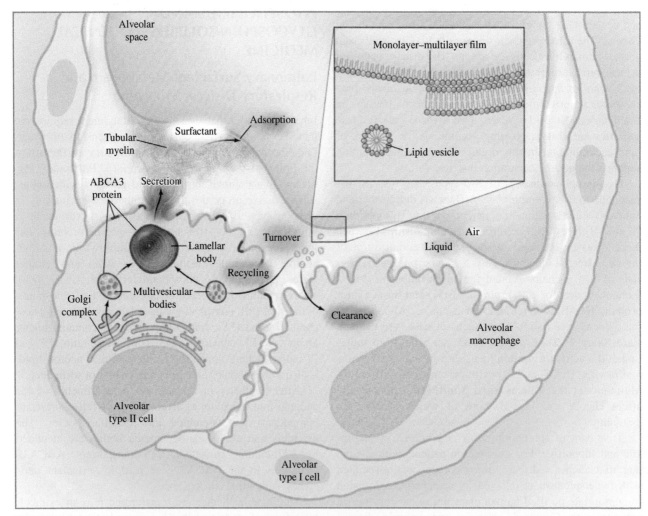

FIGURE 17.5 Diagrammatic representation of the formation and secretion of lamellar bodies from alveolar type II cells. The assembly and secretion are mediated by ATP-binding cassette transporter A3 (ABCA3). Surfactant is taken up by alveolar II cells for reuse and also cleared by macrophages. *[Reproduced with permission from Hallman, M. Lung surfactant, respiratory failure, and genes. N. Engl. J. Med. 350 (2004) 1279.]*

structure similar to that of SP-A. The importance of one surfactant protein is demonstrated in neonates who are born with an inherited deficiency of SP-B. Infants with an SP-B deficiency require ventilator support and extracorporeal membrane oxygenation. However, almost all die during the first year of life due to progressive respiratory failure. Lung transplantation is the only therapy by which the SP-B-deficient infants can be saved from death.

Surfactant biosynthesis is developmentally regulated. The capacity for the fetal lung to synthesize surfactant occurs relatively late in gestation. Although the type II cells are identifiable at 20–22 weeks of gestation, the secretion of surfactant into the amniotic fluid occurs during 30–32 weeks of gestation. Thus, for the infant, a consequence of prematurity is **respiratory distress syndrome** (**RDS**), which is a leading cause of neonatal morbidity and mortality in developed countries. The synthesis of surfactants is regulated by factors that include glucocorticoids, thyroid hormones, prolactin, estrogens, androgens, catecholamine (functioning through β-adrenergic receptors and cAMP), growth factors, and cytokines. Glucocorticoids stimulate lung maturation; thus, glucocorticoid therapy in women in preterm labor prior to 34 weeks of gestation can significantly decrease the incidence of RDS in premature neonates.

Thyroid hormones also accelerate fetal lung maturation. Fetal thyroid hormone levels may be increased by antenatal administration of thyrotropin-releasing hormone (TRH), a tripeptide that crosses the placental barrier, stimulates fetal pituitary production of thyroid-stimulating hormone (TSH), and increases fetal thyroid hormone production (Chapter 31). This indirect method of enhancement of fetal thyroid hormone production is utilized because thyroid hormones do not readily cross the placental barrier. Insulin delays surfactant synthesis, and so fetal hyperinsulinemia in diabetic mothers may increase the incidence of RDS even in the full-term infant. Androgen synthesized in the fetal testis is the probable cause of a slower onset of surfactant production in male fetuses. Administration of synthetic or natural pulmonary surfactants intratracheally to preterm infants improves oxygenation and decreases pulmonary morbidity. This result is achieved when administration is prophylactic or after the onset of RDS.

In adults, a severe form of lung injury can develop in association with sepsis, pneumonia, and injury to the lungs due to trauma or surgery. This catastrophic disorder is known as **acute respiratory distress syndrome** (**ARDS**), which has a mortality rate of more than 40%. A characteristic of ARDS is a massive influx of activated neurophils, which damage both vascular endothelium and alveolar epithelium, and result in massive pulmonary edema and impairment of surfactant function. Neutrophil proteinases (e.g., elastase) break down surfactant proteins. A potential therapeutic strategy in ARDS involves administration of both surfactant and antiproteinases (e.g., recombinant α_1-antitrypsin).

Biochemical Determinants of Fetal Lung Maturity

The need for surfactant production does not become essential until birth because no air–liquid interface exists in the alveoli *in utero*, and fetal oxygen needs are met by maternal circulation. The pulmonary system, including surfactant production, is among the last of the fetal organ systems to attain functional maturity. Since preterm birth is associated with significant neonatal morbidity and mortality due to inadequate oxygen supply to an immature pulmonary lung system, the assessment of antenatal fetal lung maturity is necessary to develop a therapeutic strategy in the management of a preterm infant. The biochemical determinants are measured primarily in the amniotic fluid obtained by amniocentesis.

In a normal pregnancy, the lung is adequately developed by about the 36th or 37th week. Biochemical changes occurring during this period of gestation can be used to evaluate fetal lung maturity when early delivery is planned. Evaluation procedures include the foam stability test, measurement of the lecithin to sphingomyelin ratio (L/S) and phosphatidylglycerol (PG) level, and lamellar body count.

A number of factors (such as hypoxia and acidosis) depress phospholipid synthesis, and administration of glucocorticoids to mothers accelerates the rate of fetal lung maturation. The fetal lung undergoes an abrupt transition from a P_{O_2} of about 20 mmHg to a P_{O_2} of 100 mmHg. This change from a hypoxic to a relatively hyperoxic condition may lead to increased production of potentially cytotoxic O_2 metabolites such as superoxide radical ($O_2^{\bullet}$), hydrogen peroxide (H_2O_2), hydroxyl radical ($^{\bullet}OH$), singlet oxygen ($^1O_2$), and peroxide radical ($ROO^{\bullet}$). These are called **reactive oxygen species** (**ROS**). The antioxidant enzyme system consists of superoxide dismutase, glutathione peroxidase, and catalase. In addition to these enzymes, other potential antioxidants are vitamin E, ascorbate, β-carotene, and thiol compounds (e.g., glutathione, cysteine). Infants born prematurely are particularly susceptible to deficiency of both surfactant and antioxidant defense. Administration of surfactant and the antioxidant enzymes using liposome technology has potential application in the management of ROS. Administration of surfactant to the lungs of very premature infants through an endotracheal tube has reduced morbidity and mortality from RDS.

CHOLESTEROL

Cholesterol (3-hydroxy-5,6-cholestene) is a steroid and contains the carbon skeleton of cyclopentanoperhydrophenanthrene, which consists of three six-membered rings and a five-membered ring. It is also a monohy-droxyalcohol and contains a double bond between C_5 and C_6:

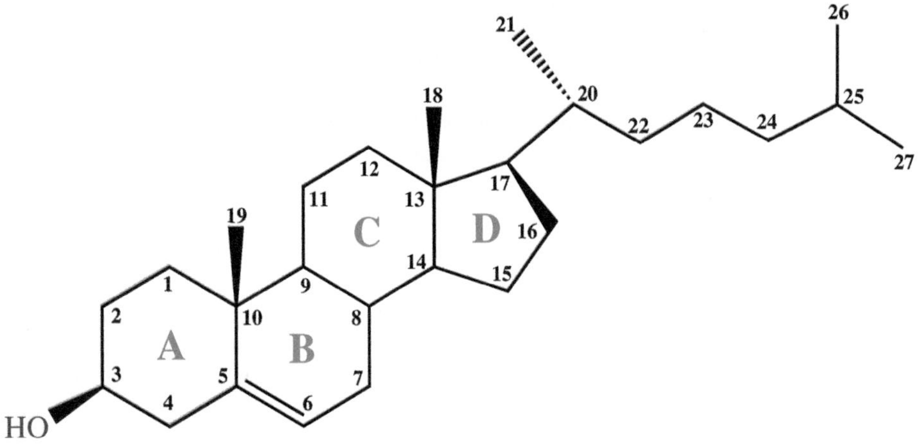

groups by a post-translational lipid modification process are required for membrane association and function of proteins such as p21[ras] and G-protein subunits.

Cholesterol

Adults normally synthesize approximately 1 g of cholesterol and consume about 0.3 g per day. Dietary cholesterol is primarily derived from foods of animal origin such as eggs and meat. Plants, yeasts, and fungi contain sterols that are structurally similar to cholesterol (sitosterols and ergosterols) but are poorly absorbed by the human intestinal tract. A rare inherited autosomal sterol storage disorder (**sitosterolemia**) is caused by defects in the ATP-binding cassette family of transporters that mediate cholesterol efflux. Treatment consists of diets low in plant sterol content with added cholestyramine to enhance sterol excretion (Chapter 18). In intestinal mucosal cells, most of the absorbed cholesterol is esterified with fatty acids and incorporated into chylomicrons that enter the blood through the lymph. After chylomicrons unload most of their triacylglycerol content at the peripheral tissues, chylomicron remnants are rapidly taken up by the liver (Chapter 18). The routing of nearly all of the cholesterol derived from dietary sources to the liver facilitates steroid homeostasis in the organism, since the liver is the principal site of cholesterol production. Although the intestinal tract, adrenal cortex, testes, skin, and other tissues can also synthesize cholesterol, their contribution is minor.

Cholesterol biosynthesis begins with isoprenoids in a multistep pathway. The end product, cholesterol, and the intermediates of the pathway participate in diverse cellular functions. The isoprenoid units give rise to dolichol, CoQ, heme A, isopentenyl-tRNA, farnesylated proteins, and vitamin D (in the presence of sunlight and 7-dehydrocholesterol). Dolichol is used in the synthesis of glycoproteins, CoQ is used in the mitochondrial electron transport chain, and attachments of farnesyl and geranyl-geranyl

Cholesterol has several functions including involvement in membrane structure, modulation of membrane fluidity and permeability, steroid hormone and bile acid synthesis (where it serves as a precursor), the covalent modification of proteins, and formation of the central nervous system in embryonic development. This last role of cholesterol was discovered through mutations and pharmacological agents that block cholesterol biosynthesis.

Cholesterol biosynthesis can be divided into six stages:

1. Conversion of acetyl-CoA to 3-hydroxy-3-methylglutaryl coenzyme-A (HMG-CoA);
2. Conversion of HMG-CoA to mevalonate, the rate-limiting step in cholesterol biosynthesis;
3. Conversion of mevalonate to isoprenyl pyrophosphates with loss of CO_2;
4. Conversion of isoprenyl pyrophosphates to squalene;
5. Conversion of squalene to lanosterol; and
6. Conversion of lanosterol to cholesterol.

The biosynthetic reactions involve a series of condensation processes and are distributed between the cytosol and microsomes. All of the carbons of cholesterol are derived from acetyl-CoA: 15 from the "methyl" and 12 from the "carboxyl" carbon atoms. Acetyl-CoA is derived either from mitochondrial oxidation of metabolic fuels (e.g., fatty acids), in which case it is transported to the cytosol as citrate (Chapter 16), or from activation of acetate (e.g., derived from ethanol oxidation) by cytosolic acetyl-CoA synthase (Chapter 16). All of the reducing equivalents are provided by NADPH.

CONVERSION OF ACETYL-CoA TO HMG-CoA

In the cytosol, three molecules of acetyl-CoA are condensed to HMG-CoA through successive action of thiolase and HMG-CoA synthase, respectively (Figure 17.6). HMG-CoA synthase is under transcriptional regulation by the sterol end products.

HMG-CoA is also synthesized in mitochondria by the same sequence of reactions, but yields the ketone bodies acetoacetate, D(−)-β-hydroxybutyrate, and acetone (Figure 17.7). Mitochondrial HMG-CoA also arises from oxidation of leucine (Chapter 15), which is ketogenic. Although HMG-CoA derived from leucine is not utilized in mevalonate synthesis, the carbon in leucine can be incorporated into cholesterol by way of acetyl-CoA. Thus, two distinct pools of HMG-CoA exist: one mitochondrial and involved in the formation of ketone bodies, and the other extramitochondrial and involved in synthesis of isoprenoid units.

Conversion of HMG-CoA to Mevalonate

Conversion of HMG-CoA to mevalonate, a two-step reduction reaction, is the rate-limiting step in cholesterogenesis. Cytosolic HMG-CoA is reduced by NADPH to

mevalonate by HMG-CoA reductase through the production of an enzyme-bound aldehyde intermediate:

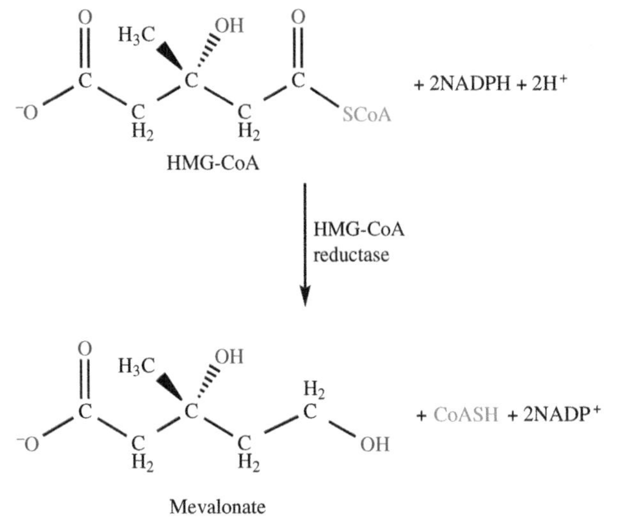

HMG-CoA reductase is an integral protein of the endoplasmic reticulum. Its C-terminal segment contains the catalytic site, which is located in the cytosol. HMG-CoA reductase is the primary target for regulation of nonsterol isoprenoid derivatives and cholesterol. HMG-CoA reductase is regulated both by short-term and long-term biochemical processes. Long-term regulation is dependent on the rate of gene transcription. Short-term regulation occurs during low cellular energy levels and consists of phosphorylation−dephosphorylation of HMG-CoA reductase and also ER-mediated proteolysis. Phosphorylation decreases its activity, whereas dephosphorylation increases it (Figure 17.8). During energy deprivation with high AMP/ATP ratios, AMP allosterically activates AMP-kinase, which, in turn, inhibits HMG-CoA reductase by phosphorylation. AMP-kinase, in addition to AMP, is also regulated by upstream protein kinases. Inactive phosphorylated HMG-CoA reductase is converted to an active state by dephosphorylation catalyzed by phosphoprotein phosphatase. Activated AMP-kinase also inhibits acetyl-CoA carboxylase, the rate-limiting enzyme of fatty acid synthesis, which is also an energy-consuming process (Chapter 16).

Cyclic AMP levels also regulate HMG-CoA reductase. Regulation by cAMP arises by way of activation of protein kinase A, which, in turn, converts phosphoprotein phosphatase inhibitor-1 (PPI-1) to its active form by phosphorylation. Activated PPI-1 inhibits phosphoprotein phosphatase and thus maintains HMG-CoA reductase in the inactive state (Figure 17.8). Elevation of plasma glucagon level (e.g., during fasting) activates cAMP production and reduces cholesterol biosynthesis. Short-term regulation of HMG-CoA reductase also occurs by degradation. During cholesterol excess, it is degraded by endoplasmic reticulum-associated proteolysis.

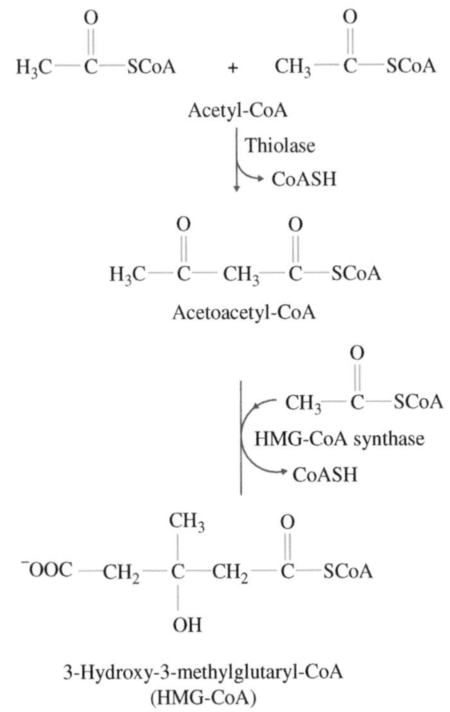

FIGURE 17.6 Biosynthesis of HMG-CoA.

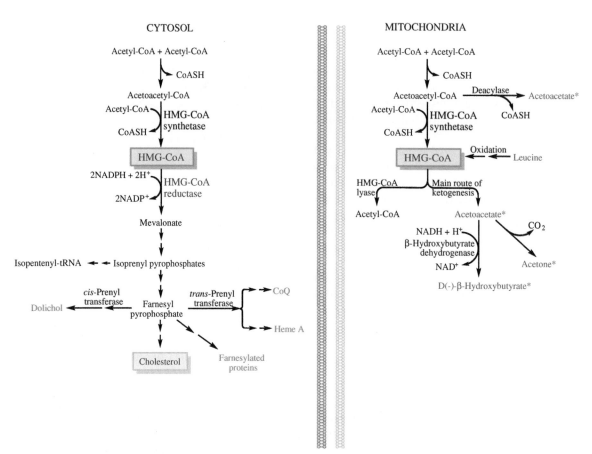

FIGURE 17.7 Mitochondrial and cytosolic biosynthesis and utilization of HMG-CoA in the liver. The molecules indicated by an asterisk are the ketone bodies. Acetoacetate and β-hydroxybutyrate (after conversion to acetoacetate) are metabolized in extrahepatic tissues. Acetone is excreted in the lungs. Note the cytosolic multifunctional isoprenoid pathway for cholesterol biosynthesis. The double arrow indicates a multistep pathway.

The synthesis of HMG-CoA reductase is regulated by transcription of the reductase gene. The regulation involves a complex of three proteins that are bound to the endoplasmic reticulum. The protein complex consists of sterol responsive element-binding protein (SREBP), SREBP cleavage-activating protein (SCAP), and a sterol-sensing protein known as INSIG1, which is an insulin-induced gene product. During low intracellular cholesterol levels, SREBP-SCAP dissociates from INSIG1 and migrates to the Golgi apparatus. In the Golgi, SREBP is converted to an active transcription factor (TF) by proteolysis. TF is translocated to the nucleus, where it binds to sterol regulatory element (SRE) and the complex initiates transcription of the HMG-CoA reductase gene. TF also coordinately promotes the transcription of genes required for low-density lipoprotein (LDL) receptors, which internalize LDL to provide intracellular cholesterol and NADPH enzymes used in *de novo* cholesterol biosynthesis. When there are excess intracellular cholesterol levels, SREBP proteolysis ceases and the transcription of genes involved in cholesterol biosynthesis is curtailed.

HMG-CoA reductase activity is also inhibited by oxygenated sterols (e.g., 27-hydroxycholesterol) but not in enucleated cells, indicating involvement of the nucleus. The oxygenated sterols are synthesized in mitochondria and may repress the HMG-CoA reductase gene or activate genes for enzymes that degrade the reductase. A rare familial sterol storage disease, **cerebrotendinous xanthomatosis**, is characterized by accumulation of cholesterol (and its reduced product cholestanol) in every tissue, especially in the brain, tendons, and aorta. This causes progressive neurological dysfunction, tendon xanthomas, premature atherosclerosis, and myocardial infarction.

Patients can develop cholesterol gallstones from a defect in bile acid synthesis. The defect is in mitochondrial C_{27}-steroid 27-hydroxylase. In these patients, the reduced formation of normal bile acids, particularly chenodeoxycholic acid, leads to the upregulation of the rate-limiting enzyme 7-α-hydroxylase of the bile acid synthetic pathway (discussed later). This leads to accumulation of 7-α-hydroxylated bile acid intermediates that are not normally utilized.

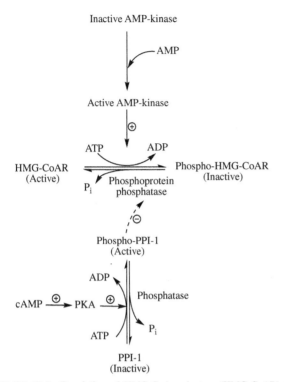

FIGURE 17.8 Regulation of HMG-CoA reductase (HMG-CoAR) by phosphorylation–dephosphorylation. Active AMP-kinase converts HMG-CoAR to an inactive state by phosphorylation. cAMP activation of protein kinase A (PKA) leads to phosphorylation of phosphoprotein phosphatase inhibitor-1 (PPI-1), converting it to its active state. Active PPI-1 inhibits phosphoprotein phosphatase and thus maintains HMG-CoA in the inactive state. $\oplus$ = stimulation; $\ominus$ = inhibition.

The overall scheme for regulation of HMG-CoA reductase by oxysterol may be constructed as follows. Cholesterol delivered to the cells via low-density lipoprotein (Chapter 18) is converted to oxygenated sterol derivatives in the mitochondria, followed by their release into the cytoplasm. Oxygenated sterols are then translocated to the nucleus by binding to oxysterol-binding protein. Once in the nucleus, the oxygenated sterols effect the appropriate gene modification, which eventually leads to decreased HMG-CoA reductase activity.

HMG-CoA reductase is inhibited competitively by structural analogues. These compounds are commonly known as **statins** and are used pharmacologically in cholesterol reduction, which can reduce the risk for **coronary artery disease** and stroke (Chapter 18). Statins inhibit HMG-CoA reductase at a much lower concentration (1 μM) compared to the K_m for HMG-CoA (10 μM). The structures of clinically effective statins are shown in Figure 17.9. Naturally occurring statins are found in a dietary supplement known as **cholestin**, which is obtained from rice fermented in red yeast. In China, red yeast rice has been used as a coloring and flavoring agent. Cholestin's safety and effectiveness as a hypocholesterolemic agent awaits long-term clinical studies.

Despite inhibition of HMG-CoA reductase by statins, cells compensate by increasing enzyme expression several fold. However, the total body cholesterol is reduced by 20%–40% due to increased expression of LDL receptors after statin administration. This enhances LDL (the major cholesterol-carrying lipoprotein) clearance from serum with a net reduction of serum cholesterol (Chapter 18). Individuals who lack functional LDL receptors (homozygous familial hypercholesterolemia; Chapter 18) do not benefit from statin therapy. However, statin therapy is useful in the treatment of heterozygous familial hypercholesterolemia. Since HMG-CoA reductase plays a pivotal role in the synthesis of many products vital for cellular metabolism, inhibitors of the enzyme may have toxic effects. Monitoring of liver and muscle function may be necessary to detect any toxicity of statin drug therapy. A decreased risk of bone fractures with statin therapy has been observed in subjects 50 years of age or older who are being treated for hypercholesterolemia. The mechanism of action of statins in bone metabolism may involve inhibition of prenylation of signaling proteins found on osteoclast cell membrane (Chapter 35).

Independent of the hypocholesterolemic effect, statins have beneficial anti-inflammatory properties, presumably linked to their inhibition of isoprenoid biosynthesis.

Conversion of Mevalonate to Isoprenyl Pyrophosphate

Isoprenyl pyrophosphates are synthesized by successive phosphorylation of mevalonate with ATP to yield the 5-monophosphate, 5-pyrophosphate, and 5-pyrophospho-3-monophospho derivatives. This last compound is very unstable and loses the 3-phosphate and the carboxyl group to yield isopentenyl pyrophosphate (IPPP), which is isomerized to 3,3-dimethylallyl pyrophosphate (DMAPP). These reactions, catalyzed by cytosolic enzymes, are shown in Figure 17.10.

Patients with severe forms of inherited mevalonate kinase deficiency exhibit mevalonic aciduria, failure to thrive, developmental delay, anemia, hepatosplenomegaly, gastroenteropathy, and dysmorphic features during neonatal development. These clinical manifestations underscore the importance of the formation of isoprenyl pyrophosphates not only for sterol synthesis, but also for the nonsterol isoprene compounds dolichol, CoQ, heme A, isopentenyl-tRNA, and farnesylated proteins.

Condensation of Isoprenyl Pyrophosphate to Form Squalene

IPPP, a nucleophile (by virtue of its terminal vinyl group), and DMAPP, an electrophile, undergo condensation with

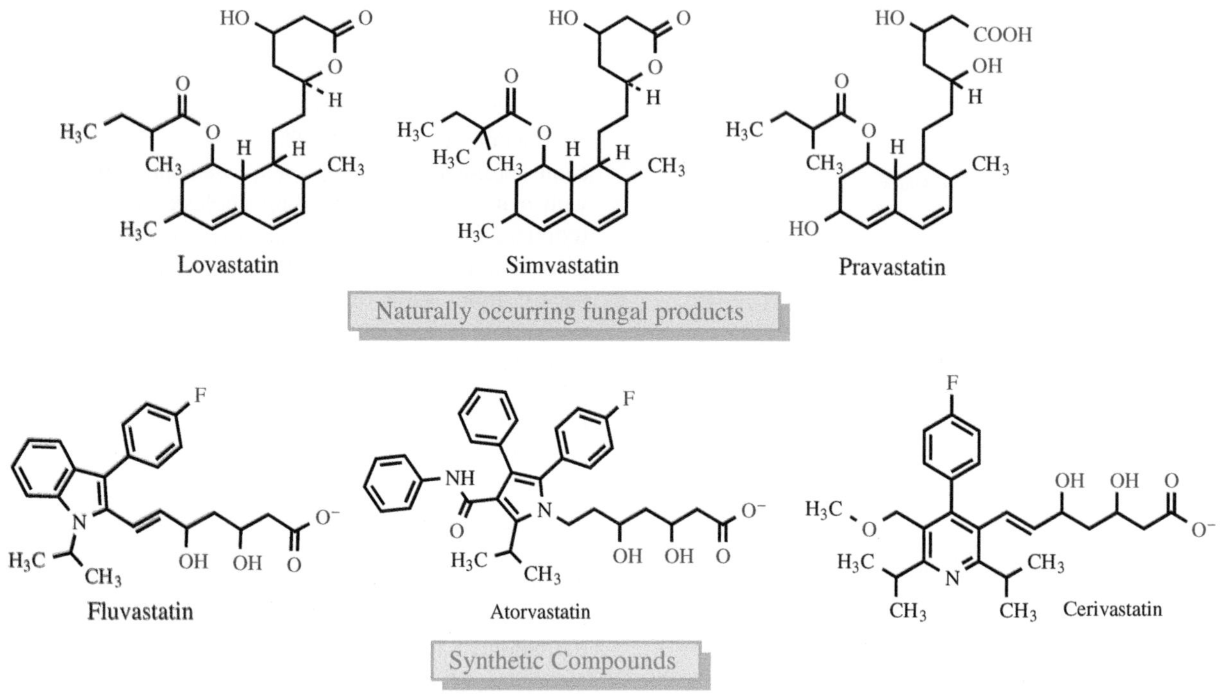

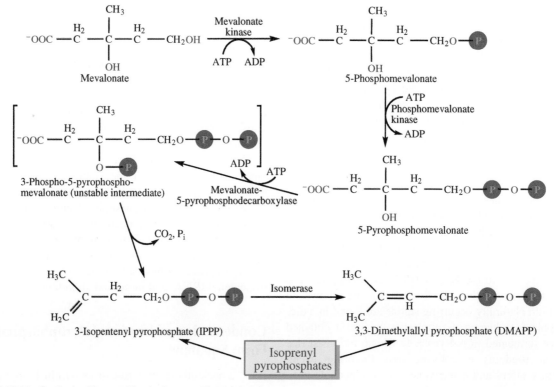

FIGURE 17.9 Structures of HMG-CoA reductase inhibitors (statins). (a) Naturally occurring fungal products (lovastatin, simvastatin, and pravastatin), (B) synthetic compounds (fluvastatin, atorvastatin, and cerivastatin).

FIGURE 17.10 Synthesis of isoprenoid units from mevalonate.

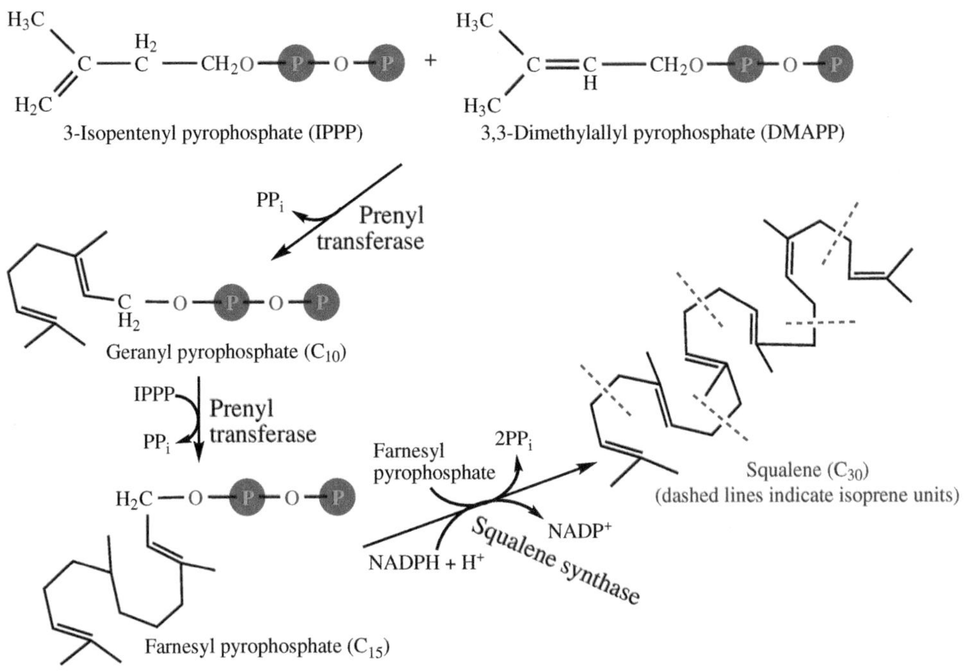

FIGURE 17.11 Synthesis of squalene from isomeric pentenyl pyrophosphates.

elimination of pyrophosphate to yield geranyl pyrophosphate (an electrophile), which condenses with a molecule of IPPP to yield a farnesyl pyrophosphate and pyrophosphate. These reactions are probably catalyzed by the same cytosolic enzyme complex. Two molecules of farnesyl pyrophosphate then condense head-to-head to form squalene by the action of microsomal squalene synthase (Figure 17.11).

The farnesyl pyrophosphate generated in this pathway is also used in the farnesylation of proteins. The farnesyl group is attached to a protein via a thioether linkage involving a cysteine residue found in the C terminus. Several proteins that are modified by farnesyl groups have been identified, e.g., growth-regulating *ras* proteins (Chapter 24) and nuclear envelope proteins. Proteins attached to a geranyl-geranyl group (a 20-C isoprene unit) have also been identified. The modification of proteins by these lipid moieties increases their hydrophobicity and may be required for these proteins to interact with other hydrophobic proteins and for proper anchoring in the cell membrane. The importance of farnesylation of proteins is exemplified by the fact that inhibition of mevalonate synthesis results in a blockage of cell growth.

Conversion of Squalene to Lanosterol

This step comprises cyclization of squalene to lanosterol (the first sterol to be formed). The cyclization begins with conversion of squalene to squalene-2,3-epoxide by a microsomal mixed function oxidase that requires O_2, NADPH, and FAD (Figure 17.12). Cyclization of squalene-2,3-epoxide to lanosterol occurs through a series of

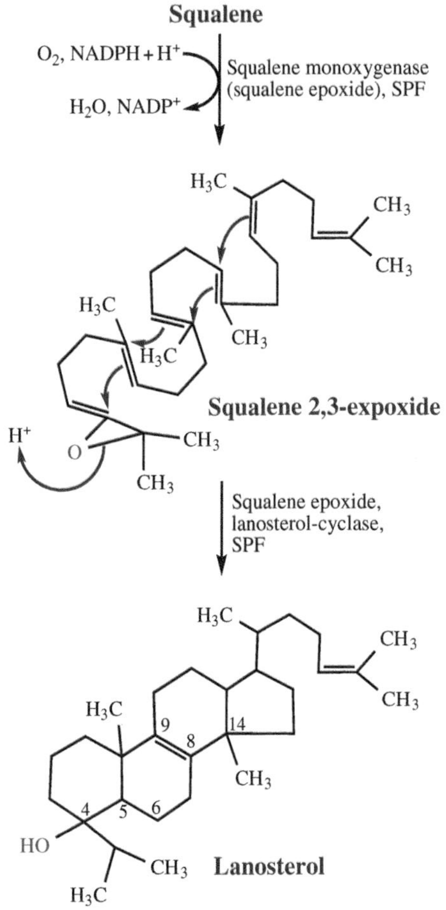

FIGURE 17.12 Cyclization of squalene to lanosterol. Supernatant protein factor (SPF), a cytosolic protein, promotes both stages of the cyclization.

concerted 1,2-methyl group and hydride shifts along the squalene chain. In both stages, the reactants are bound to supernatant protein factor (SPF), a cytosolic carrier that promotes conversion of squalene to lanosterol.

Conversion of Lanosterol to Cholesterol

Transformation of lanosterol to cholesterol is a complex, multistep process catalyzed by enzymes of the endoplasmic reticulum (microsomes). A cytosolic sterol carrier protein is also required and presumably functions as a carrier of steroid intermediates from one catalytic site to the next but may also affect activity of the enzymes. The reactions consist of removal of the three methyl groups attached to C_4 and C_{14}, migration of the double bond from the 8,9- to the 5,6-position, and saturation of the double bond in the side chain. Conversion of lanosterol to cholesterol occurs principally via 7-dehydrocholesterol and to a minor extent via desmosterol.

The importance of cholesterol biosynthesis in embryonic development and formation of the central nervous system is reflected in patients with disorders in the pathway for the conversion of lanosterol to cholesterol. Three enzyme deficiencies have been identified, those of:

1. 3β-Hydroxysteroid-Δ^{24}-reductase (also known as sterol-Δ^{24}-reductase);
2. 3β-Hydroxysteroid-Δ^8,Δ^7-isomerase (commonly known as sterol-Δ^8-isomerase);
3. 3β-Hydroxysteroid-Δ^7-reductase (also known as 7-dehydrocholesterol reductase).

Sterol-Δ^8-isomerase deficiency, known as **Conradi−Hünermann syndrome** (CDPX$_2$), is an X-linked dominant disorder. Clinical manifestations of this disorder include skeletal abnormalities, chondrodysplasia punctata, craniofacial anomalies, cataracts, and skin abnormalities. The 7-dehydrocholesterol reductase deficiency, known as **Smith−Lemli−Opitz syndrome (SLO)** is an autosomal recessive disorder occurring in about 1 in 20,000 births. Clinical manifestations of affected individuals include craniofacial abnormalities, microcephaly, congenital heart disease, malformation of the limbs, psychomotor retardation, cerebral maldevelopment, and urogenital anomalies. Measurement of 7-dehydrocholesterol in amniotic fluid during the second trimester or in neonatal blood specimens has been useful in the identification of the disorder. The sterol-Δ^{24}-reductase deficiency causes a developmental phenotype similar to SLO syndrome and is associated with accumulation of desmosterol. The inability of de novo fetal synthesis of cholesterol combined with its inadequate transport from the mother to the fetus appears to be involved in the multiple abnormalities of morphogenesis. SLO infants treated with rich sources of dietary cholesterol after birth have

shown fewer growth abnormalities. However, it is not known whether long-term dietary cholesterol supplementation can improve cognitive development, particularly since cholesterol is not transported across the blood−brain barrier.

An appreciation of the relationship between cellular cholesterol metabolism and a family of signaling molecules that participate in embryonic development is emerging. These signaling molecules are known as **hedgehog proteins**, which were initially identified in *Drosophila*. The vertebrate counterparts of hedgehog proteins participate in embryonic development, including the formation of the neural tube and its derivatives, the axial skeleton, and the appendages. The hedgehog protein is a self-splicing protein that undergoes an autocatalytic proteolytic processing, giving rise to an N-terminal and a C-terminal product. In *Drosophila*, hedgehog protein cleavage occurs between Gly-257 and Cys-258. Cholesterol is covalently attached to the carboxy terminal end of the N-terminal cleavage product. Both the autocatalytic proteolysis and intramolecular cholesterol transferase activities are located in the C-terminal portion of the hedgehog protein. The covalent modification of the N-terminal segment of the hedgehog protein is necessary for proper localization on the cell membrane at target sites to initiate downstream events (e.g., transcription of target genes). Thus, perturbations of cholesterol biosynthesis due to mutations or pharmacological agents can lead to defects in embryonic development.

Utilization of Cholesterol

Cholesterol is utilized in the formation of membranes (Chapter 9), steroid hormones (Chapters 28, 30, and 32), and bile acids. 7-Dehydrocholesterol is required for production of vitamin D (Chapter 35). Under steady-state conditions, the cholesterol content of the body is kept relatively constant by balancing synthesis and dietary intake with utilization. The major sink for cholesterol is formation of bile acids, of which about 0.8−1 g/day are produced in the liver and lost in the feces. However, secretion of bile acids by the liver is many times greater (15−20 g/day) than the rate of synthesis because of their enterohepatic circulation (Chapter 11). Cholesterol is also secreted into bile, and some is lost in feces as cholesterol and as coprostanol, a bacterial reduction product (about 0.4−0.5 g/day). Conversion of cholesterol to steroid hormones and of 7-dehydrocholesterol to vitamin D, and elimination of their inactive metabolites are of minor significance in the disposal of cholesterol, amounting to approximately 50 mg/day. A small amount of cholesterol (about 80 mg/day) is also lost through shedding the outer layers of the skin.

Bile Acids

Bile acids are 24-carbon steroid compounds. Primary bile acids (cholic and chenodeoxycholic) are synthesized in the liver from cholesterol (Figure 17.13). In human bile, about 45% is chenodeoxycholic acid, 31% cholic acid, and 24% deoxycholic acid (a secondary bile acid formed in the intestine). Early studies in rodents showed the preferred substrate to be newly synthesized cholesterol. However, whole-body turnover studies in humans using radioactive markers indicate that approximately two-thirds of bile acid is derived from HDL cholesterol delivered to the liver. Formation is initiated by 7-α-hydroxylation, the committed and rate-limiting step catalyzed by microsomal 7-α-hydroxylase. The reaction requires NADPH, O_2, cytochrome P-450, and NADPH:cytochrome P-450 reductase. Reactions that follow are oxidation of the 3-β-hydroxyl group to a 3-keto group, isomerization of the Δ^5 double bond to the Δ^4-position, conversion of the 3-keto group to a 3-α-hydroxyl group, reduction of the Δ^4 double bond, 12-α-hydroxylation in the case of cholic acid synthesis, and oxidation of the side chain. 12-α-Hydroxylase, like 7-α-hydroxylase, is associated with microsomes and requires NADPH, molecular oxygen, and cytochrome P-450. Unlike 7-α-hydroxylase, its activity does not exhibit diurnal variation. Its activity determines the amount of cholic acid synthesized. Side chain oxidation starts with 27-hydroxylation and is followed by oxidative steps similar to those of β-oxidation of fatty acids

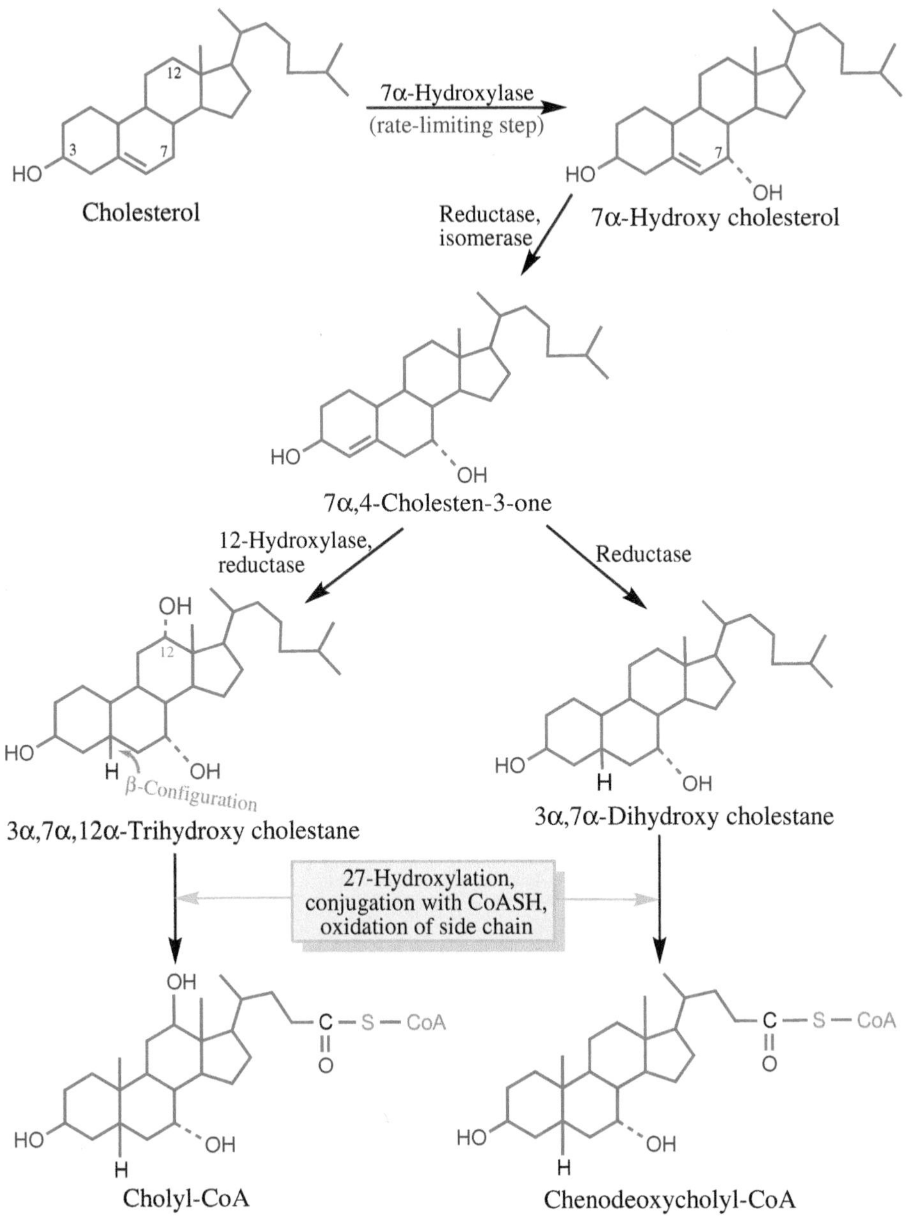

FIGURE 17.13 Formation of bile acids.

(Chapter 16). The 27-hydroxylation catalyzed by a mixed-function hydroxylase probably occurs in mitochondria and requires NADPH, O_2, and cytochrome P-450. Bile acid deficiency in cerebrotendinous xanthomatosis (see above) is due to a deficiency of 27-hydroxylase (CYP27A1). Since the substrates for bile acid formation are water insoluble, they require sterol carrier proteins for synthesis and metabolism.

Bile acids are conjugated with glycine or taurine (Figure 17.14) before being secreted into bile, where the ratio of glycine- to taurine-conjugated acids is about 3:1. Sulfate esters of bile acids are also formed to a small extent. At the alkaline pH of bile and in the presence of alkaline cations (Na^+, K^+), the acids and their conjugates are present as salts (ionized forms), although the terms **bile acids** and **bile salts** are used interchangeably. Export of bile salts in the bile from hepatocytes into the canalicular space is mediated by the bile salt-export pump (BSEP). The formation and excretion of bile is an exocrine function of hepatocytes, which is unique to the liver. BSEP is a canalicular ATP-binding cassette transporter (ABCB11) and uses ATP as an energy source. In children, ABCB11 gene mutations cause progressive intrahepatic cholestasis and end-stage liver disease, requiring orthotopic liver transplantation.

Regulation of Bile Acid Synthesis

Regulation of bile acid formation from cholesterol occurs at the 7-α-hydroxylation step and is mediated by the concentration of bile acids in the enterohepatic circulation. 7-α-Hydroxylase is modulated by a phosphorylation−dephosphorylation cascade similar to that of HMG-CoA reductase, except that the phosphorylated form of 7-α-hydroxylase is more active.

As noted earlier, the major rate-limiting step of cholesterol biosynthesis is synthesis of HMG-CoA. 7-α-Hydroxycholesterol, the first intermediate of bile acid formation, inhibits HMG-CoA reductase. The activities of 7-α-cholesterol hydroxylase and HMG-CoA reductase undergo parallel changes under the influence of bile acid levels. In the rat, they show similar patterns of diurnal variation, with highest activities during the dark period. Bile acids and intermediates do not appear to function as allosteric modifiers. In the intestines, bile acids may regulate cholesterol biosynthesis, in addition to their role in cholesterol absorption. In humans, the presence of excess cholesterol does not increase bile acid production proportionally, although it suppresses endogenous cholesterol synthesis and increases excretion of fecal neutral steroids.

Hepatic bile acid synthesis amounts to about 0.8−1.0 g/day. When loss of bile occurs owing to drainage through a biliary fistula, administration of bile acid-complexing resins (e.g., cholestyramine), or ileal exclusion, the activity of 7-α-hydroxylase increases several fold, with a consequent increase in bile acid formation. The latter two methods are used to reduce cholesterol levels in hyper-cholesterolemic patients (Chapter 18).

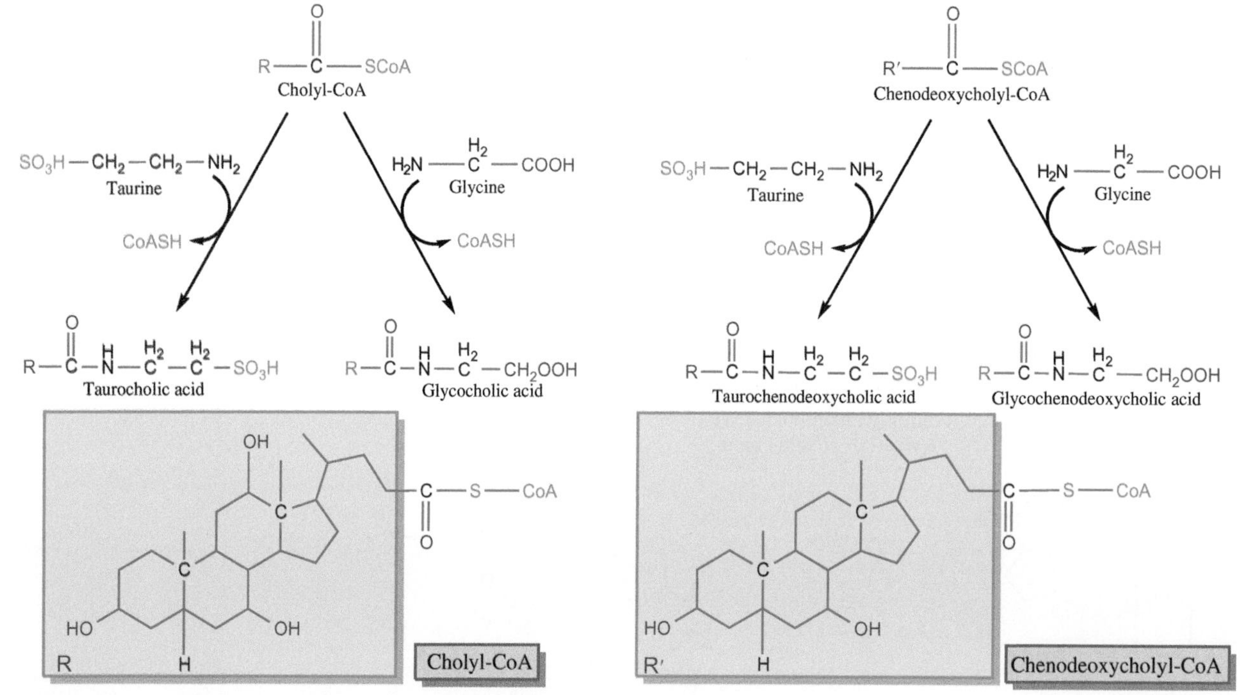

FIGURE 17.14 Conjugation of bile acids with taurine and glycine.

Disposition of Bile Acids in the Intestine and Their Enterohepatic Circulation

Bile is stored and concentrated in the gallbladder, a saccular, elongated, pear-shaped organ attached to the hepatic duct. Bile contains bile acids, bile pigments (i.e., bilirubin glucuronides; see Chapter 27), cholesterol, and lecithin. The pH of gallbladder bile is 6.9−7.7. Cholesterol is solubilized in bile by the formation of micelles with bile acids and lecithin. Cholesterol gallstones can form as a result of excessive secretion of cholesterol or of insufficient amounts of bile acids and lecithin relative to cholesterol in bile. Inadequate amounts of bile acids result from decreased hepatic synthesis, decreased uptake from the portal blood by hepatocytes, or increased loss from the gastrointestinal tract.

With ingestion of food, cholecystokinin (Chapter 11) is released into the blood and causes contraction of the gallbladder, whose contents are then rapidly emptied into the duodenum by way of the common bile duct. In the duodenal wall, the bile duct fuses with the pancreatic duct at the ampulla of Vater. Bile functions include absorption of lipids and the lipid-soluble vitamins A, D, E, and K by the emulsifying action of bile salts (Chapter 11), neutralization of acid chyme, and excretion of toxic metabolites (e.g., bile pigments, some drugs and toxins) in the feces.

The secondary bile acids, deoxycholic and lithocholic acids, are derived by 7-dehydroxylation from the deconjugated primary bile acids, cholic acid and chenodeoxycholic acid, respectively (Figure 17.15), through action of bacterial enzymes primarily in the large intestine. Minor quantities of the secondary bile acid, ursodeoxycholic (ursodiol) acid, are also produced in the intestine. It is structurally identical to chenodeoxycholic acid (chenodiol) except for orientation of the 7-hydroxyl group. In ursodeoxycholic acid, the orientation is β, whereas in chenodeoxycholic acid, the orientation is α (Figure 17.13). Synthetic ursodiol is used in the therapy of cholelithiasis (discussed later) and primary biliary cirrhosis. The main portion (>90%) of bile acids in the intestines is reabsorbed by an active transport system into the portal circulation at the distal ileum and transported bound to albumin. They are taken up by the liver, promptly reconjugated with taurine and glycine, and resecreted into bile. Both ileal absorption and hepatic uptake of bile acids may be mediated by Na^+-dependent (carrier) transport mechanisms. This cyclic transport of bile acids from intestine to liver and back to the intestine is known as **enterohepatic circulation** (Figure 17.15). About 90% of the bile acids are extracted during a single passage of portal blood through the liver. The bile acid pool in the enterohepatic circulation is 2−4 g and circulates about twice during the digestion of each meal. The amount of bile acids lost in feces is about 0.8−1.0 g/day and consists mostly of secondary bile acids (particularly lithocholic acid, the least soluble of the bile acids). The loss is made up by synthesis of an equal amount in the liver.

Bile Acid Metabolism and Clinical Medicine

In liver disorders, serum levels of bile acids are elevated, and their measurement is a sensitive indicator of liver disease. Bile acids are not normally found in urine owing to efficient uptake by the liver and excretion into the intestines. In hepatocellular disease and obstructive jaundice, however, their urinary excretion increases.

Lithocholic acid is toxic and can cause hemolysis and fever. Effects associated with hyperbile acidemia include pruritus, steatorrhea, hemolytic anemia, and further liver injury.

The main cause of **cholelithiasis** (presence or formation of gallstones) is precipitation of cholesterol in bile. Elevated biliary concentrations of bile pigments (bilirubin glucunorides) can also lead to formation of concretions known as **pigment stones** (Chapter 27). Since biliary cholesterol is solubilized by bile acids and lecithin, an excess of cholesterol along with decreased amounts of bile acids and lecithin causes bile to become supersaturated with cholesterol with the risk of forming cholesterol stones. The limits of solubility of cholesterol in the presence of bile salts and lecithin have been established by using a ternary phase diagram. If the mole ratio of bile salts and phospholipids to cholesterol is less than 10:1, the bile is considered to be lithogenic (stone forming), but this ratio is not absolute. The pattern of food intake in Western nations consists of excess fat, cholesterol, and an interval of 12−14 hours between the evening meal and breakfast. This leads to a fasting gallbladder bile that is saturated or supersaturated with cholesterol. In individuals who have gallstones, the problems associated with production of lithogenic bile appear to reside in the liver and not in the gallbladder. When the activities of HMG-CoA reductase and 7-α-hydroxylase were measured in the liver of patients who had gallstones and compared with those of controls, cholesterogenesis was found to be increased in the patients with gallstones, and bile acid synthesis was reduced. Under these conditions, the ratio of cholesterol to bile acids secreted by the liver is increased. The gallbladder provides an environment where concentration of the lithogenic bile triggers formation of gallstones. A genetic predisposition to formation of cholesterol gallstones is common in certain groups (e.g., 70% of Native American women 30 years of age or older have cholesterol gallstones). In general, the incidence of gallstones in women is three times higher than in men, and this proportion increases with age. Obesity and possibly multiparity are also associated with formation of gallstones. Any disorder of the ileum (e.g., **Crohn's disease**) or its resection

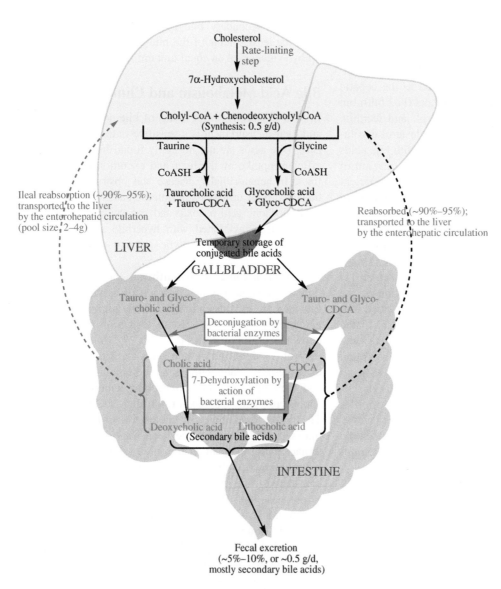

FIGURE **17.15** Formation, enterohepatic circulation, and disposition of the bile acids. CDCA = chenodeoxycholic acid.

can result in gallstones owing to impaired absorption of bile acids and depletion of the bile acid pool. Agents that increase biliary secretion of cholesterol (e.g., estrogen, oral contraceptives, and clofibrate) and those which prevent bile acid reabsorption in the intestines (e.g., cholestyramine) are also predisposing factors.

Cholelithiasis is frequently treated by surgical removal of the gallbladder (cholecystectomy). Oral treatment with ursodeoxycholic (ursodiol) and chenodeoxycholic (chenodiol) acids has effectively solubilized gallstones in a number of patients. These bile acids apparently reduce HMG-CoA reductase activity, thereby lowering cholesterol levels while enriching the bile acid pool. The increased bile acid to cholesterol ratio in the bile apparently aids in solubilizing the stones already present. However, the

effect of chenodiol is transient, and therapy may have to be long term. In addition, the treatment appears to be promising only in patients who have radiolucent gallstones and functioning gallbladders. Since chenodiol raises serum transaminase levels, the significance of which is not clear, the measurement of other liver function tests may be necessary. However, ursodiol has few side effects and is effective at a lower dosage. Two other nonsurgical treatments undergoing clinical evaluation are extracorporeal shock-wave lithotripsy to fragment gallstones and direct infusion into the gallbladder with the solvent methyl tert-butyl ether to dissolve cholesterol stones. Oral treatment with ursodiol in conjunction with extracorporeal shock-wave lithotripsy is more effective than lithotripsy alone.

CLINICAL CASE STUDY 17.1 Gaucher's Disease, Type 1

Synopsis

This clinical case was abstracted from: E.C. Larsen, S.A. Connolly, A.E. Rosenberg, Case 20-2003: A nine-year-old girl with hepatosplenomegaly and pain in the thigh, N. Engl. J. Med. 348 (2003) 2669–2677.

Synopsis

A nine-year-old girl was admitted to the hospital with an enlarged liver and spleen (hepatosplenomegaly), skeletal pain, and a history of epistaxis. Imaging and histologic bone-marrow biopsy, aspirate specimen analysis, and bone scan studies revealed a pathologic infiltrative process of histiocytes (tissue macrophages) into the organs. Based on these studies, a clinical diagnosis of Gaucher's disease was confirmed by measurement of leukocyte β-glucocerebrosidase (also known as β-glucosidase) levels and genetic studies. Since the subject did not have any neurological problems, her diagnosis was Type 1 Gaucher's disease. She was treated with recombinant β-glucocerebrosidase, which was specifically targeted to macrophages. This treatment resolved many of the subject's clinical problems.

Teaching Points

1. Gaucher's disease is due to the accumulation of glucocerebroside in the macrophages due to a defect in the lysosomal catabolic enzyme. It is an autosomal recessive disease, and its most common form is non-neuropathic type 1 disease.
2. The therapy consists of administering macrophage-targeted missing enzyme β-glucocerebrosidase.
3. The recombinant glucocerebrosidase is targeted to macrophages by its mannose-terminated oligosaccharide side chain. The recognition sites on the carbohydrate receptor sites of the macrophage internalize the mannose-terminated enzyme.
4. The recombinant enzyme therapy is clinically ineffective in the treatment of Gaucher's disease with central nervous system involvements because it cannot penetrate the blood–brain barrier.

CLINICAL CASE STUDY 17.2 Tay–Sachs Disease

This case was abstracted from: K.S. Krishnamoorthy, F. Eichler, O. Rapalino, M.P. Frosch, Case 14-2014: an 11-month-old girl with developmental delay, N. Engl. J. Med. 370 (2014) 1830–1841.

Synopsis

An 11-month-old daughter of a nonconsanguineous couple was referred to an outpatient neurology clinic for developmental delay. Perinatal history was unremarkable except for induction of labor at 38 weeks' gestation due to oligohydramnios, but otherwise the patient appeared normal and was meeting developmental milestones during the first 6 months of life. At 8 months old, the patient could no longer sit herself up, reach for objects, or feed herself, and she appeared less alert. On examination at 11 months old, she was found to have multiple abnormalities, most notably an exaggerated startle response, increased muscle tone due to spasticity and rigidity, and the developmental skills of a 6-month-old infant. Magnetic resonance imaging (MRI) of the brain showed mild diffuse hypomyelination in the deep and subcortical supratentorial white matter. Metabolic screening tests revealed very low activity of leukocyte hexosaminidase A, consistent with the diagnosis of Tay–Sachs disease. Genetic testing confirmed the diagnosis, revealing a mutation of the hexosaminidase A gene (HEXA). DNA screening of the patient's family was not performed. The patient received palliative care. As the disease progressed, she developed muscle spasms, seizures, and sleep and respiratory difficulties. She died at 28 months of age. An autopsy revealed gross, microscopic, and ultrastructural findings consistent with a storage disease limited to the neurons.

Teaching Points

1. The approach to evaluating developmental delay and spasticity in a child begins with consideration of nonprogressive causes and metabolic or degenerative causes. The main nonprogressive disorder is cerebral palsy, which is a syndrome of cerebral and motor impairment due to characteristic lesions in the brain. Cerebral palsy occurs in the early stages of development.
2. Metabolic or degenerative causes may occur either in the early or late stages of development. Disorders that manifest in the neonatal period are amino acid and organic acid disorders, urea-cycle disorders, and peroxisomal disorders. Disorders that manifest in early and late infancy are lysosomal storage disorders (Tay–Sachs disease, Krabbe's disease, Canavan's disease, and metachromatic leukodystrophy), mitochondrial disorders (Leigh's disease), glucose transporter 1 (GLUT-1) deficiency, and congenital glycosylation disorders.
3. Red flags for metabolic or degenerative disorders include failure to thrive, failure to reach normal developmental milestones, movement disorders, seizures, vision or hearing impairment, cognitive decline, development of spasticity, intermittent encephalopathy, seizures, and consanguinity.
4. Tay–Sachs disease is an autosomal recessive disorder caused by a lack of hexosaminidase A, which is a

(Continued)

CLINICAL CASE STUDY 17.2 (Continued)

lysosomal enzyme required for the degradation of ganglioside monosialic 2 (GM_2) in the neurons. Impairment or deficiency of this lysosomal enzyme results in abnormal myelin development and excessive accumulation of GM_2 in the neurons. The result is progressive weakness and loss of motor skills within the first year of life. A pathognomonic sign of Tay–Sachs disease is early excessive startle, which is caused by increased neuron excitation due to meganeurite formation at the axon hillock and a marked increase in the number of spinous processes protruding from the axons.

5. Enzymatic analysis is recommended for the diagnosis of Tay–Sachs disease, and genome testing should be used to confirm the diagnosis, to discriminate between pseudodeficiency, and to provide genetic counseling.

CLINICAL CASE STUDY 17.3 Infantile Krabbe Disease (Globoid-Cell Leukodystrophy)

This clinical case was abstracted from: K.S. Krishnamoorthy, F.S. Eichler, N.A. Goyal, J.E. Small, M. Snuderl, Case 3-2010: a 5-month-old boy with developmental delay and irritability, N. Engl. J. Med. 362 (2010) 346–356.

Synopsis

A 5-month-old boy was referred to a pediatric neurologist due to abnormal neurologic findings and developmental delay. The infant's global delay in development was attributed to difficulty in appropriate feeding accompanied by irritability and hypertonia. Several clinical and radiologic studies were performed. The cause of the rapid neurologic deterioration of this infant was considered to be a lysosomal storage disorder. The findings of brain MRI were suggestive of Krabbe leukodystrophy. This diagnosis was confirmed by the infant's decreased leukocyte galactosylceramidase activity and genetic studies. The infant's disease rapidly progressed and led to death at 9 months of age. Necropsy studies supported the diagnosis of Krabbe disease.

Teaching Points

1. Krabbe disease is due to deficiency of galactocerebrosidase, which leads to accumulation of galactosylceramides and psychosine. These abnormalities affect the growth and development of myelin.

2. Krabbe disease is an autosomal recessive disorder, occurs in about 1 in 100,000 births, and has no effective treatment.

Chapter 18

Lipids III: Plasma Lipoproteins

Key Points

1. Lipoproteins consist of triglycerides, cholesterol, phospholipids, and proteins.

2. Proteins serve as structural components, enzyme activators or inactivators, and receptor recognition molecules.

3. Five major lipoproteins are chylomicrons, very-low-density lipoprotein (VLDL), intermediate-density lipoprotein (IDL), low-density lipoprotein (LDL), and high-density lipoprotein (HDL).

4. Chylomicrons are synthesized in the gastrointestinal tract from dietary lipid. They contain apolipoprotein B-48 (apo B-48; see also Chapter 11). They are transported to the blood via lymphatic channels and are transformed to remnant chylomicrons after the removal of the triglycerides as free fatty acids by endothelial lipoprotein lipase activated by apo CII. Remnant chylomicrons are taken up by the hepatocytes and metabolized via B-48 and apo E receptors. Deficiency of apo CII or endothelial lipoprotein lipase causes hyperchylomicronemia.

5. VLDL, which is synthesized in the hepatocytes, contains apo B-100 and is progressively transformed into IDL and LDL by endothelial lipoprotein lipase and lecithin cholesterol acyl transferase (LCAT).

6. LDL supplies cholesterol for peripheral tissues for metabolic purposes. It is internalized by receptor-mediated endocytosis, which regulates intracellular cholesterol homeostasis. The liver is one of the predominant tissues that metabolize LDL and other apo B-100-containing lipoproteins. Receptor-mediated endocytosis consists of (a) LDL receptor (LDLR), (b) recognition of LDLR by apo B-100, (c) a protein required for clathrin-mediated internalization known as LDL-receptor-adapter protein (LDLRAP1), and (d) a proprotein convertase (PCSK9) whose normal function is to reduce LDLR protein levels. Genetic defects in (a), (b), (c), and (d) cause familial hypercholesterolemia (FH) leading to coronary heart disease (CHD). Defects in loss of function in (a), (b), and (c) and gain of function in (d) cause FH, which is rare. More common forms of hypercholesterolemia are caused by polygenic defects and can also be affected by secondary causes (e.g., diabetes, hypothyroidism, nephrotic syndrome).

7. Nascent HDL is synthesized in the liver and in GI cells. It undergoes maturation in blood circulation by LCAT enzyme activated by apo A-I, which converts cholesterol to cholesteryl esters, and by acquisition of other apolipoproteins. Nascent HDL acquires its cholesterol from peripheral tissues and macrophages transported by ATP-binding cassette (ABC) protein A1. Mature HDL also acquires cholesterol mediated by ABCG1 or SR-B1 transporters. Cholesteryl esters are transported from HDL to VLDL and LDL by cholesterol ester transfer protein (CETP). HDL cholesterol is delivered to the liver via SR-B1 receptor for its conversion to bile acids and its eventual removal. This pathway of cholesterol transport from peripheral tissues to the liver mediated by HDL is known as reverse cholesterol transport (RCT). Thus, the plasma HDL-C level is an indicator RCT.

8. The defects in lipoprotein disorders in the vast majority of the population with high total plasma cholesterol and LDL-C and low HDL-C levels associated with CHD risk are not completely known. CHD, the major cause of morbidity and mortality, is also associated with other risk factors (diabetes, obesity, cigarette smoking, hypertension, family history of CHD, age, and male sex). Thus, CHD can occur in the absence of lipoprotein abnormalities.

9. The treatment and management of lipoprotein abnormalities require a systematic multistep approach. These steps include lifestyle modifications (diet and exercise) and, if necessary, the use of pharmacologic approaches. The drugs used are (a) statins-HMG-CoA reductase inhibitor: inhibits cholesterol synthesis, reduces LDL-C levels, and causes a modest increase in HDL-C levels, and has desirable anti-inflammatory effects; (b) bile acid sequestrants: promotes bile acid elimination in the GI tract, reduces LDL-C levels; (c) nicotinic acid: agonist for GiPCR, inhibits adipocyte lipase, decreases LDL-C, VLDL-triglyceride, and Lp(a) levels, and increases HDL-C; (d) gemfibrozil (a fibric acid derivative): an activator for peroxisome proliferator-activated receptor-alpha (PPARα), a nuclear transcription factor that increases synthesis of lipoprotein lipase, decreases VLDL levels, and causes a modest elevation in HDL-C levels; and (e) Ezetimibe: inhibits cholesterol absorption from the GI tract and reduces LDL-C levels. All of these pharmacological agents have undesirable side effects. Combination drug therapy is used when monotherapy fails to achieve therapeutic goals. Recent studies for lowering plasma cholesterol levels in high-risk CHD subjects recommend primarily using only statin therapy.

N. V. Bhagavan and C.-E. Ha: Essentials of Medical Biochemistry, Second edition. DOI: http://dx.doi.org/10.1016/B978-0-12-416687-5.00018-X

Lipids, by virtue of their immiscibility with aqueous solutions, depend on protein carriers for transport in the bloodstream and extracellular fluids. Fat-soluble vitamins and free fatty acids are transported as noncovalent complexes. Vitamin A is carried by retinol-binding protein and free fatty acids by plasma albumin. However, the bulk of the body's lipid transport occurs in elaborate molecular complexes called **lipoproteins**.

STRUCTURE AND COMPOSITION

Lipoproteins are often called **pseudomicellar** because their outer shell is in part composed of amphipathic phospholipid molecules. Unlike simple micelles, lipoproteins contain **apolipoproteins**, or **apoproteins**, in their outer shell, and a hydrophobic core of **triacylglycerol** (also called **triglycerides**) and **cholesteryl esters**. Unesterified, or free, cholesterol contains a polar hydroxyl group and can be found as a surface component, and in the region between the core and surface (Figure 18.1). Most lipoproteins are spherical. However, newly secreted high-density lipoproteins (HDLs) from the liver or intestine are discoid and require the action of lecithin-cholesterol acyltransferase (LCAT) in plasma to expand their core of neutral lipid and become spherical. The hydrophobic core of the low-density lipoprotein (LDL) molecule may contain two concentric layers: one of triacylglycerol and another of cholesteryl ester.

The apolipoproteins are distinct physically, chemically, and immunochemically, and have important roles in lipid transport and metabolism (Table 18.1). They have specific structural domains in accordance with their individual metabolic functions. Amino acid substitutions or deletions in critical domains result in functional abnormalities. The apoproteins share a common structure in the form of an amphipathic helix, in which the amino acid residues have hydrophobic side chains on one face of the helix and hydrophilic polar residues on the other. The hydrophilic face is believed to interact with the polar head groups of the phospholipids, while the hydrophobic residues interact with their fatty acid portions.

In lipoproteins, the laws of mass action govern the interactions of lipids and most apoproteins, so that as the affinities between surface components change during lipoprotein metabolism, apoproteins may dissociate from one particle and bind to another. In fact, all of the apoproteins, with the possible exception of apoprotein B (apo B), can change their lipoprotein associations. The reason for the unique behavior of apo B remains a mystery. All lipoproteins are transport proteins and may be divided into two classes according to the composition of their major core lipids. The principal triacylglycerol carriers are chylomicrons and very-low-density lipoproteins (VLDLs), whereas most cholesterol transport occurs via LDLs and HDLs.

These four major groups of plasma lipoproteins can be separated and characterized by electrophoresis and ultracentrifugation (Table 18.2). Each group is heterogeneous and can be subdivided on the basis of variation in apoprotein and lipid compositions. The groups share several apoprotein components, e.g., apo B occurs in chylomicrons, LDLs, and VLDLs. Some apoproteins belong to families of polypeptides, of which one or another may

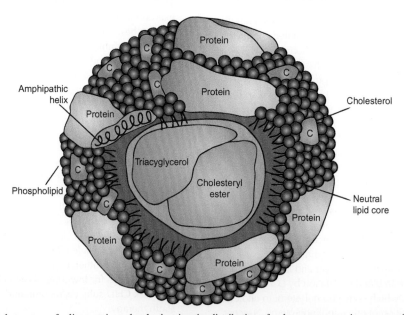

FIGURE 18.1 Generalized structure of a lipoprotein molecule showing the distribution of polar components in an outer shell composed of free cholesterol, phospholipids, and amphipathic proteins, and in an inner core composed of neutral triacylglycerols and cholesteryl esters. Phospholipids are oriented with polar head groups toward the aqueous environment and hydrophobic tails toward the neutral core, analogous to their positioning in the outer leaflet of the typical cell membrane.

TABLE 18.1 Properties of Human Apolipoproteins*

Apolipoprotein	Plasma Concentration (mg/dL) (approximate)	Molecular Weight	Major Lipoprotein Carrier	Biological Function in Addition to Structural and Transport Role
A-I	100–150	29,016	HDL	Activates LCAT
A-II	30–50	17,414	HDL	Enhances HL activity
A-IV	15	44,465	Chylomicron, HDL	Activator of LPL and LCAT
B-100	80–100	512,723	VLDL, LDL	Necessary for hepatic endogenous triacylglycerol secretion; binding to cell receptors
B-48	<5	241,000	Chylomicron	Necessary for intestinal exogenous triacylglycerol secretion; binding to hepatic receptors
C-I	7	6630	Chylomicron, VLDL	Activates LCAT
C-II	4	8900	Chylomicron, VLDL	Activates LPL
C-III	12	8800	Chylomicron, VLDL	Inhibits triglyceride hydrolysis by LPL and HL
D	5	29,000	HDL	Multilipid ligand binding protein
E_{2-4}	3–7	34,145	VLDL, IDL, HDL, Chylomicron	Binding to specific receptors
Apo(a)	10	300,000–800,000	LDL, HDL	Competes with plasminogen for its receptors and may inhibit thrombolysis

LCAT = Lecithin-cholesterol acyltransferase; LPL = lipoprotein lipase; VLDL = very-low-density lipoprotein; LDL = low-density lipoprotein; HDL = high-density lipoprotein; triglyceride = triacylglycerol (TG); HL = hepatic lipase.

TABLE 18.2 Physical Data for the Major Types of Lipoprotein

Lipoprotein (Lp)	Hydrated Density (g/mL)	S_f*	Position of Migration in Paper or Agarose Electrophoresis in Comparison with Serum Globulins
Chylomicrons	<0.95	>400	Origin
VLDL	0.95–1.006	20–400	α_2 (migrates ahead of β-globulin to pre-β position)
LDL	1.006–1.063	0–20	β (with β-globulin)
HDL_2	1.063–1.125	–	α_1 (with α_1-globulin)
HDL_3	1.125–1.210	–	
Lp(a)	1.040–1.090	–	Slow pre-β-LP

The rate at which lipoprotein floats up through a solution of NaCl of specific gravity 1.063 is expressed in Svedberg flotation units or S_f; one S_f unit = 10 − 13 cm/sec/dyne/g at 26°C. S_f can be thought of as a negative sedimentation constant.

predominate in a particular lipoprotein group. For example, apo B, when formed in chylomicrons synthesized in intestinal mucosa, is apo B-48, whereas that in VLDL from the liver is apo B-100. The designations B-48 and B-100 reflect the relative molecular masses of these proteins, B-48 being 48% of the mass of B-100. Intestinal apo B contains 2152 of the 4563 amino acids that make up the hepatic form. Oligonucleotide hybridization studies have shown that the intestinal and hepatic genes are identical. The shortened intestinal form of apo B is produced by a single nucleotide substitution of uracil for cytosine at position 6457 in the nucleotide sequence of the apo B mRNA. This changes the codon CAA (glutamine) to the termination codon UAA. The tissue-specific apo B mRNA editing, which consists of site-specific deamination of cytosine to uracil, is mediated by a multicomponent cytidine deaminase. This unique form of mRNA editing eliminates the carboxy terminal portion of apo B to form the B-48 of chylomicron. Since the carboxy terminal portion of the apo B sequence contains the apo B-binding domain, this deletion ensures a distinct metabolic routing of the chylomicron particle.

In a test tube of plasma or serum collected from a non-fasting individual and allowed to stand overnight, chylomicrons appear at the top surface in a milky layer because of their low density (*d*, 0.95). Sequential ultracentrifugation is used to obtain various lipoprotein fractions.

Classification of lipoproteins by electrophoresis and hydrated density is shown in Table 18.2. Based on their density, the individual lipoproteins are further subdivided and may possess different physiological functions (Table 18.3).

Other lipoproteins are lipoprotein (a) and lipoprotein-X, which are implicated in disease. Lipoprotein (a) [Lp(a)] is a variant of LDL and occurs in the plasma of normal subjects in the density range 1.040−1.090 g/mL. Lp(a) is similar to LDL in lipid composition. Apolipoprotein (a) [apo(a)] is bound to apo B-100 by a disulfide linkage. Apo (a) is a large glycoprotein that exhibits size heterogeneity among individuals in a range of M.W. 300,000−800,000.

The exact physiological function of Lp(a) is not known. However, apo(a) and plasminogen share extensive sequence homology. Both Lp(a) and plasminogen possess multiple tandem repeats of triple loop structural motifs, known as **kringles** (Chapter 34). Type 2 kringle IV is found in apo(a) and varies from 11 to greater than 50 tandem repeats. The exact number of repeats is an inherited

property and defines different isoforms of Lp(a). The apo (a) gene is close to the plasminogen gene (6q2.6−q2.7). Plasma Lp(a) is synthesized primarily in the liver.

Due to the structural similarity between apo(a) and plasminogen, it is thought that Lp(a) interferes with fibrinolysis and thus promotes thrombosis. Normally, formation of a blood clot (thrombus) and its dissolution by fibrinolysis are carefully regulated, and the two processes are in a delicate balance (Chapter 34). The clot is cleaved and solubilized by the serine protease plasmin, which is derived from its precursor plasminogen by proteolysis. Lp (a) exerts its antifibrinolytic activity by competitive inhibition, it interferes with the binding of plasminogen to endothelial cells and monocytes, and it binds to fibrin, which is required for formation of plasmin (Chapter 34). Elevated homocysteine levels (homocysteinemia) promote fibrin binding to Lp(a), thus further preventing the formation of plasmin and subsequent impaired thrombolysis.

Several prospective studies have shown that an excess plasma level of Lp(a) is a risk factor and an independent predictor of coronary heart disease. Survivors of myocardial infarcts have higher Lp(a) levels than controls. In individuals with Lp(a) levels greater than 30 mg/dL (0.78 mM/L), the relative risk for coronary heart disease is 2- to 3.5-fold higher than that of controls. Population studies have shown that plasma Lp(a) levels in Africans are several-fold higher compared with Asian and Caucasian populations. Lp(a) levels are elevated in end-stage renal disease. In postmenopausal women, Lp(a) is increased, but estrogen replacement therapy can lower Lp (a) levels. Most lipid-lowering drugs (except nicotinic acid) do not affect plasma Lp(a) concentration.

Lipoprotein-X (Lp-X) is an abnormal lipoprotein found in patients with obstructive liver disease or LCAT deficiency. Lp-X floats in the density range of LDL and has the same electrophoretic mobility as LDL. The composition of Lp-X differs from that of LDL, and it does not react with antisera to LDL. The major apoproteins of Lp-X

TABLE 18.3 Composition of Lipoproteins (Percent of Mass)

Protein	Major Apoproteins	Phospholipid	Cholesterol	Cholesterol Ester	Triacylglycerol	Diameter (nm)
Chylomicrons	2 (A-I, A-II, A-IV, B-48)	3	2	3	90	75−1200
VLDL	6 (B-100, C, E)	14	4	16	60	30−80
IDL	18 (B-100, E)	22	8	32	20	25−35
LDL	21 (B-100)	22	8	42	7	18−28
Lp(a)	21 [B-100, apo(a)]	20	8	45	6	25−30
HDL$_2$	37 (A-I, A-II)	32	5	20	6	9−12
HDL$_3$	54 (A-I, A-II)	26	2	15	3	5−9

isolated from patients with LCAT deficiency are albumin, apo C, and apo A. Lp-X also contains small amounts of apo D and apo E. Lp-X from patients with obstructive liver disease has been reported to lack apo A-I, a powerful activator of LCAT. The lipid constituents of Lp-X are cholesterol (almost entirely unesterified) and phospholipids. In electron microscopy, negatively stained Lp-X preparations appear as stacks of disk-like structures (**rouleaux**).

METABOLISM

Plasma lipoproteins are in a dynamic state. Their continuous synthesis and degradation are accompanied by rapid exchanges of lipid and protein components between the different lipoprotein classes. Major sites of plasma lipoprotein synthesis are the intestine and liver. Synthesis takes place in rough and smooth endoplasmic reticulum. The necessary components are triacylglycerols, cholesterol (and cholesteryl esters), phospholipids, and apoproteins. Four enzymes at various points in the metabolic cycle play important roles in the delivery, storage, and mobilization of lipoprotein lipids.

1. **Lipoprotein lipase** (**LPL**) is a glycoprotein that belongs to a family of serine esterases that includes hepatic lipase and pancreatic lipase. LPL is synthesized by many types of parenchymal cells in the body but is concentrated mainly in muscle and adipose tissue. After its synthesis in the parenchymal cells, LPL is secreted and translocated to the luminal surface of endothelial cells lining the vascular beds, where it is bound to heparan sulfate. Purified LPL is a dimer in the active state and loses activity when it dissociates into monomers. Each subunit of LPL contains binding sites for glycosaminoglycans and apo C-II, both of which promote dimerization of LPL. Apo C-II is required for the activation of LPL *in vivo* and is synthesized in the liver. In the plasma, apo C-II cycles between HDL, triacylglycerol-rich lipoprotein, chylomicrons, and VLDL.

   LPL is the main enzyme involved in the processing of chylomicrons and VLDL by hydrolysis of fatty acids from triglycerides. Phospholipids may also serve as substrates for LPL. Apo C-III inhibits the activation of LPL by apo C-II *in vitro*, albeit at high levels, but the physiological importance of inhibition is not understood.

   In the postprandial state, elevated serum insulin increases LPL activity in adipose tissue (but not in muscle) and promotes fuel storage as triacylglycerols. In the postabsorptive state, serum insulin levels decrease, causing a decrease in LPL activity in adipose tissue. However, LPL activity in muscle remains high or increases, releasing fatty acids from VLDL particles for use as fuel. In general, LPL has different functions in different tissues. In cardiac and skeletal muscle, it provides energy; in white adipose tissue, it stimulates triacylglycerol storage; in brown adipose tissue, it stimulates thermogenesis; and in lactating breast tissue, it stimulates triacylglycerol synthesis for milk production. LPL is also involved in surfactant synthesis in the lungs, and in phospholipid and glycosphingolipid synthesis in the brain.

2. **Hepatic lipase**, like LPL, is synthesized in the parenchymal cells of liver and is localized on the membrane surfaces of the hepatic endothelial cells of blood vessels, bound to heparan sulfate. It also occurs on the endothelial cells of the blood vessels of the adrenal glands and gonads. While the role of hepatic lipase is not completely understood, it continues the lipolysis of VLDL and IDL in their stepwise conversion to LDL. Hepatic lipase also hydrolyzes phospholipids and HDL-triacylglycerol. It performs these functions both on the endothelial cell surface and within endosomes, as lipoproteins are endocytosed into cells via receptor-mediated endocytosis. Hepatic lipase activity responds positively to androgens and negatively to increasing levels of estrogen.

3. **Lecithin-cholesterol acyltransferase** (**LCAT**) is a glycoprotein synthesized in the liver. LCAT circulates in the plasma with HDL, LDL, apo D, and cholesteryl ester transfer protein (CETP). It is activated by apo A-I. LCAT catalyzes the transfer of long-chain fatty acids from phospholipids to cholesterol, forming cholesteryl esters and permitting the storage and transport of cholesteryl esters in the lipoprotein core. Substrates for LCAT, namely phospholipids, are transported from triglyceride-rich lipoproteins (chylomicrons and VLDL) to HDL. This transfer is facilitated by a plasma phospholipid transfer protein (PLTP). Cholesteryl esters are exchanged between lipoproteins, a process mediated by CETP, a plasma glycoprotein that is synthesized in the liver. There is a net transfer of cholesteryl esters from HDL to LDL and VLDL with an exchange of triacylglycerol. The VLDL is transformed into LDL in the circulation (discussed later). The cholesterol content of LDL is increased, thus promoting its atherogenicity.

   HDL is antiatherogenic and removes cholesterol from peripheral cells and tissues for eventual transport to hepatocytes and excretion in the bile directly or after conversion into bile acids. The efflux of cholesterol from peripheral cells is mediated by the ATP-binding casette (ABC) transporter protein (discussed later). The flux of cholesterol transport from extrahepatic tissues (e. g., blood vessel wall) toward the liver for excretion is known as the **reverse cholesterol transport pathway**. In contrast, the forward cholesterol pathway involves the transport of cholesterol from the liver to the peripheral cells and tissues via the VLDL → IDL → LDL pathway. However, it should be noted that the liver plays a major role in the removal of these lipoproteins. Thus, the system of reverse cholesterol transport consisting of LCAT, CETP, and their carrier lipoproteins is critical for maintaining cellular cholesterol homeostasis. The role of CETP is exemplified in clinical studies involving patients with polymorphic forms of CETP that promote cholesteryl ester transfer from HDL to LDL, conditions that increase plasma CETP levels. These patients exhibit an increased risk of coronary heart disease. Homozygous CETP deficiency, found in some Japanese people, results in increased HDL cholesterol levels. There is anecdotal evidence of longevity in this group of individuals. This information led to the development of a therapeutic agent known as **torcetrapib** that inhibits CETP. In clinical trials, although CETP inhibition by torcetrapib increased plasma HDL cholesterol levels, the drug treatment was associated with increased frequency of undesirable cardiovascular events and mortality.

4. **Acyl-CoA:cholesterol   acyltransferase   (ACAT)** esterifies free cholesterol by linking it to a fatty acid. ACAT is an intracellular enzyme that prepares cholesterol for storage in hepatic parenchymal cells. The roles of these enzymes in lipoprotein metabolism will become more apparent in the discussion of the origin and fate of each class of lipoprotein.

Chylomicrons

After partial hydrolysis in the gut, dietary fatty acids, monoacylglycerols, phospholipids, and cholesterol are absorbed into the mucosal enterocytes lining the small intestine (Chapter 11). Once within the cell, the lipids are re-esterified and form a lipid droplet within the lumen of the smooth endoplasmic reticulum. These droplets consist of triacylglycerol and small amounts of cholesteryl esters and are stabilized by a surface film of phospholipid. At the junction of the smooth and rough endoplasmic reticulum, the droplet acquires apoproteins B-48, A-I, A-II, and A-IV, which are produced in the lumen of the rough endoplasmic reticulum in the same way as other proteins bound for export. The lipoprotein particle is then transported to the Golgi stacks where further processing yields **chylomicrons**, which are secreted into the lymph and then enter the blood circulation at the thoracic duct.

Synthesis and secretion of chylomicrons are directly linked to the rate of dietary fat absorption. When fat is absent from the diet, small chylomicrons with a diameter of about 50 nm are secreted at a rate of approximately 4 g of triacylglycerol per day. On a high-fat diet, the mass of lymphatic triacylglycerol transport may increase 75-fold, owing partly to greater production of chylomicrons but primarily to a dramatic increase in the size of the particles, which may have diameters of 1200 nm, and a 16-fold increase in the amount of triacylglycerol within their core.

In the circulation, chylomicrons undergo a number of changes (Figure 18.2). First, they acquire apo C and apo E from plasma HDL in exchange for phospholipids. Next, hydrolysis of triacylglycerols by endothelial LPL begins (e.g., in adipose tissue and skeletal muscle). Progressive hydrolysis of triacylglycerol through diacylglycerol and monoacylglycerol to fatty acids and glycerol ensues. The released fatty acids cross the endothelium and enter the underlying tissue cells, where they undergo re-esterification to form triacylglycerol for storage (in adipocytes) or are oxidized to provide energy (e.g., in muscle). The earlier acquisition of apo C-II by chylomicrons is important, because this apoprotein is an essential activator of LPL. Chylomicron triacylglycerol has a half-life of about 5 minutes in the circulation. LPL is most active after meals, when it is stimulated by elevated levels of plasma insulin. A low-affinity form (high K_m) of LPL, found principally in adipose tissue, is most active when triacylglycerol levels in plasma are high; it promotes lipid storage after meals. A high-affinity (low K_m) form of LPL predominates in the heart (and striated muscle tissues) and is active when triacylglycerol levels are low, as in the postabsorptive state. The high-affinity enzyme thus hydrolyzes triacylglycerols into fatty acids at sites where they will be required for energy production. Mammary gland LPL is a high-affinity type that facilitates uptake of fatty acids to promote milk-fat synthesis during lactation.

As the core triacylglycerols of a chylomicron are depleted, often reducing its diameter by a factor of 2 or more, the surface components are also modified. A substantial portion of the phospholipids and of apo A and C is transferred to HDL. The C apoproteins thus cycle repeatedly between newly produced chylomicrons and HDL. The **chylomicron remnant** is consequently rich in cholesteryl esters and apo B-48 and E.

Catabolism of chylomicron remnants may be viewed as the second step in the processing of chylomicrons. After the loss of apo C-II and other C and A apoproteins, LPL no longer acts upon the remnants. Chylomicron remnants are rapidly removed from the capillary surface by uptake into liver parenchymal cells via **receptor-mediated endocytosis**. Another receptor, known as the **LDL receptor-related protein (LRP)**, may also function in chylomicron uptake. Chylomicron remnants are transported into the lysosomal compartment where acid lipases and proteases complete their degradation. In the liver, fatty acids released this way are oxidized or are reconverted to triacylglycerol, which is stored or secreted as VLDL. The cholesterol may be used in membrane synthesis, stored as cholesteryl ester, or excreted in the bile either unchanged or as bile acids.

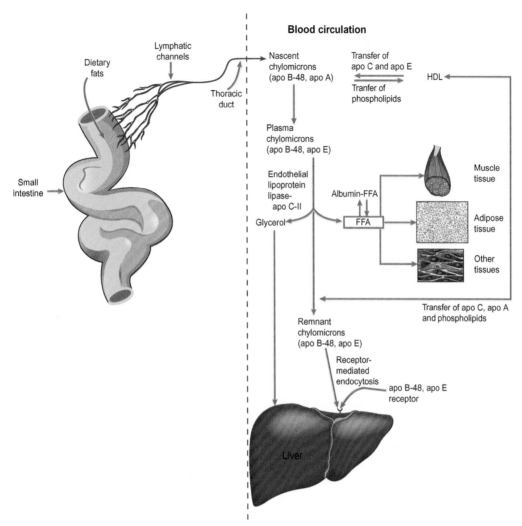

FIGURE 18.2 Steps involved in the metabolism of chylomicrons—the exogenous pathway of lipid transport. FFA = free fatty acids.

Very-Low-Density Lipoproteins and Formation of Low-Density Lipoproteins

VLDLs are produced by the parenchymal cells of the liver from lipid and apoprotein constituents in a way similar to that of chylomicron formation in enterocytes. However, while the triacylglycerol core of chylomicrons is derived exclusively from absorbed dietary lipids, the fatty acids required for the formation of VLDL triacylglycerol are derived from

1. Stored fat released from adipose tissue as fatty acids;
2. Conversion of carbohydrates to fatty acids in the liver; and
3. Hydrolysis of lipoprotein triacylglycerols on capillary endothelia and in the liver.

Synthesis of apoprotein B-100 is required for VLDL secretion. The amount of VLDL secreted by the liver is extremely variable and can be affected in a number of ways. A primary determinant of VLDL output is the flux of free fatty acids entering the liver. The liver responds to an increase in free fatty acids by synthesis of more and larger VLDLs. If saturated fatty acids predominate in the formation of triacylglycerol, the VLDL particles will be more numerous and smaller than if polyunsaturated fatty acids predominate. This finding may be related to the reduction in plasma cholesterol levels that result from elevating the proportion of polyunsaturated fats in the diet. The surface-to-volume ratio is smaller in the larger VLDLs. Since cholesterol is primarily a surface component in these particles, its secretion would be reduced. If the fatty acid flux to the liver is large, the rate of secretion of triacylglycerols in VLDLs can become saturated. The resulting triacylglycerol deposition in cytoplasmic droplets is seen in "fatty liver."

A high-carbohydrate diet results in a substantial elevation of plasma VLDL concentrations. A high-cholesterol diet alters the composition of VLDL, with cholesterol esters substituting for triacylglycerol as core components, and leads to a marked increase in apo E synthesis.

Like nascent chylomicrons, newly secreted VLDL undergoes changes in the plasma (Figure 18.3). Nascent VLDL acquires apo C and E from HDL. In chylomicrons, the apo B is B-48, whereas in VLDL only B-100 is found. LDL contains exclusively apo B-100, indicating that VLDL rather than chylomicrons is the principal precursor of LDL. In some species (e.g., the rat), most VLDL remnants, like chylomicron remnants, are rapidly taken up by the liver. In humans, as the core triacylglycerols are removed and the C apoproteins are lost, approximately half of the VLDL is rapidly removed by the liver via the apo B-100−apo E receptor pathway. The rest remains in circulation as VLDL remnants. Since some of these remnants have a density of between 1.006 and 1.019, they are called IDLs and are analogous to chylomicron remnants. Thus, the liver plays a major role in clearing lipoprotein remnants. The remaining IDLs are subjected to further catabolism by hepatic lipase. In VLDL and IDL, the cholesterol is converted to cholesteryl esters by LCAT. Thus, LDL is a cholesteryl-ester rich particle containing almost exclusively apo B-100.

Because LDL is the principal plasma-cholesterol carrier and its concentration in plasma correlates positively with the incidence of coronary heart disease, LDL is the most intensively studied plasma lipoprotein. Production in humans, via the pathway VLDL→IDL→LDL, accounts for all of the LDL normally present. However, in familial hypercholesterolemia or on a high-cholesterol diet, the VLDL that is produced is higher in cholesterol content,

smaller in size, and within the LDL density range (1.019−1.063 g/mL).

As IDL loses apo E and is converted to LDL with apo B-100 as its sole apoprotein, the residence of LDL in plasma increases from several hours to 2.5 days. This long-lived, cholesterol-rich LDL serves as a source of cholesterol for most tissues of the body. Although most cells can synthesize cholesterol under normal conditions, most endogenous production occurs in the liver and intestine, from which cholesterol is distributed to peripheral tissues by LDL. This metabolic pathway provides an efficient balance between endogenous production and dietary intake of cholesterol.

Low-Density Lipoprotein Uptake

The distribution and delivery of cholesterol to peripheral tissues is mediated by binding of LDL to specific receptors on the plasma membrane of target cells (Figure 18.4).

The number of LDL receptors on the cell membrane depends on the degree of accumulation of intracellular cholesterol, which downregulates transcription of the LDL receptor gene. The population of LDL receptors may be reduced 10-fold by this mechanism.

Cholesterol released inside cells is incorporated into an intracellular pool, which is used for membrane synthesis and for reactions that require a sterol nucleus (e.g., formation of steroid hormones or bile acids). Cholesterol (or its metabolite 26-hydroxycholesterol) suppresses its own synthesis by inhibiting two sequential enzymes of mevalonate synthesis: 3-hydroxy-3-methylglutamyl coenzyme

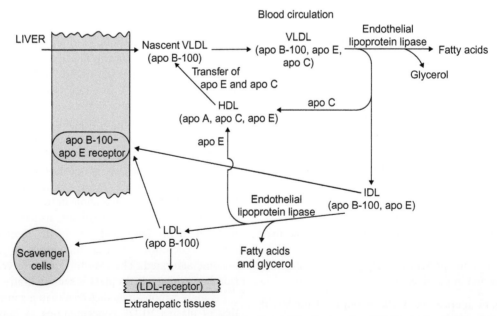

FIGURE 18.3 Conversion of very-low-density lipoproteins (VLDLs) to low-density lipoproteins (LDLs), via intermediate-density lipoproteins (IDLs) (the endogenous pathway of lipid transport). LCAT = lecithin−cholesterol acyltransferase.

A (HMG-CoA) synthase and HMG-CoA reductase, which is the rate-limiting enzyme of cholesterol biosynthesis (Chapter 17). This cholesterol also regulates the number of receptor sites on the cell membrane and, therefore, the cellular uptake of LDL. Genes for cholesterol synthesis undergo coordinated induction or repression with regard to the transcription of the respective mRNAs (Chapter 17). Some of the cholesterol is esterified (with oleate and palmitoleate) into cholesteryl esters by microsomal ACAT, whose activity is stimulated by cholesterol.

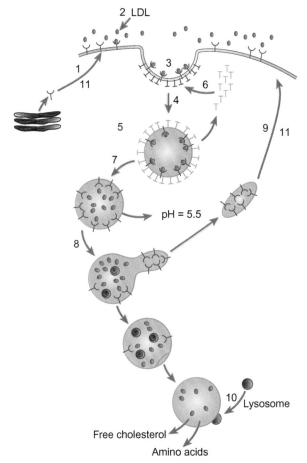

FIGURE 18.4 Cellular uptake and metabolism of LDL. (1) LDL receptor (LDLR) synthesis and maturation in the ER and Golgi apparatus and incorporation of LDLR in the plasma membrane. (2) LDLR binds to LDL via apo B-100. (3) LDLR-LDL cluster in clathrin-coated pits. (4) Clathrin-coated pits are internalized assisted by LDL-receptor-adapter protein (LDLRAP1). (5) and (6) Internalized endosomes shed clathrin with the formation of uncoated endosomes, and clathrin is returned to the pits to be reused. (7) ATP-dependent proton pumps located in the endosomal membrane lower the intravesicular pH, resulting in the separation of ligand and receptor. (8) Multivesicular bodies form as receptors are segregated into finger-like projections. (9) Most receptors are returned to the cell surface. (10) Multivesicular bodies fuse with primary lysosomes to release cholesterol and amino acids into the cytosol. (11) LDLRs are normally downregulated with a decrease in their levels by proprotein convertase subtilisin/kexin type 9 (PCSK9), in an incompletely understood mechanism.

Cholesteryl esters of LDL contain linoleate, whereas those produced by ACAT contain predominantly oleate and palmitoleate. In hepatocytes, an increase in cholesterol levels can lead to the increased activity of cholesterol 7-α-hydroxylase, the rate-limiting enzyme in the synthesis of bile acids (Chapter 17).

Thus, LDL-derived cholesterol meets the cellular requirements for cholesterol and prevents its overaccumulation by inhibiting *de novo* cholesterol synthesis, suppressing further entry of LDL, and storing unused cholesterol as cholesteryl esters or exporting it from the liver as bile acids or other sterol-derived products. About 75% of high-affinity LDL uptake occurs in the liver.

Despite this elaborate regulatory system, cells can accumulate excessive amounts of cholesteryl esters when the plasma LDL saturates the high-affinity, receptor-mediated LDL uptake process. Under these conditions, LDL enters cells by a nonspecific endocytic process known as **bulk-phase pinocytosis**. This mechanism seems to play no role in regulation of *de novo* synthesis of cholesterol and leads to its excessive accumulation, with pathological consequences (e.g., atherosclerosis). Abnormally high plasma LDL levels cause scavenger cells (e.g., macrophages) to take up lipids, which results in xanthoma in the tendons and skin.

High-Density Lipoproteins

HDLs are secreted in nascent form by hepatocytes and enterocytes (Figure 18.5). Loss of surface components, including phospholipids, free cholesterol, and protein from chylomicrons and VLDL as they are acted on by lipoprotein lipase, may also contribute to the formation of HDL in plasma. Discoid, nascent HDL is converted to spherical, mature HDL by acquiring free cholesterol from cell membranes or other lipoproteins. This function of HDL in peripheral cholesterol removal may underlie the strong inverse relationship between plasma HDL levels and the incidence of coronary heart disease. After esterification of HDL surface cholesterol by LCAT, which is activated by apo A-I, HDL sequesters the cholesteryl ester in its hydrophobic core. This action increases the gradient of free cholesterol between the cellular plasma membrane and HDL particles. Cholesteryl esters are also transferred from HDL to VLDL and LDL, utilizing the cholesteryl ester transfer protein CETP (Figure 18.6).

Removal of cholesterol from cells requires an active transport system involving an ATP-binding cassette (ABCA1) transporter. ABCA1-transporter is a member of a superfamily of proteins involved in energy-dependent transport of several substances across cell membranes. It is activated by protein kinases via phosphorylation. The role of ABCA1-transporter in cholesterol efflux is exemplified by an autosomal recessive disease known as **Tangier**

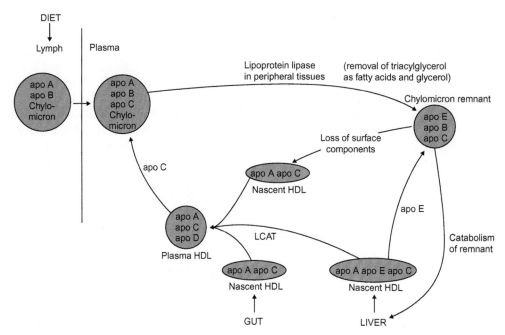

FIGURE 18.5 Sources of HDL. HDL is derived from *de novo* production in the intestinal mucosa and liver, as well as from breakdown of chylomicrons and possibly VLDL. Nascent HDL is discoid when first formed and becomes spherical in plasma as the formation and storage of cholesteryl ester in its cores (via LCAT) lowers the surface-to-volume ratio. Nascent chylomicrons pick up apo C from plasma HDL, which activates lipoprotein lipase. After losing a substantial portion of their triacylglycerols to peripheral tissues via lipoprotein lipase, chylomicrons recycle most of their apo C back to HDL and then gain apo E, which mediates their hepatic uptake as chylomicron remnants.

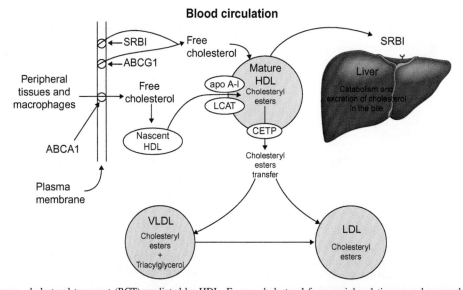

FIGURE 18.6 Reverse cholesterol transport (RCT) mediated by HDL. Excess cholesterol from peripheral tissues and macrophages is transformed via an ATP-binding cassette (ABC) A1 transporter to discoid nascent HDL, as well as to spherical HDL mediated by ABCG1 and the scavenger transporters (SR BI). Cholesterol is converted to cholesteryl esters by LCAT activated by apo A-I. HDL is internalized by the liver via SR BI transporters, and cholesterol is converted to bile acids. In the blood circulation, HDL cholesteryl esters are transferred to apo B-100 containing lipoproteins (VLDL and LDL) by cholesteryl ester transfer protein (CETP), limiting HDL's role in RCT. Defects in ABCA1 (e.g., Tangier disease) lead to massive accumulation of cholesterol. LCAT = Lecithin cholesterol acyl transferases; apo A-I = apolipoprotein A-I.

disease, in which mutations in the gene encoding the ABCA1-transporter lead to accumulation of cholesterol esters in the tissues with almost complete absence of HDL cholesterol (discussed later). Other inherited diseases caused by mutations in ABC proteins are **cystic fibrosis** (Chapter 11), **early onset macular degeneration** (Chapter 36), **sitosterolemia** (discussed later), **adrenoleukodystrophy**, **Zellweger syndrome**, and **progressive**

familial intrahepatic cholestasis. Cholesterol efflux from macrophages also occurs, mediated by another ATP-binding cassette transporter, ABCG1, and a scavenger receptor BI (SRBI). ABCG1 and SRBI transfer cholesterol to mature HDL, unlike ABCA1. ABCA1 and ABCG1 expressions are regulated by heterodimer LXR/RXR nuclear receptors. Peroxisome proliferator—activated receptor (PPAR)-α agonists (e.g., fibrates) increase plasma HDL cholesterol levels by upregulating nuclear receptors.

Under normal physiological conditions, HDL exists in two forms: HDL_2 ($d = 1.063 - 1.125$ g/mL) and HDL_3 ($d = 1.125 - 1.210$ g/mL). Fluctuations in plasma HDL levels have been principally associated with changes in HDL_2. This fraction is often found in much higher concentrations in females and may be associated with their reduced risk for atherosclerotic disease. Clinically, the cholesterol fraction of total HDL ($d = 1.063 - 1.210$) is commonly measured, and low values are frequently associated with increased risk of coronary heart disease (discussed later).

The primary determinant of HDL cholesterol level in human plasma appears to be the cholesterol efflux mediated by ABCA1-transporter (also called **cholesterol efflux regulatory protein**). A defect in this protein causes Tangier disease, which is rare, and the more common familial HDL deficiency. Polymorphisms in the ABCA1-transporter protein may lead to variations in HDL cholesterol levels, increasing the risk for premature coronary artery disease. The rate of apo A-I catabolism is regulated by the size and composition of the particle on which it resides. Smaller lipid-depleted particles have a higher rate of catabolism. In the absence of cholesterol efflux from cells, nascent HDL is not formed, and apo A-I is rapidly cleared from the plasma. Modifications in HDL size are the result of the action of enzymes, which esterify cholesterol (LCAT), transfer cholesteryl esters (CETP), or hydrolyze HDL lipids (hepatic lipase, lipoprotein lipase).

Lipoproteins and Coronary Heart Disease (Also Known as Coronary Artery Disease)

Numerous studies with prospective epidemiological and monogenic defects have shown a strong direct relationship between plasma levels of cholesterol and LDL cholesterol (LDL-C) and an inverse relationship between HDL cholesterol with coronary heart disease (CHD). The optimal and suboptimal plasma levels of cholesterol, LDL-C, and HDL-C for adults are provided in Table 18.4. For children and adolescents, the desirable levels of total cholesterol and LDL cholesterol are <170 and <110 mg/dL, respectively. Based on these studies, lowering LDL-C levels and increasing HDL-C levels (to promote reverse cholesterol transport, discussed earlier) by means of lifestyle changes (exercise and diet) and, if necessary, drug therapy, are

TABLE 18.4 CHD Risk Assessment Using Lipoprotein-Associated Cholesterol and Total Cholesterol (mg/dL)

LDL Cholesterol	Risk Category
<100	Optimal
100–129	Near or above optimal
130–159	Borderline high
160–189	High
≥190	Very high
Total Cholesterol	
<200	Desirable
200–239	Borderline high
≥240	High
HDL Cholesterol	
<40	High
≥60	Low

Levels in adults (from Executive summary of the third report of the National Cholesterol Education Program and Treatment of High Blood Cholesterol in Adults (Adult treatment panel III). JAMA 285(19):2486–2497. Table 18.2, p. 2487, 2001). To convert cholesterol to mmol/L, divide by 38.7.

recommended. It should be emphasized that decreased or elevated LDL-C levels are not always accompanied by either normal or abnormal cardiovascular events, respectively. This observation is also applicable to the inverse relationship between HDL-C and CHD. Although CHD may not exhibit any undesirable symptoms, acute coronary disease, due to ischemia, can lead to myocardial infarction (heart attack). Its diagnosis is discussed in Chapter 7.

CHD is the leading cause of morbidity and mortality in developed nations and has many risk factors in addition to suboptimal lipid parameters. The CHD risk factors include

1. Obesity (visceral obesity accompanied by metabolic syndrome; Chapter 20);
2. Little physical activity;
3. Family history of CHD;
4. Diabetes mellitus;
5. Smoking;
6. Hypertension;
7. Male sex;
8. Age (>40 years for men and >45 years for women).

The Fredrickson classification of lipoprotein disorders is based on abnormalities of concentrations of the major classes of lipoproteins in plasma (Table 18.5). Lipoprotein phenotyping is carried out by measuring total cholesterol and triacylglycerol levels, and cholesterol levels in LDL and HDL. However, the phenotypes described this

TABLE 18.5 Phenotyping Based on Hyperlipoproteinemia

Phenotype	Lipoprotein Present in Excess
I	Chylomicrons
IIa	LDL
IIb	LDL + VLDL
III	β-VLDL
IV	VLDL
V	Chylomicrons + VLDL

way do not adequately represent the current understanding of the pathophysiology underlying these disorders. The disorders can be characterized on the basis of the underlying defects in apolipoproteins, enzymes, or cellular receptors.

Hyperlipoproteinemias can be primary (genetic) or secondary. Some causes of secondary hyperlipoproteinemias include diabetes mellitus, hypothyroidism, nephrotic syndrome, uremia, ethanol abuse, primary biliary cirrhosis, and intake of oral contraceptives. Primary disorders can be due to a single-gene defect or to a combination of genetic defects. The latter type, which are known as **multifactorial** or **polygenic hyperlipoproteinemias**, are affected by secondary disorders and environmental insults such as a diet high in saturated fat and cholesterol, obesity, and ethanol abuse. In general, the polygenic hyperlipoproteinemias are more common and exhibit lower plasma lipid levels than do the single-gene defects.

Hyperlipidemias

Hypertriacylglycerolemias (Hypertriglyceridemias)

Lipoprotein lipase (LPL) deficiency is inherited as an autosomal recessive disorder. **Hyperchylomicronemia** is present from birth. Upon fat ingestion, triacylglycerol levels may rise to 5000−10,000 mg/dL. Chylomicron levels are greatly elevated but not VLDL levels (**type I hyperlipoproteinemia**). Type I hyperlipoproteinemia can also be due to a deficiency of apo C-II, an essential cofactor for LPL, or to the presence of an inhibitor of LPL in the plasma. Hyperchylomicronemia may be accompanied by an increase in VLDL (type V hyperlipoproteinemia) in which LPL is not absent but may be defective. Occasionally, this disorder is linked to low levels of apo C-II.

Hypercholesterolemias

Four genetic defects cause the phenotype known as **familial hypercholesterolemia (FH)** (Table 18.6). In FH, LDL cholesterol (LDL-C) accumulates due to defects in hepatic LDL uptake in tissues, leading to premature CHD. LDL receptor defects underlie about 80% of FH cases and are caused by more than 800 different mutations. LDL receptor protein mutations are inherited as an autosomal dominant trait. The heterozygous form occurs in approximately 1 in 500 individuals, and their serum LDL cholesterol levels are between 320 and 500 mg/dL, contributing to atherogenesis.

Three other less common causes of FHs include defects in the apolipoprotein B-100 (apo B-100), LDL-receptor-adapter protein (LDLRAP1), and the serine endoprotease enzyme proprotein convertase subtilisin/Kexin type 9 (PCSK9). It should be noted that mutations in LDLR, apo B-100, and LDLRAP-1 are loss-of-function mutations, whereas in PCSK9 the mutation is a gain-of-function mutation. All four types of FH cause the same phenotype, showing elevated plasma levels of LDL cholesterol (see Figure 18.4 and Table 18.6). Secondary hypercholesterolemia may occur in hypothyroidism from delayed clearance of LDL, the result of downregulation of LDL receptors. In the nephrotic syndrome, overproduction of VLDL results either in elevated levels of VLDL itself or in elevated levels of LDL due to an increase in conversion. Elevated LDL cholesterol has also been reported in porphyria, anorexia nervosa, and Cushing's syndrome.

TABLE 18.6 Inherited Defects Causing Familial Hypercholesterolemias (also see Figure 18.4)

Mutant Protein	Inheritance Characteristics	Change in Metabolic Function
1. LDL Receptor (LDLR)	Autosomal dominant (with a gene-dosage effect)	Loss
2. Apo B-100 (ligand for LDLR)	Autosomal dominant	Loss
3. LDLR adapter protein (LDLRAP required for clathrin-mediated internalization of LDLR by serine)	Autosomal dominant	Loss
4. Downregulation of LDLR by serine endoprotease proprotein convertase (PCSK9)	Autosomal dominant	Gain

Hypolipidemias

In **abetalipoproteinemia** (acanthocytosis, **Bassen—Kornzweig syndrome**), lipoproteins containing apo B (LDL, VLDL, and chylomicrons) are absent. This experiment of nature indicates that apo B is absolutely essential for the formation of chylomicrons and VLDL (and hence LDL). Clinical findings include retinitis pigmentosa, malabsorption of fat, and ataxic neuropathic disease. Erythrocytes are distorted and have a "thorny" appearance due to protoplasmic projections of varying sizes and shapes (hence the term *acanthocytosis*, from the Greek *akantha*, meaning "thorn"). Serum triacylglycerol levels are low (see Clinical Case Study 18.1).

HDL-deficiency occurs in the mutant protein apo A-I$_{Milano}$, which is the result of a substitution of cysteine for arginine at residue 173 of apo A-I. This mutation apparently results in enhanced uptake and hepatic catabolism of HDL. No evidence of coronary artery disease, corneal opacification, xanthomas, or hepatosplenomegaly has been reported. Plasma concentrations of HDL and apo A-I are greatly reduced. The mode of inheritance is autosomal codominant.

Tangier disease is an autosomal recessive disorder. In the homozygous state, it is characterized by the absence of plasma HDL cholesterol and deposition of cholesterol esters in the reticuloendothelial system with hepatosplenomegaly, enlarged tonsils with yellowish-orange color, enlarged lymph nodes, and peripheral neuropathy. Patients with Tangier disease, despite low levels of HDL cholesterol, are not uniformly at risk for coronary artery disease. The molecular basis for Tangier disease is a defect of cholesterol efflux from cells. The defect has been localized to mutations in a specific ABCA1 transporter protein. Defective export of cholesterol from cells, particularly from tissue macrophages, causes accumulation of cholesteryl esters. Plasma apo A-I levels in Tangier disease undergo rapid clearance since nascent HDL is not made due to a lack of cholesterol efflux from the cells.

Atherosclerosis and Coronary Heart Disease

Atherosclerosis (*athero* = fatty and *sclerosis* = scarring or hardening) of the coronary and peripheral vasculature is the leading cause of morbidity and mortality worldwide. Lesions (called **plaque**) initiated by an injury to the endothelium cause a thickening of the intima of arteries, occlusion of the lumen, and compromised delivery of nutrients and oxygen to tissues (**ischemia**). Atherosclerotic lesions primarily occur in large- and medium-sized elastic and muscular arteries and progress over decades. These lesions cause ischemia, which can result in infarction of the heart (myocardial infarction) or brain (stroke), as well as abnormalities of the extremities. The proximate cause of occlusion in these pathological conditions is thrombus formation.

The normal artery wall is composed of three layers: the tunica intima, media, and adventitia respectively (Figure 18.7). On the luminal side, the tunica intima contains a single layer of endothelial cells. These cells permit passage of water and other substances from the blood into tissue cells. On the peripheral side, the intimal layer is surrounded by a fenestrated sheet of elastic fibers (the internal elastic lamina). The middle portion of the intimal layer contains various extracellular components of the connective tissue matrix and fibers, and occasional smooth cells, depending on the type of artery, and the age and sex of the subject.

The tunica media is composed of diagonally oriented smooth muscle cells. They are surrounded by collagen, small elastic fibers, and glycosaminoglycans (proteoglycans). Most cells are attached to one another by a specific junctional complex, which are arranged as spirals between the elastic fibers and support the arterial wall.

The tunica adventitia is separated from the tunica media by a discontinuous sheet of elastic tissue (the external elastic lamina) and consists of smooth muscle cells, fibroblasts, collagen fibers, and glycosaminoglycans. This layer is supplied with blood vessels to provide nutrients.

Atherosclerosis affects mainly the intimal layer. Lesions occur because of

1. Proliferation of smooth muscle cells
2. Lipid accumulation in smooth muscle cells and scavenger cells (resident macrophages and migrant plasma monocytes)
3. Connective tissue formation.

Perhaps the most satisfying hypothesis for the formation of atherosclerosis lesions is that of "response to injury" in which lesions are precipitated by some form of injury to endothelial cells. The injury may be caused by elevated plasma levels of LDL and modified LDL (oxidized LDL), free radicals (e.g., caused by cigarette smoking), diabetes mellitus, hypertension-induced shear stress, and other factors that lead to focal desquamation of endothelial cells such as elevated plasma homocysteine levels, genetic factors, infectious microorganisms (e.g., herpes viruses, *Chlamydia pneumoniae*), or combinations of these factors. The endothelial response to injury, which eventually leads to atherosclerosis, involves platelets, leukocytes, and a variety of cytokines, e.g., vasoactive molecules, adhesion molecules, and growth factors. Once the subendothelial region is exposed, circulating platelets and monocytes become attached and release factors with chemostatic and mitogenic properties. Platelet-derived growth factor (PDGF), for example, binds with high affinity to specific binding sites on the surface of smooth muscle cells. PDGF induces DNA synthesis, thereby initiating the proliferation of smooth muscle cells. It also increases binding and degradation of LDL. Monocytes enter the arterial wall at the site of injury and produce a mitogenic factor that appears to be as potent as PDGF.

Inflammation plays a significant role in the pathogenesis of atherothrombosis [1–4]. An acute-phase systemic inflammatory marker, C-reactive protein (CRP), is elevated in the plasma of patients with acute ischemia or myocardial infarction. In a prospective study consisting of apparently healthy men, individuals with high-baseline plasma high-sensitive CRP levels had a higher risk of myocardial infarction or ischemic stroke than men in the lower quartile of CRP values (<1 mg/dL low-, $1-3$ mg/dL moderate-, and >3 mg/dL high-risk categories). The use of aspirin, an anti-inflammatory and antiplatelet drug, reduced the risk of myocardial infarction among men in the highest quartile of plasma CRP levels. These data suggest that the plasma CRP level in apparently healthy men can predict myocardial infarction and ischemic stroke and is an independent marker not altered by plasma lipid levels, body mass index, smoking, or blood pressure. Statins also decrease CRP levels in addition to their hypocholesterolemic effect (refer to Clinical Case Studies 7.1 through 7.3 for elaboration on the aspects of biomarkers and myocardial injury).

Oxidative modification of LDL (e.g., lipid peroxidation) occurs primarily in the arterial intima and is a prerequisite for LDL uptake by macrophages. This uptake by macrophages is not regulated like LDL uptake by receptor-mediated endocytosis, and the uptake involves a scavenger receptor. This unregulated uptake leads to the accumulation of LDL so that macrophages become foam cells. The foam cells form yellow patches on the arterial wall that are called **fatty streaks** (discussed later). Since oxidized LDL is an important, and possibly an essential, component in the pathogenesis of atherosclerosis, the inhibition of LDL oxidation by a variety of antioxidants (e.g., vitamins A, C, and E) has been considered in the treatment of coronary artery disease. However, a cause-and-effect association between intake of antioxidant vitamins and prevention of coronary artery disease has not been determined.

Prostaglandins and thromboxanes play major roles in platelet adhesion and aggregation to damaged intima (Chapters 16 and 34). Normal endothelium synthesizes prostacyclin (PGI_2), which inhibits platelet aggregation. In damaged endothelium, platelets

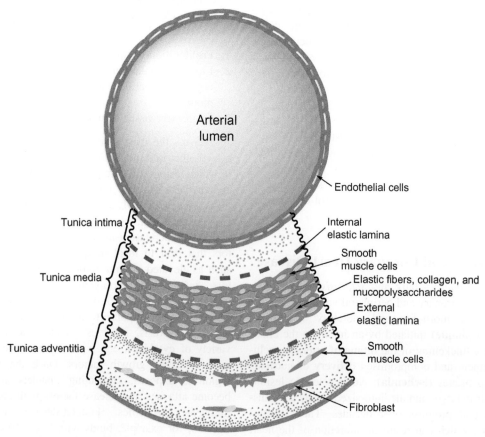

FIGURE 18.7 Cross-sectional view of an artery.

aggregate and form thromboxanes. Thromboxanes and endoperoxide (which has increased owing to lack of utilization in the synthesis of PGI_2) accelerate platelet aggregation and lead to the formation of a thrombus. Aspirin is an inhibitor of cyclooxygenase, which initiates the synthesis of prostaglandins and thromboxanes (Chapter 16). Aspirin therapy is beneficial in the prevention of heart attacks and strokes because it interferes with the function of platelets. However, aspirin may increase the risk of hemorrhagic stroke. Thus, the benefits and risks of aspirin therapy in patients with known coronary artery disease or history of stroke need to be assessed before the initiation of aspirin therapy.

Epidemiological studies have established that a reduction in plasma total cholesterol levels reduces the risk of coronary heart disease (CHD). Several studies show that therapy with cholesterol-reducing HMG-CoA reductase inhibitors (called **statins**) in patients with coronary artery disease and elevated cholesterol levels reduces the risk of both fatal and nonfatal heart attacks. Cardiovascular benefits from statin therapy have been observed even among patients with "normal" cholesterol levels. These data and the significant decreases in CHD mortality in industrialized nations in recent years, which were accompanied by decline in total plasma cholesterol levels, suggest that although a large fraction of an individual's plasma cholesterol level may be genetically determined, sociocultural determinants can strongly influence susceptibility to heart disease.

The relationship between the development of premature CHD and LDL cholesterol is well established. Epidemiological data have also suggested an inverse correlation between HDL cholesterol and premature CHD. At any given level of LDL cholesterol, the probability of CHD increases as the level of HDL cholesterol decreases. These observations suggest that a lowered HDL cholesterol level (or HDL deficiency) is an independent risk factor for premature CHD. The cardioprotective effect of HDL cholesterol may involve removal of cholesterol from the peripheral tissues and its transport to the liver for removal. As discussed earlier, the efflux of cholesterol from peripheral cells is mediated by ABCA1-transporter. Defects and polymorphisms of this protein are determining factors in the plasma HDL cholesterol levels. Animal studies have shown that the regulation and expression of the ABCA1 protein plays a significant role in cholesterol efflux and reverse cholesterol transport. Furthermore, the ABCA1 protein present in the intestinal cell membrane mediates the cholesterol efflux into the lumen, thus preventing cholesterol absorption. Potential mechanisms of the antiatherogenic effects of HDL cholesterol include

1. Inhibition of conversion of LDL to oxidized LDL, which is preferentially taken up by the tissue macrophages;
2. The prevention of adhesion of monocytes to the endothelium; and
3. The prolongation of the half-life of prostacyclin produced by endothelial cells to promote vasodilatory effects.

The inhibition of LDL oxidation by HDL has been attributed to both an HDL-associated and calcium-dependent enzyme known as **paraoxonase** and to a phospholipid fraction. HDL cholesterol and triacylglycerol levels are inversely correlated across various populations. Decreased HDL levels are associated with the male sex, obesity, physical inactivity, cigarette smoking, and hypertriacylglycerolemia. Increased HDL levels are associated with the female sex and regular vigorous exercise (e.g., running). The incidence of atherosclerosis in premenopausal women is significantly lower than in men. This difference has been attributed to estrogen. In fact, in postmenopausal women with a decline in estrogen levels, the risk of atherosclerosis rises.

The approach taken in the management of individuals with low HDL cholesterol levels (Table 18.4) is similar to that of patients in whom LDL cholesterol levels are high (discussed later). First, it is essential to exclude secondary causes of lipid abnormalities, which include acute illness, chronic diseases, and some commonly prescribed drugs (Table 18.7). Second, it is important to initiate lifestyle changes that include weight loss in obese individuals, dietary modifications (e.g., reduced saturated fat intake, including *trans*-fatty acids), exercise, avoidance of tobacco, and moderate alcohol consumption. Individuals who are homozygous for a slow-oxidizing alcohol dehydrogenase gene (ADH3) have higher HDL cholesterol levels and a lower risk of myocardial infarction. In the absence of positive results with lifestyle therapy, pharmacological agents for increasing HDL cholesterol levels are used, including niacin, and estrogen in postmenopausal women. Modest increases in HDL cholesterol levels are also observed with HMG-CoA reductase inhibitors and fibric acids.

Another factor that regulates HDL cholesterol levels is the plasma level of cholesteryl ester transfer protein (CETP). CETP, a hydrophobic glycoprotein (M.W. 741,000), facilitates the transfer of cholesteryl esters in HDL and triacylglycerols in LDL and VLDL (see earlier discussion).

In CETP deficiency due to a point mutation ($G \rightarrow A$) in a splice donor site that prevents normal processing of mRNA, the plasma HDL cholesterol levels of affected individuals are markedly high, with decreased LDL cholesterol. In affected families there was no evidence of premature atherosclerosis and, in fact, there was a trend toward longevity. These observations support the role of CETP and the antiatherogenic property of HDL. However, although pharmacological inhibition by torcetrapib increased plasma HDL cholesterol levels, the treatment produced adverse cardiovascular events (discussed earlier). Not all factors that elevate HDL levels may be protective against atherosclerosis. Because hepatic lipase deficiency causes an increase in triglyceride-enriched HDL (and IDL), it is associated with premature atherosclerosis. Examples of low plasma HDL cholesterol levels that are not associated with coronary artery disease are also known. In Tangier disease, despite low plasma or HDL cholesterol levels, coronary artery disease is an inconsistent finding. In individuals with apo A-I$_{Milano}$, despite low HDL cholesterol levels, longevity has been observed in affected individuals.

Elevated Lp(a) is a major independent risk factor for atherosclerosis in patients with familial hypercholesterolemia. This risk is independent of levels of LDL, HDL, age, sex, and smoking habits, and is primarily dependent on genetic factors. The exact mechanism of Lp(a) acceleration of atherosclerosis is not understood but may be attributable to its potential inhibition of blood clot dissolution caused by its structural similarity to plasminogen (see above and Chapter 34).

TABLE 18.7 Examples of Secondary Causes That Can Alter Plasma Lipoprotein Levels

Surgical procedures
Acute Illnesses
Viral illness
Burns
Myocardial infarction
Acute inflammatory disorders
Chronic Conditions
Diabetes mellitus
Thyroid disease
Uremia
Nephrotic syndrome
Liver disease
Commonly Prescribed Drugs
Diuretics
Progestins
Androgens
β-Adrenergic blocking agents

Lipid-Lowering Methods

Since hypercholesterolemia (in particular, LDL cholesterol) increases the risk of CHD, it seems reasonable to lower cholesterol levels in patients whose levels put them at risk. Before treatment, other risk factors, such as hypertension, cigarette smoking, obesity, and glucose intolerance need to be evaluated and corrected. Disorders that exacerbate hyperlipoproteinemia (e.g., chronic ethanol abuse, hypothyroidism, diabetes mellitus) need to be treated before lipid-lowering measures are taken (discussed earlier, Table 18.7). In uncomplicated hyperlipoproteinemia, the first step is to reduce dietary intake of cholesterol and saturated fats and to establish an ideal body weight.

If dietary therapy is unsuccessful, drug therapy should be employed. Five classes of drugs are available for treatment of hyperlipoproteinemias. Their effects are due to decreased absorption or production, or enhanced removal of lipids. They include statins, nicotinic acid, gemfibrozil, cholestyramine, and ezetimibe.

1. Statins (Chapter 17): Inhibit cholesterol biosynthesis, lower LDL-C levels, and have anti-inflammatory properties.
2. Nicotinic acids: Inhibit adipocyte lipolysis by stimulation of Gi-protein-coupled receptors followed by decreasing cAMP levels and increasing HDL-C levels.
3. Gemfibrozil: Activates lipoprotein lipase, leading to rapid turnover of VLDL. This ligand for the nuclear transcription factor PPAR-α causes modest elevation of HDL-C.
4. Cholestyramine: Bile acid sequestrant in the GI tract and its removal lowers LDL-C.
5. Ezetimibe: Inhibits intestinal cholesterol absorption and lowers LDL-C; used in the treatment of sitosterolemia due to defects in ABCG5 and ABCG8.

A novel method of lowering plasma LDL cholesterol is by raising LDL receptor levels. Since PCSK9 degrades LDL receptors, its inactivation by using specific PCSK9 antibodies reduces plasma LDL cholesterol levels (Figure 18.4). This therapeutic approach to lowering LDL cholesterol levels was devised based on the observation of individuals with loss of function PCSK9 mutations, which led to lowered LDL cholesterol levels and a reduced risk of CHD [5,6].

Recent guidelines emphasize using mainly statin therapy to lower plasma cholesterol levels. [7,8] These guidelines are summarized in reference [7] with examples of clinical cases. It should be noted that one of the undesirable side effects of statins is rhabdomyolysis. This has been attributed to statin-associated coenzyme Q_{10} biosynthesis. Recall that statins inhibit an early step in the biosynthesis of not only cholesterol but also a number of other essential metabolites. See case studies listed in references [9–12].

CLINICAL CASE STUDY 18.1 Familial Hypo β-lipoproteinemia

This clinical case was abstracted from: U. Turk, G. Basol, B. Barutcuoglu, F Sahin, S. Habif, P Tarugi, O. Bayindir, A 54-year-old diabetic man with low serum cholesterol, Clin. Chem. 58(5) (2012) 826−830.

Synopsis

A diabetic 54-year-old subject on routine serum lipid measurements revealed markedly decreased values of serum cholesterol, LDL cholesterol, and apolipoprotein B. All of the reminder studies were not significant except for elevated serum glucose and HbA$_{1C}$ values and the presence of serum IgA(κ) monoclonal protein. Further studies revealed that IgA(κ) monoclonal gammopathy was considered as monoclonal gammopathy of undetermined significance (MGUS) and not related to low-cholesterol values. The MGUS aspect is discussed in Chapter 15, and MGUS was considered relevant to the subject's low serum lipid values. After excluding secondary causes of hypo β-lipoproteinemia (HBL), genetic studies revealed the subject is heterozygous for familial HBL (FHBL).

Teaching Points

1. HBL can be caused by several nongenetic factors, and it is essential to exclude these possibilities before a genetic cause is considered.
2. FHBL is an autosomal codominant disorder accompanied by a mutation in the apolipoprotein B gene, which results in truncated apo-B lipoprotein.
3. Whereas the homozygous FHBL is accompanied by severe clinical manifestations such as malabsorption, retinitis pigmentosa, or neurologic symptoms, the heterozygous subjects may have protective cardiovascular effects due to decreased serum lipid parameters.

REQUIRED READING

[1] R.E. Gerszten, A.M. Tager, The monocyte in atherosclerosis—should I stay or should I go now? N. Engl. J. Med. 366 (2012) 1734−1736.

[2] F.K. Swirski, M. Nahrendorf, Leukocyte behavior in atherosclerosis, myocardial infarction, and heart failure, Science 339 (2013) 161−166.

[3] G.J. Randolph, Proliferating macrophages prevail in atherosclerosis, Nat. Med. 19 (2013) 1094−1095.

[4] P. Libby, Mechanisms of acute coronary syndromes and their implications for therapy, N. Engl. J. Med. 368 (2013) 2004−2013.

[5] S.G. Young, L.G. Fong, Lowering plasma cholesterol by raising LDL receptors—revisited, N. Engl. J. Med. 366 (2012) 1154−1155.

[6] T. Kosenko, M. Golder, G. Leblond, W. Weng, T.A. Legace, Low density lipoprotein binds to proprotein convertase subtilisin/kexin type-9 (PCSK9) in human plasma and inhibits PCSK9-mediated low density lipoprotein receptor degradation, J. Biol. Chem. 288 (2013) 8279−8288.

[7] J.F. Keaney, G.D. Curfman, J.A. Jarcho, A pragmatic view of the new cholesterol treatment guidelines, N. Engl. J. Med. 370 (2014) 275−278.

[8] N.J. Stone, J.G. Robinson, A.H. Lichtenstein, D.C. Goff, D.M. Lloyd-Jones, S.C. Smith, et al., Treatment of blood cholesterol to reduce atherosclerotic cardiovascular disease risk in adults: synopsis of the 2013 American College of Cardiology/American Heart Association Cholesterol Guideline, Ann. Intern. Med. 160 (2014) 339−343.

[9] A.A. Manoukian, N.V. Bhagavan, T. Hayashi, T.A. Nestor, C. Rios, A.G. Scottolini, Rhabdomyolysis secondary to lovastatin therapy, Clin. Chem. 36 (1990) 2145−2147.

[10] W.S. David, D.A. Chad, A. Kambadakone, E.T. Hedley-Whyte, Case 7-2012: a 79-year-old man with pain and weakness in the legs, N. Engl. J. Med. 366 (2012) 944−954.

[11] A.B. Goldfine, Statins: is it really time to reassess benefits and risks? N. Engl. J. Med. 366 (2012) 1752−1757.

[12] B.A. Diebold, N.V. Bhagavan, R.J. Guillory, Influences of lovastatin administration on the respiratory burst of leukocytes and the phosphorylation potential of mitochondria in guinea pigs, Biochim. Biophys. Acta 1200 (1994) 100−108.

Chapter 19

Contractile Systems

Key Points

1. Introduction
   a. Motility: requires two classes of protein, cytoskeletal and motor.
   b. Chemomechanical transduction: the controlled transformation of the chemical energy of nucleoside triphosphates into mechanical work.
2. Review of the structure and genesis of muscle
   a. Muscle cells form by the fusion and elongation of myoblasts.
      i. Structure of muscle cells: myofibrils, sarcomeres, costameres, and sarcoplasmic reticulum
3. Detailed structure of myofibrils
   a. Thin myofilaments
      i. Thin filament backbones are polymers of G-actin.
      ii. Thin filaments also contain the regulatory proteins troponin and tropomyosin.
   b. Thick myofilaments
      i. Thick filaments are mainly polymers of myosin of the Myosin II family.
      ii. Myosin II is a hexamer with a rigid tail and two globular heads.
4. Innervation of muscle
   a. The group of muscle fibers innervated by axon branches from a single motor neuron (a muscle unit) is the basic functional unit of skeletal muscle.
   b. Cardiac muscle cells do not receive motor innervation.
5. Mechanism of contraction
   a. Overview
      i. The sliding filament model and the cross-bridge hypothesis explain contraction.
   b. Excitation/contraction coupling
      i. Action potentials cause proteins in the t-tubules and in the sarcoplasmic reticulum (SR) membrane to interact to release Ca^{2+} from SR.
   c. Activation of contraction
      i. Binding of Ca^{2+} to troponin reverses the inhibition of actin binding to myosin exerted by tropomyosin.
   d. Cross-bridge cycling
      i. When tropomyosin inhibition is reversed, myosin heads and actin spontaneously bond.
      ii. Actin/myosin binding alters the neck region of the myosin heads, exerting force between the thick and thin filaments.

   iii. ATP binding dramatically reduces myosin affinity to actin, allowing detachment.
   iv. The ATP subsequently hydrolyses, restoring myosin's affinity for actin.
   v. The cross-bridge cycle repeats, producing force and filament sliding, so long as adequate concentrations of ATP and Ca^{2+} are maintained.
   e. Relaxation
      i. Return to the resting state requires resequestration of Ca^{2+} in the SR.
      ii. By two mechanisms, the Ca^{2+} spike terminates itself.
6. Diversity and plasticity in skeletal muscle
   a. Most muscle fibers can be classified by their twitch characteristics, which myosin is present, and metabolic profile.
   b. Muscle fiber diversity is due to differential expression of multiple genes for contractile and regulatory proteins in the various fiber types.
   c. The fiber type distribution has several determinants, especially the innervation.
7. Energy supply in muscle
   a. A high rate of ATP production is required to sustain muscular work.
   b. AMP-dependent kinase (AMPK) is a key sensor of energy demand.
   c. Aerobic fibers have high activities of the enzymes for fatty acid oxidation.
   d. Fast (type IIx) fibers rely on glycolytic rather than oxidative metabolism.
   e. Energy transfer from mitochondria to myofibrils is mediated by the phosphocreatine shuttle.
8. Regulation of smooth and cardiac muscle
   a. Increase in $[Ca^{2+}]_i$ initiates contraction in smooth and cardiac muscle.
   b. Smooth muscle is more dependent than skeletal upon entry of extracellular Ca^{2+}.
   c. Smooth muscle behavior differs greatly from one tissue to another.
   d. Phosphorylation of myosin light chains by myosin light chain kinase (MLCK) is a key regulator of smooth muscle contraction.
   e. Excitation/contraction coupling in myocardium is mediated mainly by calcium-induced calcium release (CICR).

N. V. Bhagavan and C.-E. Ha: Essentials of Medical Biochemistry, Second edition. DOI: http://dx.doi.org/10.1016/B978-0-12-416687-5.00019-1

9. Regulation of skeletal myogenesis by mTOR
   a. Mammalian (or mechanistic) target of rapamycin (mTOR) is a master regulator of skeletal myogenesis. Activation of mTOR pathways leads to muscle cell growth, differentiation, and survival. It is upregulated in diseases of proliferation (cancer).
   b. Inhibition of mTOR depends on the site of binding.
10. Regulation of skeletal muscle by myostatin
    a. Myostatin is a powerful negative regulator of skeletal muscle mass. It suppresses muscle growth. Myostatin plays an important role in muscle-wasting diseases.
    b. Therapeutic inhibition of myostatin has been shown beneficial in the treatment of cachexia in many muscle-wasting diseases such as cancer, AIDS, and COPD.
    c. There is cross-talk between skeletal muscle and adipose tissue such that the actions of myostatin decrease peripheral tissue fatty acid oxidation and may also play a role in morbid obesity and insulin resistance.
11. Inherited diseases of muscle
    a. Genetic diseases of myocardium are diagnosed much more often than genetic disease of skeletal muscle.
    b. Duchenne and Becker's muscular dystrophy are due to defects in or deletion of the gene for dystrophin.
    c. Defects in ion channel genes cause channelopathies, characterized by instability of membrane potential, and sometimes degenerative syndromes.
    d. Metabolic disorders can affect any aspect of substrate storage or utilization, and can cause degenerative syndromes or dynamic syndromes.
    e. Mitochondrial mutations impair ATP production.
12. Nonmuscle systems: actin
    a. When a nascent actin filament is bound at one end and growing at the other, growth can generate movement, such as extension of filopodia.
    b. Drugs that sever and/or cap actin filaments block processes such as phagocytosis and cytokinesis.
13. Nonmuscle systems: cilia
    a. Cilia are hair-like projections whose function is to propel fluid or other cells past the cell surface by an asymmetric bending motion.
    b. Cilia contain microtubules that are polymers of globular a and b tubulin.
    c. Cilia feature nine pairs of microtubules forming a cylinder, with two singlet tubules in the center.
    d. Motor proteins, called dynein, exert force between adjacent pairs of tubules.
    e. Chemomechanical transduction by dynein—tubule systems is generally similar to myosin—actin systems.
    f. Microtubules form the mitotic spindle and the scaffold for axonal transport.
    g. Antineoplastic drugs such as the taxanes and vinca alkaloids disrupt assembly or disassembly of microtubules, blocking cell division.
14. Immotile cilia syndrome
    a. Defects in genes coding for any proteins needed for synthesis and assembly of microtubules can cause absent or nonfunctioning cilia.
    b. Affected individuals have frequent/chronic bronchial, sinus, and ear infections.
    c. Kartagener's syndrome is the combination of situs inversus with the recurrent infections mentioned previously.

INTRODUCTION

All cells produce movement internally, and many are capable of motility or of changing shape. Cells specialized for changing the dimensions or shape of anatomical structures or for movement of body parts with respect to each other are called **muscle cells**.

Motility usually requires two classes of protein: cytoskeletal and motor. The ability of a cell to hold or to change shape and to move organelles within it depends on the existence of a cytoskeleton comprising **actin filaments** (microfilaments), **microtubules**, and **intermediate filaments** capable of transmitting force. Actin filaments and microtubules contain predominantly **actin** and **tubulin**, respectively.

Intermediate filaments are structural proteins not directly involved in motion. The most abundant of these in muscle cells is desmin, which performs the functions described under "Sarcomeres and Costameres" later in this chapter. Actin, tubulin, and several of the intermediate filament proteins are found in all cells.

Chemomechanical transduction also requires motor proteins. Chemomechanical transduction is the controlled transformation of the chemical energy of nucleoside triphosphates into mechanical work. In addition to the actin filaments and microtubules, the "motor" proteins, myosin and dynein or kinesin, are needed for chemomechanical transduction. These large proteins are actin- or tubulin-activated ATPases, which (in effect) capture some of the energy of hydrolysis of ATP in the form of mechanical energy. Several other proteins are associated with this process, including regulatory proteins that control contractile activity and enzymes involved in maintaining high-energy phosphate supply. This chapter discusses muscle, an actin/myosin filament system, and cilia, a microtubule/dynein system.

MUSCLE SYSTEMS

Review of the Structure and Genesis of Muscle

Mammals have four cell types specialized for contraction: skeletal muscle, cardiac muscle, smooth muscle, and myoepithelial cells (Table 19.1). Skeletal and cardiac muscle contract with more force and much more speed,

TABLE 19.1 Comparison of Four Types of Mammalian Cells Specialized for Contraction

Cell Type	Structure	Contractile Properties	Function
Skeletal muscle	Long syncytial, multinucleated cells; orderly arrangement of myosin and actin filaments gives striated appearance; each fiber is directly innervated by a motor neuron	Rapid, powerful contractions; can shorten to 60%–80% of resting length; contraction is initiated by the central nervous system under voluntary control	Movement of the bony parts across joints
Cardiac muscle	Similar to skeletal muscle but extrinsic innervation is only at the specialized nodal pacemakers; the action potential is conducted from cell to cell via gap junctions (nexuses)	Similar to skeletal muscle but contraction is initiated by automatic firing of pacemaker cells; contraction is slower and more prolonged than in skeletal muscle	Movement of blood by repetitive rhythmic contractions; beats about 3 billion times during a normal lifetime
Smooth muscle	Elongated, tapering cells; mononuclear; no striations; occur singly, in small clusters, or in sheets enclosing organs; innervated by local plexuses and extrinsically by autonomic nerves	Slow contractions under involuntary control; can shorten to 25% of resting length	Control of shape and size of hollow organs such as the digestive, respiratory, genital, and urinary tracts, and the vascular system
Myoepithelia	Basket-shaped, mononuclear cells surrounding the acini of exocrine glands; have cytoplasmic fibrils resembling smooth muscle; derived from ectoderm rather than mesoderm	Contraction stimulated by hormones (e.g., oxytocin) and presumably by autonomic nerves; may have noncontractile functions such as pressure transduction in the renal cortex	Contraction to expel contents of exocrine glands (salivary, sweat, mammary); form the dilator muscle of the iris; may be pressure transducers in juxtaglomerular cells

shorten less, and consume much more ATP during maintenance of tension than smooth muscle or myoepithelium. Characteristics of these cell types vary considerably with the functional roles of the tissues in which they occur. However, the fundamental mechanism of contraction is the same in all, and we will use skeletal muscle as our model.

Origin of Muscle Cells

Muscle cells form by the fusion and elongation of numerous precursor cells called **myoblasts**. Some stem cell precursors of myoblasts remain in an adult animal, located between the sarcolemma and basement membrane of mature muscle cells, and these are called **satellite cells** in this setting. Since each myoblast contributes its nucleus to the muscle cell thus formed, skeletal muscle fibers are all multinucleated, the longest having 200 or more nuclei. The skeletal muscles of the torso and limbs arise from the mesoderm of the somites, while those of the head arise from the mesoderm of the somitomeres, which contribute to the branchial (pharyngeal) arches.

Several growth and differentiation factors have been identified, and they play a role in committing embryonic stem cells to the muscle cell lineage and influence the rate and extent of their proliferation and differentiation. These include four growth factors of the helix–loop–helix (HLH) family of DNA-binding transcription factors, called **MyoD**, **Myf5**, **Mrf4**, and **myogenin**. In adult animals, IGF-I (see Chapter 29) is an important stimulator of satellite cell proliferation and differentiation. There are also inhibitory factors such as myostatin, the absence of which causes substantially greater muscle mass in animals with the corresponding gene defect. Pharmacologic inhibition of myostatin has therefore been recently studied for the treatment of muscle-wasting diseases, and will be discussed later in the chapter. The best-studied myogenic factors and their actions are listed in Table 19.2.

Structure of Skeletal Muscle

Individual muscle cells, or fibers, are elongated, roughly cylindrical, and usually unbranched, with a mean diameter of $10-100\,\mu m$. The plasma membrane of muscle fibers is called the **sarcolemma**, and fibers are surrounded by structural filaments of the extracellular matrix, which are often described as forming a basement membrane.

Within each fiber is a longitudinal network of tubules called the **sarcoplasmic reticulum**, or **SR**, analogous to the endoplasmic reticulum of other cells. Release of Ca^{2+} from the SR is the key step in coupling the sarcolemma action potential to activation of contraction. SR membranes

TABLE 19.2 Factors Known to Influence Muscle Development

Regulatory Factor	Role in Development	Comments
myoD	Can initiate the differentiation of stem cells to myoblasts; may stimulate proliferation of myoblasts	Either myoD or myf 5 must be present: deletion of both prevents myoblast formation
myf 5	Same as myoD	
myogenin	Required for terminal differentiation of myoblasts, myoblast fusion, and development of myotubes	Deletion of myogenin gene results in accumulation of myoblasts instead of formation of muscle cells
mrf 4	Required for development of normal amount of myofibrils; may be needed for maintenance of mature muscle cells	Deletion results in weak, fragile fibers
myostatin	Inhibits proliferation of myoblasts and formation of myotubes; plays major role in terminating the myogenic program in small animals, lesser role in large animals	Defect in myostatin gene in cattle explains the unusual degree of muscularity in Belgian Blue and Piedmontese breeds

bear large numbers of Ca^{2+} pump proteins whose role is to pump Ca^{2+} into the lumen of the SR, where it is sequestered until the fiber is stimulated. Crossing the SR perpendicular to the cell axis are **transverse tubules** (**T-tubules**), which are invaginations of the sarcolemma. They conduct the action potential into the interior of the cell, which triggers the release of Ca^{2+} from SR.

Muscle cells contain many mitochondria, which are often present as reticulum-like structures extending longitudinally in the fiber (most of them near the sarcolemma). These provide much of the high-energy phosphate needed to power contraction and to operate the Ca^{2+} pumps that control the cytosolic calcium concentration. Different types of skeletal muscle fibers differ considerably in the extent and organization of both their SR and mitochondria.

Muscle fibers contain long cylindrical myofibrils (typically $1-2\,\mu m$ in diameter) aligned longitudinally and comprising interpenetrating arrays of thin myofilaments ($6-7\,nm$ in diameter) and thick filaments ($15-16\,nm$ in diameter). These structures are the contractile apparatus of the fiber.

Myofibrils

Longitudinal sections of skeletal and cardiac muscle exhibit alternating light and dark bands (striae) visible by light microscopy and are called **striated muscle**. This banded pattern is due to regions of overlap alternating with regions of nonoverlap of the thick and thin filaments along the length of the myofibrils. Thick filaments are myosin polymers, and thin filaments are actin polymers. In cross-sections, these filaments are arranged in interpenetrating hexagonal arrays. Between the thick and thin filaments, cross-bridges may form. In addition to thick and thin filaments, the third major filament is **titin**.

The unique arrangement of these three filaments forms contractile regions called **sarcomeres**, which repeat along the length of the myofibril.

Sarcomeres and Costameres

The sarcomere is the basic structural and functional unit of the fibril. It is bordered by a Z-band on each end with adjacent I-bands, and there is a central M-line with adjacent H-bands and partially overlapping A-bands. The Z-band (or Z-disk) is a dense fibrous structure made of actin, α-actinin, and other proteins. Thin filaments (or actin filament) are anchored at one end at the Z-band. Titin is anchored to both the Z-band and the M-line. Thick filaments are anchored in the middle of the sarcomere at the M-line. The I-band is the region on either side of a Z-disc that contains only thin filaments and titin. This partial overlap in filaments makes the A-band darker at its ends, leaving a light area in the middle (H-band) where there is no overlap with the light bands. A key clue to the mechanism of contraction was the finding that during contraction the H and I bands shorten, while the A bands do not.

The sarcolemma has more-or-less regularly spaced electron-dense patches or rings of large membrane-spanning multiunit proteins. These are analogous to desmosomes, and the proteins therein are structural proteins linking the meshwork of cytoskeletal filaments that tie the myofibrils together to the extracellular matrix. The complex of cytoskeletal actin and IFs and the membrane proteins is called a **costamere**.

Desmin, an intermediate filament protein, forms a network from one Z-disk to the next across the myofibril. These networks are linked by actin, via the protein dystrophin, to the sarcolemma. This costamere structure helps hold the sarcomeres in register and plays an important role in transmitting force produced in the myofibrils to adjacent myofibrils and, via the transmembrane components of the costamere, to the extracellular matrix. This mechanism is the major means of force transmission from myofibril to tendon. Duchenne and Becker's muscular dystrophy are due to mutations in, or deletion of, the dystrophin gene.

Thin Myofilaments

Thin myofilaments are polymers of globular actin (G actin) molecules (M.W. 42,000). Although the actin gene has apparently undergone considerable duplication, normally only six actin genes are expressed in man. Three, called α-**actins**, are the actins found in sarcomeric thin filaments, each in a different type of muscle. Skeletal G-actin has N^3-methylhistidine at position 73. Myosin also contains methylhistidine. Other proteins contain methylhistidine (e.g., acetylcholinesterase) but are quantitatively insignificant compared to actin and myosin. Consequently, urinary methylhistidine excretion has been used as a marker for contractile protein turnover and as an indirect indicator of muscle mass.

G-actin has four domains arranged in a "U," such that the tertiary structure has a cleft. This structure is stabilized by the binding of MgATP or MgADP in the cleft (called an **ATPase fold**). Binding of MgATP (or MgADP) holds the cleft closed and stabilizes the molecule. ATP-G-actin spontaneously polymerizes into filaments of F-actin in solutions of physiologic ionic strength. Neither hydrolysis of the bound ATP nor other structural or catalytic proteins is required for polymerization. Each of the four G-actin domains binds noncovalently with an adjacent G-actin, such that each G-actin is thought to interact with four other G-actins in the growing filament. The four inter-actin sites are specific, such that all the G-actins in a filament have the same orientation; i.e., all have the cleft pointing toward the same end of the filament. Accordingly, the whole filament has a polarity.

G-actin is very highly conserved, both across actin genes within a species and across species. One implication of this is that variation in contractile properties between muscle types must be mainly due to the motor protein (myosin) or to the various regulatory proteins.

Thin filaments also contain two regulatory proteins called **tropomyosin** and **troponin**. Tropomyosin (M.W. 68,000) is a coiled-coil α-helical heterodimer. The proportions of α and β chains present, and which α and β chain genes are expressed, vary with the type of muscle fiber. Tropomyosin binds to F-actin along the grooves formed by the spiraling of the filament and is arranged end-to-end. Tropomyosin normally inhibits interaction

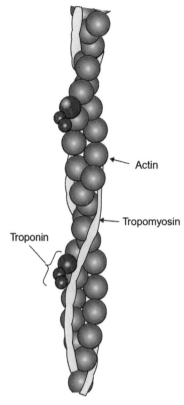

FIGURE 19.1 Model of a thin filament. Two tropomyosin filaments, each composed of two subfilaments, wind around the actin chain. They block the binding of the globular heads of myosin molecules (in thick filaments) to the actin molecules. Troponin consists of three polypeptides and binds to both actin and tropomyosin.

between actin and myosin by interfering with a projection on the myosin head.

Near one end of each tropomyosin is a molecule of **troponin** (M.W. approximately 76,000), a trimer of noncovalently bound subunits called **troponins C, I, and T** (**TnC, TnI, and TnT**) (Figure 19.1). These bind to each other, to tropomyosin, and to actin. TnC (M.W. approximately 18,000,159 amino acids) is structurally and functionally similar to calmodulin. Each TnC molecule can cooperatively bind up to four Ca^{2+} ions. Binding of Ca^{2+} to TnC (during the calcium spike induced by depolarization of the cell) reverses the tropomyosin inhibition of actin−myosin binding. Phosphorylation of TnT and TnI by a cAMP-dependent protein kinase increases actomyosin ATPase activity in cardiac muscle but has little functional significance in skeletal muscle.

Thick Myofilaments

Thick myofilaments are made primarily of myosin. Myofibrillar myosin is one member of the multigene myosin II subfamily in the myosin superfamily. Of the thirteen types of myosin that have been described, seven

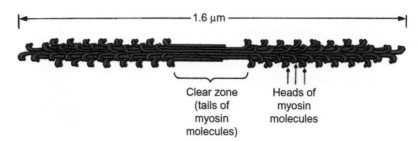

FIGURE 19.2 Organization of myosin in striated-muscle thick filaments. Filament formation begins with tail-to-tail (antiparallel) binding of myosin molecules, with subsequent parallel binding of myosin molecules to the ends of the initial nucleus, leaving the central clear zone. There are approximately 400 myosin molecules per striated thick filament.

occur in mammalian species. Most are involved in intracellular transport processes. The muscle type, myosin II, is a large hexamer (M.W. approximately 475,000) comprising two **myosin heavy chains**, or **MHCs** (M.W. approximately 200,000) and four **myosin light chains**, or **MLCs** (M.W. approximately 16,000 to 20,000). These are noncovalently linked. The size and properties of the light chains vary by tissue.

The physiologically active form of muscle myosin is the thick myofilament (Figure 19.2) formed by aggregation of myosin tails to create a bundle with the heads sticking outward. Head pairs are spaced 60° apart angularly around the filament, which corresponds to the spacing of the thin filaments in a hexagonal pattern around the thick filaments. Filament assembly depends on characteristics of the myosin tail, especially the LMM, in which much of the AA sequence exhibits a heptapeptide repeat, the sequence of which favors the formation of hydrogen and ionic bonds between adjacent LMM chains. About 400 myosin molecules aggregate to form one skeletal muscle thick filament, the number varying with species and the specific muscle.

Another set of filaments, sometimes called the **third filament system**, helps to stabilize sarcomere structure and accounts for most passive tension. A very long protein called **titin** or **connectin** holds the thick filament array centered in the sarcomere.

Another very long protein, called **nebulin**, is associated with the thin filaments. Nebulin has a molecular weight of about 700,000. Nebulin extends from either side of the Z-discs along the entire length of the thin filaments, typically 1 μm long. There are several observations that have led to the suggestion that nebulin serves as a template for thin filament assembly.

Myofibrils Contain Numerous Accessory Proteins

In the center of the A band, the thick filaments are linked by a network of proteins forming the M-line. The pattern of cross-linking of these proteins, such as **myomesin**,

skelemin, and **M-protein**, and their ability to bind the thick filament protein myosin, probably explain the hexagonal arrangement of the thick filaments. Cytosolic creatine kinase activity also localizes extensively to the M-line region and may be a property of one of the myosin cross-linking M-line proteins. Another thick filament protein, **myosin-binding protein C** (called **MyBP-C**), seems to play a role in formation of thick filaments and in binding titin (above) to the thick filament. Mutations in the gene for MyBP-C account for 15%−20% of cases of **familial hypertrophic cardiomyopathy**.

Muscle Function: Mechanism and Control of Contraction

With few exceptions, a vertebrate skeletal muscle fiber is innervated at one region along its length by a branch of an axon from a spinal or brain stem α-motor neuron (α-MN). Each α-MN innervates 10 or fewer fibers in small muscles used for finely graded movements, up to 2000 or more in large limb muscles. The group of fibers innervated by a single α-MN is called a **muscle unit**, and the muscle unit and the α-MN driving it form a motor unit. The muscle unit is the basic functional unit of skeletal muscle, since normally all fibers in a muscle unit are stimulated when their α-MN is active. In contrast, cardiac muscle contracts in response to action potentials generated spontaneously in the heart and transmitted from cell to cell through gap junctions. Contraction of the heart is modified, but not initiated, by its innervation.

Mechanism of Muscle Contraction: Overview

Our current understanding of the mechanism of contraction is reflected in the **sliding filament model** and the **cross-bridge hypothesis**. In the sliding filament model, the shortening of sarcomeres responsible for contraction in muscle is due to the sliding of the thick and thin filaments past one another due to interactions between them, such that the thin filaments are pulled toward the center of the thick filament arrays. Movement between filaments is believed to be driven by interaction between the myosin heads and the thin filaments in which the binding of myosin to actin, or **cross-bridge** formation, triggers changes in the myosin head structure that create mechanical tension. This is the **cross-bridge hypothesis**. For significant shortening to occur by this mechanism, cross-bridges must repeatedly form, create tension, be broken, and reform. This is called **cross-bridge cycling**. The speed of sarcomere shortening, then, depends on the mean cross-bridge cycling rate.

Mechanism of Contraction: Excitation/Contraction Coupling

The primary intracellular event in the activation of contraction is an increase in $[Ca^{2+}]_i$ in the sarcoplasm, from about $0.05\ \mu mol/L$ at rest to about $5-10\ \mu mol/L$ during repetitive stimulation. In skeletal muscle, nerve action potentials release acetylcholine (Ach) onto the postsynaptic membrane, depolarizing it and initiating an action potential that propagates along the sarcolemma and down the T-tubules. This depolarization results in Ca^{2+} release from the sarcoplasmic reticulum. The concentration ratio from SR to sarcoplasm is 10^4 to 10^5, so that the release of Ca^{2+} requires only that the SR conductance to Ca^{2+} be increased. The connection between depolarization and Ca^{2+} release is called **excitation–contraction** (or **E–C**) **coupling**. There are circumstances in which contraction occurs in the absence of action potentials.

E–C coupling involves interaction between a protein in the T-tubular membrane called the **DHP receptor** (or **DHPR**) and one on the SR membrane of the terminal cistern immediately apposed to it called the **ryanodine receptor**, or **RyR**. These are named for their binding to dihydropyridines and the plant alkaloid ryanodine, respectively. The DHPR is a voltage-gated Ca^{2+} channel. It comprises a voltage sensor (the $\alpha 1$ subunit), which is similar to the voltage-dependent fast Na^+ channel, and four other proteins. RyR occurs in clusters of four, each containing a Ca^{2+} channel. These have been shown to be ligand-gated Ca^{2+} channels, for which an operative ligand is moderate $[Ca^{2+}]_i$. This led to the hypothesis of **calcium-induced calcium release**, or **CICR**, to explain E–C coupling: Ca^{2+} entering through the DHPR would trigger opening of the RyR, thus serving as "trigger calcium." However, since skeletal muscle can contract in very low Ca^{2+} media, entry of external Ca^{2+} through DHPR is of limited importance. The skeletal muscle DHPR has a low Ca^{2+} conductance and slow kinetics.

It is now widely believed that, in skeletal muscle, a depolarization-induced change in the structure of the DHPR $\alpha 1$ subunit directly influences the RyR in such a

way as to markedly increase its conductance for Ca^{2+}. Since the resulting increase in $[Ca^{2+}]_i$ can open other RyR channels, this produces a surge of Ca^{2+} release. In contrast, cardiac muscle expresses a different DHPR channel subunit than skeletal muscle. The cardiac DHPR has much higher conductance and faster kinetics than the skeletal muscle type, and so admits much more extracellular calcium during the action potential. Thus, CICR does play an important role in cardiac muscle. In both cases, relaxation requires resequestering of the Ca^{2+}, as described later.

Dihydropyridines, such as the drugs **nifedipine**, **amlodipine**, and **nimodipine**, block the DHPR (and other closely related) Ca^{2+} channels. These channels are also blocked by **phenylalkylamines** (e.g., **verapamil**) and **benzothiazepines** (e.g., **diltiazem**). Since Ca^{2+} entry through these channels is not required for E–C coupling in skeletal muscle, these drugs have little effect in this tissue. Vascular smooth muscle, however, is very dependent on entry of extracellular Ca^{2+} for contraction, and in this tissue these drugs significantly reduce tension. Myocardium, which depends partially on CICR for E–C coupling, is intermediate in sensitivity to these drugs.

Three RyR genes are expressed in humans; RyR1 predominates in skeletal muscle and RyR2 in heart. Smooth muscle also has predominantly RyR1 and RyR2. Ryanodine at low concentration ($<10\ \mu M$) binds to RyR in the open state and holds these channels open, producing sustained contraction, but at high concentration, ryanodine binds to a lower affinity site that blocks the channels. **Dantrolene sodium** blocks RyR1 channels in skeletal muscle and is a **direct-acting muscle relaxant**, while producing little effect on the RyR2 channels in cardiac muscle. **Neomycin** and other **aminoglycosides** also reduce skeletal muscle SR Ca^{2+} release by binding to the low-affinity ryanodine site.

SR also contains calcium-binding proteins and structural proteins. **Calsequestrin** is a Ca^{2+}-binding protein (M.W. 44.000), especially abundant at the terminal cisternae. One calsequestrin binds about 40 Ca^{2+} ions, with a mean K_m of about 1 mM. There are three other proteins with lower capacity but higher affinity for Ca^{2+} than calsequestrin. All of these buffer the $[Ca^{2+}]$ in the SR lumen, increasing the SR Ca^{2+} capacity and limiting the $[Ca^{2+}]$ gradient against which the ATPase has to work.

Mechanism of Contraction: Activation of Contraction

Ca^{2+} binding to troponin C is the key event in activation of contraction. As described previously, depolarization leads to transient Ca^{2+} release from SR. As sarcoplasmic $[Ca^{2+}]$ rises, Ca^{2+} binds to TnC. Tropomyosin normally

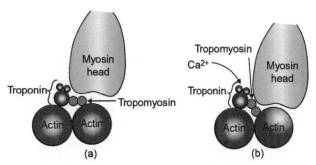

FIGURE 19.3 Schematic illustration of the interaction of myosin and actin. (a) In the absence of Ca^{2+}, tropomyosin prevents the binding of myosin heads to actin. (b) When the $[Ca^{2+}]$ rises, Ca^{2+} binds to a subunit of troponin, which causes the tropomyosin to shift slightly into the groove of the actin filament. The shift in position of tropomyosin allows the myosin heads to bind to actin. Lowering of the $[Ca^{2+}]$ level results in reversal of these events.

sterically blocks the interaction between myosin heads and actin. The troponin complex controls the behavior of tropomyosin in skeletal muscle, and as TnC saturates with Ca^{2+}, it reverses the tropomyosin inhibition of myosin binding to actin. This seems to occur by a small movement of the tropomyosin induced by the dimensional change that occurs in TnC upon Ca^{2+} binding, thus relieving a steric block, as indicated schematically in Figure 19.3. The result is that myosin heads are able to contact actin, with formation of active cross-bridges and generation of tension. The initiation of cross-bridge formation by calcium is called **activation of contraction**. SR reuptake of calcium post-stimulus lowers sarcoplasmic $[Ca^{2+}]$, desaturating the TnC and allowing the tropomyosin inhibition of cross-bridge formation to be reasserted.

Mechanism of Contraction: Cross-Bridge Cycling

The myosin neck structure allows the myosin heads to be in very close proximity to the actin. When the tropomyosin-mediated inhibition of contraction is reversed, the myosin heads interact with actin. Myosin binds initially via loop regions near the catalytic site, followed by progressively more extensive hydrogen bonding. In this process, a large surface area on each protein is removed from interaction with cell water during formation of the bond between them. In cases in which proteins bind without changes in conformation, such interaction would be associated with an energy change of 60−80 kJ/mol of bonds, a number roughly twice as great as that actually observed for binding of myosin to actin. Apparently, as the myosin moves into this tightly bound configuration, energy is captured in the form of deformation within the myosin head, which applies force (5−10 pico newtons) to the head's attachment to the thick filament (the neck region). Much, probably most, of this

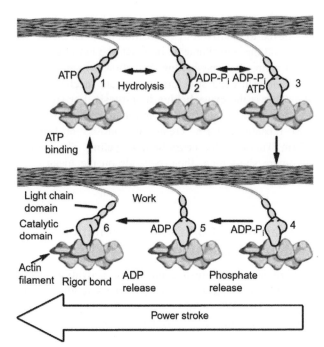

FIGURE 19.4 Schematic summary of the events of the cross-bridge cycle. At step 1, ATP binding allows the release of myosin from actin at the end of the prior power stroke. Ordinarily, ATP hydrolyzes while myosin is still attached to actin, the resulting M-ADP-P_i readily detaching from actin. In this state, myosin can reattach to actin, forming progressively stronger bonds as P_i and then ADP are released, resulting in the tightly bound rigor state. The free energy of binding of ATP to the myosin changes the conformation of its actin binding site in such a way that the affinity for actin is drastically reduced. At rest, most myosin heads will be in the M-ADP-P_i state, ready to enter stage 3 in this diagram when circumstances allow. *[Modified from R. Cooke, Modulation of the actomyosin interaction during fatigue of skeletal muscle, Muscle Nerve 36 (2007) 756−777.]*

energy is available to do mechanical work as the filaments slide past one another (the power stroke illustrated schematically in Figure 19.4), with myosin ending up very firmly bound to actin. Subsequent binding of MgATP to myosin provides the energy input to alter myosin's actin binding site to a low-affinity state, permitting detachment. ATP binding lowers the affinity of myosin for actin by a factor of about 10^4. Hydrolysis of the ATP then occurs, with little change in the free energy of the system. Subsequent events release P_i and ADP from the ATPase site.

Thus, in the cross-bridge cycle, myosin is bound with high affinity alternately to actin and to ATP, during which the ATP is hydrolyzed. Thus, the overall result is the conversion of energy of hydrolysis of ATP (about 50−60 kJ/mol under physiological conditions) to work (and heat), a process called **chemomechanical transduction**. The efficiency of this process in mammalian skeletal muscle is approximately 60%.

ATP hydrolysis normally occurs after transition of the actomyosin-ATP (A-M-ATP) complex to a weakly bound state. The complex A-M-ADP-P_i may remain weakly bound until dislodged by movement of the filaments. The released M-ADP-P_i has moderate affinity for actin, and upon reattachment, forming A-M-ADP-P_i, the phosphate release step occurs. This creates a state called **A-M*-ADP** that is the high-affinity state associated with initiating the power stroke. As the structural changes produced by increasingly tight binding to actin produce strain in the myosin head and therefore force and movement, the affinity of the ATPase site for ADP decreases, releasing the ADP. Thus, at the end of the power stroke, cross-bridges are typically in the A-M state (called the **rigor state**), their most tightly bound state, in which they will remain unless ATP is available to bind to the ATPase site and lower the affinity of the actin binding site. This model of events is illustrated schematically in Figure 19.4. In normal circumstances, it is almost impossible to deplete ATP to the point where a large proportion of myosin heads form rigor bonds, but it does happen in severely ischemic muscle and postmortem (rigor mortis). So long as $[Ca^{2+}]_i$ remains high, this cycle will continue, provided that adequate [ATP] and other appropriate conditions of the internal environment are maintained.

Since hydrolysis of one ATP occurs for every (or almost every) cross-bridge cycle in each sarcomere of each myofibril, the ATP consumption associated with contraction can be enormous, up to 100 times the resting level. There must necessarily be metabolic specializations to meet this peak demand, and to do so quickly. Energy metabolism is discussed in Chapters 13 to 16, and some muscle-specific aspects are discussed later, under "Energy Supply in Muscle."

Relaxation

Return to the resting state requires resequestration of the Ca^{2+}. This must occur rapidly even though Ca^{2+} is an operative ligand for RyR. It is accomplished by the ATP-dependent Ca^{2+} pump of SR (sarcoplasmic/endoplasmic reticulum Ca^{21}-ATPase, or **SERCA**), a "P-class" ion pump. In SR, this pump apparently occurs as single subunits coupled to a smaller protein called **phospholamban**. SERCA transports two Ca^{2+} units for each ATP hydrolyzed, and there is a lot of it. When not inhibited by phospholamban, it has very high affinity for Ca^{2+} on its sarcoplasmic face, allowing this protein to pump the sarcoplasmic $[Ca^{2+}]$ to the 0.1 μM range and below. Phospholamban inhibits the activity of the pump. The Ca^{2+} spike causes phosphorylation of phospholamban (by a CaM kinase), which disrupts this inhibition and increases the Ca^{2+}-ATPase activity 10- to 100-fold, thus greatly accelerating resequestration. Meanwhile, high $[Ca^{2+}]_i$ results in binding of Ca^{2+} to low-affinity Ca^{2+} binding sites on RyR, which, unlike the high-affinity sites,

closes the Ca^{2+} release channel. By these two mechanisms, the calcium spike, in effect, turns itself off. Myocardial phospholamban is a target for protein kinase A, and so SERCA activity in heart is increased by β-adrenergic stimulation, thereby shortening mechanical systole. Mutations in the phospholamban gene causing continuously depressed SERCA activity have been found to be one cause of dilated cardiomyopathy.

Diversity and Plasticity in Skeletal Muscle

Skeletal muscle fibers have differing mechanical and metabolic properties, and there are fiber classification schemes based on these differences. All fibers within a muscle unit are similar with respect to twitch characteristics, as these are largely determined by the innervation, and all fibers in a muscle unit are innervated by the same α-MN. However, even within a muscle unit, there may be appreciable inter-fiber differences in metabolic profile. Consequently, the following classifications should be viewed as simplified categories, not precise descriptions of populations of fibers.

Contractile Properties

Most fibers can be classified as **slow-twitch** (**ST**) or **fast-twitch** (**FT**). FT fibers have a higher twitch tension, shorter time to peak tension, shorten faster, and have a shorter relaxation time than ST, and are used for movements requiring high speed or power. ST fibers are used for sustained activity and control of posture. The speed of shortening is related to the ATPase activity of the myosin, which is mainly determined by which MHC is present, and modified by which MLCs are present, while the twitch tension and duration parameters are related to the extent of the SR, and so these two are not directly connected. However, fiber histology and myosin gene expression have some controlling influences in common, so that speed of contraction and twitch duration tend to vary inversely. In FT fibers, SR accounts for roughly 15% of the cell volume, versus only 3%–5% in ST fibers. The rates of Ca^{2+} release during E−C coupling, the peak sarcoplasmic $[Ca^{2+}]$ achieved in a twitch, and the reuptake rate of Ca^{2+} are all 2−4 times greater in FT than in ST fibers, resulting in the observed differences in twitch characteristics.

pH Dependence of Myosin ATPase Activity

The myosins in muscle differ in their inhibition by high or low pH, and can be classified on this basis. Serial thin sections of muscle specimens can be incubated in at least two different pHs with ATP and Ca^{2+} and stained for phosphate liberation. In the Brooke and Kaiser classification scheme, fibers that are light at pH >10 are type I; those that are dark are type II. This staining pattern is reversed at pH <4.4. Fibers that are light at this very acid pH but moderately dark at pH 4.6 are called type IIx, while those that are light at pH <4.4 and at pH 4.6 are type IIa. Since the speed of shortening depends on which myosin is present, the relationship between the Brooke and Kaiser fiber typing and the FT versus ST classification is quite consistent: type I fibers are slow; IIx are the fastest; and IIa are intermediate in speed. However, more sophisticated methods reveal subpopulations of ST and FT myosin types. Thus, the I, IIa, IIx scheme has been useful clinically and in analysis of muscle performance but does not convey the near continuum of myosin ATPase activity actually present across fibers.

Metabolic Profile

Muscle fibers can be classified as **high-glycolytic, high-oxidative**, or **intermediate**. Microanalytic techniques enable assessment of activity of enzymes in the energy pathways of cells, such as succinate dehydrogenase, glyceraldehyde-3-phosphate dehydrogenase, and adenylate kinase (myokinase). By staining serial sections alternately for myosin ATPase typing and enzyme profiling, it has been found that generally type I fibers are high in oxidative enzyme activity and low in glycolytic activity, while type IIx fibers are the reverse. Type IIa fibers are moderately high in both oxidative and glycolytic enzymes. Hence, type I fibers are often called **SO**, or **slow oxidative**, while type IIx are called **FG**, or **fast glycolytic** fibers. Type IIa are called **FOG**, or **fast oxidative–glycolytic**. However, the Brooke and Kaiser typing and energy pathway profiles reflect quite different specializations of the cell, and there is heterogeneity of enzyme profile within a muscle unit.

Multigene Families Encoding Muscle Proteins

Muscle fiber diversity is mainly due to the existence of multiple genes for contractile and regulatory proteins. So far as we know, multiple isoforms of all the myofibrillar proteins exist. These are encoded by families of genes in probably all mammalian species. Expression of these genes tends to be tissue-specific or fiber-type-specific, and for many there are fetal, adult, and (for some) neonatal isoforms. The significance of having multiple genes is that muscle can alter its functional characteristics to adapt to changing functional demand by selectively altering which genes are expressed.

Differentiation of Fiber Types and Muscle Plasticity

The fiber type distribution within a muscle is determined by a combination of genetic and developmental factors, and by the pattern of recruitment of the muscle unit. Due to this latter influence, the twitch and enzyme characteristics of muscle fibers are somewhat malleable, being influenced by training (especially endurance training) or by detraining (as in immobilization). Fiber type expression is also influenced by thyroid hormones, with T_4 and especially T_3 promoting the FT expression program, and so fiber type is not immutable.

There is a highly significant familial aggregation of enzyme profile, but the heritability of ST versus FT fiber type is still not well quantified. An example of developmental influence is the faster speed of shortening (when normalized for fiber length) in muscles of the head and neck derived from the mesoderm of the pharyngeal arches compared to those derived from the mesoderm of the somites. When normalized for fiber length, the extraocular muscles have the highest speed of shortening.

In general, in the mammalian embryo, the activity-independent pattern of fiber differentiation is predominantly the FT gene program, driven by thyroid hormone. The subsequent emergence and maintenance of the ST program is dependent on slow motoneuron activity, meaning small, low-threshold α-MNs that are very frequently recruited, but discharge at low action potential frequency. That is, the "default" program is FT gene expression, and ST fibers revert to FT when denervated or immobilized, whereas ST gene expression depends on chronic low-frequency stimulation. In mature human muscle fibers, extensive endurance training can shift fiber types from IIa to I or from IIx to IIa, but not from IIx to type I. Low-volume high-intensity training tends to do the opposite, albeit less markedly. Spinal cord injury, however, eventually results in almost complete disappearance of type I fibers.

The best-established pathway by which activity-dependent type I expression is mediated is the **Cn-NFAT** signaling mechanism. Cn is **calcineurin**, a Ca^{2+}/calmodulin-dependent phosphatase that acts on the NFAT family of transcription factors, causing them to translocate into the nucleus and activate transcription of ST-associated genes. Cyclosporin A inhibits Cn, inducing a partial slow to fast transition in rat ST muscles.

Control of the metabolic gene program is less clear, but studies reported in the last few years have provided some insight. The tentative conclusion is that development and maintenance of a slow oxidative fiber type is mediated at least in part by the frequently recurrent elevations in Ca^{2+} and AMP associated with slow motoneuron innervation, and possibly by the increased diffusion of fatty acids into muscle due to increases in plasma free fatty acids induced by prolonged exercise.

Energy Supply in Muscle

During muscle activation, ATP consumed by the myofibrils must be replaced by aerobic and/or anaerobic resynthesis. The magnitude of the task is increased by the augmentation of the SR Ca^{2+}-ATPase activity that also takes place. SERCA activity increases more than 10-fold during contraction. Na^+/K^+-ATPase activity also increases due to the ion fluxes accompanying repeated action potentials. The resulting ATP turnover is 100 or more times greater than at rest. The strategy for meeting these demands varies by fiber type. Type I fibers are generally oxidative. Their ATP resynthesis is largely dependent on β-oxidation of fatty acids and on glycolysis and the TCA cycle. Type I fibers usually have only modest activity of glycolytic enzymes and myokinase, and normally do not produce pyruvate at rates much in excess of the rate at which they can oxidize it. Type IIx fibers are metabolically the mirror image of type I, and type IIa is an intermediate fiber type metabolically.

AMPK (AMP-activated protein kinase, or AMP-dependent kinase) is a key sensor of energy demand. As described in Chapter 20, increases in AMP concentration activate AMPK, which initiates phosphorylation cascades that ramp up glycolysis and fatty acid oxidation. This process is also partially dependent on Ca^{2+} (via CaM kinases), which increases immediately on stimulation. As in many other tissues, AMPK is a metabolic on-switch for energy pathways, and its role is more dramatic in muscle because of the large and rapid increases in ATP synthesis that are required in this tissue (Chapter 20).

Aerobic energy production offers many important advantages over the low-oxidative high-glycolytic strategy of type IIx fibers. Most prominently, oxidative metabolism can utilize lipid and amino acids in addition to the glucose/glycogen on which anaerobic glycolysis depends. This enables a vastly greater volume of work to be performed aerobically than anaerobically. A typical value for glucose stored as glycogen in muscle is about 430 g, and about 70 g in liver, equivalent to 2100 kcal total. Lipid stores, in adipose tissue, muscle, and elsewhere, are typically 5.5 kg in an active young adult male, equivalent to almost 52,000 kcal, and this is in a relatively lean person. The derivation of TCA cycle intermediates from some amino acids and the conversion of glucogenic AAs to glucose enable protein to also serve as an energy source. For these reasons, the energy substrate available to the slow oxidative fibers is almost inexhaustible, and, due to their low actomyosin ATPase activity, replenishment of ATP by oxidative phosphorylation can keep up with ATP hydrolysis. Thus, SO fibers are fatigue-resistant, while fast glycolytic fibers that rely on glycolysis are quite susceptible to fatigue. The intermediate FOG fibers are much more fatigue-resistant than FG fibers, but less so than SO fibers.

Aerobic fiber types have high activities of the enzymes for β-oxidation of fatty acids, and their mitochondria also have abundant carnitine palmitoyl acyltransferases and associated substances. The peak rate of ATP production attainable by this pathway is about half the maximum rate achievable by aerobic glycolysis, but during prolonged moderate intensity exercise, β-oxidation is the major energy pathway. Most of the fatty acid oxidized is derived from lipolysis in adipose tissue during exercise. Adipocyte lipolysis depends, at least in part, on activation of lipoprotein lipase by catecholaminergic stimulation and other hormonal changes accompanying exercise, and takes many minutes (20 to 40) to fully accelerate. Utilization of lipid already present in the plasma as protein-bound fatty acids or as triglyceride can begin promptly, and is not dependent on hormonal mechanisms. Activation of capillary lipoprotein lipase (e.g., by heparin or methyl xanthines) measurably increases muscle lipid oxidation in endurance exercise.

The rate of lipid oxidation as a function of exercise intensity is an inverted "U"-shaped curve, increasing to a maximum at 60%−70% of maximum aerobic power, and decreasing to about 75% of this rate at 85% of maximum aerobic power. As the power requirement approaches the limits of β-oxidation, the flux through glycolysis increases rapidly.

Type IIx fibers rely on glycolytic rather than oxidative metabolism for energy. In these fibers, glucose derived from muscle glycogen and from glycogenolysis and gluconeogenesis in liver is split to pyruvate at rates far greater than the rate at which pyruvate can be oxidized. Although producing only 2 ATP per glucose molecule (3 per glucosyl unit from intramuscular glycogen), these fibers can generate such a high flux through glycolysis that the peak ATP production rate is twice the maximum rate of oxidative ATP production in type I fibers. Aerobic and anaerobic glycolysis is discussed in Chapter 12.

Lactic Acid

Lactate synthesis is a "necessary evil." In glycolysis, conversion of glyceraldehyde-3-phosphate to 1,3-bisphosphoglycerate by glyceraldehyde 3-phosphate dehydrogenase (G3PD) requires reduction of NAD to NADH. The NAD pool is small, and without rapid oxidation of NADH back to NAD, glycolysis would quickly stop at the G3PD step. Extramitochondrial oxidation of NADH back to NAD is accomplished by conversion of pyruvate to lactate via lactate dehydrogenase (LDH). Since the maximal reported rates of appearance of lactate in blood are only one-tenth the FG fiber production rate, it is clear that maximal production rate far exceeds the efflux capacity and that high-intensity exercise must necessarily produce high intramuscular lactate concentration. Intramuscular [lactate] as high as 45−50 mMol/kg of cell water has been reported in humans. Intracellular pH in muscle fibers is about 7.0, and decreases 0.4 to 0.8 pH units during intense exercise.

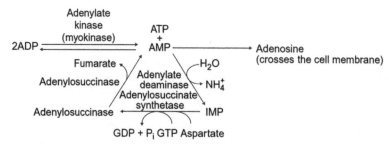

FIGURE 19.5 Role of purine nucleotides in muscle energy metabolism. The conversion of AMP to IMP prevents loss of adenosine from the cell.

Lactate efflux from muscle occurs mainly by carrier-mediated lactate-proton cotransport and by simple diffusion of undissociated lactic acid. The former probably accounts for 50%–90% of the lactate efflux, depending on fiber type and pH. There are three or more muscle lactate transporters. However, the K_m for lactate has not been found to differ significantly between fiber types, being around 30 mM. The high K_m implies that a lot (perhaps most) of the lactate produced will be retained during exercise, regardless of fiber type. Retained lactate maintains an NAD/NADH ratio, which stimulates oxidation and ensures subsequent oxidation or gluconeogenesis in the muscle.

The fate of the lactate produced is primarily oxidative (55%–70%), with as much as 90% of the labeled carbon in tracer lactate showing up as CO_2 during active recovery from exercise. Lactate released from muscle enters exclusively oxidative tissues in which lactate is low (e.g., heart, diaphragm); then it is converted back to pyruvate and oxidized. Retained lactate is oxidized to generate the ATP needed to replenish phosphocreatine stores following exercise and to restore normal distribution of Ca^{2+}, Na^+, and K^+. In liver and kidney, about 5%–15% is used for gluconeogenesis and subsequent glycogenesis. The balance is converted to alanine, glutamate, or other substances.

Protein Metabolism

Protein metabolism contributes 2%–3% of the energy requirement in exercise of a few minutes duration and rises to as much as 12% after several hours of physical work. Protein metabolism and nitrogen excretion are described in Chapter 15. Replenishment of TCA cycle intermediates such as α-ketoglutarate (derived from glutamate) or oxaloacetate (from aspartate or asparagine) is probably important to offset the loss of TCA cycle intermediates from mitochondria over time. This role of protein catabolism in supporting glucose and lipid oxidation, called **anaplerosis**, may be more important than its direct contribution to energy supply.

The Myokinase Reaction and the Purine Nucleotide Cycle

The purine nucleotide cycle is also involved in muscle energy production. During intense stimulation or when oxygen supply is limited, the high-energy bond of ADP is used to synthesize ATP, the **myokinase reaction**, as indicated in Figure 19.5. The resulting AMP can dephosphorylate to adenosine and diffuse out of the cell. Conversion of AMP to IMP via **adenylate deaminase** and then to adenylosuccinate helps sustain the myokinase reaction, especially in FG fibers, by reducing accumulation of AMP. It may also reduce the loss of adenosine from the cell since nucleosides permeate cell membranes, whereas nucleotides do not. In the heart, AMP accumulation is typically due to ischemia, and the adenosine released as a result is a potent coronary vasodilator. Accordingly, myocardium has less adenylate deaminase than skeletal muscle. However, in myocardial infarction, dilation downstream from the thrombus does no good. Large losses of adenosine from myocardium occurring due to delays in treatment are dangerous because the decrease in the total adenosine pool means that a normal ATP charge cannot occur despite subsequent therapy.

Phosphocreatine Shuttle

Energy transfer between mitochondria and the myofibrillar ATPases is mediated by phosphocreatine. The phosphocreatine shuttle is illustrated schematically in Figure 19.6. Synthesis of ATP in mitochondria is closely coupled to that of phosphocreatine. Since the reaction

$$ATP + Creatine \rightarrow ADP + Phosphocreatine$$

is energetically favored when the ATP:ADP ratio is high, phosphocreatine is rapidly formed from ATP releasing ADP, which stimulates mitochondrial respiration. Phosphocreatine diffuses from the mitochondria to the various sites of energy utilization, where creatine kinase (CK) reverses this reaction to form ATP and creatine. Creatine diffuses back to the mitochondria for rephosphorylation. Since creatine elicits formation of ADP in mitochondria, the phosphocreatine hydrolysis occurring with muscle activity

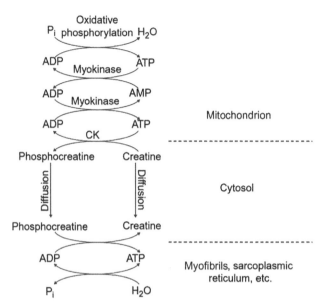

FIGURE 19.6 Phosphocreatine shuttle. A myokinase (adenylate kinase) cascade between oxidative phosphorylation and creatine kinase (CK) has been postulated. A similar cascade may exist at the myofibrillar ATPase site.

helps stimulate respiration. Cellular adenine nucleotides are compartmentalized by their very low diffusibility, with pools in the mitochondria, and at the myofibrils, SR, and other sites of energy utilization. CK is located at these sites. Phosphocreatine is much smaller and less charged, and therefore much more mobile in cells, than ATP.

Phosphocreatine also functions as an intracellular energy store. ATP inhibits mitochondrial respiration, limiting the maximum achievable concentration of ATP to about 5 mM, but phosphocreatine can accumulate to roughly 30 mM without inhibiting respiration and ATP synthesis. Phosphocreatine provides a reserve of immediately available energy that can be used for brief bursts of activity, as in throwing or jumping, which is able to cover the energy needs for a few seconds at the beginning of sprint-type activity while glycolysis is accelerating.

Creatine kinase (CK) is a dimer of subunit M and B (Chapter 7). Its cardiac muscle isoenzyme CK(MB) accounts for about 15% of the total CK, and its level in the serum has been used to assess acute myocardial infarction.

In recent years, however, more specific and sensitive cardiac troponin assays have replaced the use of CKMB. Measurement of serum muscle enzymes such as aldolase and CK remains a useful tool for quantitatively assessing skeletal muscle damage.

Muscle Fatigue

Muscle fatigue is not related to lactate accumulation. Early findings that decreasing intracellular pH, consequent to lactate accumulation, impaired activation, and

cross-bridge cycling turned out to be an artifact of performing the studies at room temperature. At 37°C to 40°C, $[H^+]$ has little effect on contraction. However, HPO_4^{2-} accumulation has marked effects on cross-bridge formation and recycling, and on the free energy of $\sim P$ hydrolysis, and $[HPO_4^{2-}]$ can increase markedly due to hydrolysis of phosphocreatine. Intramuscular $[HPO_4^{2-}]$ may rise from 2 to 30 mM during fatiguing contractions.

Regulation of Smooth and Cardiac Muscle

In skeletal, cardiac, and smooth muscle, contraction is initiated primarily by an increase in cytoplasmic $[Ca^{2+}]$. However, the main differences in histology and function of these muscle types are associated with great variety in how contraction is controlled. Smooth muscle especially differs from the model presented for skeletal muscle in the control of cytoplasmic $[Ca^{2+}]$.

Regulation of Skeletal Myogenesis by mTOR

Mammalian (or mechanistic) target of rapamycin (mTOR) is a master regulator of skeletal myogenesis, by controlling multiple stages of the myofiber formation process [1,2]. Activation of mTOR pathways leads to muscle cell growth. mTOR is a serine/threonine-specific protein kinase that has two different complexes: mTORC1 (contains raptor, which confers rapamycin-sensitive functions) and mTORC2 (contains rictor, which confers rapamycin-insensitive functions). mTOR has many functions, which include control of cell growth, proliferation, differentiation, and survival. In many cancers and proliferative disorders, mTOR signaling has been found to be upregulated, leading to the development of many drugs that target this enzyme for the treatment of cancer. Rapamycin is a macrolide antibiotic that binds noncompetitively to mTOR and inhibits some of its functions, depending on rapamycin sensitivity. The mTORC1 complex is rapamycin sensitive, and its signaling regulates translation, ribosome biogenesis, autophagy, glucose metabolism, and cellular response to hypoxia. The mTORC2 complex is rapamycin insensitive, and its signaling regulates cell survival and organization of the actin cytoskeleton. Given its sensitivity to rapamycin, mTORC1, when inhibited, would therefore be expected to be an ideal target for suppression of myogenic differentiation in skeletal muscle development. However, a recent study revealed unexpected results with knockout of raptor in mTORC1: enhancement of myoblast differentiation. Conversely, overexpression of raptor was found to suppress myoblast differentiation, suggesting that raptor has a negative function in the regulation of myogenesis. These results suggested the possibility that blockade of the mTORC1 complex in the treatment of cancer may actually lead to more aggressive lesions, a concern that has led to the development of novel drugs which can target other domains of mTOR, such as mTOR kinase inhibitors that bind to the mTOR kinase domain. Another study found that a complete depletion of either raptor or rictor had a synergizing effect with myostatin in the inhibition of myogenic differentiation. See "Regulation of Skeletal Muscle by Myostatin."

Regulation of Skeletal Muscle by Myostatin

First identified in 1997, myostatin has been studied as a powerful negative regulator of skeletal muscle mass in mammals [3−5]. Myostatin is involved in muscle homeostasis by suppressing muscle growth. In animal studies, elevated serum myostatin levels are seen in animals with diseases associated with muscle loss. On the other hand, animals with natural loss-of-function mutations of the myostatin gene, like the Belgian Blue cattle breed, have dramatically increased skeletal muscle mass. In humans, myostatin plays an important role in muscle atrophy seen in injury, infection, and muscle-wasting diseases such as cancer, AIDS, and chronic obstructive pulmonary disease. Some of myostatin's well-recognized effects on the molecular level include inhibition of myoblast proliferation, upregulation of ubiquitin-proteasomal activity, and downregulation of the IGF-Akt pathway. Newly studied roles of myostatin include adipocyte proliferation, cardiomyocyte homeostasis, and glucose metabolism.

Myostatin is synthesized in the muscle, and to a lesser extent in adipose tissue, then released into circulation. *In vitro* studies suggest that myostatin is produced primarily in the nucleus of myotubes and then released into the cell cytoplasm upon atrophic stimulation by dexamethasone. In the cytoplasm, myostatin remains in an inactive state as a propeptide. It is released from the cell in an autocrine or paracrine manner, remaining in its inactive state, or binding with other proteins to form inactive complexes. In the serum, myostatin becomes activated when cleaved into an amino-terminal propeptide and an active carboxy-terminal region. Once activated, myostatin binds with high affinity to the activin type IIB receptor (Acvr2b, or ActRIIB, or ARIIB) located on skeletal muscle. Binding activates an intracellular cascade of events, which includes the TGF-β signaling pathway and regulation of target gene expression.

Myostatin has various targets for gene expression: inhibition of myogenic regulatory factors responsible for myogenesis; decreased expression of structural myofibrillar protein genes and myogenic transcription factors; activation of proteolysis and upregulation of ubiquitin-associated genes; downregulation of the IGF-I/PI3K/AKT hypertrophy pathway; increased expression of atrophy-related genes; decreased myoblast proliferation in C2C12 cells; arrest of myoblasts in G1 phase of cell cycle; and inhibition of DNA and protein synthesis. The clinical result is a decrease in skeletal muscle mass, so the inhibition of myostatin to increase skeletal muscle mass is currently being studied for the treatment of cachexia related to muscle-wasting diseases.

Myostatin also plays an important role in adipose tissue, as there is cross-talk between skeletal muscle and adipose tissue in the regulation of total and lean body weight. In myostatin-deficient mice, there is a dramatic increase in muscle mass as expected, but there is also a dramatic reduction in fat stores and inhibition of adipogenesis. Therefore, it is suggested that myostatin decreases peripheral tissue fatty acid oxidation and may play a role in morbid obesity and insulin resistance. Therefore, inhibition of myostatin results in decreased adipocyte size, as demonstrated by a study in which myostatin prevented obesity in mice. Other therapeutic implications for myostatin inhibition may include preventing hepatic steatosis and improving glucose tolerance. Refer to references [3−5] for further reading.

Inherited Diseases of Muscle

Many disorders of skeletal and cardiac muscle due to genetic defects have been described. Taken as a group, the familial cardiomyopathies are the most common, with a combined prevalence of about 1 in 300. By comparison, the most common muscular dystrophy, Duchenne muscular dystrophy (or DMD), has a prevalence of roughly 1 in 3500 male births. It is not known whether genetic defects specific to myocardium are actually more common than those affecting skeletal muscle, or whether cardiac defects are simply more serious due to the incessant and vital nature of cardiac work, and thus more likely to be diagnosed. Most of these genetic defects affect neither skeletal nor cardiac muscle because there are cardiac-specific forms of almost all sarcomeric proteins and some enzymes. These disorders fall into the following broad categories: cardiomyopathies, muscular dystrophies, channelopathies (including the myotonias), metabolic diseases, and mitochondrial gene defects. The following paragraphs discuss only a few of these.

Hypertrophic cardiomyopathy (HCM) is a syndrome characterized by dyspnea and chest pain associated with decreased diastolic compliance and (often) outflow obstruction, and left ventricular hypertrophy, usually without dilation. More than 100 different mutations have been found in HCM patients in genes coding for β-MHC, essential and regulatory MLC, TnT, TnI, α-tropomyosin and myosin-binding protein C. Most of these are transmitted as autosomal dominant traits with variable penetrance.

Those cases due to defects in β-MHC and TnT are the most severe and have the worst prognosis, and heavy chain mutations are the most common type. About 15%–20% of HCM are due to mutations in chromosome 11p11.2, which codes for myosin-binding protein C (MyBP-C; see "Myofibrils Contain Numerous Accessory Proteins," earlier).

Dilated cardiomyopathy, like hypertrophic cardiomyopathy, can arise from mutations in many genes. Truncating mutations of the gene that encodes the sarcomere protein titin, *TTN*, are a common cause of dilated cardiomyopathy, responsible for 25% of familial cases and 18% of sporadic cases. However, in a study comparing subjects with and without *TTN* truncating mutations as a cause of their dilated cardiomyopathy, there was no significant difference in disease progression [6], which was measured by age at diagnosis, left ventricular end-diastolic measurements, ejection fraction, rates of cardiac transplantation, implantation of a left ventricular assist device, or death from cardiac causes. Refer to reference [6] for further reading.

Duchenne muscular dystrophy, or **DMD**, is a muscle degenerative disease due to a recessive mutation in the X-chromosome. The mutations involve defects in, or even complete deletion of, the gene coding for **dystrophin**. Dystrophin is a large molecule (M.W. 426,000), and the gene coding it is huge, over 2 million bases. Dystrophin is involved in attaching the cytoskeleton to the extracellular matrix, or ECM; i.e., dystrophin binds to webs of actin filaments and to transmembrane glycoproteins, which, in turn, bind to ECM proteins such as laminin and agrin. This is an important role in muscle, where dystrophin is required to transmit force from the myofibrils through the costamere structure of the fiber to the ECM and, ultimately, the tendons of the muscle (see "Myofibrils" earlier in this chapter). Similarly, dystrophin allows forces applied to the sarcolemma via the ECM to be borne by the cytoskeleton and myofibrils rather than the sarcolemma, which cannot support tension by itself. Lack of dystrophin results in mechanical stresses in muscle, tearing holes in the sarcolemma, which causes sustained high $[Ca^{2+}]_i$ and activation of Ca^{2+}-dependent proteases such as calpains. This leads to focal destruction in the fiber, migration of PMNs into the site, and either remodeling or total destruction of the fiber. **Becker's muscular dystrophy** (**BMD**) is allelic to DMD. Dystrophin may also play a role in localizing Ach receptors to the motor end-plate, and there is nitric oxide synthase (NOS) activity associated with dystrophin. NO increases phosphorylation of AMP-activated kinase, an important metabolic regulator. Consequently, the mechanism of muscle damage in DMD may be more complicated than loss of the mechanical function that dystrophin performs.

Channelopathies are a varied group of rare hereditary disorders due to defects, usually point mutations, in genes for ion channel proteins. Most of these affect voltage-gated channels. Two that do not are **Thomsen's disease** (autosomal dominant myotonia congenita) and **Becker's disease** (autosomal recessive generalized myotonia) which are both due to mutations in a gene coding for a skeletal muscle Cl-channel called **ClC-1**. For an explanation of why loss of Cl-conductance causes this problem and why the inheritance can be either dominant or recessive, see the discussion of Thomsen's and Becker's in the recommended reading for this chapter.

Most channelopathies have some features in common. There are paroxysmal attacks of myotonia or paralysis, migraine or ataxia, precipitated by physiological stressors. They are often suppressed by membrane-stabilizing agents such as mexiletine and by acetazolamide.

Metabolic disorders of muscle include those of glycogen storage, substrate transport and utilization, and electron transport chain and ATP metabolism. Some produce dynamic syndromes with symptoms occurring primarily during exertion, some cause degenerative syndromes, and some produce both. A few are discussed next.

Degenerative Syndromes

Acid maltase is a lysosomal enzyme not in the energy pathway of the cell, so its deficiency does not produce dynamic symptoms in muscle. Also called α-1,4-glucosidase, acid maltase hydrolyzes the α-1,4 bonds in glycogen. Since lysosomes degrade glycogen along with other macromolecules in the normal process of cellular turnover, this deficiency causes marked accumulation of glycogen in lysosomes. This leads to a vacuolar degeneration of muscle fibers. Heart and other tissues are also affected. Deficiency of **debranching enzyme** (also called amylo-1,6-glycosidase) is another disorder of glycogen metabolism. Since phosphorylase cannot act at or near branch points (see Chapter 14), lack of debranching enzyme results in great accumulation of limit dextrins in muscle, liver, heart, and leukocytes, with swelling and functional impairment. Since this enzyme is in the energy pathway, its lack causes dynamic symptoms, but more importantly, it causes a vacuolar degeneration. **Carnitine deficiency** causes a disorder of lipid metabolism. Carnitine is derived both from the diet and from ε-N-trimethyllysine produced by catabolism of methylated proteins including myosin, and is required for the transport of fatty acids across the mitochondrial membranes (see Chapter 16). If any enzymes or cofactors required for carnitine synthesis are deficient or defective, carnitine deficiency may develop if dietary intake is insufficient. This limits the energy supply available from β-oxidation and causes a lipid storage myopathy.

Dynamic Syndromes

Myophosphorylase deficiency is the classic example of a carbohydrate-related dynamic syndrome. Affected persons are not able to mobilize glycogen and so cannot perform high-intensity work and must rely much more extensively on lipid metabolism. Several other defects of glycolysis produce similar symptoms. All are characterized by inability to do anaerobic work and to produce lactate during ischemic exercise, which is the basis for the customary screening test for these disorders. In addition to reduced anaerobic work capacity, the low flux through glycolysis reduces maximal muscle power output and also maximal aerobic power. In phosphorylase deficiency, muscular performance can often be improved by glucose infusion, while patients with other defects of glycolysis are dependent on lipid metabolism and show little or no improvement with glucose infusion.

Mitochondrial Mutations

Human mitochondrial DNA (mtDNA) codes for 2 rRNAs and 22 tRNAs used in mitochondrial protein synthesis, and for 3 subunits of cytochrome c oxidase (also called **complex IV**), 7 subunits of NADH-CoQ reductase (also called **complex I**), 2 subunits of F_0ATPase (part of the ATP synthase complex, also called **complex V**), and the cytochrome b subunit of $CoQH_2$-cytochrome c reductase (also called **complex III**). Oxidative phosphorylation is discussed in Chapter 13. All of these are components of the respiratory chain or part of ATP synthase, or are required for their synthesis, and so are essential to aerobic ATP synthesis. mtDNA is inherited almost exclusively from the mother.

Mutations in mitochondrial DNA are much more frequent (perhaps one million times more) than in nuclear DNA, because mitochondria produce free radicals while lacking DNA ligase and other repair enzymes. When mtDNA mutations are present in a functionally significant fraction of mitochondria in oocytes, the mutations are transmitted to the offspring, where they may lead to functionally significant impairment of oxidative phosphorylation in various tissues. The most severe problems arise when these mutations are abundant in obligate aerobic tissues such as brain, heart, kidney, and to a lesser extent, muscle. Since the distribution of the zygote's mutated mitochondria to the daughter cells in the first several cell divisions is random, it is not predictable which tissues in the offspring will have the greatest enrichment of mutant mtDNA. Thus, siblings with the same mitochondrial defect can have different symptoms (or no significant symptoms), depending on which tissues end up enriched in mutant mtDNA. Thus, while it may be clear that there is a familial tendency, the genetics are distinctly non-Mendelian.

NONMUSCLE SYSTEMS

Actin

Actin is present in all eukaryotic cells, where it serves structural and mobility functions. Most movement associated with microfilaments requires myosin. The myosin to actin ratio is much lower in nonmuscle cells, and myosin bundles are much smaller (10−20 molecules rather than about 400), but the interaction between myosin and actin in nonmuscle cells is generally similar to that in muscle. As in smooth muscle, myosin aggregation and activation of actin−myosin interaction are regulated primarily by LC phosphorylation. Myosins involved in transporting organelles along actin filaments are often activated by Ca^{2+}-CaM.

Actin filaments are relatively permanent structures in muscle, whereas in nonmuscle cells microfilaments may be transitory, forming and dissociating in response to changing requirements. The contractile ring that forms during cell division to separate the daughter cells and the pseudopodia formed by migrating phagocytes comprise transient actin filaments. Belt desmosomes in epithelial cells and microvilli on intestinal epithelial cells comprise relatively permanent filaments.

If nascent actin filaments anchored to the cytoskeleton grow toward the cell membrane and continue to grow at the membrane, they will create a projection of membrane and cytoplasm in the direction of growth, especially if the filaments are connected by short linking proteins into rigid bundles. In this way, microvilli extend from the surfaces of many cell types. Projection of the acrosomal process through the zona pellucida is driven in this way, as are the extension of filopodia and lamellipodia at the leading edges of migrating cells.

The structure and properties of actin filaments can be regulated by controlling the transformation of G-actin to F-actin or the length of the F-actin filaments, and by modulating the aggregation of actin filaments into bundles or three-dimensional arrays. Several proteins affect the state of actin filaments. Proteins that bind actin monomers reduce polymerization. Gelsolin, villin, and other proteins affect actin polymerization and actin filament elongation by capping the growing filament and blocking elongation. Some accelerate nucleation, perhaps by binding to and stabilizing dimers and trimers. Many capping proteins can also sever actin filaments without depolymerizing them. Although nucleation and severing increase the number of free ends available for growth, the net effect of these proteins is a greater number of short actin filaments and an increased concentration of monomeric actin.

Cytochalasins inhibit cellular processes that require actin polymerization and depolymerization (e.g., phagocytosis, cytokinesis, clot retraction) and also act by severing and capping actin filaments. Actin filaments can be stabilized by **phalloidin**, derived from the poisonous mushroom *Amanita phalloides*. Assembly of actin filaments into bundles (as in microvilli) and three-dimensional networks is accomplished by specific cross-linking proteins.

Cilia

Tubulin and microtubules occur in all plant, animal, and prokaryotic cells, and participate in a number of essential processes. As is the case for actin filaments, microtubules occur in highly organized, relatively permanent forms such as cilia and flagella, and as transient cytoplasmic structures.

Cilia are hair-like cell surface projections, typically 0.2−0.3 μm in diameter and 10 μm long, found densely packed on many types of cells. In eukaryotes, they move fluid past cells by a characteristic whip-like motion (Figure 19.7). Eukaryotic flagella are basically elongated cilia, one or two per cell, that propel the cell through a fluid medium. Flagellar movement is wave-like and distinct from ciliary beating, although the microtubular arrangement and mechanism of movement are quite similar in the two structures.

Figure 19.8 shows a cross-section of a cilium, showing the "9 + 2" arrangement of microtubules found in the **axoneme** (core) of cilia and flagella. Nine asymmetric doublet microtubules are arranged in the periphery, and a symmetric pair of singlet tubules is in the center. Attached to the A microtubules of each doublet are dynein arms extending toward the B microtubule of the adjacent doublet. The protein **tektin**, a highly helical protein about 48 nm long, runs along the outside of each doublet where the A and B microtubules join. It is regarded as a structural rather than a regulatory protein, despite the similarity to tropomyosin. Three types of links preserve the axoneme structure. The central singlets are linked by structures, called **inner bridges**, like rungs on a ladder, and are wrapped in a fibrous structure called the **inner sheath**. Adjacent outer doublets are joined by links of **nexin** spaced every 86 nm. Outer doublets are joined to central singlets by **radial spokes**, which have a relatively globular end, or head, that interacts with the central singlet. The radial spokes are complex structures, with numerous constituent proteins. They are arranged in pairs every 96 nm along the microtubules. The inner sheath comprises primarily thin protein arms extending from the

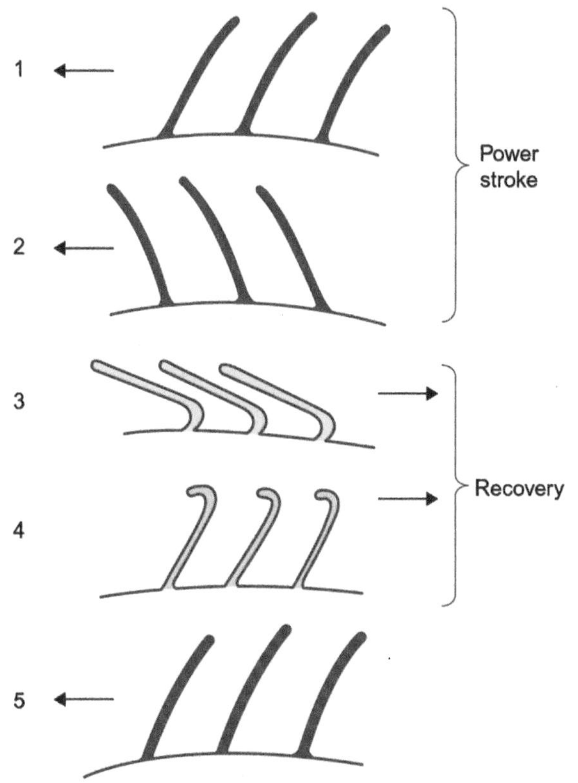

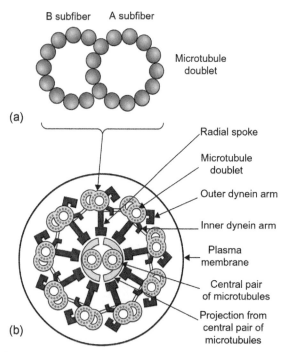

FIGURE 19.8 (a) A microtubule doublet consisting of an A subfiber with 13 protofilaments of tubulin and a B subfiber with 11 protofilaments. (b) A cilium contains 9 microtubule doublets and a central pair of microtubules. The dynein arms provide for sliding of one doublet along another during the beating of the cilium. The functions of radial spokes and the projections from the central pair of microtubules have not been as well defined.

FIGURE 19.7 Cycle of ciliary stroke shown for three cilia beating in synchrony. (1) Beginning of the power stroke. The cilia are straight and stiff. (2) Half completion of the power stroke. (3) Start of recovery stroke. The cilia are flexible and, by bending, reduce frictional drag. (4) Near completion of the recovery stroke. (5) Start of the next power stroke.

central microtubules at roughly 14 nm intervals, and the spoke heads may interact with these.

Cilia grow from **basal bodies**, one of several types of microtubule organizing centers in cells. Each basal body contains nine fused triplets of microtubules that act as nucleation centers for the growth of microtubules down the axoneme. Each triplet contains one complete A microtubule, which is continuous with the A microtubule of the axonemal doublet, and an incomplete B microtubule, which is continuous with the B microtubule of the doublet. A second incomplete microtubule, the C microtubule, is fused to the B microtubule but does not extend beyond the basal body. The basal body does not have central singlets. There are numerous proteins beside the α- and β-tubulin of the microtubules in the basal bodies, which are thought to be involved in controlling polymerization, stabilizing the axoneme structure, and securing the basal body to the cytoskeleton. These include a third type of tubulin, γ-tubulin, which is believed to form a ring that serves as the nucleus from which the tubules grow.

Microtubules are constructed of **protofilaments** (Figure 19.9), the axis of each being parallel to the axis of the microtubule. In cilia, the A microtubule and the central singlets have 13 protofilaments, while the B microtubule has 10 protofilaments of its own and shares 3 with the A microtubule, as in Figure 19.8a. The protofilaments are chains of heterodimers of α- and β-tubulin arranged end to end. The strongest bonds in the structure are clearly along the axis of the filament. In depolymerizing conditions, the dissociating ends of the microtubules have a frayed appearance due to the protomers separating from one another.

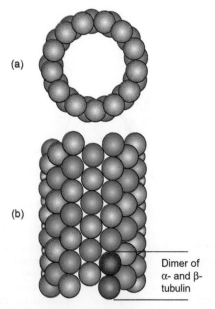

(a)

(b)

Dimer of α- and β-tubulin

FIGURE 19.9 Drawing of a microtubule. (a) Cross-sectional view of the 13 protofilaments. (b) Longitudinal view.

The **tubulins** are **globular proteins** with a mean diameter of 4 nm. They are about 450 AAs long (M.W. 50,000). The human genome contains 15–20 genes for both the α and β forms. Of these, probably five or six of each are expressed in a tissue-, developmental-, or cell cycle-dependent manner; the rest are pseudogenes. There is typically about 90% homology between the various tubulins for both the α and β forms, and although some additional diversity can be created by post-translational modification, the tubulins are functionally quite alike.

Tubulin is similar to actin and myosin in its ability to spontaneously polymerize *in vitro*. Like actin, there is a critical concentration, C_c, of αβ dimers at which equilibrium between association and dissociation occurs, and the microtubule, like actin, has a polarity and undergoes polymerization and depolymerization more rapidly at one end (+end) than at the other (−end). The heterodimers polymerize α and β all along the protomer, so that the protomer has a polarity, α at one end, β at the other, and all the protomers associate with the same polarity. So one end of a microtubule is ringed with α-tubulin and the other with β-tubulin (Figure 19.10). The αβ heterodimers each bind two molecules of GTP. The α-tubulin site binds GTP irreversibly and does not hydrolyze it because β-tubulin shields the site from water, but the β-tubulin site binds GTP reversibly and hydrolyzes it to GDP after incorporation into the protomer, and especially after additional dimers have been added. This site is called the **exchangeable site** because GTP can displace GDP from it. Microtubules, then, tend to be capped with GTP dimers but may be capped with GDP dimers if the microtubule has shortened, exposing GDP-tubulin, or when the microtubule has not grown for such a long time that most GTP has hydrolyzed. When the terminal dimers at the (+) end of the microtubule have GTP at the exchangeable site, the structure grows at dimer concentrations above C_c and slowly depolymerizes below C_c. When the (+) end dimers have GDP, however, depolymerization is rapid below C_c. This raises the surrounding dimer concentration and accelerates the growth of those microtubules that are elongating. Thus, microtubules are inherently unstable unless restrained in a structure like cilia by numerous accessory proteins.

Dynein is the motor protein in cilia and flagella. Dynein arms are arranged in two groups called **inner** and **outer dynein arms**, spaced at 24 nm intervals along the A microtubule. Dynein is a large family of microtubule-based motor proteins. Axonemal dyneins are multimers of (very) heavy chains, light chains, and intermediate chains. Each heavy chain has a large globular region with two stalks extending from it. One stalk connects to a cluster of proteins forming a "base" that attaches to the A microtubule. The other stalk projects 10 nm toward the B microtubule, and forms a small head that binds to it. The heavy

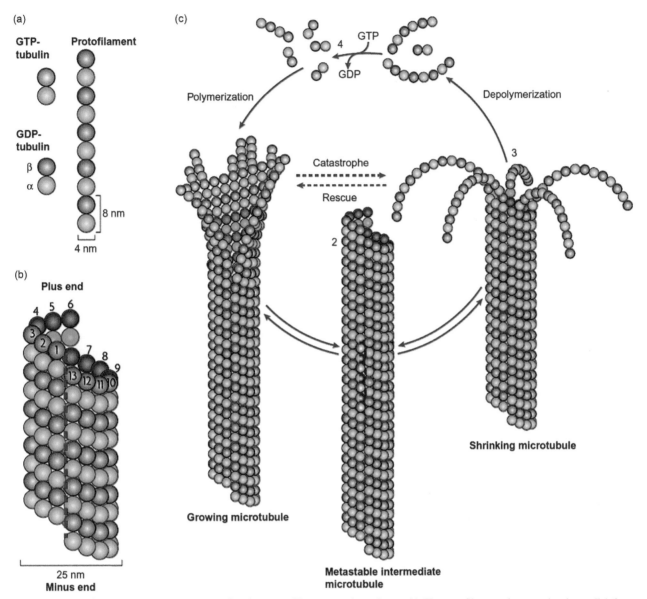

FIGURE 19.10 αβ Tubulin dimers assemble α-to-β to form protofilaments, as shown in part (a). The protofilaments then associate in parallel, forming the nascent tubule, as indicated in part (b). Note that one end of the tubule has α units only and the other has β units only, so that the whole microtubule has a polarity. Polymerization and depolymerization depend on the binding, hydrolysis, and exchange of guanine nucleotide on the β tubulin. GTP on the α tubulin is not hydrolyzed and does not exchange with GDP. When GTP has had time to hydrolyze on the β tubulin in the terminal dimers, the microtubule depolymerizes much faster at dimer concentrations below the critical concentration than it does when the nucleotide is still GTP (c3), yielding dimers and short protomers. Reassembling, or rescuing, the microtubule (c3 to c1) requires GTP hydrolysis. *[From A. Akhmanova, M.O. Steinmetz, Tracking the ends: a dynamic protein network controls the fate of microtubule tips. Nature Reviews Mol. Cell Biol. 9 (2008) 309–322. Nature Publishing Group. Figure 1, p. 310.]*

chain is quite large, about 4500 AAs long with a molecular weight of more than 540,000. It has ATPase activity, especially when bound to the B microtubule, which is believed to be localized to the head, near the tubulin-binding site. Most of the smaller chains are clustered at the base and are believed to form a fixed attachment of the dynein complex to the A microtubule, and they probably play a role in regulating the activity of the dynein, although not much is known about the regulatory mechanisms. As with myosin, there are multiple dynein genes, and the axoneme contains eight or more different heavy chains. The outer dynein arms have two or three heavy chains; the inner arms have two.

The scheme of chemomechanical transduction by dynein–tubule systems is similar in a general way to that in myosin–actin systems, but the structure of dynein is very different from that of myosin. Force generation begins by formation of a dynein cross-bridge to the

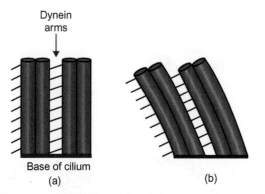

FIGURE 19.11 Sliding of one microtubule along another in an intact cilium causes the cilium to bend, creating a power stroke.

adjacent B microtubule. This is followed by a conformational change in the dynein that pulls the two microtubules past each other. Dynein is a (−) directed motor; i.e., it tries to pull its base toward the (−) end of the adjacent microtubule, which is toward the basal body in cilia and flagella. Free energy of hydrolysis of ATP is required to release the dynein head and allow the dynein bridge to recycle. As in the case of actomyosin ATPase, it is thought that the coupling of ATP hydrolysis to mechanical movement is effected at the product release step, and that this step is rate-limiting.

As the microtubules are all fixed at one end, the only way they can slide past one another is to bend, as illustrated in Figure 19.11. Activation of the dynein seems normally to occur from the base to the outer tip. How the pattern of activation is controlled or established, and what determines the direction of ciliary or flagellar bending is not understood.

Drugs Affecting Microtubules

Microtubule systems are used in the cell for many other functions, such as transport of organelles and vesicles, separating genetic material on the mitotic spindle, and other motile events of the cell cycle. Substances that interfere with microtubule growth or turnover or with microtubule interaction with motor proteins will disrupt these functions. The classic example of such a drug is **colchicine**, an alkaloid derived from the autumn crocus (*Colchicum autumnale*). Colchicine in high concentration causes cytosolic microtubules to depolymerize. In low concentrations, it does not produce this effect but binds to tubulin dimers. The tubulin−colchicine complex, even at quite low concentrations, can add on to the end of a growing (or at least stable) microtubule and block further reactions at that end. Only one or two colchicine−tubulin units at the end of a microtubule prevent any further addition or removal of dimers at that end. In cells that are replicating, this freezes the cell at metaphase. Drugs that produce such an effect are useful as **antineoplastic agents**. The well-known effect of colchicine in gouty arthritis is probably due to inhibiting the migration of granulocytes into the inflamed area by interfering with a microtubule-based component of their motility.

Vinca alkaloids (vincristine, vinblastine, vinorelbine) are derived from the periwinkle plant (*Vinca rosea*). These agents work by binding to tubulin at a site different from colchicine or paclitaxel. They block polymerization, which prevents the formation of the mitotic spindle, and are used as antineoplastic agents. **Taxanes** produce a stabilization of microtubules similar to colchicine, but by a different mechanism, and also halt cells in metaphase. **Paclitaxel** (**Taxol**) and **docetaxel** (**Taxotere**) are the taxanes commonly used clinically. Paclitaxel was originally derived from the bark of the Pacific yew. Taxanes disrupt several microtubule-based functions as completely as inhibitors of polymerization, emphasizing the importance of assembly/disassembly balance in microtubule function.

Paclitaxel also binds to, and inhibits the function of, a protein called Bcl-2. The protein Bcl-2 is an inhibitor of one or more pathways involved in mediating apoptosis, so that paclitaxel's interference with this function promotes apoptosis, in addition to its microtubule-related inhibition of cell division.

Immotile Cilia Syndrome

Defects in proteins needed for normal assembly and functioning of microtubules can cause cellular dysfunction. Several inherited disorders of this type that cause dyskinetic or completely immotile cilia and flagella have been identified. Kindreds have been identified in whom dynein arms, radial spokes, central sheath, or one or both central singlets are missing or defective. These disorders may result from a mutation in a gene needed for one of the anoxemic structures, or in a regulatory gene controlling assembly of the microtubule system in cilia and flagella.

Affected individuals manifest bronchiectasis and chronic sinusitis. Because the cilia of the respiratory epithelium are defective, mucociliary clearance is reduced or absent, leading to frequent pneumonia, colds, and ear infections. These defects were originally described in conjunction with situs inversus (lateral transposition of the thoracic and abdominal viscera), but only about half the people with immotile cilia have situs inversus. Association of the three abnormalities (bronchiectasis, sinusitis, situs inversus) is known as **Kartagener's syndrome**. These defects also cause infertility in males because the sperm are immotile. Affected females have nearly normal fertility despite the probable lack of ciliary activity in the oviducts. The function of microtubules other than in cilia and flagella is presumed to be normal; otherwise, cell division could not occur.

Although this chapter intentionally emphasizes cilia and flagella, other microtubule-based systems are also essential and deserve mention. Transport of organelles and vesicles within cells uses both microtubule- and actin-based systems. The actin-based system uses myosin as the motor protein, but, except for stress fibers and other mechanisms producing large changes in the shape of the cell, the myosin is not myosin II as in muscle, but is myosin I or V. The microtubule motors are cytosolic dynein, kinesin, and kinesin-related proteins.

Kinesins are another large family of motor proteins. Cytosolic kinesin is a tetramer of two heavy chains of unit M.W. 124,000, and two light chains of unit M.W. 64,000. The heavy chains have a head region and a tail,

similar to myosin, except that the tail is largely globular rather than a rod. There are structural similarities between myosin and kinesin heads, and in key functional groups, such as the helices flanking the ATPase site, the sequence homology is high. The light chains are associated with the tail. Typically, kinesin is a (+) directed motor; i.e., it tries to pull whatever it is attached to toward the (+) end of the microtubule. In an axon, that would be away from the cell body. Cytosolic dynein, which is smaller and simpler than axonemal dynein, is a (−) directed motor. In most kinesins, the motor region is at the N-terminal half of the molecule, but in some it is in the C-terminal region. Most of these latter kinesins are, like dynein, (−) directed.

CLINICAL CASE STUDY 19.1 Neuromuscular Junction Disorders

Vignette 1: Botulism

This clinical case was abstracted from: R.E. Scully, E.J. Mark, W.F. McNeely, S.H. Ebeling, L.D. Phillips, Case 22-1997: a 58-year-old woman with multiple cranial neuropathies, N. Engl. J. Med. 337 (1997) 184–190.

Synopsis

A 58-year-old woman presented to the hospital with a 2-day history of worsening diplopia, dysarthria, and difficulty with swallowing food. Her neurologic exam was significant for multiple cranial neuropathies such as bilateral ptosis, weakness of left eye abduction, marked adduction of the left eye when attempting upward gaze, preserved convergence, intact facial sensation but bifacial weakness, tongue weakness, slight dysmetria in the left arm, and absent deep tendon reflexes in the arms and ankles. Sensation, strength, and tone of the extremities were within normal limits. The patient's history was negative for consumption of home-canned foods, although further questioning revealed that 24 hours prior to the onset of her symptoms, the patient had opened a heat-sealed glass jar of homemade tomato-based spaghetti sauce, which she found to be rancid but had tasted its contents before discarding it. Her stool culture came back positive for *Clostridium botulinum* type B toxin. The discarded jar was later recovered, and an anaerobic culture of its contents was positive for *C. botulinum*.

Teaching Points

1. The *C. botulinum* neurotoxin causes weakness at the level of the neuromuscular junction by blocking neuromuscular transmission in cholinergic nerve fibers.
2. *C. botulinum* is a gram-positive, rod-shaped, obligate anaerobe, which is ubiquitous in nature, often found in soil and water. The bacterium is able to survive in the environment by producing spores, whose germination under anaerobic conditions causes a potent exotoxin to be synthesized and released. There are eight *C. botulinum* toxins, of which only types A, B, E, and rarely F cause

human disease. There are three main clinical presentations of *C. botulinum* infection: infant botulism, foodborne botulism, and wound botulism.

3. The toxin is absorbed by open wounds or mucosal surfaces (such as the mucous membranes of the buccal cavity, stomach, and respiratory system), as it is unable to penetrate intact skin. As in the preceding case, the toxin enters the body upon ingestion of contaminated food containing preformed *C. botulinum* toxin. The toxin remains intact within the lumen of the gut because it is not denatured by digestive enzymes. Although the toxin is large (150 kD), it is able to penetrate the intestinal epithelial wall by transcytosis or by damaging the epithelial barrier (in some animal studies). The most important site of *C. botulinum* toxin absorption occurs at the small intestine. After penetrating the intestinal wall, the toxin enters the lymphatic system and then bloodstream, where it is hematogenously spread to the cholinergic nerve fibers of predominantly the nervous, gastrointestinal, endocrine, and metabolic systems.

4. When the toxin arrives at the synaptic cleft of cholinergic nerve fibers, it is taken up into vesicles within the axon terminal. The toxin enters the body as a single 150-kD polypeptide chain, but once taken up into the vesicle, it is cleaved into two chains: a 100-kD chain and a 50-kD chain. The heavy chain binds specifically to neuronal cells and aids in the intracellular penetration of the toxin, while the light chain binds to and cleaves SNARE proteins (soluble N-ethylmaleimide sensitive fusion attachment protein). SNARE proteins are located on the presynaptic nerve terminal and on the surface of synaptic vesicles. SNARE proteins normally facilitate the fusion between nerve terminal membranes and vesicle membranes, leading to the release of neurotransmitters into the synaptic cleft. However, when SNARE proteins are cleaved by *C. botulinum* toxin, acetylcholine remains trapped in the vesicles at the axon terminal. The end result is a marked reduction

(Continued)

CLINICAL CASE STUDY 19.1 (Continued)

in the amount of acetylcholine available for release when a motor neuron is depolarized; thus, neuromuscular transmission is blocked and muscle weakness results.

5. The most common clinical features of botulism are bulbar palsy (impairment of cranial nerves IX–XII due to a lower motor neuron lesion), facial weakness, and ophthalmoparesis. Respiratory weakness, limb weakness, and depressed or absent reflexes may also occur. These symptoms may develop within 1–3 days of consuming food contaminated with the toxin. In adults, most cases of botulism follow ingestion of food stored in improperly sterilized cans or bottles, either at home or commercially, while honey consumption may be implicated in infant botulism.

6. Because botulinum toxin causes muscle weakness at the level of the neuromuscular junction, it has been used in the treatment of spasticity caused by stroke, multiple sclerosis, cerebral palsy, and spinal cord injury, which have been hypothesized to be caused by overactivity of monosynaptic muscle-stretch reflexes and/or hypertonia. The localized effects of botulinum toxin make it more favorable than systemic treatments like oral drug therapy or intrathecal injections. The therapeutic use of botulinum toxin also includes cosmetic treatment of wrinkles like "frown lines" or "crow's feet," which result from spasticity of certain facial muscles.

Vignette 2: Myasthenia Gravis

This clinical case was abstracted from: N. Venna, R.G. Gonzalez, L.R. Zukerberg, Case 39-2011: a woman in her 90s with unilateral ptosis, N. Engl. J. Med. 365 (2011) 2413–2422.

Synopsis

An independent and fully functioning 90-year-old woman presented to the emergency department with sudden onset of ptosis of the left eyelid 4 days prior. During her admission, she was diagnosed with dysfunction of the levator palpebrae superioris muscle, but then developed a wide-based gait. During rehabilitation and reconditioning over the next few weeks, her condition rapidly worsened to include widespread muscle paralysis as evidenced by coughing, choking, dysarthria, dysphonia, and dysphagia (weakness of the oropharyngeal, palatal, and vocal-cord musculature), head drop (weakness of the cervical paraspinal muscles), and difficulty breathing with consequent respiratory failure and hypercapnia (weakness of the diaphragm and intercostal respiratory muscles). The patient's clinical presentation suggested involvement of the brain stem, cerebellum, or peripheral nervous system; however, only comfort measures were administered according to the patient's wishes, and she died approximately 6–7 weeks after the onset of her symptoms.

A paraneoplastic antibody panel showed the presence of antibodies to acetylcholine receptors, as well as antibodies to striational muscle proteins. An autopsy revealed the final diagnosis of thymoma and its associated paraneoplastic syndromes of myasthenia gravis, giant-cell polymyositis of the heart and skeletal muscle, and mild brainstem encephalitis.

Teaching Points

1. Myasthenia gravis is relatively rare, but it is the most common postsynaptic neuromuscular junction disorder. It occurs when autoimmune antibodies bind to acetylcholine receptors (AChRs) at the postsynaptic neuromuscular junction of skeletal muscles, causing an increased turnover of receptors via cross-linking with neighboring receptors, absorption, and destruction. Antibody binding to AChRs also activates complement, which leads to destruction of the postsynaptic membrane. The autoantibodies may also bind to muscle-specific tyrosine kinase (MuSK), thus blocking nerve impulse transmission by reducing acetylcholine content and interrupting acetylcholine recycling. The end result is decreased transmission of nerve impulses, which is clinically exhibited by muscle weakness.

2. Because of its postsynaptic mechanism of action, myasthenia gravis is a disorder of fatigable weakness, a hallmark feature in which certain muscle groups become weak after repeated usage.

3. AChR autoantibodies originate from the thymus. About 60% of patients with AChR antibody-positive myasthenia gravis have an enlarged thymus, and about 10% have a thymoma.

4. As in the preceding case, patients with myasthenia gravis who are in respiratory distress may be experiencing a myasthenic crisis or cholinergic crisis, both of which require immediate airway management. Additional treatment involves addressing any exacerbating factors (the most common is infection), removing antibodies from circulation with the aid of plasmapheresis, diminishing the activity of the disease with immunotherapy, or removing the site of antibody formation with surgical thymectomy.

Supplemental References

[1] L.P. Rowland, Stroke, spasticity, and botulinum toxin, N. Engl. J. Med. 347 (2002) 382–383.

[2] A. Brashear, M.F. Gordon, Intramuscular injection of botulinum toxin for the treatment of wrist and finger spasticity after a stroke, N. Engl. J. Med. 347 (2002) 395–400.

[3] Y. Fujinaga, Interaction of botulinum toxin with the epithelial barrier, J. Biomed. Biotechnol. 2010 (2010) 974943.

[4] Y. Li, Y. Arora, K. Levin, Myasthenia gravis: newer therapies offer sustained improvement, Cleve. Clin. J. Med. 80 (2013) 711–721.

CLINICAL CASE STUDY 19.2 Congenital Muscular Dystrophy

This clinical case was abstracted from: R.H. Brown, P.E. Grant, C.R. Pierson, Case 35-2006: a newborn boy with hypotonia, N. Engl. J. Med. 355 (2006) 2132−2142.

Synopsis

Upon initial examination of a male neonate born full term to a mother with an uncomplicated pregnancy and delivery, a weak cry and decreased tone in the arms were noted. Over the next 2 days in the newborn nursery, his hypotonia worsened, and he developed a high-pitched cry and difficulty with feeding, which resulted in choking, oxygen desaturation, and bradycardia. Family history was unremarkable for neurologic or muscular diseases; however, the patient's father had problems with feeding and walking during infancy but was otherwise well with normal childhood development. The patient's neurologic exam was significant for diffuse hypotonia, flaccid extremities, and minimal to absent reflexes. Laboratory studies revealed a markedly elevated serum creatinine kinase level, and MRI of the brain showed an area of polymicrogyria (abnormal folding of the gyri) at the right temporal lobe. Histologic examination of a biopsy of the quadriceps muscle showed signs of dystrophic disease and decreased levels of merosin. The diagnosis of congenital muscular dystrophy with merosin deficiency was made. The patient continued to have hypotonia requiring permanent tracheostomy, gastrostomy, and continuous mechanical ventilation.

Teaching Points

1. In evaluating a "floppy infant," one should consider four broad categories: the central nervous system (CNS), peripheral nerves (sensory and motor neurons), the neuromuscular junction, and the muscle. Signs of CNS involvement include seizure, hemiparesis, and visual abnormalities. Peripheral nerve involvement may cause flaccid weakness, decreased deep tendon reflexes, or sensory deficits. The types of neuromuscular junction disorders are congenital myasthenia, neonatal myasthenia gravis, and infantile botulism. The three types of skeletal muscle diseases to be considered are congenital myopathies, congenital muscular dystrophy, and a type of neonatal myotonic dystrophy.

2. As seen in this case, the hallmark of a congenital muscular dystrophy is significant weakness at birth, markedly elevated serum creatinine kinase levels, and evidence of dystrophy on muscle biopsy. Polymicrogyria on MRI of the brain is also highly suggestive of a congenital muscular dystrophy. There are several types of congenital muscular dystrophies caused by different gene mutations (refer to case reference for detailed list).

3. Merosin, also known as laminin-2 (LAMA2), is a large extracellular glycoprotein expressed in the basal lamina of striated muscles and peripheral nerves. The role of merosin is to maintain the cell membrane and extracellular matrix relationship, the cell-to-cell interactions during normal development and differentiation, and the integrity of tissue after differentiation. Merosin deficiency caused by loss-of-function mutations of the LAMA2 gene result clinically in skeletal muscle atrophy and weakness, although there are several other gene mutations and protein deficiencies that can result in a congenital muscular dystrophy (refer to case reference for detailed list).

Supplemental Reference

[1] V. Allamand, P. Guicheney, Merosin-deficient congenital muscular dystrophy, autosomal recessive, Eur. J. Hum. Genetics 10 (2002) 91−94.

REQUIRED READING

[1] S. Basaria, S. Bhasin, Targeting the skeletal muscle-metabolism axis in prostate-cancer therapy, N. Engl. J. Med. 367 (2012) 965.

[2] Y. Ge, J. Chen, Mammalian target of rapamycin (mTOR) signaling network in skeletal myogenesis, J. Biol. Chem. 287(52) (2012) 43928−43935.

[3] J.E. Dominique, C. Gerard, Myostatin regulation of muscle development: molecular basis, natural mutations, physiopathological aspects, Exp. Cell Res. 312 (2006) 2401−2414.

[4] B. Elliot, D. Renshaw, S. Getting, R. Mackenzie, The central role of myostatin in skeletal muscle and whole body homeostasis, Acta Physiol. 205 (2012) 324−340.

[5] J.M. Argiles, M. Orpi, S. Busquets, F.J. Lopez-Soriano, Myostatin: more than just a regulator of muscle mass, Drug Discov. Today 17 (2012) 702−708.

[6] D.S. Herman, L. Lam, M.R.G. Taylor, L. Wang, P. Teekakirikul, D. Christodoulou, et al., Truncations of titin causing dilated cardiomyopathy, N. Engl. J. Med. 366 (2012) 619−628.

ENRICHMENT READING

[1] H. Morita, H.L. Rehm, A. Menesses, B. McDonough, A.E. Roberts, R. Kucherlapati, et al., Shared genetic causes of cardiac hypertrophy in children and adults, N. Engl. J. Med. 358 (2008) 1899−1908.

[2] H.A. Maluli, A.B. Meshkov, A short story of the short QT syndrome, Cleve. Clin. J. Med. 80 (2013) 41.

[3] F.C. Ernste, A.M. Reed, Idiopathic inflammatory myopathies: current trends in pathogenesis, clinical features, and up-to-date treatment recommendations, Mayo Clin. Proc. 88 (2013) 83−105.

[4] L.M. Ballou, R.Z. Lin, Rapamycin and mTOR kinase inhibitors, J. Chem. Biol. 1 (2008) 27−36.

Chapter 20

Perturbations of Energy Metabolism: Obesity and Diabetes Mellitus

Key Points

1. Metabolism encompasses numerous chemical transformations, and they comprise anabolism and catabolism.

2. The main sources of energy for metabolism are carbohydrates and lipids. Carbohydrates and lipids (triglycerides) provide about 4.1 kcal and 9.4 kcal of energy per gram, respectively.

3. In the body, energy derived from food, in part, is released as heat and also used for the synthesis of ATP.

4. Energy requirements are determined by basal metabolic rate, assimilation of food, physical activity, and physiologic state. Maintenance of desirable body weight at any age is dependent on the balance between energy intake and energy output.

5. AMP-activated protein kinase (AMPK) plays a central role in energy homeostasis. Allosteric activation of AMPK by high intracellular concentrations of AMP coupled with ATP consumption during energy expenditure promotes ATP synthetic pathways. Energy-regulating functions of AMPK also include transcriptional reprogramming of key enzymes of energy homeostasis and mitochondrial biogenesis.

6. A surrogate marker for assessing obesity is body mass index (BMI = kg/m^2). The optimal BMI for adults is ≤ 25.

7. Energy intake (feeding behavior) is regulated at the hypothalamic level by the integration signals arriving from energy acquisition, dissipation, and storage. The signals are many and redundant.

8. Excess energy intake is stored as triglycerides (fat) in the white adipose tissue cells (simply known as adipose tissue cells) and leads to obesity (BMI > 30). Brown adipose tissue does not accumulate fat and is primarily involved in the heat generation due to uncoupling of the mitochondrial oxidative phosphorylation during oxidation. Its role in adult humans is limited.

9. Obesity occurs in epidemic proportions and is associated with increased morbidity and mortality. Both heritable risk factors and eating behavior are known to play roles in the development of obesity. Interventions include behavioral (decreased food intake and increased physical activity), pharmacologic, and bariatric surgery. Both behavioral and pharmacological interventions are often temporary and moderate.

10. Perturbations of glucose homeostasis are responsible for diabetes mellitus syndrome. Plasma glucose levels are regulated by hormones secreted from the pancreas (insulin and glucagon), the adrenal cortex and medulla, gastrointestinal tract (incretins), and the pituitary. Diabetes has multiple etiologies and is a major cause of macrovascular and microvascular diseases.

11. Type 1 diabetes is due to destruction of β-cells of the pancreas and requires life-long exogenous insulin administration. The more common global epidemic type 2 disease is due to inadequate supply and tissue resistance to the physiological effects of insulin.

12. Insulin secretion from β-cells of the pancreas is primarily regulated by plasma levels of glucose and also a few amino acids and hormones. Biologic actions of insulin at the target sites (muscle, adipose tissue, and liver) are multiple and are mediated by cell surface receptors coupled to tyrosine-specific protein kinase activity.

13. The treatment and management of type 2 diabetes are multifaceted and require behavioral modifications to correct overeating and inactivity, administration of therapeutic agents that promote insulin secretion and the action of insulin at the target site, and exogenous insulin administration. These prompt therapeutic approaches are required to forestall premature microvascular and macrovascular complications. The management of diabetes also requires the assessment of fluctuations of plasma glucose levels by measuring $HgbA_{1c}$ levels and renal function by measuring urinary albumin concentration (microalbuminuria).

This chapter gives an overview of energy metabolism, its regulation, and its perturbations. Obesity and diabetes mellitus (type 2) are prevalent disorders worldwide and are examples of energy metabolism disorders.

N. V. Bhagavan and C.-E. Ha: Essentials of Medical Biochemistry, Second edition. DOI: http://dx.doi.org/10.1016/B978-0-12-416687-5.00020-8

ENERGY METABOLISM

To carry out the body's essential functions (e.g., growth, repair, pregnancy, lactation, physical activity, maintenance of body temperature), food must be consumed and utilized, and body constituents synthesized. The term *metabolism* encompasses the numerous chemical transformations that occur within the human body. Metabolism comprises **anabolism** and **catabolism**. Anabolism involves the synthesis of new molecules, usually larger than the reactants, and is an energy-requiring process. Catabolism involves degradation processes, usually the breaking down of large molecules into smaller ones, and is an energy-yielding process. **Intermediary metabolism** refers to all changes that occur in a food substance beginning with absorption and ending with excretion. Two main sources of energy for metabolism are carbohydrates and fats (lipids). Proteins have less importance as a source of energy.

Basal metabolism, determined by the **basal metabolic rate (BMR)**, is an expression of the body's vital energy needs during **physical**, **emotional**, and **digestive rests**. It represents the energy required for the maintenance of body temperature, muscle tone, the circulation of blood, the movement of respiration, and the glandular and cellular activities of a person who is awake and not involved in physical, digestive, or emotional activities. The resting metabolic energy requirements are closely related to lean body mass (as the body fat increases, the energy requirement decreases). Muscle, endocrine glands, and organs such as the liver are metabolically more active (i.e., consume a larger amount of oxygen per unit weight) than adipose tissue and bone.

The BMR for women (who have a greater proportion of fat per total body weight than men) is lower than that for men. In both sexes, the development of muscle tissue as a result of increased exercise increases the metabolic rate and the basal metabolic needs. In normal young men and women, the average body fat constitutes about 12% and 26% of total body weight, respectively; if these values are exceeded by more than 20% and 30% of the average values, respectively, they indicate the presence of obesity. The BMR changes with age. It is high at birth, increases up to 2 years of age, and then gradually decreases, except for a rise at puberty. This general decline in the BMR (approximately 2% per decade after age 21 years) is proportional to the reduction in muscle tissue (or lean body mass) and the accompanying increase in body fat. BMR is also affected by menstruation, pregnancy, and lactation.

The BMR is altered in a number of pathological states. It is increased in **hyperthyroidism**, fever (approximately 12% elevation for each degree Celsius above normal body temperature), **Cushing's syndrome**, tumors of the adrenal gland, anemia, leukemia, polycythemia, cardiac insufficiency, and injury. BMR is decreased in hypothyroidism, starvation, malnutrition, hypopituitarism, hypoadrenalism (e.g., **Addison's disease**), and anorexia nervosa.

To calculate energy requirements in an individual, one needs to take into account BMR, physical activity (muscular work), age, sex, height, and weight. Maintenance of desirable body weight at any age depends on the balance between energy intake and energy output or other requirements. During growth, pregnancy, or recovery from illness, energy demands are greater: hence, intake should be higher than output. **Obesity** and **cachexia** (a disorder characterized by general physical wasting and malnutrition) are extreme examples of problems affecting the energy stores of the body.

Mediators of Feeding Behavior

The mediators that are involved in maintaining energy balance and eating behavior are many, and their actions are complex. The major anatomic location of integration of signals for energy storage and dissipation is the hypothalamus (Figure 20.1).

The biochemical mediator **AMP-activated protein kinase (AMPK)** plays a central role in energy metabolism. AMPK is a heterotrimeric protein complex and consists of a catalytic α-subunit and regulatory β- and γ-subunits. During energy consumption coupled to the expenditure of ATP during anabolic processes, exercise, and other normal functions, AMP levels rise. The high intracellular concentrations of AMP allosterically activate AMPK, which, in turn, promotes ATP synthesis by activating key regulatory enzymes in the catabolic pathways (Chapters 12 and 16).

The energy regulatory function of AMPK also includes transcriptional reprogramming of key enzymes involved in energy expenditure and synthesis and of mitochondrial biogenesis. Metformin, a therapeutic agent used in the management of diabetes mellitus type 2, activates AMPK, which inhibits the transcription of key regulatory hepatic enzymes required for gluconeogenesis (discussed later).

The major energy reservoir of the body is triglyceride (also called **triacylglycerol**), which is stored in the white adipose tissue. The synthesis and breakdown of triglycerides are discussed in Chapter 16. The body also contains brown adipose tissue, which is involved in thermogenesis in small mammals and human neonates. Recent studies have revealed that brown adipose tissue is also active in adults (Chapter 13). Brown adipose tissues are rich in mitochondria, hence the color. They produce heat by uncoupling oxidative phosphorylation (Chapter 13). In this chapter, adipocyte refers to a cell derived from white adipose tissue only.

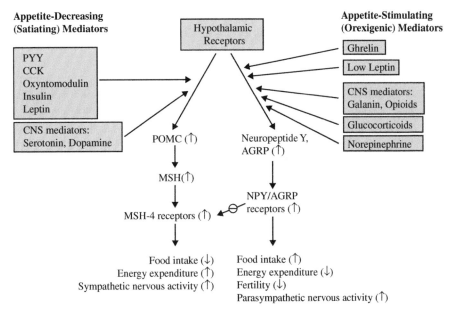

FIGURE 20.1 Integration of molecular circuitry of appetite regulation signals in the hypothalamus. Signals that stimulate the synthesis of neuropeptide Y (NPY) and agouti-related peptide stimulate appetite, whereas signals that stimulate the production of α-melanocyte-stimulating hormone (MSH) due to increased synthesis and proteolysis of pro-opiomelanocortin (POMC) depress appetite. NPY/AGRP and POMC neurons are reciprocally regulated. PYY = polypeptide YY; CCK = cholecystokinin; AGRP = agouti-related peptide; (↑) = increased; (↓) = decreased; ⊖ = inhibited.

One of the most significant mediators of the energy store in the adipose tissue is **leptin** (from the Greek *leptos*, meaning "thin"). Leptin is a protein of 167 amino acid residues that is synthesized in adipocytes. It is secreted in a pulsatile manner, in a nyctohemeral rhythm, and in proportion to the extent of the adipose tissue.

Leptin's synthesis is increased by insulin, glucocorticoids, and estrogens and is decreased by β-adrenergic agonists. Leptin's action on energy metabolism is mediated by receptors in many cells, and it binds specifically to a receptor in the hypothalamus.

The action of leptin involves at least two pathways. During starvation and weight loss, adipose tissue is decreased, with consequent low levels of leptin. The low level of leptin leads to the production of neuropeptide Y (NPY), which is synthesized in the arcuate nucleus of the hypothalamus and is transported axonally to the paraventricular nucleus. Neuropeptide Y binds to its receptor and functions as a potent appetite stimulant. The overall effect is increased appetite, decreased energy expenditure and temperature, decreased reproductive function (infertility), and increased parasympathetic activity.

An opposite set of events occurs when the leptin levels rise: decreased production of NPY and activation of the POMC pathway (Figure 20.1). The inhibition of synthesis of NPY by leptin is mediated by activation of the Janus kinase signal transduction (JAK–STAT) pathway (Chapter 28). The overall biological response of elevated leptin levels includes decreased appetite, increased energy expenditure, and increased sympathetic nervous activity (Figure 20.1).

OBESITY

The first law of thermodynamics states that the amount of stored energy equals the difference between energy intake and energy expenditure. The principal method of storage of energy is that of triglycerides in adipose tissue. Energy stores are essential for survival during times of energy deprivation.

Energy stores in an adult are maintained at a relatively constant level throughout life. However, even a small daily imbalance in energy intake over long periods of time will have a significant effect on energy storage. For example, suppose a nonobese adult's energy intake exceeds expenditure by about 1% daily for one year. This amounts to an excess of 9000 kcal and corresponds to a weight gain of about 2.5 lb (1.15 kg) per year. The chemical energy of 3500 kcal is equivalent to 1 lb (0.45 kg) of adipose tissue. Weight gain in most people is attributable to overconsumption of palatable, energy-dense foods (e.g., lipids), and a sedentary lifestyle. Childhood obesity is a risk factor for obesity in adulthood.

The overweight condition and obesity are defined using the body mass index (BMI), which is calculated as weight in kilograms divided by the square of height in meters (BMI = kg/m^2). A BMI of between 25 and 29.9 is considered as overweight and ≥30 is considered obese. Body weight is determined by a balance between energy intake and energy expenditure. The energy expenditure is required to maintain basal metabolic functions, absorption and digestion of foods (thermic effect of food), physical activity and, in children, linear growth and development. If there is a net excess of energy intake, BMI increases, and this will eventually lead to an overweight condition and ultimately to obesity and morbid obesity. In a population with stable genetic factors, an increase in obesity is primarily attributable to consumption of excess foods with high fat content and decreased physical activity. Obesity is a worldwide health problem. The prevalence of obesity-associated morbidity depends on the location of fat distribution in the body. Intra-abdominal or visceral fat deposits are associated with higher health risks than gluteofemoral adipose tissue fat accumulation. It is a risk factor for development of diabetes mellitus, hypertension, obstructive sleep apnea, and heart disease, all of which cause decreased quality of life and life expectancy.

In adults, the elevated BMI due to increased adiposity has various health risks such as diabetes and cardiovascular diseases. However, in an elderly man aged 55 and older and a woman over the age of 65, the BMI may be less useful in ascertaining good health, and obesity that is not related to abundance of central fat may be protective from all-cause mortality. Furthermore, muscle mass in the elderly is inversely associated with optimal health [1].

Feeding behavior and energy balance are regulated by a complex set of short-term and long-term physiological signals (Figure 20.1). Mechanisms that lead to obesity involve interactions between genetic, environmental, and neuroendocrine factors. It should be emphasized, however, that many myths and misinformation about obesity have prevailed in the general population. In reference [2], Tables 1 and 2 list the myths and facts about obesity.

In humans, the pathways for regulation of appetite and food consumption have been verified by identifying mutations in genes that participate in regulation of body weight. The regulators of conversion of preadipocytes to adipocytes also determine the adipose tissue mass. One of the regulators is a transcription factor known as **peroxisome proliferator-activated receptor-γ** (PPAR-γ). Thiazolidinediones are hypoglycemic agents (discussed later) that increase insulin sensitivity and also are synthetic ligands for PPAR-γ. Phosphorylation of PPAR-γ at ser114 makes the receptor less active in converting preadipocytes to adipocytes. A mutation converting proline to glutamine at position 115 blocks phosphorylation at the serine site, maintains the protein in the active state, and leads to obesity.

Leptin therapy has corrected obesity in a child with congenital leptin deficiency. In obese individuals, the presence of circulating high levels of leptin has been attributed to resistance or some other defect in the leptin receptors. This apparent paradox of high leptin levels associated with obesity is analogous to insulin resistance seen in type 2 diabetes mellitus. **Prader−Willi hereditary syndrome** of obesity is caused by the acquisition of chromosome 15

with deletions or silencing of several genes in the long arm of the chromosome from the patient's father or with acquisition of two copies of the maternal alleles instead of one maternal and one paternal allele (uniparental maternal disomy). This is a most prevalent form of dysmorphic genetic obesity (1 in 10,000−20,000 births). Patients who acquire the same chromosome with deletions from the mother or with uniparental disomy have completely different abnormal phenotypic manifestations. This syndrome is known as **Angelman syndrome**. Prader−Willi and Angelman syndromes are genomic imprinting in the causation of human diseases. Although several monogenic disorders of obesity (e.g., leptin and leptin receptor gene) are known, in the vast majority of obese patients, the molecular defects remain unknown.

The treatment of obesity, particularly with BMI >35, is challenging. It requires behavioral modifications, dietary restrictions, exercise, pharmacotherapy, and surgical intervention (bariatric surgery). The appetite suppressants sibutramine, a serotonin reuptake inhibitor, and orlistat (Chapter 11), which prevent absorption of lipids from the GI tract, are used as pharmacotherapy. The effects of these drugs on weight reduction are modest. Bariatric surgery, which consists of modifying the anatomy of the GI tract to restrict food absorption, has been used for long-term treatment of morbid obesity.

One of the most commonly used types of bariatric surgery, known as the **Roux-en-Y gastric bypass procedure**, consists of connecting the jejunum to a small pouch of proximal stomach (Figure 20.2). This bypass procedure severely limits absorption of food and promotes rapid weight loss. Patients who undergo bariatric surgery require adaptations in food intake to compensate for a smaller stomach and continual monitoring for micronutrient (vitamins and minerals) deficiencies (see the clinical study in reference [3]).

In addition to weight loss in subjects with gastric bypass surgery, a serendipitous benefit has been the amelioration of glucose intolerance due to type 2 diabetes mellitus. This beneficial effect has been attributed to increased production of incretin hormones by L-cells of the distal intestine due to rapid passage of the food from the stomach to the intestine. Incretin hormones (glucagon-like-peptide [GLP-1], glucose-dependent insulinotropic peptide [GIP], and peptide YY; Chapter 11) promote insulin secretion from pancreatic β-cells and their growth, and appetite suppression. Gastric reduction and bypass procedures may also reduce the action of the appetite-stimulating hormone ghrelin [4].

Diabetes Mellitus

Diabetes mellitus is a syndrome consisting of several metabolic disorders, all of which are characterized by

Roux-en-Y gastric bypass

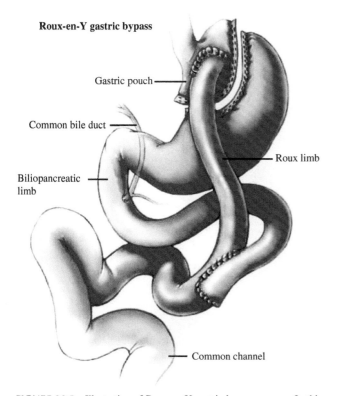

FIGURE 20.2 Illustration of Roux-en-Y gastric bypass surgery. In this procedure, small-intestinal segment jejunum is connected to the gastric pouch, bypassing significant food absorptive areas. *[Reprinted with permission from E. J. Maria, Bariatric surgery for morbid obesity. New Engl. J. Med., 356 (2007) 2179.]*

TABLE 20.1 Constituents of Endocrine Pancreas

Cell Type	Hormone	Structure
α	Glucagon	29 amino acids
β	Insulin	Two chains: A = 21 amino acids B = 30 amino acids
	Amylin	37 amino acids
δ	Somatostatin	Two forms: S-14 = 14 amino acids S-28 = 28 amino acids
F	Pancreatic polypeptide	36 amino acids

of cells; α- and δ-cells are located at the periphery and β-cells at the center. Close intercellular communication exists between different cell types known as **paracrine signaling**. The islet appears to function as an integrated unit rather than as four independent types of cell.

Insulin

Structure and Synthesis

The structures of insulin and proinsulin are shown in Figure 20.3. Proinsulin is a single polypeptide chain of 86 amino acids that permits correct alignment of three pairs of disulfide bonds. Insulin is composed of an A chain of 21 amino acids and a B chain of 30 amino acids, the chains being held together by two disulfide bonds. A third disulfide bond is present within the A chain. Human insulin differs from porcine insulin by a single amino acid and from beef insulin by three amino acids. These substitutions do not significantly affect activity—hence the use of bovine and porcine insulin in clinical therapy. Human insulin and its short-acting and long-acting analogues have been synthesized by recombinant technology for clinical use (discussed later).

Insulin biosynthesis resembles that of other export proteins. Biosynthetic steps consist of gene transcription, processing and maturation of precursor mRNA, translation of mature mRNA, translocation of preproinsulin to cisternae of the endoplasmic reticulum with removal of the N-terminal leader or signal sequence of 23 amino acids, folding of proinsulin with pairing of three specific disulfide bridges, packaging of proinsulin into secretory granules, and conversion of proinsulin to equimolar amounts of insulin and connecting peptide (C-peptide) by site-specific enzymatic cleavages.

hyperglycemia (discussed further later in chapter). It is one of the most common causes of premature death due to macrovascular (atherosclerosis) and microvascular (nephropathy, neuropathy, and retinopathy) diseases.

The pancreatic endocrine hormones insulin, glucagon, and somatostatin, and insulin's counter-regulatory hormones epinephrine, cortisol, and growth hormone, orchestrate the maintenance of optimal plasma glucose levels and thus regulate energy metabolism. This section reviews the biochemical properties of pancreatic hormones.

Endocrine Pancreas and Pancreatic Hormones

The endocrine pancreas is pivotal in metabolic homeostasis and is an integral component of metabolic regulation. It is composed of $1-2$ million islets of Langerhans scattered throughout the organ and comprising $2\%-3\%$ of its total weight. The islets contain at least four cell types (α, β, δ, and F) that secrete glucagon, insulin, somatostatin, and pancreatic polypeptide, respectively (Table 20.1). The α- and β-cells make up 20% and 75% of the islets, respectively. Most islets have a characteristic distribution

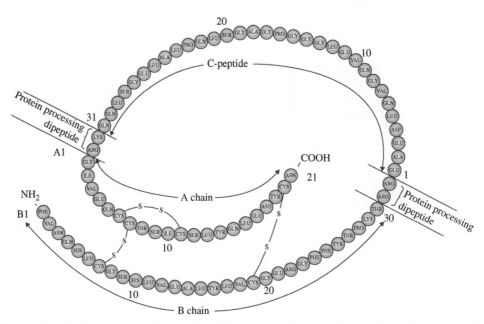

FIGURE 20.3 Structure of insulin. Insulin is derived from proinsulin by cleavage at the dipeptides Arg—Arg and Lys—Arg to give A and B chains held together by disulfide bonds. In the pig, B30 is Ala. In the cow, A8 is Ala, A10 is Val, and B30 is Ala. Bovine and porcine insulins are used extensively in clinical practice.

Conversion of proinsulin to insulin and C-peptide in secretory granules involves site-specific cleavages at the Arg—Arg and Lys—Arg sequences (Figure 20.3); these serve as signals for proteolytic processing of many other proteins. Cleavage occurs at the C-terminal end of each pair by trypsin-like enzymes and is followed by removal of the basic residues by a carboxypeptidase B-like enzyme.

Insulin monomers undergo noncovalent dimerization by formation of antiparallel β-pleated sheet associations between monomers involving the C-terminal portion of the B chain. Three insulin dimers subsequently self-associate to form hexamers in the presence of Zn^{2+}. The Zn^{2+} hexameric array of insulin probably gives the β-cell granule its unusual morphologic characteristics.

Zn^{2+} is released when insulin is secreted. Conversion of proinsulin to insulin in the secretory granule is not complete, and some proinsulin is also released on secretion of insulin. Proinsulin has less than 5% of the biological activity of insulin. The C-peptide has no physiological function, but assay of C-peptide in the serum helps distinguish between endogenous and exogenous sources of insulin.

Like genes for other proteins, the insulin gene may undergo mutation and produce an abnormal product. This process may be suspected in individuals who exhibit **hyperinsulinemia** without hypoglycemia or evidence of insulin resistance. Three abnormal insulins have been documented: one in which Ser replaces Phe at B24; another in which Leu replaces Phe at B25; and a third in which the Lys—Arg basic amino acid pair is replaced by Lys—X (X = nonbasic amino acid). The former two

mutations occur in the nucleotide sequence that codes for an invariant tetrapeptide sequence (B23—B26), and hence the abnormal insulins have greatly diminished biological activity. The third mutation yields an abnormal form of proinsulin in which the C-peptide remains attached to the A chain after processing. It is expressed as an autosomal dominant trait (**familial hyperproinsulinemia**). Affected individuals show no signs of insulin resistance and may or may not exhibit mild hyperglycemia.

Secretion

Multiple factors regulate insulin secretion from the pancreatic β-cells. Glucose, amino acids, glucagon, acetylcholine, incretins, and β-adrenergic agents stimulate insulin secretion, whereas somatostatin and α-adrenergic agents exert inhibitory influences. Most notable of the insulin secretagogues is glucose. The sequence of events that lead to insulin secretion by the β-cells as the threshold of blood glucose increases is shown in Figure 20.4.

Normal fasting plasma glucose is maintained between 70 and 105 mg/dL (3.89—5.83 mmol/L). Glucose enters the β-cells by means of **glucose transporters** GLUT1 and GLUT2 (Chapter 12). GLUT1 is a constitutive glucose transporter, and GLUT2 is a low-affinity glucose transporter capable of sensing the glucose concentration in β-cells. Glucose transport is not a rate-limiting step in the β-cell glucose metabolism because the transporters are present in greater abundance relative to physiological rates of glucose entry. After glucose enters the β-cells, it

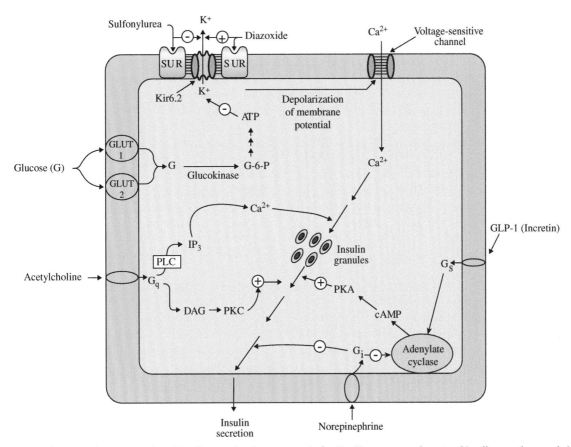

FIGURE 20.4 Diagrammatic representation of insulin secretion from pancreatic β-cells. The sequence of events of insulin secretion coupled to glucose entry into β-cells consists of glucokinase action, ATP production, inhibition of the ATP-sensitive K^+ channel, membrane depolarization, Ca^{2+} influx, and insulin release. Neurotransmitters acetylcholine and norepinephrine stimulate and inhibit insulin secretion via trimeric G-proteins G_q and G_i, respectively. Glucagon-like peptide-1 (GLP-1, incretin) promotes insulin release via the G-protein G_s. Sulfonamides and diazoxide have direct effects on sulfonylurea receptors (SURs); the former promote insulin release and the latter inhibit insulin release. $+$ = stimulation; $-$ = inhibition. Other abbreviations are given in the text.

is converted to glucose-6-phosphate by glucokinase, which is an isoenzyme of hexokinase. Glucokinase has a low affinity for glucose, is the rate-limiting step for glucose metabolism, and is not affected by feedback inhibition. The glucose-sensing device that allows rapid and precise quantitation of the ambient glucose level appears to be glucokinase, which is ultimately responsible for glucose-regulated insulin secretion.

The oxidation of glucose-6-phosphate via glycolysis, the TCA cycle, and the electron transport system coupled to oxidative phosphorylation leads to increased ATP levels (or ATP/ADP ratio) in the β-cell. The elevated ATP levels cause a closure of the ATP-sensitive K^+ channel, leading to inhibition of K^+ efflux and depolarization of the β-cell membrane potential. The ATP-sensitive K^+ channel consists of two different types of protein subunits, namely, the sulfonylurea receptor (SUR) and a potassium channel (Kir6.2). SUR is the regulatory subunit and belongs to the family of ATP-binding cassette proteins. Kir6.2 subunits participate in the actual conduction of K^+. It has been suggested that the overall functional K^+ channel is an octameric complex composed of equal numbers of SUR and Kir6.2 subunits. The depolarization of β-cell membrane that occurs with the closure of the ATP-sensitive K^+ channel activates the voltage-sensitive Ca^{2+} channel, causing Ca^{2+} influx, which ultimately leads to release of insulin from stored insulin granules. The exocytosis of insulin involves both phosphorylation and exocytosis-related ATPases. Along with insulin and C-peptide, a 37 amino acid peptide known as **amylin** is also secreted during exocytosis. Amylin is obtained by proteolytic processing of an 89 amino acid precursor molecule. Amylin, also known as **islet amyloid polypeptide**, has a number of physiologic actions that function **synergistically with insulin**. It delays gastric emptying, decreases food intake, and inhibits secretion of glucagon. Its anorectic effect is mediated by the hypothalamus (see previous section). A synthetic analogue of amylin, pralintide, is used in the management of diabetes (discussed later).

Mechanisms that alter the level of cytosolic Ca^{2+} in β-cells, other than glucose metabolism, also affect insulin release (Figure 20.4). For example, a regulatory protein associated with the ATP-sensitive K^+ channel, when occupied with sulfonylurea, inhibits K^+ efflux, causing insulin secretion. Sulfonylureas are drugs used in the management of type 2 diabetes mellitus (discussed later). Diazoxide has an opposite effect to that of sulfonylureas. It either prevents the closing, or prolongs the opening time of, the ATP-sensitive K^+ channel, resulting in the inhibition of insulin secretion and thus hyperglycemia. Somatostatin inhibits Ca^{2+} influx and causes diminished insulin secretion. Acetylcholine causes elevation of cytosolic Ca^{2+} followed by insulin secretion caused by activation of G_q-protein, phospholipase C-inositol triphosphate-Ca^{2+}, and protein kinase C (Chapter 28). Norepinephrine and epinephrine depress insulin secretion by binding at the α-adrenergic receptor sites and by inhibiting adenylate cyclase mediated by the activation of the inhibitory G-protein (G_i). This leads to the inhibition of cAMP production and results in decreased activity of protein kinase A. Decreased protein kinase A levels determine exocytosis-related phosphorylation required for insulin secretion. Release of epinephrine during stress signals the need for catabolic rather than anabolic activity. Depression of insulin secretion during exercise or trauma is also associated with epinephrine (catecholamine) secretion. A gastrointestinal hormone known as **glucagon-like peptide (GLP-1)** promotes insulin secretion via G-protein activation of the adenylate cyclase–cAMP–protein kinase A system. Gastrointestinal hormones that regulate insulin secretion may act in a feed-forward manner to signal gastrointestinal activity and metabolic fuel intake. Pancreatic glucagon, simply known as **glucagon**, stimulates secretion of insulin while somatostatin depresses it. Thus, various regulatory inputs to the β-cells are integrated to maintain secretion of optimal quantities of insulin and to maintain glucose homeostasis. The coordinated activity of three pancreatic hormones (insulin, glucagon, and somatostatin) is essential for fuel homeostasis. Furthermore, the action of insulin is opposed by glucagon and by other counter-regulatory hormones, namely epinephrine, cortisol, and growth hormone. All of these hormones correct hypoglycemia by maintaining adequate levels of glucose in tissues, such as brain, which is primarily dependent on glucose as a fuel source.

Some amino acids also function as secretagogues of insulin secretion. An example is leucine, which stimulates the release of insulin by allosteric activation of glutamate dehydrogenase. Glutamate dehydrogenase, a mitochondrial enzyme, converts glutamate to α-ketoglutarate by oxidative deamination (Chapter 15). Glutamate dehydrogenase is positively modulated by ADP and negatively modulated by GTP. The α-ketoglutarate is subsequently oxidized to provide ATP, which blocks the ATP-sensitive K^+ channel, eventually causing insulin release.

The overall glucokinase–glucose sensor mechanism as the primary regulator of glucose-controlled insulin secretion of β-cells has been substantiated by identifying mutations that affect human glucokinase. Both gain-of-function and loss-of-function mutants of glucokinase are known. The activating glucokinase mutation (Val455Met) with increased affinity for glucose results in hyperinsulinism with fasting hypoglycemia. Other mutations that impair glucokinase activity have been identified. These defects result in hyperglycemia and diabetes mellitus, known as **maturity-onset diabetes of the young (MODY)**, which also occurs as a result of mutations in genes that encode hepatocyte nuclear factors 1α, 4α, 1β, and insulin promoter factor 1.

Rare mutations, either gain-of-function or loss-of-function in the SUR/Kir6.2 components of the K^+ channel, can result in hyperglycemia or hypoglycemia, respectively. In the gain-of-function mutations, the K^+ channels remain open, leading to neonatal diabetes mellitus. The opposite set of events occur in loss-of-function mutations in which K^+ channels remain closed, causing a continual secretion of insulin leading to neonatal hypoglycemia.

Abnormalities can also result from defects in glutamate dehydrogenase function. Conversion of glutamate to α-ketoglutarate by glutamate dehydrogenase provides substrates for ATP production, and the enzyme is inhibited by GTP at an allosteric site. The importance of the sensitivity of glutamate dehydrogenase to inhibition by GTP is illustrated in patients with hyperinsulinism with episodes of hypoglycemia. All of these patients carry mutations affecting the allosteric domain of glutamate dehydrogenase and result in insensitivity to GTP inhibition. This causes increased α-ketoglutarate production followed by ATP formation and consequent insulin secretion. In addition to hyperinsulinemia, these patients also have high levels of plasma ammonia (**hyperammonemia**) that result from hepatic glutamate dehydrogenase activity. In liver mitochondria, glutamate is converted to N-acetylglutamate, which is the required positive allosteric effector of carbamoyl-phosphate synthetase. Carbamoyl-phosphate synthetase catalyzes the first step in the conversion of ammonia to urea (see Chapter 15 for ammonia metabolism). Increased activity of glutamate dehydrogenase causes a depletion of glutamate required for the synthesis of N-acetylglutamate. The decreased level of N-acetylglutamate impairs ureagenesis in the liver and is accompanied by hyperammonemia.

Biological Actions of Insulin

Insulin affects virtually every tissue. Insulin is an anabolic signal and promotes fuel storage in the form of glycogen and triacylglycerols, while inhibiting the breakdown of these two fuel stores. Insulin also promotes protein synthesis while inhibiting its breakdown. The regulation of expression of several genes, either positively or negatively, is mediated by insulin. The genes involved in the expression of enzymes that participate in fuel storage (e.g., hepatic glucokinase) are induced; those that encode catabolic enzymes (e.g., hepatic phosphoenolpyruvate carboxykinase) are inhibited. The principal effect of insulin on blood glucose is its uptake into muscle and adipose tissue via the recruitment and translocation of glucose transporter 4 (GLUT4).

The mechanism of action of insulin is complex and can be divided into three parts. The first part is the binding of insulin to its receptor on the cell membrane; the second consists of postreceptor events; and the third consists of biological responses.

Insulin Receptor

The insulin receptor is derived from a single polypeptide chain that is the product of a gene located on the short arm of chromosome 19. The proreceptor undergoes extensive post-translational processing consisting of glycosylation and proteolysis. The proteolytic cleavage yields α (M.W. 135,000) and β (M.W. 95,000) subunits that are assembled into a heterotetrameric complex ($\alpha_2\beta_2$). The subunits are held together by both disulfide linkages and noncovalent interactions. The heterotetrameric receptor is incorporated into the cell membrane with the α-subunits containing the insulin-binding domain projecting into the extracellular space and the β-subunit with its transmembrane subunit projecting into cytosolic space (Figure 20.5). The intracellular portion of the β-subunit has a tyrosine-specific protein kinase (tyrosine kinase) activity. Tyrosine kinase activity is initiated by insulin binding to α-subunits and results in autophosphorylation of at least six tyrosine residues in β-subunits. Other protein substrates are also targets of a series of phosphorylations catalyzed by the activated β-subunit tyrosine kinase. One of the key proteins that is phosphorylated is insulin receptor substrate-1 (IRS-1), which has many potential tyrosine phosphorylation sites, as well as serine and threonine residues. Phosphorylated IRS-1 interacts and initiates a cascade of reactions involving other

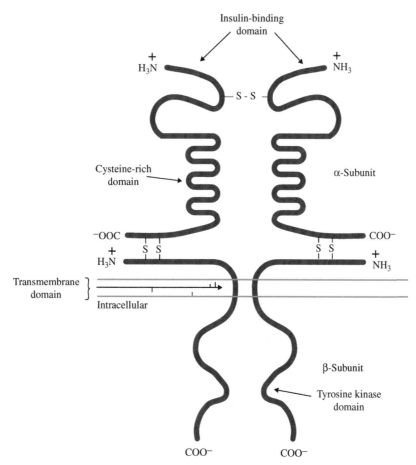

FIGURE 20.5 Model of the insulin receptor. The receptor contains two α- and two β-subunits, held together by disulfide linkages. The α-subunits are entirely extracellular and contain the insulin-binding site. The β-subunits have transmembrane and intracellular domains. Both autophosphorylation and tyrosine kinase activity reside in the β-subunits and are markedly enhanced on insulin binding.

proteins. Phosphotyrosine motifs of IRS-1 have binding sites for SH2 domains (**src homology domain 2**) contained in signal-transducing proteins such as phosphatidylinositol-3-kinase (PI-3-kinase). SH2 domains are conserved sequences of about 100 amino acids, but the role of PI-3-kinase is not understood.

Insulin-mediated IRS-1 phosphorylation also leads to activation of ras proteins. The ras proteins are proto-oncogene products that mediate signaling pathways from cell membrane receptors, regulation of cellular proliferation, differentiation, or apoptosis. Ras proteins are post-translationally modified by farnesylation at a conserved cysteine residue in the C-terminal CAAX motif; the tripeptide AAX is removed by proteolysis, and the newly exposed C terminus is methylated. The ras protein is functionally active when GTP is bound and inactive when GDP is bound. The ras-GTP/GDP cycle is regulated by GTPase-activating proteins and guanine nucleotide exchange factor proteins (Chapter 28). The pathway of IRS-1 to activation of ras protein involves a number of proteins such as growth factor receptor binding protein-2 (GRB-2). Activated ras protein stimulates a cascade of serine/threonine kinases affecting transcriptional activation of specific genes. These pathways also involve proteins with SH2 domains. In summary, responses due to insulin binding to its receptor involve a cascade of phosphorylation and dephosphorylation steps that lead to gene regulation.

One of the major effects of insulin is to promote glucose transport from the blood into muscle and adipose tissue cells. This is accomplished by recruiting and localizing GLUT4 receptor molecules in the plasma membrane. In the absence of insulin, GLUT4 remains in the intracellular vesicles. Insulin-mediated GLUT4 trafficking to the cell membrane is constitutive and multicompartmental; also, it involves the participation of several proteins. Impairment in any of the steps of GLUT4 transport can cause insulin resistance and diabetes mellitus.

The importance of the biological actions of insulin on target cells is emphasized by defects in any of the five steps involved in receptor function. These steps are analogous to the scheme described for LDL receptor gene defects (Chapter 18). Mutations can lead to

1. Impaired receptor biosynthesis;
2. Impaired transport of receptor to the cell surface;
3. Decreased affinity of insulin binding;
4. Impaired tyrosine kinase activity; and
5. Accelerated receptor degradation.

Examples of disorders caused by mutations in the insulin receptor gene are **leprechaunism** and **type A insulin resistance**. A severe form of leprechaunism is due to mutations in both alleles of the insulin receptor gene. These patients exhibit insulin resistance, intrauterine growth retardation, and many other metabolic abnormalities. Patients with type A insulin resistance exhibit insulin resistance, acanthosis nigricans, and hyperandrogenism. The latter two have been ascribed to toxic effects of insulin on the skin and ovaries. Insulin resistance is also associated with **hypoandrogenism** and **polycystic ovary disease syndrome**.

Insulin is catabolized (inactivated) primarily in the liver and kidney (and placenta in pregnancy). Liver degrades about 50% of insulin during its first passage through this organ. An insulin-specific protease and glutathione-insulin transdehydrogenase are involved. The latter reduces the disulfide bonds with separation of A and B chains, which are subjected to rapid proteolysis.

Glucagon

Glucagon is a single-chain polypeptide of 29 amino acids that shares structural homology with secretin, vasoactive intestinal polypeptide (VIP), and gastric glucagon-like peptide-1 (Chapter 11). The synthesis of glucagon in the pancreatic α-cells involves a higher molecular weight precursor. Glucagon secretion is inhibited by glucose and stimulated by arginine and alanine. Depressed plasma glucagon levels result in depressed hepatic glucose output at times when glucose is available by intestinal absorption. Amino acid-stimulated glucagon secretion counteracts the effects of the coincidental secretion of insulin, which otherwise would provoke hypoglycemia. Secretion of glucagon from the α-cells and somatostatin from the δ-cells is regulated by the other pancreatic hormones.

Figure 20.6 illustrates the coordination of islet hormone secretion. This coordination may occur by the product of one cell type regulating secretion by the others and by direct intercommunication known as **paracrine signaling** between cells (of the same or different types). Glucagon secretion is also stimulated by **gastrin** and **cholecystokinin**, presumably as an anticipatory response to intestinal fuel absorption. Stress and catecholamines provoke glucagon secretion, which increases output of hepatic glucose and adipocyte triglyceride lipolysis, releasing fatty acids to provide the extra energy requirements.

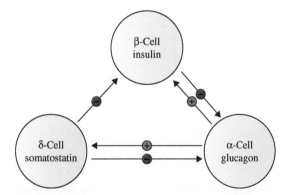

FIGURE 20.6 Paracrine regulation of islet hormonal secretion. ⊕ = stimulation; ⊖ = inhibition.

The primary targets for glucagon are the liver and adipose tissue. It stimulates hepatic glycogenolysis, gluconeogenesis, and ketogenesis, and inhibits glycogen synthesis. Glucagon stimulates adipocyte lipolysis to provide fatty acids to tissues for which glucose is not the obligatory fuel. In both liver and adipose tissues, glucagon activates adenylate cyclase, increases cAMP levels, activates cAMP-dependent protein kinase, and increases phosphorylation of the controlling enzymes in the various metabolic pathways. The initial step in the glucagon signaling pathway involves its binding to the receptor on the cell membrane and activation of the G-protein complex (Chapter 28).

The major function of insulin and glucagon is to maintain fuel homeostasis. Since glucagon opposes the actions of insulin, the [insulin]/[glucagon] ratio *determines* the metabolic fate of fuels, due to the induction or repression of appropriate enzymes. Enzymes induced by a high [insulin]/[glucagon] ratio and repressed by a low ratio are glucokinase, citrate cleavage enzyme, acetyl-CoA carboxylase, β-hydroxy-β-methylglutaryl-CoA (HMG-CoA) reductase, pyruvate kinase, 6-phosphofructo-1-kinase, 6-phosphofructo-2-kinase, and fructose-2,6-bisphosphatase. The metabolic implications of these actions are discussed later.

Somatostatin

The two forms of somatostatin, S14 and S28, are single-chain polypeptides of 14 and 28 amino acids, respectively. Somatostatin is synthesized in the δ-cells of the islets, the gut, the hypothalamus, and several other areas of the brain. Its actions appear to be primarily paracrine in nature (i.e., on other cell types in the immediate vicinity of its secretion). Somatostatin in the islets blocks insulin and glucagon secretion; in the pituitary gland, it inhibits growth hormone and thyroid-stimulating hormone (TSH) release; in the gut, it blocks secretion of gastrin and motilin, inhibits gastric acid and pepsin secretion, and suppresses gallbladder contraction. The action of somatostatin on the gastrointestinal tract leads to decreased delivery of nutrients to the blood. Analogues of somatostatin are used therapeutically in the treatment of neuroendocrine tumors.

Pancreatic Polypeptide

Pancreatic polypeptide, a peptide of 36 amino acids, is secreted in response to fuel ingestion and potentially affects pancreatic exocrine secretion of bicarbonate and proteolytic zymogens.

DIABETES MELLITUS

Diabetes mellitus, as mentioned previously, is a syndrome consisting of a group of metabolic diseases all of which are characterized by hyperglycemia. The development of diabetes comprises several pathogenic processes involving autoimmune destruction of pancreatic β-cells with consequent insulin deficiency and resistance to insulin action. The abnormalities of insulin deficiency or its action on target tissues lead to derangements in carbohydrate, lipid, and protein metabolism. The symptoms of hyperglycemia consist of polyuria, polydipsia, polyphagia, weight loss, blurred vision, and susceptibility to infections. The acute life-threatening complications of diabetes are **ketoacidosis** in the absence of insulin therapy or the **nonketotic hyperosmolar coma** (Chapter 37). Long-term complications of diabetes are development of retinopathy with potential loss of vision, nephropathy with end-stage renal disease, neuropathy with a potential for foot infections and amputation, and macrovascular disease such as atherosclerosis.

Type 1 Diabetes Mellitus

The type 1 diabetes mellitus disorder is primarily caused by autoimmune destruction of pancreatic β-cells. However, in some patients, evidence of autoimmunity may not be present, and the cause of the destruction of the β-cells may be undetermined (idiopathic). Type 1 diabetics are insulin-dependent and prone to ketoacidosis. The anti-β-cell antibodies may be directed against many different antigens, but the main ones are insulin, glutamic acid decarboxylase (GAD), and insulin storage granule membrane proteins (islet antigen-2 and zinc transporter 8). The antibodies appear in the plasma during the preclinical stage; in high-risk individuals, the presence of the autoantibodies, e.g., antibodies to GAD, can have value in predicting the development of diabetes mellitus. The inheritance of type 1 diabetes mellitus does not follow simple Mendelian inheritance, and it appears to be a polygenic disease with environmental factors playing a role in its initiation. It has been proposed that some infectious agents (e.g., coxsackie virus) may trigger the autoimmune destruction of β-cells by molecular mimicry.

Type 2 Diabetes Mellitus

Type 2 diabetes mellitus is the most prevalent form of diabetes and is characterized by both an insulin secretion defect and insulin resistance. Maturity-onset diabetes of the young (MODY), attributable to mutations of the glucose kinase gene (discussed earlier), may also be classified as type 2 diabetes mellitus. Obesity is a contributory factor and may predispose to insulin resistance with eventual development of type 2 diabetes mellitus.

Gestational Diabetes Mellitus

Glucose intolerance that is first recognized during pregnancy is classified as gestational diabetes mellitus

(GDM). The prevalence of GDM varies among ethnic groups. Studies of predominantly white women have shown that the prevalence of GDM is between 4% and 7%. GDM is associated with an increase in maternal and fetal neonatal morbidity. The clinical risks may include macrosomia (excessive fetal growth) associated with complications of labor and delivery, jaundice, respiratory distress syndrome, hypocalcemia, and polycythemia. Thus, proper diagnosis (discussed later) with appropriate intervention and treatment is required during the antepartum period of GDM. The causes of macrosomia in GDM are attributed to multiple disturbances that occur in both maternal and fetal metabolism. Fetal hyperinsulinemia due to increased delivery of maternal glucose may be one of the contributory factors. Two additional risks associated with GDM have been identified in epidemiological studies:

1. Persons exposed to GDM *in utero* have an increased risk of obesity; and
2. GDM increases the risk of development of type 2 diabetes mellitus between 5 and 16 years after the index pregnancy.

Metabolic Syndrome

Metabolic syndrome is a constellation of metabolic disorders and consists of abdominal (central) obesity, hyperglycemia, dyslipidemia, and hypertension. Metabolic syndrome increases the risk of development of type 2 diabetes mellitus, insulin resistance, and cardiovascular disorders (heart disease, stroke). Insulin resistance initiates a vicious cycle in which, as the target tissues become unresponsive to insulin, the pancreatic β-cells secrete even more insulin in order to maintain normal levels of plasma glucose. The origin of insulin resistance at the target tissues involves at least two mutually but not exclusive mechanisms: production of inflammatory cytokines and accumulation of fat as diacylglycerols in the liver and muscle. The release of inflammatory cytokines (e.g., TNF-α) and decreased secretion of anti-inflammatory cytokines (e.g., adiponectin) by the adipose tissue are implicated in the development of insulin resistance. The mechanism of ectopic accumulation of fat in the liver and muscle that may lead to insulin resistance is supported by insulin resistance that occurs in lipodystrophies. Deficiency or complete absence of adipocytes in lipodystrophies leads to accumulation of fat in the liver and muscle. Thiazolidinedione, used in the treatment of type 2 diabetes, ameliorates insulin resistance as one of its actions by promoting the conversion of preadipocytes to adipocytes (discussed earlier). It is thought that the newly formed adipocytes store fat, preventing its accumulation in the liver and muscle.

Specific Mutations that Cause Diabetes Mellitus

This group of diabetes mellitus disorders is relatively uncommon and includes mutations in genes that affect various aspects of β-cell function, including mitochondrial DNA and the insulin receptor gene; these disorders were discussed earlier.

Endocrinopathies

Excessive production of insulin counter-regulatory hormones can cause diabetes mellitus. Examples include excessive production of growth hormone (**acromegaly**), cortisol (**Cushing's syndrome**), epinephrine (**pheochromocytoma**), and glucagon (**glucagonoma**).

Miscellaneous Causes of Diabetes Mellitus

Miscellaneous causes of diabetes mellitus include diseases of the exocrine pancreas (e.g., **cystic fibrosis, hemochromatosis, pancreatitis**) and drug-induced causes (e.g., thiazides, glucocorticoids).

Diagnosis of Diabetes Mellitus

As diabetes mellitus is a heterogeneous group of metabolic disorders characterized by hyperglycemia, the diagnosis involves first excluding other causes of hyperglycemia. For example, all of the endocrinopathies that have other pathognomonic features, such as acromegaly, Cushing's syndrome, glucagonoma, and pheochromocytoma, should be identified with appropriate diagnostic procedures. After excluding all such causes of hyperglycemia, the diagnosis of type 1 or type 2 diabetes mellitus should be considered. The diagnostic criteria for diabetes mellitus are given in Table 20.2.

The diagnosis of gestational diabetes mellitus consists of two parts. A screening glucose challenge test is performed between 24 and 28 weeks of gestation after the oral administration of 50 g of glucose. This test is performed at any time of the day and irrespective of food intake. If the plasma glucose after the 50 g challenge is ≥140 mg/dL (7.8 mmol/L), the test is considered positive and requires a diagnostic test consisting of oral ingestion of 100 g of glucose after fasting for 8 hours, followed by a 3-hour oral glucose tolerance test (Table 20.3).

After the diagnosis of diabetes mellitus has been made and with the initiation of appropriate therapy (discussed later), assessment of other biochemical parameters is necessary in the management phase of the disorder to maintain the fasting blood glucose level as close to normal as possible and to prevent long-term complications. These

TABLE 20.2 Criteria for the Diagnosis of Diabetes Mellitus

1. $A_{1C} \geq 6.5\%$. The test should be performed in a laboratory using a method that is NGSP certified and standardized to the DCCT assay.*

or

2. FPG ≥ 126 mg/dL (7.0 mmol/L). Fasting is defined as no caloric intake for at least 8 hours.*

or

3. 2-h plasma glucose ≥ 200 mg/dL (11.1 mmol/L) during an OGTT. The test should be performed as described by the World Health Organization, using a glucose load containing the equivalent of 75 g anhydrous glucose dissolved in water.*

or

4. In a patient with classic symptoms of hyperglycemia or hyperglycemic crisis, a random plasma glucose ≥ 200 mg/dL (11.1 mmol/L).

*In the absence of unequivocal hyperglycemia, criteria 1−3 should be confirmed by repeat testing.
FPG = fasting plasma glucose; OGTT = oral glucose tolerance test. The diagnostic criteria are based on the report by the American Diabetes Association on the diagnosis and classification of diabetes mellitus. Diabetes Care 33, supplement 1, S62, 2010.

TABLE 20.3 Screening and Diagnosis Scheme for Gestational Diabetes Mellitus (GDM)

Plasma Glucose	50 g Screening Test	100 g Diagnostic Test
Fasting	–	95 mg/dL
1 hour	140 mg/dL	180 mg/dL
2 hour	–	155 mg/dL
3 hour	–	140 mg/dL

Screening for GDM may not be necessary in pregnant women who meet all of the following criteria: <25 years of age; normal body weight; no first-degree relative with diabetes; and not Hispanic, Native American, Asian American, or African American. The 100 g diagnostic test is performed on patients who have a positive screening test. The diagnosis of GDM requires any two of the four plasma glucose values obtained during the test to meet or exceed the values shown above. (To convert values for glucose to mmol/L, multiply by 0.05551.) The diagnostic criteria are based on the report by the American Diabetes Association on Gestational Diabetes Mellitus. Diabetes Care 23 (Suppl 1), S77, 2000.

biochemical tests include measurement of a stable form of glycosylated hemoglobin (hemoglobin A_{1c}), determination of urine albumin excretion rate, measurement of serum fructosamine levels, and self-monitoring of blood glucose levels. Hemoglobin A_{1c} levels are used to assess average glucose control over a 2- to 3-month period, since the red blood cell's lifespan is about 120 days. Entry of glucose into red blood cells depends only on the prevailing plasma glucose concentration. The glycation of hemoglobin A, the predominant form in adult humans, is a nonenzymatic continuous process and occurs throughout the 120-day lifespan of the red blood cells. In hemoglobin A_{1c}, glucose is incorporated via an N-glycosidic linkage into the N-terminal amino group of valine of each β-chain. In diabetes, HbA_{1c} levels are used in identifying individuals with risk of

developing the disorder and in the diagnosis as well. However, it should be noted that several factors affect HbA_{1c} levels [5]. These include hemoglobinopathies plus hemolytic and iron deficiency anemia.

One of the chronic complications of diabetes mellitus is diabetic nephropathy, which leads to end-stage renal disease. An initial biochemical parameter of diabetic nephropathy in the asymptomatic state is a persistent increase in urine albumin excretion rate between 20 and 200 μg/min (or 30−300 mg/d). This degree of albumin loss in the urine, called **microalbuminuria**, is a harbinger of renal failure and other complications of diabetes. It is important to identify individuals with microalbuminuria because, with appropriate therapeutic intervention, loss of renal function can be attenuated. Therapeutic interventions include blood glucose control, treatment for high blood pressure if present, and the inhibition of angiotensin II formation by use of angiotensin-converting enzyme (ACE) inhibitors or angiotensin II receptor blockers.

Chronic Complications of Diabetes Mellitus

Chronic complications of diabetes mellitus stem from elevated plasma glucose levels and involve tissues that do not require insulin (e.g., lens, retina, peripheral nerve) for the uptake and metabolism of glucose. In these tissues, the intracellular level of glucose parallels that in plasma. The chronic complications, which cause considerable morbidity and mortality, are atherosclerosis, microangiopathy, retinopathy, nephropathy, neuropathy, and cataracts. The biochemical basis of these abnormalities may be attributed to increased tissue ambient glucose concentration and may involve the following mechanisms: nonenzymatic protein glycation increased production of sorbitol, and decreased levels of *myo*-inositol.

Management of Diabetes Mellitus

The primary goal in the management of all types of diabetes mellitus (type 1, type 2, and GDM) is to maintain near-normal plasma glucose levels in order to relieve symptoms (polydipsia, polyuria, polyphagia), and to prevent both acute and chronic complications. Glycemic control is assessed by monitoring blood glucose levels (by oneself and in clinical settings), hemoglobin A_{1c}, fructosamine, and microalbuminuria. In type 1 diabetes mellitus due to β-cell destruction, administration of insulin is required throughout the person's lifetime. There are many insulin preparations that differ in duration of action. The goal of subcutaneous insulin therapy is to maintain near-normal plasma concentrations of insulin consistent with plasma glucose levels. Regular insulin associates as hexamers at the injection site, and plasma levels peak between 2 and 4 hours with duration of action between 12 and 14 hours. Structural changes in the insulin molecule have yielded insulin analogues with short-acting and long-acting pharmacokinetic profiles (Table 20.4). The short-acting insulins do not undergo aggregation, are rapidly absorbed, and reach peak values between 1 and 2 hours. The duration of long-acting insulins, glargine and detemir, is between 22 and 30 hours without any peaks. Insulin glargine is long lasting because it is less soluble due to its neutral isoelectric pH, aggregates at the injection site, and is released slowly. Insulin detemir is absorbed slowly because it aggregates and binds to plasma albumin via its fatty acid substitution, leading to protracted action. In the management of type 2 diabetes mellitus, endogenous insulin levels are augmented to maintain glycemic control with the administration of the synthetic analogue of gastrointestinal hormone, glucagon-like peptide-1. Another method of treatment to promote incretins action is to use dipeptidyl protease inhibitors (DPPs), which retard the inactivation of incretins in the plasma (see Chapter 11). A synthetic analogue of amylin, pramlintide, is also used with insulin to restore normal glycemic levels. The side effects of insulin therapy include hypoglycemia and weight gain.

The management of type 2 diabetes mellitus is based on behavioral changes with lifestyle modifications and pharmacological measures. Diet, exercise, and weight loss (in the obese) are the cornerstones of treatment in maintaining euglycemia. Physical activity stimulates insulin sensitivity, whereas physical inactivity leads to insulin resistance. In the absence or failure of lifestyle and behavioral measures in the management of type 2 diabetes mellitus, pharmacological therapy is employed. Pharmacological agents used in the treatment of type 2 diabetes mellitus include **insulins, sulfonylureas, α-glucosidase inhibitors, biguanidines**, and **thiazolidinediones**. Sulfonylureas bind to specific β-cell plasma membrane receptors and cause closure of the ATP-dependent K^+ channel with accompanying depolarization. The depolarization leads to influx of Ca^{2+} followed by insulin secretion. A benzoic acid derivative (repaglinide) is structurally similar to sulfonylurea without the sulfur group and belongs to the class **meglitinides**. This drug inhibits a different K^+ channel than the sulfonylureas in promoting insulin secretion and has rapid-onset and short-acting actions. An example of an α-glucosidase inhibitor is acarbose, a nitrogen-containing pseudotetrasaccharide. Acarbose inhibits the breakdown of carbohydrates into glucose by inhibiting intestinal brush-border α-glucosidase and pancreatic amylase (Chapter 11), and bringing about a reduction in postprandial hyperglycemia. Metformin belongs to the class of biguanidines, is chemically related to isoamylene guanidine present in the French lilac (*Galega officinalis*), and was used to treat polyuria in diabetes. Its mechanism of action in carbohydrate metabolism includes inhibition of hepatic mitochondrial complex I, elevation of AMP concentration, inhibition of adenylate cyclase with resultant reduction of cAMP levels, and protein kinase A activity (PKA). The decreased PKA activity results in the inhibition of glucagon-mediated glucose production pathways [6,7]. An adverse effect of metformin is lactic acidosis.

Rosiglitazone is an example of the thiazolidinedione class of antidiabetic agents. It enhances insulin sensitivity in liver, muscle, and adipose tissue. It promotes the conversion of non-lipid-storing preadipocytes to mature adipocytes (discussed earlier) with increased insulin sensitivity. The cellular target for the action of rosiglitazone, which functions as a transcription factor, is the nuclear receptor known as **peroxisome proliferation−activated receptor-γ** (PPAR-γ). Upon binding with the drug, PPAR-γ is activated and forms a heterodimer with retinoid X receptor (RXR). The PPAR-γ/RXR heterodimer regulates the

TABLE 20.4 Human Insulin Analogues

Name	Structural Changes
Short-acting Insulins	(occur in the B-chain)
Lispro	Proline at position 28 and lysine at position 29 interchanged
Aspart	Proline at position 28 is replaced by aspartate
Glulisine	Asparagine at position 3 is replaced by lysine and lysine at position 29 is replaced with glutamate
Long-acting Insulins	
Glargine	Asparagine at position 21 is replaced by glycine of A-chain and by addition of 2 arginines at position 30 of the B-chain
Detemir	Threonine at position 30 of the B-chain has been deleted and a fatty acid (myristic acid) acyl group has been added to lysine at position 29 of the B-chain

transcription of genes encoding proteins that control metabolic pathways (e.g., acetyl-CoA synthetase, lipoprotein lipase, GLUT4, mitochondrial uncoupling protein). An adverse effect of troglitazone is hepatic toxicity; periodic monitoring of liver function by measurement of serum alanine aminotransferase is recommended. The drug is also contraindicated in some heart diseases. When a diabetic patient becomes refractory to monotherapy, combination therapy (e.g., sulfonylurea and insulin) is used to achieve glycemic control.

Therapeutic agents used in the treatment of diabetes mellitus type 2 are listed in Table 20.5. A new class of

antidiabetic drugs to be added to the list is the inhibitor of renal sodium−glucose cotransporter 2 (SGLT2). The inhibition of SGLT2 reduces glucose reabsorption in the proximal convoluted tubule leading to increased glucose loss with consequent reduction in plasma glucose levels [8,9]. Since bariatric surgery in morbidly obese subjects normalizes plasma glucose levels (discussed earlier), consideration has been given to use of this type of surgical procedure in obese diabetic patients [10,11].

Recent guidelines for lifestyle interventions and oral pharmacological treatment for type 2 diabetes are provided in references [12] and [13], respectively.

TABLE 20.5 Pharmacologic Agents for Glycemic Control in Patients with Type 2 Diabetes

Class	Agent (Brand Name)	Expected Reduction in Glycated Hemoglobin Level (%)	Advantages	Disadvantages	Cost
Oral					
Biguanide	Metformin[§] (Glucophage)	1.0−2.0	Extensive clinical experience; hypoglycemia rare; improved lipid profile; decreased cardiovascular disease events; some weight loss in most patients	Gastrointestinal intolerance; lactic acidosis rare (avoid in patients at increased risk, such as men with a serum creatinine level of ≥1.5 mg/dL and women with a serum creatinine level of ≥ 1.4 mg/dL); vitamin B_{12} deficiency	Low (generic)
Sulfonylurea[‡]	Glyburide (Diabeta), glipizide (Glucotrol),[¶] gliclazide (Diamicron),[¶] glimepiride (Amaryl)	1.0−1.5	Extensive clinical experience	Hypoglycemia; less durability; weight gain	Low (generic)
Meglitinide	Nateglinide (Starlix), repaglinide (Prandin)	0.5−1.0	Short duration of action, hepatic clearance, glucose-dependent postprandial action	Low efficacy, hypoglycemia in some patients, weight gain	High
Thiazolidinedione	Rosiglitazone (Avandia), pioglitazone (Actos)	0.5−1.4	Hypoglycemia rare, more durable effect than that of metformin or sulfonylurea, improved lipid profile, some evidence of beneficial effect on coronary atherosclerosis (with pioglitazone)[‡]	Edema, heart failure, weight gain, increased risk of long-bone fractures and potential risk of bladder cancer and cardiovascular events (with rosiglitazone)*; use of rosiglitazone highly restricted	
DPP-IV inhibitor	Saxagliptin (Onglyza), linagliptin (Tradjenta), vildagliptin (Galvus),[¶] sitagliptin (Januvia)	0.5−0.8	Hypoglycemia rare, infrequent side effects	Less efficacy than GLP-1-receptor agonists, angioedema, unknown long-term safety, risk of pancreatitis	High

(Continued)

TABLE 20.5 (Continued)

Class	Agent (Brand Name)	Expected Reduction in Glycated Hemoglobin Level (%)	Advantages	Disadvantages	Cost
Alpha-glycosidase inhibitor	Miglitol (Glyset), voglibose (Volix),[¶‡] acarbose (Precose)	0.5–0.9	Decreased level of postprandial glucose, hypoglycemia rare, possible decrease in risk of cardiovascular disease events**	Flatulence, diarrhea	Moderate
Bile acid sequestrant	Colesevelam (Welchol)	0.5	Lowering of LDL cholesterol level; hypoglycemia rare	Gastrointestinal side effects, including constipation; low efficacy; only approved agent in class	High
D2 dopamine-receptor agonist	Bromocriptine, rapid release (Cycloset)	0.5	Hypoglycemia rare	Low efficacy; gastrointestinal side effects, including nausea; fatigue; dizziness; rhinitis; only rapid-release agent approved	High
Injectable					
GLP-1-receptor agonist	Exenatide (Byetta), exenatide once weekly (Bydureon), liraglutide (Victoza)	0.5–1.5	Hypoglycemia rare, weight loss in most patients; possible protective cardiovascular effects	Nausea and vomiting; risk of pancreatitis, thyroid C-cell hyperplasia, and tumors (with liraglutide and weekly exenatide); unknown long-term safety	High
Amylin analogue	Pramlintide (Symlin)	0.5–1.0	Weight loss in most patients, control of postprandial glycemia	Nausea and vomiting, modest effect, hypoglycemia with insulin use, unknown long-term safety	High
Insulin	Short-acting: human insulin (Novolin R or Humulin R), aspart (NovoLog), glulisine (Apidra), lispro (Humalog); long-acting: neutral protamine Hagedorn (Novolin N or Humulin N), detemir (Levemir), glargine (Lantus); mixed insulin preparations	1.0–2.5	Large effect in all patients	Hypoglycemia, weight gain	Moderate to high

[§]The use of metformin in a trial involving obese patients with newly recognized type 2 diabetes was associated with a reduction in cardiovascular disease events.
[‡]In an observational study, the use of sulfonylureas was associated with decreased renal function.
[¶]This agent is not available in the United States.
[‡]In randomized studies, pioglitazone, as compared with glimepiride, was associated with a reduction in cardiovascular disease outcomes and a decrease in the progression of coronary atherosclerosis.
[*]The use of rosiglitazone is highly restricted in the United States (and has been discontinued in Europe), predominantly on the basis of a meta-analysis.
[**]The use of this agent in patients with impaired glucose tolerance has been associated with reduced cardiovascular disease events.
Reproduced with permission from F. Ismail-Beigi, Glycemic management of type 2 diabetes mellitus, N. Engl. J. Med. 366 (2012) 1324–1315, Table 1. Pharmacologic Agents for Glycemic Control in Patients with Type 2 Diabetes.

REQUIRED READING

[1] P. Srikanthan, A.S. Karlamangla, Muscle mass index as a predictor of longevity in older adults, Am. J. Med. 127 (2014) 547−553.

[2] K. Casazza, K.R. Fontaine, A. Astrup, L.L. Birch, A.W. Brown, M.M. Bohan Brown, et al., Myths, presumptions, and facts about obesity, N. Engl. J. Med. 368 (2013) 446−454.

[3] J.F. Merola, P.P. Ghoroghchian, M.A. Samuels, B.D. Levy, J. Loscalzo, Clinical Problem Solving. At a loss, N. Engl. J. Med. 367 (2012) 67−72.

[4] S. Ikramuddin, J. Korner, W. Lee, J.E. Connett, W.B. Inabnet III, C.J. Billington, et al., Roux-en-Y gastric bypass vs intensive medical management for the control of type 2 diabetes, hypertension, and hyperlipidemia, JAMA 309 (2013) 2240−2249.

[5] T. Higgins, HbA$_{1c}$—an analyte of increasing importance, Clin. Biochem. 45 (2012) 1038−1045.

[6] R.H. Unger, E.D. Berglund, J.F. Hebener, A.D. Cherrington, Dissecting the actions of widely used diabetes drugs, Nat. Med. 19 (2013) 272−273.

[7] R.A. Miller, Q. Chu, J. Xie, M. Foretz, B. Viollet, M.J. Birnbaum, Biguanides suppress hepatic glucagon signalling by decreasing production of cyclic AMP, Nature 494 (2013) 256−260.

[8] I.B Hirsch, M.E. Molitch, Clinical decisions. Glycemic management in a patient with type 2 diabetes, N. Engl. J. Med. 369 (2013) 1370−1372.

[9] D. Vasilakou, T. Karagiannis, E. Athanasiadou, M. Mainou, A. Liakos, E. Bekiari, et al., Sodium−glucose cotransporter 2 inhibitors for type 2 diabetes, Ann. Intern. Med. 159 (2013) 262−274.

[10] P. Zimmet, K.G. Alberti, Surgery or medical therapy for obese patients with type 2 diabetes? N. Engl. J. Med. 366 (2012) 1635−1636.

[11] P.R. Schauer, S.R. Kashyap, K. Wolski, S.A. Brethauer, J.P. Kirwan, C.E. Pothier, et al., Bariatric surgery versus intensive medical therapy in obese patients with diabetes, N. Engl. J. Med. 366 (2012) 1567−1576.

[12] E.S. Schellenberg, D.M. Dryden, B. Vandermeer, C. Ha, C. Korownyk, Lifestyle interventions for patients with and at risk for type 2 diabetes, Ann. Intern. Med. 159 (2013) 543−551.

[13] A. Qaseem, L.L. Humphrey, D.E. Sweet, M. Starkey, P. Shekelle, Oral pharmacologic treatment of type 2 diabetes mellitus: a clinical practice guideline from the American College of Physicians, Ann. Intern. Med. 156 (2012) 218−231.

Chapter 21

Structure and Properties of DNA

Key Points

1. Two different types of nucleic acids, DNA and RNA, contain different nitrogenous bases: A, T, G, C, U, and modified bases.

2. DNA forms double-stranded helices; RNA has a single-stranded structure.

3. Polar phosphate backbones and hydrophobic base pairing stabilize DNA double helical structures.

4. Double- and single-stranded DNA and RNA exhibit different optical properties, measured spectrophotometrically at 260 nm.

5. Enzymes are involved in supercoiling of DNA strands and participate in regulation of gene expression.

6. Prokaryotes have circular DNAs, and eukaryotes have linear DNAs. DNA molecules in eukaryotic cells are tightly packed into chromatin structures, and chromatin exists in two different conformations: euchromatin and heterochromatin.

7. Linear DNAs in eukaryotes require unique end structures called telomeres, and telomerase extends the telomere structures.

8. The structural changes of DNA without base sequence changes through mutation affect regulation of gene expression, and heritable changes in DNA structure such as DNA methylation and histone modification are called epigenetic changes. The study of epigenetic changes is called epigenetics.

9. The sequence of nucleotides in a DNA sample can be determined by using the dideoxy chain termination method.

10. Restriction enzymes recognize specific DNA sequences and cut DNA molecules at specific sites.

11. PCR amplification produces multiple copies of identical DNA sequences.

12. Hybridization of DNA, RNA, and protein probes provide valuable tools to identify specific DNA sequences.

13. Recombinant DNA technologies are used to study genes and gene products.

INTRODUCTION

Nucleic acids store/carry the genetic information and participate in its decoding into a variety of cellular proteins. Nucleic acids include two different types: deoxyribose nucleic acids (DNA) and ribose nucleic acids (RNA). DNA stores the genetic information as long sequences of bases, and RNA plays a major role in the expression of the genetic information. In this chapter, the structure and function of DNA and its role as an information storage molecule are discussed. Also, the chapter discusses how the genetic information can be analyzed by using various molecular techniques such as DNA sequencing, polymerase chain reaction (PCR), and recombinant DNA technologies.

Discovery of DNA as the Genetic Material

The identity of the genetic material was not known in the early 1900s even though scientists had shown that deoxyribonucleic acid (DNA) was present in the nuclei of cells. DNA was chemically less complex than protein. Since DNA consists of only 4 different bases, whereas proteins contain 20 different amino acids, most scientists weren't convinced that DNA was the molecule carrying the information for specific and diverse traits exhibited by all living organisms. Rather, they thought that proteins were more suitable molecules to carry genetic information. In 1928, a British medical officer, Frederick Griffith, discovered a "transforming principle" that was responsible for converting an avirulent R strain of *Streptococcus pneumoniae* into a virulent S strain (Figure 21.1).

In 1944, a Canadian physician and bacteriologist, Oswald T. Avery, conducted an experiment using purified DNA to transform an avirulent bacterial R strain into a virulent S strain. This experiment showed that DNA was the transforming principle (Figure 21.2). However, not everyone was convinced regarding DNA being the genetic material, and many doubted that proteins were completely destroyed by proteinases in the experiment.

In 1952, a bacteriologist and geneticist named Alfred Hershey and his assistant, Martha Chase, performed a series of experiments with bacteriophages and their host bacteria *Escherichia coli*, which also indicated that DNA was the genetic material. In their experiments, S^{35}-labeled

N. V. Bhagavan and C.-E. Ha: Essentials of Medical Biochemistry, Second edition. DOI: http://dx.doi.org/10.1016/B978-0-12-416687-5.00021-X

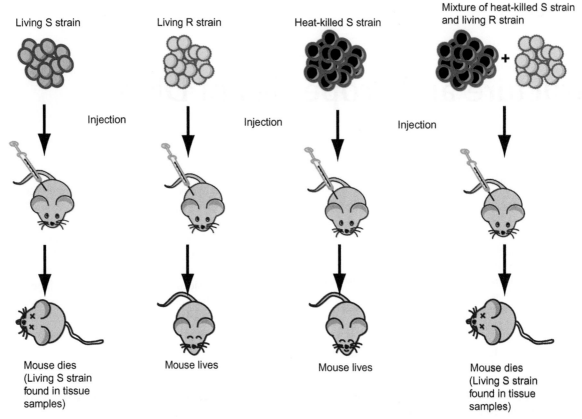

FIGURE 21.1 The Frederick Griffith experiment that discovered the "transforming principle." In this experiment, Griffith showed that a nonlethal strain (R strain) of *Streptococcus pneumoniae* was transformed by a heat-killed virulent strain (S strain) of the bacteria. Griffith hypothesized that the "transforming principle" from the heat-killed S strain converted a virulent R strain into a virulent S strain, which killed mice in the experiment.

protein and P^{32}-labeled DNA were used to show that DNA entered the cell and was the genetic material (Figure 21.3).

In 1953, James Watson and Francis Crick, working at the Cavendish Laboratories in the UK, published a structure for DNA that consisted of two strands twisted around each other (Figure 21.4). They suggested that the two strands of DNA form a right-handed helix and that the strands are wound together in a chemically antiparallel manner. Their model of the DNA structure is now known as the **B-form of DNA**. The model also suggested how genetic information is stored and replicated faithfully from one generation to the next. We now understand in great detail how DNA stores genetic information, how cells decipher the genetic information, how proteins are synthesized, and how changes in DNA lead to inherited diseases.

Components of DNA

DNA is a polymer of deoxyribonucleotides; there are four different nucleotides found in DNA. Each nucleotide of DNA consists of three major components: (1) a five-carbon deoxyribose sugar; (2) a phosphate group attached to the 5′ carbon of deoxyribose; and (3) a heterocyclic nitrogenous base attached to the 1′ carbon of the sugar (Figure 21.5). The four nucleotides differ according to the base attached to the deoxyribose sugar. A nucleotide without a phosphate group (base plus deoxyribose sugar) is called a **nucleoside**. The four different bases in DNA are adenine (A), guanine (G), cytosine (C), and thymine (T). C and T [uracil (U) in RNA molecules] are pyrimidines; G and A are purines (Figure 21.6).

Base Pairing of Bases in DNA Double Helixes

Bases found in one DNA strand are always paired with the bases in the opposite DNA strand. A always pairs with T; G always pairs with C (Figure 21.7). Since purines (A and G) always pair with pyrimidines (C and T) and pyrimidines pair with purines, any DNA molecule must contain equal numbers of purines and pyrimidines. These base pairing relationships are called **Chargaff's rules** after Erwin Chargaff, the chemist who discovered that the molar concentration of A always equals the molar content of T; the molar content of G equals the molar content of C. Thus, [A] = [T] and [G] = [C]. Also, [A] + [G] = [T] + [C] and [A] + [G]/[C] + [T] = 1.0.

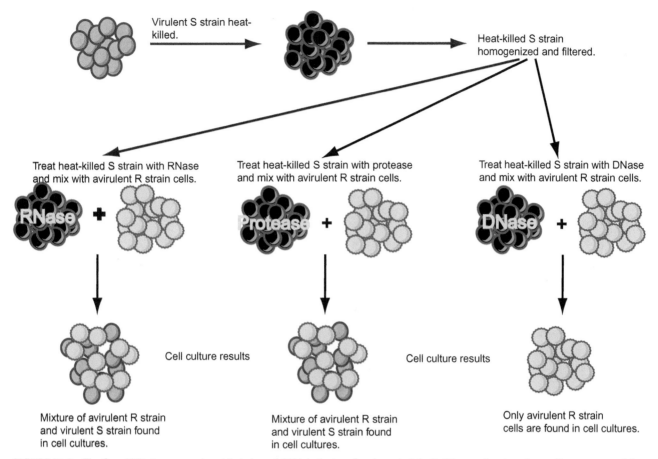

FIGURE 21.2 The Oswald T. Avery experiment that showed DNA is the transforming principle. In this experiment, various cell components of the virulent S strain were injected into living avirulent R strain cells. Only DNA from heat-killed S strain cells transformed the avirulent R strain; the conclusion was that DNA is responsible for the transformation of R strain into S strain.

However, overall base composition of DNA is variable among different organisms. Thus, [A] + [T] ≠ [G] + [C]. A−T base pairs are joined by two hydrogen bonds, and G−C base pairs are joined by three hydrogen bonds. Therefore, G−C pairs require more energy to separate compared to A−T pairs. As a result, G−C-rich DNA sequences are more stable than A−T-rich DNA sequences.

Phosphodiester Backbone of DNA

Polymerized DNA strands are formed by linking the 5′ phosphate group to the 3′ hydroxyl group of adjacent deoxyribose sugars through a $3′ \rightarrow 5′$ phosphodiester bond (Figure 21.8). These phosphodiester bonds form the backbone of DNA strands and are responsible for the characteristic negative charges on DNA molecules that stabilize DNA double helices by interacting with the surrounding water molecules. The hydrophobic bases form pairs through shared hydrogen bonds. The DNA strands are always extended by forming new phosphodiester bonds at the 3′ hydroxyl group of the last nucleotide and, therefore,

DNA sequences are written as $5′ \rightarrow 3′$ sequences. For example, the DNA sequence GAATTC is actually $5′$-pGpApApTpTpC$_{OH}$-$3′$.

Base Analogues

Base analogues are molecules that can substitute for normal bases in nucleic acids. Usually, substitution of a base analogue will result in altered base pairings and structural changes that affect DNA replication and transcription of genes. Examples of base analogues include 5-bromouracil, 2-aminopurine, 6-mercaptopurine, and acycloguanosine (Figure 21.9). Since 5-bromouracil can pair with either adenine or guanine, it also affects base pairing during DNA replication, which leads to mutations. An analogue of adenine, 2-aminopurine, also causes mutations in a similar way since it can pair with either T or C. 5-Bromouracil is used to treat neoplasms in the form of its nucleoside, 5-bromo-2-deoxy-uridine. 6-Mercaptopurine is an analogue of hypoxanthine and inhibits purine nucleotide synthesis and metabolism. 6-Mercaptopurine is used to treat acute

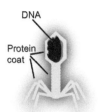

DNA

Protein
coat

Bacteriophage protein
coat labeled
with radioactive sulfur

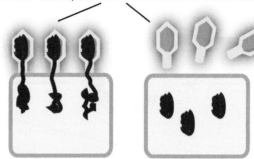

Radioactive labeled protein coats remain outside the bacterial cell

Bacteriophage shells were removed
from the host cells by using a blender
and separated by using a centrifuge.
Labeled protein coats were found
in the supernatant portion, not in
the pellet of bacterial cells.

**Conclusion: Protein is not the genetic material that directed the production
of new bacteriophages in the host cell**

DNA

Protein
coat

Bacteriophage DNA
labeled with
radioactive phosphorus

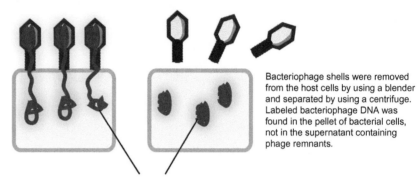

Bacteriophage shells were removed
from the host cells by using a blender
and separated by using a centrifuge.
Labeled bacteriophage DNA was
found in the pellet of bacterial cells,
not in the supernatant containing
phage remnants.

Radioactive phosphorus labeled phage DNA injected into the host cell

**Conclusion: DNA is the genetic material that directed the production
of new bacteriophages in the host cell**

FIGURE 21.3   The Alfred Hershey and Martha Chase experiment that showed DNA is the genetic material. In this experiment, Hershey and Chase showed that radio-labeled DNA of bacteriophage entered host bacterial cells and directed the production of new bacteriophages.

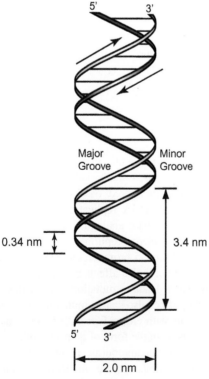

FIGURE 21.4   DNA structure presented by Watson and Crick.

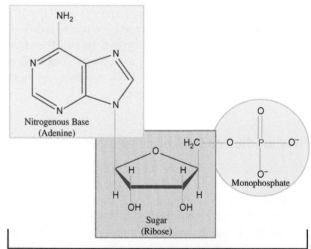

Adenosine Monophosphate (AMP)

FIGURE 21.5   Structure of the nucleotide in DNA and RNA. Each nucleotide in DNA and RNA contains a five-carbon deoxyribose sugar, a phosphate group attached to the 5′ carbon of deoxyribose sugar, and a heterocyclic nitrogenous base. Adenine monophosphate (AMP) is shown in this figure.

leukemia. Acycloguanosine (also called acyclovir) is an analogue of guanine and is one of the most commonly used antiviral drugs. It is an effective treatment for herpes simplex virus infection. Also, base analogues can inhibit DNA synthesis by inhibiting certain enzymes. For example, 5-fluorouracil is a thymine analogue that inhibits thymidylate synthases, which is a key enzyme in thymine synthesis. Because it inhibits DNA synthesis, 5-fluorouracil is used in the treatment of some cancers, and it belongs to the family of drugs called **antimetabolites**.

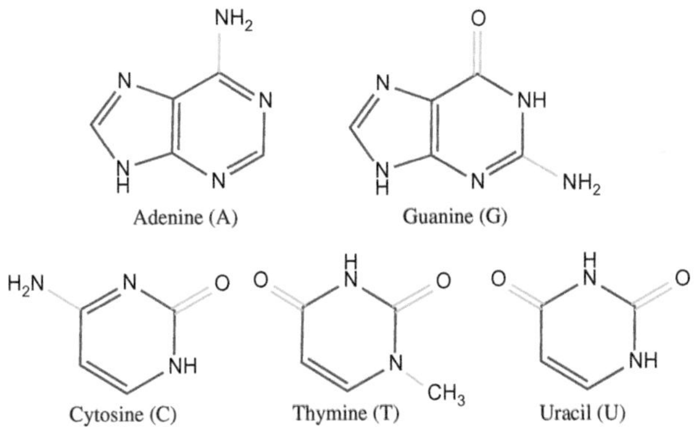

Modifications of DNA Bases

A methyl group from S-adenosyl-methionine (SAM) can be transferred to cytosine in DNA to form 5-methylcytosine, which is frequently found in heterochromatin regions of chromosomes (discussed later). Heterochromatin contains both DNA that is tightly folded and genetically inactive genes, as compared to genes located in chromatin regions of chromosomes.

Since deamination of 5-methylcytosine (m5C) converts it to the thymine base, methylation of cytosine in a certain DNA region involves higher mutation rates than in other areas of DNA (Figure 21.10). In addition to m5C, bacterial DNA contains N4-methylcytosine (m4C) and N6-methyladenine (m6A). The methylated cytosines in bacteria are involved in self-defense against infections. When bacteriophages inject their DNA into host bacteria, bacterial nucleases recognize and destroy the foreign DNA. However, bacterial DNA is immune from attacks by endogenous enzymes because bacterial DNA is methylated in specific patterns. The roles of m6A are numerous, including functions in DNA replication, repair, expression, and transposition. Also, glycosylated uracil (beta-D-glucosyl-hydroxymethyl-uracil) present in the unicellular eukaryote *T. brucei* is thought to be involved in repression of gene expression.

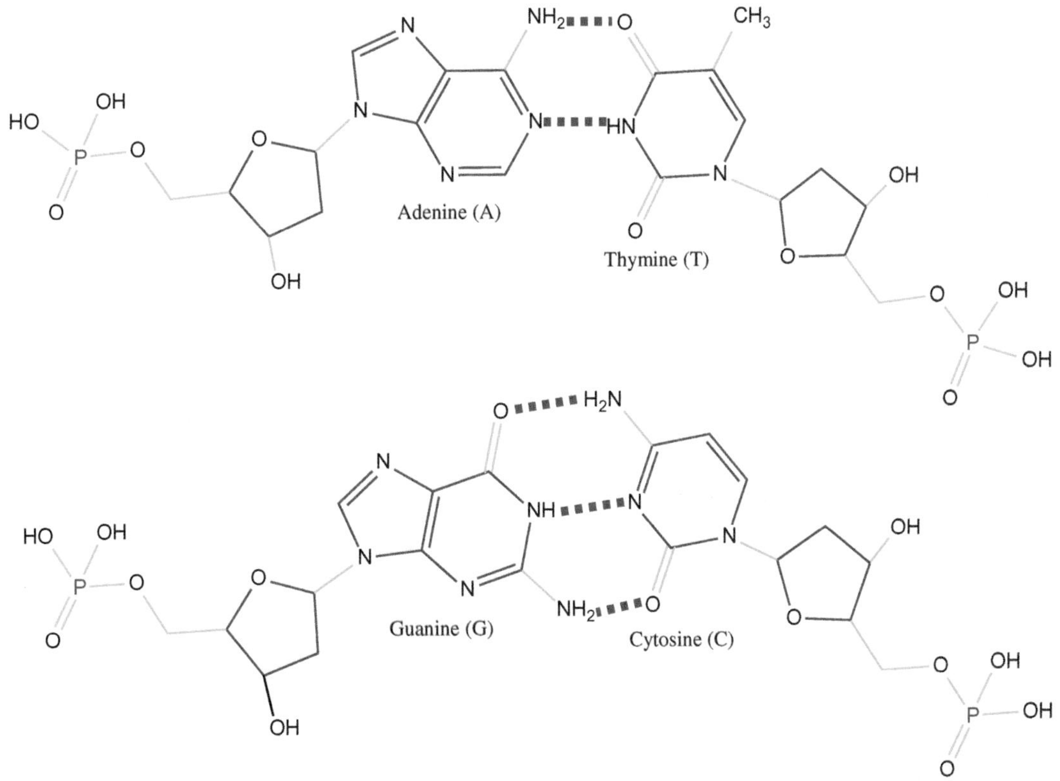

FIGURE 21.6 Four bases found in DNA (A, G, T, C) and RNA (A, G, U, C).

FIGURE 21.7 The normal pairing of bases in DNA. Adenine on one strand of DNA pairs with thymine on the complementary strand, and guanine pairs with cytosine.

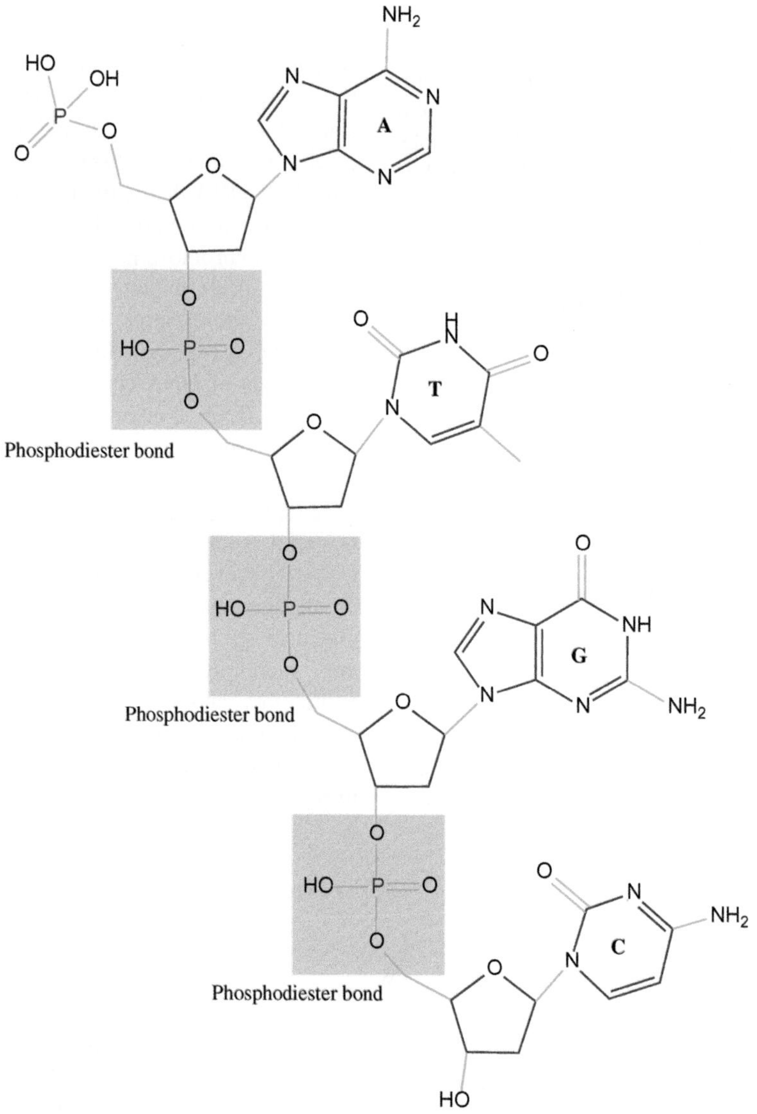

FIGURE 21.8 Phosphodiester bond linkage of DNA. Nucleotides in DNA are linked together by phosphodiester bonds.

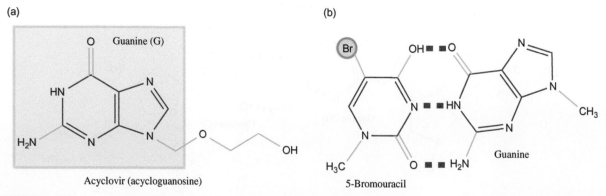

FIGURE 21.9 DNA base analogues. Structures of (a) acyclovir, a guanine derivative; and (b) 5-bromouracil paired with a normal guanine base are shown in this figure.

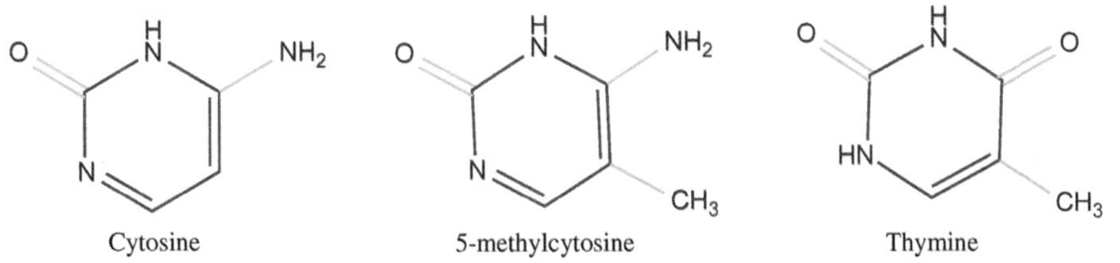

FIGURE 21.10 Structural comparisons of cytosine, methylcytosine, and thymine are shown.

The formation of 5-methylcytosine is primarily catalyzed by DNA methyltransferases (DNMT). Three different types of DNMTs have been identified in mammals based on the chemical reactions they catalyze. DNMT1, the main methyltransferase, is involved in both the methylation of newly synthesized DNA at the DNA replication fork and the maintenance of genomic methylation patterns. DNMT2 catalyzes methylation in embryonic stem cells and may be involved in RNA methylation. DNMT3A and DNMT3B mediate DNA methylation at CpG sites within promoter sequences. DNMT3L is a catalytically inactive regulatory factor for DNMT3A and DNMT3B, and is involved in genetic imprinting through methylation. It stimulates DNMT3A and DNMT3B by increasing their binding affinity for their substrate, methyl donor S-adenosyl methionine, and the unmethylated DNA sites.

It is well known that more than half of all human genes are transcribed from CpG-rich promoter sequences (called **CpG islands**) and that most of these sequences are unmethylated, which provides further evidence that methylation of cytosine suppresses gene expression. Studies have shown that methylation of DNA bases regulates gene expression by recruiting various methylated DNA binding proteins and thus plays a major role in development. For example, X-chromosome inactivation in female mammals involves methylated cytosine, resulting in impaired expression of the fragile X mental retardation 1 (FMR1) gene. FMR1 codes for FMR protein (FMRP), which is an RNA-binding protein highly expressed in the brain. FMRP regulates the expression of genes involved in Fragile X syndrome by binding to their mRNAs. The role of DNA methylation in the regulation of gene expression is discussed in more detail in Chapter 24.

DNA methylation plays major roles in the growth and development of an organism, and abnormal DNA methylation is associated with the pathogenesis of many diseases and cancers. The pattern of DNA methylation is influenced by environmental factors such as nutrition. Nutrients such as folic acid, methionine, and vitamins B_6 and B_{12}, which are involved in the formation of methyl group donor S-adenosyl methionine, affect DNA methylation and thereby regulate the expression of certain genes. As a result, a lack of these nutrients could disrupt normal metabolism and lead to certain disorders, including cardiovascular diseases and cancers. DNA methylation and histone modifications do not involve changes in DNA base sequences. However, these changes can be transferred to the next generation; the study of heritable changes in DNA structure is called **epigenetics**.

Furthermore, the pattern of DNA methylation is different depending on whether the DNA is inherited from the paternal or maternal side. Normal human development requires one set of genes inherited from each parent, and the process of selecting a specific allele to be expressed for a given gene locus is called **genetic imprinting**. Genetic imprinting requires methylation of DNA bases. If there is an error in genetic imprinting, it will lead to genetic disorders such as Beckwith−Wiedemann syndrome (BWS), Prader−Willi syndrome (PWS), and Angelman syndrome (AS). These disorders are discussed in more detail in later chapters (see also Clinical Case Study 21.1).

Determination of Base Sequences in DNA

The most widely used method of DNA sequencing is dideoxy chain termination, which was developed by Frederick Sanger. In this method, single-stranded DNA molecules are amplified by an *in vitro* polymerase chain reaction (PCR) system, which contains four deoxynucleotides (dNTPs) and four di-deoxynucleotides (ddNTPs), each labeled with four different fluorescent dye molecules. Also, included are DNA polymerase for DNA chain extension and short pieces of synthetic DNA as primers for the amplification reaction. ddNTPs are similar to dNTPs, except for the missing oxygen atom at the 3′-carbon of the deoxyribose sugar. In ddNTPs, the 2′ and 3′ carbon atoms do not contain hydroxyl groups. Since there is no hydroxyl group at the 3′ position of the sugar, DNA chain extension cannot occur at this end of the molecule. Therefore, whenever ddNTPs, complementary to the template DNA base sequence, are incorporated into the new DNA fragment being synthesized, chain extension will be stopped. The newly synthesized DNA strands containing strands of different lengths are separated by different sizes through gel electrophoresis.

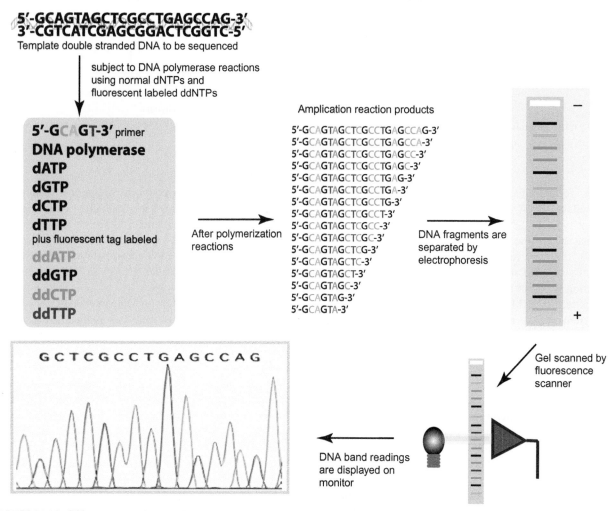

FIGURE 21.11   DNA sequencing by modified Sanger method, which uses dideoxynucleotides (ddNTPs) to stop chain elongation at the site of their incorporation. The four ddNTPs are labeled with fluorescent dyes and used in the reading of DNA sequence from the pattern of DNA bands in the gel.

The DNA fragments on the gel emit different fluorescent signals, which are examined by a laser scanning device (Figure 21.11).

Given that the methylation of cytosine residues at CpG islands plays an important role in the regulation of gene expression, several techniques have been developed to determine DNA methylation patterns, which provide valuable information on the functions of human genes and their associations with certain diseases. Such techniques include the bisulfite genomic sequencing, methylation-sensitive restriction endonuclease analysis, and methylated DNA immunoprecipitation methods. Bisulfite genomic sequencing is the most widely used method to determine methylated cytosine residues in DNA. Bisulfite converts cytosine to uracil. However, methylated cytosine residues are resistant to bisulfite treatment. Therefore, the methylation pattern of a DNA

sequence can be determined by comparing the sequence of bisulfite-treated DNA to the sequence of untreated DNA. The methylation-sensitive restriction endonuclease method uses two isoschizomers that recognize and digest the same DNA sequence, with one enzyme able to digest methylated DNA sequences (methylation insensitive) and the other unable to (methylation sensitive). Examples of such isoschizomers include *SmaI* (methylation sensitive)/*XmaI* (methylation insensitive) and *HpaII* (sensitive)/*MspI* (insensitive). In methylated DNA immunoprecipitation analysis, specific antibodies against methylated CpG islands are used to identify methylated DNA sequences. All three methods—bisulfite genomic sequencing, methylation-sensitive restriction endonuclease analysis, and methylated DNA immunoprecipitation—can be used in combination to enhance accuracy of the analysis.

Physical and Chemical Structure of DNA

Unique structural features of DNA include the following:

1. Double-stranded helix;
2. Antiparallel strands;
3. Complementary base sequences;
4. Hydrogen bonds between DNA strands.

In a DNA double helix, two strands are coiled around each other, creating its distinctive helical structure. The base sequence in one strand is complementary to the sequence of the other strand because of the unique A−T, G−C base pairings. The two antiparallel strands are joined together by hydrogen bonds between bases. Also, the hydrophobic nature of the bases and the hydrophilic nature of the phosphodiester backbone help stabilize the double helical structure by positioning hydrophilic groups to the outside and hydrophobic base pairs to the inside. The two DNA strands coil around each other in antiparallel directions and form a right-handed double helix.

Since the N-glycosidic bonds linking sugars and bases of one DNA strand are not directly opposite the glycosidic bonds of other strands, each base pairing creates a major groove on one side and minor groove on the other side of the double helix (Figure 21.12). The major groove provides space for DNA-binding proteins that perform a variety of functions, including regulation of gene expression. These proteins bind to DNA by establishing hydrogen bonds with the exposed bases.

Conformational Changes of the DNA Double Helix

Three different conformations of DNA exist in nature: A-, B-, and Z-forms (Figure 21.13). The DNA structure Watson and Crick described in their 1953 *Nature* paper is now known as the B-form of DNA, which is the naturally occurring form in cells and under laboratory physiological conditions. In its B-form, the helix makes one complete turn per every 10 bases, and the distance between adjacent base pairs is 0.34 nm. The B-form of DNA may change to the A-form in a high salt or denatured alcohol solution. The A-form of DNA is still a right-handed helix, but one complete turn of the helix in the A-form requires 11 base pairs. The distance between two adjacent base pairs in the A-form is 0.23 nm. Therefore, the A-form is more compact than the B-form, and its bases are tilted 20° relative to the helical axes. The major groove of A-form DNA is narrower, and the minor groove is wider than those of B-form.

The Z-form of DNA is quite different from the other two forms, as it is a left-handed helix and base pairs alternate in a dinucleotide repeating structure such as an alternating purine−pyrimidine sequence (5′pCGCGCGCGCG3′OH).

This repeating sequence favors the Z-form of DNA. The major and minor grooves of the Z-form show little difference in depth and width. Since CG-rich sequences are usually found in promoter sequences, the biological role of the Z-form of DNA may include regulation of gene expression. Also, it is known that Z-DNA regions may be formed to relieve torsional stress due to DNA loops that form during transcription. The alternating C−G base pairs introduce a zig-zag structure into the DNA backbone; one turn of Z-DNA requires 12 base pairs, and the distance between adjacent base pairs is 0.38 nm. Z-DNA forms in a high salt solution or in a solution of divalent cations.

Intercalating Agents

A DNA double helix can be distorted by intercalating agents that are inserted between the stacked base pairs (Figure 21.14). Intercalating agents are hydrophobic heterocyclic ring molecules that resemble the ring structure of base pairs, and include ethidium bromide, acridine orange, and actinomycin D. Insertion of these agents distorts the DNA double helix, thereby interfering with DNA replication, transcription, and repair. Such DNA distortions often result in mutations, so intercalating agents are also mutagens. Ethidium bromide is a fluorescent molecule that emits lights in the visible spectrum when DNA (loaded with ethidium bromide) is examined under ultraviolet light. It is widely used in the laboratory to visualize DNA or RNA fragments separated by gel electrophoresis. Acridine orange is a fluorescent dye also used to visualize DNA and RNA fragments. It can be used in combination with ethidium bromide to differentiate live cells from ones undergoing apoptosis. Actinomycin D interferes with both DNA replication and transcription by intercalation between bases. It is most commonly used as a chemotherapeutic (anticancer) agent to treat gestational trophoblastic neoplasia, Wilms' tumor, and rhabdomyosarcoma. It is also an antibiotic agent but is not commonly used due to its toxicity.

DNA Supercoils

If DNA is circular or if both ends of linear DNA are tied to cellular structures, adding or removing twists in DNA may induce DNA supercoiling (Figure 21.15). Adding more twists to a typical relaxed DNA double helix leads to positive supercoiling, and removing extra twists causes a negative supercoiling. Supercoiling takes place when DNA helices relieve torsional stress from overwinding or underwinding. Overwinding leads to positive supercoiling, whereas underwinding leads to negative supercoiling. Supercoiling can be expressed mathematically by using the terms *twist* and *writhe*. Twist is the number of helical turns in the DNA, and writhe is the number of times a DNA helix

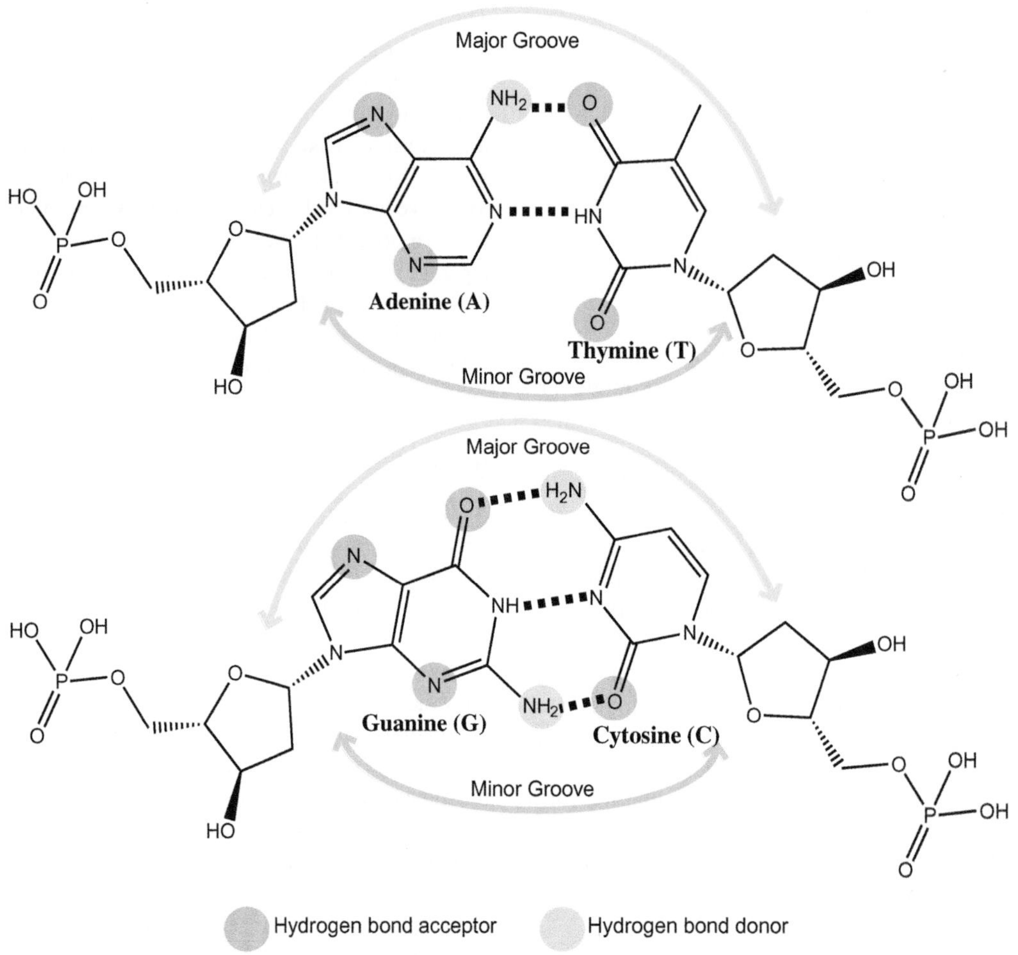

Major Groove

Adenine (A)

Thymine (T)

Minor Groove

Major Groove

Guanine (G)

Cytosine (C)

Minor Groove

Hydrogen bond acceptor   Hydrogen bond donor

FIGURE 21.12   The minor groove and major groove of the DNA double helix. Potential hydrogen donors and acceptors in the grooves are indicated that enable certain regulatory proteins to recognize specific DNA base sequences.

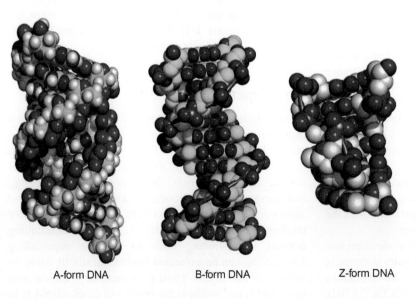

FIGURE 21.13   Three different conformations of DNA structures are shown in this figure.

A-form DNA   B-form DNA   Z-form DNA

crosses over on itself. Thus, supercoiling = twist + writhe. Packaging of DNA into the tiny nucleus of a cell during cell division requires supercoiling, since it reduces the size and volume of DNA so that it can be packaged into

chromosomes. However, growing cells require a relaxed form of DNA, which corresponds to the negative supercoiled state. In a negatively supercoiled state, DNA binding sites are exposed to many DNA binding proteins such as DNA polymerase, RNA polymerase, and other transcription factors. The degree of supercoiling in cells is maintained by the action of two important enzymes: topoisomerases I and II. Topoisomerase I changes the degree of supercoiling by cutting one strand of DNA, letting the broken strand pass through the other strand, and reannealing the ends (Figure 21.15). Topoisomerase II modifies supercoiling by cutting both DNA strands, rotating around the broken strands, and reannealing the ends. This enzyme requires hydrolysis of ATP to complete the reaction. Bacterial topoisomerase II is also called **DNA gyrase**. DNA topoisomerases are essential enzymes in DNA replication. Any mutation that inactivates these enzymes also prevents DNA replication and cell division. Therefore, inhibitors for DNA gyrase activity such as novobiocin and nalidixic acid are effective antibiotics. Also, inhibitors of eukaryotic topoisomerases such as doxorubicin and etoposide are used as chemotherapeutic agents.

Denaturation of DNA

The two antiparallel strands of a DNA double helix can be separated when hydrogen bonding between bases on opposite strands is disrupted by denaturing agents (Figure 21.16). DNA denaturing agents include pH, ionic strength, and heat. Increasing the temperature of a solution containing DNA causes DNA strands to separate. The temperature at which half of the double-stranded DNA

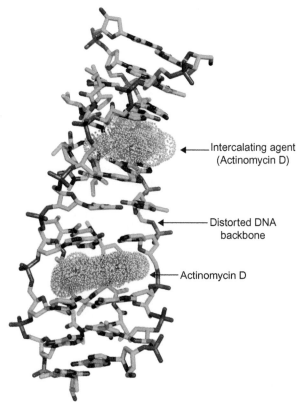

Intercalating agent
(Actinomycin D)

Distorted DNA backbone

Actinomycin D

FIGURE 21.14 Distorted backbone of DNA double helix by an intercalating agent, actinomycin D.

FIGURE 21.15 Different states of a covalent DNA circle. (a) A nonsupercoiled circular DNA having 36 turns of the helix. (b) An underwound covalent circle DNA having only 32 turns of helix. (c) The nucleotide in (b), but with four superhelical turns to eliminate the underwinding.

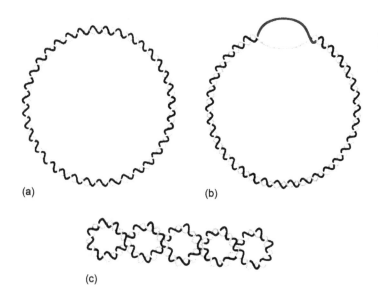

(a)

(b)

(c)

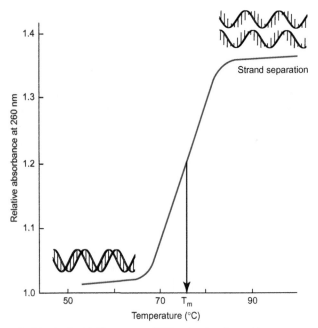

FIGURE 21.16 Melting curve of DNA showing the melting temperature (T_m) and possible molecular conformations for various degrees of melting.

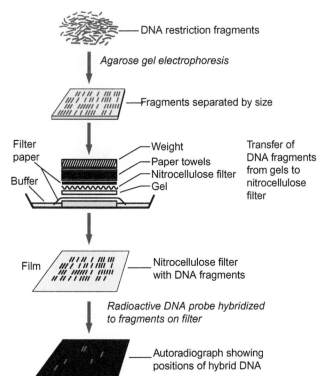

FIGURE 21.17 Procedure for a Southern blot. This analysis allows detection of a specific fragment of DNA by hybridization of a radioactive probe with a complementary DNA sequence.

molecules are converted to single-stranded DNA is defined as the melting temperature of the DNA molecule (T_m). The T_m depends on the length and the specific sequence of a DNA molecule; the greater the length of DNA, the higher the T_m. Also, the higher the GC content of a DNA molecule, the higher its T_m. Lowering the ionic strength also can break hydrogen bonding between bases, causing melting of the DNA. Lowering the ionic strength of a solution also lowers the T_m. In addition, high or low pH weakens hydrogen bonds and causes DNA denaturation. If the pH of a solution is higher than 11.5 or lower than 2.3, DNA will denature spontaneously. The degree of DNA denaturation can be measured spectrophotometrically since the optical density (OD) of a solution of DNA at 260 nm almost doubles as double-stranded DNA is converted into single-stranded DNA. This phenomenon of increasing OD_{260} is called a **hyperchromic shift** (Figure 21.16).

Renaturation of DNA

Denatured DNA strands reassemble into a double-stranded DNA helix when denaturing agents are removed. This process is called **renaturation** or **annealing** of DNA. The rate of DNA renaturation depends on the sequence of bases and the length of DNA. Short and highly repetitive DNA sequences renature faster than long, nonrepetitive DNA molecules. Differences in DNA renaturation rates are used to measure the frequency of specific base sequences, to locate specific sequences, and to analyze species of RNA that anneal to DNA.

Since small DNA fragments anneal to the complementary base sequences faster than larger DNA fragments, small DNA probes are used to detect specific DNA sequences in large DNA molecules. Such hybridization experiments are called **Southern blots**, named after the scientist who developed the analytical technique. In Southern blotting, small DNA probes are used to identify specific DNA molecules that contain complementary base sequences (Figure 21.17). In Northern blotting, DNA probes are used to detect specific RNA molecules with base sequences complementary to DNA probes.

CHROMOSOMES AND CHROMATIN

DNA molecules in eukaryotic cells are linear and tightly packaged into chromosomes (Figure 21.18). The total length of human DNA in chromosomes in a nucleus is 1.8 m, and the diameter of cell nucleus is 6 μm. Therefore, a great degree of DNA packaging is required to fit this large amount of DNA into the chromosomes in the nucleus. Packaging of DNA starts with the formation of DNA-protein complexes called **histone complexes**. The first level of DNA packaging involves the formation of nucleosome beads in which negatively charged DNA strands wrap around positively charged histone proteins. The next step in DNA packaging is the coiling of nucleosome beads into

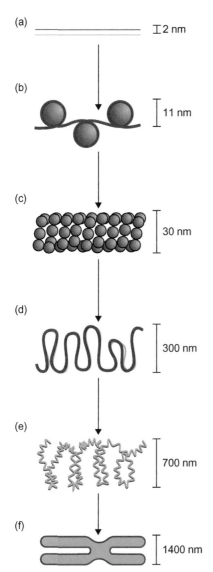

FIGURE 21.18 Hypothetical stages in the condensation of DNA with chromatin to form a chromosome. (a) Double-stranded DNA. (b)–(e) Formation of nucleosome beads and fibers consisting of histones and condensed DNA to form (f) a metaphase chromosome.

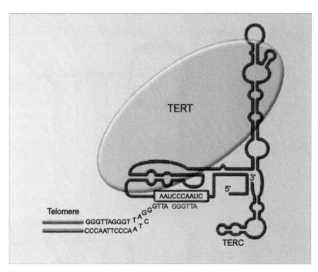

FIGURE 21.19 Structure and function of telomerase. Telomerase consists of two subunits: telomerase reverse transcriptase (TERT) and the telomerase RNA component (TERC). *[Reprinted with permission from S.E. Artandi, Telomeres, telomerase, and human disease, N. Engl. J. Med. 355(12) 1195, Figure 1.]*

30 nm nucleoprotein fibers, which are also called **chromatin fibers**. Further looping of these 30 nm fibers leads to the formation of condensed chromosomes. DNA packaging takes place during the prophase of mitosis, during which DNA must be condensed prior to being separated into the daughter cells. In the cell nucleus, two different forms of chromatins are recognized: **euchromatin** and **heterochromatin**. Euchromatin has less densely packed DNA and consists of active genes that are expressed in the cell. Heterochromatin consists of more tightly packed DNA; expression of genes in this region is limited since the tightly packed DNA is not accessible to transcription factors and polymerases required for expression of genes. Bacteria, which are prokaryotes, contain circular DNA molecules and utilize different DNA packaging processes. Also, prokaryotes do not have a true nucleus or nuclear membrane and have one or few chromosomes. In *E. coli*, the circular chromosomal DNA molecule is approximately 1400 μm long and is packaged into a cell whose length is about 2–3 μm. Therefore, some form of packaging of DNA is required to fit the large molecule into the tiny cell. Prokaryotic DNA packaging involves supercoiling of DNA by topoisomerases and histone-like proteins that bind to the DNA and make it more compact. Unlike circular DNA in prokaryotes, linear DNA sequences in eukaryotes need special structures called **telomeres** at both ends of each DNA molecule to protect the end sequences from getting degraded or fused with other DNA sequences. Telomeres in human DNA have unique repeated DNA sequences of TTAGGG and the length of telomeres range from 5 kb to 15 kb. Telomere sequences are shortened during each cell division, since DNA polymerases require short RNA primers to initiate DNA replication at the 3′ end of DNA molecules (discussed in Chapter 22). However, the TTAGGG repeat sequences can be added to each telomere structure by an enzyme called **telomerase** (Figure 21.19). Each somatic cell division event shortens telomeres up to 100 base pairs, and since most somatic cells do not contain telomerase, those cells will die after a certain number of cell divisions, depending on the length of telomeres. However, cancer cells and germ-line cells contain telomerase, and therefore, they are immortal. Telomerase is a reverse transcriptase, since it uses a short RNA strand (5′-CUAACCCUAAC-3′) as a template to synthesize DNA (TTAGGG) repeat sequences at the end of chromosomes, lengthening telomere structures (Figure 21.19).

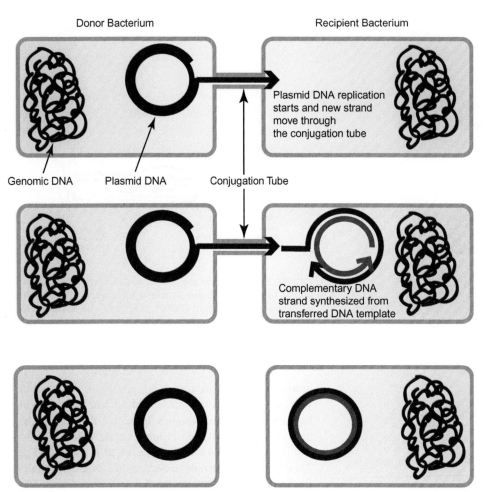

Donor Bacterium

Recipient Bacterium

Plasmid DNA replication
starts and new strand
move through
the conjugation tube

Genomic DNA   Plasmid DNA   Conjugation Tube

Complementary DNA
strand synthesized from
transferred DNA template

Separation of two bacteria takes place

FIGURE 21.20 Transfer of plasmid DNAs by the conjugation process is shown in this figure.

Human cells contain 46 chromosomes (two copies of 22 somatic chromosomes and a pair of sex chromosomes) in the cell nucleus. **Aneuploidy**, a condition in which cells contain an abnormal number of chromosomes, is responsible for many genetic disorders, such as Turner syndrome (1 X chromosome, total 45 chromosomes), Down's syndrome (3 copies of chromosome 21, total 47 chromosomes), and Klinefelter syndrome (3 sex chromosomes, XXY, total 47 chromosomes). In Down's syndrome, the most common aneuploidy in humans (1 per 1000 births), genes in the third chromosome 21 disturb normal cell functions, resulting in physical and mental disabilities. Aneuploidy can result from errors in meiosis and mitosis during gametogenesis. In some cases, chromosomal errors in mitosis after conception result in **mosaicism**, in which two or more genotypes of cells exist in the same individual.

Plasmid DNAs

Prokaryotes contain another distinctive class of circular double-stranded DNA molecule that is separate from bacterial chromosomal DNA. **Plasmids** are widespread among bacteria, and their size varies from a few genes to more than a hundred genes. Plasmid genes may provide antibiotic resistance to bacteria through mechanisms that involve inactivating drugs, making changes in drug binding sites, providing alternative metabolic pathways, and affecting drug absorption into the cell. A single cell may have just one copy of a plasmid or multiple copies. Plasmid DNA can be transferred to other bacteria through the process called **conjugation**, which involves physical cell-to-cell contact (Figure 21.20). Conjugation is an important tool that allows bacteria to exchange genes, which may be beneficial to the survival of certain bacteria. Plasmid genes transported through conjugation may protect the cell against certain antibiotics or enable cells to digest certain organic molecules otherwise unavailable for bacterial metabolism. Plasmid DNAs have played major roles in the advancement of biotechnology by providing useful tools, such as in DNA cloning vector, transformed cell screening, and *in vitro* protein expression (Figure 21.21). Human genes coding for proteins such as

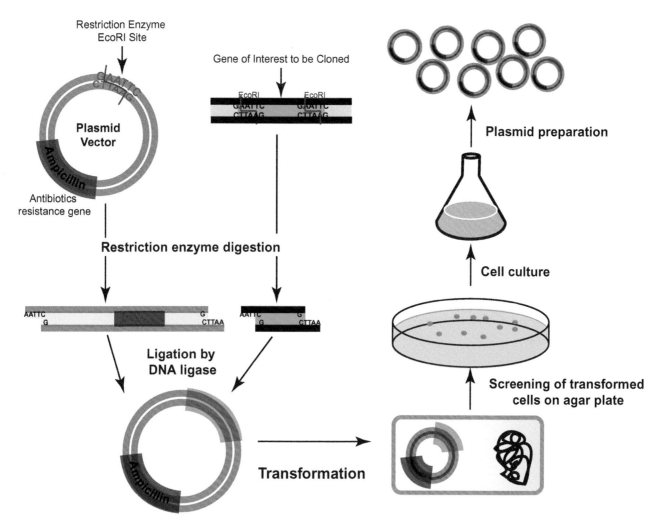

FIGURE 21.21 DNA cloning by using plasmid DNA and host bacterial system. The gene of interest is inserted into a bacterial plasmid, which carries a gene for ampicillin antibiotic resistance by using restriction enzyme digestion and DNA ligase. The resulting plasmids containing recombinant DNA fragments are introduced into bacterial cells by transformation. Only transformed cells with recombinant plasmid are able to grow on cell culture media containing an antibiotic, ampicillin.

human insulin and growth hormone have been introduced into bacteria through plasmid DNA technologies and produce the human proteins for pharmaceutical purposes.

Repetitive DNA Sequences

There are two major types of repetitive sequences found in cells: **tandem repeats** and **interspersed repeats**. A tandem repeat is a series of consecutive DNA sequences found in genomes. Three subclasses are defined: **satellites**, **minisatellites**, and **microsatellites**. Satellite DNAs are repeated sequences found near centrosomes, telomeres, and in Y chromosomes that have a repeat length of 171 bp. Satellite DNA regions can occupy from 100 kb to over 1 Mb. Alphoid DNA, the human centromeric repeat located at the centromere of all chromosomes, is an example of satellite DNA. Minisatellites have repeated sequences that range from 9 bp to 80 bp and are located in noncoding regions of DNA. Minisatellites, also

known as the **variable number of tandem repeats (VNTR)**, are the most highly variable sequence element in the human genome, and used for DNA fingerprinting analysis in forensic science. Microsatellites consist of 1 to 6 bp repeat units and can extend up to 150 bp. Like minisatellite DNA, the number of repeats in microsatellites, also referred to as **short tandem repeats (STR)**, may be different among individuals and therefore can be used for DNA fingerprinting analysis (Figure 21.22). The 6 to 50 CGG repeats found in Fragile X syndrome are an example of a microsatellite.

Unlike tandem repeats, which are repeated in one area of DNA, interspersed repeat sequences are scattered throughout the whole genome. They can be grouped into **short interspersed nuclear elements (SINEs)**, which constitute up to 13% of the human genome, and **long interspersed nuclear elements (LINEs)**, which account for up to 21% of the human genome. The specific *Alu* sequence is an example of a SINE.

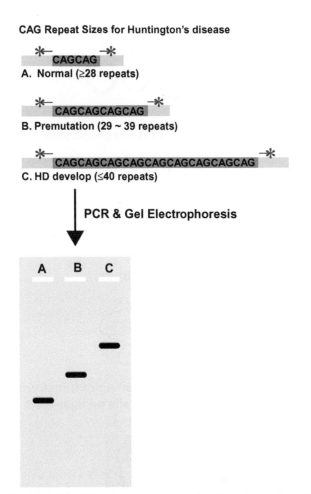

FIGURE 21.22 DNA fingerprint analysis based on the number of repeated sequences of certain alleles. Analysis of a variable number of tandem repeats (VNTR) to identify the number of triplet (CAG) sequence repeats on the Huntington gene is shown in this figure.

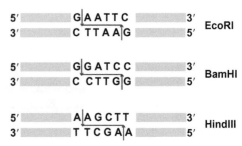

FIGURE 21.23 Palindromic DNA sequence recognized by restriction enzymes.

whole 3 billion bases of a complete human genome. These BACs were cut into small fragments of 2 kb in size. These small fragments were subjected to DNA sequencer analysis by computer programs to obtain the complete DNA sequence cloned in each BAC clone.

Enzymes in Recombinant DNA Technology

The discovery of **restriction enzymes** (restriction endonucleases) from bacteria created a new field called **genetic engineering**, and provided scientists with an important tool for the manipulation of DNA. Restriction enzymes recognize specific DNA sequences, usually 4 to 8 bases long, that are **palindrome**; i.e., the bases in one strand are the same as the bases in the other strand read in the opposite direction (Figure 21.23). Different restriction enzymes recognize different DNA sequences and cleave the double-stranded DNA, leaving either overlapping (sticky) or blunt ends, depending on the specificity of the restriction enzymes. Sticky ends are very useful in joining two different DNA molecules into one, since sticky ends contain complementary bases that can be paired. Base pairing of single-stranded ends brings two different DNA molecules together; another key enzyme in recombinant DNA technology called **DNA ligase** establishes covalent bonds between strands of two DNA fragments.

DNA Cloning Vectors

DNA cloning is a process of making identical DNA copies in large quantities by using bacteria to amplify DNA segments spliced into plasmids or with the polymerase chain reaction (PCR). In bacteria or yeast, a DNA cloning vector (plasmid) is required to insert DNA of interest into the host cells. There are five different vectors available for DNA cloning experiments, depending on the size of the DNA to be amplified (Table 21.1). The most widely used vector for DNA cloning is plasmid, which carries up to 10 kb of DNA.

RECOMBINANT DNA TECHNOLOGY

Recombinant DNA means recombining DNA fragments from different sources; a number of different techniques are required to join two different DNA molecules. Techniques such as DNA purification, digestion of DNA fragments by enzymes, and amplification of specific DNA sequences by PCR and DNA cloning are used in recombinant DNA technology. Recombinant DNA technology has been widely used to study many aspects of nucleic acids. The human genome project (HGP) serves as an example of how recombinant DNA technology is used to study DNA. In 2003, the HGP was completed and provided the complete sequence of a human genome. At first, genomic DNA was isolated, cut into small pieces by restriction enzymes, and assembled into a bacterial artificial chromosome (BAC) for the production of multiple copies of identical DNA molecules for sequencing. About 20,000 different BACs were created to cover the

TABLE 21.1 Various Vectors Used to Carry DNA Inserts for Cloning

Types of Vectors	Host Cells	Size Limit of DNA Insert	Features
Plasmid	Bacteria	0.1–10.0 kilobases (kb)	The most widely used vector system, easy to use
Lambda phage	Bacteria	8.0–20.0 kb	Higher transformation efficiency and carries larger insert, but more difficult to use than plasmid
Cosmid	Bacteria	35.0–50.0 kb	Combination of plasmid and Cos sites of lambda phage; also has high transformation efficiency and carries larger insert than does lambda
Bacterial Artificial Chromosome (BAC)	Bacteria	75.0–300.0 kb	Carries large DNA inserts; used to sequence the genome of organisms (e.g., human genome project); shows higher transformation efficiency than YAC
Yeast Artificial Chromosome (YAC)	Yeasts (Eukaryotes)	100.0–2000.0 kb	Carries very large DNA inserts, but transformation efficiency is low

Polymerase Chain Reaction (PCR)

PCR is an *in vitro* method that amplifies a specific fragment of DNA. This technique, which was developed by Kary B. Mullis in 1983, revolutionized the study of DNA and its applications. It is one of the most widely used techniques in molecular biology and biotechnology. PCR amplification of DNA has many advantages over cell-based cloning techniques. The most significant advantages are sensitivity and selectivity. PCR amplifies specific regions of DNA in a sample up to 10 kb in size in only a few hours and can use minute DNA samples. A single copy of double-stranded DNA can be amplified into 34 billion copies in a few hours after 35 cycles of PCR reaction. PCR reactions consist of 30–40 repeats of the same PCR cycle. Each cycle includes three major steps: **denaturation**, **annealing**, and **extension** (Figure 21.24). The first step of a PCR reaction is the separation of the double-stranded DNA into single-stranded DNA by heating the sample to 94°C–98°C. The next step is annealing; the heated DNA sample is cooled to allow two single-stranded DNA primers to anneal to the sample DNA. The short single-stranded oligonucleotides are complementary to sites on the target DNA. The final step is DNA chain extension, which requires the action of a heat-stable DNA polymerase to extend primers paired to each end of the DNA target sequences. The PCR technique has many applications, including identification of single nucleotide polymorphism (SNP) sequence, DNA and RNA quantifications, and forensic analysis of DNA samples obtained from crime scenes or suspects.

Nucleic Acid Hybridization

Nucleic acid hybridization is a process used to identify specific DNA sequences. Specific DNA probes are

Steps of Polymerase Chain Reaction (PCR)

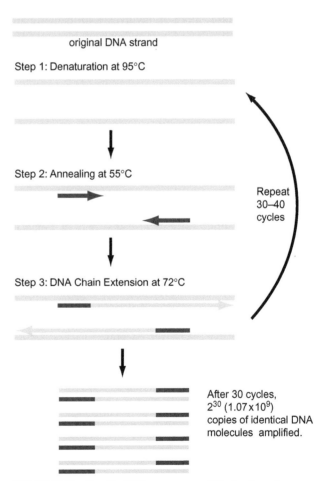

original DNA strand

Step 1: Denaturation at 95°C

Step 2: Annealing at 55°C

Repeat 30–40 cycles

Step 3: DNA Chain Extension at 72°C

After 30 cycles, 2^{30} (1.07×10^9) copies of identical DNA molecules amplified.

FIGURE 21.24 Specific DNA sequence amplification by polymerase chain reaction (PCR) technique.

denatured and annealed to sample DNA that has also been denatured. Probes used in hybridization reactions are usually chemically synthesized DNA or RNA that has been labeled with a fluorescent dye or radioactive isotope such as [32]P. There are two different types of nucleic acid hybridization techniques generally used, which are called **Northern blotting** and **Southern blotting** (Figure 21.17). Southern blotting is used to identify specific sequences on target DNA using labeled probes complementary to a DNA sequence in the sample, and Northern blotting is used to identify a specific RNA molecule in a mixture of different RNA by using labeled DNA probes. Another type of blotting technique is called **Western blotting**, which uses the same principles of hybridization but is used for identifying specific protein molecules using protein probes labeled with a fluorescent dye or with a radioisotope such as [35]S. The most widely used protein probes for Western blots are antibodies, either monoclonal or polyclonal. These are used to detect target proteins called **antigens**. A combination of Southern blot and Western blot techniques is also used in the characterization of specific DNA-binding proteins, and this technique is called a **Southwestern blot**, since it involves hybridization of both protein and DNA. Short regions of target DNA sequences are labeled and serve as probes for hybridization reactions. Many DNA-binding proteins can be characterized by this method.

Applications of Recombinant DNA Technology

Recombinant DNA technology has been widely used in the diagnosis of diseases using recombinant proteins, in the detection of normal genetic variations, in the detection of genetic variations causing diseases, and in prenatal screening for genetic diseases.

Examples of recombinant DNA applications in clinical medicine include the test for the genes causing sickle cell disease, Huntington's disease, and dozens of other Mendelian disorders. Sickle cell disease is an inherited disease caused by a point mutation in the beta hemoglobin gene. To detect the sickle cell mutation, the **restriction fragment length polymorphism (RFLP)** method is used. A single nucleotide substitution mutation (A→T) in the sixth codon changes the sixth amino acid of the beta hemoglobin from glutamic acid to valine, creating a hydrophobic patch responsible for sickling of red blood cells by polymerization of hemoglobin molecules. This single nucleotide change causes a loss of an extra restriction enzyme site for MstII, and therefore, when the MstII enzyme is used on amplified globin DNA, the sickle cell mutation can be identified (Figure 21.25).

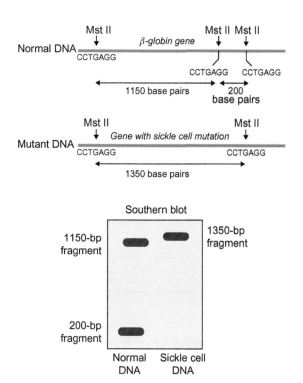

FIGURE 21.25 Detection of sickle cell anemia by PCR and RFLP analysis.

To test for Huntington's disease mutations, the variable number of tandem repeats (VNTR) method is used to identify variable numbers of repeat sequences on the Huntington gene located on chromosome No. 4. Huntington's disease is a fatal neurological disorder that results from degeneration of brain cells caused by Huntingtin protein, which is a product of a CAG triplet sequence repeat in the Huntington gene. The number of CAG triplet nucleotide sequence repeats for normal people ranges from 10 to 35 repeats; people with Huntington's disease carry 36 to 120 CAG repeats in the gene. A *HindIII* restriction enzyme site is closely linked to the Huntington gene, and digestion of DNA with the *HindIII* enzyme provides an estimate of the number of CAG repeats on the gene (Figure 21.22).

Recombinant DNA technology has also been used in forensic science to identify a criminal suspect using DNA samples obtained from various biological sources at the crime scene; this test is called **genetic fingerprinting**. Since each individual contains unique DNA repeating sequences (minisatellites or microsatellites), analysis of a number of these repeats at a given gene location provides an identification tool for investigators to identify a specific individual matching DNA fingerprint. This method can also be used in paternity tests, maternity tests, and identification of human remains (Figure 21.26).

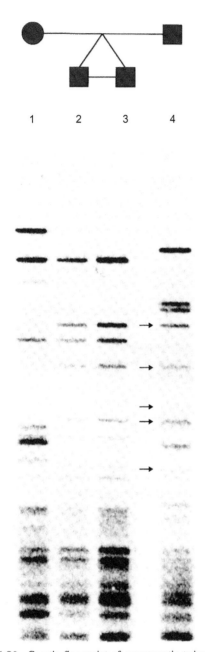

FIGURE 21.26 Genetic fingerprint of monozygotic twins (columns 2 and 3) and their parents (columns 1 and 4). The arrows indicate hypervariable sequences inherited from the father. Similar hypervariable sequences can be identified from the mother. *Reproduced with permission from A.J. Jeffreys, V. Wilson, S.L. Thein, Individual-specific "fingerprints" of human DNA, Nature 316 (1985) 76.*

CLINICAL CASE STUDY 21.1 Beckwith—Wiedemann Syndrome

This clinical case was abstracted from: T. Zhang, X. Xie, D. Xu, W. Lu, C. Dong, R. Liu, et al., Beckwith—Wiedemann syndrome: first epigenetic confirmed case report in China, Clin. Genet. 84 (2013) 603—604.

Synopsis

A baby girl was delivered at 28 weeks of gestation with a weight of 1.35 kg (>75th percentile) and height of 41 cm (>50th percentile). She exhibited an umbilical hernia, ear pits, facial nevus flammeus, macroglossia (abnormally large tongue), transient neonatal hypoglycemia, visceromegaly (abnormally large liver and ovary), and pulmonary stenosis. At 5 months of age, she was diagnosed with Beckwith—Wiedemann syndrome (BWS) and had a partial glossectomy (removal of part of the tongue).

A baby boy was delivered by caesarean section at 37 weeks of gestation with a weight of 4.23 kg (>97th percentile) and height of 52 cm (>97th percentile). He presented gigantism, macroglossia, visceromegaly (liver and kidneys), hemihyperplasia, umbilical hernia, patent ductus arteriosus, and pulmonary hypertension. At one month, he was diagnosed with BWS.

Lymphocytic DNA samples from both patients and their parents were purified and subjected to methylation status analysis at the imprinting centers (ICs) of the 11p15 region by the methylation-specific multiplex ligation-dependent probe amplification (MS-MLPA) method. The female patient exhibited loss of methylation at the KCNQ1OT1 promoter and hypermethylation at the H19DMR domains in the 11p15 region.

Teaching Points

1. Beckwith—Wiedemann syndrome (BWS) is a genetic disorder with the incidence of 1 in 13,700 newborns, and the affected individuals show macrosomia and unusual overgrowth. Typical symptoms of BWS include umbilical hernia, macroglossia, visceromegaly, exomphalos, gigantism, hemihyperplasia, earlobe creases or pits, nevus flammeus, midfacial hypoplasia, hepatomegaly and nephromegaly.

2. Children with BWS are at increased risk of developing tumors, and therefore, measurement of alpha-fetoprotein and abdominal ultrasound imaging are recommended at every 3-month follow-up. The male patient has a higher tumor development risk than the

(Continued)

CLINICAL CASE STUDY 21.1 (Continued)

female patient, since he exhibited hypermethylation at the H19DMR region.

3. The main cause of BWS is abnormal regulation of gene expression at the chromosomal 11p15.5 region. Cells contain two copies of genes inherited from both parents, and for some genes, only one of the two, either the paternal copy or the maternal copy, is expressed through the process called genomic imprinting. Genomic imprinting involves methylation of gene regulatory sequences and abnormalities of imprinted genes at the 11p15.5 region, in particular, H19 and CDKN1C, are associated with BWS.

Supplemental References

[1] S. Azzi, W.A. Habib, I. Netchine, Beckwith—Wiedemann and Russell—Silver syndromes: from new molecular insights to the comprehension of imprinting regulation, Curr. Opin. Endocrinol. Diabetes Obes. 21 (2014) 30—38.

[2] K.J. Jacob, W.P. Robinson, L. Lefebvre, Beckwith—Wiedemann and Silver—Russell syndromes: opposite developmental imbalances in imprinted regulators of placental function and embryonic growth, Clin. Genet. 84 (2013) 326—334.

ENRICHMENT READING

[1] J.D. Watson, F.H. Crick, Molecular structure of nucleic acids: a structure for deoxyribose nucleic acid, Nature 171 (1953) 737—738.

[2] M. Meselson, F.W. Stahl, The replication of DNA in *Escherichia coli*, Proc. Natl Acad. Sci. U.S.A. 44 (1958) 671—682.

[3] S.A. Grigoryev, C.L. Woodcock, Chromatin organization—the 30 nm fiber, Exp. Cell Res. 318 (2012) 1448—1455.

[4] J.C. Venter, M.D. Adams, E.W. Myers, P.W. Li, R.J. Mural, G.G. Sutton, et al., The sequence of the human genome, Science 291 (2001) 1304—1351.

[5] M.J. Bamshad, S.B. Ng, A.W. Bigham, H.K. Tabor, M.J. Emond, D.A. Nickerson, et al., Exome sequencing as a tool for Mendelian disease gene discovery, Nat. Rev. Genet. 12 (2011) 745—755.

[6] C.M. Rivera, B. Ren, Mapping human epigenomes, Cell 155 (2013) 39—55.

[7] V.K. Rakyan, T.A. Down, S. Beck, Epigenome-wide association studies for common human diseases, Nat. Rev. Genet. 12 (2011) 529—541.

[8] S.E. Artandi, Telomeres, telomerase, and human disease, N. Engl. J. Med. 355 (2006) 1195—1197.

[9] X. Deng, J.B. Berletch, D.K. Nguyen, C.M. Disteche, X chromosome regulation: diverse patterns in development, tissues and disease, Nat. Rev. Genet. 15 (2014) 367—378.

Chapter 22

DNA Replication, Repair, and Mutagenesis

Key Points

1. General features of DNA replication are:
   a. DNA replication is semi-conservative;
   b. DNA replication is bidirectional;
   c. DNA replication is semi-discontinuous.
2. Enzymes in DNA replication include:
   a. DNA polymerases' multiple activities: polymerization of nucleotides, proofreading 3′ to 5′ exonuclease, and 5′ to 3′ exonuclease activity for DNA repair and nick translation;
   b. The reverse transcriptase enzyme can synthesize DNA from an RNA template;
   c. DNA gyrase and helicase catalyze unwinding of DNA supercoils and double helices;
   d. Primases are RNA polymerases and synthesize a short piece of primer sequence required by DNA polymerases to replicate DNA molecules;
   e. DNA ligase is required to seal the nicks in double-stranded DNA;
   f. Telomerases are required to extend telomere structure of eukaryotic linear DNA sequences of chromosomes in certain types of cells.
3. Three major stages of DNA replication are initiation, elongation, and termination. Initiation of DNA replication occurs at specific sites called origins in prokaryotes. Initiation of eukaryotic DNA replication occurs at multiple initiation sites. Elongation of DNA replication proceeds in two opposite directions. For the leading strand, elongation occurs by continuous replication; for the lagging strand, elongation occurs by discontinuous replication. Okazaki fragments are synthesized during discontinuous, lagging strand DNA replication. Termination of DNA replication occurs at special sites opposite the origin sites in prokaryotes.
4. Rolling circle replication in bacteriophages can generate many copies of concatameric DNA from a single circular template DNA molecule.
5. Consequences of DNA mutation include somatic mutations, which induce cell changes mostly affecting individual cell lines, and germline mutations, which cause inheritable traits (e.g., diseases) or in rare cases lead to evolution of new species.
6. Small-scale mutations include base substitution, base deletion, and base insertion. Large-scale mutations include chromosomal mutations such as translocations, inversions, deletions, and nondisjunctions. The Ames test is used to determine whether a certain chemical is potentially carcinogenic.
7. There are three broad categories for repair types: direct reversal of damage, excision of damaged region followed by precise replacement, and damage tolerance. Excision repair includes mismatch repair, base excision repair, and nucleotide excision repair. Damage tolerance repair includes homologous recombination and SOS repair.
8. Genetic stability is accomplished by a highly accurate DNA replication system and DNA repair system whenever DNA is damaged or mutations are introduced.

INTRODUCTION

The genetic information encoded in DNA in every cell must be correctly maintained and copied from one cell generation to the next to inherit the physiological characteristics of parent organisms that were acquired over millions years of evolutionary processes. In the case of human beings, approximately 6×10^9 base pairs in each diploid cell must be copied prior to its undergoing mitosis (somatic cell) or meiosis (germ cell).

If exactly the same DNA sequences are copied from one generation to the next generation without any errors or changes, parent and progenitor organisms would have exactly the same genetic makeup, and diversity of various species could not be established. However, changes in the genetic information are routinely introduced through mutations in germline cells or errors during DNA replication. These changes introduced into DNA molecules are the sources of genetic diversity on which natural selection acts and which fuels biological evolution.

Therefore, the fidelity of DNA replication and delicate DNA repair systems of cells are two major contributors toward establishing genetic stability. On the other hand, DNA replication and repair systems allow a low frequency of genetic changes to occur during each round of replication, which establishes the diversity of living organisms. In this chapter, the basic mechanisms of DNA replication, mutation, and repair systems in prokaryotes and eukaryotes are discussed.

N. V. Bhagavan and C.-E. Ha: Essentials of Medical Biochemistry, Second edition. DOI: http://dx.doi.org/10.1016/B978-0-12-416687-5.00022-1

GENERAL FEATURES OF DNA REPLICATION

DNA Replication Is Semi-Conservative

Three possible models of DNA replication can produce the exact replica of genetic materials; these are conservative, semi-conservative, and dispersive DNA replication mechanisms (Figure 22.1). In 1953, James D. Watson and Francis F. H. C. Crick suggested a hypothetical mechanism of DNA self-duplication based on their previously published DNA structure. Their model described how base sequences of DNA are copied to generate identical duplicates, and how one strand serves as a template for the synthesis of the other strand. This mode of DNA replication is known as **semi-conservative replication**.

Meselson and Stahl proved this mode of DNA replication experimentally in 1958 by labeling DNA with a heavy isotope of nitrogen, (^{15}N). They labeled DNA in *Escherichia coli* by growing cells for several generations in medium containing $^{15}NH_4Cl$. Cells were then transferred to a medium containing $^{14}NH_4Cl$ and continued to grow exponentially. After one generation of growth in $^{14}NH_4Cl$ medium, the DNA was extracted from a sample of cells and analyzed by cesium chloride density centrifugation. All DNA was of intermediate (hybrid) density; one strand contained ^{15}N and the newly synthesized strand contained ^{14}N. After two generations of growth in ^{14}N medium, 50% of the DNA contained only ^{14}N (light DNA) and 50% was hybrid. This important experiment confirmed the prediction of the Watson–Crick model for DNA replication.

DNA Replication Is Bidirectional

Another unifying principle of DNA replication is that each molecule of DNA has one or more specific origins of replication, i.e., sequences where DNA synthesis begins. In bacteria that carry their genetic information in a circular DNA molecule, only one point of origin of replication exists. Plasmid DNA also contains a unique origin of replication. Origins of replication can be moved from one location in DNA to another or from one DNA molecule to another, just as genes can be moved by genetic engineering techniques.

In both bacterial and plasmid DNA, replication is initiated at the single unique origin and proceeds in both directions along the DNA molecule to a termination site that is located on the opposite end of the DNA molecule. Thus, replication of DNA is **bidirectional** (Figure 22.2). The ability of DNA to replicate in both directions means those cells can replicate their entire DNA in a fairly short period of time. DNA strands are polymerized by prokaryotic DNA polymerases at a speed of up to 1000 bases/second/molecule of enzyme. *E. coli* has a single genome containing over 4.6×10^6 base pairs, and therefore, it takes about 40 minutes to complete total genome duplication. For eukaryotes, even

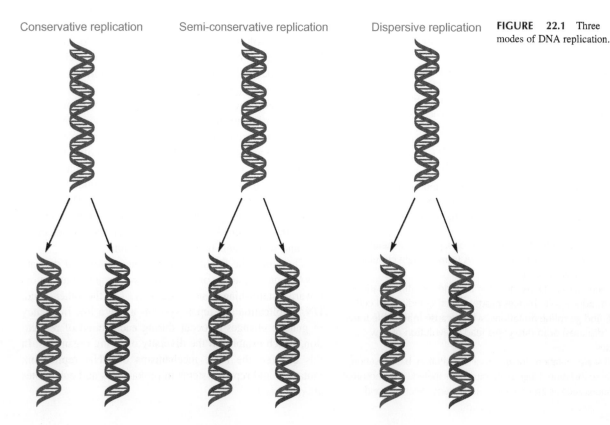

Conservative replication   Semi-conservative replication   Dispersive replication

FIGURE 22.1 Three possible modes of DNA replication.

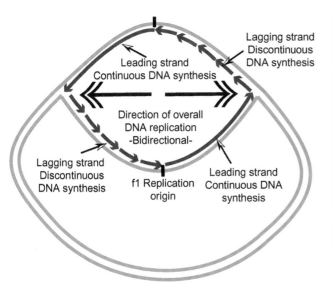

FIGURE 22.2 Bidirectional and semi-discontinuous DNA replication on a prokaryotic replication fork.

TABLE 22.1 Comparison of Parameters of DNA Replication in *E. coli* and Human Cells

DNA	*E. coli*	Human
Amount of DNA	$\sim 4.6 \times 10^6$ bp	$\sim 3.4 \times 10^9$ bp
Rate of replication at each replication fork	~ 1000 bases/second	~ 50 bases/second
Number of origins of replication	one	10^3 to 10^4
Time for cell division	~ 40 minutes	~ 24 hours

TABLE 22.2 Properties of *E. coli* DNA Polymerases

Property	Polymerase		
	I	II	III
Molecular weight	105,000	90,000	130,000
Molecules/cell	~ 400	~ 100	~ 10
Nucleotides/second	~ 20	~ 5	~ 1000
3′ exonuclease activity	yes	yes	no
5′ exonuclease activity	yes	no	no
Biological activity	RNA primer excision, DNA repair	SOS DNA repair?	Replicase

though they can begin synthesizing DNA from multiple origins, DNA polymerase processes nucleotide incorporation at a much slower speed, only up to 50 bases/second/molecule of enzyme. As the eukaryote human genome consists of approximately 3.4×10^9 base pairs, to complete DNA replication, if there were only one replication fork (replication origin), it would take approximately 40 days. However, eukaryotes finish DNA replication in several hours by employing multiple replication forks (localized regions of separated DNA strands) (Table 22.1).

DNA Replication Is Semi-Discontinuous

DNA polymerases, the enzymes that synthesize new DNA, can only add nucleotides to a 3′-OH group and therefore extend a DNA strand in the 5′ to 3′ direction. Because of the antiparallel nature of the DNA strands, at each replication fork, synthesis of one new strand of DNA (the leading strand) can be continuous, but synthesis of the other strand (the lagging strand) occurs by discontinuous replication.

At each replication fork, one DNA strand has a free 3′-OH group; the other has a free 5′-PO_4^{2-} group. Since DNA polymerase can add nucleotides only in a 5′ to 3′ direction, synthesis of the leading strand is continuous; synthesis of the lagging strand is discontinuous.

Short fragments of DNA are synthesized in the replication fork on the lagging strand in a 5′ to 3′ direction; these are called **Okazaki fragments**. These fragments are joined in the replication fork by the enzyme DNA ligase, which can form a phosphodiester bond at a single-strand break in DNA. DNA ligase joins a 3′-OH at the end of one DNA fragment to the 5′-monophosphate of the adjacent fragment (Figure 22.2).

THE ENZYMOLOGY OF DNA REPLICATION

The synthesis of DNA is a complex process because of the need for faithful replication, the necessity for high enzyme specificity, and topological constraints related to DNA coiling and interactions with other substances. Approximately 20 different enzymes are involved in bacteria to replicate DNA. In addition to polymerization reactions, DNA replication requires accurate initiation, termination, and proofreading to eliminate errors.

DNA Polymerases

Three DNA polymerases have been characterized in *E. coli* and are designated polymerase I, II, and III (Table 22.2). Although present in very low concentrations in the cell (10−20 molecules/cell for pol III vs. 400 molecules/cell for pol I), polymerase III is the polymerase that elongates both strands of the bacterial DNA in the single replication fork. Polymerase I (pol I) is primarily a DNA repair enzyme and is responsible for excision of the short RNA primer that is required to initiate DNA synthesis on both the leading and lagging strands of DNA during replication. Pol I can also remove mismatched base pairs during replication and fill in gaps in single-stranded DNA that is joined in a double helix. The function of DNA polymerase II is not clear, but it is believed to be involved in DNA repair.

All DNA polymerases select the nucleotide substrates that are complementary to the base of a template DNA strand and catalyze the formation of phosphodiester bonds to the 3'-OH end of growing chains. Therefore, the substrates for DNA polymerases are the four deoxyribonucleotides (dATP, dGTP, dCTP, and dTTP) and a single-stranded template DNA. The overall chemical reaction catalyzed by all DNA polymerases is

$$Poly(nucleotide_n) - 3'OH + dNTP \rightarrow$$
$$Poly(nucleotide_{n+1}) - 3'OH + PP_i$$

in which PP_i represents pyrophosphate cleaved from the dNTP.

No DNA polymerase can catalyze the reaction between two free nucleotides, even if one has a 3'-OH group and the other a 5'-phosphate. Polymerization can occur only if the nucleotide with the 3'-OH group is hydrogen bonded to the template strand. The oligonucleotide that is required for DNA polymerase to catalyze chain elongation is called a **primer**, which is a short ribonucleic acid (RNA) sequence. When an incoming nucleotide is joined to a primer, it supplies another free 3'-OH end, so that the growing strand itself serves as primer for the next polymerization reaction. Since polymerization occurs at only 3'-OH, DNA chain growth always proceeds in the 5'→3' direction. All known polymerases (both DNA and RNA) catalyze DNA or RNA chain growth in the 5'→3' direction.

DNA Polymerase I

DNA polymerase I is a single polypeptide chain with 928 amino acids and molecular weight of 109 kDa. It has three sites, which provide three distinct catalytic activities: 3' to 5' exonuclease, 5' to 3' exonuclease, and 5' to 3' polymerase. Occasionally, DNA polymerase adds a nucleotide to the 3'-OH terminus that can not hydrogen bond to the corresponding base in the template strand. Such a nucleotide, added in error, is removed from the primer as a result of the 3' to 5' exonuclease activity of DNA polymerase I. This exonuclease activity is called the **proofreading** or **editing** function of DNA polymerase I. Pol I also has a unique 5' to 3' exonuclease activity that is required for its DNA repair function. This activity is directed against a base-paired strand and consists of stepwise removal of nucleotides one by one from the 5'-terminus. Furthermore, the nucleotides removed can be either of the deoxyribonucleotides or ribonucleotides. The 5' to 3' exonuclease activity can be coupled to the polymerization activity to displace DNA strands. However, the main function of the 5' to 3' exonuclease activity is to remove ribonucleotide primers that are used in DNA replication.

Pol I can add nucleotides to a 3'-OH group at a single-strand (a nick) in a double helix. This activity results from the ability of pol I both to recognize a 3'-OH group anywhere in the helix and to displace the base-paired strand ahead of the available 3'-OH group. This reaction is called **strand displacement** (Figure 22.3). Under certain conditions, the displacement reaction does not occur and instead the 5' to 3' exonuclease acts on the strand that would otherwise be displaced, removing one downstream nucleotide for each nucleotide added to the 3' side of the nick. Thus, the polynucleotide of the nick moves along the strand; this reaction is called **nick translation** (Figure 22.3). It is an important laboratory procedure for producing labeled DNA that can be used as probes; simply by carrying out the reaction in the presence of radioactive- or fluorescence-labeled nucleotides, an unlabeled DNA molecule with nicks in both strands can be converted to a labeled molecule. DNA probes are used for many diagnostic and forensic purposes (Chapter 21).

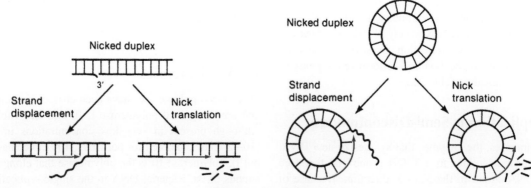

FIGURE 22.3 Strand displacement and nick translation on linear and circular DNA molecules. In nick translation, a nucleotide is removed by an exonuclease activity for each nucleotide added. The growing strand is shown in color.

DNA Polymerase III

DNA polymerase I plays an essential role in the replication process in *E. coli*, but it is not the enzyme that catalyzes the overall DNA polymerization of the replicating strands. The enzyme that accomplishes this is a less abundant enzyme: polymerase III (pol III). The simplest form of polymerase III, called the **core protein**, is 165 kD in size and consists of three polypeptide subunits, α, ε, and θ. All subunits are required for the basic DNA replication. Alpha subunits catalyze the formation of the phosphodiester bond, and the ε subunits are responsible for the 3′ to 5′ exonuclease activity and are required for proofreading. The θ subunits stabilize ε subunits to promote their exonuclease activity and are required for the high fidelity of DNA replication. The complete DNA polymerase III (holoenzyme) is the primary polymerase in *E. coli* for DNA replication; it consists of at least 10 different subunits (including those of the core protein) and provides high processing ability for the enzyme. DNA polymerase III in DNA replication forms a dimeric holoenzyme complex consisting of two identical DNA polymerase III core enzymes to direct leading-strand and lagging-strand DNA replications (Figure 22.4). Pol I and III holoenzymes are both essential for *E. coli* replication. Pol III can complete an entire strand replication of the *E. coli* genome without dissociating from the strand.

Eukaryotic DNA Polymerases

At least 19 polymerizing enzymes have been isolated from many mammalian cells (Table 22.3), and five polymerases (α, β, γ, δ, and ε) are involved in maintaining the integrity of the genome. Also, three of these polymerases (pol α, pol δ, and pol ε) function in leading-strand and lagging-strand synthesis during DNA replication. Pol α, which is a heterotetrameric enzyme, is the major polymerase of mammalian cells; it is found in the nuclei and is analogous to *E. coli* pol III. This multi-subunit enzyme has a core (4−5 subunits)

that is responsible for polymerization and a holoenzyme form that possesses additional subunits and activities. The pol α holoenzyme possesses a DNA primase activity, which synthesizes RNA primer for DNA replication. DNA pol α is associated with the initiation complex at the origin site and starts DNA replication by synthesizing a short RNA primer (10 nucleotides) and extending it up to 20−30 DNA nucleotides. This RNA−DNA hybrid fragment is further elongated by pol δ or ε on both the leading and lagging strands. Pol δ is thus the main polymerase in both the lagging and leading DNA strand synthesis in eukaryotes, analogous in function to *E. coli* pol III. Pol β is a nuclear polymerase that is involved in base excision and repair, analogous in function to *E. coli* pol I. Pol γ is found in mitochondria and is responsible for replication of mitochondrial DNA.

Other Enzymes in DNA Replication

DNA replication not only requires enzymes for adding nucleotides to the growing chains, but also requires enzymes for unwinding the parental double helix. The unwinding of the DNA helix is accomplished by **helicases**. These enzymes hydrolyze ATP and utilize the free energy of hydrolysis for unwinding the helix. As a helicase advances, it leaves single-stranded regions, and accessory proteins called **single-stranded DNA-binding proteins** (**SSB proteins**) bind tightly to these regions in order to prevent the single-stranded regions from reannealing. As a helicase creates two single-stranded regions, an RNA polymerase called **primase** binds to the regions and synthesizes small RNA primers (up to 12 bases long) for DNA polymerases to add nucleotides, since DNA polymerases can not initiate new DNA synthesis without a preceding 3′-OH group. DNA polymerases extend RNA primers only from 5′ to 3′ directions on both DNA templates (Figure 22.4). On the leading strand of DNA, DNA polymerase III continues the sequence extension to the

TABLE 22.3 Properties of Eukaryotic DNA Polymerases

Property	Polymerase				
Location in Cell	α **Nucleus**	β **Nucleus**	γ **Mitochondria**	δ **Nucleus**	ε **Nucleus**
Associated primase	Yes	No	No	No	No
3′ exonuclease	No	No	Yes	Yes	Yes
Sensitivity to aphidicolin	High	Low	Low	High	Low
Biological activity	Replication (primase activity, replication initiator)	DNA repair (Base excision repair)	Mitochondrial DNA replication	Replication (Main polymerase at the leading and lagging strand)	Replication (Leading and lagging strand)

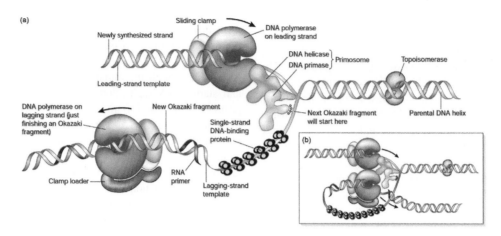

FIGURE 22.4 Bacterial DNA replication fork. (a) Simplified overview. The leading strand is initiated with an RNA primer and then elongated with a DNA polymerase, pol III. The lagging strand is synthesized in a series of fragments, each initiated by primase, which inserts short sequences of RNA primers that will be elongated by pol III. The RNA primers are removed by another DNA polymerase, pol I, and the breaks in the sugar-phosphate backbone are sealed by DNA ligase. (b) A more realistic three-dimensional illustration. *[Reproduced with permission from B. Alberts, DNA replication and recombination, Nature 421 (2003) 431–435.]*

end of the DNA sequence. However, on the lagging strand, DNA polymerase III can extend the strand only to 1000 nucleotides at one time in the direction of 5′ to 3′ from the replication forks, since the direction of the lagging-strand synthesis proceeds in the 3′ to 5′ direction (Figure 22.4). As a replication fork moves along a circular helix, coiling of the daughter molecules around one another causes the individual polynucleotide strands of the unreplicated portion of the molecule to become wound more tightly, i.e., overwound. Thus, advancement of the replication forks causes positive supercoiling (Chapter 21) of the unreplicated portion. This additional coiling can be removed by a topoisomerase that introduces negative supercoiling. There are two different types of topoisomerases, **topoisomerases I** and **II**, based on their mode of action. Topoisomerase I introduces negative supercoils by causing single-strand breaks and re-ligates strands without ATP hydrolysis. Topoisomerase II type enzymes introduce negative supercoiling by making double-strand breaks and re-ligated strands. Topoisomerase II reactions are powered by ATP hydrolysis. *E. coli* topoisomerases are called **DNA gyrases**. *In vitro* replication of circular DNA can proceed only if DNA gyrase or a similar topoisomerase is present in the reaction mixture.

Three Steps of DNA Replication: Initiation, Elongation, and Termination

Initiation

The initiation of DNA replication in *E. coli* occurs at specific sites called **replication origins**. A replication initiator protein called **DnaA** binds to the origin sequence, unwinds the AT-rich origin sequence, and provides binding space for

DnaB (DNA helicase) and **DnaC** to the site. The association of DnaB and DnaC, which is a regulator of DnaB, is required for DnaB to be loaded onto the DNA replication fork in *E. coli*. Recall that prokaryotes have circular genomic DNA and have a single origin sequence (Figure 22.5), whereas the linear genomic DNA of eukaryotes has multiple origin sequences. For example, the human genome contains up to approximately 10,000 origins. Typically, two replication complexes are formed at this single origin site in prokaryotes, and DNA replication proceeds in both directions. Therefore, two replication forks are formed that move along the DNA and make this DNA replication bidirectional.

Elongation

During the elongation stage of DNA replication, which is the addition of deoxynucleotides to the preceding 3′-OH group of RNA primers, lagging-strand and leading-strand synthesis involve different processes. For leading-strand elongation, the first RNA primer synthesized by primase at the 3′ end of the DNA template is extended by DNA polymerase III until it reaches the terminus. However, the lagging-strand elongation requires many enzymes working in coordination to extend the strand 3′ to the 5′ direction. Lagging-strand replication involves the synthesis of Okazaki fragments and their joining together by the actions of DNA polymerase I and DNA ligase. The RNA portion of Okazaki fragments is removed by the 5′ to 3′ exonuclease function of DNA polymerase I, and the deleted portions of Okazaki fragments are replaced with DNA sequences by the polymerase function of DNA polymerase I. DNA polymerase I, however, cannot join Okazaki fragments together covalently. It is DNA ligase that seals the nicks between

Okazaki fragments (Figure 22.6). The Okazaki fragments in prokaryotes are 1000–2000 bases long, whereas in eukaryotes they are only 100–200 bases long.

Termination

Termination of DNA replication in prokaryotes occurs at specific sequences called **replication termini** (*Ter*) located at the opposite end of circular genomic DNA or plasmid to the origin site. The replication terminator protein (*Tus*) binds to the *Ter* site and induces replication fork arrest, which stops DNA replication. In *E. coli*, there are 10 *Ter* sites, and binding of *Tus* proteins to these sites prevents replication forks from passing through by inhibiting helicase activity. Whereas prokaryotic cells contain circular DNA,

eukaryotic cells contain linear DNA. Therefore, termination of DNA replication in eukaryotes occurs when each replication fork meets another replication fork or reaches the linear ends of chromosomes.

Fidelity of DNA Replication

The precise copying of genetic information during DNA replication provides genetic stability to living organisms. Errors introduced during DNA replication may compromise the survival of the organism. The fidelity of DNA replication is accomplished mainly by the proofreading function of DNA polymerases. The $3'$ to $5'$ exonuclease activity of DNA polymerases is the key player in achieving accurate DNA replication. The α-subunit of *E. coli* DNA polymerase III holoenzyme catalyzes nucleotide polymerization reactions, which establishes Watson–Crick complementary base pairing to the template strand. If a wrong base is added to the growing chain of DNA, then the ϵ-subunit of DNA polymerase III, which is a $3'$ to $5'$ exonuclease, removes it and ensures that correct base pairing takes place. The accuracy of DNA polymerization during DNA replication is about one error per 10^6 nucleotide incorporations. Additionally, after the completion of DNA replication, DNA repair enzymes and systems check the DNA constantly to minimize the introduction of any errors in base pairing.

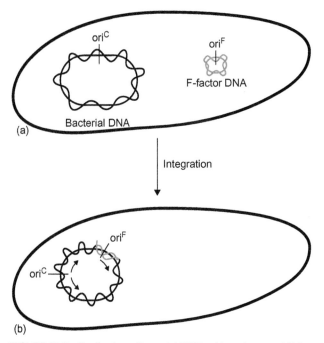

(a)

Integration

(b)

FIGURE 22.5 Replication of bacterial DNA with an integrated F factor. There are two origins of DNA replication, ori^C and ori^F. Replication is bidirectional from ori^C and unidirectional from ori^F. Somehow the cell solves this topological dilemma when cell division occurs.

Rolling Circle DNA Replication

Rolling circle DNA replication is used by bacteriophages to generate multiple copies of circular DNA. Unique features of rolling circle DNA replication include the following (1) synthesis of the primer is not necessary since template DNA gets extended by DNA polymerases; (2) the leading strand is covalently linked to the template; (3) DNA replication continues many rounds without termination, generating multiple copies of DNA, called **concatamers**; and (4) the template strand of the leading strand never separates from the circular part of the DNA molecule

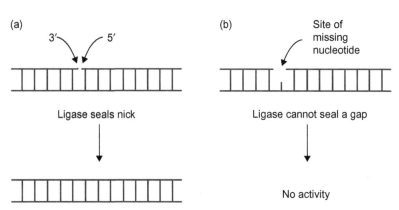

FIGURE 22.6 Action of DNA ligase. (a) A nick having a $3'$-OH and a $5'$-P terminus sealed. (b) If one or more nucleotides are absent, the gap cannot be sealed by DNA ligase.

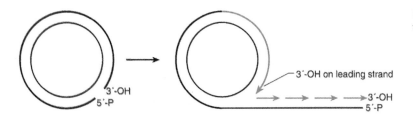

FIGURE 22.7 Rolling-circle replication. Newly synthesized DNA is shown in color.

(Figure 22.7). The newly synthesized concatameric DNA is cleaved and circularized into one unit copy of the genome for each bacteriophage offspring produced.

Reverse Transcriptase

Reverse transcriptase is an RNA-dependent DNA polymerase that was discovered in many retroviruses such as human immunodeficiency virus (HIV) and avian myeloblastosis virus (AMV) in 1970. The reverse transcriptase catalyzes the conversion of RNA template molecules into a DNA double helix and provides a very useful tool for molecular biology research. Reverse transcriptases are commonly used to produce **complementary DNA (cDNA) libraries** from various expressed mRNAs and are also used to quantify the level of mRNA synthesis when combined with the polymerase chain reaction technique, called RT-PCR. Reverse transcriptase contains three enzymatic activities: (1) RNA-dependent DNA polymerase, (2) RNase H, and (3) DNA-dependent DNA polymerase. First, RNA-dependent DNA polymerase synthesizes a DNA strand complementary to the RNA template. Then RNase H removes the RNA strand from the RNA–DNA hybrid double helix. Then the DNA-dependent DNA polymerase completes double-stranded DNA synthesis. Unlike other DNA polymerases, reverse transcriptase lacks a proofreading capability and therefore has high error rates during DNA synthesis, up to one error in 2000 base incorporations. The high error rates of viral reverse transcriptases provide selective advantage for their survival in the host system.

Eukaryotic DNA Replication

The mechanism of eukaryotic DNA replication is similar to that of prokaryotic DNA replication. However, eukaryotic DNA replication requires special consideration due to differences in DNA sizes, unique linear DNA end structures called **telomeres**, and distinctive DNA packaging that involves complexes with histones. Unlike prokaryotes, most eukaryotes are multicellular organisms, except for the unicellular eukaryotes such as yeast, flagellates, and ciliates. Therefore, DNA replication in eukaryotes is a highly regulated process and usually requires extracellular signals to coordinate the specialized cell divisions in different tissues of multicellular organisms. External signals are delivered to

cells during the G_1 phase of the cell cycle and activate the synthesis of cyclins. Cyclins form complexes with cyclin-dependent kinases (CDK), which, in turn, stimulate the synthesis of S phase proteins such as DNA polymerases and thymidylate synthase. These complexes prepare cells for DNA replication during the S phase.

Initiation of DNA replication in eukaryotes begins with the binding of the origin recognition complex (ORC) to origins of replication during the G_1 phase of the cell cycle. The ORC complex then serves as a platform for forming much more complicated pre-replicative complexes (pre-RCs). Formation of pre-RCs involves the assembly of cell division cycle 6p (Cdc6p) protein, DNA replication factor Cdt1p, mini-chromosome maintenance complex (Mcm 2p-7p), and other proteins. Pre-RCs formed during the G_1 phase are converted to the initiation complex during cell cycle transition from G_1 to S by the action of two kinases: cyclin-dependent kinase (CDK) and Dbf4-dependent kinase (DDK). Formation of an initiation complex, which includes helicase activity, unwinds the DNA double helix at the origin site (Figure 22.4). The DNA polymerase α-primase complex synthesizes the first primer. It initiates DNA replication on the leading strand and Okazaki fragments on the lagging strand. In addition to the polymerase α-primase, two DNA polymerases, δ and ε, are required for DNA replication. Polymerase δ is the major polymerase in leading-strand synthesis; polymerases δ and ε are the major polymerases in lagging-strand synthesis. This is similar to the DNA polymerase I and III in the lagging-strand synthesis of prokaryotes. In eukaryotes, Okazaki fragments generated during lagging-strand synthesis are shorter than those in *E. coli* (up to 200 bases in eukaryotes versus up to 2000 bases long in *E. coli*). Also, eukaryotic DNA replication is initiated by forming many replication forks at multiple origins to complete DNA replication in the time available during the S phase of a cell cycle.

Two key structural features of eukaryotic DNA that are different from prokaryotic DNA are the presence of histone complexes and telomere structures. Histones are responsible for the structural organization of DNA in eukaryotic chromosomes. The positive charge of histones, due to the presence of numerous lysine and arginine residues, is a major feature of the molecules, enabling them to bind the negatively charged phosphate backbones of DNA. Pairs of four different histones (H2A, H2B, H3, and H4) combine to form

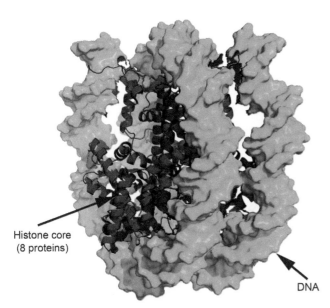

Histone core
(8 proteins)

DNA

FIGURE 22.8 Structure of a nucleosome. DNA is looped around a core of eight histone proteins (pairs of four different histone proteins) and connected to adjacent nucleosomes by linker DNA and another histone (H1, not shown).

an eight-protein bead around which DNA is wound. This bead-like structure is called a **nucleosome** (Figure 22.8). A nucleosome has a diameter of 10 nm and contains approximately 200 base pairs. Each nucleosome is linked to an adjacent one by a short segment of DNA (linker) and another histone (H1). The DNA in a nucleosome is further condensed by the formation of thicker structures called **chromatin fibers**, and ultimately DNA must be condensed to fit into the metaphase chromosome that is observed at mitosis. Despite the dense packing of DNA in chromosomes, it must be accessible to regulatory proteins during replication and gene expression. At a higher level of organization, chromosomes are divided into regions called **euchromatin** and **heterochromatin**. Transcription of genes seems to be confined to euchromatin regions, while DNA in heterochromatin regions is genetically inactive.

During DNA replication, the histone complexes of nucleosomes are separated; the leading strand retains the old histones. The lagging strand remains free of histone complexes while new histones are made and assembled. Since histones have greater affinity for double-stranded DNA, newly synthesized histone octamers are quickly added as the lagging strand is polymerized.

Since DNA in eukaryotic chromosomes is a linear molecule, problems arise when replication comes to the ends of the DNA. Synthesis of the lagging strand at each end of the DNA requires a primer so that replication can proceed in a 5′ to 3′ direction. This becomes impossible at the ends of the DNA, and the portion of RNA primer at the 5′ end of both leading and lagging strands is lost each time a chromosome is replicated. Thus, at each mitosis of a somatic cell,

the DNA in its chromosomes becomes shorter and shorter. To prevent the loss of essential genetic information during replication, the ends of DNA in chromosomes contain special structures called **telomeres**. Human telomeres are repeated end sequences of $(TTAGGG)_n$ and have typical sizes of 15–20 kb at birth. At each round of DNA replication, the telomere sequences of eukaryotic chromosomes are shortened. This is the case for normal somatic cells, and the number of DNA replications/cell divisions is linked to the timing of cell death. However, germline and cancer cells contain enzymes called **telomerases** to extend the 5′ end of lagging strands (Figure 22.9). The extension of telomere sequences by telomerases in these cells contributes to their immortality. Human telomerase is a reverse transcriptase that contains a short stretch of RNA sequence, AUCCCAAUC. This short stretch of RNA serves as a template for telomere extension and plays a major role in leading strand extension; when DNA replication is completed, telomerase binds to the 3′ end of the leading strand. This establishes base pairing with the short stretch of RNA sequence the telomerase carries and adds a 6-nucleotide sequence (GGTTAG) to the 3′ end of the strand (Figure 22.9). After leading-strand extension on the 3′ end by the telomerase is completed, DNA polymerase α completes the extension of the 5′ end of lagging strand and DNA ligase seals the nick on the lagging strand left by DNA polymerase α. Since up to 90% of tumors contain telomerases, which confer their immortality, telomerase inhibitors are being tested as a cancer therapy.

Inhibitors of DNA Replication

Inhibitors of DNA synthesis are used in the laboratory and in treatment of bacterial, viral, and neoplastic diseases. They can be divided into three main classes:

1. Those that prevent or reduce the synthesis of precursors (bases, nucleotides);
2. Those that affect either the template or the priming ability of the growing strand; and
3. Those that act directly on polymerases or other enzymes needed for replication.

Inhibitors acting on DNA polymerases and other enzymes of DNA synthesis include **acyclovir**, **aphidicolin**, **aminocoumarin**, and **camptothecins**. Acyclovir (acycloguanosine) inhibits the DNA polymerase of the *Herpes simplex* virus after conversion into acycloguanosine triphosphate. Aphidicolin reversibly inhibits eukaryotic DNA polymerases α and δ, and thereby blocks DNA replication and thus the cell cycle at the early S stage. It also inhibits viral DNA polymerases. Aminocoumarin antibiotics such as novobiocin, clorobiocin, and coumermycin inhibit DNA gyrase, which catalyzes ATP-dependent DNA supercoiling in bacteria.

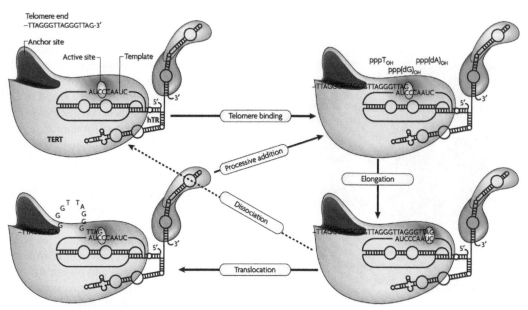

FIGURE 22.9 Steps in telomere extension by telomerase. Telomerase is a reverse transcriptase consisting of two protein components (telomerase reverse transcriptase and dyskerin) and an RNA template. First, telomerase binds to the telomere sequence at the end of chromosomes and adds six nucleotides (GGTTAG) to the telomere, which are complementary to the RNA template it carries. Next, in the translocation step, the telomerase complex moves by six nucleotides along the extended telomere sequence for another round of telomere sequence extension. The telomere synthesis terminates when the enzyme dissociates from telomere sequence. *[Reproduced with permission from C.B. Harley, Telomerase and cancer therapeutics, Nature Reviews Cancer 8 (2008) 167–179.]*

Camptothecins such as irinotecan and topotecan inhibit topoisomerase I enzymes, which catalyze ATP-independent DNA supercoiling. Camptothecins exert their inhibitory effects by trapping the topoisomerase I coupled to DNA strand breaks, which causes DNA damage in the target cells. Since normal cells can repair DNA damage, in contrast to cancer cells, which usually lack a cellular DNA repair system, camptothecins are used in the treatment of lung, ovarian, and colorectal cancers. The camptothecins are also used in the treatment of myelomonocytic syndromes, chronic myelomonocytic leukemia (CMML), acute leukemia, and multiple myeloma.

Intercalating agents such as acridines, ethidium bromide, anthracyclines, and actinomycin D induce unwinding, lengthening, and stiffening of DNA double-stranded structures. Intercalation inhibits binding of DNA-binding proteins such as DNA polymerases, RNA polymerases, topoisomerases, and other enzymes. Intercalating agents can induce multiple mutations during DNA replication and thereby inhibit cell division, and thus are lethal to actively dividing cells.

Bleomycin and **neocarzinostatin** are used as antibiotic and antineoplastic agents, since they bind tightly to DNA and cause DNA strand breakages. Bleomycin is used to treat squamous cell carcinoma, lymphomas, and testicular carcinoma. Alkyl sulphonates, nitrosoureas, and nitrogen mustards are **alkylating agents** that cross-link DNA strands,

thereby inducing DNA damage by preventing denaturation of the DNA duplex, and therefore are used as both antibiotics and anticancer drugs. **Platinum** and **gold compounds** bind to DNA and induce conformational changes that affect their functions, and thus, these metal ions are used to treat cancers.

DNA MUTATIONS AND DNA REPAIR

The genetic stability of living organisms is mainly accomplished by an accurate DNA replication system and a DNA repair system. Any change introduced into the base sequence of DNA is called a **mutation**. The most common changes are a substitution, an addition, a rearrangement, or a deletion of one or more bases. A mutation does not necessarily give rise to a mutant phenotype. Mutations introduced into somatic cells can cause disease or cell death, but these are confined to the affected individual. However, mutations introduced into germline cells can result in much more serious consequences, as they can be the causes of inheritable diseases.

Types of Mutations

There are two different categories of mutations: **large-scale** and **small-scale**. Small-scale mutations refer to a single or a few base changes, and include **base substitution**,

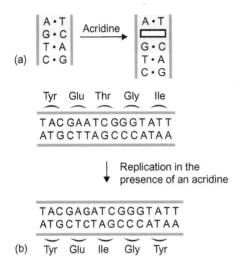

FIGURE 22.10 Frameshift mutation caused by intercalating substances (e.g., acridine). (a) Separation of two base pairs (shown as a box) by an intercalating agent. (b) A base addition resulting from replication in the presence of an acridine. The codon frame change is induced, and the change in amino acid sequence read from the upper strand in groups of three bases is also shown.

base deletion, and **base insertion**. Base substitution mutations are further divided into **transition** and **transversion** mutations, depending on the types of base changes. Purine to purine changes are called transition, and purine to pyrimidine or pyrimidine to purine mutations are called transversion. Also, depending on the consequences of a mutation, base substitutions are classified as **silent**, **missense**, and **nonsense mutations**. Silent mutation refers to base changes that do not induce amino acid sequence changes in the affected gene product. Missense mutation refers to changes that convert an amino acid codon to a different amino acid codon. Nonsense mutation or chain termination mutation refers to the base changes that result in termination of protein synthesis by introducing a new stop codon. These mutations generate one of the three stop codons: UAA, UAG, or UGA (Chapter 23). Deletion and insertion of bases cause codon frame shifts, which give rise to changes in the reading frame of the base sequence of a gene and result in the synthesis of a completely different protein. These mutations are called **frameshift mutations** (Figure 22.10). Three base insertions or deletions (whole codon) do not change codon frames for protein synthesis, but may cause disease such as Fragile X-syndrome, Huntington's disease, and myotonic dystrophy (Table 22.4).

Large-scale mutations usually refer to chromosomal mutations, which include **translocations**, **inversions**, **deletions**, and **nondisjunctions** of chromosomes. Translocations refer to the interchanges of large segments of chromosomal DNA. Inversions refer to inversion of chromosomal segments to the opposite orientation through chromosomal rearrangements. Deletions of chromosomal segments could also be caused by chromosomal rearrangements during meiosis. Nondisjunction of chromosomes occurs when pairs of chromosomes fail to separate properly during cell divisions; nondisjunctions result in genetic diseases such as Down's, Turner's, and triple-X syndromes.

Mutagen and Mutagenesis

Physical agents and chemical reagents that cause mutations are called **mutagens** and include environmental chemicals and ionizing radiations. The process that produces a mutation is called **mutagenesis**. Mutagenesis can be induced by a variety of chemicals, but also happens spontaneously as the result of faulty DNA replication and base modifications such as tautomerization and deamination of bases. A biological assay called the **Ames test** is used widely to assess mutagenic potential of chemical substances. This test is named after its developer, Bruce Ames, and is based on the assumption that any substance that is mutagenic for the test strain of bacteria may also be a carcinogen that causes cancer in eukaryotes. In the Ames test, a strain of *Salmonella typhimurium* with a mutated histidine gene, for which the amino acid histidine has to be supplied in culture media for survival, is cultured with the suspected mutagen to be tested. If the suspected mutagen reverses the mutations on the defective histidine gene and produces a normally functioning gene (back mutation), thus enabling the strain to survive in histidine-deficient culture medium, the suspected chemical compound is deemed to be a carcinogen. Although some chemicals that passed the Ames test have caused cancers in laboratory animals, its low cost and ease in testing make it an invaluable tool for screening substances in our environment for possible mutagenicity/carcinogenicity.

Types of DNA Damage

In addition to mutations, various agents and environmental factors cause damage to DNA. Examples are spontaneous depurination, base modification, interstrand cross-linking, DNA–protein cross-linking, and strand breaking. Loss of bases from a DNA double helix is called **depurination** (loss of purine bases) or **depyrimidination** (loss of pyrimidine bases); this happens because the glycosidic bonds linking deoxyribose with DNA bases are labile. If this is not corrected, the DNA strand cannot be replicated. Another type of DNA damage is **deamination** of DNA bases. Three bases (A, G, C) of DNA contain amino groups, and deamination can happen at neutral pH. Deamination of cytosine to uracil, in particular, happens at a higher rate in cells (approximately 10^2/cell/day) and

TABLE 22.4 Inherited Diseases Involving Repeated Trinucleotides that are Present in an Excessive Number of Copies*

Disease	Trinucleotide Repeat	Symptoms
Fragile X syndrome	**CGG** 5−54 unaffected 60−230 carriers 230−4000 affected	Mental retardation
Huntington's disease	**CAG** 11−30 unaffected 36−121 affected	Chorea, progressive dementia, death
Myotonic dystrophy	**CTG** 5−35 unaffected 50−4000 affected	Muscle weakness, myotonia
Spinocerebellar ataxia Type 1	**CAG** 19−36 unaffected 40−81 affected	Ataxia, dysarthria, rigidity, abnormal eye movements
Friedreich's ataxia	**GAA** 7−22 unaffected 200−900 affected	Ataxia, dysarthria, hypertrophic cardiomyopathy, diabetes
Machado−Joseph disease	**CAG** 12−37 unaffected 61−84 affected	Spasticity, dystonia

*The extent of the repeated triplet nucleotide usually determines the severity of the disease and the age of onset. These diseases may be caused by the slippage of the DNA polymerase during replication that increases the length of the repeated triplet from generation to generation. The patterns of inheritance are: Fragile X syndrome (X-linked); Huntingdon's disease; myotonic dystrophy; spinocerebellar ataxia Type 1; Friedreich's ataxia; Machado−Joseph disease.

can introduce transition mutations during subsequent DNA replication ($C \rightarrow U \rightarrow T$). Since DNA contains no U, deaminated C can be recognized by the DNA repair system and corrected before DNA replication. UV radiation from sunlight can cause two adjacent pyrimidine bases to form pyrimidine dimers in DNA, e.g., thymine dimers. Thymine dimers are the most frequently formed pyrimidine dimers caused by UV irradiation and can lead to the development of skin cancers if not corrected by the cellular repair system. Other modifications of DNA bases are caused by reactive oxygen species formed in cellular oxidative metabolism, ionizing radiation (X-rays, gamma rays), and environmental chemical substances. These DNA modifying agents also cause damage to DNA by establishing various cross-links between DNA strands and proteins, thus blocking DNA replication.

Types of DNA Repair

DNA can be damaged by external agents and by replication errors. Since maintenance of the correct base sequence of DNA and of daughter DNA molecules is essential for hereditary fidelity, repair systems have evolved that restore the correct base sequence. Most cells employ three broad types of systems to repair damage and mutations introduced by various agents:

1. Direct reversal of DNA damage;
2. Excision repair systems; and
3. Damage tolerance repair system.

Direct Reversal of DNA Damage

A pyrimidine dimer can be repaired by **photoreactivation**. Photoreactivation is a light-induced (300−600 nm) enzymatic cleavage of a thymine dimer to yield two thymine monomers. It is accomplished by **photolyase**, an enzyme that acts on dimers contained in single- and double-stranded DNA.

The enzyme−DNA complex absorbs light and uses the photon energy to cleave specific C−C bonds of the cyclobutyl thymidine dimer. Photolyase is also active against cytosine dimers and cytosine−thymine dimers, which are also formed by UV irradiation but much less frequently. Photolyases are found in bacteria, fungi, plants, and many vertebrates, but are not found in placental mammals. Another example of direct reversal of a damage repair system is **O^6-methyl-guanine-DNA methyltransferase (MGMT)**. It is an important suicide enzyme that transfers the O^6-methyl group of modified bases to its cysteine residue, which permanently inactivates the enzyme. MGMT plays a pivotal role in DNA repair against O^6-alkylating endogenous metabolites and exogenous toxic mutagens. However, MGMT in cancer cells is responsible for resistance to alkylating anticancer drug therapy targeting the O^6 position of guanine. (See Clinical Case Study 24.1.)

Excision Repairs

DNA damage that has happened to one strand of DNA can be accurately corrected by excision DNA repair systems. When the other, intact, DNA strand is used as template, the damaged sites are excised and replaced with new DNA by specific enzymes. All organisms employ at least three types of excision repair systems: mismatch repair, base excision repair, and nucleotide excision repair.

Mismatch DNA repair systems correct base mismatches introduced during DNA replication despite the polymerase's proofreading ability. In bacteria, mismatches of bases in newly synthesized DNA strands are recognized by a repair system, since parental template strands of DNA are methylated. Newly synthesized DNA strands are not methylated immediately after DNA synthesis, and therefore, a mismatched base on the unmethylated DNA strand is changed to the base complementary to the base of the methylated strand. Three key components are required for mismatch repair systems in *E. coli*: MutS, MutL, and MutH proteins. If a T−G mismatch was introduced in DNA, MutS protein scans the DNA sequence and binds to the mismatched base. Then MutH protein, which is an endonuclease, recognizes the methylated DNA sequence of GATC on the parental strand, and MutL links MutH and MutS. The linked MutH protein nicks the opposite strand of the methylated A base on the parental strand. Then, the helicase, UvrD, unwinds DNA from the nick and proceeds past the mismatched nucleotide. Exonucleases cut away the nicked GATC sequence, and single-strand binding proteins stabilize the resulting single strand. The excised DNA region is resynthesized by DNA polymerase III, and DNA ligase seals the nick to form double-stranded DNA. The end result of the mismatch repair system is a T−G mismatch correction corresponding to the methylated parental strand sequence (Figure 22.11). In eukaryotes, a DNA single base mismatch is recognized by MSH2-MSH6 heterodimer proteins, and MLH1, PMS2, and EXO1 proteins serve the function of MutL in prokaryotes. However, the counterpart of prokaryotic MutH protein has not been identified in eukaryotes.

Base excision repair (**BER**) systems handle a wide variety of individual base damage caused by oxidation,

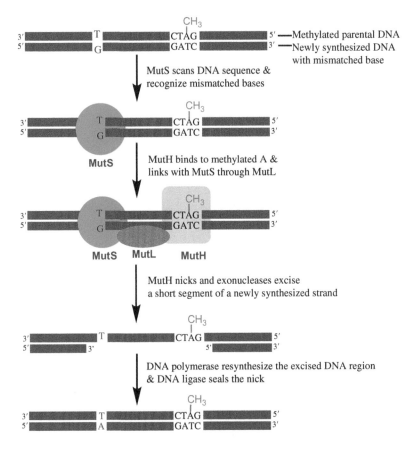

FIGURE 22.11 Prokaryotic mismatch repair. In mismatch repair, a pair of non-hydrogen-bonded bases (e.g., G:T) within a helix is recognized by MutS, and a polynucleotide segment of the daughter strand is excised, thereby removing one member of the unmatched pair. The resulting gap is filled in by pol III, and then the final seal is made by DNA ligase.

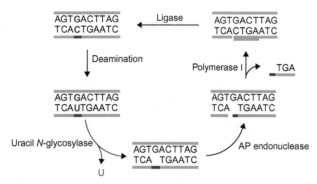

FIGURE 22.12 Scheme for base excision repair (BER). BER repairs incorrect bases (e.g., U) and damaged bases (e.g., deaminated C, methylated A).

alkylation, and deamination. Three major steps are involved in BER. First, the damaged base is recognized and removed by an appropriate DNA N-glycosylase. The DNA N-glycosylases remove the damaged bases out of the DNA strand by cutting glycosidic bonds between deoxyribose sugar and heterocyclic bases, creating an apurinic or apyrimidinic (AP) site. Next, another enzyme called AP endonuclease creates a nick by cleaving the sugar—phosphate backbone at the AP site, creating a 3'-OH terminus adjacent to the AP site. The gap at the AP site is then filled by the action of DNA polymerase and DNA ligase (Figure 22.12).

In addition to BER, all organisms adopt **nucleotide excision repair** (**NER**) systems to preserve genomic stability and overcome the consequences of DNA damaging agents. The NER system is effective in removing bulky DNA damage caused by UV light, oxidative chemicals, reactive oxygen species, cross-linking agents, and intercalating antineoplastic drugs. NER involves several steps: damage recognition, removal of damaged bases, and new DNA synthesis at the site. The *E. coli* NER system requires four proteins: UvrA, B, C, and D. UvrA dimer recognizes DNA damage and binds to the damaged site with UvrB. Then the UvrA dimer is replaced by UvrC to form a UvrBC complex. The UvrBC complex cleaves a 5th phosphodiester bond at the 3' side and an 8th phosphodiester bond at the 5' side of the damaged DNA site. Then UvrD (helicase) unwinds the DNA to release the damaged DNA strand and expose the single-stranded region. DNA polymerase I fills the excised regions, and DNA ligase seals the nick, completing the repair process (Figure 22.13). NER in eukaryotes is similar to the prokaryotic system. However, the eukaryotic NER system is much more complicated and involves more proteins. For example, the human NER system includes 17 proteins, which collaborate to repair DNA damage. The human disease xeroderma pigmentosum (XP) is caused by defects in the NER system. Investigations of XP patients have led scientists to identify the NER components: XPA to XPG. In eukaryotes, XPA protein binds to the damaged DNA site. XPB and XPD proteins are helicases to unwind the damaged DNA duplex. XPG cleaves

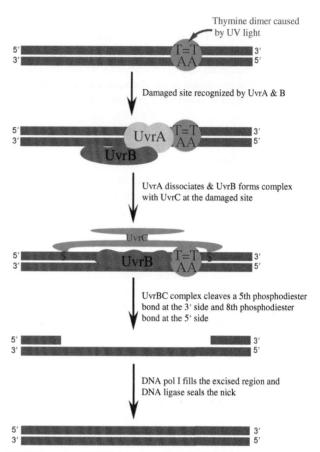

FIGURE 22.13 Nucleotide excision repair of a thymine dimer.

the 3' side of the damaged strand, and the XPF/ERCC1 complex cleaves the 5' side of the damaged strand, generating a single-stranded region of 24—32 bases covering the damaged site. The single-stranded portion is filled with correct base sequences by DNA polymerase δ/ε along with regular DNA replication complexes.

Damage Tolerance

SOS (Save Our Ship) repair includes a bypass system that allows DNA chain growth across damaged segments at the cost of fidelity of replication. It is an error-prone process; even though intact DNA strands are formed, the strands are often altered.

The SOS repair system is activated when cells are overwhelmed by UV damage. The SOS repair system causes DNA polymerases to bypass the damaged site, inserting DNA bases without proper base pairing and proofreading. At first, the UV-damaged DNA region is recognized by RecA proteins, and their binding to single DNA strands causes inactivation of LexA protein, which is a repressor protein of several SOS response genes. LexA inactivation leads to the activation of up to 43 genes, and one of these gene products causes DNA polymerase V to bypass the damaged site. As DNA polymerase bypasses the damaged site without proofreading, the introduction of mutations would occur in the repaired DNA strands. This error-prone repair is the main cause of UV-induced mutagenesis.

DNA double-strand breaks are repaired by two main mechanisms: **homologous recombination (HR)** and **nonhomologous end joining (NHEJ)**. The HR repair system is more accurate than the NHEJ system, since it uses the homologous DNA sequence on the sister chromatid to repair the damaged region. In HR repair, Rad51 protein searches the homologous copy of the damaged DNA on the sister chromatid, and the DNA damage is repaired by copying the DNA sequence from the sister chromatid (Figure 22.14). However, the NHEJ repair system does not require an identical copy of the DNA sequence to repair the DNA damage. Instead, the system joins two ends of the broken DNA double helixes by DNA ligase IV/Xrcc4 complex. First, Ku70/80 heterodimer protein recognizes double-strand breaks and recruits the protein kinase DNA-PKcs. The DNA-PKcs autophosphorylate themselves to prevent possible nuclease attack and also to promote ligase reactions; compared to HR, NHEJ is a much more error-prone repair mechanism.

Human Diseases of DNA Repair Deficiency

DNA repair systems are crucial in maintaining genomic stability, and defects in these systems result in serious disease states. One of the most studied diseases due to DNA repair defects is **xeroderma pigmentosum (XP)**. XP is caused by mutations in seven genes (XPA−XPG), all of which are involved in the nucleotide excision repair system. These genes are transmitted in an autosomal recessive manner. Affected individuals exhibit severe light sensitivity and pigmentation irregularities (see Clinical Case Study 22.1).

Ataxia telangiectasia is a rare genetic disorder of childhood transmitted in an autosomal recessive manner. This disease is characterized by neurological problems and shows an increased sensitivity to ionizing radiation such as X-rays and gamma rays, due to defects in the ATM genes located on chromosome 11q22.23.

Hereditary nonpolyposis colon cancer (HNPCC) is caused by mutations of one of five genes: hMSH2, hPMS1, MSH6, hMLH1, and hPMS2. These gene products are involved in the DNA mismatch repair system, and defects in any of these DNA repair genes predispose individuals to colon cancer, as well as to other cancers, due to microsatellite instability, which is a hallmark of HNPCC.

Fanconi's anemia (FA), a lethal aplastic anemia chiefly affecting the bone marrow, is caused by a defective DNA repair system. Genetic mutations in 16 genes are implicated in FA, and cells from affected persons cannot repair interstrand cross-links or DNA damage induced by X-rays. It is inherited as an autosomal recessive trait (see Clinical Case Study 22.2). One of the genes affected in FA is breast cancer susceptibility gene 2 (*BRCA2*), and its product, a tumor suppressor BRCA2 protein, interacts with other FA proteins in order to repair DNA cross-links. Two BRCA proteins, BRCA1 and BRCA2, are known to play major roles in homologous recombination and DNA interstrand cross-links repairs. Certain mutations in these genes are linked to the pathogenesis of breast and ovarian cancers. Mutation-screening tests on these genes are available and can be used with family history assessments to predict an increased risk of cancer development.

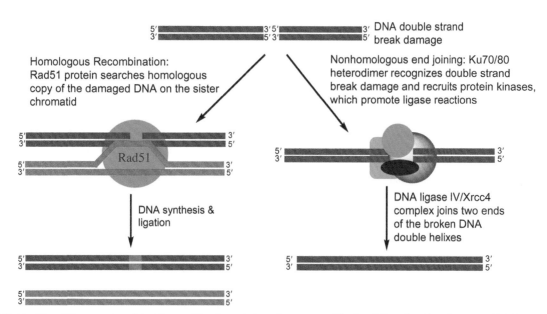

FIGURE 22.14 Two different modes of double-strand break repair: homologous recombination (HR) and nonhomologous end joining (NHEJ).

Werner syndrome is a very rare genetic disease and is inherited as an autosomal recessive trait. This disease is caused by mutations in DNA helicase genes. Affected individuals exhibit an aging rate about seven times greater than the normal rate. Because of the accelerated aging, a 10-year-old child with this disease will have similar respiratory and cardiovascular conditions to those of a 70-year-old.

Bloom's syndrome is a rare autosomal recessive disorder caused by a mutation in the gene *BLM*, which codes for DNA helicase that is involved in maintaining genomic stability. Affected individuals exhibit genetic instability in the form of increased frequencies of DNA chain breaks and interchanges.

Cockayne's syndrome is a rare autosomal recessive disease, characterized by growth retardation, large sunken eyes, and a thin prominent nose. Affected individuals exhibit a prematurely aged appearance. It is caused by mutations in the *ERCC6* and *ERCC8* genes, which are involved in the DNA repair system.

Retinoblastoma (Rb) is a rare type of cancer that develops in the retina and is caused by a mutation in the tumor-suppressor retinoblastoma (RB1) gene, which is involved in the inhibition of normal cell cycle at the first gap phase (G_1). When DNA damage is detected by certain cellular DNA sensing mechanisms, retinoblastoma proteins bind to various transcription factors and thus prevent the expression of their target genes, which is required for normal cell cycle progression. If the DNA damage is fixed, the retinoblastoma proteins will be phosphorylated by cyclin-dependent kinases (CDK), and the normal cell cycle will proceed into the synthesis (S) phase. The dysfunctions of retinoblastoma proteins result in excessive cell growth and thus lead to the development of cancerous tumors in the eye.

Certain types of **glioblastoma**, a cancerous brain tumor, are characterized by an unusual hypermethylation of the O^6-methylguanine-DNA methyltransferase (MGMT) gene promoter region. Hypermethylation of certain gene promoters results in the suppression of gene expression; therefore, inhibition of MGMT expression will lead to dysfunction of the direct reversal of the DNA damage repair system mediated by MGMT (see Clinical Case Study 24.1).

A new drug called **olaparib** (trade name Lynparza) was recently approved by the U.S. Food and Drug Administration (FDA) as a chemotherapeutic agent designed to treat advanced BRCA-mutated ovarian cancer. Olaparib is an inhibitor of **poly ADP-ribose polymerase** (PARP), an enzyme involved in DNA strand break repairs. PARP's roles in DNA repair are to recognize DNA strand breaks by binding to the damaged site, and to transfer poly ADP-ribose groups from NAD^+ to target proteins associated at the damaged site, which include histones and PARP itself. The poly ADP-ribosylation of DNA-binding proteins by PARP decreases their affinity toward DNA strands and therein relaxes chromatin assembly, leading to the repair of the damaged DNA through the recruitment of various DNA repair molecules. Conversely, the inhibition of PARP by olaparib makes DNA single-strand breaks irreparable and may lead to DNA double-strand breaks, which can be repaired by homologous recombination in normal cells. However, in BRCA-mutated ovarian cancer cells, homologous recombination is dysfunctional, and the accumulation of DNA strand breaks will lead to cancer cell death, which provides selective efficacy as a cancer treatment.

CLINICAL CASE STUDY 22.1 Xeroderma Pigmentosum: An NER Defect

This clinical case was abstracted from: S. Christen-Zach, K. Imoto, S.G. Khan, K.S. Oh, D. Tamura, J.J. Digiovanna, et al., Unexpected occurrence of xeroderma pigmentosum in an uncle and nephew, Arch. Dermatol. 145 (11) (2009) 1285–1291.

Synopsis

A 17-week-old boy showed severe sun sensitivity after being outside in a stroller for 45 minutes. The concerned parents reported that the boy's maternal uncle had been diagnosed with xeroderma pigmentosum after developing similar symptoms in infancy. Laboratory testing showed that cultured fibroblasts from the patient were hypersensitive to UV light in comparison to normal cells. Analysis of genomic DNA from the patient and his maternal uncle showed two different mutant alleles (compound heterozygote) at the XPA gene. DNA sequencing identified two mutations: a deletion of T at position 288 (c.288delT) and a deletion of 5 bases (CTTAT) (c.349_353del) in the XPA gene. DNA analysis of family members revealed that the mother and maternal grandmother of the patient carried one mutation (c.288delT), and the father and both grandfathers carried the other mutation (c.359_353del) in the XPA gene. Although the patient was not born of consanguineous marriage, both parents belong to a large close-knit community of the same ancestry.

Teaching Points

1. This case illustrates xeroderma pigmentosum, an autosomal recessive genetic disease that is characterized by severe UV light sensitivity due to nucleotide excision repair (NER) defects.
2. The NER system repairs UV-induced DNA damage caused by cross-linking of pyrimidine dimers and consists of seven XP proteins (XPA-G). Mutations in genes encoding these proteins can lead to xeroderma pigmentosum, and about 50% of all XP cases are caused by mutations in genes XPA and XPC.

References
[1] S.G. Khan, K.S. Oh, S. Emmert, K. Imoto, D. Tamura, J.J. Digiovanna, XPC initiation codon mutation in xeroderma pigmentosum patients with and without neurological symptoms, DNA Repair 8 (2009) 114–125.
[2] J.H.J. Hoeijmakers, DNA damage, age, and cancer, N. Engl. J. Med. 361 (2009) 1475–1485.

CLINICAL CASE STUDY 22.2 Fanconi's Anemia: A DNA Cross-link Repair Defect

This clinical case was abstracted from: A.X. Zhu, A.D. D'Andrea, D.V. Sahani, R.P. Hasserjian, Case 13-2006: a 50-year-old man with a painful bone mass and lesions in the liver, N. Engl. J. Med. 354 (2006) 1828–1837.

Synopsis

A 50-year-old man with a history of esophageal squamous cell carcinoma was referred to an oncologist regarding lytic bone lesions and multiple hepatic masses identified by a CT scan. A CT scan of the patient's chest, abdomen, and pelvis confirmed multiple masses in the liver, and a CT-guided, core needle biopsy of the ulnar lytic lesion showed metastatic adenocarcinoma different from the patient's previous squamous cell carcinoma. Ten years earlier, the patient had been diagnosed with Fanconi's anemia after he and his family were tested for the disease due to his brother's aplastic anemia case. Through immunohistochemistry tests for the hepatocyte-specific Hep Par 1 antigen and carcinoembryonic antigen, a diagnosis of metastatic hepatocellular carcinoma was made. The patient received chemoembolization with use of cisplatin, doxorubicin, and mitomycin in combination with ethiodized oil and trisacryl gelatin microspheres. However, 2 months later, the patient died of fever and sepsis.

Teaching Points

1. Fanconi's anemia (FA) is a rare genetic disorder following autosomal recessive inheritance. Studies have identified 16 FA genes, and mutations in these genes account for the majority (95%) of FA cases reported.

2. FA genes play a significant role in the DNA repair of interstrand cross-links caused by cross-linking agents such as X-rays, cisplatin, mitomycin, etc. Patients with defective FA genes display hypersensitivity to cross-linking agents and therefore show the two clinical characteristics: ineffective hematopoiesis and cancer susceptibility. The D1 gene, one of the 16 FA genes, is identical to the gene for *BRCA2*. Homozygous mutations in the *BRCA2* gene are linked to FA. BRCA1 and BRCA2 proteins are central to various DNA repair systems, and defects in these proteins are associated with an increased risk of breast, ovarian, and pancreatic cancers.

3. The diagnostic test for FA is the DEB test, which uses chemical cross-linking agents such as cisplatin and diepoxybutane (DEB) because cells from FA patients exhibit chromosomal instability upon exposure to these chemicals. Primary peripheral-blood lymphocytes or skin fibroblasts from the patient are used for the DEB test.

References

[1] A.R. Venkitaraman, Tracing the network connecting BRCA and Fanconi anaemia proteins, Nat. Rev. Cancer 4 (2004) 266–276.
[2] R.M. Brosh, S.B. Cantor, Molecular and cellular functions of the FANCJ DNA helicase defective in cancer and in Fanconi anemia, Front. Genet. 372 (2014) 1–14.

ENRICHMENT READING

[1] P.M.J. Burgers, Polymerase dynamics at the eukaryotic DNA replication fork, J. Biol. Chem. 284 (2009) 4041–4045.

[2] A.S. Boyer, S. Grgurevic, C. Cazaux, J.S. Hoffmann, The human specialized DNA polymerases and non-B DNA: vital relationships to preserve genome integrity, J. Mol. Biol. 425 (2013) 4767–4781.

[3] S.A. Khan, Plasmid rolling-circle replication: highlights of two decades of research, Plasmid 53 (2005) 126–136.

[4] B. Schumacher, G.A. Garinis, J.H.J. Hoeijmakers, Age to survive: DNA damage and aging, Trends Genet. 24 (2008) 77–85.

[5] P. Modrich, Mechanisms in eukaryotic mismatch repair, J. Biol. Chem. 281 (2006) 30305–30309.

[6] P.J. McKinnon, DNA repair deficiency and neurological disease, Nat. Rev. Neurosci. 10 (2009) 100–112.

[7] C. Thomas, A.V. Tulin, Poly-ADP-ribose polymerase: machinery for nuclear processes, Mol. Asp. Med. 34 (2013) 1124–1137.

[8] A.F. Swindall, J.A. Stanley, E.S. Yang, PARP-1: friend or foe of DNA damage and repair in tumorigenesis? Cancers 5 (2013) 943–958.

[9] M. Rouleau, A. Patel, M.J. Hendzel, S.H. Kaufmann, G.G. Poirier, PARP inhibition: PARP1 and beyond, Nat. Rev. Cancer 10 (2010) 293–301.

Chapter 23

RNA and Protein Synthesis

Key Points

1. Transcription is the process in which the sequence of bases in DNA is converted into a complementary sequence of bases in RNA. Translation is the process by which the codon sequence of mRNA is converted into the amino acid sequence of a protein by tRNAs, ribosomes, and many enzymes.

2. There are three major types of RNA in the cell: mRNA, tRNA, and rRNA. rRNA is the most abundant RNA in the cell and is a major component of the ribosome. rRNA provides the binding sites for mRNA and tRNA for protein synthesis.

3. RNA polymerases recognize DNA promoter sequences and initiate RNA synthesis. In prokaryotes, a single RNA polymerase is responsible for the synthesis of mRNA, tRNA, and rRNA, whereas in eukaryotes different RNA polymerases are required for synthesis of the individual RNA species. RNA polymerase I is used for rRNA synthesis, RNA polymerase II for mRNA synthesis, and RNA polymerase III for tRNA synthesis.

4. Eukaryotic mRNA synthesis involves post-transcriptional modifications such as 5′-capping, splicing, and 3′-polyadenylation; prokaryotic mRNA synthesis does not.

5. Ribosomal RNA and ribosomal proteins are assembled to form ribosomes, the required structures for the translation of mRNA codons into amino acid sequences. The prokaryotic ribosome is a 70S molecule consisting of a 50S subunit and 30S subunit. The 50S subunit contains 5S and 23S rRNAs associated with ribosomal proteins, and the 30S subunit contains 16S rRNAs associated with ribosomal proteins. The eukaryotic ribosome is an 80S molecule consisting of a 60S subunit and a 40S subunit. The 60S larger subunit contains 5S, 5.8S, and 28S rRNAs complexed with ribosomal proteins; the 40S smaller subunit consists of 18S rRNAs combined with ribosomal proteins.

6. Transfer RNAs carry specific amino acids that are determined by the tRNA anticodon sequence. The amino acid is attached to the CCA at the 3′ end of the tRNA. tRNAs also contain modified bases such as ribosylthymine, pseudouridine, and inosine that are important for their function.

7. Rifampin inhibits prokaryotic RNA polymerase, and amanitin inhibits eukaryotic RNA polymerase II. Actinomycin D blocks both eukaryotic and prokaryotic RNA synthesis by binding to the DNA template, thereby inhibiting the initiation of RNA synthesis by RNA polymerases.

8. The genetic code is universal and consists of 64 codons. The 61 codons correspond to 20 amino acids, and therefore, more than one codon is used to specify a amino acid except for tryptophan (UGG) and methionine (AUG). AUG is the start codon; and UGA, UAA, and UAG are stop codons.

9. An enzyme called aminoacyl tRNA synthetase is responsible for recognizing and attaching a specific amino acid to the tRNA carrying a corresponding anticodon. Cells have 20 different aminoacyl tRNA synthetases; each recognizes a specific amino acid and tRNA molecule.

10. Protein synthesis has three distinctive stages: polypeptide chain initiation, chain elongation, and chain termination. First, the smaller subunit of the ribosome recognizes the start codon (AUG) and with the aid of initiation factor(s) causes assembly of the initiation complex with the tRNA^{Met}, the mRNA, and the larger ribosomal subunit. Second, the tRNA binds to the A site of the ribosome, and with the help of the first elongation factor and by the action of a peptidyl transferase, peptide chain elongation takes place. This process occurs on the 23S rRNA in prokaryotes. After an amino acid is transferred from the A site tRNA to the P site tRNA, the ribosome moves a distance of three bases on the mRNA to position the next codon at the A site. This shift occurs with the aid of a second elongation factor. Finally, when translocation of the ribosome along the mRNA reaches the stop codon, no tRNA can occupy the A site and chain termination occurs. Release factors induce the release of the protein attached to tRNA from the P site of the ribosome.

11. In prokaryotes, transcription and translation are coupled, and, therefore, the rate of protein synthesis is determined by the balance between the rates of RNA synthesis and degradation. In eukaryotes, transcription and translation are not coupled, and, therefore, many reactions are required to modify RNA structure to prolong the lifespan of the mRNAs—e.g., capping and poly(A) tailing.

12. Prokaryotic protein synthesis inhibitors are used as antibiotics, since they selectively inhibit prokaryotic translation, not eukaryotic protein synthesis. These antibiotics include aminoglycosides, tetracycline, streptomycin, and erythromycin. Chloramphenicol, puromycin, and fusidic acid inhibit both eukaryotic and prokaryotic protein syntheses.

N. V. Bhagavan and C.-E. Ha: Essentials of Medical Biochemistry, Second edition. DOI: http://dx.doi.org/10.1016/B978-0-12-416687-5.00023-3

13. Collagen biosynthesis involves intracellular and extracellular modifications to produce mature, functional fibrous proteins. Intracellular modifications involve hydroxylation of proline and lysine, glycosylation of hydroxylysine, and triple helix formation. Extracellular modifications include conversion of procollagen to collagen, self-assembly into fibrils, oxidative deamination of lysyl and hydroxylysyl residues, and the formation of cross-linking between adjacent collagen molecules. Ehlers–Danlos syndrome, osteogenesis imperfecta, and scurvy are major examples of collagen disorders.

INTRODUCTION

The information for making proteins resides in the sequence of bases in the DNA. Converting the information contained in genes into proteins involves two complex processes. **Transcription** is the first step in which the sequence of bases in a gene is converted into a comple-

synthesized in them, even though both types of cells contain exactly the same genetic information. Cellular differentiation is due to differential gene expression. A tumor cell is one that has lost the ability to regulate the expression of its genetic information correctly and thus generally grows in an unregulated manner, as opposed to normal cells whose growth is regulated.

The flow of information in all cells is principally from DNA to RNA to protein. This description, formulated by Francis H. C. Crick shortly after the discovery of the structure of DNA, is known as the **central dogma** of molecular biology. Information also can flow from DNA to DNA, in both cells, and among viruses that infect cells, and from RNA to RNA during the replication of RNA viruses such as the polio virus. Permissible information transfer from RNA to DNA occurs only in the case of retroviruses such as human immunodeficiency virus (HIV). The only information transfer that is prohibited by the central dogma is from protein to RNA or to DNA. The permitted information transfers in cells (infected or uninfected) is summarized here:

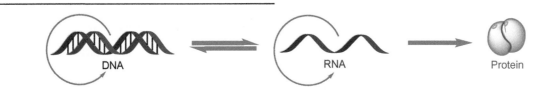

mentary sequence of bases in a molecule of RNA. Three chemically similar but functionally quite different types of RNA work together to convert the sequence of bases in DNA into the sequence of amino acids in a protein: **messenger RNA (mRNA)** carries the genetic information contained in a gene; **transfer RNA (tRNA)** and **ribosomal RNA (rRNA)** are also transcribed from genes but are used to convert the information in the sequence of bases in an mRNA molecule into the corresponding sequence of amino acids in a protein. **Translation** is the process by which an mRNA is "read" by tRNAs, ribosomes (complex structures consisting of rRNAs and ribosomal proteins), and numerous enzymes. Many other species of RNA have been identified and shown to be involved in many regulatory and catalytic roles in cells. These RNAs include small interfering RNA (siRNA), microRNA (miRNA), piwi-interacting RNA (piRNA), small nuclear RNA (snRNA), small nucleolar RNA (snoRNA), and long noncoding RNA (lncRNA).

Each type of cell is programmed to synthesize only those proteins necessary for its particular cellular functions (Chapter 24). The difference between a neuron and a liver cell is determined by the kinds of proteins that are

The synthesis of a protein in a human cell can be broadly outlined as follows. An mRNA molecule is transcribed from an antisense strand of DNA in the nucleus. The other strand of the double-stranded DNA is called "sense" because it has the identical nucleotide sequence of the mRNA, which will be translated into the amino acid sequence of a protein. The mRNA is processed by splicing out nontranslatable segments of nucleotides (**introns**) and rejoining the translatable segments (**exons**). Chemical modifications are made to both ends of the mRNA molecule, **capping** and polyadenylation to create the **poly A tail**, described later. Only the mature mRNA generated from a series of successful post-transcriptional modifications is transported into the cytoplasm. The mRNA is translated in the cytoplasm to produce the encoded protein. It is in the cytoplasm that groups of three bases in the mRNA (**codons**) are recognized by **anticodons** in the specific tRNAs that are carrying each particular amino acid.

The **genetic code** consists of 64 different codons that specify all 20 amino acids as well as codons that function to initiate and terminate translation. More than one codon may specify the same amino acid, which is called **degeneracy** of the genetic code. Finally, every organism, from bacteria to humans, uses the same codons to specify the same amino acids; this is why the genetic code is said to be universal.

Amino acids attached to their specific tRNAs are joined together in the order determined by the mRNA sequence on ribosomes with the associated enzymes, i.e., translated into a protein. Each amino acid is joined to its neighbor by a peptide bond. The specific amino acid sequence of a protein specifies its three-dimensional structure. Some proteins require the help of **chaperones** to fold into a functional configuration. When synthesis of a polypeptide chain is completed on a ribosome, it is released from the ribosome and may join with one or more similar or different polypeptides to constitute a functional protein.

STRUCTURE OF RNA

RNA is a single-stranded polyribonucleotide containing the nucleosides adenosine, guanosine, cytosine, and uridine. Unlike DNA, RNA has a 2′ hydroxyl group attached to a ribose sugar. Because of this extra hydroxyl group, RNA is less stable than DNA and more prone to chemical reactions that lead to breakdown. RNA molecules in general have a much shorter lifespan compared to that of DNA. Roughly one-third to one-half of the ribonucleotides in RNA are engaged in intrastrand hydrogen bonds, with single-stranded segments interspersed between double-stranded regions that contain up to about 30 base pairs. The base pairing produces conformations that are important to the function of the particular RNA molecules.

Ribosomal RNA (rRNA)

Ribosomal RNA is the major structural component of the ribosome and has sequence complementarity to regions of mRNA with which it can interact. Ribosomes are assembled in the **nucleolus**, a distinctive structure in the nucleus of a eukaryotic cell that serves as a factory for ribosomes, tRNAs, and various noncoding RNAs. The nucleolus is also the assembly site for telomerase and other RNA−protein complexes. Ribosomes are a complex of up to 80 ribosomal proteins, which are synthesized in the cytoplasm and transported into the nucleus for assembly, and several different ribosomal RNA molecules, three in prokaryotic ribosomes and four in eukaryotic ribosomes. For historical reasons, each class is characterized by its sedimentation coefficient, which represents a typical size. For prokaryotes, the three *Escherichia coli* rRNA molecules are used as size standards; they have sedimentation coefficients of 5S, 16S, and 23S. Eukaryotic rRNA molecules are generally larger, and there are four of them. Rat liver rRNA molecules are used as standards; the S values and the average number of nucleotides (in parentheses) are 5S (120), 5.8S (150), 18S (2100), and 28S (5050). The eukaryotic 5.8S species corresponds functionally to the prokaryotic 5S species; no prokaryotic rRNA molecule corresponds to the eukaryotic 5S rRNA. The eukaryotic 5.8S, 18S, and 28S rRNAs are products of the post-transcriptional processing of a 47S pre-rRNA molecule in the nucleolus. The 5S rRNA is produced in a different region of the nucleus, not in the nucleolus.

Transfer RNA (tRNA)

Transfer RNA molecules range in size from 73 to 93 nucleotides and have very specific secondary and tertiary structures. Each carries a covalently attached amino acid at the 3′ end and a complementary base sequence at the mRNA recognition end (anticodon) that interacts with the mRNA. Since tRNAs function as amino acid carriers, they are named by adding a superscript that designates the amino acid carried, e.g., tRNA^{Ala} for alanine tRNA. All the known tRNA molecules contain extensive double-stranded regions and form a cloverleaf structure in which open loops are connected by double-stranded stems. By careful comparison of the sequences of more than 200 different tRNA molecules, common features have been found, and a "consensus" tRNA molecule consisting of 76 nucleotides arranged in a cloverleaf form has been defined (Figure 23.1). By convention, the nucleotides are numbered 1 to 76, starting from the 5′-P terminus. The standard tRNA molecule has the following features:

1. The 5′-P terminus is always base-paired, which probably contributes to the stability of tRNA.
2. The 3′-OH terminus is always a four-base single-stranded region having the base sequence XCCA-3′-OH, in which X can be any base. This is called the **CCA** or **acceptor stem**. The adenine in the CCA sequence is the site of attachment of the amino acid to the tRNA; this is carried out by the cognate synthetase.
3. tRNA contains more than 50 different, post-transcriptionally "modified" bases. A few of these, **dihydrouridine (DHU)**, **ribosylthymine (rT)**, **pseudouridine** (ψ), and **inosine (I)** are found in particular regions and are functionally important. The precursor tRNA (pre-tRNA) molecules are produced by RNA polymerase III in eukaryotes, and certain pre-tRNA molecules contain introns that are spliced out by endonuclease and ligase. Modifications of tRNA bases happen in both the nucleus and the cytoplasm.
4. tRNA has three large single-stranded loops. The **anticodon loop** contains seven bases, 32−38. The base at position 34 is associated with wobble base pairing, base modifications at this position playing an important role in such pairing. The loop containing bases 14−21 is called the **D loop** (or the **DHU loop**); it is not constant in size in different tRNA molecules. The loop containing bases 54−60 almost always contains the sequence TψC and is called the **TψC loop**.

5. Four double-stranded regions called **stems** (or **arms**) often contain GU base pairs. The names of the stems match the corresponding loop.

6. Another loop, containing bases 44−48, is also present. In the smallest tRNAs, it contains 4 bases, whereas in the largest tRNA molecule, it contains 21 bases. This highly variable loop is called the **extra arm**.

Other Noncoding RNAs

In addition to tRNA, mRNA, and rRNA, eukaryotic cells transcribe other noncoding RNAs, which do not carry genetic information responsible for protein synthesis. These RNAs include **small nuclear RNA (snRNA)**, **small nucleolar RNA (snoRNA)**, **micro RNA (miRNA)**, **small (or short) interfering RNA (siRNA)**, **piwi-interacting RNA (piRNA)**, and **long noncoding RNA (lncRNA)**.

Small nuclear RNA molecules are synthesized by RNA polymerases II and III. These snRNAs play a major role in pre-mRNA splicing reactions as spliceosomes, which are complexes of snRNAs and other nuclear proteins (discussed later). They are found in the nucleus, and their sizes range from 100 to 300 nucleotides.

Small nucleolar RNAs are found in the nucleolar region of the nucleus and participate in rRNA processing important for ribosome assembly, e.g., methylation and pseudouridylation of bases in rRNA. snoRNAs are also produced by RNA polymerases II and III.

MicroRNAs and siRNAs are regulatory RNAs that inhibit gene expression by binding to their target mRNA sequences. Their sizes are from 20 to 30 nucleotides, and they are synthesized by RNA polymerase II. miRNAs are synthesized by processing the single-stranded 5′-capped and 3′-polyadenylated precursor miRNAs. miRNAs repress gene expression by binding to the target mRNA and thereby inhibiting protein translation. They do not have to be exactly complementary to their targets; therefore, a single miRNA may interact with many different target mRNAs. siRNAs are synthesized by processing double-stranded RNA (dsRNA) precursors, and inhibit gene expressions by binding to the specific target RNA, but with 100% accuracy in complementary base pairing. The endogenous or exogenous dsRNA activates an intracellular ribonuclease called **dicer**, which generates small dsRNA fragments (21−25 base pairs). These small dsRNA fragments (siRNA) participate in the formation of a multiprotein complex called **RNA-induced silencing complex (RISC)** that uses single-stranded siRNA (or miRNA) to recognize complementary mRNA, induce mRNA degradation, and thereby inhibit translation. siRNA repression of gene expression is thought to have been developed by cells as a defense mechanism against viral double-stranded RNA invasions.

piRNAs are small noncoding RNAs, and their size ranges from 24 to 32 nucleotides in length. The primary action of piRNA is the repression of transposons—a short

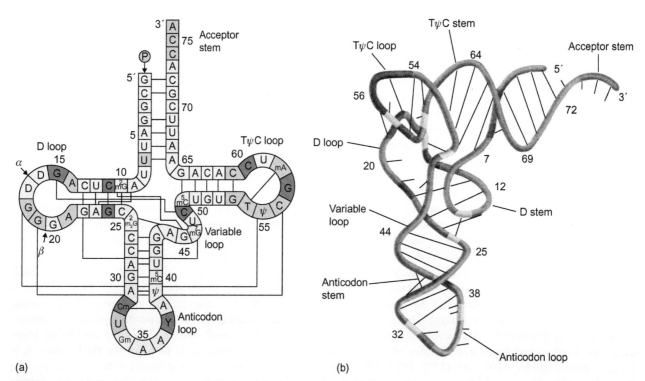

(a)   (b)

FIGURE 23.1 (a) The cloverleaf structure of tRNA with its bases numbered. (b) Schematic diagram of the three-dimensional structure of yeast tRNA^{Phe}.

DNA sequence that can jump from one location to another in the genome—to protect germline cells by forming complexes with piwi proteins. In addition to transposon silencing, piRNA/piwi protein complexes play an important role in developmental and epigenetic gene regulation by directing chromatin modification in the nucleus.

lncRNAs are more than 200 nucleotides in length, and many lncRNAs have both a 5′ cap structure and poly A tails, much like mRNA. lncRNAs are responsible for regulating a multitude of genes at different developmental stages by guiding chromatin-modifying complexes to specific target sites on chromatin.

MESSENGER RNA

Messenger RNA molecules in prokaryotic and eukaryotic cells are similar in some structural aspects, but also differ significantly. All messenger RNAs contain the same four nucleotides—A, C, G, and U—and utilize the codon AUG to initiate translation of a polypeptide and the codons UAG, UGA, and UAA to terminate translation. Prokaryotic mRNAs are polycistronic (polygenic) and usually carry information for the synthesis of several polypeptides from a single mRNA. The triplet codons in prokaryotic mRNA are transcribed from the template (antisense) strand of DNA and are subsequently translated continuously from the $5′\text{-PO}_4^{3-}$ end of the mRNA to the 3′-OH end. Since prokaryotic DNA is not separated from the cytoplasm by a nuclear membrane, translation begins on mRNA molecules before transcription is completed. Thus, transcription and translation are coupled in prokaryotes. Synthesis of each polypeptide chain in a polycistronic mRNA is determined by an AUG initiation codon and one or more stop codons that release the finished polypeptide from the ribosome.

Eukaryotic mRNAs differ from prokaryotic mRNAs in several respects (Figure 23.2). Eukaryotic genes invariably contain information for only a single polypeptide, but each gene may consist of millions of nucleotides because eukaryotic genes contain introns and exons.

The mRNA that is transcribed (primary transcript) is processed in several ways:

1. The **introns** (intervening sequences) are spliced out of the primary transcript, and the **exons** (expressed sequences) are joined together. The splicing reactions and removal of introns from the primary transcript are carried out by **small nuclear ribonucleoproteins (snRNPs)**.
2. While transcription is in process, the 5′ end of the mRNA is capped with a 7-methyl guanine nucleotide (m^7Gppp) by mRNA guanylyltransferase.
3. After the primary transcript is complete, a poly A tail ($-\text{AAA}_n\text{A}_{OH}$) is added to the 3′ terminus.
4. Other modifications of the primary transcript are possible; they include **alternative splicing**, which produces mRNAs with different sets of exons, and **RNA editing**, in which bases are modified or changed in the original transcript.
5. The functional mRNA is transported to the cytoplasm, where translation occurs on ribosomes either free in the cytosol or bound to the endoplasmic reticulum of the cell.

ENZYMATIC SYNTHESIS OF RNA

The basic chemical features of the synthesis of RNA are the following (Figure 23.3):

1. The precursors of RNA synthesis are the four ribonucleoside 5′-triphosphates (rNTPs): ATP, GTP, CTP, and UTP. The ribose portion of each NTP has an −OH group on both the 2′ and the 3′ carbon atoms.
2. In the polymerization reaction, a 3′-OH group of one nucleotide reacts with the 5′-triphosphate of a second nucleotide; a pyrophosphate is removed and a phosphodiester bond is formed. This reaction is catalyzed by RNA polymerases. There is one type of RNA polymerase present in prokaryotic cells, and three different types of RNA polymerases are present in eukaryotic cells.

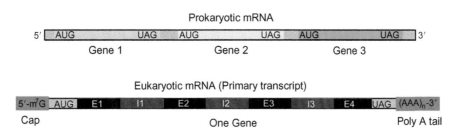

FIGURE 23.2   Structures of prokaryotic and eukaryotic primary transcripts (mRNAs). Prokaryotic mRNAs are polygenic, do not contain introns or exons, and are short-lived in the cell. Eukaryotic primary transcripts are called pre-mRNA or heterogeneous nuclear RNA (hnRNA), are monogenic, and contain introns and exons. The introns are later spliced out by splicing reactions. Eukaryotic mRNAs usually have a longer lifespan than prokaryotic mRNAs. I = intron, E = exon.

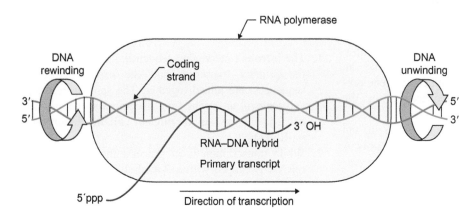

FIGURE 23.3 Model of RNA synthesis by RNA polymerase from the sense strand of DNA. See text for details.

3. The sequence of bases in an RNA molecule is determined by the base sequence of the DNA template strand. Each base added to the growing end of an RNA chain is chosen by base pairing with the appropriate base in the template strand; thus, the bases C, T, G, and A in a DNA strand cause incorporation of G, A, C, and U, respectively, in the newly synthesized RNA molecule. The RNA is complementary to the template DNA strand.

4. The RNA chain grows in the $5' \rightarrow 3'$ direction, which is the same as the direction of chain growth in DNA synthesis. The RNA strand and the DNA template strand are also antiparallel.

5. RNA polymerases, in contrast to DNA polymerases, can initiate RNA synthesis without primer.

6. Only ribonucleoside 5'-triphosphates participate in RNA synthesis, and the first base to be laid down in the initiation event is a triphosphate. Its 3'-OH group is the point of attachment of the subsequent nucleotide. Thus, the 5'-end of a growing RNA molecule terminates with a triphosphate. In tRNAs and rRNAs, and in eukaryotic mRNAs, the triphosphate group is removed.

E. coli RNA polymerase consists of five subunits: two identical α subunits and one each of β, β', and σ subunits. The total molecular weight is 465,000—it is one of the largest enzymes known. The σ subunit is easily dissociated from the enzyme and, in fact, does so shortly after polymerization is initiated. The term **core enzyme** describes the σ-free unit ($\alpha_2\beta\beta'$); the complete enzyme is called the **holoenzyme** ($\alpha_2\beta\beta'\sigma$). In this chapter, the name **RNA polymerase** is used when **holoenzyme** is meant. Several different RNA polymerases exist in eukaryotes and are described later in this chapter. An *E. coli* cell contains 3000–6000 RNA polymerase molecules. In eukaryotes, the number of RNA polymerase molecules varies significantly with cell type and is greatest in cells that actively make large quantities of protein, e.g., secretory cells.

CG	TATAAT	GTGTGG
GG	TACGAT	GTACCAC
AG	TAAGAT	ACAAATC
GT	GATAAT	GGTTGC
CT	TATAAT	GGTTAC
CG	TATGTT	GTGTGG
GC	TATGGT	TATTTC
GT	TTTCAT	GCCTCC
AG	GATACT	TACAGCC
TG	TATAAT	AGATTC
GG	CATGAT	AGCGCCC
GC	TTTAAT	GCGGTA

FIGURE 23.4 Segments of the coding strand of conserved regions from various genes showing the common sequence of six bases (Pribnow box). The start point for mRNA synthesis is indicated by the heavy letters. The "conserved T" is underlined.

PROKARYOTIC TRANSCRIPTION

The first step in prokaryotic transcription is the binding of RNA polymerase to DNA at a particular region called a **promoter**—a specific nucleotide sequence that, for different RNA classes, ranges in length from 20 to 200 nucleotides. In bacteria, a promoter is divided into subregions called the **−35 sequence** (8–10 bases long) and the −10 or **Pribnow box** (6 bases). The bases in the −35 sequence are quite variable, but the Pribnow boxes of all promoters are similar (Figure 23.4). Both the −35 sequence and the Pribnow box interact with RNA polymerase to initiate RNA synthesis.

Following RNA polymerase binding to a promoter, a conformational change occurs such that a segment of the DNA is unwound and RNA polymerase is positioned at the polymerization start site. Transcription begins as soon as the RNA polymerase−promoter complex forms and an appropriate nucleotide is bound to the enzyme.

The bactericidal drug **rifampin** binds to bacterial RNA polymerases and is a useful experimental inhibitor of initiation of transcription. It binds to the β subunit of RNA polymerase, blocking the transition from the chain initiation phase to the elongation phase; it is an inhibitor of chain initiation but not of elongation. **Actinomycin D** also inhibits initiation but does so by binding to DNA. These drugs have limited clinical use because of their toxicity.

After several nucleotides have been added to the growing chain, RNA polymerase holoenzyme changes its structure and loses the σ subunit. Thus, most elongation is carried out by the core enzyme, which moves along the DNA, binds a nucleoside triphosphate that can pair with the next DNA base, and opens the DNA helix as it moves. The open region extends over about 30 base pairs. Chain elongation does not occur at a constant rate but slows down or stops at various points along the DNA molecule. In some cases, the open region may have a regulatory function.

Termination of RNA synthesis occurs at specific base sequences within the DNA molecule. Many prokaryotic termination sequences have been determined, and most have the following three characteristics (Figure 23.5):

1. An inverted-repeat base sequence containing a central nonrepeating segment; the sequence in one DNA strand would read ABCDEF-XYZ-F′E′D′C′B′A′, in which A and A′, B and B′, and so on, are complementary bases. The RNA transcribed from this segment is capable of intrastrand base pairing, forming a stem and loop;
2. A sequence having a high G + C content;
3. A sequence of AT pairs in DNA (which may begin in the stem) that results in a sequence of 6−8 Us in the RNA.

Termination of transcription includes the following steps:

1. Cessation of RNA elongation;
2. Release of newly formed RNA; and
3. Release of the RNA polymerase from the DNA.

There are two kinds of termination events in prokaryotes: those that are self-terminating (dependent on the DNA base sequence only, termination factor independent) and those that require the presence of a termination protein called **Rho** (termination factor dependent). Both types of events occur at specific but distinct base sequences. Rho is a hexameric protein of identical subunits that bind to RNA. The action of Rho is through its helicase activity. It binds to a DNA:RNA hybrid region of a transcription bubble, the site where new RNA synthesis is taking place. By unwinding the hybrid, rho releases the newly synthesized RNA from the bubble and thereby terminates the transcription. In rho independent termination, inverted repeat DNA sequences are copied to RNA sequences, which then form hairpin loop structures due to the inverted repeat sequences. These hairpin loops cause RNA polymerase to stop the polymerization and terminate the transcription process. The transcription cycle of *E. coli* RNA polymerase is shown in Figure 23.6.

Various drugs inhibit chain elongation. **Cordycepin** (3′-deoxyadenosine) is converted to a 5′-triphosphate form and then acts as a substrate analogue. It blocks chain elongation since it lacks the required 3′-OH group.

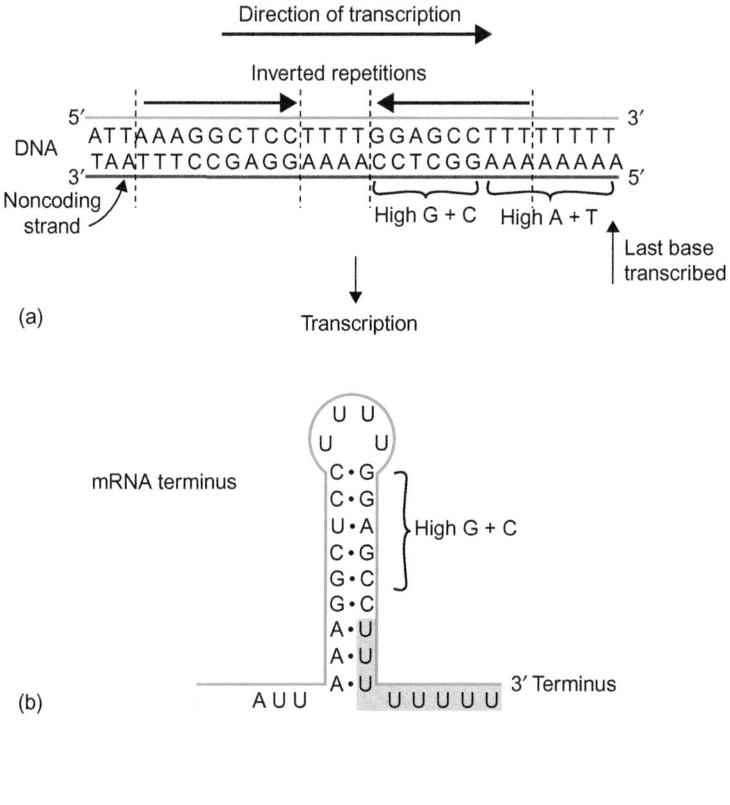

FIGURE 23.5 Factor-independent termination of transcription. Base sequence of (a) the DNA of the *E. coli* trp operon at which transcription termination occurs, and (b) the 3′ terminus of the mRNA molecule. The inverted-repeat sequence is indicated by reversed arrows. The mRNA molecule is folded to form a stem-and-loop structure.

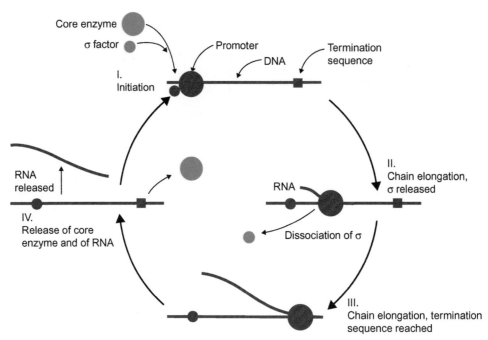

FIGURE 23.6 Transcription cycle of *E. coli* RNA polymerase showing dissociation of the σ subunit shortly after chain elongation begins; dissociation of the core enzyme during termination; and reformation of the holoenzyme from the core enzyme and the σ subunit. A previously joined core enzyme and a subunit will rarely become rejoined; instead, reassociation occurs at random.

Lifetime of Prokaryotic mRNA

All mRNA molecules are subject to degradation by RNases, which is an essential aspect of the regulation of gene expression. Proteins are not made when they are not needed, and the rate of protein synthesis is determined by a balance between the rates of RNA synthesis and degradation. The half-life of a typical prokaryotic mRNA molecule is only a few minutes, so constant production of a bacterial protein requires continued transcription. In contrast, eukaryotic mRNA molecules have a lifetime of hours to days, because they have a protective cap and poly A tails. Also, bacteria must adapt to a rapidly changing environment, whereas multicellular eukaryotic cells receive a constant supply of nutrients that maintain a uniform environment.

TRANSCRIPTION IN EUKARYOTES

The chemistry of transcription in eukaryotes is the same as in prokaryotes. However, the promoter structure and the mechanism for initiation are strikingly different.

Eukaryotic RNA Polymerases

Eukaryotic cells contain three classes of RNA polymerases, denoted I, II, and III. The RNA polymerases are distinguished by their requirements for particular ions and by their sensitivity to various toxins. All are found in the nucleus. Minor RNA polymerases are found in mitochondria and

chloroplasts. Polymerase I molecules are located in the nucleolus and are responsible for synthesis of 5.8S, 18S, and 28S rRNA molecules. Polymerase II synthesizes all RNA molecules destined to become mRNA and various noncoding RNAs such as snoRNA, miRNA, siRNA, snRNA, and lncRNAs. Polymerase III synthesizes 5S rRNA, tRNAs, and some snRNAs. Polymerases II and III are inhibited by **amanitin**, the toxic product of *Amanita phalloides* mushrooms, and are identified by their sensitivity to this substance. **Rifampin**, a powerful inhibitor of bacterial RNA polymerase, is inactive against the eukaryotic nuclear polymerases but inhibits mitochondrial RNA polymerases, although it requires higher concentrations than those needed to inhibit bacterial polymerases. Rifampin is used with other drugs (e.g., isoniazid) in treating tuberculosis. **Actinomycin D** is a general inhibitor of eukaryotic transcription by virtue of its binding to DNA. It has been useful in the treatment of childhood neoplasms (e.g., Wilms' tumor) and choriocarcinoma. However, it inhibits rapidly proliferating cells of both normal and neoplastic origin and hence produces toxic side effects.

RNA Polymerase II Promoters

The structure of eukaryotic promoters is more complex than that of prokaryotic promoters. DNA sequences, located hundreds of base pairs (bp) upstream of or downstream from the transcription start site, control the rate of initiation. Furthermore, initiation requires numerous specific proteins (**transcription factors**) that bind to particular DNA sequences. Without the transcription factors, RNA polymerase II cannot bind to a promoter. However, RNA polymerase II itself is not a transcription factor. The complexity of initiation may derive, in part, from the fact that eukaryotic DNA is in the form of chromatin, which is inaccessible to RNA polymerases. To promote transcription in eukaryotes, various transcription factors interact with the chromatin structure and induce structural changes in it that enable RNA polymerases to initiate transcription. The transcription factors form complexes and interact with specific DNA sequences such as promoters, enhancers, and RNA polymerases to mediate the initiation of transcription.

Many RNA polymerase II promoters have the following features (Figure 23.7):

1. The initiator sequence (Inr), YYANWYY, encompasses the transcription start site. The Inr is the most commonly found promoter in genes and provides a binding site for transcription factor II D, along with the TATA box.
2. A sequence, TATAWAAR, about 30 bp upstream of the transcription start site (TSS), is known as the **TATA box** and is the binding site for the TATA box binding protein of TFIID.
3. The B recognition element (BRE) has the consensus sequence SSRCGC and is located immediately upstream of the TATA box at 35 bp upstream of the TSS. The BRE provides a binding site for transcription factor II B (TFIIB).
4. The downstream core promoter element (DPE) has the consensus sequence RGWYVT and is located 30 bp downstream from the TSS. The DPE provides a binding site for TFIID.
5. A common sequence is present in the −75 region, GGCCAATCT, called the **CAAT box**. The CAAT box provides a binding site for CAAT enhancer-binding proteins (C/EBP). Another element recognized in a few promoters is the **GC box**, GGGCGG, which is located in the −90 region and serves as a binding site for the Sp1 transcription factor.

Note: The term *upstream* refers to regions in the DNA that are to the left (or 5′) of the transcription start site of a gene; the term *downstream* refers to regions to the right (or 3′) of the gene. Regulatory regions affecting gene expression may be located close to the transcription start site, thousands of bases or more upstream of or downstream from the gene, or in introns.

According to the IUPAC nucleotide code, Y = C or T, W = A or T, V = A or C or G, S = G or C, R = A or G, N = any base.

RNA Polymerase III Promoters

RNA polymerase III promoters differ significantly from RNA polymerase II promoters in that they are located downstream from the transcription start site and within the transcribed segment of the DNA. For example, in the 5S RNA gene of the South African toad (*Xenopus laevis*), the promoter is between 45 and 95 nucleotides downstream from the start point. Thus, the binding sites on RNA polymerase III are reversed with respect to the transcription direction, as compared with RNA polymerase II. That is, RNA polymerase II reaches forward to find the start point, and RNA polymerase III reaches backward. In fact, RNA polymerases can slide in either direction along a DNA template; however, they can only synthesize RNA molecules in a 5′ → 3′ direction.

Eukaryotic mRNA Synthesis

Eukaryotic mRNA used in protein synthesis is usually about one-tenth the size of the primary DNA transcript. This size reduction results from excision of the noncoding regions, **introns**. After excision, the coding fragments are rejoined by RNA splicing enzymes (Figure 23.8).

Capping and Polyadenylation

A **5′ cap** is added to the 5′ end of newly synthesized mRNA using the modified nucleotide, 7-methylguanosine. Capping occurs shortly after initiation of synthesis of the mRNA and precedes other modifications that protect the mRNA from degradation by RNases. The poly(A) terminus is synthesized by a nuclear enzyme, poly(A) polymerase. However,

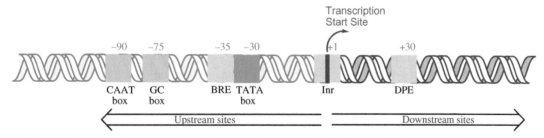

FIGURE 23.7 Sequences found in and near some RNA polymerase II promoters. Only the TATA box is represented in many promoters (some promoters lack this sequence). The CAAT box occurs much less frequently, and the GC box has only occasionally been observed. Upstream sites are very common but are not considered to be part of the promoter.

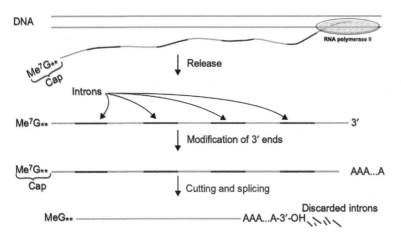

FIGURE 23.8 Schematic drawing showing production of eukaryotic mRNA. The primary transcript is capped before it is released. Then its 3′-OH end is modified, and finally the intervening regions are excised. Me^{7}G = 7-methylguanosine. **Two nucleotides whose riboses may be methylated.

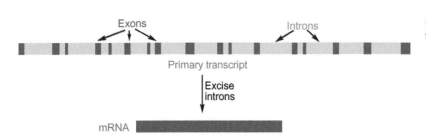

FIGURE 23.9 Diagram of the conalbumin primary transcript and the processed mRNA.

adenylate residues are not added directly to the 3′ terminus of the primary transcript. Instead, RNA polymerase transcribes past the recognition site for addition of poly(A). After synthesis of the complete RNA, endonucleolytic cleavage occurs at the poly(A) recognition site in the RNA, and poly(A) is added. A sequence, AAUAAA, located 10−25 bases upstream of the poly(A) addition site, is required for enzyme recognition of the site of polyadenylylation.

Splicing of RNA in Eukaryotes

Unmodified RNA molecules synthesized first from the template DNA strand are called **precursor mRNA (pre-mRNA)** or **heterogeneous nuclear RNA (hnRNA)**. This mRNA contains untranslated intervening sequences (**introns**). Introns interrupt the coding sequence and are excised from the pre-mRNA. In the processing of pre-mRNA in higher eukaryotes, the amount of discarded RNA ranges from 30% to nearly 90% of the pre-mRNA. The remaining coding segments (**exons**) are joined together by splicing enzymes to form translatable mRNA molecules. The excision of the introns and the formation of the final mRNA molecule by joining of the exons are called **RNA splicing** and occur in the nucleus like other pre-mRNA modifications, such as capping and poly(A) tailing. The 5′ segment (the cap) of the pre-mRNA is never discarded and hence is always present in the completely processed mRNA molecule; the 3′ segment is also usually retained.

Thus, the number of exons is usually one more than the number of introns. The number of introns per gene varies considerably. Furthermore, within different genes, the introns are distributed differently and have many sizes (Figure 23.9). Introns are usually longer than exons.

Base sequence studies of the regions adjacent to several hundred different introns indicate that common sequences can be found at each end of an intron. The sites where cutting occurs are always 5′ to GU and 3′ to AG. The rule is that the base sequence of an intron **begins with GU and ends with AG**. In addition to the GU−AG sequence, another important site called the **branch sequence** is located in the middle of the introns; it is required in the splicing reaction.

Introns are excised one by one in a specific order determined by the conformation of pre-mRNA, and ligation (splicing) occurs before the next intron is excised; thus, the number of different nuclear RNA molecules present at any instant is huge. Translation does not occur until processing is complete.

Spliceosome and Ribozymes

Different snRNPs are found in eukaryotic cells that function in removing introns from pre-mRNA. The association of small RNAs, nuclear proteins, and the introns that they attach to is referred to as a **spliceosome**. Small nuclear RNAs (snRNAs, designated U1 through U6) provide specificity to different spliceosomes so that they recognize different classes of introns through base pairing.

Introns are differentiat according to their three-dimensional structures, and each class of intron is spliced out by a different mechanism.

The splicing reaction is remarkably precise: cuts are made at unique positions in transcripts that contain thousands of bases. The fidelity of the excision and splicing reaction is extraordinary, for if an error of even one base were made, the correct reading frame would be destroyed. Such fidelity is achieved by recognition of particular sequences by snRNAs of spliceosomes. Small nuclear RNAs are usually associated with proteins and therefore called **small nuclear RNA protein particles (snRNP)**. U1 snRNP of the spliceosome recognizes the 5′-GU splicing sequence of introns and initiates splicing by base pairing. The 2′-OH group of adenine from the branch site of the intron attacks the G phosphodiester bond at the 5′ splice site of an intron. This forms a lariat structure and exposes the 3′-OH group of the exon. The 3′-OH of the exon then attacks the 5′-phosphate group of first guanine base of the next exon, displacing the 3′ end of the intron lariat structure and results in two ligated exons. The whole splicing process is orchestrated on the spliceosome complex (Figure 23.9). After completion of the splicing reactions, mRNA is transported out to the cytoplasm by shuttle proteins.

There are two types of introns, group I and II introns, that remove themselves from pre-mRNA by a self-splicing reaction without help of spliceosomes.

Group I introns were originally discovered in ciliated protozoa and subsequently were found in fungi, bacteriophages, and some other organisms. The RNA itself in a group I intron has catalytic activity, but it is not a true enzyme because it functions only once. The nucleotides in the intron that is spliced out are recycled in the cell.

Group II introns are removed from RNA by a self-splicing reaction that proceeds through an intermediate structure called a **lariat**. The removal of group II introns from RNAs also results in the splicing together of exons on either side of the intron. The ability of group II introns to specifically bind to a 5′ exon has led to their use as reagents to construct novel RNA molecules. Chemical derivatives of group II introns have been constructed that can carry out the reverse of the splicing reaction. When these introns insert themselves into RNA, they can be used to shuffle sequences or to link one RNA molecule to another.

Group II introns, which are found in bacteria, plant organelles, yeast, and fungi, also are capable of reintegrating into DNA after being excised from an RNA molecule. Group II introns encode a multifunctional intron-encoded protein (IEP) that has reverse transcriptase activity, RNA splicing activity, and DNA endonuclease activity. These three enzymatic activities allow the intron to be excised from RNA, copied into an RNA−DNA heteroduplex, and inserted into DNA at specific target sites recognized by the IEP−intron complex. Recently, group II introns have been modified so that they can be targeted to specific genes in any DNA molecule. This novel molecular mechanism raises the possibility that suitably engineered group II introns can be used therapeutically in gene therapy and possibly to inactivate viruses such as HIV by integration and disruption of essential genes.

Ribonuclease P (**RNase P**) consists of both a protein and an RNA component that has catalytic activity. RNase P functions in eukaryotic cells to process the 5′ end of precursor tRNA molecules. RNase P can be directed to cleave any RNA molecule when the target is complexed with a short complementary oligonucleotide called an **external guide**.

Ribozymes refer to catalytic RNA molecules that recognize specific target sequences in other RNA molecules. This activity and specificity make ribozymes potentially useful therapeutic agents. For example, a ribozyme could be directed to silence the expression of a deleterious gene by destroying the mRNA transcript before it can be expressed. Alternatively, ribozymes might be constructed that would inactivate the mRNA or oncogenes or the expression of genes in RNA viruses such as HIV. Laboratory studies have supported the feasibility of ribozyme therapy, but its application in clinical practice is still far in the future.

Alternative Splicing

Different splicing of the same pre-mRNA can lead to different mRNA and therefore result in the synthesis of different proteins. This type of RNA splicing is called **alternative splicing**. An example of alternative splicing in eukaryotic cells is the sex determination mechanism of *Drosophila melanogaster*. In *D. melanogaster*, development as a female requires the expression of a gene called *sxl* (sex lethal). In the female fly, synthesis of Sxl protein, the *sxl* gene product, is activated. Sxl protein, which is an RNA-binding, splicing-regulatory protein, activates female-specific protein syntheses, such as transformer 1 and 2 proteins (Tra1 and Tra2) and female-specific doublesex (Dsx) proteins, by alternative splicing. These proteins, in turn, suppress expression of male genes and activate female differentiation. In the absence of Sxl protein in the male fly, these female-specific proteins are expressed as truncated, nonfunctional proteins through different splicing patterns.

RNA Editing

Besides alternative splicing, organisms can synthesize different proteins from the same gene by modifying bases in the RNA through the process called **RNA editing**. An example of RNA editing in humans is apo B-48 synthesis in intestinal cells. Apolipoprotein B-48 (apo B-48) is a key protein component of chylomicrons, which are assembled in the small intestine to transport exogenous cholesterols and the triacylglycerols (also called **triglycerides**).

Apo B-48 is the product of the apo B gene. In the liver, this gene is transcribed into mRNA, which is responsible for the apo B-100 protein consisting of 4564 amino acids. Apo B-100 is essential for the transport of endogenous cholesterol and triacylglycerols in the blood. However, in the small intestine, the same size mRNA is used to produce apo B-48 protein consisting of 2153 amino acids (48% of apo B-100, thus named as apo B-48). In the intestine, an additional step in pre-mRNA processing (in addition to capping, splicing, and polyadenylation) takes place: enzymatic modification of the C nucleotide in codon 2153. A zinc-dependent enzyme called **cytidine deaminase** present in the intestine is responsible for the conversion of glutamine codon (CAA) to a stop codon (UAA) through deamination of cytosine.

GENETIC CODE

"Universal" Genetic Code

Production of proteins from mRNA requires translation of a base sequence into an amino acid sequence. The collection of base sequences (codons) that correspond to each amino acid and signal for initiation and termination of translation is the genetic code. The code consists of

64 triplets of bases (Table 23.1). The codons are written with the 5′ terminus of the codon at the left. The following features of the code should be noted:

1. Sixty-one codons correspond to 20 different amino acids.
2. The codon AUG has two functions. It corresponds to the amino acid methionine when AUG occurs within a coding sequence in the mRNA, i.e., within a polypeptide chain. It also serves as a signal to initiate polypeptide synthesis with methionine for eukaryotic cells, but with N-formylmethionine for prokaryotic cells. How the protein synthesizing system distinguishes an initiating AUG from an internal AUG is discussed later. The codon GUG also has the same two functions, but it is only rarely used in initiation. Once initiation has occurred at an AUG codon, the reading frame is established and the subsequent codons are translated in order.
3. Three codons—UAA, UAG, and UGA—do not represent an amino acid but serve as signals to terminate the growing polypeptide chain. The codon UGA is also used to code for selenocysteine in the synthesis of selenoproteins. When the element selenium is not present in the cell during selenoprotein synthesis, UGA codon serves as a stop codon.

TABLE 23.1 Universal Genetic Code

First Position (5′ end)	Second Position				Third Position (3′ end)
	U	C	A	G	
U	Phe	Ser	Tyr	Cys	U
	Phe	Ser	Tyr	Cys	C
	Leu	Ser	Stop	Stop	A
	Leu	Ser	Stop	Trp	G
C	Leu	Pro	His	Arg	U
	Leu	Pro	His	Arg	C
	Leu	Pro	Gln	Arg	A
	Leu	Pro	Gln	Arg	G
A	Ile	Thr	Asn	Ser	U
	Ile	Thr	Asn	Ser	C
	Ile	Thr	Lys	Arg	A
	Met	Thr	Lys	Arg	G
G	Val	Ala	Asp	Gly	U
	Val	Ala	Asp	Gly	C
	Val	Ala	Glu	Gly	A
	Val	Ala	Glu	Gly	G

The boxed codons are used for initiation. GUG is very rare.

4. Except for methionine (AUG) and tryptophan (UGG), most amino acids are represented by more than one codon. The assignment of codons is not random; with the exception of serine, leucine, and arginine, all synonyms (codons corresponding to the same amino acid) differ by only the third base. For example, GGU, GGC, GGA, and GGG all code for glycine.

The fidelity of translation is determined by two features of the system:

1. Attachment of the correct amino acid to a particular tRNA molecule by the corresponding aminoacyl synthetase; and
2. Correct codon—anticodon pairing.

The former results from the specificity of interaction of the enzyme, the amino acid, and the tRNA molecule; the latter is ensured by base pairing.

A striking aspect of the code is that, with few exceptions, the third base in a codon appears to be unimportant; i.e., XYA, XYB, XYC, and XYD are usually synonymous. This finding has been explained by the base-pairing properties of the anticodon—codon interaction. In DNA, no pairs other than GC and AT are possible because the regular helical structure of double-stranded DNA imposes steric constraints. However, since the anticodon is located within a single-stranded RNA loop, the codon—anticodon interaction does not require formation of a structure with the usual dimensions of a double helix. Molecular modeling of the interaction indicates that the steric requirements are less stringent at the third position of the codon, a feature called **wobble**. The wobble hypothesis allows for inosine (I), a nucleoside in which the base is hypoxanthine and which is often found in anticodons, to base-pair with A, U, or C. In addition, the wobble hypothesis allows for U to base-pair with G. This explains how a tRNA molecule carrying a particular amino acid can respond to several different codons. For example, two major species of yeast tRNA^{Ala} are known; one responds to the codons GCU, GCC, and GCA and has the anticodon IGC. Recall that the convention for naming the codon and the anticodon always has the 5′ end at the left. Thus, the codon 5′-GCU-3′ is matched by the anticodon 5′-IGC-3′. The other tRNA^{Ala} has the anticodon CGC and responds only to GCG.

The function of the three stop codons derives from the fact that no tRNA exists that has an anticodon which can pair with the stop codons except for the codon UGA, which is sometimes used for selenoprotein synthesis under certain conditions.

Genetic Code of Mitochondria

The genetic code of mitochondria is not the same as the "universal" code and thus has implications for the evolution of mitochondria. Human mitochondria contain a set of tRNA molecules that are not found elsewhere in the cell and a circular DNA molecule containing 16,569 base pairs. This DNA molecule encodes some mitochondrial enzymes (Chapter 13) and is the template for synthesis of all mitochondrial tRNA and rRNA molecules. Human mitochondrial DNA sequences contain the genes for 12S and 16S ribosomal RNA, 22 different tRNA molecules, three subunits of the enzyme cytochrome oxidase (whose amino acid sequence is known), cytochrome b, and several other enzymes.

The differences between the universal code and mitochondrial code are striking in that most are in the initiation and termination codons. That is, in mammalian mitochondria,

1. UGA codes for tryptophan and not for termination.
2. AGA and AGG are termination codons rather than codons for arginine.
3. AUA and AUU are initiation codons, as is AUG. Both AUA and AUG also code for methionine. AUU also codes for isoleucine, as in the universal code.
4. AUA codes for methionine (and initiation, as shown in item 3) instead of isoleucine.

The number of mammalian mitochondrial tRNA molecules is 22, fewer than the minimum number (32) needed to translate the universal code. This is possible because in each of the four-fold redundant sets (e.g., the four alanine codons GCU, GCC, GCA, and GCG), only one tRNA molecule (rather than two, as explained previously) is used. In each set of four tRNA molecules, the base in the wobble position of the anticodon is U or a modified U (not I). It is not yet known whether this U is base-paired in the codon—anticodon interaction or manages to pair weakly with each of the four possible bases. For those codon sets that are doubly redundant (e.g., the two histidine codons CAU and CAC), the wobble base always forms a G·U pair, as in the universal code. The structure of the human mitochondrial tRNA molecule is also different from that of the standard tRNA molecule [except for mitochondrial tRNA^{Leu}UUX (X = any nucleotide)]. Numerous diseases and clinical abnormalities are caused by mitochondrial mutations discussed in Chapter 13.

ATTACHMENT OF AMINO ACID TO tRNA MOLECULE

When an amino acid is covalently linked to a tRNA molecule, the tRNA is said to be **aminoacylated** or **charged**. The notation for a tRNA charged with serine is seryl tRNA (tRNA^{Ser}). The term **uncharged tRNA** refers to a tRNA molecule lacking an amino acid.

Aminoacylate is accomplished in two steps, both of which are catalyzed by an aminoacyl tRNA synthetase.

In the first (or activation) step, an aminoacyl AMP is generated in a reaction between an amino acid and ATP:

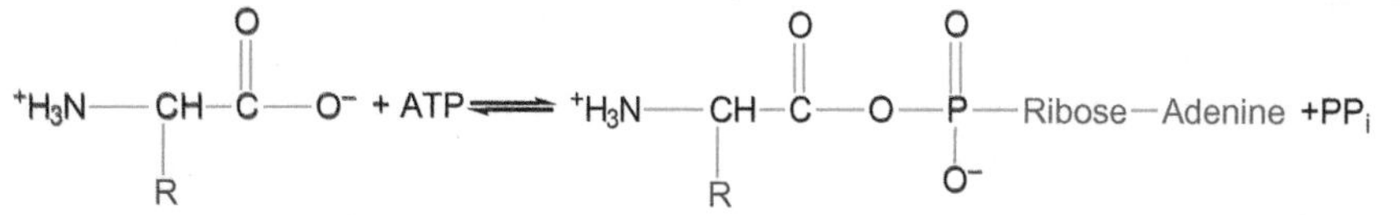

In the second (or transfer) step, the aminoacyl AMP reacts with tRNA to form an aminoacylated tRNA and AMP:

Aminoacyl AMP + tRNA $\rightleftharpoons$ aminoacyl tRNA + AMP

As in DNA synthesis, the reaction is driven to the right by hydrolysis of pyrophosphate, so the overall reaction is

Amino acid + ATP + tRNA + H$_2$O → aminoacyl tRNA + AMP + 2P$_i$

Usually, only one aminoacyl synthetase exists for each amino acid. However, for a few amino acids specified by more than one codon, more than one synthetase does exist.

INITIATOR tRNA MOLECULES AND SELECTION OF INITIATION CODON

The initiator tRNA molecule in prokaryotes, tRNA^{fMet}, has several properties that distinguish it from all other tRNA molecules. One feature is that the tRNA^{fMet} is first aminoacylated with methionine, and then the methionine is modified. Aminoacylation is by methionyl tRNA synthetase, which also charges tRNA^{fMet}. However, the methionine of charged tRNA^{fMet} is immediately recognized by another enzyme, tRNA methionyl transformylase, which transfers a formyl group from N^{10}-formyltetrahydrofolate (fTHF) to the amino group of the methionine to form N-formylmethionine (fMet) (as follows).

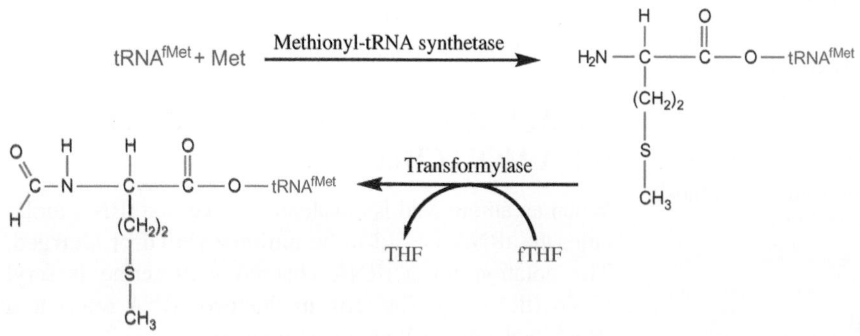

Transformylase does not formylate methionyl tRNA^{Met} because tRNA^{Met} and tRNA^{fMet} are structurally different.

Eukaryotic initiator tRNA molecules differ from prokaryotic initiator molecules in that the methionine does not undergo formylation. In eukaryotes, the first amino acid in a growing polypeptide chain is Met, not fMet. The codon for both kinds of tRNA molecules in eukaryotes is AUG, just as for prokaryotes.

The sequence AUG serves as both an initiation codon and the codon for methionine. Since methionine occurs within protein chains, some signal in the base sequence of the mRNA must identify particular AUG triplets as start codons. In eukaryotes, initiation usually begins at the triplet nearest to the 5′ terminus cap of the mRNA molecule; i.e., no particular signal is used (although sequences do play a role in the efficiency of initiation). Presumably, the ribosome, initiation factors, and the mRNA interact at the 5′ terminus, and the 40S ribosomal subunit scans down the mRNA sequence until an AUG is encountered. This AUG determines the reading frame. When a stop codon is encountered, the ribosome and the mRNA dissociate. Only one unique polypeptide is translated from a particular eukaryotic mRNA molecule.

In prokaryotes, the situation is quite different. At a fixed distance upstream from each AUG codon used for initiation is the sequence AGGAGGU (or a single-base variant). This sequence, known as the **Shine−Dalgarno sequence**, is complementary to a portion of the 3′-terminal sequence of one of the RNA molecules in the ribosome (in particular, the 16S rRNA in the 30S ribosome subunit). Base pairing between these complementary sequences orients the initiating AUG codon on the ribosome. Thus, initiation in prokaryotes is restricted to an AUG on the 3′ side of the Shine−Dalgarno sequence.

Many prokaryotic mRNA molecules are polycistronic and contain coding sequences for several polypeptides. Thus, a polycistronic mRNA molecule must possess a series of start and stop codons for each polypeptide. The mechanism for initiating synthesis of the first protein molecule in a polycistronic mRNA is the same as that in a monocistronic mRNA. However, if a second protein in a polycistronic mRNA is to be made, protein synthesis must reinitiate after termination of the first protein. This is usually accomplished by the start codon of the second protein being so near the preceding stop codon that reinitiation occurs before the ribosome and the mRNA dissociate. Otherwise, a second initiation signal is needed.

RIBOSOMES

A ribosome is a multicomponent structure that serves to bring together a single mRNA molecule and charged tRNA molecules so that the base sequence of the mRNA molecule is translated into an amino acid sequence. The ribosome also contains several enzymatic activities needed for protein synthesis. The protein composition of the prokaryotic ribosome is well known, but less is known about the composition of eukaryotic ribosomes. However, both share the same general structure and organization. The properties of the bacterial ribosomes are constant over a wide range of species, and the *E. coli* ribosome has been analyzed in great detail and serves as a model for all ribosomes.

Chemical Composition of Prokaryotic Ribosomes

A prokaryotic ribosome consists of two subunits. The intact particle is called a **70S ribosome** because of its S (sedimentation) value of 70 Svedberg units. The subunits, which are unequal in size and composition, are termed 30S and 50S subunits. The 70S ribosome is the form active in

protein synthesis. At low concentrations of Mg^{2+}, ribosomes dissociate into ribosomal subunits. At even lower concentrations, the subunits dissociate, releasing individual RNA and protein molecules. The composition of each subunit is as follows:

- 30S subunit: one 16S rRNA molecule + 21 different proteins
- 50S subunit: one 5S rRNA molecule + one 23S rRNA molecule + 34 different proteins

The proteins of the 30S subunit are termed S1, S2, ..., S21 (in this context, S is for "small" subunit); one copy of each is contained in the 30S subunit. The proteins of the 50S subunit are denoted by L (for "large" subunit). Each 50S subunit contains one copy of each protein molecule, except for proteins L7 and L12, of which there are four copies. One protein, L26, is not considered to be a true component of the 50S subunit. Most ribosomal proteins are very basic proteins (positively charged), containing up to 34% basic amino acids. This basicity probably accounts in part for their strong association with the RNA, which is acidic (negatively charged). The amino acid sequences of most of the *E. coli* ribosomal proteins are known. A summary of the structure of the *E. coli* 70S ribosome is shown in Figure 23.10.

Ribosomes are Ribozymes

The ribosomal proteins provide the structural framework for the 23S (28S in eukaryotes) rRNA that actually carries out the **peptidyltransferase** reaction. Thus, the 23S rRNA is an enzyme, and the ribosome is a **ribozyme**. Although the activity of the 23S rRNA is dependent on the ribosomal proteins, the 50S subunit that is substantially deproteinized can still perform the peptidyltransferase reaction.

A crucial feature of the peptidyltransferase reaction is a particular adenine that is conserved in rRNAs extracted

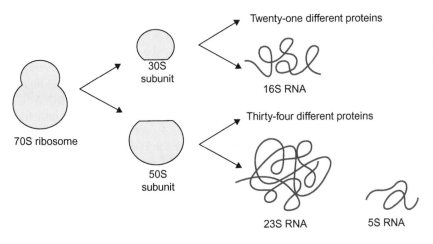

FIGURE 23.10 Dissociation of a prokaryotic ribosome. The configuration of two overlapping circles is used throughout this chapter, for the sake of simplicity.

Twenty-one different proteins

16S RNA

30S subunit

70S ribosome

Thirty-four different proteins

50S subunit

23S RNA

5S RNA

from the large ribosomal subunits of thousands of different organisms from all three kingdoms. Sequence analysis shows that this adenine base ($_{2451}A$) is always present in the active site of the ribosome and acts as a general acid−base catalyst by deprotonating the nucleophilic amine, as follows:

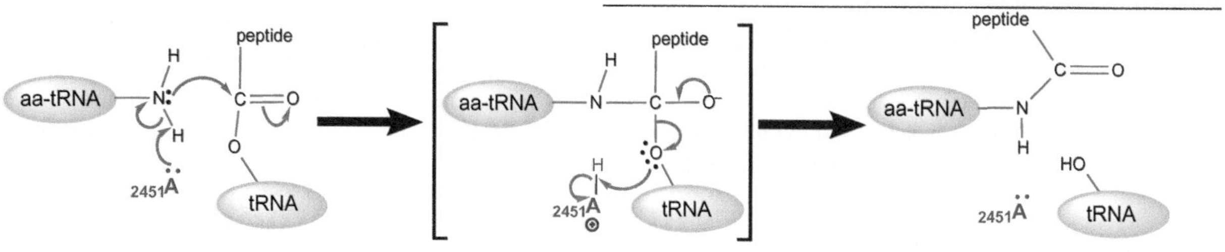

The positive charge on $_{2451}A$ in the tetrahedral intermediate is probably transferred to adjacent tRNA nucleotides that undergo tautomeric shifts. The evidence that ribosomal RNAs perform enzymatic functions critical to protein synthesis lends weight to the evolutionary significance of RNA in primordial chemical reactions that eventually led to the development of cells.

Chemical Composition of Eukaryotic Ribosomes

The basic features of eukaryotic ribosomes are similar to those of bacterial ribosomes, except for their larger size. They contain a greater number of proteins (about 80) and an additional RNA molecule. A typical eukaryotic ribosome has a sedimentation coefficient of about 80S and consists of two subunits: 40S and 60S. These sizes may vary by as much as 10% from one organism to another, in contrast to the homogeneity of bacterial ribosomes. The components of the subunits of eukaryotic ribosomes are as follows:

40S subunit: one 18S rRNA molecule + 30−35 proteins
60S subunit: one 5S, one 5.8S, and one 28S rRNA molecule + 45−50 proteins

PROTEIN SYNTHESIS

Protein synthesis has three stages:

1. Polypeptide chain initiation;
2. Chain elongation; and
3. Chain termination.

The main features of the initiation step are binding of mRNA to the ribosome, selection of the initiation codon, and binding of the charged tRNA bearing the first amino acid. In the elongation stage, there are two steps: joining of adjacent amino acids by peptide bond formation and moving the mRNA and the ribosome with respect to one another so that codons are translated successively (translocation). In the termination stage, the completed polypeptide dissociates from the ribosomes, which are then available to begin another cycle of synthesis.

Every polypeptide has an amino terminus and a carboxyl terminus. In both prokaryotes and eukaryotes, synthesis begins at the amino terminus. For a protein having the sequence H_2N−Met−Trp−Asp . . . Pro−Val−COOH, the fMet (or Met) is the initiating amino acid, and Val is the last amino acid added to the chain. Translation of mRNA molecules occurs in the $5' \rightarrow 3'$ direction.

Stages of Protein Synthesis

The mechanisms of protein synthesis in prokaryotes and eukaryotes differ slightly in detail, but the prokaryotic mechanism is used as a general model:

1. *Initiation.* Protein synthesis in bacteria begins by the association of one 30S subunit (not the 70S ribosome), an mRNA, a charged tRNA^{fMet}, three protein initiation factors (IF-1, 2, 3), and guanosine $5'$-triphosphate (GTP). These molecules make up the 30S pre-initiation complex. Association occurs at an initiator AUG codon, whose selection was described earlier. A 50S subunit joins the 30S subunit to form a 70S initiation complex (Figure 23.11). This joining process requires hydrolysis of the GTP contained in the 30S pre-initiation complex. There are two tRNA binding sites on the 50S subunit: the **aminoacyl (A) site** and the **peptidyl (P) site**. Each site consists of several L proteins and 23S rRNA. The 50S subunit is positioned in the 70S initiation complex such that the tRNA^{fMet}, which was previously bound to the 30S pre-initiation complex, occupies the P site of the 50S subunit. Positioning tRNA^{fMet} in the P site fixes the position of the anticodon of tRNA^{fMet} so that it can pair with the initiator codon in the mRNA. In this way, the reading frame is unambiguously defined.

2. *Binding of a second tRNA to the A site.* The A site of the 70S initiation complex is available to any tRNA

molecule whose anticodon can pair with the codon adjacent to the AUG complex initiation codon. However, entry to the A site by the tRNA requires the activity of a protein called **elongation factor EF−Tu** (Figure 23.12). EF−Tu binds GTP and an aminoacyl tRNA, forming a ternary complex, aminoacyl-tRNA-EF−Tu−GTP. This ternary complex enters the A site for the codon-dependent placement of the aminoacyl tRNA at the A site. During this process of placement,

GTP is hydrolyzed to GDP and P^i, and EF−Tu is released. EF−Tu−GTP is regenerated from EF−Tu−GDP through the interaction of another elongation factor protein, EF−Ts, and GTP.

3. *Chain elongation (formation of the first peptide bond).* After a charged tRNA is positioned in the A site, a peptide bond between fMet and the amino acid in the A site forms. The peptide bond is formed by the peptidyltransferase activity, whose active site resides in the

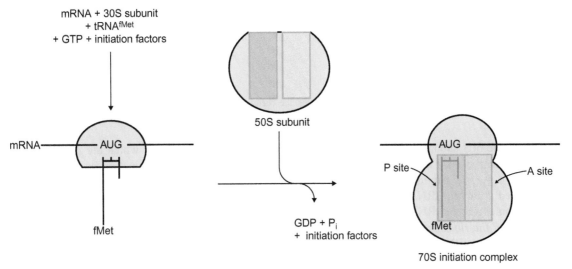

FIGURE 23.11 Early steps in protein synthesis: in prokaryotes, formation of the 30S pre-initiation complex and 70S initiation complex.

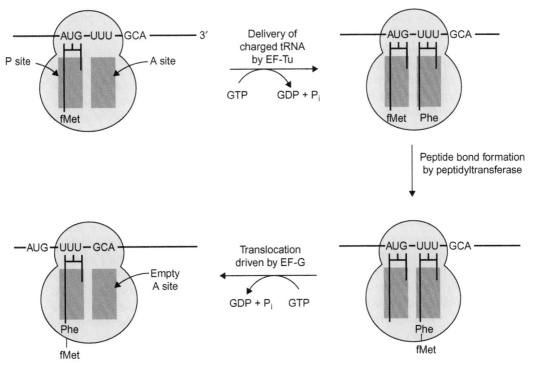

FIGURE 23.12 Elongation phase of protein synthesis: binding of charged tRNA, peptide bond formation, and translocation. Note that this phase consumes 2 GTP during each cycle of chain elongation.

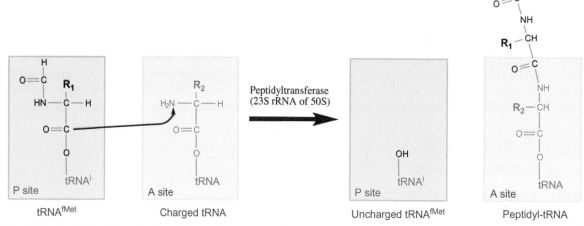

FIGURE 23.13 Peptide bond formation in prokaryotes. In eukaryotes, the chemistry is the same, although the initiating tRNA is not tRNA^{fMet} but tRNA$^{Met}_{init}$.

23S rRNA of the 50S subunit (28S rRNA of the 60S subunit in eukaryotes), as previously discussed (Figure 23.13). As the peptide bond is formed, the Met is cleaved from the tRNA^{fMet} in the P site by another ribosomal protein, tRNA deacylase.

4. *Translocation of the ribosome.* After the peptide bond forms, an uncharged tRNA occupies the P site, and a dipeptidyl tRNA is in the A site. At this point three events, which together make up the translocation step, occur: the deacylated tRNA^{fMet} leaves the P site; the peptidyl tRNA moves from the A site to the P site; and the ribosome moves a distance of three bases on the mRNA to position the next codon at the A site. The translocation step requires the activity of another elongation protein, EF—G, and hydrolysis of GTP to provide the energy to move the ribosome. Thus, the total amount of energy expended in the synthesis of one peptide bond comes from the hydrolysis of four high-energy phosphate bonds (equal to about 30 kcal = 7.5 × 4). Recall that the synthesis of each molecule of aminoacyl tRNA consumes two high-energy phosphate bonds (ATP→AMP + PP^i). (In these calculations, the one GTP molecule expended in the formation of the initiation complex is not included because of its small contribution to the overall synthesis of a polypeptide.)

5. *Refilling of the A site.* After translocation has occurred, the A site is again available for a charged tRNA molecule with a correct anticodon. When this occurs, the series of elongation reactions is repeated.

6. *Termination.* When a chain termination codon is reached, no aminoacyl tRNA is available that can fill the A site, so chain elongation stops. Since the polypeptide chain is still attached to the tRNA occupying the P site, release of the protein is accomplished by

release factors (RFs), which are proteins that, in part, respond to chain termination codons. In the presence of release factors, peptidyltransferase catalyzes the reaction of the bound peptidyl moiety with water rather than with a free aminoacyl tRNA (Figure 23.14). Thus, the polypeptide chain, which has been held in the ribosome solely by the interaction with the tRNA in the P site, is released from the ribosome.

7. *Dissociation of the 70S ribosome.* The 70S ribosome dissociates into 30S and 50S subunits, which may start synthesis of another polypeptide chain.

Role of GTP

GTP, like ATP, is an energy-rich molecule. Generally, when such molecules are hydrolyzed, the free energy of hydrolysis is used to drive reactions that otherwise are energetically unfavorable. This does not seem to be the case in protein synthesis. The reaction sequence indicates that GTP facilitates binding of protein factors either to tRNA or to the ribosome. Furthermore, hydrolysis of GTP to GDP always precedes dissociation of the bound factor. Comparison of the structure of the free factor and the factor—GTP complex indicates that the factor undergoes a slight change in conformation when GTP is bound. The function of GTP is to induce a conformational change in a macromolecule by binding to it. Since it is easily hydrolyzed by various GTPases, the use of GTP as a controlling element allows cyclic variation in macromolecular shape. When GTP is bound, the macromolecule has an active conformation, and when the GTP is hydrolyzed or removed, the molecule resumes its inactive form. GTP plays a similar role in hormone activation systems (Chapter 28).

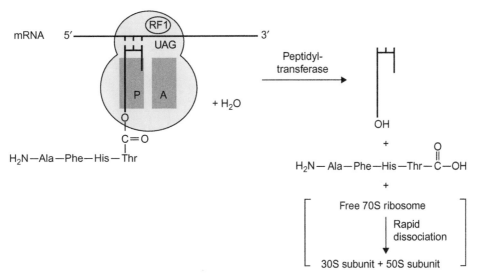

FIGURE 23.14 The steps of termination of protein synthesis.

Post-Translational Modification of Proteins

The protein molecule ultimately needed by a cell often differs from the just synthesized polypeptide chain. Modification of the synthesized chain occurs in several ways:

1. In prokaryotes, fMet is rarely retained as the NH_2-terminal amino acid. In roughly half of all proteins, the formyl group is removed by the enzyme **deformylase**, leaving methionine as the NH_2 terminal amino acid. In prokaryotes and eukaryotes, the fMet, methionine, and possibly a few more amino acids are often removed. Their removal is catalyzed by a hydrolytic enzyme: an aminopeptidase. This hydrolysis may occur as the chain is being synthesized or after the chain is released from the ribosome. The choice of deformylation versus removal of fMet usually depends on the identity of the adjacent amino acids. Deformylation predominates if the second amino acid is arginine, asparagine, aspartic acid, glutamic acid, isoleucine, or lysine, whereas fMet is usually removed if the adjacent amino acid is alanine, glycine, proline, threonine, or valine.

2. Newly created NH_2 terminal amino acids are sometimes acetylated, and amino acid side chains may also be modified. For example, in collagen, a large fraction of the proline and lysine residues are hydroxylated. Phosphorylation of serine, tyrosine, and threonine occurs in many proteins. Various sugars may be attached to the free hydroxyl group of serine or threonine to form glycoproteins or to the amido group of asparagine. Finally, a variety of prosthetic groups such as heme and biotin are covalently attached to some proteins.

3. Two distant sulfhydryl groups in two cysteines may be oxidized to form a disulfide bond.

4. Polypeptide chains may be cleaved at specific sites. For instance, chymotrypsinogen is cleaved to the digestive enzyme chymotrypsin by removal of four amino acids from two different sites. In some cases, the uncleaved chain represents a storage form of the protein that can be cleaved to generate the active protein when needed. This is true of many mammalian digestive enzymes; e.g., pepsin is formed by cleavage of pepsinogen. An interesting precursor is a huge protein synthesized in animal cells infected with some viruses; this viral protein molecule is cleaved at several sites to yield different active proteins and hence is called a **polyprotein**. Some cellular enzymatic systems are also formed by cleavage of polyproteins.

Coupled Transcription and Translation

After about 25 amino acids have been joined together in a polypeptide chain in prokaryotes, the AUG initiation site of the mRNA molecule becomes exposed, and a second initiation complex then forms. The overall configuration is of two 70S ribosomes moving along the mRNA at the same rate. When the second ribosome has moved a distance similar to that traversed by the first, a third ribosome is able to attach. This process of movement and reinitiation continue until the mRNA is covered with ribosomes separated by about 80 nucleotides. This large translation unit is called a **polyribosome** or **polysome** (Figure 23.15). Polysomes greatly increase the overall rate of protein synthesis, since 10 ribosomes traversing a single segment of mRNA clearly can

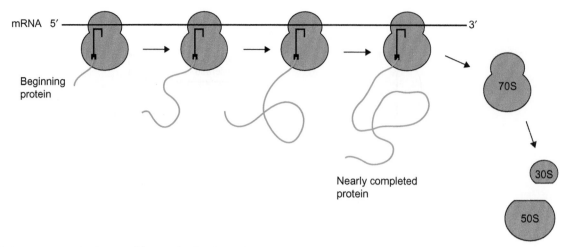

FIGURE 23.15 Bacterial polysomes. Diagram showing the relative movement of the 70S ribosome and the mRNA, and growth of the protein chain.

make 10 times as many polypeptides per unit of time as can a single ribosome. In prokaryotes, transcription and translation are coupled; this cannot occur in eukaryotes, since transcription occurs in the nucleus and translation in the cytoplasm.

Endoplasmic Reticulum

In most eukaryotic cells, two major classes of ribosomes exist: attached ribosomes and free ribosomes. The attached ribosomes are bound to an extensive cytoplasmic network of lipoprotein membranes called the **endoplasmic reticulum**. The **rough endoplasmic reticulum** consists of bound ribosomes; the **smooth endoplasmic reticulum** is devoid of ribosomes. There is no structural difference between a free and an attached ribosome, and attachment of ribosomes to the membrane occurs after synthesis of particular proteins begins.

Most endoplasmic reticulum membranes enclose large, irregularly shaped, discrete regions of the cell called **cisternae**. In this sense, the membrane system has an inside and an outside, and ribosomes are bound only to the outside. Cells responsible for secreting large amounts of a particular protein (e.g., hormone-secreting cells) have an extensive endoplasmic reticulum. Most proteins destined to be secreted by the cell or to be stored in intracellular vesicles, such as lysosomes (which contain degradative enzymes) and peroxisomes (which contain enzymes for eliminating hydrogen peroxide), are synthesized by attached ribosomes. These proteins are primarily found in the cisternae of the endoplasmic reticulum. In contrast, most proteins destined to be free in the cytoplasm are made on free ribosomes.

The formation of the rough endoplasmic reticulum and the secretion of newly synthesized proteins through the membrane are explained by the signal hypothesis (Figure 23.16). The basic idea is that the signal for attachment of the ribosome to the membrane is a sequence of very hydrophobic amino acids near the amino terminus of the growing polypeptide chain. When protein synthesis begins, the ribosome is free. Then the amino terminal hydrophobic amino acids (called **signal peptide**) interact with the **signal recognition particle** (SRP), and the signal peptide—SRP complex directs the association of the large ribosomal subunit to SRP receptor protein on the membrane surface. As protein synthesis continues, the protein moves through the membrane to the cisternal side of the endoplasmic reticulum. A specific protease termed the **signal peptidase** cleaves the amino terminal signal sequence.

Intracellular compartmentation of newly synthesized proteins is a complex process, and disorders in this process lead to severe abnormalities (discussed later). In summary, the following events take place. When the predominant portion of the protein is synthesized and has migrated in the endoplasmic reticulum, the signal sequence is removed by proteolytic cleavage. The finished protein, which is sequestered on the cisternal side of the endoplasmic reticulum, undergoes many post-translational modifications, one of which is the addition of a series of carbohydrate residues to form glycoproteins (Chapter 9). Glycoproteins are transported to the Golgi complex, where further modification of the carbohydrate residues occurs. The finished glycoproteins are then packaged into lysosomes, peroxisomes, or secretory vesicles. Secretory vesicles fuse with the plasma membrane, discharging their contents into the extracellular fluid.

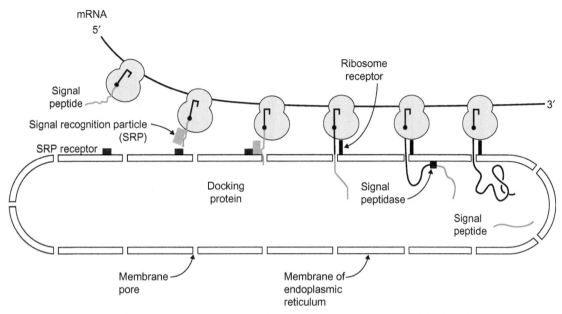

FIGURE 23.16 Signal hypothesis for the synthesis of secretory and membrane proteins. Shortly after initiation of protein synthesis, the amino-terminal sequence of the polypeptide chain binds a signal recognition protein, which then binds to a docking protein. The signal peptide is released from the signal recognition protein as the ribosome binds to a ribosome receptor, which is adjacent to a pore. Translation continues with the signal peptide passing through the pore. Once through the endoplasmic reticulum, the signal sequence is excised by the signal peptidase within the vesicle. When protein synthesis is completed, the protein remains within the vesicle, and the ribosome is released.

Defects in compartmentation processes may lead to mislocation of the proteins and cause deleterious effects. Many proteins that are destined for various organelles (e.g., mitochondria) are not routed through the endoplasmic reticulum and Golgi complex. These proteins are made on cytosolic polysomes. They possess presequences at amino terminal ends that target them to receptors on appropriate organelles. Import of proteins into cell organelles is aided by **chaperone proteins** (Chapter 4).

Compartment Disorders

Two defects of compartmentation are known. One, **mucolipidosis** or **I-cell disease** (Chapter 10), is characterized by mislocalization of lysosomal enzymes (acid hydrolases). These glycoproteins lack mannose-6-phosphate residues, and consequently they are secreted into extracellular fluids instead of being sequestered into lysosomal vesicles. The mannose-6-phosphate serves as a marker that allows the enzymes to bind with mannose-6-phosphate receptors that direct the enzymes into the vesicles to be packaged into lysosomes. The lysosomal enzymes in the extracellular fluid bring about indiscriminate destruction of tissues.

The second example of an compartmentation defect is α_1-**antitrypsin deficiency** (also known as an α_1-proteinase

inhibitor deficiency). The defect in this disorder is almost exactly the opposite of that in I-cell disease; i.e., I-cell disease is a disorder of a defect in (intracellular) retention of proteins, and α_1 antitrypsin deficiency is a defect in the secretion of a protein. α_1-Antitrypsin (α_1-AT) consists of a single polypeptide chain of 394 amino acid residues with three oligosaccharide side chains, all of which are attached to asparagine residues. It is synthesized in hepatocytes and secreted into the plasma, where it has a half-life of 6 days. α_1-AT is a defense protein, and its synthesis and release into the circulation increase in response to trauma and inflammatory stimulus. Thus, it is known as an **acute-phase reactant**. α_1-AT functions by forming a tight complex with the active site of the target enzyme that inhibits enzyme's proteolytic activity. The enzyme—inhibitor complexes are rapidly cleared by the reticuloendothelial cells. Thus, α_1-AT is a suicide protein. α_1-AT is a broad-spectrum serine proteinase inhibitor (Chapter 6). Its principal action is to inhibit leukocyte (neutrophil) elastase. This function appears to be most important in maintaining the integrity of the elastic fibers for the elastic recoil of normal lung tissue. The turnover rate of mature elastin is extremely low, and if it is destroyed, replacement is severely limited. Thus, the risk of development of a lung disease (**emphysema**) characterized by dilatation of air

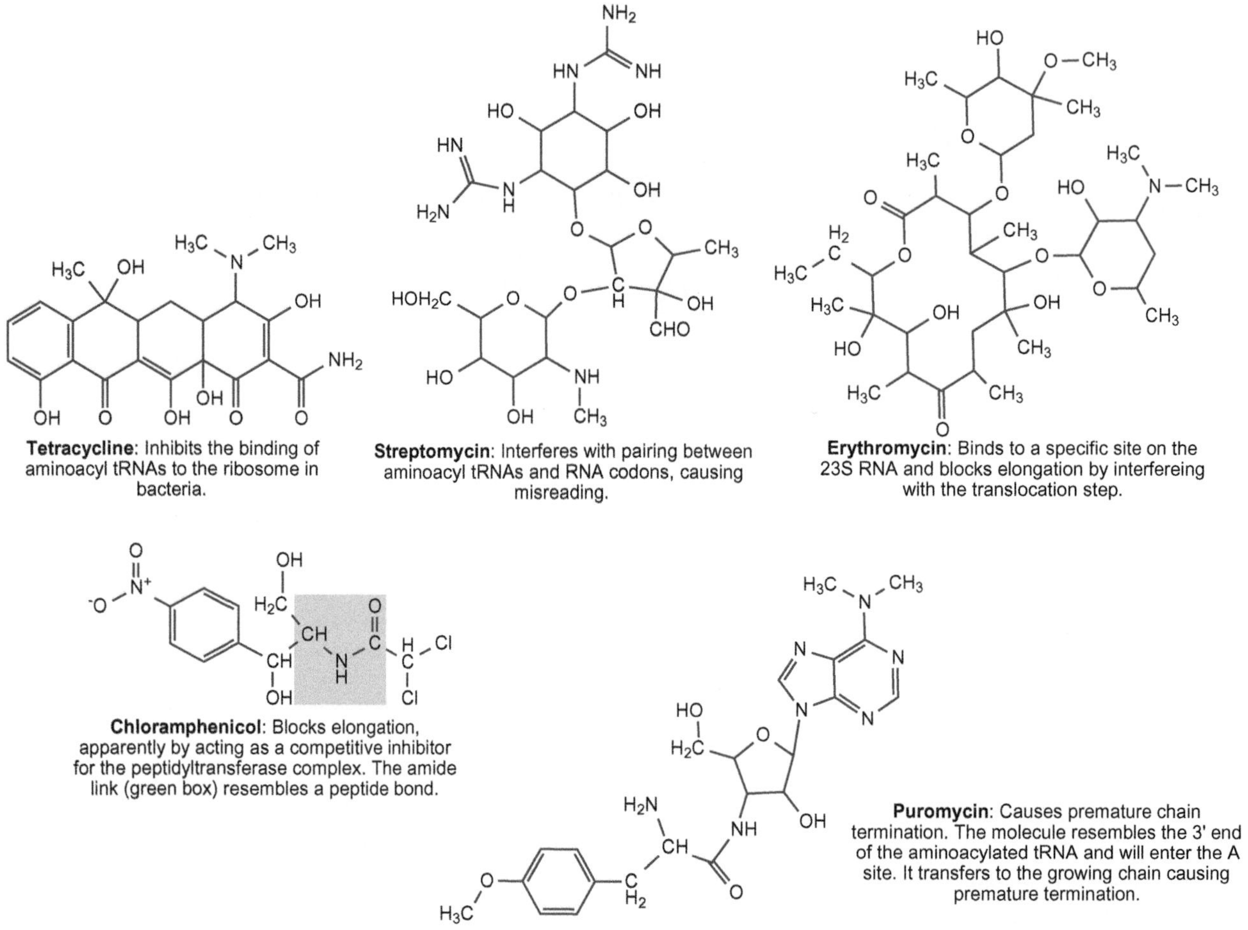

Tetracycline: Inhibits the binding of aminoacyl tRNAs to the ribosome in bacteria.

Streptomycin: Interferes with pairing between aminoacyl tRNAs and RNA codons, causing misreading.

Erythromycin: Binds to a specific site on the 23S RNA and blocks elongation by interfereing with the translocation step.

Chloramphenicol: Blocks elongation, apparently by acting as a competitive inhibitor for the peptidyltransferase complex. The amide link (green box) resembles a peptide bond.

Puromycin: Causes premature chain termination. The molecule resembles the 3' end of the aminoacylated tRNA and will enter the A site. It transfers to the growing chain causing premature termination.

FIGURE 23.17 Antibiotic inhibitors of protein synthesis. The solid lines in ring structures without any designation of an atom represent hydrogen.

spaces distal to the terminal bronchiole and destruction of bronchiole walls is high in α_1-AT deficiency.

Inhibitors of Protein Synthesis and Related Disorders

Many antibacterial agents (antibiotics) have been isolated from fungi. Antibiotics are used both clinically and as reagents for unraveling the details of protein, RNA, and DNA synthesis.

Antibiotics that have no effect on eukaryotic translation either fail to penetrate the cell membrane (which is quite common) or do not bind to eukaryotic ribosomes. The differences in effectiveness of antibiotics on bacterial and mammalian cells *in vivo* are the basis of their usefulness as therapeutic agents for bacterial infections (Figure 23.17).

Some antibiotics are active against both classes of cells. One example is **chloramphenicol**, which inhibits peptidyltransferase in both bacterial and mitochondrial ribosomes,

although eukaryotic cytoplasmic ribosomes are unaffected. Such a drug may be clinically useful if a concentration range can be maintained in the patient in which the antibacterial action is substantial but toxic effects on host cells are minimal. However, because of the potential for toxicity, such antibiotics are used only in serious infections when other drugs fail.

Three antibiotic inhibitors that act only on prokaryotes are **streptomycin**, **tetracycline**, and **erythromycin**. Other antibiotics act primarily on eukaryotic cells (Table 23.2). Eukaryotic protein synthesis can also be inhibited by toxins of bacterial origin. An example is the toxin produced by *Corynebacterium diphtheriae* bacteria carrying a lysogenic bacteriophage. Uninfected bacteria do not elaborate the toxin, and, therefore, acquisition of the phage is essential for the toxic effect. **Diphtheria** is an acute infectious disease usually localized in the pharynx, larynx, and nostrils, and occasionally in the skin. Initially, the cytotoxic effects are restricted to those tissues immediately adjacent to the bacterial growth. With increased bacterial growth,

TABLE 23.2 Inhibitors of Protein Synthesis in Eukaryotes

Inhibitor	Action
Abrin, ricin	Inhibits binding of aminoacyl tRNA
Anisomycin	Inhibits peptidyl transferase on the 80S ribosome
Diphtheria toxin	Catalyzes a reaction between NAD^+ and EF-2 to yield an inactive factor; inhibits translocation
*Chloramphenicol	Inhibits peptidyltransferase of mitochondrial ribosomes; is inactive against cytoplasmic ribosomes
*Puromycin	Causes premature chain termination by acting as an analogue of charged tRNA
*Fusidic acid	Inhibits translocation by altering an elongation factor
Cycloheximide	Inhibits peptidyltransferase
Pactamycin	Inhibits positioning of $tRNA^{fMet}$ on the 40S ribosome
Showdomycin	Inhibits formation of the $eIF2\text{-}tRNA^{fMet}\text{-}GTP$ complex
Sparsomycin	Inhibits translocation

Also active on prokaryotic ribosomes.

production of toxin increases and it is disseminated to the blood, lymphatics, and other organ systems (e.g., cardiovascular and nervous systems), where it brings about destructive changes.

The pathogenesis of diphtheria is due entirely to the toxin's inhibition of protein synthesis in the host cells. Diphtheria toxin consists of a single polypeptide chain (M.W. 63,000) with two intramolecular disulfide linkages. Toxin binds to the outer surface of susceptible cells at specific sites and enters by receptor-mediated endocytosis. The toxin is cleaved proteolytically into a small fragment A (M.W. 21,000), and a larger fragment B (M.W. 42,000), during its internalization. Fragment A is the biologically active part of the molecule and presumably penetrates the cell membrane with the aid of fragment B.

On entry into the cytoplasm, fragment A catalyzes the ADP ribosylation of the elongation factor, EF-2, leading to its inactivation and the interruption of protein synthesis. The ADP—ribose group is donated by NAD^+. The ADP ribosylation reaction, catalyzed by the toxin, is specific for EF-2 of eukaryotic cells; other proteins of eukaryotic and bacterial cells are not substrates. This specificity is due to an unusual amino acid residue in EF-2, **diphthamide**, which is the acceptor of the ADP ribosyl group. Diphthamide derives from the post-translational

modification of histidine. The acute symptoms are treated with antitoxin. The bacteria, which are gram-positive, succumb to a variety of antibiotics, including penicillin. Diphtheria is effectively prevented by immunization with toxoid (inactivated toxin) preparations.

ADP ribosylation is also involved in the action of **cholera** toxin and in certain pathogenic strains of *E. coli* (Chapter 11). Cholera toxin catalyzes the ADP ribosylation of the Gsα subunit of heterotrimeric G proteins. This activates adenylate cyclase, which catalyzes the formation of cAMP from ATP. The cAMP formed stimulates secretion of water and electrolytes from intestinal epithelial cells. Thus, patients infected with *Vibrio cholerae* secrete enormous quantities (up to 20 L/d) of water and electrolytes. Without adequate and prompt replacement, death can ensue owing to dehydration and electrolyte imbalance. Interestingly, in the case of **pertussis** toxin, ADP-ribosylation leads to inactivation of the Giα subunit of heterotrimer G proteins, whereas with cholera toxin, ADP ribosylation causes activation of the protein.

A group of plant **lectins**, such as abrin, ricin, and modeccin, are highly toxic to eukaryotic cells. Their mode of action consists of inhibition of protein synthesis by enzymatically inactivating the EF-2 binding region of the 60S ribosomal subunit, whereas the diphtheria toxin inactivates the EF-2 protein itself. Ricin is isolated from castor beans and has a molecular weight of 66,000. Like most plant and bacterial toxic proteins, ricin contains two polypeptide chains with two different but complementary functions. The A chain possesses enzymatic activity and is responsible for toxicity, and the B chain, which is a lectin, binds to galactose-containing glycoproteins or glycolipids on the cell surface. The A and B chains are linked by a labile disulfide linkage. On binding of the ricin molecule to the carbohydrate receptors of the cell surface via the B chain, the A chain enters the cytoplasm, presumably by receptor-mediated endocytosis, where it inhibits protein synthesis by irreversible inactivation of the 60S ribosomal subunit. Toxins have been used to develop highly selective cytotoxic agents targeted against specific cells. For example, ricin A has been coupled to agents that selectively bind to cell-surface membrane components. The selective agents may be monoclonal antibodies (Chapter 33), hormones, or other cell-surface ligands. These conjugates act as selective cytotoxic agents and may have potential therapeutic applications (e.g., in the treatment of cancer). Potential clinical application for diphtheria toxin may be impossible because of the prevalence of the diphtheria antitoxin in human populations.

Collagen Biosynthesis and Its Disorders

Collagen occurs in several genetically distinct forms and is the most abundant body protein; most of the body scaffolding is composed of collagen (Chapter 11). Its structure is uniquely suited for this structural role. It is a fibrillar protein but also exists in a nonfibrillar form in the basement membrane. The basic structural unit of collagen, **tropocollagen**, consists of three polypeptide chains. Each polypeptide chain has the general formula $(-Gly-X-Y-)_{333}$. Some of the amino acid residues at the X and Y positions are proline, hydroxyproline, lysine, hydroxylysine, and alanine. Collagen, which is a glycoprotein, contains only two types of carbohydrate residues: glucose and galactose linked in O-glycosidic bonds to hydroxylysyl residues.

In collagen, each polypeptide chain is coiled into a special type of a rigid, kinked, left-handed helix, with about three amino acid residues per turn. The three helical polypeptides, in turn, are wrapped around each other to form a right-handed triple-stranded superhelix that is stabilized by hydrogen bonding. The collagen molecules associate in an ordered quarter-staggered array to give rise to microfibrils, which, in turn, combine to give fibrils. Covalent cross-linkages occur at various levels of collagen fiber organization and provide great mechanical strength. Collagen biosynthesis is unusual in that it consists of many post-translational modifications (Figure 23.18). The unique post-translational reactions of collagen biosynthesis are

1. Hydroxylation of selected prolyl and lysyl residues;
2. Glycosylation of certain hydroxylysyl residues;
3. Folding of procollagen polypeptides into a triple helix;
4. Conversion of procollagen to collagen;
5. Self-assembly into fibrils;
6. Oxidative deamination of ε-amino groups of strategically located lysyl and hydroxylysyl residues to provide reactive aldehydes. These form cross-linkages between polypeptide chains of the same molecule as well as between the adjacent molecules that gives strength and stability to the fibrils.

The first three processes take place inside the cell, whereas the last three are extracellular modifications.

Collagen disorders can result from primary defects in the structure of procollagen or collagen or from secondary changes that affect collagen metabolism. Collagen disorders are both acquired and inherited. **Ehlers—Danlos syndrome** and **osteogenesis imperfecta** are examples of inherited primary collagen diseases; **scurvy** and various fibrotic processes (e.g., pulmonary fibrosis and cirrhosis) are examples of secondary collagen diseases. The mechanisms that lead to collagen diseases are

1. Aberration in the control mechanisms that alter the balance between synthesis and degradation. This imbalance can lead either to excessive deposition or to depletion of collagen.

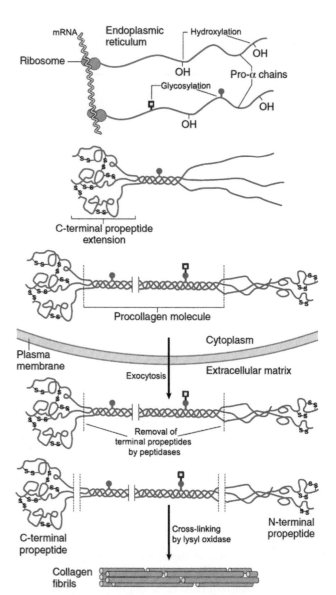

FIGURE 23.18 Collagen biosynthesis. Intracellular processes include hydroxylation of prolyly and lysyl residues, glycosylation of certain hydroxylysyl residues with glucose and/or galactose, and formation of procollagen triple helixes. Removal of terminal peptides by aminopeptidase and carboxypeptidase, formation of collagen fibrils, and extensive cross-link formation by lysyl oxidase occur in the extracellular matrix. *[Reproduced with permission from J.W. Pelley, Elsevier's integrated review biochemistry, 2nd Ed., (Chapter 20. Tissue Biochemistry, page 185, Figure 20-8).]*

2. Synthesis of structurally altered collagen due to defects at the level of DNA transcription, RNA processing, translation, or post-translational modifications.
3. Imbalance in the relative rates of synthesis of genetically distinct collagens.
4. Abnormalities in the packing of collagen molecules into a fiber or in the interaction of collagen fibers with other extracellular components of the connective tissue.

Intracellular Post-Translational Modifications

Hydroxylations of Selected Prolyl and Lysyl Residues

The hydroxylation reactions are catalyzed by three enzymes: prolyl 4-hydroxylase (usually known as **prolyl hydroxylase**), prolyl 3-hydroxylase, and lysyl hydroxylase. These enzymes are located within the cisternae or rough endoplasmic reticulum; as the procollagen chains enter this compartment, the hydroxylations begin. All three enzymes have the same cofactor requirements: ferrous iron, α-ketoglutarate, molecular oxygen, and ascorbate (vitamin C). The reducing equivalents required for the hydroxylation reaction are provided by the decarboxylation of equimolar amounts of α-ketoglutarate to succinate and carbon dioxide:

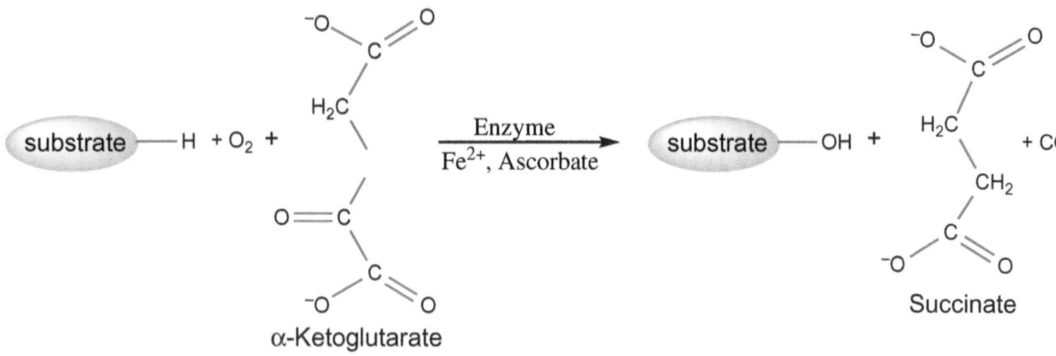

These reactions are catalyzed by two specific enzymes: hydroxylysyl galactosyltransferase and galactosylhydroxylysyl glucosyltransferase. The first enzyme catalyzes the transfer of galactose from UDP−galactose to hydroxylysyl residues, and the second enzyme transfers a glucose from UDP−glucose to galactosylhydroxylysyl residues. Both enzymes require a divalent cation, preferably manganese. The substrate has to be in the nonhelical conformation, and the glycosylation ceases when the collagen propeptides fold into a triple helix. Thus, both hydroxylation and glycosylation must occur before triple-helix formation, which is an intracellular process.

Deficiency of galactosylhydroxylysyl glucosyltransferase has been seen in the members of kindred with a dominantly inherited disease known as **epidermolysis bullosa simplex**. This disease belongs to a group of

Since ascorbate is required for hydroxylation reactions, its deficiency causes accumulation of the collagen polypeptides in the endoplasmic reticulum and their eventual secretion. However, these proteins lack the proper modifications and therefore cannot be used for assembly of collagen fibrils. Thus, the role of ascorbate in the hydroxylation of collagen is one of its major physiological roles (see also Chapter 36). **Scurvy**, a connective tissue disorder, is due to vitamin C deficiency in the diet. Most pathological changes of scurvy can be attributed to failure to synthesize collagen. Scurvy patients have defective blood vessels and poor intravascular support, leading to frequent hemorrhages, defective formation of bone and teeth, and poor wound healing. All of these manifestations can be corrected by administration of vitamin C.

Glycosylation of Hydroxylysyl Residues

The glycosylation occurs as the N-terminal ends of the polypeptide chains move into cisternae at specific hydroxylysyl residues. The carbohydrates found are galactose and the disaccharide, glucosylgalactose.

inherited disorders characterized by blister formation in response to minor skin trauma. Further investigation of this disorder may provide information about the role of glycosylation of collagen.

Formation of Disulfide Linkages and Assembly of Procollagen Polypeptides into a Triple Helix

The propeptide regions of procollagen polypeptides contain cysteine residues that can form both intra- and interchain disulfide linkages. The assembly of procollagen polypeptides into a triple helix appears to have two requirements:

1. The 4-hydroxylation of at least 100 prolyl residues per proα chain;
2. The formation of interchain disulfide linkages in the C-terminal propeptide.

Translocation and Secretion of Procollagen

After the procollagen polypeptides are assembled into a triple helix, they are secreted by the classical route. They pass through the smooth endoplasmic reticulum and the Golgi complex, where they are packaged into membranous vesicles and secreted into the extracellular space by exocytosis. This process requires ATP and may involve microtubules and microfilaments. Prevention of the formation of a triple helix (e.g., lack of 4-hydroxylation due to vitamin C deficiency) leads to the accumulation of nonhelical propolypeptides within the cisternae of the rough endoplasmic reticulum and a delayed rate in its secretion.

Extracellular Post-Translational Modification

Conversion of procollagen to collagen requires at least two proteases: a procollagen aminoprotease and a procollagen carboxyprotease. The former catalyzes removal of the N-terminal propeptide, and the latter catalyzes removal of the C-terminal propeptide. The two enzymes are endopeptidases, function at neutral pH, require a divalent cation such as Ca^{2+}, and show a preference for the helical conformation.

Formation of Collagen Fibrils from Ordered Aggregation of Collagen Molecules

The collagen molecules formed by removal of the propeptides spontaneously assemble into fibrils. At this stage, the fibrils are still immature and lack tensile strength, which is acquired by cross-linking. The initial step in cross-link formation is the oxidative deamination of α-amino groups in certain lysyl and hydroxylysyl residues catalyzed by **lysyl oxidase**. The enzyme is a copper-dependent (probably cupric) protein, and the reaction requires molecular oxygen and pyridoxal phosphate for full activity. Only native collagen fibrils function as substrates:

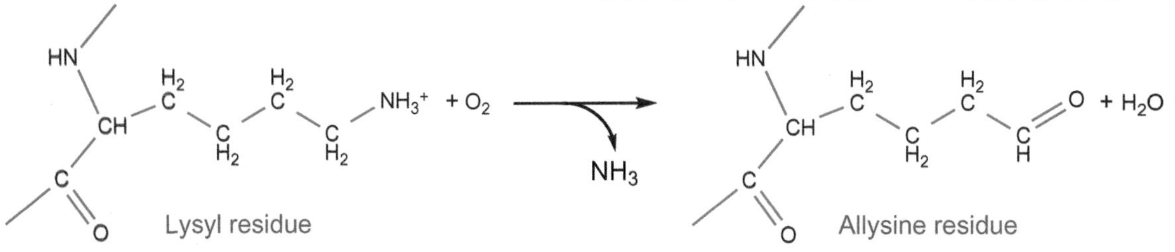

characterized by **Ehlers—Danlos syndrome type IX** and some forms of **cutis laxa**. Type IX Ehlers—Danlos syndrome patients exhibit extreme extensibility of the skin, "cigarette-paper" scarring, and easy bruisability. In cutis laxa, the skin is loose and inelastic and appears to be too large for the surface it covers. Some affected individuals exhibit deficiency of lysyl oxidase, presumably limited to skin. The animal counterpart of lysyl oxidase deficiency is seen in several variants of aneurysm-prone mice. All of these disorders are X-linked.

Lysyl oxidase undergoes irreversible inactivation when exposed to certain nitriles. β-Aminopropionitrile ($H_2N\text{-}CH_2\text{-}CH_2\text{-}CN$), which is found in certain peas (e.g., *Lathyrus odoratus*), inhibits lysyl oxidase, presumably by binding covalently to the enzyme. Ingestion of these peas results in **lathyrism**, which is characterized by multiple defects in collagen- and elastin-containing tissues. In experimental animals, lathyrism is produced by administering β-aminopropionitrile during their active growth period. Lysyl oxidase is also irreversibly inhibited by carbonyl reagents (e.g., hydroxylamine) and copper-chelating agents.

Impaired cross-linking can also occur as a result of copper deficiency, since lysyl oxidase is a copper-dependent enzyme. In experimental animals, copper deprivation causes skeletal and cardiovascular abnormalities. These abnormalities are similar to a human disorder known as **Marfan's syndrome**, an autosomal dominant trait whose molecular defect is due to mutations in the *FBN1* gene, which code for fibrillin-1 protein. A genetic defect affecting the gastrointestinal absorption of copper, known as **Menkes' (kinky-hair) syndrome**, exhibits neurological, connective tissue, pigmentary, and hair abnormalities. These defects can be

Lysyl residue ⟶ Allysine residue

In a similar reaction, the hydroxyallysine residue is formed from hydroxylysine residues. The aldehyde groups react spontaneously with each other or with amino groups, with formation of intra- and intermolecular linkages. Lysyl oxidase is also involved in the formation of elastin cross-linkages.

Several collagen disorders result from defects in the formation of cross-links (Chapter 11). The cross-linking disorders can be due to a hereditary deficiency of lysyl oxidase, inhibition of lysyl oxidase, deficiency of copper, defects in the formation of cross-links, or defects in their stabilization. The genetic deficiency of lysyl oxidase is

explained by deficient activities of various copper-requiring enzymes (Chapter 35). The connective tissue changes are attributed to the deficiency of lysyl oxidase.

Cross-linking can be inhibited by agents that can react with the aldehyde groups of allysyl and hydroxylysine residues. **Penicillamine** reacts with the aldehyde groups, forming a thiazolidine complex and rendering the aldehyde groups unavailable for cross-link formation. Penicillamine is an effective chelator of copper, thereby inhibiting lysyl oxidase. Penicillamine is used therapeutically in the treatment of disorders involving copper accumulation (**Wilson's disease**) or in lead and mercury poisoning.

CLINICAL CASE STUDY 23.1 Disorders of Collagen Biosynthesis

Vignette 1: Ehlers–Danlos Syndrome

This clinical case was abstracted from: D. Buettner, S.H. Fortier, Ehlers–Danlos syndrome in trauma: a case review, J. Emerg. Nurs. 35 (2) (2009) 169–170.

Synopsis

An 18-year-old man was brought to the emergency room with a head injury. He was unconscious and pulseless. Emergency room physicians were told that the patient had a history of diagnoses with Ehlers–Danlos syndrome (EDS), a connective tissue disorder. CPR and other resuscitation efforts such as thoracotomy, open cardiac massage, and internal cardiac defibrillation were performed but failed to revive the patient. The autopsy report confirmed that the patient had EDS type IV, known as the vascular or arterial-ecchymotic type, which is the most serious of the six EDS types. The autopsy result indicated that the patient had bilateral hemothoraces and a large left retroperitoneal hematoma due to thoracic aortic dissection.

Teaching Points

1. Ehlers–Danlos syndrome is a group of genetic disorders caused by defects in collagen biosynthesis.
2. There are six major types of EDS: classical, hypermobility, vascular, kyphoscoliosis, arthrochalasis, and dermatosparaxis.

References
[1] C.W. Chen, S.W. Jao, Ehlers–Danlos syndrome, N. Engl. J. Med. 357 (2007) e12.
[2] A.E. Sareli, W.J. Janssen, D. Sterman, S. Saint, R.E. Pyeritz, What's the connection? N. Engl. J. Med. 358 (2008) 626–632.

Vignette 2: Osteogenesis Imperfecta

This clinical case was abstracted from: A.M. Barnes, W. Chang, R. Morello, W.A. Cabral, M. Weis, D.R. Eyre, et al., Deficiency of cartilage-associated protein in recessive lethal osteogenesis imperfecta, N. Engl. J. Med. 355 (2006) 2757–2764.

Synopsis

The first child of consanguineous parents was born at 35 weeks' gestation; prenatal ultrasonography showed severe micromelia of the arms and legs. The newborn baby had a wide-open anterior fontanel, eye proptosis, and white sclerae. He also had pectus excavatum and rhizomelic limb shortening, and a radiographic survey showed over 20 fractures of the long bones and ribs. He died at 10 months of age. mRNA samples were purified and quantified from the patient's fibroblasts and tested for cartilage-associated protein (CRTAP), which forms a complex with prolyl 3-hydroxylase 1. Prolyl 3-hydroxylase 1 is required for hydroxylation of the proline residue in the 986 position of type 1 collagen. CRTAP mRNA was not detected from the patient's fibroblasts, and genomic DNA sequencing of exons of the CRTAP gene revealed a homozygous mutation in the splice donor site of exon 1. The parents of the patient were heterozygous for this mutation. This type of mutation induces premature termination of protein synthesis, since stop codons existed in the retained intron 1 sequence.

Teaching Points

Classic osteogenesis imperfecta is an autosomal dominant genetic disease caused by mutations in type 1-collagen genes. An autosomal recessive form of this disorder can be caused by mutations introduced into the CRTAP gene, which leads to the dysfunction of prolyl 3-hydroxylase required for the hydroxylation of prolyl 3-hydroxylation of type 1 collagen.

Reference
[1] F. Rauch, F.H. Glorieux, Osteogenesis imperfecta, Lancet 363 (2004) 1377–1385.

Vignette 3: Scurvy

This clinical case was abstracted from: J.M. Olmedo, J.A. Yiannias, E.B. Windgassen, M.K. Gornet, Scurvy: a disease almost forgotten. Int. J. Dermatol. 45 (2006) 909–913.

Synopsis

A 77-year-old woman presented with fatigue, bruising, gingival bleeding, and anemia. She had been eating only bread, olive oil, and red meat for 2 years because she believed that her food allergies were due to all types of fruits and vegetables. She also reported that recently she had experienced occasional nosebleeds, bleeding of the gums while brushing her teeth, and soreness of the tongue. On the day of her clinic visit, she noticed bright red blood in her stool. Laboratory tests reported normocytic anemia with a hemoglobin of 8.8 g/dL, hematocrit of 26.4%, and mean corpuscular volume of 86.1 fL. Iron studies, prothrombin time, and international normalized ratio were within normal ranges. Values for vitamin B_{12} and folate were within normal ranges, whereas vitamin C was not detected. A diagnosis of vitamin C deficiency scurvy was made, and the patient was given 100 mg of oral vitamin C three times a day for 2 weeks.

Teaching Points

Vitamin C, ascorbic acid, is required for the hydroxylation of prolyl and lysyl residues of collagen biosynthesis, which are catalyzed by prolyl and lysyl hydroxylases. Vitamin C is also required for the biosynthesis of carnitine and norepinephrine, metabolism of tyrosine, and amidation of peptide hormones, and it also facilitates iron absorption by converting ferric state iron (Fe^{3+}) to ferrous state iron (Fe^{2+}).

Reference
[1] M. Weinstein, P. Babyn, S. Zlotkin, An orange a day keeps the doctor away: scurvy in the year 2000, Pediatrics 108 (2001) e551–555.

ENRICHMENT READING

[1] F. Muller, L. Tora, Chromatin and DNA sequences in defining promoters for transcription initiation, Biochim. Biophys. Acta 2013 (1839) 118–128.

[2] A.C.M. Cheung, P. Cramer, A movie of RNA polymerase II transcription, Cell 149 (2012) 1431–1437.

[3] P.G. Higgs, N. Lehman, The RNA world: molecular cooperation at the origins of life, Nat. Rev. Genet. 16 (2015) 7–17.

[4] T.R. Cech, J.A. Seitz, The noncoding RNA revolution—trashing old rules to forge new ones, Cell 157 (2014) 77–94.

[5] K.V. Morris, J.S. Mattick, The rise of regulatory RNA, Nat. Rev. Genet. 15 (2014) 423–437.

[6] D.N. Wilson, Ribosome-targeting antibiotics and mechanisms of bacterial resistance, Nat. Rev. Microbiol. 12 (2014) 35–48.

Chapter 24

Regulation of Gene Expression

Key Points

1. In prokaryotes and eukaryotes, not all genes are expressed at any given time. The expression of genes is regulated mainly at the transcriptional initiation level.

2. Regulation of gene expression in prokaryotes is tied to the availability of nutrients in the environment and clusters of genes called operons. Operons code for enzymes required for the same metabolic pathway that are concurrently expressed by positive and negative regulation. For lactose metabolism, genes for β-galactosidase, lactose permease, and galactose transacetylase are expressed in response to the presence of lactose in the medium.

3. In positive regulation, effector molecules bind to a promoter sequence and initiate transcription. In negative regulation, an inhibitor binds to the operator and inhibits transcription.

4. Regulation of gene expression in eukaryotes is much more complicated than in prokaryotes mainly due to the larger genome size, the association of DNA with histone complexes, the existence of intracellular compartments, and the uncoupled processes of transcription and translation.

5. Regulation of the tryptophan operon is regulated by intracellular tryptophan concentrations and by inhibition of transcriptional initiation and attenuation.

6. Bacteriophages adopt a temporal transcriptional regulation mechanism to express genes on different time schedules. Bacteriophage λ can choose either lytic or lysogenic life cycles, depending on the condition of the host cells through regulation of gene expression.

7. Regulation of gene expression in eukaryotes is much more complicated than in prokaryotes mainly due to the larger genome size, the association of DNA with histone complexes, the existence of intracellular compartments, and uncoupled processes of transcription and translation.

8. Euchromatin comprises loosely packed DNA regions and is transcriptionally active chromatin. Heterochromatin comprises tightly packed DNA regions and is transcriptionally inactive chromatin.

9. Eukaryotic regulation of gene expression can be regulated at many different levels:
   a. Chromatin structure modification by DNA methylation and histone modifications;
   b. Transcriptional initiation by various transcription factors and hormone response elements;
   c. Primary transcript processing such as capping, splicing, and polyadenylation;
   d. Transport of the processed mRNA from the nucleus to the cytoplasm;
   e. Stability control of mRNA transported into the cytoplasm such as iron-response elements in the 5′ or 3′ flanking sequences of mRNAs;
   f. Downregulation of translation of mRNA (protein synthesis) by miRNAs and siRNAs; and
   g. Post-translational modifications of protein(s).

10. Transcription factors have unique structural motifs that facilitate their binding to DNA and can be classified into distinct groups based on their structural similarities: zinc finger, leucine zipper, helix—turn—helix and helix—loop—helix and homeodomain.

11. Epigenetic control of gene expression refers to the inherited pattern of specific gene expression not coded by DNA sequences. Examples of epigenetic control include X chromosome inactivation and genetic imprinting phenomena.

INTRODUCTION

Prokaryotic cells such as *Escherichia coli* can regulate their gene expression to adapt to the changes in available nutrients in the surrounding environment. Their regulation of gene expression occurs mainly through regulation of the rate of mRNA synthesis, since their transcription and translation is coupled. For example, in a lactose rich medium, *E. coli* activates expression of the gene clusters called **operons** that are required for the processing of lactose, e.g., β-galactosidase and permease. However, eukaryotic multicellular organisms regulate their gene expression through five major levels: modification of chromatin structures, the number of mRNA molecules synthesized (transcriptional), processing of the mRNA (post-transcriptional), the rate of mRNA translation (translational), and modifications of the proteins for proper function (post-translational regulation). Eukaryotic regulation of gene expression is thus much more complex than in prokaryotic gene expression and involves a variety of

N. V. Bhagavan and C.-E. Ha: Essentials of Medical Biochemistry, Second edition. DOI: http://dx.doi.org/10.1016/B978-0-12-416687-5.00024-5

447

factors and elements. In this chapter, the numerous mechanisms and factors by which regulation of gene expression is achieved in prokaryotes and eukaryotes are described.

REGULATION OF mRNA SYNTHESIS

In prokaryotes, mRNA synthesis can be controlled simply by regulating initiation of transcription. In eukaryotes, mRNA is formed from a primary transcript followed by a series of processing events (e.g., splicing, capping, polyadenylation). Eukaryotes regulate not only transcription initiation, but also the various later stages of processing.

An important aspect of mRNA regulation is determined by the degradation of mRNA molecules; i.e., translation can occur only as long as the mRNA remains intact. In bacteria, mRNA molecules have a lifetime of only a few minutes, and continued synthesis of mRNA molecules is needed to maintain synthesis of the proteins encoded in the mRNAs. In eukaryotes, the lifetime of mRNA is generally quite long (hours or days), thereby enabling a small number of transcriptional initiation events to produce proteins over a long period of time.

Metabolic pathways normally consist of a large number of enzymes; in some cases, the individual enzymes are used in a particular pathway and nowhere else. In these cases, it may be efficient to regulate expression of either all or none of the enzymes in the pathway. In bacterial systems, all enzymes of the pathway are encoded in a single polycistronic mRNA molecule called an operon, and synthesis of the single mRNA produces all the enzymes. In eukaryotes, common signals for transcription of different genes may be used, or the primary transcript can be differentially processed to yield a set of mRNA molecules, each of which encodes one protein. In eukaryotes, regulation of synthesis of the primary transcripts simultaneously regulates synthesis of all the gene products.

In prokaryotes, gene expression fluctuates in response to the environment because bacteria must be able to respond rapidly to a changing environment. However, due to the differentiation of cells of the higher eukaryotes, changes in gene expression are usually irreversible—for example, in the differentiation of a muscle cell from a precursor cell. Eukaryotes can change the number of copies of a gene during differentiation and, in this way, regulate the level of gene expression.

GENE REGULATION IN PROKARYOTES

Several common patterns of transcriptional regulation have been observed in bacteria. The particular pattern depends on the type of metabolic activity of the system being regulated. For example, in a catabolic (degradative) pathway, the concentration of the substrate for the first enzyme in the pathway often determines whether all the enzymes in the pathway are synthesized. In contrast, the final product is often the regulatory substance in a biosynthetic pathway. In the simplest mode, the absence of an end product stimulates transcription, and the presence of an end product inhibits it. Even for a gene in which a single type of protein is synthesized from monocistronic mRNAs, the protein may "autoregulate" itself in the sense that the transcriptional activity of the promoter is determined directly by the concentration of the protein. The molecular mechanisms for each regulatory system vary considerably but fall into one of two major categories: negative regulation or positive regulation.

In **negative regulation**, an inhibitor, which keeps transcription turned off, is present in the cell; and an anti-inhibitor, i.e., an inducer, is needed to turn the system on. In **positive regulation**, an effector molecule (which may be a protein, small molecule, or molecular complex) activates a promoter that initiates transcription. Negative and positive regulation are not mutually exclusive, and some degradative pathways, such as the utilization of lactose in *E. coli*, are both positively and negatively regulated.

Lactose Operon

The prototype for negative regulation is the system in *E. coli* for metabolizing lactose (lac) (Figure 24.1). The key chemical reaction carried out by the lac system is a cleavage of lactose to galactose and glucose. This reaction enables bacteria to utilize lactose as a carbon source when glucose is not available. The regulatory mechanism of the lac system, known as the **operon model**, was the first system described in detail, and it defines most of the terminology and concepts in current use. The principal regulatory features of the *lac* operon are as follows:

1. The products of two genes are required for lactose utilization. The *lacZ* gene encodes an enzyme, β-galactosidase, that degrades lactose; the *lacY* gene encodes a protein, lactose permease, needed to transport lactose and concentrate it within the cell. A third gene, *lacA*, encodes an enzyme, thiogalactoside transacetylase, which transfers an acetyl group to β-galactoside during lactose metabolism.
2. Lac enzymes encoded in the *lac* operon are transcribed into a single polycistronic mRNA molecule (lac mRNA). Enzyme levels are regulated by controlling the synthesis of that mRNA.
3. Synthesis of lac mRNA is determined by the activity of a specific regulatory protein: the lac repressor. This protein is encoded in the *lacI* gene, which is transcribed in a monocistronic mRNA molecule distinct from lac mRNA. When the repressor is active and bound to the operator sequence of *lac* operon, lac mRNA is not made. This is the negative regulation of the *lac* operon.

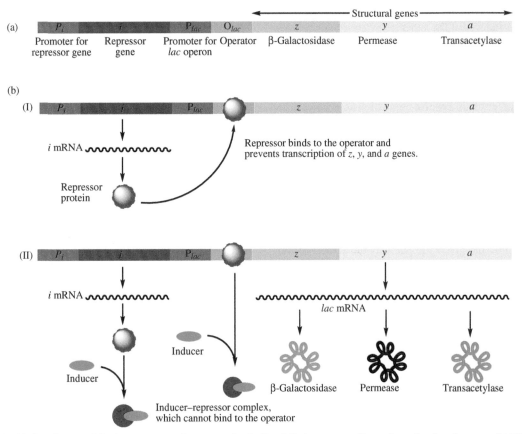

FIGURE 24.1 (a) Genetic map of the *lac* operon, not drawn to scale; the P and O sites are actually much smaller than the genes. (b) Diagram of the lac operon in (I) repressed and (II) induced states. The inducer alters the shape of the repressor, so the repressor can no longer bind to the operator.

Mutants have been isolated that produce inactive repressor proteins; these cells make lac mRNA continuously (constitutive synthesis), even in the absence of lactose. Actually, some lac mRNA is always made at a level of about one or two transcription events per generation, because repression is never complete. This basal level synthesis is responsible for a very small amount of the proteins present in the absence of lactose.

4. A sequence of bases (in the DNA)—the operator *laco*—is adjacent to the *lac* operon promoter and binds the lac repressor. When the repressor protein is bound to the operator, binding of RNA polymerase to the promoter is prevented by steric interference and initiation of transcription of lac mRNA does not occur. A mutation in the operator, which eliminates repressor binding, also leads to constitutive mRNA synthesis.

5. An inducer (a small effector molecule) is needed to initiate transcription of lac mRNA. The inducer of the *lac* operon, which is allolactose, [β-D-galactopyranosyl-(1→6)-β-D-glucopyranose] is a structural isomer of lactose formed by basal synthesis of β-galactosidase. β-galactosidase has two enzymatic functions: 1. hydrolysis of lactose to glucose and galactose, 2. isomerization of lactose to allolactose. Allolactose binds to the repressor and alters its three-dimensional structure such that it is unable to bind to the operator. Thus, in the presence of lactose or other inducers, the operator is unoccupied and the promoter is available for initiation of mRNA synthesis. It is common to refer to inactivation of the repressor by an inducer as derepression.

Thus, the overall regulatory pattern is as follows: a bacterium growing in a medium without lactose does not make the *lacZYA* product because the repressor that is present is bound to the operator and prevents synthesis of lac mRNA. The cell grows by utilizing whatever other carbon source is available such as glucose. If lactose is added (and if glucose is absent, see below), the basal level lacZYA protein brings the lactose into the cell and converts it to allolactose. As a result, the repressor is inactivated. RNA polymerase binds to the promoter, and synthesis of lac mRNA begins. Lac mRNA synthesis continues until the lactose is exhausted, at which time the inactive repressor would be reactivated and repression reestablished.

The *lac* operon is also positively regulated, presumably because of the role of glucose in general metabolism. The function of β-galactosidase is to generate glucose by cleaving lactose, so if both glucose and lactose are available, the cell can use glucose, and there is no reason for the *lac* operon to be induced. (The other cleavage product, galactose, is also converted to glucose by enzymes of the galactose operon.) Indeed, when glucose and lactose both are present in the medium, little lac mRNA is made because glucose indirectly inhibits RNA polymerase from binding to the lac promoter. This positive regulation is accomplished by **cyclic AMP**, a regulatory molecule in both prokaryotes and eukaryotes. Cyclic AMP (cAMP) is synthesized from ATP by the enzyme adenylate cyclase, and its concentration is regulated by glucose metabolism. In a bacterial culture that is starved of glucose, the intracellular concentration of cAMP is very high. If a culture is growing in a medium containing glucose, the cAMP concentration is very low. The observation that in a medium containing a carbon source that cannot enter the glycolytic pathway the cAMP concentration is high (Table 24.1), suggests that some glucose metabolite or derivative is an inhibitor of adenylate cyclase.

Cyclic AMP does not act directly as a regulator but is bound to a protein called the **cAMP receptor protein** (**CRP**), which forms a complex with cAMP (cAMP−CRP). This complex is active in the lac system and in many other operons involved in catabolic pathways. The cAMP−CRP complex binds to a base sequence in the DNA in the lac promoter region and stimulates transcription of lac mRNA; thus, cAMP−CRP functions as a positive regulator. The cAMP−CRP complex is not needed for binding of RNA polymerase to the lac promoter because a free promoter binds the enzyme. Thus, when glucose is absent, the cAMP concentration is high and there is enough cAMP−CRP to allow an active transcription complex to form; if glucose is present, cAMP levels are low, no cAMP−CRP is available, and minimal transcription of the *lac* operon occurs. Thus, the *lac* operon is independently regulated both positively and negatively, a feature common to many carbohydrate utilization operons (Figure 24.2).

Tryptophan Operon

The tryptophan (*trp*) operon is responsible for the production of the amino acid tryptophan, whose synthesis occurs in five steps, each requiring a particular enzyme. In *E. coli*, these enzymes are translated from a single polycistronic mRNA. Adjacent to the enzyme coding sequences in the DNA are a promoter, an operator, and two regions called the **leader** and the **attenuator** (Figure 24.3). The leader and attenuator sequences are transcribed. Another gene (*trpR*) encoding a repressor is located some distance from this gene cluster.

TABLE 24.1 Concentration of Cyclic AMP in Cells Growing in Media Having the Indicated Carbon Sources

Carbon Source	cAMP Concentration
Glucose	Low
Glycerol	High
Lactose	High
Lactose + glucose	Low
Lactose + glycerol	High

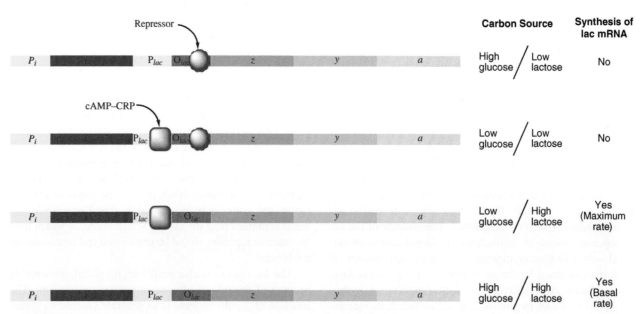

FIGURE 24.2   Three states of the *lac* operon showing that lac mRNA is made only if cAMP−CRP is present and repressor is absent.

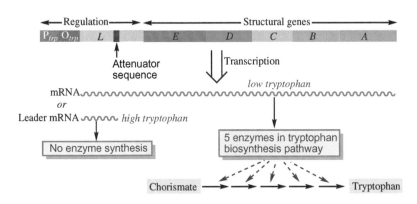

FIGURE 24.3 *Escherichia coli trp* operon. For clarity, the regulatory region is enlarged with respect to the coding region. The proper size of each region is indicated by the number of base pairs. L is the leader.

Regulation of the *trp* operon is determined by the concentration of tryptophan; when adequate tryptophan is present in the growth medium, there is no need for tryptophan biosynthesis. Transcription is turned off when a high concentration of tryptophan is present, and is turned on when tryptophan is absent. The regulatory signal is the concentration of tryptophan itself. In contrast to lactose, tryptophan is active in repression rather than induction.

The *trp* operon has two levels of regulation: an on–off mechanism and a modulation system. The protein product of the *trpR* gene (the trp aporepressor) cannot bind to the operator, in contrast to the lac repressor. However, if tryptophan synthesis is present, the aporepressor and the tryptophan molecule join together to form an active repressor complex that binds to the operator. When the external supply of tryptophan is depleted (or reduced substantially), the operator becomes exposed, and transcription begins. This type of on–off mechanism—activation of an aporepressor by the product of the biosynthetic pathway—has been observed in other biosynthetic pathways.

When the *trp* operon is de-repressed, which is usually the case unless the concentration of tryptophan in the medium is very high, the optimal concentration of tryptophan is maintained by a modulating system in which the enzyme concentration varies with the concentration of tryptophan. This modulation is affected by

1. Premature termination of transcription before the first structural gene is reached; and
2. Regulation of the frequency of premature termination by the concentration of tryptophan.

Located between the 5′ end of the trp mRNA molecule and the start codon of the *trpL* gene is a 162-base segment called the **leader** (a general term for such regions). Within the leader is a sequence of bases (bases 123 through 150) with regulatory activity. After initiation of mRNA synthesis, most mRNA molecules are terminated in this region (except in the complete absence of tryptophan), yielding a short RNA molecule consisting of only 40 nucleotides and terminating before the structural genes of the operon. This region in which termination occurs is

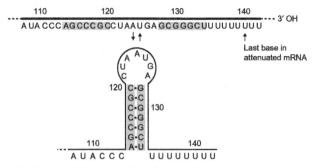

FIGURE 24.4 Terminal region of the trp leader mRNA (right end of L in Figure 24.3). The base sequence given is extended past the termination site at position 140 to show the long stretch of Us. The bases (colored lines) form an inverted repeat sequence that could lead to the stem-and-loop configuration shown (segments 3–4 in Figure 24.5).

a regulatory region called the **attenuator**. The base sequence around which termination occurs (Figure 24.4) has the usual features of a transcription termination site—namely, a possible stem-and-loop configuration in the mRNA followed by a sequence of eight AU pairs.

The leader sequence has an AUG codon that is in-phase with a UGA stop codon; together these start–stop signals encode a polypeptide of 14 amino acids. The leader sequence has an interesting feature: at positions 10 and 11 are two adjacent tryptophan codons.

Premature termination of mRNA synthesis is mediated through translation of the leader peptide. The two tryptophan codons make translation of the leader polypeptide sequence quite sensitive to the concentration of charged tRNA^{Trp}. If tryptophan is limiting, there will be insufficient charged tRNA^{Trp} and translation will pause at the tryptophan codons. Thus, biosynthesis of trp depends on two characteristics of gene regulation in bacteria:

1. Transcription and translation are coupled;
2. Base-pair formation in mRNA is eliminated in any segment of the mRNA that is in contact with the ribosome.

Figure 24.5 shows that the end of the trp leader peptide is in segment 1 and a ribosome is in contact with

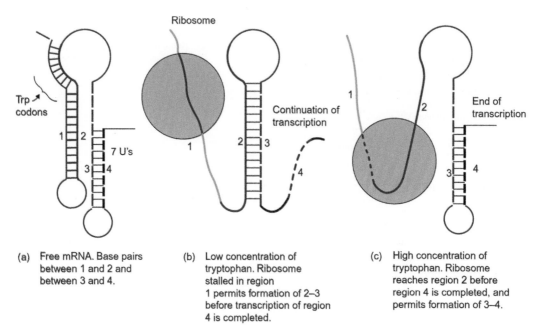

(a) Free mRNA. Base pairs between 1 and 2 and between 3 and 4.

(b) Low concentration of tryptophan. Ribosome stalled in region 1 permits formation of 2–3 before transcription of region 4 is completed.

(c) High concentration of tryptophan. Ribosome reaches region 2 before region 4 is completed, and permits formation of 3–4.

FIGURE 24.5 Model for the mechanism of attenuation in the *E. coli trp* operon.

about 10 bases in the mRNA past the codons being translated. When the final codons are being translated, segments 1 and 2 are not paired. In a coupled transcription–translation system, the leading ribosome is not far behind the RNA polymerase molecule. Thus, if the ribosome is in contact with segment 2, when synthesis of segment 4 is being completed, segments 3 and 4 are free to form the duplex region 3–4 without segment 2 competing for segment 3. The presence of the 3–4 stem-and-loop configuration allows termination to occur when the terminating sequence of seven Us is reached.

If exogenous tryptophan is not present or is present in very small amounts, the concentration of charged tRNA^{Trp} will be limiting, and occasionally a translating ribosome will be stalled for an instant at the tryptophan codons. These codons are located 16 bases before the beginning of segment 2. Thus, segment 2 will be free before segment 4 has been translated, and the 2–3 duplex will form. In the absence of the 3–4 stem-and-loop, termination will not occur, and the complete mRNA molecule will be made, including the coding sequences for the *trp* genes. Thus, once again, if tryptophan is present in excess, termination occurs and little enzyme is synthesized; if tryptophan is absent, there is no termination and the enzymes are made. At intermediate concentrations, the frequency of ribosome pausing will be such as to maintain the optimal concentrations of enzymes (Figure 24.3). This tryptophan regulatory mechanism is called **attenuation** and has been observed for several amino acid biosynthetic operons, e.g., histidine and phenylalanine. Some bacterial operons are regulated solely by attenuation without repressor–operator interactions.

Temporal mRNA Regulation in Phage Systems

The *lac* and *trp* operons are regulated by molecules that turn gene expression on and off in response to the concentration of nutrients. However, some organisms require regulation of genes with time as the key consideration. For example, in the production of phages (bacteriophages or bacterial viruses) in an infected cell, various biosynthetic activities must occur on schedule. When a bacterium is infected by a phage, the enzymes responsible for breaking open the cell and releasing progeny phages must act late in the life cycle after phage DNA replication and packaging into capsid head proteins have occurred—otherwise, the cell could burst before any phage is formed. Almost all phage systems accomplish this temporal regulation by selective and sequential transcription of particular sets of genes. This transcription gives rise to various classes of mRNA that are usually termed **early** and **late** **mRNA**. Three *E. coli* phage systems, which accomplish temporal regulation in different ways, are described next.

Phage T7 contains several promoters, but only one is recognized by *E. coli* RNA polymerase. Transcription of T7 DNA begins at this promoter. Two important proteins are encoded there by transcript (early mRNA). One phosphorylates and inactivates the *E. coli* RNA polymerase, thus preventing *E. coli* from making any bacterial mRNA. The second protein is a new phage RNA polymerase that does not recognize any *E. coli* promoters but is active at the remaining phage promoters. In the absence of *E. coli* RNA polymerase, the early mRNA is no longer made, but

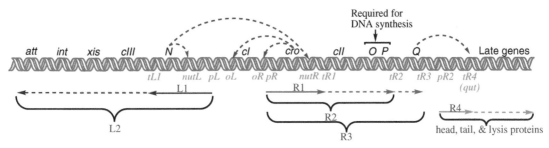

FIGURE 24.6 Genetic map of the regulatory genes of phage λ. Genes are listed above the DNA double helix; sites are below the DNA. The mRNA molecules are the dashed and solid arrows in different colors. The dashed red arrows indicate the sites of action of the N, Cro, and Q proteins.

the new T7 RNA polymerase makes the second transcript, which encodes the DNA replication enzymes and other proteins needed early in the life cycle. Transcription at the third promoter is delayed because injection of the phage DNA into the bacterium occurs very slowly. Roughly halfway through the life cycle of the phage, this promoter enters the cell, transcription occurs by T7 RNA polymerase, and the structural proteins and lysis enzymes are synthesized. Lysis does not occur until phage particles have been assembled.

Phage T4 has numerous promoters, only a few of which can be recognized by *E. coli* RNA polymerase. However, unlike T7, the late promoters are made available by successive modification of the *E. coli* enzyme. These modifications are of two types: addition of phage-encoded protein subunits and chemical modification of pre-existing subunits. Temporal regulation occurs because the gene responsible for the first modification is encoded in the first set of mRNAs, that for the second modification in the second set, and so on. To ensure that the late mRNA, which encodes the structural proteins and the lysis enzyme, is not synthesized until adequate DNA has been made by the replication system, the template for this late mRNA cannot be parental T4 DNA but must be a replica.

E. coli **phage λ** also uses *E. coli* RNA polymerase throughout its life cycle. With this phage, regulation of transcription is accomplished in two ways: modification of the positive regulatory elements, such as the cAMP−CRP complex, needed for recognition of certain promoters; specific repressors are also used to turn off expression of certain other genes. Figure 24.6 shows a portion of the genetic map of λ, which includes four regulatory genes *(cro, N, Q,* and *cII)*, three promoters (pL, pR, and pR2), and five termination sites (tLl, tRl, tR2, tR3, and tR4). Seven mRNA molecules are also shown; the L and R series are transcribed leftward and rightward, respectively, from complementary DNA strands. The genes *O* and *P* encode proteins required for λDNA replication, and the late genes encode the structural proteins of the phage and the lysis enzyme. The basic transcription sequence is as follows: two early mRNAs, Ll

and R1, are made that encode the regulatory proteins N and Cro, respectively. N is an antitermination protein and enables *E. coli* RNA polymerase to ignore certain transcription termination sites and thereby extend synthesis of these mRNAs. It acts together with a bacterial protein, NusA, by binding to a site called **nutR** in the λDNA. When RNA polymerase, which has initiated transcription at pR, reaches this site, it picks up the N−NusA complex and is thereby modified such that it is able to ignore the tR1 and tR2 terminators. A similar site, *nut*L, is present downstream from the pL promoter. Because of this antitermination effect, in time R1 is extended until the DNA replication proteins and another regulatory protein, Q, are made. Q is also an antitermination protein. A small constitutively synthesized RNA, R4, is made from the outset of the infection. The Q protein binds to a site *(qut)* downstream from the promoter for R4, causing RNA polymerase to prevent termination (antitermination), and R4 is extended to form an mRNA that encodes the head, tail, and lysis proteins. From the extended R1 mRNA, the gene-Cro protein is made. The concentration of Cro ultimately reaches a value at which the protein dimerizes, producing the active form. The Cro dimer acts as a repressor at both pL and pR; this activity turns off synthesis of early proteins that are no longer needed and prevents excess synthesis of DNA replication proteins.

Lambda, like many phages (but not T4 or T7), can engage in two alternative life cycles. In the **lytic cycle**, progeny phage particles are produced and the cell ultimately lyses, releasing the phages to the surrounding medium. In the **lysogenic cycle**, injected phage DNA is repressed and becomes inserted into the bacterial chromosome. At a later time, if the bacterial DNA is damaged sufficiently, the phage DNA is excised and a lytic cycle ensues. DNA damage, in a complicated way, leads indirectly to inactivation of the cell repressor and subsequent excision of the phage DNA and production of progeny phage. The *int* gene (Figure 24.6) encodes the enzyme called **integrase** that causes insertion of the phage DNA into the bacterial chromosome. Neither the cI repressor nor the Int protein is needed in the lytic cycle; however, both are needed in the lysogenic cycle and are coordinately regulated. The two products, cI and Int, are encoded in different mRNA molecules, but synthesis of the mRNA is initiated by a common signal. The product of the gene *cII* is a positive regulatory element. Like the cAMP–CRP complex in the *lac* operon, the *cII* product must be bound to the DNA adjacent to the promoters for the mRNAs encoding the cI and Int proteins. In this way, the choice of the lytic versus the lysogenic cycle depends on the concentration of the cII protein. If the concentration is low, neither cI repressor nor Int is made and the lytic cycle is followed; if the concentration is high, both cI repressor and Int are made and the lysogenic cycle occurs. The concentration of cII protein is regulated in response to environmental influences—e.g., increased levels of cAMP–CRP complexes due to low glucose level, and increased expression of cIII proteins that stabilize the cII.

GENE REGULATION IN EUKARYOTES

Regulation of gene expression in eukaryotes is much more complicated and involves more levels than in prokaryotes, mainly due to the larger genome size, the association of DNA with histone complexes, the existence of intracellular compartments, and the uncoupled processes of transcription and translation. Synthesis of mRNA in eukaryotes is not a simple matter of initiation at a promoter, as it is in prokaryotes, but includes several steps in which the primary transcript is converted to mRNA (Chapter 23). Control of these processing steps is also used to regulate gene expression in eukaryotes.

Various differences between the regulatory mechanisms of prokaryotes and eukaryotes are evident. First, transcription of related genes is initiated in response to a single signal, a common step in both prokaryotes and eukaryotes. Second, in eukaryotes, a single polygenic primary transcript may be differentially processed to yield a set of distinct mRNA molecules, each encoding one protein. Third, in eukaryotes, large proteins can be processed into small, active polypeptides.

Housekeeping Genes and Highly Regulated Genes

Many proteins in eukaryotic cells such as those involved in glycolysis (e.g., glucose-6-phosphate isomerase and glyceraldehyde-3-phosphate dehydrogenase; see Chapter 12) are needed continuously to maintain basal cellular functions; the genes encoding such proteins are called **housekeeping genes**. A variety of regulatory mechanisms have evolved to ensure a constant supply of

these gene products. In some instances the amount of each protein may be regulated by the strength of the promoter and ribosomal binding site. However, in housekeeping genes with strong promoters, the gene product functions as a repressor and binds to a site adjacent to the gene, thus regulating the level of transcription. This mechanism is called **autoregulation**. If the gene product is in short supply, transcription is activated; as the concentration of the product increases, the level of transcription is reduced. Due to their steady expression levels in all cells of a multicellular organism, housekeeping gene expressions are used as controls to compare the rate of other gene expressions in many experimental applications.

Unlike housekeeping genes, many eukaryotic genes are developmentally and environmentally regulated. Similarly to the way it regulates the *lac* and *trp* operons in prokaryotes, the environment that an organism is living in can affect the regulation of gene expression in eukaryotes too. Since many eukaryotes are multicellular organisms, environmental factors such as nutrition, temperature, stress, oxygen pressure, and infection can modulate the expression of various genes through the neuroendocrine signaling system. For example, insulin in a fed state and glucagon in a fasting state regulate the expression of various genes involved in energy metabolism (discussed in Chapter 20).

A well-studied example of developmentally regulated genes is **hemoglobin**, a tetrameric protein containing two α subunits and two β subunits. There are several different forms of both α and β subunits differing by only one or a few amino acids, and the forms that are present depend on the stage of development of an organism (discussed in Chapter 26). For example, the following temporal sequences show the subunit types present in humans at various times after conception:

Subunit Type	Embryonic ($<$8 wk)	Fetal (8–41 wk)	Adult (birth $\sim$)
α-like	$\zeta_2 \rightarrow \zeta_1$	α	α
β-like	ε	γ^G and γ^A	β and δ

Both the α- and β-like gene form distinct gene clusters. A property of each cluster is that the order of the genes is the order in which they are expressed in development. Two important regulatory elements for the developmental regulation of hemoglobin have been identified: (1) hypersensitivity site-40 (HS-40) for α-like globin genes on chromosome 16 and (2) locus control region (LCR) for β-like gene clusters on chromosome 11. Through complicated interactions between a variety of transcription factors, hemoglobin subunit genes are activated or silenced at the proper developmental stages.

MECHANISMS OF GENE REGULATION IN EUKARYOTES

Eukaryotic gene expression can be regulated at many different levels:

1. Chromatin structure modifications;
2. Transcriptional initiation control by transcription factors;
3. Primary transcript processing such as capping, splicing, and polyadenylation;
4. Transport of the processed mRNA from the nucleus to the cytoplasm;
5. Stability of mRNA transported into the cytoplasm;
6. Translational initiation (synthesis of protein); and
7. Post-translational modifications of protein(s).

Each of these regulatory steps is crucial to the proper functioning of the cell and the organism. As in prokaryotes, control of transcriptional initiation is the major mechanism for regulation of gene expression.

Chromatin Structure Modifications

DNA Methylation and Gene Silencing

Unlike in prokaryotes, DNA in eukaryotes is tightly packaged in the form of chromatin, as discussed in Chapter 21. In the nucleus of eukaryotic cells, there are two different types of chromatin: **euchromatin** and **heterochromatin**. These chromatin structures play a role in the regulation of gene expression. Tightly packaged DNA in heterochromatin regions is not transcriptionally active, whereas less densely packaged DNA in euchromatin regions is transcriptionally active and is accessible to many transcription factors and RNA polymerases. Therefore, through the changes in chromatin structures, gene expression can be controlled. Chromatin structures are modified by two major mechanisms: **DNA methylation** and **histone modifications**.

Most cells contain several enzymes that transfer a methyl group from S-adenosylmethionine (Chapter 15) to cytosine or adenine in DNA. These **DNA methyltransferases** (**DNMTs**) are both base-specific and sequence-specific. DNMT3A and DNMT3B, which methylate cytosine only in particular CpG sequences (CpG islands), are important for transcription. The methylation of CpG-rich promoter sequences prevents transcription of some genes. DNA methylation can directly inactivate gene expression by

1. Inhibition of transcription factor binding to methylated CpG promoter sequences; and
2. Promotion of heterochromatin formation through the recruitment of methyl-CpG-binding proteins (MeCPs) that associate with histone-modifying enzymes called **histone deacetylases** (**HDACs**). DNA methylation and

histone deacetylation together promote the formation of a more compact heterochromatin structure that prevents transcription initiation by transcription factors.

Undifferentiated and precursor cells often replicate. If methylation actually prevents gene expression in some types of cells, an inhibitory methylated site must be inherited as a methylated site in a daughter strand during DNA replication. This requires methylation of a sequence complementary (rather than identical) to a methylated site. A property of certain DNA methyltransferases (e.g., DNMT1) shows how a methylated parental strand can direct methylation of the appropriate daughter sequence; i.e., these enzymes only methylate CpG (embedded in certain surrounding sequences) and only when the CpG in the opposite strand is already methylated (Figure 24.7). In this way, the methylation pattern of parental DNA strands is inherited by the daughter strands, providing a mechanism for epigenetic inheritance. Many cancer cells exhibit aberrant DNA methylation patterns when compared to normal cells. The changed methylation pattern in these cells may lead to chromatin remodeling by the action of histone deacetylases and to abnormal gene expression. The two inhibitors of DNMTs, azacitidine and decitabine, are approved by the U.S. Food and Drug Administration (FDA) to treat myelodysplastic syndrome and acute myeloid leukemia (AML). In addition, the two U.S. FDA-approved HDAC inhibitors, vorinostat and romidepsin, are used to treat cutaneous T-cell lymphoma.

X-Chromosome Inactivation by DNA Methylation and Histone Methylation

The sex chromosome composition of human males and females is XY and XX, respectively. If cells contain more than one X-chromosome, all except one are inactivated. The cells of XX females contain a structure called a **Barr body**, which is a condensed, heterochromatic, transcriptionally inactive X-chromosome. The cells of XXY males also contain a Barr body. XXY males suffer from **Klinefelter's syndrome**; they are usually mentally retarded, sterile, and suffer from physiological and developmental abnormalities. It is thought that extra X-chromosomes must be prevented from gene expression to preserve the correct gene dosage in both males and females.

Evidence for inactivation of one X-chromosome in XX females comes from the observation that females are mosaic for X-linked alleles that are heterozygous. For example, a woman who is heterozygous for a gene that controls production of sweat glands has patches of skin that perspire and patches that do not. Cells in the patches of skin that do not perspire express the mutant allele, and the wild-type allele is silent in the inactive X-chromosome. Cells in the patches of skin that do perspire express the wild-type allele, and the mutant allele is

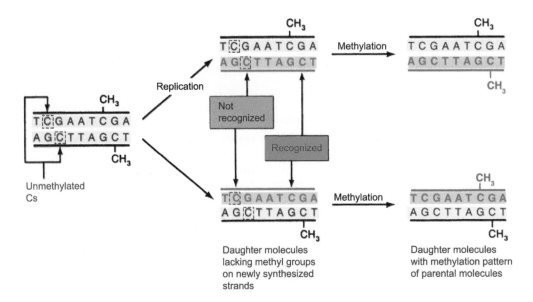

FIGURE 24.7 Mechanism by which the pattern of methylation in parental DNA is inherited in daughter molecules. The rule is that a C in a CpG sequence can be methylated only if the C in the complementary CpG sequence is already methylated. For clarity only, methyl groups have been drawn outside the sugar-phosphate strands.

silent. X-chromosome inactivation occurs early in embryonic female development, and the X-chromosome that is inactivated is selected at random. Once selected, the same X-chromosome is inactivated in subsequent cell divisions, which explains why patches of skin and other tissues differ in the expression of X-linked heterozygous genetic loci.

Inactivation of the X-chromosome is thought to occur in three phases: initiation, spreading, and maintenance. Initiation involves selection of the chromosome to be inactivated and depends on a unique genetic locus on the X-chromosome called the **X-inactivation center** (*Xic*). *Xic* includes many genes encoding long, noncoding RNAs (lncRNAs) that are involved in multiple steps of X-inactivation. One of the lncRNAs is a 19 kb RNA, **X-inactive-specific transcript** (Xist). During the initiation phase, Xist binds to Polycomb repressive complex 2 (PRC2), which mediates trimethylation of the lysine residue at position 27 of histone H3 (H3K27me3). Xist serves as a guide for PRC2 targets on the X-chromosome. Also, methylation of CpG islands at the beginning of genes in the inactive X-chromosome plays an important role in suppressing gene expression; the corresponding CpG islands in the active X-chromosome are usually unmethylated. In addition to the hypermethylation of CpG islands in the inactive X-chromosome, Xist-induced histone methylation will trigger changes in chromatin structure, converting euchromatin into the more compact form of heterochromatin that silences most genes on the X-chromosome.

Once one X-inactivation center has been triggered in a cell, the other X-chromosome is protected from inactivation by a "blocking signal." The inactivation signal spreads from the *Xic* locus, eventually inactivating almost all of the genes on that X-chromosome. However, at least 19 genes on inactive X-chromosomes have been shown to have some transcriptional activity, so inactivation is not complete. Specific maintenance mechanisms ensure that the inactive X-chromosome is clonally transmitted and that it remains inactive in subsequent cell divisions.

Genomic Imprinting and Epigenetics

It was estimated that as many as 600 autosomal genes in humans are inherited in a silent state from one parent and in a fully active state from the other parent. Such genes are said to be "imprinted" by the parents during gamete formation; this phenomenon is referred to as **genomic imprinting**, which does not change the nucleotide sequence of DNA. Rather, it is an epigenetic phenomenon in which the DNA at a particular locus is marked according to gender, and this determines whether or not the locus is expressed during embryonic development. **Epigenetics** is the term that refers to the inheritable changes in gene expression, which are not coded by the base sequences of DNA. Epigenetic control of gene expression occurs mainly during development by DNA methylation and histone modifications. Genomic imprinting and inactivation of X-chromosomes are examples of epigenetic control of expression of certain genes. Although regarded as a normal epigenetic phenomenon, two clinically different disorders are related to genomic imprinting.

Prader–Willi and **Angelman syndromes** are distinct disorders associated with multiple abnormalities and mental retardation. Both disorders are caused by mutations at the proximal end of chromosome 15 that silence one or more genes. In Prader—Willi syndrome, the paternally inherited chromosome 15 has deletions and the maternally inherited chromosome is silent. In Angelman syndrome, the opposite is the case in that the maternally inherited chromosome 15 has deletions and the paternally inherited chromosome is silent. For both syndromes, a small deletion is usually responsible for the genomic imprinting that silences the relevant gene(s). Both syndromes can also result from having both copies of chromosome 15 derive from only one parent, a condition called **uniparental disomy**. Angelman syndrome is caused by mutation in a single gene, whereas Prader—Willi syndrome is caused by mutations in more than one gene. Thus, Angelman syndrome can also be caused by mutations in the responsible gene itself, which is not the case for Prader—Willi syndrome. Diagnosis of both can be confirmed by analysis of DNA methylation in the respective genes, since methylation is the mechanism used for imprinting.

Chromatin Remodeling by Covalent Histone Modifications

Other covalent modifications of histone protein complexes can remodel the structure of chromatin. DNA double helices in chromatin tightly wrap around the histone complexes, since histones are positively charged and DNA double helix backbones are negatively charged. Therefore, modifications of histone complexes such as acetylation can result in loosening of the association between histone complexes and DNA, thereby allowing DNA-binding proteins and transcription factors to initiate gene expression. Acetylation of the basic amino acid (lysine) rich histone proteins lowers the binding affinity between DNA and histone complexes. Therefore, acetylation of histone protein **activates** and deacetylation **inhibits** gene expression in eukaryotes. Histone acetylation is catalyzed by enzymes called **histone acetyltransferases (HATs)**, and deacetylation is catalyzed by **histone deacetylases (HDACs)**.

Chromatin structure can also be modified through methylation of histone proteins. Unlike acetylation of histone complex, methylation of histone proteins on arginine and lysine residues can result in either repression or activation of gene expression. For example, methylation of lysine 4 of histone 3 (H3K4me) can lead to the activation of gene expression, and methylation of the lysine 9 and 27 residues of histone 3 (H3K9me and H3K27me) is known to inhibit gene expression through heterochromatin formation.

Other post-translational modifications of histones that affect gene expression include formylation,

phosphorylation, sumoylation, ubiquitination, ADP-ribosylation, glycosylation, and hydroxylation of various amino acids on histone proteins.

In addition to the structural changes induced by DNA methylation and histone modification, ATP-dependent **chromatin remodeling complexes** can regulate gene expression by (1) remodeling nucleosome complexes, (2) removing nucleosomes on target DNA sequences, and (3) replacing the original histone proteins with variant proteins. The remodeling complexes are thought to be recruited to their target DNA sequences through interactions with transcription activators and through recognition of histone modifications. The changes introduced in chromatin structure by remodeling complexes enable transcription initiation to proceed by making regulatory DNA elements accessible to other transcription factors.

Transcriptional Initiation by Transcription Factors

The synthesis of mRNA molecules by RNA polymerase II (RNAPII) is a multistep process requiring the interaction of numerous proteins with DNA in the region of the promoter (Chapter 23). However, before the expression of a gene can be initiated, the **transcription factor IID (TFIID)** must bind to the promoter. This is followed by the binding of other proteins called **general transcription factors (GTFs)** to DNA in the region of the promoter. A description of the functions of some human GTFs is given in Table 24.2. The various GTFs facilitate binding of RNAPII to the promoter at the correct nucleotide for initiation, destabilize the DNA at the promoter, and initiate transcription; together the GTFs and RNAPII are called the **pre-initiation complex**. Once the first phosphodiester bond has formed, transcription has been initiated.

The large number of GTFs required to initiate transcription does not completely solve the transcription problem. DNA is organized in chromatin and is wrapped around histone proteins and tightly packaged into nucleosomes (Chapter 22). These structures inhibit transcription, as well as DNA replication. A distinct class of transcription factors has been identified that modifies chromatin structure so that transcription can occur.

In addition to the GTFs that control the initiation of transcription at the TATA box and the protein factors that disassemble nucleosomes, other transcription factors are required to regulate the expression of particular genes or families of genes. Each transcription factor binds to DNA at a specific sequence, which ensures specificity (Table 24.3). The large numbers of transcription factors have certain structural motifs that facilitate their binding to DNA and, based on structural similarities, fall into five distinct groups (Figure 24.8).

TABLE 24.2 The Role of General Transcription Factors in Initiation of Gene Expression in Human Cells

Factor	Subunits*	Function
TFIID	13	Recognizes TATA box; recruits TFIIB (composed of factors TBP and TAFs)
TFIIA	3	Stabilizes TFIID binding
TFIIB	1	Orients RNAPII to start site
TFIIE	2	Recruits TFIIH (helicase)
TFIIF	2	Destabilizes nonspecific RNAPII–DNA interactions
TFIIH	9	Promotes promoter melting by helicase activity

RNAPII contains nine subunits.

TABLE 24.3 Promoter and Enhancer Controlling Elements and Transcription Factors That Bind to Them

Consensus Sequence*	Response Element	Factor
TATAAAA	TATA Box	TFIID
GGCCAATCT	CAAT Box	CTG/NF1
GGGCGG	GC Box	SP1
ATTTGCAT	Octamer	Oct1, Oct2
CNNGAANNTCCNNG	HSE	Heat shock factor
TGGTACAAATGTTCT	GRE	Glucocorticoid receptor
CAGGGACGTGACCGCA	TRE	Thyroid receptor
CCATATTAGG	SRE	Steroids

Consensus sequence is for either a promoter or enhancer element. The letter N stands for any nucleotide (A, G, C, T).

1. **Zinc finger:** The structure of the zinc finger motif (discovered in TFIIA) consists of two cysteine residues and two histidine residues separated by 12 amino acids. The two cysteine residues are separated by two amino acids, and the two histidine residues are separated by three amino acids. The cysteine and histidine residues are linked by a zinc ion, and this motif is repeated nine times in TFIIA protein. The 12 amino acids between the cysteine and histidine residues form a loop that can interact with the major groove in DNA; this loop is called the **zinc finger**.

2. **Leucine zipper:** The structure of the leucine zipper (first discovered in C/EBP, CCAAT, and enhancer-binding protein) consists of four leucine residues in an α-helical segment of the protein. Two polypeptides join to form a Y-shaped dimer whose arms can interact with the major groove of DNA. The stem of the Y-shaped structure is known as the **leucine zipper**.

3. **Helix–turn–helix (HTH):** This structure (first discovered in the λ bacteriophage repressors, cI and cro) consists of two α-helical regions of a protein joined by an amino acid sequence that allows the α-helices to turn. The two α-helices form a dyad axis of symmetry, a structure that binds tightly to the major groove of DNA.

4. **Helix–loop–helix (HLH):** This structure is similar to the helix–turn–helix in that it consists of two α-helical segments joined by a long sequence of amino acids that can form a loop. This gives the two α-helical segments more flexibility so that the proteins can fit into the large groove of DNA at some distance from one another. This may facilitate looping of the DNA to make enhancer or promoter sites more accessible.

5. **Homeodomain:** This structural domain found in *Drosophila* homeobox genes consists of 60 amino acids forming three α-helixes. The C-terminal helix has a hydrophobic region that is responsible for interacting with bases in the major groove of the DNA, and other regions of the homeodomain structure contain arginine and lysine residues that stabilize the interaction between the protein and DNA by forming hydrogen bonds with the negatively charged DNA backbone.

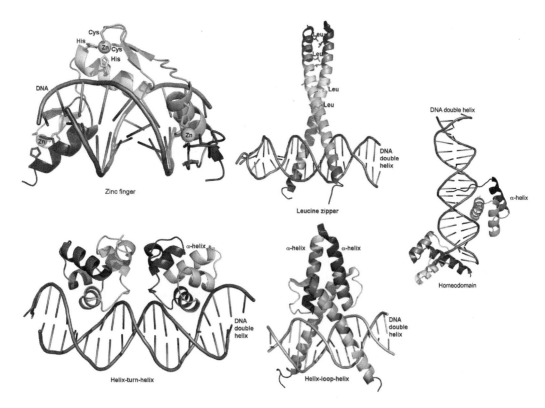

FIGURE 24.8 Structures of transcription factors that facilitate binding to DNA at specific sequences to regulate gene expression.

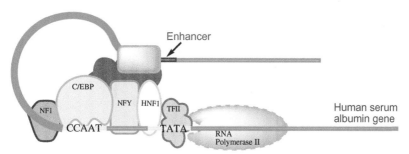

FIGURE 24.9 Expression of the human serum albumin gene is regulated by five transcription factors, four of which bind to the promoter region. TFII = transcription factor II, HNF1 = Hepatocyte nuclear factor 1, NFY = nuclear factor Y, C/EBP = CCAAT/Enhancer binding protein, NF1 = nuclear factor 1.

The serum albumin gene is one example of a gene that is regulated by several transcription factors. Although this gene is present in all tissues, it is expressed only in the liver and spleen. The gene is activated by five different transcription factors that bind DNA in a region located between the CCAAT and TATA boxes (Figure 24.9). The transcription factor NFl actually binds to the right of the CCAAT box. Binding sites for transcription factors are determined by mutating bases in the DNA one at a time and observing changes in the binding of transcription factors.

Steroid Receptors

Steroid hormones perform many functions in cells, one of which is to activate gene expression by binding to **steroid receptors**, proteins in the cytoplasm that, when activated, act as transcription factors that initiate transcription. All steroid hormones are derived from cholesterol and, as a result, have similar chemical structures. Steroid hormones differ one from another primarily in hydroxylation of particular carbon atoms and by aromatization of the steroid

A ring of the molecule. Once a steroid hormone binds to a steroid receptor protein, the complex undergoes a series of structural changes that result in the complex binding to DNA at a particular sequence called a **steroid response element (SRE)** located at some distance upstream or downstream from the promoter. The steroid hormone-receptor complexes induce changes in chromatin structure by binding to their SREs, which alters the transcriptional activities of their target genes.

Steroid receptor proteins are synthesized from a gene family that shows a high degree of homology in the DNA-binding region. All steroid receptors belong to one of five classes: androgen receptor (AR), estrogen receptor (ER), glucocorticoid receptor (GR), mineralocorticoid receptor (MR), and progesterone receptor (PR). Also, all receptors contain a zinc finger motif that, if altered by mutation, destroys the steroid receptor's function.

Regulation of RNA Processing

Initiation of transcription ultimately leads to production of a primary transcript, which in eukaryotes is processed to form an mRNA. Alternative processing patterns can yield different mRNAs. One example comes from chicken skeletal muscle in which two forms of the muscle protein myosin, LC1 and LC3, are produced. The myosin gene has two different TATA sequences that yield two different primary transcripts. These two transcripts are processed differently to form mRNA molecules encoding distinct forms of the protein (Figure 24.10).

Another type of regulation of processing involves choice of different sites of polyadenylation. One example is the differential synthesis of the hormone **calcitonin** in different tissues; another is the synthesis of two forms of the heavy chain of immunoglobulins (Chapter 33). In both cases, the differential processing includes distinct patterns of intron excision (i.e., splicing), but they are necessitated by an earlier event in which different poly (A) sites are selected from the primary transcript. That is, when the poly(A) site nearer the promoter is selected, a splice site used in the larger primary transcript is not present, so a different splice pattern results. Thus, slightly different proteins are synthesized.

The calcitonin gene consists of five exons and uses two alternative polyadenylation sites that respond to different signals in different tissues. In the thyroid,

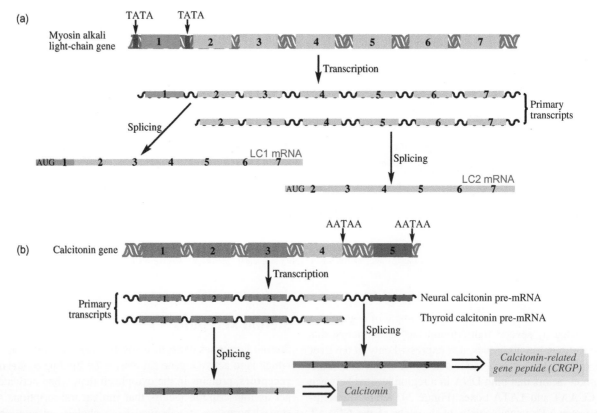

FIGURE 24.10 Alternative splicing of chicken LC1/LC3 and calcitonin genes. (a) Two distinct TATA sequences lead to production of different primary transcripts, which contain the same coding sequences. Two modes of splicing lead to the formation of distinct mRNA molecules that encode proteins having different amino terminal regions and the same carboxy terminal regions. (b) The differential expression of calcitonin genes in different tissues. Two different splicing patterns lead to the production of two different proteins, depending on the location of different polyadenylation sites.

calcitonin is produced by a signal that produces a pre-mRNA consisting of exons 1–4. The introns are then spliced out to give the mature mRNA. However, in neural tissue a different polyadenylation site is activated that produces a pre-mRNA consisting of exons 1–5 and the intervening introns. When this larger pre-mRNA is processed, the introns are spliced out, but so is exon 4, producing a mature mRNA consisting of exons 1–3 and 5. This is translated into a growth factor called **calcitonin-related gene peptide (CRGP)**. Mutations in the calcitonin gene can result in both adrenal and thyroid tumors, conditions that are classified as **multiple endocrine neoplasias**.

Alternative RNA Splicing and Editing

Most primary transcripts in eukaryotic cells derive from complete removal of all introns and complete joining of all exons. This results in only one species of mature mRNA being synthesized from each primary transcript. However, many eukaryotic genes can give rise to many different forms of mature mRNAs using different promoters and different polyadenylation sites (discussed previously), as well as by **alternative splicing**. This form of regulation of RNA processing is especially useful because a single gene can be expressed differently at various developmental stages or in different tissues of the same organism. One example is the *troponin T* gene that synthesizes the fast skeletal muscle protein. This gene consists of 18 exons; however, only 11 are found in all mature mRNAs. Five of the exons can be included or excluded, and two are mutually exclusive; if one is included, the other is excluded. Altogether, 64 different mature mRNAs can be produced by alternative splicing of the primary transcript of the *troponin T* gene.

RNA editing involves processing of RNA in the nucleus by enzymes that change a single nucleotide (insertion, deletion, or substitution). One example is the *apolipoprotein B* (apo B) gene. In the liver, this gene produces a 4536-amino acid protein (apo B-100), whereas in the small intestine, the same gene produces a 2152-amino acid protein (apo B-48). The truncated protein is identical in amino acid sequence to the first 2152 amino acids in apo B-100. This occurs because in cells of the small intestine one nucleotide (at position 6666) is edited by deamination of a cytosine residue by a sequence-specific **cytidine deaminase**. The conversion of a cytosine to uracil at this position produces a stop codon that terminates translation and produces the truncated protein. Thus, selective editing of mRNAs prior to translation in specific tissues is used to produce different proteins.

RNA Transport from the Nucleus to the Cytoplasm

Post-transcriptional modifications of pre-mRNA, such as capping, splicing, and polyadenylation, take place in the nucleus. After these modifications have been completed, the mature mRNA molecules have to be translocated into the cytoplasm, where protein synthesis occurs. During the pre-mRNA processing phase, many nuclear proteins are recruited to the mRNA being processed to form RNA-protein complexes, and only a successfully processed mRNA will form a proper mRNA–protein complex, which is required to pass through the nuclear pore complexes. Therefore, defective mRNA and other byproduct RNA molecules, such as spliced-out introns, will be subjected to degradation by a multi-subunit exosome complex specialized for RNA degradation. Mutations found in one of the exosome components (EXOSC3) are linked to a human disease called **pontocerebellar hypoplasia type 1**. Other RNA molecules are also subjected to similar transport mechanisms to get into the cytoplasm.

Translational Regulation

Translational regulation refers to the number of times a finished mRNA molecule is translated. The three ways in which translation of a particular mRNA may be regulated are

1. By the lifetime of the mRNA;
2. By the probability of initiation of translation; and
3. By regulation of the rate of overall protein synthesis.

The silk gland of the silkworm *Bombyx mori* predominantly synthesizes a single type of protein, **silk fibroin**. Since the worm takes several days to construct its cocoon, it is the amount and not the rate of fibroin synthesis that must be great; hence, the silkworm can manage with a fibroin mRNA molecule that is very long-lived. Silk production begins with chromatid amplification in which, over a period of several days, the single cell of which the gland is composed has a ploidy of 10^6. Each fibroin gene is transcribed from a strong promoter to yield about 10^4 fibroin mRNA molecules. An "average" eukaryotic mRNA molecule has a lifetime of 3 hours before it is degraded. However, fibroin mRNA survives for several days, during which time each mRNA molecule is translated repeatedly to yield 10^5 fibroin molecules. Thus, the whole unicellular silk gland makes 10^{15} molecules or 300 mg of fibroin.

Production of a large amount of a single type of protein by means of a prolonged mRNA lifetime is common in highly differentiated cells. For example, cells of the chicken oviduct, which makes ovalbumin (for egg white), contain only a single copy of the ovalbumin gene per haploid set of chromosomes, but the cellular mRNA is long-lived.

Translational and transcriptional controls are sometimes combined. For example, **insulin** (which regulates the synthesis of a large number of substances) and **prolactin** (another hormone) are required together for production of casein (milk protein) in mammary tissue. Both hormones are needed to initiate transcription, but prolactin, in addition, increases the lifetime of casein mRNA.

The synthesis of some proteins is regulated by direct action of the protein on the mRNA. For instance, the concentration of one type of immunoglobulin is kept constant by self-inhibition of translation. This protein, like all immunoglobulins, consists of two H chains and two L chains. The tetramer binds specifically to H-chain mRNA and thereby inhibits initiation of translation.

Regulation of Translation by miRNAs and siRNAs

Two classes of small noncoding RNA molecules, **microRNA (miRNA)** and **small interfering RNA**

(**siRNA**), are involved in the regulation of gene expression at post-transcriptional level. miRNAs consist of up to 25 nucleotide RNA sequences, which are partially complementary to 3′ UTR sequences of target mRNAs. Their hairpin-looped primary miRNA (pri-miRNA) transcripts are first cleaved by the action of an RNase III enzyme called **Drosha** in the nucleus, and the processed pre-miRNAs are transported into the cytosol by a transporting molecule called **exportin-5**. In the cytosol, the pre-miRNAs are further processed to double-stranded miRNA/miRNA duplexes by the catalytic action of another RNase III enzyme called **Dicer**, which removes the loop portion of double-stranded pre-miRNAs. The resulting miRNA duplexes form the multienzyme complex, **RNA-induced silencing complex (RISC)** with an endonuclease, **Argonaute 2 (AGO2)**, and other proteins (Figure 24.11). miRNAs of RISC direct the RISC to the target mRNA sequences through partial sequence complementarity for the repression of gene expression.

siRNAs are double-stranded RNA (dsRNA) molecules, up to 21 nucleotides in length. The source of siRNAs can be endogenous (transcribed from the genome in the nucleus) and exogenous (viral dsRNAs or

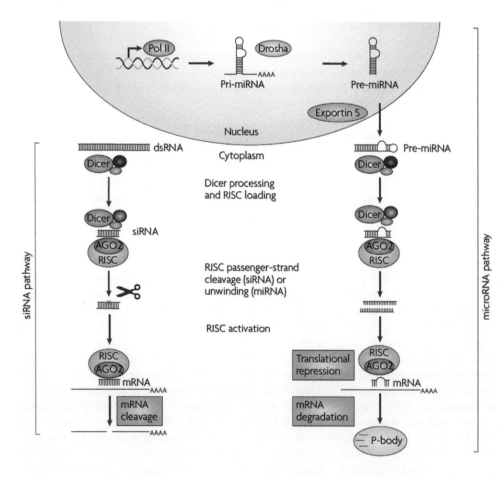

FIGURE 24.11 Mechanism of RNA interference by miRNA and siRNA in mammalian cells. *[Reproduced with permission from A. de Fougerolles, H. P. Vornlocher, J. Maraganore, J. Lieberman Interfering with disease: a progress report on siRNA-based therapeutics. Nat. Rev. Drug Discov. 6(6) (2007) 444, Figure 1.]*

experimentally prepared/injected dsRNAs to disable specific target mRNA). The long dsRNA molecules in the cytosol are processed by the previously mentioned Dicer enzyme, as in miRNA processing, which produces 2-nucleotide overhangs on 3′ ends. These processed dsRNAs are involved in the RNA interference (RNAi) pathway by forming RISC with AGO2 and other specific proteins and leading RISC to the target mRNA sequence for degradation, thereby inhibiting translation of the target mRNAs.

mRNA Stability Regulation by Iron Concentration in Cells

Iron is an important constituent of human cells and regulates many biochemical functions (Chapter 27). Iron acts as an effector molecule in the translation of several mRNAs by binding to either 31 or 51 stem-loop structures, called **iron-response elements** (**IREs**), that flank the coding sequences for several genes whose expression is regulated by iron. In particular, the mRNAs for the transferrin receptor, light and heavy chains of ferritin, an erythrolytic form of aminolevulinic acid synthetase, and a form of mitochondrial aconitase are regulated by IREs and an **IRE-binding protein** (**IREBP**). The IRE in the ferritin mRNA is in the 5′ flanking sequence. Translation of the mRNA is regulated by binding of IREBP, whose activity, in turn, is regulated by the concentration of iron in the cell. In contrast, the IRE of transferrin mRNA is in the 3′ flanking sequence; binding of IREBP to this IRE regulates turnover of the mRNA. Therefore, IREBP binding to IRE of ferritin and transferrin mRNAs will increase the rate of transferrin mRNA translation and inhibit the translation of ferritin mRNA (see Chapter 27 for details). This example illustrates the variety of ways that gene expression can be regulated under varying physiological conditions in human cells. Thus, selective editing of mRNAs prior to translation in specific tissues is used to produce different proteins.

Polyproteins

In prokaryotes, coordinated regulation of the synthesis of several gene products is accomplished by regulation of the synthesis of a single polycistronic mRNA molecule encoding all of the products. The analogue to this arrangement in eukaryotes is the synthesis of a **polyprotein**, a large polypeptide that is cleaved after translation to yield individual proteins. Each protein can be thought of as the product of a single gene. In such a system, the coding sequences of each gene in the polyprotein unit are not separated by stop and start codons but instead by specific amino acid sequences that are recognized as cleavage sites by particular proteases, i.e., protein-cutting enzymes. Polyproteins have been observed with up to eight cleavage sites; the cleavage sites are not cut simultaneously, but rather in a specific order. Use of a polyprotein serves to maintain an equal molar ratio of the constituent proteins; moreover, delay in cutting at certain sites introduces a temporal sequence of production of individual proteins, a mechanism frequently used by animal viruses.

Some polyproteins are differentially cleaved in different tissues. An example is **pro-opiomelanocortin** (**POMC**), a polyprotein that is the source of several hormones synthesized in the pituitary gland. In the anterior lobe of the pituitary, the polyprotein is cleaved to release β-lipotropin and adrenocorticotropic hormone (ACTH). In the intermediate lobe, a different pattern of cleavage forms β-endorphin and α-melanotropin.

Regulation of Protein Activity

Many enzymes contain several subunits and are regulated by a process known as **allosteric regulation**. A common arrangement in enzymes is that the binding sites for the molecule that is acted on (the substrate) and the inhibitor (which may be the product) are located on different subunits. If binding of the inhibitor prevents binding of the substrate, the information from a site on one subunit must somehow be transmitted to the other subunit. This can be accomplished by the following subunit interactions. Binding of the inhibitor molecule alters the shape of the subunit to which it is bound, resulting in changes in the reactive sites on other subunits. If the subunits remain in contact, all subunits adjoining the first will undergo a conformational change at their respective subunit interaction sites, altering, in turn, the substrate-binding site of other subunits. Proteins capable of undergoing such conformational interactions are called **allosteric proteins** (Chapter 4).

In mammalian cells, cAMP is called a **second messenger**, because it regulates the activities of many proteins. Furthermore, certain hormones and cAMP work in concert to regulate enzymatic activities. Many hormones regulate metabolic processes, such as glucose metabolism and calcium utilization, through binding to specific receptors in the cell membranes of target cells. However, many hormones are not capable of penetrating the target cells, and instead the binding of a hormone to a membrane receptor induces intracellular synthesis of cAMP and other second messengers; these second messengers then cause the desired metabolic effects (Chapter 28).

CLINICAL CASE STUDY 24.1 Methylation Status of MGMT Promoter-Glioblastoma

This clinical case was abstracted from: T.T. Batchelor, A.G. Sorenson, D.N. Louis, Case 17-2012: a 54-year-old man with visual-field loss and a mass in the brain, N. Engl. J. Med. 366 (2012) 2112–2120.

Synopsis

A 54-year-old man was admitted to a hospital due to loss of vision, dizziness, and a mass in the brain. Upon examination, nystagmus was noted. The patient's vital signs and laboratory test results, including a complete blood count, blood electrolyte levels, calcium, glucose, coagulation tests, and renal and hepatic functions, were normal. Magnetic resonance imaging (MRI) of the brain showed two adjacent masses in the left occidental and posterior parietal regions. The patient had a history of gastroesophageal reflux disease, *Helicobacter pylori* infection, and hematuria. His uncle had had an inoperable primary brain tumor, but other family members were healthy. The patient showed no lesions outside the brain. Possible diagnosis was a primary brain tumor, most likely glioblastoma, considering his age and MRI images. Craniotomy was performed in order to confirm diagnosis and remove the masses. Examination of the tumor specimen confirmed the glioblastoma diagnosis, and methylation-specific PCR was subsequently performed on DNA samples extracted from the tumor. The result showed a methylated O^6-methylguanine-DNA methyltransferase (MGMT) gene promoter. The patient received chemoradiation therapy for 6 weeks. Nineteen months after the removal procedure, the patient developed recurrent glioblastoma at a different site in the brain. He died 28 months after the initial diagnosis of glioblastoma.

Teaching Points

1. The enzyme, O^6-methylguanine-DNA methyltransferase (MGMT), is important in the direct reversal of the damage repair system, as it removes the O^6-methyl group from modified DNA bases. Excessive methylation of the MGMT promoter suppresses gene expression, thereby resulting in the dysfunction of the direct reversal of the DNA damage repair system.

2. The methylation status of the MGMT gene promoter can be assessed by methylation-specific PCR. In methylation-specific PCR, the DNA samples treated with and without sodium bisulfite will be subjected to PCR with two different sets of primers. Sodium sulfite converts unmethylated cytosines to uracil, and thus by comparing the two PCR reactions, one can determine the methylation status of the DNA samples.

References

[1] J.G. Herman, J.R. Graff, S. Myohanen, B.D. Nelkin, S.B. Baylin, Methylation-specific PCR: a novel PCR assay for methylation status of CpG islands, Proc. Natl Acad. Sci. USA 93 (1996) 9821–9826.

[2] E. Smith, T. Bianco-Miotto, P. Drew, D. Watson, Method for optimizing methylation-specific PCR, BioTechniques 35 (2003) 32–33.

ENRICHMENT READING

[1] F. Muller, L. Tora, Chromatin and DNA sequences in defining promoters for transcription initiation, Biochim. Biophys. Acta 2013 (1839) 118–128.

[2] S. Grunberg, S. Hahn, Structural insights into transcription initiation by RNA polymerase II, Trend. Biochem. Sci. 38 (2013) 603–610.

[3] H. Kwak, J.T. Lis, Control of transcriptional elongation, Annu. Rev. Genet. 47 (2013) 483–508.

[4] M. Ptashne, The chemistry of regulation of genes and other things, J. Biol. Chem. 289 (2014) 5417–5435.

[5] K.V. Morris, J.S. Mattick, The rise of regulatory RNA, Nat. Rev. Genet. 15 (2014) 423–437.

[6] M.A. Zabidi, C.D. Arnold, K. Schernhuber, M. Pagani, M. Rath, O. Frank, et al., Enhancer-core-promoter specificity separates developmental and housekeeping gene regulation, Nature (2014). Available from: http://dx.doi.org/10.1038/nature13994.

[7] T.J. Santangelo, I. Artsimovitch, Termination and antitermination: RNA polymerase runs a stop sign, Nat. Rev. Microbiol. 9 (2011) 319–329.

[8] A.N. Schechter, Hemoglobin research and the origins of molecular medicine, Blood 112 (2008) 3927–3938.

[9] K. Helin, D. Dhanak, Chromatin proteins and modifications as drug targets, Nature 502 (2013) 480–488.

[10] S.B. Rothbart, B.D. Strahl, Interpreting the language of histone and DNA modifications, Biochim. Biophys. Acta 2014 (1839) 627–643.

[11] A. Kohler, E. Hurt, Exporting RNA from the nucleus to the cytoplasm, Nat. Rev. Mol. Cell Biol. 8 (2007) 761–773.

Chapter 25

Nucleotide Metabolism

Key Points

1. Purine and pyrimidine nucleotides are synthesized both *de novo* and by reutilization of preformed bases known as salvage pathways.

2. For the purine nucleotide *de novo* biosynthetic pathway, C and N atoms are derived from glycine (C-4, C-5, and N-7), glutamine, (N-3 and N-9), aspartate (N-1), CO_2 (C-6) and tetrahydrofolate one-carbon derivatives (C-2 and C-8). The first nucleotide synthesized is IMP, and it is the precursor for AMP and GMP synthesis. The pathway is regulated at several steps. Inhibitors of this pathway serve as antineoplastic agents.

3. Conversion of ribonucleotides to their respective deoxy forms occurs exclusively at the nucleoside diphosphate level, catalyzed by a reductase. This step is inhibited by hydroxyurea and is used in the treatment of some cancers.

4. Preformed purines, hypoxanthine, and guanine derived from nucleotide turnover are reutilized (salvaged) by converting them to IMP and GMP catalyzed by HGPRT. HGPRT deficiency causes a severe clinical disorder.

5. The end product of purine catabolism is uric acid, which is relatively insoluble. Xanthine oxidase is the terminal enzyme in the pathway, and it is a molybdopterin flavoprotein. Uric acid is eliminated in the renal excretory system. Xanthine oxidase inhibitors are used in the treatment of hyperuricemic conditions such as gout and tumor lysis syndrome.

6. Adenosine deaminase (ADA) and purine nucleoside phosphorylase (PNP) catalyze sequential steps in the metabolism of ribo- and deoxyriboadenosine, and they are highly expressed in T- and B-lymphocytes. ADA deficiency and PNP deficiency inherited as an autosomal recessive disorder causes immunodeficiency diseases.

7. Unlike purine nucleotide biosynthesis, pyrimidine biosynthesis consists of, first formation of the ring structure utilizing aspartate, glutamine, bicarbonate, and ATP and, then, conversion to a nucleotide derivative. UMP is the parent compound for the synthesis of cytidine and deoxycytidine phosphates and thymidine nucleotides.

8. The donor of the methyl group for thymidylic acid (TMP) synthesis from dUMP is N^5,N^{10}-methylene tetrahydrofolate. Fluorodeoxyuridylate derived from 5-fluorouracil (5-FU) inhibits thymidylate synthetase, and methotrexate inhibits dihydrofolate reductase required for the regeneration of tetrahydrofolate. Both 5-FU and methotrexate are antineoplastic drugs.

9. In humans, the committed step in pyrimidine biosynthesis is the formation of carbamoyl phosphate (CP) catalyzed by CP synthetase II (CPS-II). CPS-II is under feedback allosteric regulation by negative (UDP, UTP) and positive (PRPP, ATP) modulators. Coordination of purine and pyrimidine nucleotide biosynthesis occurs at several steps.

A nucleotide consists of a purine or pyrimidine base, a pentose (or deoxypentose), and a phosphate. The synthesis and degradation of nucleotides are discussed in this chapter. Structural features of nucleotides are discussed in Chapter 21, and how nucleotides are used in the synthesis of DNA and RNA in Chapters 22 and 23, respectively.

Table 25.1 gives the nomenclature of purine and pyrimidine nucleosides and nucleotides. Names of purine nucleosides end in *-osine*, whereas those of pyrimidine nucleosides end in *-idine*; guanine nucleoside is guanosine and should not be confused with guanidine, which is not a nucleic acid base; thymidine is a deoxyriboside.

Nucleotides are synthesized by two types of metabolic pathways: **de novo synthesis** and **salvage pathways**. The former refers to synthesis of purines and pyrimidines from precursor molecules; the latter refers to the conversion of preformed purines and pyrimidines derived from nucleic acid turnover and by addition of ribose-5-phosphate to the base. *De novo* synthesis of purines is based on the metabolism of one-carbon compounds.

ONE-CARBON METABOLISM

One-carbon moieties of different redox states are utilized in the biosynthesis of purine nucleotides and thymidine monophosphate (also known as **thymidylate**). They are also used in the metabolism of several amino acids (particularly serine and homocysteine), the initiation of protein biosynthesis in bacteria and mitochondria by formylation of methionine, and the methylation of a variety of metabolites. These one-carbon reactions utilize coenzymes derived from **folic acid** or **folate**. Folate is a

N. V. Bhagavan and C.-E. Ha: Essentials of Medical Biochemistry, Second edition. DOI: http://dx.doi.org/10.1016/B978-0-12-416687-5.00025-7

TABLE 25.1 Nomenclature of Nucleosides and Nucleotides

Base	Nucleoside (Base-Sugar)*	Nucleotide (Base-Sugar Phosphate)†
Purines		
Adenine (6-aminopurine)	Adenosine	Adenosine monophosphate (AMP) or adenylic acid
	Deoxyadenosine	Deoxyadenosine monophosphate (dAMP)
Guanine (2-amino-6-oxypurine)	Guanosine	Guanosine monophosphate (GMP)
	Deoxyguanosine	Deoxyguanosine monophosphate (dGMP)
Hypoxanthine (6-oxypurine)	Inosine	Inosine monophosphate (IMP)
	Deoxyinosine	Deoxyinosine monophosphate (dIMP)
Xanthine (2,6-dioxypurine)	Xanthosine	Xanthosine monophosphate (XMP)
Pyrimidines		
Cytosine (2-oxy-4-aminopyrimidine)	Cytidine	Cytidine monophosphate (CMP)
	Deoxycitidine	Deoxycytidine monophosphate (dCMP)
Thymine (2,4-dioxy-5-methylpyrimidine)	Thymidine (thymine deoxyriboside)	Thymidine monophosphate (TMP) (thymine deoxyribotide)
Uracil (2,4-dioxypyrimidine)	Uridine	Uridine monophosphate (UMP)

*The sugar residue can be ribose or deoxyribose. If it is deoxyribose, it is identified as such; otherwise, it is assumed to be ribose, with the exception of thymidine, which is deoxyriboside.
†A nucleotide is a nucleoside monophosphate; the monophosphates are sometimes named as acids.

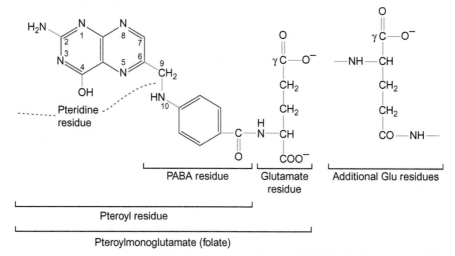

FIGURE 25.1 Structure of folic acid showing its components. The numbered parts participate in one-carbon transfer reactions. In nature, folate occurs largely as polyglutamyl derivatives in which the glutamate residues are attached by isopeptide linkages via the γ-carboxyl group. The pteridine ring structure is also present in tetrahydrobiopterin, a coenzyme that participates in the hydroxylation of phenylalanine, tyrosine, and tryptophan (Chapter 15).

vitamin required by humans (and other animals) because of their inability to synthesize it (Chapter 36).

Folic acid is the common name for **pteroylglutamic acid**, a compound consisting of a heterobicyclic pteridine, *p*-aminobenzoic acid (PABA), and glutamic acid (Figure 25.1). The combination of the first two produces pteroic acid. Humans lack the enzymes capable of synthesizing PABA or of linking pteroic acid to glutamate.

The antimicrobial activity of sulfonamides is due to their competitive inhibition of the bacterial enzyme that incorporates PABA into dihydropteroic acid (Chapter 6).

Folates have a wide biological distribution; a rich dietary source is green leaves. They occur in nature largely as oligoglutamyl conjugates (e.g., in plants, predominantly as pteroylheptaglutamate) in which the peptide linkages occur between the γ-carboxyl group of one glutamate and the α-amino group of the next (Figure 25.1). The mechanism of intestinal absorption of folate is not completely understood. However, the ingested folylpolyglutamates must be converted to folylmonoglutamate prior to absorption. The folylpolyglutamates are rapidly hydrolyzed in the intestines at neutral pH by the brush-border enzyme pteroylpolyglutamate hydrolase (also called **conjugase**) to pteroylmonoglutamate (folic acid). If the folylpolyglutamates enter the intestinal epithelial cells intact, they may be converted to folylmonoglutamate within lysosomes by lysosomal hydrolase. Folate transport in the intestine and the choroid plexus is mediated by a specific carrier, and the disorder **hereditary folate malabsorption** is associated with a defective folate carrier.

Individuals affected with hereditary folate malabsorption exhibit early onset of failure to thrive, megaloblastic anemia, and severe mental retardation. Therapy requires the administration of large doses of oral and systemic folates. Folate is reduced and converted to N^5-methyltetrahydrofolate in the intestines and secreted into the circulation.

Tissue needs for folate are met by uptake from plasma. In the plasma, about two-thirds of the folate is protein-bound. Within tissue cells, N^5-methyltetrahydrofolate transfers its methyl group to homocysteine with the formation of methionine (Chapter 15). This reaction is catalyzed by homocysteine methyltransferase, a vitamin B_{12} coenzyme-dependent enzyme, and appears to be the major site of interdependence of these two vitamins.

In the tissues, tetrahydrofolate is converted to polyglutamyl forms by an ATP-dependent synthetase. In the liver, the major form is pteroyl pentaglutamate. Reduced polyglutamyl forms, each substituted with one of several one-carbon moieties, are the preferred coenzymes of folate-dependent enzymes. Reduction of folate (F) to tetrahydrofolate (FH_4) occurs in two steps: F is reduced to 7,8-dihydrofolate (FH_2); and FH_2 is reduced to 5,6,7,8-tetrahydrofolate (FH_4). Both of these reactions are catalyzed by a single NADPH-linked enzyme, dihydrofolate reductase (Figure 25.2).

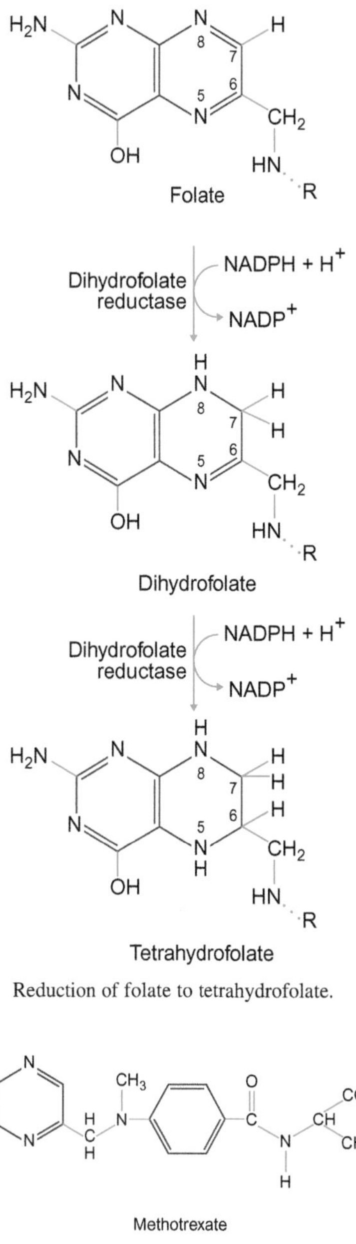

Inhibitors of Dihydrofolate Reductase

Methotrexate (Figure 25.3), a structural analogue of FH_2, is a potent inhibitor of dihydrofolate reductase and is used in chemotherapy of neoplastic disease. Methotrexate is not effective against infections from bacteria and protozoa, since these organisms are impermeable to folate and its analogues. However, methotrexate inhibits dihydrofolate reductase of both bacterial and protozoal origin in cell-free preparations. Pyrimethamine is extremely effective against protozoan (e.g., malarial parasite) infections, is ineffective against bacterial infections, and is a mild inhibitor of the mammalian enzyme. Trimethoprim is an effective inhibitor of both bacterial and protozoal enzymes, but has minimal inhibitory action against the mammalian enzyme.

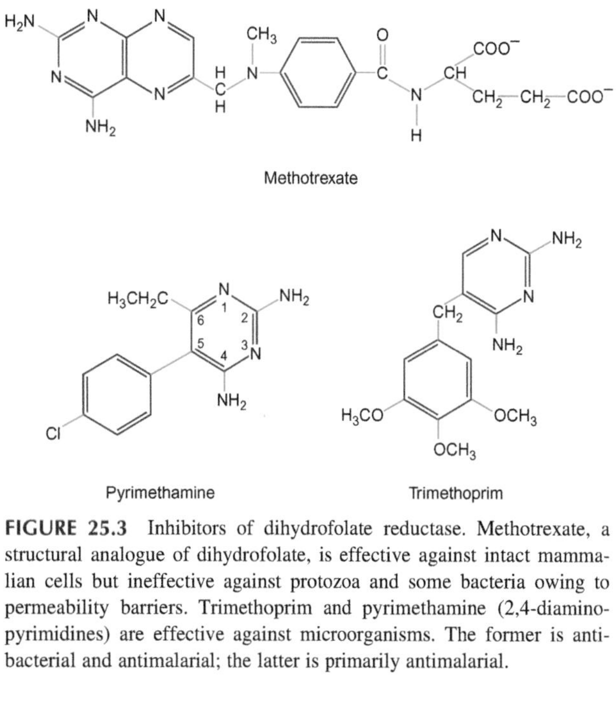

FIGURE 25.2 Reduction of folate to tetrahydrofolate.

FIGURE 25.3 Inhibitors of dihydrofolate reductase. Methotrexate, a structural analogue of dihydrofolate, is effective against intact mammalian cells but ineffective against protozoa and some bacteria owing to permeability barriers. Trimethoprim and pyrimethamine (2,4-diaminopyrimidines) are effective against microorganisms. The former is antibacterial and antimalarial; the latter is primarily antimalarial.

These selective enzyme inhibitors have been used in the treatment of bacterial and malarial infections.

Formation of One-Carbon Derivatives of Folate

There are several one-carbon derivatives of folate (of different redox states) that function as one-carbon carriers in different metabolic processes. In all of these reactions, the one-carbon moiety is carried in a covalent linkage to one or both of the nitrogen atoms at the 5- and 10-positions of the pteroic acid portion of tetrahydrofolate. Six known forms of carrier are shown in Figure 25.4. Folinic acid (N^5-formyl FH_4), also called **leucovorin** or **citrovorum factor**, is chemically stable and is used clinically to prevent or reverse the toxic effect of folate antimetabolites, such as methotrexate and pyrimethamine.

Serine, a nonessential amino acid, is the main source of one-carbon fragments. Some key processes that utilize folate-mediated one-carbon transfer reactions are as follows:

1. Synthesis of purine nucleotides (N^{10}-formyl FH_4);
2. Methylation of deoxyuridylate to thymidylate (N^5,N^{10}-methylene FH_4) (an essential step in the biosynthesis of DNA, impairment of this reaction is responsible for the main clinical manifestations of folate deficiency);
3. Synthesis of formylmethionyl-tRNA (N^{10}-formyl FH_4) required for initiation of protein synthesis in prokaryotes (Chapter 23) and in mitochondria; and
4. Conversion of homocysteine to methionine (N^5-methyl FH_4) (Chapter 15).

Inherited disorders of folate transport and metabolism include defects in folate carrier (hereditary folate malabsorption, previously discussed), deficiency of N^5,N^{10}-methylene FH_4 reductase (Chapter 15), or functional deficiency of N^5-methyl FH_4 methyltransferase due to defects of vitamin B_{12} metabolism (Chapter 36) or formiminotransferase.

Several observational studies have shown that dietary supplementation of folate in women of childbearing age can reduce the risk of fetal neural tube defects (e.g., spina bifida).

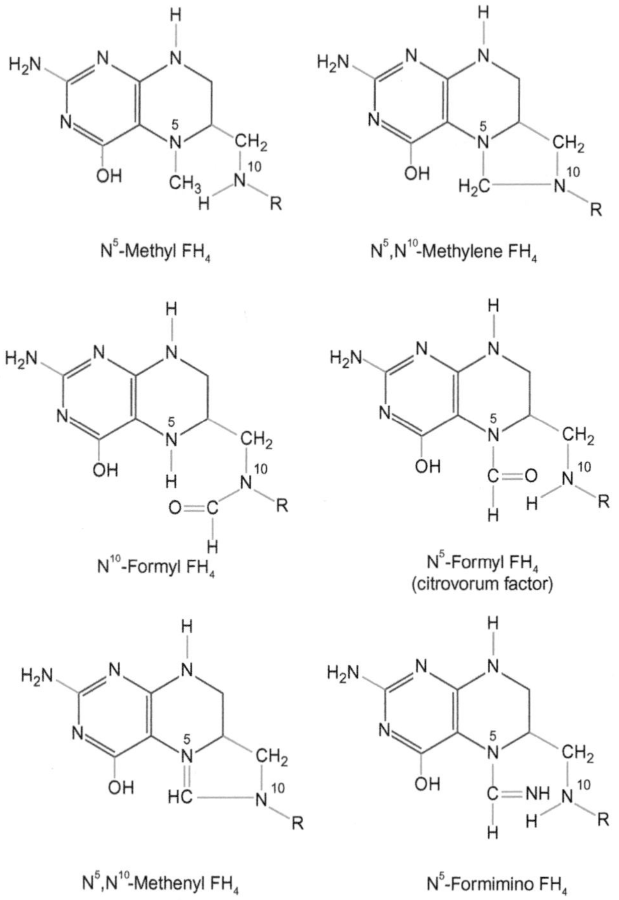

FORMATION OF 5-PHOSPHORIBOSYL-1-PYROPHOSPHATE

5-Phosphoribosyl-1-pyrophosphate (PRPP) is a key intermediate in nucleotide biosynthesis. It is required for

FIGURE 25.4 The six one-carbon substituents of tetrahydrofolate. The oxidation state is the same in the one-carbon moiety of N^5-formyl, N^{10}-formyl, and N^5,N^{10}-methenyl FH_4. N^{10}-formyl FH_4 is required for *de novo* synthesis of purine nucleotides, whereas N^5,N^{10}-methylene FH_4 is needed for the formation of thymidylic acid.

de novo synthesis of purine and pyrimidine nucleotides and the salvage pathways, in which purines are converted to their respective nucleotides via transfer of ribose 1-phosphate group from PRPP to the base, i.e.:

$$\text{Purine} + \underset{\text{(PRPP)}}{\text{P-ribose-P-P}} \rightarrow \text{purine-ribose-P} + PP_i$$

PRPP is synthesized from ribose 5-phosphate by the following reaction:

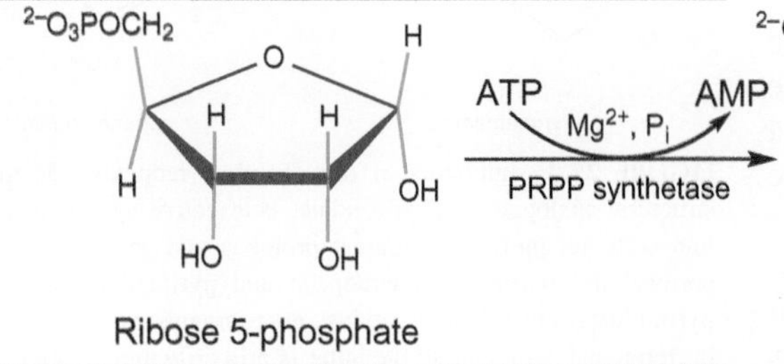

Ribose 5-phosphate   PRPP

In this reaction, the pyrophosphate group of ATP is transferred to ribose 5-phosphate; the product PRPP is a high-energy compound. PRPP synthetase has an absolute requirement for inorganic phosphate (P_i), which functions as an allosteric activator. The enzyme is inhibited by many nucleotides, the end products of the pathway for which PRPP is an essential substrate. The gene for PRPP synthetase is located on the X-chromosome. Mutations in this gene have given rise to PRPP synthetase variants with increased catalytic activity, which leads to overproduction of uric acid (discussed later, under "Gout"). The main source of ribose-5-phosphate is the pentose phosphate pathway (Chapter 14).

BIOSYNTHESIS OF PURINE NUCLEOTIDES

Purine nucleotides can be produced by two different pathways. The salvage pathway utilizes free purine bases and converts them to their respective ribonucleotides by appropriate phosphoribosyltransferases. The *de novo* pathway utilizes glutamine, glycine, aspartate, N^{10}-formyl FH_4, bicarbonate, and PRPP in the synthesis of inosinic acid (IMP), which is then converted to AMP and GMP.

De Novo Synthesis

The atoms of the purine ring originate from a wide variety of sources (Figure 25.5). C_2 and C_8 are derived from N^{10}-formyl FH_4. Glycine is incorporated entirely and provides C_4, C_5, and N_7. The α-amino group of aspartic acid provides N_1. The amide group of glutamine contributes N_3 and N_9. Carbon dioxide (or HCO_3^-) provides C_6. The stepwise synthesis of IMP is depicted in Figure 25.6. The following features of purine biosynthesis should be noted:

1. The glycosidic bond is formed when the first amino group of the purine ring is incorporated (reaction 1; Figure 25.6). Free purines or purine nucleosides are not intermediates of this pathway. In pyrimidine biosynthesis, the pyrimidine ring is formed completely before addition of ribose 5-phosphate.

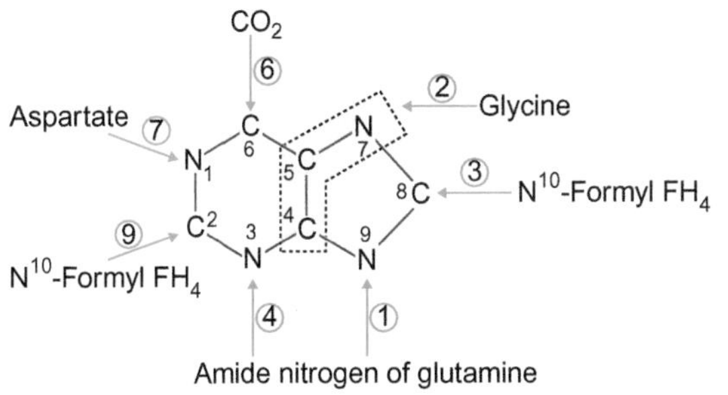

FIGURE 25.5 Sources of the atoms in *de novo* synthesis of the purine ring. The numbered reactions correspond to the reaction steps in Figure 25.6.

2. The biosynthesis is accomplished by several ATP-driven reactions. Ring formation (cyclization) is achieved by incorporating nucleophilic amino groups and electrophilic carbonyl groups at the appropriate positions.

3. The first purine nucleotide synthesized is inosinic acid (IMP), which is the precursor for the synthesis of adenylic acid (AMP) and guanylic acid (GMP) in two different pathways (Figure 25.7).

4. All the required enzymes occur in the cytosol. This is also true for enzymes of salvage pathways, nucleotide interconversion, and degradation.

5. *De novo* synthesis is particularly active in the liver and placenta. Nonhepatic tissues (e.g., bone marrow) depend on preformed purines that are synthesized in the liver and transported by red blood cells. They are very effective in salvaging the purines and exhibit little or no xanthine oxidase activity, which oxidizes free purines.

Salvage Pathways

The reutilization of purine bases, after conversion to their respective nucleotides, constitutes salvage pathways. These pathways are particularly important in extrahepatic tissues. Purines arise from the intermediary metabolism of nucleotides and the degradation of polynucleotides. Salvage occurs mainly by the phosphoribosyltransferase reaction:

$$\text{Base} + \text{PRPP} \rightleftharpoons \text{base-ribose-phosphate} + PP_i$$

Human tissue contains two phosphoribosyltransferases (Figure 25.8). Adenine phosphoribosyltransferase (APRT) catalyzes the formation of AMP from adenine. Hypoxanthine guanine phosphoribosyltransferase (HGPRT) catalyzes the formation of IMP and GMP from hypoxanthine and guanine, respectively. HGPRT also catalyzes the conversion of other purines (6-thiopurine, 8-azaguanine, allopurinol) to their respective ribonucleotides. The K_{eq} of both phosphotransferases favors formation of ribonucleotides. During salvage, only two high-energy bonds are used, whereas in *de novo* synthesis of AMP or GMP, at least six high-energy bonds are expended. Deficiencies of these enzymes are discussed later.

A quantitatively less significant salvage pathway uses purine nucleoside phosphorylase and nucleoside kinase:

$$\text{Base} + \text{ribose 1-phosphate} \underset{\text{nucleoside phosphorylase}}{\rightleftharpoons} \text{base-ribose} + P_i$$

$$\text{Base-ribose} + \text{ATP} \xrightarrow{\text{nucleoside kinase}} \text{base-ribose-phosphate} + \text{ADP}$$

The phosphorylase can catalyze the formation of inosine or deoxyinosine, and of guanosine or deoxyguanosine, but not adenosine or deoxyadenosine. However,

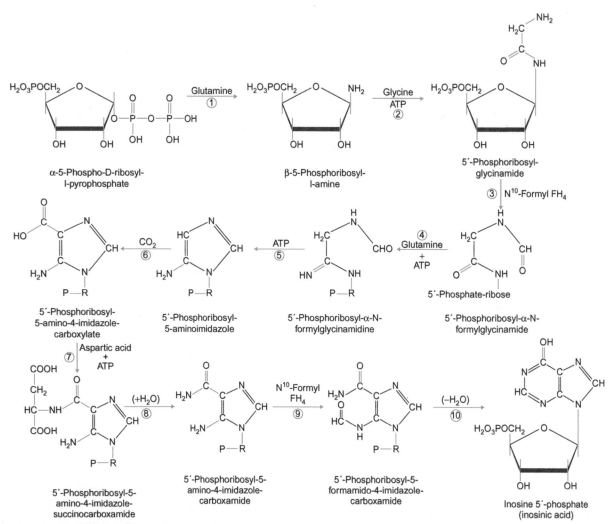

FIGURE 25.6 Biosynthesis of inosinic acid. The reactions are numbered. Enzymes: (1) amidophosphoribosyltransferase; (2) phosphoribosylglycina-mide synthetase; (3) phosphoribosylglycinamide formyltransferase; (4) phosphoribosylformylglycinamidine synthetase; (5) phosphoribosyl-aminoimidazole synthetase; (6) phosphoribosylaminoimidazole carboxylase; (7) phosphoribosylaminoimidazolesuccinocarboxamide synthetase; (8) adenylosuccinate lyase; (9) phosphoribosylaminoimidazolecarboxamide formyltransferase; (10) IMP cyclohydrolase. *The solid lines in ribose structures without any designation of an atom represent hydrogen.

the last two nucleosides can be converted to inosine and deoxyinosine by adenosine deaminase. The normal function of the phosphorylase appears to be the formation of free hypoxanthine and guanine for conversion to uric acid. Deficiency of adenosine deaminase or purine nucleoside phosphorylase results in immunodeficiency disease (Chapter 33).

In muscle, a unique nucleotide reutilization pathway, known as the **purine nucleotide cycle**, uses three enzymes: myoadenylate deaminase, adenylosuccinate synthetase, and adenylosuccinate lyase. In this cycle, AMP is converted to IMP with the formation of NH_3, and IMP is then reconverted to AMP. Myoadenylate deaminase deficiency produces a relatively benign muscle disorder, which is characterized by muscle fatigue following exercise (see "Myoadenylate Deaminase Deficiency" later in this chapter).

Nucleoside kinases specific for inosine or guanosine have been described. Adenosine kinase is widely distributed in mammalian tissues.

Dietary Purines

Purines derived from food do not participate significantly in the salvage pathways described previously; they are mostly converted to uric acid by intestinal xanthine oxidase.

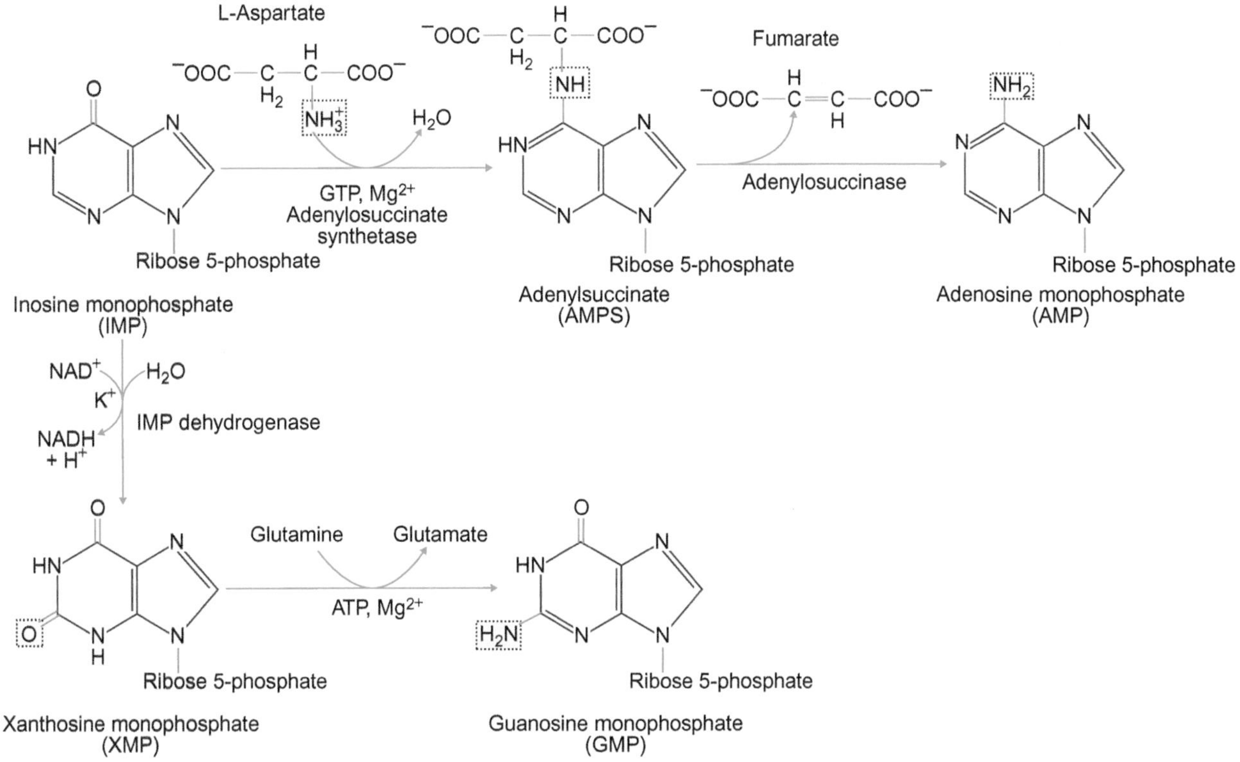

FIGURE 25.7 Biosynthesis of AMP and GMP from IMP.

CONVERSION OF NUCLEOSIDE MONOPHOSPHATES TO DIPHOSPHATES AND TRIPHOSPHATES

The triphosphates of nucleosides and deoxynucleosides are substrates for RNA polymerases and DNA polymerases, respectively. They are formed from monophosphates in two stages. Conversion to diphosphates is catalyzed by kinases. These enzymes are base-specific but not sugar-specific. ATP is the usual source of phosphate. However, in some cases, other triphosphates or dATP may be used. Typical reactions are as follows:

$$\underset{\text{(dGMP)}}{\text{GMP}} + \text{ATP} \xrightarrow{\text{guanylate kinase}} \underset{\text{(dGDP)}}{\text{GDP}} + \text{ADP}$$

$$\text{AMP} + \text{ATP} \xrightarrow{\text{adenylate kinase}} \text{ADP} + \text{ADP}$$

The diphosphates are converted to the triphosphates by the ubiquitous enzyme nucleoside diphosphate kinase. Remarkably, the lack of base or sugar specificity applies to the phosphate acceptor and the phosphate donor. Hence, the general reaction is

$$\underset{\text{(acceptor)}}{\text{dXDP}} + \underset{\text{(donor)}}{\text{dYTP}} \xrightarrow{\text{nucleoside diphosphate kinase}} \text{dXTP} + \text{dYDP}$$

Recall that the conversion of ADP to ATP occurs mostly by mitochondrial oxidative phosphorylation coupled to electron transport (Chapter 13).

FORMATION OF PURINE DEOXYRIBONUCLEOTIDES

Conversion of ribonucleotides to the deoxy form occurs exclusively at the diphosphate level. Ribonucleoside diphosphate reductase (ribonucleotide reductase) catalyzes the reaction. This enzyme is found in all species and tissues. The immediate source of reducing equivalents is the enzyme (E) itself in which two sulfhydryl groups are oxidized to a disulfide. The general reaction is

Ribonucleoside diphosphate + E(SH)$_2$

$$\xrightarrow{\text{ribonucleoside diphosphate reductase}} \text{deoxyribonucleoside}$$

diphosphate + E(S-S) + H$_2$O

In this reaction, the 2′-hydroxyl group of ribose is replaced by hydrogen, with retention of configuration at the 3′-carbon atom:

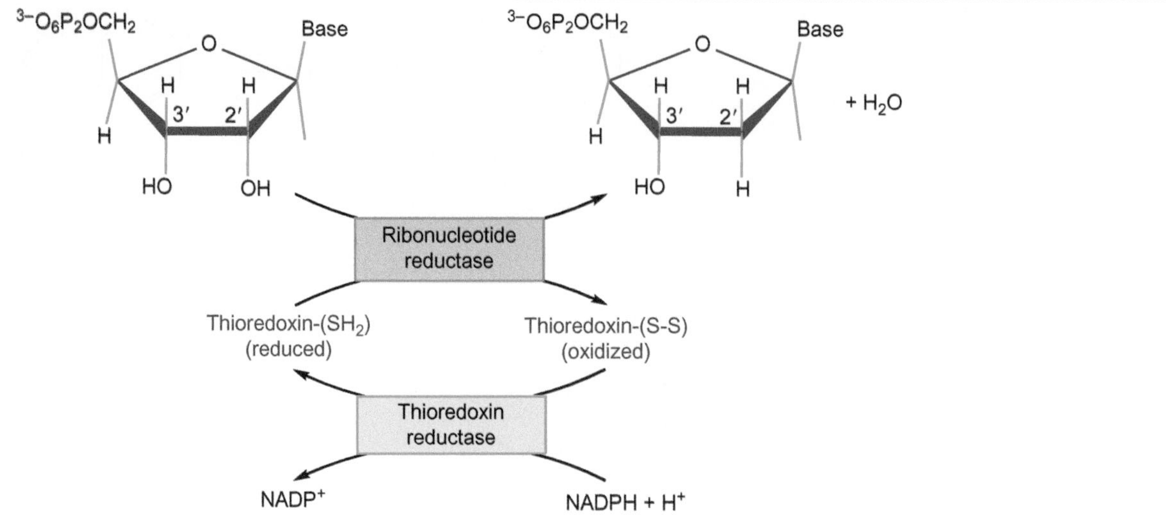

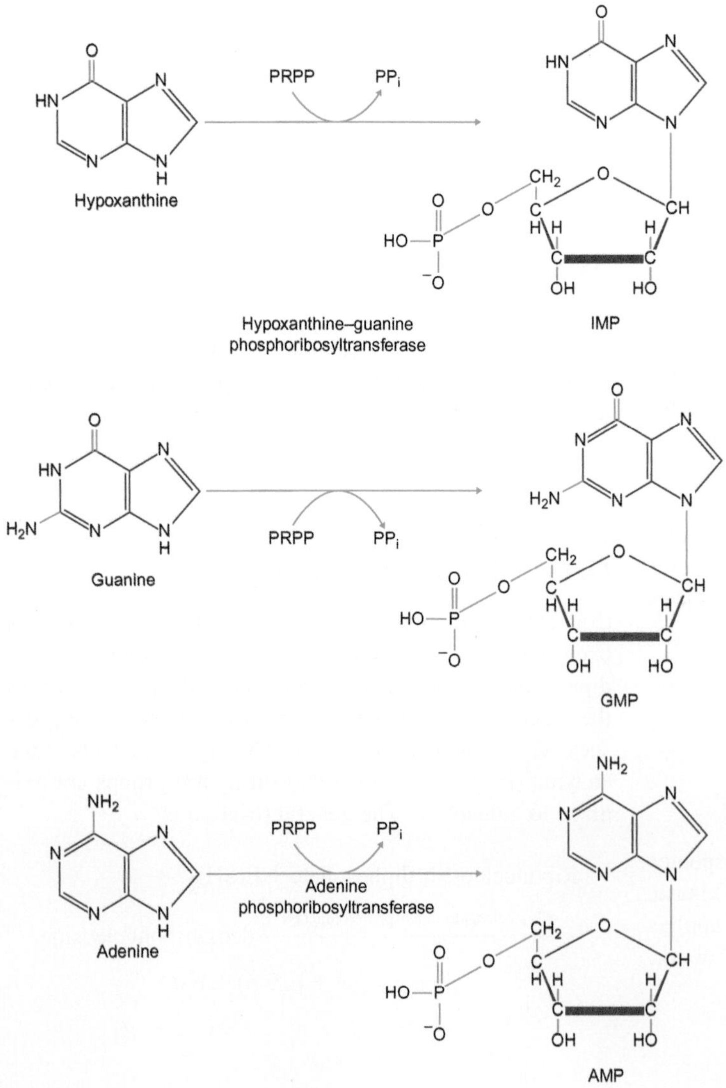

FIGURE 25.8 Salvage pathways of purine nucleotide synthesis. The preformed purines can be converted to mononucleotides in a single step, using PRPP.

Regeneration of the ribonucleotide reductase is accomplished by thioredoxin, a dithiol polypeptide (M.W. 12,000) coenzyme, which also plays a role in other protein disulfide reductase reactions. In thioredoxin, two cysteine residues in the sequence $-Cys-Gly-Pro-Cys-$ are converted to cystine disulfide. Reduced thioredoxin is regenerated by thioredoxin reductase, a flavoprotein enzyme that uses $NADPH + H^+$.

Ribonucleotide reductase consists of two subunits, B_1 and B_2, neither of which possesses catalytic function. B_1 is a dimer in which each monomer contains a substrate binding site and two types of allosteric effector binding sites. One type of effector site confers substrate specificity, and the other is regulatory. B_1 contains a pair of sulfhydryl groups that are required for catalytic activity. B_2 is also a dimer, contains one nonheme Fe(III), and has an organic free-radical delocalized over the aromatic ring of a tyrosine residue in each of its polypeptide chains. The catalytic site is formed from the interaction of B_1 and B_2. A free-radical mechanism involves the tyrosyl residues, the iron atom of B_2, and sulfhydryl groups of B_1. An antineoplastic agent, **hydroxyurea**, inhibits ribonucleotide reductase by inactivating the free radical.

Ribonucleotide reductase is regulated so as to ensure a balanced supply of deoxynucleotides for DNA synthesis. For example, if excess dATP is present, decreased synthesis of all the deoxyribonucleotides ensues, whereas ATP stimulates formation of dCDP and dUDP. Binding of TTP stimulates formation of dGDP and hence of dGTP. Binding of dGTP stimulates the formation of dADP and hence of dATP. In this way, these nucleotide effectors, by binding to various regulatory sites, tend to equalize the concentrations of the four deoxyribonucleotides required for DNA synthesis.

REGULATION OF PURINE BIOSYNTHESIS

Regulation of *de novo* purine biosynthesis is essential because it consumes a large amount of energy as well as glycine, glutamine, N^{10}-formyl FH_4, and aspartate. Regulation occurs at the PRPP synthetase reaction, the amidophosphoribosyltransferase reaction, and the steps involved in the formation of AMP and GMP from IMP.

PRPP Synthetase Reaction

PRPP synthetase requires inorganic phosphate as an allosteric activator. Its activity depends on the intracellular concentrations of the end products of pathways in which PRPP is substrate. These end products are purine and pyrimidine nucleotides (Figure 25.9).

Increased levels of intracellular PRPP enhance *de novo* purine biosynthesis. For example, in patients with HGPRT deficiency, the fibroblasts show accelerated rates

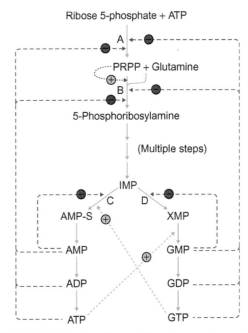

FIGURE 25.9 Feedback regulation of the *de novo* pathway of purine biosynthesis. Solid lines represent metabolic pathways, and broken lines indicate sites of feedback regulation. $\oplus$ = stimulatory effect; $\ominus$ = inhibitory effect. Regulatory enzymes: (A) PRPP synthetase; (B) amidophosphoribosyltransferase; (C) adenylosuccinate synthetase; (D) IMP dehydrogenase.

of purine formation. Several mutations of PRPP synthetase, which exhibit increased catalytic activity with increased production of PRPP, have been described in subjects exhibiting gout.

Amidophosphoribosyltransferase Reaction

The amidophosphoribosyltransferase reaction is the first and uniquely committed reaction of the *de novo* pathway and is the rate-determining step. Amidophosphoribosyltransferase is an allosteric enzyme and has an absolute requirement for a divalent cation. The enzyme is inhibited by AMP and GMP, which bind at different sites. It is also inhibited by pyrimidine nucleotides at relatively high concentrations.

Inhibition by AMP and GMP is competitive with respect to PRPP. The human placental enzyme exists in a small form (M.W. 133,000) and a large form (M.W. 270,000). The small form is catalytically active. Ribonucleotides convert the active form to the large form, whereas PRPP does the opposite. The regulatory actions of PRPP synthetase and amidophosphoribosyltransferase are coordinated. When there is a decrease in the intracellular concentration of adenine ribonucleotides, PRPP synthetase is activated. This results in increased synthesis of PRPP, which in turn converts the inactive form of amidophosphoribosyltransferase to the active form and increases production of purine nucleotides.

Regulation of Formation of AMP and GMP from IMP

In the formation of AMP and GMP from IMP, ATP is required for GTP synthesis, and GTP is needed to form ATP (Figure 25.7). In addition, adenylosuccinate synthetase is inhibited by AMP, and IMP dehydrogenase is inhibited by GMP (Figure 25.9).

INHIBITORS OF PURINE BIOSYNTHESIS

A variety of inhibitors of purine biosynthesis function at different stages and are used as antimicrobial, anticancer, and immunosuppressive agents.

Inhibitors of Folate Biosynthesis

The **sulfonamide** drugs were the first effective antibacterial agents to be employed systemically in humans. These drugs resemble *p*-aminobenzoic acid (PABA) in structure and inhibit utilization of that compound, required for the synthesis of folate in bacteria.

Inhibitors of Formation of IMP

Folate analogues, such as methotrexate (Figure 25.3), are folate antagonists. They block production of FH_2 and FH_4 by dihydrofolate reductase and lead to diminished purine biosynthesis (inhibition of reactions 3 and 9 in Figure 25.6). Methotrexate also affects metabolism of amino acids and pyrimidine (inhibition of thymidylate synthesis) and inhibits DNA, RNA, and protein synthesis. It is effective in the treatment of breast cancer, cancer of the head and neck, choriocarcinoma, osteogenic sarcoma, and acute forms of leukemia. High doses of methotrexate can be tolerated provided that the patient also receives folinic acid (N^5-formyl FH_4), which decreases damage to bone marrow and prevents the development of leukopenia and thrombocytopenia. Methotrexate also produces mucositis, gastrointestinal symptoms, and liver damage by a direct toxic effect on hepatic cells. Chronic oral administration of methotrexate in psoriasis has been associated with an increased incidence of postnecrotic cirrhosis.

The "rescue" of normal (but not tumor), cells from methotrexate toxicity by folinic acid is partly explained by differences in membrane transport. For example, osteogenic sarcoma cells (which do not respond to conventional doses of methotrexate treatment) are not rescued by folinic acid administered after methotrexate, presumably owing to the absence of transport sites for folinic acid in the neoplastic cells. The therapeutic effects of administration of methotrexate and "rescue" with folinic acid are superior to those of methotrexate alone. Resistance to methotrexate can develop from increased activity of dihydrofolate reductase, synthesis of an enzyme having a lower affinity for the inhibitor, decreased transport of the drug into tumor cells, decreased degradation of the reductase, and genetic amplification of the gene for dihydrofolate reductase.

Inhibitors of Formation of AMP and GMP

Hadacidin (N-formyl-N-hydroxyglycine) is an analogue of aspartic acid isolated from fungi.

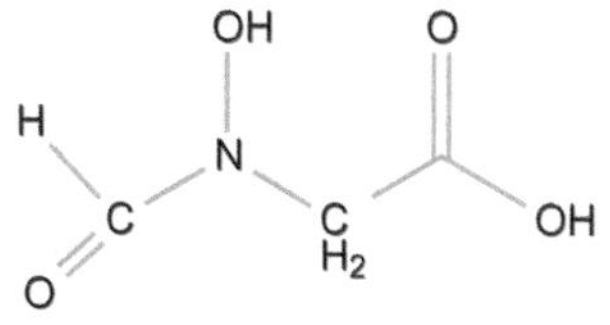

By competing with aspartate, it inhibits adenylosuccinate synthetase and hence the synthesis of AMP; it is an experimental antineoplastic agent.

Mycophenolic acid and **ribavarin monophosphate** inhibit IMP dehydrogenase and hence GMP synthesis.

Inhibitors of Multiple Steps: Purine Analogues

6-Mercaptopurine is a structural analogue of hypoxanthine and is converted to thio-IMP in the salvage pathway.

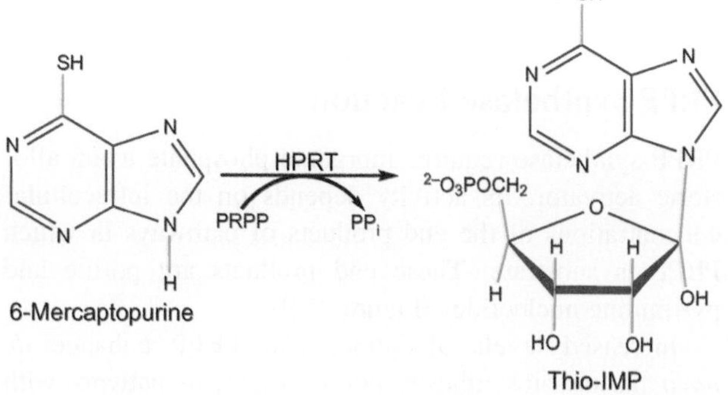

Thio-IMP prevents production of AMP and GMP by inhibiting the following reactions:

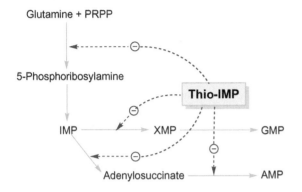

6-Mercaptopurine is used in the treatment of acute leukemias. However, resistant tumor cells develop rapidly, probably because of altered specificity or lack of phosphoribosyltransferases, so that thio-IMP (the active inhibitor) is not formed. Other mechanisms may include altered cell permeability and an increased rate of destruction of 6-mercaptopurine.

Azathioprine, which is a prodrug, is converted to 6-mercaptopurine *in vivo*. Its principal use is as an immunosuppressive agent:

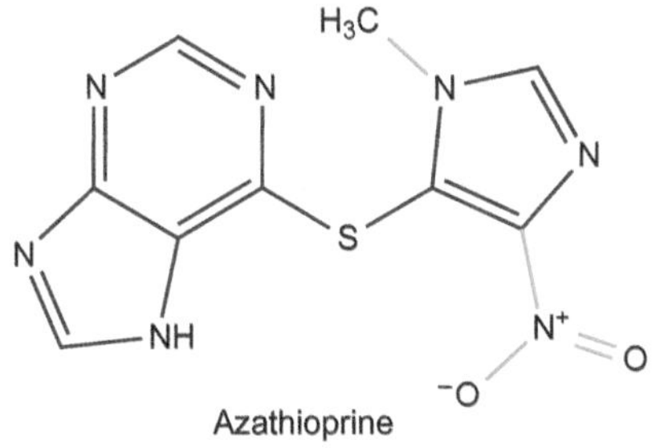

Azathioprine

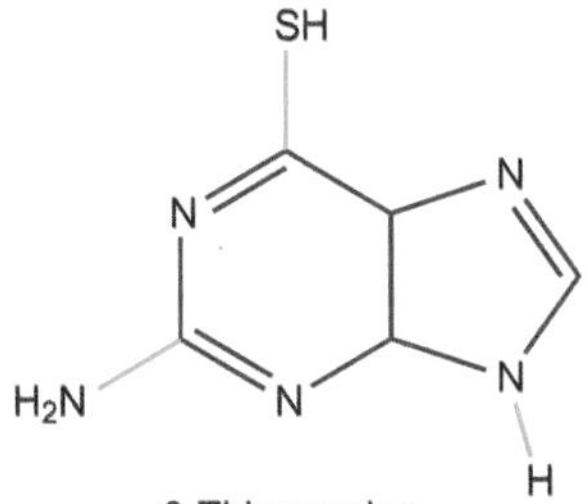

6-Thioguanine

6-Thioguanine is similar to 6-mercaptopurine in its action. The most active form is 6-thio-GMP, which inhibits guanylate kinase and, at higher concentrations, IMP dehydrogenase. Thio-IMP and thio-GMP also inhibit PRPP amidotransferase.

Thiopurines, in part, are inactivated by 5-methylation by thiopurine S-methyltransferase (TPMT). Individuals with variant alleles, which give rise to low TPMT activities

when treated with thiopurines, can have life-threatening drug-induced myelosuppression. Polymorphisms of the TPMT gene with lowered gene expression require lower doses of the drug for therapeutic management. This aspect of gene expression that affects drug metabolism is known as **pharmacogenomics**. Thiopurines are also partially inactivated by conversion to 6-thiouric acid by xanthine oxidase. Inhibitors of xanthine oxidase used in the treatment of gout and in other hyperuricemic conditions can also affect the pharmacokinetic properties of thiopurines.

Vidarabine (adenine arabinoside), an analogue of adenosine, does not interfere with purine biosynthesis but affects DNA synthesis:

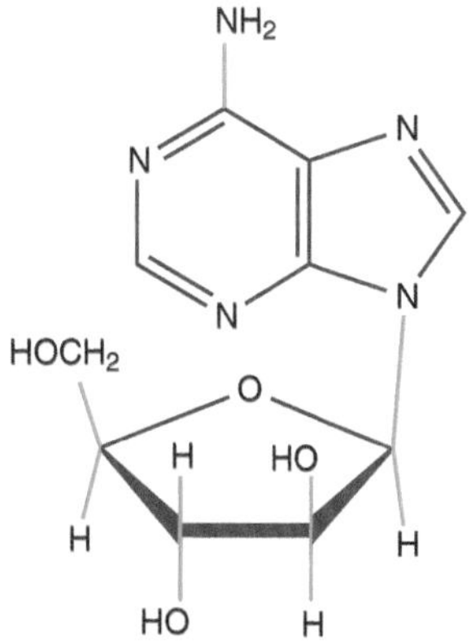

It is phosphorylated to adenine arabinoside triphosphate, which inhibits viral DNA polymerases, but not the mammalian enzymes, by competition with dATP. Originally developed as an antileukemia drug, it has proved useful in the treatment of herpes virus infections. Several other purine nucleoside analogues are used as antiviral agents; these include acyclovir, valacyclovir, ganciclovir, penciclovir, and famiciclovir. After conversion to their respective triphosphate derivatives, these drugs inhibit viral DNA polymerase. Viruses inhibited by these drugs are herpes simplex, varicella zoster, cytomegalovirus, and hepatitis B. Antiviral agents that are not purine nucleoside analogues are amantadine analogues and α-interferon.

Inhibition of Conversion of Ribonucleoside Diphosphate to Deoxyribonucleoside Diphosphate

Hydroxyurea, an antineoplastic agent, acts by destroying an essential free radical in the active center of ribonucleotide reductase:

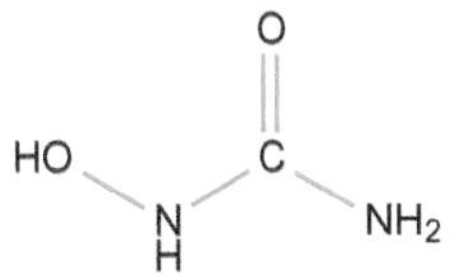

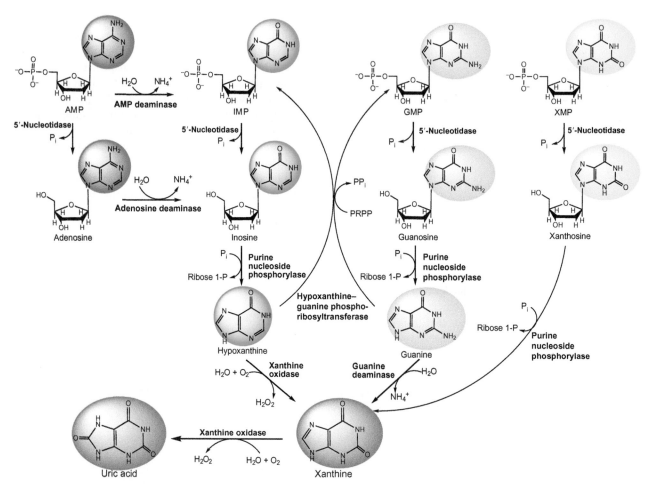

FIGURE 25.10 Pathways for the formation of uric acid from purine nucleotides.

A serendipitous finding of hydroxyurea treatment is the induction of hemoglobin F (HbF) production in red blood cells. Hydroxyurea therapy results in an increase in both the content and number of red blood cells that contain HbF (known as **F cells**). This property of hydroxyurea has been used in the treatment of **sickle cell disease** because of the antisickling effect of HbF (Chapter 26).

CATABOLISM OF PURINE NUCLEOTIDES

Degradation of purines and their nucleotides occurs during turnover of endogenous nucleic acids and degradation of ingested nucleic acids (Figure 25.10), during which most of the purines are converted to uric acid.

Degradation of purine nucleoside phosphates (AMP, IMP, GMP, and XMP) begins by hydrolysis by 5′-nucleotidase (Figure 25.10), to produce adenosine, inosine, guanosine, xanthosine, and phosphate, respectively. Adenosine is converted to inosine by adenosine deaminase (adenosine aminohydrolase). For inosine, guanosine, and xanthosine, the next step is catalyzed by purine

nucleoside phosphorylase and involves a phosphorylation and a cleavage to produce ribose 1-phosphate and hypoxanthine, guanine, and xanthine, respectively. (The enzyme also acts on deoxyribonucleosides to release deoxyribose 1-phosphate.) The pentose sugars are metabolized further or excreted. Purine nucleoside phosphorylase functions in purine salvage, and deficiency of this enzyme results in decreased cell-mediated immunity.

Hypoxanthine and guanine are converted to xanthine by xanthine oxidase and guanine aminohydrolase, respectively. Thus, all purine nucleosides produce xanthine. Xanthine oxidase is very active in the intestinal mucosa and the liver, and is present in most tissues. The products of the xanthine oxidase reaction are uric acid and hydrogen peroxide. Uric acid is excreted in the urine. In mammals other than primates, a liver enzyme, urate oxidase, converts uric acid to allantoin. In humans, uric acid, urea, and creatinine are the end products of nitrogen metabolism. All three are excreted in urine. The primary end product is urea (Chapter 15). In humans, uric acid production and excretion are a balanced process. On a daily basis, about two-thirds of the uric acid produced is excreted by the kidneys and the remainder via intestinal bacterial degradation. In chronic renal failure and hyperuricemia, extrarenal uric acid disposal is enhanced. Renal excretion of uric acid is a multicomponent bidirectional process involving glomerular filtration (100%) and proximal tubular reabsorption (98%−100%), followed by proximal tubular secretion (50%) and proximal tubular reabsorption (40%−44%). The proximal renal tubular reabsorption of urate is mediated by urate transporter-1 (URAT1). It is located in the apical membrane and transports urate from the lumen into the proximal tubular cells in exchange for anions. Ultimately, the renal clearance of uric acid is only about 7%−10% of creatinine clearance (Chapter 15).

Xanthine Oxidase Reduction

Xanthine oxidase contains FAD, nonheme iron (Fe-S), and a pterin complexed to a molybdenum cofactor. The liver and gastrointestinal tract contain large amounts of the enzyme. The enzyme is present in two forms: one with dehydrogenase activity (xanthine dehydrogenase) and the other with oxidase activity. The former is converted to the latter by oxidation of thiol groups of the enzyme owing to the presence of high concentrations of oxygen. It is active on purines, aldehydes, and pteridines. The reactions catalyzed on purines are

$$\text{Hypoxanthine} + H_2O + O_2 \rightarrow \text{xanthine} + H_2O_2$$

and

$$\text{Xanthine} + H_2O + O_2 \rightarrow \text{uric acid} + H_2O_2$$

Molybdenum, an essential cofactor, is the initial acceptor of electrons during purine oxidation and undergoes reduction from Mo^{6+} to Mo^{4+}. Deficiency of molybdenum can result in **xanthinuria**. The electrons from molybdenum are passed successively to the iron-sulfur center, to FAD, and finally to oxygen. The oxygen incorporated into xanthine and uric acid originates in water. Xanthine oxidase also yields the superoxide radical, O_2^- which is then converted to hydrogen peroxide by superoxide dismutase (Chapter 13). This may result in the formation of free radicals

$$H_2O_2 + Fe^{2+} \rightarrow Fe^{3+} + OH^- + OH^\bullet$$

which can cause tissue destruction. Thus, the xanthine oxidase reaction can cause tissue injury during reperfusion of organs that have been deprived of oxygen (e.g., ischemic myocardium, transplanted organs). The xanthine oxidase reaction is further promoted in hypoxic organs by providing increased substrates owing to the enhanced adenine nucleotide breakdown that occurs during oxygen

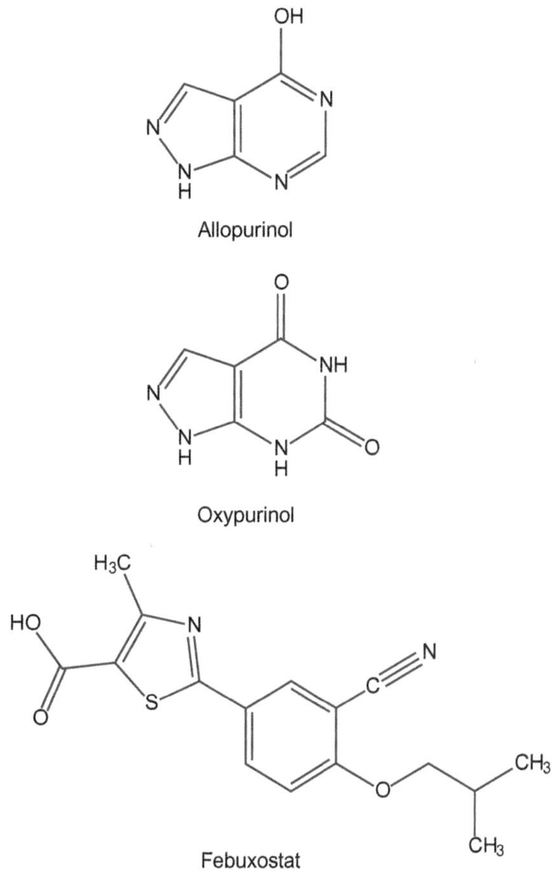

FIGURE 25.11 Structures of inhibitors of xanthine oxidase.

deprivation. Hydrogen peroxide is inactivated by catalase or by peroxidase. Xanthine oxidase is inhibited by allopurinol (a purine analogue) and febuxostat (Figure 25.11), which are used in the treatment of gout and other hyperuricemic conditions (e.g., tumor lysis syndrome).

DISORDERS OF PURINE NUCLEOTIDE METABOLISM

Several disorders affect purine metabolism. They are gout and the syndromes associated with deficiency of HGPRT, APRT, adenosine deaminase, nucleoside phosphorylase, myoadenylate deaminase, and xanthine oxidase.

Gout

Gout is a heterogeneous group of genetic and acquired diseases characterized by elevated levels of urates in blood (**hyperuricemia**) and of uric acid in urine (**uricuria**) (see Clinical Case Study 25.1, Vignette 1). Hyperuricemia in men is defined as serum urate concentration greater than 7 mg/dL (420 μmol/L), and in women 6 mg/dL (357 μmol/L). In gout, hyperuricemia is a common biochemical occurrence; however, many hyperuricemic subjects may not develop clinical gout (asymptomatic hyperuricemia). All clinical symptoms of gout arise from the low solubility of urate in biological fluids. The maximum solubility of urate in plasma at 37°C is about 7 mg/dL. However, in peripheral structures and in the extremities, where the temperature is well below 37°C, the solubility of urate is decreased. When urate is present in supersaturated solutions, crystals of monosodium urate monohydrate form easily. Deposits of aggregated crystals, known as **tophi**, in and around joints of the extremities, initiate an inflammatory foreign body reaction (acute arthritis) that involves leukocytes, complement, and other mediators (Chapter 33). This reaction causes severe pain, swelling, redness, and heat in the affected areas. Initial attacks are usually acute and frequently affect the metatarsophalangeal joint of the big toe. Tophi may also be present in subcutaneous tissues, cartilage, bone, and kidneys. Formation of urinary calculi (urolithiasis) of urate is common. Gout is potentially chronic and disfiguring.

Primary gout is a disorder of purine metabolism seen predominantly in men. The condition is multifactorial and involves genetic and nongenetic factors. Occurrence in women is uncommon; when it does occur, it is usually found in postmenopausal women. The blood urate concentration of normal men is ~ 1 mg/dL higher than that in women, but this difference disappears after menopause. Thus, in women, the postmenopausal rise in serum urate levels may increase the risk of developing gout. Gout is very rare in children and adolescents.

Primary gout may be due to overproduction or underexcretion of uric acid or to a combination of both. Frequently, siblings and other close relatives of afflicted individuals have high levels of uric acid in blood but do not develop gout, indicating that hyperuricemia is not the only factor involved. Primary renal gout is due to underexcretion caused by a renal tubular defect in uric acid transport. Primary metabolic gout is due to overproduction of purines and uric acid. The prevalence of gout is high in some populations (e.g., 10% of the adult male Maori of New Zealand). In Europe and the United States, the prevalence rate is 0.13%–0.37%.

Secondary gout develops as a complication of hyperuricemia caused by massive tumor cell death and is known as **tumor lysis syndrome** (e.g., leukemia). This type of hyperuricemia is usually associated with abnormally rapid turnover of nucleic acids. Due to release of intracellular contents, **acute tumor lysis syndrome** is also accompanied by several electrolyte abnormalities (hyperkalemia, hyperphosphatemia, and hypocalcemia) with severe clinical consequences (see Clinical Case Study 25.1, Vignette 2).

The mechanism of hyperuricemia in most individuals who have gout is unknown. Biochemical lesions that lead to hyperuricemia and may eventually lead to gout are given in Figure 25.12.

Treatment

Various drugs are used in the management of gout in three clinical situations:

1. To treat acute gouty arthritis;
2. To prevent acute attacks; and
3. To lower serum urate concentrations.

Acute gouty attacks commonly affect the first metatarsal joint of the foot. In aspirated joint fluids, birefringent urate crystals can be seen in the polarized light microscope, which is used in definitive diagnosis. The treatment of acute gout includes the administration of colchicine, nonsteroidal anti-inflammatory drugs (NSAIDs), corticosteroids, adrenocorticotropic hormone (ACTH), and analgesics. Colchicine and NSAIDs can also be used prophylactically to prevent acute attacks in patients with gout. Drugs used to lower serum urate concentrations include probenecid, sulfinpyrazone, and allopurinol.

Colchicine depolymerizes microtubules and structures (such as the mitotic spindle) consisting of microtubules. It is effective in decreasing pain and the frequency of attacks, but its mechanism of action is obscure.

Allopurinol, an analogue of hypoxanthine, inhibits xanthine oxidase and reduces formation of xanthine and uric acid. Allopurinol is also converted to allopurinol ribonucleotide by HGPRT. Uric acid production is decreased through depletion of PRPP. Allopurinol ribonucleotide also inhibits PRPP-amidotransferase allosterically.

A recently introduced drug, **febuxostat**, is a nonpurine inhibitor of xanthine oxidase (Figure 25.11). It is used in patients who exhibit undesirable toxic reaction to allopurinol therapy and in chronic renal insufficiency. Rapid removal of uric acid in serum is also accomplished by administration of recombinant rasburicase, which converts uric acid to soluble allantoins. Rasburicase is used in the treatment of hyperuricemia of acute tumor lysis syndrome (Clinical Case Study 25.1, Vignette 2).

Drugs that increase uric acid excretion in humans include **probenecid**, which is effective in the regulation of hyperuricemia and the resolution and prevention of tophi, and **sufinpyrazone**, which has similar effects. Both agents are weak organic acids and probably act as competitive inhibitors of tubular reabsorption of uric acid.

Dietary and Lifestyle Factors

Serum urate levels can be lowered by dietary and lifestyle changes. These include correction of obesity, avoidance of ethanol consumption, and avoidance of high-purine foods (e.g., organ meats).

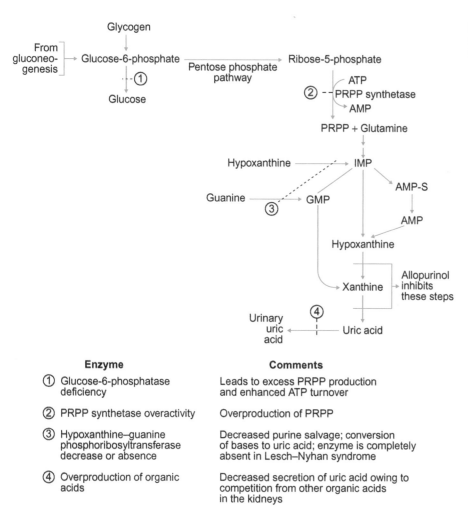

FIGURE 25.12 Composite diagram of the biochemical lesions that may lead to hyperuricemia.

Enzyme	Comments
① Glucose-6-phosphatase deficiency	Leads to excess PRPP production and enhanced ATP turnover
② PRPP synthetase overactivity	Overproduction of PRPP
③ Hypoxanthine–guanine phosphoribosyltransferase decrease or absence	Decreased purine salvage; conversion of bases to uric acid; enzyme is completely absent in Lesch–Nyhan syndrome
④ Overproduction of organic acids	Decreased secretion of uric acid owing to competition from other organic acids in the kidneys

Lesch−Nyhan Syndrome

Lesch−Nyhan syndrome is characterized by a virtual absence of HGPRT, excessive production of uric acid, and abnormalities of the central nervous system. These abnormalities include mental retardation, spasticity (increased muscle tension resulting in continuous increase of resistance to stretching), choreoathetosis (characterized by irregular, jerky, or explosive involuntary movements, and writhing or squirming, which may involve any extremity or the trunk), and a compulsive form of self-mutilation. The disorder associated with partial deficiency of HGPRT also leads to hyperuricemia but lacks the devastating neurological and behavioral features characteristic of Lesch−Nyhan syndrome. Both disorders are X-linked.

The hyperuricemia in Lesch−Nyhan patients is explained, at least in part, on the basis of intracellular accumulation of PRPP leading to increased purine nucleotide biosynthesis *de novo* and increased production of uric acid. Such patients do not usually develop gouty arthritis early in life, but do exhibit uric acid crystalluria and stone formation.

In Lesch−Nyhan patients, all tissues are devoid of HGPRT. The disorder thus can be detected by an assay for HGPRT in erythrocytes and by cultured fibroblasts. The former test has been used in the detection of the heterozygous state. HGPRT is a 217-amino acid cytosolic enzyme coded for by a single gene on the X-chromosome. Several mutations of the HGPRT gene are known.

The mechanism by which a deficiency of HGPRT causes central nervous system disorders remains unknown. Lesch−Nyhan patients do not show anatomical abnormalities in the brain. In normal subjects, HGPRT activity is high in the brain and, in particular, in the basal ganglia where *de novo* purine biosynthesis is low. This suggests the importance of the purine salvage pathway in this tissue. However, the relationship between HGPRT deficiency and neurological manifestations is not understood. The most significant abnormality identified in neurotransmitter systems is in the dopaminergic pathway (Chapter 30). There is no known treatment for the central nervous system dysfunction. Allopurinol has been used in the management of the hyperuricemia.

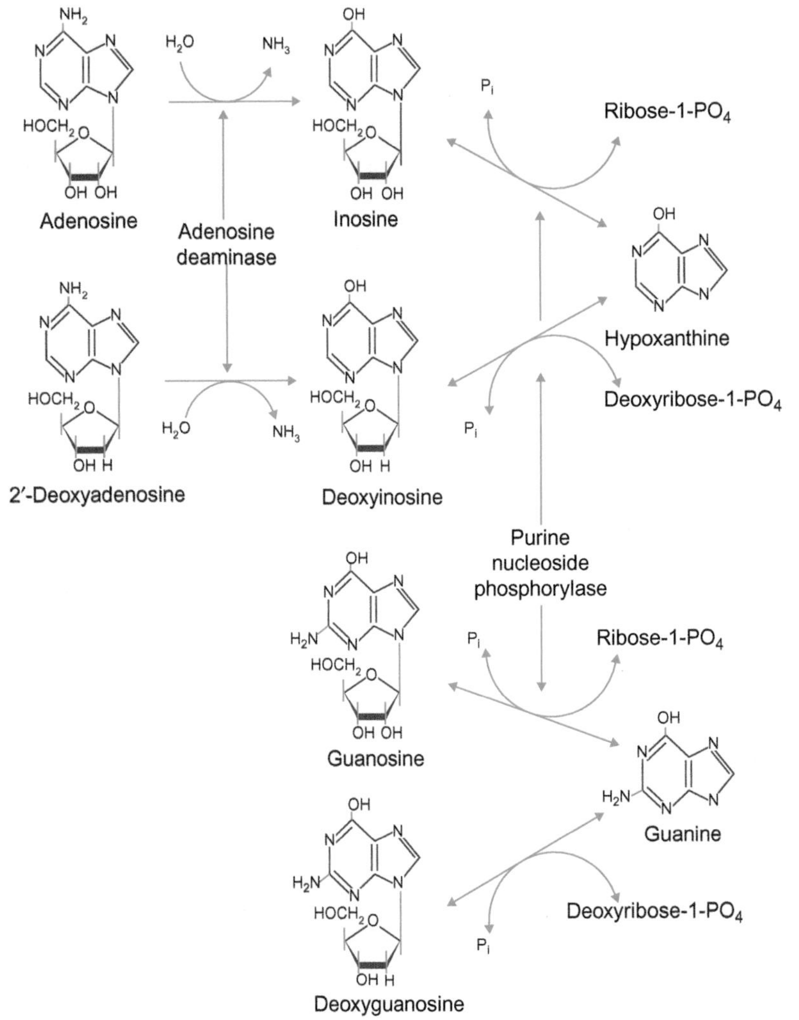

FIGURE 25.13 Reactions catalyzed by adenosine deaminase (ADA) and purine nucleotide phosphorylase (PNP). ADA and PNP participate in the purine catabolic pathway, and deficiency of either leads to immunodeficiency disease. *The solid lines in ribose and deoxyribase structures without any designation of an atom, represent hydrogen. [*Slightly modified and reproduced, with permission, from N. M. Kredich, M. S. Hershfield, Immunodeficiency diseases caused by adenosine deaminase and purine nucleoside phosphorylase deficiency. In C. S. Scriver, A. L. Beaudet, W. S. Sly, D. Valle (Eds). The metabolic basis of inherited disease, 6th ed., New York, NY: McGraw-Hill, 1989.*]

Adenine Phosphoribosyltransferase Deficiency

Adenine phosphoribosyltransferase (APRT) isolated from erythrocytes is a dimer with each subunit having a molecular weight of 19,481; the gene is located on chromosome 16. This autosomal recessive trait results in inability to salvage adenine, which accumulates and is oxidized to 2,8-dihydroxyadenine by xanthine oxidase. The main clinical abnormality is the excretion of 2,8-dihydroxyadenine as insoluble material (gravel) in the urine. These stones can be confused with those of urate on routine analysis. Thus, appropriate chemical analyses are required, particularly in the pediatric age group, to identify APRT-deficient individuals. The neurological disorders characteristic of HGPRT deficiency are not found in APRT deficiency, indicating that APRT may not play a significant role in the overall regulation of purine metabolism in

humans. APRT deficiency is treated with a low-purine diet and allopurinol.

Adenosine Deaminase Deficiency and Purine Nucleoside Phosphorylase Deficiency

Adenosine deaminase (ADA) deficiency and purine nucleoside phosphorylase (PNP) deficiency are two autosomal recessive traits that cause immune system dysfunction. The reactions catalyzed by ADA and PNP are shown in Figure 25.13. Both enzymes function in the conversion of adenosine and deoxyadenosine to hypoxanthine. PNP is also involved in conversion of guanosine and deoxyguanosine to guanine. In ADA or PNP deficiency, the appropriate substrates accumulate, along with other alternative products, resulting in toxic effects on the cells of the immune system. In ADA deficiency, deoxyadenosine is converted to dATP by salvage enzymes, which inhibits deoxyribonucleotide biosynthesis, compromising DNA biosynthesis.

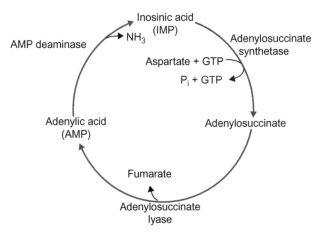

FIGURE 25.14 Purine nucleotide cycle. This cycle plays an important role in energy production in skeletal muscle during exercise.

Myoadenylate Deaminase Deficiency

Myoadenylate deaminase (or AMP deaminase) deficiency is a relatively benign muscle disorder characterized by fatigue and exercise-induced muscle aches. This disorder is presumably inherited as an autosomal recessive trait. The relationship between the exercise-induced skeletal muscle dysfunction and AMP deaminase deficiency is explained by an interruption of the purine nucleotide cycle.

The purine nucleotide cycle of muscle consists of the conversion of $AMP \rightarrow IMP \rightarrow AMP$ and requires AMP deaminase, adenylosuccinate synthetase, and adenylosuccinate lyase (Figure 25.14). Flux through this cycle increases during exercise. Several mechanisms have been proposed to explain how the increase in flux is responsible for the maintenance of appropriate energy levels during exercise. Fumarate produced in the cycle serves as an anaplerotic substrate in supporting the TCA cycle.

BIOSYNTHESIS OF PYRIMIDINE NUCLEOTIDES

Pyrimidine nucleotides, in common with purine nucleotides, are required for the synthesis of DNA and RNA. They also participate in intermediary metabolism. For example, pyrimidine nucleotides are involved in the biosynthesis of glycogen (Chapter 14) and of phospholipids (Chapter 17). Biosynthesis of pyrimidine nucleotides can occur by a *de novo* pathway or by the reutilization of preformed pyrimidine bases or ribonucleosides (salvage pathway).

De Novo Synthesis

The biosynthesis of pyrimidine nucleotides may be conveniently considered in two stages: the formation of uridine monophosphate (UMP) and the conversion of UMP to other pyrimidine nucleotides.

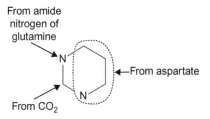

FIGURE 25.15 Sources of the pyrimidine ring atoms in *de novo* biosynthesis.

Formation of UMP

The synthesis of UMP starts from glutamine, bicarbonate, and ATP, and requires six enzyme activities. The sources of the atoms of the pyrimidine ring are shown in Figure 25.15, and the pathway is shown in Figure 25.16.

1. The pyrimidine base is formed first and then the nucleotide by the addition of ribose 5-phosphate from PRPP. In contrast, in *de novo* purine nucleotide biosynthesis, ribose 5-phosphate is an integral part of the earliest precursor molecule.
2. In the biosynthesis of both pyrimidine and urea (or arginine) (Chapter 15), carbamoyl phosphate is the source of carbon and nitrogen atoms. In pyrimidine biosynthesis, carbamoyl phosphate serves as donor of the carbamoyl group to aspartate with the formation of carbamoyl aspartate. In urea synthesis, the carbamoyl moiety of carbamoyl phosphate is transferred to ornithine, giving rise to citrulline. In eukaryotic cells, two separate pools of carbamoyl phosphate are synthesized by different enzymes located at different sites. Carbamoyl phosphate synthetase I (CPS I) is located in the inner membrane of mitochondria in the liver and, to a lesser extent, in the kidneys and small intestine. It supplies carbamoyl phosphate for the urea cycle. CPS I is specific for ammonia as nitrogen donor and requires N-acetylglutamate as activator. Carbamoyl phosphate synthetase II (CPS II) is present in the cytosol. It supplies carbamoyl phosphate for pyrimidine nucleotide biosynthesis and uses the amido group of glutamine as nitrogen donor. The presence of physically separated CPSs in eukaryotes probably reflects the need for independent regulation of pyrimidine biosynthesis and urea formation, despite the fact that both pathways require carbamoyl phosphate. In prokaryotes, one CPS serves both pathways.
3. In mammalian tissue, the six enzymes are encoded by three genes. One gene codes for a multifunctional polypeptide (Pyr 1−3) that is located in the cytosol and has carbamoyl phosphate synthetase II (Figure 25.17), aspartate transcarbamoylase, and dihydroorotase activity. Each subunit of Pyr 1−3 has a molecular weight of 200,000−220,000, and

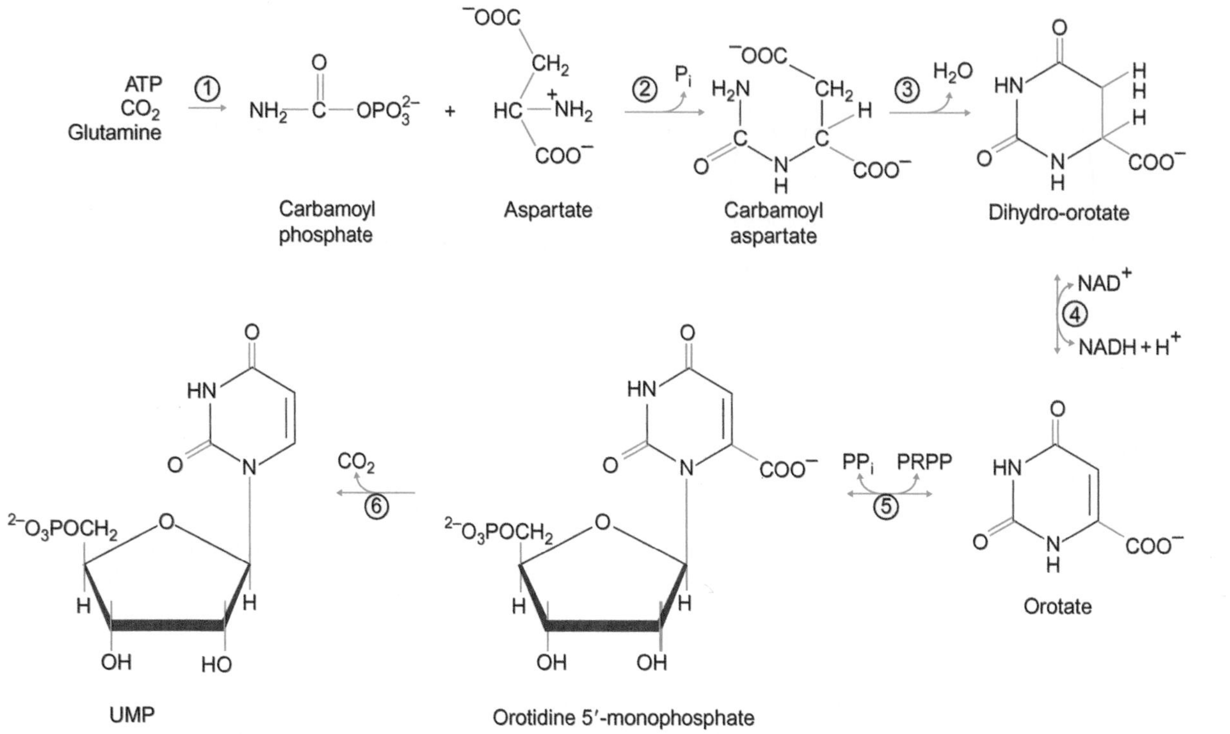

FIGURE 25.16 *De novo* pathway of uridine-5′-monophosphate (UMP) synthesis. Enzymes: (1) carbamoyl phosphate synthetase II; (2) aspartate transcarbamoylase; (3) dihydro-orotase; (4) dihydro-orotate dehydrogenase; (5) orotate phosphoribosyltransferase; (6) orotidine-5′-monophosphate decarboxylase (orotidylate decarboxylase). *The solid lines in ribose structures without any designation of an atom represent hydrogen.

the native enzyme exists as multiples of three subunits. The second gene codes for dihydro-orotate dehydrogenase, which is located on the outer side of the inner mitochondrial membrane. Dihydro-orotate, the product of Pyr 1−3, passes freely through the outer mitochondrial membrane and converts to orotate. Orotate readily diffuses to the cytosol for conversion to UMP. The third gene codes for another multifunctional polypeptide known as **UMP synthase** (Pyr 5,6). Pyr 5,6 (M.W. 55,000) contains orotate phosphoribosyltransferase and orotidylate (orotidine-5′-monophosphate) decarboxylase activity. Use of multifunctional polypeptides is very efficient, since the intermediates neither accumulate nor become consumed in side reactions. They are rapidly channeled without dissociation from the polypeptide. Other pathways in eukaryotic cells, such as fatty acid synthesis, occur on multifunctional polypeptides.

4. Conversion of dihydroorotate to orotate is catalyzed by dihydroorotate dehydrogenase, a metalloflavoprotein that contains nonheme iron atoms and flavin adenine nucleotides (FMN and FAD). In this reaction, the electrons are probably transported via iron atoms and flavin nucleotides that are reoxidized by NAD^+.

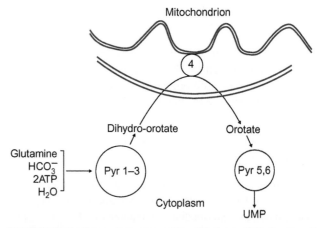

FIGURE 25.17 Schematic representation of the intracellular location of the six enzymes of UMP biosynthesis in animals. Pyr 1−3 = 1, carbamoyl phosphate synthetase II; 2, aspartate transcarbamoylase; 3, dihydro-orotase. 4 = Dihydro-orotate dehydrogenase. Pyr 5,6 = 5, orotate phosphoribosyltransferase; 6, orotidine-5′-monophosphate decarboxylase.

5. Biosynthesis of purine and pyrimidine nucleotides requires carbon dioxide and the amide nitrogen of glutamine. Both use an amino acid "nucleus"—glycine in purine biosynthesis and aspartate in pyrimidine biosynthesis. Both use PRPP as the source of ribose 1-phosphate.

6. UMP is converted to UTP by uridylate kinase and nucleoside diphosphate kinase.

$$\text{UMP} + \text{ATP} \underset{}{\overset{\text{uridylate kinase}}{\rightleftharpoons}} \text{UDP} + \text{ADP}$$

$$\text{UDP} + \text{ATP} \underset{}{\overset{\text{nucleoside diphosphate kinase}}{\rightleftharpoons}} \text{UDP} + \text{ADP}$$

Formation of Other Pyrimidine Nucleotides

UMP is the parent compound in the synthesis of cytidine and deoxycytidine phosphates and thymidine nucleotides (which are deoxyribonucleotides).

Synthesis of Cytidine Nucleotides

CTP is synthesized from UTP by transfer of the amide nitrogen of glutamine to C-4 of the pyrimidine ring of UTP. This reaction requires ATP as an energy source. See the following reaction:

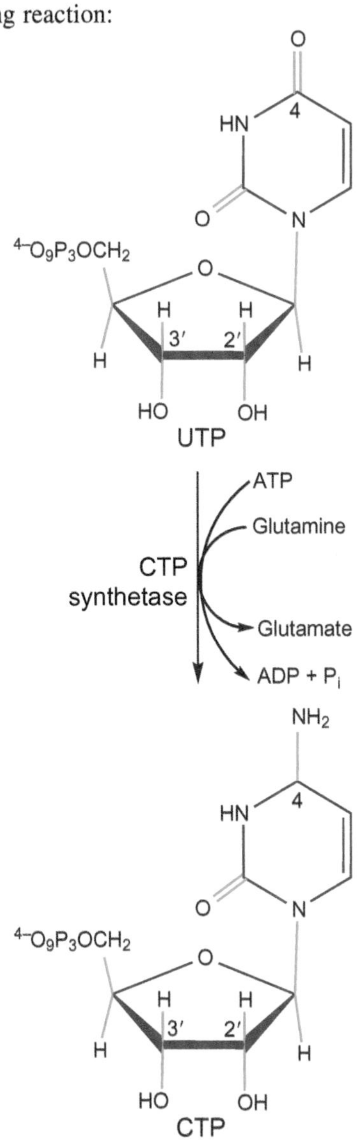

Deoxycytidine phosphates result from reduction of CDP to dCDP by a mechanism analogous to that described for purine nucleotides. Then dCDP is converted to dCTP by nucleoside diphosphate kinase.

Synthesis of Thymidine Nucleotides

De novo synthesis of thymidylic acid (TMP) occurs exclusively by methylation of the C-5 of dUMP (Figure 25.18) by thymidylate synthase. The methylene group of N^5,N^{10}-methylene FH_4 is the source of the methyl group, and FH_4 is oxidized to FH_2. For sustained synthesis, FH_4 must be regenerated by dihydrofolate reductase. Recall that deoxynucleotides are formed at the diphosphate level by ribonucleotide reductase: thus, UDP is converted to dUDP; then to dUTP; dUMP is then generated mainly by dUTPase.

$$\text{dUTP} \overset{\text{dUTPase}}{\longrightarrow} \text{dUMP} + \text{PP}_i$$

The dUTPase reaction is very important because the DNA polymerases cannot distinguish dUTP from TTP and catalyze significant incorporation of dUMP into DNA when dUTP is present. Incorporation of the base U into DNA is not deleterious because all cells contain a uracil N-glycosylase that removes U from DNA. Figure 25.19 summarizes the pathway for TTP synthesis.

Thymidylate synthase is competitively inhibited by fluorodeoxyuridylate (FUDRP), with formation of a stable ternary complex with methylene FH_4. FUDRP is generated by a salvage pathway from exogenous 5-fluorouracil (FU) or fluorodeoxyuridine (FUDR). FUDR is a useful drug in cancer chemotherapy because it inhibits TMP formation in proliferating cells. Thymidine nucleotide deficiency can also be induced by competitive inhibitors of dihydrofolate reductase, e.g., aminopterin (4-aminofolate) and methotrexate (4-amino-10-methylfolate; see Figure 25.3). 5-FU is normally inactivated by **dihydropyrimidine dehydrogenase (DPD)**, up to 85% of the administered drug. Patients with inherited deficiency of DPD treated with standard doses of 5-FU experience serious clinical consequences leading to myelosuppression and gastrointestinal and neurologic toxicity. A similar toxic manifestation due to deficiency of thiopurine metabolizing enzyme was discussed earlier. The study of the role of genetic inheritance that leads to variations in drug response is known as **pharmacogenomics**.

Salvage Pathways

Pyrimidines derived from dietary or endogenous sources are salvaged efficiently in mammalian systems. They are converted to nucleosides by nucleoside phosphorylases and then to nucleotides by appropriate kinases.

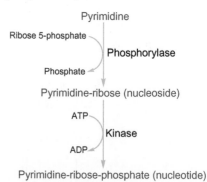

FIGURE 25.18 Synthesis of thymidylic acid (TMP). Fluorodeoxyuridylate inhibits conversion of dUMP to TMP, and methotrexate inhibits regeneration of the tetrahydrofolate coenzyme.

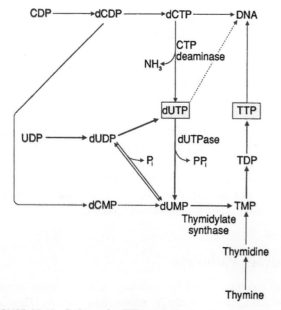

FIGURE 25.19 Pathway for TTP synthesis. The key intermediate is dUTP, which is converted to dUMP by dUTPase.

UMP can also be synthesized from uracil and PRPP by uracil phosphoribosyltransferase:

Pyrimidine

Ribose 5-phosphate ⟍
⟍ Phosphorylase
Phosphate ⟋

Pyrimidine-ribose (nucleoside)

ATP ⟍
⟍ Kinase
ADP ⟋

Pyrimidine-ribose-phosphate (nucleotide)

Pyrimidine Analogues

This group of compounds (Figure 25.20) includes several drugs that are clinically useful in the treatment of neoplastic disease, psoriasis, and infections caused by fungi and DNA-containing viruses. The mechanism of action of **5-fluorouracil**, a halogenated pyrimidine, was mentioned

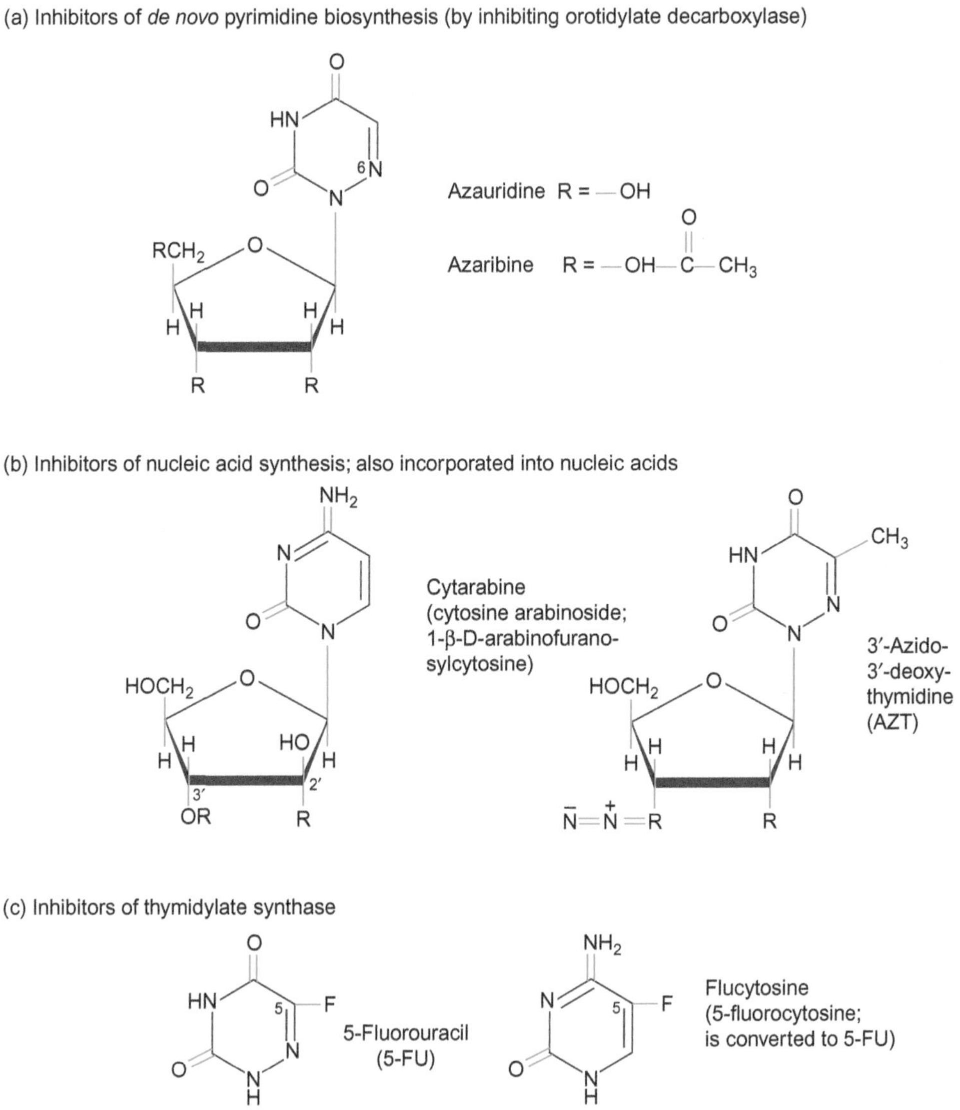

(a) Inhibitors of *de novo* pyrimidine biosynthesis (by inhibiting orotidylate decarboxylase)

Azauridine R = —OH

Azaribine   R = —OH—$\overset{\overset{\displaystyle O}{\|}}{C}$—CH$_3$

FIGURE 25.20 Pyrimidine analogues. All of these compounds require conversion to appropriate nucleotides before they become active.

(b) Inhibitors of nucleic acid synthesis; also incorporated into nucleic acids

Cytarabine
(cytosine arabinoside;
1-β-D-arabinofurano-
sylcytosine)

3′-Azido-
3′-deoxy-
thymidine
(AZT)

(c) Inhibitors of thymidylate synthase

5-Fluorouracil
(5-FU)

Flucytosine
(5-fluorocytosine;
is converted to 5-FU)

previously. An antifungal agent, **flucytosine (5-fluorocytosine)**, acts through conversion to 5-fluorouracil by cytosine deaminase in the fungal cells. **Idoxuridine (iododeoxyuridine)**, another halogenated pyrimidine derivative, is used in viral infections. It is a competitive inhibitor (via phosphorylated derivatives) of the incorporation of thymidylic acid into DNA.

Cytarabine (cytosine arabinoside) is an analogue of 2′-deoxycytidine. The 2′-hydroxyl group of the arabinose moiety is in a *trans* position with respect to the 3′-hydroxyl group (instead of in a *cis* position, as in the ribose) and causes steric hindrance to rotation of the base around the nucleoside bond. Phosphorylated derivatives of cytarabine inhibit nucleic acid synthesis as well as being incorporated into nucleic acids.

6-Azauridine and its triacetyl derivative, azaribine, inhibit pyrimidine biosynthesis. Azaribine is better absorbed than 6-azauridine, but it is converted in the blood to 6-azauridine, which undergoes intracellular transformation to 6-azauridylic acid (6-AzUMP). 6-AzUMP, a competitive inhibitor of orotidylate decarboxylase, blocks formation of UMP (Figure 25.20), resulting in high rates of excretion of orotic acid and orotidine in the urine.

5-Azacytidine, after conversion to a phosphorylated derivative, arrests the DNA synthesis phase of the cell cycle. It also affects methylation of certain bases, which leads to hypomethylation.

Azathymidine, 3′-azido-3′-deoxythymidine (AZT), after conversion to the corresponding 5′-triphosphate, inhibits viral reverse transcriptase. Hence, AZT is used in the treatment of AIDS.

Regulation of *De Novo* Pyrimidine Biosynthesis

In prokaryotic cells aspartate transcarbamoylase, an allosteric protein, is inhibited by the end product CTP and activated by ATP. In mammals, carbamoyl phosphate synthetase II (CPS II) activity of the trimeric multifunctional protein (Pyr 1–3) is the primary regulatory site. CPS II allosteric ligand UTP inhibits, whereas PRPP and ATP activate, the enzyme activity. Activation by ATP may be important in achieving a balanced synthesis of purine and pyrimidine nucleotides. During rapid proliferation of cells, either as a normal physiological process or in pathological processes (e.g., tumor growth), there is an increased demand for nucleotides required for nucleic acid synthesis. In those periods of cellular growth, CPS II activity is altered by phosphorylation at its regulatory site, leading to relief of UTP inhibition and enhancement in the stimulation of PRPP activation. The phosphorylation is mediated by mitogen-activated protein (MAP) kinase. The MAP kinase cascade is activated in response to growth factors (e.g., epidermal growth factor). PRPP is also an essential (and probably a rate-limiting) substrate for the orotate phosphoribosyltransferase reaction (reaction 5 in Figure 25.16) promoting *de novo* pyrimidine nucleotide synthesis by induction of Pyr 1–3. Another potential site of regulation is orotidine-5-phosphate decarboxylase, which is inhibited by UMP, CMP, allopurinol nucleotide, and oxypurinol.

COORDINATION OF PURINE AND PYRIMIDINE NUCLEOTIDE BIOSYNTHESIS

A balanced synthesis of pyrimidine and purine nucleotides is essential for the biosynthesis of nucleic acids in growing cells. Several mechanisms exist for coordinating nucleotide synthesis.

1. PRPP affects purine and pyrimidine nucleotide biosynthesis. PRPP formation is activated by inorganic phosphate and inhibited by several end products of pathways that use PRPP. In the purine *de novo* pathway, PRPP activates amidophosphoribosyltransferase and is the rate-limiting substrate for the enzyme. In purine salvage pathways, PRPP is the substrate for HGPRT and APRT. In the pyrimidine pathway, PRPP activates CPS II, may induce Pyr 1–3, and is a rate-limiting substrate for orotate phosphoribosyltransferase. Thus, changes in the levels of PRPP bring about concordant changes, while enhanced utilization by one pathway might result in reciprocal changes in the other. This interrelationship between the synthesis of purine and pyrimidine nucleotides was seen in a patient who had a deficiency of PRPP synthetase and exhibited orotic aciduria and hypouricemia, consistent with decreased synthesis of purine and pyrimidine nucleotides.
2. Coordination of synthesis of purine and pyrimidine nucleotides is affected by activation of CPS II by ATP.
3. Cultured mammalian cells (e.g., human lymphocytes) fail to grow when exposed to adenosine and have increased pools of ADP, ATP, and GTP; decreased pools of UDP, UTP, and CTP; and decreased incorporation of [^{14}C] orotic acid. These effects are reversed by exogenous uridine. These results suggest that adenosine inhibits the conversion of orotic acid to orotidine-5′-monophosphate. Adenosine deaminase reduces the toxic effect of adenosine by converting it to inosine. Increased levels of purine nucleotides may also inhibit PRPP synthesis. This inhibition of formation of pyrimidine nucleotides in the presence of excess purine nucleosides and nucleotides has been termed **pyrimidine starvation**.

CATABOLISM OF PYRIMIDINE NUCLEOTIDES

Pyrimidine catabolism occurs mainly in the liver. In contrast to purine catabolism, pyrimidine catabolism yields highly soluble end products.

ABNORMALITIES OF PYRIMIDINE METABOLISM

Hereditary orotic aciduria is a rare autosomal recessive trait. In this disorder, both orotate phosphoribosyltransferase and orotidine-5′-monophosphate decarboxylase activities (reactions 5 and 6 in Figure 25.16) are markedly deficient. Recall that these activities occur on the polypeptide Pyr 5,6.

Orotic aciduria is characterized by failure of normal growth and by the presence of hypochromic erythrocytes and megaloblastic bone marrow, none of which is improved by the usual hematinic agents (e.g., iron, pyridoxine, vitamin B_{12}, and folate). Leukopenia is also present. Treatment with uridine (2–4 g/d) results in marked improvement in the hematological abnormalities, in growth and development, and in decreased excretion of orotic acid. These patients are pyrimidine auxotrophs and require an exogenous source of pyrimidine just as all humans need vitamins, essential amino acids, and essential fatty acids.

Deficiency of folate or **vitamin B_{12}** can cause hematological changes similar to hereditary orotic aciduria.

Folate is directly involved in thymidylic acid synthesis and indirectly involved in the metabolic functions of vitamin B_{12}. Orotic aciduria without the characteristic hematological abnormalities occurs in disorders of the urea cycle that lead to the accumulation of carbamoyl phosphate in mitochondria (e.g., ornithine carbamoyltransferase deficiency; see Chapter 15). The carbamoyl phosphate exits from the mitochondria and augments cytosolic pyrimidine biosynthesis. Treatment with allopurinol or 6-azauridine also produces orotic acid-uria as a result of inhibition of orotidine-5′-phosphate decarboxylase by their metabolic products.

CLINICAL CASE STUDY 25.1 Hyperuricemia

Vignette 1: Gout
This clinical case was abstracted from: T. Neogi, Gout, N. Engl. J. Med. 364 (2011) 443−452.

Synopsis
A 54-year-old man with four arthritic attacks during the previous year was found to have serum urate levels of 7.2 mg/dL (428 µmol/L) on a routine laboratory test. Values of serum urate more than 6.8 mg/dL (404 µmol/L) are considered hyperuricemic. His hyperuricemia was treated with a xanthane oxidase inhibitor, allopurinol. He was also obese and had hypertension managed by hydrochlorothiazide. His serum renal function test was within the reference interval, with creatinine of 1.0 mg/dL (88 micromol/L). The concern was what course of medical options to undertake to correct his hyperuricemia. After consideration to attain normal serum urate levels, the patient's allopurinol dose was incrementally increased to achieve normal levels. In order to prevent acute attacks, daily colchicine administration was used as prophylaxis. The patient was advised to refrain from intake of alcohol and excessive consumption of meat and seafood. He was also advised to lose weight.

Teaching Points
1. Acute inflammatory gout disease is caused by monosodium urate crystals undergoing phagocytosis in the synovial joint fluid. This process leads to activation of the inflammatory response, with production and release of several inflammatory mediators, leading to severe pain. The acute attacks are treated with colchicine, glucocorticoids, and/or nonsteroidal anti-inflammatory drugs.
2. Hyperuricemia can result from overproduction of urate. Underexcretion of urate from the kidneys may also contribute to hyperuricemia.
3. Several medications may exacerbate hyperuricemia (e.g., hydrochlorothiazide).
4. For patients who cannot tolerate allopurinol, a nonpurine xanthine oxidase inhibitor like febuxostat may be considered.
5. Hyperuricemia can also be treated with uricosuric drugs: probenecid, sulfinpyrazone, and benzbromarone.
6. If conventional treatments cannot correct hyperuricemia, a pegylated porcine recombinant uricase is used, which converts uric acid to soluble allantoin. Allantoin is rapidly cleared through the renal system. Note that uricase is not present in humans, and thus, the relatively insoluble uric acid is the end product of purine catabolism.

Vignette 2: Tumor Lysis Syndrome
This clinical case was abstracted from: S.C. Howard, D.P. Jones, C.H. Pui, The tumor lysis syndrome, N. Engl. J. Med. 364 (2011) 1844−1854.

Synopsis
An 8-year-old male with cervical lymphadenopathy, malaise, and repeated vomiting was brought to the emergency department. In the ED, the patient's white blood cell count was slightly elevated, and microscopic examination revealed lymphoblasts. He was diagnosed with T-cell acute lymphoblastic leukemia. Laboratory studies revealed he had hyperphosphatemia and acute renal failure, and later he was diagnosed with acute tumor lysis syndrome. His problem was resolved after 2 months of medical therapy.

Teaching Points
1. Excessive cell destruction of tumor cells during chemotherapy causes cellular contents to be spilled into the bloodstream, causing hyperuricemia, hyperkalemia, hyperphosphatemia, and acute kidney failure. Hypocalcemia occurs secondary to hyperphosphatemia. Acute kidney failure results from deposition of calcium phosphate and urate crystals in the renal tubules, and is assessed by measurement of serum creatinine levels.
2. Management of tumor lysis syndrome and its acute toxic effects involves intravenous fluid administration, renal dialysis, and administration of rasburicase. Rasburicase is a recombinant uricase (urate oxidase), which converts uric acid to a more soluble form, allantoin. Currently, rasburicase is recommended for the treatment of gout.

Chapter 26

Hemoglobin

Key Points

1. Hemoglobin (Hb) present in red blood cells (RBCs) transports O_2 from the lungs to the tissues, and it also, along with RBC carbonic anhydrase, participates in the transport of CO_2 from the tissues to the lungs for elimination.

2. The predominant Hb after one year of age is HbA, a tetramer ($\alpha_2\beta_2$), and each subunit contains a heme group with an iron atom in the Fe^{2+} state. Cooperativity of Hb in binding with O_2 and allosteric regulatory binding properties with CO_2, H^+, Cl^-, and 2,3-DPG (2,3-bisphosphoglycerate) are based on subunit interactions. 2,3-DPG stabilizes deoxy conformation and promotes O_2 release from HbO_2. The shape of the percent (%) saturation and profile for Hb is sigmoidal.

3. Myoglobin is a monomer that provides O_2 during muscle contraction and binds more tightly to O_2 than Hb. Its saturation profile is hyperbolic. Maternal Hb ($\alpha_2\beta_2$) transfers O_2 to fetal hemoglobin ($\alpha_2\gamma_2$) due to its higher affinity for O_2 because of its decreased binding to 2,3-DPG.

4. Erythropoietin stimulates proliferation of precursor cells for the production of RBC. It is synthesized in the juxtatubular interstitial cells of the renal cortex in response to hypoxemia. Recombinant erythropoietin preparations are used in the treatment of anemia patients with chronic kidney disease who are undergoing dialysis.

5. Genetic defects of deletions and point mutations in the α- and β-globin genes lead to decreased chain synthesis, resulting in α- and β-thalassemia syndromes, respectively. These syndromes are clinically heterogeneous and cause ineffective erythropoiesis and hemolysis.

6. Sickle cell disease, which is due to point mutation leading to substitution of glutamate to valine in the β-chain at the 6th position, yields HbS. HbS polymerizes in the deoxy state and affects RBC deformability and survival and causes vaso-occlusion. The treatment is palliative; it also uses hydroxyurea, which promotes HbF synthesis, ameliorating the symptoms. Reawakening of HbF production by regulating erythroid transcription factors is under investigation as a possible therapeutic intervention in the treatment of β-thalassemia syndromes, sickle cell, and malaria diseases. Elevated Hb F levels in the erythrocytes thwarts the susceptibility of malarial parasites.

7. When Fe^{2+} of the heme group is converted to Fe^{3+}, the Hb is converted to methemoglobin, and it is not an oxygen carrier. Methemoglobinemia can occur due to defects in the enzymes, converting Fe^{3+} to Fe^{2+}, or specific amino acid substitutions in the heme pocket of the α- and β-globin chain or acquired causes. The treatment requires blood transfusion and in some cases methylene blue, which, in its reduced form, converts metHb to Hb.

8. The signal transduction of erythropoietin (EPO) is initiated by its binding to cell membrane receptors followed by activation of tyrosine kinase Jak-2 (see Chapters 28 and 29).

9. Individuals who take part in competitive endurance sports and have been administered recombinant EPO as a performance-enhancing drug to increase their oxygen-carrying capacity are banned from athletic participation.

10. The α-gene complex consists of one functional ζ gene and two functional α genes, and the β-gene complex consists of five functional genes: ϵ, Gγ, Aγ, δ, and β. In diploid cells, gene numbers are doubled. The genetic and regulatory mechanisms of β-globin gene cluster expression are being explored to stimulate HbF production. Elevated hemoglobin F production attenuates the symptoms of subjects with sickle cell diseases and β-thalassemia syndromes.

11. Polycythemia vera (primary polycythemia, a disorder of increased erythrocytes) and Hb are associated with an activating mutation (V617F) of tyrosine kinase Jak-2. The intracellular signaling of Jak-2 V617F mutant undergoes constitutive activation independent of ligand. The mutation is acquired and occurs in more than 95% of polycythemia vera patients. In about 60% of essential thrombocytosis and primary myelofibrosis, the Jak-2 V617F mutation is found. Treatment for these disorders is palliative or hematopoietic stem cell transplantation for selected patients.

Hemoglobin is the predominant protein in the red blood cell and is responsible for transporting oxygen, carbon dioxide, and protons between the lungs and tissues. The study of hemoglobin has led to detailed knowledge of how oxygen and carbon dioxide transport is accomplished and regulated, and has provided insight into the functioning of other allosteric proteins (Chapter 6). Hemoglobin is the first allosteric protein for which molecular details of allosteric effector binding and the mechanism of allosteric

N. V. Bhagavan and C.-E. Ha: Essentials of Medical Biochemistry, Second edition. DOI: http://dx.doi.org/10.1016/B978-0-12-416687-5.00026-9

action are known. Studies of the genes coding for the globin polypeptide chains have provided a better understanding of many anemias and of the regulation of expression of other eukaryotic genes. Correction of genetic defects in sickle cell anemia and thalassemia by the introduction of new genetic information into bone marrow cells (gene therapy) is being actively explored.

STRUCTURE OF HEMOGLOBINS

Globin Chains

Mammalian hemoglobins are tetramers made up of two α-like subunits (usually α) and two non-α subunits (usually β, γ, or δ). These subunits differ in primary structure but have similar secondary and tertiary structures. However, the differences in tertiary structure among them are critical to the functional properties of each subunit. Each globin subunit has associated with it, by noncovalent interaction, an Fe^{2+}−porphyrin complex known as a **heme group**. Oxygen binding occurs at the heme iron. The predominant hemoglobin present in adult erythrocytes is $\alpha_2 \beta_2$, known as **hemoglobin A_1 (HbA)**.

Each tetramer has a molecular weight of about 64,500, and each α-like and β-like chain has a molecular weight of about 15,750 and 16,500, respectively. The subunits are situated at the corners of a tetrahedron (Figure 26.1). In a tetramer, dissimilar chains are more strongly joined than similar chains.

Association involves salt bridges, hydrogen bonds, van der Waals forces, and hydrophobic interactions. About 60% of the contact between subunits does not change during reversible oxygen binding (packing contacts), while about 35% does (sliding contacts). Contacts of $\alpha\alpha$ and $\beta\beta$ form about 5% of the total intersubunit contacts, the remainder being $\alpha\beta$ contacts. In deoxyhemoglobin, there is no $\beta\beta$ contact. This information was derived principally from the X-ray crystallographic studies of Perutz, Kendrew, and coworkers and clarifies how **2,3-bisphosphoglycerate (BPG)** or **2,3-diphosphoglycerate (2,3-DPG)** regulates oxygen transport. (Note: 2,3-bisphosphoglycerate and 2,3-diphosphoglycerate are identical.)

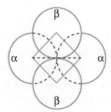

FIGURE 26.1 Arrangement of hemoglobin subunits. The four polypeptide subunits are located at the corners of a tetrahedron to give a roughly spherical structure.

The secondary structures of α- and β-chains are similar. Each chain contains helical and nonhelical segments and surrounds a heme group (Figure 26.2). Almost 80% of the amino acids exist in an α-helical conformation.

Heme Group

Heme consists of a porphyrin ring system with an Fe^{2+} fixed in the center through complexation to the nitrogens of four pyrrole rings. The porphyrin system is a nearly planar aromatic ring formed from four pyrrole rings linked by $=CH-$ (methene) bridges. The pyrrole rings are substituted so that different porphyrins are distinguished by variations in their side chains (see also Chapters 13 and 27). The heme porphyrin is protoporphyrin IX. Iron has a coordination number of six; i.e., each atom of iron can bond with six electron pairs. In heme, two coordination positions of iron are not occupied by porphyrin nitrogens. When heme is associated with a globin chain, a histidyl nitrogen (from histidine F8, important in the allosteric mechanism) bonds with the fifth coordination position of iron, while the sixth position is open for combination with oxygen, water, carbon monoxide, or other ligands. Portions of the seven or eight helices of a globin chain form a hydrophobic crevice near the surface of the subunit. Heme lies in this crevice, between helices E and F. The low dielectric constant of the crevice prevents permanent oxidation of iron (from Fe^{2+} to Fe^{3+}) by oxygen and is responsible for the reversible binding of oxygen. Evidence from electron paramagnetic resonance studies shows that, in oxyhemoglobin, the oxygen can be considered to oxidize the iron as long as the oxygen is bound. When the oxygen is released, Fe^{3+} is again reduced to Fe^{2+}. Reversible O_2 binding is a unique property of hemoglobin. If the Fe^{2+} in hemoglobin is permanently oxidized to the ferric state (as in methemoglobin), the Fe^{3+} binds tightly to a hydroxyl group, and oxygen will not bind.

FUNCTIONAL ASPECTS OF HEMOGLOBIN

Oxygen Transport

The primary function of hemoglobin is to transport oxygen from the lungs to the tissues. Hemoglobin forms a dissociable complex with oxygen:

$$\text{Deoxyhemoglobin} + 4O_2 \rightleftarrows \text{oxyhemoglobin}$$

This reaction goes to the right with an increase in oxygen pressure (as in the lungs) and to the left with a decrease in oxygen pressure (as in the tissues), in accordance with the law of mass action. Table 26.1 shows typical partial pressures of oxygen between the atmosphere and tissue mitochondria. Hemoglobin increases the solubility of oxygen in the blood 70-fold.

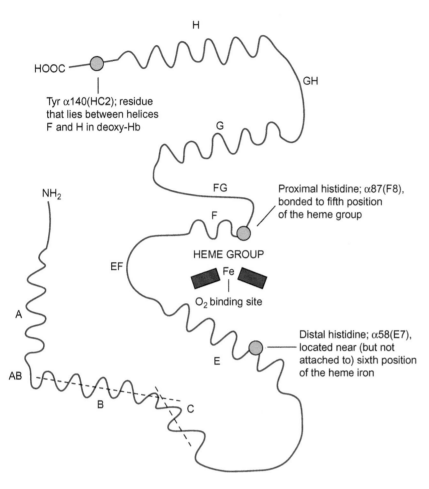

H

HOOC

Tyr α140(HC2); residue
that lies between helices
F and H in deoxy-Hb

GH

G

NH₂

FG

Proximal histidine; α87(F8),
bonded to fifth position
of the heme group

F

EF

HEME GROUP

Fe

O₂ binding site

A

Distal histidine; α58(E7),
located near (but not
attached to) sixth position
of the heme iron

AB

E

B

C

FIGURE 26.2 Secondary structure of the α-chain of human hemoglobin. The helical regions (labeled A–H, after Kendrew), N and C termini, and the histidines located near the heme group are indicated. The axes of the B and C helices are indicated by dashed lines. Note that the α-chain lacks helix D present in the β-chain. The amino acid residues are numbered by two different methods: from the N terminus of the polypeptide chain and from the N-terminal amino acid residue of each helix. Nonhelical regions are designated by the letters of helices at each end of a region.

TABLE 26.1 Partial Pressures of Oxygen

P$_{O_2}$ (mmHg)	Stage*
158	Inspired air
100	Alveolar air
95	Arterial blood
40	Peripheral capillary beds
<30 (est.)	Interstitial fluid
<10 (est.)	Inside cells in tissues
<2 (est.)	Inside mitochondria

During transport from lungs to mitochondria in tissues, where O_2 is used as a terminal electron acceptor. est. = estimated.

In Figure 26.3, the oxygen saturation percentage for hemoglobin is plotted against oxygen pressure of the gas above the surface of the hemoglobin solution. The curves, known as **binding** or **dissociation curves**, are often characterized by their P$_{50}$ value, the oxygen pressure at which the hemoglobin is 50% saturated with oxygen. For comparison, the dissociation curve for myoglobin is also shown. The hemoglobin curves are sigmoid (S-shaped), whereas the myoglobin curve is a rectangular hyperbola. This difference arises because hemoglobin is allosteric and shows cooperative oxygen-binding kinetics (Chapter 6), whereas myoglobin is not allosteric. The binding of each molecule of oxygen to hemoglobin causes binding of additional oxygen molecules. The cooperative binding of oxygen by hemoglobin is the basis for the regulation of oxygen and, indirectly, carbon dioxide levels in the body. Myoglobin is a heme protein found in high concentrations in red muscle, particularly cardiac muscle, where it functions as a storage site for oxygen. At the oxygen pressures found in tissues, the dissociation curve for myoglobin lies to the left of the hemoglobin curves; consequently, hemoglobin (Hb) can readily oxygenate myoglobin (Mb). In the equation

$$Hb(O_2)_4 + 4\,Mb \rightleftharpoons Hb + 4\,MbO_2$$

the equilibrium lies far to the right.

In the hemoglobin curves, the change from A to B to C is termed a **rightward shift**. The farther to the right a

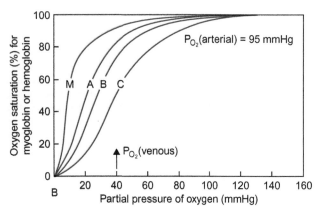

FIGURE 26.3 Oxygen dissociation curves for myoglobin and hemoglobin at several CO_2 pressures. The myoglobin plot (curve M) is very similar to the plot for a hemoglobin subunit (monomer). Curves A, B, and C are for hemoglobin at P_{CO_2} = 20, 40, and 80 mmHg, respectively. P_{O_2} (arterial) and P_{O_2} (venous) indicate normal values for oxygen tensions. During exercise, P_{O_2} (venous) will be lower, around 20–25 mmHg, because of the increased extraction of oxygen from blood by exercising muscle.

curve is shifted, the larger is the P_{50} value and the lower the oxygen affinity. In this example, the shift is due to an increase in P_{CO_2}, since pH, temperature, and 2,3-DPG concentration were held constant. However, the shift could also have been caused by an increase in temperature (as in strenuous exercise or fever), a decrease in pH (as in acidosis or exercise), an increase in 2,3-DPG concentration (see below), or a combination of these variables. For example, an increase in P_{CO_2} is usually accompanied by a decrease in pH. Substances that cause a rightward shift are called **negative allosteric effectors** or **allosteric inhibitors** of hemoglobin. They decrease the stability of the oxy form (R-state) of hemoglobin or increase the stability of the deoxy form (T-state). **Positive allosteric effectors**, or **allosteric activators**, increase the affinity of hemoglobin for oxygen. They cause a leftward shift by increasing the stability of the R-state or by decreasing the stability of the T-state. Oxygen and carbon monoxide are positive allosteric effectors.

A rightward shift decreases the saturation of hemoglobin with oxygen at any particular P_{O_2}. If the blood is initially fully saturated (arterial blood, P_{O_2} = 95 mmHg), a rightward shift would increase the amount of oxygen received by the tissues, as number of moles of oxygen per unit time. At P_{O_2} = 40 mmHg, only about 15% of the oxygen is released from hemoglobin on curve A (85% saturation at 40 mmHg), while 25% and 50% of the oxygen are released on curves B and C, respectively. On the other hand, if the arterial P_{O_2} = 45 mmHg, a rightward shift would decrease both the initial and final percentage saturations to roughly the same extent, and on a molar basis the tissues would receive almost the same amount of oxygen regardless of the position of the curve.

As blood travels from the lungs to the tissues and back again, pH, P_{O_2}, temperature, and other factors vary continually. Consequently, the curve that describes the affinity of hemoglobin for oxygen differs from one moment to the next.

A shift in the dissociation curve can be very important. In chronic conditions in which the supply of oxygen to the lungs is normal and arterial blood is saturated with oxygen but the ability of the blood to deliver oxygen to the tissues is impaired because of a low hemoglobin concentration (anemia) or a low cardiac output (cardiac insufficiency), the intraerythrocytic 2,3-DPG concentration increases. A rightward shift of the dissociation curve will not affect the percent saturation during loading in the lungs (since the upper part of the curve is quite flat), but it will decrease the percent saturation in the tissues, thereby providing a useful increase in moles of oxygen per unit time available for metabolism.

During respiration in a normal individual, P_{CO_2} is higher (and pH lower) in the tissues than in the lungs, oxygen unloading is facilitated at the tissues, and loading occurs more readily in the lungs. The decrease in oxygen saturation at constant P_{O_2} (i.e., decrease in oxygen affinity) with increasing P_{CO_2}, or decreasing pH above pH 6.3 is known as the **alkaline Bohr effect**.

Mechanism of Oxygenation

Hemoglobin has two quaternary structures that correspond to the deoxygenated (deoxy; five-coordinate iron; T-state) and the oxygenated (oxy; six-coordinate iron; R-state) forms. X-ray crystallographic studies have demonstrated that the coordination number of the iron is the crucial difference between these forms. All six-coordinate hemoglobins (oxyhemoglobin, carbon monoxyhemoglobin, and methemoglobin) are in the R-state, whereas deoxyhemoglobin is in the T-state. Binding of one molecule of oxygen to deoxyhemoglobin causes a change in the tertiary structure of the binding subunits, which results in a change in quaternary structure (T-to-R transition) that enhances the binding of oxygen to the remaining subunits. Thus, up to four molecules of O_2 per molecule of Hb can be bound.

The "machinery" for this transition is composed of the globin peptide chain, amino acid side chains, and heme. In deoxyhemoglobin, the bond between the nitrogen of histidine F8 and the iron atom of heme is tilted at a slight angle from the perpendicular to the plane of the porphyrin ring. The porphyrin ring is domed "upward" toward the histidine, with the iron at the apex of the dome. Thus, the iron is in an unfavorable position for binding of a sixth ligand, such as oxygen. This is a primary cause of the weak binding of O_2 to T-state hemoglobin (Figure 26.4).

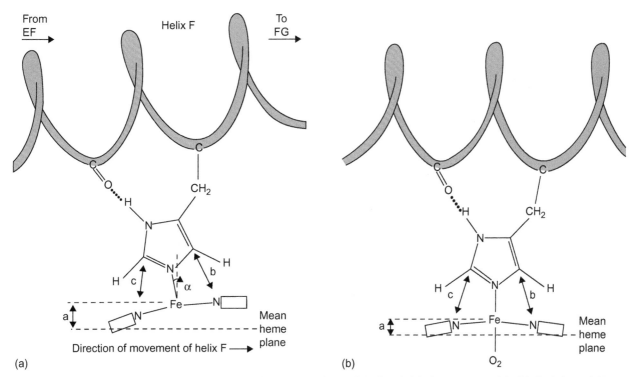

FIGURE 26.4 Schematic representation of the geometry of the unliganded (a) and the liganded (b) heme group, and the F helix in hemoglobin. The iron atom in the unliganded heme (a) is located out of the mean heme plane (distance a) and the iron and the porphyrin ring are slightly domed toward histidine. The histidine is tilted as indicated by the bond angle α ($8-10°$) and the distance b > c. When O_2 binds to the iron (b), the iron atom is closer to the heme plane [distance a decreases], the N–Fe bond becomes more nearly perpendicular ($\alpha = 1$ degree and b = c), and the heme acquires a less domed conformation. These events pull helix F across the heme group, setting off a series of reactions leading to the T-to-R transition. *[Modified and redrawn with permission from I. Geis and R. E. Dickerson.]*

Deoxyhemoglobin is in an unstrained conformation, as would be expected for a stable molecular form, despite its designation as the T ("tense") state. It is stabilized by hydrogen bonds, salt bridges, and van der Waals contacts between amino acid side chains on the same subunit and on different side chains. Breaking of many of these bonds occurs during oxygen binding, destabilizing the deoxy structure and causing the release of the Bohr protons (discussed later). Oxyhemoglobin is also stabilized by several noncovalent bonds, formation of which aids in the T-to-R transition.

Upon binding of the first molecule of oxygen to completely deoxygenated hemoglobin, strain is introduced into the molecule as a result of competition between maintenance of the stable deoxy form and formation of a stable iron–O_2 bond. "Strain" refers to "long" bonds that have higher energy than do normal length, minimal-energy bonds. When possible, the "stretched" bonds shorten to lower their energy and thereby produce movement within the molecule. In an effort to reduce the strain and return to an energetically more favorable state, the F helix slides across the face of the heme, causing the bond between histidine F8 and the iron atom to straighten toward the perpendicular (Figure 26.4b), and the heme

rotates and sinks further into the heme pocket. These movements occur in both α and β subunits but are greater in the β subunits. The change in radius of the iron on O_2 binding, originally thought to "trigger" the conformational changes, is now considered to be of minor importance in the T-to-R transition.

The movement of heme and surrounding amino acid residues accomplishes two things. First, it relieves the strain on the six-coordinated heme iron, increasing the strength of the Fe–O_2 bond (i.e., increasing oxygen affinity). Second, it strains other noncovalent interactions elsewhere in the subunit, causing some of them to break. This is the beginning of the quaternary structure transition from T-state to R-state. The energy needed for the conformational changes is provided by the energy of binding of oxygen to the heme iron and by the formation of new, noncovalent bonds typical of oxyhemoglobin. Under normal conditions of P_{CO_2} and P_{O_2} in the lungs, more energy is released by binding four oxygen molecules than is needed to break the noncovalent interactions in deoxyhemoglobin. The opposite situation prevails in extrapulmonary capillary beds.

The binding of an oxygen molecule in one subunit induces strain in another subunit and causes some noncovalent interactions to break. Thus, Hb(O$_2$), Hb(O$_2$)$_2$, and Hb(O$_2$)$_3$ represent different transient structures. Which type of subunit (α or β) binds oxygen first is unclear, although a three- to four-fold difference in oxygen affinity exists between the subunits in the tetramer. It is assumed that oxygen binds first to the α-subunit because its heme is more readily accessible.

The T-to-R transition occurs after the binding of two or three oxygen molecules, when the noncovalent interactions are too few to stabilize the deoxy form. The oxygen binding breaks the remaining noncovalent interactions that stabilize the deoxy structure, relieves the strain on the oxygenated subunits, and increases the affinity for oxygen of the unliganded subunits. The higher oxygen affinity of the oxy form and the formation of new noncovalent interactions accounts for the cooperative binding of oxygen.

The sum of the motions of individual side chains on binding and release of oxygen produces a relative reorientation of the four subunits (Figure 26.5). The motion can be considered as rotation of $\alpha\beta$ pairs, joined by $\alpha_1\beta_1$ type interactions, around the $\alpha\alpha$ axis. Movement occurs primarily at the $\alpha_1\beta_2$ interfaces. The effect is to move the β-subunits closer to each other. Therefore, a change in the tertiary structure of the subunits brings about the change in quaternary structure of the tetramer.

At pH 7.4, deoxyhemoglobin releases 0.7 mol of H$^+$ for each mole of oxygen it binds, the process being reversible. However, under physiological conditions, deoxyhemoglobin in whole blood releases only 0.31 mol of H$^+$ per mole of O$_2$ bound. This phenomenon is the alkaline Bohr effect (decrease in oxygen affinity of hemoglobin with decrease in pH or increase in P$_{CO_2}$); these H$^+$ ions are known as **Bohr protons**. An equivalent statement would be that oxyhemoglobin is a stronger acid than the deoxy form. In fact, the pK$'$ for oxyhemoglobin is 6.62, while that for deoxyhemoglobin is 8.18. This change in pK$'$ occurs because interactions of the basic side chains involved in the salt bridges are broken during the T-to-R transition. Release of the protonated side chains decreases their pKs, making it easier for the H$^+$ to separate. It is estimated that, at pH 7.4, 40% of the Bohr protons come from the C-terminal histidines on the β-chains, 25% from the N-terminal amino groups on the α-chain, and 35% from many amino acids, particularly His α122, His β143,

and Lys β144. The amount each of these groups contributes depends on pH and other conditions prevailing at the time of measurement.

A primary allosteric effector of hemoglobin in human erythrocytes, in addition to H$^+$ and CO$_2$, is the organic phosphate 2,3-DPG. Deoxyhemoglobin binds 1 mol of 2,3-DPG per mole of hemoglobin, whereas oxyhemoglobin does not. The reason for this differential binding is the location of the binding site (Figure 26.5). The negatively charged 2,3-DPG molecule fits between the β-subunits, where it binds to positively charged residues. In oxyhemoglobin, the β-subunits are close together, the cleft between them being too small for entry of the 2,3-DPG molecule. Binding of 2,3-DPG inhibits this movement of the β-chains and helps to stabilize the deoxy form. This binding is an equilibrium phenomenon, and the amount of 2,3-DPG bound depends on the concentrations of 2,3-DPG, O$_2$, CO$_2$, H$^+$, and hemoglobin. Chloride promotes H$^+$ binding to deoxyhemoglobin and acts as an allosteric effector, so its concentration must also be considered. All of these factors interact to determine the amount of oxygen bound to hemoglobin at any particular P$_{O_2}$.

Function, Metabolism, and Regulation of Organic Phosphates in Erythrocytes

2,3-DPG, ATP, inositol hexaphosphate (IHP), and other organic phosphates bind to hemoglobin and decrease its affinity for oxygen. IHP, the principal organic phosphate in avian erythrocytes, is the most negatively charged of these compounds and binds the most tightly; however, it is not found in human red cells.

2,3-DPG levels in the erythrocytes help regulate hemoglobin oxygenation (Figure 26.6). Unloading of oxygen at the P$_{O_2}$ in tissue capillaries is increased by 2,3-DPG, and

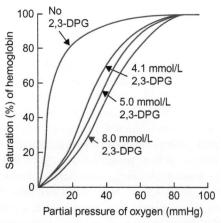

FIGURE 26.6 Effect of 2,3-bisphosphoglycerate (2,3-DPG) on the oxygen saturation curve of hemoglobin. Note that 2,3-DPG decreases the affinity of hemoglobin for oxygen.

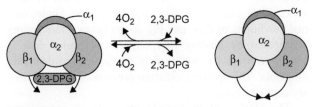

FIGURE 26.5 Relative motion of the β-subunits of hemoglobin on oxygenation and deoxygenation.

small changes in its concentration can have significant effects on oxygen release.

Fetal hemoglobin ($\alpha_2\gamma_2$) binds less tightly to 2,3-DPG and thus has a higher oxygen affinity than maternal hemoglobin ($\alpha_2\beta_2$), allowing for maternal-to-fetal oxygen transport.

In humans, 2,3-DPG is the most abundant phosphate compound in the red cell, and it is formed by rearrangement of 1,3-bisphosphoglycerate, an intermediate in glycolysis (Chapter 12).

In the blood specimens used in the treatment of transfusion therapy, 2,3-DPG levels decline significantly in storage and therefore affect oxygen delivery to the tissues. This process can be prevented by adding glucose, phosphate, and adenine to stored blood specimens. The safety of blood transfusions is also improved by washing of red blood cells and leukoreduction [1].

Carbon Dioxide (CO_2) Transport

In addition to carrying oxygen from the lungs to the tissues, hemoglobin helps to move carbon dioxide from tissues back to the lungs, where it can be eliminated. This CO_2 movement is accomplished with little or no change in blood pH, owing to the buffering capacity of hemoglobin (discussed in Chapter 2).

Nitric Oxide (NO) Binding to Hemoglobin

NO binds to hemoglobin at two sites. One is involved in scavenging NO through its binding to the ferrous iron of heme. The affinity of NO at the Fe^{2+}-heme site is about 8000 times greater than oxygen affinity at the same site. The binding of NO at the second site is a reversible process and occurs at the β 93 cysteine residue with the formation of S-nitrosothiol. This interaction of NO with β 93 cysteine is linked to the binding of oxygen to hemoglobin in the lungs and also to release of NO and O_2 into the tissues. It is thought that NO reduces regional vascular resistance. In the blood, NO is also transported by binding to plasma albumin; the exact physiological role of the transport of NO by hemoglobin or albumin is not understood.

Erythropoietin

Erythropoietin is a glycoprotein hormone that regulates red blood cell production in a feedback loop manner between kidney and bone marrow based on oxygen tension. It consists of 165 amino acids and has a molecular weight of 30,000−34,000; approximately 30% is accounted for by covalently linked carbohydrate. Erythropoietin is produced by the fetal liver and, shortly after birth, production switches from the liver to the

kidney. In the fetus, erythropoietin functions in a paracrine−endocrine fashion because liver is the site of erythropoietin synthesis as well as erythropoiesis. The mechanism of this developmental switch is unclear. In the liver, erythropoietin synthesis occurs in the Ito cells and in a subset of hepatocytes. In the kidney, erythropoietin is synthesized in the peritubular interstitial fibroblast-like cells. Hypoxia is the primary physiological stimulus for the dramatic rise of erythropoietin levels, which can rise up to 1000-fold through an exponential increase in the number of erythropoietin-producing cells, as well as in the rate of synthesis. Hypoxic states can occur due to loss of red blood cells, decreased ambient oxygen tension, presence of abnormal hemoglobin with increased oxygen affinity (discussed later), and other causes that limit oxygen delivery to the tissues (e.g., chronic obstructive lung disease). The expression of the erythropoietin gene is affected by a number of other physiological and pharmacological agents in addition to hypoxic states. Transition metal ions Co^{2+}, Ni^{2+}, Mn^{2+}, and iron chelator desferrioxamine can stimulate erythropoietin gene expression. Carbon monoxide, nitric oxide, inflammatory cytokine tumor necrosis factor-α, and interleukin-1 can prevent expression of the gene. The latter is implicated as one of the factors causing anemia in chronic disease (**anemia of chronic disease**). Erythropoietin is the primary regulator of erythropoiesis and red blood cell mass. Erythropoietin action is mediated by its binding to plasma cell membrane receptors of erythroid progenitors and precursors. The signal transduction of erythropoietin (EPO) is initiated by its binding to cell membrane receptors (EPORs) followed by activation of tyrosine kinase Jak-2 (see Chapters 28 and 29). The sensitivity of erythroid progenitors to erythropoietin appears to be under developmental regulation. Colony-forming unit erythroid (CFU-E) and proerythroblasts have a peak number of erythropoietin receptors. In peripheral reticulocytes, the receptors are undetectable. Erythropoietin's function is mediated through its receptor and includes proliferation, survival, and terminal differentiation of CFU-E cells.

The mechanism of oxygen sensing and signal transduction that leads to activation of the erythropoietin gene and erythropoietin synthesis in renal cells involves the following steps. The first step in the hypoxia-induced transcription of the erythropoietin gene consists of production of oxygen free radicals (O_2^-) by cytochrome b-like flavoheme NADPH oxidase in proportion to oxygen tension. Superoxide in the presence of iron is converted to other reactive oxygen species (e.g., $OH^•$). During normal oxygen tension, superoxide and other reactive oxygen species oxidize hypoxia-inducible factor-α (HIF-α) and cause its destruction in the ubiquitin−proteasome pathway. However, at low oxygen tension, HIF-α is preserved and it forms a heterodimer with constitutively expressed HIF-β. The dimer HIF-α/HIF-β is translocated into the nucleus, where it interacts with hypoxia response elements to activate gene expression.

Recombinant human erythropoietin has been used in the correction of anemia of chronic renal failure. It has also been used in other disorders of anemia such as anemia of prematurity, anemia of inflammation, and anemia of malignancy. Several synthetic peptide-based erythropoiesis-stimulating agents including EPO-receptor activator at the target sites have been developed for the anemia of chronic disease [2−5]. Individuals who take part in competitive endurance sports and have been administered recombinant EPO as a performance-enhancing drug to increase their oxygen-carrying capacity are banned from athletic participation.

Polycythemia is a disorder characterized by an increase in the number, and in the hemoglobin content, of circulating red cells. In patients who have chronic anoxia from impaired pulmonary ventilation or congenital or acquired heart disease, the increase in plasma erythropoietin leads to secondary polycythemia. Some renal cell carcinomas, hepatocarcinomas, and other tumors, which produce physiologically inappropriate amounts of erythropoietin, may also cause secondary polycythemia. Conversely, anemia can result from renal insufficiency and from chronic disorders that depress erythropoietin production. *Polycythemia vera* (primary polycythemia) is a malignancy of erythrocyte stem cells, and it is associated with an activating mutation (V617F) of tyrosine kinase Jak-2. The intracellular signaling of Jak-2 V617F mutant can undergo constitutive activation independent of ligand. The mutation is acquired and occurs in more than 95% of polycythemia vera patients. In about 60% of essential thrombocytosis and primary myelofibrosis, Jak-2 V617F mutation is found. Treatment for these disorders is palliative or hematopoietic stem cell transplantation for selected patients.

INHERITED DISORDERS OF HEMOGLOBIN STRUCTURE AND SYNTHESIS

Although hemoglobin disorders are extremely diverse, they can be generally classified into two somewhat-overlapping groups:

1. Thalassemia syndromes are characterized by a decreased rate of synthesis of one or more of the globin peptides. Although there are exceptions, the globin chains produced in thalassemic states usually have a normal amino acid sequence. In **hereditary persistence of fetal hemoglobin (HPFH)**, there is a failure at birth to switch from synthesis of fetal hemoglobin (HbF) to adult hemoglobin (HbA), so the individual continues to have high levels of HbF throughout life. Although HPFH does not cause any major hematological abnormalities and is compatible with normal life, HPFH is protective against β-thalassemia syndromes, sickle cell diseases, and malaria. Thus, reawakening HbF genes has tremendous prospects in the treatment of β-globin gene disorders (e.g., β-thalassemias, sickle cell disease). In addition, high erythrocyte HbF levels affect the survivability of malarial parasites. In stimulating HbF production, genomic targets affecting transcription factors in erythroid progenitor cells are under investigation (Figure 26.7 and [6−12]).

2. Hemoglobinopathies are characterized by alterations in function or stability of the hemoglobin molecule arising from changes in the amino acid sequence of a globin chain. They include single amino acid substitutions (the largest group), insertion or deletion of one or more amino acids, drastic changes in amino acid sequence caused by frameshift mutations, combination of different pieces of two normal chains by unequal crossing over during meiosis to produce fusion hemoglobins, and an increase or a decrease in chain length from mutations that create or destroy stop codons. Frameshift mutants, chain termination mutants, and fusion polypeptides produce thalassemia-like syndromes. About 750 abnormal hemoglobins have been described, some of which cause no physiological abnormality; consequently, they are not true hemoglobinopathies.

Normal Hemoglobins

Each molecule of human hemoglobin is a tetramer of two α-like (α or ζ) and two β-like (β, γ, δ, or ε) chains. The normal hemoglobins, in order of their appearance during development, are

1. Embryonic hemoglobins:
   a. Gower I = $\zeta_2\varepsilon_2$
   b. Gower II = $\alpha_2\varepsilon_2$
   c. Portland = $\zeta_2\gamma_2$.
2. Fetal hemoglobin. Hemoglobin F = $\alpha_2\gamma_2$: the γ may be Gγ or Aγ, depending on whether glycine or alanine is present at γ136. HbF, the major oxygen carrier in fetal life, accounts for less than 2% of normal adult hemoglobin and falls to this level by the age of 6−12 months.
3. Adult hemoglobins:
   a. Hemoglobin A = $\alpha_2\beta_2$. This is the major adult type (HbA$_1$), comprising 95% of the total hemoglobin.
   b. Hemoglobin A$_2$ = $\alpha_2\delta_2$. This type accounts for less than 3.5% of the total hemoglobin. Synthesis of δ chains is low at all times.

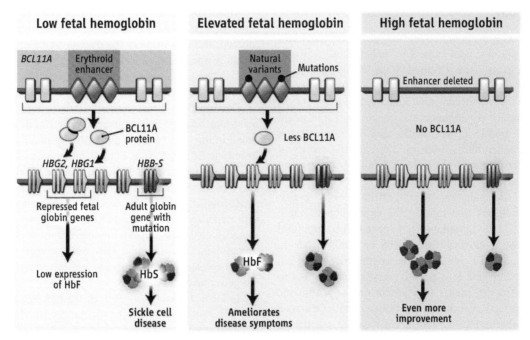

FIGURE 26.7 Effects of globin gene expression modulator, BCL11A, synthesis on HbF production. The expression of BCL11A leads to the silencing of fetal globin genes. The mutations and deletion of enhancer sequence for the BCL11A gene resulted in stimulation of HbF production. *[Reproduced with permission from R.C. Hardison, G.A. Blobel, GWAS to therapy by genome edits? Science 342 (2013) 206.]*

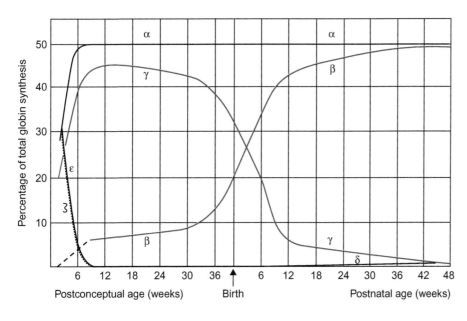

FIGURE 26.8 Changes in human globin chain concentration during development. *[Reproduced with permission from W. G. Wood, Hemoglobin synthesis during human fetal development. Br. Med. Bull. 32, (1976), 282.]*

Embryonic hemoglobins normally are not found in the fetus after the first trimester of pregnancy because synthesis of ζ- and ε-chains ceases before the tenth week of gestation (Figure 26.8). HbF has a higher affinity for oxygen than HbA, facilitating oxygen transfer from the maternal to the fetal circulation, because of the lower affinity of HbF for 2,3-DPG. Since β-chain synthesis remains low relative to γ-chain synthesis throughout intrauterine life, HbF predominates during that period. The mechanisms that control hemoglobin switching (change in expression of the globin genes during development) represent an active area of research.

Synthesis of the different hemoglobin chains is directed by different genes, at least one for each chain type. The α-gene complex and β-gene complexes are located in chromosomes 16 and 11, respectively. The α-gene complex consists of one functional ζ gene and two functional α genes, and the β-gene complex consists of five functional genes: ε, Gγ, A_Y, δ and β. In diploid cells, these gene numbers are doubled.

Thalassemias

The thalassemias are a heterogeneous group of hypochromic, microcytic anemias caused by unbalanced synthesis of globin chains (see Clinical Case Study 26.1). In Southeast Asia, the Philippines, China, the Hawaiian Islands, and the Mediterranean countries, thalassemia syndromes are relatively common and constitute a significant public health problem. Worldwide, they are probably the most common hereditary diseases. The clinical severity of these disorders ranges from mild or totally asymptomatic forms to severe anemias, causing death *in utero*. Presentation depends on the specific globin genes; acquired conditions that modify the expression of these genes; race; and inheritance of genes for structural hemoglobin variants (most often HbS or HbC; discussed later). Anemia and other characteristic hematological abnormalities are due to ineffective erythropoiesis and to precipitation of the excess free globin chains within the red cells. Ineffective erythropoiesis causes the appearance of hypochromic red cells, with a clear center and darker rim containing the hemoglobin (target cells). The precipitates form inclusions, called **Heinz bodies**, which are removed ("pitted out") by the spleen with consequent membrane damage. The resulting cells, called **poikilocytes**, have abnormal shapes and a shortened lifespan. Sequestration of abnormal cells by the spleen and the reticuloendothelial hyperplasia caused by hemolysis leads to splenomegaly.

The thalassemias are classified according to which chains are deficient. The most important types are the α- and β-thalassemias. In each type, globin chain synthesis may be reduced (α^+, β^+) or absent (α^0, β^0). By the use of probes to detect the presence or absence of specific globin genes, thalassemias due to deletion of globin genes have been distinguished from nondeletion varieties. δ-, δ-β-, and γ-δ-β-thalassemias are extremely rare.

α-Thalassemias

In order of increasing severity, the four clinical entities are

1. The **silent carrier state** (α-thalassemia-2);
2. α-**Thalassemia trait** (α-thalassemia-1);
3. **HbH(β_4) disease**; and
4. **Hydrops fetalis** with **Hb Barts** (γ_4), or **homozygous α-thalassemia** (not compatible with life; infants are stillborn).

The first three are α^+-thalassemias, and the last is an α^0-thalassemia. Since non-α-globin synthesis is not affected in α-thalassemia, the relative proportions of hemoglobins A, A_2, and F are unchanged, although all three are reduced in concentration.

These conditions are produced by deletion of one, two, three, or four copies of the α-globin gene; the more copies missing, the more severe the disease. Although deletion is probably the most common cause of the α-thalassemias (and is the *only* cause identified for α^0-thalassemias), a number of nondeletion mutations are known.

Decreased α-chain synthesis leads to formation of two abnormal hemoglobin tetramers. Hemoglobin Barts (γ_4) is found in the cord blood in association with all of the α-thalassemias. It arises from normal production of γ-chains in association with a lack of α-chains. After birth, if the infant survives, the γ-chain is replaced by the α-chain. If α-chain production is severely depressed, excess β-chains form and self-associate (β_4). Neither Hb Barts nor HbH shows cooperativity, and their oxygen-binding curves resemble that of myoglobin. Consequently, they bind oxygen very tightly, do not release it to the tissues, and are almost useless for oxygen transport.

In HbH disease, 5%−30% of the hemoglobin in adults is HbH, which is detected as a rapidly moving hemoglobin during electrophoresis at pH 8.4. In some patients, a slow-moving, minor band of Hb Constant Spring may also be detected. HbH is unstable and precipitates in older red cells and under oxidant stress (e.g., treatment of the patient with primaquine) as Heinz bodies, which are removed by the spleen. Following splenectomy, these inclusions can be seen in peripheral blood after staining with methylene blue. In erythrocytes from unsplenectomized patients, fine inclusions of precipitated HbH can be induced and visualized by staining for 1 hour with brilliant cresyl blue. Although other unstable hemoglobins also produce inclusion bodies, staining for 1 hour is usually not sufficient for their visualization. Patients with iron deficiency together with HbH disease may not show the presence of HbH.

A nondeletional form of HbH disease has been identified. In this subtype, two α-genes are deleted from one chromosome in a diploid cell, and the other chromosome contains one functional gene and one hypofunctioning gene. The net effect is higher clinical severity during infancy [13,14].

Several nondeletion α^+-thalassemias have been described. Four mutations in the normal stop codon for the α-globin gene have been identified. In these cells, the mutated stop codon specifies an amino acid, and translation continues beyond the normal point, thereby extending the α-chains by 31 additional amino acids. Synthesis is terminated by a stop codon downstream from the normal stop codon. Hb Constant Spring is relatively common in Southeast Asian populations. In heterozygotes, about 1% of the α-globin is Hb Constant Spring (25% is expected for the output of one α-globin gene), apparently because the mRNA is unstable. Because of the low rate of synthesis of Hb Constant Spring, HbH disease results when a chromosome carrying this mutation is present along with one lacking α-globin genes. Extended α-chains also result from frameshift mutations (Hb Wayne) and duplications (Hb Grady).

Other α-globin mutants have a mutation at the intron−exon boundary (mRNA splice boundary) that prevents RNA processing or a mutation in the recognition sequence needed for polyadenylation of the mRNA during RNA maturation. In other α^+-thalassemias with both α-genes present but active only at a reduced level, the defect has not been identified.

Thalassemias of the β-Globin Gene Family

In thalassemias of the β-globin gene family, there is reduced synthesis of β-chains, with or without reduced synthesis of γ- or δ-chains. An isolated decrease in γ- or δ-globin synthesis would probably be benign and likely to be detected only by chance. Hemoglobin Lepore (which consists of normal α-chains and an abnormal α-β fusion chain) is usually included with the β-thalassemias, since synthesis of normal β-chains is reduced or absent.

The β-thalassemias are more important in terms of patient suffering and expense than the α-thalassemias. Because all of the normal hemoglobins of fetal and adult life require α-globin chains for normal function, homozygous α-thalassemia (hydrops fetalis) is usually fatal *in utero* by the third trimester of pregnancy. When ζ-chain synthesis is unusually prolonged and a living fetus is born, death invariably occurs soon after delivery. In the α-thalassemias, even a single α-locus seems sufficient to preclude serious morbidity. In contrast, the β-globin chain is not needed until after birth. Except for rare instances in which γ-, δ-, and β-chain syntheses are all absent, β-thalassemic fetuses are delivered normally at term. In homozygous β-thalassemia, problems begin about 4−6 months postnatally, when γ-chain synthesis has declined and β-chain synthesis should have taken over. Thus, β-thalassemia is a crippling disease of childhood. Disorders of β-thalassemia trait are compatible with normal life.

The molecular defects of β-thalassemias and related disorders are heterogeneous; nearly 200 mutations have been identified. Many of the mutations are single-nucleotide substitutions affecting critical loci in the expression of β-globin-like genes. Examples of defects include nonsense and frameshift mutations in the exons, point mutations at intron−exon splice junctions, and mutations in the conserved sequence in the 5' region and the 3'-poly-adenylation site of the gene.

Hemoglobinopathies

The term **hemoglobinopathies** refers to hemoglobin disorders caused by normal synthesis of qualitatively abnormal globin chains. Transcription and translation of mutant genes usually proceed at a normal rate, but the products denature rapidly or function abnormally.

The most common mutations are the single amino acid substitutions caused by one nucleotide change in a globin gene. Globin chains that contain two mutations are extremely rare and appear to have resulted from a second mutation in a common mutant (usually HbS or HbC).

Hemoglobin S and Sickling Disorders

The most common, severe, and best-studied hemoglobin mutation is HbS, which causes sickle cell anemia (or disease) in homozygous individuals and sickle cell trait in heterozygous individuals. The characteristic change in shape of erythrocytes from biconcave disk to curved and sickle-like occurs at low oxygen pressures. Carriers of sickle cell trait are asymptomatic, since their red cells do not sickle unless the P_{O_2} drops below about 25 mmHg; this happens only under unusual circumstances, such as unassisted breathing at high altitudes, some severe forms of pneumonia, and occasionally during anesthesia.

The symptoms of sickle cell anemia are due to sequestration and destruction of the abnormal, sickled erythrocytes in the spleen and to the inability of many sickled cells to pass through capillaries. The normal lifespan of 120 days for red blood cells is decreased to 10−12 days. Painful vaso-occlusive crises can occur anywhere in the body and are the major complication in sickle cell anemia. These crises are caused by blockage of capillaries in the affected tissue by the deformed red cells, which cause hemostasis, anoxia, further sickling, and eventually infarction. Increased adhesion of sickled erythrocytes to capillary endothelial cells may also contribute to capillary obstruction. Crises occur only when the circulation slows or hypoxia is present because the normal circulatory rate does not keep cells deoxygenated for the approximately 15 seconds required for sickling to begin. In contrast to the splenomegaly usually seen in chronic hemolytic anemias, a small, fibrous spleen is usually seen in adults with sickle cell anemia. This autosplenectomy is due to repeated infarction of the spleen. The clinical complications of sickle cell disease are highly variable. The severe forms of the disease occur in homozygous SS disease and S/β°-thalassemia, and a milder disease occurs in double heterozygous SC disease and S/β$^+$-thalassemia. Patients who have a concurrent α-thalassemia trait or high HbF levels have a mild form of the SS disease. The latter observation has been utilized in pharmacological approaches that raise HbF levels using hydroxyurea (discussed later).

Some of the clinical consequences in SS disease include megaloblastic erythropoiesis, aplastic crisis, stroke, bone pain crisis, acute chest syndrome, and proneness to infection particularly by *Pneumococcus*, *Salmonella*, and *Haemophilus* due to hyposplenism. Prophylactic use of penicillin and antipneumococcal and *Haemophilus* vaccines has aided in the management of life-threatening infectious complications of SS disease. Neonatal screening has been used in the identification of infants with sickle cell disease so that risk of infection can be modulated by appropriate immunizations and penicillin prophylaxis. The acute chest syndrome characterized by chest pain is due to clogged pulmonary capillaries; in a small number of studies, patients have been treated with inhaled nitric oxide, which dilates blood vessels, with clinical improvement.

Sickle cell trait is present in about 8% of Black Americans and to a much greater extent (as high as 45%) in some Black African populations. The homozygous condition causes considerable morbidity and about 60,000–80,000 deaths per year among African children. HbS also occurs in some parts of India, the Arabian region, and occasionally in the Mediterranean area. The HbS mutation that occurs in eastern Saudi Arabia and India, known as the **Asian haplotype**, has different flanking DNA sequences surrounding the β-globin locus. Thus, the Asian haplotype may represent an independent occurrence of the HbS mutation that is distinct from its African counterpart. The deleterious gene likely has persisted in these populations because HbS increases resistance to malaria caused by *Plasmodium falciparum*, which was, until recently, endemic in those areas. A similar explanation has been advanced for the high frequencies of β-thalassemia and single α-locus genotypes in these regions. The biological basis of resistance to malaria has been established in laboratory experiments. As schizonts of *P. falciparum* grow in erythrocytes, they lower the intracellular pH and generate hydrogen peroxide. The lower pH promotes sickling, and the hydrogen peroxide damages cell membranes of thalassemic erythrocytes. In both cases, the erythrocyte membranes become more permeable to potassium ions; the resulting decrease in intracellular potassium kills the parasites.

The mutation in HbS replaces glutamic acid (a polar amino acid) with valine (a nonpolar residue) at position 6 of the β-chains. The solubilities of oxy- and deoxy-HbA and oxy-HbS are similar, being about 50 times that of deoxy-HbS. On deoxygenation, HbS precipitates, forming long, rigid, polymeric strands that distort and stiffen the red cells (Figure 26.9). The very high concentration of hemoglobin in the erythrocytes (340 mg/mL), giving an average intermolecular distance of about 1 nm (10 Å), minimizes the time necessary for precipitation to occur. Dilution of HbS, as in sickle cell trait (HbS/HbA heterozygotes), reduces the concentration below the point at which sickling readily occurs. Similarly, the mildness of homozygous HbS disease in populations such as those of

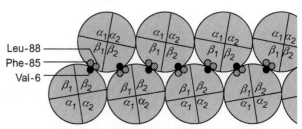

FIGURE 26.9 Structure of hemoglobin S (HbS) polymer. The valine at the β6 position of the deoxy-HbS fits into the hydrophobic pocket formed by leucine and phenylalanine at β85 and β88 of an adjacent β-chain. Because each β-chain has an "acceptor" pocket and a "donor" valine, the HbS polymer has a double-stranded, half-staggered structure. *[Reproduced with permission from S. Charache, Advances in the understanding of sickle cell anemia. Hosp. Pract. 21(2) (1986) 173. J. E. Zupko, illustrator.]*

eastern Saudi Arabia and Orissa, India, is attributed to an accompanying elevation of HbF, which reduces the effective concentration of HbS tetramers. Thus, stimulation of HbF production by gene manipulations is under active investigation (discussed earlier).

Other abnormal hemoglobins can interact with HbS and alter the course of the disease. The most common is HbC, which has a lysine in place of glutamic acid, also at β6. The gene has a frequency second only to that of HbS in Black Americans and in some Black African populations. People heterozygotic for HbC are asymptomatic, but homozygotic individuals have a mild hemolytic anemia with splenomegaly. Because of its insolubility, crystals of HbC can sometimes be seen in peripheral blood smears from homozygous individuals. As a result of the coincidental distribution of the genes for HbS and HbC, heterozygotes for both hemoglobins are not uncommon. HbSC disease has a severity intermediate between that seen in people homozygotic for HbS and those homozygotic for HbC. Unlike HbA, HbC copolymerizes with HbS. In contrast, replacement of the β6 glutamic acid by alanine (HbG-Makassar), or deletion of β6 (Hb Leiden), results in hemoglobin that neither precipitates nor interacts with HbS. Hb San Jose (Glu → Gly at residue β37) is a harmless variant.

Individuals who are heterozygous for HbS and HBO$_{Arab}$ have a hemolytic anemia of a severity similar to sickle cell anemia (see Clinical Case Study 26.2). Certain α-chain variants (e.g., Hb Memphis), when present in HbS homozygotes, can ameliorate the clinical course of the disease. The severity of HbS-β-thalassemia depends on whether the thalassemia is β^0 or β$^+$ and, if it is β$^+$, on how much normal β-chain is synthesized. Severity also depends on the HbF concentration.

A patient with heterozygosity for both HbS and Hb Quebec-chori exhibited clinical symptoms suggestive of sickle cell disease. Hb Quebec-chori, an electrophoretically silent variant at acid and alkaline pH

(β87 Thr→Ile), polymerizes with HbS with the stabilization of the polymer under hypoxic conditions, leading to sickling of red blood cells. Thus, Hb Quebec-chori provides an example of a hemoglobin that has the potential to polymerize with HbS and cause sickle cell disease in a sickle cell trait condition that is otherwise benign by itself.

Determination of the structure of crystalline HbS has shown that in the β-subunits of oxy-HbA and oxy-HbS, a "hydrophobic pocket" between helices E and F is closed; this opens in the deoxy form. In HbA, the residues at the surfaces of the globin subunits are hydrophilic (polar) and do not interact with this pocket. In HbS, however, the β6 valine is hydrophobic and fits into the hydrophobic pocket (formed by leucine and phenylalanine at β85 and β88) of an adjacent β-chain to form a stable structure. Since each β-subunit in deoxy-HbS has an "acceptor" hydrophobic pocket and a "donor" valine, linear aggregates form (Figure 26.8).

Understanding of the sickling process and of the structure of the HbS polymer provides a rational basis for ways of correcting the molecular defect. Thus, dilution of the HbS in the red cells, blockage of the interaction of the β6 valine with the hydrophobic pocket, and decrease of the deoxy-HbS oxy-HbS ratio should reduce the likelihood of sickling and the severity of the disease. These approaches have been tried but so far have failed to ameliorate the disease.

Among Bedouin Arabs and some populations of central and southern India, high HbF levels reduce the severity of sickle cell disease by inhibiting the formation of HbS polymers. This observation has led to therapeutic approaches to induce higher levels of HbF in patients with sickle cell disease. The therapeutic agents are hydroxyurea and short-chain fatty acid derivatives. Hydroxyurea is an antineoplastic agent that inhibits ribonucleotide reductase (Chapter 25). It is thought that in bone marrow, hydroxyurea selectively kills many precursor cells but spares erythroblasts that produce HbF. Hydroxyurea therapy also results in decreased circulating granulocytes, monocytes, and platelets. These changes, along with increased HbF, reduce vaso-occlusion due to a decreased propensity for sickling and adherence of red blood cells to the endothelium. The long-term toxicity of hydroxyurea due to its myelosuppressive and teratogenic effects is not known. Induction of HbF by short-chain fatty acids such as butyrate was discovered in infants of diabetic mothers who had a delay in switching from HbF to HbA in association with elevated serum levels of amino-n-butyrate. This led to studies of HbF induction with several short-chain acids, including butyrate, all of which induce production of HbF. The mechanism by which butyrate and other short-chain fatty acids affect gene expression involves transcriptional activation of a γ-globin gene at the promoter site. One of the mechanisms of regulation of transcription, which occurs at the promoter site, is due to changes in histone acetylation and deacetylation (Chapter 24). Acetylation of specific histones by acetylases allows increased binding of transcription factors to target DNA that stimulates transcription. Deacetylation by histone deacetylases of histones prevents transcription. Butyrate is an inhibitor of deacetylase, and it is thought that this action leads to induction of γ-globin gene expression. Since hydroxyurea acts by a different mechanism in the elevation of HbF, butyrate's action can be additive or synergistic. Sustained induction of HbF by pulse butyrate therapy in patients with sickle cell anemia has resulted in an up to 20% increase in HbF along with marked clinical improvement. Reawakening HbF production in erythroid cells by modulating genetic targets is under investigation.

Hemoglobin E results from the substitution of glutamic acid with lysine at the 26th residue of the β-chain; it has a high prevalence in Southeast Asia. Neither heterozygous nor homozygous states are associated with significant clinical abnormalities. The homozygous state exhibits a mild anemia, and both states show red blood cell indices resembling those of heterozygous thalassemic states, namely, hypochromic microcytosis. The latter has been attributed to activation of a cryptic splice site in exon 1 of β-globin mRNA caused by β^E mutation. This leads to abnormal splicing of β-globin mRNA, producing a less stable and ineffective mRNA. Coinheritance of the HbE gene with different forms of thalassemia genes is also known.

Screening and Prenatal Diagnosis

Neonatal screening programs are aimed at detection of the most common and most harmful hemoglobins. Screening can provide a basis for genetic counseling. Prenatal diagnosis of genetic disease often complements programs for detection of heterozygotes. If both parents are known to be carriers, a homozygous fetus can be detected early in pregnancy.

DERIVATIVES OF HEMOGLOBIN

Carbon Monoxide–Hemoglobin

Carbon monoxide, an odorless gas, has an affinity for hemoglobin that is 210 times that of oxygen. Thus, in the equilibrium reaction

$$HbO_2 + CO \rightleftharpoons HbCO + O_2 \quad K_{eq} = 2.1 \times 10^{-2}$$

the equilibrium lies far to the right. Like oxygen, CO binds to the sixth position of the heme iron. When CO and O_2 are present together in appreciable quantities, CO is bound preferentially and oxygen is excluded, effectively causing an anemic hypoxia. Even if some O_2 is bound to the hemoglobin, it cannot be released owing to the tight binding of CO and to its action as an allosteric activator, with the result that hemoglobin is trapped in the R-state.

Carbon monoxide poisoning may result from breathing automobile exhaust, poorly oxygenated coal fires in stoves and furnaces, or incomplete combustion of a carbon-containing compound. Breathing air containing 1% CO for 7 minutes can be fatal; automobile exhaust contains about 7% CO. Unconsciousness, a cherry-red discoloration of the nail beds and mucous membranes (due to large amounts of R-state carbon monoxide—hemoglobin), and spectrophotometric analysis of the blood are useful for clinical diagnosis.

Treatment of carbon monoxide poisoning consists of breathing a mixture of 95% O_2 and 5% CO_2, which will usually eliminate carbon monoxide from the body in 30—90 minutes, or breathing hyperbaric oxygen.

The enzymatic oxidation of heme produces CO and biliverdin in equimolar amounts (Chapter 27). The CO is transported via the blood to the lungs, where it is released.

Carbaminohemoglobin

Some of the CO_2 in the bloodstream is carried as carbamino compounds (Chapter 2). These compounds form spontaneously in a readily reversible reaction between CO_2 and the free α-amino groups in the N-terminal residues of the Hb chains.

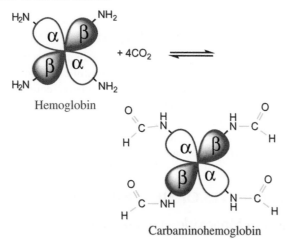

Methemoglobin

In the presence of oxygen, dissolved hemoglobin is slowly oxidized to methemoglobin, a derivative of hemoglobin in which the iron is present in the ferric (Fe^{3+}) state (see Clinical Case Study 26.3). In metHb, the ferric ion is bound tightly to a hydroxyl group or to some other

anion. The heme porphyrin (protoporphyrin IX) that contains an Fe^{3+} ion is known as **hemin**.

Methemoglobinemia

A small but constant amount of methemoglobin is produced in the red blood cells. It is reduced by specific enzymes (methemoglobin reductases) and NADH, which is generated in glycolysis (see Clinical Case Study 26.3, Vignettes 1 and 2). Reductases isolated from human red cells also use NADPH, but to a lesser extent. Inability to reduce metHb produces **methemoglobinemia** and tissue anoxia.

Hereditary methemoglobinemia may arise from the following:

1. Deficiency of NADH cytochrome-b_5 methemoglobin reductase (usually a recessive trait). MetHb values may range from 10% to 40% of the total Hb (normal = 0.5%). Treatment involves administration of methylene blue (discussed later).
2. Defects in the hemoglobin molecule that makes it resistant to reduction by metHb reductases and ascorbate or methylene blue. These hemoglobins are collectively called the **M hemoglobins**. HbM is inherited as a dominant trait, since homozygosity for an M β-chain hemoglobin would be lethal. In four types that involve a His→Tyr substitution, the phenolic hydroxyl group forms a stable complex with Fe^{3+}, making it resistant to reduction to Fe^{2+}. In HbM Milwaukee, a glutamic acid residue substitutes for a valine at position 67 near the distal histidine and forms a stable complex with Fe^{3+}.

Acquired acute methemoglobinemia is a relatively common condition caused by a variety of drugs such as phenacetin, aniline, nitrophenol, aminophenol, sulfanilamide, and inorganic and organic nitrites and nitrates. The condition is less commonly produced by chlorates, ferricyanide, pyrogallol, sulfonal, and hydrogen peroxide. These compounds appear to promote the oxidation of Hb by oxygen. Symptoms include brownish cyanosis, headache, vertigo, and somnolence. Diagnosis is based on the occurrence of brownish cyanosis and the presence of excessive amounts of metHb (measured spectrophotometrically). Treatment usually consists of removal of the offending substance and administration of methylene blue. The mechanism of action of methylene blue (MB) in converting methemoglobin to normal hemoglobin is shown in Figure 26.10. MB undergoes an oxidation—reduction cycle coupled to glucose-6-phosphate dehydrogenase (G6PD). G6PD is required for the production of NADPH, which in turn is required for the conversion of MB to its active reduced state, known as **leukomethylene blue**. It mediates the conversion of

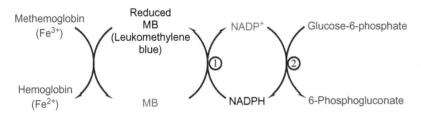

FIGURE 26.10 Role of methylene blue (MB) in converting methemoglobin to hemoglobin. NADPH generated in the glucose-6-phosphate dehydrogenase reaction (1) converts MB to its active form leukomethylene blue by NADPH-methemoglobin reductase (2). In subjects with G6PD deficiency, NADPH synthesis is impaired, limiting the clinical utility of MB.

methemoglobin to hemoglobin. Thus, MB administration to treat methemoglobinemia in G6PD-deficient individuals is contraindicated. MB as an oxidant can exacerbate the disorder in G6PD deficiency. These subjects may require a blood transfusion.

Cyanmethemoglobin

Cyanide poisoning does not cause production of cyanohemoglobinemia or cyanosis. It does produce cytotoxic anoxia by inhibiting cytochrome oxidase, thereby preventing utilization of O_2 by tissues (Chapter 13). Cyanide poisoning is detected by the characteristic odor of HCN gas (odor of bitter almonds) on the breath and by laboratory tests (absorption spectra, tests for CN^-).

Treatment of cyanide poisoning consists of diverting the cyanide into the production of cyanmetHb. First, some of the normal hemoglobin is converted to methemoglobin by intravenous infusion of a solution of $NaNO_2$ or by inhalation of amyl nitrite. Once metHb is formed, CN^- can replace OH^- at position 6 of the iron, since it has a higher affinity for Fe^{3+} than OH^-. CyanmetHb is no more toxic than metHb, and cells containing it can be eliminated by normal body processes. The cyanide bound to metHb is always in equilibrium with free CN^-, and this uncomplexed cyanide is converted to thiocyanate

(SCN^-; nontoxic) by administration of thiosulfate (Chapter 6). Hydrosulfide anion (HS^-) derived from hydrogen sulfide (H_2S) also inhibits cytochrome oxidase by binding to ferric heme. Similarly to cyanide toxicity treatment, administration of nitrites relieves the toxicity of H_2S by formation of methemoglobin followed by sulfmethemoglobin, which is chemically analogous to cyanomethemoglobin. In treatment of H_2S toxicity, there is no role for thiosulfate.

Glycated Hemoglobins

Both α- and ε-amino groups of hemoglobin form amino-1-deoxyfructose adducts on reaction with glucose. Other hexoses can give rise to similar adducts (e.g., galactose in galactosemia; Chapter 14). The main sites of *in vivo* glycation in order of prevalence are β-Val1, β-Lys66, α-Lys61, β-Lys17, and α-Val1. The adduct formed with the amino terminus of the β-chains is known as HbA_{1c}, which makes up about 4%−5.6% of the total hemoglobin in normal red blood cells. Its concentration is increased in uncontrolled diabetics who have hyperglycemia. Because HbA_{1c} accumulates within the erythrocyte throughout the cell's normal lifespan, it is used as an indicator of the success of long-term blood glucose control in diabetics (Chapter 20).

CLINICAL CASE STUDY 26.1 Thalassemia

This clinical case was abstracted from: E.J. Benz, C.C. Wu, A.R. Sohani, A 62-year-old woman with anemia and paraspinal masses, N. Engl. J. Med. 365 (2011) 648−658.

Synopsis

A 62-year-old previously healthy woman of Cambodian descent presented to the emergency department with a chief complaint of shortness of breath, exertional dyspnea, and fatigue. A chest radiograph showed paraspinal masses of the lower thorax with no acute etiology. Follow-up with her primary care physician revealed splenomegaly on physical exam. CBC showed only mild anemia but profound microcytosis. Peripheral blood smear showed hypochromia, microcytosis, and anisocytosis, with occasional elliptocytes, schistocytes, teardrop red cells, and polychromatophilic red cells. On repeat MRI of the thoracic spine, the patient's paraspinal masses appeared unchanged with

no evidence of mass effect. CT-guided needle biopsy of one of the masses revealed clusters of maturing erythroid and granulocytic precursors in the absence of cortical or trabecular bone. These findings were consistent with extramedullary hematopoiesis, which could have been due to either myeloproliferative disease with myelofibrosis, or stress erythropoiesis from chronic anemia. Bone marrow biopsy and aspirate smears showed normocellular but markedly abnormal, left-shifted erythroid hyperplasia, with normal granulocytic and megakaryocytic lineages and normal myeloid blasts. There was no fibrosis of the bone marrow. Hemoglobin-electrophoresis results suggested HbH disease (or α-thalassemia intermedia), confirmed by genetic screening, which revealed deletion mutations of three out of four α-globin genes. Treatment was not recommended due to the patient's relatively mild condition, so management of this

(Continued)

CLINICAL CASE STUDY 26.1 (Continued)

patient was instead focused on genetic counseling, namely screening for anemia in the patient's descendants and their spouses to determine the risk of severe anemia in their offspring.

Teaching Points

1. In comparison with the other various causes of anemia, thalassemia syndromes show a pattern of disproportionately low mean corpuscular volume for the degree of anemia. Iron studies should be performed to rule out concomitant iron-deficiency anemia, which also presents as a microcytic anemia.

2. Hemoglobin H disease, or α-thalassemia intermedia, occurs as a result of deletion mutations of three of four α-globin genes, causing a deficiency of the α-globin chains of hemoglobin and an excess production of β-globin chains.

3. As a result of this accumulation of unpaired and underutilized globin chains, which form polymers of β-globin (Hemoglobin H), the membranes of developing erythroblasts become damaged. This can result in clinical symptoms of anemia, splenomegaly, extramedullary hematopoietic masses, osteopenia, skeletal deformities, and cardiac disease. Patients with thalassemia may also develop cirrhosis and diabetes mellitus as a result of iron deposition from ineffective erythropoiesis.

Supplemental References

[1] E. Smitaman, A.N. Rubinowitz, Extramedullary hematopoiesis associated with β-thalassemia, N. Engl. J. Med. 362 (2010) 253.

[2] R.T. Chung, J. Misdraji, D.V. Sahani, A 43-year-old man with diabetes, hypogonadism, cirrhosis, arthralgias, and fatigue, N. Engl. J. Med. 355 (2006) 1812–1819.

[3] A. Lal, M.L. Goldrich, D.A. Haines, M. Azimi, S.T. Singer, E.P. Vichinsky, Heterogeneity of hemoglobin H disease in childhood, N. Engl. J. Med. 364 (2011) 710–718.

[4] B.J. Bain, Current concepts: diagnosis from the blood smear, N. Engl. J. Med. 353 (2005) 498–507.

[5] D. Rund, E. Rachmilewitz, Beta-thalassemia, N. Engl. J. Med. 353 (2005) 1135.

[6] V.G. Sankaran, J. Xu, R. Byron, H.A. Greisman, C. Fisher, D.J. Weatherall, et al., A functional element necessary for fetal hemoglobin silencing, N. Engl. J. Med. 365 (2011) 807–814.

CLINICAL CASE STUDY 26.2 HbS/O$_{Arab}$ disease

This case was abstracted from: E.K. O'Keeffe, M.M. Rhodes, A. Woodworth, A patient with a previous diagnosis of hemoglobin S/C disease with an unusually severe disease course, Clin. Chem. 55 (2009) 1228–1233.

Synopsis

A 17-year-old African American male was seen by a physician at a hematology clinic for treatment of sickle cell disease (e.g., pain crises, acute chest syndrome). He had previously been diagnosed at another hospital with hemoglobin sickle cell (HbS/C) disease. However, the patient's clinical condition appeared more severe than expected for HbS/C disease. A reexamination of the hemoglobin analysis by high-performance liquid chromatography and isoelectric focusing revealed an Hb profile of S/O$_{Arab}$, which may appear clinically similar to HgS/C disease. The patient was treated with hydroxyurea.

Teaching Points

1. The differential diagnosis of hemoglobinopathies and thalassemia syndromes is made based on using multiple laboratory methods (see "Supplemental Material for Enrichment" next). HbS is coinherited with other abnormal variants (e.g., C, D, E, and O$_{Arab}$).

2. The pharmacotherapy of sickle cell disease is hydroxyurea, which ameliorates the disease by increasing hemoglobin F levels.

Supplemental Material for Enrichment
Laboratory Evaluation of Hemoglobin Disorders
The differential diagnosis of hemoglobinopathies and some thalassemias is made in the laboratory. Some of the laboratory methods are described here.

Electrophoretic Procedures
Most of the **common** hemoglobin mutants can be identified by electrophoretically separating the hemoglobins on cellulose acetate at pH 8.4–8.6 and on citrate agar at pH 6.0–6.2. At pH 8.4, the hemoglobins are negatively charged and migrate toward the anode (positive electrode) in an electric field at a rate dependent on their net charge (see Figure 26.11 and Table 26.2). Separation of the hemoglobins with citrate agar as the support medium is based on net charge and adsorption of the hemoglobin to the supporting gel. The degree of adsorption may depend on hemoglobin solubility and may be due to an impurity in the agar rather than to the agar itself. Patterns obtained on citrate agar with various hemoglobins are shown in Figure 26.12. The value of using these two electrophoretic systems can be seen in the following examples:

1. Hbs S, D, and G migrate together about halfway between Hbs A and A$_2$ on cellulose acetate electrophoresis. With citrate agar gels, Hbs D and G migrate cathodically with HbA, while HbS continues to migrate anodically, separating it from the other hemoglobins.

2. Similarly, HbC comigrates with Hbs A$_2$, E, O, and C-Harlem on cellulose acetate but is readily separated from them on citrate agar.

3. On cellulose acetate at pH 8.4, HbC-Harem (two mutations, each changing the net charge on the tetramer by +2) migrates with HbC (one mutation, change of +4). These two hemoglobins are well separated from each other on citrate agar by electrophoresis.

4. While HbO$_{Arab}$ comigrates with Hb C and E at A$_2$ position at pH 8.4, it migrates between S and A at pH 6.0.

(Continued)

CLINICAL CASE STUDY 26.2 (Continued)

Two modifications of classical hemoglobin electrophoresis are also useful for studying these proteins: isoelectric focusing and electrophoresis of individual globin chains, with or without

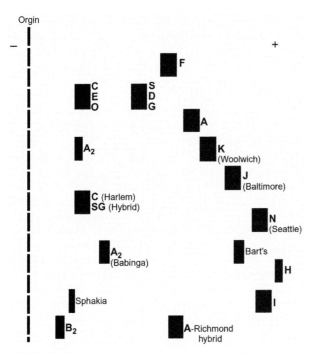

Orgin

FIGURE 26.11 Relative mobilities of some hemoglobins on cellulose acetate (pH 8.4). ($O = O_{Arab}$).

hemoglobins that are not separable by electrophoresis of the complete hemoglobin tetramers.

Nonelectrophoretic Procedures

It is apparent from the preceding discussion that many hemoglobins with different amino acid substitutions demonstrate identical electrophoretic mobilities. Methods that rely on differences other than net charge are needed to establish the identity of these hemoglobins. Definitive molecular biological techniques are employed in the evaluation of hemoglobin disorders.

HbS can be differentiated from most other hemoglobins by its **insolubility on deoxygenation**. The test is carried out in the presence of a reducing agent (sodium dithionite or sodium metabisulfite, which consumes dissolved oxygen) in an appropriate buffer. When HbS is added, the solution becomes turbid because of the precipitation of deoxyHbS. The turbidity can be quantitated spectrophotometrically and related to the amount of HbS added. When the same reagents are used with intact erythrocytes, HbS can be precipitated within the cells, causing them to sickle. The extent of sickling can be assessed by microscopic examination. Because HbC-Harlem and Hb-Bart's behave like HbS in the solubility test, they may give false positive results in tests for HbS.

HbF is identified and quantitated by its **resistance to denaturation by strong alkali**. The hemolsyate is mixed with an alkaline buffer (pH 12.8) and incubated to denature the nonfetal hemoglobins. The denatured hemoglobin is then precipitated by the addition of ammonium sulfate, and the undenatured HbF in the clear supernatant is measured by spectrophotometry.

TABLE 26.2 Amino Acid Substitutions and Net Charge Alterations in Hemoglobins Whose Relative Mobilities are Shown in Figure 26.11

Hemoglobin	Amino Acid Change*	Charge Alteration in Tetramer*
C	$Glu^6 \rightarrow Lys$	+4
E	$Glu^{26} \rightarrow Lys$	+4
O-Arab	$Glu^{121} \rightarrow Lys$	+4
C-Harlem	$Glu^6 \rightarrow Val = Asp^{73} \rightarrow Asn$	+4
S	$Glu^6 \rightarrow Val$	+2
G-San Jose	$Glu^7 \rightarrow Gly$	+2
D-Los Angeles	$Glu^{121} \rightarrow Gln$	+2
Sydney	$Val^{67} \rightarrow Ala$	0
I-Philadelphia	$Lys^{16**} \rightarrow Glu$	−4

*As compared to HbA. Hemoglobins A_2, F, Bart's, and H have multiple amino acid differences from HbA, and their electrophoretic mobilities depend on the net effect of all these differences on their net charge at pH 8.4–8.6.
**Mutation in the alpha globin gene; mutations of all other hemoglobin variants are in the beta globin gene.

(Continued)

CLINICAL CASE STUDY 26.2 (Continued)

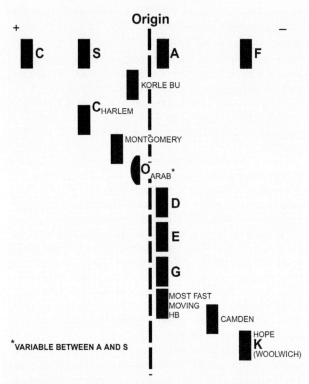

FIGURE 26.12 Relative mobilities of some hemoglobins on citrate agar (pH 6.0). The differences between this gel and cellulose acetate are the differential interactions of the hemoglobins with the gel and the changes in net charge caused by the lower pH.

HbA$_2$ can be separated from most other common hemoglobins by **anion exchange chromatography** using diethylaminonoethylcellulose. Separation results from differences in the interactions of the charged groups of the various hemoglobins with the positively charged groups on the anion exchange resin. Following separation, the HbA$_2$ can be quantitated by spectrophotometry. Although several other hemoglobins (including C, E, O, and D) are eluted with HbA$_2$, it is unlikely that one of these hemoglobins would be present in a person being tested for elevated HbA$_2$. HbA$_2$ measurement is used in the diagnosis of β-thalassemia trait, in which the HbA$_2$ is elevated to about twice the normal level (the upper normal limit is about 3%−3.5%).

HbH and many unstable hemoglobins spontaneously precipitate within the red cells, forming **Heinz bodies**, which can be detected in splenectomized patients by staining with methylene blue. Alternatively, **precipitation** of these hemoglobins can be induced and the precipitates visualized by incubation of the erythrocytes with a redox dye such as brilliant cresyl blue.

Many abnormal hemoglobins are also much more readily **denatured by heat** than normal hemoglobins. Heating an unstable hemoglobin for 30 minutes at 60°C usually causes complete denaturation, whereas HbA is hardly precipitated at

all under the same conditions. The sulfhydryl groups of the mutant globins are generally more exposed than those in normal hemoglobin, making them **more reactive toward parachloromercuribenzoate** (**PCMB**). Treatment with PCMB for several hours precipitates many mutant hemoglobins, whereas it only causes dissociation of HbA.

A sensitive method for identifying changes in the primary structure of hemoglobin is **peptide mapping** or **fingerprinting**. High-performance liquid chromatographic techniques have been used as a sensitive method to detect various hemoglobins, including fetal hemoglobins γ^A and γ^G.

Molecular biology techniques, such as Southern blot analysis and polymerase chain reaction (PCR), have been used in the diagnosis of hemoglobinopathies (see following discussion).

Thalassemia Syndromes
Thalassemia syndromes are a group of heterogeneous inherited disorders. They are relatively common in persons of Mediterranean, African, and Southeast Asian ancestry, and are characterized by an unbalanced and defective (absent or reduced) rate of synthesis of one or more globin chains of hemoglobin. Defects in α-globin chain synthesis are designated as α-thalassemia. The clinical manifestations in thalassemia syndromes occur due to inadequate hemoglobin synthesis and accumulation of unused globin subunits due to unbalanced synthesis. The former leads to hypochromic microcytic anemia and the latter leads to ineffective erythropoiesis and hemolytic anemia. The clinical severity of these disorders depends on the nature of the mutation. They range from asymptomatic mild hypochromic microcytosis (α-thalassemia silent trait) to early childhood mortality (homozygous β-thalassemia) to *in utero* death (Hb-Bart's hydrops fetalis).

β-Thalassemia: The β and β-like genes are located in chromosome 11 (see earlier discussion). In homozygous β-thalassemia syndrome, the β-globin chain synthesis is either absent (β°) or severely reduced (β$^+$). The molecular defects are numerous and include gene deletion, as well as defects in transcription, processing, transport, and translation of mRNA. The clinical symptoms manifest after 6 months of age, when the switch of synthesis from HbF to HbA occurs. Numerous clinical problems follow that require transfusion therapy with its consequent complications (e.g., iron overload; see Chapter 27). Heterozygous β-thalassemia (β-thalassemia trait) is accompanied by mild hypochromic microcytic anemia without any significant clinical abnormalities. β-Thalassemia trait can be identified by elevated HbA$_2$ ($>$3.5%) and, in some mutations, by HbF and normal serum iron studies. β-Thalassemia can also result from the presence of abnormal "δ"—β fusion polypeptides instead of normal β-polypeptide. These are known as hemoglobin Lepore.

α-Thalassemia: The α-globin gene cluster is found in chromosome 16, and each chromosome contains two α genes

(Continued)

CLINICAL CASE STUDY 26.2 (Continued)

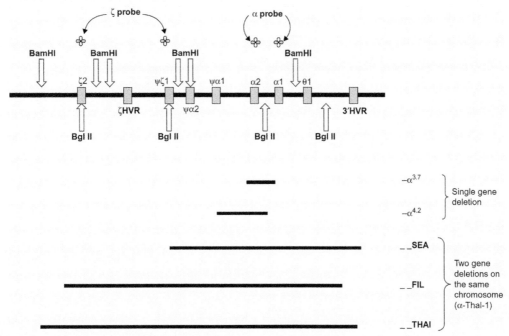

FIGURE 26.13 Restriction enzyme map of α-globin gene cluster and some of the deletions (indicated as dark lines).

(α_2 and α_1). Thus, each individual has four copies of the α gene. The genes are present in the order: 5'-ζ2-$\psi\zeta$1-$\psi\alpha$2-$\psi\alpha$1-α2-α1-θ1-3' (Figure 26.13). The ψ genes are thought to be relics of past evolutionary gene mutations. The θ1 gene does not yield a viable globin protein product, and its function is not known. α-Thalassemia is probably the most common genetic defect, and its prevalence correlates with geographic areas of malaria incidence. The clinical manifestations of α-thalassemia depend on the number of genes affected. Gene deletions are the most common defect, and point mutations account only for about 5% of α-thalassemia.

α-Thalassemia silent carrier ($\alpha\alpha/\alpha$): This genotype is caused by a single α_2-globin gene defect, primarily due to $-\alpha^{3.7}$ or $-\alpha^{4.2}$ deletion. The superscript indicates the length of deletion in kilobases of DNA. Both deletions are caused by unequal crossover points occurring at different locations. The reciprocal product of unequal crossover yields a tripled α-globin gene. Individuals who have inherited a single α-gene defect do not exhibit any hematological or clinical manifestations, and the relative proportion of hemoglobins A, A$_2$, and F are normal.

α-Thalassemia 2 ($\alpha - /\alpha -$): Inheritance of two single gene defects gives rise to clinically significant microcytosis, and the relative proportions of hemoglobins A, A$_2$, and F are normal.

α-Thalassemia-1($-- /\alpha\alpha$): Two gene deletions on the same chromosome are characterized by microcytic anemia. Different deletions have been observed. Southeast

Asian ($\alpha\alpha/^{--SEA}$) and Filipino ($\alpha\alpha/^{--FIL}$) are examples of two-gene deletions occurring in the designated geographic areas. Deletion of a regulatory element (HS-40) also results in thalassemia syndrome.

Hemoglobin H disease ($- -/-\alpha$): Compound heterozygosity for α-thalassemia-1 and α-thalassemia-2 deletions give rise to hemoglobin H (HbH) disease. These individuals exhibit hemolytic, microcytic anemia, which is not life-threatening. HbH(β_4) inclusions are identified by supravital staining with brilliant cresyl blue (discussed earlier) and as fast-moving hemoglobin in a cellulose acetate, alkaline pH 8.4 electrophoresis technique (Figure 26.11).

Hemoglobin Bart's hydrops fetalis ($- -/- -$): This occurs due to the inheritance of two α-thalassemia-1 alleles, and the offspring does not have any functional α-genes. This condition is incompatible with life and results in a stillborn or critically ill fetus with hydrops fetalis. Hemoglobin Bart's is a γ-tetramer and does not function as a normal hemoglobin.

Thalassemia syndromes also occur along with other mutations in the α-globin gene. One example is hemoglobin Constant Spring, which is a result of mutation in the stop codon that gives rise to an abnormally large α-globin chain. In rare cases, α-thalassemia is associated with mental retardation. An X-linked form of mental retardation, known as ATR-X syndrome, is a result of mutations in the XH2 gene. The XH2 gene is a member of a subgroup of the family of genes that

(Continued)

CLINICAL CASE STUDY 26.2 (Continued)

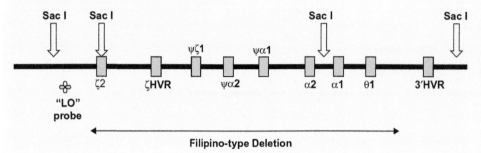

FIGURE 26.14 Restriction enzyme map of Filipino-type deletion in α-globin gene cluster.

Filipino-type Deletion

codes for proteins with many diverse functions (e.g., DNA helicase, DNA repair enzymes, putative global transcriptional regulatory proteins). It is thought that XH2 mutations result in downregulation of α-globin gene expression.

Laboratory Diagnosis of α-Thalassemia Syndromes
DNA-based analysis is utilized for definitive testing in prospective parents for assessing the risk of transmission of α-thalassemia-1 genes and in prenatal diagnosis to identify Hb-Bart's hydrops fetalis. The methods may include Southern blot analysis and PCR methodology using appropriate α-globin gene cluster probes and primers, respectively. In Southern blot analysis, DNA obtained from leukocytes or fetal cells is subjected to digestion with specific restriction endonucleases, and the resulting fragments are separated by electrophoresis.

Blotting is performed by alkaline transfer to nylon membranes and exposed to labeled (e.g., radioactive fluorescent) DNA probes. The genotypes are established by using previously defined restriction fragment length patterns. Figure 26.13 shows the map of an α-globin gene locus with restriction sites useful for diagnosis. In the Filipino α-thalassemia-1 ($\alpha\alpha/^{--FIL}$), due to a large deletion involving the entire α-globin gene locus, digestion with restriction endonuclease, BamHI, and Bgl II is not useful in identifying the deletion. Thus, Filipino α-thalassemia-1 is identified by using a probe ("LO Probe") derived from sequences of about 4 kb 5′ to the ζ-globin gene and digestion with the restriction endonuclease Sac I or Sst I (Figure 26.14).

CLINICAL CASE STUDY 26.3 Methemoglobinemia

Vignette 1: Methemoglobinemia Due to G6PD Deficiency
Synopsis
A 15-year-old male was brought to the emergency department by his father with complaints of dystonia, fatigue, nausea, and dizziness. The patient had consumed large quantities of preserved meat for several days. A comprehensive physical examination and testing revealed the following: jaundice with elevated indirect reacting serum bilirubin; elevated serum LDH; and severe anemia with 20% methemoglobin (MHb). The patient was treated with intravenous methylene blue and his symptoms deteriorated with further lowering of hemoglobin value. The methylene blue administration was discontinued, and the patient required blood transfusion therapy. It was later discovered by the attending physician that the patient had been diagnosed with glucose-6-phosphate dehydrogenase deficiency (G6PD) 8 years earlier.

Teaching Points
1. In MHb, the normal ferrous (Fe^{2+}) iron is in the ferric (Fe^{3+}) state in the heme group and is incapable of oxygen transport. MHb can occur both due to inherited (Hb M variants, erythrocyte NADH-cytochrome-b_5 reductase) and noninherited causes. In this case study, the consumption of preserved meat containing nitrites, particularly in a patient with G6PD deficiency, was thought to

be the precipitating, and exacerbating, factor in the initiation of methemoglobinemia, hemolysis (note serum LDH level is elevated), and hemolytic jaundice. *In vivo* hemolysis can also be diagnosed by elevated reticulocytes and decreased serum haptoglobin levels (hemoglobin complexes with haptoglobin, and the complex is catabolized in macrophages).

2. The preferred treatment for methemoglobinemia is methylene blue. It reduces the MHb to normal (Fe^{2+}) hemoglobin. This process requires an *in vivo* reduction of methylene blue by NADPH, which is produced by the reaction catalyzed by G6PD in the hexose monophosphate pathway, to leukomethylene blue, which is responsible for the reduction of MHb to normal Hb. Thus, in G6PD deficiency, this cycle is interrupted and methylene blue becomes toxic.

3. This case illustrates the importance of glycolysis, which is responsible for the production of NADH required for erythrocyte NADH-cytochrome-b_5-reductase, and the hexose monophosphate shunt pathway, which is required for producing NADPH and maintaining GSH levels. GSH prevents oxidative insult to red blood cells in the presence of normal hemoglobin.

Supplemental Reference
[1] P.A. Hart, B.M. Horst, G.C. Kane, 77-year-old woman with back pain and shortness of breath, Mayo Clin. Proc. 85 (2010) 176–179.

(Continued)

CLINICAL CASE STUDY 26.3 (Continued)

Vignette 2: Methemoglobinemia Due to Benzocaine Exposure

This case was abstracted from: I. Shu, P. Wang, A 70-year-old man with blue skin, Clin. Chem. 60 (2014) 595−599.

Synopsis

A 70-year-old man with hypertension, chronic obstructive pulmonary disease (COPD), and porcine aortic valve replacement presented to the hospital with dizziness and confusion. He was started on medications to treat a subarachnoid hemorrhage found on computed tomography of the head. He was also started on a variety of medications to treat his chronic conditions, which included amiodarone, atorvastatin, azithromycin, benzocaine, citalopram, clonazepam, fluticasone, levalbuterol, metoprolol, and ropinirole. Two days later the patient developed fever, atrial fibrillation, shortness of breath, and lethargy. A blood culture returned positive for gram-positive cocci. Due to concerns for bacterial endocarditis, the patient was pretreated with topical benzocaine spray in preparation for transesophageal echocardiography (TEE). Upon completion of the TEE procedure, the patient developed respiratory failure, hypotension, bradycardia, and cyanosis. A blood sample was taken for further laboratory studies, which appeared dark brown in color and raised the concern for methemoglobinemia. His methemoglobin (Met-Hb) level was found to be markedly elevated at 39.0%, which rapidly rose to 67.7% upon remeasurement 6 minutes later (reference range 0.0%−1.0%), confirming the diagnosis of methemoglobinemia. The patient was treated with intravenous methylene blue with a return of Met-Hb levels to normal.

Teaching Points

1. In this case, the patient developed acquired methemoglobinemia likely due to benzocaine exposure. Acquired methemoglobinemia is mostly caused by exposure to oxidizing toxins, including nitrates and chlorate compounds. A study of 138 acquired methemoglobinemia cases found that 42% were caused by dapsone, 4% by benzocaine, and 4% by primaquine. Upon admission, the patient in this case was treated with a benzocaine lozenge for a complaint of sore throat, and he was again treated with topical benzocaine spray for the TEE procedure.

2. Oxidizing toxins such as benzocaine should not be used in patients with G6PD-deficiency (see Clinical Case Study 26.3, Vignette 1), and should be used with caution in children, the elderly, and patients who are active smokers, as these people are at higher risk of drug-induced methemoglobinemia. Ascorbic acid can be used as an alternative treatment option.

3. Symptoms of methemoglobinemia include persistent cyanosis despite oxygen supplementation, dark-brown-colored arterial blood even after exposure to air, a very high oxygen partial pressure, and an oxygen saturation gap.

REQUIRED READING

[1] K. Grimshaw, J. Sahler, S.L. Spinelli, R.P. Phipps, N. Blumberg, New frontiers in transfusion biology: identification and significance of mediators of morbidity and mortality in stored red blood cells, Transfusion 51 (2011) 874−880.

[2] T.B. Drüeke, Anemia treatment in patients with chronic kidney disease, N. Engl. J. Med. 368 (2013) 387−389.

[3] S. Fishbane, B. Schiller, F. Locatelli, A.C. Covic, R. Provenzano, A. Wiecek, et al., Peginesatide in patients with anemia undergoing hemodialysis, N. Engl. J. Med. 368 (2013) 307−319.

[4] I.C. Macdougall, Novel erythropoiesis-stimulating agents: a new era in anemia management, Clin. J. Am. Soc. Nephrol. 3 (2008) 200−207.

[5] Z. Kiss, S. Elliot, K. Jedynasty, V. Tesar, J. Szegedi, Discovery and basic pharmacology of erythropoiesis-stimulating agents (ESAs), including the hyperglycosylated ESA, darbepoietin alfa: an update of the rationale and clinical impact, Eur. J. Clin. Pharmacol. 66 (2010) 331−340.

[6] R.C. Hardison, G.A. Blobel, GWAS to therapy by genome edits? Science 342 (2013) 206−207.

[7] D.E. Bauer, S.C. Kamran, S. Lessard, J. Xu, Y. Fujiwara, C. Lin, et al., An erythroid enhancer of *BCL11A* subject to genetic variation determines fetal hemoglobin level, Science 342 (2013) 253−257.

[8] D.E. Bauer, S.C. Kamran, S.H. Orkin, Reawakening fetal hemoglobin: prospects for new therapies for the β-globin disorders, Blood 120 (2012) 2945−2953.

[9] L. Shi, S. Cui, J.D. Engel, O. Tanabe, Lysine-specific demethylase 1 is a therapeutic target for fetal hemoglobin induction, Nat. Med. 19 (2013) 291−294.

[10] V.G. Sankaran, Targeted therapeutic strategies for fetal hemoglobin induction, Hematology Am. Soc. Hematol. Educ. Program 2011 (2011) 459−465.

[11] K.M. Musallam, A.T. Taher, M.D. Cappellinin, V.G. Sankaran, Clinical experience with fetal hemoglobin induction therapy in patients with β-thalassemia, Blood 121 (2013) 2199−2212.

[12] M.H. Steinberg, D.H. Chui, G.J. Dover, P. Sebastiani, A. Alsultan, Fetal hemoglobin in sickle cell anemia: a glass half full? Blood 123 (2014) 481−485.

[13] E.J. Benz, Newborn screening for α-thalassemia—keeping up with globalization, N. Engl. J. Med. 364 (2011) 770−771.

[14] A. Lai, M.L. Goldrich, D.A. Haines, M. Azimi, S.T. Singer, E.P. Vichinsky, Heterogeneity of hemoglobin H disease in childhood, N. Engl. J. Med. 364 (2011) 710−718.

ENRICHMENT READING

[1] F.B. Piel, D.J. Weatherall, The α-Thalassemias, N. Engl. J. Med. 371 (2014) 1908−1916.

Chapter 27

Metabolism of Iron and Heme

Key Points

1. Heme, an iron—porphyrin complex, is the prosthetic group of hemoglobin, myoglobin, cytochromes, and many other proteins.

2. Iron is required for the biosynthesis of heme and other nonheme-iron-containing proteins.

3. Acquisition of iron by entry into the enterocytes, its storage in the liver and macrophages, its utilization by erythroid cells for erythropoiesis and its requirement for other cellular functions are regulated by a number of proteins. Hepcidin, synthesized in the liver, regulates a number of steps in iron homeostasis. Loss of function in these regulatory proteins leads to hemochromatosis.

4. Iron deficiency anemia is the most prevalent nutritional disorder. It is due to inadequate dietary intake of iron and/or chronic blood loss.

5. Heme biosynthesis in the hematopoietic tissues and the liver requires eight enzymes that are present in both the cytosol and the mitochondria. The regulatory step is catalyzed by ALA synthase and is inhibited at its allosteric site by heme. Inherited disorders of porphyrin biosynthesis lead to porphyrias with various clinical manifestations.

6. Catabolism of red blood cells in the macrophages leads to the production of bilirubin, a waste product of porphyrin. Iron is sequestered and reutilized.

7. Bilirubin is transported to the liver bound to albumin, conjugated with glucuronic acid using UDP-glucuronic acid, and eliminated via the biliary system.

8. Elevated serum bilirubin levels (jaundice) can be due to liver disease, excessive hemolysis, and various genetic defects in its metabolism in the liver. Neonatal jaundice, if not corrected promptly, can cause damage to the CNS. Phototherapy reduces serum bilirubin levels in neonates.

Heme, an iron—porphyrin complex, is the prosthetic group of many important proteins. The central role of hemoglobin and myoglobin in oxygen transport and storage was discussed in Chapter 26. Heme proteins or enzymes are involved in redox reactions (e.g., cytochromes) and participate in many oxidation reactions needed for synthesis of metabolically important compounds, as well as for degradation and detoxification of waste products and environmental toxins.

Ionic forms of iron (referred to hereafter as iron) also participate in a variety of enzymatic reactions as nonheme irons, which are present as iron—sulfur clusters (e.g., mitochondrial electron transport). Fe-S cluster assembly occurs in mitochondria mediated by frataxin. In the inherited disease Friedreich's ataxia, frataxin deficiency has been observed. This autosomal recessive disease is most often caused by expansion of the GAA repeat in the first intron of the gene and is accompanied by sensory loss and impairment of gait. There are also both storage and transportable forms of iron that are bound to proteins. Under normal physiological conditions, only trace amounts of free iron exist. In the body, if iron exceeds the sequestration capacity of the iron-binding proteins present in different physiological compartments, the free iron can cause tissue damage. Cellular injury is caused by reactive oxygen species that are produced from H_2O_2 in a reaction catalyzed by iron. Thus, iron homeostasis in the body is in a delicate balance: either a deficiency or an excess of iron results in abnormalities and presents as a common cause of human diseases.

IRON METABOLISM

The total body iron of a 70 kg adult is about 4.2—4.4 g. The distribution of iron in various body compartments is shown in Table 27.1. Several specific proteins participate in the orchestration of iron metabolism (discussed later). Iron metabolism consists of the absorption of dietary iron from the gastrointestinal tract, transport in the blood, storage in the liver and macrophages, and utilization in the cells requiring synthesis of iron-containing proteins (e.g., hemoglobin in erythroblasts). Iron deficiency causes **anemia**, and iron excess causes iron accumulation diseases known as **hemochromatosis**.

N. V. Bhagavan and C.-E. Ha: Essentials of Medical Biochemistry, Second edition. DOI: http://dx.doi.org/10.1016/B978-0-12-416687-5.00027-0

Iron Absorption, Transport, Utilization, and Storage

See Figure 27.1 for an overview of iron metabolism.

Absorption

Ferrous iron is absorbed principally from the mature enterocytes lining the absorptive villi of the duodenum. The amount of iron absorption by these enterocytes is determined by the prior programming of the duodenal crypt cells based on iron requirements of the body as they undergo maturation. At the apical membrane of the enterocyte, Fe^{3+} is converted to Fe^{2+} by **ferrireductase** followed by its uptake mediated by divalent transporter 1 (DMT1). DMT1 also transports other metal ions (e.g., Cu^{2+}, Zn^{2+}, and cobalt). The internalized Fe^{2+} is either temporarily stored after conversion to Fe^{3+} as ferritin or transported across the cell for transport to the portal capillary blood circulation. The exit of iron from the enterocytes at the basolateral surface requires the participation of **HFE protein** and a copper-containing ferroxidase

TABLE 27.1 Distribution of Iron in a 70 kg Adult[1]

Circulating erythrocytes	1800 mg[2]
Bone marrow (erythroid)	300 mg
Muscle myoglobin	300 mg
Heme and nonheme enzymes	180 mg
Liver parenchyma[3]	1000 mg
Reticuloendothelial macrophages[4]	600 mg
Plasma transferrin[5]	3 mg

[1]These are approximate values. Premenopausal women have lower iron stores due to periodic blood loss through menstruation. Iron balance in the body is maintained by intestinal absorption of 1–2 mg/day and by loss of 1–2 mg/day.
[2]1 mg = 17.9 mmol.
[3]Primarily storage forms of iron.
[4]Senescent red blood cells are catabolized by the macrophages; the salvaged iron is temporarily stored and made available via transferrin for erythron and for hemoglobin synthesis.
[5]Transportable form of iron.

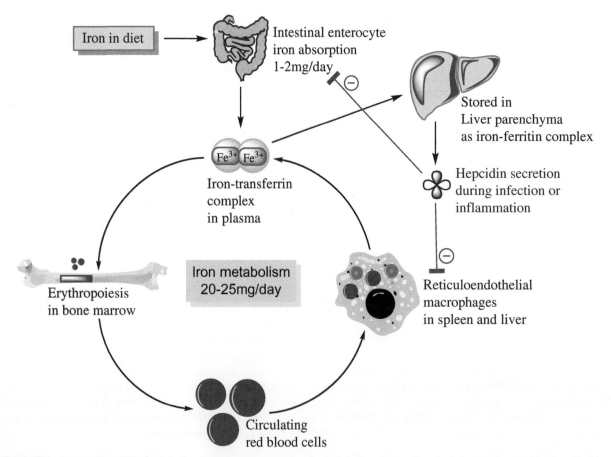

FIGURE 27.1 An overview of iron metabolism. Intestinal enterocytes uptake 1–2 mg/day of iron. Transferrin in blood carries iron to erythroid precursor cells for heme production and hepatocytes in the liver for iron storage. Hepcidin mainly released from the liver inhibits iron release from enterocytes and reticuloendothelial macrophages.

known as **hephaestin**, which converts Fe^{2+} to Fe^{3+}. Plasma **transferrin** transports iron in the ferric state to its target sites. The regulation of intestinal iron absorption is critical because iron excretion from the body is a limiting physiological process (discussed later). The small intestine is also an excretory organ for iron, since that stored as ferritin in the epithelial cells is lost when they are shed and replaced every 3−5 days. Heme iron is transported intact into the mucosal cells, and the iron is removed for further processing.

Plasma Iron Transport

Over 95% of plasma iron is in the Fe^{3+} state bound to the glycoprotein transferrin, a monomeric β_1-globulin (M.W. 80,000). Transferrin is synthesized primarily in the liver. Each molecule of transferrin can bind two Fe^{3+} ions. The binding is extremely strong under physiological conditions. Its half-life in humans is about 8 days.

The bulk of transferrin iron is delivered to immature erythroid cells for utilization in heme synthesis. Iron in excess of this requirement is stored as ferritin and hemosiderin. Unloading of iron to immature erythroid cells is by **receptor-mediated endocytosis**. The process begins in clathrin-coated pits with the binding of diferric transferrin to specific plasma membrane transferrin receptors. The next step is the internalization of the transferrin−transferrin receptor-HFE protein complex with formation of endosomes. In the endosomes, a proton pump acidifies the complex to pH 5.4, and by altering the conformation of proteins, iron is released from transferrin bound to transferrin receptors. In the acidified endosomes, DMT1 facilitates iron transport into the cytosol. Both apotransferrin (and a fraction of ironbound transferrin) and transferrin receptors are returned to cell surfaces for reuse. In this type of receptor-mediated endocytosis of transferrin−transferrin receptor complex, the endosomes do not come into contact with lysosomes. The process is therefore unlike that of low-density lipoprotein receptor-mediated internalization (Chapter 18). In the erythroid cells, most of the iron released from the endosomes is transported into mitochondria for heme synthesis in nonerythroid cells, and the iron is stored predominantly as ferritin.

Storage of Iron

Iron is stored in the apoferritin shell in the ferric state as a polynuclear hydrated ferric oxide−phosphate complex. Apoferritin is a protein shell consisting of 24 subunits of two types: a light (L) subunit (M.W. 19,000) and a heavy (H) subunit (M.W. 21,000). H chains possess ferroxidase activity and convert Fe^{2+} to Fe^{3+}.

Coordinate Regulation of Iron Uptake and Storage in Nonerythroid Cells

Iron uptake is regulated by transferrin receptors and storage of iron as ferritin, which occurs post-transcriptionally for these two proteins. The regulation maintains an optimal intracellular-transit chelatable iron pool for normal functioning in the body. The regulatory process consists of an interaction between iron regulatory elements (IREs) and iron regulatory proteins (IRPs) 1 and 2. One copy of each IRE has been identified in the 5′ untranslated region (UTR) of H and L ferritin mRNAs and five copies in the 3′ untranslated region of transferrin receptor mRNA. IRE sequences are highly conserved and have a stem−loop structure with a CAGUGN sequence at the tip of the loop. IRPs are RNA-binding proteins that bind to IREs and regulate the translation of the respective mRNAs.

When there are low levels of intracellular chelatable iron, iron storage declines due to inhibition of ferritin synthesis, and cellular entry of iron increases due to enhanced transferrin receptor synthesis. An opposing set of events occurs during intracellular chelatable iron excess or iron-replete states. Coordinated control occurs when IRP binds to IRE at the 5′ UTR of ferritin mRNAs inhibiting ferritin synthesis: simultaneously, the binding of IRP to IRE at the 3′ UTR of transferrin receptor mRNA stimulates transferrin receptor synthesis (Figure 27.2). Intracellular iron regulates the level of IRPs. During the expansion of the iron pool, IRPs are inactivated, leading to efficient translation of ferritin mRNA and rapid degradation of transferrin receptor mRNA. In iron-replete cells, IRP1 acquires iron by the formation of iron−sulfur clusters (4Fe−4S) that bind to IREs with low affinity. During iron deficiency states, IRP1 lacks a 4Fe−4S cluster and binds to IREs with high affinity. IRP1, when it possesses an iron−sulfur cluster, has aconitase activity, normally a TCA cycle enzyme (Chapter 12).

Measurement of serum ferritin levels has diagnostic utility. In iron deficiency anemia, serum ferritin levels are low; in iron storage disease, the levels are high. However, serum ferritin levels can also be elevated under many other circumstances, including liver diseases and chronic inflammatory diseases.

Alterations of Plasma Transferrin Concentration

Plasma transferrin levels are commonly measured in the evaluation of disorders of iron metabolism (discussed later). It is customary to measure transferrin concentration indirectly from the maximum (or total) iron-binding capacity (TIBC) of plasma (reference interval for adults, 250−400 μg/dL). It can also be measured directly by immunological methods (reference interval for adults, 220−400 mg/dL). **Hypertransferrinemia** (or increased TIBC) can occur with diminished body iron stores, as in iron deficiency anemia or

At low cytosolic mobile iron pool:
IRP levels increase and IRP binds to mRNA-IREs
of ferritin and transferrin receptors.

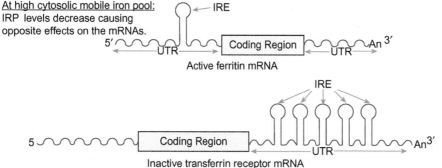

At high cytosolic mobile iron pool:
IRP levels decrease causing
opposite effects on the mRNAs.

FIGURE 27.2 Coordinated translational regulation of ferritin mRNA and transferrin receptor mRNA in nonerythroid cells. Iron regulatory proteins (IRPs) are RNA-binding proteins that bind to iron regulatory elements (IREs). IREs are hairpin structures with loops consisting of CAGUGN sequences and are located at the 5′ untranslated region (UTR) and 3′ UTR for ferritin mRNA and transfer mRNA, respectively.

during pregnancy (because of enhanced mobilization of storage iron to supply maternal and fetal demands). Hypertransferrinemia of iron deficiency is corrected by oral iron supplementation, whereas that due to pregnancy is not. Exogenous administration of estrogens (e.g., oral contraceptives) also causes hypertransferrinemia.

Hypotransferrinemia can result from protein malnutrition and accompanies hypoalbuminemia. Since transferrin has a much shorter half-life (8 days) than albumin (19 days), measurement of the transferrin level may be a more sensitive indicator of protein malnutrition than albumin measurement (see also . Chapter 15). Hypotransferrinemia also results from excessive renal loss of plasma proteins (e.g., in nephrotic syndrome).

Regulation of Iron Metabolism

Gastrointestinal Tract

The enterocyte, the first entry portal for iron absorption, plays the predominant role in the iron metabolism. In the enterocytes, iron absorption is increased during iron deficiency and decreased during the body's iron excess. The molecular circuitry signals that mediate iron absorption involve a network of signals. Hypoxia-inducible factors (HIFs), which are transcription factors, are involved in the enterocyte iron absorption. During hypoxic conditions, the HIF signaling cascade is upregulated, promoting iron entry into the enterocytes. Thus, in iron deficiency, anemia HIF promotes absorption. However, inappropriate HIF stimulation can lead to chronic iron accumulation [1]. It should also be noted that HIF upregulation stimulates erythropoietin production in renal cells, which promotes erythropoiesis in the bone marrow (see "Erythropoietin" in Chapter 26).

Liver

Hepcidin, a 25-amino acid protein containing eight cysteine residues, is synthesized in the liver [2] and plays a central role in iron homeostasis [3]. It is a negative modulator of iron levels in the body; it decreases iron absorption and release of iron from macrophages by inactivating ferroportins. The hepcidin gene is transcriptionally activated by **hemojuvelin (HJV)**, **HFE protein**, and **transferrin receptor 2 (TFR2)**. Thus, the inactivating mutations in the genes of hepcidin—HJV, HFE, and TFR2—are related to hereditary hemochromatosis. Activation of the hepcidin gene by interleukin-6 (IL-6), which is released during inflammation, results in anemia [4]. Action of IL-6 is mediated by the activation of signal transducer and activator of the transcription-3 (STAT-3) pathway. Under physiological conditions, hepcidin synthesis is positively regulated by bone morphogenetic proteins via

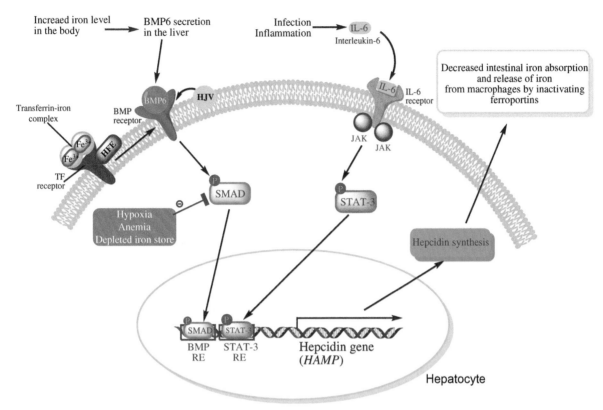

FIGURE 27.3 Hepcidin expression in the hepatocyte. The two pathways of hepcidin regulation consist of the iron homeostasis and the inflammation and infection (see text for details). HAMP = hepcidin gene, HJV = hemojuvelin, STAT-3 = signal transducer and activator of the transcription-3, BMP RE = BMP response element, STAT-3 RE = STAT-3 response element, JAK = Janus-associated kinase, TF = transferrin.

phosphorylation of downstream transcription factors. Anemia, hypoxia, and depleted iron stores inhibit the synthesis of hepcidin, promoting iron absorption (Figure 27.3).

Macrophage

Senescent red blood cells are catabolized in the macrophages (discussed later), and the iron is reclaimed and temporarily stored. The exit of Fe^{2+} from macrophages requires cell-membrane-bound **ferroportin 1** and **glycosylphosphatidylinositol (GPI)-linked ceruloplasmin**. Ceruloplasmin, like enterocyte hephaestin, is a copper-containing ferro-oxidase. It converts Fe^{2+} to Fe^{3+} for transferrin-mediated transport. Thus, copper deficiency causes anemia.

Disorders of Iron Metabolism

Iron Deficiency Anemia

Iron deficiency anemia is the most prevalent nutritional disorder [5,6]. Its cause may comprise many overlapping factors: dietary iron deficiency; absence of substances that favor iron absorption (ascorbate, amino acids, succinate); presence of compounds that limit iron absorption (phytates, oxalates, excess phosphates, tannates); lack of iron absorption due to gastrointestinal disorders (malabsorption

syndrome, gastrectomy); loss of iron due to menstruation, pregnancy, parturition, lactation, chronic bleeding from a gastrointestinal tract peptic ulceration, hemorrhoids, cancer, colonic ulceration, or hookworm infestation of the genitourinary tract (uterine fibroids); enhanced demand for growth or new blood formation; deficiency of iron transport from mother to fetus; abnormalities in iron storage; deficiencies in release of iron from the reticuloendothelial system (infection, cancer); inhibition of incorporation of iron into hemoglobin (lead toxicity); and rare genetic conditions (transferrin deficiency, impaired cellular uptake of iron by erythroid precursors). In the initial phase of depletion of the iron content of the body, the iron stores maintain normal levels of hemoglobin and other iron proteins. With exhaustion of storage iron, hypochromic and microcytic anemia becomes manifest.

The clinical characteristics of iron deficiency anemia are nonspecific and include pallor, rapid exhaustion, muscular weakness, anorexia, lassitude, difficulty in concentrating, headache, palpitations, dyspnea on exertion, angina on effort, peculiar craving for unnatural foods (pica), ankle edema, and abnormalities involving all proliferating tissues, especially mucous membranes and the nails. The onset is insidious and may progress slowly over many months or years.

Physiological adjustments take place during the gradual progression of the disorder, so that even a severe hemoglobin deficiency may produce few symptoms. Iron deficiency may affect the proper development of the central nervous system. Early childhood iron deficiency anemia may lead to cognitive abnormalities.

Individuals who have **congenital atransferrinemia** lack apotransferrin and suffer from severe hypochromic anemia in the presence of excess iron stores in many body sites, susceptibility to infection (transferrin inhibits bacterial, viral, and fungal growth, probably by binding the iron required for growth of these organisms), and retardation of growth. This condition does not respond to administration of iron. Intravenous administration of transferrin normalizes the iron kinetics. There are reports of a rare congenital defect in uptake of iron by red cell precursors that leads to severe hypochromic anemia with normal plasma iron and transferrin levels.

Microcytic anemia occurs frequently in thalassemia syndromes (Chapter 26), but these patients do not require iron supplementation unless they have concurrent iron deficiency as assessed by measurement of serum iron levels and TIBC. Iron deficiency anemia can also be assessed from the plasma ferritin concentration (which, when decreased, reflects depleted iron stores), red cell protoporphyrin concentration (increased because of lack of conversion to heme), and the number of sideroblasts in the bone marrow (which parallels iron stores). Sideroblasts are erythrocyte precursors (normoblasts) containing free ferritin—iron granules in the cytoplasm that stain blue with Prussian blue reagent. There is a close correlation between plasma iron levels, TIBC, and the proportion of sideroblasts in bone marrow. In hemolytic anemias, pernicious anemia, and hemochromatosis, the serum iron level increases and sideroblast numbers reach 70% (normal range, 30%−50% of total cells). In iron deficiency, the sideroblasts are decreased in number or absent.

Before treatment is initiated, the cause of negative iron balance must be established. Treatment should correct the underlying cause of anemia and improve the iron balance. In general, oral therapy with ferrous salts is satisfactory; however, sometimes parenteral therapy is preferred, e.g., in proven malabsorption problems, gastrointestinal disease and excessive blood loss, and for patients who cannot be relied on to take oral medication.

Iron Storage Disorders

Excessive accumulation of iron (chronic iron overload) can result from the following (see also Clinical Case Study 27.1):

1. Defective erythropoiesis (dyserythropoiesis); impaired hemoglobin synthesis leading to lack of utilization and consequent accumulation of iron in mitochondria, e.g., from inhibition of ALA synthase activity due to vitamin B_6 deficiency; inhibition of heme synthesis by lead; impairment of pyridoxine metabolism in alcoholic patients; familial sideroblastic anemias; and Cooley's anemia.
2. Repeated blood transfusions, e.g., in Cooley's anemia or sickle cell disease.
3. Hereditary hemochromatosis defects that lead to decreased hepcidin production (discussed earlier), in which there is an increased rate of absorption of iron in the presence of normal or enlarged iron stores and normal hematopoiesis.
4. High dietary iron and substances that enhance its absorption (e.g., Bantu siderosis).
5. Hereditary atransferrinemia.

In all of these disorders, the gastrointestinal tract cannot limit absorption of iron to a significant extent. Thus, the "mucosal block" responsible for keeping out unnecessary iron on a daily basis is susceptible to disruption, perhaps at more than one point. Iron overload leads to progressive deterioration in pancreatic, hepatic, gonadal, and cardiac function. Clinical manifestations include cirrhosis, diabetes mellitus, life-threatening arrhythmias, and intractable heart failure. Removal of excess iron produces clinical improvement, particularly of diabetes and congestive heart failure.

In iron storage diseases accompanied by normal erythropoiesis (e.g., hereditary hemochromatosis), removal of excess iron is accomplished by repeated bloodletting (phlebotomy). Therapeutic phlebotomy of a unit of blood (which contains about 250 mg of iron) may be performed up to three times per week. When the iron stores become depleted, reaccumulation of iron is prevented by four to six phlebotomies per year. In asymptomatic patients, periodic determination of serum ferritin provides a measure of storage of iron [7,8].

In hemochromatosis secondary to refractory anemias (e.g., Cooley's anemia, sickle cell anemia), patients require repeated blood transfusions to survive childhood and adulthood. Therapy consists of administration of iron-chelating agents. Deferoxamine, deferiprone, and deferasirox are used as iron chelators in the transfusion overload disorders [9]. In disorders of intravascular hemolysis (e.g., transfusion, sepsis, sickle cell disease), free hemoglobin and hemin are released. Both hemoglobin and hemin are toxic and produce adverse clinical conditions involving vascular, hepatic, and renal systems. Two plasma proteins, namely haptoglobin and hemopexin, function as scavenger molecules for free hemoglobin and hemin, respectively [10].

HEME BIOSYNTHESIS

The principal tissues involved in heme biosynthesis are the hematopoietic tissues and the liver. Biosynthesis

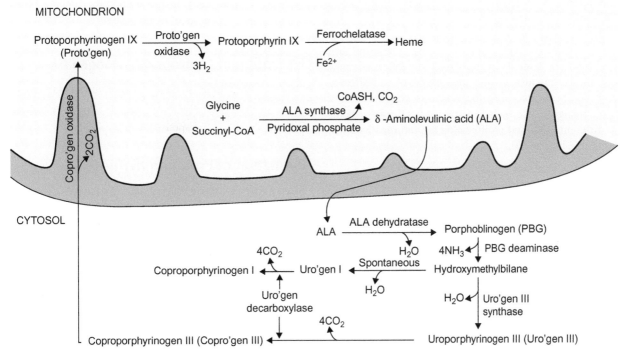

FIGURE 27.4 Biosynthetic pathway of heme. The pathway consists of eight irreversible reactions, four each in the mitochondrion and the cytosol. The primary site of regulation is the ALA synthase step.

requires the participation of eight conserved enzymes, of which four (the first and the last three) are mitochondrial and the rest are cytosolic (Figure 27.4). The reactions are irreversible. Glycine and succinate are the precursors of porphyrins. Some of these enzymes are coded by erythroid-specific and housekeeping genes [11].

Formation of δ-Aminolevulinic Acid

δ-Aminolevulinic acid (ALA) formation is catalyzed by mitochondrial ALA synthase, which condenses glycine and succinyl-CoA to ALA:

The ALA synthase is located on the matrix side of the inner mitochondrial membrane. It is encoded by a nuclear gene.

Heme synthesis also requires a functional tricarboxylic acid (TCA) cycle and an oxygen supply. The primary regulatory step of heme synthesis in the liver is apparently that catalyzed by ALA synthase. The regulatory effects are multiple. The normal end product, heme, when in excess of that needed for production of heme proteins, is oxidized to **hematin**, which contains a hydroxyl group attached to the Fe^{3+} atom. Replacement of the hydroxyl group by a chloride ion produces **hemin**. Hemin and heme inhibit ALA synthase allosterically. Induction of ALA synthase is suppressed by hemin and increased by a variety of xenobiotics (e.g., environmental pollutants) and natural steroids. In erythropoietic tissues, where the largest amount of heme is synthesized, regulation of heme biosynthesis may also involve the process of cell differentiation and proliferation of the erythron, which occurs to meet a change in the requirements for synthesis of heme. The differentiation and proliferation are initiated by erythropoietin (Chapter 26).

Formation of Porphobilinogen

Two molecules of ALA are condensed by cytosolic zinc containing ALA dehydratase to yield porphobilinogen (PBG).

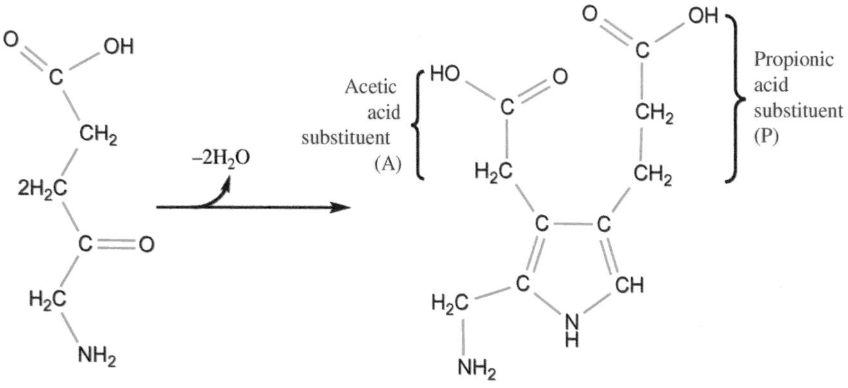

There are four zinc ions per octamer of the enzyme, and they are bound via the reduced thiol groups. Zinc is required for enzyme activity. Lead is a potent inhibitor of ALA dehydratase (see Clinical Case Study 27.2).

Formation of Uroporphyrinogen III

Uroporphyrinogen III formation occurs in the cytosol and requires the successive action of porphobilinogen deaminase (or methylbilane synthase) and uroporphyrinogen III synthase. Porphobilinogen deaminase catalyzes condensation of four porphobilinogen molecules in a symmetrical head-to-tail arrangement to form a straight-chain tetrapyrrole, hydroxymethylbilane. Uroporphyrinogen III synthase catalyzes the rearrangement of one of the pyrrole rings (ring D in Figure 27.5) to form an asymmetrical tetrapyrrole, followed by its cyclization to form uroporphyrinogen III. In the absence of uroporphyrinogen III synthase (e.g., in congenital erythropoietic porphyria), the hydroxymethylbilane

cyclizes spontaneously to the uroporphyrinogen I isomer, which is not a precursor of heme (Figure 27.5).

Formation of Coproporphyrinogen III

Cytosolic uroporphyrinogen decarboxylase catalyzes successive decarboxylation of the four acetic groups to yield four methyl groups (Figure 27.6).

Formation of Protoporphyrinogen IX

Mitochondrial coproporphyrinogen oxidase is localized in the intermembrane space and is probably loosely bound to the outer surface of the inner membrane. It catalyzes the successive conversion of propionic acid groups of ring A and ring B to vinyl groups (Figure 27.7).

Formation of Protoporphyrin IX and Heme

Both of these steps occur in mitochondria (Figure 27.8). Porphyrinogen oxidase removes six hydrogen atoms (four from methane bridge carbons and two from pyrrole nitrogens) from protoporphyrinogen to yield protoporphyrin. The oxidase has an absolute requirement for oxygen. Protoporphyrin oxidase is bound to the inner mitochondrial membrane, and its active site faces the cytosolic side of the membrane. Formation of heme is accomplished by ferrochelatase (or heme synthase), which incorporates Fe^{2+} into protoporphyrin and is inhibited by lead. Zinc can function as a substrate in the absence of iron.

DISORDERS OF HEME BIOSYNTHESIS

The porphyrias are a group of disorders caused by abnormalities in heme biosynthesis [11]. They are inherited and acquired disorders characterized by excessive accumulation and excretion of porphyrins or their precursors.

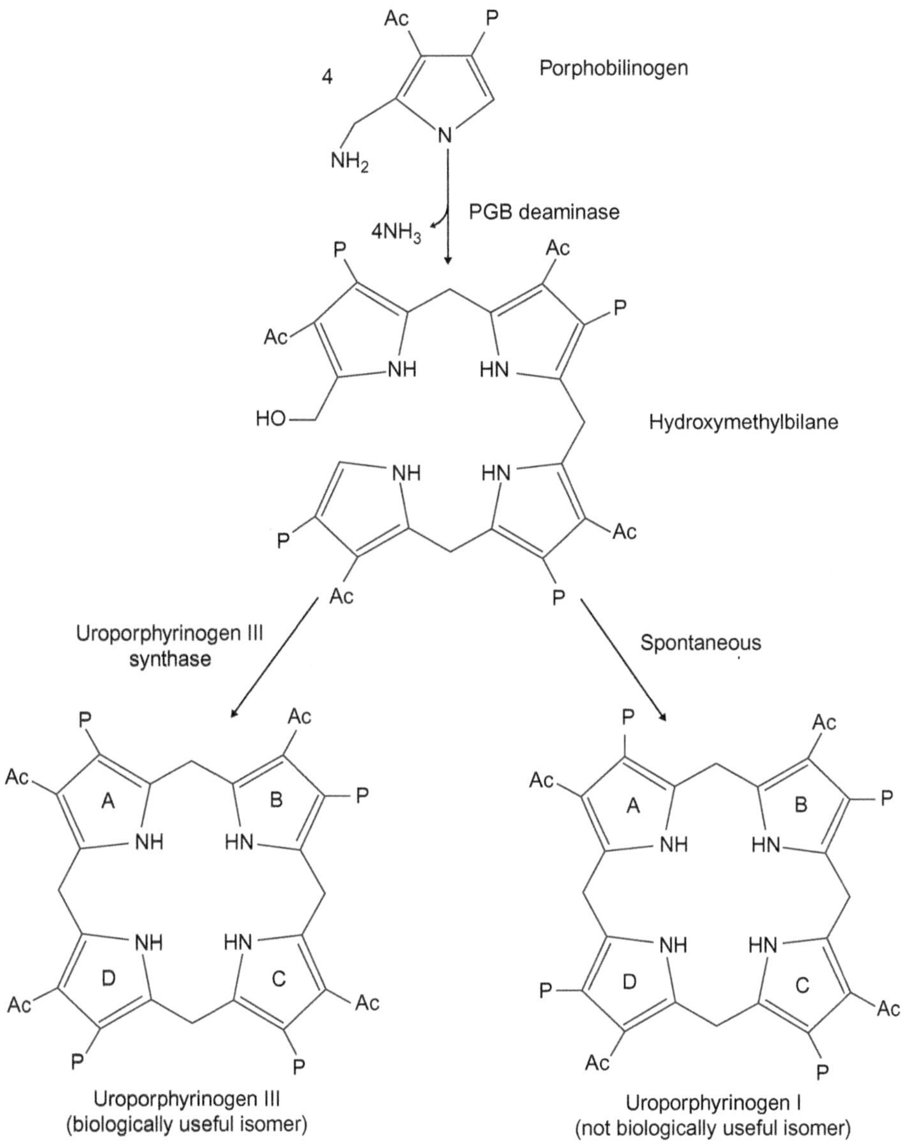

FIGURE 27.5 Synthesis of uroporphyrinogen I and III. The latter is the biologically useful isomer, and its formation requires the action of uroporphyrinogen-III synthase. $Ac = -CH_2COOH$; $P = -CH_2CH_2COOH$.

Defects in any one of the eight enzymes involved in heme biosynthesis may cause inherited porphyrin-related disorders (Figure 27.9). Porphyrins have a deep red or purple color (Greek *porphyra* = purple). Porphyrins are excreted by different routes, depending on their water solubility. For example, uroporphyrin with its eight carboxylic group substituents is more water-soluble than the porphyrins derived from it and is eliminated in the urine, whereas protoporphyrin (which contains two carboxylic groups) is excreted exclusively in bile. Coproporphyrin has four carboxylic groups and is found in bile and urine.

These disorders are associated with acute or cutaneous manifestations (or both). In the acute state, the presentation may include abdominal pain, constipation, hypertension, tachycardia, and neuropsychiatric manifestations.

Cutaneous problems consist of photosensitivity (itching, burning, redness, swelling, and scarring), hyperpigmentation, and sometimes hypertrichosis (an abnormally excessive growth of hair). Four porphyrias can manifest as acute disorders: δ-ALA dehydratase deficiency porphyria; acute intermittent porphyria; hereditary coproporphyria; and variegate porphyria.

Porphyria may be classified as hepatic or erythropoietic. However, enzyme defects are sometimes common to both tissues. Porphyrias can be induced by alcohol, stress, infection, starvation, hormonal changes (e.g., menstruation), and certain drugs. These drugs presumably precipitate acute manifestations in susceptible subjects since they are inducers of cytochrome P-450 and increase the need for synthesis of heme as they deplete the

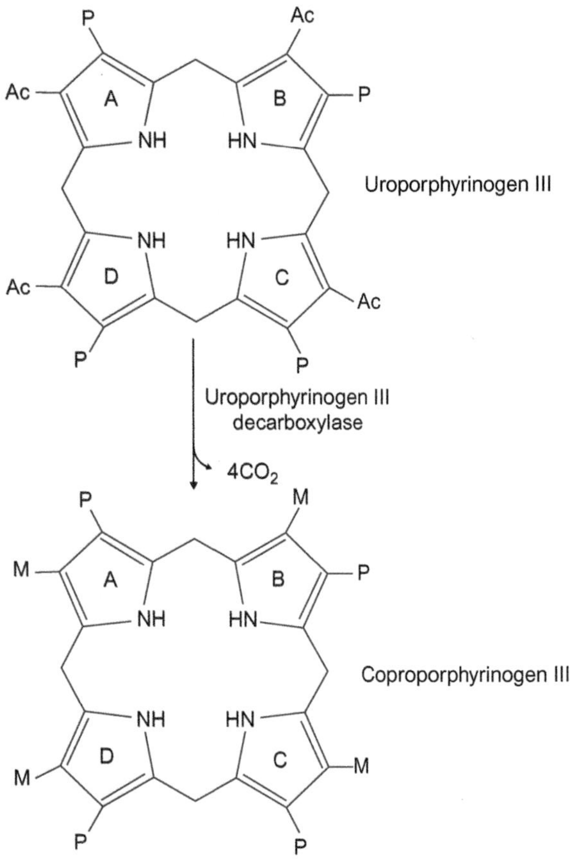

FIGURE 27.6 Formation of coproporphyrinogen III from uroporphyrinogen III. Acetic acid side chains (Ac) are decarboxylated to methyl groups (M), sequentially, starting clockwise from ring D. P = —CH₂CH₂COOH.

mitochondrial pool of free heme. Major hepatic porphyrias include **acute intermittent porphyria**, **variegate porphyria**, **hereditary coproporphyria**, and **porphyria cutanea tarda**. The principal erythropoietic porphyrias are **hereditary erythropoietic porphyria** and **erythropoietic protoporphyria**.

HEME CATABOLISM

When heme proteins are degraded in mammals, the polypeptides are hydrolyzed to amino acids while the heme groups are freed of their iron, which is salvaged, and are converted to bilirubin. After transport to the liver, bilirubin is coupled to glucuronic acid and the conjugated bilirubin is excreted into bile as the principal bile pigment. When increased production or decreased excretion of bilirubin causes increased plasma concentration, it diffuses into tissues and produces jaundice. The yellow coloration of jaundiced skin and sclerae has aroused much interest and has made bilirubin the subject of extensive research. Fractionation and quantitation of serum bilirubin are now widely used for diagnosis and prognosis of hepatobiliary disease. Bilirubin is a waste product and has no known beneficial physiological function. However, both the conjugated and the unconjugated forms of bilirubin show antioxidative properties (e.g., inhibition of lipid peroxidation). The physiological role of the antioxidative property of bilirubin is not known.

Bilirubin is a yellow-orange pigment that in its unconjugated form is strongly lipophilic and cytotoxic. It is virtually

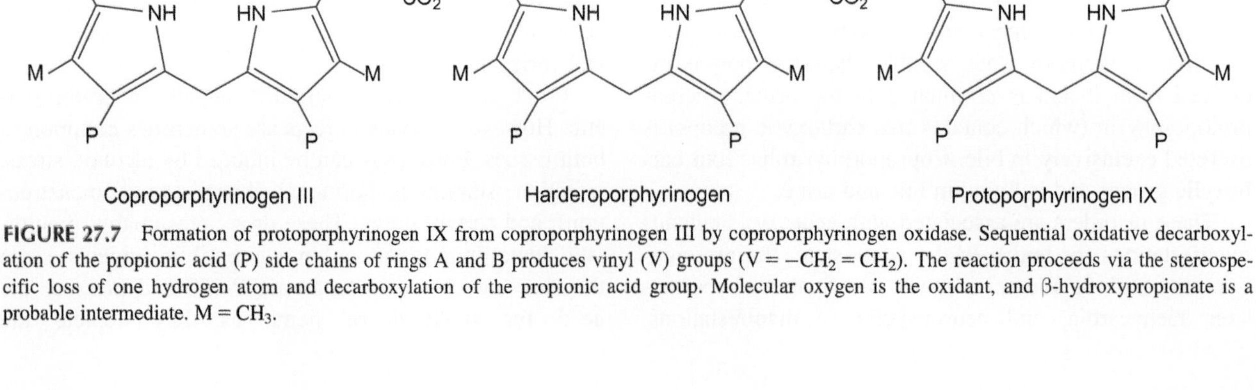

FIGURE 27.7 Formation of protoporphyrinogen IX from coproporphyrinogen III by coproporphyrinogen oxidase. Sequential oxidative decarboxylation of the propionic acid (P) side chains of rings A and B produces vinyl (V) groups (V = —CH₂ = CH₂). The reaction proceeds via the stereospecific loss of one hydrogen atom and decarboxylation of the propionic acid group. Molecular oxygen is the oxidant, and β-hydroxypropionate is a probable intermediate. M = CH₃.

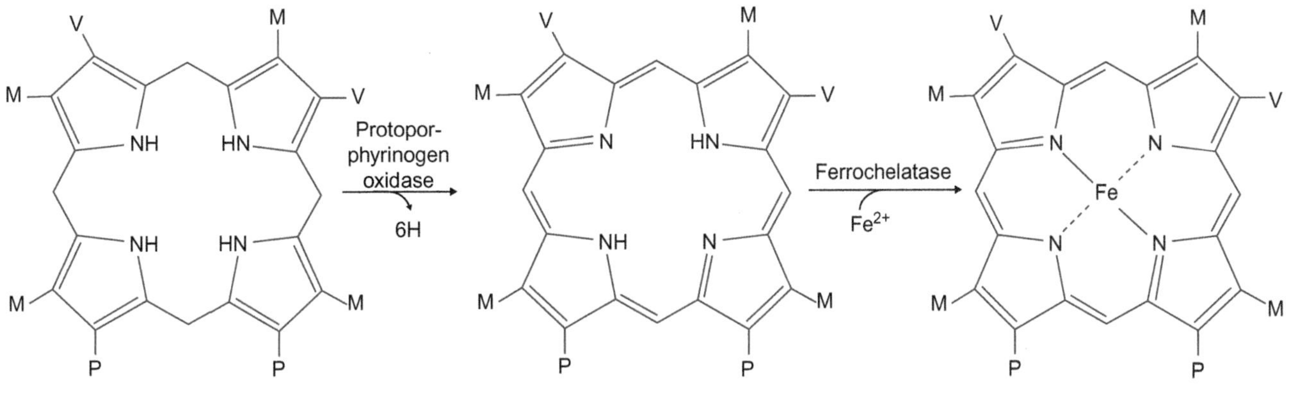

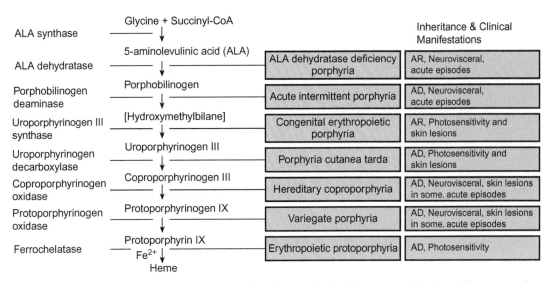

FIGURE 27.8 Formation of heme. In the reaction catalyzed by protoporphyrinogen oxidase, six hydrogens are removed and the primary electron acceptor is not known, but oxygen is required for enzyme activity. In the terminal step of heme synthesis, only Fe^{2+} is incorporated into protoporphyrin. (For key to letters see Figure 27.7.)

Enzyme	Pathway	Porphyria	Inheritance & Clinical Manifestations
ALA synthase	Glycine + Succinyl-CoA ↓		
ALA dehydratase	5-aminolevulinic acid (ALA) ↓	ALA dehydratase deficiency porphyria	AR, Neurovisceral, acute episodes
Porphobilinogen deaminase	Porphobilinogen ↓	Acute intermittent porphyria	AD, Neurovisceral, acute episodes
Uroporphyrinogen III synthase	[Hydroxymethylbilane] ↓	Congenital erythropoietic porphyria	AR, Photosensitivity and skin lesions
Uroporphyrinogen decarboxylase	Uroporphyrinogen III ↓	Porphyria cutanea tarda	AD, Photosensitivity and skin lesions
Coproporphyrinogen oxidase	Coproporphyrinogen III ↓	Hereditary coproporphyria	AD, Neurovisceral, skin lesions in some. acute episodes
Protoporphyrinogen oxidase	Protoporphyrinogen IX ↓	Variegate porphyria	AD, Neurovisceral, skin lesions in some. acute episodes
Ferrochelatase	Protoporphyrin IX Fe^{2+} ↓ Heme	Erythropoietic protoporphyria	AD, Photosensitivity

FIGURE 27.9 Heme biosynthesis pathway and the enzyme defects in various porphyrias. AD = autosomal dominant; AR = autosomal recessive.

insoluble in aqueous solutions below pH 8, but readily dissolves in lipids and organic solvents and diffuses freely across cell membranes. Bilirubin toxicity is normally prevented by tight binding to serum albumin. Only when the binding capacity of albumin is exceeded can a significant amount of unconjugated bilirubin enter cells and cause damage. Conjugated bilirubin is hydrophilic and does not readily cross cell membranes, even at high concentrations. Of the 250–300 mg (4275–5130 μmol) of bilirubin normally produced in 24 hours, about 70%–80% is derived from hemoglobin. The remainder comes from several sources, including other heme proteins (e.g., cytochromes P-450 and b_5, catalase), ineffective hemopoiesis (erythrocytes that never leave the marrow), and "free" heme (heme never incorporated into protein) in the liver. Hemoglobin heme has a lifespan equal to that of the red cell (about

125 days), whereas heme from other sources (with the exception of myoglobin, which is also quite stable) turns over much more rapidly. Hepatic P-450 enzymes have half-lives of 1–2 days. It is increased by drugs that induce hepatic P-450 oxygenases and in erythropoietic porphyria and anemias associated with ineffective erythropoiesis (lead poisoning, thalassemias, and some hemoglobinopathies).

Formation of Bilirubin

A summary of the pathway for bilirubin metabolism and excretion is shown in Figure 27.10. Release of heme from heme proteins and its conversion to bilirubin occur predominantly in the mononuclear phagocytes of liver, spleen, and bone marrow (previously known as the **reticuloendothelial system**), sites where sequestration of

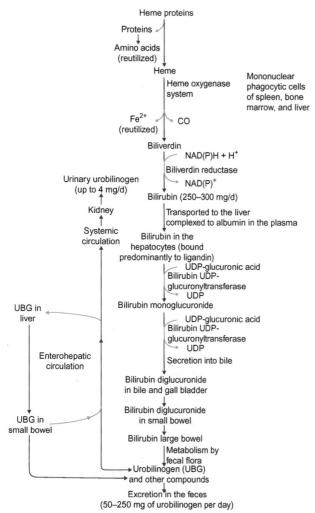

FIGURE 27.10 Catabolic pathway for the heme group from hemoproteins (predominantly hemoglobin).

aging red cells occurs. Renal tubular epithelial cells, hepatocytes, and macrophages may also contribute to bilirubin formation under some conditions. Structures of the intermediates in the conversion of heme to bilirubin are shown in Figure 27.11. The initial step after the release of heme is its binding to heme oxygenase, a microsomal enzyme distinct from the microsomal P-450 oxygenases. Heme oxygenase catalyzes what appears to be the rate-limiting step in catabolism of heme. It is induced by heme and requires O_2 and NADPH for activity. The activity of the inducible isoenzyme form of heme oxygenase is highest in the spleen, which is involved in the sequestration of senescent erythrocytes. The constitutive form of heme oxygenase is mainly localized in the liver and brain. After binding, the α-methene carbon of heme is oxidized (hydroxylated) to α-hydroxyhemin, which undergoes autoxidation to biliverdin (a blue-green pigment) with consumption of O_2 and release of iron and carbon

monoxide (derived from oxidation of the α-methene bridge). Since CO production in mammals occurs primarily by this pathway, measurement of expired CO has been used to estimate heme turnover. A potent competitive synthetic inhibitor of heme oxygenase is tin (Sn) protoporphyrin, which has potential therapeutic use in the treatment of neonatal jaundice (see later).

Biliverdin is reduced to bilirubin by NAD(P)H-dependent biliverdin reductase, a cytosolic enzyme that acts at the central methene bridge. Although both molecules have two propionic acid groups, the polarity of biliverdin is greater than that of bilirubin. Bilirubin can form six internal hydrogen bonds between the carboxylic groups, the two lactam carbonyl oxygens, and four pyrrolenone ring nitrogens, and thus prevents these groups from hydrogen-bonding with water (Figure 27.12). Esterification of the propionyl side chains of bilirubin with glucuronic acid disrupts the hydrogen bonds and increases its solubility. Phototherapy for neonatal jaundice also acts by disrupting the hydrogen-bonded structure of unconjugated bilirubin.

Hemoglobin and heme released from intravascular hemolysis or blood extravasations (e.g., subcutaneous hematomas) are bound, respectively, by **haptoglobin** and **hemopexin** to form complexes that cannot be filtered by the kidney. This action prevents renal loss of the heme iron and protects the renal tubules from possible damage by precipitated hemoglobin. Haptoglobin–hemoglobin and hemopexin–heme complexes are processed in mononuclear phagocytic cells in a way similar to that for hemoglobin. Haptoglobin and hemopexin are glycoproteins synthesized in the liver. The former is an α2-globulin and an acute-phase reactant [10].

Circulatory Transport of Bilirubin

Bilirubin formed in extrahepatic tissues is transported to the liver for excretion in bile. Since bilirubin is virtually insoluble in aqueous media, it is transported to the liver bound noncovalently to serum albumin. The bilirubin–albumin complex increases the amount of bilirubin carried per volume of plasma and minimizes the diffusion of bilirubin into extrahepatic tissues, thereby preventing bilirubin toxicity. Because of the formation of this complex, bilirubin does not normally appear in urine. Urinary bilirubin is almost invariably conjugated bilirubin (see later) and signifies the presence of a pathological process. An albumin molecule binds two molecules of bilirubin at one high-affinity site and at one to three secondary sites. Bilirubin conjugated with glucuronic acid also binds to albumin but with much lower affinity. Another form of bilirubin (probably conjugated), very tightly (probably covalently) bound to albumin, has been described. The mechanism of its formation is not known, although

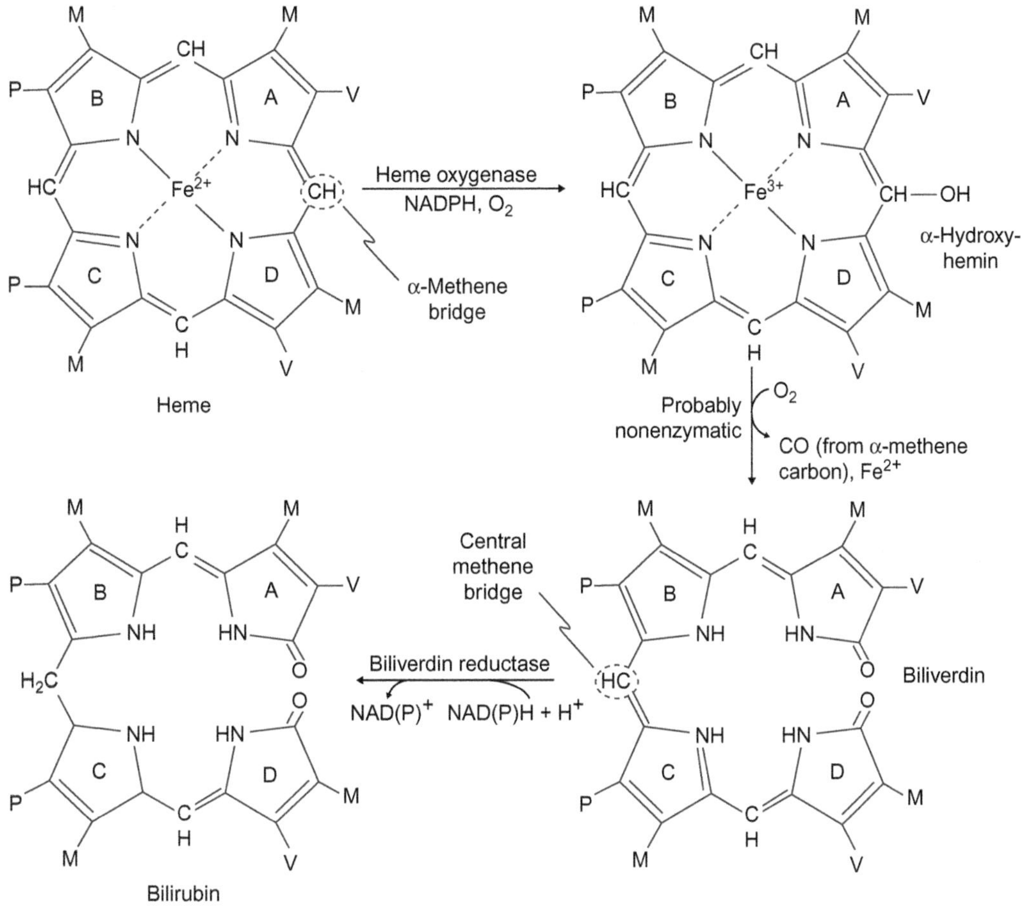

FIGURE 27.11 Conversion of heme to bilirubin in the monocytic phagocytic cells. Carbon monoxide and bilirubin are generated. The Fe^{3+} released is conserved and reutilized. Biliverdin and bilirubin are lactams. P = propionic acid; M = methyl; V = vinyl.

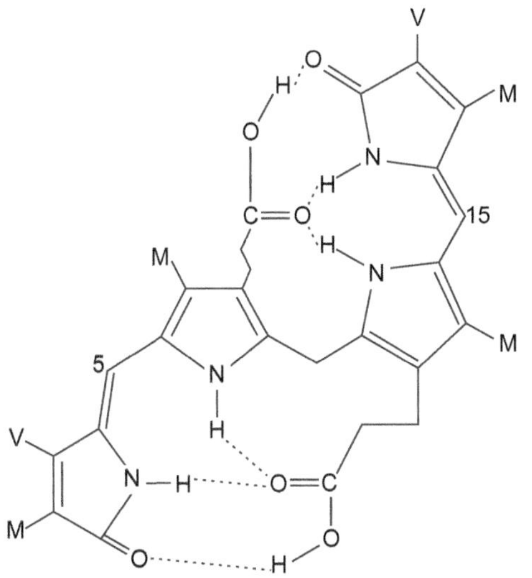

FIGURE 27.12 Conformation of bilirubin showing involuted hydrogen-bonded structure between NH/O and OH/O groups. Despite the presence of polar carboxyl groups, bilirubin is nonpolar and lipophilic. Glucuronidation disrupts hydrogen bonds and provides polar groups to yield water-soluble pigments. (For key to letters see Figure 27.11.)

blockage of biliary flow associated with an intact hepatic conjugating system releases a chemically reactive form of bilirubin into the circulation.

If the capacity of albumin to bind bilirubin is exceeded because of increased amounts of unconjugated bilirubin or decreased concentration of albumin, bilirubin readily enters extrahepatic tissues. In neonates, this can cause **kernicterus**, a serious condition associated with permanent neurological damage (see later). Bilirubin can be displaced from binding to albumin by sulfonamides, salicylates (notably aspirin), and cholangiographic contrast media. Use of these substances in jaundiced newborn infants increases the risk of occurrence of kernicterus.

Hepatic Uptake, Conjugation, and Secretion of Bilirubin

Hepatocytes take up bilirubin from the sinusoidal plasma and excrete it, after conjugation with glucuronic acid, across the canalicular membrane into the bile. The entry and exit steps and the transport of bilirubin within the cell are not completely understood. The following is a plausible interpretation of the available data.

Since binding of bilirubin to albumin is usually reversible, a small amount of free bilirubin is present in plasma in equilibrium with albumin-bound bilirubin. It is probably this free bilirubin that is taken up at a rate determined by its plasma concentration. As this free bilirubin concentration decreases, more bilirubin is released from albumin and becomes available for uptake. Alternatively, the albumin–bilirubin complex may bind to specific hepatocyte plasma membrane receptors, and thereby bilirubin is released to enter the cell. Both models are consistent with the finding that albumin does not accompany bilirubin into the hepatocyte.

The entry step seems to be carrier-mediated, is saturable, is reversible, and is competitively inhibited by sulfobromophthalein, indocyanine green, cholecystographic agents, and several drugs. Bile salts do not compete with bilirubin for hepatic uptake. After it enters hepatocytes, bilirubin is transported to the smooth endoplasmic reticulum for glucuronidation bound to a protein. Two cytosolic proteins, **Z protein** (fatty acid-binding protein) and **ligandin** (Y protein), bind bilirubin. Under normal conditions, ligandin is probably the principal hepatic bilirubin-binding protein and may serve the same protective and transport functions intracellularly as does albumin in plasma. It may also help limit reflux of bilirubin into plasma, since its affinity for bilirubin is at least five times greater than that of albumin. Z protein (M.W. 11,000) becomes important at high plasma bilirubin concentrations. The concentration of ligandin in the liver does not reach adult levels until several weeks after birth, whereas neonatal and adult levels of Z protein are the same. This lack of ligandin, together with low hepatic glucuronyltransferase activity, is the probable cause of transient, "physiological," nonhemolytic, **neonatal jaundice**.

Glucuronidation of bilirubin in the endoplasmic reticulum by UDP-glucuronyltransferase produces an ester between the 1-hydroxyl group of glucuronic acid and the carboxyl group of a propionic acid side chain of bilirubin (Figure 27.13). In bile, about 85% of bilirubin is in the diglucuronide form, and the remainder is in the monoglucuronide form. Glucuronidation increases the water solubility of several lipophilic substances. There appear to be many UDP-glucuronyltransferases in the endoplasmic reticulum, which differ in substrate specificity. (Biosynthesis of UDP-glucuronic acid was described in Chapter 14.) Secretion across the canalicular membrane into bile appears to be the rate-limiting step in hepatic bilirubin metabolism. It is probably carrier-mediated, requires energy, is saturable, and is unaffected by bile salts. Bilirubin can be made water-soluble by conversion to its isomers. These are known as **photobilirubins** and are formed when bilirubin is exposed to blue light of 400–500 nm wavelength. Photobilirubins cannot form the intramolecular hydrogen bonds characteristic of the natural isomer of bilirubin (Figure 27.12). Thus, they are more polar and readily excreted in the bile without the requirement for glucuronidation. **Lumirubin**, a structural isomer of bilirubin, is formed by light-induced intramolecular

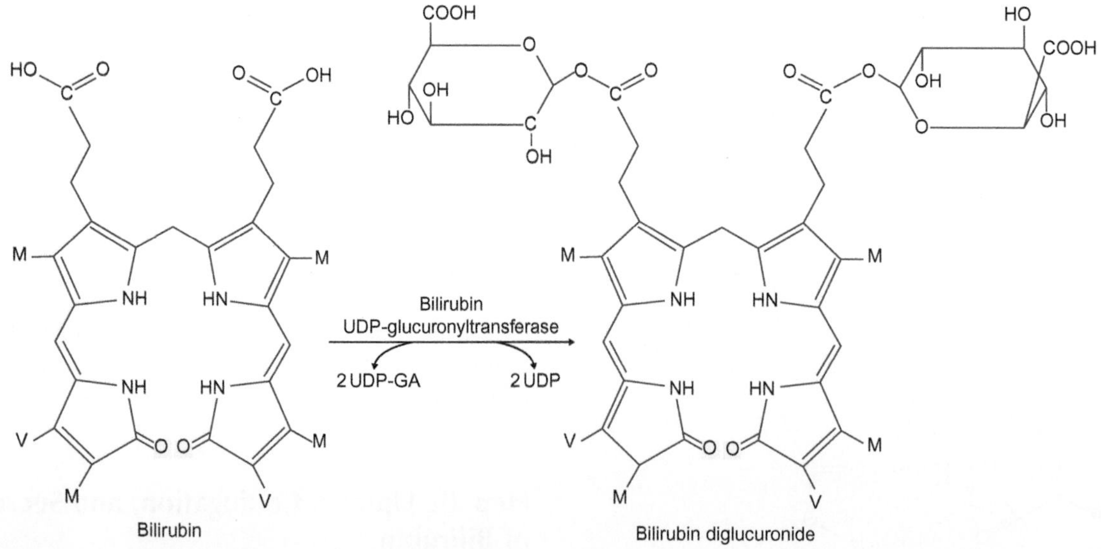

FIGURE 27.13 Formation of bilirubin diglucuronide. Glucuronidation occurs in two steps via formation of monoglucuronide. Mono- and diglucuronides are more water-soluble and less lipophilic than bilirubin. Conversion of bilirubin to water-soluble products is obligatory for excretion of bilirubin from hepatocytes. M = methyl; V = vinyl; UPD-GA = UDP-glucuronic acid.

cyclization of the vinyl side group of C-3. It contains a seven-membered ring, is stable, is polar, and is excreted without conjugation. These observations explain the mechanism of phototherapy commonly used for treatment of neonatal hyperbilirubinemia.

Bilirubin in the Intestinal Tract

Most bilirubin entering the intestine in bile is in the diglucuronide form, which is very poorly absorbed in the small and large intestines. In the lower small intestine and colon, bacteria remove glucuronic acid residues and reduce bilirubin to colorless **urobilinogen** and **stercobilinogen**. Exposure to air oxidizes these to urobilin and stercobilin, respectively (i.e., red-orange pigments that contribute to the normal color of stool and urine). Other degradation products of bilirubin are present in minor amounts in feces.

Urobilinogen is excreted mostly in the feces, but a small fraction is absorbed from the colon, enters the portal circulation, is removed by the liver, and is secreted into the bile. That which is not removed from the portal blood by the liver enters the systemic circulation and is excreted by the kidneys. Urobilinogen excretion in urine normally amounts to 1−4 mg per 24 hours, as opposed to the 40−280 mg (67−470 μmol) excreted in feces.

Lack of urobilinogen in the urine and feces indicates biliary obstruction; stools are whitish ("clay-colored") owing to the absence of bile pigment. Urinary and fecal urobilinogen excretion increases in hemolytic anemia.

Disorders of Bilirubin Metabolism

The plasma of normal subjects contains 0.1−1 mg of bilirubin per deciliter (2−17 μmol/L), mostly in the unconjugated form. Unconjugated bilirubin is known as **indirect-reacting bilirubin** and conjugated bilirubin as **direct-reacting bilirubin**.

Jaundice occurs when plasma becomes supersaturated with bilirubin (>2−2.5 mg/dL) and the excess diffuses into the skin, sclera, and other tissues. The sclera is particularly affected because it is rich in elastin, which has a high affinity for bilirubin. Reddish-yellow pigments, particularly carotene and lycopene, may give a yellowish tinge to the skin, but they do not usually produce scleral coloration. Hyperbilirubinemia may result from elevation of unconjugated or conjugated bilirubin levels.

Unconjugated Hyperbilirubinemias

Unconjugated hyperbilirubinemias result from imbalance between the rates of production of pigment and of its uptake or conjugation in the liver. Because of the large reserve capacity of the liver for conjugation and

excretion of bilirubin, increased production seldom elevates unconjugated serum bilirubin to more than 3−4 mg/dL. If a greater increase occurs, some degree of liver dysfunction probably also occurs. These disorders are usually due to decreased uptake of pigment by hepatocytes or to failure of these cells to store, transport, or conjugate bilirubin. Bilirubinuria does not accompany these disorders. Except in infancy, or when pigment gallstones form, unconjugated hyperbilirubinemias are benign.

Gilbert's syndrome may be the most common cause of mild, persistent, nonhemolytic, unconjugated hyperbilirubinemia. Serum bilirubin concentration rarely exceeds 5 mg/dL and usually fluctuates between 1.3 and 3 mg/dL. Other liver function tests are normal. The syndrome is usually asymptomatic and is detected during routine laboratory testing or examination for other diseases. Family studies suggest that Gilbert's syndrome is an autosomal dominant disorder. The unconjugated hyperbilirubinemia in Gilbert's syndrome is due to decreased UDP-glucuronyltransferase activity resulting from an insertion mutation found in the promoter region of the enzyme. The wild-type promoter $A[TA]_6TAA$ is mutated to $A[TA]_7TAA$. Mutations affecting the coding region of the enzyme, although rare, also occur.

In **Crigler−Najjar syndrome type I**, activity of hepatic bilirubin UDP-glucuronyltransferase is undetectable and bilirubin conjugates are absent from the serum, bile, and urine, but biliary secretion of sulfobromophthalein and indocyanine green is normal. The disease is apparent shortly after birth, kernicterus develops, and death commonly occurs during the neonatal period. The effectiveness of phototherapy is often transient. The enzyme is not inducible by phenobarbital. This autosomal recessive defect occurs in all races. Orthotopic liver transplantation is the definitive treatment, and it normalizes serum bilirubin levels.

Crigler−Najjar syndrome type II (Arias syndrome) is milder, usually benign, and is caused by partial deficiency of bilirubin UDP-glucuronyltransferase. Jaundice may not appear until the second or third decade of life. The monoglucuronide is the predominant pigment in bile. Phenobarbital induces the enzyme. Dominant and recessive inheritance patterns have been described. An accurate diagnosis of type I, as opposed to type II Crigler−Najjar syndrome, is essential since orthotopic liver transplantation is an important therapy for type I patients.

Conjugated Hyperbilirubinemias

Conjugated hyperbilirubinemias are due to intra- or extra-hepatic reduction in bile flow (cholestasis) with spillage of conjugated bilirubin into the bloodstream, which may occur from injury to the endothelial cells lining bile ductules or from reverse pinocytosis by the hepatocytes. Since the serum bilirubin is mostly the water-soluble glucuronide, bilirubinuria is usually present.

Abdominal tumors, gallstones, strictures, hepatitis, and cirrhosis can mechanically block the biliary canaliculi or ducts. If obstruction affects only intrahepatic bile flow, hyperbilirubinemia occurs when 50% or more of the liver is involved. Extrahepatic obstruction can also elevate serum bilirubin. Nonmechanical cholestasis can be caused by bacterial infection, pregnancy, sex steroids and other drugs, or it may be genetically determined.

In cholestasis, bile salts and bile pigments are retained and appear in the circulation, and steatorrhea and deficiencies of fat-soluble vitamins may occur. These deficiencies are often manifested as hypoprothrombinemia (from lack of vitamin K) and osteomalacia (from lack of vitamin D). The magnitude depends on the degree of obstruction. If blockage is complete, urinary urobilinogen will be absent, and the stools will have a pale, clay-like color.

Familial diseases include **Dubin–Johnson syndrome**, **Rotor's syndrome**, and **benign familial recurrent cholestasis**. All three disorders are uncommon or rare, and all are benign.

Neonatal Hyperbilirubinemia

Normal neonates are frequently hyperbilirubinemic [12,13]. Birth interrupts normal placental elimination of pigment, and the "immature" liver of the neonate must take over. Normally, serum bilirubin levels rise on the first day of life, reaching a maximum (rarely greater than 10 mg/dL) by the third or fourth day. This type is mostly unconjugated. If the placenta is functioning normally, jaundice will not be present at birth. If jaundice is present at birth, a cause other than hepatic immaturity must be sought.

The primary blocks to bilirubin metabolism are low activity of bilirubin glucuronyltransferase and a low concentration of ligandin in the liver at birth. Secretion of conjugated bilirubin into the bile is also reduced.

Hepatic immaturity may be partly due to diversion *in utero* of blood from the liver by the ductus venosus. When this channel closes shortly after birth and normal hepatic blood flow is established, concentrations of a number of substances rise within the hepatocytes and may induce enzymes needed for their metabolism. Accumulation of bilirubin in plasma may play an important role in hastening the maturation. Although the liver normally matures within 1–2 weeks after birth, hypothyroidism can prolong this process for weeks or months.

The neonate is at risk for kernicterus if the serum unconjugated bilirubin level is higher than 17 mg/dL. Kernicterus is characterized by yellow staining of clusters of neuronal cell bodies in the basal ganglia, cerebellum, and brain stem, leading to motor and cognitive deficits or death. Immaturity and perhaps hypoxia make the blood–brain barrier permeable to bilirubin and contribute to the likelihood of kernicterus. The biochemical basis of bilirubin encephalopathy is due to many causes: inhibition of RNA and protein synthesis; carbohydrate metabolism (both cAMP-mediated and Ca^{2+}-activated); phospholipid-dependent protein kinases; enzymes involved in the electron transport system; and impaired nerve conduction.

A major complicating factor can be hemolytic anemia such as that of **erythroblastosis fetalis** caused by Rh incompatibility between mother and child. The hemolysis increases the rate of bilirubin formation, which soon overwhelms the liver and produces severe jaundice and kernicterus. Sickle cell anemia has a similar effect. Congenital absence of bilirubin UDP-glucuronyltransferase (Crigler–Najjar syndrome type I) usually causes a kernicterus that is fatal shortly after birth. Inhibition of glucuronyltransferase by various drugs (e.g., novobiocin) or toxins can increase the severity of neonatal jaundice. "Breast milk jaundice" is due to the presence in breast milk of a substance (perhaps pregnane-3α,20β-diol) that inhibits bilirubin glucuronyltransferase, although the resulting unconjugated hyperbilirubinemia is seldom serious enough to cause neurotoxicity or to require discontinuation of breast-feeding. Other risk factors for pathologic hyperbilirubinemia include Gilbert's syndrome (discussed earlier) and glucose-6-phosphate dehydrogenase deficiency (Chapter 14). Conjugated hyperbilirubinemia is rare during the neonatal period. It can result from impaired hepatocellular function or extrahepatic obstruction. Hepatocellular defects can be caused by bacterial, viral, or parasitic infections, cystic fibrosis, α_1-antitrypsin deficiency, Dubin–Johnson and Rotor's syndromes, and other genetic diseases. Extrahepatic obstruction can be congenital (biliary atresia) or acquired. Treatment of neonatal jaundice is usually by phototherapy. A decrease in bilirubin production in the neonatal period can also be achieved by inhibiting the rate-limiting enzyme of bilirubin formation from heme, namely, the heme oxygenase [14]. A potent competitive inhibitor of heme oxygenase is the synthetic heme analogue tin (Sn^{4+}) protoporphyrin (see Clinical Case Study 27.3). When administered parenterally, tin protoporphyrin safely decreases bilirubin formation. Exchange transfusions also rapidly decrease plasma bilirubin levels.

CLINICAL CASE STUDY 27.1 Hereditary Hemochromatosis, an Iron Storage Disorder

This case was abstracted from: R.T. Chung, J. Misdraji, D.V. Sahani, Case 33-2006: a 43-year-old man with diabetes, hypogonadism, cirrhosis, arthralgias, and fatigue, N. Engl. J. Med. 355 (2006) 1812–1819.

Synopsis

A 43-year-old Caucasian male presented with a chief complaint of fatigue, decreased libido, and erectile dysfunction. His physical examination revealed tanned skin. Comprehensive imaging and laboratory testing showed that the subject had hypogonadotropic hypogonadism, recent onset diabetes type 2, arthralgias, and hemochromatosis. The initial diagnosis of hemochromatosis was based on highly elevated serum iron, percent iron saturation and ferritin, and aminotransferase levels. Liver biopsy revealed marked iron deposition with cirrhosis. Genetic testing showed homozygous mutation in the HFE gene, resulting in the expression of C282Y-mutant HFE protein. The patient was placed on a therapeutic phlebotomy schedule to reduce iron overload, treatment with insulin for diabetes, and testosterone supplementation for decreased libido. Lifestyle modification of avoidance of ethanol intake, iron and vitamin C supplements, and decreased consumption of red meat were implemented.

Teaching Points

1. Either iron deficiency or iron excess leads to severe diseases. The most common nutritional disorder is due to iron deficiency, which leads to hypochromic, microcytic anemia. The accumulation of iron in the body, either due to frequent blood transfusions (e.g., in the management of sickle cell disease and thalassemias) or due to genetic causes because of mutations in iron regulatory proteins, leads to a pathological condition known as hemochromatosis. The toxicity of iron is due to the production of superoxide anions and hydroxyl radicals, which inactivate proteins, lipids, and nucleic acids.

2. Iron absorption, storage, and utilization are orchestrated by gastrointestinal duodenal cells, macrophages of liver and spleen, and hepatocytes. The players are transport proteins: divalent metal transporter 1 (DMT1); ferroportin; storage protein ferritin; plasma transport protein transferrin; transferrin receptors 1 and 2 (tfR1 and 2); HFE protein; hemojuvelin; cytokines; and finally the conductor hepcidin.

3. Hepcidin is a 25-amino acid peptide consisting of four intra-disulfide bonds; it attenuates (inhibits) iron absorption and iron release from macrophages. The expression of hepatic hepcidin is regulated by iron regulatory proteins, hypoxia, and inflammatory mediators (cytokines) via signaling pathways involving bone morphogenetic proteins (BMP). Thus, mutation in iron regulatory proteins

that result in decreased hepcidin synthesis can lead to all forms of currently known genetic hemochromatosis.

4. Genetic hemochromatosis is the most common genetic disorder in populations of European ancestry, and among the types of genetic hemochromatosis, HFE mutation (C282Y) is the most common type (type 1). The homozygous C282Y mutation does not always lead to genetic hemochromatosis. Clinical penetrance of the homozygous C282Y mutation is incomplete and is probably affected by modifier genes.

5. Timely diagnosis of genetic hemochromatosis in subjects with high serum iron, percent iron saturation, and ferritin levels, and unexplained elevated serum aminotransferase levels is vital to prevent multiorgan iron damage. Initiation of iron removal by regular therapeutic phlebotomy can ameliorate symptoms. Iron removal by chelation therapy is utilized in some forms of genetic hemochromatosis associated with anemia and in transfusion-dependent secondary hemochromatosis.

6. Anemia of chronic disease that occurs in acute and chronic inflammatory immune disorders is caused by a cytokine-mediated increase of hepcidin synthesis, leading to decreased availability of iron required for heme biosynthesis. In addition to changes in iron homeostasis, anemia of chronic disease affects erythropoietin synthesis and proliferation of erythroid precursor cells. In anemia due to chronic renal disease, the primary cause has been attributed to decreased erythropoietin production in the kidneys. Thus, administration of recombinant erythropoietin is employed in correcting anemia of chronic renal disease.

Supplemental References

[1] N.C. Andrews, S. Anupindi, K. Badizadegan, Case 21-2005: a four-week-old male infant with jaundice and thrombocytopenia, N. Engl. J. Med. 353 (2005) 189–198.

[2] M.J. Flamm, Hemochromatosis: a new look at a familiar disease, Cortland Forum 20 (2007) 35–37.

[3] J.Y. Abuzetun, R. Hazin, M. Suker, J. Porter, A rare case of hemochromatosis and Wilson's disease coexisting in the same patient, JAMA 100 (2008) 112–114.

[4] A. Zhang, C.A. Enns, Iron homeostasis: recently identified proteins provide insight into novel control mechanisms, J. Biol. Chem. 284 (2009) 711–715.

[5] B.R. Bacon, R.S. Britton, Clinical penetrance of hereditary hemochromatosis, N. Engl. J. Med. 358 (2008) 291–292.

[6] K.J. Allen, L.C. Gurrin, C.C. Constantine, N.J. Osborne, M.B. Delatycki, A.J. Nicoll, et al., Iron-overloaded-related disease in HFE hereditary hemochromatosis, N. Engl. J. Med. 358 (2008) 221–230.

[7] R.E. Fleming, B.R. Bacon, Orchestration of iron hemostasis, N. Engl. J. Med. 352 (2005) 1741–1743.

[8] A. Pietrangelo, Hereditary hemochromatosis: a new look at an old disease, N. Engl. J. Med. 350 (2004) 2383–2397.

[9] D.W. Lee, J.K. Andersen, D. Kaur, Iron dysregulation and neurogeneration: the molecular connection, Mol. Interv. 6 (2006) 89–97.

[10] D.W. Swinkles, M.C.H. Janssen, J. Bergmans, J.J.M. Marx, Hereditary hemochromatosis: genetic complexity and new diagnostic approaches, Clin. Chem. 52 (2006) 950–968.

CLINICAL CASE STUDY 27.2 Lead Poisoning

This case was abstracted from: L.S. Friedman, L.H. Simmons, R.H. Goldman, A.R. Sohani, Case 12-2014: a 59-year-old man with fatigue, abdominal pain, anemia, and abnormal liver function, N. Engl. J. Med. 370 (2014) 1542–1550.

Synopsis

A 59-year-old man presented to the clinic with a 3-day history of fatigue, epigastric pain, nausea, and ankle swelling. His physical examination was unremarkable, but laboratory studies revealed anemia and elevated liver enzymes. Review of a peripheral blood smear showed microcytic anemia, polychromasia (red cell enlargement with a purplish hue), and basophilic stippling (punctate basophilic inclusions that are evenly distributed throughout the cytoplasm). He was sent home with omeprazole for the treatment of peptic-ulcer disease and bleeding ulcer. However, within one week his abdominal pain worsened, and he developed an unusual constellation of symptoms raising concerns for lead poisoning, such as behavioral changes and dysgeusia (altered sense of taste). The patient's blood lead level was found to be markedly elevated. A thorough occupational history did not reveal a definitive source for the patient's lead poisoning, except perhaps the daily usage of an Italian mug and spoon containing lead-based paint. The patient's condition was considered severe, so chelating therapy was administered (calcium disodium EDTA and 2,3-dimercaptosuccinic acid).

Teaching Points

1. The most common source of lead poisoning in the United States is from workplace exposures (e.g., lead-based paint,

building renovations). A blood lead level of 10 μg/dL or higher is considered elevated. In acute lead poisoning, blood lead levels may reach as high as 100 μg/dL or more.

2. At a blood lead level of about 55 μg/dL, heme synthesis becomes impaired. Lead is a potent inhibitor of heme synthesis by binding to the enzyme 5-aminolevulinic acid (ALA) dehydratase. Lead also inhibits the enzyme heme chelatase in the final step of heme synthesis.

3. Patients with lead poisoning may present clinically like patients with porphyria, a disorder in which there is a deficiency of ALA dehydratase. The resulting overproduction of ALA in both disorders can cause symptoms of nausea, abdominal pain, constipation, restlessness, and pain in the arms and legs.

4. Basophilic stippling is a hallmark feature of lead poisoning and sideroblastic anemia, but it may not be seen in all cases of lead poisoning. Coarse basophilic stippling is a result of impaired hemoglobin synthesis or impaired incorporation of iron into heme, causing abnormal ribosomal structure and incomplete RNA degradation, which appear as punctate inclusions evenly distributed throughout the cytoplasm.

5. The treatment of lead poisoning is administration of a chelating agent, such as calcium disodium EDTA or 2,3-dimercaptosuccinic acid. Removal of the lead source is also an important part of the treatment.

CLINICAL CASE STUDY 27.3 Hyperbilirubinemia

This case was abstracted from: G.J. Mizejewski, K.A. Pass, Sn-Mesoporphyrin interdiction of severe hyperbilirubinemia in Jehovah's Witness' newborns as an alternative to exchange transfusion, Pediatrics 108 (2001) 1374–1377.

Synopsis

Two unrelated newborn infants at the same hospital were found to have progressive hyperbilirubinemia despite phototherapy. Their hyperbilirubinemia was due to immune hemolytic disease of the newborn. Exchange transfusion therapy was considered when their plasma unconjugated bilirubin had reached a level of 19.5 mg/dL. However, the families of both infants refused exchange transfusion therapy due to religious concerns. Alternative therapy with Sn-mesoporphyrin (Sn-MP) was necessary. Sn-MP is a powerful inhibitor of heme oxygenase. Intramuscular administration of a single dose of Sn-MP led to a sustained decrease in the infants' plasma bilirubin levels and the resolution of hyperbilirubinemia.

Teaching Points

1. Bilirubin is a catabolic product of heme synthesized in the macrophages in two enzymatic steps. The first step

is catalyzed by the rate-limiting enzyme heme oxygenase. Elimination of bilirubin requires albumin-bound transport to the liver, conjugation with glucuronic acid, and eventual removal via the biliary-gastrointestinal tract.

2. Newborn unconjugated hyperbilirubinemia, a common condition, is usually treated with phototherapy. Note that the photoisomers of bilirubin are water-soluble, do not require glucuronidation, and are eliminated in the urine. Severe hyperbilirubinemia can occur due to prematurity, isoimmune hemolytic disease, glucose-6-phosphate dehydrogenase deficiency, asphyxia, acidosis, and hypoalbuminemia. Genetic defects in bilirubin glucuronidation also result in hyperbilirubinemia.

3. High plasma levels of bilirubin cause brain damage (e.g., bilirubin encephalopathy evolving into kernicterus). The precise plasma levels of bilirubin that cause abnormal neurologic manifestations are not understood (see references [2] and [3]). A determining factor for the management of hyperbilirubinemia is its progressive increase, unresponsiveness to phototherapy, and the presence of risk factors.

(Continued)

CLINICAL CASE STUDY 27.3 (Continued)

4. In the treatment of neonatal hyperbilirubinemia, exchange blood transfusion therapy is effective for rapid elimination of bilirubin. However, inhibition of its production via heme oxygenase by Sn-MP is also effective.

5. While the primary management consideration for neonatal jaundice requires incorporation of methods that decrease plasma bilirubin levels, the management of adult jaundice requires the diagnosis of diseases of the hepatic-biliary system (e.g., cancer, cirrhosis, infection), followed by their treatment.

Supplemental References

[1] M.J. Maisels, A.F. McDonagh, Phototherapy for neonatal jaundice, N. Engl. J. Med. 358 (2008) 920−928.

[2] J.F. Watchko, Neonatal hyperbilirubinemia: what are the risks? N. Engl. J. Med. 354 (2006) 1947−1948.

[3] T.B. Newman, P. Liljestrand, R.J. Jeremy, D.M. Ferriero, Y.W. Wu, E.S. Hudes, et al., Outcomes among newborns with total serum bilirubin levels of 25 mg per deciliter or more, N. Engl. J. Med. 354 (2006) 1889−1900.

[4] S.K. Moerschel, L.B. Cianciaruso, L.R. Tracy, A practical approach to neonatal jaundice, Am. Fam. Physician 77 (2008) 1255−1262.

REQUIRED READING

[1] M. Mastrogiannaki, P. Matak, C. Peyssonnaux, The gut in iron homeostasis: role of HIF-2 under normal and pathological conditions, Blood 122 (2013) 885−892.

[2] D. Meynard, J.L. Babitt, H.Y. Lin, The liver: conductor of systemic iron balance, Blood 123 (2014) 168−176.

[3] R.E. Fleming, P. Ponka, Iron overload in human disease, N. Engl. J. Med. 366 (2012) 348−359.

[4] H. Drakesmith, A.M. Prentice, Hepcidin and the iron-infection axis, Science 338 (2012) 768−772.

[5] S.R. Pasricha, H. Drakesmith, J. Black, D. Hipgrave, B. Biggs, Control of iron deficiency anemia in low- and middle-income countries, Blood 121 (2013) 2607−2617.

[6] C. Brugnara, J. Adamson, M. Auerbach, R. Kane, I. Macdougall, A. Mast, Iron deficiency: what are the future trends in diagnostics and therapeutics? Clin. Chem. 59 (2013) 740−745.

[7] N.C. Andrews, Closing the iron gate, N. Engl. J. Med. 366 (2012) 376−377.

[8] C. Camaschella, Treating iron overload, N. Engl. J. Med. 368 (2013) 2325−2327.

[9] A.V. Hoffbrand, A. Taher, M.D. Cappellini, How I treat transfusional iron overload, Blood 120 (2012) 3657−3669.

[10] D.J. Schaer, P.W. Buehler, A.I. Alayash, J.D. Belcher, G.M. Vercellotti, Hemolysis and free hemoglobin revisited: exploring hemoglobin and hemin scavengers as a novel class of therapeutic proteins, Blood 121 (2013) 1276−1284.

[11] M. Balwani, R.J. Desnick, The porphyrias: advances in diagnosis and treatment, Blood 120 (2012) 4496−4504.

[12] J.R. Ingelfinger, Bilirubin-induced neurologic damage—mechanisms and management approaches, N. Engl. J. Med. 369 (2013) 2021−2030.

[13] American Academy of Pediatrics Subcommittee on Hyperbilirubinemia, Management of hyperbilirubinemia in the newborn infant 35 or more weeks of gestation, Pediatrics 114 (2004) 297−316.

[14] S. Schulz, R.J. Wong, H.J. Vreman, D.K. Stevenson, Metalloporphyrins—an update, Front. Pharmacol. 3 (2012) 1−16.

Chapter 28

Endocrine Metabolism I: Introduction and Signal Transduction[1]

Key Points

1. Hormones and neurotransmitters are integrated, and they coordinate cellular functions in the body.
2. The physiologic response to hormones can be autocrine, paracrine, or endocrine.
3. Hormones can be amino acid-derived amines, peptides, proteins or glycoproteins, steroids, or eicosanoids.
4. The nervous and endocrine systems function in a coordinated manner to promote growth, homeostasis, and reproductive competence.
5. Feedback regulation of an endocrine system (usually negative) involves both simple feedback loops (e.g., insulin secretion regulated by plasma glucose levels) and complex feedback loops (e.g., hypothalamic, pituitary, thyroid in the secretion of thyroxine). Some of the feedback loops can be positive.
6. Hormone secretion can be pulsatile, episodic, daily, monthly, or seasonal.
7. Steroid hormones and thyroid hormones (T_3 and T_4) require transport proteins in the blood to reach the target sites.
8. Some hormones have multiple physiologic effects at a given target site (e.g., insulin), and some hormones have different physiological effects at different sites (e.g., testosterone).
9. The physiological response to a hormone (ligand) is determined by the presence of a specific receptor at the target cell. The receptors may be located on the plasma cell membrane, in the cytosol, or in the nucleus.
10. Hormone recognition and binding at their specific receptor binding site initiate the signal transduction amplification pathways, which culminates in an appropriate biological response. This chapter discusses only a few selected pathways.
11. Nuclear receptors are responsible for the action of the thyroid hormone tetraiodothyronine (T_3). T_3 binding to the nonhistone receptor proteins stimulates transcription at the target sites. However, it can also inhibit transcription of thyroid stimulating hormone (TSH) at the pituitary (a negative feedback process).
12. All steroid hormones, like T_3, mediate their action via nuclear receptors. However, glucocorticoids and aldosterone initially bind to cytosolic receptors.
13. Amine, polypeptide, and protein hormones (either growth promoting or inhibiting) initiate their action by binding to plasma membrane receptors on the cell surface.
14. Cell surface receptors can be G-protein-coupled receptors (GPCR), tyrosine kinase receptors, or guanylyl kinase receptors. GPCR pathways include the heterotrimeric G-protein-coupled adenylate cyclase−cAMP system and the G-protein-coupled phosphatidyl-inositol-Ca^{2+} pathway.
15. G-protein-coupled receptor signaling pathways affect diverse metabolic processes and are regulated by several different mechanisms involving the regulation of GTPase activity of the Gα subunit. Some bacterial toxins act by activating or inhibiting Gα subunit activity.
16. Monomeric G-proteins anchored to the inner cytoplasmic membrane also participate in the normal cellular functions when activated by external stimuli. Some of these are proto-oncogenes and, when mutated, become oncogenes that promote cancer.
17. Tyrosine kinase receptors are of two types. The first type contains an extracellular receptor domain and an intracellular domain with tyrosine kinase activity (e.g., insulin). The second type consists of the extracellular hormone-induced dimerization with resultant activation of constitutively associated tyrosine kinases known as Janus kinases (JAK) and involves cytosolic signal transducer and activator of transcription pathway (STAT). These signal amplifications of JAK−STAT pathways are involved in several other pathways of the cytokines of growth factors (e.g., growth hormone, prolactin).
18. Nonreceptor tyrosine kinases are found in the cytosol and are involved in the signal transduction pathways of normal cellular processes. Abnormalities in these pathways can lead to cancer (e.g., chronic myelogenous leukemia, BCR-ABL tyrosine kinase).
19. Specific inhibitions at the receptor sites using monoclonal antibodies or tyrosine kinase inhibitors are therapeutic targets used in the management of some cancers.

---

1. Endocrine topics not discussed in Chapters 28 through 32 that are covered elsewhere in the text are as follows: gastrointestinal hormones, Chapter 11; eicosanoids, Chapter 16; pancreatic hormones, Chapter 20; parathyroid hormone and vitamin D, Chapter 35; renin−angiotensin system and antidiuretic hormone, Chapters 30 and 37. A list of expanded acronyms appears in the appendix.

N. V. Bhagavan and C.-E. Ha: Essentials of Medical Biochemistry, Second edition. DOI: http://dx.doi.org/10.1016/B978-0-12-416687-5.00028-2

20. Guanylyl cyclases can be either bound to cell surfaces or found in the cytosol. They catalyze the conversion of GTP to cGMP, which then initiates the signal transduction pathway. Natriuretic peptides initiate their action of vasodilation and natriuresis by binding to receptor-guanylyl cyclases. Nitric oxide initiates its action of smooth muscle relaxation by binding to cytosolic guanylyl cyclase.

21. All of the cyclic nucleotide (cAMP and cGMP) mediated pathways are terminated by specific phosphodiesterases, and protein phosphorylation pathways mediated by various kinases are terminated by phosphatases.

22. Many other signal transduction pathways are known. Some of these pathways are discussed at the appropriate sections in the text. The supplemental reading list provides sources that describe the various pathways.

The survival of multicellular organisms depends on the integration and coordination of differentiated cell functions and the ability to react appropriately to internal and external influences that threaten to disrupt homeostatic conditions. These requirements are fulfilled by a form of intercellular communication in which chemical signals (messengers) released by one cell evoke a receptor-mediated response in another. There are two types of chemical messengers: neurotransmitters and hormones. Neurotransmitters convey signals from one neuron to another or from a neuron to an effector cell, travel very short distances to reach their target sites, and function within the specialized regions of synapses and junctions. Hormones are usually defined as messengers that are transported by the blood to distal target cells. Because they are released into the interstitial space and then into the blood, they are called **endocrine** (ductless; "secreted within") secretions to distinguish them from those that are released into the external environment ("exocrine" or ductal secretions).

Hormones and neurotransmitters may have evolved from the same or similar ancestral prototypes in unicellular organisms. Thus, the structure of hormones and neurotransmitters and their functions as chemical messengers may have been highly conserved during evolution. The fact that many prototypical messengers in unicellular organisms predate the appearance of nerve cells implies that neurotransmitters may be a specialized form of hormone. Chemically, there are four types of hormones:

1. Hormonal amine;
2. Peptide, protein, or glycoprotein;
3. Steroid; and
4. Eicosanoid.

Amine hormones and some agents of the second type have counterparts in the nervous system that function as neurotransmitters.

HORMONAL AMINES

Hormonal amines are derived from amino acids and, in most cases, represent simple modifications of the parent compound. All of these amines except the thyroid hormones (Chapter 31) are decarboxylated products that are synthesized both within and outwith the nervous system. Within the nervous system, they are important neurotransmitters; outside the nervous system, the cells that produce them are modified postsynaptic neurons (e.g., adrenal medulla), blood-derived cells (e.g., basophils, mast cells), or APUD[2] cells (e.g., enterochromaffin cells). Regardless of where they are released, hormonal amines exert their effects through specific receptor sites located in various parts of the body. These systemic receptors (Table 28.1) and their subtypes probably have identical counterparts in the central and peripheral nervous systems. All of these amines, except T_3, exert rapid systemic effects that usually involve smooth-muscle activity. Because the amines are hydrophilic, their receptors are located on the outer surface of target cells, and most if not all of their effects are mediated by intracellular mediators (discussed later).

PEPTIDE, PROTEIN, AND GLYCOPROTEIN HORMONES

Cells that produce peptide, protein, or glycoprotein hormones are derived embryologically from the entoderm or ectoblast (progenitor of ectoderm and neuroectoderm). More than 40 hormones have been identified, containing from three to over 200 amino acid residues, and they are discussed in various parts of the text.

All hormonal peptides and many hormonal proteins are synthesized as part of a larger molecule (**preprohormone**) that contains a leader sequence (signal peptide) at its amino terminal end. The leader sequence is removed as the nascent precursor enters the lumen of the endoplasmic reticulum, and the resultant prohormone undergoes post-translational processing after being packaged into secretory granules by the Golgi complex. Post-translational processing involves proteolytic cleavage at specific sites, usually paired basic amino acid residues (Lys−Arg, Arg−Lys, Arg−Arg, Lys−Lys), by endopeptidases within the secretory granule. In the synthesis of insulin, for example, removal of the 23-amino acid leader sequence of preproinsulin results in proinsulin, which has two pairs of basic residues (Lys−Arg and Arg−Arg) that are cleaved

---

2. APUD (Amine Precursor Uptake and Decarboxylation) cells are derived from the neuroblast (stem cells that give rise to nerve cells and neural crest cells) or the entoderm. They have the ability to synthesize and release peptide hormones and, as their name implies, take up amine precursors (e.g., dopa) and decarboxylate them, producing hormonal amines (e.g., dopamine).

TABLE 28.1 Examples of G-proteins Involved in the Propagation of Extracellular Stimuli with Physiological Effects

G Protein	Location	Stimulus	Effector	Effect
G_s	Liver	Epinephrine, glucagon	Adenylyl cyclase	Glycogen breakdown
G_s	Adipose tissue	Epinephrine, glucagon	Adenylyl cyclase	Fat breakdown
G_s	Kidney	Antidiuretic hormone	Adenylyl cyclase	Conservation of water
G_i	Heart muscle	Acetylcholine	Potassium channel	Decreased heart rate and pumping force
G_i/G_o	Brain neurons	Enkephalins, endorphins, opioids	Adenylyl cyclase, potassium channels, calcium channels	Changes in neuron electrical activity
G_q	Smooth muscle cells in blood vessels	Angiotensin	Phospholipase C	Muscle contraction, blood pressure elevation
G_{olf}	Neuroepithelial cells in the nose	Odorant molecules	Adenylyl cyclase	Odorant detection
Transducin (G_t)	Retinal rod and cone cells of the eye	Light	cGMP phosphodiesterase	Light detection (vision)

to yield insulin and the C-peptide (Chapter 20). Likewise, the endogenous opiates arise from site-specific cleavages of their respective prohormones (Chapter 29). When the cell is stimulated to secrete, all major fragments of the prohormone, active and inactive, are released by calcium-dependent exocytosis.

Although several peptide hormones have multiple anatomical sites of synthesis, there is usually only one major source of the circulating hormone. The presence of the same hormone in ancillary sites may indicate that it functions as a local hormone at those sites. For example, somatostatin produced by the hypothalamus is transported by blood to the anterior pituitary, where it inhibits the release of growth hormone (GH); somatostatin produced by the hypothalamus also functions as an inhibitory neurotransmitter when released into synapses in the central nervous system; somatostatin produced by pancreatic islet cells acts locally to inhibit the release of insulin and glucagon; and somatostatin produced by the gastrointestinal mucosa acts locally to inhibit the secretion of gastrointestinal hormones (Chapter 11). Other "brain−gut" peptides (Chapter 29) also illustrate this point, although they are not released into the general circulation in significant amounts.

Several families of peptide hormones have been described, including the opiomelanocortin family (endorphins, adrenocorticotropic hormone, melanocyte-stimulating hormone), the somatomammotropin family (growth hormone, prolactin, human placental lactogen), the glycoprotein hormones (thyroid-stimulating hormone, luteinizing hormone, follicle-stimulating hormone, human

chorionic gonadotropin), the insulin family (insulin, insulin-like growth factors, somatomedins, relaxin), and the secretin family (secretin, glucagon, glicentin, vasoactive intestinal peptide). Members in a family are related in structure and function. Thus, members of the secretin family regulate secretory activity in target cells, while those of the insulin family promote cell growth.

STEROID HORMONES

The four kinds of steroid hormones differ in structure and action; they are the androgens (C_{19}), the estrogens (C_{18}), the progestins (C_{21}), and the corticosteroids (C_{21}) (Chapters 30 and 32). All are synthesized from cholesterol and are produced by mesoderm-derived cells regulated by at least one peptide hormone. The hormonal form of vitamin D_3 [$1,25(OH)_2$-D_3] is a sterol derived from an intermediate of cholesterol biosynthesis and is discussed in Chapter 35.

Organization of the Endocrine System

The nervous and endocrine systems function in a coordinated manner to promote growth, homeostasis, and reproductive competence. The need to rapidly adjust physiological processes in response to impending disturbances in homeostasis is met by the nervous system, while the endocrine system effects the more prolonged, fine-tuned adjustments. Together, they effect appropriate organismic adaptation. Some interdependence exists between the two systems. The nervous system, for example, relies heavily on a continuous supply of glucose, the circulating levels of which are under multiple hormonal controls, while maturation of the nervous system is regulated by thyroid hormone, and maintenance of mental acuity in adults depends on the availability of both thyroid hormone and glucocorticoids. On the other hand, hormone production is generally dependent on the nervous system, albeit to varying extents. The synthesis of insulin, glucagon, calcitonin, parathyroid hormone, and aldosterone appears to require little or no neural regulation, whereas that of the hypothalamic peptides and adrenal medullary hormones is totally dependent on it. It may be significant that those hormones that are relatively independent of neural control serve to regulate the level of substances that profoundly affect nerve function (glucose, calcium, sodium, potassium).

Feedback Regulation of the Endocrine System

Hormone secretion is regulated by both negative simple feedback and complex multiple-level feedback loops. Positive feedback regulation also occurs for some hormones. These aspects are discussed along with each hormone in various sections of the text.

Additional Properties of Hormones

Secretion of hormones occurs at different time intervals to meet physiological demands. The secretions can be pulsatile or episodic, exhibit diurnal rhythm, and undergo daily, monthly, and even seasonal variations.

The transport of water insoluble hormones in blood circulation requires transport proteins that are mostly synthesized in the liver (e.g., steroid and thyroid hormones). Water-soluble hormones, such as amino acid derivatives and polypeptides, do not require any transport proteins.

Some hormones exhibit multiple effects at a given target site. Insulin, for example, promotes the uptake of glucose by skeletal muscle and adipose tissue, and also promotes several anabolic pathways. Some metabolic pathways are regulated by multiple hormones to produce an integrated response (e.g., glycogen synthesis and breakdown and gluconeogenesis; Chapter 14).

Some hormones produce different effects at different target sites. Testosterone promotes spermatogenesis in the testes but also promotes growth of the accessory sex glands. Furthermore, testosterone brings about these varying effects via transformation into dihydrotestosterone at the selected target sites (Chapter 32).

Some tumors (e.g., small cell lung, breast gynecologic, and hematologic cancers) may inappropriately produce hormones (e.g., ADH, PTH or PTHrP, ACTH or CRH, IGF-2 or insulin). These hormones exert their expected hormonal functions. However, they cause undesirable clinical manifestations (SIADH, hypercalcemia, Cushing's syndrome, and hypoglycemia). These disorders are known as **paraneoplastic endocrine syndromes**.

The functions of hormones are mediated by the presence of specific receptors at the target sites and the subsequent signal amplification systems (discussed in the next section).

Evaluating endocrine abnormalities requires determination of the plasma levels of the free hormone, its transport protein, and secondary and tertiary stimulation or inhibition by the hormones. In addition, provocative tests are used to test functioning of overall hormone secretion at different levels (e.g., pituitary, hypothalamus). All of these aspects are discussed in the following chapters.

Mechanism of Hormone Action and Signal Transduction

A target cell is equipped with specific receptors that enable it to recognize and bind a hormone selectively and preferentially. Usually, there is one type of receptor for a hormone, but in some cases more than one may exist. For example, α_1-, α_2-, β_1-, and β_2-adrenergic receptors are available for catecholamines. The number of receptors for a given hormone in a given cell varies from about 10,000 to more than 100,000.

Hormone recognition (and binding) by a receptor initiates a chain of intracellular events that ultimately lead to the effect of the hormone. Binding can be described as a lock-and-key interaction, with the hormone serving as the key and the receptor as the lock. The structural attributes of a hormone allow it to bind to its receptor and to unlock the expression of receptor function. The receptor, which frequently is a hormone-dependent regulatory protein, is functionally coupled to key enzyme systems in the cell, such that hormone binding initiates a receptor-mediated activation of enzymatic reactions, or it is functionally coupled to a region on chromatin, such that hormone binding initiates expression of one or more structural genes. Stated in another way, a hormone-responsive cell is programmed to carry out certain functions when a sensor (receptor) receives the appropriate signal (hormone).

In general, the number of receptors for a hormone determines how well the cell responds to that hormone. The following factors influence the number of receptors in a cell:

1. The genotype of the cell—determines whether the cell is capable of receptor synthesis and, if so, how much and of what type. (Some endocrinopathies involve receptor deficiency or a defect in receptor function.)
2. The stage of cellular development.

3. Hormones themselves acting as important regulators of the number of receptors in a cell; this regulation may be homologous or heterologous.

Homologous regulation occurs when a hormone affects the number of its own receptors. **Downregulation** (decrease in receptor number) is seen when a target cell is exposed to chronically elevated levels of a hormone. Downregulation involves a decrease in receptor synthesis and is a means by which a cell protects itself against excessive hormonal stimulation. Insulin, the catecholamines, GnRH, endogenous opiates, and epidermal growth factor downregulate their receptors. **Upregulation** (increase in receptor number) also occurs; prolactin, for example, upregulates its receptors in the mammary gland.

Heterologous regulation is more widespread and is a mechanism by which certain hormones influence the actions of other hormones. Some hormones diminish the production of receptors for another hormone, thereby exerting an antagonistic effect. Growth hormone, for example, causes a reduction in the number of insulin receptors. More prevalent, however, is augmentation of receptor number by a heterologous hormone. Estrogen, for example, increases the number of receptors for progesterone, oxytocin, and luteinizing hormone, while thyroid hormone increases the number of β-adrenergic receptors in some tissues.

Hormone receptors can be classified into three types on the basis of their locations in the cell and the types of hormone they bind:

1. Nuclear receptors, which bind triiodothyronine (T_3) after it enters the cell;
2. Cytosolic receptors, which bind steroid hormones as they diffuse into the cell; and
3. Cell surface receptors, which detect water-soluble hormones that do not enter the cell (peptides, proteins, glycoproteins, catecholamines). The mechanism of action of each of these receptor types is different because each is associated with different post-receptor events in the cell.

The transduction of hormone signals into cellular responses consists of second messenger systems, covalent modifications of proteins and lipids, and signal amplification systems with **interactive communications (cross-talk)**.

TYPES OF HORMONE RECEPTORS

Nuclear Receptors

Receptors for thyroid hormone (TRs), 1,25-dihydroxyvitamin D (VDRs), and retinoic acid (RARs) are called **nuclear receptors** because they are located in the nucleus

already bound to DNA (nuclear chromatin), even in the absence of their respective hormone or ligand. These nuclear receptors closely resemble the steroid hormone receptors and belong to the same "superfamily" of DNA-binding proteins. The similarities among the members of this superfamily are striking, particularly in their DNA recognition domains and in the corresponding receptor recognition segment of DNA. The receptor molecule consists of three domains: a carboxy terminal that binds the ligand (hormone/vitamin); a central DNA-binding domain (DBD); and an amino terminal that may function as a gene enhancer. The ligand-binding domain has high specificity for the ligand, and is the trigger that initiates receptor-mediated regulation of gene expression by the ligand. The receptor is bound to DNA by two "zinc fingers" (Chapter 24) in the DBD, which associates with a specific nucleotide sequence in DNA called the **response element** for the ligand. The response element is usually located "upstream" (at the $5'$ end) of a promoter for the gene that is regulated by the ligand, such that binding of the ligand to the receptor activates the response element (via the DBD), which then activates (transactivates) transcription of the gene. Transactivation of gene transcription involves the binding of RNA polymerase II to the promoter region of the gene and construction of an RNA transcript (hnRNA) from the DNA gene sequence (Figure 28.1). The hnRNA is then spliced to yield mature mRNA (messenger RNA), which translocates to the cytoplasmic compartment and becomes associated with ribosomes. Ribosomal translation of the mRNA results in the synthesis of a nascent polypeptide, the primary sequence of which is encoded in the gene that was activated by the ligand.

Steroid Hormone Receptors

Steroid hormone receptors belong to a large family of DNA-binding proteins that include receptors for thyroid hormone, retinoic acid, and 1,25-dihydroxyvitamin D. There are specific receptors for glucocorticoids (GRs), mineralocorticoids (MRs), estrogens (ERs), androgens (ARs), and progestogens (PRs), all of which are coded for by different genes. Unlike the TRs, there appears to be only one functional receptor protein for a given steroid that is encoded on a single gene. Like the TRs, the steroid receptor is a single protein with three regions (domains): a carboxy terminal region that binds the steroid specifically; a central DBD; and an amino terminal region that may function as a gene enhancer. The DBD of the steroid receptor projects two zinc fingers, which allow recognition of, and binding to, the hormone response element (HRE) of DNA.

A generic depiction of the mechanism for steroid hormone activity at a target cell is shown in Figure 28.2. The

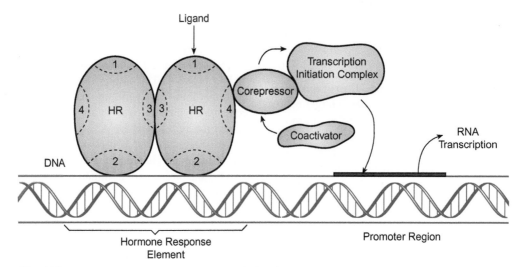

FIGURE 28.1 Thyroid hormone, 1,25-dihydroxyvitamin D, and retinoic acid receptor regulation of transcription. The hormone receptor (HR) is dimerized at site (3) and is bound to DNA at hormone response element site (2). Without the ligand, transcription is inactive due to the interaction of HR with corepressor at site (4). When the ligand (hormone) binds to HR, the bound corepressor dissociates, leading to an interaction between the coactivator and HR. These regulatory changes result in increased transcription.

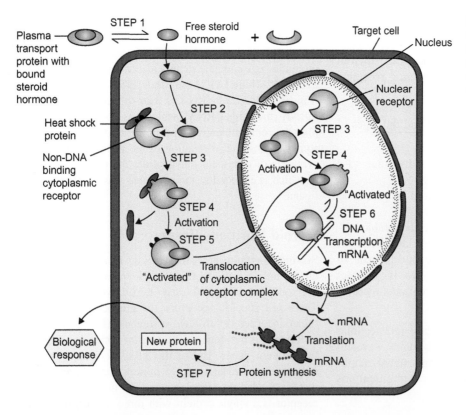

FIGURE 28.2 A generic depiction of the mechanism of steroid hormone action (see text for details). *[Reprinted with permission from A. W. Norman, G. Liwack, Hormones, 2nd ed. San Diego, CA: Academic Press, 1997, p. 40.]*

first step consists of dissociation of the hormone from the plasma transport protein and entry into the cell by diffusion across the plasma membrane. In the second step, the hormone binds with the receptors located either in the cytoplasm or in the nucleus. Receptors for glucocorticoid and aldosterone are found in the cytoplasm, and receptors for estrogen and progesterone are found in the nucleus.

Recall that receptors for thyroid hormone, retinoic acid, and 1,25-dihydroxyvitamin D are all located in the nucleus.

Activation of the receptor occurs in steps 3 and 4. The receptors exist in cells complexed with a dimer of heat-shock protein (HSP, 90 kDa), which conserves the functional status of the receptor molecule in the absence of the hormone. In addition, HSPs may function to prevent the unoccupied receptor from binding to DNA by masking the DNA-binding domain. HSPs bound to the hormone-binding domain of the receptor are displaced by the hormone which the receptor is specific for. Once formed, steroid−receptor complexes aggregate into homodimers, and in this form the receptor is said to be "activated." The activated cytoplasmic receptors are translocated via a nucleopore to the nucleus (step 5). In step 6, the activated receptor triggers (or inhibits) transcription of the hormone-regulated gene. In the event of mRNA production, the newly synthesized mRNA is translocated to the cytoplasm where it is translated (step 7). The biological effect of the hormone reflects the function of the protein (e.g., enzyme, transporter, hormone, receptor), and the duration of the effect is a function of the half-life of the mRNA (minutes to hours) and of the protein (hours to days).

The steroid receptor homodimer may also regulate gene expression by interacting with transcription factors that are associated with either the HRE or the promoter. By binding to certain transcription factors, the activated hormone receptor may

1. Stimulate gene expression by removing an inhibitory transcription factor;
2. Stimulate gene expression by facilitating the activation of the promoter by a transcription factor; or
3. Inhibit gene expression by removing a stimulatory transcription factor.

One important transcription factor that steroid receptors interact with is AP-1 (accessory/auxiliary/adapter protein-1), which promotes the expression of genes involved in cell proliferation. The growth-suppressing effect of glucocorticoids is probably mediated by the binding of the hormone−glucocorticoid receptor (GR) complex to one component of AP-1 (c-jun) which, in effect, removes the AP-1−mediated promotion of cell proliferation. Another transcriptional factor involved in steroid hormone action is the **cAMP response element binding protein** (CREB), which mediates the genomic effects of peptide hormones that utilize the cAMP second messenger system (see later). A number of steroid-regulated genes are subject to modulation by CREB and are thus subject to the regulatory influence of both steroid and peptide hormones.

CELL SURFACE RECEPTORS

Water-soluble hormones (hormonal amines, polypeptides, and proteins) are unable to penetrate the lipid matrix of the cell membrane and thus initiate their actions by binding to cell surface receptors on target cells. Three known

receptor superfamilies are coupled to a distinct set of intracellular post-receptor events. They are

1. G-protein-coupled receptors (GPCRs);
2. Receptors that initiate intracellular tyrosine kinase or guanylyl cyclase activities;
3. Ligand-gated oligomeric ion channels regulated by neurotransmitters.

G-Protein-Coupled Receptors and G-Proteins

GPCRs belong to a large family of receptor proteins, and each of these consists of seven transmembrane helixes. The intracellular C-terminal segment of a GPCR interacts with heterotrimeric GTP-binding proteins (G-proteins) to initiate signal transduction when the hormone is bound to a GPCR. Trimeric G-proteins consist of α-, β-, and γ-subunits. The specificity of the G-proteins resides in the α-subunit; their various physiologic functions are listed in Table 28.1. In two of these cases, an intracellular second messenger amplification system is coupled to a hormone−GPCR signal transduction unit. One α-subunit is coupled to the adenylate cyclase−cAMP system, and the other α-subunit is associated with the phospholipase C-phosphatidylinositol-Ca^{2+} pathway (IP_3 pathway).

Heterotrimeric G-Protein-Coupled Adenylate Cyclase−cAMP System

The effect of ligand binding to specific receptors mediated by heterotrimeric G-proteins on adenylate cyclase is shown in Figure 28.3. The G-proteins belong to a family of regulatory proteins, each of which regulates a distinct set of signaling pathways. The cell membrane receptors coupled to intracellular G-proteins share an α-helical serpentine structure spanning seven transmembranes. The α-helical segments are linked by alternating intracellular and extracellular peptide loops with the N-terminal region located on the extracellular side and the C-terminal region located on the intracellular side (Figure 28.3).

Extracellular signals detected by the membrane receptors are diverse and include hormones, growth factors, neurotransmitters, odorants, and light (Table 28.1). The next stage in the propagation of extracellular stimuli to the intracellular events via the GPCR-G-protein signal transduction unit involves adenylate cyclase, which can have either a stimulatory or an inhibitory effect on the adenylate cyclase (Figure 28.3). Binding of an appropriate ligand to R_s stimulates the adenylate cyclase, whereas binding to R_i inhibits the enzyme. The G-proteins are heterotrimeric, consisting of an α-subunit and a tightly coupled β,γ-subunit. The α-subunit is the unique protein in the trimer. It possesses sites for interaction with cell membrane receptors and the β,γ-subunit, intrinsic GTPase

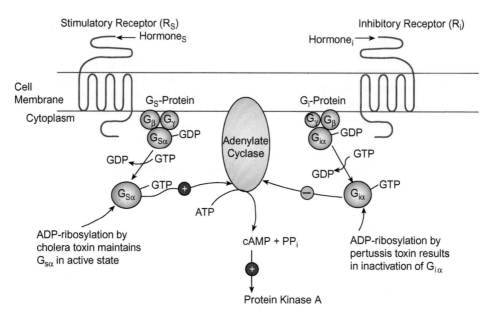

FIGURE 28.3 Dual control of adenylate cyclase activity by guanine nucleotide-binding proteins (G). Subscripts s and i denote stimulatory and inhibitory species, respectively. Plus and minus indicate activation and inactivation, respectively.

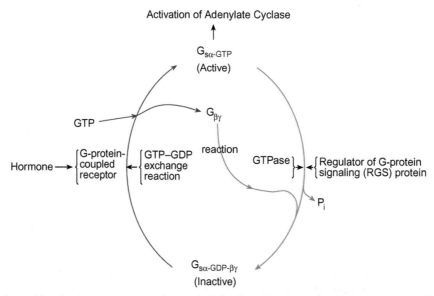

FIGURE 28.4 Activation and inactivation (or return to resting state) of signal transducer trimeric stimulatory G-protein (G_s). Hormone binding to membrane G-protein-coupled stimulatory receptor causes a GTP and GDP exchange reaction in $G_{s\alpha}$ and dissociation of $G_{\beta\gamma}$. $G_{s\alpha\text{-}GTP}$ activates adenylate cyclase and initiates the signal transduction pathway. The return of $G_{s\alpha\text{-}GTP}$ to the resting state is mediated by the intrinsic GTPase activity of the $G_{s\alpha}$ subunit and also through promotion of GTPase activity by the regulator of G-protein signaling (RGS) protein. Activation of G_i protein occurs by analogous reactions of G_s when a hormone binds to its corresponding inhibitory receptor. This leads to the formation of $G_{i\alpha\text{-}GTP}$, which inhibits adenylate cyclase. Gain in function or loss in function can result from mutations in genes coding for receptor protein, $G_{s\alpha}$, $G_{i\alpha}$, or RGS protein.

activity, and ADP ribosylation. In the unstimulated state, the G-protein is present in the heterotrimeric (α,β,γ-) form with GDP tightly bound to the α-subunit. Upon activation of the receptor by the bound ligand, GDP is exchanged for GTP on the α-subunit, followed by dissociation of $G_{s\alpha\text{-}GTP}$ from the β,γ-subunit. The α-subunit's G_s and G_i are designated $G_{s\alpha}$ and $G_{i\alpha}$. The G-protein subunits undergo post-translational covalent additions of

lipids such as palmityl, farnesyl, and geranyl groups. These hydrophobic groups linked to G-protein subunits are necessary for proper anchoring in the cell membrane.

Figure 28.4 shows the cyclic functioning of the G_s protein. Binding of a hormone to R_s permits the binding of GTP to $G_{s\alpha}$ to form a $G_{s\alpha\text{-}GTP}$ complex and the release of the G_β- and G_γ-subunits. The $G_{s\alpha\text{-}GTP}$ complex activates adenylate cyclase, which remains active as long as GTP is bound to $G_{s\alpha}$. Binding of GTP to $G_{s\alpha}$ also activates a GTPase intrinsic to $G_{s\alpha}$ that slowly hydrolyzes the bound GTP, eventually allowing reassociation and formation of the trimer and inactivation of adenylate cyclase. A similar series of events is initiated by binding of a ligand to R_i. In this instance, dissociation of the G_i trimer by formation of $G_{i\alpha\text{-}GTP}$ inhibits adenylate cyclase, allowing the β,γ-subunit to bind to $G_{s\alpha}$, thus preventing $G_{s\alpha}$ from activating adenylate cyclase. GTP hydrolysis by GTP-activated $G_{i\alpha}$-GTPase allows reformation of the G_i trimer. GDP is displaced from $G_{i\alpha}$ by binding of another molecule of GTP with reactivation of $G_{i\alpha}$. The inhibitory effect occurs because the target cell contains much more G_i than G_s. $G_{i\alpha}$ may also interact with membrane calcium transport and the phosphatidylinositol pathway. The function of $\beta-\gamma$ dimers is not completely understood, but they may participate in the regulation of downstream effectors in the signaling pathway.

The G-protein-coupled receptor signaling pathway is regulated by several different mechanisms. Negative regulation of the signaling pathway can occur at both the receptor and G-protein levels. Receptor phosphorylation by G-protein-coupled receptor kinase and **arrestin** proteins-binding can lead to the **inability** of the ligand to activate the signaling systems (e.g., the vision-enabling rhodopsin and β-adrenergic receptor$-$G-protein transducing system). Negative regulation also occurs after the activation of G-protein by promoting the GTPase activity of $G_{s\alpha}$. This is achieved by regulators of G-protein signaling (RGS) proteins, which bind to $G\alpha$ and promote GTPase activity. This process of converting $G\alpha$ from an active state to an inactive state is rapid. Transcriptional regulation and post-translational modification of G-proteins provide other means for regulation of the G-protein-coupled signaling pathway.

Both $G_{s\alpha}$ and $G_{i\alpha}$ contain sites for NAD-dependent ADP ribosylation. Cholera toxin ADP-ribosylates a specific arginine side chain of $G_{s\alpha}$ and maintains it in a permanently active state, while islet-activating protein, one of the toxins in *Bordetella pertussis*, ADP-ribosylates a specific cysteine side chain of $G_{i\alpha}$, permanently blocking its inhibitory action because it maintains $G_{i\alpha}$ in the GDP-bound inactive form. Thus, both toxins stimulate adenylate cyclase activity and lead to excessive production of cAMP. Cholera toxin causes a diarrheal illness (Chapter 11), while toxins of *B. pertussis* cause whooping cough, a respiratory disease affecting ciliated bronchial epithelium.

Response of the adenylate cyclase system to a hormone is determined by the types and amounts of various constituent proteins. Cyclic AMP production is limited by the amount of adenylate cyclase present. When all the adenylate cyclase is fully stimulated, further hormone binding to R_s cannot increase the rate of cAMP synthesis. In cells having many different R_s (adipocytes have them for epinephrine, ACTH, TSH, glucagon, MSH, and vasopressin), maximal occupancy of the receptors may not stimulate cAMP production beyond that which can be achieved by full occupancy of only a few of the receptor types. Therefore, the greatest stimulation that can be achieved by a combination of several hormones will not be simply the sum of the maximal effects of the same hormones given singly. A hormone's ability to stimulate cAMP production may depend on the cell type. For example, epinephrine causes large increases in cAMP concentration in muscle but has relatively little effect on liver. The opposite is true for glucagon (see Chapter 14). Within a particular cell type, destruction of one type of R_s does not alter the response of the cell to hormones that bind other stimulatory receptors.

In prokaryotic cells, cAMP binds to catabolite regulatory protein (CAP), which then binds to DNA and affects gene expression (Chapter 24). In eukaryotic cells, cAMP binds to cAMP-dependent protein kinase, which contains two regulatory (R) and two catalytic (C) subunits. On binding of cAMP, the catalytic subunits separate and become active:

$$\underset{\text{(inactive)}}{R_2C_2} + 4\,cAMP \rightleftharpoons 2(R\text{-}2\,cAMP) + \underset{\text{(active)}}{2C}$$

They then catalyze ATP-dependent phosphorylation of serine and threonine residues of various cell proteins, often altering the activities of these proteins. Some, such as phosphorylase kinase, become activated, whereas others, such as glycogen synthase, are inactivated (Chapter 14). Cyclic AMP-dependent protein kinase remains active (while intracellular cAMP concentration, controlled by the relative rates of cAMP synthesis by adenylate cyclase and of degradation by phosphodiesterase, is elevated above basal levels). The reaction catalyzed by the phosphodiesterase is

$$3',5'\text{-}cAMP + H_2O \rightarrow 5'\text{-}AMP$$

As cAMP concentration decreases, reassociation of catalytic and regulatory subunits of the kinase occurs. The cAMP system also derives some of its specificity from the proteins that are substrates for cAMP-dependent protein kinase. It is an indirect regulatory system, since cAMP modulates the activity of cAMP-dependent protein kinase, and the kinase, in turn, affects the activities of a variety of metabolic enzymes or other proteins. This indirectness is the basis for amplification of the hormonal signal. Activation of adenylate cyclase by binding of a hormone molecule to a receptor causes formation of many molecules of cAMP and allows activation of many molecules of the protein kinase, each of which in turn can phosphorylate many enzymes or other proteins. This "amplification cascade" accounts in part for the extreme sensitivity of metabolic responses to small changes in hormone concentrations. Finally, the response to increased cAMP concentrations within a cell usually involves several metabolic pathways and a variety of enzymes or other proteins.

Abnormalities in Initiation of G-Protein Signal

Two well-studied bacterial toxins, cholera toxin and pertussis toxin, disturb normal functioning of $G\alpha$ proteins by ADP ribosylation of specific amino acid residues (discussed earlier). Germline and somatic mutations in the genes coding for G-proteins can cause endocrine disorders in several ways. Mutations can be either activating (gain-of-function) or inactivating (loss-of-function). Examples of gain-of-function mutations that result in the ligand-independent activator of $G_{s\alpha}$ include some adenomas of pituitary and thyroid, some adenomas of adrenals and ovary, and syndromes of hyperfunction of one or more endocrine glands. In most of these disorders, a point mutation alters an amino acid in $G_{s\alpha}$, leading to inhibition of the GTP hydrolysis required for the termination of activity of $G_{s\alpha}$. An example of autonomous hyperhormone secretion and cellular proliferation of several endocrine glands is **McCune−Albright syndrome**. In this disorder, two missense mutations result in the replacement of $Arg^{201} \rightarrow His$ or $Arg^{201} \rightarrow Cys$ in the $G_{s\alpha}$ of affected endocrine glands. Some of the clinical manifestations of McCune−Albright syndrome include hyperthyroidism, polyostotic fibrous dysplasia, precocious puberty, and hyperpigmented irregularly shaped skin lesions (known as **café-au-lait lesions**). The cutaneous hyperpigmentation is caused by hyperfunctioning of melanocyte-stimulating hormone (MSH) receptors coupled to $G_{s\alpha}$ in the melanocytes.

An example of decreased function due to impaired activation or loss of $G_{s\alpha}$ and resistance to hormone action is pseudohypoparathyroidism (**Albright's hereditary osteodystrophy**). Patients with this syndrome exhibit generalized resistance to the action of those hormones dependent on $G_{s\alpha}$ for their function, despite elevated serum hormone levels. Several dysmorphic features are also characteristic of these patients.

G-protein abnormalities may also arise from mutations in genes encoding proteins that regulate G-protein signaling. Several RGSs have been identified. RGS proteins bind to G_α and accelerate the hydrolysis of GTP. The physiological responses of G-proteins regulated by RGSs are fast and include vagal slowing of the heart rate (G_i), retinal detection of photones (G_i), and contraction of vascular smooth muscle (G_q).

G-Protein-Coupled Phosphatidylinositol−Ca^{2+} Pathway

The calcium system is more complex than the cAMP system and contains a variety of mechanisms for transducing the Ca^{2+} signal into changes in cellular function. In a resting cell, the pool of cytosolic calcium is very small, and most of the intracellular Ca^{2+} is bound in mitochondria or the endoplasmic reticulum, or is bound to the plasma membrane. The cytosolic calcium ion concentration is $\sim 0.1\,\mu mol/L$, in contrast to the high concentration ($\sim 1000\,\mu mol/L$) outside cells. This 10,000-fold gradient is maintained by the plasma membrane, which is relatively impermeable to calcium, by an ATP-dependent Ca^{2+} pump (which extrudes Ca^{2+} in exchange for H^+) in the plasma membrane, and by "pump-leak" systems in the endoplasmic reticulum and inner mitochondrial membranes. A change in the permeability of any of these membranes alters the calcium flux across the membrane and causes a change in cytosolic calcium concentration. This change occurs in the calcium-mediated second messenger system. Plasma membranes regulate cytosolic calcium concentration by their role as a diffusion barrier and by containing the ATP-dependent Ca^{2+} pump and receptors for extracellular messengers. Activation of some of these receptors increases the Ca^{2+} permeability of the plasma membrane and thus the level of cytosolic calcium. In other cases, the activated receptor produces one or more second messengers that increase cytosolic Ca^{2+} concentration from the calcium pool in the endoplasmic reticulum. Calcium then becomes a "third messenger." The endoplasmic reticulum is the source of Ca^{2+} in the initial phase of cell activation by some hormones in many systems. Since the plasma membrane can bind Ca^{2+}, the rise in cytosolic calcium following binding of some extracellular messengers may be due to release from this intracellular calcium pool.

Mechanism of the Calcium Messenger System

Most extracellular messengers that cause a rise in cytosolic Ca^{2+} concentration also increase the turnover of phosphoinositides in the plasma membrane. The pathway is initiated by the binding of extracellular messengers to specific receptors located on the plasma membrane and activating G-proteins (e.g., G_q). The activated $G_{q\alpha \cdot GTP}$-protein, in turn, activates phospholipase C, which catalyzes hydrolysis of phosphatidylinositide 4,5-bisphosphate, initiating the intracellular second-messenger effects (Figure 28.5). The action of phospholipase C results in the formation of two products, inositol 1,4,5-trisphosphate (IP_3) and diacylglycerol (DAG), which function as intracellular messengers (Figure 28.6).

IP_3 causes a rise in cytosolic Ca^{2+} by opening the channels of the endoplasmic reticulum, which contains the

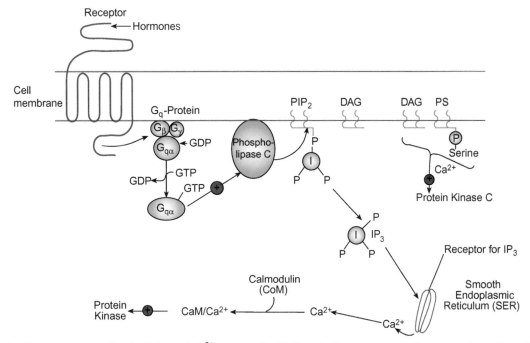

FIGURE 28.5 G_q protein-coupled phosphatidylinositol-Ca^{2+} pathway. The binding of a hormone at a specific receptor site results in the activation of G-protein, which, in turn, activates phospholipase C via $G_{q\alpha\text{-}GTP}$-protein. The action of phospholipase C on phosphatidylinositol 4,5-bisphosphate (PIP_2) yields inositol trisphosphate (IP_3) and diacylglycerol (DAG), which, along with phosphatidylserine (PS), activates protein kinase C. IP_3 binds to receptors on smooth endoplasmic reticulum (SER), releasing Ca^{2+}, which, in turn, activates another set of protein kinases. $\oplus$ = Activation.

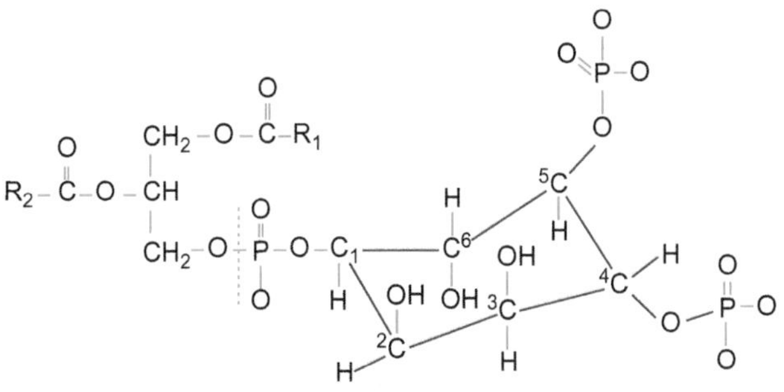

FIGURE 28.6 Structure of phosphatidylinositol 4,5-bisphosphate. Hydrolysis at the dotted line is catalyzed by phospholipase C and releases diacylglycerol and inositol 1,4,5-triphosphate, which are the intracellular second messengers.

intracellular reservoir of calcium. The cytosolic Ca^{2+} can also be increased by opening the channels of the plasma membrane mediated by cAMP. Ca^{2+} complexed to its binding proteins (e.g., calmodulin) diacylglycerol and phosphatidylserine activates protein kinase C. Active protein kinase C phosphorylates serine and threonine residues of downstream protein kinases of the signaling pathway to produce eventual cellular responses. Protein phosphatases convert the active kinases to inactive kinases. Diacylglycerol released due to the phospholipase C action usually contains arachidonic acid in the second position and is a source of synthesis of eicosanoids. Membrane phospholipids also provide arachidonate due to the action of phospholipase A_2 (discussed in Chapter 16).

Tumor-promoting phorbol esters such as 12-O-tetradecanoylphorbol-13-acetate (an ingredient in croton oil that promotes skin tumor production by carcinogens) can directly activate protein kinase C, mimicking DAG and bypassing the receptors. When activated by phorbol esters, protein kinase C phosphorylates cytosolic myosin light-chain kinase (Chapter 19) and other proteins, particularly those related to secretion and proliferation.

Monomeric Guanine Nucleotide-Binding Proteins with GTPase Activity

Monomeric G-proteins, like the trimeric G-proteins discussed previously, also propagate incoming messages transmitted through receptors to signal amplification systems. Several distinct families of monomeric G-proteins have been identified. They include products of the **Ras, Rho,** and **Rab gene families**, and elongation factor EF-Tu required for protein synthesis (Chapter 23). The Ras and Ras-like proteins are anchored to the inner surface of the plasma membrane via the lipid moieties of farnesyl or geranylgeranyl groups attached to their C-terminus (Figure 28.7). They consist of 129 amino acid residues and have a molecular weight of 23 kDa. These monomeric G-proteins function as molecular switches, and their activities are turned on when bound to GTP and turned off when the GTP is hydrolyzed to GDP due to G-protein's intrinsic GTPase activity. The GDP–GTP exchange and GTP

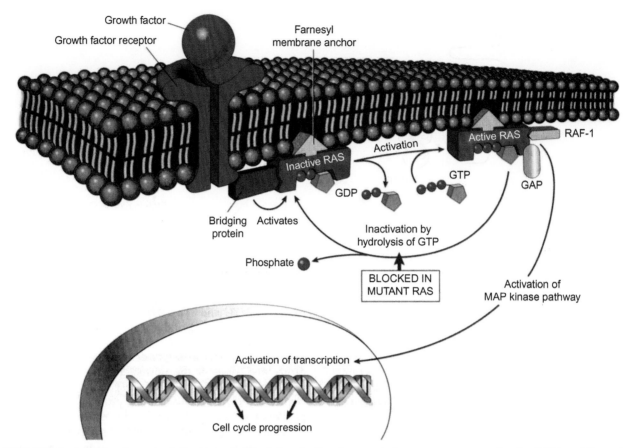

FIGURE 28.7 Activation of monomeric Ras G-protein. Stimulation of cell-surface growth factor receptors converts inactive GDP bound Ras to the activated GTP bound state. Activated Ras recruits downstream signaling pathways involving several proteins with the mitogen-activated protein (MAP) kinase pathway. *[Reproduced with permission from V. Kumar, A. K. Abbas, N. Fausto, Robbins and Cotran pathologic basis of disease, 7th ed. Philadelphia, Elsevier and Saunders, 2005, p. 297, Figure 7-32.]*

hydrolysis are regulated by auxiliary proteins. There are many functions of Ras and Ras-like proteins when activated by appropriate stimuli. They include cell proliferation, differentiation, and apoptosis.

Some tumor-promoting viruses contain monomeric G-proteins that show homology with mammalian G-proteins. These genes are known as **proto-oncogenes**, with normal growth and differentiation functions. An example of G-protein found in the rat sarcoma virus has been named the Ras gene. Mutations in the proto-oncogenes convert them to oncogens, which can lead to cancer. Impaired GTPase activity of G-proteins due to their mutations or to mutations in any of the auxiliary regulatory proteins can potentially lead to cancer.

Receptors That Initiate Tyrosine Kinase (TK) Activity

Both of these kinases can either be receptor bound, can be found in the cytosol, or can occur in other cytoplasmic organelles of the cell. TK receptors (TKR) are transmembrane proteins, and they possess binding sites at their extracellular domains for polypeptide hormones and growth factors (ligands). When the ligands bind to their specific TKRs, their catalytic intracellular kinase domain becomes active and mediates appropriate biological responses. Insulin receptor is an example of a TKR.

Receptors for Insulin and Growth Factors

Insulin exerts its effects by altering the state of phosphorylation of certain intracellular enzymes by a mechanism that does not involve cAMP but that requires specific binding to surface receptors with tyrosine kinase activity. Insulin exerts acute (minutes), delayed-onset (hours), and long-term (days) effects entirely through a single receptor. The insulin receptor is a transmembrane heterotetramer consisting of one pair each of two dissimilar subunits linked by disulfide bridges (discussed in Chapter 20).

Receptors for Growth Hormone and Prolactin

Growth hormone (GH) and prolactin (PRL) belong to the "helix bundle peptide" (HBP) hormone family, which includes human placental lactogen (hPL), erythropoietin

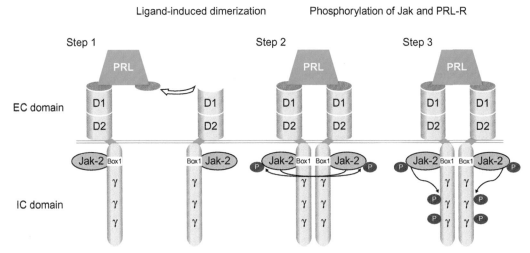

FIGURE 28.8 Mechanism of prolactin receptor activation. Step 1: Activation of prolactin receptor consists of ligand-induced sequential receptor homodimerization driven by the two binding sites of prolactin. Step 2: In the intracellular domain of the homodimer of the ligand−receptor complex, a tyrosine kinase (known as Janus kinase 2 [Jak-2]) is activated. Step 3: Jak-2 kinase causes autophosphorylation and phosphorylation of the receptor. γ = tyrosine residues, sites of phosphorylation, D = domains, IC = intracellular, EC = extracellular. *[Reproduced with permission from M. E. Freeman, B. Kanyicska, A. Lerant, G. Nagy, Prolactin: structure, function, and regulation of secretion, Physiological Rev. 80 (2000) 1530.]*

(EPO), and many interleukins and cytokines. All members of this family have similar mechanisms of action; they utilize membrane receptors that are straight-chain glycoproteins with extracellular, transmembrane, and intracellular domains. The membrane receptors for this group of hormones undergo dimerization when bound with their respective receptors. Each dimerized receptor is bound to one molecule of hormone. The intracellular events of the ligand-mediated activation of the receptor consist of activation of a constitutively associated tyrosine kinase known as **Janus kinase 2** (**Jak-2**). Jak-2 transphosphorylates itself and phosphorylates the intracellular domains of the receptor. Figure 28.8 illustrates the mechanism of prolactin receptor activation.

The signal transduction of Janus kinases is mediated via the phosphorylation of signal transducer and activator transcription (STAT) proteins. Thus, the primary substrates for JAKs are intracytoplasmic STAT proteins that eventually induce the transcription of target genes. Specific phosphatases render the system inactive, serving as an off switch for gene expression. Mutations causing either gain-of-function or loss-of-function of JAKs or STAT proteins can lead to serious metabolic consequences such as cancer. For example, the gain-of-function mutation of substitution of phenylalanine to valine (V617F) in the specific domain known as JH2 of Jak-2 makes the enzyme constitutively active, leading to disorders of polycythemia vera and myeloproliferative diseases (Chapter 26). Loss-of-function of a STAT protein results in growth hormone insensitivity, thus affecting growth-promoting and metabolic actions of growth hormone (Chapter 29).

Nonreceptor Tyrosine Kinases

Nonreceptor kinases lack transmembrane domains and are found in the intracellular parts of the cell. An example of a nonreceptor tyrosine kinase is the product of proto-oncogene c-ABL, which regulates cell survival and proliferation. Aberrant c-ABL tyrosine kinase activation occurs due to the translocation of the c-ABL proto-oncogene located at the distal tip of chromosome 9 into the breakpoint cluster region (BCR) of chromosome 22, resulting in the chimeric gene BCR-ABL. This results in the hematologic cancer **chronic myelogenous leukemia**. A new strategy of targeted therapy is used in the treatment of these cancers instead of traditional cytotoxic drugs. An orally active tyrosine kinase inhibitor, imatinib, is used in the treatment of chronic myelogenous leukemia and other related cancers (e.g., gastric intestinal stromal tumors). Multikinase inhibitors (e.g., sorafenib) of multiple overactive molecular pathways are used in the treatment of some cancers (e.g., hepatocellular carcinoma).

Monoclonal antibodies against receptor tyrosine kinases or its ligands have been used as targeted therapies in the treatment of some cancers. An example is inhibition of overamplified HER-2 receptor tyrosine kinase in breast cancer with humanized monoclonal antibody (Chapter 32).

Signal Transduction Mediated via the Guanylyl Cyclases

Guanylyl cyclase catalyzes the conversion of GTP to cGMP (guanosine 3′-5′-phosphate). The biological effects are initiated by cGMP binding to downstream signaling molecules. Guanylyl cyclase can either be plasma membrane bound or present in the cytosol. The membrane-bound enzyme has a hormone recognition extracellular membrane recognition site and a cytosolic domain with guanylyl cyclase activity. Natriuretic peptides (ANP and BNP) released from the heart in response to myocardial stretch due to increased blood volume cause natriuresis and vasodilation by binding to receptor guanylyl cyclase (Chapter 37). An example of activation of cytosolic guanylyl cyclase, which is a heme-containing enzyme, is nitric oxide (Chapter 15).

ENRICHMENT READING

[1] A.M. Spiegel, G proteins—the disease spectrum expands, N. Engl. J. Med. 368 (2013) 2515–2516.

[2] M.A. Nesbit, F.M. Hannan, S.A. Howles, V.N. Babinsky, R.A. Head, T. Cranston, et al., Mutations affecting G-protein subunit α_{11} in hypercalcemia and hypocalcemia, N. Engl. J. Med. 368 (2013) 2476–2486.

[3] J.J. O'Shea, S.M. Holland, L.M. Staudt, JAKs and STATs in immunity, immunodeficiency, and cancer, N. Engl. J. Med. 368 (2013) 161–170.

[4] W. Warsch, C. Walz, V. Sexl, JAK of all trades: JAK2-STAT5 as novel therapeutic targets in *BCR-ABL1* chronic myeloid leukemia, Blood 26 (2013) 2167–2175.

[5] J. Berlin, Beyond exon 2—the developing story of *RAS* mutations in colorectal cancer, N. Engl. J. Med. 369 (2013) 1059–1060.

[6] J. Douillard, K.S. Oliner, S. Siena, J. Tabernero, R. Burkes, M. Barugel, et al., Panitumumab-FOLFOX4 treatment and *RAS*

mutations in colorectal cancer, N. Engl. J. Med. 369 (2013) 1023–1034.

[7] A. Kikuchi, H. Yamamoto, A. Sato, Selective activation mechanisms of *Wnt* signaling pathways, Trends Cell Biol. 19 (2009) 119–129.

[8] K.M. Cadigan, M. Peifer, Wnt signaling from development to disease: insights from model systems, Cold Spring Harb. Perspect. Biol. 1 (2009) 1–23 a002881.

[9] R.V. Amerongen, R. Nusse, Towards an integrated view of Wnt signaling in development, Development 136 (2009) 3205–3214.

[10] R. Baron, M. Kneissel, Wnt signaling in bone homeostasis and disease: from human mutations to treatments, Nat. Med. 19 (2013) 179–192.

[11] L. Wang, M.S. Lawrence, Y. Wan, P. Stojanov, C. Sougnez, K. Stevenson, et al., SF3B1 and other novel cancer genes in chronic lymphocytic leukemia, N. Engl. J. Med. 365 (2011) 2497–2506.

[12] Y. Ge, J. Chen, Mammalian target of rapamycin (mTOR) signaling network in skeletal myogenesis, J. Biol. Chem. 287 (2012) 43928–43935.

[13] M.M. Awad, R. Katayama, M. McTigue, W. Liu, Y. Deng, A. Brooun, et al., Acquired resistance to crizotinib from a mutation in *CD74-ROS1*, N. Engl. J. Med. 368 (2013) 2395–2401.

[14] J.M. Goldman, Ponatinib for chronic myeloid leukemia, N. Engl. J. Med. 367 (2012) 2148–2149.

[15] S. Panjarian, R.E. Lacob, S. Chen, J.R. Engen, T.E. Smithgall, Structure and dynamic regulation of Abl kinases, J. Biol. Chem. 288 (2013) 5443–5450.

[16] A.T. Shaw, D. Kim, K. Nakagawa, T. Seto, L. Crino, M. Ahn, et al., Crizotinib versus chemotherapy in advanced *ALK*-positive lung cancer, N. Engl. J. Med. 368 (2013) 2385–2394.

Chapter 29

Endocrine Metabolism II: Hypothalamus and Pituitary

Key Points

1. The endocrine systems of the hypothalamus and pituitary (hypophysis) have an integrated anatomical and functional relationship.

2. The afferent signals received by the hypothalamus are transmitted to the pituitary by two hypothalamic–pituitary axes: (1) portal venous blood circulation, which transports hypothalamic (hypophysiotropic) hormones to target sites at the anterior lobe of the pituitary, and (2) hypothalamic nerve tracts in which neuro-hormones are synthesized and then stored in the posterior lobe of the pituitary.

3. The hypophysiotrophic hormones are all peptides, except for dopamine, which is synthesized from tyrosine. These hormones either inhibit or stimulate the synthesis and release of hormones from specific cells of the anterior pituitary. The pituitary hormones belong to three families: somatomammotropin (GH and PRL), the glycoproteins (LH, FSH, and TSH), and opiomelanocortin (ACTH, β-endorphin, and related peptides).

4. The anterior pituitary hormones regulate growth (GH), milk production (PRL), and secretions of hormones from the adrenal glands (Chapter 30), thyroid gland (Chapter 31), and the glands of the reproductive system (Chapter 32).

5. The neurohypophyseal hormones, antidiuretic hormone (ADH), and oxytocin regulate water balance (Chapter 37) and milk production (Chapter 32), respectively.

6. The brain contains several peptides that share homology with gastrointestinal hormones and affect eating behavior (Chapters 11 and 20).

7. The perception of pain in the CNS is regulated by peptides known as endogenous opioids; they are derived from post-translational processing of three prohormones: pro-opiomelanocortin (POMC), proenkephalin A, and proenkephalin B.

8. Growth hormone (GH) is synthesized in pituitary somatotroph cells and is the most abundant of the anterior pituitary hormones. It is a nonglycosylated globular protein and consists of 191 amino acid residues (M.W. 22 kDa). Its secretion is pulsatile and is stimulated by GHRH and ghrelin, and inhibited by somatostatin.

9. The effect of growth hormone on growth and the metabolic activity of somatic cells consists of both direct and indirect actions. The indirect actions are mediated by insulin-like growth factors (IGFs) synthesized and released from the liver.

10. The action of growth hormone at the target site is initiated by its binding to specific cell surface receptors followed by activation of intracellular Janus tyrosine kinase-signal transducing activators of transcription (JAK-STAT), leading to a protein phosphorylation cascade (Chapter 28).

11. Growth hormone deficiency leads to short stature and can occur due to defects in its synthesis by the pituitary cells, whereas growth hormone insensitivity occurs due to mutations of the GH receptor or in the STAT proteins of the signal-transducing system. Recombinant human GH is used to treat primary GH deficiency, and recombinant IGF-I is used to treat growth hormone insensitivity disorders.

12. Growth hormone–secreting tumors result in gigantism during childhood.

13. Prolactin is secreted by pituitary lactotroph cells and inhibited by the neurotransmitter dopamine originating from the hypothalamus. It is a globular protein of 199 amino acid residues. Its action at the target site involves the JAK-STAT pathway. Its functions include mammary gland development and milk production.

14. Hyperprolactinomas (e.g., pituitary lactotroph adenoma), which can cause hypogonadism, are treated with dopamine agonists, radiotherapy, or surgery.

HYPOTHALAMUS

The hypothalamus is a small region of the brain in the ventral aspect of the diencephalon. In the adult human, it is about 2.5 cm in length and weighs about 4 g. Ventromedially, it surrounds the third ventricle and is continuous with the infundibular stalk of the pituitary (hypophysis). The cone-shaped region of the hypothalamus, the **median eminence**, consists mainly of axonal fibers from hypothalamic neurons, which either terminate in the median eminence or continue down into the posterior lobe of the pituitary. It is perfused by a capillary network (primary plexus) derived from the carotid arteries.

N. V. Bhagavan and C.-E. Ha: Essentials of Medical Biochemistry, Second edition. DOI: http://dx.doi.org/10.1016/B978-0-12-416687-5.00029-4

Blood from the primary plexus is transported by portal vessels (hypophyseal portal vessels) to another capillary network (secondary plexus) in the anterior lobe of the pituitary (adenohypophysis) (Figure 29.1).

The hypothalamus contains a high density of nerve cell bodies clustered into nuclei which participate in a variety of functions, such as integrating neurons and endocrine systems. Neurons in each of these nuclei tend to send their axons to the same regions in the form of tracts. These nuclei innervate the median eminence, other hypothalamic nuclei, the posterior pituitary, and various structures in the extrahypothalamic central nervous system. Many of the hypothalamic neurons are presumably monoaminergic as they synthesize and release the neurotransmitter amines norepinephrine, serotonin, or dopamine. Many are also peptidergic as they synthesize and release neuropeptides. At least 12 hypothalamic neuropeptides have been identified.

The hypothalamus is an important integrating area in the brain. It receives afferent signals from virtually all parts of the central nervous system (CNS) and sends efferent fibers to the median eminence, the posterior pituitary, and certain areas of the central nervous system. In the median eminence, a number of peptidergic fibers terminate in close proximity to the primary plexus, which transports their neuropeptide hormones via the portal blood flow to the anterior pituitary. Since these neuropeptide hormones affect the function of the anterior pituitary cells, they are called **hypophysiotropic**. Two hypothalamic nuclei, the paraventricular nucleus (PVN) and the supraoptic nucleus (SON), consist of two populations of neurons that differ in size: the parvicellular (small-celled) neurons and the magnocellular (large-celled) neurons. The parvicellular neurons of the paraventricular nucleus produce a peptide (see later discussion of ACTH) that is transported by axoplasmic flow to the median eminence and then released into the hypophyseal portal blood. The magnocellular neurons of the paraventricular and supraoptic nuclei send their long axons directly into the posterior pituitary (neurohypophysis), where their dilated nerve endings are closely positioned next to capillaries that drain the tissue. Peptide hormones synthesized in these magnocellular neurons are transported to the neurohypophysis by axoplasmic flow and then released into the

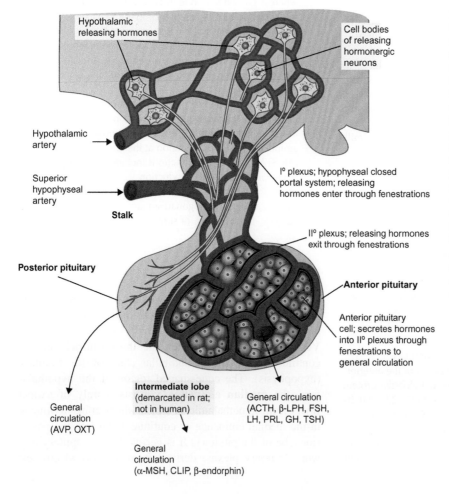

FIGURE 29.1 Hypothalamic—pituitary regulatory system. Hypothalamic releasing hormones and release-inhibiting hormones are synthesized in various neurons. These hypothalamic hormones are transported to the anterior pituitary via a portal venous system, and the anterior pituitary target cells either release or inhibit the release of specific hormones into the general circulation. The posterior pituitary hormones are synthesized and packaged in cell bodies in the hypothalamus and then are transported to nerve endings in the posterior pituitary. Afterward they are released following appropriate stimuli. The explanation for the abbreviations appears in the text. CLIP = corticotrophin-like intermediate lobe peptide; LPH = lipotrophic hormone; MSH = melanocyte-stimulating hormone. *[Slightly modified and reproduced with permission from A. W. Norman, G. Litwack, Hormones, 2nd ed., New York: Academic Press, 1997, p. 89.]*

TABLE 29.1 Hypophysiotropic Hormones of the Human Hypothalamus

Hormone	No AA^2 (M.W.)	Gene (Human)	Origin^1 (Production Site)	Functions	Mechanism of Action
Gonadotropin-releasing hormone GnRH (LHRH)	10 (1182)	8p21–q11.2	PON, OVLT	Stimulates LH and FSH secretions	$Gq_{\alpha\text{-PLC}\beta Ca^{2+}\text{-PKC}}$
Corticotropin-releasing hormone (CRH)	41 (4758)	8q13	PVN_{parv}	Stimulates POMC synthesis and processing ($>$ ACTH secretion)	$G_{s\alpha\text{-AC}}$ activation
GH-releasing hormone (GHRH)	44 (5040)	20	TIDA (arcuate, VMN)	Stimulates GH secretion	$G_{s\alpha\text{-AC}}$ activation
Somatostatin (SS) (SIRH)	14 (1638)	3q28	PeriVN, PVN_{parv}, PON	Inhibits GH secretion; also inhibits TSH secretion	$G_{i\alpha\text{-AC}}$ inhibition
Thyrotropin-releasing hormone (TRH)	3 (362)		PVNparv, PeriVN	Stimulates TSH secretion; also stimulates PRL release	$Gq_{\alpha\text{-PLC}\beta Ca^{2+}\text{-PKC}}$
Prolactin-inhibiting hormone ((PIH); dopamine)	n.a. (153)	n.a.	TIDA (arcuate, VMN)	Inhibits PRL release; also indirectly inhibits LH, FSH release	$G_{i\alpha\text{-AC}}$ inhibition (D_2)
Prolactin-releasing factor (PRF)	unknown			Stimulates PRL release	

^{1}PeriVN = periventricular nucleus; PVN_{parv} = parvicellular paraventricular nucleus; PON = preoptic nucleus; OVLT = organum vasculosum of the lamina terminalis; VMN = ventromedial nucleus; TIDA = tuberoinfundibular dopaminergic system; LHRH = luteinizing hormone-releasing hormone; POMC = pro-opiomelanocortin
^{2}AA = amino acid residue; M.W. = molecular weight; n.a. = not applicable.

general circulation. These neuropeptide hormones are called **neurohypophyseal**. Other hypothalamic neurons, called **neuroregulatory**, make synaptic contact with other neurons, and their neuropeptide products function as neurotransmitters or as neuromodulators. The properties and functions of hypophysiotropic hormones are summarized in Table 29.1.

Neurohypophyseal Peptide Hormones

The magnocellular neurons of the paraventricular and supraoptic nuclei synthesize antidiuretic hormone (ADH, vasopressin, or arginine vasopressin) and oxytocin, neuropeptide hormones that are released into the general circulation in significant amounts (Table 29.2). These neuropeptides are synthesized in separate cells and transported by axoplasmic flow to nerve endings in the neurohypophysis. Cholinergic signals to the magnocellular neurons stimulate the release of ADH and oxytocin via nerve impulse propagation along the axon and Ca^{2+}-dependent exocytosis of secretory granules. Release mechanisms for ADH and oxytocin are under separate control.

ADH is a nonapeptide that contains an intrachain disulfide bridge. ADH is synthesized as part of a

TABLE 29.2 Neurohypophyseal Hormones

Hormone	Structure
Antidiuretic hormone (ADH; arginine vasopressin)	S—S bridge between Cys^1 and Cys^6; Cys^1—Tyr^2—Phe^3—Gln^4—Asn^5—Cys^6—Pro^7—Arg^8—Gly^9—NH$_2$
Oxytocin	S—S bridge between Cys^1 and Cys^6; Cys^1—Tyr^2—Ile^3—Gln^4—Asn^5—Cys^6—Pro^7—Leu^8—Gly^9—NH$_2$

precursor glycoprotein (propressophysin), whose gene is located on chromosome 20. Propressophysin is packaged into secretory granules and transported to the neurohypophysis. During transit, propressophysin is split by proteases in the granule wall yielding ADH and a "carrier" polypeptide called **neurophysin II**, or **nicotine-sensitive neurophysin** (NSN), whose release is stimulated by nicotine. On neural stimulation, ADH and NSN are released in equimolar amounts. NSN has no known biological activity. In the collecting ducts of the kidney, ADH acts by a G_s-protein coupled cAMP-mediated mechanism to increase the permeability of ductal cells to water by mobilizing aquaporin proteins to the apical membrane, which prevents diuresis (see also Chapter 37). At relatively high concentrations, ADH is a potent vasoconstrictor. High concentrations are attained after massive hemorrhage, when constriction of blood vessels prevents further blood loss. Major stimuli for ADH release are (1) an increase in plasma osmolality, monitored by hypothalamic osmoreceptors; and (2) a decrease in blood volume, monitored by baroreceptors in the carotid sinus and in the left atrial wall. Afferent fibers to the vasomotor center in the hindbrain sense a fall in blood volume, and diminished noradrenergic signals from this center to the hypothalamic magnocellular neurons stimulate ADH release. Because norepinephrine inhibits ADH release, a decrease of this inhibition results in stimulation of ADH.

Oxytocin is a nonapeptide that differs from ADH at the amino acids in positions 3 and 8. It is synthesized in separate magnocellular neurons from a gene that codes for pro-oxyphysin, the precursor of oxytocin and its "carrier" (neurophysin I, or estrogen-sensitive neurophysin). Neurons in the paraventricular and supraoptic nuclei synthesize pro-oxyphysin, which is packaged and processed during transit to the neurohypophysis. The principal action of oxytocin is ejection of milk from the lactating mammary gland ("milk let-down"), and it also participates in parturition (Chapter 32). Oxytocin is released by a neuroendocrine reflex mechanism and stimulates contraction of estrogen-conditioned smooth muscle cells. The mechanism of action of oxytocin does not involve cAMP but may involve regulation of increased intracellular Ca^{2+}. The oxytocin receptor belongs to a seven-membrane-spanning receptor family.

Neuroregulatory Peptides

Neuronal projections from the hypothalamus to other regions of the brain relay important output information that influences blood pressure, appetite, thirst, circadian rhythm, behavior, nociception (pain perception), and other factors. Although many of these neurons release neurotransmitter amines at synapses, some of them are known to release neurotransmitter peptides. These include, among others, peptides that closely resemble hormones formed in the gastrointestinal system, as well as endogenous opiates.

Brain—Gut Peptides

Gut hormone-like neurotransmitters have been detected in the brain and are believed to function in a manner that complements their counterparts in the gastrointestinal tract (see Chapter 11). For example, cholecystokinin (CCK) functions in the gut to promote digestion by acting on the gallbladder and exocrine pancreas. The gut-derived CCK is a large peptide (33 amino acid residues), the last 8 of which confer biological activity. In the brain, a smaller CCK with the same 8 carboxy terminal amino acids functions as a neurotransmitter for appetite suppression. Thus, in the brain and in the gut, CCK influences some facet of eating. Gastrin, a 17-residue hormone in the gut that stimulates gastric acid secretion and that has the same five carboxy terminal residues as CCK, has also been detected in the brain, where it is believed to have an effect on appetite.

Endogenous Opiates

Opioid neurotransmitters in the brain are peptides that modulate pain perception and/or the reaction to perceived pain; they include the enkephalins, endorphins, dynorphins, and neoendorphins. All exert their effects by binding to specific types of opiate receptors that are located in various parts of the CNS, but particularly in those regions that function in pain perception. Derivatives of opium, the extract of the poppy *Papaver somniferum*, exert their analgesic and psychological effects through these opiate receptors. The endogenous opiates include β-endorphin, the enkephalins (met-enkephalin and leu-enkephalin), the dynorphins, and the neoendorphins. All are peptides varying in size from 5 to 31 amino acid residues. All have in common an amino terminus consisting of either of two pentapeptide sequences: Tyr—Gly—Gly—Phe—Met (the met-enkephalin sequence) or Tyr—Gly—Gly—Phe—Leu (the leu-enkephalin sequence). The fundamental endogenous opioid peptides are the pentapeptides, met-enkephalin and leu-enkephalin, which function as neurotransmitters in the CNS. All of the known endogenous opioids are derived from three different prohormones: pro-opiomelanocortin (POMC); proenkephalin A; and proenkephalin B (prodynorphin).

PITUITARY GLAND (HYPOPHYSIS)

The pituitary is a small, bilobed gland connected to the base of the hypothalamus by the infundibular process (pituitary stalk). Embryologically, it derives from Rathke's pouch (buccal epithelium) and the infundibulum (neuroectoderm). The former gives rise to the anterior lobe (anterior pituitary or adenohypophysis). The latter is a projection of the hypothalamus and gives rise to the posterior lobe (posterior pituitary or neurohypophysis). In many species, an "intermediate lobe" also exists, but in humans, this is rudimentary and apparently nonfunctional. The pituitary in an average-sized adult weighs only about 0.5 g, 75% of which is anterior pituitary.

TABLE 29.3 Somatomammotropin Family

Hormone	Site of Secretion	Chemistry	Location of Gene	Function	Regulation of Secretion
Growth Hormone (GH)	Anterior pituitary (somatotroph)	191 aa (22 kDa)	17q22-q24	Stimulates protein synthesis in many tissues; exerts protein-sparing effects; stimulates production of insulin-like growth factors (IGFs)	Primarily stimulated by GHRH and inhibited by somatostatin
Prolactin (PRL)	Anterior pituitary (lactotroph)	198 aa (23 kDa)	6p22.2-q21.3	Stimulates growth and protein synthesis in breast; promotes milk secretion during lactation	Primarily inhibited by PIH (dopamine); stimulated by PRF and TRH
Placental Lactogen (hPL)	Placenta (syncytiotrophoblast)	191 aa (22 kDa)	17q22-q24	Exerts protein-sparing effect in maternal tissues; promotes fetal growth by stimulating production of fetal IGF-I	Progesterone-stimulated placental growth

The anterior pituitary is not innervated by nerve fibers from the hypothalamus but is well vascularized by the portal blood that drains the median eminence. The portal blood flows primarily from the median eminence to the anterior pituitary; however, some vessels may transport blood in the opposite direction (retrograde flow). The posterior pituitary consists mainly of nerve endings of hypothalamic magnocellular neurons and contains no portal connections with the hypothalamus; its vascular connections are largely independent of those in the anterior pituitary. The hormones of the posterior pituitary (neurohypophyseal hormones) have been discussed previously.

The anterior pituitary has five endocrine cell types, each of which produces different hormones. In humans, at least seven are produced: growth hormone (GH); prolactin (PRL); luteinizing hormone (LH); follicle-stimulating hormone (FSH); thyroid-stimulating hormone (TSH); adrenocorticotropic hormone (ACTH, corticotropin); and β-endorphin. β-Lipotrophic hormone (β-LPH) is also secreted, but it serves mainly as a precursor of β-endorphin and is not regarded as a hormone. In species that possess a prominent intermediate lobe, the pituitary also secretes α- and β-melanocyte-stimulating hormones (MSH), which affect skin coloration.

Anterior pituitary hormones are classified into three families: the somatomammotropin family (GH and PRL); the glycoprotein hormones (LH, FSH, and TSH); and the opiomelanocortin family (ACTH, β-endorphin, and related peptides). These three families appear to have evolved from three separate ancestral polypeptides, because homologous members of each family occur in other parts of the body. For example, human placental lactogen (hPL) is somatomammotropin, human chorionic gonadotrophin (hCG) is a glycoprotein hormone, and the brain and gut produce substances related to endorphins (enkephalins and dynorphin).

Somatomammotropin

Members of the somatomammotropin family are single-chain proteins with two or three intrachain disulfide bridges. Their molecular weight is about 20,000, and they function mainly to promote protein synthesis (Table 29.3). The hypothalamus produces both a stimulating and an inhibiting hormone to regulate secretion of the adreno-hypophyseal hormone.

Growth Hormone (Somatotropin)

Human growth hormone consists of 191 amino acid residues (M.W. 22,000) and contains two disulfide bridges. It is a globular protein and does not undergo glycosylation. It exhibits extensive sequence homology with prolactin (76%) and placental lactogen (94%). The gene encoding growth hormone (GH) is on chromosome 17, q22-q24, and its expression is regulated by a transcription factor (Pit-1/GHF-1) that is promoted by hormones that stimulate GH synthesis (e.g., GHRH, glucocorticoids, thyroid hormone). GH is the most abundant hormone in the human pituitary gland, averaging about 6 mg per adult gland. Growth hormone secretion is pulsatile, with about 10 pulses per day; its basal blood level averages 2 ng/mL in adults and 6 ng/mL in preadolescent and adolescent boys. GH is cleared from circulation with a half-life of about 25 minutes in lean adults, but is cleared more rapidly in obese subjects. GH is inactivated mainly by the liver but also by the kidney. About 40% of the hormone is bound to **GH-binding protein (GHBP)**, a fragment of the GH receptor, which serves to prolong the half-life of GH 10-fold. Its primary structure is species-specific and antigenic.

Actions of GH

The actions of GH are initiated by its binding to the membrane receptor of target cells. The GH and PRL receptors are single-membrane-bound proteins that belong to class I of the cytokine receptor superfamily (Chapter 28). Each receptor contains an extracellular, a transmembrane, and an intracellular domain. Receptor activation begins when one molecule of growth hormone (with its two binding sites) binds to the first receptor, followed by binding to the second receptor, resulting in receptor dimerization. The hormone—receptor complex activates a tyrosine kinase known as **Janus kinase 2 (Jak-2)** that is associated with the proximal region of the intracellular domains of the dimerized receptor (Figure 29.2). Two Jak-2 kinases transphosphorylate each other and also phosphorylate tyrosine residues of the receptor. Subsequent downstream single transduction pathways are many and involve signal transducer and activator of transcription (STAT) proteins, activation of mitogen-activated protein (MAP) kinase, and phosphoinositide-3-kinase (PI kinase) pathways.

GH promotes the transport and incorporation of amino acids into skeletal muscle, cardiac muscle, adipose tissue, and liver, and is responsible for the proportional growth of visceral organs and lean body mass during puberty. Linear growth during childhood requires the presence of GH. GH acts directly on cartilage tissue to promote the endochondral growth that results in skeletal growth. Although GH has a direct effect on chondrocyte stem cells, the growth-promoting effect of GH is due to its stimulation of the chondrocytes to produce insulin-like growth factor I (IGF-I, discussed later), which then acts locally to stimulate cellular replication in the distal proliferative zone of the epiphyseal plate. Thus, the growth-promoting effect of GH can be abolished by blocking the IGF-I receptor, and can be duplicated by exogenous IGF-I treatment in the absence of GH. The importance of GH is in its ability to stimulate IGF-I production within bone cartilage, an ability that is unique to GH. This explains why GH deficiency results in growth retardation, despite the fact that bone cartilage has the ability to produce IGF-I.

GH exerts a "protein-sparing" effect by mobilizing the body's energy substrates, such as glucose, free fatty acids, and ketone bodies, in the same tissues in which it stimulates protein synthesis. GH elevates blood glucose levels by lowering glucose uptake by skeletal muscle by means of inhibiting hexokinase activity and desensitizing the tissue to the actions of insulin. In addition, GH increases the activity of hepatic glucose-6-phosphatase, and thereby increases glucose secretion. GH promotes lipolysis in adipocytes, possibly by increasing the

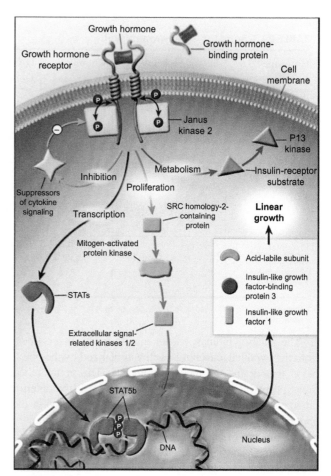

FIGURE 29.2 Mechanism of growth hormone (GH) action at the target cell receptor site. GH binds to the extracellular domain of the specific receptor and activates the intracellular phosphorylation cascade initiated by Janus kinase 2. See text for details. *[Reproduced with permission from E. A. Eugster, O. H. Pescovitz, New revelations about the role of STATs in stature. N. Engl. J. Med. 349 (2003), 1111.]*

synthesis of hormone-sensitive lipase (HSL), and by increasing ketogenesis in the liver. These protein-sparing effects of GH are diabetogenic and explain how GH functions as an insulin antagonist (Chapter 20).

Regulation of GH Release

The protein-anabolic and protein-sparing actions of GH require the metabolic effects of insulin, glucagon, cortisol, and thyroid hormone in the unstressed individual. These actions depend on fine-tuned control of GH release, which is achieved mainly by substrate feedback to the hypothalamus (Figure 29.3).

A fall in blood glucose level stimulates GH release, whereas a rise inhibits release. GH release also occurs in response to certain amino acids, the most potent being L-arginine, with a latency period of about 30 minutes. Circulating hormones exert their effects at the level of the

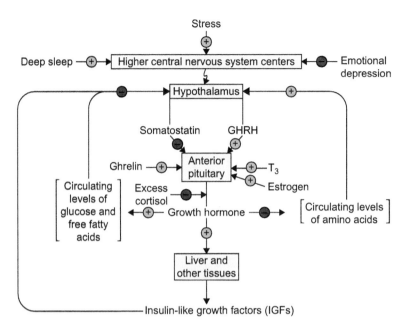

FIGURE 29.3 Regulation of growth hormone secretion in humans. + = stimulation, − = inhibition.

hypothalamus or the pituitary. GH influences its own secretion by way of IGFs that exert a negative feedback effect at the median eminence. However, it is not known if this feedback involves a decrease in GHRH, an increase in somatostatin, or both. Other hormones promote the synthesis and release of GH at the level of the anterior pituitary. While estrogen promotes an increase in somatotroph numbers and GH mRNA levels, androgens increase and IGF-I decreases the somatotroph response to GHRH. Glucocorticoids and thyroid hormone act in concert to stimulate GH gene expression. However, pharmacological concentrations of glucocorticoids strongly inhibit GH release in response to GHRH (Figure 29.3).

Superimposed on this fine regulation of GH secretion by substrates and hormones is the coarse regulation by higher brain centers that operate on an open-loop, substrate-independent basis. These influences are dramatic and result in a several-fold increase in GH levels during stress and during deep sleep. GH is one of several hormones released in response to stress (see later). It is also one of the few released during deep sleep (EEG stages III and IV) irrespective of the time of day. The amount of GH released during deep sleep is substantial, accounting for about 75% of the daily output of GH. In children who are unable to achieve deep sleep because of emotional disturbances, the absence of sleep-induced GH results in growth retardation.

Insulin-Like Growth Factors

Insulin-like growth factors (IGFs) are GH-dependent polypeptide hormones that promote cell replication in most mesenchymally derived tissues, and are responsible for the growth-promoting effects of GH. There are two IGFs, I and II, both of which are 7 kDa proteins that resemble proinsulin in structure. The IGFs exert insulin-like biological effects when tested in insulin bioassay systems *in vitro* and account for the **nonsuppressible insulin-like activity** (**NSILA**) in plasma that had been described before the discovery of the IGFs. The similarities and differences of IGFs with insulin and among themselves are summarized in Table 29.4.

Six serum proteins, produced mainly in the liver, that bind IGFs in the circulation have been identified. They are designated IGF-binding proteins (IGFBPs) 1 through 6. IGFBP-1 appears to retard target tissue uptake of IGF-I, while IGFBP-3 appears to enhance it; the latter accounts for most of the IGF found in blood. About 80%−90% of the IGF-I in circulation is bound to IGFBP-3, along with an 88 kDa acid-labile subunit (ALS). This complex confers protection on the IGF and prolongs its half-life to 12−15 hours. About 5% of the IGF is unbound, and the remaining 5%−15% is bound to IGFBP-1, -2, or -4 as smaller complexes.

Unlike insulin and other peptide hormones that are released from storage granules, IGFs are released as they are produced. Most tissues produce IGFs in small amounts that are sufficient for local (paracrine/autocrine) effects and do not contribute significantly to the circulating pool of IGFs. The highest concentration of IGF-I and -II is found in the circulation and is primarily derived from the liver. GH and insulin are positive regulators of IGF-I synthesis in the nonfasting state. During mild starvation when GH levels increase, there is a decline in hepatic production of IGF-I due, in part, to the decline in insulin. Refeeding, which causes insulin to rise and GH to fall, promotes IGF-I production only if the diet is adequate in calorie

TABLE 29.4 Insulin and Insulin-Like Growth Factors

	Insulin (Ins)	IGF-I	IGF-II
M.W.	5734	7649	7471
Site of production	β-cell, pancreatic islet	Many sites	Many sites
Major source of circulating form	β-cell, pancreatic islet	Liver	Liver
Regulation of secretion	Glucose, amino acids, catecholamines	GH, insulin, nutritional status	GH
GH-dependence	−	+ + + +	+
Production rate	2 mg/day	10 mg/day	13 mg/day
Plasma levels (adults)	<5 ng/mL	200 ng/mL	700 ng/mL
Characteristics in circulation	Pulsatile	Nonpulsatile	Nonpulsatile
	Variable	Steady, nonvariable	Steady, nonvariable
Binding proteins in serum	None	3 (types 1, 3, 4)	6 (types 1−6)
Half-life	10 minutes	12−15 hours	15 hours
Target tissues	Liver, muscles, adipose, skin, connective tissue, bone	Muscles, connective tissue, cartilage, bone	Muscles, connective tissue, cartilage, bone
Receptor type	Plasma membrane heterotetramer ($\alpha 2\beta 2$)	Plasma membrane heterotetramer ($\alpha 2\beta 2$)	IGF-I receptor; plasma membrane IGF-II/man-6-P receptor (monomeric)
	$Ins > IGF\text{-}I_{100}$	$IGF\text{-}I > II_{100} \gg Ins_{1000}$	$IGF\text{-}II \gg I_{500}$
Regulation of receptor density	Ins inhib	IGF-I inhib	IGF-I inhib (IGF-IR)
	GH inhib	Starvation inhib	Starvation inhib (IGF-IR); insulin, IGF-I stim (IGF-IIR)

inhib = inhibits; stim = stimulates; ins = insulin.

and protein content. Note that circulating IGF-I does not mediate the growth-promoting effect of GH but is an important feedback regulator of GH. The indispensability of IGF-I for linear growth and the fact that it mediates the effect of GH are firmly established. IGF-I, locally produced in endochondral bone in response to GH, promotes skeletal growth by stimulating clonal chondrocyte expansion of the distal proliferative zone of the epiphyseal plate.

Under physiological conditions, IGF-I is a relatively minor regulator of fuel homeostasis. It does not mediate the effects of GH on intermediary metabolism. In fact, many of the effects of IGF-I resemble those of insulin, not GH. IGF-I promotes the production and actions of erythropoietin and is responsible for the increased packed volume of blood that results from elevated GH levels.

Disturbances in GH and IGF

Only rarely has a condition of IGF deficiency or excess been described that is not accompanied by a disturbance in GH production or responsiveness (see Clinical Case Studies 29.1 and 29.2). Thus, a GH deficiency results in IGF-I deficiency, whereas a GH excess results in IGF-I excess. GH deficiency in children, due to defects of the GH gene and its related genes, results in reduced growth rate and growth retardation (dwarfism) due to secondary IGF-I deficiency. Insensitivity to GH action, which is accompanied by phenotypic features consistent with GH deficiency, occurs due to either defects in GH receptors at the target site or defects in the post-GHR signal transduction pathways. These disorders are characterized by normal or high levels of serum GH and low levels of IGF-I and IGF binding protein-3. The main cause of GH resistance is a genetic defect in the growth hormone receptor (GHR); the resultant condition is known as **Laron-type dwarfism**. This is an autosomal recessive disorder characterized by normal-to-high levels of serum growth hormone and low levels of IGF-I. Analysis of the GHR gene in Laron-type dwarfism has revealed point mutations, deletions, and splicing defects. Adults who are deficient primarily in GH have decreased lean body mass, increased adiposity, and are at increased risk of cardiovascular disease.

The STAT5b mutation is an example of how a post-receptor molecular abnormality causes dwarfism due to insensitivity to GH action (see also Clinical Case Study 29.2).

Growth hormone deficiency in children can be treated by administration of recombinant GH or, in some cases, growth hormone-releasing hormone. Recombinant IGF-I is used in the treatment of children with GH insensitivity syndrome. Potential targets for recombinant GH use that are undergoing clinical investigation include children with idiopathic short stature, persons with wasting syndrome associated with human immunodeficiency virus infection, critically ill patients, and sick elderly individuals.

GH hypersecretion in children causes increased growth rate and can result in **gigantism**. However, GH excess occurs infrequently in childhood. GH excess occurs most frequently in the middle-aged and leads to **acromegaly** (akros = extremities, plus megas = large), a condition in which the cartilaginous tissues proliferate, resulting in distorted overgrowth of the hands, feet, mandibles, nose, brow, and cheek bones. Because epiphyseal cartilage is absent in long bones, there is no gain in height. Acromegaly promotes insulin resistance and results in cardiovascular complications. Resistance to the effects of GH results in a condition similar to GH deficiency.

Excess production of GH due to somatotroph adenomas may be treated by surgical resection, irradiation, or in some cases with somatostatin analogues (e.g., octreotide) that suppress GH secretion, or by a combination of these treatments.

GH has been abused by athletes to enhance their physical performance for its purported anabolic and metabolic effects. GH has also been touted as an anti-aging drug. Studies have not substantiated any of the presumed beneficial effects of GH. Recombinant and endogenous GH are chemically identical 22 kDa proteins. Normal serum contains endogenous 22 kDa GH protein and other molecular species of GH (20 kDa GH and their dimers and multimers). Since the injected recombinant 22 kDa GH increases its proportion relative to other normal GH species in the serum, the determination of the proportions of 22 kDa protein with its other forms by sensitive immunochemical analysis has been used in the detection of GH doping in sports.

Prolactin

Human prolactin (PRL) contains 199 amino acid residues (M.W. 23,500) and three intramolecular disulfide bridges. In healthy adults, the anterior pituitary releases very little PRL under nonstressed conditions, primarily because PRL release is under hypothalamic inhibition. This inhibition is exerted by dopamine, or prolactin-inhibitory hormone (PIH). The hypothalamus also secretes PRF, but the major regulator appears to be PIH.

The biological actions of prolactin are initiated by its membrane receptor, followed by the associated intracellular signal transduction pathways. These actions are similar to growth hormone (discussed previously; also see Chapter 28). In humans, the function of PRL may be restricted to promotion of lactation, but there is some evidence that PRL suppresses gonadal function in females. There is no consensus regarding a physiological role for PRL in males. During late pregnancy, the maternal pituitary releases increasing amounts of PRL in response to rising levels of estrogen, a stimulator of PRL synthesis. Elevated levels of PRL stimulate milk production in the mammary gland (Chapter 32). After parturition, PRL promotes milk secretion via a neuroendocrine reflex that involves sensory receptors in the nipples (Chapter 32). In mammary tissue, it binds to alveolar cells and stimulates the synthesis of milk-specific proteins (casein, lactalbumin, and lactoglobulin) by increasing production of their respective mRNAs. Accordingly, there is a lag of several hours before this effect of PRL is seen. In the liver, PRL stimulates the synthesis of its own receptors. PRL receptors occur in the mammary gland, liver, gonads, uterus, prostate, adrenals, and kidney.

Disturbances in Prolactin

In women, **hyperprolactinemia** is often associated with amenorrhea, a condition that resembles the physiological situation during lactation (lactational amenorrhea). Excess PRL may inhibit the menstrual cycle directly, by a suppressive effect on the ovary, or indirectly, by decreasing the release of GnRH (Chapter 32). In men, hyperprolactinemia is not associated with altered testicular function, but is often accompanied by diminished libido and impotence. This finding suggests that PRL may serve an important role in regulating certain behavior patterns. **Prolactinomas** are the most common hormone-secreting pituitary adenomas and, in addition to the previously mentioned clinical characteristics, the patient may exhibit visual field defects. Prolactinoma may be treated by surgery, radiotherapy, or pharmacotherapy. This last method consists of using a dopamine D_2 receptor agonist such as **bromocriptine** or **cabergoline** to suppress prolactin secretion. Hyperprolactinemia can also occur due to loss of function mutations in the prolactin receptor, causing downstream signaling dysfunctions in JAK2-STAT5 receptor pathways (see reference 1 for Clinical Case Study 29.1).

The Opiomelanocortin Family

All members of this family are derived from a single pro-hormone, pro-opiomelanocortin (POMC). The prohormone molecule contains three special peptide sequences: enkephalin (opioid), melanocyte-stimulating hormone (MSH), and corticotropin (ACTH). After translation, POMC is processed by proteases contained in tissue. In the adenohypophyseal corticotrophs, there is no further processing of ACTH and β-endorphin, whereas in the intermediate lobe of other species, ACTH can be processed to form α-MSH; γ-LPH can be processed to form β-MSH. The human hypothalamus and/or brain may be capable of forming α- and β-MSH from POMC, since these peptides appear to exert central effects.

Inherited defects in POMC production result in deficiency or a complete lack of secretion of ACTH, MSH (α-, β-, γ-), and β-endorphin. These patients exhibit adrenal insufficiency, red hair pigmentation, and early-onset obesity that have been related to a deficiency of α-MSH production. Animal studies have shown that α-MSH regulates food intake by activation of melanocortin receptor-4. In mice, a POMC gene defect also leads to adrenal deficiency, altered pigmentation, and obesity with loss of a significant portion of the animal's excess weight when treated with a stable α-MSH preparation. Other genetic causes of obesity include defects in genes encoding prohormone convertase-1 (i.e., inability to convert POMC to various peptides), leptin, and leptin receptor (Chapter 20).

ACTH is a polypeptide of 39 residues, the first 24 of which are required for corticotropic activity and which do not vary among species. $ACTH_{1-24}$ has been synthesized and is used for diagnostic purposes. Because it contains the MSH sequence in residues 6−9 (His−Phe−Arg−Trp), ACTH has intrinsic melanocyte-stimulating activity and can thus cause skin darkening if present in high concentrations. In normal adults, the pituitary contains about 0.25 mg of ACTH, and the basal level of ACTH in blood is approximately 50 pg/mL.

ACTH acts mainly on the cells of the zona fasciculata of the adrenal cortex to stimulate the synthesis and release of cortisol (Chapter 30). It also stimulates the secretion of adrenal androgens from the zona reticularis. Binding of ACTH to receptors activates the formation of cAMP, which mediates cortisol formation and secretion, as well as protein synthesis. Deficiency of ACTH leads to a reduction in the size and activity of adrenocortical cells in the inner two zones.

ACTH secretion is regulated to ensure constant, adequate levels of cortisol in the blood. Corticotrophs in the pituitary are under tonic stimulation by CRH and are modulated by the negative feedback effect of blood cortisol, which inhibits POMC synthesis by repression of the POMC gene. Although cortisol also exerts an effect on the hypothalamus and related areas in the brain (limbic system, reticular formation), this activity may involve the regulation of emotion and behavior and not regulation of CRH release. The regulation of ACTH and cortisol is illustrated in Figure 29.4. Secretion of POMC in the intermediate lobe is not regulated by CRH and glucocorticoids because the intermediate lobe is poorly vascularized and contains no glucocorticoid receptors. However, the intermediate lobe is rich in dopaminergic fibers so that dopamine agonists (ergocryptine) decrease—and antagonists (haloperidol) increase—the synthesis and release of POMC-derived peptides.

β-Endorphin is a 31-amino acid polypeptide released together with ACTH. When it is introduced into the third ventricle of the brain, it produces dramatic behavioral changes, but it does not when injected systemically. Thus,

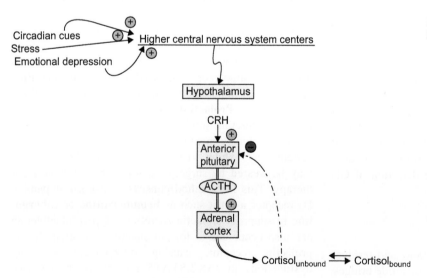

FIGURE 29.4 Regulation of ACTH secretion in humans. + = stimulation, − = inhibition.

the function of circulating β-endorphin remains unclear. The CNS and gastrointestinal effects of the hormone are probably induced by that secreted locally. Circulating β-endorphin may act in conjunction with enkephalins released by the adrenal medulla to produce stress analgesia.

Human N-terminal fragment (hNT) and **γ-melano-cyte-stimulating hormone (γ-MSH)** are released with ACTH and β-endorphin. γ-MSH, a fragment of hNT, may be the putative aldosterone-stimulating hormone of the pituitary. hNT and γ-MSH may also be major stimulators of adrenocortical proliferation, whereas ACTH is the regulator of adrenocortical steroidogenesis.

Pituitary Independent Cutaneous POMC Production and Ultraviolet-Induced Formation of Melanin

Ultraviolet (UV) radiation induces skin pigmentation (sun tanning) by stimulating production of melanin. This process is mediated by the following sequence of events. UV damage to keratinocytes leads to activation of transcription factor tumor suppressor protein, p53, which upregulates the pro-opiomelanocortin (POMC) gene leading to multicomponent precursor POMC production. Enzymatic post-translational processing of POMC yields three biologically active peptides: adrenocorticotropic hormone (ACTH), α-melanocyte-stimulating hormone (MSH), and opioid peptide β-endorphin. MSH, by binding to melanocortin-1 receptors located on the cell membrane of melanocytes, stimulates the production of melanin, which is packaged in melanosomes, which, in turn, are subsequently transported out and deposited in keratinocytes. The cutaneous production of POMC and its processing is independent of anterior pituitary processing. The addictive potential for indoor tanning among young individuals may be attributed to the release of β-endorphin. A beneficial action of UV is the production of vitamin D (Chapter 35). It should be emphasized that UV radiation, whether due to indoor skin tanning or exposure to excessive sunlight, is a risk factor for melanoma and nonmelanoma cutaneous cancers.

Glycoprotein Hormones

The glycoprotein hormones may have evolved from a single ancestral polypeptide (Table 29.5). In the anterior pituitary, the thyrotrophs produce thyrotropin, and the gonadotrophs produce LH and FSH. Each glycoprotein hormone is composed of two dissimilar subunits (α and β) that are glycosylated and noncovalently bound. Synthesis of β-subunits is rate-limiting. Glycosylation takes place post-translationally on the endoplasmic reticulum membranes at about the time when disulfide bridges form in each subunit. "Mature" subunits then dimerize and are packaged in secretory granules.

The α-subunits of all pituitary glycoprotein hormones are identical but show minor differences from the α-subunit of hCG. The β-subunits of all these hormones are different and confer hormonal specificity (discussed in the ensuing chapters).

TABLE 29.5 Glycoprotein Family

Hormone	Site of Secretion	Peptide Length	Location of Gene	Carbohydrate (%)	Function	Regulation of Secretion
Thyroid-stimulating hormone (TSH)	Anterior pituitary (thyrotroph)	TSHα = 89aa TSHβ = 112aa	6q21.1-23 1p22	16.2 (1% sialic acid)	Stimulates thyroid growth and development	Stimulated by TRH and inhibited by thyroid hormone and somatostatin
Luteinizing hormone (LH)	Anterior pituitary (gonadotroph)	LHα = 89aa LHβ = 112aa	6q21.1-23 19q12.32	15.7 (1% sialic acid)	Stimulates steroidogenesis in gonads	Stimulated by GnRH and inhibited by estrogen
Follicle-stimulating hormone (FSH)	Anterior pituitary (gonadotroph)	FSHα = 89aa FSHβ = 115aa	6q21.1-23 11p13	18.2 (5% sialic acid)	Stimulates gametogenesis and estrogen formation in gonads	Stimulated by GnRH and inhibited by estrogen
Chorionic gonadotropin (hCG)	Placenta (syncytiotrophoblast)	hCGα = 89aa hCGβ = 112aa	6q21.1-23 19q13.32	31.0 (12% sialic acid)	Stimulates steroidogenesis in gonads and placenta	Autocrine?

CLINICAL CASE STUDY 29.1 Short Stature in Childhood

This case was abstracted from: D.B. Allen, L. Cuttler, Short stature in childhood—challenges and choices, N. Engl. J. Med. 368 (2013) 1220–1228.

Synopsis

An 11½-year-old boy was brought to his primary care doctor for concerns of short stature and slowed growth rate. Over the past 2 years, the patient went from the 3rd percentile to just below the 1st percentile for height. The patient had no symptoms, and review of systems was unremarkable. At birth, the patient's size was normal, and throughout childhood he had an otherwise normal development with no medical problems. Family history was negative for any growth-restricting conditions. His mother's height was 5 feet (152 cm), and his father's height was 5 feet 6 inches (167 cm). Physical examination was normal with appropriate prepubertal development. Laboratory studies revealed normal complete blood count (CBC), erythrocyte sedimentation rate (ESR), thyrotropin, tissue transglutaminase antibody, growth hormone levels, and insulin-like growth factor I (IGF-I). His bone age was 9 years, with a predicted adult height of 5 feet 5 inches (165 cm). This correlated very closely with the predicted adult height based on midparental height calculations, making a pathologic cause of this patient's markedly short stature unlikely, especially given his otherwise healthy condition. Further studies for hormonal, renal, inflammatory, immune, and hematologic disorders were not pursued. The patient was given the diagnosis of idiopathic short stature due to familial short stature and/or constitutional delay of growth and puberty (CDGP). He was managed with reassurance and observation, with the option to treat electively with oxandrolone (an anabolic steroid) or recombinant human growth hormone in the event of psychological distress caused by his short stature.

Teaching Points

1. The most common causes of short stature are familial short stature, constitutional delay of growth and puberty (CDGP), or both. Other causes of short stature are failure to thrive (due to inadequate energy intake [calories], caloric losses from malabsorption, or excess caloric needs due to chronic disease), intrinsic shortness (due to genetic syndromes like Turner or Silver–Russell, congenital disorders like intrauterine growth restriction and bone dysplasia, or acquired growth restriction from spinal irradiation), delayed growth, or attenuated growth (due to endocrine disorders, severe chronic disease, or medications like stimulants or glucocorticoids).

2. Short stature in childhood should be thoroughly evaluated when there is a severe height deficit (<1st percentile for age), an abnormally slow growth rate (<10th percentile for bone age), a significant discrepancy between predicted and midparental height, or an abnormality in body proportions.

3. Treatment with recombinant human growth hormone can safely increase growth rate and adult height, but it is expensive and has undetermined long-term benefits and risks.

4. There is inadequate evidence linking growth-enhancing therapy with long-term psychosocial benefits, and there appears to be no harm in short stature; therefore, most children with short stature are reasonably managed with observation and reassurance.

5. For those children with psychological distress or impairment as a result of their short stature, low-dose oxandrolone is an elective treatment that effectively accelerates growth. Oxandrolone is a synthetic testosterone analogue that preserves or restores muscle mass in patients with cachexia, AIDS-associated myopathy, alcoholic hepatitis, and other pathologic delays in growth like Turner syndrome. It is also being studied for improving body composition and strength in severe burn patients.

Supplemental References

[1] M.G. Jeschke, C.C. Finnerty, O.E. Suman, G. Kulp, R.P. Micak, D.N. Herndon, The effect of oxandrolone on the endocrinologic, inflammatory, and hypermetabolic responses during the acute phase postburn, Ann. Surg. 246 (2007) 351–362.

[2] L.E. Cohen, Idiopathic short stature, JAMA 311 (2014) 1787–1796.

CLINICAL CASE STUDY 29.2 Growth Hormone Insensitivity

This clinical case was abstracted from: E.M. Kofoed, V. Hwa, B. Little, K.A. Woods, C.K. Buckway, J. Tsubaki, et al., Growth hormone insensitivity associated with STAT5b mutation, N. Engl. J. Med. 349 (2003) 1139–1147.

Synopsis

A 7-year-old female offspring of a consanguineous married couple presented to an endocrine center for evaluation of growth failure and attenuation in weight gain. She was born at 33 weeks of gestation with respiratory difficulties, and was noted to have poor weight gain and growth failure throughout her childhood, but no abnormalities had been found. At present, she also had immunodeficiency syndrome and an accompanying pulmonary disorder caused by an opportunistic infection by *Pneumocystis carinii*. Endocrinological testing revealed that the patient had elevated serum growth hormone levels and low level of insulin-like growth factor I (IGF-I). Furthermore, growth hormone therapy did not improve the growth rate. Molecular studies showed that the subject's growth hormone resistance was due to a homozygous missense mutation in the STAT5b gene, which is involved in the postreceptor growth hormone signaling cascade.

(Continued)

CLINICAL CASE STUDY 29.2 (Continued)

Teaching Points

1. This case illustrates that the subject's growth retardation was a result of perturbations in the growth hormone (GH) intracellular signaling system. Actions of GH at the target cell sites are initiated by its binding to specific cell surface receptor sites, followed by activation of intracellular Janus tyrosine kinase–signal transducing activators of transcription (JAK-STAT) leading to a protein phosphorylation cascade.

2. In the liver, GH promotes the synthesis of insulin-like growth factor I (IGF-I), which, in turn, mediates the growth-promoting and anabolic effects. IGF-I consists of 70 amino acid residues (M.W. 7649 Da) and is transported bound primarily to a binding protein, IGFBP-3. The binding of IGF-I to IGFBP-3 increases the half-life of IGF-I by preventing its rapid renal clearance. Thus, serum levels of IGF-I are used as a screening test for either GH deficiency or excess.

3. Growth hormone deficiency can occur because of (1) lack of production of GH from the pituitary due to mutations in genes of the GHRH receptor, GH, or transcription factor; (2) mutations in the target cell's extracellular component of the GH receptor (Laron syndrome); or (3) mutations in the intracellular signaling system involving STAT5b.

4. Primary GH deficiency is treated with recombinant GH and Laron syndrome with recombinant IGF-I.

5. STAT5b deficiency also explains the subject's immunodeficiency because it is also involved in the activation of T-cells (see Chapter 33).

Supplemental References

[1] E.A. Eugster, O.H. Pescovitz, New revelations about the role of STATs in stature, N. Engl. J. Med. 349 (2003) 1110–1112.
[2] J.A. Schlechte, Prolactinoma, N. Engl. J. Med. 349 (2003) 2035–2041.
[3] S.P. Taback, H.J. Dean, E. Elliott, Management of short stature, West J. Med. 176 (2002) 169–172.
[4] S. Melmed, Acromegaly, N. Engl. J. Med. 355 (2006) 2558–2573.
[5] M. Saugy, N. Robinson, C. Saudan, N. Baume, L. Avois, P. Mangin, Human growth hormone doping in sport, Br. J. Sports Med. 40(Suppl. I) (2006) i35–i39.
[6] M.E. Molitch, Medication-induced hyperprolactinemia, Mayo Clin. Proc. 80(8) (2005) 1050–1057.

REQUIRED READING

[1] P.J. Newey, C.M. Gorvin, S.J. Cleland, C.B. Willberg, M. Bridge, M. Azharuddin, et al., Mutant prolactin receptor and familial hyperprolactinemia, N. Engl. J. Med. 369 (2013) 2012–2020.

Chapter 30

Endocrine Metabolism III: Adrenal Glands

Key Points

1. Adrenal glands consist of two embryologically, histologically, and functionally distinct endocrine glands: cortex and medulla, the former surrounding the latter.

2. The adrenal cortex is divided into three zones: glomerulosa, fasciculata, and reticularis. All zones produce zone-specific steroid hormones: aldosterone exclusively in glomerulosa, and primarily cortisol in fasciculata, and androgens and estrogens in the reticularis.

3. The synthesis of steroid hormones begins with cholesterol, which is obtained mainly from circulating LDL.

4. The synthesis of the steroid hormones is compartmentalized due to subcellular localization of enzymes located in mitochondria and endoplasmic reticulum. With the exception of β-hydroxysteroid dehydrogenase, all of the other enzymatic reactions are mediated by the cytochrome P-450 (CYP) family of mixed-function oxidases. The rate-controlling step for steroidogenesis is the importation of cholesterol into the mitochondria.

5. Anterior pituitary hormone, corticotropin (ACTH), promotes glucocorticoid (cortisol) and androgen synthesis by activating adenylate cyclase. The synthesis of ACTH is regulated by the hypothalamic hormone corticotropic-releasing hormone (CRH). The hypothalamic–pituitary–adrenal axis is under feedback control.

6. Aldosterone synthesis and release are promoted by angiotensin II and extracellular K^+. Angiotensin II synthesis is regulated by the renin–angiotensin system.

7. Cortisol secretion is pulsatile, diurnal, and stimulated by stress and inflammatory conditions. The actions of cortisol are initiated by steroid-mediated gene regulation transduction pathways (Chapter 28). The physiological actions of cortisol affect virtually all organ systems of the body.

8. All of the inherited defects of cortisol biosynthesis are autosomal recessive, resulting in congenital adrenal hyperplasia due to the decreased cortisol production, with resultant elevated ACTH secretion, and varying pathology depending on the deficiency of the end product and accumulation of the intermediates.

9. Neonatal screening procedures for the inherited defects of adrenal steroids are utilized in their detection and treatment.

10. Synthetic steroids that possess higher activities than cortisol and aldosterone are used therapeutically in adrenal insufficiency and inflammatory diseases.

11. Primary adrenal deficiency (Addison's disease) is mostly due to autoimmune-mediated destruction of adrenal cortices. A prompt treatment with supplementation of corticosteroids and mineralocorticoids is required. Hypercortisolism can occur due to corticotropin-secreting adenomas (Nelson's disease) and adenomas of adrenal cortices (Cushing syndrome).

12. Epinephrine and norepinephrine (known as catecholamines) are synthesized from tyrosine in the chromaffin cells of the adrenal medulla, which is a specialized sympathetic ganglion. Their synthesis occurs in the cytosol and in granules.

13. Catecholamine synthesis and release into blood circulation are stimulated acutely by sympathetic stimulation such as during trauma, stress, and hypoglycemia. Catecholamines initiate their biological response by binding to structurally related transmembrane cell surface receptors, which then affect G-proteins, which leads to either an increase or decrease of cyclic AMP levels.

14. The physiological effects of catecholamines are rapid and widespread: mobilization of fuels (glucose and fatty acids) for energy production and regulation of the cardiovascular system.

15. Excessive production of catecholamines in tumors of the adrenal medulla (pheochromocytoma) causes several metabolic abnormalities including hypertension and hypermetabolism.

INTRODUCTION

The adrenal glands, a pair of well-vascularized glands positioned bilaterally above the cranial poles of the kidney, consist of two embryologically, histologically, and functionally distinct regions. The outer region of each (**adrenal cortex**) accounts for about 80% of the weight of the gland, is derived from the celomic epithelium of the urogenital ridge (mesodermal), and produces steroid hormones. The inner core of each gland (**adrenal medulla**) is derived from the neural crest (neuroectodermal), represents a modified sympathetic ganglion that has assumed an endocrine function, and synthesizes and secretes catecholamines and enkephalins. Neural supply, by preganglionic sympathetic

N. V. Bhagavan and C.-E. Ha: Essentials of Medical Biochemistry, Second edition. DOI: http://dx.doi.org/10.1016/B978-0-12-416687-5.00030-0

fibers from the splanchnic nerve, is mainly to the adrenal medulla. Blood is supplied mainly to the adrenal cortex via the inferior phrenic, celiac, and renal arteries and is drained from the adrenal medulla.

The adult adrenal cortex is divided into three zones histologically: the outermost **zona glomerulosa**, the middle **zona fasciculata**, and the inner **zona reticularis**. All three zones are exclusive steroid producers; together, they can produce all classes of steroid hormones. Under normal conditions, major products are 11-hydroxylated C_{21} steroids (corticosteroids) (Table 30.1). The two types of corticosteroids are the **mineralocorticoids**, which regulate sodium and potassium levels, and the **glucocorticoids**, which regulate carbohydrate metabolism. The major mineralocorticoid aldosterone (4-pregnen-11β,21-diol-3,18,20-trione), is produced exclusively by the zona glomerulosa, which contains the unique mitochondrial enzyme aldosterone synthase (CYP11B2). The major glucocorticoid in humans is cortisol (or hydrocortisone) (4-pregnen-11β,17α,21-triol-3,20-dione), which is produced exclusively by the zona fasciculata (major) and zona reticularis (minor). These two inner zones contain 17α-hydroxylase (CYP17), but lack enzymes to convert corticosterone to aldosterone. Regional differences in steroid production exist because of the distribution of enzymes (Figure 30.1).

Aldosterone and cortisol are not the only adrenal steroids with mineralocorticoid and glucocorticoid activities. Corticosterone (4-pregnen-11β,21-diol-3,20-dione), a glucocorticoid about one-fifth as active as cortisol, has little mineralocorticoid activity relative to aldosterone; 11-deoxycorticosterone (DOC) (4-pregnen-21-ol-3,20-dione) has about one-tenth the mineralocorticoid activity of aldosterone but is devoid of glucocorticoid activity (Table 30.1). Both are intermediates in the biosynthesis of aldosterone (Figures 30.1 and 30.2). However, under normal conditions, circulating corticosterone and DOC originate mainly from the zona fasciculata. The sodium-retaining action of DOC is normally obscured by the pronounced effect of aldosterone. However, in certain disease states (e.g., CYP17 deficiency), the levels of DOC can result in hypokalemia and hypertension in the absence of aldosterone.

SYNTHESIS OF CORTICOSTEROIDS

Synthesis of corticosteroids begins with cholesterol, which in most instances is acquired from circulating low-density lipoprotein (LDL) and also from high-density lipoprotein (HDL). The steroidogenic pathway begins in the mitochondria. The unesterified cholesterol is transported from the cytosol to the matrix side of the mitochondrial membrane; this process is mediated by a 37 kDa phosphoprotein known as steroidogenic acute regulatory (StAR) protein. The import of cholesterol from cytosol to mitochondria is the rate-limiting step in steroidogenesis. Trophic stimuli such as ACTH in the zona fasciculata and zona reticularis, and elevated cytosolic Ca^{2+} in the zona glomerulosa, cause increased synthesis of the StAR protein within minutes. Other proteins in addition to StAR are also involved in cholesterol transfer across the mitochondrial membrane.

A number of enzymes in the steroidogenic pathway are cytochrome P-450 (CYP) proteins that require NADPH, electron-transfer systems, and molecular oxygen. The CYP enzymes are located in the mitochondria and in the endoplasmic reticulum (Figure 30.2), and they consist of different electron-transfer systems. The mitochondrial electron-transfer system requires adrenodoxin (an iron−sulfur protein) and adrenoreductase (a flavoprotein), whereas the microsomal system utilizes CYP-450 reductase. These enzymes catalyze the cleavage of carbon−carbon bonds and hydroxylation reactions. The CYP-450 enzymes are distinguished by specific notations. For example, CYP11A (also known as cholesterol desmolase, side chain cleavage enzyme, and P-450scc) is a mitochondrial enzyme and catalyzes the conversion of cholesterol (a C_{27} compound) to pregnenolone, a C_{21} steroid. CYP21 refers to the hydroxylase that catalyzes the hydroxylation at carbon 21. The pathways of steroidogenesis are shown in Figure 30.3. Formation of active corticosteroids in each zone of the adrenal cortex involves a series of transformations of the cholesterol molecule (steroidogenesis). The enzymes for steroidogenesis are distributed in the mitochondria and in the endoplasmic reticulum (Figure 30.2). Conversion of cholesterol to pregnenolone, the initial step, is regulated by ACTH and angiotensin II (see later). Pregnenolone in the zona glomerulosa is converted to aldosterone by transformations that involve formation of DOC (Figure 30.1). In the zona fasciculata, pregnenolone is directed mainly at the formation of cortisol and involves 17α-hydroxylation (Figures 30.1 and 30.3). In the zona reticularis, pregnenolone is utilized mainly for the formation of the androgen dehydroepiandrosterone (DHEA) and its sulfate (DHEAS), which also requires 17α-hydroxylation (Figures 30.1 and 30.3). DHEAS is synthesized from the addition of sulfate group at the 3-hydroxyl group. Phosphoadenosine phosphosulfate (PAPS) is sulfate donor (Chapter 15) and catalyzed by sulfotransferases. This hydroxylation is required for synthesis of cortisol and androgens (and also estrogens), but not for aldosterone. Since the zona glomerulosa lacks 17α-hydroxylase, it cannot synthesize cortisol or androgens.

The zona reticularis of the adrenal cortex secretes substantial amounts of DHEA and DHEAS daily, equaling or exceeding the amount of cortisol. Most of the DHEA and all of the DHEAS come from the adrenals. Negligible amounts of testosterone, dihydrotestosterone (DHT), and

TABLE 30.1 Relative Activity of Natural and Synthetic Corticosteroids

Steroid	Structure	Relative Bioactivity	
		Glucocorticoid	**Mineralocorticoid**
Natural			
Cortisol	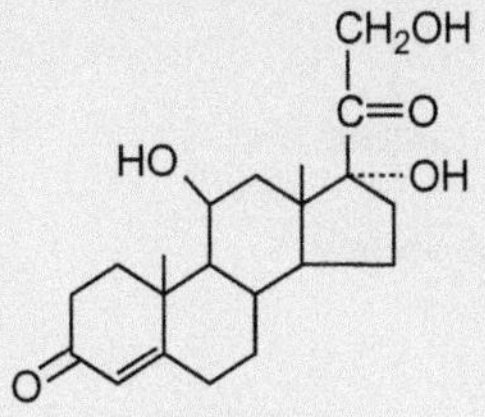	100	100
Corticosterone		20	200
11-Deoxycorticosterone		0	2000
Aldosterone	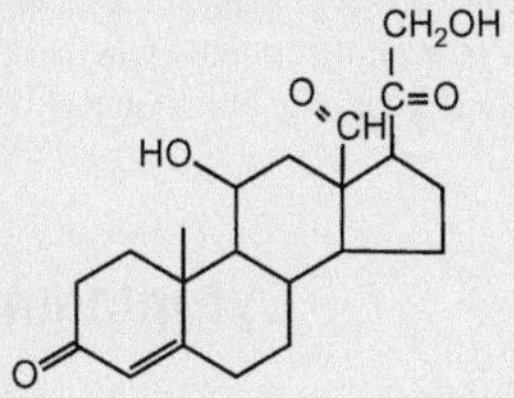	10	40,000
Synthetic			
Prednisolone (Δ^1-cortisol)	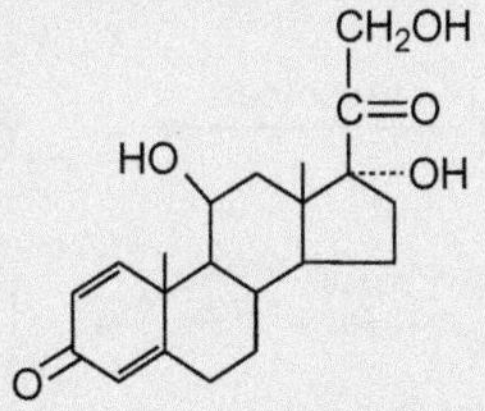	400	70
Methylprednisolone (6α-methylprednisolone)	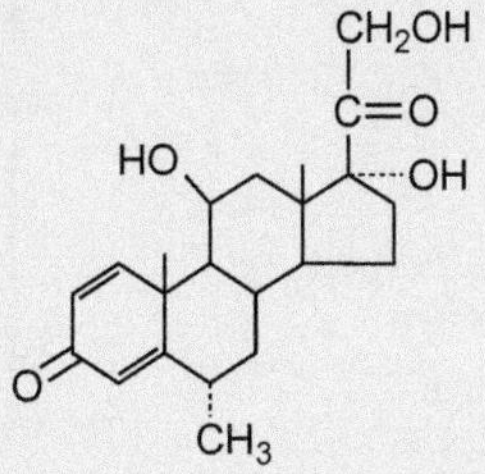	500	50
Dexamethasone (9α-fluoro-6α-methylprednisolone)	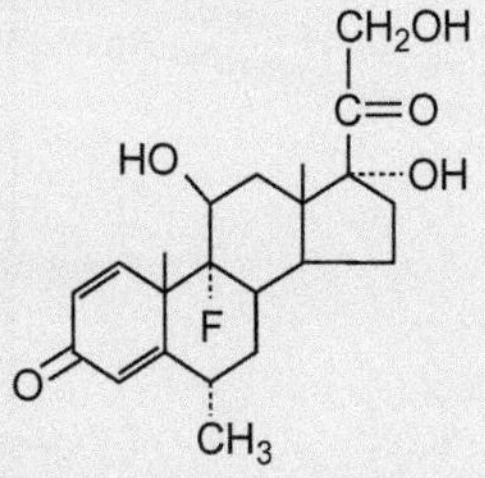	3000	0
Fludrocortisone (9α-fludrocortisol)	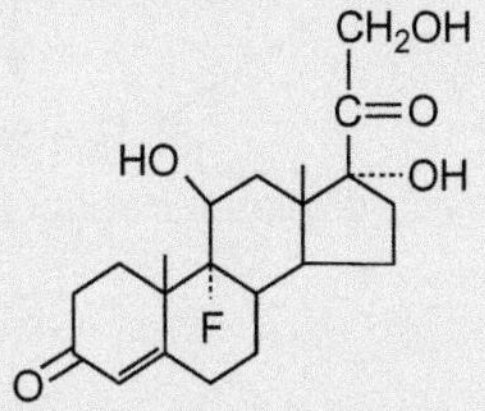	1000	40,000

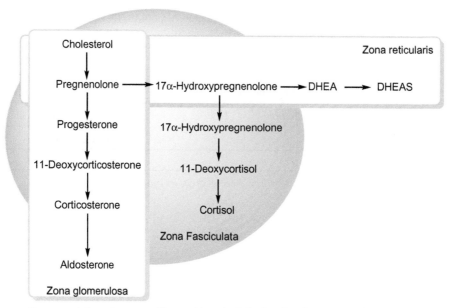

FIGURE 30.1 Diagrammatic representation of zone-specific steroid synthesis in the adrenal cortex. All three zones can be sites of convertsion of cholesterol into pregnenolone, but because of the presence/absence of certain enzymes, the flow of pregnenolone is preferentially directed to the synthesis of the steroid for which the particular zone is specific. It has not been determined whether the zona fasciculata produces dehydroepiandrosterone (DHEA) and dehydroepiandrosterone sulfate (DHEAS) in physiologically significant amounts.

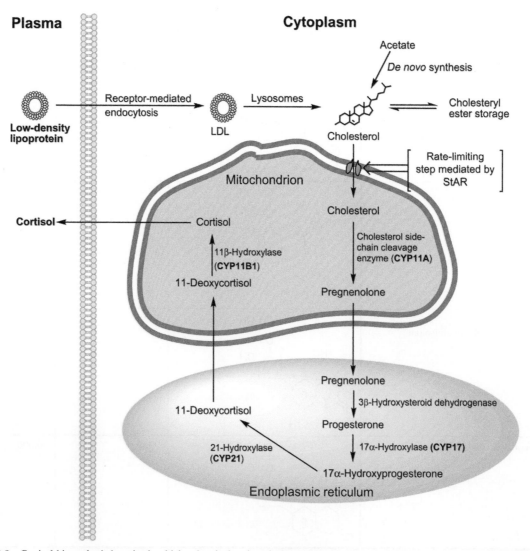

FIGURE 30.2 Cortisol biosynthesis by mitochondrial and endoplasmic reticulum enzymes of the adrenal cortex. The rate-limiting step in steroidogenesis is the import of cholesterol from cytoplasm to mitochondria, which is mediated by the steroidogenic acute regulatory protein (StAR).

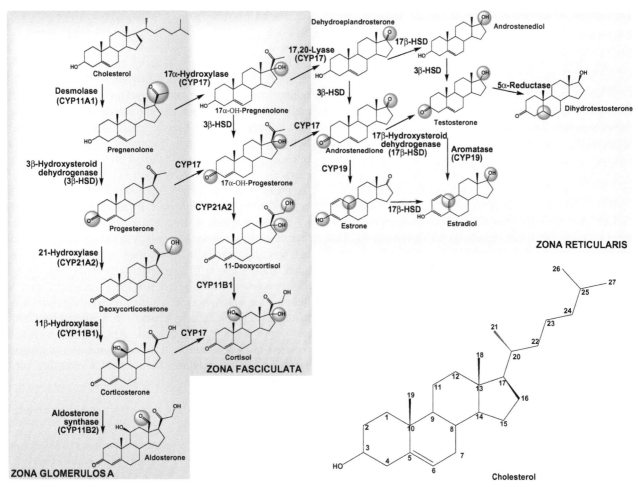

FIGURE 30.3 Biosynthesis of steroid hormones from cholesterol. The different pathways occur to varying extents in the adrenals, gonads, and placenta. The systematic names for cytochrome P-450 enzymes, namely CYP followed by a number, are given in parentheses. CYP11B2 and CYP17 possess multiple enzyme activities.

estradiol are secreted by the cortex. However, DHEA and, to a lesser extent, DHEAS undergo conversion to estradiol in skeletal muscle and adipose tissue. They can also be converted to testosterone. The adrenal cortex accounts for about two-thirds of the urinary 17-ketosteroids, which are a measure of androgen production. This steroidogenic versatility makes the adrenal cortex an important organ in certain disease states (see later).

Regulation of Corticosteroid Secretion

The zona glomerulosa and zona fasciculata differ in their content of enzymes, major secretory products, and regulators of their functions. They can be regarded as separate endocrine tissues.

Regulation of Aldosterone Secretion

The major regulators of aldosterone secretion are the renin−angiotensin system and extracellular potassium

ions (K^+). The former is sensitive to changes in intravascular volume and arterial pressure, whereas the latter is an aldosterone-regulated ion that feeds back to reduce aldosterone synthesis (simple negative feedback). Aldosterone secretion is also influenced (but not regulated) by ACTH and, directly and indirectly, by natriuretic peptides (ANP and BNP; Chapter 37) secreted by the heart.

Potassium ions exert a direct and stimulatory effect on aldosterone secretion that is independent of the renin−angiotensin system. A small increase in serum potassium levels elicits a rise in serum aldosterone levels, whereas a small decrease in serum potassium levels results in reduced levels of serum aldosterone. Aldosterone, by promoting the excretion of potassium, lowers the serum potassium level and thereby completes the negative feedback circuit. Potassium ions are believed to act by causing depolarization of the zona glomerulosa cells, allowing calcium ions to enter the cells through voltage-gated calcium channels. This increase in intracellular calcium ions fuels the calcium-dependent intracellular events triggered by angiotensin II (see later) and would explain why the angiotensin II effect on aldosterone secretion is augmented in hyperkalemic states and attenuated in hypokalemic states.

The **renin—angiotensin system** consists of components derived from precursor molecules produced by the liver and kidney. The juxtaglomerular apparatus of the kidney is a specialized region of the afferent arteriole that releases **renin**, an aspartyl protease (M.W. 42,000), in response to several stimuli (see later). Renin, derived from an inactive precursor (prorenin) by the action of prorenin processing enzyme (Figure 30.4), acts on the Leu—Val bond at the amino terminus of a circulating α_2-globulin called **angiotensinogen**, or renin substrate, secreted by the liver. The liberated amino terminal segment is an inactive decapeptide called angiotensin I. The physiologically important fate of this decapeptide is realized during passage through capillary beds in the lung and other tissues, where a converting enzyme, a dipeptidyl carboxypeptidase present in endothelial cells, splits off the carboxy terminal dipeptide (His—Leu) to yield the highly active octapeptide **angiotensin II** (Figure 30.4). Angiotensin II exerts sodium-dependent arteriolar constriction, and it acts directly on the zona glomerulosa to stimulate the secretion of aldosterone. Both of these effects are mediated by membrane-bound receptors coupled, via a G_s-subunit, to the phospholipase C-phosphatidylinositol 4,5-bisphosphate (PLC-PIP$_2$) complex, which utilizes calcium ions and protein kinase C (PKC) as intracellular effectors (Chapter 28). The effect of angiotensin II on aldosterone secretion involves an activation of both CYP11A and CYP11B2 (aldosterone synthase) activities in the zona glomerulosa and is influenced by dietary sodium intake. A low-sodium diet enhances the aldosterone response to angiotensin II, whereas a high-sodium diet tends to attenuate it.

Because renin release leads to the formation of angiotensin II and secretion of aldosterone, and because, under normal conditions, the concentrations of angiotensinogen and converting enzyme are not rate-limiting, any factor that influences the release of renin influences the secretion of aldosterone. Two important regulators of renin release are the mean renal arterial blood pressure and the extracellular fluid volume (blood volume). A decrease in the mean renal arterial blood pressure is sensed by baroreceptors in the juxtaglomerular apparatus, which responds by releasing renin. A decrease in blood volume, such as that caused by hemorrhage or by standing from a reclining position, results in diminished venous return to the heart. Baroreceptors in the atrial walls of the heart signal this change via cranial nerves to the vasomotor center in the medulla oblongata, which then relays the information to the juxtaglomerular apparatus via β-adrenergic fibers, and renin output is stimulated. Aldosterone, which is secreted in response to renin release, promotes potassium excretion and extracellular fluid retention, and thus inhibits further release of renin (Figure 30.5).

Regulation of Cortisol Secretion

The principal regulator of cortisol secretion is ACTH (see Chapter 29), the release of which is regulated by CRH and by circulating unbound cortisol. ACTH is also known as corticotropin. The balance of the effects of CRH and cortisol on the anterior pituitary maintains fairly constant

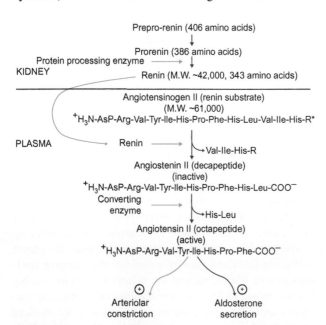

FIGURE 30.4 Renin—angiotensin system. R* is the remainder of the amino acid sequence.

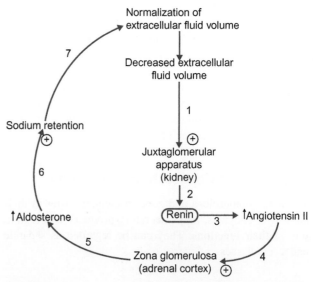

FIGURE 30.5 Regulation of renin release. Numbers 1−7 indicate the steps involved in the secretion of renin by the juxtaglomerular apparatus and its action. $\oplus$ = stimulatory; $\ominus$ = inhibitory; ↑ = increased production.

ACTH stimulation of the adrenals and circulating levels of cortisol. This closed-loop feedback system is superseded by neural signals from higher regions of the brain in non-steady-state conditions (e.g., circadian rhythm, stress).

ACTH exerts effects on the zona fasciculata by binding to cell membrane receptors following an elevation of intracellular cAMP via the G_s protein system (Chapter 28). It increases the activities of cholesterol esterase and cholesterol side chain cleavage enzyme (CYP11A), thereby stimulating production of pregnenolone. The cholesterol side chain cleaving enzyme is a cytochrome P-450 enzyme present in mitochondria. ACTH promotes the uptake of low-density lipoprotein (LDL) by increasing the number of LDL receptor sites. The adrenal cortex, although capable of *de novo* cholesterol synthesis, normally depends on LDL for its cholesterol requirements. By increasing the size of the pregnenolone pool, ACTH promotes the flow of steroids through the major pathways of the zona fasciculata. It also stimulates DHEA and DHEAS secretion by the zona reticularis, and aldosterone secretion from the zona glomerulosa. ACTH is not an important regulator of aldosterone secretion because there is no feedback communication between ACTH release and the zona glomerulosa.

Metabolism of Corticosteroids

Steroid hormones are hydrophobic molecules that affect diverse regions of the body through the bloodstream. Because of their very limited solubility in water, steroid hormones are rendered soluble either by binding to proteins or by conjugation. Steroids that are conjugated, either with sulfate or glucuronic acid, are unable to enter target tissues because of their water soluble (hydrophilic) property; they are usually destined for excretion by the kidney. Although a few tissues (e.g., placenta) have sulfatases that can liberate a steroid from its sulfoconjugated state, most tissues are lacking in this ability and are therefore unresponsive to the high levels of conjugated steroids in the blood. The physiologically important means by which steroid hormones are solubilized for transport in blood is the use of certain serum proteins that solubilize steroids. The liver produces three serum proteins that serve to solubilize steroid hormones: albumin, a nonspecific adsorber of hydrophobic substances that binds steroids with low affinity; corticosteroid-binding globulin (CBG, also called **transcortin**), which binds C_{21} steroids but has the greatest affinity for cortisol; and testosterone−estradiol-binding globulin (TeBG, also called **sex hormone-binding globulin**, SHBG) which binds steroids of the androstane (C_{19}) and estrane (C_{18}) series but has greater specificity for androgens than estrogens. Due to high-affinity binding, steroids are not accessible for tissue uptake while bound to either CBG or TeBG. Therefore, the greater the fraction of the steroid that is bound to these proteins, the longer the half-life of the steroid.

Synthetic Corticosteroids

The corticosteroids are ineffective when taken by mouth because of inactivation in the liver. Synthetic corticosteroids have been developed that are relatively resistant to hepatic inactivation and are active when taken orally (Table 30.1). Certain structural modifications enhance glucocorticoid activity while suppressing mineralocorticoid activity; other modifications enhance both.

Biological Actions of Aldosterone

Mechanism of Action

Aldosterone exerts its effect by binding to type I corticosteroid receptors in the cytoplasm, translocating to the nucleus, and binding to an acceptor site in the chromatin, resulting in gene activation and synthesis of a specific protein (see Chapter 28). Aldosterone induces the synthesis of aldosterone-induced protein (AIP), which is involved in transcellular Na^+,K^+-ATPase regulation.

The type I corticosteroid receptor (mineralocorticoid receptor) binds cortisol and aldosterone with equal affinity. Because the circulating level of cortisol normally exceeds that of aldosterone by about 1000-fold, activation of the receptor by aldosterone would probably not occur were it not for the presence of a cortisol-inactivating enzyme in cells responsive to aldosterone. This enzyme, 11β-HSD, catalyzes the conversion of cortisol to cortisone, a metabolite that is not recognized by the receptor. Inhibition or absence of this enzyme leads to excessive aldosterone-like effects due to receptor activation by cortisol, a condition referred to as "apparent mineralocorticoid excess" (AME). AME can result from ingestion of large amounts of licorice, which contains glycyrrhetinic acid, an inhibitor of 11β-HSD.

Physiological Effects of Aldosterone

The major effect of aldosterone is on the distal tubules of nephrons, where it promotes sodium retention and potassium excretion. Under the influence of aldosterone, sodium ions are actively transported out of the distal tubular cell into blood, and this transport is coupled to passive potassium flux in the opposite direction. Consequently, intracellular $[Na^+]$ is diminished and intracellular $[K^+]$ is elevated. This intracellular diminution of $[Na^+]$ promotes the diffusion of sodium from the filtrate into the cell, and potassium diffuses into the filtrate. Aldosterone also stimulates sodium reabsorption from salivary fluid in the salivary gland and from luminal fluid in the intestines, but these sodium-conserving actions are of minor importance.

Biological Actions of Cortisol

Mechanism of Action

Cortisol and the synthetic glucocorticoids (GCs) bind to the type II corticosteroid receptor (glucocorticoid receptor, GR), and the resultant GC–GR homodimer activates glucocorticoid response elements (GREs) in chromatin, which ultimately leads to the genomic effects characteristic of GCs (Chapter 28). However, not all of the biological effects of GCs can be explained by this mechanism. In exerting its antimitogenic effects, GCs interfere with mitogen-stimulated AP-1 transactivation of late response genes by binding of c-jun by the GC-GR complex. Thus, no induction of a GC-responsive gene is required for this effect.

Not all of the GC effects involve binding of GC to the GR. Elevated levels of cortisol and synthetic GCs are known to exert rapid-onset, nongenomic effects that are seen within minutes, and do not involve the GR or a change in gene expression. Such nongenomic effects, which include the rapid suppression of ACTH release, the inhibition of exocytosis in inflammatory cell types, and the strong inhibition of growth hormone (GH) release, are believed to involve the direct binding of the extracellular GC–CBG complex to cell membrane receptors.

Physiological Effects of Cortisol

Normal circulating levels of cortisol, including the circadian early morning rise and the moderate elevations after meals and minor stress, help to sustain basic physiological (vegetative) functions. Large amounts of cortisol released in response to major stresses enable the individual to withstand, or cope with, the metabolic, cardiovascular, and psychological demands of the situation. Cortisol (and other glucocorticoids) promotes the conservation of glucose as an energy source in several ways:

1. Cortisol induces and maintains the activity of all of the gluconeogenic enzymes in the liver. By increasing hepatic formation of glucose, cortisol promotes its conversion to hepatic glycogen. Insulin reduces cortisol-stimulated gluconeogenesis but potentiates its effect on glycogenesis. Glucagon, on the other hand, augments the gluconeogenic action of cortisol while inhibiting its effect on glycogen deposition.
2. Cortisol inhibits glucose utilization in peripheral tissues, such as skeletal muscle, adipose tissue, bone matrix, lymphoid tissue, and skin, by inhibiting glycolysis and promoting the use of fatty acids. This action is modulated by insulin and thyroid hormones but is potentiated by growth hormone (GH).
3. Cortisol promotes the liberation of fatty acids from adipose tissue by inducing and maintaining the synthesis of hormone-sensitive lipase (HSL), an effect supported by GH. The actual activity of HSL is controlled by those hormones that trigger its phosphorylation (glucagon, catecholamines) or dephosphorylation (insulin, prostaglandin E).
4. Cortisol, by inhibiting glucose utilization in peripheral tissues, exerts a mild antianabolic effect on these tissues. This effect diminishes their rate of amino acid incorporation, thus making more amino acids available for metabolism for hepatic protein synthesis and gluconeogenesis.

Cortisol has an important permissive influence on the cardiovascular system by conditioning many components of the system to respond maximally to regulatory signals. In the absence of GC, inadequacy of cardiovascular response can result in circulatory collapse and death, whereas in GC excess, increased cardiovascular responsiveness may result in hypertension. Cortisol is required for the vascular smooth muscle to respond to the vasoconstrictor effect of catecholamines (norepinephrine), but the mechanism is incompletely understood. The hypotension in GC deficiency is normalized by intravenous cortisol treatment. Cortisol enhances the positive inotropic effect of catecholamines on the heart, possibly by promoting coupling of β-adrenergic receptors to adenylate cyclase. Elevated levels of cortisol increase both cardiac output and stroke volume, whereas cortisol deficiency has the opposite effect. Cortisol (and GCs) increases the density of β_2-adrenergic receptors in vascular smooth muscle and promotes the vasodilating effect of epinephrine in some vascular beds.

Cortisol is an important modulator of the renin–angiotensin system. It stimulates the synthesis of angiotensinogen (renin substrate) by the liver, probably by prolonging the half-life of the angiotensinogen mRNA; it also stimulates the synthesis of angiotensin-converting enzyme (ACE) in vascular endothelial cells by promoting the expression of the ACE gene. By these actions, cortisol increases the magnitude of the renin–angiotensin response. Concurrently, however, cortisol promotes the expression of the natriuretic peptide (ANP) genes in cardiac muscle cells and thus allows for better modulation of a renin–angiotensin overshoot.

Cortisol maintains the reactivity of the reticular activating system, the limbic system, and areas of the thalamus and hypothalamus to sensory "distress" signals and to endogenous opiates. Cortisol thus has an important role in behavioral and neuroendocrine responses to stress. The presence of glucocorticoids appears to be necessary for the stress analgesia that is attributed to endogenous opiates. Cortisol promotes tissue responsiveness to catecholamines and induces adrenal medullary phenylethanolamine-N-methyltransferase (PNMT), which converts norepinephrine to epinephrine (see later). Thus, even before it reaches the general circulation, cortisol promotes cardiovascular performance by promoting the formation of a cardiotropic hormone.

These central, metabolic, and cardiovascular effects of cortisol are accentuated when large amounts of cortisol are released in response to severe stress. Although the plasma cortisol concentration attained during stress often increases 10-fold, this hypercortisolism is relatively acute and ephemeral, and the effects do not resemble those seen after chronic excesses of cortisol or synthetic glucocorticoids.

Pharmacological Effects of Glucocorticoids

When plasma glucocorticoid levels are chronically elevated, whether because of hyperactivity of the adrenal cortex or administration or consumption of synthetic glucocorticoids, some of the "physiological" effects become exaggerated, while other effects not normally seen appear. Protein catabolism is enhanced in skeletal muscle, skin, bone matrix, and lymphoid tissues by inhibition of protein synthesis and of cellular proliferation (DNA synthesis). Glucose utilization is severely inhibited and hepatic gluconeogenesis enhanced, which can lead to muscle weakness and atrophy, thinning and weakening of the skin, osteoporosis, diminished immunocompetence (from destruction of lymphocytes in lymphoid tissues), increased susceptibility to infections, and poor wound healing. These are classical features of **Cushing's syndrome** and **Cushing's disease**, in which the adrenal cortex secretes supraphysiological amounts of cortisol.

Chronic excess of glucocorticoids leads to elevated levels of glucose and free fatty acids in blood. Because these effects indicate inadequate counteraction by insulin, they are diabetogenic and are associated with an abnormal glucose tolerance test. Although glucocorticoid excess stimulates lipolysis, which leads to hyperlipidemia, body fat is not depleted. In fact, a form of obesity ("central obesity") involving redistribution of body fat to the abdomen, upper back, and face appears to be characteristic of glucocorticoid toxicity.

Chronic glucocorticoid excess during the period of growth (e.g., the peripubertal period) leads to suppression of cellular proliferation and production of growth-promoting hormones and results in stunted growth. Proliferation of fibroblasts and other cell types required for longitudinal bone growth and for somatic growth in general is severely affected. The function of these cell types is also inhibited, and adequate formation of tissue matrices is not possible. Release of GH and formation of IGFs, 5′-deiodinase activity, and release of TSH and ACTH are all inhibited. Synthetic glucocorticoids can exert a long-lasting repression of POMC synthesis in the anterior pituitary. After cessation of chronic glucocorticoid treatment, ACTH levels do not return to normal before 2−3 months, and cortisol secretion resumes only after an additional 6 months. Thus, withdrawal of exogenous steroids results in a state of adrenocortical deficiency for 8−9 months, during which time the imposition of stressful stimuli may have undesirable consequences.

High local concentrations of glucocorticoids inhibit or diminish inflammatory and allergic reactions. Cell-mediated inflammation (of joints, bursae, etc.) and allergic reactions (IgE-induced) are caused by release of agents designed to combat infection (see Chapter 33). Thus, chemotaxis of neutrophils and other invasive cells to the affected area is followed by release of lysosomal enzymes (e.g., collagenase), histamine, prostaglandin E_2 (PGE_2), superoxide anion radicals, and other mediators of inflammation that cause tissue destruction and vascular permeability changes (see Chapter16). Glucocorticoids counteract the inflammatory response by

1. Inhibiting phospholipase A_2 activity, thereby decreasing the synthesis of PGE, and of the potent chemotactic substance leukotriene B_4;
2. Stabilizing membranes of the lysosomes and secretory granules, thereby inhibiting the release of their contents; and
3. Acting directly on the capillary endothelium to render it less permeable.

This endothelial effect and the permissive effect of glucocorticoids on catecholamine-induced vasoconstriction inhibit edema and swelling. Use of glucocorticoids to reduce inflammation due to bacterial infection should not be undertaken without concurrent use of antibiotics.

Adrenal Androgen, Dehydroepiandrosterone

The zona reticularis is the innermost layer of the cortex, about equal in size to the zona glomerulosa in the adult. Although it shares similarities with the zona fasciculata, both histologically and functionally the zona reticularis should be regarded as a distinct entity because it has certain features that are not found in the other zones. One distinct feature is the relatively late appearance and growth of this zone (**adrenarche**) and its functional decline in late adulthood (**adrenopause**). Another unique feature is that the zona reticularis is the exclusive site of DHEAS formation in the adult, owing to the presence of a steroid sulfotransferase that attaches a sulfate to DHEA.

Disturbances in Adrenocortical Function
Deficiency

Primary adrenocortical insufficiency (Addison's disease) is a condition in which secretion of all adrenal steroids diminishes or ceases owing to the deterioration of adrenocortical function. The adrenal cortex atrophies as a result of an infectious disease or autoimmune disorder (see Clinical Case Study 30.1). Circulating levels of aldosterone and cortisol decrease, whereas those of renin and ACTH increase. ACTH levels increase enough to produce darkening of the skin, a hallmark of Addison's disease. Aldosterone deficiency encourages potassium retention and sodium loss and leads to hyperkalemia, hyponatremia hypovolemia, and hypotension. Deficiency of cortisol renders the tissues more sensitive to insulin and, hence, hypoglycemia may develop. In addition, cortisol deficiency diminishes responsiveness of tissues to catecholamines, particularly in vascular smooth muscles, which do not contract adequately in response to α-adrenergic stimulation. All of these changes can lead to circulatory collapse.

More specific adrenocortical deficiencies occur when only specific steroidogenic enzymes are affected. If production of cortisol, which is the exclusive feedback suppressor of ACTH release, is deficient, hypersecretion of ACTH with overstimulation of the adrenal cortex (hyperplasia) results. This condition is called **congenital adrenal hyperplasia**. The most common enzyme deficiency is that of 21-hydroxylase (CYP21), and neonatal diagnosis of this disorder is accomplished by measuring 17-hydroxyprogesterone. The incidence of 21-hydroxylase deficiency varies from 1:10,000 to 1:18,000 live births. The deficiency can be severe or mild, giving rise to a continuum of clinical symptoms; a deficiency in the level of 21-hydroxylase causes a lack of production of mineralocorticoids and glucocorticoids (Figure 30.3). Two steroid precursors, progesterone and 17-hydroxyprogesterone, accumulate and are diverted to the synthesis of androgens. The decreased production of cortisol removes the negative feedback on ACTH production by the pituitary, resulting in excess production in ACTH and overstimulation of the adrenal glands. *In utero*, female fetuses exposed to high levels of androgen show masculinization (virilization) and ambiguous genitalia. In contrast, male newborns with 21-hydroxylase deficiency appear normal at birth.

In infants with classic congenital adrenal hyperplasia as a result of 21-hydroxylase deficiency, the mineralocorticoid deficiency causes a rapidly evolving "salt wasting" crisis due to renal loss of Na^+ and accumulation of K^+. These changes result in hyponatremia, hyperkalemia, hyperreninemia, and hypovolemic shock. The electrolyte disturbances are treated with fluid and Na^+ replacement, and administration of both mineralocorticoid and glucocorticoid. The administration of glucocorticoid also corrects hypoglycemia. Because this life-threatening disorder is treatable, newborn screening programs have been developed. They consist of assaying 17α-hydroxyprogesterone from a filter paper dried blood spot. In milder, attenuated forms of 21-hydroxylase deficiency, virilization may be the sole clinical manifestation. In young women, this condition may cause hirsutism and menstrual irregularities.

Prenatal diagnosis and fetal therapy of affected female fetuses may be desirable since corrective surgery of external genitalia is not always optimal. Oral administration of a synthetic glucocorticoid (dexamethasone, which crosses the placenta) to the mother during pregnancy can limit the virilization of a female fetus. Dexamethasone is used instead of cortisol because cortisol does not readily cross the placental barrier.

Primary hypercortisolism (Cushing's syndrome) is usually due to an autonomous adrenocortical tumor and is characterized by low levels of circulating ACTH. Virtually all of the effects that have been described for glucocorticoid excess occur in this syndrome. Skeletal muscle, skin, bone, and lymphoid tissue exhibit protein loss, which can lead to muscle weakness, fragility of the skin, osteoporosis, and diminished immunocompetence. Chronic excess of cortisol encourages lipolysis in adipose tissues and favors fat deposition in the face and trunk regions, resulting in the characteristic "moon face" and central obesity. Because cortisol is an insulin antagonist, some glucose intolerance may develop. Removal of the adrenal tumor corrects this syndrome, but it also creates adrenocortical insufficiency, which requires exogenous glucocorticoid treatment. In the case of a unilateral tumor, the other adrenal gland atrophies in Cushing's syndrome, because ACTH secretion is chronically suppressed. Recovery of ACTH and cortisol secretion following removal of the tumor may take several months.

Overproduction of cortisol by hyperplastic adrenal cortices can result from oversecretion of ACTH (**secondary hypercortisolism**) due to a defect at the level of the pituitary or median eminence or to ectopic production of ACTH. The former condition is known as **Cushing's disease** and the latter as **ectopic ACTH syndrome**. In both conditions, the adrenal cortex functions normally and secretes cortisol in proportion to the level of ACTH. Diagnosis of Cushing's disease consists of establishing loss of the normal diurnal pattern of cortisol secretion, and an elevated 24-hour urinary free-cortisol level. If these determinations are equivocal, additional provocative tests are utilized (e.g., dexamethasone suppression test). See required reading reference 1 for the clinical case study on Nelson's syndrome, which is a variant of Nelson's disease.

Cushing's disease, which is more prevalent than ectopic ACTH syndrome, can be distinguished from the latter by its glucocorticoid suppressibility. Large doses of dexamethasone administered for three consecutive days suppress ACTH secretion. In ectopic ACTH syndrome, malignant transformation in some tissues (notably the lung) induces, in that tissue, the synthesis and release of ACTH (a paraneo-plastic endocrine syndrome). Because the transformed tissue is unresponsive to cortisol, ACTH and cortisol levels continue to rise and cannot be depressed by dexamethasone. A novel therapeutic approach to treat corticotropin-secreting adenomas is through the use of a somatostatin analogue known as pasireotide, since these adenomas express somatostatin receptors [2].

ADRENAL MEDULLA

The adrenal medulla is a modified sympathetic ganglion that lacks postsynaptic axonal projections. As such, it can be regarded as the endocrine component of the sympathetic division of the autonomic nervous system. Like its sympathetic ganglia counterparts, the adrenal medulla receives numerous preganglionic cholinergic fibers from the spinal cord that transmit autonomic efferent signals from the brain stem and higher centers. Unlike its sympathetic ganglia counterparts, which release norepinephrine into synaptic junctions at target cells, the adrenal medulla releases mainly epinephrine into the systemic circulation. Cells of the adrenal medulla are often referred to as "chromaffin cells" because they contain "chromaffin granules," electron-dense membrane-bound secretory vesicles with an affinity for chromic ions. Chromaffin granules contain catecholamines ($\sim 20\%$), various proteins ($\sim 35\%$), ATP (15%), lipids ($\sim 20\%$), calcium ions, ascorbic acid, and other substances; they are the adrenal medullary counterparts of secretory vesicles in ganglion cells.

Regulation of Release

The catecholamine content of mature chromaffin granules has been estimated to consist of 80% epinephrine (E),

16% norepinephrine (NE), and 4% dopamine (DA), although the percentage of E depends on the rate of cortisol production (discussed later). On cholinergic stimulation and depolarization of the adrenal medullary cells, the intracellular Ca^{2+} ion concentration increases, promoting fusion of chromaffin granules with the plasma membrane. This leads to exocytosis of all soluble granule constituents, including the catecholamines, ATP, dopamine β-hydroxylase, calcium ions, and chromogranins.

Adrenal medullary cells have plasma membrane receptors for acetylcholine (ACh) of the neuronal nicotinic subtype (N_N). These receptors are cation channels that span the plasma membrane and are activated by ACh to rapidly increase Na^+ and K^+ permeabilities (Na^+ influx rate $\sim 5 \times 10^7$ ions/s), causing the cells to depolarize and release their catecholamines by exocytosis. The cholinergic stimulation of exocytosis is accompanied by an activation of tyrosine hydroxylase activity within the adrenal medullary cell, and this promotes the biosynthesis of more catecholamines. This cholinergic activation of tyrosine hydroxylase involves phosphorylation of the enzyme, presumably by a $Ca^{2+}-$calmodulin-dependent protein kinase.

Synthesis of Epinephrine

Epinephrine (adrenaline) (Figure 30.6) is synthesized from tyrosine by conversion of tyrosine to 3,4-dihydroxyphenylalanine (dopa) by tyrosine-3-monooxygenase (tyrosine hydroxylase) in the cytosol. The mixed-function oxidase requires molecular oxygen and tetrahydrobiopterin, which is produced from dihydrobiopterin by NADPH-dependent dihydrofolate reductase. In the reaction,

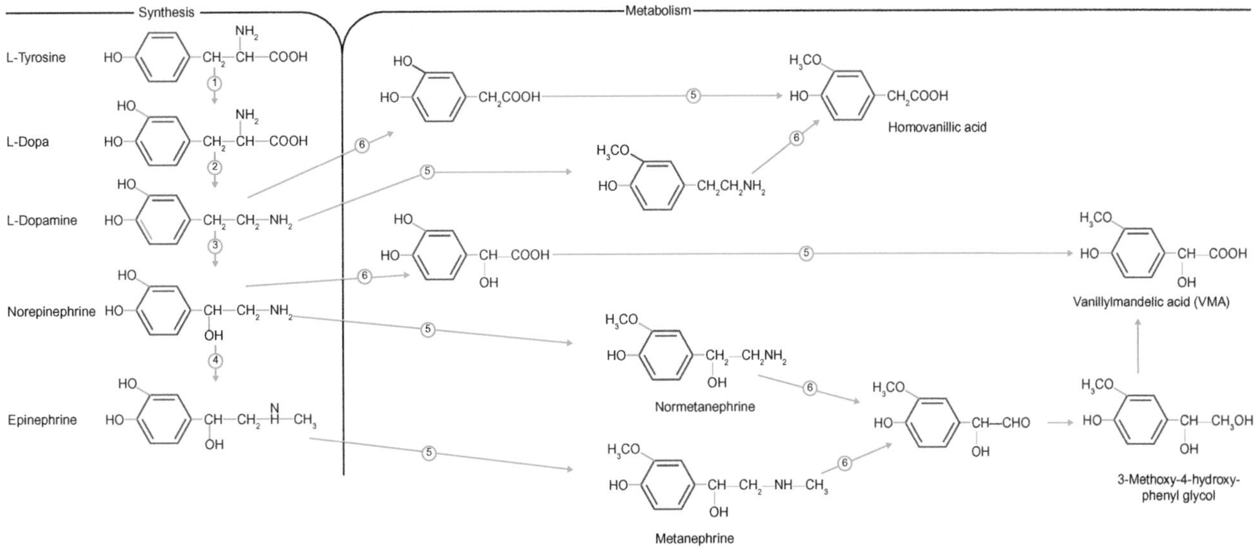

FIGURE 30.6 Synthesis and metabolism of catecholamines. Arrows indicate molecular conversions catalyzed by specific enzymes. Bold arrows indicate major (preferred) pathways. Enzymes: (1) tyrosine hydroxylase; (2) aromatic L-amino acid decarboxylase; (3) dopamine β-monooxygenase; (4) S-adenosyl-L-methionine:phenylethanolamine-N-methyltransferase (PNMT); (5) catechol-methyltransferase; (6) monoamine oxidase.

tetrahydrobiopterin is oxidized to dihydrobiopterin, which is reduced to the tetrahydro form by NADH-dependent dihydropteridine reductase. These reactions are similar to the hydroxylations of aromatic amino acids (phenylalanine and tryptophan), in which an obligatory biopterin electron donor system is used.

The rate-controlling step of catecholamine synthesis is the tyrosine hydroxylase reaction, for which the catecholamines are allosteric inhibitors. The enzyme is activated by the cAMP-dependent protein kinase phosphorylating system. α-N-Methyl-p-tyrosine is an inhibitor of this enzyme and is used to block adrenergic activity in pheochromocytoma (see below):

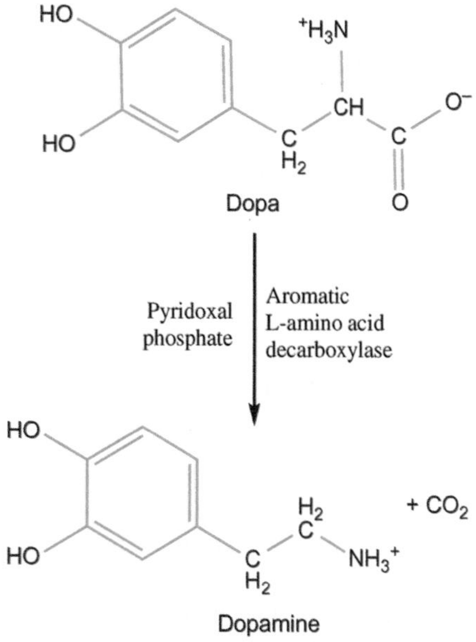

Dopa

Pyridoxal phosphate | Aromatic L-amino acid decarboxylase

Dopamine

Dopa is decarboxylated to 2-(3,4-dihydroxyphenyl) ethylamine (dopamine) by the aromatic L-amino acid decarboxylase, a nonspecific cytosolic pyridoxal phosphate-dependent enzyme also involved in the formation of other amines (e.g., 5-hydroxytryptamine).

Tyrosine can be decarboxylated to tyramine by the aromatic L-amino acid decarboxylase of intestinal bacteria. Tyramine, which is present in large amounts in certain foods (e.g., aged cheeses, red wines), is converted to the aldehyde derivatives by monoamine oxidase (MAO). However, individuals who are receiving MAO inhibitors for the treatment of depression can accumulate high levels of tyramine, causing release of norepinephrine from sympathetic nerve endings and of epinephrine from the adrenal medulla. This results in peripheral vasoconstriction and increased cardiac output, which can lead to hypertensive crises that cause headaches, palpitations, subdural hemorrhage, stroke, or myocardial infarction.

In dopaminergic neurons, dopamine is not metabolized further but is stored in presynaptic vesicles. In noradrenergic neurons and in the adrenal medulla, dopamine enters the secretory granules and is further hydroxylated to norepinephrine. This reaction is catalyzed by dopamine-β-monooxygenase, a copper protein that requires molecular oxygen and ascorbate or tetrahydrobiopterin:

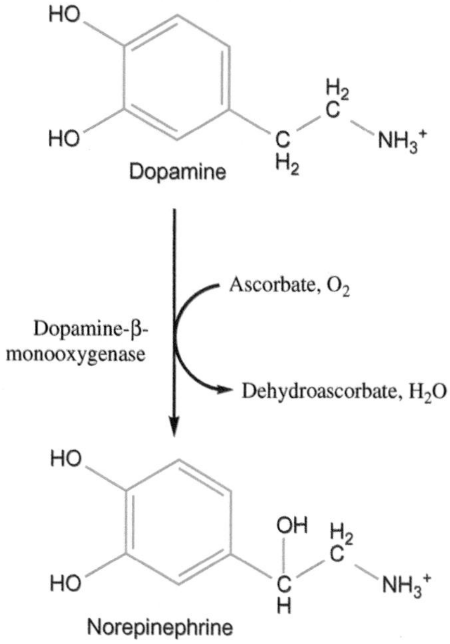

Dopamine

Dopamine-β-monooxygenase | Ascorbate, O_2 → Dehydroascorbate, H_2O

Norepinephrine

Norepinephrine diffuses into the cytosol, where it is converted to epinephrine by methylation of its amino group. This reaction, in which the methyl group is donated by S-adenosylmethionine, is catalyzed by S-adenosyl-L-methionine:phenylethanolamine-N-methyltransferase (PNMT). Epinephrine enters the granules and remains there until it is released:

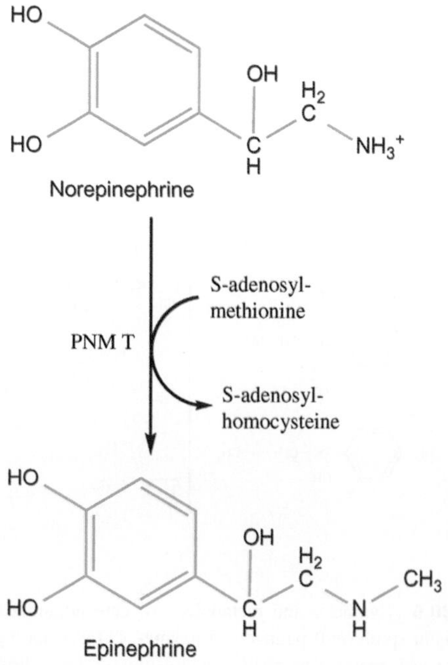

Norepinephrine

PNMT | S-adenosyl-methionine → S-adenosyl-homocysteine

Epinephrine

Catecholamine neurotransmitters in the central nervous system are synthesized within that tissue because they cannot cross the blood−brain barrier. However, dopa readily crosses the blood−brain barrier and promotes catecholamine synthesis. Thus, in disorders involving a deficiency of catecholamine synthesis, administration of dopa may have beneficial effects. In **Parkinson's disease**, in which deficiency of dopamine synthesis affects nerve transmission in the substantia nigra of the upper brain stem, administration of dopa leads to some symptomatic relief. Parkinsonism is a chronic, progressive disorder characterized by involuntary tremor, decreased motor power and control, postural instability, and muscular rigidity.

Regulation of Catecholamine Secretion

Regulation of Synthesis

Dopamine (**DA**) and **norepinephrine** (**NE**) are allosteric inhibitors of tyrosine hydroxylase and regulate catecholamine synthesis when the adrenal medulla is quiescent (unstimulated). Continuous stimulation of the adrenal medulla (as during prolonged stress) promotes tyrosine hydroxylase activity primarily because the turnover of DA and NE is rapid. Tyrosine hydroxylase activity is also regulated by cAMP and by cholinergic nerve activity. The enzyme is active when phosphorylated. Tonic cholinergic impulses maintain the activity of tyrosine hydroxylase, whereas chronic, intense cholinergic stimulation increases the activity of tyrosine hydroxylase and dopamine β-hydroxylase by increased protein synthesis.

Epinephrine production is catalyzed by PNMT, which is induced by glucocorticoids (cortisol). The venous drainage of the adrenal cortex, which contains very high concentrations of newly released cortisol, bathes the adrenal medulla before entering the general circulation.

Regulation of Release

The main regulator of catecholamine release from the adrenal medulla is cholinergic stimulation, which causes calcium-dependent exocytosis of the contents of the secretory granules. Exocytosis of the granular content releases epinephrine (E), NE, DA, dopamine β-hydroxylase, ATP, peptides, and chromaffin-specific proteins that are biologically inert. The amounts of DA and NE released are minor in comparison with that of E. Of the total catecholamine content in the granules, approximately 80% is E, 16% is NE, and the remainder is mostly DA.

Metabolism of Catecholamines

Two enzymes responsible for the inactivation of catecholamines are present in most tissues but are particularly abundant in the liver. **Catechol-O-methyltransferase** (**COMT**) is a cytosolic, Mg^{2+}-dependent enzyme that catalyzes methoxylation of catecholamines at the hydroxyl group at position 3. COMT utilizes S-adenosylmethionine as the methyl donor and usually initiates inactivation. **Monoamine oxidase** (**MAO**), a mitochondrial enzyme that oxidizes the amino side chain of catecholamines, acts generally (but not invariably) on methoxylated catecholamines. About 70% of the total output of urinary catecholamines is 3-methoxy-4-hydroxymandelic acid (also called **vanillylmandelic acid**, or **VMA**). Unmodified catecholamines represent 0.1%−0.4% of the total.

Biological Actions of Catecholamines

Catecholamines exert their effects through specific receptors on the target cell surface [3,4]. However, the effects elicited depend on the type or subtype of receptor with which they interact. There are three types of catecholamine receptor: dopamine, α-adrenergic, and β-adrenergic. Each of these consists of at least two or more subtypes, which differ with respect to ligand affinity, tissue distribution, postreceptor events, and drug antagonists. Their biological effects are rapid and are summarized in Table 30.2.

Disturbances in Adrenal Medullary Function

Because there is considerable overlap in the functions of the sympathetic nervous system and the adrenal medulla, elimination of the adrenal medulla would be tolerated as long as the autonomic nervous system remained functionally intact (see Clinical Case Study 30.2). However, overproduction of adrenal medullary hormones would be disruptive. Such catecholamine excesses are seen in patients with tumors of the adrenal medullary chromaffin cells and/or tumors of chromaffin tissue located outside the adrenal gland. These tumors are called **pheochromocytomas**, and, although their incidence is rare, the pathophysiology should be understood for full appreciation of adrenergic catecholamine functions under normal conditions. One site at which pheochromocytomas often develop is the organ of Zuckerkandl, a collection of pheochromocytes at the bifurcation of the aorta. Familial predisposition to pheochromocytoma can occur due to activating mutations of the *RET* proto-oncogene that cause multiple endocrine neoplasia type II (Chapter 35) and mutations in the von Hippel−Lindau tumor suppressor gene [5]. Familial and a few sporadic cases of pheochromocytoma and paragangliomas (tumors of the extra-adrenal sympathetic ganglia) are associated with loss of function mutations in succinate dehydrogenase subunit B. Complex II (succinate dehydrogenase) is a part of TCA cycle and mitochondrial electron transport. This

TABLE 30.2 The Catecholamine Receptors

Receptor Subtype[1]	Relative Affinity[2]	Subtype-Specific Agonist	Subtype-Specific Antagonist	Postreceptor Events	Tissue	Physiologic Ligand[3]	Effect
D1	DA≫E=NE		Metoclopramide	G_s; AC stimulation; increased cAMP	Vascular smooth muscle (kidney, coronary, mesenteric)	DA	Relaxation (low levels)
D2	DA≫E=NE	Bromocriptine		G_s; AC inhibition; decreased cAMP	Lactotrope (anterior pituitary)	DA	Inhibition of prolactin release
α₁[4]	E≳NE≫I	Methoxamine; Phenylephrine	Phentolamine; Phenoxybenzamine	G_q;PLC:IP$_3$:DAG; Ca^{2+}-Protein kinase	Vascular smooth muscle (skin)	NE	Constriction
					Vascular smooth muscle (arterioles)	NE	Constriction
					Vascular smooth muscle (abdominal viscera except liver)	NE	Constriction
					Intestinal smooth muscle	NE	Relaxation
					Intestinal sphincters	NE	Contraction
					Urinary smooth muscle (ureter, sphincter of bladder)	NE	Increased ureter tone, contraction of sphincter
					Iris (radial muscle)	NE	Contraction [mydriasis (pupil dilation)]
					Skin (pilorector muscle)	NE	Contraction (pilorection)
					Skin (sweat gland; palms of hands; soles of feet)[5]	NE	Secretion
					Hypothalamus	NE	Stimulation of ADH release
α₂[6]	E≳NE≫I	Clonidine	Yohimbine	G_i;AC inhibition; Decreased cAMP	β-cells of islets of Langerhans	NE, E	Inhibition of insulin secretion
					Vascular smooth muscle (skin)	E	Constriction (vasoconstriction)
					Vascular smooth muscle (veins)	NE, E	Constriction
					Hypothalamus	NE	Decreased SRIH and/or increased GRH release (results in increased GH release from pituitary)

Receptor	Agonist order	Agonist	Antagonist	Mechanism	Tissue	Agonist	Response
β₁[7]	I > E = NE	Dobutamine	Metoprolol; atenolol	Gₛ:AC stimulation; Increased cAMP	Heart (SA node, VA node, His–Purkinje system, ventricles)	NE, E	Increased firing rate; increased conduction velocity; increased contractility; increased heart rate; increased cardiac output
					Juxtaglomerular apparatus of kidney	NE	Stimulation of renin release
					Adipose tissue (white)	E	Lipolysis
β₂	I > E ≫ NE	Terbutal; salbutamol; soterenol	Butoxamine	Gₛ:AC stimulation; increased cAMP	Vascular smooth muscle (skeletal muscle)	E	Relaxation (vasodilation)
					Vascular smooth muscle (coronary)	E	Relaxation (vasodilation)
					Bronchiolar smooth muscle	E	Relaxation (bronchodilation)
					Skeletal muscle	E	Glycogenolysis; increased contractility?
					α-cells of islets of Langerhans	NE, E	Stimulation of glucagon release
					Intestinal smooth muscle		
					Genitourinary smooth muscle		
					Liver	E	Glycogenolysis; gluconeogenesis
β₃	I = NE > E, low affinity receptor	Fenoterol	None[8]	Gₛ:AC stimulation; increased cAMP	White adipose tissue (visceral > subcutaneous)	NE, E	Lipolysis
					Brown adipose tissue	NE (and E)	Thermogenesis

[1] α₁ and α₂ each have three subtypes; information about these can be found in D. B. Bylund, Subtypes of α₁- and α₂-adrenergic receptors. *FASEB J 6* (1992), 832–839.

[2] DA = dopamine; E = epinephrine; NE = norepinephrine; I = isoproterenol (a nonselective β-adrenergic agonist that stimulates all three types of β-adrenoceptors).

[3] NE = norepinephrine as neurotransmitter released from sympathetic nerve endings at synaptic junctions, acting on postjunctional adrenergic receptors; E = plasma-borne epinephrine that binds to extrajunctional adrenergic receptors that are not associated with nerve terminals.

[4] Localized entirely postsynaptically.

[5] Most of the sweat glands are innervated by **cholinergic** sympathetic fibers that release ACh at synaptic junctions and respond to sympathetic stimulation with an increase in secretion (increased sweating). The exceptions are the sweat glands in the palms of the hands and the soles of the feet, which are innervated by adrenergic fibers that release NE at synaptic junctions. Emotional excitement activates these adrenergic nerves and causes sweating in the hands and feet ("adrenergic sweating"). Emotional excitement usually has no effect on sweating in other regions because the cholinergic sympathetic nerves are mainly controlled by the thermoregulatory center of the hypothalamus and are not affected by behavioral reactiveness. Note, however, that all sweat glands have α₁ adrenoceptors and are therefore capable of responding to plasma epinephrine (and NE?) with an increase in sweat secretion. This would result in more generalized sweating.

[6] Located both pre- and postsynaptically.

[7] Primarily involved in neuronal functions.

[8] Metoprolol is a nonselective antagonist.

observation suggests a link between mitochondrial dysfunction and tumorigenesis. Pheochromocytoma is usually characterized by intermittent to permanent hypertension with potentially life-threatening consequences. The biochemical diagnosis in patients suspected of pheochromocytoma consists of measuring epinephrine, norepinephrine, and their metabolites—namely, metanephrine, normetanephrine, dihydroxyphenylglycol, and vanillylmandelic acid in a 24-hour urine specimen. Measurement of plasma levels of total free metanephrines has a very high sensitivity for detecting pheochromocytoma [6]. However, elevated plasma metanephrine levels can occur during medication with, for example, tricyclic antidepressants and serotonin−norepinephrine reuptake inhibitors, which give rise to false-positive results [7]. Tumors of the pheochromocytes can be surgically removed.

CLINICAL CASE STUDY 30.1 Primary Adrenal Deficiency (Addison's Disease)

This case was abstracted from: K. Visrodia, R. Shvashankar, A. T. Wang, 58-year-old woman with progressive nausea and fatigue, Mayo Clin. Proc. 89 (2014) 1004−1008.

Synopsis

A 58-year-old woman was eventually presented to the emergency department with acute illness consisting of nausea, vomiting, fatigue, and weakness. Initially, she was diagnosed as hypothyroid because her serum TSH was elevated. However, despite normal T_4 and T_3 serum values, treatment with levothyroxine worsened her clinical condition.

The subject's laboratory studies revealed hyponatremia, hyperkalemia, and low plasma cortisol levels. Based on all of these studies, a diagnosis of primary adrenal insufficiency was made. This diagnosis was also supported by very high levels of serum corticotropin (ACTH), which occurs due to loss of negative feedback from cortisol.

Resolution of the case: The subject underwent appropriate replacement therapy with glucocorticoid and mineralocorticoid therapy, and her clinical condition and laboratory studies were resolved. The replacement therapy is life-long; therefore, the patient wears a medical bracelet identifying her condition.

Teaching Points

1. Autoimmune adrenalitis, which promotes the destruction of adrenal cortex, is the most common cause of primary adrenal deficiency.

2. A prompt diagnosis and therapeutic intervention of primary adrenal deficiency are essential to prevent fatal adrenal crisis.

3. The hyperpigmentation of the subject is due to elevated corticotropin, which stimulates melanocyte receptors.

CLINICAL CASE STUDY 30.2 Pheochromocytoma

This case was abstracted from: A.S. Davison, S.J. Pattman, R. D.G. Neely, R. Bliss, S.G. Ball, A 12-cm mass with no symptoms and unremarkable laboratory results, Clin. Chem. 59 (2013) 1561−1566.

Synopsis

A 60-year-old man with a past medical history of hypertension controlled with an angiotensin receptor blocker presented to his primary care physician with abdominal discomfort and flu-like symptoms. He was found to have an elevated serum alkaline phosphatase level of 131 U/L (reference interval 35−120). An abdominal ultrasound revealed a 12-cm cystic mass in the area of the left kidney, and a CT scan suggested that the mass was located on the left adrenal gland. Plasma-free metanephrines showed only borderline increases in normetanephrine and metanephrine, which were not diagnostic of pheochromocytoma. However, because the patient was not taking any medications known to cause a physiological increase in plasma-free metanephrines, and because of the size of the mass, the patient was started on phenoxybenzamine, an α-blocker, in preparation for surgical resection of the adrenal mass. After the surgery, histologic examination of the mass revealed cystic pheochromocytoma with intratumoral hemorrhage and degeneration. The patient's plasma-free metanephrine levels returned to normal after the surgery, and his hypertension resolved. Genetic testing revealed no germline mutations known to cause pheochromocytoma, supporting the diagnosis of sporadic disease.

Teaching Points

1. Pheochromocytoma (PCC) and paraganglioma (PGL) are rare neuroendocrine tumors of the chromaffin cells, usually found in the adrenal medulla in about 80% of cases. Because the adrenal gland normally secretes catecholamines in response to stress, a tumor of the adrenal gland may present clinically with headaches, palpitations, diaphoresis, and severe or life-threatening hypertension. Catecholamine excess may also cause glucose intolerance as manifested by diabetes mellitus (refer to Chapter 20).

2. The adrenal gland secretes catecholamines, mainly epinephrine, norepinephrine, and dopamine, in response to stress. The metabolites of these catecholamines are vanillylmandelic acid, metanephrine, and normetanephrine, which are then excreted from the body in the urine.

(Continued)

CLINICAL CASE STUDY 30.2 (Continued)

Measurement of plasma-free metanephrines or urine-fractionated metanephrines is the recommended screening test for PCC/PGL because these tests are not subject to the fluctuating levels seen for catecholamines.

3. Concentrations of catecholamine metabolites should be interpreted in conjunction with imaging results. Radiologic features like a thickened wall, lack of internal septa, or persistent enhancement after contrast media usage may indicate an unusual pathology like cystic PCC, which is associated with minimally increased or normal plasma-free metanephrine levels.

4. Several drugs can cause elevated metanephrine levels by altering catecholamine production, leading to false positive results for PCC. These include tricyclic antidepressants, selective serotonin reuptake inhibitors, tetracyclines, nonspecific alpha-blockers, calcium channel blockers, and beta-blockers.

Supplemental References

[1] S.G. Waguespack, T. Rich, E. Grubbs, A.K. Ying, N.D. Perrier, M. Ayala-Ramirez, A current review of the etiology, diagnosis, and treatment of pediatric pheochromocytoma and paraganglioma, J. Clin. Endocrinol. Metab. 95 (2010) 2023–2037.

[2] S.G. Sheps, N.S. Jiang, G.G. Klee, J.A. van Heerden, Recent developments in the diagnosis and treatment of pheochromocytoma, Mayo Clin. Proc. 65 (1990) 88–95.

[3] J.J. Hines, D.L. Kessler, Unexplained remission of long standing severe diabetes mellitus, Ann. Intern. Med. 55 (1961) 314–316.

[4] P.G. Balestrieri, S. Spandrio, G. Romanelli, G. Giustina, Diabetes mellitus as presenting feature in extra-adrenal pheochromocytoma, Acta Diabetol. Lat. 27 (1990) 261–265.

REQUIRED READING

[1] N.A. Tritos, P.W. Schaefer, T.D. Stein, Case 40-2011: a 52-year-old man with weakness, infections, and enlarged adrenal glands, N. Engl. J. Med. 365 (2011) 2520–2530.

[2] A. Colao, S. Petersen, J. Newell-Price, J.W. Findling, F. Gu, M. Maldonado, et al., A 12-month phase 3 study of pasireotide in Cushing's disease, N. Engl. J. Med. 366 (2012) 914–924.

[3] J.J. Rilstone, R.A. Alkhater, B.A. Minassian, Brain dopamine-serotonin vesicular transport disease and its treatment, N. Engl. J. Med. 368 (2013) 543–550.

[4] J.E. Ahlskog, Parkinson disease treatment in hospitals and nursing facilities: avoiding pitfalls, Mayo Clin. Proc. 89 (2014) 997–1003.

[5] N.M. Neary, K.S. King, K. Pacak, Drugs and pheochromocytoma—don't be fooled by every elevated metanephrine, N. Engl. J. Med. 364 (2011) 2268–2270.

[6] D. Astuti, F. Latif, A. Dallol, P.L.M. Dahia, F. Douglas, E. George, et al., Gene mutations in the succinate dehydrogenase subunit SDHB cause susceptibility to familial pheochromocytoma and to familial paraganglioma, Am. J. Hum. Genet. 69 (2001) 49–54.

[7] G. Eisenhofer, D.S. Goldstein, M.W. Walther, P. Friberg, J.W.M. Lenders, H.R. Keiser, et al., Biochemical diagnosis of pheochromocytoma: how to distinguish true- from false-positive test results, J. Clin. Endocrinol. Metab. 88 (2003) 2656–2666.

Chapter 31

Endocrine Metabolism IV: Thyroid Gland

Key Points

1. The thyroid gland consists of two lobes connected by an isthmus; it is adjacent to both sides of the trachea.

2. The gland consists of spherical follicles, which are formed by a single layer of epithelial endocrine cells. These cells are responsible for the synthesis of thyroid hormones. The lumen of the follicle contains storage form thyroid hormones consisting of colloidal material. The thyroid gland also contains another type of endocrine cell; these cells are scattered around the follicles. These cells, known as parafollicular cells and also called C-cells, secrete calcitonin, which is involved in calcium metabolism.

3. The synthesis and secretion of thyroid hormones—mostly T_4 (L-isomer) and small amounts of T_3—are under the regulation of the cerebral cortex–hypothalamus–anterior pituitary–thyroid–peripheral tissues axis. The hypothalamus releases thyrotropin-releasing hormone (TRH), stored in the median eminence, into hypothalamic–pituitary portal vein. TRH causes anterior pituitary thyrotropes to release thyroid-stimulating hormone (TSH) by the G-protein-mediated intracellular second messenger Ca^{2+}, diacylglycerol and phosphatidylinositol, pathway.

4. TSH stimulates many processes of the thyroid gland, including the growth of the gland and synthesis and secretion of the thyroid hormones. The effects of TSH on the follicular cells are mediated by production of second messengers, Ca^{2+}, phosphatidylinositol, and c-AMP by the stimulation of $G_Q\alpha$- and $G_s\alpha$-proteins, respectively.

5. The synthesis of T_4 and T_3 in the thyroid follicular cells consists of the following steps:

    a. Active transport of I^- mediated by Na^+/I^- symporter coupled to Na^+, K^+-ATPase.

    b. Eventually, iodide ion transportation to the follicular lumen by a pendrin-mediated active transport system located in the apical membrane.

    c. An obligatory H_2O_2 (peroxide) generating system involving molecular oxygen (O_2) and NADPH, required for conversion of an I^- ion into an active form.

    d. Organification of an I^- ion into the selected tyrosine residues of thyroglobulin (TG) followed by coupling of the iodinated tyrosine residues to form either T_4-TG or T_3-TG, both reactions catalyzed by thyroid peroxidase. The main product is T_4-TG.

6. Thyroid peroxidase is a hemoprotein and is bound to apical plasma membrane of thyroid follicular cells with its catalytic domain facing the colloidal space of the lumen. Iodinated-TG is stored within the follicle as colloidal material.

7. The release of thyroid hormones from the follicular cells to the capillary blood occurs by retrieval of iodinated-TG from the colloidal droplets into the follicular cell by endocytosis, fusion of droplets with lysosomes, proteolysis of iodinated-TG by lysosomal proteases and release of T_4, T_3, and other iodinated tyrosines. T_4 and T_3 are released in to the capillary blood. The metabolically inactive iodinated tyrosine derivatives are deiodinated, and the iodide that is reclaimed is used in thyroid hormone biosynthesis.

8. T_4 and T_3 are transported in the blood bound to proteins, primarily thyroxine-binding globulin, and to a lesser extent to transthyretin and albumin. The free T_4 and free T_3, which are biologically active forms, constitute only 0.03% and 0.3% of the total, respectively. Thus, the presence of a large reservoir of T_4 and T_3 provides a buffer against acute changes in thyroid function as well as the changes in the concentrations of the binding proteins. T_4 has a larger pool size and has a half-life of about 7 days, whereas T_3 has a smaller pool size and a half-life of 1 day.

9. T_4 is a prohormone. In the peripheral tissues, deiodination of T_4 in the outer ring 5'-position yields T_3, which is biologically active, whereas the deiodination in the inner ring 5-position gives rise to an inactive product known as reverse T_3. These reactions are catalyzed by deiodinases, which contain selenocysteine at their active sites. Deiodinases are under metabolic regulation.

10. Numerous metabolic and developmental functions of thyroid hormones are mediated by binding to nuclear receptors that interact with transcription factors leading to changes in gene expression. Thyroid hormones are essential to energy homeostasis.

11. Thyroid hormone deficiency has many causes: iodide deficiency in the diet, thyroid agenesis, defects in synthesis, loss of thyroid tissue (e.g, autoimmune disease, radiation, thyroid surgery). Neonatal hypothyroidism has serious clinical consequences of mental retardation and growth abnormalities (cretinism). Timely diagnosis and supplementation of T_4 corrects the clinical abnormality.

12. Excess thyroid hormone production gives rise to a hypermetabolic state affecting a number of organs including the heart by reinforcing catecholamine hormone

N. V. Bhagavan and C.-E. Ha: Essentials of Medical Biochemistry, Second edition. DOI: http://dx.doi.org/10.1016/B978-0-12-416687-5.00031-2

action via the β-adrenergic receptors. One form of hyperthyroidism is an autoimmune disease and is mediated by autoantibodies produced against the TSH receptor on the follicular cell. Thyroid hormone synthesis is inhibited by thionamide-containing drugs propylthiouracil (PTU) and methimazole. PTU also inhibits the conversion of T_4 to T_3 in the peripheral tissues. The abnormal cardiovascular manifestations of thyroid hormones are treated with β-adrenergic antagonists. Excess iodide administration inhibits thyroid hormone synthesis (Wolff–Chaikoff effect) and is used in certain hyperthyroid states.

13. Both deteriorating conditions at either end of thyroid function—hypo (myxedema coma) or hyper (thyrotoxic storm) thyroidism—are medical emergencies and require prompt treatment strategies.

INTRODUCTION

The thyroid gland consists of two lobes connected by an isthmus, positioned on the ventral surface of the trachea just below the larynx. It receives adrenergic fibers from the cervical ganglion and cholinergic fibers from the vagus; it also is profusely vascularized by the superior and inferior thyroid arteries. Histologically, thyroid tissue is composed of numerous follicles lined by a single layer of epithelial cells around a lumen filled with proteinaceous material called **colloid** (Figure 31.1).

The follicle cells, called **thyrocytes**, produce the thyroid hormones and are derived from the entodermal pharynx. Interspersed between follicles are specialized amine precursor uptake and decarboxylation (APUD) cells

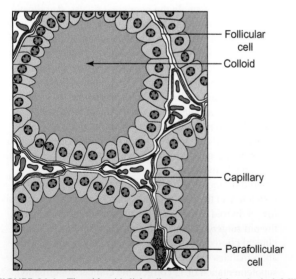

FIGURE 31.1 Thyroid epithelial cells are arranged in spherical follicles. The central portion of the follicles contains colloidal material made of thyroglobulin-coupled thyroid hormones. Parafollicular cells are also endocrine cells and produce calcitonin, which is involved in calcium metabolism.

derived from the neural crest, called **C-cells** or **parafollicular cells**. These cells produce calcitonin (CT), a polypeptide hormone that participates in Ca^{2+} homeostasis (discussed in Chapter 35).

THYROID HORMONE SYNTHESIS

Two substrates are required in the synthesis of thyroid hormones. The intrinsic substrate is **thyroglobulin (Tg)**, a homodimeric glycoprotein (M.W. 669,000) synthesized in the rough endoplasmic reticulum of the thyrocytes and secreted into the follicular lumen by exocytosis. Tg contains 134 tyrosyl residues, only 25–30 of which undergo iodination and only 4 of which become a hormonogenic segment of the molecule.

The extrinsic substrate is elemental **iodine**, present in food as inorganic iodide. Iodide is readily absorbed via the small intestine; it is almost entirely removed from the general circulation by the thyroid and kidney. Iodide that is taken up by salivary and gastric glands is secreted in salivary and gastric fluids, and is returned to the plasma iodide pool by intestinal reabsorption. A small amount of iodide is taken up by the lactating mammary gland and appears in milk. The synthesis and release of thyroid hormone involves a number of steps (Figure 31.2).

1. *Uptake of Iodide.* Iodide (I^-) is actively taken up by the follicular cells at the basolateral membrane against electrical and concentration gradients. The active uptake of I^- is mediated by the Na^+/I^- symporter, an intrinsic membrane protein with approximately 13 transmembrane segments. The Na^+ and I^- bound symporter releases both Na^+ and I^- on the cytoplasmic side. The empty symporter then returns to its original conformation, exposing binding sites on the external surface of the cell. The internalized Na^+ is pumped out of the cell by the ATP-dependent Na^+, K^+-ATPase (ouabain sensitive) to maintain an appropriate ion gradient. In the normal gland, the limiting step of thyroid hormone synthesis is uptake of I^-. Both I^- and thyroid-stimulating hormone (TSH) regulate the Na^+/I^- symporter function. TSH promotes I^- uptake, whereas excess I^- decreases I^- uptake (discussed later). Anions of similar charge and ionic volume (e.g., ClO_4^-, BF^-, TcO_4^-, SCN^-) compete with I^- for transport in the follicular cell.

2. *Activation and Organification of Iodide.* These processes take place in the thyroid follicular lumen. Iodide enters the lumen by a pendrin-dependent active process. Iodide that enters thyrocyte lumen is "activated" by oxidation that is catalyzed by thyroperoxidase (TPO). Hydrogen peroxide is required and is supplied by an H_2O_2-generating system that may be an NADPH (NADH) oxidase system similar to that of

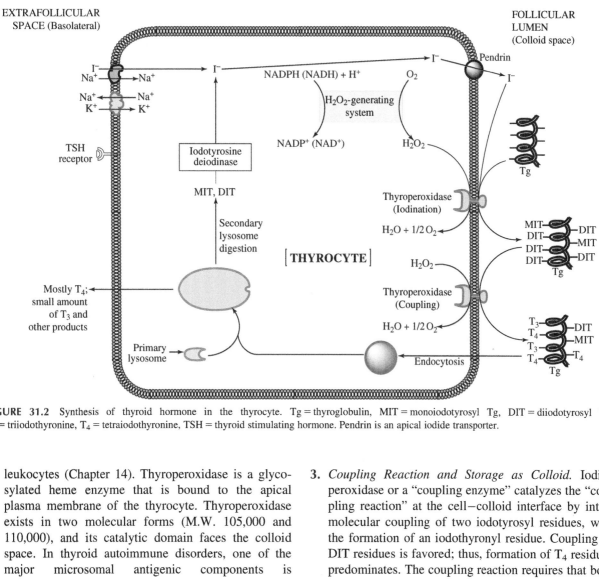

EXTRAFOLLICULAR
SPACE (Basolateral)

FOLLICULAR
LUMEN
(Colloid space)

FIGURE 31.2 Synthesis of thyroid hormone in the thyrocyte. Tg = thyroglobulin, MIT = monoiodotyrosyl Tg, DIT = diiodotyrosyl Tg, T_3 = triiodothyronine, T_4 = tetraiodothyronine, TSH = thyroid stimulating hormone. Pendrin is an apical iodide transporter.

leukocytes (Chapter 14). Thyroperoxidase is a glycosylated heme enzyme that is bound to the apical plasma membrane of the thyrocyte. Thyroperoxidase exists in two molecular forms (M.W. 105,000 and 110,000), and its catalytic domain faces the colloid space. In thyroid autoimmune disorders, one of the major microsomal antigenic components is thyroperoxidase.

In vitro, iodide peroxidase catalyzes the reaction $2H^+ + 2I^- + H_2O_2 \rightarrow I_2 + 2H_2O$. However, *in vivo* the enzyme probably forms an enzyme-bound iodonium ion $(E-I^+)$ or a free radical of iodine. These activated derivatives iodinate the tyrosyl residues of Tg ("organification of iodide") by the action of thyroperoxidase and H_2O_2. Thionamides (e.g., thiourea, thiouracil, propylthiouracil, methimazole) inhibit this organification of iodide without affecting iodide uptake. The first position iodinated is position 3, which forms monoiodotyrosyl-Tg (MIT-Tg). The second is position 5, which forms diiodotyrosyl residues (DIT-Tg). Iodination is accompanied by structural changes (e.g., cystine formation) and includes a conformational change in the Tg molecule. Activation and organification occur at the cell–colloid interface where thyroperoxidase activity is prevalent.

3. *Coupling Reaction and Storage as Colloid.* Iodide peroxidase or a "coupling enzyme" catalyzes the "coupling reaction" at the cell–colloid interface by intramolecular coupling of two iodotyrosyl residues, with the formation of an iodothyronyl residue. Coupling of DIT residues is favored; thus, formation of T_4 residues predominates. The coupling reaction requires that both substrates be iodinated and that one of the substrates be DIT-Tg. For this reason, no T_0, T_1, or T_2 residues are formed. Under normal conditions, less than half of the iodotyrosyl residues in Tg undergo coupling such that of the total iodinated residues in Tg, 49% are MIT, 33% are DIT, 16% are T_4, 1% is T_3, and a trace amount is rT_3. The ratio of T_4 to T_3 is ≥ 10 in a typical American diet, but it rises or falls with dietary iodide content. The coupling reaction appears to be the most sensitive to inhibition by the thionamides (propylthiouracil, methimazole), being inhibited at doses that do not inhibit the organification reactions. As a final note, although proteins other than Tg are iodinated in nonthyroid tissues that take up iodine (e.g., salivary gland, mammary gland, intestinal mucosa), no coupling of the iodinated residues occurs. Thus, the requirements for the coupling reaction may be more stringent than previously believed.

4. *Processing of TG and Release of Thyroid Hormone.* Thyroglobulin, with its tetra-iodothyronyl residues, is a pre-prohormone stored in the follicular lumen.

When follicle cells are stimulated by TSH or TSI (see later), their luminal border ingests colloid (and hence Tg) into endocytotic vesicles. These vesicles fuse with lysosomes to form phagolysosomes, which, during transit toward the basolateral surface of the cell, hydrolyze Tg and release T_4, DIT, MIT, and a small amount of T_3. These substances are released into the cytosol, and T_4 and T_3 diffuse into the blood. MIT and DIT do not enter the circulation in significant amounts because they are rapidly deiodinated by iodotyrosine deiodinase, an enzyme complex that contains ferredoxin, NADPH:ferredoxin reductase, and a deiodinase containing flavin mononucleotide (FMN). This reaction promotes recycling of iodide within the follicle cell. Some of the iodide, however, diffuses into plasma and constitutes the daily "iodide leak," estimated to be about 16 μg/day. By mechanisms that are not clear, a small amount of intact Tg also leaks out of the thyroid and can be found circulating in plasma in normal individuals. Tg leakage at a rate of 100 μg/day has been reported in euthyroid individuals, and a concentration of 15−25 μg Tg per liter of serum is considered to be normal. The route of Tg leakage is by way of the lymphatic drainage and is increased when the gland is excessively stimulated. Increased entry of Tg into the circulation may result in an immune response because of the antigenic nature of this glycoprotein.

The release of thyroid hormone following endocytosis of colloid is inhibited by iodide and is known to be reduced with high dietary intake of iodine. This inhibition is due to the inverse relationship between the iodide content of Tg and the digestibility of iodo-Tg by lysosomal peptidase; that is, poorly iodinated Tg is more readily digested than richly iodinated Tg.

Regulation of Thyroid Hormone Synthesis

Thyroid-Stimulating Hormone

Thyroid-stimulating hormone (TSH) (also called **thyrotropin**), which is secreted by the pituitary, plays a central role in the regulation of growth and function of the thyroid gland. TSH receptors are functionally coupled to G-proteins; thus, the extracellular stimulus by TSH is transduced into intracellular signals mediated by a number of G-proteins. Activation of G_S-protein results in the stimulation of the adenylate cyclase−cAMP−protein phosphorylation cascade. Other G-proteins coupled to TSH-receptor activation include G_Q-protein, which mediates the phospholipase C-phosphatidylinositol 4,5-bisphosphate-Ca^{2+} signaling pathway (see Chapter 28 for a detailed discussion).

The TSH receptor is a glycoprotein that consists of three major domains: a long amino-terminal extracellular segment that confers binding specificity; a transmembrane segment; and a carboxy-terminal intracytoplasmic segment that mediates G-protein interactions (see Chapter 28). TSH-receptor activation promotes iodide trapping, organification of iodide, and endocytosis and hydrolysis of colloid by a mechanism that involves cAMP.

Thyroid disorders consisting of either hypofunction or hyperfunction can result from constitutive inactivating or activating mutations, respectively, in the TSH receptor. Similarly, $G_S\alpha$-inactivating (loss-of-function) or activating (gain-of-function) mutations can result in **hypothyroidism** and **hyperthyroidism**. An example of the former is **Albright hereditary osteodystrophy**, and the latter is **McCune−Albright syndrome**. Both syndromes are discussed in Chapter 28. In addition, the $G_S\alpha$ mutations can also affect function of other polypeptide hormones that initiate their action via cell surface $G_S\alpha$ receptors (e.g., parathyroid hormone, gonadotropins). All of the effects of TSH on the thyroid gland are exaggerated in hyperthyroidism due to the excessive stimulation of the TSH receptor, whether or not the ligand is TSH. For example, **Graves' disease** of the thyroid gland is due to a thyroid-stimulating antibody that binds the TSH receptor and, in so doing, activates it. This **thyroid-stimulating immunoglobulin (TSI)** or **long-acting thyroid stimulator (LATS)** is an expression of an autoimmune disease that may be hereditary, although in some cases it may be the result of a normal immune response. For example, certain strains of bacteria contain a membrane protein that is homologous to the TSH receptor and that elicits an immune response in infected individuals. The antibodies generated resemble TSI. Thus, some cases of hyperthyroidism may be due to cross-recognition across phylogenetic lines (molecular mimicry).

Human chorionic gonadotropin (hCG), a placental hormone, has some TSH-like activity. hCG synthesis is initiated during the first week after fertilization and reaches its highest level near the end of the first trimester, after which levels decrease. The action of hCG on thyrocytes may help to meet the increased requirement for thyroxine during pregnancy. The total thyroxine pool is increased in pregnancy due to elevated levels of thyroxine-binding globulin. However, the serum free T_4 level, which represents the biologically active form, remains the same compared to nonpregnant status. During the first trimester, the fetal requirement for thyroxine is met by maternal circulation until the fetal pituitary−thyroid axis becomes functional late in the first trimester. The maternal−fetal transport of thyroxine that occurs throughout pregnancy minimizes fetal brain abnormalities in fetal hypothyroidism (discussed later). Maternal hypothyroidism (e.g., due to women

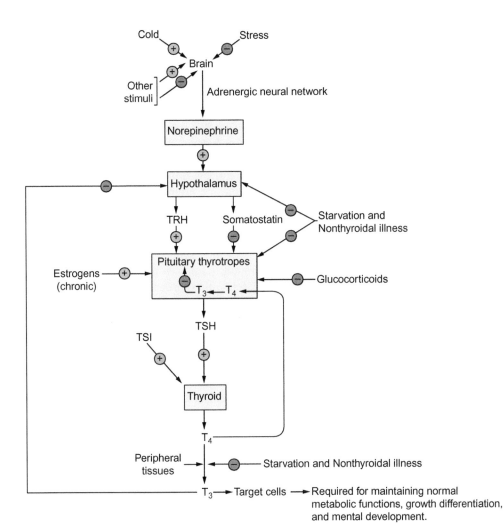

FIGURE 31.3 Schematic representation of the major steps in the regulation of thyroid hormone secretions and metabolism at five levels, namely, brain, hypothalamus, pituitary thyrotropes, thyroid, and peripheral tissues.

consuming iodine-deficient diets) is associated with defects in the neurological functions of offspring.

Release of TSH from the anterior pituitary is stimulated by hypothalamic thyrotropin-releasing hormone (TRH) and inhibited by thyroid hormone. These opposing signals to the anterior pituitary determine the magnitude of TSH secretion. Many other factors influence secretion of TSH and thyroid hormone (Figure 31.3). For example, starvation reduces conversion of T_4 to T_3 in peripheral tissues but not in the anterior pituitary; consequently, TSH secretion is not suppressed but circulating T_3 levels fall. Stress, somatostatin, estrogens, and cold exposure (in infants) also affect TSH release, but except for cold stress in infants, they are not regarded as important regulators of TSH and thyroid hormone secretion.

Iodide

The thyroid gland is capable of adjusting its synthetic activity to the supply of iodine. When the iodide supply is low, the thyroid makes maximal use of iodide; when the iodide supply is abundant, the thyroid defends itself against hormone overproduction by reducing iodide uptake. The thyroid gland autoregulates its iodide supply by an internal feedback mechanism that controls the intraglandular handling of iodide and the thyroid response to TSH. The exact nature of this feedback is not known, but the size of the organified iodide pool is believed to have an inhibitory effect on the rate of iodide uptake, intraglandular T_4/T_3, the completeness of thyroglobulin hydrolysis, and the responsiveness to TSH.

This autoregulatory mechanism is not prominent when iodide intake is within the normal range but becomes physiologically significant when there is a moderate deficiency or excess of iodide. On an ordinary diet in which the iodine intake ranges between 300 μg and 1000 μg/day, T_4 production increases linearly with increased consumption of iodine. With lower iodine intake (e.g., 50–150 μg/day), T_4 production increases due to increased efficiency in iodide processing and to a greater response to TSH; in fact, although plasma TSH levels are within the normal range, a mild enlargement of the gland is not an uncommon consequence (e.g., nontoxic goiter). On the other hand, with a diet enriched in iodine (e.g., $1-2$ mg/day), the increase in T_4 production as a function of iodine consumption is subnormal, due to negative feedback of the organified iodide pool on T_4 production and the thyroid response to TSH; as a result, plasma TSH levels are maintained within the normal range.

Abnormal thyroid function is seen at the extremes of iodine intake: at the lower end, less than 50 μg/day; at the upper end, 5 mg/day or more. At the lower end, frank hypothyroidism occurs because there is an insufficient supply of iodine to meet the T_4 production demands, despite the efficiency with which the limited iodide is used. In this condition, the iodide deficiency creates an organified iodide pool consisting of MITs and a paucity of DITs, and, consequently, little or no coupling occurs because substrate requirements are not met. Products of Tg digestion are mainly MITs, small amounts of DITs, and possibly some T_3 and T_4. The decrease in thyroid hormone increases output of TSH, which, in its futile attempt to increase T_4 production, creates glandular overgrowth due to thyrocyte hypertrophy and colloid accumulation ("hypothyroid goiter," "colloid goiter," or "endemic goiter"). At the upper end, further increase in iodide intake causes a shutdown of thyroid gland function and acutely reduces the thyroid response to TSH; consequently, there is a paradoxical decline in T_4 production as a function of iodide consumption accompanied by a dramatic reduction in vascular supply to the gland. This phenomenon is called the **Wolff–Chaikoff effect** and is observed only in thyroid glands that are stimulated (e.g., with TSH or in hyperthyroidism). Thus, the rapid but transient inhibitory effect of thyroid hormone release by elevated plasma iodide concentration is therapeutically efficacious in the management of severe forms of hyperthyroidism (e.g., thyrotoxicosis) and in preoperative thyroidectomy of hyperthyroid subjects.

TRANSPORT AND METABOLISM OF THYROID HORMONES

Thyroid hormones in the blood are mainly bound to serum proteins. The major thyroid hormone-binding protein in blood is **thyroxine-binding globulin** (TBG). TBG exhibits a high affinity for T_4 ($K_d = 50$ pM) and a moderate affinity for T_3 ($K_d = 500$ pM). Approximately 75% of circulating T_4 is bound to TBG, 15% to transthyretin, and 10% to albumin. Unbound thyroxine, also known as **free thyroxine** (FT_4), is the biologically active form. About 0.03% of T_4 and about 0.3% of T_3 is unbound, so that only a small fraction is accessible for target cell uptake.

The albumin-bound hormone dissociates rapidly in the vicinity of target cells. The cellular uptake of T_4 and T_3 is mediated by carrier-mediated processes at stereospecific binding sites. At high-affinity sites, the uptake is energy-, temperature-, and often Na^+-dependent. T_3 is metabolized about six times more rapidly than T_4 and has a metabolic clearance rate (MCR) 25 times as high as that of T_4. The half-life for T_4 is 7 days, whereas for T_3 it is about 18 hours.

Mutations in a human serum albumin gene that substitutes either histidine or proline for arginine at position 218 increases its binding affinity for T_4. These mutations are expressed as an autosomal dominant trait and occur with relatively high frequency. Carriers of these mutations have high levels of total serum T_4, but their free T_4 and TSH concentrations are within the normal range. Individuals are euthyroid (normal thyroid), and their condition is known as **familial dysalbuminemic hyperthyroxinemia (FDH)**.

Two reactions account for the metabolic fate of about 80% of the T_4 in plasma: about 40% is converted to T_3 via 5′-deiodination (activation); another 40% of the T_4 is converted to rT_3 by 5-deiodination (inactivation). These two reactions are catalyzed by three enzymes designated types I, II, and III iodothyronine deiodinases (Figure 31.4). Types I and II both catalyze the 5′-deiodination reaction but differ with respect to substrate specificity, tissue distribution, and regulation. Type III is a 5-deiodinase, which catalyzes the removal of iodine from position 5 of the inner ring. All three deiodinases contain selenocysteine at their active site, and selenium is required for enzyme activity. Type I is deiodinase microsomal enzyme present in the liver, kidney, and thyroid, with specificity for thyronines bearing iodine at position 5′. It is responsible for catalyzing the formation of most of the circulating T_3 and is a high-capacity ($K_m = 1$ μM) processor of T_4. A unique feature of the type I enzyme is that it is sensitive to inhibition by **propylthiouracil (PTU)**, a known inhibitor of thyroid gland function that has little or no effect on the activity of the other deiodinases. It is also inhibited in conditions such as starvation, caloric restriction, and non-thyroid-related illnesses, which may deplete the cofactor supply and/or depress enzyme synthesis.

The type II deiodinase is less abundant and has a higher specificity and a higher affinity for T_4 than type I. The type II enzyme is present in the pituitary, brain, brown fat, and skin, and is responsible for intracellular T_3 formation that leads to a cellular response; whereas the type I enzyme produces T_3 for export to other cells, and the type II enzyme is self-serving, producing its own supply of intracellular T_3. A unique feature of the type II enzyme is that it is sensitive to inhibition by T_4 by an unknown mechanism. The activity of the type II enzyme

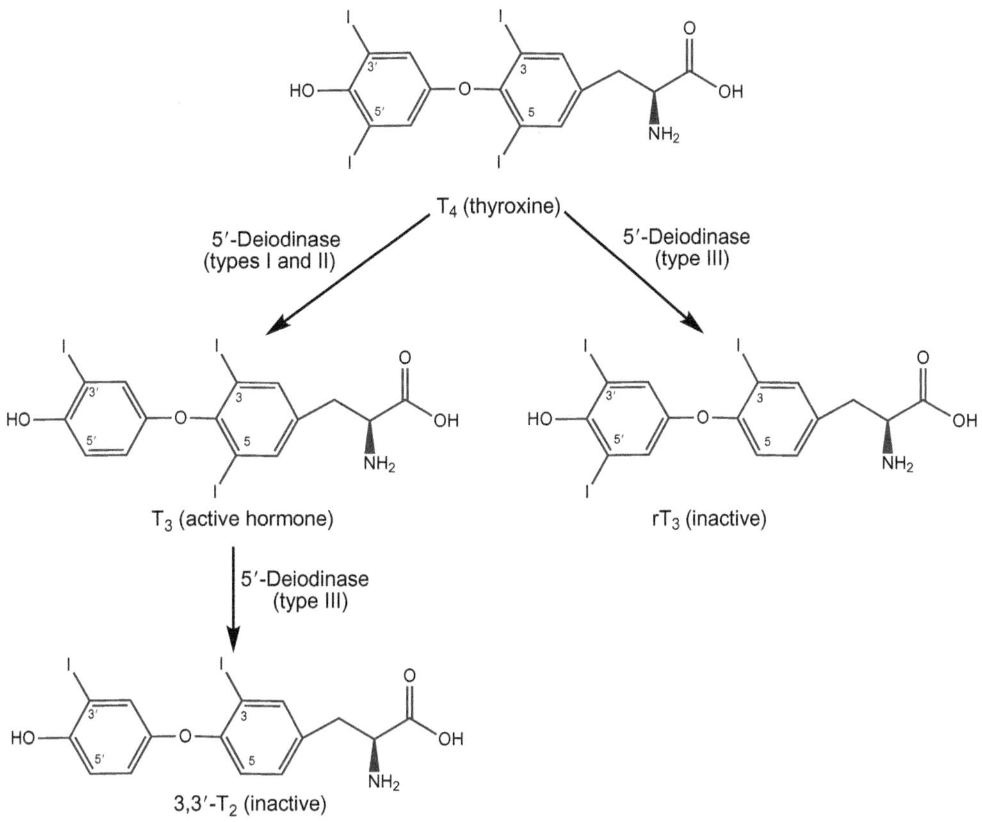

FIGURE 31.4 Activating and inactivating pathways of T$_4$.

is increased in hypothyroidism and is depressed in hyperthyroidism, whereas these conditions have the opposite effect on the activity of the other two deiodinases. The type III deiodinase is mainly involved with the degradation of T$_3$ and the inactivation of T$_4$, both involving the removal of the inner ring 5-iodide. The type III enzyme is widely distributed among tissues and is responsible for the formation of almost all (95%) of the circulating rT$_3$. Its activity is not altered by most of the factors or conditions that inhibit the activity of types I or II deiodinases, and this explains why there is a reciprocal relationship between plasma levels of T$_3$ and rT$_3$.

Not all metabolic processing of iodothyronines involves deiodination, although it is the main metabolic route for T$_4$, T$_3$, and rT$_3$, accounting for about 80% of the fate of each compound. The remaining 20% is conjugated in the liver with sulfate or glucuronide, and/or processed by oxidative deamination and decarboxylation to form thyroacetic acid derivatives with minimal biological activity.

Inorganic iodide liberated from deiodinations enters the extracellular compartment and is excreted in the urine ($\sim$488 μg/day) or stool ($\sim$12 μg/day via bile), or is reabsorbed by the thyroid gland ($\sim$108 μg/day) for resynthesis of thyroid hormones. At the average daily intake of $\sim$500 μg inorganic iodide, this amounts to no net change

in the iodide status, although the small amount of iodide that is lost through the skin and gastrointestinal tract would place the body in a negative iodide balance unless intake is adjusted accordingly.

BIOLOGICAL ACTIONS OF THYROID HORMONES

Mechanism of Action and Latency Period

All but a few of the thyroid effects that have been identified occur at the level of gene transcription, mediated by nuclear thyroid hormone receptors (Chapter 28). These effects have a longer latency period than for most steroids; some of the relatively early responses show a latency period of several hours. Three T$_3$ nuclear receptor isoforms that are members of the nuclear receptor family, named TRα_1, TRβ_1 and TRβ_2, are tissue specific, and they may also function coupled to other nuclear receptors (e.g., retinoic acid X receptor). T$_3$ binding to its receptors initiates binding of other accessory nuclear proteins that either activate or inactivate gene transcription (see Clinical Case Study 31.1 with Vignettes 1 and 2).

Physiological Effects

Cardiovascular System

Thyroid hormone enhances cardiac contractility and exerts a positive chronotropic effect on the heart, increasing the heart rate by a mechanism that may involve more than a potentiation of the β-adrenergic effect [1]. In the ventricular myocardium, thyroid hormone stimulates the synthesis of myosin heavy chain α (MHCα), while inhibiting the synthesis of MHCβ, by regulating the expression of the respective genes. The result is that myocardium with a higher MHCα content has higher calcium and actin-activated ATPase activities and increased velocity of muscle fiber shortening. Thus, the thyroid hormone-conditioned ventricle is capable of increased contractile performance.

Intermediary Metabolism

Thyroid hormone increases both lipolysis and lipogenesis, although lipogenesis is stimulated before lipolysis, due to early induction of malic enzyme (malate dehydrogenase), glucose-6-phosphate dehydrogenase, and fatty acid synthase. Thyroid hormone lowers serum cholesterol levels by maintaining low-density lipoprotein (LDL) receptor density via regulation of hepatic lipase activity, which catalyzes conversion of intermediate-density lipoproteins (IDL) to LDL, and probably also by maintaining lipoprotein lipase activity, which promotes triglyceride clearance. In liver, kidney, skeletal muscle, cardiac muscle, and adipose tissue, thyroid hormone stimulates Na^+, K^+-ATPase gene expression. Thyroid hormone is essential to energy homeostasis (Chapter 20). T_3 upregulates genes involved in energy expenditure, one of which codes for uncoupling protein (UCP2). Uncoupling proteins mediate oxidation of substrates in the mitochondria without the synthesis of ATP but coupled to thermogenesis (Chapter 13).

Growth and Maturation

Thyroid hormone stimulates the production of IGF-I directly (liver) and indirectly (via increased growth hormone, GH). In adults with hypothyroidism, basal serum GH levels are normal, but the GH responses to provocative stimuli, nocturnal GH secretion, and serum IGF-I are all subnormal. In **cretinism** (infantile hypothyroidism), linear growth is severely retarded, and the resulting dwarfism is characterized by a retention of the high upper to lower body ratio of infancy. The growth retardation in people with cretinism is primarily due to the delayed appearance of ossification centers in long bone, and secondarily to a deficiency in growth factors.

A deficiency of thyroid hormone during fetal development due to untreated or undertreated maternal hypothyroidism results in a neurological deficit in the offspring [2,3]. In congenital hypothyroidism, a normal maternal thyroid can meet the fetal requirements for thyroid hormone. However, in the postnatal period, these infants require prompt thyroid hormone replacement therapy throughout their life, beginning in the first few weeks of life. If untreated, they inevitably develop growth and mental retardation. Thyroid hormone is essential for maturational development of the CNS and is required for the development of axonal projections and myelination. One of the important effects of thyroid hormone is to promote the synthesis of myelin basic protein.

Congenital hypothyroidism in Western populations occurs in about 1 in 4000 births. Congenital hypothyroidism is caused by embryogenic defects leading to thyroid agenesis (50%–55%), dysgenesis (30%–35%), errors in the synthesis of thyroid hormone (10%–15%), and defects in the pituitary–hypothalamic axis (4%). Neonatal screening for congenital hypothyroidism is performed using blood collected on filter paper from newborns older than 12 hours. The laboratory test consists of measuring T_4, TSH, or both. Abnormal values must be reconfirmed, and additional tests include other parameters (e.g., TBG levels). Prompt and early intervention with thyroid replacement therapy prevents mental retardation and cretinism.

Reproductive System

Thyroid hormone increases total plasma androgen levels by increasing the production of testosterone-binding globulin (TeBG) by the liver. Hyperthyroidism is associated with increased plasma TeBG levels, higher total testosterone levels, but normal free testosterone levels in adult men. There is a high incidence of gynecomastia in hyperthyroid men (40%–83%) due to a higher circulating level of free estradiol in plasma. The elevated free-estradiol level may be explained by the lower affinity of estrogen for TeBG, and also by increased conversion of androgen to estrogen in hyperthyroidism.

In prepubertal boys, but not in older boys or adult men, hypothyroidism is associated with a high incidence of increased testicular growth (macroorchidism) that is not accompanied by any change in testosterone levels. Elevated follicle-stimulating hormone levels are found in most cases, but there is no testicular maturation. The mechanism of this disorder is not known.

Pharmacological Effects

The effects of thyroid hormone may persist for up to a week because of its long half-life in the circulation, its long-acting effect on the genome of the target cell, and the slow rate of recovery of some target tissues (e.g., brain).

Thyroid hormone excess depletes body energy stores and imparts hypersensitivity of tissues to catecholamines. Protein synthesis is inhibited, protein catabolism is accelerated, and the antianabolic actions of cortisol are enhanced. During the period of growth, these factors would lead to growth suppression, muscle weakness, weight loss, and depletion of liver and muscle glycogen stores. Gluconeogenesis is fueled by elevated substrate levels (amino acids, glycerol, lactate, and pyruvate), and this contributes to the blood glucose pool; however, since peripheral glucose utilization is also increased, hyperglycemia is not severe. Thyroid hormone excess also leads to depletion of triacylglycerol stores, mainly due to its action on catecholamine-induced lipolysis. Tissue consumption of free fatty acids and ketones is also increased. Thus, serum lipid levels fall, including the level of serum cholesterol. In essence, the storage forms of glucose, amino acids, and fatty acids are depleted, and these substrates are rapidly metabolized, resulting in the need for increased food intake.

Because thyroid hormone increases tissue responsiveness to catecholamines, many symptoms of thyroid hormone excess are those that characterize catecholamine excess. Effects of T_3 on cardiovascular hemodynamics consist of increased tissue thermogenesis, decreased systemic vascular resistance, decreased effective arterial filling volume, increased renal sodium reabsorption, increased blood volume, and increased cardiac inotropy and chronotropy leading to increased cardiac output. Behavioral changes (such as nervousness, restlessness, short attention span, and emotional lability) are common, some of which require a longer period for decay than the metabolic changes following normalization of thyroid hormone levels. Clearly, the central actions of thyroid hormone involve more than potentiation of adrenergic neurotransmission.

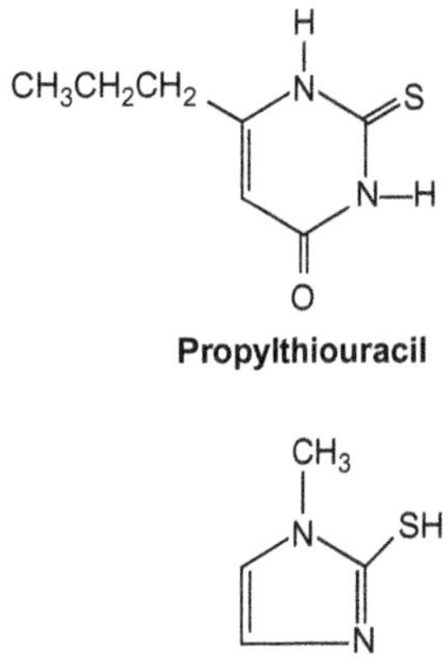

Thyroid Disorders

Disturbances in thyroid metabolism can occur at any level of the hypothalamus−pituitary thyroid−peripheral tissue axis. Several of these disorders have been discussed previously. Hyperthyroidism is more prevalent in women than men. The three most common causes of hyperthyroidism are Graves' hyperthyroidism, toxic multinodular goiter, and toxic adenoma. The clinical features of hyperthyroidism include hyperkinesis, weight loss, cardiac anomalies (e.g., atrial fibrillation), fatigue, weakness, sweating, palpitations, and nervousness [4,5; also see Clinical Case Study 31.2]. The typical biochemical laboratory parameters in primary hyperthyroidism are increased serum free T_4 and decreased serum TSH.

The antithyroid drugs propylthiouracil and methimazole are used in the management of hyperthyroidism. They are thionamides (Figure 31.5) and inhibit thyroid peroxidase-mediated iodination of tyrosine residues of

Propylthiouracil

Methimazole

FIGURE 31.5 Structure of inhibitors of thyroid hormone biosynthesis.

TABLE 31.1 Major Classes of Hypothyroidism

Loss of Functional Thyroid Tissue
Chronic autoimmune thyroiditis (Hashimoto's disease)
Idiopathic hypothyroidism
Postradioactive iodine treatment
Post-thyroidectomy
Biosynthetic Defects in Thyroid Hormonogenesis
Inherited defects
Iodine deficiency
Antithyroid agents
Hypothyrotropic Hypothyroidism
TSH deficiency
TRH deficiency
Peripheral Resistance to Thyroid Hormones

From A. W. Norman, G. Litwak, Hormones, 2nd ed., San Diego, CA: Academic Press, 1997.

Tg. Propylthiouracil also inhibits the peripheral conversion of T_4 to T_3. Radioiodine ablation therapy is also used in the treatment of hyperthyroidism [6].

Thyroglobulin is unique to thyroid follicular cells. Therefore, serum thyroglobulin levels are used as a tumor marker in the assessment of complete removal of the thyroid gland in subjects with well-differentiated thyroid carcinoma. However, the presence of auto-antibodies to thyroglobulin in the serum interferes with its quantification.

Major causes of hypothyroidism are listed in Table 31.1. The most common cause of hypothyroidism is failure of the thyroid gland; this is known as **primary**

hypothyroidism. In adults, the cause of primary hypothyroidism is often spontaneous autoimmune disease (e.g., Hashimoto's thyroiditis) or destructive therapy for hyperthyroid states (e.g., Graves' disease). In adults, hypothyroidism has an insidious onset with a broad range of symptoms. The typical biochemical laboratory parameters include the measurement of serum free T_4 and TSH, which are decreased and increased, respectively. In addition, measurement of serum autoantibodies for thyroglobulin and thyroperoxidase is also utilized.

Mild thyroid abnormalities produce **subclinical hypo- and hyperthyroidism**. In subclinical hypothyroidism, serum TSH levels are mildly elevated but FT_4 levels are normal compared to reference intervals. Subclinical hypothyroidism may be associated with abnormalities of cardiac and neuropsychological functions. In subclinical hyperthyroidism, the serum TSH levels are suppressed (decreased) with normal FT_4 and T_3 levels. It may be associated with many of the hyperthyroid symptoms that are mild and subtle. Subclinical thyroid disorders require

further clinical evaluations to exclude other diseases that mimic the symptoms of thyroid diseases (e.g., psychiatric and heart diseases).

In many non-thyroid-critical diseases (e.g., involving liver, heart, kidney, pancreas, starvation) despite a normal thyroid, the serum levels of TSH, T_4, and T_3 may be decreased. These clinical conditions have been named as euthyroid sick syndrome, which may be due to abnormalities of type I deiodinase, binding proteins, elevated cortisol levels, and drugs.

Extreme deteriorating thyroid functions, either hypo- or hyperthyroidism known as **myxedema coma** or **thyrotoxic storm**, respectively, require prompt medical intervention. In the case of myxedema coma, administration of T_4 and T_3 is required. In thyrotoxic storm, a rare life-threatening complication of hyperthyroidism, the management involves the administration of antithyroid drugs, an iodide solution to initiate Wolff–Chaikoff therapeutic modality (discussed earlier), and β-adrenergic receptor antagonist (e.g., propranolol) as an antiarrhythmic agent.

CLINICAL CASE STUDY 31.1 Hypothyroidism Due to Thyroid Hormone Resistance

Vignette 1: Loss of Function of Thyroid Hormone (T_3) Receptor at the Alpha Gene Locus
This case was abstracted from: E. Bochukova, N. Schoenmakers, M. Agostini, E. Schoenmakers, O. Rajanayagam, J.M. Keogh, et al., A mutation in the thyroid hormone receptor alpha gene, N. Engl. J. Med. 366 (2012) 243–249.

Synopsis
A 6-year-old girl presented with persistent height deficit, growth retardation, and severe constipation. Physical examination revealed bradycardia, hypotension, growth attenuation of the lower segment of her body, delayed tooth eruption, borderline-high body mass index, hypermobility of the ankle and knee, reduced muscle tone, impaired gross and fine motor coordination, and restricted adaptive behavior. Laboratory studies revealed borderline low levels of total and free thyroxine (T_4), elevated levels of total and free triiodothyronine (T_3), and normal levels of thyroid-stimulating hormone (TSH). She also had borderline low levels of insulin-like growth factor I (IGF-I), decreased basal metabolic rate, normal growth hormone tests, and increased serum sex hormone-binding globulin (indicative of increased thyroid hormone activity in the liver). The patient was started on T_4 therapy, with subsequent normalization of free T_4, free T_3, and IGF-I levels, but her low growth rate, heart rate, and blood pressure did not resolve. Her serum sex hormone-binding globulin levels also remained elevated. DNA studies revealed a nonsense mutation in the thyroid hormone receptor alpha (THRA) gene, identified as E403X mutant, which is a potent, dominant negative inhibitor of wild-type thyroid hormone receptor alpha (TRα). The patient's

mixed presentation in which the skeletal, gastrointestinal, and myocardium were affected but the liver and hypothalamic–pituitary axis were spared suggests she had differential sensitivity to thyroid hormone activity due to tissue-specific expression of different types of thyroid hormone receptors.

Vignette 2: Loss of Function of Thyroid Hormone (T_3) Receptor at the Beta Gene Locus
This case was abstracted from: J. Leey, P. Cryer, Discrepant thyroid function test results in a 44-year-old man, Clin. Chem. 59 (2013) 1703–1707.

Synopsis
A 44-year-old man with hyperthyroidism treated with propylthiouracil and radioactive iodine presented with fever, fatigue, muscle tenderness, 10-pound weight loss, and diarrhea. His mother, grandmother, and daughter also had hyperthyroidism with goiter. The patient was found to have a diffuse goiter, elevated free thyroxine (T_4) and triiodothyronine (T_3) levels, and unsuppressed thyroid-stimulating hormone (TSH). An MRI of the brain ruled out a TSH-secreting pituitary tumor. DNA analysis revealed a missense mutation in the THRB gene (thyroid hormone receptor beta), confirming the clinical diagnosis of thyroid hormone resistance.

Teaching Points
1. The thyroid gland primarily secretes the hormone thyroxine (T_4), which is relatively inactive and is converted to the active hormone triiodothyronine (T_3) by thyroxine 5'-deiodinase. Binding of T_3 with nuclear receptors accounts for the actions of thyroid hormone in the body.

(Continued)

CLINICAL CASE STUDY 31.1 (Continued)

T_3 hormone receptors are encoded by two genes: THRA and THRB (thyroid hormone receptor alpha and beta). These two genes undergo alternate splicing to create the receptor subtypes TRα1, TRβ1, and TRβ2. These receptor subtypes have differing tissue distributions. TRα1 is found predominantly in the bone, skeletal and cardiac muscle, gastrointestinal tract, and central nervous system. TRβ1 is found predominantly in the liver and kidney. TRβ2 is found predominantly in the hypothalamus, pituitary, cochlea, and retina. See Table 31.2.

2. Because different target tissues contain distinct receptor subtypes, insensitivity of one receptor subtype leads to a reduction in T_3 binding at that site, causing excess thyroid stimulating hormone production and subsequent activation of the other receptor subtypes, resulting in conflicting hypothyroid and hyperthyroid symptoms. THRB gene mutation is the most common cause of thyroid hormone resistance, and has an autosomal dominant genetic basis.

3. Surgery and radioactive iodine therapy are ineffective treatments for patients with thyroid hormone resistance. Therapy is focused on symptomatic relief, as there is no

definitive treatment for correcting the underlying gene mutation. Tachycardia, tremor, heat intolerance, and sweating may be alleviated with a β-adrenergic blocker. Some studies have found that the mutant receptor either inactivates or directly competes with wild-type receptors, suggesting that some THRB mutations can be overcome by treatment with high concentrations of T_3 [5].

Supplemental References

[1] V.K.K. Chatterjee, T. Nagaya, L.D. Madison, A. Rentoumis, J.L. Jameson, Thyroid hormone resistance syndrome: inhibition of normal receptor function by mutant thyroid receptors, J. Clin. Invest. 87 (1991) 1977–1984.

[2] S. Refetoff, R.E. Weiss, S.J. Usala, The syndromes of resistance to thyroid hormone, Endocr. Rev. 14 (1993) 348–399.

[3] T.J. Visser, Thyroid hormone transporters and resistance, Endocr. Dev. 24 (2013) 1–10.

[4] W.H. Dillmann, Cellular action of thyroid hormone on the heart, Thyroid 12 (2002) 447–452.

[5] G.A. Brent, Mechanisms of disease: the molecular basis of thyroid hormone action, N. Engl. J. Med. 331 (1994) 847–853.

[6] J.H. Oppenheimer, H.L. Schwartz, Molecular basis of thyroid hormone-dependent brain development, Endocr. Rev. 18 (1997) 462–475.

TABLE 31.2 Tissue Distribution and Function of T_3 Receptor

THRA Gene	THRB Gene	
TRα1 receptor	**TRβ1 receptor**	**TRβ2 receptor**
Tissue location: • Bone • Skeletal muscle • Cardiac muscle • Gastrointestinal tract • Central nervous system	Tissue location: • Liver • Kidney • Brain	Tissue location: • Hypothalamus • Pituitary • Cochlea • Retina
Functions: • Positive chronotropic effect [5] • Enhances cardiac contractility [5] • Stimulates transcription of myosin heavy chain α [5] • Reduces systemic vascular resistance [5]	Functions: • Stimulates lipogenesis and lipolysis [5] • Mediates weight loss [5] • Thermogenesis [5] • Stimulates production of insulin-like growth factor [5]	Functions: • Facilitates normal brain development [6] • Increases growth hormone synthesis [5]
Gene mutation or knockout: • Bradycardia [4] • Decreased cardiac contractility [4] • Low body temperature [5] • Increases systemic vascular resistance [5] • Prolonged diastolic relaxation of the heart [5] • Prolonged relaxation of deep-tendon reflexes [5] • Diminished colonic motility [1] • Retarded bone age [1]	Gene mutation or knockout: • Increased serum intermediate-density and low-density lipoprotein [5] • Decreased cholesterol and triglyceride clearance [5] • Decreased hepatic lipase activity [5]	Gene mutation or knockout: • Diminished synthesis of mRNAs for myelin-associated proteins [6] • Cognitive deficits [1] • Deafness [5]

CLINICAL CASE STUDY 31.2 Hyperthyroidism Due to Graves' Disease

This case was abstracted from: G.A. Brent, Graves' disease, N. Engl. J. Med. 358 (2008) 2594–2605.

Synopsis

A 23-year-old woman presented with palpitations, loose stools, weight loss, and increased irritability. On physical exam, she was tachycardic and hypertensive, and she appeared anxious with a symmetrically enlarged, firm, non-tender thyroid gland and audible thyroid bruit. Serum thyrotropin levels were markedly decreased to 0.02 μU/mL (reference range 0.35 to 4.50), and free thyroxine (T_4) levels were increased to 4.10 ng/dL (reference range 0.89 to 1.76). Further investigation revealed the diagnosis of Graves' disease.

Teaching Points

1. Fifty to eighty percent of patients with hyperthyroidism have underlying Graves' disease, in which autoimmune IgG antibodies bind to and activate the G-protein-coupled thyrotropin receptor of the anterior pituitary gland. Thyrotropin receptor activation leads to increased thyroid hormone production and subsequent hypertrophy and hyperplasia of the thyroid gland. Thyroid function tests may reveal elevated serum T_4 and T_3 levels, with suppression of serum thyrotropin levels.

2. Clinical symptoms of hyperthyroidism due to Graves' disease include weight loss, tremor, heat intolerance, sleep difficulty, frequent bowel movements, irritability, proximal muscle weakness, and menstrual irregularity. Older patients are less likely than younger patients to have tachycardia and tremor, but more often will have weight loss or depression (apathetic hyperthyroidism) or cardiovascular manifestations like atrial fibrillation.

3. Patients with Graves' disease may be treated with antithyroid drugs such as propylthiouracil or methimazole (thioamides), radioiodine therapy, or surgical thyroidectomy.

REQUIRED READING

[1] I.M. Grais, J.R. Sowers, Thyroid and the heart, Am. J. Med. 127 (2014) 691–698.

[2] G.A. Brent, The debate over thyroid-function screening in pregnancy, N. Engl. J. Med. 366 (2012) 562–563.

[3] J.H. Lazarus, J.P. Bestwick, S. Channon, R. Paradice, A. Maina, R. Rees, et al., Antenatal thyroid screening and childhood cognitive function, N. Engl. J. Med. 366 (2012) 493–501.

[4] E.P. Rhee, J.A. Scott, A.S. Dighe, Case 4-2012: a 37-year-old man with muscle pain, weakness, and weight loss, N. Engl. J. Med. 366 (2012) 553–560.

[5] R.S. Bahn, Mechanisms of disease: Graves' ophthalmopathy, N. Engl. J. Med. 362 (2010) 726–738.

[6] D.S. Ross, Radioiodine therapy for hyperthyroidism, N. Engl. J. Med. 364 (2011) 542–550.

Chapter 32

Endocrine Metabolism V: Reproductive System

Key Points

1. In the early embryonic state, the gonads of both males and females are morphologically identical.

2. After about 8 weeks of gestation, the product of a specific gene called SRY (for "sex-determining region Y") in the Y-chromosome (SRY) inhibits the action of DAX-1 (dosage-sensitive sex reversal, adrenal hypoplasia congenital, critical region on the X-chromosome, gene 1), a gene located in the X-chromosome that results in the development of testes. In the absence of the Y-chromosome, the X-chromosome directs the development of fetal gonads into ovaries.

3. In females, during the early embryonic state, methylation randomly and permanently inactivates one of the X-chromosomes, forming a nondescript mass of inert DNA known as a Barr body. However, about 15% of the genes escape inactivation, and the degree of inactivation may vary from one female to another.

4. The regulation of the reproductive system in females and males is determined by external environmental cues and the interaction between internal signals located in specific brain centers, the hypothalamus, anterior pituitary, gonads, and peripheral tissues. These interrelationships are commonly known as the hypothalamic−pituitary−gonadal axis and are integrated and programmed by several hormones.

5. The gonads contain several specific classes of cells that function in gametogenesis and hormone production. Gonadotropin-releasing hormone (GnRH) synthesized and released by hypothalamic neurons enters the hypothalamic−pituitary portal system and activates specific anterior pituitary cell plasma membrane receptors promoting the synthesis and release of follicle-stimulating hormone (FSH) and luteinizing hormone (LH). The pulsatile and episodic secretion of GnRH maintains an optimal balance of FSH and LH secretion during various physiologic conditions. GnRH, FSH, and LH secretions are under feedback regulation by the product of end organs (ovaries and testes).

6. In the testes, Sertoli cells are interspersed among spermatogenic cells, whereas Leydig cells reside more distant from germ cells and are separated by a basement membrane in the interstitium of the testes outside the seminiferous tubules.

    a. Sertoli cells have multiple functions, such as synthesizing androgen-binding protein (ABP) and many other proteins necessary for spermatogenesis as well as estradiol from testosterone provided by Leydig cells. Their functions are stimulated by FSH via G-protein-coupled receptors. Protein hormones synthesized in the Sertoli cells include inhibin, follistatin, and activin. Inhibin suppresses FSH release by the pituitary gonadotrophs, and it also functions as a paracrine hormone. FSH release is stimulated by activin and inhibited by follistatin, which is an activin-binding protein.

    b. The synthesis and secretion of testosterone occur in the Leydig cells under the stimulation of LH. Testosterone is produced by and released from Leydig cells, and these processes are stimulated by LH. The biochemical pathway for testosterone synthesis is the same as that used in the adrenal cortex.

    c. Testosterone and other androgens have multiple tissue-specific reproductive and nonreproductive functions based on receptor specificity and the presence of key enzymes. Testosterone functions mainly in the testes, pituitary, and muscle tissues. Aromatization of testosterone by aromatase yields estrogen in adipose tissue, liver, CNS, skin, and hair. Reduction of testosterone by 5 α-reductase yields dihydrotestosterone in the prostate, scrotum, penis, and bone, resulting in their respective biological effects.

7. a. In the ovaries, the two principal endocrine cell types are granulosa and theca cells (also known as interstitial cells). In follicles, granulosa cells encompass germ cells, whereas theca cells are located distant from the germ cells and are separated by a basement membrane. They are under the influence of different microenvironments. Both types of cells mediate steroid and protein hormone synthesis under the regulation of FSH and LH. The principal sex steroid hormones are estradiol and progesterone. The biochemical pathways for the synthesis of steroid hormones are the same as those that exist in the adrenal cortex. Both theca and granulosa cells mediate steroid hormone synthesis. Under LH stimulation, theca cells synthesize and secrete testosterone, which diffuses into granulosa cells, where it is converted to estradiol by aromatization supported by FSH. Following ovulation, transformed granulosa and

N. V. Bhagavan and C.-E. Ha: Essentials of Medical Biochemistry, Second edition. DOI: http://dx.doi.org/10.1016/B978-0-12-416687-5.00032-4

theca cells, known as luteal cells, synthesize and release progesterone. Three polypeptide hormones are produced in the ovaries: inhibin, activin, and follistatin. Inhibin and follistatin suppress the secretion of FSH, whereas activin increases the secretion of FSH.

b. Several interacting processes that are highly regulated are involved in female reproduction. Puberty and the development of secondary sexual characteristics are initiated by many hormonal signals. GnRH secretion stimulates the production of FSH and LH, which, in turn, stimulate estrogen production by the follicles in the ovaries.

c. Only a small number of ova (about 400) undergo maturation and are available for fertilization by sperm on a periodic basis. The periodic and controlled maturation of the oocyte in the ovary, followed by its release and coupled with the physiologic changes that occur in the uterus, are known as the menstrual cycle. The menstrual cycle occurs over about 28 days and is coordinated and orchestrated by several hormones derived from the brain, pituitary, and ovary. Endocrine abnormalities can cause menstrual cycle disorders, leading to either the absence of menstruation (amenorrhea) or irregular menstruation (oligomenorrhea). Ovulation is the process in which the ovum is released from the mature follicle (Graffian follicle). The postovulatory follicle undergoes many biochemical and morphological changes to become a corpus luteum (yellow body). The granulosa and theca cells of the corpus luteum are known as lutein cells. The corpus luteum is an endocrine organ that produces progesterone and estradiol, which are required for the implantation of the fertilized eggs (zygote) in the uterus. In the absence of fertilization of the ovum, the corpus luteum atrophies and transforms into a nonfunctional structure known as the corpus albicans. These changes cause an abrupt loss of progesterone and estrogen that leads to the tissue destruction of the endometrium, the thickened lining of the uterus. The discharge of endometrial cells along with blood is known as menstruation.

d. Pregnancy is initiated when the fertilized ovum is implanted on the uterine wall followed by the formation of the placenta from fetal trophoblasts. Initially, the placenta secretes human chorionic gonadotropic (hCG) hormone, which maintains the function of the corpus luteum so that it continues to produce steroid hormones as well as protein hormones, which resemble hypothalamic and pituitary hormones. Parturition requires the action of steroid and protein hormones as well as prostaglandins.

e. Estrogen has many actions, including both reproductive and nonreproductive functions. Its action is mediated by the presence of the isoforms of the estrogen receptors (α, β, or both) in target tissue and differences in estrogen receptor conformations upon ligand binding and interaction with transcriptional coregulatory proteins. Ultimately, DNA-bound estrogen receptors at specific sites regulate either the positive or negative transcription of specific genes. Estrogen has shown to be a risk factor for breast and endometrial carcinogenesis. In postmenopausal women with estrogen-positive breast cancers, inhibition of estrogen formation from testosterone by aromatase via specific inhibitors has provided therapeutic benefits. Selective estrogen receptor modulators demonstrate both tissue-specific agonistic and antagonistic effects on estrogen action and also have been used therapeutically in the management of breast cancer.

SEX DETERMINATION

The sex of an individual is the outcome of genetic determinants (genotypic sex) and of hormonal effects that confer structural features characteristic of a given sex (phenotypic sex). Both genetic and hormonal determinants normally operate at only two phases of life: during fetal development and at puberty. Genotypic sex is either homogametic (XX) or heterogametic (XY). In humans and other mammals, homogametic (XX) pairing programs ovarian development and oocyte formation (oogenesis), while heterogametic (XY) pairing leads to testicular development and spermatogenesis. In females, one of the X-chromosomes is randomly inactivated permanently during early embryonic life when the embryo consists of fewer than 200 cells. The inactive X-chromosome is known as a **Barr body**. The inactivation process occurs by an epigenetic modification of methylation of DNA, a process initiated by the X-inactivation-specific transcript gene (XIST). However, some genes escape inactivation in the inactivated X-chromosome. Thus, both X-chromosomes, although one of them is mostly inactivated, participate in the female phenotype. This phenomenon is illustrated in **Turner syndrome**, in which the reduced complement of genes that are normally expressed from both X-chromosomes gives rise to abnormal phenotypes.

In both genotypes, the embryonal gonads develop from the epithelium and stroma of the urogenital ridge, a thickening of the celomic (ventromedial) aspect of the mesonephros that emerges at about the second or third week of pregnancy. Both the epithelium and stroma of the urogenital ridge are derived from the intermediate mesoderm of the embryo; however, invading this structure at about week 4 are primordial germ cells from the yolk sac, which take residence in association with the epithelial cells of the developing gonad and replicate. During the germ cell invasion, the epithelial cells of the gonads undergo proliferation and begin entering the stromal spaces as cord-like projections, called **primary sex cords**. Until about week 6, the gonads are undifferentiated and uncommitted; that is, the structure has the potential to develop into either ovaries or testes.

In the presence of a Y-chromosome, the SRY gene (sex-determining region of the Y-chromosome) antagonizes the action of the DAX-1 (dosage-sensitive sex reversal, adrenal hypoplasia critical region, on chromosome X, gene 1) gene. Mutations in the SRY gene can result in an XY female with gonadal dysgenesis. Testis differentiation may also require the participation of a gene located on chromosome 17 (SOX9 gene). The precise interaction between SOX9 and SRY in gonadal development is not yet understood; however, both genes code for DNA-binding proteins. Mutations in the SOX9 (SRY-box 9) gene can result in sex reversal in XY individuals with **camptomelic dysplasia**. The latter is a severe form of skeletal dysplasia associated with dysmorphic features and cardiac defects.

After the fetal gonads are converted into testes by SRY, testosterone and Müllerian-inhibiting substance (MIS) control subsequent sexual differentiation. MIS causes the regression of the Müllerian duct, which would otherwise develop into female genital tracts. Testosterone promotes the development of the Wolffian duct into male internal genitalia, including epididymis, vas deferens, and seminal vesicles.

Testosterone also causes differentiation of the urogenital sinus into the prostate gland and masculinizes the external genitalia; however, these tissues must first convert testosterone into dihydrotestosterone (DHT), a reaction catalyzed by cytoplasmic 5α-reductase type 2. The effects of testosterone on the Wolffian duct and urogenital sinus are dependent on the presence of androgen receptors and 5α-reductase, which are expressed by genes on the X-chromosome and chromosome 2, respectively, and are present in both genotypic sexes. This explains why exposure of the fetus to androgens during the critical period may lead to masculinization of the internal and external genitalia of genotypically female fetuses. It also explains why lack of expression of either gene can lead to feminization in genotypically male fetuses.

The gonad has two interrelated functions: gametogenesis and hormone production. Gametogenesis depends on gonadal hormone production, which is influenced by gametogenesis. Both processes are controlled by luteinizing hormone (LH) and follicle-stimulating hormone (FSH), which are released in response to pulsatile secretion of hypothalamic gonadotropin-releasing hormone (GnRH) and to the levels of circulating gonadal hormones. Early pulsatile secretion of GnRH leads to precocious puberty (see Clinical Case Study 32.3). In addition, gametogenesis is regulated by the paracrine actions of gonadal hormones.

Male and female fertility are radically different. The obvious difference between the sexes in gametogenesis is the formation of ova (ootids) in the female and of sperm (spermatozoa) in the male. Spermatogenesis becomes operational from about the time of puberty and continues throughout life. The process takes about 74 days. Oogenesis occurs in three phases. The initial phase involves proliferation of the stem cells (oogonia) and occurs only in fetal life. The second phase involves the first

maturational division (formation of the secondary oocyte) and occurs about the time of ovulation (see later). The third phase involves the second and final maturational division (formation of ova) and occurs at fertilization. Because the number of oogonia is determined before birth, the number of fertilizable ova that can be produced in a woman's lifetime is limited and is greatly reduced by normal degenerative processes (atresia), so that of the estimated seven million oogonia in the fetal ovaries, only about 300–500 ova ultimately develop. Unlike the continuous generation of sperm in the testes, the ovaries generally produce only one secondary oocyte every 22–28 days. The fertile lifetime in the average woman is about 35–40 years.

Ovaries and testes produce identical steroid hormones, but the amounts and their patterns of secretion are different (Table 32.1). Although testosterone is present in both males and females, its level in the male is about 18 times that in the female; conversely, circulating levels of estradiol in the female are about 3–15 times those in the male. In either sex, the major sex steroid originates in the gonads, whereas in the opposite sex this steroid is generated in substantial amounts by the adrenal cortex or by peripheral conversion of another steroid. For example, almost all of the circulating estradiol in the female comes from the ovaries, whereas in the male only about one-third comes from the testis, the rest being generated by peripheral conversion of androgenic precursors. Finally, the pattern of gonadal steroid secretion is different. In the male, the secretion of gonadal steroids is fairly constant, although minor fluctuations occur as a result of circadian rhythms. In the adult female, gonadal steroid secretion undergoes dramatic, cyclic changes at about monthly intervals. These cyclic changes, which are dictated by processes regulating oogenesis, are referred to as the menstrual cycle.

TESTES

The overall regulation of the male reproductive system is shown in Figure 32.1.

Regulation of Spermatogenesis: Sertoli–Neuroendocrine Axis

Sertoli cells are epithelial cells that line the seminiferous tubules of the testes. At their basal aspects, these cells form the basement membrane and tight junctions that make up the highly selective "blood–testes barrier," which normally prevents entry of immune cells into the lumen. Their function is to provide nutritional and hormonal support for cells undergoing spermatogenic transformation (i.e., spermatocytes and spermatids). For this reason, they are often referred to as "nurse cells." Sertoli cells require FSH stimulation for their maintenance and specifically for production of **androgen-binding protein (ABP)**. ABP, which has a high affinity for

TABLE 32.1 Calculating Levels and Origin of Sex Steroids

Steroid	Adult Men				Adult Women				
	Plasma Level (ng/dL)	Approximate Contribution (%) of			Cycle Phase*	Plasma Level (ng/dL)	Approximate Contribution (%) of		
		Testes	Adrenal						Periphery
Ovary				Adrenal					Periphery
Testosterone (T)	700	90	5	5 (A, DHEA)	—	40	60	10	30 (A, DHEA)
Dihydrotestosterone (DHT)	40	20	—	80 (T, A)	—	20	—	—	100 (A, T)
Androstenedione (A)	115	60	30	10 (DHEA, T)	—	160	60	30	10 (DHEA)
Dehydroepiandrosterone (DHEA)	640	10	85	5 (DHEAS)	—	500	10	85	5 (DHEAS)
DHEA sulfate (DHEAS)	260,000	—	100	—	—	130,000	—	100	—
Estradiol (E$_2$)	3	30	—	70 (T, A)	EF	10 ⎱			
					LF	50 ⎰ 95	—		5 (A, DHEA, E$_1$)
					ML	20			
Progesterone (P)	30	70	30	—	F	100	70	30	—
					L	1100	95	5	—

*EF = early follicular phase; LF = late follicular phase; ML = midluteal phase; F = follicular phase; L = luteal phase.

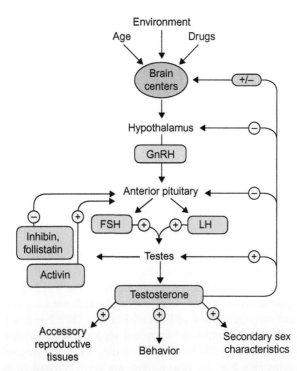

FIGURE 32.1 An overview of the regulation of the male reproductive system. Plus indicates positive, and minus indicates negative, regulation. *[Reproduced from R.A. Rhodes, G.A. Tanner, Medical physiology, 2nd ed. Philadelphia, PA: Lippincott Williams & Wilkins, 2003.]*

testosterone, maintains high levels of testosterone in the seminiferous tubules and thereby helps maintain spermatogenesis. FSH also stimulates aromatase activity in the Sertoli cell, thus stimulating conversion of testosterone into estradiol. This FSH-sensitive Sertoli cell activity accounts for most of the testicular output of estradiol.

Regulation of Testicular Steroidogenesis: Leydig–Neuroendocrine Axis

Leydig cells (interstitial cells) are situated in the vascularized compartments outside the seminiferous tubules in the testes. Leydig cells are steroidogenic cells, converting cholesterol to pregnenolone under regulation by LH. The major steroid product of the Leydig cells is testosterone, which accounts for most of the steroid output by the adult testes. The preferred pathway is the Δ^4 pathway, although detectable amounts of Δ^5 steroids (dehydroepiandrosterone, androstenediol) are secreted (Figure 32.2). LH acts at the cholesterol side chain cleavage step. LH release is held in check by the level of non-TeBG-bound testosterone in the circulation, although the principal inhibitor is DHT (TeBG = testosterone–estrogen binding globulin). Because the anterior pituitary contains 5α-reductase activity, it converts testosterone to DHT and responds to DHT by diminishing the output of LH. LH release is also inhibited by testosterone via inhibition of

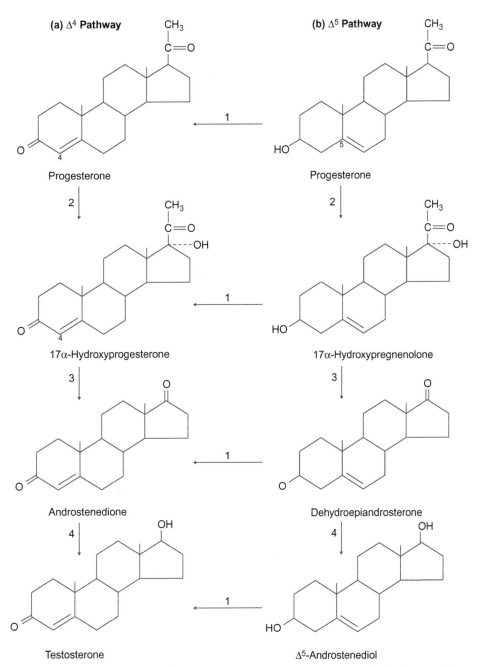

FIGURE 32.2 Testosterone biosynthetic pathways. (a) Δ^4 Pathway (preferred pathway in human testes). (b) Δ^5 Pathway (all enzymes located in the endoplasmic reticulum). Enzymes: (1) 3β-hydroxy-steroid dehydrogenase and Δ^5, Δ^4-isomerase; (2) 17α-hydroxylase (CYP17); (3) C_{17-20}-lyase; (4) 17β-hydroxysteroid dehydrogenase.

GnRH release. In the preoptic hypothalamus, testosterone can be converted to DHT or estradiol, and may thereby inhibit the pulse frequency (DHT) or pulse height (estradiol) of GnRH released at the median eminence.

Secretion of testosterone is fairly constant and consistent in adult men, although higher levels occur in the morning and lower levels in late evening. This circadian rhythm is not accompanied by changes in the levels of LH; thus, it appears to be an intrinsic testicular rhythm. Testosterone levels decline by 10−15% between the ages of 30 and 70 years, accompanied by a reduction in tissue responsive to androgenic stimulation.

Metabolism of Testosterone

Testosterone in plasma exists in two fractions: TeBG-bound (44%) and non-TeBG-bound (56%). The plasma level of TeBG in adult males is ~ 25 nM, which approximates the normal testosterone level in adult men (~ 22 nM); thus, an increase in testosterone production or a decrease in TeBG level will result in a higher level of testosterone available for tissue uptake.

In androgen target tissues that lack 5α-reductase, testosterone activates the androgen receptor (AR)−androgen response element (ARE) mechanism directly and induces the expression of androgen-dependent genes (Chapter 28). After dissociating from the AR, testosterone may be metabolized by the target cell or make its way to the liver for inactivation. The liver modifies three parts of the testosterone structure that are important for hormonal activity: oxidation of the 17β-OH group (to produce a 17-keto derivative); saturation of the Δ^4 in ring A (to produce an androstane derivative); and reduction of the 3-keto-group (to produce a 3α-hydroxylated derivative (Figure 32.3).

In androgen target tissues that contain 5α-reductase, testosterone is converted to DHT, which activates the AR−ARE sequence leading to an androgenic response. DHT undergoes inactivation when it dissociates from AR; the major metabolites (except in the liver and kidney) are the reduced, hydroxylated metabolites 3α-androstenediol and 3β-androstenediol, which are then conjugated to glucuronic acid for renal excretion. These and other polar compounds account for about 30%−50% of the testosterone metabolites excreted daily. 5α-Reductase deficiency causes male pseudohermaphroditism. 5α-Reductase inhibitors are used to treat benign prostatic hyperplasia.

In tissues that contain aromatase (CYP19), testosterone can be converted to estradiol, which may exert an effect *in situ* if the tissue is an estrogen-responsive one, and/or may return to plasma for distribution to estrogen target tissues. Unlike the androgenic steroids, the only major site of estrogen inactivation appears to be the liver; therefore, estrogen theoretically can be recycled until it is transported to the liver, which itself is an estrogen-responsive tissue. Testosterone-derived estradiol accounts for a small percentage (0.3%) of the testosterone metabolized.

Testosterone is known to exert some important biological effects in the liver and kidney, which are major testosterone-inactivating tissues in the body. About 50% of the testosterone is removed from plasma with each passage through the liver. The main metabolites of testosterone in the liver are etiocholanolone, androsterone, and epiandrosterone, collectively referred to as the 17-ketosteroids (17KS), all of which are released as glucuronide conjugates for excretion by the kidney. This route accounts for 50%−70% of the total testosterone metabolized daily.

An epimer of testosterone, epitestosterone (17α-hydroxylated testosterone), is produced by the testes and excreted as such in the urine in amounts approximately equal to that of testosterone (T:epiT $\sim$ 1:1). Epitestosterone is biologically inactive, but is not a metabolite, and is believed to be produced only by the gonads; thus, it is used as a gonadal steroid marker. In women, the ratio of T to epiT is also normally 1:1. Urinary T:epiT is useful in monitoring abuse of anabolic steroids by athletes because the ratio increases when any exogenous testosterone derivative is used.

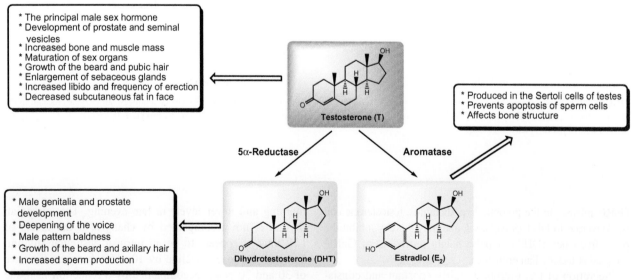

FIGURE 32.3 Multiple biological effects of testosterone (T). These effects are mediated by testosterone, dihydrotestosterone, and estradiol (E_2).

Biological Effects of Androgens

The biological effects of androgenic hormones can be of two types: (1) reproductive (androgenic), i.e., promoting the primary and secondary sexual characteristics of a male; and (2) nonreproductive, which includes anabolic effects. Both types of effects are mediated by the same ARs; therefore, testosterone and DHT are capable of exerting either type of effect. However, under physiological conditions, a critical factor that determines which hormone is active in a given tissue is the presence or absence of 5α-reductase. As underscored previously, pharmacological treatment with any steroid that can activate the ARs will evoke both androgenic and anabolic responses.

FEMALE REPRODUCTIVE SYSTEM

The overall regulation of the female reproductive system is shown in Figure 32.4.

Menstrual Cycle

The menstrual cycle consists of the **follicular phase** and the **luteal phase**, each lasting about 2 weeks. The follicular phase is a period of ovarian follicular growth (folliculogenesis) that results in ovulation (release of secondary

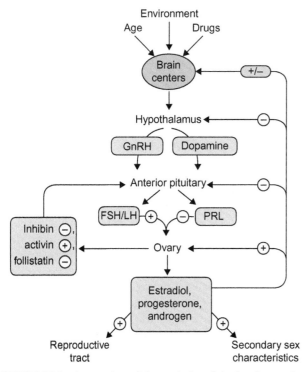

FIGURE 32.4 An overview of the regulation of the female reproductive system. Plus indicates positive, and minus indicates negative, regulation. *[Reproduced from R. A. Rhodes, G. A. Tanner, Medical physiology, 2nd ed. Philadelphia, PA: Lippincott Williams & Wilkins, 2003.]*

oocyte). This phase is dominated by estrogen produced by the growing follicle itself. During the follicular phase, the uterine endometrium is stimulated by estrogen to proliferate and to synthesize cytosolic receptors for progesterone. Follicular production of estrogen depends on FSH and LH. The luteal phase is progesterone-dominated. The follicle that ruptures at ovulation becomes the corpus luteum, which produces progesterone and estradiol. During this period, the uterine endometrium becomes secretory under the influence of progesterone. About 5 days into the luteal phase, the endometrium is ready to accept a blastocyst for implantation; in the absence of fertilization, however, the corpus luteum degenerates after about 12 days, steroid production ends, and the endometrium deteriorates (menstruation). The first day of menstruation is the first day of the menstrual cycle.

Endocrine Control of Folliculogenesis

Two types of endocrine cells are associated with the ovarian follicle. One is the granulosa cell, which resides within the follicle and is encased by the basal lamina. Like the Sertoli cell in the testes, the granulosa cell is perfused by plasma transudate, not blood, and possesses FSH-sensitive aromatase activity; thus, FSH stimulates the formation of estrogen by the granulosa cell. Unlike Sertoli cells, however, granulosa cells proliferate in response to estrogen, and this proliferation is inhibited by androgens. The nonendocrine function of granulosa cells is to promote growth of oocytes by conditioning the follicular fluid. The other type of follicular cell with endocrine function is the theca interna cell. Such cells are positioned along the outer border of the basal lamina; thus, they are located outside the follicle and are perfused by blood. Theca interna cells, like the Leydig cells of the testes, respond to LH with production of androgen from cholesterol. Unlike Leydig cells, they produce mainly androstenedione, although some testosterone is also produced. Theca interna cells provide androgens for estrogen production by granulosa cells.

Hormonal Control of Follicle Growth

During the follicular phase, the ovarian follicle (pre-antral follicle) grows by pronounced proliferation of granulosa cells. During the second half of the follicular phase, the follicle accumulates fluid, which leads to formation of an antrum (antral follicle).

In the pre-antral follicle, LH stimulates the internal cells to produce androgens (mainly androstenedione), which diffuse through the basal lamina into the granulosa cell compartment. The granulosa cells, under the influence of FSH, aromatize the androgens into estrogens (mainly estradiol), which act directly on the granulosa

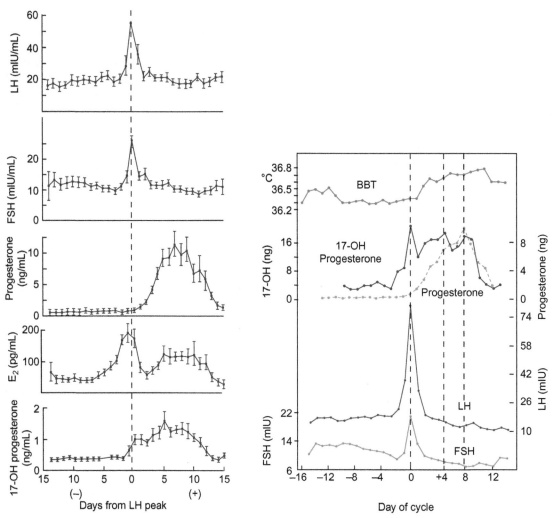

FIGURE 32.5 Hormonal and physiological changes during the menstrual cycle. BBT = basal body temperature.

cells to stimulate proliferation. This causes the follicle to grow and to accumulate fluid, and leads to the formation of an antrum. Although the intrafollicular concentration of estradiol is sufficient to stimulate proliferation of granulosa, it is not high enough to enter the general circulation. For this reason, the level of circulating estradiol does not increase during the first half of the follicular phase (Figure 32.5), and thus the release of LH and FSH is not inhibited but remains at a fairly constant level.

In the antral stage of follicular growth, production of estradiol by granulosa cells increases owing to the acquisition of LH receptors by the granulosa cells. Induction of LH receptors is brought about by the combined effects of FSH and estradiol, and enables the granulosa cells to begin producing estradiol from pregnenolone. Thus, the formation of estradiol by aromatization of androgens derived from the theca interna is augmented by estradiol synthesized *de novo*. The increased estradiol pool causes marked acceleration in follicle growth and spillage of

estradiol into the general circulation. Consequently, a relatively rapid increase in circulating estradiol levels is seen during the last 5 or 6 days of the follicular phase, which exerts an initial negative feedback on the release of FSH (Figure 32.5). This rise in estradiol levels is a critical cue for the neuroendocrine system because it indicates that the ovarian follicle is ready for ovulation. This message is in the form of an approximately three-fold increase in estradiol levels, which stimulates the release of a large surge of LH and FSH. In contrast to the negative feedback effect that estradiol normally exerts on the release of gonadotropins, the very high levels of estradiol presented over 2 to 3 days exert a positive feedback effect. Two mechanisms may operate in this positive feedback: high levels of estradiol for 2 or 3 days may (i) increase the sensitivity of the pituitary to GnRH or (ii) increase the release of GnRH from the hypothalamus (or both).

The latter mechanism is believed to involve the formation of catecholestrogen (2-hydroxyestradiol) after uptake by the median eminence, which, because of its abundance, competes with hypothalamic norepinephrine for inactivation by catechol-O-methyltransferase (COMT). This inactivation results in a higher norepinephrine content in the median eminence, which favors GnRH release (see Clinical Case Study 32.3).

The large bolus of LH released (preovulatory LH surge) in response to positive feedback by estradiol induces ovulation in about 1 day, probably by stimulating production of granulosa plasminogen activator, which leads to the formation of plasmin, an enzyme that may be responsible for the digestion of the basal lamina and, consequently, for the rupture of the follicle (ovulation).

Hormonal Control of Luteal Function

After ovulation, the granulosa cells proliferate in response to the preovulatory LH surge, and the theca interna cells and perifollicular blood vessels invade the cavity of the collapsed follicle. Under the influence of LH, the granulosa and invasive theca cells differentiate into luteal cells, which are characterized by their high lipid content. These luteal cells are steroidogenic and produce large amounts of progesterone and moderate amounts of estradiol. The ruptured follicle thus becomes the corpus luteum. Morphogenesis of the corpus luteum is not complete until about 1 to 4 days after ovulation, and luteal production of progesterone and estradiol gradually increases to a maximum about 6 to 7 days after ovulation. Thus, there is an interval of about 3 days after ovulation during which levels of circulating estradiol are reduced (Figure 32.5), and this interval is required for proper transport of the ovum through the fallopian tube into the uterus. Exposure to high levels of estrogen during this interval would lead to expulsion of the ovum or to blockage of ovum transport. The rise in levels of progesterone and of estradiol during the first week of the luteal phase is required for the endometrium to become secretory in preparation for implantation and pregnancy. The corpus luteum has a lifespan of about 12 days; it can synthesize steroids autonomously without extra-ovarian hormonal stimulation. Although the corpus luteum has receptors for LH, release of LH (and FSH) during the luteal phase is strongly inhibited by the potent negative feedback effect of progesterone and estradiol. Thus, if fertilization and implantation do not occur, the corpus luteum degenerates (luteolysis), and its production of progesterone and estradiol rapidly declines (Figure 32.5). Withdrawal of progesterone and estradiol during luteolysis results in deterioration of the endometrium and its shedding (menstruation). If fertilization and implantation occur, secretion of chorionic gonadotropin (hCG) (Chapter 29) by the implanting blastocyst stimulates the corpus luteum to continue producing progesterone, and luteolysis is prevented.

The regularity of the menstrual cycle in women of reproductive age can be affected by anatomical defects of the uterus or vagina, or by functional or structural defects in the hypothalamic−pituitary−ovarian axis that affect hormonal secretions. Complete cessation of menses (for more than 6 months) is known as **amenorrhea,** and a reduction in the frequency is known as **oligomenorrhea.** Physiologic states of amenorrhea include prepuberty, pregnancy, lactation, and postmenopause. A common pathologic cause of amenorrhea can result from a reduction in the secretion of GnRH from the hypothalamic neurons into the hypothalamic−hypophyseal portal system, with a corresponding reduction in the secretion of FSH and LH from the pituitary. Decreased GnRH can occur due to weight loss, anorexia nervosa, excessive exercise, debilitating diseases, or psychological trauma.

Pregnancy

Human Chorionic Gonadotropin (hCG)

After fertilization of the ootid in the fallopian tube, the zygote develops into a blastocyst by the time it enters the uterine cavity (about 4 days after ovulation). (See Clinical Case Study 32.1.) By the seventh day after ovulation, the blastocyst is completely embedded in the endometrium, and its outer trophoblast has differentiated into an inner, mitotically active cytotrophoblast and an outer, nonmitotic syncytiotrophoblast. The syncytiotrophoblast begins producing hCG from about the eighth day after ovulation and continues to produce increasing amounts of hCG until maximal levels are attained at about 6−8 weeks of pregnancy. hCG stimulates the corpus luteum to continue producing progesterone from about the eighth day after ovulation and thus protects the corpus luteum from luteolysis. As mentioned previously, hCG promotes expression of the male genotype in the fetus by stimulating testosterone production by fetal Leydig cells from about the eighth week of pregnancy. Moreover, hCG may be the stimulus for the initial development of the fetal zone of the adrenal cortex, which appears at about the sixth week. Lastly, and perhaps most importantly, hCG stimulates steroidogenesis in syncytiotrophoblasts at 4−5 weeks of pregnancy by inducing CYP11A (side chain cleavage enzyme), 3β-hydroxysteroid dehydrogenase (3β-HSDH/iso), and CYP19 (aromatase). These enzymes are critical for the maintenance of pregnancy because the placenta assumes the role of the ovaries as the major generator of progesterone and estrogen after the sixth week of pregnancy.

The implantation of the fertilized ovum at a site other than the endometrium is known as **ectopic pregnancy.** Patients with ectopic pregnancy frequently exhibit abdominal pain with amenorrhea. By far the most common cause is impaired fallopian tube function. The tubal pathology can result from pelvic infection, endometriosis (occurrence of endometrial tissue outside the uterus), or previous surgery. The implanted embryo commonly dies in the early weeks of pregnancy. Tubal rupture and the associated bleeding require emergency care. Thus, early diagnosis is essential and consists of serial measurements of serum levels of hCG using antibodies specific for the β-subunits of hCG (β-hCG) and serum progesterone and pelvic ultrasonography. In ectopic pregnancy, serum β-hCG levels are lower than those observed at the same gestational age in a normal pregnancy. A serum progesterone level of 25 ng/mL (79.5 nM/L) or higher excludes ectopic pregnancy with 97.5% sensitivity. The treatment for ectopic pregnancy consists of either surgery to remove the embryonic tissue or medical treatment with methotrexate, a folic acid antagonist. Methotrexate inhibits both *de novo* purine and pyrimidine nucleotide biosynthesis (Chapter 25), blocking DNA synthesis and cell multiplication of the actively proliferating trophoblasts. Serum β-hCG levels are used as a tumor marker for germ cell cancers. hCG mimics the actions of LH, and it can induce ovulation in the ovaries and testicular testosterone production. Thus, it is clinically used in the treatment of undescended testes, and infertility of hCG is also used in the treatment of testicular atrophy and decreased libido in males who have abused the banned anabolic androgenic steroids (AAS). AAS inhibits, the hypothalamic—pituitary—gonadal axis with resultant decreased *in vivo* production of testosterone. AAS abuse also causes gynecomastia due to increased production of estrogen from androgens.

Placental Steroids

The absence of progesterone is incompatible with the gravid pregnant state. Progesterone acts in three ways to maintain pregnancy:

1. It maintains placental viability, thus ensuring adequate exchange of substances between maternal and fetal compartments;
2. It maintains perfusion of the decidua basalis (maternal placenta), presumably by inhibiting the formation of vasoconstrictive prostaglandins; and
3. It diminishes myometrial contractility, possibly by increasing the resting membrane potential or by inhibiting the formation of prostaglandins E_1 and $F_{2\alpha}$ (Chapter 16).

The placenta is the major site of conversion of cholesterol to progesterone after the sixth week; however, it is not capable of synthesizing cholesterol from acetate. Moreover, the placenta lacks CYP17 and CYP21, which explains why the major steroid it produces is progesterone. Although the placenta cannot process cholesterol beyond progesterone, it contains some important steroid-modifying enzymes that enable it to produce substantial amounts of estrogen.

Small amounts of estrogen are required for the maintenance of pregnancy because estrogen maintains tissue responsiveness to progesterone. The placenta forms estrogen from androgenic steroids. The main androgen used is dehydroepiandrosterone (DHEA), derived mainly from the fetal compartment in the form of DHEAS (DHEA sulfate).

Human Placental Lactogen

Human placental lactogen (hPL) is produced by the syncytiotrophoblast from about the time when production of hCG begins to diminish. Production of hPL is proportional to placental growth, and its level reflects placental well-being. hPL exerts GH-like effects in both fetal and maternal compartments. In the fetus, it promotes the formation of insulin-like growth factor and growth factors believed to promote growth of most, if not all, fetal tissues. In the mother, hPL promotes nitrogen retention and utilization of free fatty acids, and it creates a state of mild insulin resistance that benefits the fetus because it increases the availability of maternal glucose for fetal consumption. In the mother, hPL also exerts a prolactin-like effect and, with estrogen and progesterone, promotes ductal and alveolar growth in the mammary gland during the third trimester of pregnancy. However, prolactin is believed to be more important than hPL in this effect.

The placenta lacks the key enzyme necessary for formation of estrogens from cholesterol (CYP17), and relies on androgenic precursors from the maternal and fetal compartments. The main androgen used comes from the fetal zone of the fetal adrenal; this is DHEAS, which is also taken up and metabolized by fetal liver into 16α-hydroxy-DHEAS. The placenta converts DHEAS into estrone (E1) and estradiol (E2), and processes 16β-hydroxy-DHEAS into estriol (E3). Estrogens enter the maternal circulation and appear in maternal urine as conjugated estrogens.

Parturition

During late gestation, rising levels of estrogen are thought to increase the synthesis of oxytocin in hypothalamic magnocellular neurons to induce oxytocin receptors in the myometrium and to increase myometrial contractility by lowering the membrane potential. During this period, relaxin from the decidua softens the cervix for impending delivery. At term, an unidentified triggering factor in fetal urine stimulates uterine production of PGE_1 and $PGF_{2\alpha}$, which lead to myometrial contraction, mainly in the fundus, and encourage movement of the infant through the cervical canal. Dilation of the cervix by the infant stimulates the release of oxytocin by neuroendocrine reflex, further stimulating uterine contractions, which, in turn, causes more cervical stretching and a positive feedback to oxytocin release, until delivery is complete.

Lactation

During mid- and late-pregnancy, growth and differentiation of mammary glands are promoted by the combined effects of estrogen, progesterone, and prolactin (or hPL). In the presence of prolactin or hPL, estrogen stimulates branching of the mammary ducts and increases their growth. Estrogen also stimulates production of lactogenic receptors in mammary ductal cells and acts on the anterior pituitary to stimulate prolactin secretion. In the presence of prolactin or hPL, progesterone acts on the estrogen-conditioned mammary gland to promote differentiation of the terminal ductal buds into alveoli and, consequently, into lobules. Synthesis and secretion of milk occur in these lobules. Lactogenesis begins during the third trimester of pregnancy and involves synthesis of the milk-specific proteins casein, lactalbumin, and lactoglobulin. The primary regulator of lactogenesis is prolactin, although the participation of additional hormones is needed (i.e., insulin, cortisol, thyroid hormone). Lactation involves release of milk into the alveolar lumen. Lactation is inhibited by estrogen and, thus, is held in check by the high levels of estrogen during pregnancy; the withdrawal of estrogen after birth triggers lactation.

Biological Effects of Estrogens

The biological effects of estrogen are many. Estrogen is the main determinant of female reproductive function, bone maintenance, and cardioprotection. Its effect on the brain includes reproductive behavior and function, learning, and memory. The mechanism of action of estrogen in the improvement of cognitive function is not understood, but it is thought that estrogen delays the onset of Alzheimer's disease. Estrogen is a risk factor for breast cancer, and in rodent studies it has been shown that estrogen and its catechol metabolites are carcinogens. Breast cancer consists of several tumor subtypes with different natural history and requires different individual treatment strategies in addition to, or in place of, conventional chemotherapy. In postmenopausal women with ER positive breast cancer, estrogen deprivation is a key therapeutic approach. This is accomplished by the inhibition of aromatase, which is responsible for the conversion of androgens to estrogens in the malignant breast cancer cells. Structures of the therapeutically used aromatase inhibitors are shown in Figure 32.6.

As estrogen mediates its biological effects via different receptors, selective inhibition of estrogen action in the

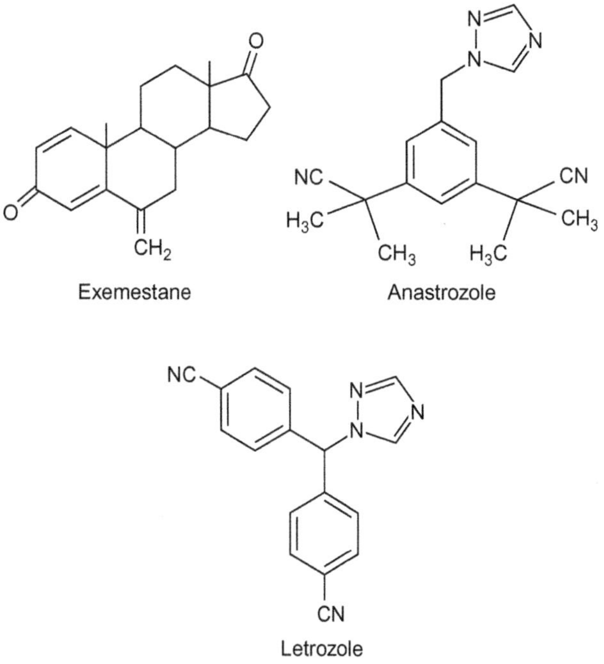

FIGURE 32.6 Structures of orally active third-generation aromatase inhibitors. Exemestane can be used in chemoprevention in postmenopausal women with high-risk breast cancer.

target tissue (e.g., breast), yet preserving its normal physiologic functions in other tissues, has been used clinically in the management of breast cancer. This group of compounds with agonist and antagonist properties against estrogens is known as **selective estrogen receptor modulators** (discussed later).

In 20%−30% of invasive breast cancers, overexpression of human epidermal growth factor receptor type 2 (referred to as HER2/neu) is associated with tumor promotion. HER2 is a member of a family of four (HER1−4) transmembrane tyrosine kinase receptors, and their normal cell function includes growth, differentiation, adhesion, migration, and other cellular functions. In overexpressed HER2, the extracellular domains in the absence of a ligand adopt a fixed ligand-activated conformation. Once activated, the signal-transduction cascade promotes cell proliferation mediated by the downstream signaling events of the RAS−MAPK pathway and inhibits apoptosis via the phosphatidylinositol 39-kinase-AKT-mammalian target of rapamycin (mTOR) pathway. Current treatment strategies for overexpressed HER2 breast cancers include the use of a humanized monoclonal antibody (e.g., trastuzumab) against HER2 and tyrosine kinase inhibitors (e.g., lapatinib). Examples of familial syndromes associated with the risk of developing breast cancer are *BRCA1* and *BRCA2* gene mutations. These genes encode DNA excision repair proteins (Chapter 22). BRCA1 protein also has other functions including transcriptional regulation.

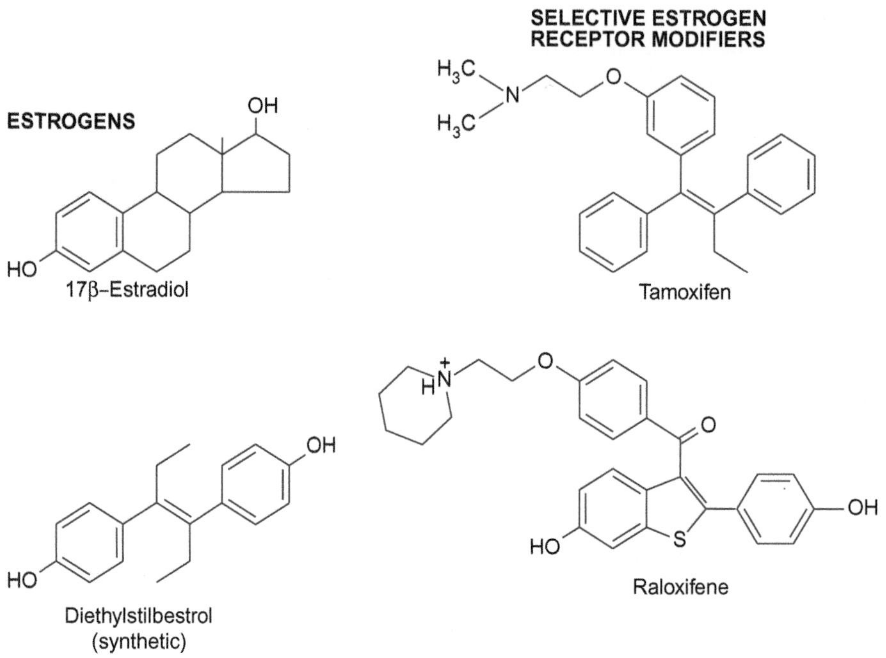

FIGURE 32.7   Structures of estrogens and selective estrogen receptor modifiers.

Selective Estrogen Receptor Modulators (SERM)

As discussed earlier, estrogens have tissue selectivity, in part, based on the type of estrogen receptor and the DNA-bound transcription complex present in the target cells. Drugs are being developed to mimic (agonist) or antagonize (antagonist) the effect of estrogen (Figure 32.7). An example of a synthetic estrogen is diethylstilbestrol. Tamoxifen and raloxifene are mixed agonists and antagonists of estrogen activity, and are named SERMs (also known as **designer estrogens**). Tamoxifen antagonizes the action of estrogen in breast tissue and is used in the treatment of ER-positive breast cancer. However, tamoxifen is an estrogen agonist in the uterus and increases the risk of endometrial cancer. In bone and in the cardiovascular system, tamoxifen has beneficial effects similar to those of estrogen. Raloxifene is used in the treatment of osteoporosis as an antiresorptive agent in postmenopausal women but does not exhibit estrogen-related adverse effects on the endometrium. Raloxifene has a similar effect on the serum lipid profile compared to estrogen; however, neither tamoxifen nor raloxifene increases the serum concentration of HDL cholesterol. Many phytoestrogens contain nonsteroidal precursors of estrogenic substances (Figure 32.8), which may affect estrogen-sensitive tissues if ingested in significant amounts. However, indications are that the intake must be very high for this to occur.

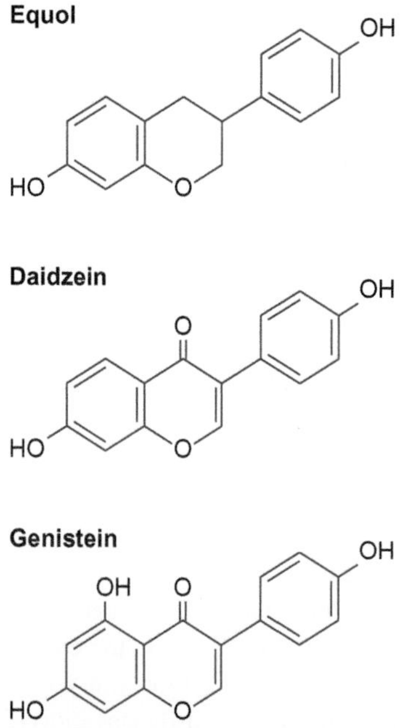

FIGURE 32.8   Structures of phytoestrogens. These are diphenolic compounds found in plants, such as soybeans, sprouts of cloves, and alfalfa.

Biological Effects of Progesterone

The human progesterone receptor (PR) gene on chromosome 11 encodes a single PR protein that is homologous to the other steroid hormone receptors, but binds progesterone preferentially. Like the ER, PR is predominantly found in the nucleus; however, it shuttles between the nucleus and the cytoplasm. PR exists in a stable form bound to heat-shock proteins (hsp), but is activated when progesterone (P) displaces the hsp and binds to the hormone-binding domain. The P−PR complex dimerizes and binds to two half-sites of progesterone response elements (PREs) in chromatin; this promotes transcription of a progesterone-sensitive gene. The PRE is identical to the glucocorticoid response element and androgen response element, but the significance of three hormones sharing the same response element in DNA is not known. PR production is stimulated by estrogen and in some tissues is downregulated by progesterone. Clinical cases for both delayed and precocious puberty are presented in Clinical Case Studies 32.2 (Vignettes 1 and 2) and 32.3, respectively.

- **Endometrium:** Following adequate priming of the endometrium by estradiol, progesterone increases glycogen deposition in the glandular epithelium and stimulates secretory function of the glands from about day 4 after ovulation. Progesterone also stimulates fluid accumulation in the endometrial stroma, with maximal edema occurring at about the time of implantation (day 7). Progesterone promotes the decidual reaction in the stroma in response to the implanting blastocyst and stimulates the production of decidual prolactin (dPRL) by the decidual cells. In the absence of conception, progesterone causes predecidualization of the stroma from about day 9 in a process of terminal (irreversible) differentiation that leads to degeneration of the tissue when progesterone levels decline. Progesterone is required for pregnancy to continue, in part because it keeps the spiral arteries patent and promotes maternal blood flow to the placenta. Progesterone is also involved in the pathogenesis of endometriosis, a condition in which endometrial tissue grows in ectopic sites (most commonly in the peritoneal cavity); treatment with danazol, a synthetic androgen that antagonizes progesterone, reduces growth of the tissue.
- **Myometrium:** Progesterone increases the resting membrane potential of the myometrium and thereby reduces its contractility. Progesterone antagonizes the effect of estrogen by inhibiting synthesis of ER and blocks $PGF_{2\alpha}$-induced contraction of the myometrium. Accordingly, the administration of an antiprogesterone causes a rise in myometrial ER and increased responsiveness to $PGF_{2\alpha}$-induced contractions. There is evidence that myometrial contractility is promoted by the formation of gap junctions between muscle cells; progesterone presumably inhibits gap junction formation and therefore inhibits myometrial contractions.

- **Mammary Gland:** In the presence of PRL, progesterone stimulates lobuloalveolar development in the breast of pubertal girls, but only after the tissue has been stimulated by estrogen. As discussed previously, estrogen promotes the growth of the ducts and induces the synthesis of progesterone receptors in the tubular epithelium. During the luteal phase of the menstrual cycle and during pregnancy, progesterone (in the presence of estrogen and PRL) stimulates maximal proliferative growth of the breast mainly by alveolar (glandular) growth. In breast cancer cells, progesterone reduces the formation of estrogen from androgenic precursors but increases the production of some of the autocrine growth factors (e.g., TGF-α).
- **Central Nervous System:** In addition to its inhibitory effect on GnRH secretion, progesterone causes thermogenesis during the menstrual cycle and pregnancy. In a normal menstrual cycle, the oral basal temperature in most women increases by about 0.5°F around the time of ovulation and remains elevated until shortly before the onset of menses. In pregnant women, the oral basal temperature remains elevated to term. An anesthetizing and EEG-altering effect of progesterone administered to laboratory rodents and the increase in ventilatory drive (hyperventilation) stimulated by progesterone may also reflect CNS actions of the steroid.
- **Immunosuppression and Anti-inflammatory Effects:** In addition to its ability to suppress myometrial and endometrial prostaglandin formation, progesterone suppresses T-lymphocyte proliferation and interleukin-8 synthesis, and increases prostaglandin dehydrogenase activity, all of which contribute to preventing maternal rejection of the implanting conceptus (an allograft).

Pharmacological Enhancement of Fertility

Female infertility due to inadequate gonadotropin stimulation of the ovary has been successfully treated by two approaches:

1. Increasing gonadotropin levels by supplying exogenous gonadotropins; and
2. Increasing gonadotropin levels by stimulating their release from the pituitary.

   Exogenous Gonadotropin Treatment: Women who are infertile because of a deficiency in gonadotropins are given exogenous FSH for 7−12 days to promote folliculogenesis followed by a single large dose of hCG to induce ovulation. This has resulted in ovulation in the majority of patients (∼90%) and pregnancy in about half. However, about 17%−20% of the patients have multiple births due to the secretion of more than one follicle.

Stimulation of Endogenous Gonadotropin Secretion: In women who are infertile because of a neuroendocrine system that is overly sensitive to negative estrogen feedback, an antiestrogen is used to reduce the intensity of the negative feedback. By reducing the intensity of estrogen feedback, normal gonadotropin secretion should resume and lead to ovulation. Clomiphene (50 mg) is taken daily for 5 days to block the estrogen effect in the neuroendocrine system; in the majority of patients ($\sim 80\%$), ovulation occurs about a week later, and about 30%–40% of the women conceive. The incidence of multiple births by this method ($\sim 8\%$) is higher than in untreated patients ($\sim 1\%$) but lower than with gonadotropin treatment. In some laboratories that conduct *in vitro* fertilization, the two protocols (exogenous gonadotropins and clomiphene) are combined for a greater yield of secondary oocytes.

Pharmacological Prevention of Pregnancy (Female Contraception)

Currently, the most effective method to prevent conception is by oral steroid hormone treatment (Figure 32.9) to inhibit or modify the cyclic changes in endogenous reproductive hormones. Two effective types of oral contraceptive are used with proven success: one to prevent ovulation and the other to prevent implantation.

The most effective means of preventing ovulation is by the oral administration of ovulation-inducing gonadal steroid derivatives at dosages that will prevent cyclic changes in LH and FSH secretion. This method has proven to be more than 99% effective over the past 40 years and is currently used by more than 50 million women worldwide. The treatment protocol is of four different types, all of which involve the use of synthetic steroid derivatives that have long biological half-lives. The first three protocols involve estrogen + progestin combinations taken for 3 weeks, followed by no treatment for 1 week.

1. Fixed combined dose of estrogen + progestin daily ("combination pill");
2. Biphasic type: fixed dose estrogen + two different doses of progestin;
3. Triphasic type: fixed dose estrogen + three different doses of progestin (stepwise increase at 1-week intervals); and
4. "Mini" pill, containing a progestin only; taken daily without interval of withdrawal.

Among these, the combination pill is the most widely used and is probably the most effective at a success rate of >99%, whereas the mini pill is the least effective at $\sim 95\%$ because it does not always prevent ovulation. The rationale for varying the dose of progestin was that reducing the total amount of steroid would reduce the risk of myocardial infarction, hypertension, and stroke, all of which were subsequently shown not to be altered by varying the dose of progestin. The rationale for the progestin-only protocol was to eliminate the risk of endometrial cancer and breast cancer associated with estrogen; however, some of the health benefits of estrogen (e.g., increased HDL, osteoprotection) are eliminated and some side effects (e.g., breakthrough bleeding) are introduced.

In women using the combination pill, the plasma levels of LH, FSH, estradiol, and progesterone were found to be greatly suppressed and unchanged throughout the 3-week period on the pill. Folliculogenesis did not occur during treatment, but there was no evidence of increased atresia; when electing to bear a child, former pill users experienced normal menstrual cyclicity, pregnancy, and lactation. Long-term observations indicated that pill users experienced menopause within the normal age range and had no health problems that could be attributed to the long-term use of the pill, although recent epidemiological studies suggest that combination pill users may be at increased risk of developing breast cancer prior to age 45.

Postcoital pharmacologic prevention of implantation can be achieved by oral administration of a synthetic estrogen (25 mg diethylstilbestrol) twice daily for 5 days. This "morning after pill" treatment stimulates fallopian tube contractions during the period of conceptus travel toward the uterus, such that the conceptus is propelled into the uterus prematurely and is resorbed, or is trapped within the fallopian tube because of spastic contraction of the isthmus. Although effective, the high-dose estrogen produces nausea, vomiting, and menstrual problems. Alternatively, oral administration of a combination pill (50 μg ethinylestradiol + 0.5 mg norgestrel), two tablets within 72 hours postcoitus and two tablets 12 hours later, is also effective and does not produce undesirable side effects. This treatment does not involve an alteration in fallopian tube activity, but the treatment may serve to convert the endometrium into a less-receptive tissue for blastocyst implantation.

Another postcoital treatment is the use of a progesterone antagonist, mifepristone (RU-486), which apparently works by reversing the effects of progesterone on the endometrium, and thereby interferes with implantation. Administration of the drug 2 days after the midcycle LH surge prevents implantation by a direct effect on the endometrium and has no effect on the circulating levels of gonadotropins or on luteal function.

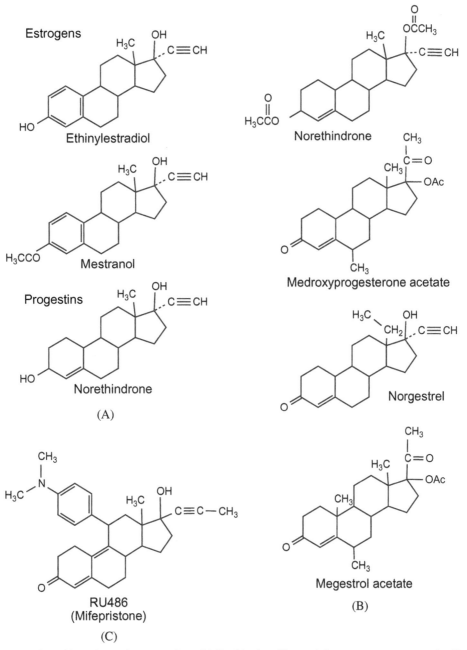

FIGURE 32.9 Structures of steroids used as oral contraceptives. (A) Combination pills containing an estrogen + progestin. (B) The progestin-only pill. (C) RU-486 or mifepristone, which can be utilized as an abortifacient. *[Reproduced with permission from S.W. Norman and G. Litwack, Hormones, 2nd Ed. San Diego: Acedemic Press, 1997].*

CLINICAL CASE STUDY 32.1 Familial Infertility Due to Absence of Zona Pellucida

This case was abstracted from: H.-L. Huang, C. Lv, Y.-C. Zhao, W. Li, X.-M. He, P. Li, et al., Mutant ZP1 in familial infertility, N. Engl., J. Med. 370 (2014) 1220–1226.

Synopsis

This case study describes a consanguineous family consisting of four sisters with the diagnosis of primary infertility, despite normal maturity and more than 12 months of unprotected sexual intercourse. One of the infertile sisters, who was 32 years old and served as a proband for this study, underwent further clinical and molecular investigations. The morphological studies of oocytes obtained from the subject revealed the absence of insoluble zona pellucida (ZP). The molecular studies revealed a homozygous deletion of the gene ZP1, due to a frameshift mutation. This mutation resulted in a premature stop codon, causing production of a truncated protein unable to form the zona pellucida around the ova, which accounted for the subject's infertility.

Teaching Points

1. The oocyte produces four zona pellucida (ZP) glycoproteins: ZP1, ZP2, ZP3, and ZP4. These glycoproteins form an insoluble, meshlike matrix around the oocyte, in which repeating ZP2–ZP3 units are cross-linked by ZP1. The role of the zona pellucida is to support early development of the oocyte, facilitate gamete recognition during fertilization, prevent polyspermia, and protect early embryos prior to implantation. Absence or dysfunction of the zona pellucida leads to sterility.

2. Gene knockout of either ZP2 or ZP3 results in complete absence of the zona pellucida, whereas gene knockout of ZP1 results in two types of eggs: one with a normal zona pellucida and one with a swollen and loosely organized zona pellucida. This abnormal formation of the zona pellucida may be due to intracellular sequestration of ZP proteins and disruption of their trafficking to the cell surface.

3. In this case, consanguinity was suggested to have played a role in the autosomal recessive inheritance of a homozygous ZP mutation. The parents of the affected women were first cousins, both with heterozygous genotype (+ / −). Of their seven offspring who underwent genetic testing, four were infertile (homozygous affected genotype − / −), one produced a child (heterozygous genotype + / −), and the remaining two did not attempt to have children (homozygous affected genotype − / −). Their grandparents had a heterozygous genotype (+ / −).

Supplemental Reference

[1] P.M. Wassarman, Zona pellucida glycoproteins, J. Biol. Chem. 283 (2008) 24285–24289.

CLINICAL CASE STUDY 32.2 Delayed Puberty

Vignette 1: Constitutional Delay of Growth and Puberty

This case was abstracted from: M.R. Palmart, L. Dunkel, Delayed puberty, N. Engl. J. Med. 366 (2012) 443–453.

Synopsis

A 14-year-old boy with an unremarkable past medical history was brought to his primary care physician for concerns of delayed pubertal development. His height of 57.5 inches (146 cm) was less than the 3rd percentile for age, and his weight of 82 pounds (37 kg) was at the 3rd percentile. Family history was significant for late pubertal development in his mother and father. The patient's predicted adult height based on midparental height calculations was 65.8 inches (167 cm). Physical examination revealed Tanner stage 1 pubic hair and prepubertal-sized testes (see case reference for a diagram of Tanner stages). His presentation suggested the diagnosis of constitutional delay of growth and puberty (CDGP). Further evaluation was recommended to rule out pathologic causes of delayed puberty, which included growth rate monitoring, bone age analysis, biochemical analyses, measurement of serum luteinizing hormone (LH), measurement of serum follicle-stimulating hormone (FSH), measurement of serum insulin-like growth factor 1 (IGF-1), and measurement of serum testosterone. The recommended treatment for this patient's CDGP was supportive therapy and close monitoring of growth.

Teaching Points

1. Delayed puberty is defined as the absence of testicular enlargement in boys, or the absence of breast development in girls, at an age that is 2 to 2.5 standard deviations greater than the mean age of pubertal development in the population (usually 14 years of age in boys and 13 years of age in girls).

2. The most common cause of delayed puberty in both sexes is constitutional delay of growth and puberty (CDGP). The cause of CDGP is unknown, but there is usually a family history of delayed puberty. CDGP is a diagnosis of exclusion and can be made only after underlying causes of delayed puberty have been ruled out. Treatment with growth hormone, anabolic steroids, or aromatase inhibitors is not recommended to induce the appearance of secondary sexual characteristics; however, low-dose sex steroids may be administered electively to alleviate the psychosocial difficulties associated with short stature and pubertal delay. Refer to Chapter 29 and Clinical Case Study 29.1.

Vignette 2: Estrogen Resistance

This case was abstracted from: S.D. Quaynor, E.W. Stradtman, H.G. Kim, Y. Shen, L.P. Chorich, D.A. Schreihofer, L.C. Layman, Delayed puberty and estrogen resistance in a

(Continued)

CLINICAL CASE STUDY 32.2 (Continued)

woman with estrogen receptor α variant, N. Engl. J. Med. 369 (2013) 164–171.

Synopsis

A 15-year-old female presented to her primary care physician with absent breast development, primary amenorrhea, and intermittent lower abdominal pain. The patient was enrolled in a study and, at 17 years old, was found to have Tanner stage 1 breast development, Tanner stage 4 pubic hair, severe facial acne, and a body mass index of 16.6. Laboratory studies revealed a serum estradiol level 10 times the normal value, a normal follicular phase, normal luteinizing hormone levels (LH), and normal follicle-stimulating hormone levels (FSH). Her bone age was 11 to 12 years. Ultrasonography revealed a small uterus, no endometrial stripe, and significantly enlarged multicystic ovaries. MRI of the brain was normal. Her clinical

presentation suggested estrogen resistance. A trial of exogenous estrogens was started, but after 5 months of high-dose estrogen therapy, her breasts remained at Tanner stage 1. DNA sequencing revealed a homozygous missense mutation of the *ESR1* gene (estrogen receptor α) as a cause of her profoundly estrogen-resistant state.

Teaching Points

1. Other causes of delayed puberty are hypergonadotropic hypogonadism (Turner's syndrome, gonadal dysgenesis), permanent hypogonadotropic hypogonadism (tumors or infiltrative diseases of the hypothalamus or pituitary), and functional hypogonadotropic hypogonadism (systemic illness, anorexia, excessive exercise, hypothyroidism). In Vignette 2, the patient's delayed puberty was caused by estrogen resistance due to abnormal estrogen receptors.

CLINICAL CASE STUDY 32.3 Precocious Puberty

This case was abstracted from: J. Carel, J. Leger, Precocious puberty, N. Engl. J. Med. 358 (2008) 2366–2377.

Synopsis

A 6-year-old girl presented to her pediatrician with unexpected breast development for her age. Her height was in the 97th percentile for her age, and her body mass index was 17.9. The patient's growth velocity was increased for her age. She had Tanner stage 3 breast development and Tanner stage 2 pubic hair development. Her presentation suggested the diagnosis of progressive central precocious puberty. Further evaluation was recommended to confirm the diagnosis, which included bone age assessment, measurement of serum luteinizing hormone levels, measurement of serum estradiol levels, MRI and ultrasound of the pelvis to exclude tumors, and GnRH (gonadotropin-releasing hormone)-agonist stimulation tests. The recommended treatment for this patient's central precocious puberty was a GnRH agonist.

Teaching Points

1. The initiation of puberty results from activation of the hypothalamic–pituitary–gonadal axis and requires the coordinated pulsatile secretion of GnRH, which is regulated by various stimulatory and inhibitory factors. Early pulsatile GnRH activation due to hypothalamic tumors or idiopathic etiologies leads to progressive central precocious puberty. In an attempt to elucidate the idiopathic etiologies of progressive central precocious puberty, there have been numerous efforts to identify genes associated with central precocious puberty.

2. There are only two rare gene mutations known to cause central precocious puberty: a gain-of-function mutation of the gene kisspeptin-1 (KISS1), and a gain-of function mutation of the gene for its receptor KISS1R (also known as G-protein-coupled receptor 54, or GPR54). KISS1 and GPR54 are powerful stimulators of the central reproductive axis. The GPR54 receptor is located at the GnRH neurons of the hypothalamus, and its activation results in intracellular signaling that leads to GnRH release. KISS1 (also known as metastin, a metastasis suppressor gene) encodes kisspeptin proteins that bind and activate GPR54. A gain-of function mutation of the GPR54 gene was found to prolong intracellular signaling, and a gain-of-function mutation of the KISS1 gene was found to increase the bioavailability of kisspeptin via increased resistance to degradation. Both mutations result in a greater amplitude pulse of GnRH, which leads to early initiation of puberty [3].

3. A recent study found another gene mutation associated with central precocious puberty: a loss-of-function mutation of the makorin ring-finger 3 gene (MKRN3). The mechanism of action of MKRN3 protein is not yet well understood, but animal studies suggest that MKRN3 is a powerful inhibitor of the central reproductive axis. MKRN3 protein is abundantly expressed in the developing brain and is involved with ubiquitination and cell signaling. A frameshift mutation of the MKRN3 gene was found to cause the production of abnormal proteins that prematurely activated pulsatile secretion of GnRH, thus initiating

(Continued)

CLINICAL CASE STUDY 32.3 (Continued)

puberty. MKRN3 mutations are also known contribute to Prader–Willi syndrome [1].

4. Precocious puberty can also result from peripheral causes (e.g., gonadal and adrenal tumors). Treatment for peripheral or GnRH-independent precocious puberty may require aromatase inhibitors and selective estrogen receptor modifiers (SERMs).

5. The treatment begins with administration of GnRH agonist, which forms a depot providing a continuous supply. This causes desensitization of pituitary gonadotropins with the inhibition of luteinizing hormone (LH) and follicle-stimulating hormone (FSH) release.

6. Many cases of idiopathic precocious puberty may not require any treatment because of their lack of progression.

Supplemental References

[1] A.P. Abreu, A. Dauber, D.B. Macedo, S.D. Noel, V.N. Brito, J.C. Gill, et al., Central precocious puberty caused by mutations in the imprinted gene MKRN3, N. Engl. J. Med. 368 (2013) 2467–2475.
[2] I.A. Hughes, Releasing the brake on puberty, N. Engl. J. Med. 368 (2013) 26.
[3] L.G. Silveira, S.D. Noel, A.P. Silveira-Neto, A.P. Abreu, V.N. Brito, M.G. Santos, et al., Mutations of the KISS1 gene in disorders of puberty, J. Clin. Endocrinol. Metab. 95(5) (2010) 2276–2280.

ENRICHMENT READING

[1] J.L. Ross, C.A. Quigley, D. Cao, P. Feuillan, K. Kowal, J.J. Chipman, et al., Growth hormone plus childhood low-dose estrogen in Turner's syndrome, N. Engl. J. Med. 364 (2011) 1230–1242.

[2] L. Cuttler, R.L. Rosenfield, Assessing the value of treatments to increase height, N. Engl. J. Med. 364 (2011) 1274–1276.

[3] A.K. Topaloglu, J.A. Tello, L.D. Kotan, M.N. Ozbek, M.B. Yilmaz, S. Erdogan, et al., Inactivating KISS1 mutation and hypogonadotropic hypogonadism, N. Engl. J. Med. 366 (2012) 629–635.

[4] H. Huang, C. Lv, Y. Zhao, W. Li, X. He, P. Li, et al., Mutant ZP1 in familial infertility, N. Engl. J. Med. 370 (2014) 1220–1226.

[5] D.J. Handelsman, Mechanisms of action of testosterone: unraveling a Gordian knot, N. Engl. J. Med. 369 (2013) 1058–1059.

[6] J.S. Finkelstein, H. Lee, S.M. Burnett-Bowie, J.C. Pallais, E.W. Yu, L.F. Borges, et al., Gonadal steroids and body composition, strength, and sexual function in men, N. Engl. J. Med. 369 (2013) 1011–1022.

[7] S. Caburet, V.A. Arboleda, E. Llano, P.A. Overbeek, J.L. Barbero, K. Oka, et al., Mutant cohesin in premature ovarian failure, N. Engl. J. Med. 370 (2014) 943–949.

[8] J.D. Iams, Prevention of preterm parturition, N. Engl. J. Med. 370 (2014) 254–261.

[9] R.D. Nerenz, E.S. Jungheim, C.M. Lockwood, Premenopausal amenorrhea: what's in a number? Clin. Chem. 60 (2014) 29–34.

[10] D. Bode, D.A. Seehusen, D. Baird, Hirsutism in women, Am. J. Physician 85 (2012) 373–380.

[11] S. Harrison, N. Somani, W.F. Bergfeld, Update on the management of hirsutism, Cleve. Clin. J. Med. 77 (2010) 388–398.

[12] J.L. Shea, E.P. Diamandis, B. Hoffman, Y.M. Dennis, J. Canick, D. van den Boom, A new era in prenatal diagnosis: the use of cell-free fetal DNA in maternal circulation for detection of chromosomal aneuploidies, Clin. Chem. 59 (2013) 1151–1159.

[13] J. Cha, X. Sun, S.K. Dey, Mechanisms of implantation: strategies for successful pregnancy, Nat. Med. 18 (2012) 1754–1767.

[14] L.A. Hughes, J.D. Davies, T.I. Bunch, V. Pasterski, K. Mastroyannopoulou, J. MacDougall, Androgen insensitivity syndrome, Lancet 380 (2012) 1419–1428.

[15] K.M. Pantalone, C. Faiman, Male hypogonadism: more than just a low testosterone, Cleve. Clin. J. Med. 79 (2012) 717–725.

[16] S. Basaria, S. Bhasin, Targeting the skeletal muscle–metabolism axis in prostate-cancer therapy, N. Engl. J. Med. 367 (2012) 965–967.

[17] N.E. Davidson, T.W. Kensler, "MAPping" the course of chemoprevention in breast cancer, N. Engl. J. Med. 364 (2011) 2463–2464.

[18] P.E. Goss, J.N. Ingle, J.E. Ales-Martinez, A.M. Cheung, R.T. Chlebowski, J. Wactawski-Wende, et al., Exemestane for breast-cancer prevention in postmenopausal women, N. Engl. J. Med. 364 (2011) 2381–2391.

[19] M.R. Palmert, L. Dunkel, Delayed puberty, N. Engl. J. Med. 366 (2012) 443–453.

Chapter 33

Immunology

Key Points

1. The "Introduction" section will enable the student to
   a. Compare innate and adaptive aspects of the human immune response.
   b. Describe the concept of immunologic memory and its importance in immunity.
2. The "Components of the Immune System" section will enable the student to
   a. Describe the means by which cells of the immune system are differentiated.
   b. Describe functions and properties of the major cell types and organs associated with the human immune system.
3. The "Communication within the Immune Response" section will enable the student to
   a. Describe how immune cells communicate through cytokines and cell surface molecules.
   b. Describe the function of the major histocompatibility complex in the immune response.
   c. Describe the importance of the exogenous and endogenous pathways in antigen presentation.
4. The "Complement System and Inflammation" section will enable the student to
   a. Define the role of the complement cascade in the host immune response.
   b. Specify three distinct mechanisms in the initiation of cleavage of C3.
   c. Explain the function of Toll-like receptors and their importance in the host immune response.
   d. Describe inflammation in terms of its beneficial role in the immune response.
   e. Describe why and how phagocytes migrate to and from the circulation to the tissues.
   f. Explain the importance of opsonization in phagocytosis of microbes.
5. The "Antibodies" section will enable the student to
   a. Describe the structure of the antibody molecule.
   b. Describe how an antibodies function is related to its structure.
6. The "B-cell Development and Antibody Diversity" section will enable the student to
   a. Describe how B-cell development results in B-cell diversity.
   b. Describe the mechanism of VDJ rearrangement in relationship to antibody diversity.

7. The "T-cell Development and T-cell Receptor Diversity" section will enable the student to
   a. Describe the structure and function of the T-cell receptor.
   b. Describe the mechanism of TCR gene rearrangement and its importance in T-cell diversity.
   c. Explain thymic selection and its importance in T-cell diversity.
8. The "Adaptive Immune Response: Specific Antibody Response" section will enable the student to
   a. Describe the role of T cells in the production of antigen-specific antibodies.
   b. Explain the mechanism and importance of somatic hypermutation on antibody diversity.
   c. Describe the role of memory B cells in the anamnestic response.
9. The "Adaptive Immune Response: Cell-Mediated Immune Response" section will enable the student to
   a. Describe the major cell types and their role in generating a cell-mediated immune response.
   b. Explain the role of cross-presentation in cell-mediated immune responses.
   c. Describe how cytotoxic T cells target cells for apoptosis.

INTRODUCTION

The human immune system is the integration of a number of organs, tissues, cells, and cell products that interact to maintain a body's homeostasis as a result of the wide array of organisms that we encounter on a daily basis. This system is the basis for the immune response that serves to protect us from our encounters with the outside world. The general characteristics of this immune response are specificity, adaptability, and memory. One successful outcome of our immune response to an organism or its products is immunity. For the immune response to be able to induce this resistance to disease, an elaborate multilayer interaction between many different immune mechanisms must come into play each time the host encounters the specific organism. Characteristics of this immune response therefore must include diversity and the ability to adapt, so as to respond to several different

N. V. Bhagavan and C.-E. Ha: Essentials of Medical Biochemistry, Second edition. DOI: http://dx.doi.org/10.1016/B978-0-12-416687-5.00033-6

pathogens at one time as well as respond to previously seen microorganisms or new microorganisms. Specificity of the immune response is critical to recognition, as it is profoundly important for the response to be able to discriminate between self and non or altered self so that the immune response's effector functions can target the threat with little damage to surrounding tissue. Immunity therefore can be thought of as the ability to "remember" organisms previously encountered and react promptly to prevent disease associated with that specific organism.

Memory of a previous interaction between host and microbe is a unique characteristic of the immune response. This immunologic memory allows for a rapid and specific response against an organism. Finally, the need for control of the immune response is key to an appropriate response. So we can think of the essential steps to an appropriate immune response as (1) recognition of an invading microbe as nonself, (2) activation of the key specific effector cells and molecules for the response, (3) mediation of an appropriate response to remove the microbe, and (4) modulation of the response once the microbe is no longer present.

Any substance, from molecule to organism, that is capable of being recognized by the immune system, is called an **antigen**. The degree to which an antigen induces an immune response is termed **antigenicity** or **immunogenicity** [1].

WHAT GENERATES THE IMMUNE RESPONSE?

As mentioned previously, the immune system is a complex multilayered network of molecules, cells, and tissues that interact on a local and a systemic level. One can think of the immune response as beginning on an external surface of the host with the epithelium providing a mechanical barrier. The physiologic environment of the skin with its population of normal flora organisms and low pH presents a formidable barrier to colonization by potential pathogenic organisms. In the lungs, coughing and sneezing mechanically eject pathogens and other irritants from the respiratory tract. The flushing action of tears and urine also mechanically expels pathogens, while mucus secreted by the respiratory and gastrointestinal tract serves to trap and entangle microorganisms.

Chemical barriers also protect against infection. The skin and respiratory tract secrete antimicrobial peptides such as the β-defensins. Enzymes such as lysozyme and phospholipase A in saliva, tears, and breast milk are also antibacterials. In the stomach, gastric acid and proteases serve as powerful chemical defenses against ingested pathogens. Cells such as neutrophils and macrophages engulf or phagocytose potential pathogens while natural

TABLE 33.1 Overview of Innate and Adaptive Immune Responses

Innate Immune Response	Adaptive Immune Response
Present prior to exposure to the threat	Requires startup time to activate response
Functions early in the response	Activated if innate immunity has failed to eliminate the threat
Responds to components common to multiple threats	Responds to a unique component of threat
Goal: Control and/or eliminate the threat while preparing for an appropriate adaptive response	Goal: Eliminate the threat and establish long-term memory

killer (NK) cells induce killing of altered infected host cells. Collectively, these initial nonspecific immune mechanisms of dealing with potential pathogens and altered or infected host cells have traditionally been termed the **innate immune response**. The innate immune response is characterized by its rapid nonspecific recognition and response to potential pathogens and for its activation of the adaptive immune response. The adaptive immune response is an orchestrated interaction of specialized cells and soluble molecules that initiate a specific response against a targeted threat. Unlike the innate response, adaptive immunity requires a period of time before it can mount a finely tuned specific response to the targeted organism, but in doing so, it also generates a memory component that gives the immune system the ability to remember the best way to deal with that specifically targeted organism. It is this memory component that allows the host to mount a vigorous, rapid targeted response should it encounter that specific organism again (Table 33.1, Figure 33.1).

COMPONENTS OF THE IMMUNE SYSTEM

The cells of the immune response as well as erythrocytes and platelets are derived from a common hematopoietic stem cell (HSC) in the bone marrow. The two main lineages of immune cells are the myeloid and the lymphoid lineage (Figure 33.2). From the common myeloid progenitor (CMP) in the myeloid lineage arise the monocytes/macrophages and the granulocytes, while from the common lymphoid progenitor (CLP) in the lymphoid lineage arise the T lymphocyte (T cell), B lymphocyte (B cell), and natural killer (NK) cell

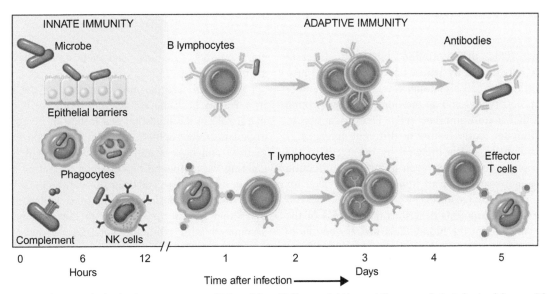

FIGURE 33.1 The innate and adaptive immune response. *[Adapted from V. Kumar, Robbins and Cotran: pathologic basis of disease, 5th ed., 1995 (Ch. 6, Diseases of immunity, Fig. 1, p. 194).]*

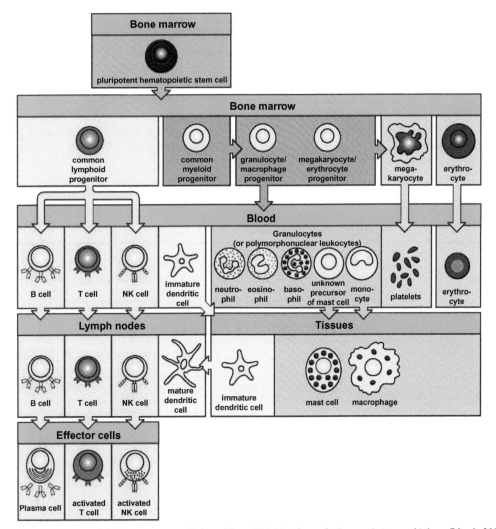

FIGURE 33.2 Cellular components of the immune system. *[Adapted from K.M. Murphy et al., Janeway's immunobiology, 7th ed., 2007 (Ch. 1 Fig. 3, p. 4).]*

populations. This differentiation into the different immune cells is driven by various cytokines, small proteins important for cell signaling, released by a number of different cell types such as the bone marrow stromal cells and other mature immune cells. The production of all these immune cells from the different lineages is under tight control and is dictated by the needs of the host. Such physiologic states such as stress or infection can greatly increase any or all immune cells needed for the host to regain homeostasis.

The term **hematopoiesis** describes the process of the generation of cells in the blood. This process begins in the early embryo and continues throughout the life of an individual. Since most mature blood cells are short-lived, new blood cells continually arise from hematopoietic stem cells and become committed to either the myeloid or lymphoid lineage of development. Historically, the phenotype of the different cell types has been based on staining patterns of these cells, but with standardization of the nomenclature for cell surface molecules by the clusters of differentiation designation (CD1, CD2, etc.), cells can now be defined based on what molecules are present on their cell surface (Table 33.2).

The granulocytes, which arise from the myeloid lineage, are a group of immune cells characterized by the presence of intracellular granules and named according to appearance under various histological stains. Hence, basophils possess basic staining granules, eosinophils possess eosin-staining granules, and neutrophils have neutral staining granules.

Neutrophils are the most numerous of the leukocytes in a healthy host and are actively motile cells that play a major role in the ingestion and killing of invading microorganisms. Neutrophils are rapidly produced in the bone marrow and have a half-life of only 6 to 8 hours after entering the peripheral blood. Upon activation and migration in tissues, these cells can survive for 1−2 days.

Eosinophils, basophils, and mast cells are granulocytes that play a special role in the immune response associated with epithelial surfaces of the respiratory, gastrointestinal, and urogenital tracts. Mast cells play an important role in inflammation by releasing the contents of their granules that contain numerous inflammatory mediators.

Produced by the bone marrow, monocytes circulate in the bloodstream for about 1 to 3 days before moving into tissues throughout the body. Once in the tissues, monocytes mature into macrophages and then engulf and digest cellular debris. If organisms are also present, they can stimulate other immune cells to respond to the potential pathogen. Macrophages (MΦ) have various names depending on the tissue they reside in: liver—Kupffer cells, kidney—mesangial cells, lung—alveolar macrophages; or brain—microglial cells. Once in the tissues, these macrophages can have a lifespan measured in years.

Cells of the lymphocytic line are agranulocytic and are composed mainly of T cells, B cells, and NK cells.

TABLE 33.2 Cellular Components of the Immune System

Lymphoid Cells		
Cellular Component	**Principal Functions**	**Characteristic Receptors**
B cells (B lymphocytes) Mature in bone marrow, reside in lymph nodes, spleen, GI, and respiratory tract mucosa	Recognition of antigens, become antibody-secreting plasma cells, act as antigen-presenting cells (APC); some B cells differentiate into memory cells for rapid recognition of previously encountered antigens.	Antigen receptors are surface IgM (monomeric IgM), IgD MHC class II. Express CD3, CD19, CD20, CD21 (complement receptor), CD32 (Fcγ receptor II), CD35 (complement receptor), CD40 and CD45 on surface.
T cells (T lymphocytes) Mature in thymus	Regulation of adaptive immune response.	T-cell receptors (TCRs), which recognize specific antigen/MHC complexes on the surface of cells. All T cells express CD3 on their surface.
T helper (T$_H$) cells T$_H$1 T$_H$2 T$_H$17	T-cell subclass that releases necessary cytokines for cell stimulation and proliferation. T$_H$1: regulation of cellular immunity via IL-2 and IFN-γ. T$_H$2: stimulates eosinophil-rich inflammation via IL-4 and -5. T$_H$17: mucosal immunity, inflammation.	TCRs recognize antigen/MHC class II complex on APCs. Express CD3 and CD4.

(Continued)

TABLE 33.2 (Continued)

Lymphoid Cells		
Cellular Component	**Principal Functions**	**Characteristic Receptors**
T follicular helper cells (T_{FH})	Key cell type required for the formation of germinal centers (GC) and the generation of long-lived serological memory.	Chemokine receptor CXCR5.
T regulatory cells (Tregs)	Suppress B- and T-cell responses to antigen stimulation.	TR, CD3, CD25, Foxp3.
Cytotoxic T lymphocytes (CTLs)	Cytotoxic against tumor cells and virus-infected host cells. Destroy infected cells through perforin-dependent permeability changes. Produce TNF-α and TNF-β. Promote apoptosis.	TCRs recognize antigen/MHC class I complex on cells. Express CD3 and CD8.
Natural killer cells (NK cells)	Granular lymphocytes that recognize and destroy virus-infected and cancer cells via ADCC (antibody-dependent cell-mediated cytotoxicity). Exclude destruction of cells that express MHC I. Stimulated by IL-2, IFN-γ, TNF-α. Create pores that permit transfer of proteins which promote apoptosis of infected cells.	Nitric oxide (NO) Ig, CD3 or TCRs. Express CD16 (FcγIIR), i.e. surface receptors that bind via Fc portion of IgGs. Express CD56 on surface. CD16 and CD56 used for identification in cell sorting.

Myeloid Cells		
Cellular Component	**Principal Functions**	**Characteristic Receptors**
Neutrophils Mature in bone marrow	First cells recruited to sites of inflammation, dominant phagocytic granulocyte, able to mediate ADCC, major defense against pyogenic bacteria, use respiratory burst to kill cells.	Adhere to foreign cells via CD11b/CD18. Neutrophils are attracted to sites of complement activation by C5a, a chemotactic peptide from C5.
Basophils	Blood "analog" of tissue mast cell, stimulated by complement C3.	Possess Fc receptors that bind IgE, causing degranulation and histamine release.
Mast cells Mature in bone marrow, resident in tissue	Tissue granulocytes that produce cytokines and release heparin, histamine.	Fc receptors that bind IgE, causing degranulation and histamine release.
Eosinophil	Attack parasites too large for phagocytosis, possess granules that contain toxic proteins, adhere to foreign organisms via C3b, release proteins and enzymes that damage foreign organism membranes, kill with active oxygen burst. Engage in IgE-mediated ADCC.	High-affinity receptors for IgE, bind via Fc region of antibody.
Monocytes Mature in bone marrow, found in blood	Mononuclear phagocyte in blood, are precursors of tissue macrophage. Process protein antigens and present peptides to T cells via MHC I and II. Engage in IgG-mediated ADCC.	Express MHC class II surface proteins, CD11b, CD14, FcγRI (CD64), and FcγRII (CD62).
Macrophages Develop from monocytes	Tissue phagocytes (APCs) that process antigens for presentation to T cells, possess MHC receptors; principal attack is on bacteria, viruses, and protozoa that enter cells. Engage in IgG-mediated ADCC.	Express MHC class II surface proteins, CD14, CD68, Toll-like receptors, C3b receptors, FcγRI (CD64), and FcγRII (CD62).
Dendritic cells	Found in skin (Langerhans cells) in T-cell areas of lymph nodes and spleen. Recognize and endocytose foreign carbohydrate antigens that are not found in vertebrate animals. Act as APCs to CD 28-bearing T cells, stimulated by IFN-α, TNF-α.	Express MHC class II surface proteins, CD11c, Toll-like receptors, and FcγRI (CD64).

Functionally, both T and B cells act in an antigen-specific manner, whereas NK cells lack this antigen specificity but react to missing or altered host cell surface molecules.

T cells play a central role in the initiation, propagation, and control of antigen-specific immune responses. They are distinguished from B and NK cells by expression of the T-cell receptor (TCR) molecules on their cell surface. A number of different subsets of T cells have been described based on expression of certain cell surface molecules, such as $CD4^+$ T cells and $CD8^+$ T cells. Furthermore, immune function and patterns of protein secretion can further subdivide T-cell populations into $CD4^+$ T helper 1 (T_H1), $CD4^+$ T helper 2 (T_H2), $CD4^+$ T helper 17 (T_H17), $CD4^+$ T follicular helper (T_{FH}), $CD4^+$ T regulatory (Treg), and $CD8^+$ cytotoxic T cells.

The function of a B cell is to synthesize and secrete antibodies in response to a specific antigen. There are two different subsets of B cells—B1 cells and B2 cells—that differ in their developmental origin, cell surface marker expression, and their immune function. B1 cells predominate in the developing fetus, spontaneously secrete antibodies, and operate under a differentiation program that is unique from that of conventional B2 cells, which become the predominant cells in adults. Upon stimulation with antigen, B cells develop into plasma cells, which function as antibody secreting "factories" or memory cells, which are maintained by the immune system to be able to respond quickly following a subsequent exposure to the same antigen.

Dendritic cells (DCs) are bone marrow-derived and are found in most tissues of the body. They are the most effective and efficient at presenting antigens to other cells of the immune system. As with the lymphocytes, dendritic cells are composed of several subsets that express different cell surface molecules and signaling proteins. Dendritic cells in the lymph nodes are called **interdigitating cells,** and in the skin, they are called **Langerhans cells**.

While the immune system interacts and encompasses the entire host, certain tissues are integral to the development and maintenance of an immune response. Traditionally, these tissues have been divided into the primary and secondary lymphoid tissues.

The primary (or central) lymphoid tissues, the bone marrow, and the thymus are the sites for the generation and early maturation of lymphocytes. Hematopoiesis occurs in the bone marrow, producing all the cells associated with the immune response, but the thymus is required for T cells to fully mature.

The secondary (peripheral) lymphoid tissues are located throughout the body and include the lymph nodes, spleen, tonsils, and the mucosa-associated lymphoid tissue (MALT). The function of the secondary lymphoid tissues is for the collection and interaction of antigen with the cells of the immune system. It is at these sites that the immune response will decide on which effector mechanism to use to deal with the foreign antigen [2,3].

COMMUNICATION WITHIN THE IMMUNE RESPONSE

The immune system is a complex system composed of numerous different cell types located throughout the body. Coordinating an effective immune response requires active communication between the cells of the immune response so that each player knows when and how to contribute to the overall response. A miscommunication between cells, such as might occur if a signal is provided at an inappropriate time or is not shut off appropriately, can result in dire consequences for the host. Because of this, usually more than one signal is required to activate or inhibit a cell to perform its function in an immune response. This multiple signaling mechanism helps ensure that the proper cells function at the right time and for the right length of time during the immune response (Figure 33.3). In general, cells of the immune system communicate by direct cell-to-cell contact and through the production of soluble factors that bind cognate receptors expressed at the surface of target cells. Examples of molecules involved in direct cell-to-cell contact include cell surface proteins, the major histocompatibility complex (MHC), T-cell receptors, and numerous co-stimulatory molecules (CD28, B7).

Cytokines are some of the most important molecules produced by cells in order to communicate and orchestrate the immune response. As with hormones in the endocrine system, cytokines can produce a wide range of different effects on many different cells throughout the body. Cytokines may act systemically throughout the host in an endocrine fashion, in a paracrine manner on only cells in close proximity, or even on the cell that produced it in an autocrine manner. The functionality of cytokines can be described as being (1) pleiotropic, having different effects on different cells; (2) redundant, having many different cytokines with the same function; (3) synergistic, in that when the cytokines act together, they increase the effects of one another; or (4) antagonistic, in which one negates the function of the other (Figure 33.4). All of this, and the fact that cytokines act only through specific cytokine receptors on the cells, shows how important these molecules are to an effective and nonharmful immune response.

Cytokines have been called **chemokines, interleukins**, or **lymphokines**, depending on their targets and observed function. Previously, cytokines that targeted leukocytes were called interleukins (IL), but the term *interleukin* is now used to describe any cytokine with no bearing on its presumed function or target (Table 33.3).

The major histocompatibility complex (MHC) is a set of highly polymorphic evolutionarily related genes that encode for proteins that are essential players in the processing and binding of antigenic peptides during the immune response. The function of MHC proteins is to

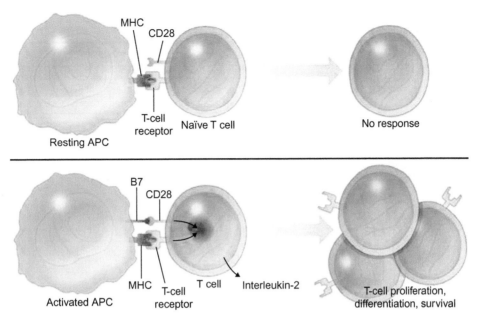

FIGURE 33.3 Multiple signals are required for immune cell function. *[Adapted from A.H. Sharpe, A.K. Abbas. T-cell co-stimulation—biology, therapeutic potential, and challenges. N. Engl. J. Med. 355(10) (2006) 974.]*

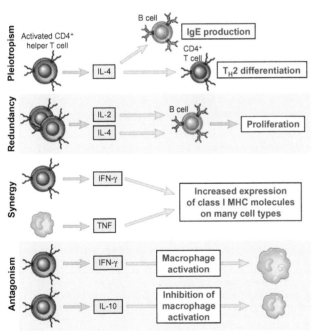

FIGURE 33.4 Functionality of cytokines. *[Adapted from A.K. Abbas, A.H. Lichtman, Cellular and molecular immunology, 6th ed., 2007 (Ch. 12, Cytokines, Fig. 2, p. 270.)]*

present a wide range of self and foreign protein fragments to different subsets of T cells. The MHC genes are also expressed co-dominantly, meaning that an individual expresses the alleles inherited from each parent. In this way, the number of MHC molecules that bind peptide for presentation to T cells can be maximized. In humans,

these genes are located on Chromosome 6, and the gene products are referred to as **human leukocyte antigens (HLAs)** and are categorized into a number of different classes based on their chemical structure, tissue distribution, and function within the immune response [4].

HLA Class I Antigens

HLA class I gene products are membrane-bound glycoproteins found on the surface of all nucleated cells. The structure of these molecules consists of a polymorphic α heavy chain with three extracellular domains, two of which form the peptide-binding region, a single transmembrane domain. A nonpolymorphic light β-chain called **β_2 microglobulin**, while not encoded by a gene in the MHC region, is important for the transport of the HLA molecule to the surface of the cell. In humans, there are three separate genes for HLA class I, which are referred to as HLA-A, HLA-B, and HLA-C genes; and their respective gene products are called **HLA-A, HLA-B**, and **HLA-C** proteins (Figure 33.5). These genes are expressed codominantly within a cell, meaning that each cell will express six different HLA class I genes: one HLA-A, HLA-B, and HLA-C each from the maternal chromosome 6, and one HLA-A, HLA-B, and HLA-C each from the paternal chromosome 6. The function of all of these HLA class I molecules is to bind polypeptide fragments that have been processed from intracellular proteins found within the cell's cytosol and present the resulting HLA/ polypeptide complex to cytotoxic CD8$^+$ T cells.

TABLE 33.3 Selected Cytokines of Biologic Importance

Cytokine	Sources	Biologic Effects
IL-1	Activated macrophages	Pleiotropic cytokine involved in various immune responses, inflammatory processes, and hematopoiesis. Induces fever through the hypothalamus, the "endogenous pyrogen."
IL-2	Helper CD4 T cells	Major inducer of proliferation for T and B cells.
IL-3	Activated T cells, mast cells	Involved in a variety of cell activities such as cell growth, differentiation, and apoptosis.
IL-4	T_H2 cells, mast cells, eosinophils, basophils	On resting B cells, macrophages, increases MHC II expression. On activated B cells, stimulates proliferation and differentiation and induces antibody class switch to IgE. Along with IL-10, it decreases activated macrophage activity. Drives naïve T cells to become T_H2 cells.
IL-5	T_H2 cells, mast cells, eosinophils	Stimulates proliferation and differentiation of activated B cells and induces antibody class switch to IgA. Very important in stimulating the growth and differentiation of eosinophils.
IL-6	T_H2 cells, monocytes, macrophages	Induces synthesis of acute-phase proteins by hepatocytes. Acts on proliferating B cells to promote differentiation into plasma cells and stimulate antibody secretion.
IL-7	Thymus, intestinal epithelial cells,	Essential role in early T-cell and B-cell development.
IL-8	Macrophages, endothelial cells	Major mediator of the inflammatory response, promotes chemotaxis and margination by neutrophils.
IL-9	T cells, B cells, macrophages	Acts as a regulator of a variety of hematopoietic cells by stimulating cell proliferation and preventing apoptosis.
IL-10	T_H2 cells, macrophages	Pleiotropic effect on immunoregulation and inflammation. Downregulates the expression of T_H1 cytokines, MHC class II, and co-stimulatory molecules on macrophages. Enhances B-cell survival, proliferation, and antibody production.
IL-11	Stromal cells	Induction of synthesis of acute phase proteins, enhances stimulation of hematopoietic cell proliferation by IL-3 and T-cell development of B cells.
IL-12	Macrophages, dendritic cells	Induces T_H1 cell development, activates NK and cytotoxic T cells.
IL-13	T_H2 cells	Involved in B-cell maturation, differentiation; promotes IgE isotype switching.
IL-17	Activated T cells	Induces and mediates proinflammatory responses. Drives naïve T cells to become T_H17 cells.
Interferon γ (IFN-γ)	T_H1, cytotoxic T cells	Activates macrophages. Drives naïve T cells to become T_H2 cells. Induces antibody class switch to IgG antibodies.
Tumor necrosis factor α (TNF-α)	Macrophages, NK cells, B cells, T cells, mast cells	Activates monocytes, PMN cells; promotes inflammation.
Transforming growth factor β (TGF-β)	Platelets, macrophages, lymphocytes	Immunosuppressive.
GM-CSF	T cells, macrophages, B cells, mast cells	Promotes growth of granulocytes and monocytes.

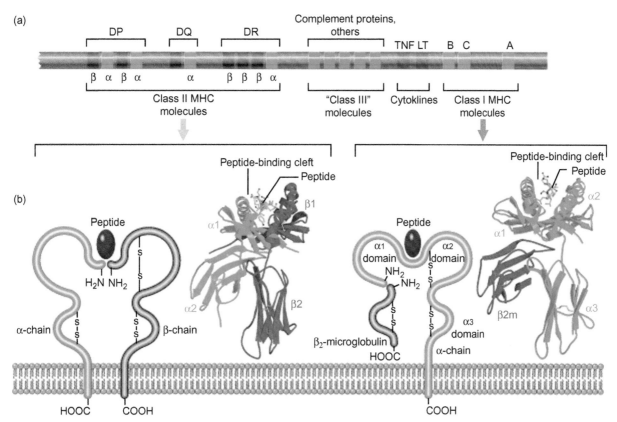

FIGURE 33.5 The HLA complex and the structure of HLA molecules. (a) The location of genes in the HLA complex is shown. (b) Schematic diagrams and crystal structures of class I and class II HLA molecules. [*Adapted from V. Kumar, Robbins and Cotran: pathologic basis of disease, 7th ed., 1995 (Ch. 6, Diseases of immunity, Fig. 9, p. 203.)*]

The complex process of generation of polypeptide fragments, their association with HLA class I molecules within the cells, and then their transport to the cell surface is known as the **endogenous pathway of antigen expression**. Large proteins within the cell cytoplasm are routinely degraded into short peptides by proteolytic complexes known as **proteasomes**, and subsequently become associated with transport proteins called **transporters associated with antigen processing**, or **TAPs**, encoded by the *TAP1* and *TAP2* genes. The TAP1 and TAP2 proteins interact to form a channel, which ferries the polypeptide fragments from the cytoplasm into the endoplasmic reticulum. Once inside the endoplasmic reticulum, these polypeptides bind to a newly synthesized HLA class I α heavy chain. This polypeptide/HLA class I α heavy chain complex then associates with a β_2-microglobulin to form a stable molecule that is transported to the cell surface for presentation to $CD8^+$ cytotoxic T cells.

HLA Class II Antigens

HLA class II gene products are also membrane-bound glycoproteins, but unlike HLA class I molecules, they are found on only a limited set of cells. These cells are collectively referred to as antigen-presenting cells and include dendritic cells, activated macrophages, B lymphocytes, thymic epithelial cells, and activated endothelial cells. The structure of this molecule consists of a heterodimer consisting of noncovalently polymorphic peptides called α- and β-chains. The extracellular portions of both α- and β-chains contain two domains ($\alpha1$, $\alpha2$ and $\beta1$, $\beta2$), and each chain has a transmembrane component. It is the interaction of the $\alpha1$ and $\beta1$ domains that creates the peptide-binding region of the HLA class II molecule. The class II molecules are coded for in a region called **HLA-D**, also located on chromosome 6. There are three subregions—HLA-DP, HLA-DQ, and HLA-DR—and each subregion encodes one α-chain and one β-chain (Figure 33.5). As with HLA class I, the genes for HLA class II are highly polymorphic and are also codominantly expressed on the surface of antigen-presenting cells.

HLA class II molecules bind polypeptides produced by the exogenous pathway. Proteins from outside antigen-presenting cells are internalized by a process known as **endocytosis** and are placed in endocytic vacuoles. Like the two chains of HLA class I molecules, the HLA class II molecules are manufactured and then brought together; but unlike HLA class I molecules, HLA class II molecules are not loaded with the small polypeptides in the endoplasmic reticulum. Instead, the α- and β-chains associate with another protein, the invariant chain, which acts to block the peptide-binding region and inhibits premature loading of polypeptide. Enclosed in membranous vesicles, the complexes of HLA class II molecules and invariant chains intersect with endocytic vesicles with their cargo of exogenous peptides. The HLA class II transporting vesicles and the endocytic vesicles fuse to form the HLA class II compartment in which proteases degrade the exogenous proteins into small polypeptides of 10–20 amino acids in length. Within the HLA class II compartment, the small exogenous polypeptides outcompete the invariant chain for space into the antigen-binding region, and the polypeptide-laden HLA class II molecules are then exported to the surface of the cell, where they interact with $CD4^+$ helper T cells.

Nonclassical HLA Antigens

The HLA complex also includes genes for certain nonclassical HLA proteins that are structurally similar to HLA class I or class II, but are not polymorphic and play different roles in human immune response. A few examples of nonclassical class I molecules are HLA-G, which forms a dimer with β_2-microglobulin that acts to control immune responses at the fetal-maternal interface; and HLA-E, which binds peptides derived from the signal sequences of the classical HLA-A, -B, and -C molecules, creating complexes that are recognized by receptors on natural killer (NK) cells. Recognition of these HLA-E complexes inhibits NK cells from attacking normal cells, so that they kill only cells that fail to express HLA class I proteins, as is often seen when cells become infected by viruses or undergo malignant transformation. Examples of nonclassical molecules encoded in the class II region include HLA-DM and HLA-DO, both of which are heterodimers that localize to endocytic vesicles and help promote polypeptide loading onto class II molecules.

THE COMPLEMENT SYSTEM AND INFLAMMATION

The complement system is an enzyme cascade that had been traditionally described as a mechanism for augmenting (or complementing) the biologic activities in an antibody response. The complement system is now appreciated as being one of the most important humoral systems to recognize the conserved patterns present on potential pathogens and altered or damaged host cells. It is not just the complement system's ability to recognize potential pathogens, but it is its ability to translate that recognition into an appropriate innate or adaptive immune response that makes it a key component in a successful

immune response. In general, this is accomplished by components that have recognized pathogens or altered self antigens directly so as to activate cell-bound receptors and/or initiate cleavage of complement factors that acquire the ability to bind to complement receptors and/or regulators on other cells in the immune system. Also of importance is the expression pattern of the complement receptors on different immune cells, as this can impact the magnitude of the generated immune response. Therefore, it is the interaction of this direct action of recognition and of the cleavage fragments with distinct cell-bound receptors/regulators that can direct the immune response toward an innate or an adaptive immune response.

The proteins of the complement system designated C1 through C9 circulate in the blood and surrounding tissues in an inactive form. In response to the recognition of some common molecular components of microorganism, they become sequentially activated, proceeding in a cascade where the binding of one protein component promotes the binding of the next protein component in the cascade.

There are three initiation pathways for the induction of the complement cascade: the classical complement pathway, the lectin pathway, and the alternative complement pathway. While each of these pathways is activated by a slightly different mechanism, they all ultimately initiate a key process in the complement system: the cleavage of C3 to C3a and C3b. Activation of the classical complement pathway is initiated by IgM and/or IgG antibodies binding to antigen and forming antigen–antibody complexes on the surface of potential pathogens. The lectin pathway does not require antibody for complement activation but instead uses a molecule called **mannose-binding lectin** (**MBL**). MBL binds to mannose-containing polysaccharides (mannans) that are found on the surface of various bacterial cells and initiates the complement cascade. And finally there is the alternative complement pathway, which does not rely on a pathogen-binding protein, in contrast with the other two pathways. In the alternative pathway there is a low level of cleavage of C3 into C3a and C3b by enzymes in the blood. If there is no pathogen present, the C3a and C3b protein fragments are quickly degraded. However, if there is a nearby pathogen, some of the C3b binds to the surface of the organism and continues the complement cascade. Production of C3b continues the cascade to form C5a and C5b, which, in turn, begins formation of membrane pore-forming complex known as the **membrane attack complex** (**MAC**).

Components produced by the initiation of any of the complement pathways are responsible for a number of important immune functions, such as the induction and maintenance of inflammation secondarily due to C3a's and C5a's (known as **anaphylatoxins**) ability to trigger degranulation of histamine and other vasoactive substances from tissue mast cells. C3a and C5a are also able to chemotactically attract phagocytes to the area of complement activation and then, via C3b attachment, enhance attachment or opsonize the targeted antigens to phagocytes for engulfment and removal. Finally, the MAC can cause direct lysis of gram-negative bacteria and targeted human cells by means of its ability to induce pore formation in plasma membranes (Figure 33.6).

Because of the relative nonspecific initiation of the complement system, there is the potential for damage to

host tissues as well as potential pathogens. This requires the complement system to be tightly regulated by a number of different complement control proteins that are divided into soluble inhibitors that inhibit complement progression such as factor H, factor I and C1 inhibitor, and membrane-bound factors that are designed to protect individual host cells such as complement receptor 1 (CR1), membrane cofactor protein (MCP), decay accelerating factor (DAF), and CD 59 (Table 33.4).

Many cells in the immune system such as macrophages, dendritic cells, and B cells use the same sort of

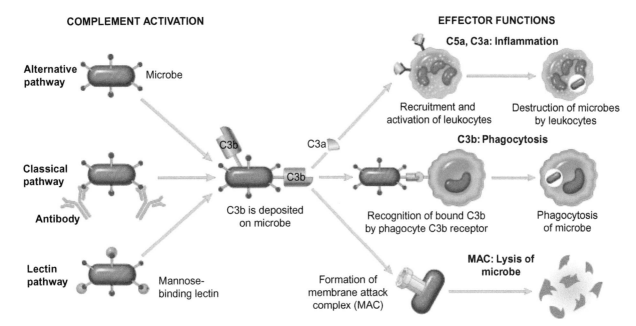

FIGURE 33.6 The complement system. *[Adapted from V. Kumar, Robbins and Cotran: pathologic basis of disease, 7th ed., 1995 (Ch. 2, Acute and chronic inflammation, Fig. 14, p. 65).]*

TABLE 33.4 Selected Complement Regulatory Proteins

Regulatory Protein	Fragment Target	Function
C1 inhibitor	C1 complex	Inhibits spontaneous activation
Factor I	C3b, C4b	Inactivates C3b, C4b with cofactors listed below
Factor H	C3b	Cofactor in the cleavage of C3b by factor I
Complement receptor 1 (CR1)	C3b, C4b	Cofactor in the cleavage of C3b or C4b by factor I; mediates phagocytosis of opsonized antigens
Membrane cofactor protein (MCP)	C3b, C4b	Cofactor in the cleavage of C3b or C4b by factor I
Decay-accelerating factor (DAF)	C4b2a, C3bBb	Inhibits activation of C3 and C5
CD 59	C9	Inhibits MAC formation by interfering with C9

nonspecific recognition cues as in the lectin pathway of the complement system. The receptors on these cells recognize molecules and structures constantly associated with potential pathogens, but not the host, and are called **pattern recognition receptors (PRRs)**. PRRs tend to recognize molecules not associated within the human host such as lipopolysaccharides (LPS) and flagellin from the gram-negative organisms; peptidoglycan and lipoteichoic acids from the gram-positive organisms; bacterial unmethylated CpG DNA and double-stranded RNA (dsRNA) from viruses; and glucans from fungal cell walls. Collectively, these nonhuman molecules are called **pathogen-associated molecular patterns (PAMPs)**. The PRRs include, among others, a family of cell surface-expressed proteins called **Toll-like receptors (TLRs)** and a set of intracellular proteins called the **nucleotide-binding oligomerization domain proteins (NOD-like receptors, NLRs)**. TLRs and NLRs have been shown to be crucial in not only a host immune response against potential microbial pathogens but also in homeostasis of the host's normal microbial flora. TLRs have been the best characterized, and it is known that, upon binding to their appropriate microbial molecular pattern, TLRs activate intracellular signaling pathways within the host cell that upregulate the production of proinflammatory cytokines that trigger inflammation and drive the development of an adaptive immune response.

In terms of the immune response, inflammation is a protective response elicited by foreign material or cell injury, which serves to destroy, dilute, or wall off (sequester) both the injurious agent and the injured tissue. The classical signs of inflammation, as described by Celsus more then two centuries ago, are heat (*calor*), pain (*dolor*), redness (*rubor*), swelling (*tumour*) and loss of function (*functio laesa*). Physiologically, this response involves a complex series of events, including dilatation of arterioles, capillaries, and venules that results in increased permeability and blood flow, leakage of blood vessel fluids, including plasma proteins, and migration of leukocytes into the inflammatory site. Since most of the players in the immune response are located in the blood, inflammation is a means by which defense molecules and leukocytes can leave the circulation and enter the tissue around the injured or infected site. Therefore, inflammation is a key player in the immune response, but an inappropriate, excessive, or prolonged inflammatory response can cause harm.

A variety of chemical inflammatory mediators released from a number of different leukocytes interact to both initiate and modulate inflammation. These include activated macrophages that release proinflammatory cytokines such as IL-1, IL-6, and TNF-α. All three of these cytokines are known as **endogenous pyrogens**, as they induce fever by acting on the hypothalamus, and induce

the production of acute phase proteins such as C-reactive protein (CRP) and mannose-binding protein by the liver and increased leukocyte production (hematopoiesis) in the bone marrow, leading to leukocytosis. Mast cells in the connective tissue as well as basophils, neutrophils, and platelets that have left the blood due to injured capillaries can release vasodilators such as histamine, kinins, leukotrienes, and prostaglandins. Certain products of the complement pathways (C5a and C3a) can bind to mast cells and trigger release of their vasoactive agents. Also, damaged tissues can activate the coagulation cascade and produce inflammatory mediators like bradykinin.

Some major physiological events which occur during an inflammatory response that are important for the immune response include immediate vasoconstriction of the blood vessels leading away from the site of initiation, vasodilation to increase the volume of blood flow at the site, and increased vascular permeability resulting in an influx of immune molecules and cells from the circulation into tissues.

The benefits of inflammation as a result of this increased permeability include increased tissue concentrations of such molecules as clotting factors that activate the coagulation cascade, causing fibrin clots to form to localize the infection, stop the bleeding, and act as a chemoattractant for leukocytes; antibodies that help to identify and block the action of microbes through a variety of methods; and proteins of the complement pathway that mark microorganisms for removal by phagocytic leukocytes (C3b) and act as a chemoattractant to attract phagocytes (C5a), the MAC inserting into the membrane of the cell and allow ions, water, and other small molecules to freely pass through the pore. As a result, the targeted cell will not be able to maintain osmolality, thus resulting in cell death, which stimulates more inflammation, provides nutrients to feed the cells of the inflamed tissue, and releases lysozyme. Lysozyme degrades peptidoglycan and β-defensins, which build pores in the cytoplasmic membranes of many bacteria, and transferrin, which binds Fe^{3+}, thereby depriving microbes of needed iron.

Another very important aspect of this vascular change as a consequence of inflammation is the attraction and activation of leukocytes such as neutrophils and macrophages to the site. It is the leukocyte interactions with the vascular endothelium and the complex processes that include the capture of free-flowing leukocytes from the circulation and subsequent leukocyte rolling, arrest, firm adhesion, and ensuing migration through the vascular wall (diapedesis) that play a key role in both the innate and adaptive immune response. The process of leukocyte extravasation begins after recognition of nonself molecules, and subsequent activation macrophages and vascular endothelial cells release proinflammatory cytokines such as TNF-α and IL-1. These proinflammatory cytokines induce the local vascular endothelial cells to increase their surface expression of adhesion molecules such as selectins and to produce chemokines such as IL-8 that are bound to the vascular wall also (selectins are molecules found on leukocytes that bind to carbohydrates expressed on endothelial cells and other leukocytes. During diapedesis, selectins enable the circulating leukocytes to slow their flow and roll along the inner blood vessel wall). The surface expression of the selectins on the vascular cells causes the circulating leukocytes to begin to attach to the adhesion molecule and roll along the inner blood vessel wall. Chemokine receptors on the surface of the rolling leukocytes interact with the surface bound IL-8 and trigger the expression of another family of adhesion molecules called **integrins** on the surface of the leukocytes. These integrins enable the slowly rolling leukocytes to bind tightly to adhesion molecules such as intercellular adhesion molecules (ICAMs) and vascular cell adhesion molecules (VCAMs) on the luminal surface of the vascular endothelial cells. Once the leukocyte binds tightly to the activated endothelial cells, they flatten out and migrate across the endothelial layer to leave the circulation and enter the tissue underneath. Chemokines and other chemotactic factors continue to attract the leukocytes to the area of the injury where leakage of fibrinogen and plasma fibronectin from the inflammatory response for scaffolding aid the migration and retention of leukocytes at the injury site.

Once the leukocytes reach the area and encounter the intruding microorganisms, a mechanism central to the immune response begins phagocytosis. Phagocytosis is the process of engulfing material by a cell to form an internal vacuole or phagosome. A number of different cells in the immune system such as neutrophils, macrophages, dendritic cells, and B cells are able to engulf and degrade material via phagocytosis. The process of phagocytosis begins with recognition of the invading organism by the phagocyte. If the organism has been opsonized with an antibody such as IgG or C3b fragments from the activation of complement, then the phagocyte binding of the organism to the phagocyte is greatly enhanced; indeed, phagocytic binding cannot occur without opsonization of the particle (Figure 33.7). The phagocyte engulfs the opsonized microbe, which then winds up enclosed in a phagocytic vesicle called a **phagosome**. Lysosomes, then fuse with the phagosome to form a structure termed a **phagolysosome**.

There are two major killing systems at the disposal of macrophages and neutrophils: the oxygen-dependent system and the oxygen-independent system. The oxygen-dependent system is the more important of the two, as the system generates large quantities of powerful oxidizing molecules collectively known as **reactive oxygen species (ROS)**. The cytoplasmic membrane of phagocytes

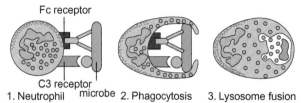

Normal antimicrobial action

FIGURE 33.7 Opsonization and phagocytosis. [*Adapted from D. Male et al., Immunology, 7th ed., 2006 (Ch. 24, Hypersensitivity-Type IV, Fig. 2, p. 345).*]

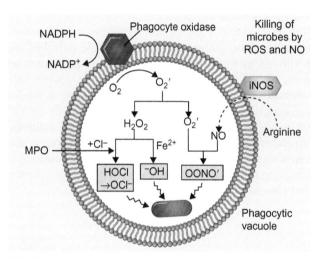

FIGURE 33.8 Production of microbicidal reactive oxygen intermediates within phagocytic vesicles. [*Adapted from V. Kumar, Robbins and Cotran: pathologic basis of disease, 7th ed., 1995 (Ch. 2, Acute and chronic inflammation, Fig. 9, p. 60).*]

contains the enzyme oxidase molecules located in the phagolysosome membrane, which converts oxygen into superoxide anion (O_2^-), which can combine with water to form hydrogen peroxide (H_2O_2) and hydroxyl (OH) radicals. In neutrophils, but not macrophages, the hydrogen peroxide can combine with chloride (Cl^-) ions after added enzyme myeloperoxidase (MPO) fuses primary granules to form hypochlorous acid (HOCl) and singlet oxygen. In macrophages, hydrogen peroxide can combine with nitric oxide (NO) to form peroxynitrite radicals (Figure 33.8).

In the oxygen-independent system of killing, some lysosomes contain defensins; lysozyme; lactoferrin, which deprives bacteria of needed iron; cathepsin G, a protease that causes damage to microbial membranes; and various other digestive enzymes that exhibit antimicrobial activity by breaking down proteins, lipids, and carbohydrates.

While the overall effect of an inflammatory response is to recruit various cells and components that have a direct role in the defense of the actual site of microbial invasion against the intruding microorganism, inflammation is also an important player in initiating the generation of the adaptive immune response [5−7].

ANTIBODIES

Immunoglobulins or antibodies are proteins that account for approximately 15%−20% of the total protein portion of human serum. The function of these molecules in the immune response is to bind a specific antigen and then signal other molecules or cells to respond to that specific antigen. Antibodies produced during the immune response initiate a wide variety of biological activities to defend the host from potential pathogens. Antibodies by themselves have the ability to neutralize and aggregate microbes and toxins so as to prevent bacteria, viruses, and exotoxins from binding to receptors on host cells. In concert with other molecules and cells of the immune system, antibodies can coat or opsonize microbes that aid phagocytic cells in internalizing the microbes and NK cell killing by antibody-dependent cellular cytotoxicity (ADCC). Antibodies also can activate the classical complement pathway, resulting in the direct lysis of the microbe, opsonization of the microbe, and induction of inflammation (Figure 33.9).

The immunoglobulin proteins are divided into five different classes (isotypes)—IgA, IgD, IgE, IgG, and IgM—based on their polypeptide structure. Two of these immunoglobulin classes can be further divided into subclasses—IgA1 and IgA2; IgG1, IgG2, IgG3, and IgG4—again, based on their polypeptide structure. The relative proportion of each class of immunoglobulin in the serum is as follows: ∼80% IgG, ∼15% IgA, ∼10% IgM with IgE and IgD both accounting for <0.1%.

The general structure of the immunoglobulin molecule consists of four polypeptide chains, two heavy chains (H chain) of ∼50 kDa each, and two light chains (L chain) of ∼25 kDa in size. There are five classes of heavy chains, μ, γ, α, ε, and δ, which form the five antibody isotypes IgM, IgG, IgA, IgE, and IgD respectively. For the light chains, there are only two types, kappa (κ) and lambda (λ), and they can combine with any of five classes of heavy chains. However, approximately 60% of antibody molecules use kappa chains and 40% use lambda chains. Immunoglobulin molecules can also be divided into two different regions based on structure and functionality. These regions are designated as the constant and variable regions. On the immunoglobulin heavy chain, one can find regions with relatively constant amino acid sequences that are designated as C_H1, C_H2, and C_H3. Also on the immunoglobulin heavy chain is a region noted for its variable amino acid

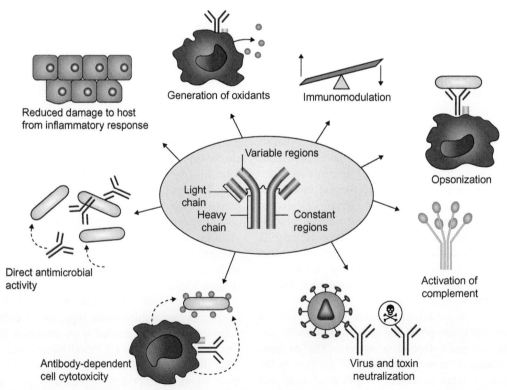

FIGURE 33.9 Biologic function of antibodies. *[Adapted from A. Casadevall, E. Dadachova, L.A. Pirofski. Passive antibody therapy for infectious diseases. Nat. Rev. Microbiol. 2(9) (2004) 699, Fig. 1).]*

sequence called the **variable region** (**V$_H$**). Within this variable region are three short highly variable amino acid sequences that are called the **hypervariable regions**. The immunoglobulin light chains consist of one constant region C$_L$ and one variable region V$_L$, and as with the heavy chain variable region, there are also three hypervariable regions. When two heavy and two light chains of the immunoglobulin molecule are folded into a tertiary structure resembling the letter *Y*, the hypervariable regions come together to form the walls of the antigen-binding cleft. These stretches of protein sequence are also called **complementarity-determining regions** (**CDRs**). The region between the C$_H$1 and C$_H$2 domains is called the **hinge region** because it allows some flexibility between the two arms of the Y-shaped immunoglobulin molecule. Digestion of an IgG molecule with the proteolytic enzyme papain cleaves the immunoglobulin at the hinge region to generate two functionally distinct fragments: the fragment antigen binding (Fab) portion and the fragment crystallizable (Fc) portion. The two Fab fragments generated are the antigen-binding regions and are each composed of one constant and one variable domain from each of the heavy and the light chains. The Fc portion corresponds with the paired C$_H$2 and C$_H$3 domains and is the part of the immunoglobulin molecule that interacts with other effector molecules and cells of the immune system (Figure 33.10). It is the Fc fragment that accounts for the functional differences between heavy-chain isotypes. Depending on the class and subclass of the immunoglobulin, the biological activities of the Fc portion include the ability to activate the complement pathway; bind to phagocyte mast cells, eosinophils, or NK cells; and determine the tissue distribution of the immunoglobulin. Two classes of immunoglobulin, IgM and IgA, have a slightly more complex structure. IgM is a pentamer consisting of five monomeric IgM molecules connected at their Fc portions by a *J*, or joining, chain. Secretory IgA is a dimer of two monomeric IgA molecules connected at their Fc portions by a J chain and stabilized with a secretory component (Figure 33.11).

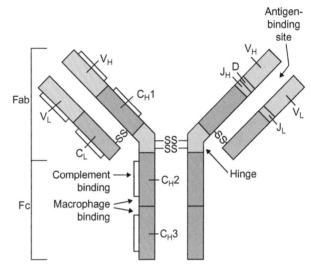

FIGURE 33.10 Schematic view of typical immunoglobulin structure. *[Adapted from K. Barrett et al., Ganong's review of medical physiology, 23rd ed., 2009 (Ch. 3, Immunity, infection & inflammation, Fig. 10, p. 72).]*

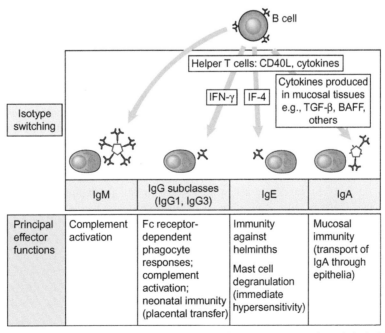

	IgM	IgG subclasses (IgG1, IgG3)	IgE	IgA
Principal effector functions	Complement activation	Fc receptor-dependent phagocyte responses; complement activation; neonatal immunity (placental transfer)	Immunity against helminths Mast cell degranulation (immediate hypersensitivity)	Mucosal immunity (transport of IgA through epithelia)

FIGURE 33.11 Immunoglobulin isotypes. *[Adapted from A.K. Abbas, A.H. Lichtman, Basic immunology, 3rd ed., 2009 (Ch. 7, Humoral immune responses, Fig. 10, p. 142).]*

Different immunoglobulin classes trigger different effector mechanisms through interaction of immunoglobulin Fc regions with specific Fc receptors (FcRs) on immune cells. The different types of Fc receptor are all classified based on the class of antibody that they recognize: Fc-gamma receptors (FcγR) bind IgG and Fc-alpha receptors (FcαR) bind IgA; those receptors that bind IgE are called **Fc-epsilon receptors (FcεR)**. Unbound antibodies do not bind efficiently to Fc receptors, whereas antigen−antibody complexes do bind efficiently (IgE−FcRε is an exception). Therefore, the effector cell that expresses the Fc receptor will not be activated in a nondirected manner.

The development of stable cell lines from the fusion of immortalized myeloma cells with B cells from immunized mice gave researchers the ability to produce large amounts of antigen-specific antibodies. These monospecific or monoclonal antibodies have opened a whole new area of drug therapy (drug names ending in -mab) based on their ability to specifically target and block cancerous cells, infectious agents, or components of the immune system. Also, when these monoclonal antibodies are coupled with radioisotopes or cytotoxins, the resulting conjugated monoclonal antibody allows for direct killing of targeted cells. Because monoclonal antibodies are murine in nature and in time cause the human recipient to mount a vigorous immune response to the murine antibody, newer therapeutic monoclonal antibodies are either chimeric, meaning they have the murine antigen-binding variable region grafted on to a human constant region (these drugs end in -ximab); humanized, meaning they have the murine hypervariable region fused to a human antibody (these drugs end in -zumab); or fully human monoclonal antibodies (these drugs end in -mumab). Table 33.5 lists the currently available therapeutic monoclonal antibodies [8,9].

B-CELL DEVELOPMENT AND ANTIBODY DIVERSITY

The human immune response must be able to provide antibodies against any and all non-self antigens encountered by the host, yet the human genome doesn't encode a unique antibody gene for each possible antigen. B cells create this vast antibody diversity by randomly selecting from a number of genes that are then spliced together to encode for a unique heavy and light chain combination.

Cells destined to become B cells are formed in the bone marrow at a fairly constant rate of $\sim 10^9$ cells per day. The changes that these precursor cells go through to finally become mature B cells involve a specific series of somatic gene rearrangements associated with the production of immunoglobulins. The expression of immunoglobulins and other cell surface proteins has allowed for the description of distinct stages of B-cell differentiation.

The heavy and light chains that make up an immunoglobulin molecule are reflected in the heavy and light gene arrangement shown in Figure 33.12. The heavy chain locus is composed of many different Variable (V) genes, Diversity (D) genes, and Joining (J) genes that will be important in producing the wide range of antigen binding required of antibodies. Also located on the heavy chain locus are several Constant (C) genes that encode for the heavy chain constant regions. There are two light chain loci (κ or λ), both of which contain multiple V and J genes, but unlike the heavy chain locus, contain no D genes. Again, following the V and J genes are C genes associated with the light chain constant regions.

The first immunoglobulin gene rearrangement for production of the immunoglobulin heavy chain begins in the pro-B-cell developmental stage. This first recombination event to occur is between any one of the approximately 25 D and 6 J gene segments within the human heavy chain locus. The second recombination event occurs between any one V gene and the previously DJ gene segments. Finally the VDJ gene segment is spliced to the Cμ constant region (Figure 33.12). A number of different proteins are required for these recombination processes; a few, such as the proteins from the recombination-activating genes 1 and 2 (RAG1 and RAG2), have been well characterized. RAG1 and RAG2 help initiate VDJ recombination by cleaving and helping to remove the DNA from between the joining V, D, and J gene segments.

While this combinational diversity generated from the VDJ rearrangement is great, it is not unlimited. Other mechanisms, which can increase the diversity of immunoglobulins by orders of magnitude, come into play during the recombination event. Enzymes such as exonucleases can remove several nucleotides from the cut ends of the gene segments, whereas other enzymes such as terminal deoxynucleotidyl transferase (TdT) can randomly add nucleotides, called **N additions**, to the joint between the two cut ends. This random addition or subtraction of nucleotides in the recombined VDJ gene segment makes the newly assembled coding regions for the immunoglobulins unique and is called **junctional diversity**. Due to the random nature of VDJ recombination and the possibility of a cell producing a nonfunctional heavy chain, the pre-B cell pairs the newly formed heavy chain with a surrogate light chain and expresses the pair on its cell surface. If the heavy chain successfully binds the surrogate light chain and is displayed on the pre-B-cell surface, a signal is given to halt any somatic rearrangement of the heavy chain locus on the other chromosome. This allelic exclusion will prevent the pre-B cell from making more than one heavy chain.

Once the heavy chain locus is turned off from further recombination, the pre-B cell begins the rearrangement of

TABLE 33.5 Therapeutic Monoclonal Antibodies

Year of Approval	Generic Name	Trade Name	Monoclonal Antibody Type	Target
1986	Muromonab-CD3	Orthoclone OKT3	Murine	anti-CD3
1994	Abciximab	ReoPro	Chimeric	anti-gpIIb/IIIa
1997	Rituximab	Rituxan	Chimeric	anti-CD20
1997	Daclizumab	Zenapax	Humanized	anti-CD25
1998	Basiliximab	Simulect	Chimeric	anti-CD25
1998	Palivizumab	Synagis	Humanized	anti-RSV
1998	Infliximab	Remicade	Chimeric	anti-TNF-α
1998	Trastuzumab	Herceptin	Humanized	anti-HER2
2000	Gemtuzumab ozogamicin	Mylotarg	Humanized	anti-CD33 cytotoxin labeled calicheamicin
2001	Alemtuzumab	Campath	Humanized	anti-CD52
2002	Ibritumomab tiuxetan	Zevalin	Murine	anti-CD20; radiolabeled yttrium-90
2002	Adalimumab	Humira	Human	anti-TNF-α
2003	Omalizumab	Xolair	Humanized	anti-IgE
2003	Tositumomab-I131	Bexxar	Murine	anti-CD20 radiolabeled either yttrium-90 or indium-111
2003	Efalizumab	Raptiva	Humanized	anti-CD11a
2004	Cetuximab	Erbitux	Chimeric	anti-EGF receptor
2004	Bevacizumab	Avastin	Humanized	anti-VEGF
2004	Natalizumab	Tysabri	Humanized	anti-α4-integrin
2006	Ranibizumab	Lucentis	Humanized	anti-VEGF
2006	Panitumumab	Vectibix	Human	anti-EGF receptor
2007	Eculizumab	Soliris	Humanized	anti-complement C5
2008	Certolizumab pegol	Cimzia	Humanized	anti-TNF-α
2009	Golimumab	Simponi	Human	anti-TNF-α
2009	Canakinumab	Ilaris	Human	anti-IL-1β
2009	Ustekinumab	Stelara	Human	anti-IL-12/23
2009	Ofatumumab	Arzerra	Human	anti-CD20
2010	Denosumab	Prolia	Human	anti-RANKL
2011	Belimumab	Benlysta	Human	anti-BAFF
2011	Ipilimumab	Yervoy	Human	anti-CTLA-4
2011	Brentuximab vedotin	Adcetris	Chimeric	anti-CD30 drug conjugated
2012	Raxibacumab	ABthrax	Human	anti-PA of *Bacillus anthracis*
2013	Pertuzumab	Perjeta	Humanized	anti-HER2
2013	Ado-trastuzumab emtansine	Kadcyla	Humanized	anti-HER2 drug conjugated

HER2: human epidermal growth factor receptor 2, RSV: respiratory syncytial virus, EGF: epidermal growth factor, VEGF: vascular endothelial growth factor, RANKL: receptor activator of nuclear factor kappa-B ligand, BAFF: B-cell activating factor, CTLA-4: cytotoxic T-lymphocyte antigen 4, PA: protective antigen

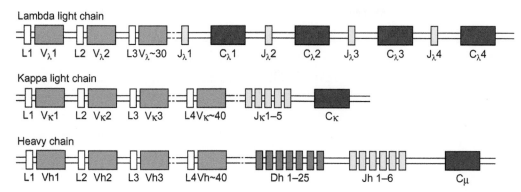

FIGURE 33.12 Immunoglobulin gene arrangements. *[Adapted from K.M. Murphy et al., Janeway's immunobiology, 7th ed., 2007 (Ch. 4, Fig. 4, p. 147).]*

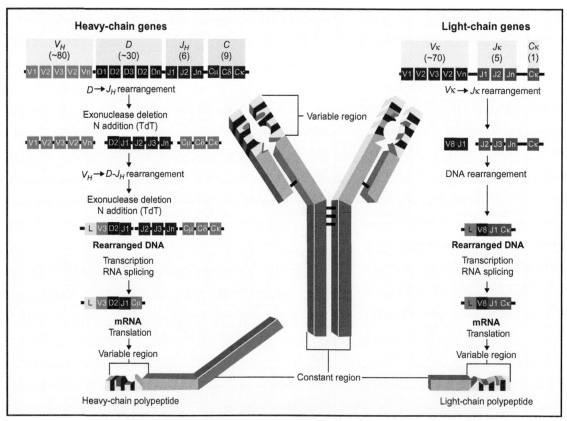

FIGURE 33.13 Production of heavy and light immunoglobulin chains. *[Adapted from R.S. Schwartz, Shattuck lecture: Diversity of the immune repertoire and immunoregulation. N. Engl. J. Med. 348(11) (2003) 1018, Fig. 1).]*

light chain locus. The mechanism for rearrangement of light chain loci is the same as for heavy chain loci, with the exception that light chain genes have only V, J, and constant gene segments, but no D segments. The pre-B cell begins to rearrange the κ chain locus first. If that VJ rearrangement does not produce a functional light chain, it tries a VJ rearrangement of the other κ allele and finally the λ-chain alleles. Since both the heavy and light chains contribute to the antigen-binding site, the range of

immunoglobulin diversity is further expanded by the possibility of different light chains associating with a heavy chain. Once both the heavy and light chains are successfully rearranged to form a functional IgM that is expressed on the cell surface, the lymphocyte becomes known as an **immature B cell** (Figure 33.13).

Because of the random nature of immunoglobulin production, the host must remove any immature B cells that produce any autoreactive antibodies that might injure the host. Immature B cells at this stage are still in the bone marrow, and any antigens they encounter in this environment are likely to be self antigens. If the immature B cell's surface-bound IgM is cross-linked by self antigen, that immature B cell is induced to undergo apoptosis in a process known as **clonal deletion**. There is one possible rescue for an autoreactive immature B cell, called **receptor editing**. If sufficient levels of RAG1 and RAG2 proteins remain in the immature B cell, it can attempt to rearrange a new light chain locus in the hope that the new IgM will no longer react with self antigens.

Once a functional, nonautoreactive IgM is produced, the immature B cells begin to express both IgM and IgD isotypes with identical antigen specificity. This is accomplished by alternative mRNA splicing. One mRNA is transcribed that encodes the VDJ segment and both the Cμ segment and the Cδ segment. This mRNA transcript is then spliced to remove the Cμ exons, which are translated into IgD, or the Cδ exons are removed and the mRNA is translated into IgM. Co-expression of both these immunoglobulin isotypes on the surface of the B cell renders it "mature" and ready to respond to antigen.

The cytoplasmic tail of the membrane-bound immunoglobulin (MBI) does not have catalytic or signaling activity. Instead, the MBIs are associated with two invariant chains: Igα and Igβ. It is the aggregation or conformational changes in the MBI molecule that cause changes in the Igα and Igβ heterodimer that allow it to signal the inside of the cell that antigen has been bound by the MBI on the outside of the cell. This complex of MBI, Igα and Igβ, is known as the **B-cell receptor** (BCR) complex (Figure 33.14) [10].

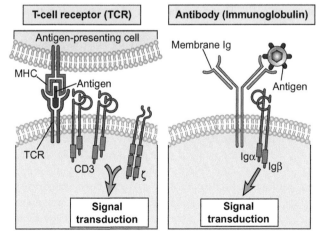

FIGURE 33.14 T-cell receptor and B-cell receptor signaling. *[Adapted from A.K. Abbas, A.H. Lichtman, Cellular and molecular immunology, 6th ed., 2007 (Ch. 7, Antigen receptors and accessory molecules of T lymphocytes, Fig. 2, p. 138).]*

T-CELL DEVELOPMENT AND T-CELL RECEPTOR DIVERSITY

T cells are lymphocytes that are produced in the bone marrow but require a journey through the thymus for differentiation and selection before they mature. These T cells are distinguished from the other two types of lymphocytes, B cells and NK cells, by the presence of either CD4 or CD8 surface molecules and a special cell surface receptor called the **T-cell receptor** (TCR), whose function is to recognize antigens bound to MHC molecules presented from antigen-presenting cells. The surface of each T cell also displays thousands of identical TCRs.

There are two different types of T cells based on the polypeptides that make up their TCRs. Alpha/beta (αβ) T cells have their TCRs composed of a heterodimer of one alpha (α)-chain with one beta (β)-chain. Both αβ-chains have a structure containing a variable (V) region and a constant (C) region not unlike what is found in immunoglobulin structures. Within the V regions, there lie three hypervariable regions or the complementarity-determining regions (CDRs) that make up the antigen−MHC-binding site. Gamma/delta (γδ) T-cell TCRs are also composed of a heterodimer, but this time with a gamma (γ)-chain paired with a delta (δ)-chain. The following discussion of T cells concerns αβ T cells, as our understanding of the function of γδ T cells in the host immune response is less clear.

The antigen−MHC-binding aspect of the TCR presents a risk to the host if their T cells recognize self-peptide/self-MHC complexes and mount an autoimmune response against the self peptide. The host's immune response deals with this potential problem by requiring all T cells to go through a process of selection as they journey through the thymus. This thymic selection process selects the useful, neglects the useless, and destroys the harmful in the T-cell population. T-cell precursors, like all immune cells, are formed in the bone marrow. Those cells that migrate to the thymus have neither a complete TCR nor expression of CD4 or CD8 and are called **double-negative pro T cells**. Once these precursor cells reach the thymus, they begin to randomly generate a TCR and synthesize both CD4 and CD8 cell surface markers and so are called **double-positive immature T cells**. Most of these immature T cells (>95%) will produce a TCR that does not bind to any of the self-peptide−MHC molecules presented on the surface of the thymus cells. Unless these immature T cells can try again with new randomly generated TCRs, these cells will die by apoptosis. Some of those remaining immature T cells whose TCRs have bound a self-peptide antigen presented by an MHC class II molecule stop expressing CD8 and become CD4+ T cells. Other immature T cells that have bound a self-peptide antigen presented by an MHC class I molecule stop expressing CD4 and become CD8$^+$ T cells. Both of these populations of immature T cells are said to have undergone positive selection. However, now those cells whose TCRs have bound very strongly to self peptide/MHC are destroyed by apoptosis. Those few immature T cells whose TCRs have bound self peptide/MHC at an affinity below the threshold that induces apoptosis then leave the thymus and become mature CD4 or CD8 T cells. It is this process of negative selection that is important in eliminating T cells that might otherwise mount an autoimmune response.

Generation of the TCR involves many of the same concepts and processes that are used by the lymphocyte to generate diversity in immunoglobulins. The first step is the random rearrangement of the V, D, J, and C region genes of the β-chain of the TCR in a way very similar to that of the heavy chain rearrangement needed for immunoglobulin synthesis. In fact, the same enzymes are used for both. Again, as in the immunoglobulin light chain rearrangement, the TCR α-chain is generated by a random VJ gene segment recombination event. Along with the random nucleotide additions and deletions at the gene segment recombination sites, this helps account for the great diversity in specificity of the T-cell receptor (Figure 33.15).

As with membrane-bound immunoglobulin, the TCR does not have catalytic or signaling activity. Instead, the TCRs are associated with two accessory molecules, CD3 and the zeta (ζ) chains, which form what is known as the **T-cell receptor complex** (Figure 33.14) [11].

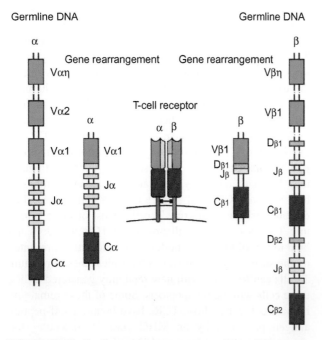

FIGURE 33.15 T-cell receptor rearrangements. *[From K.M. Murphy et al., Janeway's immunobiology, 7th ed., 2007 (Ch. 4, Fig. 10, p. 157).]*

T-CELL ACTIVATION AND DIFFERENTIATION

T cells contribute to the host immune response in two major ways. One set of T cells is vital in orchestrating the roles of other players in the immune system. Chief among these are the effector or helper T cells (T$_H$ cells). Typically, these T$_H$ cells are identifiable by their cell surface expression of the CD4 cell marker. T$_H$ cells are essential for activating a wide variety of cells such as B cells, macrophages, natural killer cells, and other T cells. There are also at least four different types or subpopulations of effector CD4 T cells based on the cytokines they produce and their role within the immune response (Figure 33.16). T$_H$1 cells produce cytokines such as IFN-γ and IL-12 that promote cell-mediated immunity, produce antibodies that promote phagocytosis, and block the production of T$_H$2 cells. T$_H$2 cells produce cytokines such as IL-4, IL-5, and IL-13 that recruit and activate a variety of cell types such as eosinophils and mast cells as well as promote other immune responses at mucosal surfaces against extracellular pathogens, particularly helminthes. T$_H$17 cells are associated with cytokines IL-17 and IL-23 and appear to play an important role in the recruitment of neutrophils in response to extracellular bacterial and fungal pathogens. T$_{FH}$ cells are a specific subset of T$_H$ cells characterized by expression of numerous molecules such as the master transcription factor B-cell lymphoma 6 (Bcl-6) and the C-X-C chemokine receptor type 5 (CXCR5). The reported key function of T$_{FH}$ cells is to provide help to B cells to support their activation, expansion, and differentiation, as well as the formation of the germinal center (GC) within lymph nodes.

Another subset of effector T cells act to regulate the immune response by inhibiting the action of other immune cells. These regulatory T (Treg) cells function to modulate the activities of a wide variety of cellular components of the immune systems. Treg cells use appropriate immunomodulatory mechanisms that can effectively inhibit the targeted cell population.

The second way in which T cells contribute to the host's immune response is by directly attacking those host cells that are infected or malignant. These cells, known as **cytotoxic** or **killer T cells**, usually express the CD8 cell surface marker, along with a TCR. In addition to killing host cells that have been infected by intracellular pathogens such as viruses or have been transformed by cancer, they are responsible for the rejection of tissue and organ grafts. The CD8 molecule, like the CD4 molecule, determines which MHC molecule, class I or class II, the T cell will recognize but not how the T cell will behave.

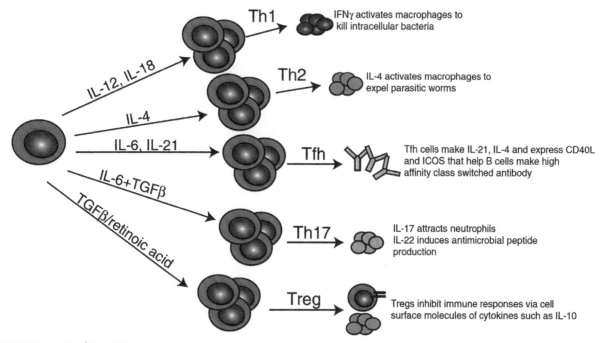

FIGURE 33.16 CD4$^+$ T-cell lineages. *[From A.S. McKee, M.K. MacLeod, J.W. Kappler, P. Marrack. Immune mechanisms of protection: can adjuvants rise to the challenge? BMC Biol. 12(8) (2010) 37, Fig. 2.]*

THE ADAPTIVE IMMUNE RESPONSE: SPECIFIC ANTIBODY RESPONSE

A mature B cell is covered in cell-surface-bound immunoglobulins of a single specificity. For this B cell to become activated to produce and secrete antibodies, it must come into contact with the specific antigen to which it binds; plus it must receive help from an antigen-specific T helper cell. This is a very important feature of antibody production, as it allows only those antibodies to be made that will be of benefit to the host immune response. This antigen-specific activation of both the B- and T-cell populations is known as **clonal selection**. A critical difference between B cells and T cells is how each interacts with its specific antigen. A B cell can recognize its specific antigen in a soluble form by using membrane-bound immunoglobulin. A T cell, however, recognizes its specific antigen as a peptide fragment presented on an MHC molecule by an antigen-presenting cell through its TCR. So for a specific antibody to be made, the correct B cell, T cell, and antigen-presenting cell must interact with the specific antigen but also with each other. It is the secondary lymphoid tissues such as the lymph nodes, spleen, and Peyer's patches where all these players can interact to initiate antibody production (Figure 33.17).

Antigen-presenting cells such as macrophages and dendritic cells having encountered non-self antigens such as those from an invading microorganism are phagocytosed, processed, and then presented on MHC class II molecules on the surface of the cell. The antigen-presenting cells can then enter the lymphatic system vessels, where they migrate to the regional lymph nodes. As the antigen-presenting cells enter the lymph node, they have matured and are now able to present the processed non-self-antigen MHC class II complex to an ever-changing population of CD4 T cells that are passing through the lymph node. When a CD4 T cell recognizes the non-self antigen MHC class II complex through its unique TCR, this serves as a first signal for activating the T cell. The affinity between CD4 and the MHC class II molecule keeps the T cell and the antigen-presenting cell closely bound together during the antigen-specific activation process. A second signal involving interaction of co-stimulatory molecules, such as B7, CD40, and T-cell activating cytokines with their corresponding partners' receptor ligands on the CD4 T cell, is also necessary for full T-cell activation. This activation signal leads to T-cell proliferation and differentiation into either T_H1 or T_H2 helper T cells (Figure 33.18).

B cells that have not been previously exposed to antigen, also known as **naïve B cells**, can be activated either with T-cell help in a thymus-dependent manner or without T-cell help in a thymus-independent manner. Most proteins are T-cell-dependent antigens, as they activate B cells in a strongly T-cell-dependent manner, whereas carbohydrates and lipids tend to activate B cells in a T-cell-independent manner. In either case, B-cell recognition of antigen is the first step in B-cell activation.

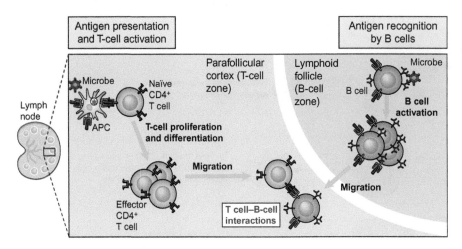

FIGURE 33.17 Antigen recognition by T and B cells in the lymph node. *[From A.K. Abbas, A.H, Lichtman, Basic immunology, 3rd ed., 2009 (Ch. 7, Humoral immune responses, Fig. 7, p. 139).]*

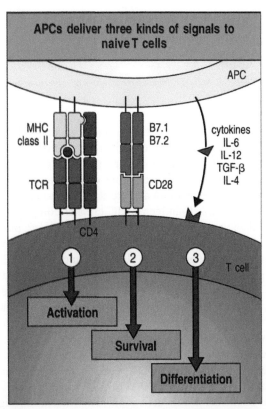

FIGURE 33.18 Antigen-presenting cell activation of CD4 T cells. *[Adapted from K.M. Murphy et al., Janeway's immunobiology, 7th ed., 2007 (Chap. 8, Fig. 19, p. 345).]*

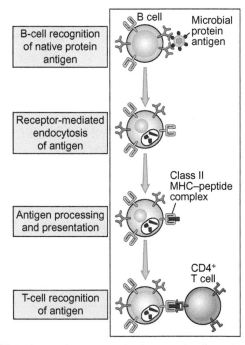

FIGURE 33.19 Antigen presentation by naïve B cells. *[Adapted from A.K. Abbas, A.H. Lichtman, Basic immunology, 3rd ed., 2009 (Ch. 7, Humoral immune responses, Fig. 8, p. 140).]*

For a T-cell-dependent activation of a naïve B cell, the first signal occurs when the membrane-bound immunoglobulin on the B-cell surface binds its specific epitope of a thymus-dependent antigen. Once the antigen is bound, the B cell engulfs the antigen and digests it so that it can present the peptides on its surface in association with an MHC class II molecule.

This process is analogous to the way antigens are processed by other antigen-presenting cells such as macrophages and dendritic cells. These events activate the naïve B cell to produce increased amounts of MHC class II molecules, co-stimulatory molecules such as B7, and receptors for T-cell-derived cytokines. The activated B cell can now interact with helper T cells such as T$_{FH}$ cells that will promote antibody production (Figure 33.19). The TCR on the helper T cell binds to the antigen MHC class II complex being presented by the B cell. But here again, a second signal from surface-bound co-stimulatory molecules, such as CD28 and CD40L on the T$_H$ cell, must

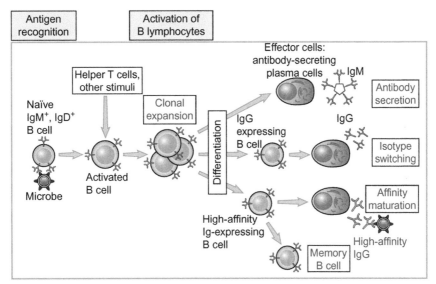

FIGURE 33.20 Consequences of T-cell help on activated B cells. *[Adapted from A.K. Abbas, A.H. Lichtman, Basic immunology, 3rd ed., 2009 (Ch.7, Humoral immune responses, Fig. 1, p. 132).]*

bind the B7 and CD40 molecules on the activated B cell to induce the cognate T_H cell to produce cytokines such as IL-2, IL-4, IL-5, and IL-6. It is the release of these T_H cell cytokines that directs the activated B cell to proliferate and differentiate into antibody-secreting plasma cells and memory cells (Figure 33.20).

As B cells proliferate after T-cell help, they undergo a process that can fine-tune the immunoglobulin binding to a non-self antigen known as **affinity maturation**. The reason is that the immunoglobulin variable gene locus of B cells undergoes an extremely high rate of somatic mutation that is at least $10^3 - 10^4$-fold greater than the normal rate of mutation in the rest of the cell's genome. This somatic hypermutation creates an opportunity for selection of variant B cells that possess an enhanced ability to recognize and bind a specific non-self antigen through its immunoglobulin. The longer and more tightly the antigen binds to the B cell through its surface-bound immunoglobulin, the greater the chance that B cell has of surviving and replicating. This allows the binding of the antibody to a non-self antigen to be improved on over time.

Another outcome of T-cell help is isotype or class switching of the antigen-specific antibody. This process allows the activated B cell to produce different antibody classes such as IgA, IgE, or IgG. Since only the constant region of the immunoglobulin heavy chain changes during class switching, the variable gene used remains unchanged. This gives the progeny of a single activated B cell the ability to produce antibodies that are specific for the same antigen, but with the ability to produce the effector function appropriate for each antigenic challenge. Fine-tuning of this T-cell help in the form of different combinations and amounts of these cytokines

produced by the T cells enables the activated B cell to switch the class of antibodies being produced so as to have the appropriate biologic function (Figure 33.11).

Many plasma cells migrate to the bone marrow, where they may continue to secrete antibodies for months or years after the antigen has been eliminated. But during the proliferation and differentiation that arises during B-cell activation, some of the activated B cells stop proliferating and do not secrete antibody, but instead enter the circulation to become long-lived memory B cells. Memory B cells are capable of immunologic memory of the original antigen, or what is called an **anamnestic response**. If that same antigen again is detected by the host immune system while the memory B cells are still present, these memory B cells will initiate a rapid, heightened secondary response that is adequate to deal with the antigen (Figure 33.21).

Naïve B cells can also be activated in a thymus-independent manner. Since this activation does not result from T-cell help, the consequences of that help—affinity maturation, class switching, and memory differentiation—are not seen in thymus-independent antigen response. That means the antibody response generated is, generally, a short-lived IgM that does not give rise to a memory response [12].

THE ADAPTIVE IMMUNE RESPONSE: CELL-MEDIATED IMMUNE RESPONSE

Cell-mediated immunity (CMI) is that arm of the immune response that does not involve antibodies but rather incorporates the activation of macrophages and NK cells, enabling them to destroy intracellular pathogens, the production of antigen-specific CD8 cytotoxic T lymphocytes (CTLs), and the release of various cytokines that

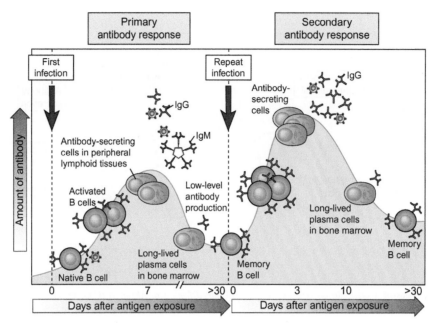

FIGURE 33.21 Primary and secondary antibody responses. *[Adapted from A.K. Abbas, A.H. Lichtman, Cellular and molecular immunology, 6th ed., 2007 (Ch. 10, B-cell activation and antibody production, Fig. 2, p. 218).]*

influence the function of other cells involved in both the adaptive and innate immune responses in response to a non-self antigen (Figure 33.22). Cell-mediated immunity is directed primarily at removing virus-infected cells, but is also a very important player in defending against fungi, protozoans, intracellular bacteria, and cancers. It also plays a major role in transplant rejection.

As in antibody production, helper T cells, particularly T_H1 cells, are key players in initiating, propagating, and controlling the cell-mediated immune response. In the lymph node, naïve CD4 T cells come into contact with antigen-presenting cells that are displaying non-self-antigen MHC class II complexes on their surface. When a naïve CD4 T cell recognizes the non-self antigen MHC class II complex through its unique TCR, this serves as a first signal for activating the T cell. A second signal involving interaction of co-stimulatory molecules, such as B7, CD40, and T-cell activating cytokines with their corresponding ligands on the CD4 T cell, is also necessary for full T-cell activation. Here, cytokines elaborated from the antigen-presenting cell such as IL-12 drives the differentiation of the naïve CD4 T cell into becoming a T_H1 helper T cell (Figure 33.23).

The now differentiated T_H1 cell can, in turn, begin to activate macrophages, thus enabling them to better kill intracellular pathogens. The term **activated macrophages** is used to describe those macrophages possessing characteristics such as enhanced MHC class II cell surface expression resulting in increased antigen presentation; increased surface expression of co-stimulatory molecules; increased killing of intracellular organisms; tumor cell lysis; and maximal secretion of inflammatory mediators

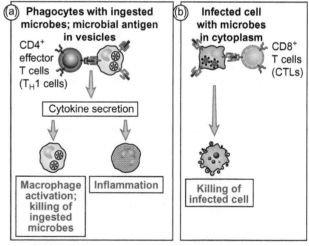

FIGURE 33.22 Cell-mediated immune responses. *[Adapted from A.K. Abbas, A.H. Lichtman, Basic immunology, 3rd ed., 2007 (Ch. 6, Effector mechanisms of cell-mediated immunity, Fig. 1, p. 114).]*

including TNF-α, IL-1, IL-12, reactive oxygen intermediates (ROI), and nitric oxide (NO). This activation of macrophages is initiated by the recognition of a specific non-self antigen MHC class II complex on the macrophage cell surface by the TCR of a T_H1 cell. Again, a second signal is necessary and, in this case, it is the interaction of CD40-CD40L to induce the T_H1 cell to secrete cytokines, particularly IFN-γ.

Activated macrophages are capable of increased uptake and killing of organisms, but they also are critical for the induction and activation of another cell type

Overview of CD4⁺ T cell-mediated immunity to bacteria and fungi

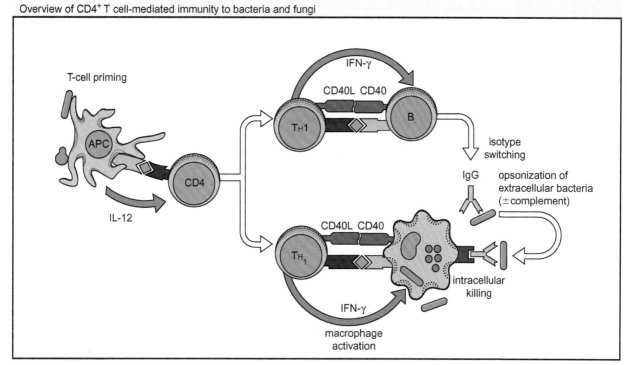

FIGURE 33.23   Activation of T$_H$1 cells. *[Adapted from D. Male et al., Immunology, 7th ed., 2006 (Ch. 14, Immunity to viruses, Fig. 11, p. 252).]*

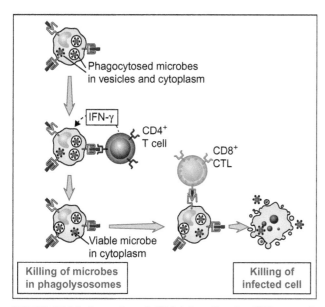

FIGURE 33.24   Interaction of CD4 helper T cells, CD8 cytotoxic T cells, and macrophages in the cell-mediated immune response. *[From A.K. Abbas, A.H. Lichtman, Basic immunology, 3rd ed., 2007 (Ch. 6, Effector mechanisms of cell-mediated immunity, Fig. 11, p. 126).]*

important in cell-mediated immunity, the CD8 cytotoxic T cell (Figure 33.24).

Stimulation of naïve CD8 cytotoxic T cells with antigen and co-stimulation results in proliferation but relatively weak effector functions (Figure 33.25a). Development of

effector function can be driven by help provided from CD4 T$_H$1 cells in the form of cytokines such as IL-2 and IFN-γ either directly to the naïve CD8 cytotoxic T cells (Figure 33.25b) or indirectly by enhancing the ability of antigen-presenting cells (Figure 33.25c). CD8 cytotoxic T cells, however, express TCRs that, along with CD8, recognize specific nonself-peptides bound to MHC class I molecules. It is the affinity of CD8 for the MHC class I molecule that binds the cytotoxic T cell and its target cell tightly together during antigen-specific activation. Since only antigen-presenting cells have both MHC class I and MHC class II on their cell surface, they play a key role in cytotoxic T-cell activation.

It is the mechanism of cross-presentation that allows antigen-presenting cells to take up, process, and present extracellular antigens with both MHC class I molecules to CD8 cytotoxic T cells and MHC class II molecules to CD4 T$_H$1 cells. The intracellular mechanisms of cross-presentation are still unclear, but endocytosed proteins are transported out of the endosome and into the cytoplasm, where they are processed by the proteasome into peptides, transported by the TAP transporter into the endoplasmic reticulum, associated with MHC class I molecules, and transported to the cell surface, where they can be detected by specific CD8 T cytotoxic cells.

Once the CD8 cytotoxic T cells have become activated, they are ready to enter the circulation and kill infected host cells via a programmed cell death mechanism known as

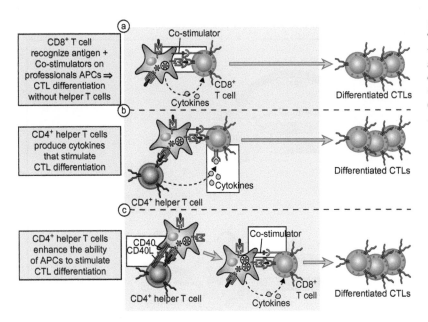

FIGURE 33.25 Activation of naïve CD8 cytotoxic T cells. (a) No CD4 T-cell help in CD8 T-cell activation. (b) Direct CD4 T-cell help in CD8 T-cell activation. (c) Indirect CD4 T-cell help in CD8 T-cell activation. *[From A.K. Abbas, A.H. Lichtman, Cellular and molecular immunology, 6th ed., 2007 (Ch. 9, Activation of T lymphocytes, Fig. 4, p. 193).]*

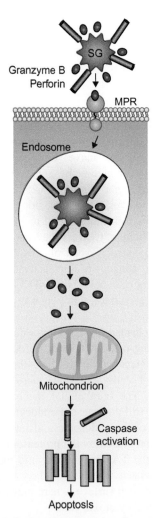

FIGURE 33.26 Induction of apoptosis in cells targeted by CD8 cytotoxic T cells using the perforin/granzyme pathway. *[Adapted from J.A. Trapani, M.J. Smyth, Functional significance of the perforin/granzyme cell death pathway. Nat. Rev. Immunol. 2(10) (2002) 738, Fig. 1).]*

apoptosis. The CD8 cytotoxic T cells induce apoptosis in their target cells by means of two different mechanisms. The first and most important is by the granzyme/perforin pathway. This pathway is activated when the TCRs and CD8 of a circulating cytotoxic T-cell encounter the non-self antigen/MHC class I complex on the surface of the virus-infected cell or tumor cell, which sends a signal that triggers the release of the contents of the T cells' intracellular granules. These granules contain two membrane-pore-forming proteins, perforin and granulysin, and a family of serine proteases known as **granzymes**, complexed together with a proteoglycan called **serglycin (SG)**. The exact role each of these components plays in the induction of apoptosis is still debated, but it is known that certain granzymes initiate the caspase cascade to DNA fragmentation and cell death (Figure 33.26).

Activated CD8 cytotoxic T cells can also trigger apoptosis of targeted cells through Fas/FasL interactions. Activated CD8 cytotoxic T cells have a cell surface receptor called **Fas ligand** or **FasL** that can bind to the death receptor Fas found on the surface of most cell types. This FasL/Fas interaction triggers the activation of caspase enzymes, which leads to degradation of cytoskeletal structural proteins and fragmentation of the target cells' chromosomal DNA.

The ability of the diverse elements of the human immune system to correctly identify and properly react to the wide array of organisms that we encounter on a daily basis is one of the marvels of nature. Our ability to understand these complex and elaborate multilayer interactions between many different immune mechanisms will be the cornerstone in the progress of medicine for generations to come [13].

CLINICAL CASE STUDY 33.1 Complement Deficiency

This clinical case was abstracted from N. Daskas, K. Farmer, R. Coward, M. Erlewyn-Lajeunesse, Meningococcal disease associated with an acute post-streptococcal complement deficiency. Pediatr. Nephrol. 22(5) (2007) 747–749.

Synopsis

A 15-year-old Japanese male was admitted with suspected bacterial meningitis. The patient appeared to be acutely ill with spiking fever (39°C), chills, myalgias, severe headache, and vomiting. His past medical history was remarkable for two previous hospitalizations for positive blood and cerebral spinal fluid (CFS) cultures of *Neisseria meningitidis* at age 2 and 13 years. There was no family history of recurrent meningococcal disease. The patient received ceftriaxone intravenously for 7 days.

Complement studies revealed the total hemolytic complement activity (CH_{50}; antibody-sensitized sheep erythrocytes as target) was less than half of the normal level. Complement assays showed that values of Clq, C2, factor B, C4, and C5 were all present at normal amounts, but the C9 component was undetectable. At the time of discharge, the patient was immunized with pneumococcal polysaccharide vaccine, *Hemophilus influenzae* B conjugate vaccine, and the tetravalent meningococcal polysaccharide vaccine (serogroups A, C, Y, and W135). The patient and his family were educated about the risk of recurrence of meningococcal disease and were advised to seek early medical attention in the case of fever.

Teaching Points

The complement system consists of a collection of circulating and membrane-bound proteins that play a vital role in the innate host immune response. The membrane attack complex (MAC) is composed of the terminal complement components (C5b, C6, C7, C8, and C9). The MAC forms a pore-like structure in cell membranes, resulting in cell lysis. This mechanism is particularly important for the complement-mediated lysis of gram-negative organisms.

The prevalence of deficiencies of terminal complement components C5, C6, C7, C8, and C9 varies greatly geographically and ethnically, but all are typically associated with recurrent systemic infections caused by *Neisseria gonorrhoeae* and *N. meningitidis*. Infection with *N. meningitidis* in these individuals frequently involves unusual rare serotypes (W135 and Y). Individuals with terminal complement component deficiency generally present with recurrent meningococcal disease after the age of 10 years.

CLINICAL CASE STUDY 33.2 Defects in Intracellular Killing by Phagocytes

This clinical case was abstracted from: R. Fehon, S. Mehr, E. La Hei, D. Isaacs, M. Wong, Two-year-old boy with cervical and liver abscesses. J. Paediatr. Child Health. 44(11) (2008) 670–672.

Synopsis

A 2-year-old white male was referred to the Clinical Immunology service with the chief complaint of recurrent skin and soft tissue infections. The patient's past medical history was significant for recurrent perianal abscesses beginning at 4 months of age. Due to his poor response to oral antibiotics, at 9 months of age, he underwent a surgical procedure to incise, drain, and repair multiple perianal fistulas associated with abscess progression. At 1 year of age, he had a right cervical lymphadenitis due to *Staphylococcus aureus* that was again treated with incision and drainage and a course of IV cefazolin. A cellulitis of his left elbow that grew *Serratia marcescens* upon culture was the patient's last significant infection in the last 3 months. The WBC count, hemoglobin level, and erythrocyte sedimentation rate were normal. Screening tests for serum antibody levels, anergy skin testing, and serum complement component levels were normal. A sample of the patient's blood was sent to a neutrophil research laboratory with results that demonstrated no reduction of nitroblue tetrazolium (NBT). Testing of the patient's mother demonstrated two populations of neutrophils in which one population reduced NBT (a normal reaction) and a population that did not reduce NBT. A diagnosis of chronic granulomatous disease was made, and the patient was placed on subcutaneous injections of gamma interferon (IFN-γ) and daily doses of oral trimethoprim-sulfamethoxazole.

Teaching Points

This case illustrates the importance of not only phagocytoses but also of the proper functioning of intercellular killing mechanisms for the control of normal flora organisms. Chronic granulomatous disease (CGD) is the name given to a collection of inherited genetic disorders that disrupt the NADPH oxidase complex and that are characterized by the inability of the phagocyte to produce microbicidal reactive superoxide anion and associated metabolites such as hydrogen peroxide. While ingestion of bacteria, degranulation, and phagolysosome formation are normal in CGD patients, regardless of the exact genetic form of CGD, the end result is that superoxide cannot be produced, and in turn, bacterial killing is ineffective. The presence of recurrent, indolent, pyogenic infections due to *Staphylococcus aureus* and gram-negative bacteria, such as *Pseudomonas, Klebsiella,* and *Serratia,* beginning in the first year of life is pathognomonic for CGD. This disease may be inherited as either X-linked-recessive (as in this case) or as autosomal-recessive. A standard screening test for CGD is the nitroblue tetrazolium (NBT) test, in which neutrophils are incubated with NBT and examined microscopically for color changes. Normal neutrophils will readily reduce NBT (which appears to color the cells blue) due to their ability to produce O_2^-; CGD patients' neutrophils will show little or no color changes in their cells. Aggressive prophylactic antibiotic therapy and IFN-γ therapy have been very effective in limiting the frequency of infections in these patients. Bone marrow transplants can cure the disease but are limited by matching donors.

CLINICAL CASE STUDY 33.3 IgA Deficiency

This clinical case was abstracted from: T. Nishikawa, Y. Nomura, Y. Kono, Y. Kawano, Selective IgA deficiency complicated by Kawasaki syndrome. Pediatr. Int. 50(6) (2008) 816–818.

Synopsis

A 28-year-old female was admitted to the hospital for multiple trauma related to an automobile accident. Due to moderately severe blood loss, the patient was transfused with one unit of typed and cross-matched whole blood. Almost immediately upon infusion, the patient began to show signs of urticaria, erythema, cutaneous flushing, and bronchoconstriction (wheezing and dyspnea). The transfusion was stopped, antihistamines and bronchodilators were administered, and the patient's vital signs closely monitored. After the transfusion reaction, a quantitative immunoglobulin enzyme-linked immunosorbent assay (ELISA) revealed that the patient had normal levels of 1100 mg/dL of IgG (normal range 751–1560 mg/dL) and 76 mg/dL of IgM (normal range 46–304 mg/dL), but an IgA level of less than 1.0 mg/dL (normal range 76–390 mg/dL). Further laboratory investigations revealed she produced anti-IgA antibodies that reacted with both IgA_1 and IgA_2 isotypes. Before discharge, the patient was advised to wear an identification bracelet to prevent inadvertent plasma or IVIG administration, which could lead to anaphylaxis.

Teaching Points

IgA deficiency is defined clinically in the United States as a serum IgA level <7 mg/dL in males or females 4 years of age or older, but with normal serum levels of both IgG and IgM. While selective absence of IgA is the most common of antibody deficiency disorders, it is also the one antibody deficiency that is most commonly clinically asymptomatic. This case illustrates the clinical implications of IgA deficiency even in asymptomatic individuals. IgA-deficient individuals are recognized to be at an increased risk for developing anaphylactic reactions after therapeutic blood products. Allergic responses to blood products containing IgA are secondary to anti-IgA antibodies developed by these individuals. It is currently unclear what leads to the production of anti-IgA antibodies in individuals with IgA deficiency.

Allergic responses can range from a less severe anaphylactoid reaction characterized by urticaria, periorbital swelling, dyspnea, and/or perilaryngeal edema to the very severe as in bronchospasm, hypotension, and shock, culminating on rare occasions in death.

CLINICAL CASE STUDY 33.4 Impaired Development of T Cells and B Cells

This clinical case was abstracted from: R.P. Katugampola, G. Morgan, R. Khetan, N. Williams, S. Blackford, Omenn's syndrome: lessons from a red baby. Clin. Exp. Dermatol. 33(4) (2008) 425–428.

Synopsis

A 3-week-old male presented to his pediatrician with an erythematous scaly rash of his scalp and was treated for atopic dermatitis. He had one episode of otitis media at the age of 2 months that was treated with antibiotics and resolved on follow-up examination. At 5 months of age, he developed a fever of 39°C and was admitted to the hospital for evaluation of fever of unknown origin. On physical examination, the patient's height and weight were noted to be in the <10th percentile for his age. A severe erythematous macular-papular rash was observed throughout his body with numerous serous sanguineous weeping lesions on his scalp. A small amount of mucoid diarrhea was noted in his diaper. Hematological and immunological laboratory results are shown below. Further immunologic testing showed the T-cell repertoire was restricted, with oligoclonal T cells present upon analysis of the T-cell receptor β (TCR-β) variable region by RT-PCR.

Total WBC	4280 cells/mm³	6000–17,500 cells/mm³
Absolute eosinophil count	4667 cells/mm³	300 cells/mm³
Absolute neutrophil count	9130 cells/mm³	1000–8500 cells/mm³
Absolute lymphocyte count	433 cells/mm³	5400–7200 cells/mm³
Absolute T-cell count	199 cells/mm³	2400–6900 cells/mm³
Absolute B-cell count	256 cells/mm³	700–2500 cells/mm³
Absolute NK-cell count	410 cells/mm³	100–1000 cells/mm³
IgG	1–5 mg/dL	480–1152 mg/dL
IgM	33 mg/dL	41–227 mg/dL
IgA	3 mg/dL	14–107 mg/dL
IgE	26 mg/dL	1–17 mg/dL

With consent of his parents, the patient underwent a conditioning, or preparative, regimen in preparation for a matched, unrelated bone marrow transplant. Within 2 weeks of the conditioning regimen, the patient's erythematous rash improved and was completely resolved by 1 month. Three

(Continued)

CLINICAL CASE STUDY 33.4 (Continued)

months after immunological reconstitution, the patient was doing well at home.

Teaching Points

This case illustrates that mutations in the *RAG1* or *RAG2* genes can cause impairment in V(D)J recombination, resulting in arrest in T-cell and B-cell development and maturation, and the eventual development of humoral and cellular immunodeficiency. Classically, Omenn's syndrome has the typical features of a severe combined immunodeficiency (SCID) and is characterized by early onset of erythroderma, hepatomegaly, failure to thrive, protracted diarrhea, hypereosinophilia, high IgE levels, low to absent B-cell numbers, and normal to low T-cell numbers. T cells in individuals with Omenn's syndrome exhibit a very narrow TCR repertoire. The presence of these T cells with their limited TCR repertoire that may target autoantigens present in host tissues may explain the rashes

and diarrhea associated with Omenn's syndrome. Initially, Omenn's syndrome was first associated with either *RAG1* or *RAG2* mutations; however, multiple additional genetic causes have been identified; theyinclude *AIRE*, *IL-7Rα*, and *ARTEMIS* gene mutations. However, in some patients, no specific genetic mutation can be identified. *RAG1* and *RAG2* are genes that encode specific proteins that form a complex required for V(D)J recombination in T and B cells to produce the broad array of immunoglobulin and T-cell receptors. Mutations within either *RAG1* or *RAG2* that lead to a complete lack of functional recombinase activity result in an absence of these receptors and, as a consequence, lead to an absence of mature T and B cells. Bone marrow transplantation is currently the only therapeutic option for Omenn's syndrome, as patients seldom survive longer than the first few months of life.

REQUIRED READING

[1] D.F. Kelly, A.J. Pollard, E.R. Moxon, Immunological memory: the role of B cells in long-term protection against invasive bacterial pathogens, JAMA 294(23) (2005) 3019–3023.

[2] S. Alvermann, C. Hennig, O. Stüve, H. Wiendl, M. Stangel, Immunophenotyping of cerebrospinal fluid cells in multiple sclerosis: in search of biomarkers, JAMA Neurol. 71(7) (2014) 905–912.

[3] T.T. Hansel, S.L. Johnston, P.J. Openshaw, Microbes and mucosal immune responses in asthma, Lancet 381(9869) (2013) 861–873.

[4] O. Osborn, J.M. Olefsky, The cellular and signaling networks linking the immune system and metabolism in disease, Nat. Med. 8(3) (2012) 363–374.

[5] S. Sethi, F.C. Fervenza, Membranoproliferative glomerulonephritis—a new look at an old entity, N. Engl. J. Med. 366(12) (2012) 1119–1131.

[6] G. Mihai, M.G. Netea, J.W.M. van der Meer, Immunodeficiency and genetic defects of pattern-recognition receptors, N. Engl. J. Med. 364 (2011) 60–70.

[7] C. Iadecola, J. Anrather, The immunology of stroke: from mechanisms to translation, Nat. Med. 17(7) (2011) 796–808.

[8] A. Vaglio, J. Zwerina, IgG4-related disease, N. Engl. J. Med. 366 (17) (2012) 1646.

[9] J.B. Bice, E. Leechawengwongs, A. Montanaro, Biologic targeted therapy in allergic asthma, Ann. Allergy Asthma Immunol. 112(2) (2014) 108–115.

[10] A. Palumbo, K. Anderson, Multiple myeloma, N. Engl. J. Med. 364 (2011) 1046–1060.

[11] J.E. Smith-Garvin, G.A. Koretzky, M.S. Jordan, T cell activation, Annu. Rev. Immunol. 27 (2009) 591–619.

[12] C. King, New insights into the differentiation and function of T follicular helper cells, Nat. Rev. Immunol. 9(11) (2009) 757–766.

[13] B. Rehermann, Pathogenesis of chronic viral hepatitis: differential roles of T cells and NK cells, Nat. Med. 19(7) (2013) 859–868.

ENRICHMENT READING

[1] R.M. Anthony, L.I. Rutitzky, J.F. Urban Jr., M.J. Stadecker, W.C. Gause, Protective immune mechanisms in helminth infection, Nat. Rev. Immunol. 7(12) (2007) 975–987.

[2] V.L. Brandt, D.B. Roth, V(D)J recombination: how to tame a transposase, Immunol. Rev. 200 (2004) 249–260.

[3] S. Carpenter, L.A. O'Neill, How important are toll-like receptors for antimicrobial responses? Cell. Microbiol. 9(8) (2007) 1891–1901.

[4] A. Casadevall, E. Dadachova, L.A. Pirofski, Passive antibody therapy for infectious diseases, Nat. Rev. Microbiol. 2(9) (2004) 695–703.

[5] A. Casadevall, L.A. Pirofski, Antibody-mediated regulation of cellular immunity and the inflammatory response, Trends Immunol. 24(9) (2003) 474–478.

[6] A. Corthay, A three-cell model for activation of naïve T helper cells, Scand. J. Immunol. 64(2) (2006) 93–96.

[7] P.J. Delves, I.M. Roitt, The immune system. First of two parts, N. Engl. J. Med. 343(1) (2000) 37–49.

[8] P.J. Delves, I.M. Roitt, The immune system. Second of two parts, N. Engl. J. Med. 343(2) (2000) 108–117.

[9] R.N. Germain, T-cell development and the CD4–CD8 lineage decision, Nat. Rev. Immunol. 2(5) (2002) 309–322.

[10] S. Gordon, Alternative activation of macrophages, Nat. Rev. Immunol. 3(1) (2003) 23–35.

[11] C.M. Grimaldi, R. Hicks, B. Diamond, B-cell selection and susceptibility to autoimmunity, J. Immunol. 174(4) (2005) 1775–1781.

[12] H. Hawlisch, J. Köhl, Complement and toll-like receptors: key regulators of adaptive immune responses, Mol. Immunol. 43(1–2) (2006) 13–21.

[13] P.E. Jensen, Recent advances in antigen processing and presentation, Nat. Immunol. 8(10) (2007) 1041–1048.

[14] J. Klein, A. Sato, The HLA system. First of two parts, N. Engl. J. Med. 343(10) (2000) 702–709.

[15] J. Klein, A. Sato, The HLA system. Second of two parts, N. Engl. J. Med. 343(11) (2000) 782–786.

[16] A. Lanzavecchia, N. Bernasconi, E. Traggiai, C.R. Ruprecht, D. Corti, F. Sallusto, Understanding and making use of human memory B cells, Immunol. Rev. 211 (2006) 303–309.

[17] J. Lieberman, The ABCs of granule-mediated cytotoxicity: new weapons in the arsenal, Nat. Rev. Immunol. 3(5) (2003) 361–370.

[18] A.D. Luster, Chemokines—chemotactic cytokines that mediate inflammation, N. Engl. J. Med. 338(7) (1998) 436–445.

[19] C.S. Ma, E.K. Deenick, M. Batten, S.G. Tangye, The origins, function, and regulation of T follicular helper cells, J. Exp. Med. 209(7) (2012) 1241–1253.

[20] M.M. Markiewski, J.D. Lambris, The role of complement in inflammatory diseases from behind the scenes into the spotlight, Am. J. Pathol. 171(3) (2007) 715–727.

[21] M.A. Mascelli, H. Zhou, R. Sweet, J. Getsy, H.M. Davis, M. Graham, et al., Molecular, biologic, and pharmacokinetic properties of monoclonal antibodies: impact of these parameters on early clinical development, J. Clin. Pharmacol. 47(5) (2007) 553–565.

[22] R. Medzhitov, Recognition of microorganisms and activation of the immune response, Nature 449(7164) (2007) 819–826.

[23] M.F. Mescher, J.M. Curtsinger, P. Agarwal, K.A. Casey, M. Gerner, C.D. Hammerbeck, Signals required for programming effector and memory development by CD8 + T cells, Immunol. Rev. 211 (2006) 81–92.

[24] W.M. Nauseef, How human neutrophils kill and degrade microbes: an integrated view, Immunol. Rev. 219 (2007) 88–102.

[25] D. Nemazee, Receptor editing in lymphocyte development and central tolerance, Nat. Rev. Immunol. 6(10) (2006) 728–740.

[26] M. Noack, P. Miossec, Th17 and regulatory T cell balance in autoimmune and inflammatory diseases, Autoimmun. Rev. 13(6) (2014) 668–677.

[27] J. Ollila, M. Vihinen, B cells, Int. J. Biochem. Cell Biol. 37(3) (2005) 518–523.

[28] A. Pichlmair, C. Reis e Sousa, Innate recognition of viruses, Immunity. 27(3) (2007) 370–383.

[29] J.H. Russell, T.J. Ley, Lymphocyte-mediated cytotoxicity, Annu. Rev. Immunol. 20 (2002) 323–370.

[30] R.S. Schwartz, Shattuck lecture: diversity of the immune repertoire and immunoregulation, N. Engl. J. Med. 348(11) (2003) 1017–1026.

[31] W.D. Shlomchik, Graft-versus-host disease, Nat. Rev. Immunol. 7 (5) (2007) 340–352.

[32] A.H. Sharpe, A.K. Abbas, T-cell costimulation—biology, therapeutic potential, and challenges, N. Engl. J. Med. 355(10) (2006) 973–975.

[33] Y. Takahama, Journey through the thymus: stromal guides for T-cell development and selection, Nat. Rev. Immunol. 6(2) (2006) 127–135.

[34] K. Takeda, S. Akira, Toll-like receptors in innate immunity, Int. Immunol. 17(1) (2005) 1–14.

[35] J.A. Trapani, M.J. Smyth, Functional significance of the perforin/granzyme cell death pathway, Nat. Rev. Immunol. 2(10) (2002) 735–747.

[36] D Vestweber, Adhesion and signaling molecules controlling the transmigration of leukocytes through endothelium, Immunol. Rev. 218 (2007) 178–196.

[37] C.T. Weaver, L.E. Harrington, P.R. Mangan, M. Gavrieli, K.M. Murphy, Th17: an effector CD4 T cell lineage with regulatory T cell ties, Immunity 24(6) (2006) 677–688.

[38] J.M. Woof, D.R. Burton, Human antibody-Fc receptor interactions illuminated by crystal structures, Nat. Rev. Immunol. 4(2) (2004) 89–99.

Chapter 34

Biochemistry of Hemostasis

Key Points

1. Hemostatic response is rapid, local, and temporary (normal physiological hemostasis).
2. Coagulation cascade reactions occur via two *interdependent* pathways; the extrinsic and intrinsic pathways both produce thrombin and result in fibrin formation.
3. Blood circulation to tissues after hemostasis involves clot digestion and dissolution by action of fibrinolytic system components.
4. Action of the coagulation proteases is controlled *in vivo* through inactivation by inhibitory proteins that circulate in the blood.
5. Anticoagulant therapy is used to reduce formation of a clot, and fibrinolytic therapy is used to eliminate formed clots.
6. Therapeutic fibrinolysis is obtained from activation of plasminogen by administration of activators of plasminogen.
7. Molecular explanations for hemostatic function are provided by coagulation factor structures and reactions and the formation of specific reaction complexes that localize the reactions and make the local reactions extremely fast.
   a. How the new direct inhibitors of thrombin and factor Xa act and how they differ from naturally occurring inhibitors
8. Thrombosis—hemostatic reactions at sites NOT associated with hemorrhage-threatening injury.
   a. Contact system reactions and thrombosis versus hemostasis
9. *In vitro* the coagulation pathways are defined by the laboratory tests used to evaluate hemostatic system function.
10. Key case studies illustrate hemostatic components as markers of a variety of diseases.

Hemostasis is the spontaneous arrest of blood loss from ruptured blood vessels. Organisms with pressurized circulatory systems require a hemostatic system that responds rapidly to an injury and that is localized to the injury site. The structure that forms to prevent excessive blood loss, the hemostatic plug, is spatially constrained so as not to

occlude the ruptured blood vessels that provide nutrients to the surrounding tissues. Because blood must continue to flow adjacent to the hemostatic plug, it must not create a prolonged disruption of blood flow. The normal hemostatic plug is not permanent, and thus, hemostasis must be integrated with a mechanism for its removal as the injury is repaired and the wound heals.

HEMOSTATIC RESPONSE OCCURS IN TWO PHASES

In human and other vertebrate animals, hemostatic response can be conceptually divided into two interdependent phases: primary and secondary hemostasis. Primary hemostasis involves cellular components, conveniently divided into two groups: blood vessel subendothelial cells and blood platelets. The secondary phase involves the plasma coagulation proteins, the coagulation cascade. Primary and secondary hemostasis act synergistically to prevent excessive blood loss without starving the surrounding tissues of nutrients transported to them from blood.

Primary hemostasis begins with blood vessel contraction (vasoconstriction) because vasoconstrictor substances are released at the site of the injury. Platelets present in the blood recognize and adhere to collagen fibers and other macromolecular substances exposed on the subendothelium, and thus, the site of the primary hemostatic response is localized to the injury site. The adherent platelets immediately undergo dramatic changes in their structure to form a hemostatic plug, initiate secondary hemostasis, and prevent extravasation without blood vessel occlusion.

Secondary hemostasis includes the processes by which the hemostatic plug is consolidated as the result of a sequence of the reactions that involve protein clotting factors in plasma. This sequence, or "cascade," of reactions is generally described as **blood clotting**. The clotting reactions ultimately produce the protease thrombin, which transforms the circulating protein fibrinogen into a fibrin

N. V. Bhagavan and C.-E. Ha: Essentials of Medical Biochemistry, Second edition. DOI: http://dx.doi.org/10.1016/B978-0-12-416687-5.00034-8

meshwork that surrounds the platelets in the plug and mechanically reinforces it.

The plasma clotting system consists of three "subsystems." A "procoagulant subsystem" provides the rapid, localized response to the injury and, because of the enmeshing of the platelets by fibrin, a hemostatic plug that is spatially constrained and mechanically stable. An "anticoagulant subsystem" modulates two of the key reactions of the procoagulant system, prothrombin and factor X activation, by inactivating cofactor proteins (factors Va and VIIIa) that are critical components in making the procoagulant reactions rapid and local to the injury site. The "anticoagulant subsystem," through inhibitors of the clotting proteases, acts to shut down the process. The "fibrinolytic subsystem," which is proteolytic digestion by plasmin of the fibrin that reinforced the hemostatic plug, is responsible for the temporary nature of the plug. Digestion of the fibrin occurs after tissue repair has commenced and hemorrhage is no longer a threat.

Primary Hemostasis: Platelets and Hemostatic Plug Formation

Primary hemostasis is dominated by processes that initiate transformation of the quiescent, discoid, resting platelets into their active forms. Activated platelets secrete the contents of their storage granules in response to stimuli that trigger cell membrane receptor activation and initiate intracellular signaling pathways. The platelet subcellular organelles fuse with the outer membranes and release their contents into the area surrounding the injury site. ADP and serotonin released from the dense granules recruit additional platelets that spread on the exposed subendothelial cell surfaces and adhere to these surfaces. Coagulation system proteins present in the platelet α-granules are released and provide ligands that, upon binding to receptors on different platelets, cause the activated platelets to aggregate to form the platelet plug. At the same time as granule release, arachidonic acid from the phospholipids of the platelet membranes is transformed into prostaglandins E_2 and H, which are then further transformed into thromboxane A_2, the most potent platelet-activating and -aggregating agent. Aspirin targets the synthesis of thromboxane and is the basis for the prevention of thrombosis.

Recognition of subendothelial and smooth muscle cell membrane surface components by the platelets occurs through receptors on the external surface of the platelet membrane. Collagen, present in the subendothelial space, binds to the platelet receptor *GpIa/IIa*. The plasma proteins, von Willebrand factor, fibrinogen, and vitronectin provide bridging molecules between the exposed endothelial surface and the platelets. Von Willebrand factor binds to the platelet via *GpI*; fibrinogen binds via the receptor *GpIIb/IIIa*. Another platelet protein, glycocalicin, binds the platelets via the receptor, *GpIb/IX*. At the same time that adhesion of the platelets to the newly exposed surface occurs, the procoagulant subsystem provides thrombin to activate the platelets. This occurs through a unique receptor mechanism that requires proteolytic action by thrombin and transmembrane signaling via a G-protein-coupled mechanism.

During the activation of platelets, phosphatidyl serine and phosphatidyl ethanolamine, normally facing inward toward the cytoplasm, are catalytically flipped by the platelet phospholipid translocating protein (scramblase or flippase) so that their head groups are now exposed to the extracellular environment. The modified external membrane surface becomes "procoagulant"; i.e., it provides a surface onto which the plasma coagulation factors can bind and promote the rapid formation of thrombin.

Platelet Receptors and Platelet Granule Release Reactions in Hemostasis

Several hemorrhagic disease states are identified from defects in platelet membrane receptors: coagulation factor receptors that bridge the primary and secondary hemostatic responses. The platelet receptor for fibrinogen and fibrin, *GpIIb/IIIa*, is defective in Glanzmann's thrombasthenia. Bernard–Soulier syndrome is caused by a defect(s) in *GpIb*, causing impaired binding of von Willebrand factor (vWF). Von Willebrand factor, a circulating plasma protein, is responsible for a common but generally mild bleeding disorder: von Willebrand disease (vWD). The prevalence of defects in von Willebrand factor may be as high as 1%, although not all defects lead to evident disease. Specific granules within the platelet, α-granules, contain coagulation factors that are also found in circulating plasma. Release of these proteins from the granules at the site of platelet adhesion aids the localization of the response to the injury site [1,2].

The dense granules of platelets contain polymers of phosphate [3], novel substances that promote activation of the coagulation system proteins in the same way as membrane surfaces, but as two- rather than three-dimensional surfaces.

Secondary Hemostasis: Clotting Factors and the Coagulation Cascade

Secondary hemostasis, or "blood clotting," refers to the reactions of the blood clotting factors that are circulating in the plasma. These reactions do not require platelets but do require phospholipid bilayer surfaces. Clotting culminates in the transformation of blood from a readily flowing fluid to a gel. The sequence of reactions comprising the coagulation system (procoagulant reactions) was

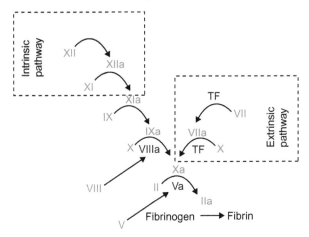

FIGURE 34.1 Classical "cascade" or "waterfall" model for coagulation, 1964. This model represents the ordered sequence of *in vitro* transformations of precursor molecules from their inactive forms to their catalytically active forms. The figure has been corrected to place Factors V and VIII outside the linear sequence; this differs from the original proposals. In all figures, protease precursors are orange, proteases are dark green, cofactor proteins are yellow, and activated cofactor proteins are yellow green. Tissue factor (TF) is shown in light blue.

described as a "cascade" [4] or "waterfall" in 1964 [5]. The essential difference between these two proposals is that in the cascade hypothesis the amplification of the initial stimulus took place at each stage during an ordered sequence of transformations of precursor molecules from their inactive forms to their catalytically active forms, providing a context in which the hemorrhagic deficiencies, which had been primarily identified in patients with bleeding disorders, could be understood [6]. The classical cascade is shown in Figure 34.1.

The "classical" clotting cascade depiction of the procoagulant does not represent the additional properties of the procoagulant subsystem that have been discovered since its initial presentation [7]. Our current understanding of the procoagulant process is more completely represented by Figures 34.2A and 2B. The occurrence of the reactions on the surface of membrane lipids from damaged cells, from activated platelets, or from exogenous sources in laboratory tests is generally recognized. Figure 34.2A illustrates the formation of activation complexes on cell surfaces and provides a basis for the very large increases in the rates of reactions that occur within the activation complexes. The contact component of coagulation is associated with

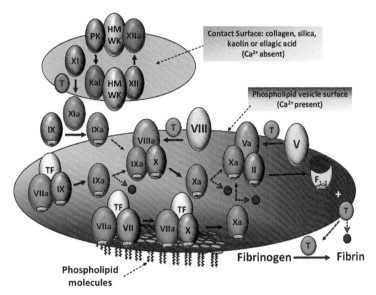

FIGURE 34.2A A contemporary model for the procoagulant subsystem of hemostasis. Transformations of protease precursors to active proteases are organized on the surfaces to which the proteins bind. In each stage of the sequence, a protease precursor, a protease from the preceding complex, and an activated cofactor protein form a noncovalent reaction complex on the membrane surface (shown as a blue "island"). Dissociation of the protease formed in one complex (the product protease) permits that protease to diffuse to the next complex (shown by dashed arrows) and catalyze the transformation of a different precursor protein into an active protease. Thrombin does not bind to the surface of the membrane; it dissociates and diffuses so that it can convert the soluble fibrinogen into fibrin. Vitamin K-dependent proteins are prothrombin and Factors VII, IX, and X (shown with Gla domains near the membrane surface). The unactivated cofactor proteins, tissue factor, and Factors V and VIII are shown in yellow. Red dashed arrows pointing to a red dot indicate dissociation of proteases from the surface and/or reaction complexes—a condition that renders the proteases susceptible to inactivation by protease inhibitors in the blood (see Figure 34.8).

The contact phase reactions (shown on the gray "island") involve two protease precursors, Factor XII and prekallikrein, and one cofactor protein, plasma high molecular weight kininogen (HMWK). Reactions occur on a surface provided from exogenous sources.

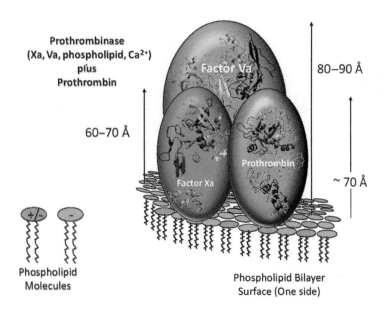

Prothrombinase
(Xa, Va, phospholipid, Ca²⁺)
plus
Prothrombin

60–70 Å

80–90 Å

~ 70 Å

Phospholipid
Molecules

Phospholipid Bilayer
Surface (One side)

FIGURE 34.2B Expanded view of an activation complex, **prothrombinase** with overlaid cartoon structures from X-ray crystallography Prothrombin (**4OO3**), Factor Xa (composite of **2W26** and **1WHE**), and Factor Va (**1SDD**) coordinated from the Protein Data Bank (http://www.rcsb.org). Gla domains are indicated by stick representation of the glutamate residues in crystal structures.

abnormal clotting function tests, but its components are not associated with bleeding disorders. Several lines of evidence implicate the contact factors with thrombosis but by as yet unknown mechanisms.

HEMOSTATIC SYSTEM FACTORS AND PROTEIN STRUCTURE/FUNCTION RELATIONSHIPS

The 20 plasma proteins (clotting factors) that participate in blood clotting can be divided into four categories based on their functions: (1) protease precursors, (2) cofactor proteins, (3) protease inhibitors, and (4) other proteins not included in the first three categories.

Protease Precursors

Protease precursors, also called **zymogens** or **proenzymes**, become catalytically active upon specific proteolytic cleavage. Extensive structural similarities (sequence homologies) are found among all the protease precursors. The regions of the molecules that express protease activity after activation are found in the C-terminal one-half to one-third of each molecule (protease domain). The N-terminal regions provide recognition sites for the cofactor proteins, cell membrane lipid surfaces, and other specific protein interactions.

The amino acid sequences and the three-dimensional structures of the protease domains are homologous to the pancreatic serine proteases, chymotrypsin and trypsin. Primary specificity, i.e., the amino acid residue providing the peptide bond that is hydrolyzed, is trypsin-like; that is recognition is of arginyl and lysyl residues in the protein substrates.

Regulatory sites, distinct from the active site, are present in the coagulation protease domains. These sites, designated

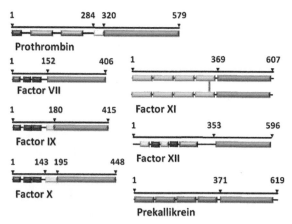

FIGURE 34.3 Bar diagrams for the zymogen (proenzyme) forms of the coagulation Factors II (prothrombin), VII, IX, X, XI, XII, and prekallikrein. Domains and motifs are color-coded: Gla domains (blue), kringles (orange), protease (green), EGF domains (magenta), apple domains (cyan), fibronectin domains (blue-green), and activation peptides (light yellow). The numbers indicate the cleavage sites that transform the precursors into active enzymes. The white bar in prothrombin is an activation peptide that remains covalently attached as the thrombin A chain; the active site-containing chain of thrombin is green.

exosites, are involved in the binding of protein substrates and effector molecules that alter the catalytic efficiency of the proteases and their susceptibility to inactivation (see the section describing the anticoagulant subsystem).

The N-terminal domains of the hemostatic protease precursors and proteases are constructed of several different protein structural motifs. These motifs, in various combinations and by placement in different positions, provide the building blocks of the non-protease-forming regions. Bar diagrams illustrating the structural similarities among the proteases are shown in Figure 34.3. Numbers on the bars are cleavage sites in the polypeptide chain sequence.

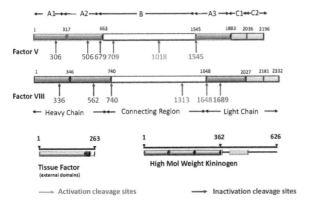

--→ Activation cleavage sites   ——→ Inactivation cleavage sites

FIGURE 34.4 Bar diagrams for the cofactor proteins, Factor V and Factor VIII. Factor V and Factor VIII coagulant (not the von Willebrand factor carrier of Factor VIII) contain six distinct structural domains. The three A domains, A1 and A2 at the N-terminal end of the polypeptide chain, are separated from the A3 domain by a highly glycosylated B domain. The two C domains are at the C-terminal end of the molecule. The A domain sequences are homologous to the A domains of ceruloplasmin. Both Factor Va and Factor VIIIa act as catalysts in the activation of prothrombin and Factor X, respectively. Activation sites are indicated by green arrows; inactivation sites by red arrows. In Factor Va, complete inactivation requires cleavage of Arg^{306}.

Tissue factor is a membrane-spanning protein (only the domains are shown) and high molecular weight kininogen is a cofactor in the activation of prekallikrein and Factor XII in the contact system and is the source of bradykinin (yellow). Domains shown in rose and cyan are not found in the other coagulation proteins.

Cofactor Proteins

Cofactor proteins, after activation, bind both the protease (enzyme) and the proenzyme (substrate). They bind to lipid membranes, either to the surfaces or, in the case of integral proteins (tissue factor and thrombomodulin), span the cell membrane. Cofactor proteins enhance the specificity of the reactions and increase the rate of activation of the proenzymes. Schematic structures for the cofactor proteins Factors V and VIII, are shown in Figure 34.4. Numbers on the bars are cleavage sites.

Recombinant Factor VIII preparations, with structural modifications to alleviate immunogenicity of plasma-derived Factor VIII that is administered therapeutically to hemophiliacs (Factor VIII deficiency), are improving the hemostatic capability of hemophiliacs, without the almost universal development of antibodies to plasma Factor VIII that otherwise occurs too frequently [8,9].

Protease Inhibitors and Protease Inactivation

Protease inhibitors (proteins) that circulate in plasma irreversibly inactivate the activated clotting factors after they have "essentially completed" their actions. With two exceptions—α-2 macroglobulin and tissue factor pathway inhibitor, all of the inhibitors of the procoagulant, anticoagulant, and fibrinolytic proteases are **serpins**, an acronym for *ser*ine *p*rotease

inhibitors, a group of highly homologous protease inhibitors. The irreversible inactivation of the activated factors is achieved by the protease inhibitors acting as suicide substrates; i.e., they react and inactivate (kill) the protease. The most important inhibitor of the procoagulant proteases is antithrombin. The anticoagulant heparin acts by catalyzing protease inactivation by antithrombin and some other serpins.

The anticoagulant system components—Protein C, Protein S, and thrombomodulin—act to produce the activated form of Protein C (Ca). Protein Ca proteolytically inactivates Factors Va and VIIIa.

Other Proteins of the Hemostatic System

The other proteins are fibrinogen; the protransglutaminase (Factor XIII); a metalloprotease (ADAMTS13); and one adhesion/carrier protein (von Willebrand factor). These are unique, structurally diverse proteins.

Fibrinogen

Fibrinogen is the precursor to the spontaneously polymerizing fibrin monomer. Polymerized fibrin comprises the gelatinous clot, and thus, fibrinogen and fibrin are best categorized as structural proteins. Fibrin strands, as noted previously, provide the reinforcement necessary for an adequate hemostatic plug. Fibrin can also act as a "surface" for fibrinolytic system proteins (see Figure 34.5).

Factor XIII (Plasma Protransglutaminase)

Factor XIII, after activation to Factor XIIIa, is a transglutaminase that catalyzes the formation of covalent cross-links in fibrin. In the absence of Factor XIIIa action, the fibrin polymer is unstable and physiologically inadequate. Factor XIII exists in two forms: as a tetramer in plasma, a_2b_2; and as a dimer, a_2, in the platelet granules. Factor XIIIa catalyzes the formation of "isopeptide," ε-(γ-glutamyl) lysine bonds between the ε-amino groups of Lys residues and γ-carboxamido groups of Gln residues of fibrin monomers in polymerized fibrin.

Von Willebrand Factor

Von Willebrand factor (vWF) is a multi-subunit protein that serves both to anchor the platelets to the subendothelial collagen and as a carrier protein for Factor VIII in plasma. The circulating vWF is the largest protein of the hemostatic system. Circulating vWF is made of protomeric units that are ~250 kDa dimers of 125 kDa building-block vWF polypeptide chains. Recognition sites on the von Willebrand factor polypeptide chain enable it to bind to the platelet receptors *GpIb* and *GpIIb/IIIa* and to collagen in the extracellular matrix. Because of the very large size of vWF and the action of ADAMTS13, a protease that cleaves vWF, vWF multimers present in plasma vary in size from 500 kDa to >15,000 kDa. The distribution of

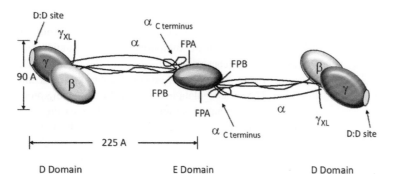

FIGURE 34.5 Fibrinogen and polymerizing fibrin monomers. The fibrinogen molecule is a 340,000 Da protein consisting of three pairs of polypeptide chains: two Aα chains, two Bβ chains, and two γ chains. The A and B designations refer to the two A peptides and two B peptides cleaved from fibrinogen by thrombin to produce the self-polymerizing fibrin monomer (Fn_m), the building block of the fibrin blood clot. The polypeptide chains are linked together by disulfide bridges between CySH residues of the chains.

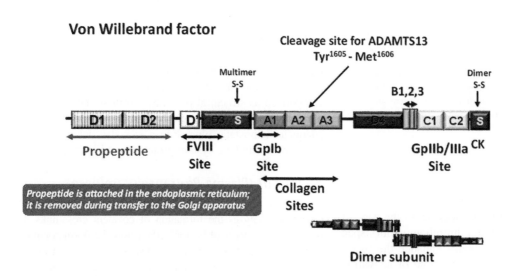

FIGURE 34.6 A bar diagram representation of the von Willebrand factor (vWF). The various domains and the functions that they express are marked on the diagram. The cleavage site by the vWF processing enzyme, ADAMTS13 (Tyr^{1605}-Met^{1606}), is responsible for the formation of the multiplicity of different molecular weight forms of vWF. At bottom right is a diagram showing the cross-linking of protomers to form the dimeric building block of the multimeric vWF in the circulating blood.

sizes leads to varied phenotypic expression of vWD and is used in categorizing the types of vWD (Figure 34.6).

ADAMTS13

ADAMTS13 (A Disintegrin-like Metalloprotease domain with ThromboSpondin type 1 motifs) is a metalloprotease that cleaves the large vWF molecules that "bridge" activated platelets in small platelet aggregates. Excessive action of ADAMTS13 increases the risk of bleeding. Too little action from ADAMTS13, however, as would occur in a hereditary deficiency or in situations in which antibodies to the ADAMTS13 interfere with its action or result in its removal from the circulation, increases the risk of thrombosis [10].

Membrane Phospholipid Surfaces

The membranes of damaged endothelia, subendothelial smooth muscle cells, and activated platelets are

components of the hemostatic system. They provide surfaces onto which the proteases, protease precursors, and cofactor proteins bind. The formation of complexes of protease, protease precursor, and cofactor protein is a thermodynamically favorable reaction and is triggered by the exposure of the inner membrane surfaces upon injury. The inner membrane phospholipids are phosphatidyl serine and other negatively charged phospholipids. The coordinated binding of the coagulation proteins to the phospholipids exposed as a result of cell membrane rupture serves to localize the coagulation reactions at the injury site and, via the platelet membrane changes, with the hemostatic plug.

The Formation of Fibrin: Clotting and the Process of Fluid-to-Gel Transformation

Thrombin's proteolytic cleavage of four peptide bonds in the 6-polypeptide chain fibrinogen molecule converts it to fibrin. This results in transformation of blood from a freely flowing fluid into a gel-like clot. This is the final step of the procoagulant subsystem. Two A fibrinopeptides are removed from the amino-terminal ends of the Aα chains, and two B fibrinopeptides are removed from the amino-terminal ends of the Bβ chains. The cleavage products, fibrin monomers (Fn_m), self-associate to form the "rope-like strands" of fibrin that create the gelatinous clot.

Cross-links formed by the action of Factor XIIIa create the stable form of fibrin. Multiple cross-links are formed among α-chains of several different fibrin monomers. This creates a molecular species-designated α-polymer. Two adjacent γ-chains from two different fibrin monomers also are cross-linked to form a species-designated γ−γ dimer. Fragments of fibrin after action by plasmin are used as diagnostic markers of thrombosis. Noncovalently associated fibrin is physiologically unsatisfactory because the dissociation of the fibrin results in recurrent bleeding.

Fibrinogen is an acute phase reactant,[1] and thus, its concentration is substantially increased in several clinical situations, particularly those associated with inflammation. When the fibrinogen concentration is increased, the action of thrombin on fibrinogen is faster—a consequence of greater binding of fibrinogen to thrombin (see Chapter 6).

REACTIONS OF THE COAGULATION CASCADE: THE PROCOAGULANT SUBSYSTEM

Rapid, localized proenzyme activation required for adequate hemostatic response occurs only in a complex of protease, protease precursor, and cofactor protein assembled on the surface of a platelet or a damaged cell membrane. With protease and proenzyme alone, the reactions are 10^5 times slower. Expressed in terms of the same amount of product formed in the two situations, a 10^5 decrease represents the difference between 1 minute and about 6 months to form the same amount of product! The reaction complexes of the coagulation cascade [7,11] are shown in Figure 34.2A; the complex involved in prothrombin activation (prothrombinase) is shown in Figure 34.2B.

The first stage in the formation of complex is a reversible, noncovalent association of protease, cofactor protein (strictly, activated cofactor protein), a protease precursor, and a membrane surface to form the activation complex. Second, irreversible proteolytic action in the complex converts the protease precursor into an active protease, which then diffuses away to the subsequent complex where it participates in the next step in the cascade (Figure 34.2A).

Third, after the activation phase, the residual protease dissociates from the complex and is inactivated by the protease inhibitors present in the blood; proteases in the activation complexes are protected from inactivation. Fourth, complete dissociation is assured by proteolytic inactivation of the activated cofactor proteins (Va and VIIIa) by activated Protein C. In fact, the inactivation of Va and VIIIa precludes reassembly of the complexes. In the initial stages of the secondary hemostatic response, the first two "steps" predominate. As the hemostatic plug becomes consolidated by fibrin reinforcement, the third and fourth steps predominate.

The Final, Common Pathway of Clotting: Activation of Prothrombin by Prothrombinase

The activation of prothrombin to thrombin and the conversion of fibrinogen to fibrin are frequently described as the final, common pathway of blood coagulation (Figure 34.2A). This designation, the result of the convergence of the "intrinsic" and "extrinsic" pathways with the formation of Factor Xa, is common in coagulation literature. An expanded picture of the prothrombin activation complex, prothrombinase, is shown in Figure 34.2B; the protein secondary structures within the ellipsoidal cartoons are from actual three-dimensional structures of the prothrombinase components. After proteolytic cleavage, Fragment 1-2 retains the Gla domain and the two kringles of the prothrombin molecule (see Figure 34.3); however, thrombin is "freed" to diffuse to act on fibrinogen. Although there is an explosive increase in the reaction rate[2] (>100,000 times), only a small fraction (<<1%) of the total prothrombin available (~1.5 μM) is required to be converted to thrombin to transform fibrinogen to fibrin.

An interesting mutation in the gene for prothrombin, a G-to-A transition in the 3′ untranslated region at nucleotide 20210, results in an elevated concentration of prothrombin in the circulation (>115% of average concentration in individuals without bleeding disorders) [12]. It is not known how the mutation causes the elevated prothrombin levels, but the defect is associated with a twofold increase in the risk of thrombosis.

Initiating Coagulation: The Extrinsic Pathway, Injury, and Tissue Factor Exposure

The reactions of the coagulation system that are initiated *in vitro* by the addition of tissue homogenates

---

1. Acute phase reactants are plasma proteins that undergo large increases in synthesis and thus become concentrated in plasma in response to acute inflammation caused by surgery, myocardial infarction, and infections.

2. Amplification in a sequence of enzyme-catalyzed reactions might be interpreted to imply that more of the product is formed at each stage than in the previous stage. However, in the coagulation "cascade," the principal amplification is of reaction rate, not quantity of product formed.

(thromboplastin) are grouped under the title "extrinsic pathway." The reaction pathway that is followed *without* tissue homogenate being added was designated the "intrinsic pathway" because it was assumed that no extrinsic components were necessary.[3] The two pathways are not independent, as is evident by the common components, but the terms continue to be used both because of their historical roots and because the commonly performed *in vitro* coagulation tests can be related to them. See Figures 34.1 and 34.2A.

Initiation of the Procoagulant Subsystem and Activation of Factor VII

Initiation of the hemostatic response upon traumatic injury occurs at the site of injury and involves a component, tissue factor, present in the subendothelium. The first complex to form is almost certainly the "extrinsic" Factor X activation complex, tissue factor, Factor VIIa, and Factor X (Figures 34.1 and 34.2A). Factor VIIa binds to tissue factor, an integral membrane protein, when it is exposed upon injury of the blood vessel wall. The phospholipid bilayer surface is provided by the damaged cell membranes. *In vivo*, Ca^{2+} is present in the blood. *In vitro*, Ca^{2+} is added only because the anticoagulant solutions used for blood collection employ substances such as citrate ion, EDTA, or oxalate to bind Ca^{2+} to prevent activation complex formation.

A recombinant factor, VIIa, has been very successful in treating hemophiliacs with factor VIII antibodies and patients with other serious bleeding episodes [13].

Activation of Factor X by Factor VII(VIIa)

Factor X is activated by a single peptide bond cleavage, $Arg^{51}-Ile^{52}$ in the heavy chain of the pro-enzyme, whereas the light chain and its Gla domain remain attached and thus promote association with the cell membrane phospholipids.

The Intrinsic Pathway: *In Vitro* versus *In Vivo* Processes

Activation of Factor X

Factor IXa, Factor VIIIa (after activation of Factor VIII by thrombin), and the phospholipid surface comprise the "intrinsic pathway" activator of Factor X. Factor VIII, as described previously, is activated prior to its becoming functional in the activation of Factor X. Circulating Factor VIII is bound to von Willebrand factor as a carrier and as a

stabilizer. Because of the binding of vWF to collagen in the subendothelium and to the receptor *GpIb* on platelets, Factor VIII is preferentially associated with the hemostatic plug at the injury site. This association provides an additional mechanism distinct from Factor VIII binding to the membrane lipids for localization. Some patients who appear to be Factor VIII deficient (having classical hemophilia) actually possess defective von Willebrand factor and can be treated by increasing the quantity of vWF rather than of Factor VIII. An explanation for the initial origin of Factor VIIa remains ambiguous.

The Xa is structurally and functionally the same whether formed via the extrinsic or intrinsic pathway activation complexes.

Activation of Factor IX by Factor VIIa and Tissue Factor

Factor IX can be activated by the complex containing Factor VIIa, tissue factor, phospholipid bilayer surface, and Ca^{2+} ions. This "crossover" reaction for Factor IX activation, although perhaps not strictly a reaction of the "intrinsic pathway," contributes about half of the Factor IXa that is formed in situations in which tissue factor is present. Factor IX is activated as the result of the cleavage of two peptide bonds, Arg^{145} and $Arg^{180}-Ile^{181}$, to form a two-chain enzyme. The chains are linked by a disulfide bond, and thus, the Gla domain remains associated with the protease domain and maintains phospholipid association capability.

Activation of Factor IX by Factor XIa

In the classical reaction of the "intrinsic pathway," Factor IXa is formed from Factor IX by the action of Factor XIa (the protease). The *in vivo* "surface" for this reaction is believed to be a glycosaminoglycan, a sugar polymer to which Factor IX and Factor XIa can bind. In this situation, the "surface" is actually a two-dimensional polymer that acts like a chain between the molecules and on which they can migrate toward or away from each other. Reaction between Factor XIa and IX on two-dimensional surfaces occurs more rapidly than if the proteins were free in solution.

The Contact Phase of the *In Vitro* Intrinsic Pathway of Coagulation

Figure 34.2A shows that two other reactions exist (reactions associated with the gray surface); these reactions are not part of *in vivo* hemostasis but are likely linked to thrombosis. These reactions, the "contact phase" or "contact system," are considered important only for *in vitro* clotting. Individuals with deficiencies in the components of the "contact phase"

---

3. It is now known that the glass test tube provides a surface upon which proteins bind and initiate clot formation in a series of reactions that do not require Ca^{2+}.

do not suffer from bleeding. (Note that thrombotic plugs are less stable in the absence of Factor XII.)

The contact phase reactions are important for *in vitro* coagulation function testing; they underlie the classical "activated partial thromboplastin time" test. This test is so named because it initially was believed to involve nothing extrinsic to blood plasma; the glass vessel in which the blood was contained was unrecognized as a participant in the reactions. Historically, the partial thromboplastin was considered to be generated from plasma proteins alone and augmented by phospholipid, a platelet substitute. Factor XIIa, however, is suspected to be involved in hemostasis, although enigmatically because it can activate Factor VII *in vitro* in plasma cooled for storage. Analogous to the reactions of the clearly physiologically important components of blood clotting, high molecular weight kininogen acts as a cofactor protein in the activation of prekallikrein by Factor XIIa and, in a reciprocal fashion, the activation of Factor XII by kallikrein. High molecular weight kininogen is a source of bradykinin, a highly potent vasoconstrictor.

The recent identification of the components of contact activation in the risk for thrombosis is yet to be firmly established but is a provocative observation that may clarify some of the "mysteries" of blood clotting in hemostasis and as agents leading to thrombosis [14].

SHUTTING OFF COAGULATION: THE ANTICOAGULANT SUBSYSTEM

The procoagulant subsystem of blood clotting consists of a series of irreversible chemical reactions. Peptide bond cleavages convert the precursors of the coagulation proteases and cofactor proteins into catalytically active forms. Regulation of coagulation, unsurprisingly, consists of opposing irreversible chemical reactions, additional proteolysis to inactivate the cofactor proteins, and irreversible reaction with inhibitors to inactivate the procoagulant proteases [15].

Activation of Protein C and Inactivation of Factors Va and VIIIa

The cofactor protein-enhanced protease activations of the procoagulant subsystem are opposed by the inactivation of the cofactor protein. The very large increases in the rates of activation of prothrombin and Factor X, in the presence of Factors Va and VIIIa, respectively, make the hemostatic response rapid. However, if such rates were to continue unabated, extension of the hemostatic plug into the blood vessel could occlude the vessel and result in ischemia and death to the adjacent cells and tissues. If the pathologically extended hemostatic plug is formed in the venous system

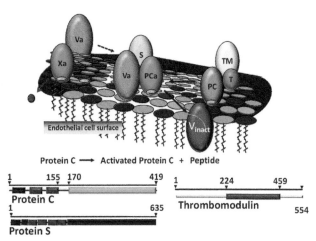

FIGURE 34.7 Anticoagulant subsystem activation of Protein C occurs adjacent to the injury site; inactivation occurs on the exposed surface at the injury site. Protein C is activated through cleavage by thrombin at Arg[169]–Ile[170]. Protein S is vitamin K-dependent protein that functions as a cofactor protein. Thrombomodulin is an integral membrane protein. Red dashed arrows pointing to a red dot indicate dissociation of proteases from the surface and/or reaction complexes—a condition that renders the proteases susceptible to inactivation by protease inhibitors in the blood (see Figure 34.8). See Figure 34.3 for motif and domain color coding.

(red thrombus) and is freed from the vessel wall, the clot can move to the lungs, creating a pulmonary embolus [16].

In the anticoagulant subsystem, Protein C is activated by thrombin in a complex with thrombomodulin to produce activated Protein C, the protease that inactivates Factors Va and VIIIa [17]. The binding of thrombin to thrombomodulin transforms thrombin to a protease that efficiently cleaves Protein C. Whereas the hemostatic reactions that prevent blood loss at the injury site are associated with the ruptured blood vessels, thrombomodulin is on the endothelium, where it is appropriately located to prevent clot expansion and vessel occlusion (Figure 34.7).

Risk for thrombosis arises when inactivation of Factors Va and VIIIa is impaired. In the phenomenon called **activated Protein C resistance**, a mutation in the Factor V gene, G to A at nucleotide 1691, results in the replacement of the normal Arg residue at position 506 in the heavy chain of Factor Va by a Gln residue. Individuals carrying this mutation, called **Factor V (Leiden)** [18], are at increased risk of venous thrombosis and venous thromboembolism. The poorer ability of activated Protein C to cleave Gln[506] also slows the cleavage of Arg[306] in Factor Va. A mutation at Arg[306], Factor V (Cambridge), also confers activated Protein C resistance [19] (see also Figure 34.4).

The Factor V (Leiden) mutation is very common in individuals of Western European origin. The prevalence is approximately 5% in Caucasians in the Western Hemisphere; the mutation is very rare in Africans and is absent in Asians [20].

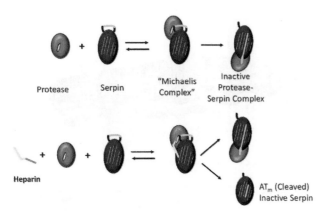

FIGURE 34.8 Protease inactivation by serpins. Protease inactivation occurs by reaction between protease and inhibitor, e.g., antithrombin. The protease is a stoichiometric reactant in this instance; it is not a catalyst. The protease is indicated by green, the inhibitor by red, and the inactivated protease by gray. Stripes on the inhibitor represent the β-sheets.

Irreversible Inactivation of Proteases: Protease Inhibitors

Protease inactivation occurs by stoichiometric reaction between a protease and an inhibitor, which results in the formation of a "covalent" ester bond between the reactive site residue of the inhibitor and the active site serine residue of the protease [21]. The name given these inhibitors, serpins (see Figure 34.8), reflects their reaction with serine proteases. The proteases, thrombin, Factor Xa, and Factor IXa are all inactivated by antithrombin, while Factor VIIa, Factor Xia, and Factor XIIa are inhibited at a much lower rate. Other serpins can inactivate procoagulant proteases; heparin cofactor II can inactivate thrombin; and α-1 protease inhibitor can inactivate Factor Xa. An altered α-1 protease inhibitor (α-1 protease inhibitor Pittsburgh), in which an Arg has replaced Ala at the inhibitor reactive site, is a good inhibitor of thrombin. The serpin inhibitor C1 inactivates Factor XIIa.

Because inactivation of Factors Va and VIIIa by activated Protein C promotes the dissociation of the proteases, cofactor protein inactivation facilitates protease inactivation.

Mechanism of Action of Heparin as a Therapeutic Anticoagulant

The inactivation of procoagulant proteases can be catalyzed by glycosaminoglycans, which are sulfated polysaccharide molecules found on the surface of the normal endothelial cells and in the basophilic granules of mast cells [22].

Heparin is a polymer of repeating disaccharide "building blocks." All heparins bind to antithrombin; however, heparin molecules that contain a unique pentasaccharide sequence (high-affinity heparins) bind with particularly high affinity and cause a conformation change in antithrombin that makes it react more rapidly with the target protease. The magnitude of the heparin-catalyzed increase in the rates of protease inactivation by antithrombin depends on the molecular weight of the high-affinity heparin molecules. The higher the heparin molecular weight, the greater the rate of protease inactivation; an upper limit occurs at a molecular weight of ∼20,000. Thrombin inactivation requires heparin molecules of sufficient molecular weight to permit the formation of a "tethered" thrombin−heparin−antithrombin complex. Factor Xa inactivation is much less sensitive to the molecular weight of the heparin molecules and is the only protease that can be efficiently inactivated by the very low molecular weight high-affinity pentasaccharide (Figure 34.9).

Inhibitors of the Contact Phase Proteases

Several protease inhibitors inactivate the proteases of the contact phase. Among the serpins are C-1 inactivator, α-1 protease inhibitor, and antithrombin. The target proteases for these inhibitors are Factor XIIa, kallikrein, and Factor XIa. The molecular mechanisms are the same as those described for the procoagulant and fibrinolytic system proteases (Figure 34.8).

Another inhibitor present in plasma, α-2 macroglobulin, inhibits by a completely different mechanism from the serpins. It entraps the proteases within a "cavity" that is created by the four subunits of the α-2 M molecule, preventing the protease acting on a protein substrate. The protease active site is not blocked; the entrapped protease still cleaves low molecular weight substrates.

CLOT REMOVAL: THE FIBRINOLYTIC SUBSYSTEM

Clot Dissolution: Fibrinolysis

The fibrinolytic subsystem degrades the fibrin in the hemostatic plug by proteolytic digestion. The system is activated as a result of the formation of the fibrin itself by the association of plasminogen and plasminogen activators with the fibrin. Fibrinolysis is responsible for the temporary nature of the fibrin clot, which is removed after wound repair has occurred. Recently, recombinant plasminogen activator, t-PA, has become available and is a highly effective therapeutic biopharmaceutical for treating thrombosis [23].

Plasminogen Activation

Two different proteases are responsible for the physiological activation of plasminogen: tissue-type

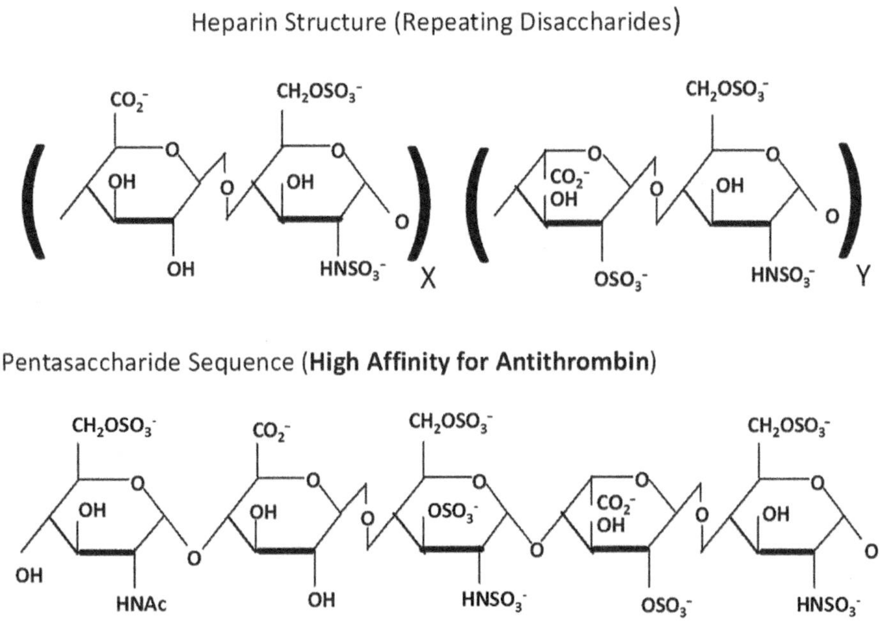

Heparin Structure (Repeating Disaccharides)

FIGURE 34.9 Structure of heparin. Heparin is a polymer of repeating disaccharide units that contain one uronic acid and one hexosamine residue. The uronic acid residues may be either glucuronic acid or iduronic acid, monosaccharide acids that differ in their stereochemistry. The hexosamine residue is exclusively glucosamine. Both uronic acid and glucosamine residues can be modified by O and N sulfation and the glucosamine residue by N acetylation.

Pentasaccharide Sequence (High Affinity for Antithrombin)

plasminogen activator (t-PA) and urinary plasminogen activator (u-PA). Tissue-type PA is the principal activator of plasminogen and is synthesized primarily in endothelial cells [24].

Activation of plasminogen occurs as a result of the binding of both plasminogen and t-PA to fibrin. This results in a nearly 400 times decrease in the K_m for the proteolytic cleavage of plasminogen by t-PA. Plasminogen binds to fibrin via lysyl residues exposed on fibrin fibrils and as a result of plasmin proteolysis of Lys-X peptide bonds, which increases the number of exposed Lys residues to which the plasminogen and t-PA can bind and further promotes activation. The activation of plasminogen by t-PA and u-PA is opposed by two protein inhibitors: plasminogen activator inhibitor 1, or PAI-1, and plasminogen activator inhibitor 2, or PAI-2 (see Figure 34.10). PAI-1 inactivates t-PA very rapidly but does so less rapidly when the t-PA is bound to fibrin. Higher than normal concentrations of PAI-1 have been associated with thrombosis, indicating the necessity for balance between t-PA activation of plasminogen and t-PA inactivation by PAI-1.

A plasma procarboxypeptidase (procarboxypeptidase U) or thrombin-activatable fibrinolysis inhibitor (TAFI) after activation removes the lysyl residues from the fibrin degradation products produced by plasmin. This reaction reduces the enhanced fibrin-dependent plasminogen activation by t-PA and also results in the loss of "protection" of t-PA that is bound to fibrin from inactivation by PAI-1.

Plasminogen can also be activated rapidly by the formation of a "stoichiometric" complex between the plasminogen molecule and proteins from several strains of

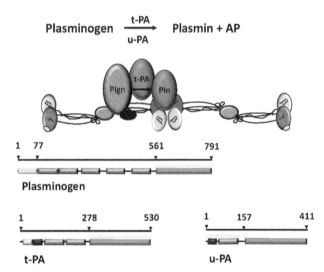

FIGURE 34.10 Plasminogen activation. Plasminogen is converted to plasmin as the result of cleaving a single peptide bond, $Arg^{561}-Val^{562}$. Two different proteases are responsible for the physiological activation of plasminogen: tissue-type plasminogen activator (t-PA) and urinary plasminogen activator (u-PA). Motifs and domains are color-coded as shown in Figure 34.3.

hemolytic *Streptococcus* and *Staphylococcus*. The "streptokinase-plasmin" complex (SK-plasmin) converts plasminogen to plasmin and also digests fibrin by itself. In contrast to normal plasmin, "streptokinase-plasmin" is not inhibited by α-2 antiplasmin. Streptokinase–plasmin (ogen) is used therapeutically in fibrinolytic therapy to remove thrombi, but it is very immunogenic and hence the preference for recombinant t-PA for fibrinolysis [25].

Fibrin degradation products

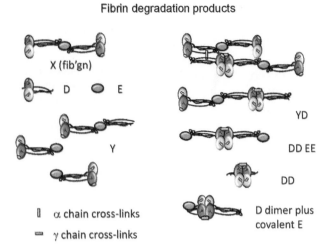

⫾ α chain cross-links

▬ γ chain cross-links

FIGURE 34.11 Products of fibrinolysis. Products of plasmin action are used to identify ongoing fibrinolysis and to distinguish fibrinolysis from fibrinogenolysis. The presence of larger fibrin fragments that result from Factor XIIIa catalyzed cross-linking is the basis for the distinction between fibrinogenolysis and fibrinolysis. Detection of D-dimer (DD) is used in detecting intravascular thrombi.

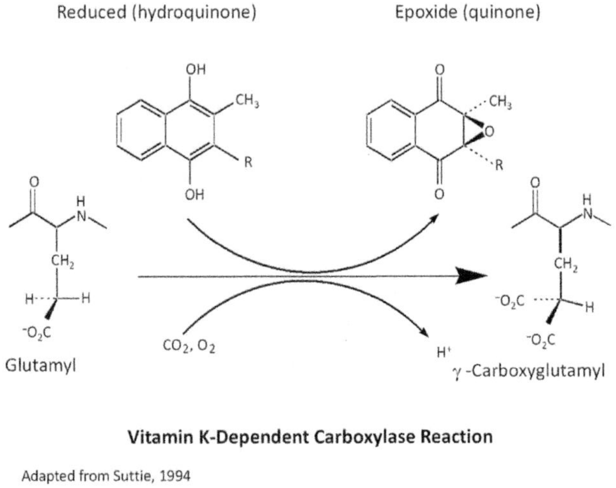

Vitamin K-Dependent Carboxylase Reaction

Adapted from Suttie, 1994

FIGURE 34.12 Vitamin K action in Glu carboxylation. Formation of Gla from Glu requires abstraction of an H ion from the γ position of Glu in the proteins that is followed by the addition of a carboxyl group from CO_2. The proteins to be carboxylated are marked by a propeptide that serves as a signal to the vitamin K-dependent carboxylase.

Degradation of Fibrin (Products of Fibrinolysis)

The principal substrate of plasmin is fibrin, and except under pathological situations, e.g., disseminated intravascular coagulation (DIC), the concentrations of the products of plasmin action are usually less than 10 μg/mL (less than 0.5% of the circulating fibrinogen). Because of the opportunities for artifactual generation of fibrin degradation products (FDPs) during the collection of the blood sample and the common use of serum, the true concentrations of FDPs in normal plasma are probably <50 ng/mL. Fibrin degradation products are illustrated in Figure 34.11. D-dimer is a marker of fibrinolysis of venous blood clots. (Fibrin degradation products also inhibit the polymerization of fibrin.)

In disseminated intravascular coagulation (DIC), fibrinogen can be digested by the fibrinolytic enzyme plasmin. This leads to bleeding, even though clotting is also occurring.

VITAMIN K, ORAL ANTICOAGULANTS, AND THEIR MECHANISMS OF ACTION

Vitamin K acts in the liver as an enzyme cofactor in the reactions that add an additional carboxyl group to particular Glu residues to form Gla (γ-carboxyglutamic acid) in the vitamin K-related coagulation factors. Vitamin K is a substituted naphthoquinone. A methyl group at the 2 position and a prenyl (or phytyl) side chain of varying length are required for the vitamin to be active. The two primary sources of vitamin K in humans are green plants and intestinal flora. The active form of the vitamin is the reduced hydroquinone form, not the quinone form found in the diet. In the vitamin K-dependent carboxylase reaction, the vitamin is converted from the hydroquinone to an epoxide (Figure 34.12). This stereospecific reaction requires O_2 and the hydroquinone for the formation of a Glu anion. Reaction of the Glu anion with CO_2 leads to the formation of the γ-carboxy Glu.

In the absence of vitamin K, or in the presence of antagonists of vitamin K action, uncarboxylated vitamin K-dependent proteins are synthesized and secreted from the liver [26]. These proteins do not bind Ca^{2+} normally and do not bind to the surface of the phospholipids in the cell membranes and importantly have physiologically significant coagulant activity.

The physiological expression of oral anticoagulant action is an increase in the time required for clotting in the prothrombin time assay. The slowing of all of the reactions that lead to the formation of thrombin is the direct result of the reduced concentrations of the γ-carboxylated vitamin K-related proteins in the reaction complexes on the membrane surface. The effects of oral anticoagulant blockage on the carboxylation reaction are common to all vitamin K-related proteins.

Action of Warfarin and other Vitamin K Antagonists

Vitamin K antagonists were first identified by K. P. Link while he was investigating a hemorrhagic disease in cattle. The compound responsible for the hemorrhagic disease, a

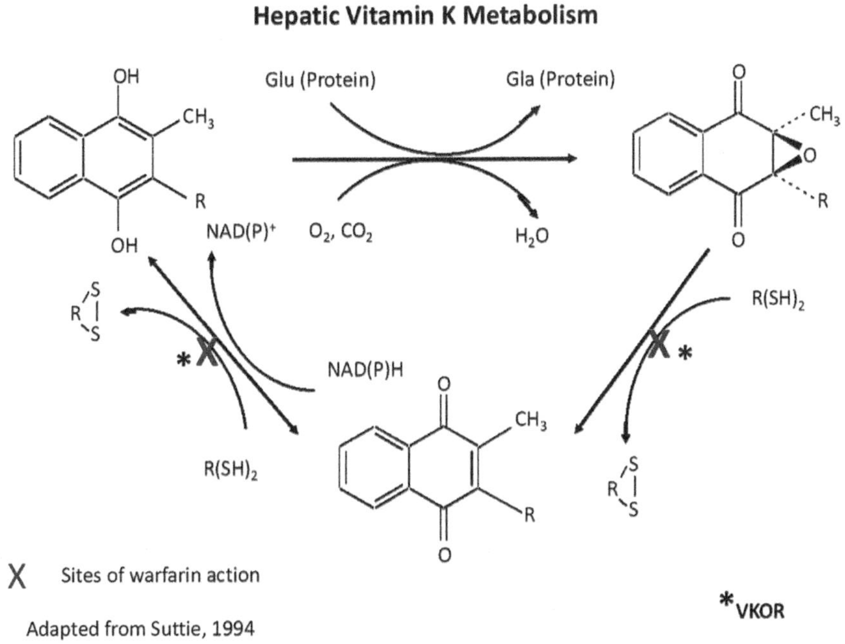

Hepatic Vitamin K Metabolism

X Sites of warfarin action

Adapted from Suttie, 1994

*VKOR

FIGURE 34.13 Hepatic vitamin K metabolism. Oral anticoagulant drugs act indirectly on the process of Glu carboxylation of the vitamin K-dependent proteins. The vitamin K antagonists block the reduction of the reaction intermediate, vitamin K epoxide, that results in the accumulation of vitamin K epoxide and other nonfunctional forms of vitamin K. Without cycling of the vitamin K-related reaction intermediates in the cycle shown, a depletion of functional vitamin K occurs.

product of microbial metabolism (spoiling) of sweet clover, was identified as 3,3'-methyl-bis-4-(hydroxycoumarin), dicoumarol. A synthetic, clinically useful compound with the ability to function like dicoumarol is warfarin [1-(4'-hydroxy-3'-coumarinyl)-1-phenyl-3-butanone]. This compound is sold in the United States under the name Coumadin™. Other vitamin K antagonists are used widely in Europe.

The oral anticoagulant drugs act *indirectly* on the process of Glu carboxylation of the vitamin K-dependent proteins in the liver [27]. The vitamin K antagonists prevent the reduction of a reaction intermediate, vitamin K−epoxide. This is illustrated in Figure 34.13.

Vitamin K antagonists are processed in the liver by the cytochrome P-450 (CYP) systems that metabolize many other drugs. Drug interactions commonly alter the effectiveness of oral anticoagulant drugs. This requires that oral anticoagulant therapy be monitored frequently during the early stages of therapy and at regular intervals after stable anticoagulation; a balance is achieved between unwanted hemorrhage and inadequately reduced risk of thrombosis. Warfarin therapy must be monitored biweekly to monthly and is the principal use for the prothrombin time test. Dietary influences, e.g., high consumption of vitamin K-rich foods such as leafy green vegetables, can alter the efficacy of vitamin K antagonists. Although highly effective, warfarin is problematic for patients because of the narrow range in which risk of clotting and risk of bleeding is in balance. Warfarin is the "top ranked" drug on the USP list of adverse drug reactions.

New anticoagulants—active site-directed inhibitors of thrombin and Factor Xa—are now available for prevention of thrombus formation in atrial fibrillation [28−32]. These new antithrombotic drugs (e.g., dabigatran, rivaroxaban, apixaban, and others being developed) are intended to replace warfarin. However, whereas the effect of warfarin can be reversed by administration of vitamin K if excessive bleeding occurs, agents to reverse the effects of the new direct inhibitors are not yet available.

HEMOSTATIC SYSTEM DYSFUNCTION: BLEEDING AND THROMBOSIS

Bleeding disorders can be both hereditary and acquired. Laboratory diagnosis of bleeding risk is identified by a prolonged time for a clot to form in an *in vitro* clotting test. Prolonged clotting times can be described as indicating a factor deficiency,[4] with more complex tests employed to identify the specific factor. Classical hemophilia (due to a deficiency of FVIII, hemophilia A) and FIX deficiency (hemophilia B) are the most common

---

4. Deficiency is generally understood to mean a diminished ability to participate in the clotting cascade as evidenced by a prolonged time to clot relative to the normal clotting time. Normal is defined as the time for a sample of a plasma pool to lead to clot formation in one of the general clotting function tests (prothrombin time, PT, and activated partial thromboplastin time, aPTT) or in specific clotting factor tests. It may be the result of a decrease in concentration of a clotting protein or a mutation that results in the clotting factor being less efficient in the reactions of the clotting cascade.

bleeding disorders. Defects in vWF can result in impaired hemostasis; however, because vWF involves interaction with collagen *in vivo* and with platelets, vWF deficiencies are not reflected in clotting time measurements. Other procoagulant factor deficiencies are rare [33].

Thrombosis, an unwanted hemostatic response when there is no threat of hemorrhage from an injury, is categorized as arterial or venous. Whereas a hemostatic plug forms *in vivo*, in response to a blood vessel injury and risk of hemorrhage, when rupture of an atheromatous plaque occurs in an artery wall, particularly a coronary artery, it provides a nidus for a thrombus [34], an unwanted "hemostatic plug" called a **white thrombus**. Infarction, death to the surrounding tissues because of oxygen starvation, can occur; or pieces of the thrombus may be "torn" away and moved to other organs, e.g., the brain, to cause a thrombotic stroke [35,36].

Venous thrombosis results when a clot forms in a vein (red thrombus comprising fibrin and entrapped red blood cells) in regions of stasis (very low blood flow), such as at valves in veins. If a venous thrombus dislodges and is carried to the lungs, it produces a pulmonary embolus, which is an occlusion of blood flow within the lungs that can be fatal.

Dysfunction of the hemostatic system [37,38] that is not hereditary is generally *not* attributable to a single cause. Deficiencies in antithrombin and the mutated FV Leiden are the two most prevalent deficiencies that increase the risk of thrombosis. Heterozygous individuals—for example, with half of the normal amount of antithrombin or of Protein C—do not necessarily present with thrombosis, nor do individuals with half the amount of a procoagulant factor show an increased risk of bleeding. Rather a combination of defects in more than one component is required or injuries of such large magnitude that the entire hemostatic system is overwhelmed. Each defect can cumulatively increase the risk of thrombosis at some time during an individual's lifetime. Thrombosis is a principal risk of death, increasing with age, dramatically so for smokers even though the damage from smoking is not directly related to coagulation.

LABORATORY ASSESSMENT OF COAGULATION SYSTEM FUNCTIONS

Sample collection and sample handling are particularly important in coagulation factor testing [39,40]. The venipuncture, or capillary puncture for some devices, is a blood vessel injury that elicits a hemostatic response. When a tourniquet is employed to facilitate venipuncture,

which is almost universally the case, the tourniquet should be in place for the shortest time possible. Plasminogen activator (t-PA) is released from the vessel in response to the compression and activates plasminogen. Badly collected samples from extensive tourniquet compression can be rendered unclottable because of the fibrinogenolysis that occurs from plasmin action. Normally, all abnormal coagulation system functional tests should be reconfirmed.

Prothrombin Time

The prothrombin time (PT), as proposed by Quick, is the most commonly performed coagulation function test. It is used to monitor oral anticoagulant therapy and also as a preoperative screening test to warn of possible bleeding risk in patients with a personal or family history of bleeding. Measured clotting times are extremely dependent on the animal and tissue source and the quality of the thromboplastin used. Variability can be expected because of the assay's dependence on the number of tissue factor molecules and the quantity of membrane surface provided by the thromboplastin. To improve the comparability of prothrombin time measurements, thromboplastins are standardized by comparison with an international reference thromboplastin. Standardized thromboplastins are described by a number, the International Standardization Index (ISI), that relates the thromboplastin to the international standard. Although this effort has improved comparisons of oral anticoagulant therapy monitoring, it is still difficult to accurately compare the results of different laboratories and different thromboplastin manufacturers.

Further improvement in measuring oral anticoagulation therapy has been achieved by relating the ratio of the prothrombin time of the patient to the mean value for "normal" individuals who are not on oral anticoagulant therapy. This ratio, called the **International Normalized Ratio (INR)** [41], better relates the anticoagulation assays from different laboratories. Monitoring oral anticoagulant therapy is critical to maintaining a balance between the risk of thrombosis and the risk of bleeding.

Activated Partial Thromboplastin Time

Intrinsic pathway components are measured in the activated partial thromboplastin time (APTT). The designation "activated" refers to the performance of the test in two stages. In the first stage, the citrate-anticoagulated plasma is preincubated with the surface activator (see "The Contact Phase of the *In Vitro* Intrinsic Pathway of Coagulation, Figures 34.1 and 34.2A) to form activated

Factor XI. In the second stage, Ca^{2+} is added to initiate the activation of Factor IX and the remaining reactions of the "intrinsic" pathway. The clotting time is determined from the time of addition of Ca^{2+}. The most common uses of the APTT are screening for possible Factor VIII or Factor IX deficiencies and for monitoring heparin therapy. Because the rate of activation of Factor IX and the clotting time are very dependent on the Factor XIa, the APTT is sensitive to deficiencies in Factor XII, prekallikrein, high molecular weight kininogen, and Factor XI.

The APTT can be shorter than normal. This suggests a defect in Factor V, called Factor V (Leiden), in which mutations result in slower cleavage of Factor Va by activated Protein C. A modified version of the APTT that employs Factor V-deficient plasma in the second stage of the procedure is used to indicate the possibility of this mutation; confirmation is by a nucleic acid sequence-based test.

Thrombin Time

Purified thrombin is added to plasma samples, and the time for clotting is measured. This test, the thrombin time (TT), primarily reflects the concentration of fibrinogen. However, it also reflects the ability of the fibrinopeptides to be cleaved and the polymerization of fibrinogen. Separation of these three contributions requires a quantitative measurement of the concentration of fibrinogen that is not related to time for clot formation. The thrombin time can be prolonged if heparin is present in the patient's plasma sample because it will promote preferential inactivation of the added thrombin by antithrombin and reduce the amount of thrombin that can act on fibrinogen.

Specific Factor Assays

Specific factor assays are variations on the APTT or PT tests. In the APTT and PT, dilutions of the patient's plasma are made into hereditarily deficient or depleted "substrate" plasma. The assays are then performed in the usual way. The clotting times are compared with those obtained from dilutions of pooled normal plasma, commonly 1 to 10, 1 to 20, 1 to 50, and 1 to 100. A graph of the logarithm of the clotting time (Y-axis) versus the logarithm of the concentration as percentage of normal (X-axis) is used to determine the amount of the factor activity in the patient's plasma. The normal pooled plasma is conventionally assigned a value of 100% activity. Many variations exist for specific factor assays, e.g., the venom of Russell's viper (*Daboia russelli*, previously known as *Vipera russelli*), and phospholipids may be substituted for

thromboplastin in a PT-like assay. An enzyme in Russell's viper venom rapidly and relatively specifically activates Factor X. In conjunction with Factor X-deficient "substrate" plasma, this provides a specific Factor X assay.

Immunoglobulin inhibitors to Factor VIII and Factor IX are found in patients with hemophilia A and B, respectively, as a consequence of their treatment with Factor VIII and Factor IX preparations to arrest bleeding. Evidence for inhibitors can be seen from APTT assays using specific factor-deficient substrate plasmas. This evidence is obtained by making dilutions of the patient's plasma and observing that higher dilutions give higher levels of activity for the factor in the patient's plasma. This reflects the simultaneous dilution of the antibody inhibitor and the decreased inhibition at the higher dilution.

Assay of Heparin

Heparin therapy may be monitored by its increases in the clotting time in the APTT although this method for measuring heparin is difficult to standardize. Heparin is more specifically assayed by its effect on Factor Xa inactivation by antithrombin. Such Factor Xa-based heparin assays usually employ purified Factor Xa as a reagent and Factor X-deficient substrate plasma as the source of antithrombin. The prolongation of the clotting time that results from the heparin in the patient's plasma is compared with pooled normal plasma that is known to be free of heparin. Many variations of this heparin assay are available. Heparin assays can use thrombin rather than Factor Xa; however, the low molecular weight heparins are not reliably measureable in thrombin-based assays.

Inadequacy of assays can be exploited; an example of relevance to heparin was adulteration that resulted in the death of more than 100 recipients of the adulterated product [42].

Thrombin Generation

Automated methods for measurement of total thrombin formation are now available; a variation of the thrombin generation test (TGT) monitors thrombin formed using specific fluorogenic substrates for thrombin. This procedure shows sensitivity to the concentrations of procoagulant factors and protease inhibitors including heparin, and the action of activated Protein C; and it provides information about the clotting reactions not available for elapsed time clotting tests [43]. Tables 31.1–31.4 provide more information.

TABLE 34.1 Platelet Substructures

Subcellular Component	Contents	Function(s) and Disorders
Alpha granules		Large (200–500 nm), granule (~40 per platelet)
		Disorders: α-storage pool disease (gray platelet syndrome)
	Fibrinogen	Platelet source of fibrinogen, released upon platelet activation
	wWF	Platelet source of von Willebrand factor, released upon platelet activation
	PGDF	Platelet-derived growth factor; widely used growth factor
	Thrombospondin(s)	Multifunctional group of proteins initially identified from platelets
	P-selectin	Cell adhesion molecule, found in platelets, on endothelial and other cells
	Platelet factor 4 (PF4)	Cationic (positively charged) protein that binds to heparin to neutralize its ability to catalyze protease inactivation
	β-Thromboglobulin	Chemokine with multiple activities, e.g., megakaryocyte maturation
	Platelet factor 3 (PF3)	Broad, (test method) defined term for platelet procoagulant activity. Disorders: Scott syndrome (phospholipid scramblase deficiency?) Stormorken syndrome (amino phospholipid translocase defect?)
	Factor V	Procoagulant, after activation supports local prothrombin activation
Dense (δ) granules		Small, dense, by electron microscopy; sparse (<10 per platelet)
	Serotonin	Disorder: δ-storage pool disease
	ADP, ATP	Adenosine diphosphate, adenosine triphosphate
	Ca, Mg	Divalent cation mediators of Gla protein interactions with phospholipids
	Polyphosphates	Polymers of phosphate that act as two-dimensional surfaces in hemostasis
Microtubular system	Actin/myosin	Cytoskeleton components, responsible for platelet retraction
Dense tubular system	Cyclooxygenase (COX1)	Generation of thromboxane A2, site of action of NSAIDs, aspirin
Open canalicular system		Provide access for granule contents to platelet exterior

TABLE 34.2 Platelet Components and Component Functions[†]

Common/Group Name	Structure, Function(s), and Disorders
Integrins[‡]	
GpIb-IX-V	Membrane-spanning glycoprotein subunits: GpIbα—binds vWF(via A1 region), binds integrin αMβ2, P-selectin, thrombin (via thrombin exosites) to activate PAR1, interacts with platelet cytoskeletal proteins; GpIbβ—binds cytoskeletal proteins; GpIX—binds vWF; GpV—binds collagen
Subunit stoichiometry: (2GpIbα, 2GpIbβ, 2GpIX, GpV), GpV = (integrin α5β3)	Disorders: Bernard–Soulier syndrome (heterozygous carrier asymptomatic); mutations in GpIbα are hypersensitive to vWF
GpIc-IIa, (GpIIa)	Fibronectin receptor
GpIa-IIa-VI (integrin α2β1)	Collagen receptor
GpIIb/IIIa	αIIbβ3 binds vWF (via C1 region), principal receptor for fibrinogen (cross-links platelets within the plug)
GpIIb = integrin αIIbβ3)	Disorder: Glanzmann's thrombasthenia (heterozygous carrier asymptomatic)
	Antagonists: abciximab (Reopro™), eptifibatide (Integrelin™), tirofiban (Aggrastat™), clopidogrel (Plavix™)
Integrin αvβ3	Receptor for vitronectin

(Continued)

TABLE 34.2 (Continued)

Common/Group Name	Structure, Function(s), and Disorders
GpVI	Binds to exposed collagen, upregulates integrin synthesis
P-selectin (PADGEM), GMP (granule membrane protein) 140	Granule membrane receptor, *after* platelet activation, P-interacts with vWF on endothelium, (recruits) monocytes, neutrophils to hemostatic plug
SNARE Proteins	
	Mediate granule-membrane interactions, fusion of granule membranes with cell membrane
Other Adhesion Molecules	
PECAM-1	Involved in cell migration; function is not exclusively related to platelets; may inhibit platelet collagen interaction
platelet/endothelial cell adhesion molecule	
Protease-Activated Receptors (PARs)	
PAR1	Activation of platelets; receptor is activated by cleavage by low concentrations of thrombin; exposes the ligand, the "tethered" peptide; activation signal transmitted via G protein
PAR4	Responds only to high concentrations of thrombin, adjunct to PAR1?
Platelet Small Molecule Receptors, Agonists, and Antagonists	
P_2Y_{12}	ADP receptor, Agonist: ADP, induces aggregation, Antagonist—clopidogrel (Plavix) looks like ADP receptor deficiency
PIP_2, PA	Phospholipids; act in initiating secretion by granules; via arachidonic acid and prostaglandin (thromboxane A2)
	Antagonist: aspirin
Epinephrine	Platelet-aggregating agent (used for *in vitro* testing)

[†]*Because of the multi-subunit nature of the receptors, several subunits are found in receptors that are associated with different ligands.*
[‡]*Group name is based on the property of mediating a connection between cells and the extracellular matrix (ECM). Integrins are noncovalent dimers of α- and β-subunits. There are more than 18 different α-subunits and 8 β-subunits. Varying combinations of these subunits give rise to the different integrins. Only a few of the integrins are found on platelet membranes.*
Normal platelet count (normal range) $150-450 \times 10^9 \, L^{-1}$, lifespan 8–10 days.

TABLE 34.3 Components of the Hemostatic System: Procoagulant Subsystem Components: Plasma Coagulation Factors

Factor Number[i]	Common Name[ii]	Function(s)
I	Fibrinogen	Precursor to fibrin monomer, associates to form fibrin polymers, i.e., clot
II[iii]	**Prothrombin**	Precursor to thrombin
III[iv]	Thromboplastin	Tissue homogenate that contains tissue factor and phospholipids; products of recombinant tissue factor and synthetic lipids are also available
III[iv]	**Tissue factor (TF)**	Transmembrane cofactor protein exposed upon tissue injury; interacts with F VIIa to initiate clotting
IV	Ca^{2+}	Required for vitamin K protein to bind to phospholipids, active conformations in other proteins
V	**Proaccelerin**	Cofactor protein that, after activation, catalyzes conversion of prothrombin to thrombin by F Xa in prothrombinase
	Accelerator globulin (Ac-G)	
VI	Abandoned[v]	(*Discovered to be the activated form of F V*)

(*Continued*)

TABLE 34.3 (Continued)

Factor Number[i]	Common Name[ii]	Function(s)
VII	**Proconvertin**, plasma thromboplastin component (PTC)	Precursor to protease (VIIa) that, with tissue factor, converts F X to F Xa; protease responsible for initiation of the clotting process via the "extrinsic pathway"
VIII	Antihemophilic factor (AHF), antihemophilic globulin (AHG)	Cofactor protein that, after activation, catalyzes conversion of F X to F Xa by F IXa in complex; sometimes called tenase
IX	Christmas factor, antihemophilic factor B	Precursor to protease (IXa) that, with F VIIIa, converts F X to F Xa; in the clotting process via the "intrinsic pathway"
X	Stuart–Prower factor (autoprothrombin III)	Precursor to protease (Xa) that, with F Va, converts prothrombin to thrombin; a reaction of the "final common pathway" of coagulation
XI	Plasma thromboplastin antecedent (PTA)	Precursor to protease (XIa) that converts F IX to F IXa; a reaction of the "intrinsic pathway" of coagulation
XII	Hageman factor	Precursor to protease (XIIa) that, with HMWK, converts prekallikrein to kallikrein *in vitro*; a reaction of the "intrinsic pathway" of coagulation; converts F VII to F VIIa in refrigerated plasma.
XIII	Fibrin stabilizing factor, plasma transglutaminase	Precursor of plasma transglutaminase that catalyzes the formation of cross-links between fibrin monomers in fibrin to create a stable fibrin meshwork around platelets; activation is catalyzed by thrombin
	Laki–Lorand factor, fibrinoligase	
	Prekallikrein, Fletcher factor	Precursor to kallikrein; *in vitro* converts F XII to F XIIa and F XI to F XIa; both activation reactions are catalyzed by HMWK
	High mol wt kininogen	A protein cofactor that enhances the activation of F XII to F XIIa and F XI to F XIa
	Fitzgerald, Flaujac, Williams factor	
	von Willebrand factor	Adhesive protein that anchors platelets to collagen exposed upon tissue injury
	ADAMTS13 (*A d*isintegrin and *m*etalloprotease with *t*hrombospondin domains)	Metalloprotease, cleaves very large vWF to produce lower molecular weight polymers

Anticoagulant Subsystem Components

Common Name[ii]	Function in Anticoagulant Subsystem
Cofactor Protein Inactivation	
Protein C (PC)	Precursor to protease, PCa (APC) that cleaves F Va and F VIIIa to inactivate them
Protein S (PS)	Cofactor protein that catalyzes inactivation of F Va and F VIIIa by APC
Thrombomodulin (CD141)	Cofactor protein that catalyzes PC activation by thrombin
Thrombin	Protease cleaving PC to product APC
Protein C inhibitor	Plasma inhibitor (SERPIN) that inactivates APC
Proteinase Inactivation	
Antithrombin (antithrombin III)	Plasma protein inhibitor (serpin) that inactivates thrombin, F Xa, F IXa, F VIIa (F XIa and F XIIa, but inefficiently)—reaction is catalyzed by heparin, which contains a *unique* pentasaccharide sequence
Heparin cofactor II (Leuserpin)	Plasma protein inhibitor (serpin) that inactivates thrombin—reaction catalyzed by heparin, dermatan sulfate, and other polyanions
Tissue factor pathway inhibitor (TFPI, LACI, EPI)	Plasma protein inhibitor that modulates activation of F X by F VIIa/TF (Kunitz type inhibitor)
α-1 Proteinase inhibitor (α-1 PI) α-1 Antitrypsin	Plasma protein inhibitor (serpin) that inactivates thrombin, F Xa and other proteases; primary target protease is leukocyte elastase
α-2 Macroglobulin (α-2 M)	Plasma protein inhibitor that captures proteases and prevents protease action on protein substrates; low molecular weight synthetic substrates are unaffected

(Continued)

TABLE 34.3 (Continued)

Fibrinolytic System Components	
Common Name[ii]	Function in Fibrinolysis

Fibrinolytic System Components	
Common Name[ii]	Function in Fibrinolysis

Proteinase Precursors	
Plasminogen	Precursor of the fibrinolytic protease plasmin that digests the fibrin after wound healing is underway
Tissue plasminogen activator (**t-PA**)	Protease from the endothelium that converts plasminogen to plasmin
Urinary plasminogen activator (**scu-PA**)	Protease that converts plasminogen to plasmin
Proteinase Inhibitors	
α-2 Antiplasmin (α-2 AP)	Plasma protein inhibitor that rapidly inactivates plasmin that has dissociated from fibrin
α-2 Plasmin inhibitor	
α-2 Macroglobulin (α-2 M)	Plasma protein inhibitor that captures proteases, including plasmin and prevents protease action on protein substrates; low molecular weight synthetic substrates are unaffected
Plasminogen activator inhibitor 1 (PAI-1)	Plasma protein inhibitor that inactivates t-PA
Plasminogen activator inhibitor 2 (PAI-2)	Plasma protein inhibitor that inactivates t-PA
Plasminogen activator inhibitor 3 (PAI-3), Protein C inhibitor	Plasma protein inhibitor that inactivates t-PA; preferentially inactivates APC

[i]Plasma coagulation factors are all designated by Roman numerals, platelet factors by Arabic numerals.
[ii]Common names may be of primarily historical significance. The factor nomenclature was devised to reduce the number of synonyms for the same component which had made description of the process of coagulation unnecessarily awkward.
[iii]Color (orange) indicates that the factor is a protease precursor or a cofactor protein and, when the Roman numeral is marked (magenta), a vitamin K-dependent protein; serpin protease inhibitors (red).
[iv]Factor III is used as a synonym for thromboplastin, the mixture of tissue factor and phospholipid.
[v]Discovered to be the activated form of Factor V after the numeral had been assigned.

TABLE 34.4 Anticoagulant Subsystem Components

Common Name	Function in Anticoagulant Subsystem
Cofactor Protein Inactivation	
Protein C (PC)	Precursor to protease, PCa (APC) that cleaves F Va and F VIIIa to inactivate them
Protein S (PS)	Cofactor protein that catalyzes inactivation of F Va and F VIIIa by APC
Thrombomodulin (CD141)	Cofactor protein that catalyzes PC activation by thrombin
Thrombin	Protease cleaving PC to product APC
Protein C inhibitor	Plasma inhibitor (serpin) that inactivates APC
Proteinase Inactivation	
Antithrombin (antithrombin III)	Plasma protein inhibitor (serpin) that inactivates thrombin, F Xa, F IXa, F VIIa (F XIa and F XIIa, but inefficiently)—reaction is catalyzed by heparin, containing a *unique* pentasaccharide sequence
Heparin cofactor II (leuserpin)	Plasma protein inhibitor (serpin) that inactivates thrombin—reaction catalyzed by heparin, dermatan sulfate, and other polyanions
Tissue factor pathway inhibitor (TFPI, LACI, EPI)	Plasma protein inhibitor that modulates activation of F X by F VIIa/TF (Kunitz type inhibitor)
α-1 Proteinase inhibitor (α-1 PI) α-1 Antitrypsin	Plasma protein inhibitor (serpin) that inactivates thrombin, F Xa, and other proteases; primary target protease is leukocyte elastase
α-2 Macroglobulin (α-2 M)	Plasma protein inhibitor that captures proteases and prevents protease action on protein substrates; low molecular weight synthetic substrates are unaffected

CLINICAL CASE STUDY 34.1 Platelet Defects

Platelet defects are principally associated with primary hemostasis and are commonly identified from platelet counts, platelet histological examination, and/or specialized receptor binding measurements. The bleeding time is a measure of primary hemostasis, and is primarily sensitive to platelet count.

Coagulation Factor Deficiencies

Coagulation factor deficiencies (defects) principally affect secondary hemostasis and are most commonly reflected in abnormal results in clotting time-based assays. Assay methods for assessing the function of the hemostatic system are designed to emphasize the particular aspects, primary or secondary, of pathways. *In vivo*, the convenience of such separation does not exist. Estrogen therapy has been linked to increased risk for thrombosis. There are no changes in components of the procoagulant, anticoagulant, or fibrinolytic system that can be unambiguously linked to an increase in risk of thromboembolic disease sometimes associated with estrogen therapy.

A combined deficiency of both Factors V and VIII has long perplexed investigators. The gene for FV is autosomal (chromosome 1q 23) but for Factor VIII is sex-linked (X-chromosome). Both cofactor proteins are found in reduced amounts, but the proteins that were present were fully active. Two genes that code for proteins that are required for the packaging and transporting of Factors V and VIII from the endoplasmic reticulum to the Golgi apparatus are actually defective, and the combined deficiency is related to transport, not actual function of the molecules.

The most prevalent mutations in antithrombin result in weakened binding of heparin. Higher concentrations (doses) of heparin used as a therapeutic agent would thus be indicated for such patients. Current laboratory methods do not commonly identify this situation, however. Chemical equilibrium is the basis for dosage adjustment in patients with heparin-binding mutations.

Many mutations have been identified in fibrinogen. Classic examples are (1) fibrinogen Detroit, in which Arg 19 is replaced by Ser in the A chain, resulting in an abnormally slow rate of cleavage by thrombin and fibrinogen, even though this is not the site of cleavage; and (2) fibrinogen Paris, in which cross-linking of fibrin is impaired. Differences in protein conformation not only explain allostery, but also explain reactivity differences caused by genetic mutations that are not necessarily the reactive amino acid residue.

A mutation in Factor XIII (Leu34Val) has been suggested to reduce risk for venous thrombosis. The decreased extent of cross-linking of fibrin provides the biochemical explanation.

Mutations in ADAMTS13 can lead to excessively large aggregates of vWF-bridged platelet thrombi that can cause microvascular occlusion. More dramatically, antibodies to ADAMTS13 are responsible for thrombotic thrombocytopenic purpura (TTP) and idiopathic thrombocytopenic purpura (ITP). Quaternary structure differences can be caused by post-processing defects, not in the principal protein, but in an enzyme that processes it.

Antibody formation occurs in some patients with hemophilia A (F VIII deficiency), thus rendering administration of F VIII ineffective. These IgG inhibitors of F VIII can be bypassed by administering large amounts of recombinant activated F VII (rFVIIa) Novoseven™. Branched pathways provide an opportunity for compensation for defects that are not solved by simply providing more F VIII because the F VIII is recognized as a foreign protein and antibody formation to it occurs.

Recombinant t-PA (Ateplase™) is used to digest thrombi in coronary arteries; it is arguably superior to other thrombolytic agents because of the mechanism that localizes t-PA to fibrin that does not exist with other plasminogen activators such as streptokinase.

A deficiency of all vitamin K-dependent proteins exists that is related to defects in the gene for VKOR, the vitamin K epoxide reductase (oxidoreductase) gene.

Some individuals possess a variant of cytochrome P-450, CYP2C9, that makes them very sensitive to warfarin. To reduce the risk of over-anticoagulation, the FDA has approved a genetic test to identify this variant prior to beginning oral anticoagulant therapy. Expression of genetic defects is indirect and illustrates the challenge of extrapolating from gene defects to phenotype defects.

Common rodenticides are vitamin K antagonists and are frequently labeled as "superwarfarins" because they have very long half-life. A common rodenticide, brodifacoum (3-(3-(4'-bromobiphenyl-4-yl)-1,2,3,4-tetrahydro-1-naphthyl)-4-hydroxy-coumarin), can be lethal to humans as well as many other animals. The same mechanism of action is insufficient to fully evaluate the effects of particular drugs; pharmacological properties that are related the steady-state concentrations of the drug may be more important.

Laboratory Diagnosis
Vignette 1: Fibrinogen
Testing on a plasma sample from a patient gave a PT of 12.9 s (normal 10.5 s–12.2 s) and an APTT of 38 s (normal 20 s–32 s). A TT was determined and the clotting time was 15 s (normal 10 s–12 s). The patient's plasma sample was incubated with an enzyme that digests heparin. After this treatment, the TT was 14.7 s. The concentration of fibrinogen was determined by a method that weighs the fibrin after clotting. The fibrinogen was 340 mg/dL, a value higher than the normal range. What might explain these results?

Teaching Point
Fibrinogen proteolysis or fibrin monomer polymerization may be impaired as a result of a mutation that alters the fibrinogen molecule.

Vignette 2: Factor XIII
A 60-year-old female is referred because of bleeding that has been a problem since infancy. Childbirth (multiple) and surgery have required treatment with cryoprecipitate and other

(Continued)

CLINICAL CASE STUDY 34.1 (Continued)

plasma protein fractions that contain fibrinogen, von Willebrand factor, Factor VIII, Factor XIII, and many other plasma proteins. All ordinary coagulation functions tests were determined and found to be normal. Her platelet count was measured and found to be at the high end of the normal range. The addition of plasmin to the plasma clot resulted in almost immediate dissolution (lysis) in contrast to ~10 minutes for a control clot formed from pooled normal plasma. Plasmin was not detectable in a freshly collected plasma sample from the patient. After addition of Ca^{2+} and incubation for an hour, the patient's plasma clot was mixed with an equal volume of 8 M urea, and the clot dissolved. A plasma clot prepared in the same manner from normal plasma did not dissolve upon addition of 8 M urea. What is the likely cause of the patient's bleeding disorder and her laboratory findings?

Teaching Point
Factor XIII deficiency: there is no expected effect on any of the clotting time measurements. The susceptibility to plasmin suggests an unstabilized clot. The solubility in urea is diagnostic for un-crosslinked fibrin.

Vignette 3: Activated Protein C Resistance
A 30-year-old male with recurrent thromboses is referred for diagnosis. The PT and APTT results determined on his freshly drawn plasma sample are 10.1 s (normal 10 s−12 s) and 18 s (normal 20 s−32 s). Which components of the hemostatic system would you initially suspect as being responsible for the thrombosis and the test results? Measurements of antithrombin in the presence and absence of heparin are normal. What would you consider candidate factors to explain the thrombosis? If both Protein C and Protein S were normal, what would you suspect? If assays for the inactivation of Factor V by activated Protein C were normal, which of the known factors in the anticoagulant subsystem would you ascribe his thrombosis to?

Teaching Point
The components are all mentioned in Vignette 3. Because of the current attention on APC resistance, almost certainly this test would be performed first. Antithrombin is indicated as being measured first because of its simplicity compared to Protein C and Protein S measurements. If there is no APC resistance, thrombomodulin defects might be considered.

Vignette 4: Antibiotic-Induced Vitamin K Deficiency
A 4-year-old female was brought to the "urgent care" department of your hospital after a minor automobile accident in which glass shards had caused some cuts. Continued bleeding from the cuts was noted from a pile of blood-spotted tissues collected during her 3-hour wait in the emergency room! What questions would you ask her and/or her parents about the bleeding? If the parents indicated that she had been receiving high doses of antibiotics for an infection during the last few weeks, what coagulation test would you expect to be abnormal?

Teaching Point
A history of bleeding by the girl, parents, or other relatives should be explored. If there were none, given the patient's age and any information about infectious diseases in the community, use of antibiotics might be asked. The PT could be prolonged if the girl had used antibiotics. This is the result of the antibiotics killing the intestinal bacteria that produce vitamin K from dietary sources of phylloquinones.

Vignette 5: Hemophilia
A young African American male patient is brought to you because of a hemarthrosis sustained after twisting his knee while running. His PT is normal, but his APTT is 51 s (normal 20 s−32 s). Mixing equal volumes of the patient's plasma and normal plasma shortens the APTT to 25 s. Specific factor assays show a normal Factor IX activity, but a Factor VIII activity of ~ 25% of that in pooled normal plasma. Why was his PT normal? What else, prior to the Factor IX and VIII assay results, might have been responsible for the prolonged APTT?

Teaching Point
Factor VIII is not involved in the PT, i.e., "extrinsic" pathway. Before the normal Factor IX result was obtained, Factor XII deficiency might cause a prolonged APTT. Because the patient was not under treatment for anything, heparin would be unlikely.

Vignette 6: Von Willebrand Disease
A 7-year-old male is brought to you because of recurrent nose bleeds and bleeding after minor injuries associated with ordinary child's "horse play." There is no history of any type of bleeding disorder in the four brothers in the mother's family; she is the only girl. The PT and the APTT tests were performed, and both were at the low end (shorter clotting times) of the normal ranges. A platelet count was made and found to be normal. What hemostatic system components might be responsible for the bleeding but not be detectable in the two tests for which results had been obtained? What test would you suggest to complete the diagnosis?

Teaching Point
Von Willebrand factor and Factor XIII should be tested. Because Factor XIII is easier, it might be done first.

Von Willebrand's Disease and vWF Multimer Patterns
An example of the pattern of multimers of vWF is shown in Figure 34.14. Differences in the distribution of the various multimers account for many of the manifestations of von Willebrand's disease [1,2].

The two references cited are important for understanding the heterogeneity and diagnosis of bleeding tendencies associated with vWF and ADAMST13.

Vignette 7: α-1 Antitrypsin
A 10-year-old boy had a severe lifelong hemorrhagic disorder that had necessitated more than 50 hospitalizations.

(Continued)

CLINICAL CASE STUDY 34.1 (Continued)

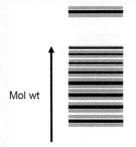

Von Willebrand's Disease and vWF multimer patterns

FIGURE 34.14 Severity of von Willebrand's disease in some cases is related to the distribution of molecular sizes of the multimeric vWF and their relative amounts. In this schematic representation of the banding pattern on gel electrophoresis of vWF in plasma, darker, broader bands indicate amount of the variant.

Laboratory examination showed prolonged bleeding, clotting, partial thromboplastin, prothrombin, and thrombin times. These findings were due to a potent inhibitor of the thrombin–fibrinogen reaction. This inhibitor was similar to heparin in that it acted immediately and did not interfere with the coagulant activities of certain venoms. It differed from heparin in not being adsorbed to barium citrate or neutralized by protamine sulfate. The inhibitory effect was found in the α-1 globulin fraction. It was identified immunologically and functionally as a double-bonded α-1 antitrypsin of a previously unreported phenotype. The inhibitory effects were depressed by trypsin and heterologous anti-α-1 antitrypsin [3].

Teaching Points

This example is noted in the text under "Irreversible Inactivation of Proteases: Protease Inhibitors." The paper presents a classical molecular diagnosis of a bleeding disorder due to a single mutation that results in a gain of function/specificity.

Vignette 8: Laboratory-Created Artifacts

A plasma sample was obtained from a female patient for routine coagulation testing; i.e., there was no history of bleeding. The PT was 11.9 s (normal 10.5 s–12.2 s). The APTT was 80 s (normal 20 s–32 s). A sample was sent to a reference laboratory specializing in coagulation factor testing. At the reference laboratory, the PT was 11.6 s and the APTT was 43 s. Because of this discrepancy, blood was drawn at the reference laboratory and a new plasma sample prepared. In a second set of tests performed at the reference laboratory, the PT was 10.9 s and the APTT was 94 s. What might account for the discrepancy between the results obtained on the original plasma sample in the two laboratories?

Teaching Point

If the initial sample were not centrifuged appropriately to remove the platelets, platelet fragmentation and microparticle formation could have occurred during transport to the reference laboratory. The membrane fragments could artifactually shorten the APTT.

Supplemental References

[1] J.E. Sadler, U. Budde, J.C.J. Eikenboom, E.J. Favaloro, F.G.H. Hill, L. Holmberg, et al., Update on the pathophysiology and classification of von Willebrand disease: a report of the Subcommittee on von Willebrand Factor, J. Thromb. Haemost. 4(10) (2006) 2103–2014

[2] W.L. Nichols, M.E. Rick, T.L. Ortel, R.R. Montgomery, J.E. Sadler, B.P. Yawn, et al., Clinical and laboratory diagnosis of von Willebrand disease: a synopsis of the 2008 NHLBI/NIH guidelines, Am. J. Hematol. 84(6) (2009) 366–370.

[3] J.H. Lewis, R.M. Iammarino, J.A. Spero, U. Hasiba, Antithrombin Pittsburgh: an alpha1–antitrypsin variant causing hemorrhagic disease, Blood 51(1) (1978) 129–137.

REQUIRED READING

[1] J. Yip, Y. Shen, M.C. Berndt, R.K. Andrews, Primary platelet adhesion receptors, IUBMB Life 57(2) (2005) 103–108.

[2] D.M. Goodman, A.E. Burke, E.H. Livingston, JAMA patient page. Bleeding disorders, JAMA 308(14) (2012) 1492.

[3] J.H. Morrissey, Polyphosphate multi-tasks, J. Thromb. Haemost. 10(11) (2012) 2313–2314.

[4] R.G. Macfarlane, An enzyme cascade in the blood clotting mechanism, and its function as a biochemical amplifier, Nature 202 (1964) 498–499.

[5] E.W. Davie, O.D. Ratnoff, Waterfall sequence for intrinsic blood clotting, Science 145(3638) (1964) 1310–1312.

[6] M.P. Esnouf, R.G. Macfarlane, Enzymology and the blood clotting mechanism, Adv. Enzymol. Related Areas Mol. Biol 30 (1968) 255–315.

[7] C.M. Jackson, Y. Nemerson, Blood coagulation, Annu. Rev. Biochem. 49 (1980) 765–811.

[8] D.W. Scott, K.P. Pratt, C.H. Miao, Progress toward inducing immunologic tolerance to factor VIII, Blood 121(22) (2013) 4449–4456.

[9] A.D. Shapiro, Long-lasting recombinant factor VIII proteins for hemophilia A, Hematology Am. Soc. Hematol. Educ. Program (2013) 37–43.

[10] A. Arning, M. Hiersche, A. Witten, G. Kurlemann, K. Kurnik, D. Manner, et al., A genome-wide association study identifies a gene network of ADAMTS genes in the predisposition to pediatric stroke, Blood 120(26) (2012) 5231–5236.

[11] J.C. Whisstock, Assembling the machinery of coagulation, Blood 122(16) (2013) 2773–2774.

[12] R.F. Franco, M.D. Trip, H. ten Cate, A. van den Ende, M.H. Prins, J.J. Kastelein, et al., The 20210 G-->A mutation in the 3'-untranslated region of the prothrombin gene and the risk for arterial thrombotic disease, Br. J. Haematol. 104(1) (1999) 50–54.

[13] U. Hedner, Recombinant coagulation factor VIIa: from the concept to clinical application in hemophilia treatment in 2000, Semin. Thromb. Hemost. 26(4) (2000) 363–366.

[14] G.J. Miller, M.P. Esnouf, A.I. Burgess, J.A. Cooper, J.P. Mitchell, Risk of coronary heart disease and activation of factor XII in middle-aged men, Arterioscl. Thromb. Vasc. Biol. 17(10) (1997) 2103–2106.

[15] I. Bjork, S.T. Olson, Antithrombin. A bloody important serpin, Adv. Exp. Med. Biol. 425 (1997) 17–33.

[16] C.T. Esmon, N.L. Esmon, B.F. Le Bonniec, A.E. Johnson, Protein C activation, Methods Enzymol. 222 (1993) 359–385.

[17] C.T. Esmon, W.G. Owen, The discovery of thrombomodulin, J. Thromb. Haemost. 2(2) (2004) 209–213.

[18] R.M. Bertina, P.H. Reitsma, F.R. Rosendaal, J.P. Vandenbroucke, Resistance to activated protein C and factor V Leiden as risk factors for venous thrombosis, Thromb. Haemost. 74(1) (1995) 449–453.

[19] D. Williamson, K. Brown, R. Luddington, C. Baglin, T. Baglin, Factor V Cambridge: a new mutation (Arg306–>Thr) associated with resistance to activated protein C, Blood 91(4) (1998) 1140–1144.

[20] T. Koster, F.R. Rosendaal, H. de Ronde, E. Briet, J.P. Vandenbroucke, R.M. Bertina, Venous thrombosis due to poor anticoagulant response to activated protein C: Leiden Thrombophilia Study [see comments], Lancet 342(8886–8887) (1993) 1503–1506.

[21] I. Bjork, S.T. Olson, J.D. Shore, Molecular mechanisms of the accelerating effect of heparin on the reactions between antithrombin and clotting proteinases, In: D.A. Lane, U. Lindahl (Eds.), Heparin: chemical and biological properties, clinical applications, CRC Press, Inc., Boca Raton, 1989, pp. 229–255.

[22] M.F. Scully, V. Ellis, V.V. Kakkar, Localization of heparin in mast cells, Lancet 2(8509) (1986) 718–719.

[23] H.R. Lijnen, D. Collen, Interaction of plasminogen activators and inhibitors with plasminogen and fibrin, Semin. Thromb. Hemost. 8(1) (1982) 2–10.

[24] M. Hanss, D. Collen, Secretion of tissue-type plasminogen activator and plasminogen activator inhibitor by cultured human endothelial cells: modulation by thrombin, endotoxin, and histamine, J. Lab. Clin. Med. 109(1) (1987) 97–104.

[25] M. Verstraete, R.W. Brower, D. Collen, A.J. Dunning, J. Lubsen, P.L. Michel, et al., Double-blind randomised trial of intravenous tissue-type plasminogen activator versus placebo in acute myocardial infarction, Lancet 2(8462) (1985) 965–969.

[26] C.M. Jackson, J.W. Suttie, Recent developments in understanding the mechanism of vitamin K and vitamin K-antagonist drug action and the consequences of vitamin K action in blood coagulation, Prog. Hematol. 10 (1977) 333–359.

[27] J.-K. Tie, D.-Y. Jin, D.L. Straight, D.W. Stafford, Functional study of the vitamin K cycle in mammalian cells, Blood 117(10) (2011) 2967–2974.

[28] M.M. Samama, The mechanism of action of rivaroxaban—an oral, direct Factor Xa inhibitor—compared with other anticoagulants, Thromb. Res. 127(6) (2011) 497–504.

[29] A. Banerjee, D.A. Lane, C. Torp-Pedersen, G.Y. Lip, Net clinical benefit of new oral anticoagulants (dabigatran, rivaroxaban, apixaban) versus no treatment in a 'real world' atrial fibrillation population: a modelling analysis based on a nationwide cohort study, Thromb. Haemost. 107(3) (2012) 584–589.

[30] A. Gomez-Outes, A.I. Terleira-Fernandez, M.L. Suarez-Gea, E. Vargas-Castrillon, Dabigatran, rivaroxaban, or apixaban versus enoxaparin for thromboprophylaxis after total hip or knee replacement: systematic review, meta-analysis, and indirect treatment comparisons, BMJ 344 (2012) e3675.

[31] S.Z. Goldhaber, A. Leizorovicz, A.K. Kakkar, S.K. Haas, G. Merli, R.M. Knabb, et al., Apixaban versus enoxaparin for thromboprophylaxis in medically ill patients, N. Engl. J. Med. 365 (23) (2011) 2167–2177.

[32] L.A. Castellucci, C. Cameron, G. Le Gal, M.A. Rodger, D. Coyle, P.S. Wells, et al., Efficacy and safety outcomes of oral anticoagulants and antiplatelet drugs in the secondary prevention of venous thromboembolism: systematic review and network meta-analysis, BMJ 347 (2013) f5133.

[33] M. Greaves, H.G. Watson, Approach to the diagnosis and management of mild bleeding disorders, J. Thromb. Haemost. 5(Suppl 1) (2007) 167–174.

[34] Z.S. Kaplan, S.P. Jackson, The role of platelets in atherothrombosis, Hematology Am. Soc. Hematol. Educ. Program (2011) 51–61.

[35] S.M. Bates, I.A. Greer, S. Middeldorp, D. Veenstra, A.-M. Prabulos, P.O. Vandvik, et al., VTE, thrombophilia, antithrombotic therapy, and pregnancy: Antithrombotic therapy and prevention of thrombosis, 9th ed.: American College of Chest Physicians Evidence-Based Clinical Practice Guidelines, Chest 141(2 Suppl) (2012) e691S–736S.

[36] S.M. Bates, R. Jaeschke, S.M. Stevens, S. Goodacre, P.S. Wells, M.D. Stevenson, et al., Diagnosis of DVT: antithrombotic therapy and prevention of thrombosis, 9th ed.: American College of Chest Physicians Evidence-Based Clinical Practice Guidelines, Chest 141(2 Suppl) (2012) e351S–418S.

[37] B.J. Hunt, Bleeding and coagulopathies in critical care, N. Engl. J. Med. 370(22) (2014) 2153.

[38] B.J. Hunt, Bleeding and coagulopathies in critical care, N. Engl. J. Med. 370(9) (2014) 847–859.

[39] A.H. Kamal, A. Tefferi, R.K. Pruthi, How to interpret and pursue an abnormal prothrombin time, activated partial thromboplastin time, and bleeding time in adults, Mayo. Clin. Proc. 82(7) (2007) 864–873.

[40] P.M. Mannucci, A. Tripodi, Hemostatic defects in liver and renal dysfunction, Hematology Am. Soc. Hematol. Educ. Program (2012) 168–173.

[41] S. Kitchen, I. Jennings, T.A. Woods, I.D. Walker, F.E. Preston, Two recombinant tissue factor reagents compared to conventional thromboplastins for determination of international normalised ratio: a thirty-three-laboratory collaborative study. The Steering Committee of the UK National External Quality Assessment Scheme for Blood Coagulation, Thromb. Haemost. 76(3) (1996) 372–376.

[42] R. Sasisekharan, Z. Shriver, From crisis to opportunity: a perspective on the heparin crisis, Thromb. Haemost. 102(5) (2009) 854–858.

[43] R. Al Dieri, B. de Laat, H.C. Hemker, Thrombin generation: what have we learned? Blood. Rev. 26(5) (2012) 197–203.

ENRICHMENT READING

[1] C.M. Jackson, Y. Nemerson, Blood coagulation, Annu. Rev. Biochem. 49 (1980) 765–811.

[2] K.G. Mann, R.J. Jenny, S. Krishnaswamy, Cofactor proteins in the assembly and expression of blood clotting enzyme complexes, Annu. Rev. Biochem. 57 (1988) 915–956.

[3] C.T. Esmon, Interactions between the innate immune and blood coagulation systems, Trends Immunol. 25(10) (2004) 536–542.

[4] H.R. Roberts, D.M. Monroe, J.A. Oliver, J.Y. Chang, M. Hoffman, Newer concepts of blood coagulation, Haemophilia 4(4) (1998) 331–334.

Chapter 35

Mineral Metabolism

Key Points

1. Normal function of the human body requires both large (macro) and trace (micro) quantities of minerals. Many of these have been discussed throughout the text and require continued integration and understanding with normal and abnormal functions. Some are discussed in this chapter.

2. Calcium and phosphorus have numerous metabolic roles in energy metabolism, signal transduction, and bone homeostasis.

3. Parathyroid hormone and the vitamin D endocrine system, acting on bone, kidneys, and intestines, orchestrates bone formation, resorption, and remodeling. Osteocytes and osteoblasts of bone participate as a part of the integrated endocrine system. A number of other hormones such as thyroid hormone, estrogens, cortisol, calcitonin, and growth hormone also participate in bone homeostasis. Vitamin D also has non-bone-related metabolic functions.

4. Disorders of bone include rickets osteomalacia, osteoporosis, and Paget's disease, and they require appropriate diagnosis followed by proper treatment.

5. Copper, zinc, selenium, and molybdenum are essential cofactors for many enzymes, and defects in their metabolism lead to abnormal clinical manifestations.

The chemical elements—exclusive of the common elements carbon, hydrogen, nitrogen, oxygen, and sulfur—necessary for the normal structure and function of the body are collectively known as **minerals** and their study as **bioinorganic chemistry**. The minerals can be classified as **macrominerals** and **trace elements**. Macrominerals are present in large amounts and include sodium, potassium, chloride, phosphate, calcium, and magnesium. Trace elements are needed in very small amounts. Improvement in the sensitivity of analytical methods has increased the number of known essential trace elements, and the list is likely to grow. Certain other elements have no known biological function and are toxic.

Iron, the central element in oxygen transport and utilization, is discussed in Chapter 27. Iodine, a constituent of thyroid hormones, is discussed in Chapter 31. Sodium, potassium, and chloride, which are important for maintaining proper osmolality and ionic strength, and for generating the electrical membrane potential, are discussed in Chapter 37. Most of this chapter is devoted to the metabolism of calcium and phosphorus because of their importance in the skeleton and other body systems. Because of its chemical and biological relationship to calcium, magnesium is also covered. The trace elements are discussed with emphasis on those which have a known biochemical function.

CALCIUM AND PHOSPHORUS

Distribution and Function

Calcium is the fifth most abundant element on earth and the principal extracellular divalent cation in the human body. A healthy, 70 kg adult body contains 1–1.25 kg of calcium (25–33 g/kg of fat-free tissue), while a 3.5 kg newborn contains about 25 g of calcium. About 95%–99% of body calcium is in the skeleton as hydroxyapatite crystals. The remainder is in the extracellular fluid and is exchangeable with that in periosteal fluid, bone-forming surfaces, and soft tissues. Skeletal calcium is slowly exchangeable with extracellular fluid calcium, and the skeleton is thus a reservoir of calcium. The steady-state extracellular and periosteal fluid concentrations of calcium depend, in large part, on the balance between bone formation and bone resorption, which are regulated by a number of hormones.

The plasma concentration of calcium is kept remarkably constant throughout life at about 8.8–10.3 mg/dL (2.20–2.58 mmol/L). The normal serum calcium concentration is maintained by the integrated actions of parathyroid hormone (PTH), vitamin D endocrine system, calcitonin, and cytokines such as transforming growth factor β and interleukin-6. The principal target sites for these hormones in the regulation of both calcium and phosphorus homeostasis are the gastrointestinal tract, kidney, and bone. Based on physiological signals, the gastrointestinal tract regulates absorption, the kidney regulates reabsorption and excretion, and bone regulates accretion and mobilization of calcium and phosphorus. Abnormal serum calcium concentration has deleterious physiological

N. V. Bhagavan and C.-E. Ha: Essentials of Medical Biochemistry, Second edition. DOI: http://dx.doi.org/10.1016/B978-0-12-416687-5.00035-X

effects on diverse cellular processes involving muscular, neurological, gastrointestinal, and renal systems.

Three forms of calcium are in equilibrium in serum: non-diffusible calcium bound primarily to albumin; diffusible complexes of calcium with lactate, bicarbonate, phosphate, sulfate, citrate, and other anions; and diffusible ionized calcium (Ca^{2+}). Ionized calcium accounts for approximately half the total serum calcium, and nondiffusible and complexed calcium account for 45% and 5%, respectively. Ionized calcium is the physiologically active form; its concentration is regulated by the parathyroid gland. A decrease in serum ionized calcium can cause **tetany** (involuntary muscle contraction) and related neurological symptoms, regardless of the total serum calcium concentration.

Although ionized and protein-bound calcium are in equilibrium, release from the protein-bound fraction is slow, and changes in plasma protein (especially albumin) concentration result in parallel changes in total plasma calcium. A decrease in serum albumin of 1 g/dL results in a decrease of about 0.8 mg/dL in total serum calcium. The equilibrium among the three forms of serum calcium is affected by changes in blood pH. Thus, at pH 6.8 (acidosis), about 54% of serum calcium is in the ionized form, whereas at pH 7.8 (alkalosis), only 38% is ionized. The most accurate and sensitive method of determining ionized calcium concentration is using an ion-specific electrode.

Besides being the principal component of teeth and skeleton, calcium is essential for blood coagulation (Chapter 34), muscle contraction (Chapter 19), secretion of digestive enzymes (Chapter 11), secretion and action of many hormones (Chapter 28), and other body systems. Calcium is intimately involved in the signal transduction pathway (Chapter 28) and in contractile systems (Chapter 19).

About 80% of the phosphate in the body is combined with calcium as hydroxyapatite in the skeleton. The remainder is present in many organic compounds as phosphate esters and anhydrides, e.g., nucleic acids, nucleoside triphosphates (particularly ATP), membrane phospholipids, and sugar-phosphate metabolites. Phosphate is distributed fairly equally between extracellular and intracellular compartments. Inorganic phosphate is a substrate in oxidative phosphorylation (Chapter 13), in glycogen breakdown (Chapter 14), in the formation of 1,3-bisphosphoglycerate from glyceraldehyde-3-phosphate (Chapter 12), in conversion of nucleosides to free base- and sugar-phosphate (Chapter 25), and in several other reactions.

Bone Structure, Formation, and Turnover

The human body contains 206 bones whose size and shape are highly diverse; many contain joints at their ends that connect them to adjacent bones [1,2]. The functions of bone include maintenance of external form, provision of a structural framework for attachment of muscles, weight-bearing support, and protection of internal organs. In addition, the interior medullary cavity of the bone is filled with soft, pulpy material known as **bone marrow** that houses the hematopoietic system. The cells involved in bone formation and resorption arise in the hematopoietic system. There are two basic types of bone: compact (or cortical) and cancellous (or trabecular). The outer layer is compact bone. It is dense, solid, and responsible for mechanical and protective functions. The interior part contains the cancellous bone, which has the appearance of sponge-like or honeycombed structures. Because of its large surface area, cancellous bone contains bone-forming cells and is the site of mineral-requiring bone formation.

The skeleton is the body's principal reservoir of calcium and phosphorus. Contrary to its appearance, bone is a dynamic tissue, and calcium and phosphate are continuously deposited and released. Bone is a modified connective tissue consisting of a cellular component, an organic matrix, and an inorganic (mineral) phase. Its cells are osteoblasts, osteoclasts, osteocytes, and osteoprogenitor cells. The last are a type of mesenchymal cell that can differentiate into any of the other three types and to which the other types can revert.

During bone formation, osteoblasts secrete tropocollagen, mucopolysaccharides, sialoproteins, and lipids to form the organic matrix. When this matrix matures into an insoluble, fibrillar network (osteoid), mineralization begins with a nucleation step, followed by precipitation of calcium and phosphate from the surrounding interstitial fluid. The initial deposits are amorphous and have the composition of brushite ($CaHPO_4 \cdot 2H_2O$). This mineral changes to hydroxyapatite, a hard, crystalline compound of approximate composition $Ca_{10}(PO_4)_6(OH)_2$. The incorporation of fluoride ions into bones and teeth increases the ratio of crystalline to amorphous calcium phosphate, which increases the hardness of the mineral. The calcium phosphates are quite insoluble; their precipitation may be enhanced by an increase in the $Ca^{2+} \times PO_4^{3-}$ ion product, perhaps by the action of alkaline phosphatase on sugar phosphates and pyrophosphates in the bone matrix. Alkaline phosphatase may regulate bone mineralization by hydrolysis of pyrophosphate, which is a potent inhibitor of mineralization *in vitro*. This enzyme is localized in osteoblasts, and its activity is increased in sera of patients afflicted by rickets, osteomalacia, and hyperparathyroidism, all of which are associated with increased osteoblastic activity.

As mineralization progresses, osteoblasts become surrounded by growing bone and differentiate into osteocytes, which reside in individual lacunae in the bone. These lacunae communicate with each other via canaliculi, exchanging substrates and metabolites. Osteocytes nourish the bone, which is a living, highly vascularized tissue. The bone cells account for 2%−3% of mature bone volume. Bone mass is about 65% mineral and 35% organic matrix.

Bone turnover (remodeling) is a dynamic, continuous process. The adult skeleton is renewed about every 10 years. The remodeling process is tightly regulated and is coupled to the resorption of old and defective bone and formation of new bone. The remodeling is accomplished by the formation of temporary anatomical structures known as **basic multicellular units**. In these units, multinucleated osteoclasts located in the front resorb the existing bone, and the osteoblasts coming from the rear carry out bone formation. The attachment of osteoclasts to the bone surface occurs at specific target sites consisting of integrin receptors that recognize specific bone matrix proteins. Osteoclast attachment is mediated by mechanical stimuli or release of chemotactic substances from the damaged bone. Resorption of the bone requires hydrogen ions, lysosomal enzymes, and collagenase, which are secreted through the ruffled borders containing microvilli of osteoclasts. The low pH is responsible for the solubilization of the mineral component of the bone. The requisite hydrogen ions are derived from organic acids and H_2CO_3 formed locally by hydration of CO_2 by carbonic anhydrase present in the osteoclasts. The importance of carbonic anhydrase in the production of H^+ and bone resorption is evident in **osteopetrosis** (discussed later).

Osteoblasts recruited to the site of the erosion cavity carry out bone formation. During bone matrix synthesis, osteoblasts become lining cells or osteocytes, and some undergo apoptosis. Thus, in bone remodeling, regulators of apoptosis of osteoclasts and osteoblasts play a major role. For example, increased production of cytokines, namely interleukin-1, interleukin-6, and tumor necrosis factor that occurs due to estrogen deficiency, leads to increased bone resorption to a level greater than that of bone formation and causes **osteoporosis** (discussed later). Osteocytes comprise more than 90% of bone cells. The osteocyte−canalicular system plays an important role in activating the bone remodeling process by functioning as a transducer that detects microfractures or other flaws in the bone structure. Osteocytes also undergo apoptosis with increasing age.

Both osteoblasts and osteoclasts are derived from osteoprogenitor cells originating in the bone marrow. Osteoblast precursors are pluripotent mesenchymal stem cells, and the osteoclast precursors are hematopoietic cells of the monocyte−macrophage lineage. The development of osteoblasts and osteoclasts is regulated by several growth factors and cytokines whose responsiveness, in turn, is modulated by systemic hormones.

The formation of osteoclasts, activation, and their survival require **a protein** secreted by osteoblastic stromal cells in response to 1,25-(OH)$_2$D, PTH, PGE$_2$, and IL-11. This protein is a specific ligand (L) that binds to receptors located on the osteoclast precursor cells, initiating the formation of multinucleated osteoclasts. The receptor is known as **receptor activator of nuclear factor-κB (RANK)** and its ligand as RANKL. The RANK−RANKL system is responsible for bone resorption. It is regulated by osteoprotegerin, an endogenous ligand that competes for binding to RANK with RANKL. Administration of monoclonal antibodies (**denosumab**) to RANKL as a means of targeted therapy for osteoporosis to postmenopausal women with low bone mineral density (BMD) increases their BMD (Figure 35.1).

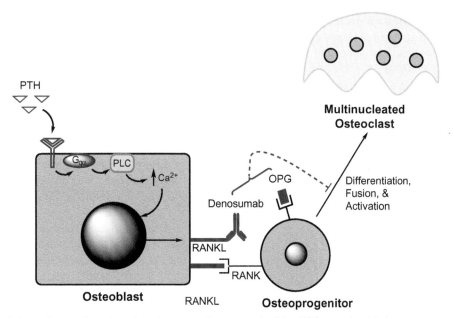

FIGURE 35.1 Regulation of osteoclast formation (see text for more details). PTH = parathyroid hormone, PLC = phospholipase C, OPG = osteoprotegerin, ⊥ = inhibition.

Bone morphogenetic proteins (BMPs) are also signaling molecules involved in the formation of bone that govern other developmental functions. These proteins belong to the transforming growth factor β superfamily. BMPs regulate many steps in the formation of new bone, such as mobilization of progenitor cells and their differentiation and proliferation into chondrocytes and osteoblasts. Thus, BMPs have therapeutic implications in the enhancement of osteoblast differentiation and bone formation. For example, *in vitro* and rodent studies have shown that one particular BMP (BMP-2) promotes bone formation. HMG-CoA reductase inhibitors that decrease hepatic cholesterol biosynthesis also activate the promoter of the BMP-2 gene, thereby promoting new bone formation. HMG-CoA reductase inhibitors (statins) cause decreased prenylation of proteins, such as GTP-binding proteins, and induction of osteoclast apoptosis. (For a discussion on the mevalonate–cholesterol multifunctional pathway and HMG-CoA reductase inhibitors, see Chapter 17.) Population-based, case-controlled epidemiological studies have shown that the **statin** used by elderly individuals is associated with a decreased risk of bone fractures (e.g., hip fracture). Some **bisphosphonates** that are used clinically as potent antiresorptive agents may also affect prenylation of proteins by inhibiting enzymes in the mevalonate–cholesterol biosynthetic pathway more distal to the HMG-CoA reductase catalyzed step (e.g., farnesyl–pyrophosphate synthase).

The transformation of precursor cells to osteoclasts is stimulated by parathyroid hormone (PTH), thyroxine, growth hormone, and vitamin D metabolites, whereas calcitonin, estrogens, and glucocorticoids inhibit the formation of osteoclasts. Osteoblast formation is promoted by calcitonin, estrogen, growth hormone, inorganic phosphate, and mechanical stress, and is antagonized by PTH and vitamin D metabolites. The actions of some of these agents are mediated by cAMP.

Calcium and Phosphate Homeostasis

Four primary factors influencing calcium and phosphate homeostasis are diet, vitamin D and its metabolites, PTH, and calcitonin. Table 35.1 lists other hormones known to affect homeostasis of these elements by interacting with one or more of these factors.

Calcium and Phosphate in the Diet

Sufficient dietary calcium acid phosphate must be absorbed to support growth (including during pregnancy) and replace minerals lost from the body. Phosphates are present in adequate quantities in a wide variety of foods, and it is very unlikely that hypophosphatemia results from inadequate dietary phosphorus. Phosphate is absorbed from the small intestine with Ca^{2+} as a counterion, and by an independent process that requires vitamin D metabolites. Normal daily intake of phosphate is about 800–1500 mg. Phosphate is highly conserved by the body, and obligatory losses are minimal. Serum phosphate concentration is maintained within a narrow range (reference interval: 2.5–5.0 mg/dL or 0.80–1.60 mmol/L) by phosphate intake and urinary phosphate excretion. Thus, in phosphate homeostasis, the kidneys have a primary role. The filtered phosphate at the glomeruli is reabsorbed in the proximal tubules by Na^+/PO_4^{3-} cotransporters. The transport is driven by the intracellular Na^+ gradient, which is maintained by Na^+/K^+-ATPase. Both PTH and a hormone produced by the bone and connective tissue, known as **fibroblast growth factor (FGF) 23** and **klotho system**, inhibit urinary phosphate reabsorption. Thus, FGF-23 and the klotho system participate in phosphate homeostasis [3–7]. The participation of FGF-23 as a part of the bone endocrine system is presented in Figure 35.2. (Recommended reading references [3–5] present clinical cases.)

Calcium homeostasis is profoundly dependent on diet and intestinal absorption. **Rickets** and postmenopausal **osteoporosis** are related to inadequate intestinal absorption of calcium. In the adult, an unavoidable loss of about 300 mg of Ca^{2+} per day occurs in urine, feces, and sweat. During pregnancy and lactation, there is calcium deposition in the fetus and calcium loss in milk. About 80% of the 25 g of calcium present in a full-term fetus is deposited during the last trimester of pregnancy. Human breast milk contains 30 mg of Ca^{2+} per deciliter (7.5 mmol/L), and about 250 mg of calcium per day is lost in milk during lactation. Increased calcium intake is recommended for older women because of the occurrence of osteoporosis. Administration of vitamin D metabolites together with calcium is indicated for older women.

Calcium is absorbed both actively and passively throughout the small intestine and, to a small extent, in the colon. The active transcellular transport occurs in the duodenum and the passive paracellular process takes place in the jejunum and ileum. The chemical gradient and the sojourn time of the food passing through the intestine determine the movement of calcium that occurs by a passive process. The absorption of calcium in the colon becomes nutritionally significant under conditions of small intestine resection.

The active, saturable calcium transport consists of three steps: uptake by the brush-border cell membrane, diffusion through the cytoplasm, and extrusion at the basolateral surface where calcium is transferred to the portal blood circulation. The first step of calcium uptake is not energy-dependent, the second step of transcellular calcium movement is thought to be a rate-limiting process, and the third step of calcium extrusion from the

TABLE 35.1 Factors That Can Influence Calcium and Phosphate Homeostasis

Factor	Pathological Effects on Calcium and Phosphorus Metabolism*	Mechanism/Site of Action[†]
Vitamin D	Excess (vitamin D toxicosis) causes hypercalciuria and hypercalcemia, leading to urolithiasis and soft tissue (especially renal) calcification; insufficiency causes rickets and osteoporosis.	Active form 1,25-$(OH)_2$D binds to vitamin D receptor (VDR) located in the nuclei of target cells, causing either stimulation or repression of specific gene-encoding proteins required for mineralization and remodeling of bone. VDR belongs to the steroid/thyroid hormone receptor superfamily and is expressed in many tissues.
Parathyroid hormone (PTH)	Primary hyperparathyroidism causes hypercalcemia, hypophosphatemia, and increased urinary cAMP; hypoparathyroidism causes hypocalcemia and hyperphosphatemia, often with soft tissue calcification, tetany, and convulsions.	Binds to cell surface receptors and activates G_s-protein-coupled adenylate cyclase and G_q-protein coupled to phospholipase C; increases bone mineralization and activity of renal 1α-hydroxylase; in kidney, reabsorption of Ca^{2+} increases and reabsorption of phosphate decreases.
Parathyroid hormone-related protein (PTHrP)	Predominant factor responsible for hypercalcemia of malignancy.	PTHrP functions through the activation of PTH receptors; it has paracrine functions; it regulates rate of cartilage differentiation and increases placental calcium transport; it may also have specific PTHrP receptors.
Calcitonin	Neither deficiency nor excess of calcitonin is known to have any pathological effects; plasma calcitonin is increased in medullary carcinoma of the thyroid.	Binds to cell surface receptors in bone and kidney, increasing intracellular cAMP; may also function by activation of phospholipase C signal transduction pathway, inhibiting osteoclast activity; decreases Ca^{2+} release from bone and stimulates Ca^{2+} and phosphate excretion in kidneys; generally antagonistic to PTH.
Magnesium	Hypermagnesemia decreases PTH secretion; mild hypomagnesemia increases PTH secretion; severe hypomagnesemia (<0.5 mmol/L) inhibits PTH secretion, even with hypocalcemia.	Needed to maintain parathyroid responsiveness to serum Ca^{2+}.
Diet	Inadequate Ca^{2+} intake produces rickets or osteomalacia; hypophosphatemia rarely (if ever) results from dietary inadequacy; excess dietary Ca^{2+} causes hypercalciuria and risk of urolithiasis.	Ca^{2+} and inorganic phosphate are absorbed in the small bowel by active 1,25-$(OH)_2$D-requiring processes; Ca^{2+} decreases PTH secretion and may decrease 1α-hydroxylase activity; inorganic phosphate decreases 1α-hydroxylase activity.
Estrogens	Decreased estrogens are probably responsible for postmenopausal osteoporosis.	Estrogens increase renal 1α-hydroxylase activity; promote osteoblast differentiation.
Glucocorticoids[‡]	Osteoporosis occurs in Cushing's disease and in patients treated with glucocorticoids for immunosuppression; glucocorticoids are used in the treatment of hypercalcemia.	Decrease intestinal Ca^{2+} absorption; may antagonize 1,25-$(OH)_2$D or PTH; may directly stimulate parathyroids; may have direct effects on intestine and bone independent of PTH and 1,25-$(OH)_2$D.
Growth hormone/ insulin-like growth factors[‡]	Absence during bone growth causes slow growth and short stature (dwarfism).	Growth and development related to calcium and vitamin D metabolism.
Insulin[‡]	Decreased bone mass in insulin-dependent diabetes; in rats, decreased insulin lowers intestinal Ca^{2+} absorption, intestinal calcium-binding proteins, and serum 1,25-$(OH)_2$D.	Probably increases conversion of 25-(OH)D to 1,25-$(OH)_2$D.
Prolactin[‡]	No pathological effects are known.	In chicks, prolactin administration increases renal 1α-hydroxylase activity and serum 1,25-$(OH)_2$D concentration.
Thyroid hormones[‡]	Bone abnormalities are seen in hyper- and hypothyroidism.	Affects vitamin D metabolism.

*Effects observed in humans unless otherwise noted.
[†]Results are from animal studies; some of these mechanisms may also explain effects observed in humans.
[‡]These hormones have other major roles in the body, and their effects on calcium and phosphate homeostasis are generally of secondary importance.

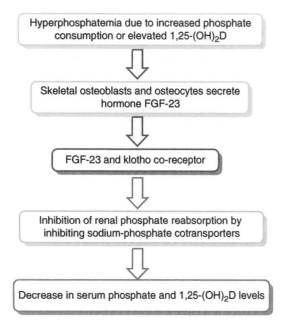

FIGURE 35.2 Regulation of phosphate homeostasis by FGF-23. The skeleton functions as an endocrine organ. Phosphate homeostasis is integrated with PTH and 1,25(OH)$_2$D. PTH reduces renal phosphate reabsorption and 1,25(OH)$_2$D increases the intestinal phosphate absorption.

enterocyte is an energy-dependent process. This latter process is mediated by Ca^{2+}-ATPase, which forms a transmembrane segment, and the enzyme undergoes phosphorylation-induced conformational changes. This transcellular active calcium transport requires 1α,25-dihydroxyvitamin D [1,25-(OH)$_2$D], or calcitriol, which is responsible for inducing the synthesis of enterocyte calcium-binding proteins that promote absorption and transcellular movement, as well as enhancing the number of Ca^{2+}-ATPase pumps. Increased physiological need during pregnancy, lactation, and growth enhances calcium absorption. The molecular mechanism by which this occurs is not understood. Phosphate is also absorbed in the small intestine by an active process, with maximal absorption occurring in the middle of the jejunum. 1,25-(OH)$_2$D also mediates phosphate absorption.

Intestinal calcium absorption is influenced by dietary factors. Lactose and other sugars increase water absorption, thereby enhancing passive calcium uptake. The effect of lactose is especially valuable because of its presence in milk, a major source of calcium. Lactose also increases absorption of other metal ions. This effect may contribute to the incidence of lead poisoning (plumbism) among young inner-city children exposed to high dietary levels of both lead and lactose.

Calcium absorption is reduced by high pH; complexing agents such as oxalate, phytate, free fatty acids, and phosphate; and shortened transit times. These factors are probably of clinical importance only when associated with vitamin D deficiency, marginal calcium intake, or malabsorption disorders. Absorption is also reduced by increased intake of protein, fat, and plant fiber; increasing age; stress; chronic alcoholism; immobilization (e.g., prolonged hospitalization); and drugs such as tetracycline, thyroid extract, diuretics, and aluminum-containing antacids.

As intestinal absorption of calcium increases, urinary calcium excretion also increases. When the latter exceeds 300 mg/day, formation of calcium phosphate or calcium oxalate stones (**urolithiasis**) may occur. Hypercalciuria may result from decreased reabsorption of calcium due to a renal tubular defect or from increased intestinal absorption of calcium. Hypercalciuria may be due to an intrinsic defect in the intestinal mucosa or secondary to increased synthesis of 1,25-(OH)$_2$D in the kidney. Disordered regulation of 1,25-(OH)$_2$D synthesis is relatively common in idiopathic hypercalciuria. Treatment usually includes increased fluid intake, decreased Na$^+$ and protein intake, and Ca^{2+} below the recommended intake for age and sex. Increased vitamin D intake, hyperparathyroidism, and other disorders can also cause hypercalciuria and urolithiasis.

In some forms of steatorrhea, calcium, which normally binds to and precipitates oxalate in the intestine, binds instead to fatty acids, producing increased oxalate absorption and hyperoxaluria. Even though urinary calcium is decreased under these conditions, the concentration of urinary oxalate may be elevated sufficiently to cause precipitation of calcium oxalate crystals. Stone formation can be exacerbated by a diet that contains foods rich in oxalate, such as rhubarb, citrus fruits, tea, and cola drinks. Increased oxalate absorption can be reduced by calcium administered with meals as a water-soluble salt.

Vitamin D Metabolism and Function

Although rickets was first described in the mid-1600s, it was not until the 1920s that deficiency of vitamin D was recognized as its cause [8−12; also see Clinical Case Study 35.1]. Despite its designation as a vitamin, dietary vitamin D is needed only if a person receives inadequate exposure to sunlight. Normally, vitamin D$_3$ is synthesized in the skin by irradiation of 7-dehydrocholesterol (Figure 35.3). Sufficient exposure to ultraviolet radiation can cure rickets.

The principal compound formed in the skin is **cholecalciferol** (vitamin D$_3$). The critical step requiring irradiation is the breaking of the 9,10 bond in the sterol B ring to form a secosterol. A secosterol occurs when one of the rings of the steroid skeleton cyclopentanoperhydrophenanthrene has undergone carbon−carbon breakage. Ring opening is accomplished by light of wavelength from 290 to 320 nm (ultraviolet-B radiation) with a maximal effect

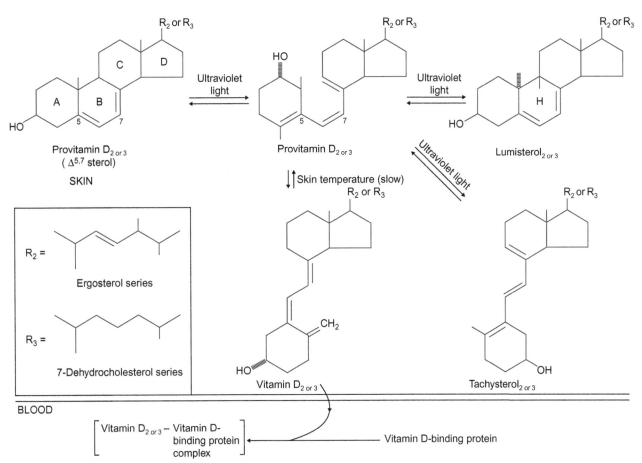

FIGURE 35.3 Conversion of provitamin D (ergosterol or 7-dehydrocholesterol) to vitamin D in the skin and its transport in the blood complexed with vitamin D-binding protein (also known as transcalciferin, Gc-protein, and group-specific component).

at 297 nm. 7-Dehydrocholesterol is present at high concentration in the stratum spinosum and stratum basale of the epidermis. Thermal isomerization of previtamin D_3 to vitamin D_3 occurs slowly, over 2–3 days. The vitamin D_3 formed diffuses gradually through the basal layers of the skin into the circulation. Excessive exposure to solar radiation causes conversion of previtamin D_3 to tachysterol$_3$ and lumisterol$_3$, which are biologically inactive. Melanin can reduce the formation of previtamin D_3 by absorbing part of the solar radiation. This effect may partly explain the greater susceptibility of dark-skinned children to rickets. The concentration of 7-dehydrocholesterol in skin is reported to decrease with increasing age; if so, inadequate synthesis of vitamin D_3 may contribute to senile osteoporosis. The recommended daily allowance (RDA) for vitamin D is 400 IU. However, some studies recommend daily supplementation of 800–1000 IU to achieve option vitamin D nutritional states.

Ergocalciferol (vitamin D_2) is formed by irradiation of ergosterol, a plant sterol common in the diet. It differs from cholecalciferol in the side chain attached to the D ring (Figure 35.3). Irradiation of ergosterol is an

important commercial method for the synthesis of vitamin D_2, which is used for enriching cows' milk. The practice of fortifying milk and milk products with vitamin D has nearly eliminated rickets as a major disease of infancy and childhood in industrialized countries. The metabolism and biological activity of vitamin D_2 in humans are identical to those of vitamin D_3. For this reason, the subscripts 2 and 3 are omitted from vitamin D and its metabolites.

In the liver, vitamin D is hydroxylated to 25-hydroxyvitamin D [25-$(OH)_2$D], the principal circulating metabolite of vitamin D (Figure 35.4). In the kidney, hydroxylation at position 1 yields 1,25-$(OH)_2$D. This metabolite has the highest specific activity of the naturally occurring metabolites. 24,25-$(OH)_2$D is also synthesized by renal mitochondria and in other tissues in relatively large amounts in animals with adequate intake of vitamin D, calcium, and phosphorus. These conditions are opposite to those that favor synthesis of 1,25-$(OH)_2$D. Consequently, the serum concentration of 24,25-$(OH)_2$D varies inversely with that of 1,25-$(OH)_2$D.

There are two hepatic vitamin D 25-hydroxylases: the major one in the mitochondria and the other in the smooth

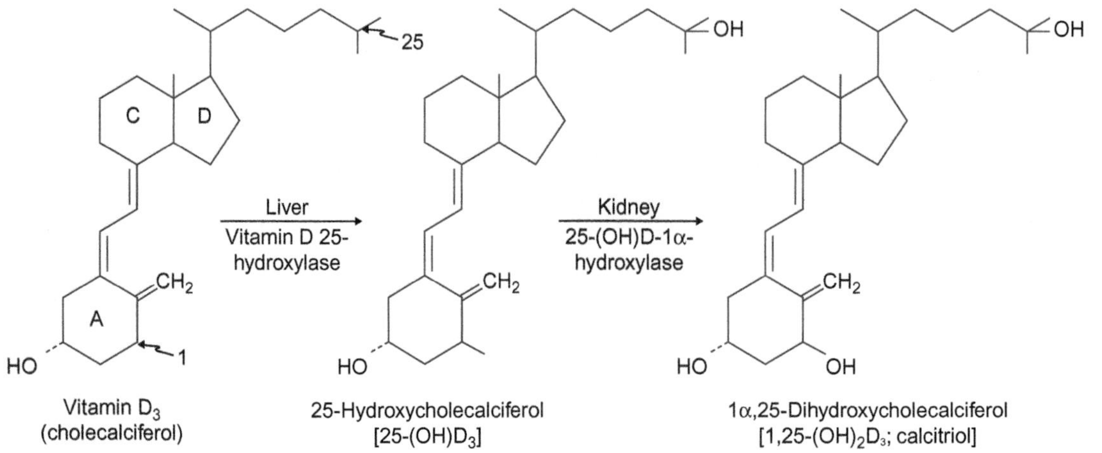

FIGURE 35.4 Activation of vitamin D_3, successive hydroxylations in the liver and the kidneys.

endoplasmic reticulum (Figure 35.4). Both require NADPH and molecular oxygen. The microsomal enzyme appears to be a P-450 mixed-function oxidase. 25-Hydroxylase activity also occurs in intestine, kidney, and lung. 25-Hydroxylase is apparently regulated only by availability of its substrate, leading to a high plasma concentration of 25-(OH)D and a low concentration of vitamin D. 25-(OH)D-1α-hydroxylase is found in the inner mitochondrial membrane of the cells lining the proximal convoluted renal tubules. It is a mixed-function oxidase that requires molecular oxygen, a flavoprotein, a ferredoxin, and a cytochrome P-450 for activity. Placental tissue and macrophages contain 1α-hydroxylase activity.

Hypercalcemia that occurs in some patients with chronic granulomatous diseases, such as tuberculosis, sarcoidosis, and silicosis, is presumably due to the increased 25-(OH)D-1α-hydroxylase activity found in the inflammatory cells in the granulomas. The hydroxylase activity does not appear to be under the usual tight feedback control by serum calcium levels.

Renal osteodystrophy [due to decreased synthesis of 1,25-(OH)$_2$D secondary to kidney failure] is treatable with synthetic 1,25-(OH)$_2$D (also known as **calcitriol**) or 1α-(OH)D. These compounds are also useful in other renal disorders such as hypoparathyroidism and vitamin D-dependent rickets. Deficiency of renal 1α-hydroxylase has been found in patients with **hereditary vitamin D-dependent rickets type I**, in which the serum concentration of 1,25-(OH)$_2$D is low. Patients respond to treatment with 1,25-(OH)$_2$D.

Synthesis of 1,25-(OH)$_2$D is the principal control point in vitamin D metabolism. Although the serum concentrations of vitamin D and 25-(OH)D$_3$ show seasonal and other types of variation, serum concentration of 1,25-(OH)$_2$D remains constant, owing to feedback control of its

synthesis. Activity of renal 1α-hydroxylase is increased by PTH, hypocalcemia (both through PTH and directly), and hypophosphatemia. Calcitonin has no effect on the activity of 1α-hydroxylase. Growth hormone, estrogens, androgens, prolactin, and insulin may also influence the activity of 1α-hydroxylase indirectly (Table 35.1). Thyroxine and glucocorticoids are necessary for bone mineralization, and indirectly influence calcium homeostasis and synthesis of 1,25-(OH)$_2$D (Table 35.1). Glucocorticoids appear to act as antagonists of 1,25-(OH)$_2$D, reducing intestinal calcium absorption independently of any effect on vitamin D metabolism. Because of the high plasma concentration of 25-(OH)D with a longer half-life of 2 to 3 weeks compared to the short half-life of circulating 1,25-(OH)$_2$D (1−5 hours), small changes in activity of the 1α-hydroxylase rapidly change the plasma concentration of 1,25-(OH)$_2$D. Deviations from normal serum concentrations of calcium and phosphate are rapidly corrected by this mechanism. The nutritional status of vitamin D is determined by measuring serum concentration of total 25-(OH)D (also known as **calcidiol**) levels.

The main target tissues for 1,25-(OH)$_2$D are bone, intestine, and kidney. In the intestine, absorption of dietary calcium and phosphorus is increased. In bone, resorption is accelerated. In the kidney, 1,25-(OH)$_2$D is localized in the nuclei of the distal convoluted tubule cells [recall that synthesis of 1,25-(OH)$_2$D occurs in the proximal tubular cells], and reabsorption of calcium and phosphate is increased. In each case, the changes brought about by 1,25-(OH)$_2$D act on the kidney to reduce production of 1,25-(OH)$_2$D, thereby forming a regulatory feedback loop.

On entry into target cells, 1,25-(OH)$_2$D migrates into nuclei, where it binds with high affinity to the vitamin D receptor (VDR), which is a member of the nuclear receptor family. The binding of 1,25-$(OH)_2$D, which functions as a ligand to the nuclear receptor VDR, is analogous to the initiation of biochemical action by steroid–thyroid hormones (Chapters 30 and 31). The binding of the ligand to VDR brings about conformational changes so that the retinoid X receptor (RXR) can combine with it, forming a heterodimer complex. The ligand–VDR–RXR complex recruits additional coactivators and binds to vitamin D response elements (VDREs) in the promoter regions of target genes. The specific binding occurs via the two zinc finger modules of the DNA-binding domain of the receptors. The complex interacts with the general transcription apparatus of 1,25-$(OH)_2$D-responsive genes. The target genes can be either upregulated or downregulated. An example of upregulation is calcium-binding proteins (CaBPs), which facilitate intestinal calcium absorption. Inhibition of PTH gene expression in the parathyroid glands is an example of downregulation.

The vitamin D receptor is widely distributed not only in the target tissues, but also in many other tissues such as thyroid, skin, adrenal, liver, breast, pancreas, muscle, prostate, and lymphocytes, and in numerous malignant cells. Thus, the nonskeletal actions of vitamin D include regulation of proliferation, differentiation, and immunomodulation. Low levels of 25-(OH)D appear to be a risk factor and are associated with some chronic inflammatory diseases and cancer. Serum levels for 25-(OH)D in ng/mL of <10, $10-24$, $25-80$, >80 are considered deficiency, insufficiency, sufficiency, and toxic level, respectively.

An autosomal recessive disease known as **hereditary vitamin D-resistant rickets**, **type II** is caused by defects in the gene for the vitamin D receptor that render it nonfunctional. Thus, this knockout of VDR function illustrates its importance. Patients with this disease have high circulating levels of 1,25-$(OH)_2$D (unlike type I disease, which is due to 1α-hydroxylase deficiency, discussed earlier), severe rickets, and alopecia. The alopecia may be due to defective VDR in the hair follicles. Type II patients do not benefit from 1,25-$(OH)_2$D therapy, and may require both intravenous calcium therapy and high-dose oral calcium administration.

In bone, a CaBP called **osteocalcin** contains 49 amino acids (M.W. 5500–6000). Its synthesis is stimulated by 1,25-$(OH)_2$D. Osteocalcin contains four residues of γ-carboxyglutamic acid, which require vitamin K for their synthesis and are important as binding sites for calcium (Chapter 34).

1,25-$(OH)_2$D increases reabsorption of phosphate in the kidney and intestinal absorption of phosphate. In the intestine, phosphate is absorbed as a counterion with Ca^{2+} and also by a calcium-independent route. Phosphate flux through both pathways is increased by 1,25-$(OH)_2$D but more slowly than in calcium transport. The major

excretory route for vitamin D is bile. Vitamin D metabolites may undergo conjugation in the liver prior to secretion.

Parathyroid Hormone

Humans usually have four parathyroid glands, located behind the thyroid. They were frequently removed during thyroidectomy, leading to hypocalcemia. The parenchyma of the glands is composed of **chief cells** and **oxyphil cells**. The chief cells are more numerous and are responsible for the production of PTH. The oxyphil cells have no known function.

Parathyroid hormone (PTH) is a polypeptide of 84 amino acids (M.W. 9500). Full hormonal activity is present in the N-terminal 34-residue peptide [PTH(1–34)]. Removal of the first two residues eliminates biological activity, even though PTH(3–34) binds well to PTH receptors. PTH can be secreted, sequestered in an intracellular storage pool, or degraded within the parathyroid gland. Secretion is thought to occur by exocytosis, although the number of secretory granules is inadequate to maintain the observed rate of sustained release of PTH. It appears that most proPTH is rapidly degraded at a rate that parallels changes in extracellular Ca^{2+} concentration. Chronic hypocalcemia leads to hypertrophy and hyperplasia of the parathyroid glands, whereas chronic hypercalcemia has the opposite effect. At normal levels of calcium, a decrease in Mg^{2+} stimulates PTH secretion, while an increase in Mg^{2+} inhibits it. Although the effects of Mg^{2+} and Ca^{2+} are qualitatively similar, Ca^{2+} is two to four times as potent as Mg^{2+}. During low levels of serum Ca^{2+}, Ca^{2+}-sensing receptors located on cell membranes of parathyroid cells are stimulated and promote PTH release by an initial activation of G_qα-phospholipase C-inositol triphosphate pathway. In humans, PTH secretion is suppressed by severe hypermagnesemia (three times normal), even when accompanied by hypocalcemia. Although moderate hypomagnesemia stimulates PTH output, extremely low levels of Mg^{2+} severely decrease the secretion and peripheral effectiveness of PTH, possibly because of the need for Mg^{2+} or several enzymes' activity. 1,25$(OH)_2$D suppresses PTH gene expression. Target tissues for PTH are bone and kidney. PTH affects intestinal calcium absorption only indirectly, by regulating 1,25-$(OH)_2$D. The action of PTH on target cells is mediated by its binding to a distinct family of plasma membrane G-protein-coupled receptors, followed by stimulation of adenylate cyclase and cAMP production. G_sα-inactivating (loss-of-function) or activating (gain-of-function) mutations can cause hypoparathyroidism (Albright hereditary osteodystrophy) or hyperparathyroidism (McCune–Albright syndrome), respectively. Parathyroid hormone-related protein also binds to the same receptor in initiating its biological actions (discussed later).

In bone, PTH promotes resorption and new bone formation. Bone dissolution predominates at high concentrations of PTH, while formation is more important at physiological levels. PTH also stimulates calcium release independently of resorption and more rapidly than it stimulates resorption. This pathway is probably most important in short-term regulation of the serum calcium level. Calcium released in this way may come from a pool of soluble calcium in the extracellular fluid of bone.

PTH promotes bone resorption by increasing the number of osteoclasts indirectly, by binding initially to osteoblasts. PTH-stimulated osteoblasts express RANK ligand (RANKL) on their cell membrane, which, in turn, interacts with RANK expressing osteoclast and their precursor cells. Thus, RANKL−RANK interaction results in signaling events stimulating the number of osteoclasts and their activity (discussed previously) accompanied by bone resorption (Figure 35.1). Since PTH stimulates osteoblastic activity, low levels of PTH promote bone formation (discussed later).

PTH increases the activity of renal 25-(OH)D-1α-hydroxylase.

Normally, 65%−80% of filtered calcium and 85%−90% of filtered phosphate are reabsorbed, mainly in the proximal tubule. The daily loss of 700−800 mg of phosphate is balanced by dietary intake (discussed earlier). Fine-tuning of calcium excretion is accomplished by PTH in the distal convoluted tubules and collecting ducts. Phosphate excretion is regulated by PTH in the proximal tubules. Elevation of the PTH level increases reabsorption of calcium and decreases reabsorption of phosphate from the tubules. This phosphaturic action opposes the phosphate-sparing action of 1,25-$(OH)_2$D.

Urine concentrations of cAMP are normally 100 times higher than those in plasma or cytoplasm. Urinary cAMP is formed by the action of PTH on the renal tubules. In hypocalcemia due to pseudohypoparathyroidism, reabsorption of Ca^{2+} from the collecting ducts is impaired because the cells do not respond to PTH. Serum levels of PTH are high, but the urine contains very little cAMP. The defect may be in the receptors for PTH or distal to the receptor. PTH may also affect tubular reabsorption by cAMP-independent mechanisms.

Thus, a complex relationship exists among serum Ca^{2+} and phosphate, PTH, and vitamin D and its metabolites. Release of PTH in response to low serum Ca^{2+} directly mobilizes calcium from bone and increases synthesis of 1,25-$(OH)_2$D, which, in turn, mobilizes skeletal Ca^{2+} and causes increased intestinal calcium absorption. These effects raise the serum Ca^{2+} level sufficiently to reduce PTH secretion. The effect of PTH on the kidneys occurs within minutes, whereas the effects of PTH on

bone and (indirectly) on the intestine take hours and days, respectively. An increase in serum phosphate acts in a way qualitatively similar to that of hypocalcemia to release PTH, increase excretion of phosphate in the proximal tubules, and decrease intestinal phosphate absorption. These events are mediated predominantly by the decrease in serum calcium that accompanies a rise in phosphate concentration. In addition, phosphate may inhibit 25-(OH)-D-1α-hydroxylase.

Primary hyperparathyroidism results from hyperplasia, adenoma, or carcinoma of the parathyroid glands and from ectopic production of the hormone by squamous cell carcinoma of the lung or by adenocarcinoma of the kidney [13]. In about 10% of hyperparathyroidism, hyperplasia or tumors of the parathyroid glands occur due to familial disorders known as **multiple endocrine neoplasias (MEN)**. MEN syndromes consist of three subtypes (I, IIA, IIB), and are characterized by hyperplasia or tumors (or both) involving two or more endocrine glands in the same individual. MEN syndromes are inherited as an autosomal dominant trait; however, they may also arise sporadically.

Secondary hyperparathyroidism develops whenever hypocalcemia occurs. This condition frequently arises from chronic renal failure or intestinal malabsorption. In chronic renal failure, the hypocalcemia is secondary to hyperphosphatemia caused by the inability of the diseased kidneys to excrete phosphate. The loss of renal tissue also decreases 1α-hydroxylase activity, which leads to decreased intestinal calcium absorption.

Hypoparathyroidism is characterized by hypocalcemia, hyperphosphatemia, and a decrease in circulating PTH. Ectopic calcification is common. The most common form is due to inadvertent removal of, or damage to, the parathyroid glands during thyroid gland surgery or surgery to remove malignant tumors in the neck. The congenital absence of the parathyroids combined with thymic agenesis is known as **diGeorge's syndrome**. This condition is usually fatal by age 1−2 years because of hypocalcemia and immunodeficiency. **Familial hypoparathyroidism** may be an autoimmune disease, but its inheritance is complex and the presence of circulating antiparathyroid antibodies does not always correlate well with the occurrence of the disease.

Parathyroid Hormone-Related Protein

A second member of the parathyroid hormone family, **parathyroid hormone-related protein** (PTHrP), is quite similar to PTH in amino acid sequence and protein structure. Like PTH, it activates the parathyroid hormone receptor, causing increased bone resorption and renal tubular calcium reabsorption. Increased serum concentrations of PTHrP are the predominant cause of hypercalcemia in cancer patients with solid tumors. This observation led to its discovery and to the elucidation of its many cellular functions in normal tissues. In contrast to PTH, which is expressed only in parathyroid glands, PTHrP is detected in many tissues in fetuses and adults; it is found in epithelia, mesenchymal tissues, endocrine glands, and the central nervous system. This protein is also the principal regulator of placental calcium transport to the fetus.

Calcitonin

Parafollicular cells (C cells) scattered throughout the thyroid gland synthesize, store, and secrete calcitonin (thyrocalcitonin). These cells are derived from neural crest cells that fuse with the thyroid gland. In nonmammalian vertebrates, they remain together as discrete organs, i.e., ultimobranchial bodies. Continuous secretion of calcitonin occurs in eucalcemia, while hypercalcemia and hypocalcemia modulate this secretion. Gastrin, pentagastrin, cholecystokinin, and glucagon stimulate release of calcitonin, but there is no evidence that they are physiological regulators.

Procalcitonin (a 116-amino acid peptide), by intracellular proteolytic processing, is converted to active calcitonin (a 32-amino acid peptide). Under normal conditions the procalcitonin is not secreted; however, during severe bacterial infection and sepsis, its serum levels are increased to high levels due to its release from many tissues. Serum levels of procalcitonin are used as a surrogate marker for bacterial infection.

The principal target organ for calcitonin is bone, but renal excretion of calcium and phosphate is also directly affected. In bone and kidney, calcitonin activates adenylate cyclase by binding to a distinct class of G-protein-coupled receptors. Calcitonin may also exert effects on cytosolic Ca^{2+} and IP_3 levels by activation of the phospholipase C signal transduction pathway. Secretion of calcitonin is stimulated by hypercalcemia, but the effect of the hormone on calcium transport appears to be secondary to increased phosphate uptake by target cells. The number and activity of osteoclasts are decreased, and urinary excretion of hydroxyproline is decreased. Calcitonin may also inhibit release of calcium from the extracellular fluid calcium pool, but it increases calcium and phosphate excretion by renal tubules. Some tubular cells respond to calcitonin, PTH, and vasopressin, while others respond only to one or two of these hormones. In general, the actions of calcitonin in kidney and in bone are antagonistic to those of PTH. Calcitonin decreases secretion of gastrin and of gastric acid, and inhibits bile flow.

The exact biological role of calcitonin remains elusive. Neither deficiency nor excess seems to produce pathological changes. Calcitonin may prevent hypercalcemia during childhood when calcium intake is high, or it may be important in adults during periods of hypercalcemia or high calcium intake.

Calcitonin is a useful marker for medullary carcinoma of the thyroid, which occurs both sporadically and as a dominantly inherited disease. In this type of tumor, the plasma concentration of calcitonin is 1–1000 mg/mL (normal concentration ranges from undetectable to 0.05 ng/mL). Also, urinary hydroxyproline excretion is decreased. Ectopic secretion of calcitonin also occurs from several types of pulmonary tumor, in addition to other hormones.

Calcitonin is used in the treatment of **Paget's disease** (osteitis deformans), a chronic disorder characterized by increased bone remodeling, normocalcemia, and normophosphatemia, frequent episodes of hypercalciuria leading to stone formation, and elevation of serum alkaline phosphatase and urinary hydroxyproline levels. The disease does not appear to be primarily a derangement of calcium metabolism. Calcitonin reduces the levels of serum alkaline phosphatase and urinary hydroxyproline, and may relieve other symptoms of the disease as well. Bisphosphonates, especially etidronate disodium, also reduce bone resorption in this disease. Various cancers are accompanied by hypercalcemia and may respond to treatment with calcitonin. Salmon calcitonin, the type most frequently used clinically, is 25–100 times more potent compared to human calcitonin.

Disorders of Calcium and Phosphorus Homeostasis

Several disorders have already been discussed; the following are additional disorders of calcium and phosphorus homeostasis [17–26; see also Clinical Case Studies 35.2 and 35.4]. In **rickets** in children and **osteomalacia** in adults, there is failure of mineralization of osteoid, with consequent "softening" of the bones. In rickets, there is defective mineralization of bone and of the cartilaginous matrix of the epiphyseal growth plates. In osteomalacia, since the epiphyseal plates are closed, only bone is affected. The most common cause is deficiency of vitamin D or, more rarely, deficiency of 1,25-$(OH)_2$D. Characteristic findings are eucalcemia or hypocalcemia, hypophosphatemia, hypocalciuria, and elevated serum alkaline phosphatase. Hypocalcemia or hypophosphatemia due to inadequate dietary intake or excess urinary loss of calcium or phosphate, respectively, causes identical lesions.

The most common metabolic bone disease is **osteoporosis**, which results from many environmental factors such as poor diet, smoking, alcohol consumption, and lack of exercise. However, recent evidence indicates that genetic variation may be the most important factor in the determination of bone mass, development of osteoporosis, and the risk of fracture. Osteoporosis can occur at any age. Poor diet and lack of exercise early in life contribute to the development of osteoporosis later in life, particularly in genetically susceptible individuals. Thus, the seeds of osteoporosis are sown early in life. Lack of intake of calcium and vitamin D in adequate quantities and an insufficient amount of weight-bearing exercise (e.g., walking and running but not swimming) lead to failure in achieving peak bone mass in late adolescence and early adulthood. Under normal conditions, skeletal mass is a balance between bone formation, bone resorption, bone cell proliferation, and apoptosis. It is only recently that it has been recognized that the skeleton is a delicately balanced regenerating tissue, regulated as precisely as the destruction and synthesis of blood cells.

An estimated 75 million people are affected by osteoporosis to some degree in the United States. Osteoporosis is a systematic skeletal disease characterized by bone mass and microarchitectural deterioration with a consequent increase in bone fragility and susceptibility to fracture. Operationally, osteoporosis can be defined as a certain level of bone mineral density, measured by dual-energy X-ray absorptiometry. Other biochemical markers used in the assessment of monitoring therapy are provided in Table 35.2.

Osteoporosis, a skeletal disorder, requires both increased resorption and one or more defects in bone formation. During adolescence, the rate of bone formation is higher than the rate of bone resorption; in older persons, the rate of bone resorption is greater than that of bone formation. Recent research indicates that peak bone mass, skeletal structure, and metabolic activity are all determined by a large number of different genes interacting with environmental factors; thus, osteoporosis is a polygenic-multifactorial disease. Candidate genes that may play a role in the development of osteoporosis include vitamin D receptor, estrogen receptor, transforming growth factor β, interleukin-6, collagen type I genes, and collagenase.

Osteoporosis is a common disorder in postmenopausal women. Estrogen or other drugs are used to prevent osteoporosis in postmenopausal women. Bisphosphonates and calcitonin act as antiresorptive drugs. Actually, these drugs decrease the rates of both bone resorption and bone formation, but they affect the remodeling cycle so that there is a net increase in bone mineral density of 5%–10%. Both etidronate and alendronate (bisphosphonates, Figure 35.5) are used in the treatment of osteoporosis in postmenopausal women and provide a treatment option if estrogen replacement therapy is contraindicated. It is thought that bisphosphonates are incorporated into bone matrix and incapacitate osteoclasts upon entry during resorption. Promotion of osteoclast apoptosis through inhibition of the mevalonate–cholesterol biosynthetic pathway, which leads to loss of G-protein prenylation, is also a possible mechanism of action. Rare and potential complications of long-term bisphosphate therapy, which are atypical fractures and osteonecrosis due to suppression of normal bone turnover, can occur. The use of HMG-CoA reductase inhibitors (statins), which also inhibit G-protein prenylation, decreases the risk of fractures in the elderly (discussed earlier). Statins, independent of their lipid-lowering activity, can improve endothelial function,

TABLE 35.2 Biochemical Markers of Bone Formation and Bone Resorption

Markers of Bone Formation (measured in serum)
1. Bone-specific alkaline phosphatase: rich in osteoblasts
2. Osteocalcin: major noncollagen protein of bone matrix
3. Amino- and carboxyl-terminal procollagen 1 extension peptides: byproduct of collagen biosynthesis
Markers of Bone Resorption
Serum Marker
1. Bone-specific acid phosphatase: a lysosomal enzyme of osteoclasts
Urine Markers (all of these are collagen breakdown products)
1. N-telopeptide and C-telopeptide
2. Pyridinium cross-links (pyridinoline and deoxypyridinoline)
3. Post-translational modification of lysine and hydroxylysine residues of collagen
4. Hydroxylysine glycosides
5. Hydroxyproline

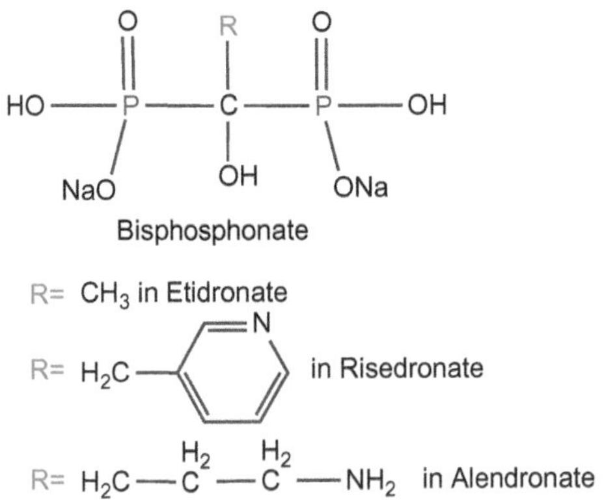

FIGURE 35.5 Structures of bisphosphonates. These antiresorptive therapeutic agents are characterized by a geminal bisphosphonate bond.

including the formation of new blood vessels (angiogenesis). The mechanism of action involves increased formation of nitric oxide and nitric oxide-dependent pathways. These pathways are linked to the activation of *Akt*-protooncogene/protein kinase mediated cellular processes.

Calcitonin therapy results in decreased bone resorption. Osteoclasts have calcitonin receptors, and calcitonin inhibits their activity. Sodium fluoride stimulates bone formation by unknown mechanisms. In women with osteoporosis, fluoride therapy produced increased bone mineral density but no reduction in the rate of vertebral fractures. Other drugs, known as **selective estrogen receptor modulators** (raloxifene, droloxifene, idoxifene, and levormeloxifene), may provide an alternative to estrogen replacement therapy (Chapter 32). Subcutaneous administration of low doses of recombinant PTH(1−34), known as teriparatide, does not affect serum calcium concentration but promotes bone formation and increases mineral density. This anabolic action of PTH is probably mediated by decreasing osteoblast apoptosis. An emerging novel therapy is the use of monoclonal antibodies (denosumab) against RANKL, which inhibits osteoclast formation (discussed earlier). Many pharmacological agents, as an undesirable effect, promote osteoporosis either by inhibiting osteoblasts or promoting osteoclast activities or both. Some examples are as follows:

1. Glucocorticoids (as anti-inflammatory agent)
2. Thyroxine (used in the treatment of hypothyroidism)
3. Aromatase inhibitors (used in breast cancer)
4. Anticonvulsants
5. Antiretrovirals
6. Proton pump inhibitors.

Osteogenesis imperfecta, perhaps the most common hereditary disease of bone, is due to a defect in collagen formation (see Clinical Case Study 23.1, vignette 2). **Osteopetrosis** (marble bone disease) is marked by the formation of abnormally dense and condensed bone. It is

a genetically, biochemically, and clinically heterogeneous disease. However, the underlying mechanism for the various types of osteopetrosis is a failure in bone resorption due to defects in osteoclasts. In some forms of osteopetrosis, providing normal osteoclastic precursors by bone marrow transplantation has yielded clinical improvement. Stimulation of osteoclast formation, activity with 1,25-$(OH)_2D$, or recombinant interferon-γ also has yielded modest clinical improvement.

One form of osteopetrosis is caused by a deficiency of one of the isoenzymes of carbonic anhydrase (carbonic anhydrase II). This is an autosomal recessive disorder, and four different mutations in the structural gene of carbonic anhydrase II have been identified. As discussed earlier, the acidic environment necessary for resorption is generated by the action of carbonic anhydrase II. Patients with carbonic anhydrase II deficiency also exhibit renal tubular acidosis and cerebral calcification. The cause of renal tubular acidosis is due to the failure of reclamation of bicarbonate from the glomerular filtrate because of the deficiency of carbonic anhydrase in the renal tubular cells (see Chapter 37 for the role of carbonic anhydrase in the reclamation of filtered bicarbonate in the kidney).

Three genes coding for isoenzymes of carbonic anhydrase I, II, and III, all belonging to the same gene family, are clustered on chromosome 8q22. These are all zinc metalloenzymes, soluble, and monomeric, and have molecular weights of 29,000. Isoenzyme I is found primarily in erythrocytes (Chapters 2 and 37) and isoenzyme III in skeletal muscle. There are yet other isoenzymes coded for by different genes.

Disorders that cause systemic calcification of the blood vessels can lead to complications of thrombosis and ischemia. This clinical syndrome is known as **calciphylaxis**, and may occur in end-stage renal disease patients undergoing dialysis and in other hypercalcemic states. The mechanism of abnormal calcification is not known. In some patients, parathyroidectomy may be beneficial in normalizing plasma Ca^{2+} and phosphate levels.

Hyperphosphatemia, often seen in renal failure, produces no specific signs or symptoms. In contrast, chronic **hypophosphatemia** can produce weakness, bone pain, congestive cardiomyopathy, dizziness, and hemolytic anemia. Severe hypophosphatemia is potentially fatal and is usually due to hyperphosphaturia, shifting of phosphate from extracellular to intracellular fluid (as in electrolyte and pH imbalances), or diminished intestinal absorption of phosphate. Iatrogenic hypophosphatemia may occur in diabetic ketoacidosis. The immediate goals of therapy are to normalize glucose metabolism and to restore fluid and electrolyte balance. Since metabolism of glucose requires phosphate, rapid entry of glucose into insulin-dependent tissues as a result of insulin therapy shifts phosphate to the intracellular fluid. This effect can be exacerbated by intravenous rehydration with phosphate-poor fluids.

Rehydration and hyperalimentation of alcoholics (who are often phosphate depleted due to poor diet, diarrhea, and vomiting) can lead to hypophosphatemia. Concurrent hypomagnesemia can increase phosphaturia.

MAGNESIUM

Magnesium is the fourth most abundant cation in the body after sodium, potassium, and calcium, and is the second most abundant cation in intracellular fluid after K^+. Mg^{2+} is needed in many enzymatic reactions, particularly those in which $ATP^{4-}Mg^{2+}$ is a substrate. Magnesium binds to other nucleotide phosphates and to nucleic acids and is required for DNA replication, transcription, and translation. The DNA helix is stabilized by binding of histones and magnesium to the exposed phosphate groups. In green plants, chlorophyll—a magnesium—porphyrin complex similar to heme—is vital for photosynthesis.

A normal 70 kg adult body contains about 25 g of magnesium (11–18 mmol/kg wet weight). Bone contains 60%–65% (about 45 mmol/kg wet weight) of the body's Mg^{2+}, complexed with phosphate. Of the other tissues, liver and muscle contain the highest concentrations of Mg^{2+}, approximately 7–8 mmol/kg wet weight. Only 1% of total body magnesium is in extracellular fluid. The skeletal and extracellular fluid Mg^{2+} pools probably exchange freely with each other, but not with the intracellular pool, which remains stable even when there are large fluctuations in the level of serum Mg^{2+}. Thus, the plasma concentration does not accurately reflect total body stores of magnesium.

The kidney appears to be the main organ responsible for maintaining plasma Mg^{2+} concentration within normal limits. Most of the serum Mg^{2+} that is filtered at the glomerulus is reabsorbed, and only about 3%–5% is excreted in the urine. Urinary excretion varies with plasma Mg^{2+} concentration. PTH enhances the tubular reabsorption of Mg^{2+}, whereas aldosterone decreases it.

Hypermagnesemia suppresses PTH secretion, whereas hypomagnesemia, if not too severe, stimulates PTH secretion or inhibits it at very low plasma Mg^{2+} concentrations. The suppressive effect of hypomagnesemia on PTH secretion occurs even in hypocalcemia. The hypocalcemia that accompanies marked hypomagnesemia can be corrected by magnesium repletion. Both hypomagnesemia and hypocalcemia cause tetany. At plasma Mg^{2+} concentrations of 20 mg/dL, anesthesia and paralysis of peripheral neuromuscular activity occur; they can be reversed by intravenous administration of calcium.

Hypermagnesemia occurs in acute or chronic renal failure, in hemodialysis, and in women receiving magnesium sulfate for treatment of pre-eclampsia. The clinical manifestations resemble the effects of curare. At serum Mg^{2+} levels of 2.5–5.0 mmol/L, cardiac conduction is affected, and at concentrations above 12.5 mmol/L, cardiac arrest occurs in diastole. Hypomagnesemia can occur in steatorrhea, alcoholism, diabetic ketoacidosis, and many other disorders. Tetany usually occurs at serum Mg^{2+} concentrations below 1 mmol/L.

ESSENTIAL TRACE ELEMENTS

The biochemical roles of several trace elements such as iron, selenium, cobalt, and iodine have been discussed elsewhere in the text. In the following sections, copper, zinc, molybdenum, and selenium metabolism are discussed.

Copper

Copper is necessary, together with iron, for hematopoiesis, probably partly because it is needed for the synthesis of ferroxidase (ceruloplasmin). Many enzymes require

TABLE 35.3 Examples of Copper-Containing Enzymes and Their Functions

Enzyme	Functional Significance
1. Cytochrome c oxidase	Terminal enzyme of mitochondrial electron transport; oxidative phosphorylation
2. Superoxide dismutase	Inactivation of reactive oxygen species; antioxidant defense
3. Ceruloplasmin	Ferroxidase; iron metabolism
4. Tyrosinase	Synthesis of melanin
5. Dopamine β-monooxygenase	Synthesis of norepinephrine and epinephrine
6. Lysl oxidase (also called protein-lysine 6-oxidase)	Required for cross-linking of collagen and elastin; maturation of collagen
7. Peptidylglycine monooxygenase	Required for removal for carboxyl terminal residue and α-amidation; processing and maturation of neuroendocrine and gastrointestinal peptide hormones
8. Amine oxidases	Deamination of primary amines

copper for activity. Examples of some of the copper enzymes and their functions are given in Table 35.3. Mitochondrial iron uptake may be blocked by deficiency of a cuproprotein, perhaps cytochrome oxidase. Several inherited diseases involving abnormalities in copper metabolism (Wilson's disease, Menkes' syndrome) or copper enzymes (X-linked cutis laxa, albinism) occur in humans and in several animal species. The oxidation state of copper in biological systems is $+1$ or $+2$. The average adult human body contains $70-100$ mg of copper. The highest concentrations (in decreasing order) are in liver, brain, heart, and kidney. Muscle contains about 50% of total body copper. Of the remainder, about one-fifth is in the liver ($3-11\,\mu g/g$ wet weight). Human erythrocytes contain $1.0-1.4\,\mu g$ of copper per milliliter, of which more than 60% is in superoxide dismutase. Normal serum contains $200-400$ mg/L of ceruloplasmin, which is a copper containing protein..

Copper is absorbed from food in the upper small intestine. The absorption is primarily dependent on the quantity of the copper present in the diet. High intake of zinc diminishes copper absorption by inducing metallothionein formation in the mucosal cells. Metallothioneins, due to their high affinity for copper, bind it preferentially, and the bound copper is lost during the sloughing of cells from the villi. Copper accumulation in patients with **Wilson's disease** can be reduced by giving oral zinc acetate, which decreases absorption (discussed later). Absorbed copper is transported to the portal blood, where it is bound to albumin (and probably transcuprein), amino acids, and small peptides. Copper binds to albumin at the N-terminal tripeptide (Asp–Ala–His) site. The recently absorbed copper is taken up by the liver, which plays a central role in copper homeostasis.

Copper occurs in many foods; particularly good sources are liver, kidney, shellfish, nuts, raisins, and dried legumes. Copper deficiency due to diet is rare except in malnutrition and in children with chronic diarrhea. It occurs in total parenteral nutrition with fluids low in copper, particularly following intestinal resection and in patients who receive large amounts of zinc to improve wound healing or for management of sickle cell anemia.

About 40%–70% of ingested copper is absorbed. The appearance in plasma of ingested copper with a peak occurring 1.5–2.5 hours after eating suggests that absorption begins in the stomach and proximal small intestine. Absorption may be inversely related to the metallothionein content of the mucosa. Although pinocytosis occurs in the intestines of human infants, their inability to absorb copper in Menkes' syndrome suggests that physiologically important absorption is by an active process, even at an early age. Absorption is enhanced by complex formation with L-amino acids and small peptides. Newly absorbed copper is taken up by the liver in a saturable

transport process and secreted into plasma as ceruloplasmin (0.5–1.0 mg/day).

Copper is excreted in bile in an amount roughly equal to daily absorption (~ 1.7 mg/day). Toxic levels in liver occur in primary biliary cirrhosis, Indian childhood cirrhosis (a familial, probably genetic, disease, limited to Asia), and other liver diseases in which bile flow is disrupted. Toxicity is rare unless there is pre-existing liver disease or ingestion of large amounts of copper salts. Biliary copper is in a form that cannot be readily reabsorbed. Some loss also occurs in urine and sweat. Renal copper reabsorption may be important for copper homeostasis. Approximately 0.5 mg of copper is lost during each menstrual period. During lactation, copper loss averages 0.4 mg/day. The biological functions of copper are listed in Table 35.3. Like iron, copper is intimately involved in adaptation to an aerobic environment. The proteins contain copper bound directly to specific side chains.

Wilson's disease is an autosomal recessive disease of copper metabolism (see Clinical Case Study 35.3). It has a prevalence of 1 in 30,000 live births in most populations. The disease has a highly variable clinical presentation. It is characterized by impairment of biliary copper excretion, decreased incorporation of copper into ceruloplasmin, and accumulation of copper in the liver and, eventually, in the brain and other tissues. Biochemical findings include low serum ceruloplasmin, high urinary copper excretion, and high hepatic copper content. Some patients have normal serum ceruloplasmin levels, and heterozygous individuals do not consistently show reduced levels of this protein.

The genetic defect in Wilson's disease resides in the long arm of chromosome 13 and the gene codes for a copper-transporting ATPase. Thus far, more than 40 different mutations have been found. The defect in ATPase results in reduced biliary excretion of copper from the hepatocytes. The end result may lead to hepatic failure. Cu-ATPase belongs to a family of ubiquitous proteins that are involved in the translocation or movement of cations such as H^+, Na^+, K^+, Ca^{2+}, and other metal ions (discussed in the appropriate areas of the text). This family of ATPases is designated as P-type ATPase. The P designation stems from the fact that, during the catalytic cycle, an invariant aspartic acid residue undergoes phosphorylation by ATP and dephosphorylation by the phosphatase domain that results in a conformational change promoting cation transfer. Another common feature of P-type ATPase is that all variants possess an ATP-binding domain at the carboxy terminus. Because intestinal copper absorption is unaffected, there is a net positive copper balance. Normal hepatic copper uptake ensures that copper accumulation occurs first in the liver. As the disease progresses, nonceruloplasmin serum copper increases, and copper deposits occur in various tissues, e.g., Descemet's membrane in the cornea (Kayser–Fleischer rings), basal ganglia (leading to lenticular degeneration), kidney, muscle, bone, and joints. Cultured fibroblasts from patients exhibit increased copper uptake.

Penicillamine, a chelating agent, solubilizes copper and other heavy metals and promotes their excretion in urine. Triethylenetetramine, another copper-chelating agent, may be of particular value in patients who are or who are becoming sensitive to penicillamine. As discussed earlier, since excessive zinc prevents copper absorption, oral intake of zinc acetate is used in the management of Wilson's disease.

Menkes' syndrome (Menkes' steely-hair syndrome) is a rare, X-linked recessive disorder in which infants have low levels of copper in serum and in most tissues except kidney and intestine, where the concentration is very high. They also have greatly reduced plasma ceruloplasmin levels. Hair of the affected infants has a characteristic color and texture (*pili torti*, "twisted hair"). It appears tangled and dull, has an ivory or grayish color, and is friable. Weakness and depigmentation of hair and defects in arterial walls (leading to aneurysms) are explained by loss of activity of copper-dependent enzymes (Table 35.3). Cerebral dysfunction may be due to a disturbance in energy metabolism or neurotransmitter synthesis secondary to decreased activity of cytochrome oxidase and dopamine β-hydroxylase.

The defect in copper metabolism is generalized, and the degenerative changes cannot be reversed by parenteral copper administration. However, parenteral copper administration corrects serum ceruloplasmin and hepatic copper deficiencies. Other tissues take up copper administered parenterally, but activities of their copper enzymes are not normalized. The failure of postnatal treatment is due to the fact that many of the deleterious effects occur *in utero*.

The molecular defect in **Menke's syndrome**, like that in Wilson's disease, resides in a P-type ATPase. The gene for the enzyme is located on the X-chromosome. Wilson's disease is characterized by defective biliary excretion; in Menkes' syndrome, the defect is a failure to transport copper to the fetus during development, as well as failure to absorb copper from the gastrointestinal tract after birth.

Aceruloplasminemia is an autosomal recessive disorder characterized by progressive neurodegeneration and accumulation of iron in the affected parenchymal tissues. Iron accumulation in this disorder is consistent with ceruloplasmin's role as a ferroxidase in iron metabolism (discussed in Chapter 27).

Zinc

Zinc is an essential nutrient in animals. The discovery that zinc deficiency is the cause of parakeratosis in pigs was the first demonstration of the practical importance of zinc in animal nutrition. More than 100 zinc enzymes are

known, many of which are in the liver and include examples from all six classes of enzymes. In these enzymes, zinc may have a regulatory role or be required for structure or for catalytic activity. Zinc is not readily oxidized or reduced from its usual oxidation state of $+2$ and is not involved in redox reactions.

Table 35.4 lists some zinc-containing enzymes important for mammalian metabolism. Zinc has an important regulatory function in fructose-1,6-bisphosphatase, and has structural catalytic and regulatory roles in DNA and RNA polymerases. Many of the transcription factors, namely, DNA-binding proteins, contain zinc finger motifs consisting of Zn^{2+} bound to four cysteine or two cysteine and two histidine residues in a coordinate covalent linkage (Chapter 24). Thymidine kinase is also a zinc enzyme. Zinc is present in gustin, a salivary protein secreted by the parotid glands that may be necessary for the proper development of taste buds. A common sign of zinc deficiency is hypogeusesthesia. Gustin is structurally similar to nerve growth factor (NGF) isolated from male mouse submaxillary glands and other sources, which contains 1 mol of zinc per mole of protein. The metal prevents autocatalytic activation of NGF. Along with copper, zinc is required for activity of superoxide dismutase (Table 35.3).

Human zinc deficiency was first reported in boys from Iran and Egypt who were small in stature, hypogonadal, and mildly anemic. The zinc content of their hair, plasma, and erythrocytes was below normal range. Etiological factors included high dietary phytate content, loss of zinc through sweating, and geophagia (the practice of eating earth and clay, which bind zinc and other metals). All showed clinical improvement following oral supplementation with zinc.

Acrodermatitis enteropathica (AE) is an autosomal recessive disorder characterized by defective absorption of zinc; thickened, ulcerated skin on the extremities and around body orifices; alopecia; diarrhea; growth failure; and abnormalities in immune function. The concentration of zinc in serum and hair is reduced. AE responds to large amounts of oral zinc.

Molybdenum

In humans, molybdenum participates at the catalytic site of sulfite oxidase and xanthine oxidase. Molybdenum occurs as pterin-based cofactor and undergoes oxidation—reduction reactions. Molybdenum deficiency can lead to progressive neurological defects and mortality. The toxic effects have been attributed to accumulation of sulfite arising from the catabolism of sulfur-containing amino acids and lipids, due to deficient action of sulfite oxidase.

TABLE 35.4 Some Mammalian Zinc Proteins

Enzyme	Source	Molecular Weight	Zn Atoms/ Protein Molecule	Functions/Comments
Alcohol dehydrogenase	Human liver	87,000	4	Requires $2NAD^+$; two zinc atoms used for catalysis—function of the other two is not known but may stabilize enzyme
Alkaline phosphatase	Human placenta	125,000	~3	Unknown function; role of Zn^{2+} unknown but needed for activity
Leucine aminopeptidase	Porcine kidney	300,000	6	N-terminal exopeptidase; Zn binds at two sites: one for catalysis and the other for regulation
5′-AMP aminohydrolase	Rat muscle	290,000	2	–
Carbonic anhydrase B	Human erythrocyte	26,600	1	Catalyzes reversible hydration of CO_2; needed for CO_2 transport, buffering of ECF, gastric and renal acid secretion; Zn is catalytic
Carboxypeptidase A	Bovine pancreas	34,500	1	C-terminal exopeptidase; involved in digestion of proteins; Zn functions in catalysis as a Lewis acid
Glutamate dehydrogenase	Bovine liver	1,000,000	2–4	–
α_2-Macroglobulin	Human serum	840,000	3–8	–
Malate dehydrogenase	Bovine heart	40,000	1	–
Metallothionein	Human renal cortex	10,500	3	Usually also contains 4–5 Cd; may serve as source of Zn
Rhodanese (sulfur transferase)	Bovine liver	37,000	2	Metabolizes SCN^-

Selenium

Selenium is an essential trace element that has an important role in oxidation–reduction reactions in the body [27]. The primary inorganic form of selenium is sodium selenite, which is found in natural water and soil. Sodium selenite is inactive and requires organification to exert its biological actions. Sodium selenite is taken up and then incorporated into plants as an amino acid complex that contains sulfur (Se-methylselenocysteine). In humans, the organified form of selenium is selenocysteine, the 21st amino acid to be discovered (see Chapter 3). Proteins that contain this amino acid are referred to as selenoproteins. The incorporation of selenocysteine during selenoprotein synthesis requires the codon UGA; otherwise, in the absence of selenocysteine, the codon UGA serves as a stop codon. Of the 25 selenoproteins identified in humans, many have unknown functions, but the glutathione peroxidase family is well understood as antioxidant enzymes. Glutathione peroxidases reduce lipids to lipid alcohols, and hydrogen peroxidases to water, while oxidizing glutathione to glutathione disulfide, thereby preventing oxidative injury. Patients who live in geographic areas that lack selenium may develop selenium deficiency, which results in Keshan disease. Keshan disease was identified in 1935 in the Keshan province of China, characterized by rapidly progressive and fatal cardiomyopathy. It was not until 1962 that the connection between Keshan disease and selenium deficiency was made, due to morphologic similarities between Keshan disease and white muscle disease in sheep that lived in areas with low selenium levels in the soil. The absence of antioxidant selenoproteins was found to involve a variety of proatherothrombotic mechanisms in the development of cardiomyopathy, including inactivation of nitric oxide leading to increased platelet-dependent thrombosis, promotion of endothelial dysfunction and structural abnormalities of the myocardium, exacerbation of ischemia–reperfusion injury in cardiomyocytes, increased risk of cardiovascular events in patients with coronary artery disease, and increased risk of stroke.

CLINICAL CASE STUDY 35.1 Vitamin D Insufficiency

This clinical case is about vitamin D insufficiency in a healthy postmenopausal woman without any overt symptoms of skeletal disease and was derived from: C.J. Rosen, Vitamin D insufficiency, N. Engl. J. Med. 364 (2011) 248–254.

Synopsis

A healthy postmenopausal woman aged 61 years on a skeletal risk factor assessment was found to have a bone mineral density T-score of −1.5. A score of −1.5 is suggestive for a marginally decreased mineral density. Her serum 25-hydroxyvitamin D (25(OH)D) was 21 ng/mL, which is at the suboptimal level. She is 5 ft. 2 in. tall and weighs 130 lbs. She has no family history of hip fractures.

Recommendation by the physician:

1. Implementation of an exercise regimen;
2. Supplemental calcium intake of 1200 mg/day;
3. Regarding vitamin D supplementation, institution of daily supplementation of 600 IU (800 IU for individuals with risk factors).

It should be noted that vitamin D nutritional status is assessed by quantitation of serum 25(OH)D.

Teaching Points

1. Severe vitamin D deficiency [>10 ng/mL of serum 25 (OH)D] results in bone fractures, pain, and weakness of muscles. This condition requires prompt vitamin D supplementation.
2. Levels of vitamin D insufficiency considered are in the range of 10–30 ng/mL, and this condition requires either calcium or vitamin D or both, depending on the risk factors for fracture and age.
3. Regular exposure to sunlight rapidly induces production of vitamin D in the skin.
4. Intestinal inflammatory diseases and medications such as phenobarbital, phenytoin, and glucocorticoids can cause decreased serum 25(OH)D levels.
5. The association between several chronic diseases and low 25(OH)D have been implicated in observational studies. However, the mechanisms for these observations are not understood and require large-scale randomized controlled trials.

CLINICAL CASE STUDY 35.2 Disorders of Calcium Metabolism

Vignette 1: Hypercalcemia Due to Primary Hyperparathyroidism

This clinical case was abstracted from: C. Marcocci, F. Cetani, Primary hyperparathyroidism, N. Engl. J. Med. 365 (2011) 2389–2397.

Synopsis

A routine laboratory study of a 62-year-old woman revealed a serum calcium level of 10.8 mg/dL (reference range 8.4–10.4 mg/dL). Her serum parathyroid hormone (PTH) value was elevated at 70 pg/mL (reference range 11–54 pg/ mL). The patient also had hypertension controlled with an angiotensin receptor blocker. Her ratio of calcium to creatinine clearance was 0.025, and her bone mineral density was decreased. She was negative for any history of kidney stones and had an unremarkable family history. A recommendation of parathyroidectomy was made after ultrasonography and delayed-phase planar sestamibi scan of the right thyroid lobe showed a single enlarged parathyroid gland, later identified as an adenoma. Postoperative follow-up studies would include periodic measurement of serum calcium, 25-hydroxy vitamin D, and bone mineral density.

Teaching Points

1. Serum calcium levels are tightly regulated by parathyroid hormone (PTH). PTH acts on the bone and kidney to raise serum calcium levels when they are low. When high serum calcium levels are detected by calcium-sensing receptors on the surface of parathyroid hormone cells, PTH release is inhibited. The target tissues for PTH are the bone and kidney. In the bone, PTH releases calcium from a pool of soluble calcium stored in the extracellular fluid of bone. In the kidney, PTH increases the synthesis of 1,25-$(OH)_2$D, which mobilizes skeletal calcium and increases intestinal absorption of calcium. Abnormally elevated PTH can therefore cause elevated levels of serum calcium (hypercalcemia).

2. The most common cause of hypercalcemia is primary hyperparathyroidism (1 in 1000 with increased age). A major frequent cause of primary hyperparathyroidism is a single adenoma of the glands. Risk factors include head and neck irradiation, lithium therapy, and hyperplasia of the glands. Patients with hypercalcemia should be evaluated for use of drugs (e.g., lithium, thiazides), renal disease, and various malignancies. Malignancy-related hypercalcemia may be a cause of parathyroid hormone-related protein, which does not cross-react with serum parathyroid hormone determination. Rarely, ectopic parathyroid hormone production can occur in some tumors.

3. Symptoms of primary hyperparathyroidism include nephrolithiasis (formation of renal stones; see Vignette 2), decreased bone density, fatigue, anxiety, cognitive impairment, and hypertension. All patients with symptomatic primary hyperparathyroidism should be treated with parathyroidectomy, as there is no definitive medical

(Continued)

CLINICAL CASE STUDY 35.2 (Continued)

treatment. Preoperative treatment to reduce serum calcium levels may include saline infusion, loop diuretics, or bisphosphonates.

Supplemental References

[1] F.J. Fernandez-Fernandez, P. Sesma, Primary hyperparathyroidism, N. Engl. J. Med. 366 (2012) 860−861.

[2] N. Franceschini, A. Kshirsagar, Cinacalcet HCl: a calcimimetic agent for the management of primary and secondary hyperparathyroidism, Expert Opin. Investig. Drugs 8 (2003) 1413−1421.

Vignette 2: Calcium Kidney Stones Due to Hypercalciuria

This case was abstracted from: E.M. Worcester, F.L. Coe, Calcium kidney stones, N. Engl. J. Med. 363 (2010) 954−963.

Synopsis

A 43-year-old man underwent evaluation for recurrent kidney stones, for which he had been taking daily potassium citrate for the past 9 years. Chemical analysis of the kidney stones revealed a composition of 80% calcium oxalate and 20% calcium phosphate. The 24-hour urine calcium level was elevated at 408 mg (reference range 25−300 mg/24-hour specimen in males). The 24-hour urine oxalate level was within normal limits, and urine pH was 5.6 (reference range 4.5 to 8). In the absence of systemic disease, further workup for metabolic abnormalities suggested hypocitraturia and hypercalciuria as a cause of his stone formation. He was started on a thiazide-type diuretic, and he was advised to increase fluid intake and decrease dietary intake of sodium and protein.

Teaching Points

1. There are many inherited and systemic diseases associated with calcium kidney stones, but most stones have an unknown etiology (idiopathic).

2. Kidney stones form when urinary calcium oxalate or calcium phosphate concentrations exceed their solubility in urine. This condition, in which the solution contains more dissolved material than can be dissolved by the solvent under normal conditions, is called supersaturation, and is the driving force in kidney stone formation. Supersaturation is expressed as a ratio of the concentration of urinary calcium oxalate or calcium phosphate to its solubility. Crystals dissolve at supersaturation levels less than 1 but precipitate at supersaturation levels greater than 1.

3. There are several conditions that alter supersaturation, based on the type of stone. For example, calcium phosphate supersaturation rapidly increases when urine pH rises from 6 to 7, whereas calcium oxalate supersaturation remains unaffected by urine pH. Certain urinary substances can also affect the rate of stone formation, but at present the only substance that can be modified in clinical practice is citrate, which retards the growth of calcium crystals. Stone formation can also result from anatomic abnormalities that cause urinary stasis, or from metabolic abnormalities that affect the excretion of calcium and oxalate.

4. The most common metabolic abnormality found in patients with recurrent calcium kidney stones is familial or idiopathic hypercalciuria (see Vignette 3).

5. Kidney stones do not require removal or fragmentation unless they cause persistent pain, obstruction, infection, or serious bleeding. Otherwise, treatment focuses on pain control with opioid analgesics and nonsteroidal anti-inflammatory agents, and aiding in the passage of the stones with α_1-adrenergic receptor blockers or calcium-channel blockers. A study showed that α_1-adrenergic receptor blockers decrease the force of ureteral contraction, thereby allowing stone expulsion. Calcium channel blockers were also found to be effective in increasing the rate of stone expulsion.

6. The formation of recurrent kidney stones may be prevented by lowering urinary calcium and oxalate excretion and raising urine volume, thus decreasing supersaturation. Increasing fluid intake to a target urine volume of 2 to 2.5 liters per day was found to significantly reduce the recurrence of stone passage. Another study found that a diet low in animal protein, sodium, and oxalate, but with normal calcium intake, resulted in reduced stone formation. It is not recommended to restrict calcium from the diet because it may reduce bone mineral density and increase the rate of fracture. Thiazide-type diuretics were found to decrease urine calcium excretion and significantly reduce the recurrence rates of calcium stones.

Supplemental Reference

[1] M. Lipkin, O. Shah, The use of alpha-blockers for the treatment of nephrolithiasis, Rev. Urol. 8 (2006) S35−S42.

Vignette 3: Familial Hypocalciuric Hypercalcemia

This case was abstracted from: C. Marcocci, S. Borsari, E. Pardi, G. Dipollina, T. Giacomelli, A. Pinchera, F. Cetani, Familial hypocalciuric hypercalcemia in a woman with metastatic breast cancer: a case report of mistaken identity. J. Clin. Endocrinol. Metab. 88(11) (2003) 5132−5136.

Synopsis

A 45-year-old woman with metastatic breast cancer was found to have hypercalcemia. She was diagnosed with hypercalcemia secondary to malignancy and was treated with bisphosphonates. However, her serum calcium levels did not return to normal. Further investigation revealed a mildly elevated serum calcium level of 10.9 mg/dL (reference range 8.9−10.1 mg/dL), a normal parathyroid hormone (PTH) level of 24.4 pg/mL (reference range <65 pg/mL), a 24-hour urinary calcium level of 160 mg (reference range 25−300 mg/24-hour specimen), and a markedly decreased calcium/creatinine clearance ratio of 0.007 (reference range 0.01−0.02). These findings suggested the diagnosis of familial hypocalciuric hypercalcemia. Genetic studies revealed a missense mutation of the patient's calcium-sensing receptor gene. Diagnostic testing of the patient's family members revealed a similar mutation among those who also had laboratory values consistent with

(Continued)

CLINICAL CASE STUDY 35.2 (Continued)

familial hypocalciuric hypercalcemia. The patient was expected to have a benign course and did not receive treatment, but she underwent regular monitoring of serum calcium levels given her history of metastatic breast cancer.

Teaching Points

1. Familial hypocalciuric hypercalcemia is a heterogeneous disorder with three types (1, 2, and 3), caused by various germline loss-of-function mutations that affect the calcium-sensing receptor (CASR), resulting in abnormal receptor signaling, increased serum calcium levels, and decreased urinary calcium levels. See Table 35.5.

2. Familial hypocalciuric hypercalcemia type 1 is caused by a loss-of-function mutation of the calcium-sensing receptor gene (*CASR*), resulting in diminished sensitivity of the parathyroid gland to calcium. The result is uninhibited transcription and release of PTH, which raises serum calcium levels and causes hypercalcemia. Hypocalciuria in this disease is due to the mutant CASR located in the kidney, leading to increased calcium reabsorption and decreased calcium excretion in the urine.

3. A recent study showed that familial hypocalciuric hypercalcemia type 2 is caused by a loss-of-function mutation of

the gene that encodes G-protein subunit α_{11}. G-proteins (guanine nucleotide-binding proteins) are heterotrimeric complexes composed of an α-subunit (bound to guanosine triphosphate), a β-subunit, and a γ-subunit. The role of G-proteins is to couple with a wide variety of receptors to facilitate receptor interaction with downstream effectors, resulting in the regulation of many cellular processes. G-protein subunit α_{11} (Gα_{11}) is involved in calcium-sensing receptor signaling. Loss-of-function mutations of its gene (*GNA11*) results in diminished G-protein-coupled receptor signaling in response to calcium binding.

4. Familial hypocalciuric hypercalcemia type 3 is caused by altered calcium-sensing receptor endocytosis, due to a mutation of adaptor-related protein complex 2, sigma 1 subunit (AP2S1).

Supplemental References

[1] M.A. Nesbit, F.M. Hannan, S.A. Howles, V.N. Babinsky, R.A. Head, T. Cranston, et al., Mutations affecting G-protein subunit α_{11} in hypercalcemia and hypocalcemia, N. Engl. J. Med. 368 (2013) 2476–2486.

[2] A.M. Spiegel, G-proteins—the disease spectrum expands, N. Engl. J. Med. 368 (2013) 2515–2516.

[3] P. Glendenning, Summary statement from a workshop on asymptomatic primary hyperparathyroidism: a perspective for the 21st century, J. Clin. Endocrinol. Metab. 87 (2002) 5353–5361.

TABLE 35.5 Types of Familial Hypocalciuric Hypercalcemia

	Loss-of-Function Gene Mutation	Effect on Calcium-Sensing Receptor (CASR)
Type 1	CASR (calcium-sensing receptor)	Diminished sensitivity of CASR to calcium
Type 2	GNA11 (G-protein subunit α_{11})	Decreased G-protein-coupled CASR signaling
Type 3	AP2S1 (adaptor-related protein complex 2, sigma 1 subunit)	Altered CASR endocytosis

CLINICAL CASE STUDY 35.3 Wilson's Disease, a Disorder of Copper Metabolism

This case was abstracted from: N. Houchens, G. Dhaliwal, F. Askari, B. Kim, S. Saint, The essential element, N. Engl. J. Med. 368 (2013) 1345–1351.

Synopsis

A 21-year-old woman was being evaluated for a 10-day history of progressive fatigue, weakness, light-headedness, exertional dyspnea, dark urine, and syncope. Past medical history and family history were noncontributory. Her only medication was an oral contraceptive. On physical examination, she was febrile and tachycardic with scleral icterus, but otherwise had normal findings. Laboratory studies revealed leukocytosis, profound anemia, and elevated aminotransferases. Analysis of

peripheral blood smear revealed signs of hemolysis. The patient required multiple erythrocyte transfusions for refractory anemia but was discharged upon achieving hemodynamic stability after receiving prednisone therapy for a presumed autoimmune hemolytic anemia. Three months later, the patient returned with recurrent dark urine, fatigue, and jaundice. Ultrasound of the abdomen revealed a nodular liver with splenomegaly. Ophthalmologic slit-lamp examination revealed Kayser–Fleischer rings, and a 24-hour urinary copper level was markedly elevated at 4703 μg (reference range <55). Genetic testing revealed a mutation of the gene encoding adenosine triphosphatase copper transporter (*ATP7B*). She was diagnosed with Wilson's disease and was treated with

(Continued)

CLINICAL CASE STUDY 35.3 (Continued)

plasmapheresis, hemodialysis, and liver transplantation. Pathological examination of the explanted liver revealed copper deposits with a concentration of 526 µg per gram of liver (reference range 10−35). The patient's symptoms resolved.

Teaching Points

1. Wilson's disease is an autosomal recessive disorder characterized by ineffective copper metabolism in the liver. It is caused by a loss-of-function mutation of the gene that encodes an adenosine triphosphatase copper transporter (*ATP7B*) required for copper excretion into bile. ATP7B is also required to incorporate copper into ceruloplasmin precursors. Ceruloplasmin is the major copper-carrying protein in the serum. Failure to excrete copper results in its accumulation in the liver, and when released into the plasma, it is deposited in the brain, eyes, skin, and kidney.

2. Copper deposition may lead to the generation of reactive oxygen species that cause hepatic parenchymal injury, hepatocellular damage, and elevated levels of aminotransferases. In severe disease, hepatic necrosis releases massive amounts of copper into the plasma, causing oxidative damage and dysfunction to red blood cells.

3. Neurologic manifestations of Wilson's disease include movement disorders (dystonia, tremor, ataxia), dysarthria, dysphagia, and memory loss. Psychiatric abnormalities (psychosis, depression, behavioral disturbances) have also been reported.

4. Ocular deposition of copper results in Kayser−Fleischer rings (golden-brown rings at the corneoscleral junction) or sunflower cataracts (radiating, multicolored central opacities).

5. Untreated Wilson's disease is fatal. Treatment includes chelating agents like penicillamine or trientine that induce urinary copper excretion; zinc for maintenance therapy; or tetrathiomolybdate, which is an experimental drug that decreases copper absorption and deposition by forming stable copper−albumin complexes. Treatment for Wilson's disease must be continued indefinitely to prevent copper reaccumulation. In severe liver disease, transplantation may be required.

CLINICAL CASE STUDY 35.4 Paget's Disease, a Disorder of Bone Mineralization

This case was abstracted from: S.H. Ralston, Paget's disease of bone, N. Engl. J. Med. 368 (2013) 644−650.

Synopsis

A 73-year-old man presented with a 5-year history of low back pain, which was worse with standing. His symptoms progressed over the past year to include pain in the buttocks and legs while walking. He was taking acetaminophen with no relief of his symptoms. The neurologic examination was normal. Laboratory studies revealed an elevated total serum alkaline phosphatase level of 350 U/L (reference range 40−125). Radiographs of the spine revealed coarse trabecular patterning of the lumbar and thoracic vertebrae with expansion of the vertebral bodies. These characteristic radiographic findings led to the diagnosis of Paget's disease, with spinal stenosis. A trial of risedronate (an oral bisphosphonate) or zoledronic acid (an intravenous bisphosphonate) therapy was recommended.

Teaching Points

1. Paget's disease (osteitis deformans) is caused by increased and disorganized bone remodeling. Bone remodeling is a tightly regulated process that requires the resorption of old and defective bone by osteoclasts, followed by the formation of new bone by osteoblasts. There are several gene mutations associated with Paget's disease, all of which affect osteoclast differentiation and function.

2. Although the minerals calcium and phosphorus comprise the inorganic phase of bone, the process of bone mineralization and remodeling depends mainly on the organic phase of bone (osteoblasts and osteoclasts). Paget's disease therefore does not appear to be primarily due to derangements in calcium or phosphate, which is also suggested by normal serum calcium and phosphate levels seen in patients with Paget's disease. However, a rise in serum alkaline phosphatase is seen (a bone isoenzyme; alkaline phosphatase is also elevated in cholestasis; see Chapter 7). Serum alkaline phosphatase regulates bone mineralization by hydrolyzing pyrophosphate (a potent inhibitor of mineralization *in vitro*), and serves as a marker for increased osteoblastic activity.

3. The most common symptom of Paget's disease is localized bone pain, usually of the axial skeleton (pelvis, femur, lumbar spine, skull, and tibia). It is a disease that predominantly affects older people, rarely occurring before the age of 55 years. Infection is a potential trigger.

4. Bisphosphonates are the first-line treatment for localized bone pain due to increased metabolic activity caused by Paget's disease. Salmon calcitonin has also been shown to be very effective in the treatment of pain related to Paget's disease.

REQUIRED READING

[1] C. Scholtysek, J. Katzenbeisser, H. Fu, S. Uderhardt, N. Ipseiz, C. Stoll, et al., PPARβ/δ governs Wnt signaling and bone turnover, Nat. Med. 19 (2013) 608–613.

[2] D. Nikitovic, J. Aggelidakis, M.F. Young, R.V. Lozzo, N.K. Karamanos, G.N. Tzanakakis, The biology of small leucine-rich proteoglycans in bone pathophysiology, J. Biol. Chem. 287 (2012) 33926–33933.

[3] Q.H. Meng, E.A. Wagar, Severe hypophosphatemia in a 79-year-old man, Clin. Chem. 60 (2014) 928–932.

[4] O.R. Hamnvik, C.B. Becker, B.D. Levy, J. Loscalzo, Wasting away, N. Engl. J. Med. 370 (2014) 959–966.

[5] C. Bergwitz, M.T. Collins, R.S. Kamath, A.E. Rosenberg, Case 33–2011: A 56-year-old man with hypophosphatemia, N. Engl. J. Med. 365 (2011) 1625–1635.

[6] M. Pi, L.D. Quarles, Novel bone endocrine networks integrating mineral and energy metabolism, Curr. Osteoporos. Rep. 11 (2013) 391–399.

[7] J. Donate-Correa, M.M. de Fuentes, C. Mora-Fernandez, J.F. Navarro-Gonzalez, Pathophysiological implications of fibroblast growth factor-23 and Klotho and their potential role as clinical biomarkers, Clin. Chem. 60 (2014) 933–940.

[8] M.F. Holick, Vitamin D deficiency, N. Engl. J. Med. 357 (2007) 266–281.

[9] T.D. Thacher, B.L. Clarke, Vitamin D insufficiency, Mayo Clin. Proc. 86 (2011) 50–60.

[10] K.A. Kennel, M.T. Drake, D.L. Hurley, Vitamin D deficiency in adults: when to test and how to treat, Mayo Clin. Proc. 85 (2010) 752–758.

[11] M.F. Holick, Bioavailability of vitamin D and its metabolites in black and white adults, N. Engl. J. Med. 369 (2013) 2047–2048.

[12] C.E. Powe, M.K. Evans, J. Wenger, A.B. Zonderman, A.H. Berg, M. Nalis, et al., Vitamin D-binding protein and vitamin D status of black Americans and white Americans, N. Engl. J. Med. 369 (2013) 1991–2000.

[13] C. Marcocci, F. Cetani, Primary hyperparathyroidism, N. Engl. J. Med. 365 (2011) 2389–2397.

[14] M. Tonelli, N. Pannu, B. Manns, Oral phosphate binders in patients with kidney failure, N. Engl. J. Med. 362 (2010) 1312–1324.

[15] Y.H. Lien, Phosphorus: another devil in our diet? Am. J. Med. 126 (2013) 280–281.

[16] M.P. Whyte, C.R. Greenberg, N.J. Salman, M.B. Bober, W.H. McAlister, D. Wenkert, et al., Enzyme-replacement therapy in life-threatening hypophosphatasia, N. Engl. J. Med. 366 (2012) 904–913.

[17] D.C. Bauer, Calcium supplements and fracture prevention, N. Engl. J. Med. 369 (2013) 1537–1543.

[18] F.S. van Dijk, M.C. Zillikens, D. Micha, M. Riessland, C.L.M. Marcelis, C.E. de Die-Smulders, et al., *PLS3* mutations in X-linked osteoporosis with fractures, N. Engl. J. Med. 369 (2013) 1529–1536.

[19] H.D. Nelson, E.M. Haney, T. Dana, C. Bougatsos, R. Chou, Screening for osteoporosis: an update for the U.S. Preventive Services Task Force, Ann. Intern. Med. 153 (2010) 99–111.

[20] D.J. Prockop, New targets for osteoporosis, N. Engl. J. Med. 367 (2012) 2353–2354.

[21] M. McClung, S.T. Harris, P.D. Miller, D.C. Bauer, K.S. Davison, L. Dian, et al., Bisphosphonate therapy for osteoporosis: benefits, risks, and drug holiday, Am. J. Med. 126 (2013) 13–20.

[22] C.M. Laine, K.S. Joeng, P.M. Campeau, R. Kiviranta, K. Tarkkonen, M. Grover, et al., WNT1 mutations in early-onset osteoporosis and osteogenesis imperfecta, N. Engl. J. Med. 368 (2013) 1809–1816.

[23] C.B. Becker, Sclerostin inhibition for osteoporosis—a new approach, N. Engl. J. Med. 370 (2014) 476–477.

[24] M. Mannstadt, A.E. Lin, L.P. Le, Case 24–2014: a 27-year-old man with severe osteoporosis and multiple bone fractures, N. Engl. J. Med. 371 (2014) 465–472.

[25] M. Nestle, M.C. Nesheim, To supplement or not to supplement: The U.S. Preventive Services Task Force recommendations on calcium and vitamin D, Ann. Intern. Med. 158 (2013) 701–702.

[26] V.A. Moyer, Vitamin D and calcium supplementation to prevent fractures in adults: U.S. Preventive Services Task Force recommendation statement, Ann. Intern. Med. 158 (2013) 691–696.

[27] J. Loscalzo, Keshan disease, selenium deficiency, and the selenoproteome, N. Engl. J. Med. 370 (2014) 1756–1760.

ENRICHMENT READING

[1] L.A. Allen, A.V. Ambardekar, K.M. Devaraj, J.J. Maleszewski, E.E. Wolfel, Missing elements of the history, N. Engl. J. Med. 370 (2014) 559–566.

[2] W.M. Konstantopoulos, M.B. Ewald, D.S. Pratt, Case 22–2012: A 34-year-old man with intractable vomiting after ingestion of an unknown substance, N. Engl. J. Med. 367 (2012) 259–268.

[3] F.P. Guengerich, Thematic minireview series: metals in biology 2013, J. Biol. Chem. 288 (2013) 13164.

[4] H.M. Girvan, A.W. Munro, Heme sensor proteins, J. Biol. Chem. 288 (2013) 13194–13203.

[5] C. Gherasim, M. Lofgren, R. Banerjee, Navigating the B$_{12}$ road: assimilation, delivery, and disorders of cobalamin, J. Biol. Chem. 288 (2013) 13186–13193.

[6] R.R. Mendel, The molybdenum cofactor, J. Biol. Chem. 288 (2013) 13165–13172.

[7] Y. Hu, M.W. Ribbe, Biosynthesis of the iron-molybdenum cofactor of nitrogenase, J. Biol. Chem. 288 (2013) 13173–13177.

[8] M.A. Farrugia, L. Macomber, R.P. Hausinger, Biosynthesis of the urease metallocenter, J. Biol. Chem. 288 (2013) 13178–13185.

[9] F.P. Guengerich, Thematic minireview series: metals in biology 2012, J. Biol. Chem. 287 (2012) 13508–13509.

[10] J.M. Argüello, D. Raimunda, M. González-Guerrero, Metal transport across biomembranes: emerging models for a distinct chemistry, J. Biol. Chem. 287 (2012) 13510–13517.

[11] C.C. Philpott, Coming into view: eukaryotic iron chaperones and intracellular iron delivery, J. Biol. Chem. 287 (2012) 13518–13523.

[12] C. Correnti, R.K. Strong, Mammalian siderophores, siderophore-binding lipocalins, and the labile iron pool, J. Biol. Chem. 287 (2012) 13524–13531.

[13] Y. Nicolet, J.C. Fontecilla-Camps, Structure-function relationships in [FeFe]-hydrogenase active site maturation, J. Biol. Chem. 287 (2012) 13532–13540.

[14] J.D. Aguirre, V.C. Culotta, Battles with iron: manganese in oxidative stress protection, J. Biol. Chem. 287 (2012) 13541–13548.

[15] V. Hodgkinson, M.J. Petris, Copper homeostasis at the host-pathogen interface, J. Biol. Chem. 287 (2012) 13549–13555.

Chapter 36

Vitamin Metabolism

Key Points

1. Vitamins are a heterogeneous group of organic compounds; are generally not synthesized in the body; and are required in catalytic amounts to maintain growth, reproduction, and homeostasis by participating in several metabolic pathways.
2. They are divided into fat-soluble (vitamins A, D, E, and K) and water-soluble (B-complex and vitamin C) vitamins.
3. Metabolic functions of many of these vitamins are discussed throughout the text. This chapter summarizes and recapitulates the functions of vitamins A, E, thiamine, riboflavin, pyridoxine, cobalamin (vitamin B_{12}), folic acid, niacin, pantothenic acid, biotin, and ascorbic acid (vitamin C). Clinical aspects of hyper- and hypovitaminosis for each of these are discussed.

INTRODUCTION

The word **vitamin** is used to describe any of a heterogeneous group of organic molecules that are needed in small quantities for normal growth, reproduction, and homeostasis, but that the human body is unable to synthesize in adequate amounts. The group includes the fat-soluble vitamins (A, D, E, and K) and the water-soluble vitamins (B-complex and C). Vitamins are generally needed in catalytic quantities and do not function as structural elements in the cell. Other organic compounds are not synthesized in the body but are required for maintenance of normal metabolism, such as essential fatty acids (Chapter 16) and essential amino acids (Chapter 3). These substances are needed in relatively large quantities, serve as nonregenerated substrates in metabolic reactions, and are used primarily as structural components in lipids and proteins. Vitamins discussed in other chapters include vitamin D (Chapter 35) and vitamin K (Chapter 34). All the B vitamins function as cofactors or precursors for cofactors in enzyme-catalyzed reactions and are discussed in appropriate chapters. Additional properties and less-well-defined actions are reviewed here.

Vitamin deficiency is caused by nutritional inadequacy, or may result from malabsorption, effects of pharmacological agents, and abnormalities of vitamin metabolism or utilization in the metabolic pathways. Thus, in biliary obstruction or pancreatic disease, the fat-soluble vitamins are poorly absorbed despite adequate dietary intake, because of steatorrhea. Absorption, transport, activation, and utilization of vitamins require the participation of enzymes or other proteins whose synthesis is under genetic control. Dysfunction or absence of one of these proteins can produce a disease that is clinically indistinguishable from one caused by dietary deficiency. In vitamin-dependent or vitamin-responsive disorders, use of pharmacological doses of the vitamin can overcome the blockage sufficiently for normal function to occur.

Vitamin deficiency can result from treatment with certain drugs. Thus, destruction of intestinal microorganisms by antibiotic therapy can produce symptoms of vitamin K deficiency. Isoniazid, used to treat tuberculosis, is a competitive inhibitor of pyridoxal kinase, which is needed to produce pyridoxal phosphate. Isoniazid can produce symptoms of pyridoxine deficiency. To prevent this, pyridoxine is often incorporated into isoniazid tablets. Methotrexate and related folate antagonists act by competitively inhibiting dihydrofolate reductase (Chapter 25).

Excessive intake of vitamins A and D produces **hypervitaminosis**. Vitamin D toxicosis was discussed in Chapter 35; vitamin A toxicosis is discussed later in this chapter. The toxicity of high doses of vitamin B_6 is also covered later in the chapter.

Certain vitamins can be synthesized by humans in limited quantities. Niacin (also known as nicotinic acid or vitamin B_3) can be formed from tryptophan (Chapter 15). This pathway is not active enough to satisfy all the body's needs; however, when one is calculating the RDA for niacin, 60 mg of dietary tryptophan is considered equivalent to 1 mg of dietary niacin. In Hartnup's disease (Chapter 15), a rare hereditary disorder in the transport of monoaminomonocarboxylic acids (e.g., tryptophan), a pellagra-like rash may appear, suggesting that over a long period of time dietary intake of niacin is insufficient for metabolic needs. This pattern also occurs in carcinoid syndrome, in which much tryptophan is shunted into the synthesis of 5-hydroxytryptamine. Vitamin D is

N. V. Bhagavan and C.-E. Ha: Essentials of Medical Biochemistry, Second edition. DOI: http://dx.doi.org/10.1016/B978-0-12-416687-5.00036-1

synthesized in the skin, provided radiant energy is available for the conversion (Chapter 35):

7-Dehydrocholesterol $\xrightarrow{\text{photons}}$ cholecalciferol (vitamin D_3)

Physiological age-related changes in the elderly can affect nutritional status. Decreased active intestinal transport and atrophic gastritis impair the absorption of vitamins and other nutrients. Reduced exposure to sunlight can lead to decreased vitamin D synthesis. Many drugs may impair both appetite and absorption of nutrients. Some examples of unfavorable drug–nutrient interactions are drugs that inhibit stomach acid production (e.g., omeprazole); drugs that reduce vitamin B_{12} absorption; anticonvulsant drugs (e.g., barbiturates, phenytoin, primidone) that act by inducing hepatic microsomal enzymes which accelerate inactivation of vitamin D metabolites and aggravate osteoporosis (Chapter 35); interference with folate metabolism by antifolate drugs (methotrexate) used in the treatment of some neoplastic diseases; and vitamin B_6 metabolism affected by isoniazid, hydralazine, and D-penicillamine. Examples of negative impacts of vitamins on drug action are vitamin B_6-dependent action of peripheral conversion of L-3,4-dihydroxyphenylalanine (L-dopa) to L-dopamine, which is mediated by aromatic L-amino acid decarboxylase and prevents L-dopa's transport across the blood–brain barrier; also, ingestion of large amounts of vitamin K-rich foods or supplements, and action of warfarin on anticoagulation (Chapter 34). L-dopa is the metabolic precursor of L-dopamine and is used in the treatment of Parkinson's disease (Chapter 30). Thus, L-dopa is administered along with a peripherally acting inhibitor of aromatic L-amino acid decarboxylase (e.g., carbidopa).

FAT-SOLUBLE VITAMINS

The fat-soluble vitamins share many properties, despite their limited chemical similarity. They are absorbed into the intestinal lymphatics, along with other dietary lipids, after emulsification by bile salts. Lipid malabsorption accompanied by steatorrhea usually results in poor uptake of all the fat-soluble vitamins. Deficiency disease (except in the case of vitamin K) is difficult to produce in adults because large amounts of most fat-soluble vitamins are stored in the liver and in adipose tissue. The fat-soluble vitamins are assembled from isoprenoid units; this fact is apparent from examination of the structures of vitamins A, E, and K; cholesterol, the precursor of vitamin D, is derived from six isoprenoid units (Chapter 17). Specific biochemical functions for vitamins A, D, and K are known, but a role for vitamin E, other than as a relatively nonspecific antioxidant, remains elusive.

Vitamin A

Nutrition and Chemistry

The role of vitamin A in vision is fairly well understood, but the other functions of vitamin A are only beginning to be elucidated (e.g., mucus secretion, maintenance of the integrity of differentiated epithelia and of the immune system, growth, and reproduction). Loss of night vision (nyctalopia) is an early sign of vitamin A deficiency, and clinical features of well-developed deficiency include epidermal lesions, ocular changes, growth retardation, glandular degeneration, increased susceptibility to infection, and sterility.

Natural and synthetic compounds with vitamin A activity and inactive synthetic analogues of vitamin A are collectively termed **retinoids**. The most biologically active, naturally occurring retinoid is all-*trans* retinol, also called **vitamin A** (Figure 36.1). The β-ionone ring is required for biological activity. Vitamin A_1 exists free or as retinyl esters of fatty acids (primarily palmitic acid) in foods of animal origin including eggs, butter, cod liver oil, and the livers of other vertebrates. In most dietary retinol, the four double bonds of the side chain are in the *trans* configuration. They are readily oxidized by atmospheric oxygen, inactivating the vitamin. They can be protected by antioxidants such as vitamin E.

Several provitamins are present in yellow and dark green, leafy vegetables and fruits, such as carrots, mangoes, apricots, collard greens, and broccoli. They are collectively known as the **carotenoid pigments** or **carotenes**. The most widely occurring and biologically active is β-carotene. Other nutritionally important carotenoid pigments are cryptoxanthin, a yellow pigment found in corn, and α- and γ-carotenes. Other carotenoids, such as lycopene (the red pigment of tomatoes) and xanthophyll, lack the β-ionone ring essential for vitamin A activity.

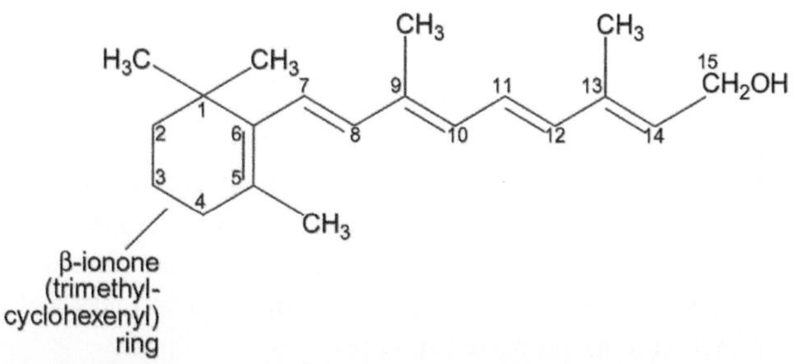

FIGURE 36.1 Chemical structure of all-*trans* retinol (vitamin A_1), the most active form a vitamin A. Oxidation of C_{15} to an aldehyde or an acid produces retinaldehyde (retinal) and retinoic acid, respectively. The *cis–trans* isomerization of the double bond between C_{11} and C_{12} occurs during functioning of retinaldehyde in vision.

Absorption, Transport, and Metabolism

Retinyl esters are hydrolyzed in the intestinal lumen by pancreatic carboxylic ester hydrolase, which also hydrolyzes cholesteryl esters. In mucosal cells, retinol is re-esterified, mostly with long-chain fatty acids, by acyl-CoA:retinol acyltransferase. The retinyl esters are incorporated into chylomicron particles and secreted into the lacteals. In humans and rats, 50% of the retinol is esterified with palmitic acid, 25% with stearic acid, and smaller amounts with linoleic and oleic acids. These esters are eventually taken up by the liver in chylomicron remnants. β-Carotene is cleaved in the intestinal mucosa to two molecules of retinaldehyde by β-carotene-15,15′-dioxygenase, a soluble enzyme. Other provitamins are also activated on cleavage by this enzyme. Retinaldehyde is then reduced to retinol by retinaldehyde reductase, using either NADH or NADPH.

In the liver, retinyl esters are hydrolyzed and re-esterified. More than 95% of hepatic retinol is present as esters of long-chain fatty acids, primarily palmitate. In an adult receiving the RDA of vitamin A, a year's supply or more may be stored in the liver. More than 90% of the body's supply of vitamin A is stored in the liver. The hepatic parenchymal cells are involved in its uptake, storage, and metabolism. Retinyl esters are transferred to hepatic fat-storing cells (also called **Ito cells** or **lipocytes**) from the parenchymal cells. The capacity of these fat-storing cells may determine when vitamin A toxicosis becomes symptomatic. During the development of hepatic fibrosis (e.g., in alcoholic liver disease), vitamin A stores in Ito cells disappear, and the cells differentiate to myofibroblasts. These cells appear to be the ones responsible for the increased collagen synthesis seen in fibrotic and cirrhotic livers.

Retinol is released from the liver and transported in plasma bound to retinol-binding protein (RBP), which is synthesized by hepatic parenchymal cells. Less than 5% circulates as retinyl esters. Retinol-binding protein from human plasma is a monomeric polypeptide (M.W. 21,000) that has a single binding site. Transfer of retinol into cells may be mediated by cell surface receptors that specifically recognize RBP. After binding and releasing its vitamin A, RBP appears to have decreased affinity for prealbumin (discussed later) and is rapidly filtered by the kidney and degraded or excreted.

The retinol—RBP complex circulates as a 1:1 complex with prealbumin (**transthyretin**; M.W. 55,000), which also functions in the transport of thyroid hormones. Interaction between transthyretin and RBP is quite specific (dissociation constant $10^{-6}-10^{-7}$ mol/L).

The retinol—RBP—transthyretin complex (RTC) (M.W. 76,000) migrates electrophoretically as an α-globulin. Normal concentrations of RBP and transthyretin in human plasma are 40—50 and 200—300 μg/mL, respectively.

Transthyretin, a glycoprotein with a high tryptophan content, is synthesized in the liver and migrates electrophoretically ahead of albumin. It has a half-life of 2 days and has a small pool size. These two properties make it a sensitive indicator of nutritional status (Chapter 15). Its plasma level decreases in protein—calorie malnutrition, liver disease, and acute inflammatory diseases. Oxidation of retinol produces retinoic acid (tretinoin). The reaction is irreversible. Retinoic acid enters the portal blood, is transported bound to albumin, and is not stored to any great extent.

Function

Vitamin A and its derivatives retinal and retinoic acid have many essential biological roles in such processes as vision, regulation of cell proliferation and differentiation, and as morphogenetic agents during embryonic development and differentiation. Both natural and synthetic analogues (known as **retinoids**) possess varying degrees of biological activity ascribed to vitamin A. Other than the role of retinal in vision, retinoids exert their biological actions by binding to specific nuclear receptors, such as transcription factors that modulate gene expression. The retinoid receptors belong to two subfamilies: retinoic acid receptors (RARs) and retinoid X receptors (RXRs). Both the RARs and RXRs contain three isotypes—α, β, and γ—encoded by separate genes. Thus, the retinoid receptor family includes RARα, RARβ, RARγ, RXRα, RXRβ, and RXRγ, which are members of the steroid and thyroid hormone superfamily of receptors (Chapter 28). All of these receptors possess at least two domains: a ligand-binding domain and a DNA-binding domain. The DNA-binding domain of the receptor recognizes retinoic acid response elements within the target gene via the two zinc finger binding motifs.

The receptors form dimers of various combinations and permutations within the superfamily, and each dimer may perform a specific biological function by binding to specific DNA response elements. Activation of retinoid receptors can lead to inhibition of cell proliferation, induction of differentiation, and induction of apoptosis during normal cell development. The importance of RARs in cell proliferation and differentiation is illustrated by acute promyelocytic leukemia (PML), which is a subtype of acute myeloid leukemia. PML is characterized by abnormal hypergranular promyelocytes, a cytogenetic translocation, and a bleeding disorder secondary to consumptive coagulopathy and fibrinolysis. The bleeding problems are presumably initiated by procoagulant phospholipids present in the leukemic cells (Chapter 34).

The cytogenetic hallmark of PML consists of a balanced reciprocal translocation between the long arms of chromosomes 15 and 17, designated as t(15;17)

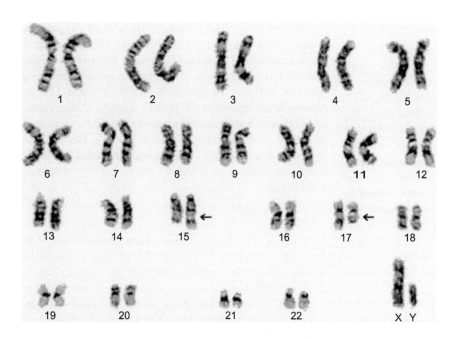

FIGURE 36.2 Karyotype of a promyelocytic leukemic cell, with balanced reciprocal translocation between chromosomes 15 and 17, t(15;17). The rearranged chromosomes are identified by arrows. *[Courtesy of Dr. Mark H. Bogart.]*

(Figure 36.2). The translocation results in two recombinant chromosomes: one abnormally long $15q^+$ and one shortened $17q^-$. This translocation fuses the PML gene located on chromosome 15 with the RARα gene located on chromosome 17. The two chimeric genes formed are PML-RARα and RARα-PML. Of these, the PML-RARα gene is transcriptionally active and produces a PML-RARα protein. This protein, which is an oncoprotein, is responsible for the pathogenesis of PML by interfering with the normal functions of PML and RARα genes in differentiation of myeloid precursors. The normal PML gene product has growth suppressor activity. Treatment with all-*trans* retinoic acid has produced complete remissions in PML by promoting the conversion of leukemic blast cells into mature cells. The reason is that the PML-RARα oncoprotein maintains responsiveness to retinoic acid's biological actions. Retinoic acid therapy may also augment the degradation of the PML-RARα fusion protein. Transgenic mice with a PML-RARα fusion gene develop PML, which supports the pathogenic role of the fusion protein in human PML.

More than 1500 retinoids have been synthesized. Two in particular, 13-*cis*-retinoic acid (isotretinoin) and etretinate (Figure 36.3), have generated considerable interest as agents for the treatment of dermatological disorders. 13-*Cis*-retinoic acid inhibits sebum production, is the drug of choice for treatment of severe cystic acne, and is useful in the treatment of other forms of acne. Etretinate is used in Europe for treatment of psoriasis and related disorders.

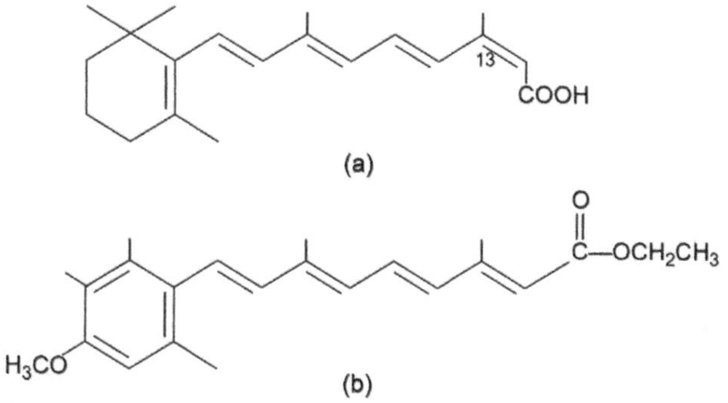

Hypo- and Hypervitaminosis A

Serious vitamin A deficiency is a major public health problem; where poverty and poor nutrition are common,

FIGURE 36.3 Structures of 13-*cis*-retinoic acid (isotretinoin) (a) and etretinate (b), synthetic retinoids used as pharmacological agents.

many persons suffer with active corneal involvement. This condition leads to permanent blindness in about 50% of cases; many of those affected are preschool children. Night blindness occurs early in vitamin A deficiency. There is a reduction in rhodopsin concentration, followed by retinal degeneration and loss of photoreceptor cells. Degenerative changes may be due to instability of free opsin or may indicate an additional nutritive function for retinaldehyde (or retinoic acid) in retinal cells. The degenerative changes of retinitis pigmentosa, a relatively common, inherited cause of blindness, closely resemble those of vitamin A deficiency.

Ingestion of vitamin A in high excess of the RDA can cause toxicity. A daily intake of more than 7500 retinol equivalents (25,000 IU) is not recommended, and doses in excess of 3000 retinol equivalents (10,000 IU) should be used only with medical supervision. Acute toxicity after a single large dose may last several days. Chronic toxicity occurs in adults after daily doses of 7500−15,000 retinol equivalents (25,000−50,000 IU) taken for several months.

Vision and Vitamin A

Photosensitivity in various organisms is based on photoisomerization of retinaldehyde; in humans, 11-*cis*-retinaldehyde complexes with opsin to produce rhodopsin (visual purple). Absorption of a photon by the π-electron system of retinaldehyde changes it to the all-*trans* isomer; this reaction—the only light-catalyzed step in vision—is transduced to an action potential transmitted in the optic nerve. The human visual system functions over a range of light intensity. If properly dark adapted, the eye can detect a single photon. Vitamin A deficiency reduces the amount of rhodopsin in the retina, thereby increasing the minimum amount of illumination that can be detected (the visual threshold) and causing night blindness.

The retina consists of a thin layer of photoreceptor cells connected to the optic nerve via axons, synapses, and the bodies of several intermediate cells. Other structures of the eye provide the optical system for focusing light onto the retina. Visual pigments are present in the approximately 100 million rod and 5 million cone cells. The cones are less sensitive than the rods but are required for color vision. Three types of cone cells differ with respect to the wavelengths of light to which they are maximally sensitive. Comparison and integration of inputs from these three cell types by the visual cortex produce color vision. At low light levels, only the rods are active and color vision is lost. A higher pigment content and greater molar absorptivity of the pigment contribute to the higher sensitivity of rods.

Rhodopsin has a broad absorption spectrum with a maximum at 500 nm, making the rods most sensitive to green light. In the human retina, cone cell absorption maxima are 445 nm (blue), 535 nm (green), and 570 nm (red). Opsins in the three types of cone cells have different primary structures and are coded by three separate genes. Those for red- and green-sensitive opsins are on the X-chromosome, while that for blue-sensitive opsin is on an autosome. Congenital color blindness, which affects about 9% of the male population, results from the absence of one or more cone cell types or from a decrease in the amount of one of the pigments. Absence of red or green cones occurs in 2.5% of males. Absence of blue cones occurs in only 0.001% of males. Decrease in red cone pigment (protanomaly) or green cone pigment (deuteranomaly) occurs in 1.3% and 5.0%, respectively, of males. The decrease in pigment type presumably is due to a mutation in a structural gene. Absence of two or three types of cone cells is extremely rare.

Light reception and energy transduction take place in the rod outer segments, while cellular metabolism occurs

in the rod inner segment, which is rich in glycogen, mitochondria, and rough endoplasmic reticulum. The outer segment contains 500−1500 flattened membrane sacs (disks), each about 16 nm thick, which are electrically isolated from the surrounding plasma membrane and which contain rhodopsin as an integral, transmembrane protein. The disks are continually replaced from the outer segment nearest the nucleus. Phagocytosis of old disks by cells of the retinal pigment epithelium occurs at the top of the outer segment. In the rhesus monkey, the lifetime of a disk is 9−13 days. Disks are formed by sealing off invaginations of plasma membrane; the composition of their contents differs from that of cytoplasm. Blindness due to hereditary defects in regeneration of disks has been observed in rats.

Cone cells are similar in many ways to rod cells. Because cone cells lack distinct disks, their visual pigments are located on deep invaginations of the plasma membrane that resemble disks and probably are formed by a similar mechanism. Disk membranes are composed primarily of phospholipid and rhodopsin at a molecular ratio of about 70:1. The phospholipids are 40% phosphatidylcholine, 40% phosphatidylethanolamine, and 13% phosphatidylserine. The membrane has high fluidity because the lipids are extensively unsaturated, which gives rhodopsin considerable rotational and translational mobility. Average spacing of rhodopsin molecules is 5.6 nm.

In the dark, rhodopsin has a red color, which changes to pale yellow on exposure to light. This bleaching takes place within a few milliseconds after absorption of a photon with the formation of all-*trans* retinaldehyde and opsin (Figure 36.4).

In the dark, the cation (Na^+)-specific cGMP-gated channels located in the rod outer segment (ROS) are

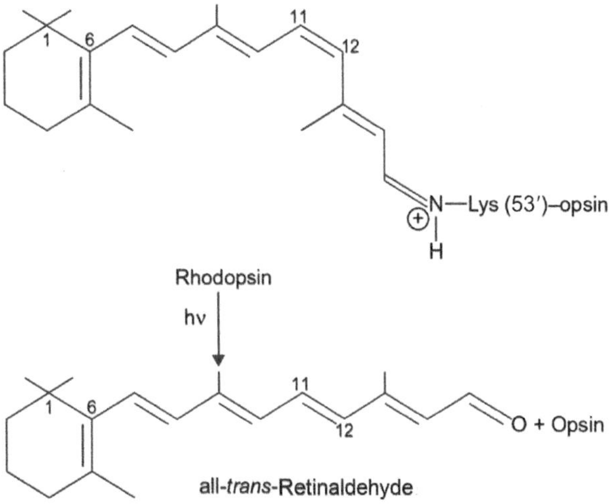

FIGURE 36.4 Absorption of light by 11-*cis*-retinaldehyde initiates the configurational change to all-*trans* retinaldehyde, culminating in hydrolysis of the bond between retinaldehyde and Lys of opsin.

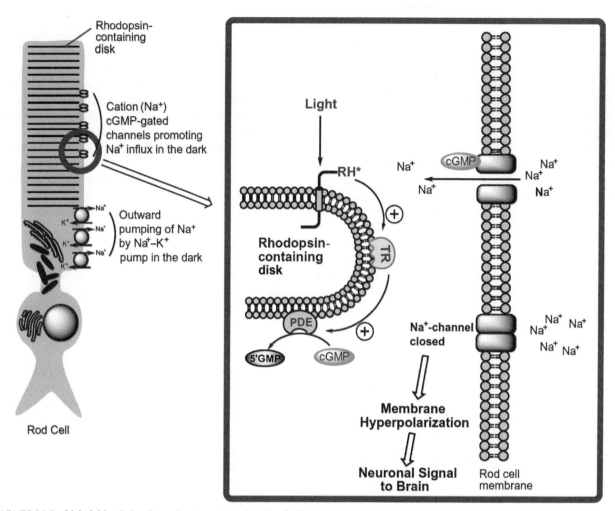

FIGURE 36.5 Light-initiated signal transduction pathway in rod cells. In the dark (left side), the steady-state level of cations is maintained by passive inward Na^+ import into the outer segment and active Na^+ export from the inner segment. In the light-initiated amplification system (right side), depletion of cGMP leads to closing of the Na^+-gated channels, followed by hyperpolarization and inhibition of neurotransmitter release (see text for details and Figure 36.6 for signal amplification). RH* = activated rhodopsin; TR = transducin; PDE = phosphodiesterase; + = stimulation.

open, thus promoting the influx of Na^+. The steady-state of cations is maintained by outward pumping of Na^+ by Na^+/K^+-ATPase pumps located in the inner segment. Exposure to light blocks cGMP-gated cation channels, causing the inside of the plasma membrane to become more negative and resulting in hyperpolarization. The signal of hyperpolarization is transmitted to the synaptic body and eventually to the brain. In the dark, because of the influx of Na^+, the rod cells are depolarized, and neurotransmitters (possibly aspartate or glutamate) are released from the presynaptic membrane of the rod cell at relatively high rates. In the hyperpolarized state, neurotransmitter release is inhibited (Figure 36.5).

The coupling of rhodopsin present in the disk membrane to cation-specific cGMP-gated channels of the plasma membrane involves signal amplification after photon absorption. The coupling mechanism (Figure 36.6) begins with the absorption of light by 11-*cis*-retinal-rhodopsin and its conversion to all-*trans* retinal and photoactivated rhodopsin (R*). In the next step, activated rhodopsin initiates an amplification process that consists of activating transducin, a signal-coupling G-protein. The active form of transducin, $T_{\alpha\text{-GTP}}$, activates cGMP-phosphodiesterase (PDE). In the second amplification process, PDE hydrolyzes cGMP with the closure of cation channels. The opening and closing of these cation channels mediated by cGMP is similar to Ca^{2+} ion channel gating by cAMP in olfaction.

Photoactivation of rhodopsin produces a conformational change that allows it to interact with and activate transducin, resulting in the replacement of GDP by GTP at the α-subunit. Depending on the lifetime of the activated rhodopsin molecule, each molecule can activate up to 500 transducin molecules. The α-subunit GTP complex of

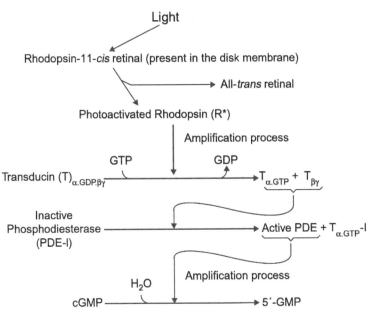

Light

Rhodopsin-11-*cis* retinal (present in the disk membrane)

⟶ All-*trans* retinal

Photoactivated Rhodopsin (R*)

Amplification process

GTP   GDP

Transducin (T)$_{\alpha.GDP.\beta\gamma}$ ⟶ T$_{\alpha.GTP}$ + T$_{\beta\gamma}$

Inactive
Phosphodiesterase ⟶ Active PDE + T$_{\alpha.GTP}$-I
(PDE-I)

Amplification process

H_2O

cGMP ⟶ 5′-GMP

Plasma membrane cation channels closure;
Decreased Ca^{2+} and Na^+ concentration; hyperpolarization

FIGURE 36.6 Mechanism for stimulus−response coupling of photon absorption to the closure of plasma membrane cation channels. I = inhibitory subunit of phosphodiesterase.

transducin disinhibits PDE by removing its inhibitor subunit (I), causing it to hydrolyze cGMP to 5′-GMP. The decrease in cGMP concentration closes the plasma membrane cation channels, decreasing Na^+ and Ca^{2+} concentrations. Some cases of retinitis pigmentosa are caused by mutations in the transducin gene that prevent it from disinhibiting PDE, leading to an accumulation of cGMP that is toxic to the retina. The transducin-GTP complex remains active as long as the GTP bound to G_α is not hydrolyzed.

The recovery and adaptation of the rod and the cone cells to the dark state begin shortly after illumination. The GTP bound to the α-subunit is hydrolyzed by the intrinsic GTPase activity of the α-subunit, followed by its reassociation with $T_{\beta\gamma}$ to form inactive $T_{\alpha.GDP.\beta\gamma}$. The formation of $T_{\alpha.GDP.\beta\gamma}$ results in the reversion of PDE to its inactive state. Photoactivated rhodopsin is inactivated by phosphorylation at multiple serine and threonine residues, and is catalyzed by rhodopsin kinase and binding of an inhibitory protein known as **arrestin**. Low Ca^{2+} levels and recoverin stimulate rhodopsin kinase activity. Prestimulus cGMP levels are attained through the activation of guanylyl cyclase by low Ca^{2+} levels, which converts GTP to cGMP.

Vitamin E

Nutrition and Chemistry

Vitamin E was crystallized and its structure determined in 1936. Eight plant vitamers are known (Figure 36.7), the most abundant and active being α-**tocopherol**.

Vitamin E vitamers are viscous, light yellow oils that are heat stable but readily degraded by oxygen or ultraviolet light. Principal sources are the vegetable oils, particularly wheat germ and salad oils. Green vegetables, beef liver, butter, milk, and eggs contain appreciable amounts. Animals presumably obtain vitamin E from plants in their diet. Fish liver oils, although rich in vitamins A and D, are devoid of vitamin E.

The RDA depends on age and increases as the amount of polyunsaturated fatty acid (PUFA) in the diet increases. However, foods that are rich in PUFA are also rich in vitamin E.

Absorption, Transport, and Metabolism

Vitamin E is absorbed as free tocopherol, along with other fat-soluble vitamins and dietary lipids. Tocopheryl acetate, the form commonly used for dietary supplementation, is hydrolyzed before absorption. Uptake requires bile salts. A selective impairment of vitamin E absorption without malabsorption of other fat-soluble vitamins has been identified; it was corrected after a large oral intake of the vitamin. Patients with chronic fat malabsorption and abetalipoproteinemia (Chapter 18) may develop vitamin E deficiency.

About 75% of the absorbed vitamin E enters the lymphatics in chylomicrons, and the rest in other lipoproteins. In plasma, vitamin E is carried by lipoproteins and erythrocytes. In humans, vitamin E is present in greatest amounts in adipose tissue, liver, and muscle. Its principal excretory route is the feces, probably by way of bile.

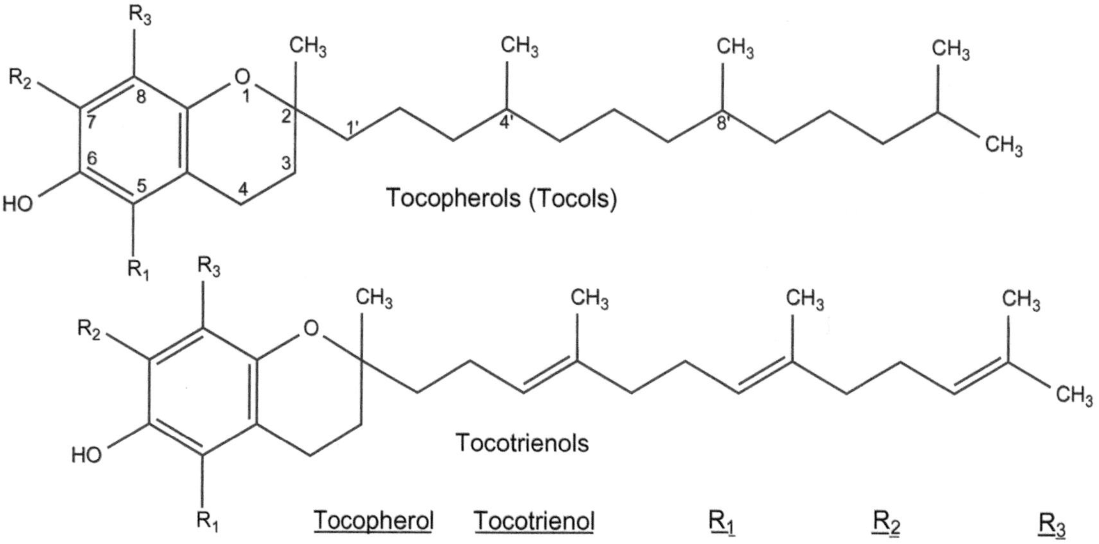

Tocopherol	Tocotrienol	R_1	R_2	R_3
Alpha-	Alpha-	—CH_3	—CH_3	—CH_3
Beta-	Beta-	—CH_3	—H	—CH_3
Gamma-	Gamma-	—H	—CH_3	—CH_3
Delta-	Delta-	—H	—H	—CH_3

FIGURE 36.7 Structures of naturally occurring plant compounds having vitamin E activity. The nucleus in each is 6-hydroxychroman. Attachment of a saturated 16-carbon chain produces the tocopherols; the tocotrienols have a 16-carbon unsaturated chain. Both groups are optically active. The tocotrienols have one chiral center at carbon 2, while the tocopherols have three, at carbons 2′, 4′, and 8′. Tocol is a tocopherol in which R_1, R_2, and R_3 are all hydrogen atoms. The tocopherols can be viewed as methylated tocols.

Function

The function most consistent with symptoms of vitamin E deficiency in animals is that of a general, membrane-localized antioxidant, which protects cellular and subcellular membranes from attack by endogenous and exogenous free radicals. In membranes, vitamin E may be located near enzyme complexes that produce free radicals, such as NADPH-dependent oxidase systems. Selenium alleviates some symptoms of vitamin E deficiency, probably through its role as a cofactor for glutathione peroxidase, which reduces peroxides generated within cells, thereby preventing formation of free radicals. Vitamin E in the plasma membrane may act as a first line of defense against free radicals, while glutathione peroxidase may be a second line of defense. Most enzymes affected by vitamin E deficiency are membrane bound or are involved in the glutathione—peroxidase system.

Vitamin E may also possess antiatherogenic properties. *In vitro* studies have shown that oxidized low-density lipoproteins (LDL) are proatherogenic (Chapter 18) and vitamin E retards LDL oxidation. Thus, it is thought vitamin E supplementation might reduce the morbidity and mortality from coronary artery disease. Non-antioxidant functions of vitamin E may involve several cellular signaling pathways (e.g., protein kinase C-initiated pathways). The signaling pathways cause inhibition of proliferation of smooth muscle cells, platelet adhesion, and aggregation and function of adhesion molecules. Vitamin E also may attenuate the synthesis of leukotrienes and increase synthesis of prostacyclin by upregulating phospholipase A_2 and cyclooxygenase. All of these actions of vitamin E may contribute to its protective properties against the development of atherosclerosis. In elucidating the various biological functions of the vitamin E complex, the specific roles played by each of the four species of tocopherols and four of trienols are not understood. Some studies have suggested that tocotrienols may be superior to tocopherols in their cardiovascular-related effects.

Hypo- and Hypervitaminosis E

Characteristic lesions of vitamin E deficiency in animals include necrotizing myopathy (inaccurately referred to as nutritional muscular dystrophy), exudative diathesis, nutritional encephalomalacia, irreversible degeneration of testicular tissue, fetal death and resorption, hepatic necrosis, and anemia. Several of these conditions are directly related to peroxidation of unsaturated lipids in the absence of vitamin E, and others can be prevented by synthetic antioxidants or vitamin E.

It is difficult to produce vitamin E deficiency in adult humans. Adult males who were depleted of vitamin E for 6 years showed no symptoms, although serum tocopherol concentrations became very low. However, their erythrocytes lysed more readily than normal when exposed to hydrogen peroxide or other oxidizing agents *in vitro*. This finding led to the use of low-plasma vitamin E and increased susceptibility of erythrocytes to oxidative hemolysis as criteria for vitamin E deficiency.

Premature infants and children with chronic cholestasis may develop spontaneous vitamin E deficiency. In premature infants, the deficiency manifests itself as increased red cell fragility and mild hemolytic anemia. It has been claimed, but not established, that these infants respond to administration of vitamin E. The anemia is not prevented by vitamin E, and only small improvements in red cell indices follow vitamin E treatment. A role has been claimed for vitamin E in prophylaxis of **retrolental fibroplasia** and **bronchopulmonary dysplasia**, two types of oxygen-induced tissue injury that occur in premature infants treated aggressively with oxygen.

Children with chronic cholestasis may exhibit a neuromuscular disorder that responds to treatment with vitamin E given parenterally or in large oral doses. Some patients show neurological signs despite normal serum levels of vitamin E, which may arise from the associated increase in levels of serum lipids, which contain vitamin E. In patients with cystic fibrosis, steatorrhea and fat malabsorption with subnormal plasma and tissue concentrations of vitamin E are common, but neuromuscular disorders such as those in chronic cholestasis do not occur.

Vitamin E deficiency occurs due to genetic defects in the formation of hepatic α-tocopherol transfer protein. This transport protein plays a central role in the liver, and one of its functions is to facilitate incorporation of α-tocopherol into nascent very-low-density lipoproteins (VLDLs). Since there is no specific transport protein for vitamin E in plasma, the delivery of vitamin E to the tissues is primarily mediated by VLDL—LDL transport mechanisms. Thus, deficiency of hepatic α-tocopherol transport protein causes low plasma levels of vitamin E with impairment of delivery to the tissues. Patients with the transport protein deficiency exhibit peripheral neuropathy and ataxia. Early and vigorous vitamin E supplementation in patients with neurological symptoms and with low plasma levels of vitamin E has yielded therapeutic benefits.

A pharmacological role for vitamin E may exist in claudication arising from peripheral vascular disease. Studies with small numbers of patients having cystic fibrosis, glucose-6-phosphate dehydrogenase deficiency, and sickle cell anemia conditions associated with decreased erythrocyte half-lives showed that many had chemical evidence of vitamin E deficiency. Administration of vitamin E supplements (400−800 IU/d) significantly increased red cell survival time. Claims that doses of vitamin E 10−20 times the RDA are beneficial for treatment of skin disorders, fibrocystic breast

disease, sexual dysfunction, cancer, baldness, and other disorders have not been substantiated.

Self-medication with high doses of vitamin E appears to be relatively nontoxic. However, in patients receiving warfarin, vitamin E in excess of 400 IU/d may further depress coagulability and produce coagulopathy.

WATER-SOLUBLE VITAMINS

The B group of vitamins and vitamin C serve as coenzymes or coenzyme precursors. The B-complex includes thiamine, riboflavin, pyridoxine, niacin, pantothenic acid, biotin, folate, and cobalamin. Inositol, choline, and para-aminobenzoic acid, usually classified as vitamin-like substances in humans, are sometimes included with the B-complex vitamins. They will be discussed briefly at the end of the chapter. The B vitamins occur in protein-rich foods and in dark green, leafy vegetables. A deficiency of one B vitamin is usually accompanied by deficiencies of others in the group and of protein. Vitamin C is the antiscorbutic factor of citrus fruit and other fresh fruits and vegetables. The symptoms of deficiency of water-soluble vitamins are similar to, and include, disorders of the nervous system and of rapidly dividing tissues, such as the gastrointestinal epithelium, mucous membranes, skin, and cells of the hematopoietic system.

Thiamine (Vitamin B₁)

Nutrition and Chemistry

Thiamine, or vitamin B_1, also called **aneurin**, is the anti-beriberi factor. The active coenzyme form is thiamine pyrophosphate (TPP) or cocarboxylase (Figure 36.8).

The principal dietary sources include fish, lean meat (especially pork), milk, poultry, dried yeast, and whole-grain cereals. Bread, cereals, and flour-based products are frequently enriched with this vitamin. Thiamine is present

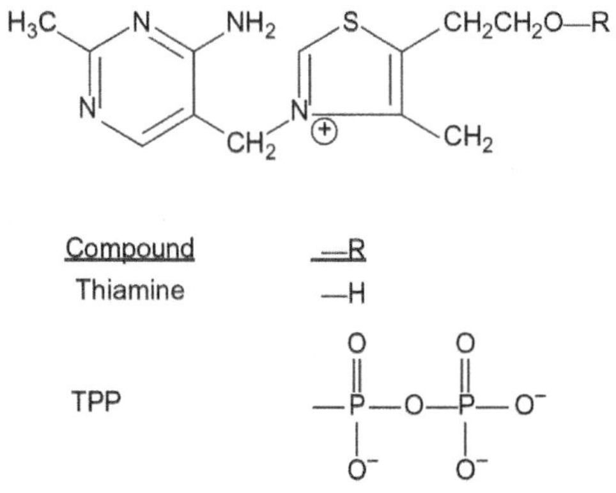

FIGURE 36.8 Structures of thiamine (vitamin B_1) and thiamine pyrophosphate (TPP). They consist of a six-membered pyrimidine ring and a five-membered thiazole ring, linked through a methylene group.

in the outer layers of rice grains, where it was first identified. Deficiency is common in Asian countries where polished rice is the principal dietary staple. The RDA depends on energy intake.

Absorption, Transport, and Metabolism

Thiamine is absorbed by a pathway that is saturable at concentrations of $0.5-1.0$ μmol/L. Oral doses in excess of 10 mg do not significantly increase blood or urine concentrations of vitamin B_1. In humans, absorption occurs predominantly in the jejunum and ileum. Some ferns, shellfish, fish, and species of bacteria contain thiaminase, which cleaves the pyrimidine ring from the thiazole ring. This enzyme causes thiamine deficiency in cattle. In plasma, thiamine is transported bound to albumin and, to a small extent, other proteins. TPP is synthesized in the liver by transfer of the pyrophosphate group from ATP by thiamine pyrophosphokinase.

Function

In humans, TPP is a coenzyme for transketolation, an important reaction in the pentose—phosphate pathway, and for the oxidative decarboxylations catalyzed by pyruvate dehydrogenase, branched-chain α-ketoacid decarboxylase, and α-ketoglutarate dehydrogenase complexes.

Hypovitaminosis

Major diseases caused by thiamine deficiency are the **Wernicke—Korsakoff syndrome**, commonly associated with chronic alcoholism, and beriberi, the classic thiamine deficiency syndrome. Both are related to diets high in carbohydrate and low in vitamins. Wernicke—Korsakoff syndrome comprises **Wernicke's encephalopathy** and **Korsakoff's psychosis**. The encephalopathy presents with ataxia, confusion, and paralysis of the ocular muscles, which are relieved by administration of thiamine. Korsakoff's psychosis is characterized by amnesia and is only slightly responsive to thiamine. These disorders reflect different stages of the same pathological process. Some areas within the brain appear to be more susceptible than others to the effects of thiamine depletion. The nature of the biochemical lesion is unknown. Thiamine deficiency in chronic alcoholics probably has multiple causes, including poor diet, an inhibitory effect of alcohol and of any accompanying folate deficiency on intestinal transport of thiamine, and reduced metabolism and storage of thiamine by the liver due to alcoholic cirrhosis.

Beriberi occurs whenever thiamine intake is less than 0.4 mg/day for an extended period of time. It occurs where polished rice is a dietary staple and, in Western society, in poor and elderly populations and alcoholics. Beriberi has wet, dry, and cardiac types, and an individual may have more than one type. "Wet" refers to pleural and peritoneal effusions and edema; "dry" refers to polyneuropathy without effusions. Cardiomyopathy is the principal feature of the cardiac type. An infantile form occurs in breast-fed infants, usually 2—5 months of age, nursing from thiamine-deficient mothers. The symptoms of beriberi remit completely with thiamine supplementation. A subclinical deficiency of thiamine occurs in hospital patients and the elderly. Deficiency of thiamine and other vitamins may contribute to a generally reduced state of health in these populations. Thiamine deficiency can be assessed by measuring blood levels. Increased blood levels of pyruvate and lactate suggest thiamine deficiency. Measurement of erythrocyte transketolase activity, which requires TPP as a coenzyme, confirms the deficiency. Four inherited disorders responsive to treatment with pharmacological doses of thiamine are known. Gastric bypass surgery, employed in the treatment of morbid obesity, can lead to risk of several nutritional deficiencies, including thiamine [1].

Riboflavin (Vitamin B$_2$)

Riboflavin, or vitamin B_2, is the precursor of flavin mononucleotide (FMN) and flavin adenine dinucleotide (FAD), cofactors for several oxidoreductases that occur in all plants, animals, and bacteria (Figure 36.9).

The FAD-requiring enzymes in mammalian systems include the D- and L-amino acid oxidases, mono- and diamine oxidases, glucose oxidase, succinate dehydrogenase, α-glycerophosphate dehydrogenase, and glutathione reductase. FMN is a cofactor for renal L-amino acid oxidase, NADH reductase, and α-hydroxy acid oxidase. In succinate dehydrogenase, FAD is linked to a histidyl residue, and in liver mitochondrial monoamine oxidase, to a cysteinyl residue. In other cases, the attachment is noncovalent, but the dissociation constant is very low.

Use of oral contraceptives may increase the dietary requirement for riboflavin. Riboflavin status can be evaluated from the activity of erythrocyte glutathione reductase, an FAD-requiring enzyme, before and after addition of exogenous FAD. Low initial activity or a marked stimulation by FAD (or both) is indicative of ariboflavinosis.

Although only synthesized in plants, bacteria, and most yeasts, riboflavin is ubiquitous in plants and animals. Good dietary sources include liver, yeast, wheat germ, green leafy vegetables, whole milk, and eggs. Riboflavin is readily degraded when exposed to light, especially at elevated temperatures, and considerable decrease in its content in foods can occur during cooking accompanied by exposure to light. The greatest amounts of flavin nucleotides are in the liver, kidney, and heart.

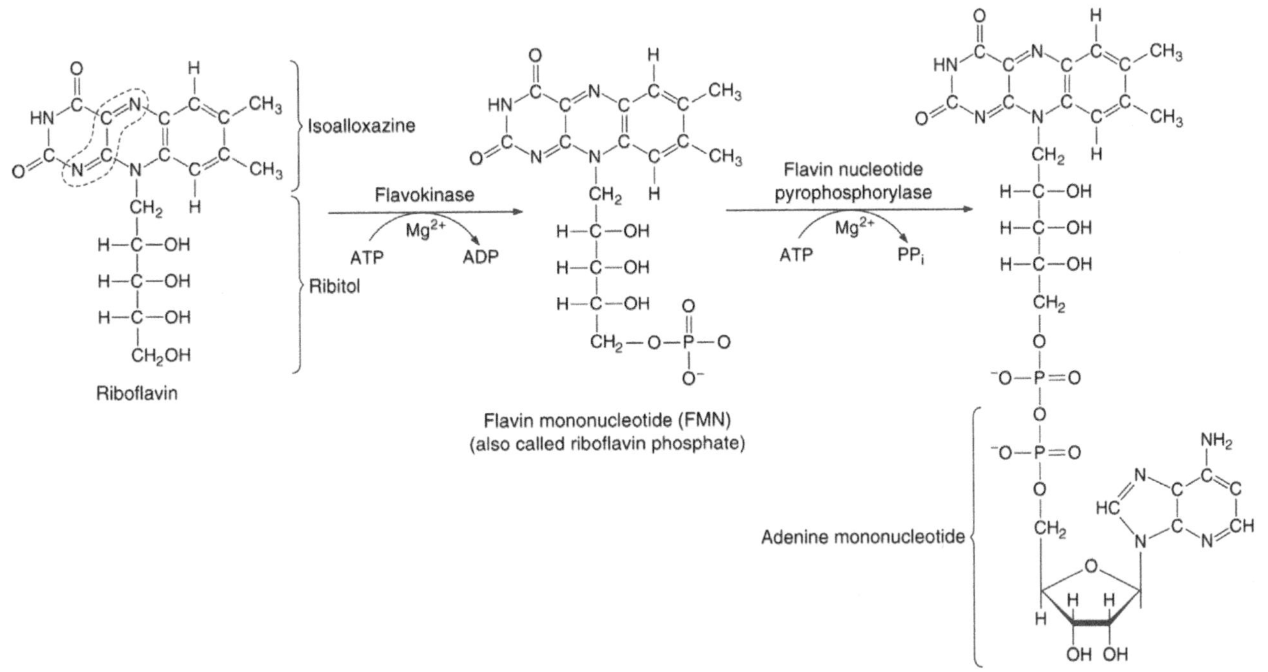

FIGURE 36.9 Synthetic pathways of converting riboflavin (vitamin B_2) to its two cofactors, FMN and FAD. Riboflavin contains an isoalloxazine ring system joined to a ribosyl group (a sugar alcohol) through a cyclic amine on the middle ring. Because of the isoalloxazine nucleus, the three compounds are yellow and exhibit a yellow-green fluorescence in aqueous solutions. FMN is not a true nucleotide because the bond between the flavin (isoalloxazine) ring and the ribityl side chain is not a glycosidic bond. Two nitrogen atoms and two carbon atoms, identified by the dashed area in the isoalloxazine ring system, participate in the oxidation–reduction reactions of the flavin coenzymes.

Dietary riboflavin is present mostly as a phosphate, which is rapidly hydrolyzed before absorption in the duodenum. In humans, the rapid, saturable absorption of riboflavin following an oral dose suggests that it is transported by a carrier-mediated pathway located predominantly in the duodenal enterocytes. The process may be sodium-dependent. Bile salts enhance absorption of riboflavin. Fecal riboflavin is derived from the intestinal mucosa and the intestinal flora. This is the predominant excretory route for the vitamin.

Signs of riboflavin deficiency include cheilosis, angular stomatitis, magenta tongue, and localized seborrheic dermatitis. Some of these conditions may be due to concurrent deficiency of other B-complex vitamins, since it is difficult to produce "pure" riboflavin deficiency in humans. No toxicity following large doses of riboflavin has been reported.

Pyridoxine (Vitamin B_6)

Three closely related compounds have vitamin B_6 activity. With the exception of glycogen phosphorylase (Chapter 14) and kynureninase, all of the pyridoxine-requiring enzymes are involved in transamination or decarboxylation of amino acids (Chapter 15).

Cobalamin (Vitamin B_{12})

Vitamin B_{12} is unusual among the vitamins because its two coenzyme forms contain an organometallic bond between cobalt and carbon [2,3]. The structure of the cobalamin family of compounds is shown in Figure 36.10. The corrin ring system of four pyrrole rings linked by three methene bridges is similar to the porphyrin ring system. The 5,6-dimethylbenzimidazole ring is sometimes replaced by 5-hydroxybenzimidazole, adenine, or similar groups. Dietary cobalamins generally contain Co^{3+} while the coenzyme forms contain Co^{1+}. Reduction of B_{12} is catalyzed by specific NADH-dependent reductases. In the coenzyme form, the group attached to the sixth position of cobalt is 5'-deoxyadenosine (adenosylcobalamin) or a methyl group (methylcobalamin). In the liver and several other tissues, about 70% of the cobalamin is present as adenosylcobalamin in mitochondria, 1%–3% as methylcobalamin in the cytoplasm, and the rest probably as hydroxocobalamin. Dietary cobalamins are released by gastric acid and bind to cobalophilins (R-proteins), derived primarily from saliva, and to intrinsic factor (IF), a glycoprotein secreted by gastric parietal cells. In the duodenum, pancreatic proteases degrade cobalophilins, allowing cobalamin to bind to IF. Transfer to IF is aided by the bicarbonate of pancreatic juice. Patients with pancreatic insufficiency may have vitamin B_{12} deficiency.

The IF−cobalamin complex binds to specific receptors on the luminal surface of ileal mucosal cells. The receptor protein is known as **cubilin**, which is tethered by the protein **amnionless**. Binding requires Ca^{2+} and occurs optimally at pH >6.6. The IF−B_{12}−receptor complex undergoes endocytosis, followed by release of B_{12}. Transfer of cobalamin to plasma is specifically to transcobalamin II (TCII). The TCII−B_{12} complex enters virtually all cells of the body via a specific cell surface receptor by receptor-mediated endocytosis (Figure 36.11).

Vitamin B_{12} Deficiency

Deficiency of vitamin B_{12} causes both neurological and hematological abnormalities; it can occur due to inadequate dietary intake and disorders associated with its absorption, transport, and conversion to its coenzyme forms. (See Clinical Case Study 36.1.)

Pernicious anemia is a megaloblastic anemia caused by malabsorption of vitamin B_{12} secondary to inadequate secretion of normal IF. The word *pernicious* reflects the unremitting and usually fatal course of the disease prior to the discovery that it could be corrected with the administration of vitamin B_{12}. Virtually all cases result from atrophic gastritis, which, in turn, may result from autoimmune attack of gastric parietal cells, leading to absence of IF and achlorhydria. Pernicious anemia may rarely be due to an inherited defect in the structure or secretion of IF. Several cases of congenital absence of IF have been described. Anemia due to inborn errors usually manifests itself early in life, while the average age for appearance of atrophic gastritis is 60 years; the latter condition is rare, but not unknown, before the age of 30.

The two reactions in mammalian systems in which cobalamins participate are conversion of L-methylmalonyl-CoA to succinyl-CoA by methylmalonyl-CoA mutase, and methylation of homocysteine to methionine by N^5-methyltetrahydrofolate homocysteine methyltransferase (Figure 36.11).

Inborn errors in the synthesis of adenosylcobalamin or of both adenosyl- and methylcobalamins have been described. They cause, respectively, methylmalonic aciduria alone or combined with homocystinuria. These disorders respond to treatment with pharmacological doses of vitamin B_{12}. Methylmalonic aciduria that does not

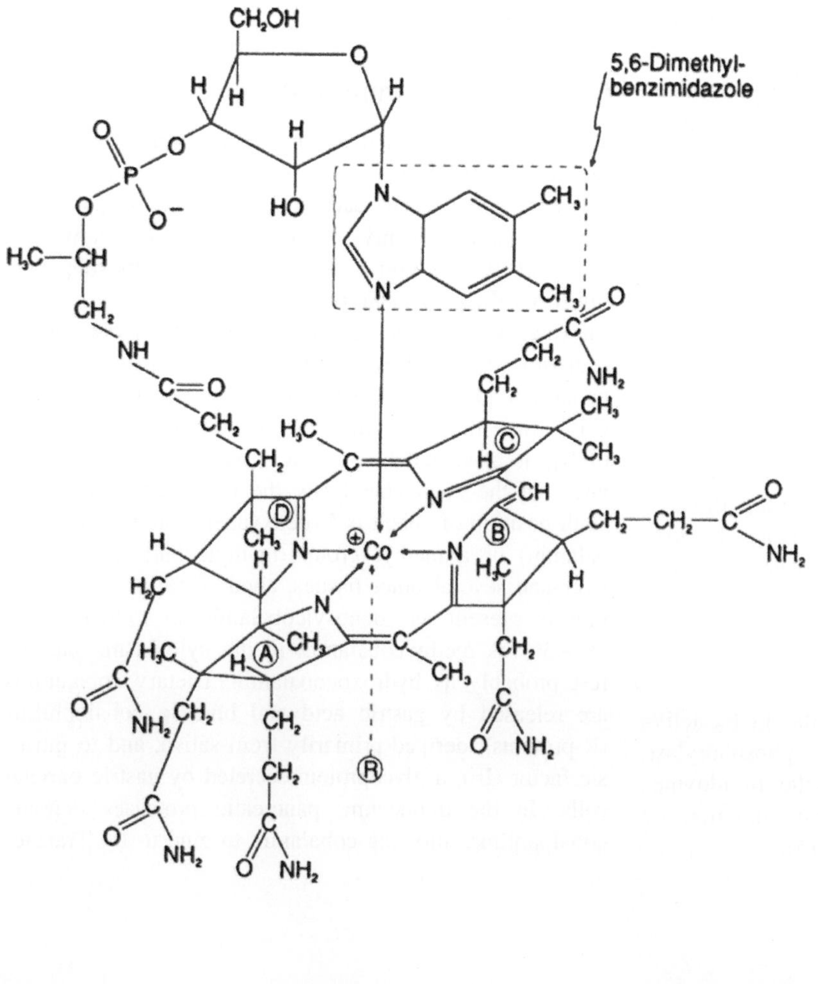

FIGURE 36.10 Structure of the cobalamin family of compounds. A through D are the four rings in the corrinoid ring system. The B ring is important for cobalamin binding to intrinsic factor. If $R = -CH_3$, the molecule is methylcobalamin. Arrows pointing toward the cobalt ion represent coordinate covalent bonds.

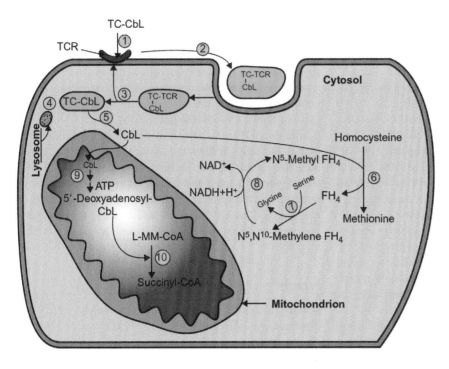

FIGURE 36.11 Steps involved in the transport of cobalamin (vitamin B_{12}), its internalization by receptor-mediated endocytosis, and its utilization. (1) Binding of cobalamin (CbL) and transcobalamin (TC, transcobalamin II) complex to specific receptors (TCRs). (2) Endocytosis of TC−CbL−TCR complex in the clathrin-coated pits. (3) Dissociation of TCR upon acidification in the endosomes and its return to the cell surface. (4) Fusion of endosomes with lysosomes, freeing of CbL and degradation of TC by proteolysis. (5) Transport of CbL to cytosol. (6) Conversion of homocysteine to methionine using N^5-methyltetrahydrofolate (FH_4) with the formation of methyl-CbL as an intermediate catalyzed by methionine synthase. (7) and (8) Steps involved in the reconversion of FH_4 to N^5-methyl FH_4 by serine hydroxymethyltransferase and N^5, N^{10}-methylene FH_4 reductase, respectively. (9) and (10) Formation of coenzyme 5-deoxyadenosyl-CbL for methylmalonyl-CoA mutase, which catalyzes the conversion of L-methylmalonyl-CoA (L-MM-CoA) to succinyl CoA.

respond to vitamin B_{12} is probably due to an abnormal methylmalonyl-CoA mutase.

Nitrous oxide (N_2O), used as an anesthetic agent, inactivates vitamin B_{12}. Subjects with marginal vitamin B_{12} stores and subjects with N^5,N^{10}-methylenetetrahydrofolate reductase deficiency (see Figure 36.11) can develop vitamin B_{12} deficiency within weeks of administration of the anesthetic.

Major clinical manifestations of vitamin B_{12} deficiency include hematologic abnormalities (e.g., megaloblastic macrocytic anemia, hypersegmentation of neutrophils) and neurological deficiency (e.g., sensory neuropathy). In addition, since dividing cells require vitamin B_{12}, all rapidly growing cells are affected. One visible sign of vitamin B_{12} deficiency is a sore mouth and tongue, and the tongue may be bald and shiny.

In vitamin B_{12} deficiency, neurological damage may sometimes occur in the absence of hematological abnormalities and may be mistaken for multiple sclerosis, diabetic neuropathy, or neuropsychiatric disorders. In some elderly subjects, malfunctioning of the gastric parietal cells leads to a reduction in both intrinsic factor and H^+ production (achlorhydria), causing vitamin B_{12} malabsorption. Two metabolites that accumulate due to reduced catalytic function of L-methylmalonyl-CoA mutase and methionine synthase as a result of vitamin B_{12} deficiency are methylmalonic acid and homocysteine. In the clinical assessment of subtle vitamin B_{12} deficiency in the presence of normal or low−normal serum levels of vitamin B_{12}, measurement of homocysteine and methylmalonic acid serum levels may prove valuable in detecting incipient vitamin B_{12}

deficiency. In folate deficiency, serum homocysteine levels are also increased, but methylmalonic acid levels remain in the normal range. In vitamin B_{12}-deficient subjects, folate therapy may normalize homocysteine levels, along with some of the hematological abnormalities, but it will not correct the neurological manifestations. Thus, inappropriate folate therapy may mask vitamin B_{12} deficiency (discussed in the next section).

Folic Acid (Pteroylglutamic Acid)

The common name **folic acid** is derived from Latin *folium*, "leaf," because this vitamin was originally isolated from dark green, leafy vegetables, such as spinach. Folic acid metabolism was discussed in Chapter 25.

A close relationship exists between the metabolism of folates and of vitamin B_{12}. Deficiency of either produces megaloblastic anemia, and symptoms of vitamin B_{12} deficiency anemia are reversed by large doses of folate. However, folate does not reverse the neurological abnormalities associated with vitamin B_{12} deficiency and may even exacerbate them. Correction of megaloblastosis by folate is not necessarily an indication of the absence of vitamin B_{12} deficiency; in fact, it could mask developing vitamin B_{12} deficiency until the advent of neurological damage. The common point in metabolism of folate and B_{12} is transmethylation of homocysteine to methionine (Figure 36.11).

Homocysteine, an amino acid associated with coronary heart disease as a risk factor, is elevated in folate deficiency. If the elevation of plasma homocysteine is due to folate deficiency, supplementation of folate corrects the plasma homocysteine level and may decrease the morbidity and mortality from atherosclerotic disease, which can lead to heart attack and stroke. Vitamin B_6 and Vitamin B_{12} deficiencies can also cause elevated plasma homocysteine levels.

Folate supplementation, in addition to a healthy diet that includes foods rich in folate, has been recommended for women before and in the early weeks of pregnancy. Folate supplementation helps prevent the majority of birth defects of the brain and spinal cord, called **neural tube defects (NTDs)**. Although the exact role of folate in the prevention of NTDs is not understood, the involvement of folate in nucleic acid and amino acid metabolism in the rapidly developing fetus is probable. Also, supplementation of dietary folate may correct metabolic defects in individuals who inherit defects in folate metabolism. A recent study indicates that as many as one in seven people carry a mutation affecting folate metabolism.

A screening test for NTDs has been developed using maternal serum to measure α-fetoprotein (AFP) levels at 15−20 weeks of gestation. AFP is synthesized in the fetal liver and yolk sac during development, and is the main fetal serum protein (analogous to serum albumin in adults). AFP appears in the amniotic fluid from fetal urination. Its amniotic fluid concentration parallels fetal serum levels, except that the amniotic fluid level is about 150 times lower in concentration. Through placental transfer, AFP also appears in maternal serum. During the screening test period of 15−20 weeks, maternal serum AFP concentration increases by about 15% per gestational week. If NTDs are present (e.g., anencephaly, open spina bifida), fetal serum leaks into the fluid compartment, raising AFP amniotic fluid levels, as well as maternal serum levels. Thus, elevated levels of maternal serum AFP are used to identify women who may be carrying a fetus with an NTD. However, because of the overlap in AFP values between unaffected and affected pregnancies, the test is not diagnostic and requires confirmatory diagnostic procedures (e.g., acetylcholinesterase and AFP levels in the amniotic fluid, and high-resolution ultrasonography).

Low levels of maternal serum AFP may also be informative in assessing other fetal abnormalities such as Down's syndrome. Some of the salient features of Down's syndrome include trisomy 21, malformations, dysmorphic features, and mental retardation. The risk of Down's syndrome increases with maternal age. At age 35, the risk at birth is 1 in 385, and at age 40, it is 1 in 105. In estimating the risk of Down's syndrome, three other biochemical serum parameters in addition to AFP are measured: unconjugated estriol (which is decreased); human chorionic gonadotropin (hCG, which is increased); and inhibin A (which is elevated). The measurement of these four maternal serum markers is used in conjunction with maternal age, twin gestation, maternal insulin-dependent diabetes mellitus, maternal weight, ethnic derivation, and smoking, to assess risk. Evaluating all of these parameters in comparison to values obtained with healthy women of comparable gestational age has increased the sensitivity and reduced false-positive rates in the assessment of Down's syndrome in the fetus.

Serum AFP levels are elevated in hepatocellular carcinomas and malignancies involving the ovaries and testes. Yolk sac tumors, which occur more frequently in the ovaries of young women and girls, and in the testes of boys, raise serum levels of AFP. Serum hCG measurement is also helpful in the diagnosis and management of germ cell tumors. Thus, hCG and AFP are also used as tumor markers.

Niacin

Niacin (nicotinic acid; pyridine-3-carboxylic acid) and nicotinamide are precursors of NAD^+ and $NADP^+$. Niacin occurs in meat, eggs, yeast, and whole-grain cereals, in conjunction with other members of the vitamin B group. A limited amount of niacin can be synthesized in the body from tryptophan, but it is not adequate to meet metabolic needs. Nicotinic acid has been used to increase serum HDL cholesterol levels in the therapy of atherosclerosis (Chapter 18).

Pantothenic Acid (Pantoyl-β-Alanine)

Pantothenic acid is ubiquitous in plant and animal tissues and especially abundant in foods rich in other B vitamins. Pantothenic acid is a precursor for the synthesis of coenzyme A (CoA, CoASH) and forms part of the "swinging sulfhydryl arm" of the fatty acid synthase complex (Chapter 16).

Biotin

Biotin is widely distributed in foods. Beef liver, yeast, peanuts, kidney, chocolate, and egg yolk are especially rich sources. The intestinal flora synthesizes biotin. Fecal excretion reflects this enteric synthesis. Total daily urinary and fecal excretion exceeds dietary intake.

Biotin deficiency occurs when large amounts of raw egg white are consumed. Egg white contains avidin, a protein (M.W. 70,000) that binds biotin strongly and specifically, preventing its absorption from the intestine. Because of the tight binding and specificity of biotin, avidin-labeled probes have been used to detect proteins and nucleic acids to which biotin has been covalently attached ("biotinylated" molecules). Avidin is a homotetramer. Each subunit contains 128 amino acids and binds one molecule of biotin. The affinity of

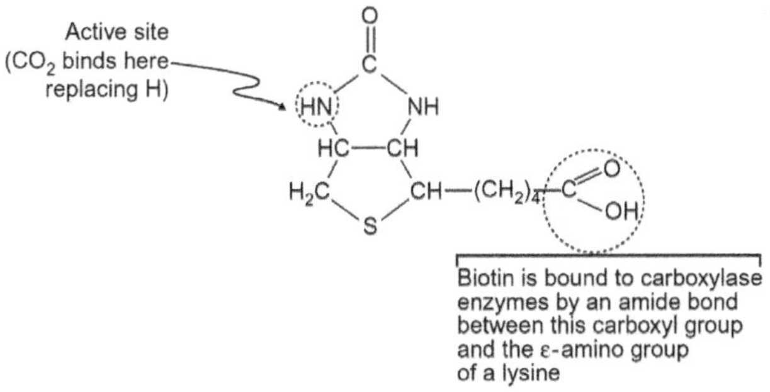

FIGURE 36.12 Structure of biotin.

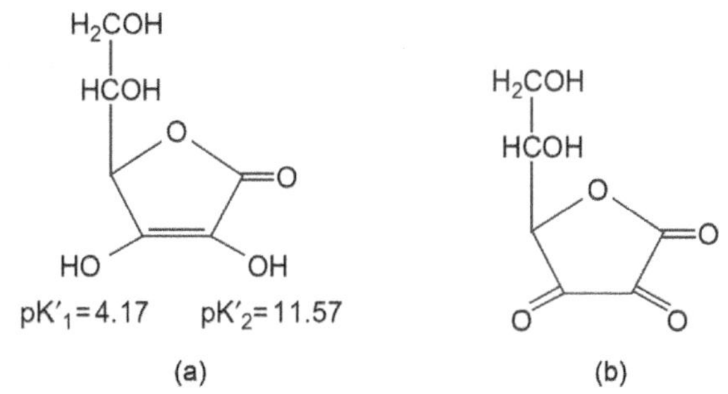

Active site
(CO₂ binds here replacing H)

Biotin is bound to carboxylase enzymes by an amide bond between this carboxyl group and the ε-amino group of a lysine

$$pK'_1 = 4.17 \qquad pK'_2 = 11.57$$

(a)

(b)

FIGURE 36.13 Structure of L-ascorbic acid (a) and L-dehydroascorbic acid (b).

avidin for biotin is abolished by heat and other denaturants. Biotin deficiency can result from sterilization of the intestine by antibiotics and from administration of biotin analogues.

The structure of biotin consists of fused imidazole and tetrahydrothiophene rings and a carboxyl-containing side chain (Figure 36.12). In oxybiotin, which can substitute for biotin in most species, the sulfur of the tetrahydrothiophene ring is replaced by oxygen, making it a tetrahydrofuran ring.

Biotin is a coenzyme for the carbon dioxide fixation reactions catalyzed by acetyl-CoA carboxylase (Chapter 16), propionyl-CoA carboxylase, pyruvate carboxylase, and β-methylcrotonyl-CoA carboxylase. Carboxylation reactions that do not require biotin are the addition of C_6 to the purine ring (Chapter 25), the formation of carbamoyl phosphate (Chapter 15), and the γ-carboxylation of glutamyl residues of several of the clotting factors, which requires vitamin K (Chapter 34). Biotin is bound to an apoenzyme by an amide linkage to a lysyl B-amino group (Figure 36.11). This binding occurs in two steps, catalyzed by holocarboxylase synthetase:

$$Biotin + ATP \rightarrow biotinyl\ 5'\text{-}adenylate + PP_i$$

$$Biotinyl\ 5'\text{-}adenylate + apocarboxylase \rightarrow$$
$$holocarboxylase + AMP$$

Most dietary biotin is bound to protein, the amide linkage being broken prior to absorption. At least eight children have been described who have multiple carboxylase deficiency with low activity of several of the biotin-requiring carboxylases; i.e., multiple carboxylase deficiency. Pharmacological doses of biotin restored the activity of carboxylases in these patients, indicating that the defect was not in the apocarboxylases. Thus, the defect is presumably in the intestinal transport system, in holocarboxylase synthetase, or in some step in cellular uptake or intracellular transport of biotin.

Proteolysis of biotin-containing enzymes releases ε-biotinyllysine, or biocytin. Biotinidase cleaves biocytin

and biotinylated peptides, resulting from degradation of endogenous carboxylases, to biotin and lysine. Thus, biotin is recycled. Deficiency of biotinidase may cause biotin deficiency, manifested clinically by neurological problems, cutaneous findings, and developmental delay. These defects can be corrected by pharmacological doses of biotin. Toxicity due to excessive consumption of biotin is not known. Biotinidase deficiency is an autosomal recessive disorder with an estimated incidence of 1 in 72,000–126,000. Many newborn screening programs for genetic diseases include testing for this enzyme. Prompt treatment with oral biotin administration of 5–20 mg/day in affected infants will prevent clinical consequences. If the treatment is delayed, neurological manifestations (e.g., hearing loss and optic atrophy) and developmental delay occur, and may not revert to normal.

Ascorbic Acid (Vitamin C)

Humans and guinea pigs lack the enzyme that converts L-gulonolactone to 2-keto-L-gulonolactone, which is required for biosynthesis of ascorbic acid, L-ascorbic acid, and L-dehydroascorbic acid (Figure 36.13), which is biologically equivalent in humans, presumably because of the ready reduction of dehydroascorbate to ascorbate in the body. Ascorbic acid (M.W. 176.1) is a six-carbon enediol lactone (ketolactone) having a configuration analogous to that of glucose. The enolic hydroxyl groups dissociate with $pK'_1 = 4.17$ and $pK'_2 = 11.57$. It is one of the strongest naturally occurring reducing agents known. Ascorbate is a specific electron donor for eight enzymes and also may participate in several nonenzymatic reactions as a reductant (Table 36.1). However, it should be emphasized that the *in vivo* role of ascorbate as a reductant in nonenzymatic reactions (based on its redox potential) is not established. Other reductants may participate or substitute for ascorbate in nonenzymatic reactions.

Major dietary sources of vitamin C are fresh, frozen, and canned citrus fruits. Other fruits, leafy green vegetables, and tomatoes are important contributors to ascorbate

TABLE 36.1 Metabolic Functions of Ascorbate

A: As Specific Electron Donor for Enzymes and Their Metabolic Role

Enzymes	Metabolic Role
Three collagen hydroxylases	Collagen biosynthesis
Two enzymes in carnitine biosynthesis	Carnitine is essential for mitochondrial fatty acid oxidation
Dopamine β-monooxygenase	Necessary for synthesis of norepinephrine and epinephrine
4-Hydroxyphenylpyruvate dehydrogenase	Participates in tyrosine metabolism
Peptidylglycine α-monooxygenase	Required for amidation of peptide hormones

B: As Potential Chemical Reductant and/or Antioxidant in Nonenzymatic Reactions

Reaction or Function	Consequence
Gastrointestinal iron absorption	Increase
Oxidative DNA and/or protein damage	Decrease
Low-density lipoprotein oxidation	Decrease
Endothelial-dependent vasodilation	Increase
Lipid peroxidation	Decrease
Oxidants and nitrosamines in gastric juice	Decrease
Extracellular oxidants from neutrophils	Decrease

intake. Human milk contains 30−55 mg/L, depending on maternal intake of vitamin C. Exposure to copper, iron, and oxygen can destroy vitamin C by oxidation. The vitamin is heat-labile, so excessive cooking will degrade it. D-ascorbate (isoascorbate or erythroascorbate), frequently used as a food preservative, has one-twentieth the biological activity of L-ascorbate.

Absorption of vitamin C from the small intestine is a carrier-mediated process that requires sodium at the luminal surface. Transport is most rapid in the ileum and resembles the sodium-dependent transport of sugars and amino acids, but the carrier is distinct for each class of compound. Some ascorbate may also enter by simple diffusion. With dietary intake less than 100 mg/day, efficiency of absorption is 80%−90%. With intake equal to the RDA, plasma ascorbate is 0.7−1.2 mg/dL, and the ascorbate pool size is 1500 mg. Scurvy becomes evident when the pool is less than 300 mg, at which point plasma ascorbate is 0.13−0.24 mg/dL. Highest tissue concentrations of ascorbate are in the adrenal gland (cortex > medulla).

Most signs of scurvy can be related to inadequate or abnormal collagen synthesis. Ascorbate enhances prolyl and lysyl hydroxylase activities. Collagen formed in scorbutic patients is low in hydroxyproline and poorly cross-linked, resulting in skin lesions, bone fractures, and rupture of capillaries and other blood vessels. The absolute amount of collagen made in scorbutic animals may also decrease independently of the hydroxylation defect. The anemia of scurvy may result from a defect in iron absorption or folate metabolism.

Ascorbate increases the activity of hydroxylases needed for the conversion of p-hydroxyphenylpyruvate to homogentisic acid (Chapter 15), synthesis of norepinephrine from dopamine (Chapter 30), and two reactions in carnitine synthesis.

CLINICAL CASE STUDY 36.1 Vitamin B₁₂ Deficiency

This clinical case was abstracted from: A. Puig, M. Mino-Kenudson, A.S. Dighe, Case 13-2012: a 62-year-old man with paresthesias, weight loss, jaundice, and anemia, N. Engl. J. Med. 366 (2012) 1626−1633.

Synopsis

A 62-year-old man with paresthesias, weight loss, and anemia was admitted to the hospital. Four months prior to admission, the patient's routine examination was unremarkable. However, thereafter he developed signs of peripheral neuropathy, changes in gait, dyspnea, glossitis, and dark urine. Significant abnormal laboratory values included MCV of

124 fL (reference range 80−100 fL), hemoglobin of 6.3 g/dL (reference range 13.5−17.4 g/dL), bilirubin of 3.4 mg/dL (reference range 0.0−1.0 mg/dL), lactate dehydrogenase (LDH) of 1404 μ/L (reference range 110−210 μ/L), haptoglobin of <6 mg/dL (reference range 16−199 mg/dL), and vitamin B₁₂ of 61 pg/mL (reference range >250 pg/mL). Serum iron studies revealed iron deficiency, normal folate levels, presence of anti-intrinsic factor antibodies, and markedly elevated serum methylmalonic acid and gastrin levels. These results led to the diagnosis of vitamin B₁₂ deficiency due to pernicious anemia. The patient's anemia was treated with packed red blood cell transfusion, his vitamin B₁₂ deficiency was treated with

(Continued)

CLINICAL CASE STUDY 36.1 (Continued)

intramuscular injection of vitamin B_{12}, and his iron deficiency was treated with administration of intravenous iron. Although his vitamin B_{12} and iron values returned to normal, his peripheral neuropathy continued to be a problem.

Teaching Points

1. The only dietary source of vitamin B_{12} (cobalamin) for humans is animal products (meat and dairy). Adequate absorption of vitamin B_{12} from the diet requires five factors: adequate dietary intake of vitamin B_{12}, liberation of vitamin B_{12} from binding to proteins via gastric acid-pepsin, liberation of vitamin B_{12} from binding to R factors (transcobalamin I) via pancreatic proteases, binding of vitamin B_{12} to intrinsic factor secreted by gastric parietal cells, and binding of vitamin B_{12} to functional CbL-IF receptors (cobalamin-intrinsic factor) of intact ileum.

2. After absorption by ileal enterocytes, vitamin B_{12} binds with transcobalamin II, and together they enter the cells where vitamin B_{12} is metabolized into coenzymes involved in the generation of succinyl-CoA (an important intermediate in the citric acid cycle) and methionine (an essential amino acid).

3. Vitamin B_{12} deficiency may be caused by gastric abnormalities (pernicious anemia, gastrectomy, bariatric surgery, gastritis, autoimmune metaplastic atrophic gastritis), small bowel disease (malabsorption syndrome, ileal resection or bypass, Crohn's disease, blind loops, fish tapeworm infection), pancreatitis, diet (strict vegans, pregnant vegetarians), medications that block absorption

(neomycin, biguanides, proton pump inhibitors, or histamine 2 receptor antagonists), or inherited transcobalamin II deficiency.

4. Vitamin B_{12} deficiency causes degeneration of the dorsal columns of the spinal cord, causing peripheral neuropathy. Vitamin B_{12} deficiency also causes macrocytic anemia, reflected by elevated MCV, hemolysis, decreased serum haptoglobin levels, and elevated lactate dehydrogenase (LDH) levels.

5. The elevated level of serum gastrin that may be seen in vitamin B_{12} deficiency is due to atrophic gastritis, in which destruction of acid-producing parietal cells results in decreased gastric acid levels, which stimulate the production of gastrin.

6. Methylmalonic acid is elevated in both macrocytic anemias due to vitamin B_{12} and folate deficiency. However, homocysteine levels are elevated only in folate deficiency and not in vitamin B_{12} deficiency. Thus, measurement of serum methylmalonic acid and homocysteine is useful in distinguishing between the two deficiencies.

Supplemental References

[1] S.P. Stabler, Vitamin B_{12} deficiency, N. Engl. J. Med. 368 (2013) 149–160.
[2] R.K. Pruthi, A. Tefferi, Pernicious anemia revisited, Mayo Clin. Proc. 69 (1994) 144.
[3] R.H. Allen, S.P. Stabler, D.G. Savage, J. Lindenbaum, Metabolic abnormalities in cobalamin (vitamin B_{12}) and folate deficiency, FASEB J 7 (1993) 1344.
[4] R. Green, L.J. Kinsella, Current concepts in the diagnosis of cobalamin deficiency, Neurology 45 (1995) 1435.

REQUIRED READING

[1] J.F. Merola, P.P. Ghoroghchian, M.A. Samuels, B.D. Levy, J. Loscalzo, At a loss, N. Engl. J. Med. 367 (2012) 67–72.
[2] H.F. Bunn, Vitamin B12 and pernicious anemia—the dawn of molecular medicine, N. Engl. J. Med. 370 (2014) 773–776.
[3] A. Puig, M. Mino-Kenudson, A.S. Dighe, Case 13-2012: a 62-year-old man with paresthesias, weight loss, jaundice, and anemia, N. Engl. J. Med. 366 (2012) 1626–1633.

Chapter 37

Water, Electrolytes, and Acid–Base Balance

Key Points

1. The concentrations of various electrolytes, pH, and water balance are determined by many interwoven systems to maintain homeostasis.
2. The composition and volume of extracellular fluid are regulated by complex hormonal and nervous system mechanisms that coordinate to control osmolality, volume, and pH.
3. Kidneys regulate acid–base balance by excreting nonvolatile acids and conserving HCO_3^-. Lungs regulate volatile acid CO_2 excretion. These functions are coordinated, and any imbalance results in metabolic and respiratory acid–base disorders.
4. The acid–base disorders are assessed by measuring arterial blood gases and pH values, venous blood electrolyte concentrations, and serum and urinary anion gap values. These disorders often occur as complex conditions.

WATER METABOLISM

Water is the most abundant body constituent; it comprises 45%–60% of total body weight (Figure 37.1). In a lean person, it accounts for a larger fraction of the body mass than in an overweight person. Since most biochemical reactions take place in an aqueous environment, the control of water balance is an important requirement for homeostasis.

Water permeates cell membranes through water channels consisting of integral membrane proteins known as **aquaporins**. Solute concentrations are regulated because of barriers imposed by membrane systems. These barriers give rise to fluid pools or compartments of different but rather constant composition (Figure 37.2).

Intracellular fluid makes up 30%–40% of the body weight, or about two-thirds of total body water. Potassium and magnesium are the predominant cations. The anions are mainly proteins and organic phosphates, with chloride and bicarbonate at low concentrations.

Extracellular fluid contains sodium as the predominant cation and accounts for 20%–25% of body weight, or one-third of total body water. It makes up vascular, interstitial, transcellular, and dense connective tissue fluid pools. Vascular fluid is the circulating portion, is rich in protein, and does not readily cross endothelial membranes. Interstitial fluid surrounds cells and accounts for 18%–20% of total body water. It exchanges with vascular fluid via the lymph system. Transcellular fluid is present in digestive juices, intraocular fluid, cerebrospinal fluid (CSF), and synovial (joint) fluid. These fluids are secretions of specialized cells. Their composition differs considerably from that of the rest of the extracellular fluid, with which they rapidly exchange contents under normal conditions. Dense connective tissue (bone, cartilage) fluid exchanges slowly with the rest of the extracellular fluid and accounts for 15% of total body water.

Movements of water occur mainly via aquaporins affected by osmosis and filtration. In osmosis, water moves to the area of highest solute concentration. Thus, active movement of salts into an area creates a concentration gradient down which water flows passively.

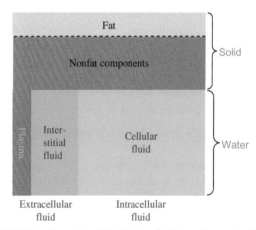

FIGURE 37.1 Proportional distribution of solids and water in a healthy adult.

N. V. Bhagavan and C.-E. Ha: Essentials of Medical Biochemistry, Second edition. DOI: http://dx.doi.org/10.1016/B978-0-12-416687-5.00037-3

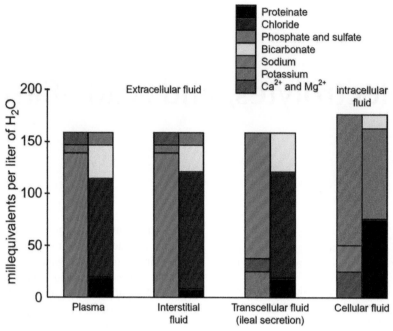

FIGURE 37.2   Composition of body fluids.

In filtration, hydrostatic pressure in arterial blood moves water and nonprotein solutes through specialized membranes to produce an almost protein-free filtrate: this process occurs in the formation of the renal glomerular filtrate. Filtration also accounts for movement of water from the vascular space into the interstitial compartment, which is opposed by the osmotic (oncotic) pressure of plasma proteins. Cells move ions (especially Na^+ and K^+) against a concentration gradient by a "sodium pump" that actively transports sodium across the plasma membranes (Chapter 11).

The kidneys are the major organs that regulate extracellular fluid composition and volume via their functional units known as **nephrons**. The average number of nephrons present in adult per kidney is about 1 million, and this number declines with the normal aging process. Low birth weight infants due to intrauterine growth retardation have a decreased number of nephrons.

Three main processes occur in the nephrons:

1. Formation of a virtually protein-free ultrafiltrate at the glomerulus;
2. Active reabsorption (principally in the proximal tubule) of solutes from the glomerular filtrate; and
3. Active excretion of substances such as hydrogen ions into the tubular lumen, usually in the distal portion of the tubule (Figure 37.3).

The normal **glomerular filtration rate (GFR)** is 100−120 mL/min; about 150 L of fluid passes through the renal tubules each day. Since the average daily urine volume is 1−1.5 L, 99% of the glomerular filtrate is reabsorbed. Approximately 80% of the water is reabsorbed in the proximal tubule, a consequence of active absorption of solutes. Reabsorption in the rest of the tubule varies according to the individual's water balance, in contrast to the **obligatory** reabsorption that occurs in the proximal tubule.

The facultative absorption of water depends on the establishment of an osmotic gradient by the secretion of Na^+ from the ascending loop and uptake by the descending loop in the loop of Henle. As a result, the proximal end of the loop is hyperosmotic (1200 mosm/kg), and the distal end is hypo-osmotic with respect to blood. The collecting ducts run through the hyperosmotic region. In the absence of antidiuretic hormone (ADH; see Chapter 29), the cells of the ducts are relatively impermeable to water. They become permeable to water in the presence of ADH, however, and the urine becomes hyperosmotic with respect to blood.

HOMEOSTATIC CONTROLS

The composition and volume of extracellular fluid are regulated by complex hormonal and nervous mechanisms that interact to control its osmolality, volume, and pH. The osmolality of extracellular fluid is due mainly to Na^+ and accompanying anions. It is kept within narrow limits (285−295 mosm/kg) by regulation of water intake (via a

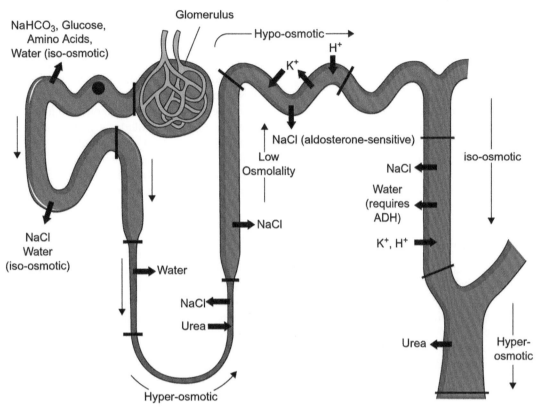

FIGURE 37.3   Principal transport processes in the renal nephron. ADH = antidiuretic hormone.

thirst center) and water excretion by the kidney through the action of ADH. The volume is kept relatively constant, provided the individual's weight remains constant to within ± 1 kg. Volume receptors sense the effective circulating blood volume, which, when decreased, stimulates the renin–angiotensin–aldosterone system and results in retention of Na^+ (Chapter 30). The increased Na^+ level leads to a rise in osmolality and secretion of ADH, with a resultant increase in water retention. Antagonistic systems exist that result in increased Na^+ excretion. Natriuretic peptides (NP), atrial natriuretic peptide (ANP), and brain natriuretic peptide (BNP, also called **natriuretic hormones**) are released by the cardiocytes of the cardiac atria and ventricles, respectively, in response to mechanical stretch caused by the plasma volume expansion. NP induces diuresis and natriuresis. These effects result from renal hemodynamic changes associated with increases in GFR and inhibition of Na^+ reabsorption from inner medullary collecting ducts. ANP and BNP are 28- and 32-amino acid peptides, respectively. They both have a single disulfide linkage and are derived from precursors by intracellular proteolysis. Some of the stimuli other than blood volume that function as secretogogues of NP include high blood pressure, elevated serum osmolality, increased heart rate, and elevated levels

of plasma catecholamines. Activation of the NP genes in cardiocytes by glucocorticoids leads to their increased synthesis. NP also regulates Na^+ and water homeostasis by different mechanisms that include inhibition of steps in the renin–angiotensin–aldosterone pathway and inhibition of ADH secretion from the posterior pituitary cells.

The mechanism of action of NP on target cells involves the formation of cGMP via the activation of plasma membrane receptor. The NP receptor itself is a guanylyl cyclase with its ligand-binding domain located in the extracellular space and its catalytic domain in the cytosolic domain. The receptor has only one membrane-spanning domain. This NP receptor-activated cGMP complex is unique and does not involve any G-proteins. A soluble cytosolic guanylyl cyclase that is activated by nitric oxide binding to the heme group of the enzyme causes vascular relaxation (Chapter 15). The intracellular formation of cGMP causes activation of cGMP-dependent protein kinases, which mediate the actions of ANP. Blood levels of BNP or its other circulatory peptide fragments are used in the diagnosis of heart failure. BNP is also used therapeutically. A combination of inhibitors of two systems that promote optimal cardiovascular function, which ameliorates neurohormonal overactivation in subjects with congestive heart failure, has been valuable as synergistic therapeutic agents. This combined therapy with the two inhibitors includes a protease inhibitor, known as **neprilysin**, that increases the plasma levels of vasoactive peptides such as natriuretic peptides and bradykinin by their inhibition of degradation, and an angiotensin II-receptor blocker [1,2].

The pH of extracellular fluid is kept within very narrow limits (7.35−7.45) by buffering mechanisms (see also Chapter 2), the lungs, and the kidneys. These three systems do not act independently. For example, in acute blood loss, release of ADH and aldosterone restores the blood volume, and renal regulation of the pH leads to shifts in K^+ and Na^+ levels.

WATER AND OSMOLALITY CONTROLS

Despite considerable variation in fluid intake, an individual maintains water balance and a constant composition of body fluids. The homeostatic regulation of water is summarized in Figure 37.4. Body water is derived from 2−4 L of water consumed daily in food and drink and 300 mL of metabolic water formed daily by oxidation of lipids and carbohydrates. Water loss occurs by perspiration and expiration of air ($\sim$1 L/day), in feces ($\sim$200 mL) (Chapter 11), and in urine (1−2 L/day).

Water balance is regulated to maintain the constant osmolality of body fluids. This osmolality is directly related to the number of particles present per unit weight of solvent. A solution that contains 1 mol of particles in 22.4 kg of water (22.4 L at 4°C) exerts an osmotic pressure of 1 atm and has an osmolality of 0.0446. Conversely, the osmotic pressure of an osmolal solution (1 mol of particles/kg of water) is 22.4 atm. In this sense,

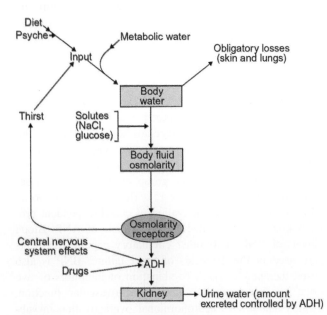

FIGURE 37.4 Regulation of osmolality in the body.

"number of particles" is roughly defined as the number of noninteracting molecular or ionic groups present. Since glucose does not readily dissociate, 1 mol dissolved in 1 kg of water (a molal solution) produces 1 mol of "particles" and has an osmolality of 1. Sodium chloride dissociates completely in water to form two particles from each molecule of NaCl so that a molal solution of NaCl is a 2 osmolal solution. Similarly, a molal solution of Na_2SO_4 or $(NH_4)_2SO_4$ is a 3 osmolal solution. In practice, the milliosmole (mosm) is the unit used.

With aqueous solutions, *osmolarity* is sometimes used interchangeably with *osmolality*. Although this practice is not strictly correct (moles of particles per liter of solution versus moles of particles per kilogram of solvent), in water at temperatures of biological interest, the error is fairly small unless solute concentrations are high (i.e., when an appreciable fraction of the solution is not water). Thus, with urine, the approximation is acceptable, whereas with serum it is not, because of the large amount of protein present. Although osmolarity is more readily measured, it is temperature-dependent, unlike osmolality. Osmolality is commonly measured by freezing point or vapor pressure depression. In terms of vapor pressure (P^v), the osmotic pressure (π) is defined as follows:

$$\pi = P^v_{\text{pure solvent}} - P^v_{\text{solution}}$$

As defined here, osmolality = π/22.4, where π is measured in atmospheres. In one instrument, solution and solvent vapor pressures are measured by the use of sensitive thermistors to detect the difference in temperature decrease caused by evaporation of solvent from a drop of pure solvent and a drop of solution. Because the rate of evaporation (vapor pressure) of the solution is lower, the temperature change will be less, and the vapor pressure difference can be calculated.

The freezing point of a solution is always lower than that of the solvent. The exact difference depends on the solvent and the osmolality of the solution. For water:

$$\text{Osmolality} = \frac{\Delta T}{1.86}$$

where ΔT is the freezing point depression in °C. Instruments that measure the freezing point of a sample are used in clinical laboratories to determine serum and urine osmolality.

Since water passes freely through most biological membranes, all body fluids are in osmotic equilibrium; consequently, the osmolality of plasma is representative of the osmolality of other body fluids. The osmotic pressure of extracellular fluid is due primarily to its principal cation Na^+ and the anions Cl^- and HCO_3^-. Taking twice the Na^+ concentration gives a good estimate of serum osmolality. Thus, normal plasma contains 135−145 mEq of Na^+/L (3.1−3.3 g/L), and normal plasma osmolality is about 270–290 mosm/kg (this corresponds to an osmotic pressure of 6.8–7.3 atm and a freezing point depression of 0.50°C–0.54°C). Glucose contributes only 5–6 mosm/kg (or 0.1 atm) to the osmotic pressure. Plasma protein contributes approximately 10.8 mosm/kg. Because of their size and general inability to pass through biological membranes, proteins are important determinants of fluid balance between intravascular and extravascular spaces. That portion of the osmotic pressure which is due to proteins is often referred to as the **oncotic pressure**.

Since many molecules in plasma interact, the measured osmolality of a sample is an effective osmolality and is lower than the value calculated from the concentrations of all the ions and molecules it contains. A solution that has the same effective osmolality as plasma is said to be **isotonic**, e.g., 0.9% saline, 5% glucose, and Ringer's and Locke's solutions. If a solute can permeate a membrane freely, then a solution of that solute will behave like pure water with respect to the membrane. Thus, a solution of urea will cause red cells to swell and burst as pure water does, because urea moves freely across erythrocyte membranes.

The osmolality of urine can differ markedly from that of plasma because of active concentration processes in the renal tubules. The membranes of renal collecting ducts show varying degrees of water permeability and permit removal of certain solutes without simultaneous uptake of water. Plasma osmolality can be calculated from the concentrations of plasma Na^+, glucose, and serum urea nitrogen:

$$\text{Osmolality} = 1.86(Na^+ \text{ mEq/L}) + \frac{\text{glucose (mg/dL)}}{18}$$
$$+ \frac{\text{serum urea nitrogen (mg/dL)}}{2.8}$$

The numerical denominators for glucose and urea nitrogen convert the concentrations to moles per liter. Such an estimated osmolality is usually 6–9 mosm less than the value determined by freezing point or vapor pressure measurements. If the latter value is much greater than the estimated value, molecules **other** than Na^+, glucose, and urea must account for the difference. Such "osmolal gaps" occur in individuals suffering from drug toxicity (alcohol, barbiturates, salicylates), acute poisoning due to unknown substances, and acidosis (keto-, lactic, or renal). Determination of osmolality is helpful in the management of patients with fluid and electrolyte disorders, e.g., chronic renal disease, nonketotic diabetic coma, hypo- and hypernatremia, hyperglycemia, hyperlipemia, burns, sequelae to major surgery or severe trauma (particularly serious head injuries), hemodialysis, or diabetes insipidus [3]. Changes of about 2% or more are detected by hypothalamic osmoreceptors (Chapter 29), which elicit a sensation of thirst and production of hypertonic urine. Under conditions of fluid restriction, urine osmolality can reach 800–1,200 mosm/kg (normal is 390–1090 mosm/kg), or three to four times the plasma levels. Decrease in plasma osmolarity (as in excessive water intake) produces urine with decreased osmolality. Water losses from skin and lungs are not subject to controls of this type.

ADH acts at the renal tubules and collecting ducts to raise cAMP levels, followed by aquaporin-mediated water uptake. Urinary levels of cAMP are increased by ADH. Factors other than plasma hypertonicity may stimulate ADH secretion. Thus, in acute hemorrhage, extracellular fluid volume drops abruptly, and ADH is secreted to increase the volume at the expense of a drop in osmolarity. Biological actions of ADH (also known as **vasopressin**) are mediated by its cell membrane receptor subtypes. Receptor subtype VIa causes vasoconstriction, VIb causes secretion of ACTH, and V2 causes water reabsorption and secretion of von Willebrand factor and factor VIII (Chapter 34).

Inappropriate ADH secretion can occur in the presence of water overload and a decline in plasma Na^+ concentration and osmolality. Fear, pain, and certain hormone-secreting tumors can cause inappropriate ADH secretion, which leads to hyponatremia and water retention. Morphine and barbiturates increase, and ethanol decreases, secretion of ADH.

In **diabetes insipidus**, due to defective ADH receptors or to diminished ADH secretion, renal tubules fail to recover the water from the glomerular filtrate. In cases of deficiency of ADH, hormone replacement by 8-lysine vasopressin or 1-deamino-8-D-arginine vasopressin (administered as a nasal spray or subcutaneously) is effective. In osmotic diuresis, e.g., in **diabetes mellitus** with severe glycosuria, the solute load increases the osmolality of the glomerular filtrate and impairs the ability of the kidney to concentrate the urine. Extracellular fluid volume in a normal adult is kept constant; body weight does not vary by more than a pound per day despite fluctuations in food and fluid intake. A decrease in extracellular fluid volume lowers the effective blood volume and compromises the circulatory system. An increase may lead to hypertension, edema, or both. Volume control centers on renal regulation of Na^+ balance. When the extracellular fluid volume decreases, less Na^+ is excreted; when it increases, more Na^+ is lost. Na^+ retention leads to expansion of extracellular fluid volume, since Na^+ is confined to this region and causes increased water retention. Renal Na^+ flux is controlled by the aldosterone–angiotensin–renin system (Chapter 30) and natriuretic peptides (discussed earlier).

ELECTROLYTE BALANCE

The major electrolytes are Na^+, K^+, Cl^-, and HCO_3^- (HCO_3^- is discussed later, under "Acid–Base Balance").

Sodium

The average Na^+ content of the human body is 60 mEq/kg, which consists of 50% extracellular fluid, 40% bone, and 10% intracellular Na^+ [4]. The chief dietary source of sodium is salt added in cooking. Excess sodium is largely excreted in the urine, although some is lost in perspiration. Gastrointestinal losses are small except in diarrhea.

Sodium balance is integrated with the regulation of extracellular fluid volume. Depletional **hyponatremia** (sodium loss greater than water loss) may result from inadequate Na^+ intake, excessive fluid loss from vomiting or diarrhea, diuretic abuse, and adrenal insufficiency (see Clinical Case Study 37.1). Hyponatremia affects extracellular fluid volume, as it occurs in congestive heart failure, uncontrolled diabetes, cirrhosis, nephrosis, and inappropriate ADH secretion. Severe hyponatremia, a hypoosmolar condition, can cause cerebral edema, resulting in neurocognitive dysfunction. The treatment may require therapy with hypertonic saline and ADH receptor antagonists (vasopressin V2 receptor antagonist, vaptans).

Hypernatremia results from loss of hypo-osmotic fluid (e.g., in burns, fevers, high environmental temperature, exercise, kidney disease, diabetes insipidus) or increased Na^+ intake (e.g., administration of hypertonic NaCl solutions, ingestion of $NaHCO_3$).

Potassium

The average K^+ content of the human body is 40 mEq/kg. K^+ occurs mainly in the intracellular spaces. It is required for carbohydrate metabolism, and increased cellular uptake of K^+ occurs during glucose catabolism. K^+ is widely distributed in plant and animal foods, the human requirement being about 4 g/day. Insulin and catecholamines promote a shift of K^+ into the cells. Excess K^+ is excreted in the urine, a process regulated by aldosterone.

Plasma K^+ plays a role in the irritability of excitable tissue. A high concentration of plasma K^+ leads to electrocardiographic (ECG) abnormalities and possibly to cardiac arrhythmia, which may be due to the lowering of the membrane potential. Low concentration of plasma K^+ increases the membrane potential, decreases irritability, and produces other ECG abnormalities and muscle paralysis.

Hyperkalemia may occur in chronic renal disease, adrenal insufficiency, and in the disorders of drugs that inhibit the renin−angiotensinogen−aldosterone axis. Treatment consists of promotion of cellular uptake of K^+ by administration of insulin with glucose and β_2-adrenergic agonists. In severe cases, ion exchange resins given orally, which bind K^+ in intestinal secretions and hemodialysis, are used to correct hyperkalemia [5].

Hypokalemia may occur with loss of gastrointestinal secretions (which contain significant amounts of K^+) and from excessive loss in the urine because of increased aldosterone secretion or diuretic therapy. Hypokalemia is usually associated with alkalosis (see Clinical Case Study 37.2 [6]).

Chloride

Chloride is the major extracellular anion. About 70% is in the extracellular fluid. The average Cl^- content of the human body is 35 mEq/kg. Chloride in food is almost completely absorbed. Plasma levels of Na^+ and Cl^-, in general, undergo parallel alterations. However, in metabolic alkalosis, chloride concentration increases.

ACID−BASE BALANCE

Normal blood pH is 7.35−7.45 (corresponding to 35−45 nmol of H^+ per liter). Values below 6.80 (160 nmol of H^+ per liter) or above 7.70 (20 nmol of H^+ per liter) are seldom compatible with life. A large amount of acid produced is a byproduct of metabolism. The lungs remove 14,000 mEq of CO_2 per day. From a diet that supplies 1−2 g of protein per kilogram per day, the kidneys remove 40−70 mEq of acid per day as sulfate (from oxidation of sulfur-containing amino acids), phosphate (from phospholipid, phosphoprotein, and nucleic acid catabolism), and organic acids (e.g., lactic, β-hydroxybutyric, and acetoacetic). These organic acids are produced by incomplete oxidation of carbohydrate and fats, and in some conditions (e.g., ketosis; see Chapter 16) considerable amounts may be produced.

The most important extracellular buffer is the carbonic acid−bicarbonate system:

$$CO_2 + H_2O \leftrightarrow H_2CO_3 \leftrightarrow HCO_3^- + H^+$$

As discussed in Chapter 2, at a blood pH of 7.4, the ratio of $[HCO_3^-]$ to $[H_2CO_3]$ is 20:1, and the system's buffering capacity can neutralize a large amount of acid. The system is independently regulated by the kidneys, which control the plasma HCO_3^- level, and by the respiratory rate, which regulates the P_{CO_2}. Protein and phosphate buffer systems also operate in plasma and erythrocytes. Proteins are especially important buffers in the intracellular fluid. The hydroxyapatite of bone also acts as a buffer.

The medullary respiratory center senses and responds to the pH of blood, perfusing it, and is the source of pulmonary control. P_{CO_2} and perhaps P_{O_2} also influence the center, together with nervous impulses from higher centers of the brain. A decrease in pH results in an increased respiratory rate and deeper breathing with a consequent increase in the respiratory exchange of gases and lowering of P_{CO_2}, which elevates the pH. Similarly, a decrease in

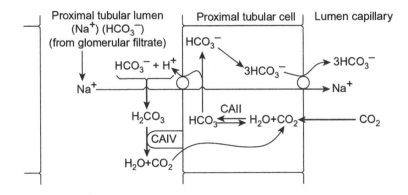

FIGURE 37.5 Reclamation of bicarbonate. The filtered Na^+ is reabsorbed by the proximal tubular cell in exchange for H^+. The filtered HCO_3^- is converted to H_2O and CO_2 catalyzed by luminal carbonic anhydrase IV (CAIV). CO_2 diffuses in the tubular cell, where it is hydrated to H_2CO_3 by carbonic anhydrase II and dissociated to H^+ and HCO_3^-. Three molecules of HCO_3^- and one of Na^+ are transported to the peritubular capillary by the basolateral cotransporter.

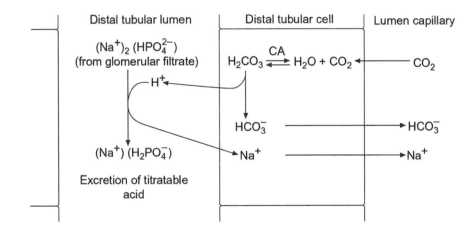

FIGURE 37.6 Hydrogen ion secretion and Na^+/H^+ exchange coupled to the conversion of HPO_4^{2-} to $H_2PO_4^-$ in the distal tubule lumen. CA = carbonic anhydrase.

respiratory rate leads to the accumulation of CO_2, an increase in P_{CO_2}, and a decrease in pH. Pulmonary responses to fluctuations in blood pH are rapid, while renal compensatory mechanisms are relatively slow. The kidneys actively secrete H^+ ions through three mechanisms:

1. Na^+/H^+ exchange;
2. Reclamation of bicarbonate; and
3. Production of ammonia and excretion of NH_4^-.

The proximal tubule is responsible for reclamation of most of the 4500 mEq of HCO_3^- filtered through the glomeruli each day. H^+ secreted into the tubules in exchange for Na^+ from the tubular fluid (an energy-dependent process mediated by Na^+/H^+-ATPase) combines with HCO_3^- to form CO_2 and water. The CO_2 diffuses into the tubular cells, where it is rehydrated to H_2CO_3 by carbonic anhydrase, and dissociates to bicarbonate and H^+. The HCO_3^- diffuses into the bloodstream, resulting in the reclamation of bicarbonate (Figure 37.5).

The formation of H_2CO_3 from H^+ and HCO_3^- in the tubular lumen is catalyzed by the membrane-bound isoenzyme of carbonic anhydrase (CAIV). CAIV is located on the brush border lining the lumen of the proximal tubules

of the kidney. The CO_2 diffuses into the tubular cells, where it is rehydrated to H_2CO_3 by a different isoenzyme of carbonic anhydrase—namely, carbonic anhydrase II (CAII). HCO_3^- is transported from the cytosol of the proximal tubular cell to peritubular capillary blood by a basolateral cotransporter, which transports three molecules of HCO_3^- and one of Na^+. An inherited CAII deficiency is an autosomal recessive disorder causing renal tubular acidosis and osteopetrosis. The latter is due to a lack of the H^+ required for bone remodeling processes (Chapter 35). The proximal tubular mechanism for reclamation of HCO_3^- becomes saturated at approximately 26 mEq HCO_3^-/L. At higher levels, HCO_3^- appears in the distal tubule and may be excreted in the urine. Under conditions of elevated P_{CO_2} secretion is more active, possibly owing to intracellular acidosis. In proximal renal tubular acidosis, HCO_3^- reabsorption is impaired and saturation occurs at a lower concentration (about 16−18 mEq/L), so plasma $[HCO_3^-]$ is low while urine pH may be high because of the presence of HCO_3^-.

Na^+/H^+ exchange may also be coupled to formation of $H_2PO_4^-$ from HPO_4^{2-} in the lumen (Figure 37.6). This coupling is of particular importance in distal tubules and acidifies the urine to a maximum pH of about 4.4.

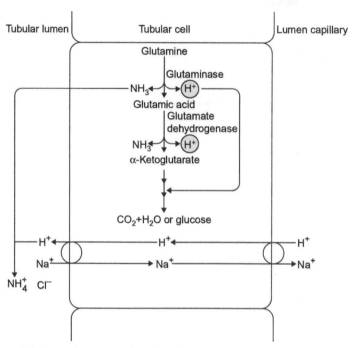

FIGURE 37.7 Formation of ammonia in the renal tubule cells from glutamine and secretion of the ammonium ion in the urine.

The proximal tubule cannot maintain an H^+ gradient of more than one pH unit between the lumen and the intracellular fluid. The $H_2PO_4^-$ present (termed the **titratable acidity**) can be measured by titrating the urine to pH 7 (pK_2 for $H_3PO_4 = 6.2$). Normal values for titratable acidity are 16–60 mEq/24 hours, depending on the phosphate load.

Generation of ammonia from glutamine and its excretion as NH_4^+ is an important mechanism for the elimination of protons, particularly during severe metabolic acidosis when it becomes a significant mode of proton excretion. Most renal glutamine is derived from muscle (Chapter 15). Glutamine provides two molecules of NH_3:

$$\text{Glutamine} + H_2O \xrightarrow{\text{glutaminase}} \text{glutamate} + NH_3 + H^+ (\leftrightarrow NH_4^+)$$

and

$$\text{Glutamate}^- \underset{\substack{\text{NADH} \quad \text{NADH} + H^+}}{\overset{\substack{\text{Glutamate} \\ \text{dehydrogenase}}}{\rightleftarrows}}$$

$$\alpha\text{-ketoglutarate}^{2-} + NH_3 + H^+ (\rightleftarrows NH_4^+)$$

The ratio of $[NH_3]/[NH_4^+]$ depends on intracellular pH; however, the NH_3 readily diffuses into the tubular lumen and forms an ammonium ion that is no longer able to pass freely through the membranes and remains "trapped" in the urine, where it is associated with the dominant counterion (Figure 37.7). Protons produced in these reactions are consumed when α-ketoglutarate is either completely oxidized or converted to glucose. Thus, they do not add to the existing "proton burden" due to severe acidosis (see also Chapters 12 and 14).

Disorders of Acid–Base Balance

Disorders of acid–base balance are classified according to their cause, and the direction of the pH change, into respiratory acidosis, metabolic acidosis, respiratory alkalosis, or metabolic alkalosis. [7,8; reference 8 also includes illustrative case studies] Any derangement of acid–base balance elicits compensatory changes in an attempt to restore homeostasis (Figure 37.8). Acidosis due to respiratory failure leads to compensatory renal changes, which lead to increased reclamation of HCO_3^-.

In the assessment of acid–base disorders, commonly measured electrolytes are serum Na^+, K^+, H^+ (as pH), Cl^-, and HCO_3^-. Other anions (e.g., sulfates, phosphates, proteins) and cations (e.g., calcium, magnesium, proteins) are not measured routinely but can be estimated indirectly, since (to maintain electrical neutrality) the sum of the cations must equal that of the anions. Serum Na^+ and K^+ content accounts for 95% of cations, and Cl^- and HCO_3^- for about 85% of anions.

$$\text{Unmeasured anions} = [Na^+] + [K^+] - [Cl^-] - [HCO_3^-]$$

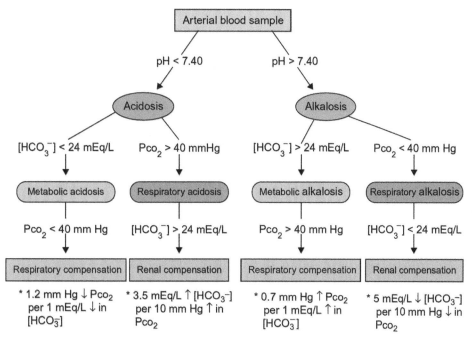

FIGURE 37.8 Classification, characteristics, and compensation of simple acid–base disorders. ↑ = increased, ↓ = decreased. *[Reproduced with permission from R. M. Berne, M. N. Levy, Principles of physiology, 3rd ed. (Maryland Heights, MO: Mosby Inc., 2000, Ch. 39, Figure 39.5, p. 479).]*

The unmeasured anion is commonly known as the **anion gap**, which is normally 12 ± 4 mEq/L. This value is useful in assessing the acid–base status of a patient and in diagnosing metabolic acidosis. Disorders that cause a high anion gap are metabolic acidosis, dehydration, therapy with sodium salts of strong acids, therapy with certain antibiotics (e.g., carbenicillin), and alkalosis. A decrease in the normal anion gap occurs in various plasma dilution states, hypercalcemia, hypermagnesemia, hypernatremia, hypoalbuminemia, disorders associated with hyperviscosity, some paraproteinemias, and bromide toxicity.

The urinary anion gap (UAG) is utilized in the diagnosis of metabolic acidosis caused by defects in renal reclamation of bicarbonate. As discussed previously, renal acid excretion requires urinary excretion of NH_4^+ and its accompanying Cl^- ions. Thus Cl^- concentration is used as a surrogate level of NH_4^+ in the UAG calculation, along with urinary Na^+ and K^+ concentrations, as follows:

$$UAG = (Na^+ + K^+) - Cl^-$$

In metabolic acidosis due to bicarbonate loss in chronic diarrhea, renal compensation leads to an increased level of NH_4^+, and therefore Cl^- excretion, resulting in a negative UAG; whereas in disorders of primary renal acidification systems, the UAG has a positive value.

Respiratory acidosis is characterized by the accumulation of CO_2, a rise in P_{CO_2} (hypercapnia or

hypercarbia), a decrease in $[HCO_3^-]/[P_{CO_2}]$, and a decrease in pH (see the Henderson–Hasselbalch equation, Chapter 2). It may result from central depression of respiration (e.g., narcotic or barbiturate overdose, trauma, infection, cerebrovascular accident) or from pulmonary disease (e.g., asthma, obstructive lung disease, infection). Increased $[H^+]$ is, in part, buffered by cellular uptake of H^+ with corresponding loss of intracellular K^+. In acute hypercapnia, the primary compensatory mechanism is tissue buffering. In chronic hypercapnia, the kidneys respond to elevated plasma P_{CO_2}, increasing the amount of HCO_3^- formed by carbonic anhydrase in the tubules and by excreting more H^+. The primary goal of treatment is to remove the cause of the disturbed ventilation. Immediate intubation and assisted ventilation also aid in improving gas exchange.

Metabolic acidosis [9,10] with increased anion gap occurs in diabetic or alcoholic ketoacidosis; lactic acidosis from hypoxia, shock, severe anemia, alcoholism, or cancer; toxicity from ingestion of salicylates, methanol, paraldehyde, or ethylene glycol; and renal failure. Lactic acid acidosis caused by deprivation of tissue oxygen, inhibition of gluconeogenesis, and some drugs and toxins is due to the accumulation of L-lactate, which is the end product of glycolysis (Chapter 12). Frequently, L-lactate (simply referred to as **lactate**) is the metabolite measured in assessing metabolic acidosis. However, D-lactate may be produced under certain clinical conditions, such as diminished colonic motility, short bowel syndrome, jejunoileal bypass, overgrowth of D-lactate producing grampositive organisms (e.g., *Lactobacillus* species, *Streptococcus bovis*). Carbohydrate malabsorption and ingestion of large amounts of carbohydrates may also exacerbate the development of D-lactate acidosis. In addition, an impairment of D-lactate metabolism may also contribute to D-lactic acidosis. D-lactate is converted to pyruvate by D-2-hydroxy acid dehydrogenase, a mitochondrial enzyme present in liver, kidney, and other tissues. The clinical manifestations of D-lactic acidosis include episodes of encephalopathy and metabolic acidosis. Since serum D-lactate is not normally a measured clinical parameter, the critical indices of suspicion of D-lactic acidosis in the clinical conditions mentioned previously include increased anion gap metabolic acidosis with negative tests for L-lactate and ketoacidosis. Metabolic acidosis with normal anion gap occurs in renal tubular acidosis, carbonic anhydrase inhibition, diarrhea, ammonium chloride administration, chronic pyelonephritis, and obstructive uropathy. In both of these groups, plasma HCO_3^- levels decrease and tissue buffering occurs by exchange of extracellular H^+ for intracellular K^+. Thus, plasma K^+ levels may increase.

Metabolic acidosis produces prompt stimulation of respiratory rate and a decrease in P_{CO_2}. This effect cannot be sustained, however, because the respiratory muscles become tired. Renal compensation is slower but can be maintained for an extended period because of induction of glutaminase.

Treatment is by correction of the cause of the acidosis (e.g., insulin administration in diabetic ketoacidosis) and neutralization of the acid with $NaHCO_3$, sodium lactate, or TRIS [tris(hydroxymethyl)aminomethane] buffer. Problems that may occur following alkali replacement therapy include development of respiratory alkalosis, particularly if the low CO_2 tension persists, and further decline in the pH of CSF, which may decrease consciousness. The alkaline "overshoot" results from resumption of oxidation of organic anions (e.g., lactate, acetoacetate) with the resultant production of bicarbonate from CO_2. Severe acidosis should be corrected slowly over several hours. Potassium replacement therapy is frequently needed because of the shift of intracellular K^+ to extracellular fluid and the loss of K^+ in the urine.

Respiratory alkalosis occurs when the respiratory rate increases abnormally (hyperventilation), leading to a decrease in P_{CO_2} and a rise in blood pH. Hyperventilation occurs in hysteria, pulmonary irritation (pulmonary embolus), and head injury with damage to the respiratory center. The increase in blood pH is buffered by plasma HCO_3^- and, to some extent, by the exchange of plasma K^+ for intracellular H^+. Renal compensation seldom occurs, because this type of alkalosis is usually transitory.

Metabolic alkalosis is characterized by elevated plasma HCO_3^- levels. It may result from the administration of excessive amounts of alkali (e.g., during $NaHCO_3$ treatment of peptic ulcer) or of acetate, citrate, lactate, and other substrates that are oxidized to HCO_3^-, and from vomiting, which causes loss of H^+ and Cl^-.

In excessive loss of extracellular K^+ from the kidneys, cellular K^+ diffuses out and is replaced by Na^+ and H^+ from the extracellular fluid. Since K^+ and H^+ are normally secreted by the distal tubule cells to balance Na^+ uptake during Na^+ reabsorption (see Figure 37.3), if extracellular K^+ is depleted, more H^+ is lost to permit reabsorption of the same amount of Na^+. Loss of H^+ by both routes causes hypokalemic alkalosis. Excessive amounts of some diuretics and increased aldosterone production can cause the hypokalemia that initiates this type of alkalosis.

In compensation, the respiratory rate decreases, raising P_{CO_2} and lowering the pH of blood. This mechanism is limited because if the respiratory rate falls too low, P_{O_2} decreases to the point where respiration is again stimulated. Renal compensation involves decreased reabsorption of bicarbonate and formation of alkaline urine. Because the urinary bicarbonate is accompanied by Na^+ and K^+, if the alkalosis is accompanied by extracellular fluid depletion, renal compensation by this mechanism may not be possible. Treatment consists of fluid and electrolyte replacement and NH_4Cl to counteract the alkalosis.

Acid−base disturbances frequently coexist with two or more simple disorders. In these settings, blood pH is either severely depressed (e.g., a patient with metabolic acidosis and respiratory acidosis) or normal. Both plasma HCO_3^- and pH may be within normal limits when metabolic alkalosis and metabolic ketoacidosis coexist, as in a patient with diabetic ketoacidosis who is vomiting. In this situation, an elevated anion gap may be the initial abnormality that can be detected in the underlying mixed acid−base disturbance. Acetylsalicylate (aspirin) toxicity causes metabolic acidosis and respiratory alkalosis. The latter is due to the stimulation of respiratory centers, resulting in hyperventilation and decreased P_{CO_2}.

CLINICAL CASE STUDY 37.1 Hyponatremia and Hyperkalemia in a Neonate

This case was abstracted from: S.P. Paul, B.A. Smith, T.M. Taylor, J. Walker, Take with a grain of salt, Clin. Chem. 59 (2013) 348−352.

Synopsis

A 5-day-old girl was admitted to the hospital for 15% loss of her original birth weight. The patient was born at term after an uncomplicated pregnancy and delivery. She was fed normal-term formula milk and was found to be mildly dehydrated on initial exam. Electrolytes were within normal limits. She was started on a feeding regimen with close monitoring of weight, but after 5 days the patient's weight remained unchanged, and her laboratory results showed hyponatremia and hyperkalemia. Further biochemical and endocrine testing revealed a markedly elevated plasma renin level of 854 mIU/L (reference range 4−190 mIU/L) and a markedly elevated serum aldosterone level of >5786 ng/L (reference range 300−2000 ng/L). These laboratory results were consistent with the diagnosis of pseudohypoaldosteronism type 1 (PHA1) as a cause of her hyponatremia and hyperkalemia.

Teaching Points

1. There are several classifications and etiologies of hyponatremia [see supplemental reference 4]. The presence of both hyponatremia and hyperkalemia in a neonate with excessive weight loss suggests a problem with sodium chloride metabolism, most commonly congenital adrenal hyperplasia due to salt-wasting 21-hydroxylase deficiency (SW21-OHD). However, SW21-OHD has other manifestations such as female virilization, genitourinary anomalies, and abnormal serum levels of 17-hydroxyprogesterone, cortisol, renin, and aldosterone. Other causes include adrenal hypoplasia; cystic fibrosis; cerebral salt-wasting syndrome; and aldosterone insensitivity secondary to urinary tract infections, pyelonephritis, or obstructive uropathy.

2. Pseudohypoaldosteronism (PHA) is a group of electrolyte imbalance disorders caused by **aldosterone resistance of the aldosterone receptor**. PHA1 is caused by mutations of the gene coding for the luminal epithelial sodium channel (coupled with the Na^+-K^+ ATPase pump) of many organs including kidney, lung, colon, and sweat and salivary glands. Receptor dysfunction and unresponsiveness to aldosterone lead to excretion of sodium (hyponatremia) and retention of potassium (hyperkalemia).

3. The electrolyte abnormalities of PHA1 may present after the patient has already developed symptoms of excessive weight loss, failure to thrive, feeding difficulties, vomiting, and dehydration within the first 2 weeks of life.

4. Treatment for PHA1 is sodium chloride replacement therapy. For the management of other causes of hyponatremia, see supplemental references 1 and 2. Severe hyponatremia requires urgent infusion of hypertonic saline to correct cerebral edema; however, rapid correction of chronic hyponatremia may result in serious neurologic injury in some cases. Patients with asymptomatic hyponatremia do not need immediate therapy to acutely correct serum sodium.

Supplemental References

[1] J.G. Verbalis, S.R. Goldsmith, A. Greenberg, C. Korzelius, R.W. Schrier, R.H. Sterns, et al., Diagnosis, evaluation, and treatment of hyponatremia: expert panel recommendations, Am. J. Med. 126 (2013) S1−S41.
[2] C. Vaidya, W. Ho, B.J. Freda, Management of hyponatremia: providing treatment and avoiding harm, Cleve. Clin. J. Med. 77 (2010) 715−726.
[3] H.J. Adrogue, N.E. Madias, Hyponatremia, N. Engl. J. Med. 342 (2000) 1581−1589.
[4] T.A. Kotchen, A.W. Cowley, E.D. Frohlich, Salt in health and disease—a delicate balance, N. Engl. J. Med. 368 (2013) 1229−1237.

CLINICAL CASE STUDY 37.2 Hypokalemia

This case was abstracted from: C.H. Ho, K.E. Lewis, J.L. Johnson, L.M. Opas, A 20-year-old woman with fatigue and palpitations, Cleve. Clin. J. Med. 81 (2014) 283−288.

Synopsis

A 20-year-old woman presented to the emergency department with fatigue and palpitations. She had no significant past medical history. She was not on any medications. She was afebrile with a blood pressure of 92/48 mmHg, heart rate of 73 bpm, and respiratory rate of 5 bpm. Physical examination was unremarkable. Electrocardiography (EKG) showed ST-segment depression, prolonged QT interval, T-wave inversion, PR prolongation, increased P-wave amplitude, and U waves. Laboratory testing revealed hypokalemic, hyponatremic, hypochloremic metabolic alkalosis with compensatory

respiratory acidosis. Further questioning revealed that the patient had an 8-year history of daily self-induced vomiting and an eating disorder. She was treated with oral and intravenous potassium chloride with normalization of serum potassium level above 3 mmol/L, and resolution of her electrocardiographic U wave, ST depression, and prolonged QT interval.

Teaching Points

1. Potassium is mostly intracellular, stored predominantly in the myocytes and hepatocytes, with only 2% of potassium located in the extracellular space. Acute hypokalemia occurs when potassium shifts from the extracellular space into the cell, in response to alkalosis, insulin secretion, or beta-adrenergic stimulation. Chronic hypokalemia can

(Continued)

CLINICAL CASE STUDY 37.2 (Continued)

occur in patients with long-standing renal or gastrointestinal loss of potassium.

2. The most common cause of hypokalemia is diuretic use (renal loss). In a patient who is not taking diuretics, protracted vomiting should be considered (gastric loss). Potassium loss may also occur in diarrhea or laxative abuse. Other differential diagnoses for hypokalemia are Liddle, Bartter, or Gitelman syndromes. See Table 37.1 for a step-wise approach to determine the cause of hypokalemia.

3. Calculating the urinary anion gap and urinary pH may help distinguish between the different causes of acidosis and hypokalemia.

4. Electrocardiographic findings for hypokalemia include ST-segment depression, prolonged QT interval, T-wave inversion, PR prolongation, increased P-wave amplitude, and U waves. Hyperkalemia is associated with prolonged PR interval, decreased P wave amplitude, widened QRS complex, and characteristic peaked T waves. Severe hypokalemia can lead to fatal arrhythmias such as ventricular tachycardia, ventricular fibrillation, and high-grade atrioventricular block.

5. Other life-threatening complications of hypokalemia include hypoventilation due to respiratory muscle weakness. Of the different formulations of potassium, the preferred treatment of hypokalemic metabolic alkalosis is administration of potassium chloride due to chloride's effect on pendrin, a chloride—bicarbonate exchanger, which increases excretion of bicarbonate at the cortical collecting duct, thereby correcting metabolic alkalosis.

TABLE 37.1 An Approach to Determining the Cause of Hypokalemia

Review history and medications

- Renal potassium loss: diuretics, cisplatin, amphotericin B
- Extrarenal potassium loss: vomiting; laxatives; cutaneous loss from burns, excessive exercise, use of "sauna suits"

Look at blood pressure

- Elevated blood pressure suggests primary or secondary hyperaldosteronism (or pseudoaldosteronism)

Look at acid—base status

- Acidosis suggests renal tubular acidosis
- Alkalosis suggests Bartter syndrome, Gitelman syndrome, or hyperaldosteronism

Check urinary potassium—creatinine ratio when the site of potassium loss is unclear

- Renal loss: ratio >15
- Extrarenal loss: ratio <15

Check plasma aldosterone—renin ratio in patients with hypertension

- Ratio >20 suggests primary hyperaldosteronism
- Ratio <10 suggests secondary hyperaldosteronism

Check urinary chloride in patients with metabolic alkalosis

- Low urinary chloride (<10 mmol/L) suggests volume contraction
- High urinary chloride (>10 mmol/L) suggests renal chloride loss from hyperaldosteronism, diuretic use, or Bartter or Gitelman syndrome

Check urinary anion gap and urinary pH in patients with normal anion gap metabolic acidosis

- A negative anion gap suggests gastrointestinal or renal loss of bicarbonate (type II or proximal renal tubular acidosis)
- A urinary pH greater than 5.5 in the setting of acidosis suggests impaired ability of the kidneys to acidify urine and raises the possibility of renal tubular acidosis

[This table was reproduced with permission from: C.H. Ho, K.E. Lewis, J.L. Johnson, L.M. Opas, A 20-year-old woman with fatigue and palpitations, Cleve. Clin. J. Med. 81 (2014) 283—288.]

REQUIRED READING

[1] M. Jessup, Neprilysin inhibition—a novel therapy for heart failure, N. Engl. J. Med. 371 (2014) 1062—1064.

[2] J.J.V. McMurray, M. Packer, A.S. Desai, J. Gong, M.P. Lefkowitz, A.R. Rizkala, et al., Angiotensin—neprilysin inhibition versus enalapril in heart failure, N. Engl. J. Med. 371 (2014) 993—1004.

[3] A.H. Ropper, Hyperosmolar therapy for raised intracranial pressure, N. Engl. J. Med. 367 (2012) 746—752.

[4] T.A. Kotchen, A.W. Cowley, E.D. Frohlich, Salt in health and disease—a delicate balance, N. Engl. J. Med. 368 (2013) 1229—1237.

[5] M.M. Sood, A.R. Sood, R. Richardson, Emergency management and commonly encountered outpatient scenarios in patients with hyperkalemia, Mayo Clin. Proc. 82 (2007) 1553—1561.

[6] H.K. Jensen, M. Brabrand, P.J. Vinholt, J. Hallas, A.T. Lassen, Hypokalemia in acute medical patients: risk factors and prognosis, Am. J. Med. 128 (2015) 60—67.

[7] J.K. Berend, A.P.J. de Vries, R.O.B. Gans, Physiological approach to assessment of acid—base disturbances, N. Engl. J. Med. 371 (2014) 1434—1445.

[8] J.L. Seifter, Integration of acid—base and electrolyte disorders, N. Engl. J. Med. 371 (2014) 1821—1831.

[9] K.S. Kamel, M.L. Halperin, Acid-base problems in diabetic ketoacidosis, N. Engl. J. Med. 372 (2015) 546—554.

[10] J.A. Kraut, N.E. Madias, Lactic acidosis, N. Engl. J. Med. 371 (2014) 2309—2319.

Index

Note: Page numbers followed by "*b*," "*f*," and "*t*" refer to boxes, figures, and tables, respectively.

A

A site, ribosome, 434–436
ABC transporters. *See* ATP-binding cassette (ABC) transporters
ABCA1 protein, 335
Abetalipoproteinemia, 154, 333
ABP. *See* Androgen-binding protein (ABP)
Abrin, 160
Absorption. *See* Digestion and absorption
Acarbose, 376
ACAT. *See* Acyl-CoA:cholesterol acyltransferase (ACAT)
ACC. *See* Acetyl-CoA carboxylase (ACC)
Aceruloplasminemia, 676
Acetaminophen, 94–95
Acetylcholine receptors (AChR), 360
Acetyl-CoA, formation from pyruvate, 176–180
Acetyl-CoA carboxylase (ACC)
  fatty acid synthesis properties, 283
  mechanism, 283
  regulation, 283
N-Acetylcysteine, 28
N-Acetylglucosamine, 133
N-Acetylneuraminic acid (Neu5Ac), 102
Acid–base balance
  buffers, 706, 708
  disorders, 708–712
Acidemia, 19
Acini, 142
Aconitate dehydratase, 180
Acridine orange, 389
Acromegaly, 374
ACTH. *See* Corticotropin (ACTH)
Actin, 5, 343, 354–355
Actinomycin D, 389, 424–426
Activated partial thromboplastin time (APTT), 650–651
Activated protein C resistance, 657
Activation energy, 60
Activation of contraction, 345–346
Acute coronary syndrome (ACS), 92
Acute promyelocytic leukemia (PML), 685–686
Acute respiratory distress syndrome (ARDS), 307
Acute-phase reactant, 439
Acyclovir, 409
Acyl-CoA, transport into mitochondrial matrix, 271
Acyl-CoA dehydrogenase, 271–274

Acyl-CoA synthase, 270
Acyl-CoA synthetase, 154
Acyl-CoA:cholesterol acyltransferase (ACAT), 326
AD. *See* Alzheimer's disease (AD)
ADA. *See* Adenosine deaminase (ADA)
ADAMTS13, 642
Addison's disease, 364, 567–568, 574*b*
  case study, 574*b*
Adenine. *See* Purine nucleotides
Adenine nucleotide translocase, 197
Adenine phosphoribosyltransferase (APRT), 469, 472*f*, 480
Adenosine deaminase (ADA), 480
S-Adenosylmethionine (SAM), 251–252
Adenylate cyclase, receptor signaling, 537–540
Adenylate deaminase, 350
Adenylate kinase, 471
ADH *See* Alcohol dehydrogenase (ADH); Antidiuretic hormone (ADH)
Adrenal gland. *See also* Aldosterone; Cortisol
  anatomy, 559–560
  catecholamine synthesis and release, 569–571
  cortex, 559–560
  corticosteroids
    metabolism, 565
    regulation of secretion, 563–565
    relative activities, 561*t*
    synthesis, 560–569
    synthetic compounds, 565
  disturbances
    cortical function, 567–569
    medullary function, 571–574
  medulla, 569–574
Adrenoleukodystrophy (ALD), 6, 276, 329–331
Affinity labeling, 76
AFP. *See* α-Fetoprotein (AFP)
Agammaglobulinemia, 264
AGO2. *See* Argonaute 2 (AGO2)
Agonist, 57
AK-STAT signaling, 543
Alanine, 22, 207–208
Alanine, gluconeogenesis, 208
Alanine aminotransferase (ALT), 94
Albright's hereditary osteodystrophy, 540, 580
Albumin, 262–263
Alcohol dehydrogenase (ADH), 335
ALD. *See* Adrenoleukodystrophy (ALD)

Aldolase, glycolysis, 169*t*, 170
Aldolase B, 217
Aldosterone
  mechanism of action, 565
  physiological effects, 565
  secretion regulation, 563–564
Alendronate, 672–673
Alkalemia, 19
Alkaline Bohr effect, 492, 494
Allopurinol, 202, 477–478
Allosteric activators, 491–492
Allosteric inhibitors, 491–492
Allosteric proteins, 463
Allosteric regulation, 80–81, 463
  kinetics of, 81–83
Allysine residues, 124–125
Alois Alzheimer, 45
ALS. *See* Amyotrophic lateral sclerosis (ALS)
Alternative splicing, 423, 429, 461
Alzheimer's disease (AD), 45–49, 46*t*
  amyloidosis, 36
  diagnosis of, 49–50
  neural cell death, 8
  protein folding defects, 5–6
Amenorrhea, 597
Ames test, 411
Amidophosphoribosyltransferase, 473
D-Amino acid oxidase, 6, 200, 232
Amino acids, metabolism
  abbreviations, 24*t*
  acidic amino acids, 24
  basic amino acids, 25
  classification, 22–28
  derivatives as drugs, 28
  electrolyte and acid–base properties, 28–29
  essential, 22, 24*t*, 229
  free amino acids, 27
  metabolism
    arginine, 244
    branched-chain amino acids, 250–251
    cysteine, 253–255
    glutamate dehydrogenase, 233
    glutamate–glutamine cycle, 233
    glycine, 247
    histidine, 250
    methionine, 251–253
    organ specificity, 234–238
    phenylalanine, 256–257
    transamination, 233
    tryptophan, 258–267
    tyrosine, 256–257